The Future of Aging

Gregory M. Fahy · Michael D. West ·
L. Stephen Coles · Steven B. Harris
Editors

The Future of Aging

Pathways to Human Life Extension

Editors
Gregory M. Fahy
Intervene Biomedical, LLC
Box 478
Norco CA 92860
USA
intervene@sbcglobal.net

Dr. Michael D. West
BioTime, Inc.
1301 Harbor Bay Parkway
Alameda CA 94502
USA
mwest@biotimemail.com

L. Stephen Coles
Institute of Molecular Biology
Department of Chemistry and Biochemistry
Young Hall
University of California
Los Angeles
CA 90095-1569
USA
scoles@ucla.edu

Steven B. Harris
Critical Care Research, Inc.
10743 Civic Center Drive
Rancho Cucamonga CA
91730
USA
sbharris1@earthlink.net

ISBN 978-90-481-3998-9 e-ISBN 978-90-481-3999-6
DOI 10.1007/978-90-481-3999-6
Springer Dordrecht Heidelberg London New York

Library of Congress Control Number: 2010927444

© Springer Science+Business Media B.V. 2010
No part of this work may be reproduced, stored in a retrieval system, or transmitted in any form or by any means, electronic, mechanical, photocopying, microfilming, recording or otherwise, without written permission from the Publisher, with the exception of any material supplied specifically for the purpose of being entered and executed on a computer system, for exclusive use by the purchaser of the work.

Cover illustration: On the cover: The small nematode worm, *C. elegans* (wavy lines), can realize some very large gains in lifespan. Compared to the standard N2DRM (wild-type) worm, worms with a strong mutation in a single gene (the *age-1 (mg44)* allele) can live 10 times longer, and can do so in excellent health. This striking result brings into question the very nature of aging, and raises the possibility of someday extending the lifespans of humans in good health as well. The latter subject is the theme taken up in this book.

Printed on acid-free paper

Springer is part of Springer Science+Business Media (www.springer.com)

This book is dedicated to the fond memory of Christopher B. Heward, Ph.D., a tireless and dedicated clinical biogerontologist and a superb laboratory scientist who worked at the interface of business and biomedicine as the President of the Kronos Science Laboratories, where he conceived and pursued a variety of research and development projects aimed variously at early detection and prevention of age-related diseases and understanding and slowing the aging process in patients. He was internationally recognized as a seminal thinker in the area of biological aging and was the author and co-author of numerous scientific articles and book chapters. He was a consummately skeptical enthusiast for interventive gerontology.

It is an extreme and tragic irony that Chris developed cancer at a relatively young age and died on January 10, 2009 after a vigorous 3-month battle. He set a magnificent example of how to live life courageously to the fullest without illusion. He will be sorely missed by all who knew him.

Preface

Biogerontology is coming of age. It now seems clear that a fundamental understanding of the molecular pathways underlying age-related pathology has the potential to translate into a net economic impact of over a trillion dollars a year in the US alone and to alleviate a massive toll of human suffering. At the same time, a tidal wave of aging post World War II baby boomers is poised to begin flooding the shores of our health care system within the coming decade. It is therefore now strategic and timely for medical research to turn its attention to the molecular and cellular mechanisms of aging and age-related disease, and the purpose of this book is to provide a broad perspective on future possibilities for mitigating aging and improving the human condition that can be envisioned today based on available knowledge and reasonable deduction.

One of the largest barriers to successful intervention in aging is likely psychological rather than technological. A widespread misperception persists that modifying human aging borders on the impossible. However, this skepticism is not supported by science. There are now numerous examples of therapies based on fundamental insights in molecular biogerontology that are in clinical development, and many more such therapies are expected to unfold over time scales ranging from the very near term to the more distant future. We call attention to many of these in this volume.

Without a doubt, there has never been a more exciting time in the history of biogerontology than the present and, just as certainly, the best is yet to come. Ten years ago, Leonard Guarente and Cynthia Kenyon expressed their growing sense of hope and optimism about the future of biogerontology in these inspiring words in an outstanding article in *Nature* (Guarente and Kenyon 2000):

> The field of ageing research has been completely transformed in the past decade.... When single genes are changed, animals that should be old stay young. In humans, these mutants would be analogous to a ninety year old who looks and feels forty-five. On this basis we begin to think of ageing as a disease that can be cured, or at least postponed.... The field of ageing is beginning to explode, because so many are so excited about the prospect of searching for – and finding – the causes of ageing, and maybe even the fountain of youth itself.

But the results that stimulated such hope in 2000 have been far surpassed in the last ten years. To give just one example, the seminal observation in the 1980s that genetic damage (the *age-1* mutation) in *C. elegans* could extend lifespan by 40–65% (Klass 1983; Johnson 1987; Friedman and Johnson 1988), which ultimately led to the modern paradigm of genetic regulation of aging, was confirmed but on a practical level dwarfed in 2007 by the discovery that stronger mutations in the same *age-1* gene can actually extend lifespan in *C. elegans* by a staggering 10-fold (Ayyadevara et al. 2008; Shmookler Reis and McEwen 2010). The finding that a genetic hierarchy controlled by one gene can in effect control 100% of the normal aging process for a time in excess of the normal maximum lifespan in at least this species is a profound development that we celebrate on the cover of the present volume as symbolic of the future potential of interventive gerontology. Nor have results with higher organisms failed to excite the interest of biogerontologists, the public, and the business community as well. The recent large investment in biogerontology by a major pharmaceutical company marks a new era of promise for intervention into human aging.

Despite all of the progress of the past decade of biogerontological research, unanimity of views on the meaning and implications of present knowledge has not yet been achieved. Different interpretations of and hypotheses about aging are still possible, and controversy continues even over the meaning of specific animal and cellular models and their eventual applicability or lack of applicability to or implications for human intervention. Accordingly, the reader will notice basic differences of opinion between some of the contributors to this book. This is as it should be, and it is hoped that some of these disagreements will be stimulating. In fact, we may be justifiably accused of deliberately stoking the fires of disagreement in some cases in the interests of promoting debate and thereby accelerating the resolution of conflicting views and stimulating new thoughts in general.

Despite some present disagreements, however, the more important truth is that modern biogerontology has developed many commonly-agreed upon broad themes that seem sufficiently reliable to justify an examination of their implications for the future of human aging, and such an examination now seems particularly timely.

Although this book is entitled "*The* Future of Aging," no one can predict *the* future in specific detail, and we do not attempt to do this here. Instead, our goal is only to provide a broader and more complete and practical sense of what the future might bring than has been available in one place heretofore.

Part One of this volume considers historical, anthropological, philosophical, ethical, evolutionary, and general evidentiary perspectives on the proposal that aging in the future may be appreciably different from aging as it is experienced today. Given often-repeated skepticism that aging either can or should be significantly modified in man, the arguments in Part One should help to establish the contrary view that aging intervention ought to be both generally feasible and desirable.

Part Two presents many specific and potentially transformative proposed interventive approaches, organized in roughly chronological order, with the earliest interventions coming first and the more distant interventions coming last. One can imagine useful interventions proceeding in a series of steps, many of which seem likely to be steps outlined in this book. The possible steps described here begin to provide a kind of tentative roadmap to the future of potential pathways to human life extension in which greater and greater control over aging is achieved over the next several decades. But it seems unlikely we will have to wait decades for major changes to become evident. In fact, in trying to decide on the sequence of chapters in Part Two, it was striking how many of these chapters, while offering powerful and paradigm-changing potential interventions, are also vying to be among the first to have their approach put into practice. And yet even as impressive as the current list of interventive approaches is, the rapid pace of ongoing developments in biogerontology ensures that the list is incomplete, a fact that underscores even more the potential that lies ahead.

It is hoped that the present glimpse into the future of human aging and the refinements to it that will follow will be of assistance to policy makers as they attempt to decide how public funds should be directed and how population aging should be addressed. Governmental projections of future human life expectancy have been consistently and significantly in error in failing to anticipate steady and even accelerating reductions in mortality rates (Horiuchi 2000; Tuljapurkar et al. 2000; Vaupel 2010), perhaps in part because of the lack of an adequately convincing argument for or any sufficiently concrete therapeutic basis for a timetable for greater future longevity. In addition, current NIH spending priorities continue to ignore the fact that learning how to successfully mitigate aging itself would be a dramatically more efficient and therefore more humane way to deal with the diseases of aging, which account for the bulk of US health care expenditures, than continuing to devote such an overwhelming share of our research resources to a piecemeal attack on these diseases. Although this point has been made many times before, perhaps the availability of a credible roadmap to a better future for aging will make the argument easier for policy makers to accept in the future.

In the end, the extent to which aging can be modified will impact all of us and our loved ones in a very personal way. This bald fact alone justifies bringing together thought leaders in the field of aging research to attempt to glimpse the future even if we are at a time in history when the full function of the human genome is not known and our present knowledge of aging remains very incomplete. Fortunately, despite the acknowledged limitations of today's knowledge, today's evidence already speaks for itself. We truly and unquestionably are learning how to intervene, in diverse ways, into aging, and it seems inevitable that this will materially affect human aging. The future of aging will begin to arrive sooner than many of us think.

We thank the editorial staff at Springer, and particularly Fabio de Castro, Marlies Vlot, and Tanja van Gaans, for their trust and for their steadfast support of our concept for this book, even in the face of some strong opposition by certain reviewers. We thank as well all of the anonymous reviewers of the concept of this book, whose comments both pro and con helped us to refine our message. We also thank Brenda Peters for her devoted, highly skillful, cheerful, and very helpful volunteer editorial assistance, and Karen Jessie of Integra India for outstanding and understanding production assistance.

Finally, we thank our many wonderful authors, who cheerfully and patiently endured an at times rather vigorous process of peer review. Their opinions, of course, remain their own!

References

Ayyadevara S, Alla R, Thaden JJ, Shmookler Reis RJ (2008) Remarkable longevity and stress resistance of nematode PI3K-null mutants. Aging Cell 7:13–22

Friedman DB, Johnson T (1988) Three mutants that extend both mean and maximum life span of the nematode, Caenorhabditis elegans, define the age-1 gene. J Gerontol 43:B102–B109

Guarente L, Kenyon C (2000) Genetic pathways that regulate ageing in model organisms. Nature 408:255–262

Horiuchi S (2000) Greater lifetime expectations. Nature 405:744–745

Johnson T (1987) Aging can be genetically dissected into component processes using long-lived lines of Caenorhabditis elegans. Proc Natl Acad Sci 84:3777–3781

Klass MR (1983) A method for the isolation of longevity mutants in the nematode Caenorhabditis elegans and initial results. Mech Ageing Develop 22:279–286

Shmookler Reis RJ, McEwan JE (2010) Methuselah's DNA: defining genes that can extend longevity. In: Fahy GM, West M, Coles LS, Harris SB (eds) The Future of Aging: Pathways to Human Life Extension. Springer, Berlin Heidelberg New York (in press)

Tuljapurkar S, Li N, Boe C (2000) A universal pattern of mortality decline in the G7 countries. Nature 405:789–792

Vaupel JW (2010) Biodemography of human ageing. Nature 464:536–542

California, USA

Gregory M. Fahy
Michael D. West
L. Stephen Coles
Steven B. Harris

Contents

Part I Introduction and Orientation

1. **Bridges to Life** 3
   Ray Kurzweil and Terry Grossman

2. **Analyzing Predictions: An Anthropological View of Anti-Aging Futures** 23
   Courtney Everts Mykytyn

3. **Towards Naturalistic Transcendence: The Value of Life and Life Extension to Persons as Conative Processes** 39
   Steven Horrobin

4. **The Ethical Basis for Using Human Embryonic Stem Cells in the Treatment of Aging** 63
   L. Stephen Coles

5. **Evolutionary Origins of Aging** 87
   Joshua Mitteldorf

6. **Precedents for the Biological Control of Aging: Experimental Postponement, Prevention, and Reversal of Aging Processes** 127
   Gregory M. Fahy

Part II The Future of Aging

7. **An Approach to Extending Human Lifespan Today** 227
   Christopher B. Heward

8. **Near Term Prospects for Ameliorating Cardiovascular Aging** ... 279
   Roger Yu, Kaveh Navab, and Mohamad Navab

9. **Near Term Prospects for Broad Spectrum Amelioration of Cancer** 307
   Zheng Cui

10. **Small Molecule Modulators of Sirtuin Activity** 331
    Francisco J. Alcaín, Robin K. Minor, José M. Villalba, and Rafael de Cabo

11	**Evolutionary Nutrigenomics** 357
	Michael R. Rose, Anthony D. Long, Laurence D. Mueller,
	Cristina L. Rizza, Kennedy C. Matsagas, Lee F. Greer, and
	Bryant Villeponteau
12	**Biological Effects of Calorie Restriction: Implications for Modification of Human Aging** 367
	Stephen R. Spindler
13	**Calibrating Notch/TGF-β Signaling for Youthful, Healthy Tissue Maintenance and Repair** 439
	Morgan Carlson and Irina M. Conboy
14	**Embryonic Stem Cells: Prospects of Regenerative Medicine for the Treatment of Human Aging** 451
	Michael D. West
15	**Maintenance and Restoration of Immune System Function** 489
	Richard Aspinall and Wayne A. Mitchell
16	**Mitochondrial Manipulation as a Treatment for Aging** 521
	Rafal Smigrodzki and Francisco R. Portell
17	**Life Extension by Tissue and Organ Replacement** 543
	Anthony Atala
18	**Telomeres and the Arithmetic of Human Longevity** 573
	Abraham Aviv and John D. Bogden
19	**Repairing Extracellular Aging and Glycation** 587
	John D. Furber
20	**Methuselah's DNA: Defining Genes That Can Extend Longevity** . 623
	Robert J. Shmookler Reis and Joan E. McEwen
21	**Reversing Age-Related DNA Damage Through Engineered DNA Repair** 641
	Clifford J. Steer and Betsy T. Kren
22	**WILT: Necessity, Feasibility, Affordability** 667
	Aubrey D.N.J. de Grey
23	**Comprehensive Nanorobotic Control of Human Morbidity and Aging** 685
	Robert A. Freitas Jr.

Appendices: Two Unusual Potential Sources of Funding for Longevity Research 807

Appendix A: SENS Foundation: Accelerating Progress Toward Biomedical Rejuvenation 809
 Michael Rae

Appendix B: The Manhattan Beach Project 827
 David Kekich

Index 843

Contributors

Francisco J. Alcaín Departamentode Biología Celular, Fisiología e Inmunología, Facultad de Ciencias, Campus Universitario de Rabanales, Edificio Severo Ochoa, Universidad de Córdoba, 14014-Córdoba, Spain, bc1altef/@uco.es

Richard Aspinall Cranfield Health, Vincent Building, Cranfield University, Cranfield, Bedfordshire, MK43 0AL, UK, r.aspinall@cranfield.ac.uk

Anthony Atala Wake Forest Institute for Regenerative Medicine, Wake Forest University School of Medicine, Medical Center Boulevard, Winston-Salem, NC 27157, USA, aatala@wfubmc.edu

Abraham Aviv The Center of Human Development and Aging, University of Medicine and Dentistry of New Jersey, New Jersey Medical School, Newark, NJ, USA, avivab@umdnj.edu

John D. Bogden Department of Preventive Medicine and Community Health, University of Medicine and Dentistry of New Jersey, New Jersey Medical School, Newark, NJ, USA, bogden@umdnj.edu

L. Stephen Coles Institute of Molecular Biology, Department of Chemistry and Biochemistry, Young Hall, University of California, Los Angeles, CA 90095-1569, USA, scoles@ucla.edu

Zheng Cui Section of Tumor Biology, Department of Pathology, Wake Forest University School of Medicine, Winston-Salem, NC 27157, USA, zheng.cui@gmail.com

Morgan Carlson Department of Bioengineering, University of California, Berkeley, CA 94720, USA, morgancarlson@gmail.com

Irina M. Conboy Irina Conboy, B108B Stanley Hall, Department of Bioengineering, UC Berkeley, Berkeley, CA 94720-1762, USA, iconboy@berkeley.edu

Rafael de Cabo Laboratory of Experimental, Gerontology, National Institute on Aging, NIH, Baltimore, MD 21224, USA, decabora@mail.nih.gov

Aubrey D.N.J. de Grey SENS Foundation, PO Box 304, El Granada, CA 94018, USA, aubrey@sens.org

Gregory M. Fahy Intervene Biomedical LLC, Box 478, Norco, CA 92860, USA, intervene@sbcglobal.net

Robert A. Freitas Jr. The Institute for Molecular Manufacturing, Palo Alto, CA, USA, rfreitas@rfreitas.com

John D. Furber Legendary Pharmaceuticals, PO Box 14200, Gainesville, FL 32604, USA, johnfurber@LegendaryPharma.com

Lee F. Greer Genescient, LLC, Irvine, CA, USA, lgreer@genescient.com

Terry Grossman Frontier Medical Institute, Denver, CO, USA, terry.grossman@comcast.net

Christopher B. Heward[†] Kronos Science Laboratory, Phoenix, AZ 85018, USA

Steven Horrobin College of Medicine, University of Edinburgh, Edinburgh, Scotland, s_horrobin@hotmail.com

David Kekich Maximum Life Foundation, Huntington Beach, CA 92648, USA, kekich@maxlife.org

Betsy T. Kren Department of Medicine, University of Minnesota Medical School, 420 Delaware Street SE, MMC 36, Minneapolis, MN, USA, krenx001@umn.edu

Ray Kurzweil Kurzweil Technologies, Inc., Wellesley, MA, USA, ray@kurzweiltech.com

Anthony D. Long Genescient, LLC, Irvine, CA, USA, along@genescient.com

Kennedy C. Matsagas Genescient, LLC, Irvine, CA, USA, kmatsagas@genescient.com

Joan E. McEwen Department of Geriatrics, University of Arkansas for Medical Sciences, Little Rock, AR 72205, USA, McEwenJoanE@uams.edu

Robin K. Minor Laboratory of Experimental Gerontology, National Institute on Aging, NIH, Baltimore, MD 21224, USA, minorro@mail.nih.gov

Wayne Mitchell Cranfield University, Cranfield Health, Vincent Building, Cranfield, Bedfordshire, MK43 0AL, UK, w.mitchell@cranfield.ac.uk

Joshua Mitteldorf Department of Ecology and Evolutionary Biology, University of Arizona, Tucson, AZ 85720, USA, josh@mathforum.org

Laurence D. Mueller Genescient, LLC, Irvine, CA, USA, lmueller@genescient.com

---

[†]Deceased

Courtney Everts Mykytyn Highland Park, 5657 Fallston Street, CA 90042, USA, cemykytyn@gmail.com

Kaveh Navab David Geffen School of Medicine, UCLA, Los Angeles, CA 90095, USA, KNavab@ucla.edu

Mohamad Navab Department of Cardiology, UCLA Medical Center, 10833 Le Conte Ave., Los Angeles, CA 90095, USA, MNavab@mednet.ucla.edu

Francisco R. Portell Gencia, Inc., Charlottesville, VA, USA, francisco@genciabiotech.com

Michael Rae SENS Foundation, 1230 Bordeaux Drive, Sunnyvale, CA 94089, USA, michael.rae@sens.org

Cristina L. Rizza Genescient, LLC, Irvine, CA, USA, crizza@genescient.com

Michael R. Rose Genescient, LLC, Irvine, CA, USA, mrose@genescient.com

Robert J. Shmookler Reis Department of Geriatrics, University of Arkansas for Medical Sciences, and Research Service, Central Arkansas Veterans Healthcare System, Little Rock, AR 72205, USA, rjsr@uams.edu

Rafal Smigrodzki Gencia, Inc., Charlottesville, VA, USA, rafal@genciabiotech.com

Stephen R. Spindler Department of Biochemistry, University of California, Riverside, CA 92521, USA, spindler@ucr.edu

Clifford J. Steer Department of Medicine, University of Minnesota Medical School, VFW Cancer Research Center, 406 Harvard Street SE, Minneapolis, MN, USA, steer001@umn.edu

José M. Villalba Departamento de Biología Celular, Fisiología e Inmunología, Facultad de Ciencias, Universidad de Córdoba, 14014- Córdoba, Spain, bc1vimoj@uco.es

Bryant Villeponteau Genescient, LLC, Irvine, CA, USA, bvilleponteau@genescient.com

Michael D. West CEO, BioTime, Inc. and Embryome Sciences, Inc., Alameda, CA, USA, mwest@biotimemail.com

Roger Yu Hospital Medicine Program, Beth Israel Deaconess Medical Center, 330 Brookline Ave, Span 2, Boston, MA 02215, USA, RogerYu.MD@gmail.com

Part I
Introduction and Orientation

Chapter 1
Bridges to Life

Ray Kurzweil and Terry Grossman

Contents

1.1 Introduction	3
1.2 Bridge One	4
1.2.1 Caloric Restriction	4
1.2.2 Exercise	7
1.2.3 Nutritional Supplementation	9
1.2.4 Benefiting from Predictive Genomics	11
1.3 General Perspectives on the Bridges to Come: Accelerating Progress in Nature and in Technology	14
1.4 Bridge Two: The Biotechnological Revolution	15
1.5 Bridge Three: Miniaturization, Nanotechnology, and Artificial Intelligence	16
1.5.1 Nanotechnology	16
1.5.2 Getting Help from Artificial Intelligence	17
1.6 Accelerating Gains in Longevity	18
1.7 Conclusions	19
References	19

1.1 Introduction

In two recent books (Kurzweil and Grossman 2004, 2009), we have previously presented an optimistic exploration of medical knowledge and tools that give readers a powerful opportunity to greatly improve their health. In a snapshot in time of today's medicine and tomorrow's potential, we showed that improving your health today will have long-lasting positive effects and make it more likely for you to benefit from the more powerful medical tools of the future.

R. Kurzweil (✉)
Kurzweil Technologies, Inc., Wellesley, MA, USA
e-mail: ray@kurzweiltech.com

Employing a "bridge" metaphor, what we call the first bridge of present-day therapies and guidance should enable many people to remain healthy long enough to take full advantage of the second bridge – the Biotechnology Revolution, which is providing us with the means to reprogram the outdated software that underlies our biology. As we learn the genetic and protein codes of our biology, we are gaining the means of turning off disease and aging while we turn on our full human potential. This second bridge, in turn, will lead to the third bridge – the Nanotechnology-AI (Artificial Intelligence) revolution. This revolution will enable us to rebuild our bodies and brains at the molecular level. In the present chapter, we briefly summarize this overview of the future of aging to provide a general context for the rest of this book.

1.2 Bridge One

Bridge one is what you can do right now to slow down aging and disease processes in order to increase your chances of remaining healthy until later bridges become manifest. We are only about 15 years from the maturation of the second bridge, and about 20–25 years from the full realization of the third bridge. It is theoretically quite feasible, therefore, even for baby boomers to remain in good health until we get to the "tipping point" in life extension. Keep in mind that bridge one is a moving frontier: our knowledge of how to stay healthy and vital is rapidly accumulating. As our knowledge increases, bridge one will essentially morph into bridge two.

The best proven method of increasing lifespan across a broad range of animal species is caloric restriction. So, let's begin by looking at modified caloric restriction as one of the cornerstones of our bridge one therapies and then have a brief look at other suggestions for heading off aging changes. By following these simple suggestions, it is our goal to help individuals live long enough and remain in good health long enough to take full advantage of the more powerful bridge two and three technologies that will lead to more profound increases in human longevity in the decades ahead.

1.2.1 Caloric Restriction

Caloric Restriction (CR) involves eating less food while still maintaining adequate nutrition. As such, CR is different from starvation or famine. CR is scientifically the best-demonstrated intervention to safely increase longevity across a broad variety of species, including C. elegans, Drosophila, and mice, and to be tested for this effect in monkeys. Compared with controls eating ad libitum, CR animals have been shown to increase their average life expectancy as well as their maximum lifespan. More than 2,000 animal studies show dramatic results across many different species (Lane et al. 2002). Consequently, we have good evidence that restricting calories slows down aging and can extend youthfulness. These experiments have not yet shown life extension in humans, but studies of humans practicing CR have shown reduction in disease and aging biomarkers similar to those seen in animal studies (Hursting et al. 2003). Free radicals cause a gradual deterioration of body tissues, particularly fragile cell membranes, or a stiffening of collagen proteins by cross-linkage, in both animals and humans, and biogerontologists attribute multiple processes associated with aging to the effects of free radical damage (Ferrari and Torres 2003). CR animals have significantly lower levels of free radicals, so they also have less free-radical damage to their cell membranes. Levels of liver enzymes that detoxify free radicals are also about 60 percent higher in low-calorie animals (Koizumi et al. 1987).

CR animals have more robust DNA-repairing enzymes. Random point mutations in DNA can lead to cancer and can accelerate other aging processes, so more effective DNA-repair enzymes would help explain the slower aging and lower rate of tumors in these animals (Licastro et al. 1988).

Of course, there is a limit to the degree to which CR can be done in an attempt to extend life. All CR animals eventually die. Because of the need to obtain adequate nutrients, it is not possible to reduce calories to one-third of normal levels and expect animals to live three times as long. It appears that, at least in the case of mammals such as rats, the optimal level of CR for longevity is about two-thirds of the calories that the animals would consume if they were eating ad libitum.

1.2.1.1 CR in Humans

A number of studies have demonstrated the potential benefit of CR for humans. For example, the people living in the Okinawa region of Japan have 40 times the number of centenarians compared with the Japanese mainland, and they experience very little serious disease before age 60 (Facchini et al. 1989; Kahn 1990). Okinawans who follow the traditional ways remain active much longer than their peers in other areas of Japan. The key difference seems to be their lower caloric intake. Extrapolating animal studies to humans, some researchers have estimated that our maximum lifespan might be extended from 120 to 180 years (Angier 1990). Yet very few of us live to 120 as it is, and others' extrapolations have been far less optimistic. The truth is that the true quantitative benefits of CR for life extension in humans are simply unknown, but we certainly know that restricting calories can improve human health and reduce many risk factors for life-limiting diseases in people. For a detailed analysis of the effects and potential benefits of CR, see Chapter 12.

The benefits of CR, whatever they may be, will apply only to your remaining life expectancy. If you are now 40 and have a remaining life expectancy of, say, 40 years, CR will only apply to that remaining period if you have not begun a CR program already. Therefore, in theory, the earlier you start CR, the greater the benefits. However, regardless of when you start, you'll achieve the health benefits of maintaining a lower weight.

One of the biggest problems of using CR for human life extension, of course, is that it isn't pleasant. But Dr. C. Ronald Kahn, Executive Director of the Joslin Diabetes Center at Harvard Medical School, and his colleagues have created a mouse model that seems to enjoy at least some of the benefits of CR without requiring CR itself. These FIRKO (Fat-specific Insulin Receptor Knock Out) mice have been genetically modified such that they lack a single gene that controls insulin's ability to allow adipocytes to store fat (Blüher et al. 2003). FIRKO mice can eat considerably more than normal mice – as much as they want – yet have 50–70 percent less body fat. They appear resistant to diabetes and remain healthier far later in life than control animals, and can live 18 percent longer. Both FIRKO mice and CR animals have significantly lower blood glucose levels (Cerami 1985), which may reduce non-enzymatic glycation reactions (for more details about the latter, see Chapter 19). Fortunately, many companies are now pursuing drugs that can block the fat insulin receptor and hopefully allow humans to gain benefits similar to those of FIRKO mice, without CR.

For now, we suggest a moderate form of CR, not as austere as the 35 percent reduction typically used in animal experiments. In Okinawa, it is customary to recite "hara hachi bu" (stomach 80 percent full) before eating. In other words, the Okinawans try to restrict calories about 20 percent, and this has translated into the longest life expectancy in the world.

We suggest the following guidelines:

Eat 20 percent fewer calories. A 140 pound sedentary woman should eat approximately 1,800 calories daily to maintain her weight. Hara hachi bu would mean eating 1,440 calories – and would lead to a weight loss of about 25 pounds (Kurzweil and Grossman 2004). A 180 pound moderately active man needs 2,700 calories to remain at the same weight. Reducing this by 20 percent would correspond to 2,160 calories per day – and gradually lead to a 35 pound weight loss.

Select foods low in caloric density. For example, choose oranges instead of orange juice. Another way to reduce calories is to eat low-starch vegetables such as broccoli and summer squash, which are filling and have relatively few calories, instead of potatoes or rice.

Focus on fiber. Another choice is foods rich in fiber, which provides bulk and texture with no digestible calories. Fiber also has health benefits by lowering cholesterol levels, improving regularity, and reducing the risks of colon cancer. Most vegetables are, of course, high in fiber. There are also many foods designed to be carbohydrate substitutes that use fiber (as well as vegetable protein) to replace the bulk and texture of starch, such as low-carbohydrate cereals and breads.

Consider calorie blockers. Another strategy for weight loss is to block the digestion of some of the food that you eat. There are some limited but promising approaches to doing this with carbohydrate and fat blockers.

Starch blockers. Precose is a prescription starch blocker taken with meals. It is mostly used in the treatment of diabetes, but delays starch absorption. There are a number of "starch blockers" available over-the-counter; however, we have had mixed results with these products in our own informal tests, which suggest that Precose may prove more effective than nonprescription starch blockers.

Fat blockers. Orlistat, available by prescription as Xenical or over-the-counter as Alli, blocks lipases. Orlistat is said to block up to one-third of fat calories. (Harrison et al. 2003). Keep in mind that key nutritional supplements are fat based (for example, EPA/DHA), so do not take fat blockers within an hour of taking oil based supplements.

More effective calorie blockers, as well as body-fat inhibitors, are in the pipeline. In the meantime, there are extensive benefits to restricting calories and maintaining a lower body weight.

A combination of moderate CR along with a diet that avoids high-glycemic-load foods (that is, foods high in simple sugars and starches) and that is generally low in fat, while emphasizing healthy fats (for example, fish oil, nuts, and the monounsaturated fat found in extra virgin olive oil) can allow you to still eat plenty of food – and therefore never be hungry.

Recent research has identified certain substances to be caloric restriction mimetics, that is, they produce at least some of the biochemical changes created by reducing calories. One CR mimetic is the prescription drug metformin, which reduces early stage insulin resistance, mimics the gene expression changes induced by CR, and extends lifespan in animals (Dhahbi et al. 2005). Another seems to be resveratrol, which is found in red wine. Mice that were fed significant amounts of resveratrol along with a high-fat diet lived as long as mice fed a normal-fat diet and were protected from the negative health consequences of the calories and fat of the high fat diet (Baur et al. 2006). You would need to supplement with about 400–700 mg a day of resveratrol to match the supplementation of these mice, but certain variations of resveratrol may be effective at much lower doses (Barger et al. 2008). For more information about the effects of resveratrol and the sirtuin pathways and their relationship to calorie restriction, see Chapter 10.

1.2.2 Exercise

Evidence from the medical literature suggests that both aerobic and resistance (strength) training exercise are associated with substantial health benefits (McGinnis and Foege 1993). All major progressive diseases such as heart disease, stroke, and type 2 diabetes are dramatically reduced with regular exercise. Our ancestors spent their lives as hunter-gatherers only a few dozen centuries ago, living a lifestyle that entailed plenty of exercise, and part of the modern epidemic of degenerative disease results from our modern-day sedentary lifestyles.

1.2.2.1 The Remarkable Benefits of Exercise

In an 8-year study published in the Journal of the American Medical Association researchers divided 13,344 participants into five fitness categories according to their exercise habits, ranging from sedentary (no regular exercise program) to high fitness (walking or running 20–30 miles per week or more). Overall death rates for the people in the moderate exercise group were 60 percent less than those of the sedentary group and there was even more benefit for the high-fitness category, particularly with respect to cardiovascular disease (Blair et al. 1989).

Despite this proven value of exercise, 72 percent of American women and 64 percent of American men do not participate in any regular physical activity. Yet it does not require a considerable expenditure of time to reap benefits. A Harvard Medical School study showed that as little as 30 minutes per day can provide significant benefit (Lee 2003).

An effective exercise program should include three forms of training: aerobic exercise, resistance training and stretching.

1.2.2.2 Aerobic Exercise

Aerobic exercise such as walking, swimming, cycling, jogging or cross-country skiing significantly lowers the risk of cardiovascular disease, cancer, and other diseases. It also provides immediate benefits such as weight loss, lower blood pressure, better sleep and mood and improved cholesterol profile (Liem et al. 2003).

An effective aerobic exercise session requires continuous, rhythmic use of the large muscles for at least 20 minutes. According to the American College of Sports Medicine (ACSM), during aerobic exercise, the goal is to keep your heart rate between 60 and 80 percent of your maximum predicted heart rate, which you can estimate as 220 minus your age [(www.acsm.org/health+fitness/pdf/fitsociety/fitsc103.pdf), p. 5].

The four phases of aerobic exercise are:

1. Warmup. Begin to warm your muscles by walking at an easy pace, about 3 miles per hour, for a few minutes.
2. Aerobic exercise. The majority of your exercise routine is the aerobic phase, in which you exercise vigorously enough to bring your heart rate into your training range. Sessions should last at least 20 minutes – and preferably 30 minutes or more – to get a training effect on your heart.
3. Cool down. Cool down by again by walking about 3 miles per hour for a few minutes. This allows your heart rate to return to normal gradually and prevents pooling of blood in your legs and feet, which can lead to lightheadedness or syncope.
4. Recovery (stretching). Muscles, tendons and ligaments tend to tighten after exercise, so it is important to lengthen these tissues by gentle stretching at the end of each session.

1.2.2.3 Resistance (Strength) Exercise

Resistance or strength training consists of exercises designed to increase strength and muscular endurance. You can use free weights, weight machines or specially designed rubber bands. For many people, rubber bands are an optimal choice with which to begin. Strength training with bands has many advantages: it is easy to learn; bands are inexpensive; they can be used at home; and they are lightweight and simple to carry with you when traveling. Free weights, on the other hand, require proper lifting technique and balance to avoid injury and to train the correct muscles. Weight machines are easier to use, but often require that you go to a gym.

The older you are, the more important it is to perform regular strength training. Sedentary individuals can lose up to 10 percent of their lean muscle mass each decade after 30 years of age, and the loss accelerates after 60 (Deschenes 2004).

The following tips should be helpful:

- Work out 2–3 times a week on non-consecutive days.
- Always exhale during the exertion phase of each exercise.
- Perform at least one set of exercises on each major muscle group.
- First work the large muscle groups, such as the chest and back, and then the smaller muscles, such as the biceps.
- Use slow, careful, controlled movements.
- Do 8–12 repetitions per set.
- Vary your program and increase either weights or thickness of bands as you progress.

1.2.2.4 Stretching

Stretching increases the range of motion in your joints. Muscles, tendons, and ligaments tend to shorten either as a result of aging or after exercise. Flexibility training or stretching helps to slow down that process. The American College of Sports Medicine (ACSM) identifies many other benefits to flexibility training, including better physical performance, better circulation, improved posture, stress relief, and enhanced coordination and balance (American College of Sports Medicine 2010). In addition, stretching may help to prevent aging of the extracellular matrix (see Chapter 19).

According to the ACSM guidelines (ACSM Fit Society Page, Spring 2002, "Enhancing your flexibility," p. 5), the following procedures should be followed.

> Warm up first to make the muscles supple and easier to stretch.
> Focus on the major muscle groups (front and back of the legs, shoulders, chest, etc.).
> Perform the stretches at least three times a week.
> Stretch muscles slowly until you feel a slight pull, not pain.
> Hold each stretch for 10–30 seconds. Don't bounce.
> Start slowly and work up.

1.2.3 Nutritional Supplementation

Many people believe that with good diet, vitamin and mineral supplementation is unnecessary. However, numerous factors contribute to a scenario in which most adults need supplementation. Current farming methods are associated with lower levels of the vitamin and mineral content in fruits and vegetables, and almost no one eats enough fresh produce to get adequate amounts of these nutrients without supplementation. Digestive function also decreases with age so that nutrient absorption and assimilation is reduced. By taking a daily multiple vitamin mix, the body can use what it needs and simply discard the rest.

In addition, some new research suggests that many people have genetic defects that can only be corrected by taking appropriate nutritional supplementation. Bruce Ames at the University of California, Berkeley has demonstrated at least 50 genetic diseases that involve defective vitamin-cofactor binding sites that can be corrected by nutritional supplementation (Ames et al. 2002). Some of these are very common and altogether affect over a billion people. As one example, methylenetetrahydrofolate reductase (MTHFR) is the enzyme that turns folic acid into its active form and helps convert homocysteine back to methionine. Up to 20 percent of people in the Caucasian and Asian populations have MTHFR polymorphisms and require much more folic acid than the RDA of 400 micrograms per day for optimal methylation. For people that carry this polymorphism, taking the RDA amount is rarely adequate to control abnormal methylation, a major cause of many serious diseases and accelerated aging (Botto and Yang 2000; Ames et al. 2002).

It is estimated that even in developed countries, much of the population consumes less than the RDA amount of one or more vitamins. In the underdeveloped world, the situation is often critical, involving multiple deficiencies. But the consequences of lacking even one nutrient can be quite serious. Deficiencies of vitamins C, B6, B12, folic acid, and the minerals iron and zinc, for example, lead to DNA damage and can cause cancer (Ames and Wakimoto 2002).

Taking nutritional supplements makes economic sense as well. A study commissioned by Wyeth Consumer Healthcare showed if all Americans over age 65 were to take a multiple vitamin daily, Medicare would save an estimated $1.6 billion over 5 years because of the resulting decrease in healthcare expenditures (Dobson et al. 2004).

Finally, silent inflammation has been increasingly found to be a factor predisposing to many serious chronic diseases. The results of a double-blind, placebo-controlled study published in the American Journal of Medicine showed that silent inflammation decreased by one-third when patients took a daily multiple vitamin (Church et al. 2003).

1.2.3.1 Basic Supplement Recommendations

The two supplements that we feel should be taken by virtually everyone over 30 are a multiple vitamin/mineral formulation and fish oil. For the former, we suggest a daily multiple vitamin formula that contains enough vitamins and minerals to meet the needs of optimal nutrition. These amounts are sometimes greater than the RDA amounts, which are primarily intended to prevent deficiency diseases like scurvy and rickets. Typically, a one-a-day daily multiple vitamin pill won't suffice, and you'll need to take a daily regimen that typically has 2–6 pills per day. This will allow each tablet or capsule to be of a comfortable size for you to swallow. It is not acceptable to take 2–6 one-a-day vitamins either. Doing so would lead to your getting excessive amounts of one vitamin or mineral and perhaps not enough of another.

In addition to eating fish several times each week, it is helpful to take supplemental fish oil, which is a rich source of the omega-3 fatty acids eicosapentanenoic acid (EPA) and docosahexaenoic acid (DHA). EPA and DHA are precursors to anti-inflammatory eicosanoids in the body that help reduce silent inflammation.

Several medical authorities have come out in support of fish oil supplementation in certain cases (Kris-Etherton et al. 2002). The American Heart Association recommends one gram of fish oil daily for patients with coronary artery disease (Kris-Etherton et al. 2002) The National Institutes of Health (NIH) considers fish oil of value in the treatment of heart patients as well, and also feels it has value in the treatment of elevated triglycerides and high blood pressure (http://www.nlm.nih.gov/medlineplus/druginfo/natural/patient-fishoil.html). There is no RDA for omega-3 fats, but NIH recommends four grams a day for healthy adults. Vegetarians can get 2.5 g of omega-3 fats from each teaspoon of flaxseed oil.

1.2.4 Benefiting from Predictive Genomics

Uncovering your personal health risks, largely hidden until now within your genes, will enable you to formulate a preventive health program that is specific to you (see also Chapter 7, by Heward). The current consensus is that there are in the vicinity of 23 thousand human genes. Your ability to obtain solid information about these has just begun, but it will increase exponentially over the next few years. Modern genomic analysis can now accomplish in minutes what once took weeks of testing. As a result, the one-size-fits-all type of medicine that physicians have been practicing will soon be replaced by individualized therapies. This will enable you to create the perfect bridge one program for yourself: a diet and exercise plan as well as nutritional supplements and prescription drugs personalized for your genome.

Predictive genomics remains a relatively new bridge one diagnostic tool, but there are already a number of commercially available genomics tests to help predict your predisposition to many serious, but preventable or modifiable, diseases, such as heart disease, Alzheimer's, or cancer. When doing genomics testing, it is important to remember that, for most genes tested, your genes merely express tendencies. Your lifestyle choices have a much larger role in determining what happens, or how your genes are expressed.

Genomics tells you your tendencies, and therapies that will be able to alter these tendencies are still in their infancy, which is another reason it is important for you to remain as healthy as possible for the next decade or two, until these new treatments are more fully evolved. These bridge one therapies and lifestyle choices will help you avoid or significantly delay irreversible physiologic changes (heart attacks, strokes, and dementia). Then you will be able to take fuller advantage of the powerful bridge two gene-based therapies. Soon after that, you'll benefit from the bridge three Nanotechnology-AI Revolution, which will mean that damage to your body that is presently irreversible will, in principle, be correctable.

1.2.4.1 Evaluating SNPs

It is estimated that each person may carry as many as one million single nucleotide polymorphisms (SNPs) [A. Braun, quoted in (Duncan 2002)]. One aspect of predictive genomics attempts to identify the most significant SNPs to determine how likely you are to be predisposed to develop a specific disease or health risk, and to evaluate the possibility that this condition might appear under particular environmental circumstances or lifestyle choices (Ames et al. 2002). Genomics testing panels are now commercially available for up to three dozen or so SNPs at a cost of less than $30 per gene, while only a few years ago, it was rare to get a genetic test done for less than $300 each. Testing is currently offered by 23andMe, Navigenics, Decode Genetics, and SeqWright. Within a few years, in accordance with Ray's Law of Accelerating Returns (Kurzweil 2005), for the same thousand dollars, you will be able to get a panel that reads your entire genome. By the end of the decade, you will probably have access to DNA chips that will test for most, if not all, of the 3.6 million SNPs currently identified. In the spring of 2007, Nobel Laureate James Watson, 79, had his entire genome sequenced by Life Sciences, Inc. of Connecticut in collaboration with the Baylor College of Medicine in Houston. The total cost of this project was more than $1 million (Davies 2007). The results of his genome were published in Nature in April of 2008 (Wheeler et al. 2008).

A SNP that can be beneficial to a person in one circumstance can be harmful under different conditions. For example, consider the "thrifty gene." This is an SNP that helps people survive on minimal calories and throughout history has given individuals a better chance of survival during periods of famine or near starvation. Today, however, this gene creates more problems than benefit, since carriers are more predisposed to obesity under conditions of excess or even merely adequate calories. Years ago, when famine was a regular occurrence, Pima Indians from the Southwestern United States, who carry copies of this gene, were better able to survive long periods of near starvation. Since this gene conferred a survival advantage, it became more prevalent in this population. But this has led to the majority of modern-day Pima Indians being overweight and having a high rate of diabetes, so it is now a disadvantage (http://diabetes.niddk.nih.gov/dm/pubs/pima/obesity/obesity.htm).

Almost all of the most common, disabling, and deadly degenerative diseases of our time, including cardiovascular disease, cancer, Type-2 Diabetes, and Alzheimer's disease, are the result of the interaction between genetic and environmental factors. Tests for some of these genes are now available. Genomics testing allows you to gain a deeper understanding of your individualized health risks and develop more specific and effective interventions.

1.2.4.2 Problems with Genomic Testing

Even if you could know all of the hundreds of thousands of SNPs you possess, like James Watson (or Craig Venter, who also had it done), the information overload would be a major problem. No one knows what to do with most of the information.

We need to wait for the bioinformatics scientists to catch up and provide us with sophisticated computer programs that can make sense of the enormous amount of data. Today's clinicians who have started to use genomics testing, therefore, limit screening to just a few of the more common polymorphisms.

In addition, not many patients would want to know that they have a genetic problem that can't be treated by any presently available therapy. For example, many women whose mothers have been diagnosed with breast cancer frequently delay testing for the BRCA1 gene, since a positive test suggests a high likelihood of developing breast or ovarian cancer themselves (Lancaster 1997). Most wait until they have had children to do the testing, since the only treatment for a positive test is prophylactic bilateral mastectomy and possible oophorectomy. As a result, most present-day testing focuses on the hundred or so SNPs that can currently be modified through interventions such as diet, lifestyle, nutritional supplements, or prescription pharmaceuticals.

1.2.4.3 The Complex Benefits of Genomic Testing: Apo E

The Apo E (Apolipoprotein E) polymorphisms are genetic markers associated with varying degrees of risk for cardiovascular disease and Alzheimer's disease and provide a good example of the benefits of genomic testing. We will first examine the specific risks and benefits associated with the different Apo E polymorphisms, and then describe how this information can prompt specific lifestyle recommendations, which can help an individual modify the real-life risks of the more dangerous genetic variants.

Apolipoprotein E comes in three alleles: E2, E3 and E4. Because of very minor differences in just one or two of their amino acids, they each differ significantly in their ability to carry fat and cholesterol in the bloodstream. Apo E2, for instance, is quite effective at clearing cholesterol from the arteries, while Apo E4 is much less efficient (Li et al. 2003).

If you possess one or two copies of the E4 allele, you may have an increased chance of having elevated cholesterol, triglycerides, and coronary heart disease (Eto et al. 1988; Singh et al. 2002). Even more important, Apo E4 is also correlated with a significantly increased risk of Alzheimer's disease (AD). If you do not have any copies of Apo E4, you have a 9 percent risk of developing AD by age 85. If you have just one copy of the gene as the E3/E4 genotype, which is carried by more than 25 percent of the population – you have a 27 percent chance of developing AD by the same age; in other words, triple the risk. But if you have two copies – the E4/E4 genotype – the risk rises to 55 percent, a sixfold increase (Kamboh 1995; Myers et al. 1996; Farrer et al. 1997). Furthermore, the average age that AD is diagnosed is much younger, depending on the number of copies of Apo E4 carried: 84 years old if you have no copies of E4, 75 years if one copy, and around 68 with two copies.

The Apo E2 gene, on the other hand, appears to confer some degree of protection against the development of Alzheimer's, and patients with at least one copy of E2 have a 40–50 percent reduction in their AD risk at least through age 80 (Kardaun et al. 2000; Sando et al. 2008). Apo E2 is not perfect, however, because some forms of heart disease are more common in patients with this allele.

The Apo E3 form is the most common – more than 60 percent of the population is E3/E3. This allele affords some protection against both heart disease and AD.

Although the Apo E4 is a significant risk factor for Alzheimer's and may be associated with other forms of dementia, most people who carry the Apo E4 gene still do not develop dementia, and about one-half of patients diagnosed with AD do not possess any copies (Slooter et al. 1998). In some studies, the proportion of patients with dementia that is attributable to the Apo E4 allele is estimated to be about 20 percent.

But if there were no benefits associated with Apo E4, it probably would have been selected out of the gene pool long ago. People who carry this gene have a much lower incidence of some serious diseases, such as age-related macular degeneration (AMD), the leading cause of blindness in the developed world (Hamdi and Keney 2003). Meanwhile, people who carry the more favorable E2 are at much higher risk of losing their vision to AMD (Friedman et al. 2007).

Free-radical damage appears to play a key role in developing the specific type of damage seen in AD (Retz et al. 1998). Consequently, when we find that an individual possesses the Apo E4 gene, we feel special efforts to limit free-radical damage should be implemented. Apo E4 carriers should begin taking aggressive free-radical damage-control measures as early in life as possible.

Predictive genomics testing is available today that can provide previously unknowable genetic information personalized to each individual. This new medical specialty is in its infancy, and, as with any new science, there are perils and pitfalls. But today's primitive, incompletely understood tests will lead rapidly to ever more sophisticated analyses. Today's limited view of the genome will ultimately lead to significant breakthroughs in personalized medicine in the years ahead.

1.3 General Perspectives on the Bridges to Come: Accelerating Progress in Nature and in Technology

The coming of predictive genomics is just one indication that information technology is increasingly encompassing everything of value. It's not just computers and electronic gadgets. It now includes the field of biology. We're beginning to understand how life processes such as maturation, disease, and aging are manifested as information processes and gaining the tools to actually manipulate and change those processes.

The acceleration of information processes is inherent in evolution, which is essentially a process of creating more sophisticated information structures. Evolutionary processes work through indirection. Evolution creates a capability, and then it uses that capability to evolve the next stage. That's why the next stage goes more quickly, and that's why the fruits of an evolutionary process grow exponentially. Today we can see this effect in the form of a doubling of capability and capacity of our information technologies such as computers and communication networks every year.

But we can also see this process going back to the dawn of biological evolution. The first paradigm shift in biological evolution, the evolution of cells, and in particular DNA (actually, RNA came first) – the evolution of an information processing backbone that would allow evolution to record the results of its experiments – took a billion years. Once DNA and RNA were in place, the next stage, the Cambrian explosion, when all the body plans of the animals evolved, went a hundred times faster. Then those body plans were used by evolution to concentrate on higher cognitive functions, which evolved in only millions of years. Biological evolution kept accelerating in this manner. Homo sapiens, our species, evolved in only a few hundred thousand years, the blink of an eye in biological evolutionary terms.

Then, again working through indirection, biological evolution used one of its creations, the first technology-creating species, to usher in the next stage of evolution, which was technology. The enabling factors for technology were a higher cognitive function with an opposable appendage, so we could manipulate and change the environment to reflect our models of what could be. The first stages of technology evolution – fire, the wheel, stone tools – only took a few tens of thousands of years.

Technological evolution has continued to accelerate. Half a millennium ago the printing press took a century to be adopted, half a century ago the first computers were designed pen on paper and were wired with screw drivers. Now computers are designed in only a few weeks' time by computer designers sitting at computers, using advanced computer-assisted design software. The computer in your cell phone today is a million times smaller and a million times cheaper than the computers that existed when the authors of this chapter started college 40 years ago, yet they are a thousand times more powerful. That's a billion-fold increase in price-performance in about 40 years, which we will do again over the next 25 years since the rate of exponential growth is itself growing faster.

These trends are rapidly bringing us closer to the advent of bridge two and bridge three.

1.4 Bridge Two: The Biotechnological Revolution

One of the profound implications of accelerating progress stems from the fact that we are coming to understand our biology as a set of information processes. We have on the order of 23,000 little software programs inside us called genes. These evolved in a different era. Thousands of years ago, it was not in the interest of our species to live past child-rearing, as food and other resources were very scarce, and life expectancy was in the twenties.

As an example of the outdated "software" that runs in our bodies, one of those programs, called the fat insulin receptor gene, says, basically, "hold onto every calorie because the next hunting season might not work out so well." Now that we live in an era of abundance, that gene underlies an epidemic of obesity, and we'd like to change that program.

We now have new technologies that for the first time allow us to reprogram the outdated software that underlies our biology. Doing this is the essence of "bridge two." RNA interference, where we put fragments of RNA inside the cell as a drug, can effectively inhibit selected genes. RNA interference is capable of effectively turning genes off by blocking the messenger RNA expressing that gene (see also Chapter 21).

When the fat insulin receptor was turned off in mice at the Joslin Diabetes Center, the mice ate ravenously yet remained slim. They didn't get diabetes or heart disease and they lived 20 percent longer: they got the benefit of caloric restriction without the restriction, and there are several pharmaceutical companies that are developing fat insulin receptor (FIR) gene inhibitors for the human market. The FIR gene is just one of the genes we'd like to turn off, and there are over a thousand drugs in the development and testing pipeline using RNA interference.

We also have effective new means of adding new genes, not just to a baby but also to a mature individual. A company that the authors advise (United Therapeutics) has a process in which it takes cells out of the body (lung cells in this case), adds a gene in-vitro, inspects that the gene got added correctly, then multiplies the cell with the new gene a million fold (using another new technology) and then injects these cells back into the body. These new cells migrate back to the lungs, and this technique has cured a fatal disease (pulmonary hypertension) in animal models and is now undergoing human trials. There are hundreds of such developments now in the pipeline.

The new paradigm of rational drug design involves a deeper understanding of the information processes underlying biology, the exact sequence of steps that leads up to a process like atherosclerosis, or cancer, or insulin resistance, and then provides very precise tools to intervene. After we design these drugs using computer-assisted design tools, we can test them using biological simulators, which are also scaling up in accuracy and scope at an exponential pace.

1.5 Bridge Three: Miniaturization, Nanotechnology, and Artificial Intelligence

1.5.1 Nanotechnology

Another exponential process is miniaturization, which will lead inexorably to nanotechnology, our "third bridge" to maximum human life extension. We've begun to demonstrate the feasibility of actually constructing devices at the molecular level that can perform useful functions. One of the biggest applications of this, again, will be in biology, where we will be able to go inside the human body and extend its range beyond the current limitations of biology.

Rob Freitas has designed a nanorobotic red blood cell, which is a relatively simple device for storing and releasing oxygen and carbon dioxide. A conservative analysis of these robotic "respirocytes" shows that if you were to replace ten percent of your red blood cells with these robotic versions, you could do an Olympic sprint for 15 minutes without taking a breath, or sit at the bottom of your pool for four hours. A robotic white blood cell has also been designed. Somewhat more complicated, it would download software to combat specific pathogens (see Chapter 23 in the present book for details and references for both of these devices).

If that sounds futuristic, we can point out that we already have at least a first generation of experimental blood cell-size devices, working to perform therapeutic functions in animals. For example, one scientist cured type I diabetes in rats with this type of nanoengineered device (Desai 2003).

Another scientist has designed millimeter-sized reservoirs, carved from silicon, that store drugs in implantable devices for smart drug delivery, with programmable release of drugs using microprocessors or wireless control (Prescott et al. 2006). And scientists at Swiss Federal Institute of Technology are developing microrobots that can perform eye surgery, controlling them with an external magnetic field (Yeşin et al. 2006). An advanced neural implant, in the research stage, is a microchip-based artificial hippocampus. It could ultimately be used to replace damaged brain tissue destroyed in an accident, during a stroke, or by neurodegenerative conditions such as Alzheimer's disease (Berger et al. 2005).

1.5.2 Getting Help from Artificial Intelligence

When will we have the software or methods of human intelligence? To achieve the algorithms of human intelligence, a grand project to reverse-engineer the brain is underway, and not surprisingly, is making exponential progress. The spatial resolution of in-vivo brain scanning in 3D volume is doubling every 1.1 years (Trajtenberg 1990; Robitaille et al. 2000; Reiser et al. 2008; Lee et al. 2005).

We're also showing that we can convert all of this data into working models and simulations of brain regions. There are already working simulations of regions of the auditory cortex (Watts 2003), visual cortex (Riesenhuber and Poggio 2002; Cadieu et al. 2007), cerebellum (Medina et al. 2000), and columns in the cerebral cortex (Markram 2006). These simulations have been tested to compare their capabilities to human performance, and they compare well. If you follow the trends in reverse brain engineering, it's a reasonable conclusion that we will have reverse-engineered all of the several hundred regions of the brain by the late 2020s.

By early in the next decade, computers won't look like today's notebooks and personal digital assistants (PDAs). They will disappear, integrated into our clothing and environment. Images will be written to our retinas from our eyeglasses and contact lenses providing full-immersion virtual reality. We'll be interacting with virtual personalities; we can see early harbingers of this already. The broad trend is to become closer to our technology. Today we can access much of human knowledge using devices we carry in our pockets. Soon these computers will move into our clothing, and then into our bodies and brains.

If we go out to 2029, there will be many doublings of capability in terms of this exponential progression of information technology. There will be about thirty doublings in the next 25 years. That's a factor of a billion in capacity and price performance over today's technology, which is already quite formidable. At the same time, we are shrinking technology by a factor of 100 in 3-dimensional volume per decade, so that's about a 100,000 fold shrinkage in size of key features over the next quarter century.

By 2029, we will have completed the reverse engineering of the brain, we will understand how human intelligence works, and that will give us new insight into ourselves. Non-biological intelligence will then combine the suppleness and subtlety of our pattern-recognition capabilities with ways in which computers have already demonstrated their superiority. Every time you use Google you can see one aspect of the power of non-biological intelligence. Machines can remember things very accurately. They can share their knowledge instantly. We can share our knowledge, too, but only at the slow bandwidth of language. As nonbiological intelligence masters the subtle and supple powers of biological intelligence, it will provide a powerful combination.

This will not be an alien invasion of intelligent machines coming from over the horizon to compete with us. It's emerging from within our civilization and will extend the power of our civilization. Even today we routinely do intellectual feats that would be impossible without our technology. In fact, our whole economic infrastructure couldn't manage without the intelligent software that's underlying it. In essence, this is what our technology has always provided: an extension of our physical and mental reach.

1.6 Accelerating Gains in Longevity

Every time we have technological gains, we make gains in life expectancy. Sanitation was a big one, antibiotics was another. These two revolutions occurred in the last two centuries; before that life expectancy was only 37. We're now in the beginning phases of the biotechnology revolution. We're exploring, understanding, modeling, simulating, and redesigning the information processes underlying biology.

Health and medicine have been hit or miss up until recently, so gains in life expectancy were linear: in recent decades, we've been adding three months every year. But with the reprogramming of genes and the simulation of biology, biology has now become an information technology and is therefore subject to "the law of accelerating returns" (Kurzweil 2005), which articulates a double exponential rate of growth in information technology. As a result, we will make exponential gains in life expectancy very soon.

These technologies will be a thousand times more powerful than they are today in a decade and a million times more powerful in two decades. According to our models, within 15 years from now, we'll be adding more than a year every year not just to infant life expectancy, but to your remaining life expectancy. That's not a guarantee of immortality, but it will change the metaphor of the sands of time running out. So, if you watch your health today, the old-fashioned way, for a little while longer, you may have a good chance of actually living to see the remarkable 22nd century.

1.7 Conclusions

We have described above a few of the steps you can take to slow down disease and aging processes so that even baby boomers can remain in good health until the maturation of the biotechnology revolution ("bridge two") about 15 years from now. Bridge one is your own personal bridge as it needs to be based on your personal genomics and health status based on tests such as lipid and other levels, weight, and other personal assessments. You don't catch diseases such as heart disease, cancer, and type II diabetes walking down the street – these conditions are the end result of processes that are decades in the making. You can find out where you are in the development of these diseases and step back from the "edge of the cliff" before you fall off.

Bridge two – now under construction – involves modeling and simulating biological processes on computers, designing highly targeted interventions and then testing them in silico. It also involves new methods to reprogram the information processes underlying biology such as RNA interference (to turn genes off), new in vitro methods of gene therapy (to add new genes to a mature individual) and methods to turn on and off proteins and enzymes. We will soon no longer be stuck with the outdated software in our bodies that evolved thousands of years ago when conditions were very different. And then will come Bridge three.

Now that biology and health has become an information technology, it has become subject to "the law of accelerating returns," which is a doubling of capability each year (Kurzweil 2005). So these technologies will be a million times more powerful in 20 years, and that will usher in a dramatic new era in extending human life and potential.

References

American College of Sports Medicine, ed. (2010) ACSM's Resource manual for Guidelines for Exercise Testing and Prescription, Lippincott, Williams and Wilkins: Philadelphia, pp. 346–347, 513–514

Ames BN, Elson-Schwab I, Silver EA (2002) High-dose vitamin therapy stimulates variant enzymes with decreased coenzyme binding affinity (increased K_m): relevance to genetic disease and polymorphisms. Am J Clin Nutr 75:616–658

Ames BN, Wakimoto P (2002) Are vitamin and mineral deficiencies a major cancer risk? Nat Rev Cancer 2:694–704

Angier N. Diet offers tantalizing clues to a long life. New York Times, April 17, 1990, sec. C

Barger JL, Kayo T, Pugh TD, Prolla TA, Weindruch R (2008) Short-term consumption of a resveratrol-containing nutraceutical mixture mimics gene expression of long-term caloric restriction in mouse heart. Exp Gerontol 43:859–866

Baur JA, Pearson KJ, Price, NL, Jamieson HA, Lerin C, Kalra A, Prabhu VV, Allard JS, Lopez-Lluch G, Lewis K, Pistell PJ, Poosala S, Becker KG, Boss O, Gwinn D, Wang M, Ramaswamy S, Fishbein KW, Spencer RG, Lakatta EG, Couteur D, Shaw RJ, Navas P, Puigserver P, Ingram DK, de Cabo R, Sinclair, DA (2006) Resveratrol improves health and survival of mice on a high-calorie diet. Nature 444:337–342

Berger TW, Ahuja A, Courellis SH, Deadwyler SA, Erinjippurath G, Gerhardt GA, Gholmieh G, Granacki JJ, Hampson R, Hasio MC, Lacoss J, Marmarelis VZ, Nasiatka P, Srinivasan V, Song D, Tanguay AR, Wills J (2005) Restoring lost cognitive function: hippocampal-cortical neural prostheses. IEEE Eng Med Biol 24:30–44

Blair SN, Kohl HW 3rd, Barlow CE, Gibbons LW (1989) Physical fitness and all-cause mortality. A prospective study of healthy men and women. JAMA 262:2395–2401

Blüher M, Kahn BB, Kahn CR (2003) Extended longevity in mice lacking the insulin receptor in adipose tissue. Science 299:572–574

Botto LD, Yang Q (2000) 5,10-Methylenetetrahydrofolate reductase gene variants and congenital anomalies: a HuGE review. Am J Epidemiol 151:862–877

Cadieu C, Kouh M, Pasupathy A, Connor C, Riesenhuber M, Poggio T (2007) A model of V4 shape selectivity and invariance. J Neurophysiol 98:1733–1750

Cerami A (1985) Hypothesis: glucose as a mediator of aging. J Am Geriatr Soc 33:626–634

Church TS, Earnest CP, Wood KA, Kampert JB (2003) Reduction of C-reactive protein levels through use of a multivitamin. Am J Med 15(115):702–707

Davies K (2007) Computational Biology: Project Jim: Watson's Genome Goes Public: 454 Baylor Scientists Present Jim Watson with His Personal Genome, Flaws, and All. BioIT-World, 6:5,16

Desai TA (2003) MEMS-based technologies for cellular encapsulation: focus on nanopore biocapsules for diabetes mellitus therapy. Am J Drug Deliv 1:3–11

Deschenes MR (2004) Effects of aging on muscle fibre type and size. Sports Med 34:809–824

Dhahbi JM, Mote PL, Fahy GM, Spindler SR (2005) Identification of potential calorie restriction mimetics by microarray profiling. Physiol Genomics 23:343–350

Dobson A, DaVanzo J, Consunji M, Gilani J, McMahon P, Sen N, Preston B (2004) A Study of the Cost Effects of Daily Multivitamins for Older Adults. The Lewin Group: Falls Church.

Duncan, DE (2002) DNA as destiny. Wired 10:183

Eto M, Watanabe K, Chonan N, Ishii K (1988) Familial hypercholesterolemia and apolipoprotein E4. Atherosclerosis. 72:123–128

Facchini A, Haaijman JJ, Labo G (eds) (1986) Immunoregulation in Aging. EURAGE: Rijswijk 53–61

Farrer LA, Cupples LA, Haines JL, Hyman B, Kukull WA, Mayeux R, Myers RH, Pericak-Vance MA, Risch N, van Duijn CM (1997) Effects of age, sex, and ethnicity on the association between apolipoprotein E genotype and Alzheimer's disease. A meta-analysis. APOE and Alzheimer Disease Meta Analysis Consortium. JAMA 278:1349–1356

Ferrari CK, Torres EA (2003) Biochemical pharmacology of functional foods and prevention of chronic diseases of aging. Biomed Pharmacother 57:251–260

Friedman DA, Lukiw WJ, Hill JM (2007) Apolipoprotein E ε4 Offers Protection Against Age-Related Macular Degeneration. Med Hypotheses 68(5):1047–1055

Hamdi HK, Keney C (2003) Age-Related Macular Degeneration: A New Viewpoint. Frontiers in Bioscience 8:e305–e314

Harrison SA, Ramrakhiani S, Brunt EM, Anbari MA, Cortese C, Bacon BR (2003) Orlistat in the treatment of NASH: a case series. Am J Gastroenterol 98:926–930

Hursting SD, Lavigne JA, Berrigan D, Perkins SN, Barrett JC (2003) Calorie restriction, aging, and cancer prevention: mechanisms of action and applicability to humans. Ann Rev Med 54:131–152

Kahn C (1990) His theory is simple: eat less, live longer. A lot longer. Longevity 2 [Oct]:60–62, 64, 66

Kamboh R (1995) Apolipoprotein E polymorphism and susceptibility to Alzheimer's disease. Hum Biol 67:195–215

Kardaun JW, White L, Resnick HE, Petrovitch H, Marcovina SM, Saunders AM, Foley DJ, Havlik RJ (2000) Genotypes and phenotypes for apolipoprotein E and Alzheimer disease in the Honolulu-Asia aging study. Clin Chem 46:1548–1554

Koizumi A, Weindruch R, Walford RL (1987) Influences of dietary restriction and age on liver enzyme activities and lipid peroxidation in mice. J Nutr 117:361–367

Kris-Etherton PM, Harris WS, Appel LJ for the Nutrition Committee (2002) Fish consumption, fish oil, omega-3 fatty acids, and cardiovascular disease. Circulation 106:2747–2757

Kurzweil R (2005) The Singularity is Near: When Humans Transcend Biology. Viking, New York, pp. 491–496

Kurzweil R, Grossman T (2004) Fantastic Voyage: Live Long Enough To Live Forever. Rodale, New York

Kurzweil R, Grossman T (2009) Transcend: Nine Steps to Living Well Forever. Rodale, New York

Lancaster JM (1997) BRCA 1 and 2—A Genetic Link to Familial Breast and Ovarian Cancer. Medscape Women's Health 2: 7

Lane MA, Mattison J, Ingram DK, Roth GS (2002) Caloric restriction and aging in primates: relevance to humans and possible CR mimetics. Microsc Res Tech 59:335–338

Lee IM (2003) Physical activity in women: how much is good enough? JAMA 290:1377–1378

Lee JKT, Sagel SS, Stanley RJ, Heiken JP (2005) Computed Body Tomography with MRI Correlation, 2nd Ed. Lippincott Williams & Wilkins, New York

Li X, Du Y, Huang X (2003) Association of apoliproprotein E gene polymorphism with essential hypertension and its complications. Clin Exp Med 2:175–179

Licastro F, Weindruch R, Davis LJ, Walford RL (1988) Effect of dietary restriction upon the age-associated decline of lymphocyte DNA repair activity in mice. AGE 11:48–52

Liem AH, Jukema JW, van Veldhuisen DJ (2003) Secondary prevention in coronary heart disease patients with low HDL: what options do we have? Int J Cardiol 90:15–21

Markram H (2006) The blue brain project. Nat Rev Neurosci 7:153–160

McGinnis JM, Foege WH (1993) Actual causes of death in the United States. JAMA 270:207–212

Medina JF, Garcia KS, Nores WL, Taylor NM, Mauk MD (2000) Timing mechanisms in the cerebellum: testing predictions of a large-scale computer simulation. J Neurosci 20:5516–5525

Myers RH, Schaefer EJ, Wilson PW, D'Agostino R, Ordovas JM, Espino A, Au R, White RF, Knoefel JE, Cobb JL, McNulty KA, Beiser A, Wolf PA (1996) Apolipoprotein E epsilon4 association with dementia in a population-based study: The Framingham study. Neurology 46:673–677

Prescott JH, Lipka S, Baldwin S, Sheppard NF Jr, Maloney JM, Coppeta J, Yomtov B, Staples MA, Santini JT Jr (2006) Chronic, programmed polypeptide delivery from an implanted, multireservoir microchip device. Nat Biotechnol 24:437–438

Reiser MF, Becker CR, Nikolaou K, Glazer G (2008) Multislice CT. Springer, Berlin, Heidelberg, New York

Retz W, Gsell W, Munch G, Rosler M, Riederer P (1998) Free radicals in Alzheimer's disease. J Neural Transm Suppl 54:221–236

Riesenhuber M, Poggio T (2002) Neural mechanisms of object recognition. Curr Opin Neurobiol 12:162–168

Robitaille P-MLAM, Abduljalil AM, Kangarlu A (2000) Ultra high resolution imaging of the human head at 8 Tesla: 2 K × 2 K for Y2K. J Comp Assist Tomogr 24:2–8

Sando SB, Melquist S, Cannon A, Hutton ML, Sletvold O, Saltvedt I, White LR, Lydersen S, Aasly JO (2008) APOE epsilon 4 lowers age at onset and is a high risk factor for Alzheimer's disease; a case control study from central Norway. BMC Neurol Apr 16;8:9

Singh PP, Singh M, Mastana SS (2002) Genetic variation of apolipoproteins in North Indians. Hum Biol 74:673–682

Slooter AJ, Cruts M, Kalmijn S, Hofman A, Breteler MM, Van Broeckhoven C, van Duijn CM (1998) Risk estimates of dementia by apolipoprotein E genotypes from a population-based incidence study: the Rotterdam Study. Arch Neurol 55: 964–968

Trajtenberg M (1990) Economic Analysis of Product Innovation: The Case of CT Scanners. Harvard University Press, Cambridge

Watts L (2003) Visualizing complexity in the brain. In: Fogel D, Robinson C, (eds) Computational Intelligence: The Experts Speak. IEEE Press/Wiley, Hoboken, pp. 45–56; http://www.lloydwatts.com/wcci.pdf

Wheeler DA, Srinivasan M, Egholm M, Shen Y, Chen L, McGuire A, He W, Chen YJ, Makhijani V, Roth GT, Gomes X, Tartaro K, Niazi F, Turcotte CL, Irzyk GP, Lupski JR, Chinault C, Song XZ, Liu Y, Yuan Y, Nazareth L, Qin X, Muzny DM, Margulies M, Weinstock GM, Gibbs RA, Rothberg JM (2008) The complete genome of an individual by massively parallel DNA sequencing. Nature 452:872–876

Yeşin KB, Vollmers K, Nelson BJ (2006) Modeling and control of untethered biomicrorobots in a fluidic environment using electromagnetic fields. Int J Robotics Res 25:527–536

Chapter 2
Analyzing Predictions: An Anthropological View of Anti-Aging Futures

Courtney Everts Mykytyn

Contents

2.1	Introduction	23
2.2	Understanding Predictions	24
	2.2.1 The Functions of Predictions	24
	2.2.2 The Requirements for Successful Predictions	25
2.3	A Context: A Brief Outline of the "Anti-Aging" Field	26
2.4	Historical Trends as a Basis for Prediction	28
2.5	Moral Obligation as a Basis for Prediction	31
2.6	Conclusion	34
References		35

2.1 Introduction

Some scholars argue that contemporary, Western, science-based societies are more oriented to the future than ever before (Giddens 1998). The burgeoning anti-aging field provides an excellent example of this assertion. Anti-aging medicine is in large part a future field because it is built upon predictions that, in different ways, assert that aging itself *can* and *should* be a target for biomedical intervention. While there is no scientific consensus that a therapy to retard or reverse aging does now, or will ever, exist, hopes and expectations for anti-aging thrive in medical, scientific and lay communities. The hope for this future is driving its pursuit: marshaling resources, fostering interest, and challenging understandings of aging, nature, and the goals of biomedicine. Thus, predictions speak not only of the future but also affect what we do and work toward today. More than rhetoric, these predictions organize the present in light of future promises.

C.E. Mykytyn (✉)
Highland Park, 5657 Fallston Street, CA 90042, USA
e-mail: cemykytyn@gmail.com

I have studied anti-aging medicine from an anthropological perspective since 1999: interviewing anti-aging practitioners, observing in clinics, attending conferences, and reading the vast array of popular and academic publications on the topic. My research is a cultural analysis of how anti-aging medicine has emerged and the ways in which aging is understood and dealt with by the medical and scientific communities. Throughout the course of this research, it has become clear that the future, or, rather, a sustainably compelling and solicitous vision of the future, is a point of contention. It is also a powerful and highly consequential vision that affects not only the future but also the "present" of aging. Thus, I argue that much of the work done in and around anti-aging has happened at the juncture between the present and the future.

My aim here is to provide an anthropological perspective on the predictions that have come to characterize anti-aging. I do not endeavor to criticize nor advocate for the science of anti-aging. I will not attempt to evaluate the efficacy or scientific merit of various claims, leaving this task to the much more qualified scientific contributors to this volume. Rather, I attempt to contextualize the futures emerging in this field and explicitly in this volume within a broader, social science understanding of predictions as themselves objects that can be analyzed (Brown and Michael 2003).

2.2 Understanding Predictions

2.2.1 The Functions of Predictions

Predictions for a particular technique or field do much of the heavy lifting of science politics and practice. In a material sense, predictions shape what we do today – mobilizing professional roles, marshaling resources, and creating networks of people joined in the imaginary (Borup et al. 2006; Fujimura 2003). They provide a powerful mechanism for prioritizing research agendas. They supply a goal to work toward and in so doing lodge a critique of the status quo. It is clear that predictions rely upon particular histories (Fortun and Fortun 2005; Brown 2003; Marcus 1995; Rescher 1998) to provide a foundation of feasibility and that predictions must come packaged within a moral argument (Blumenfeld 1999; Fischer 2005; Fortun and Fortun 2005; Franklin 2003; Rabinow 1996). Thus, predictions are cultural products as they are always grounded in what people do and how these acts are conceptualized.

The social science literature on predictions discusses at length how the "good" that predictions tender is often undermined by their failure to materialize (Brown 2003; Brown and Michael 2003). In Brown's model of early research agendas, predictions are significant in securing interest yet are often tremendously exaggerated (see also Borup et al. 2006; Franklin and Lock 2003; Fujimura 2003; Geels and Smit 2000; Hogle 2003; Konrad 2006). Exaggeration may indeed be hyperbole, though the deliberateness of any overstated claims may be difficult to assess. Nonetheless, the disillusionment that failed pursuits engender can be profound in terms of the perceived legitimacy and stability of a given field and in the lives of individuals who had high hopes dashed. Perhaps more significant, however, is the funneling of resources toward a particular set of predictions that fail to come to fruition rather than toward other avenues of scientific research.

2.2.2 The Requirements for Successful Predictions

An influential prediction – and by this, I mean a prediction that has staying power, or cultural purchase – must be grounded in feasibility. This feasibility factor often emerges as an extension of a historical trajectory: in the past we have done *this* and so *this* is likely to continue *this* way into the future. For example, the history of technological progress is a well-worn trope often employed to substantiate a given prediction by suggesting that the prediction is likely to come to fruition premised on the fact that many other phenomenal advances have happened already.

Moreover, for a prediction to have an aura of feasibility, it must also outline a roadmap for success, a sense of what must be undertaken to achieve the end goal. The roadmap and the particularly drawn history both serve to counter any science-fiction "feel" of a future prediction by grounding it in both past and present practice. Thus, predictions for the future include both a moral and scientific premise and are morally and scientifically critical in guiding contemporary work.

But these pillars of feasibility (histories and roadmaps) are not enough to maintain strategic predictions. An influential prediction is also one that can appropriate a moral authority such as that the science is "right" – or even obligatory – because it predicts a clear and serious benefit to humanity and earth. Within much of the anti-aging field and Western discourse at large, aging is constructed as painful and costly for the individual and for society. Anti-aging medicine provides a possible solution to these problems. This perspective links with the powerful ethical imperative to scientifically overcome adversity and "victimhood" to foster anti-aging predictions in the context of present-day realities of scientific funding, research and practice. Imagining a more humane future, therefore, demands its pursuit just as, conversely, imagining negative consequences demands suspension of activities leading to them.

Perhaps one of the more important jobs of a prediction or set of predictions is to foster financial investment – and financial investment itself is often linked with feasibility and promise. Investment in a research initiative is not only an investment in scientific understanding, but is also an investment in an imagined future. Leonid Gavrilov laments that while aging research often makes its way into mainstream reporting, only "10–20% of research projects are funded" (Gavrilov 2004). The American Academy of Anti-Aging Medicine (A4M), perhaps the most well-known and certainly the most controversial organization devoted in some way to anti-aging goals, employs statistics stating that less than 1% of the NIH is dedicated to the study of aging (Mitka 2002 in American Academy of Anti-Aging Medicine 2002b); therefore, the notion that no anti-aging currently exists is linked to the lack of investment rather than the lack of feasibility.

During a session regarding funding at the Integrative Medical Therapeutics for Anti-Aging Medicine™ conference, the associate director of the NIA responded to such critiques. Warner defended the NIA and situated *all* biology of aging research as, in effect, anti-aging research (Warner 2004). While citing the inconclusive work on caloric restriction and the failure to identify aging biomarkers (despite the substantial investment), Warner places the onus on scientists by asserting that the NIA only responds to the grants submitted and therefore cannot and should not be held exclusively accountable for any perceived lack of funding. This suggests that funding is also contingent upon capturing intellectual investment as well.

The Methuselah Mouse Prize, founded in part by Áubrey de Grey, is an interesting mechanism for marshaling both financial and intellectual resources. The Methuselah Mouse Prize awards researchers who have been able to demonstrate certain anti-aging research goals. While soliciting funds to support the prize from the public at large, the prize also stimulates researchers interested in the prize to seek scientific grants to carry out their research. Thus, the prize promotes the channeling of funding sources toward anti-aging research ends while simultaneously raising money for the payout as well.

In the remainder of this article, I will sketch out some of the contours of the anti-aging field and examine these two components of influential predictions: (a) issues of feasibility which rely upon historical foundations and (b) the moral arguments which underpin the mandate to pursue such a future. I will leave the work of outlining the various roadmaps to effective anti-aging interventions to the authors in the remainder of this volume. I hope to offer a convincing argument that the imagination of a given future affects the present by impressing the significance of its creation on the work of today. By attending to predictions, we can better understand not only this field via an examination of its hopes and beliefs, but also the cultural milieu in which it emerges. This should provide a useful anthropological lens through which we can explore the "culture" of anti-aging medicine and research.

2.3 A Context: A Brief Outline of the "Anti-Aging" Field

Clearly and comprehensively describing the field of anti-aging, if one could even call anti-aging a field, is extremely problematic. It seems at first logical to take a kind of census approach by outlining the field according to traditional, recognizable categories such as clinical medicine (those thousands of anti-aging physicians and health care workers who practice in anti-aging clinics or incorporate anti-aging approaches in their practices), scientific research (those scientists working in academic settings or biotech firms), for-profit businesses (those companies developing or selling anti-aging related products), and non-profit organizations (organizations dedicated to anti-aging ideas in some way). These are, however, very awkward categories as their boundaries are highly porous in both obvious and subtle ways – and have arguably become increasingly permeable since the early 1990s. Moreover, this kind of institutional geography does not well address the cultural purchase of anti-aging overall. Perhaps most problematically, however, is that this kind of field mapping glosses over the diversity of definitions and descriptions of anti-aging practices and ideologies.

"Anti-aging" means many different things to many different people (Arking et al. 2003; de Grey et al. 2002; Mykytyn 2006b), and many other people who "do" what might be called "anti-aging" work under other names (such as "longevity medicine," "age-management medicine," "rejuvenation research," etc.). Instead of attempting to describe what I am corralling together as "anti-aging" according to an institutional geography, it is perhaps more illustrative to tease out the cleavages and nuances of the various goals and practices of those individuals and groups who believe that aging can and should be a target for biomedical intervention.

The variety of anti-aging objectives can be largely distilled into those that aim to (1) compress morbidity, (2) retard or decelerate the aging process or, (3) rejuvenate or reverse the aging process and perhaps "cure" aging altogether (Post and Binstock 2004, President's Council on Bioethics 2003). The compression of morbidity (Fries 1980), whereby the period of late-onset illness (or the steep, late decline of aging) is condensed into smaller timeframes, is an approach that has long been championed by gerontology and, as such, does not present a "new" or revolutionary approach. The retardation of the aging process is an approach whereby advocates seek to push back the decline of biological aging into much later chronological ages. We may be seeing some trends suggesting that this is already taking place. Nonetheless, the idea that a 90 year old could feel like a 50 year old (Miller 2004) and live to 140 or beyond is a staggering prediction. The third and most radical approach attempts not only to postpone the physiological decline of aging indefinitely but also to reverse the process of aging itself. This is the most controversial of claims both in terms of scientific feasibility and socio-moral impact. Despite, or perhaps because of the shocking nature of such objectives, this approach tends to capture a lion's share of attention, especially with Aubrey de Grey's Strategies for Negligible Senescence (SENS) project (Mykytyn 2010). The latter of these anti-aging goals depend upon a scientific and clinical intervention focused on the process of aging itself rather than the disease model that has largely characterized biomedical approaches to late-life care (Mykytyn 2009).

Another important cleavage in the world of anti-aging rages around the question of whether or not an efficacious intervention currently exists. While the American Academy of Anti-Aging Medicine (A4M), a hugely controversial organization that continues to boast a swelling membership of physicians and health care practitioners involved in clinical anti-aging medicine (Mykytyn 2010), and the anti-aging "marketplace" captures perhaps twenty billion patient/consumer dollars a year (Freedonia 2005), many gerontologists decry this work as raising false expectations at best (Butler 2001) and profiteering at worst. In 2002, for example, *Scientific American* published a position paper signed by fifty-one gerontologists stating emphatically that "there are no lifestyle changes, surgical procedures, vitamins, antioxidants, hormones or techniques of genetic engineering available today that have been demonstrated to influence the processes of aging" (Olshansky et al. 2002). This is but one illustration of the efforts that gerontologists have made to distance themselves from those seen as "entrepreneurs" selling snake oil (Binstock 2003).

Practitioners I have interviewed argue that while their work may not be lengthening the quantity of life, it is, they claim, positively impacting its quality. More importantly, however, they believe that the anti-aging focus of their practice is a different and superior way to take care of people than what they experienced in the more traditional practices that preceded their anti-aging work (Mykytyn 2006a). Additionally, noted futurist Ray Kurzweil, for whom a luncheon was auctioned off on ebay for $4050 (Methuselah Foundation 2005), argues along somewhat similar lines in that the goal of keeping people as healthy as possible today could serve as a kind of "bridge" until a true anti-aging intervention is available (Kurzweil and Grossman 2004) (see also Chapter 1).

While many other debates around anti-aging are salient and significant – such as how to scientifically achieve an anti-aging intervention and when such could become available – the predictions of what anti-aging could mean and whether it currently exists offer a condensed introduction to the emerging ideas, practices and rhetoric of anti-aging. The innovation of present day anti-aging medicine or anti-aging science in many respects takes the shape of a new approach rather than one specific technology or advance. Targeting the *process of aging* specifically for biomedical intervention serves as the launch-pad for a new model for the study and treatment of aging (Mykytyn 2006a).

2.4 Historical Trends as a Basis for Prediction

History is a crucial component of compelling predictions. History can ground the imagination of a certain tomorrow in the seemingly solid rock of yesterday. However, history is less straightforward than it is contentious. A particularly identified history that highlights some and silences other moments or trends may lend credibility to a prediction by revealing trajectories that support the assertions of the prediction. Alternatively, another read of history may undermine the prediction by juxtaposing it with a history of failure or trends that run counter to the predictions' claims. Histories are particularly controversial in the quest for an anti-aging intervention. The histories of medical quackery, scientific revolutionaries, longevity increases, and the "boom" of biotechnological triumph provide the terrain for the bulk of this predicting.

Statistics revealing that people (primarily Westerners) are living longer than ever before on average are ubiquitous. The strong presence of this fact has birthed volumes of worry over the "graying" of society and its consequences. This demographic has also been important for many anti-aging proponents as it serves to substantiate claims for future trends. For example, the A4M asserts that "there's no reason to imagine that we won't do at least as much in the next century" (American Academy of Anti-Aging Medicine 2002a). They argue that as the numbers have been on the rise over the past century, the trajectory can only be expected to continue given the ongoing maturation of science and biomedicine. The historical gains in average life span, in this view, beget the likelihood of future successes.

However, while this application of history to the future presupposes continuing progress, a counter-perspective asserts that such increases have little to do with the likelihood of the trend's continuance. The National Institute on Aging, for example, argues that the increase in average life span is directly attributable to "improvements in sanitation, the discovery of antibiotics, and medical care" (National Institute on Aging 1996). Critics argue that using historical longevity trends is ill-advised for grounding future longevity trends since most of the predictable gains in lifespan have already been realized. The trajectory that the A4M employs is, in other words, due for a change.

The primary point upon which much of the anti-aging medicine forecasting pivots, however, is the impact of biotechnological progress. For example, noted futurist Ray Kurzweil asserts that the tempo of technical progress is "doubling every decade, and the capability (price, performance, capacity, and speed) of specific information technologies is doubling every year" (Kurzweil and Grossman 2004). For this reason, he and co-author Terry Grossman predict that with "aggressive application" of our scientific knowledge, life extension therapies will emerge over the next couple of decades. Thus, biotechnological progress should not only be included in thinking about any historical trajectory, it must be thought of within the context of its own present-day rate of growth.

The A4M also banks upon this trajectory of biotechnological progress in their "future equation." The "equation" for the future of human longevity depends upon the belief that medical and technological information doubles every three and a half years (American Academy of Anti-Aging Medicine 2000a). This equation asserts that just as we have no reason to assume that longevity trends might abate, neither should we assume that scientific discovery will slow. Drawing from a history of technical and scientific success, the A4M confidently poses a future in which these triumphs will continue at the same astounding rate and that we can, quite literally, count on it. Moreover, the A4M counters pessimism by asserting that many of the models of mortality and longevity "ignore the enormous potential for technology to function as the quantum leap accelerating the extent and achievement of scientific discovery leading to practical human immortality" (American Academy of Anti-Aging Medicine 2000b). Thus, they expect the rise of biotechnological knowledge to not only continue but to continue geometrically. The successes of scientific innovations of sanitation and antibiotics only further support the idea that science will continue to produce effective innovations that will, the A4M argues, increase longevity.

Aubrey de Grey, of the SENS Foundation, also offers a set of assertions regarding the boom of contemporary scientific advance. De Grey's notion of the "escape velocity" (de Grey and Sprott 2004) refers to science's ability to predict, cope with, and solve impending problems of aging faster than they accumulate. Marking not only science's obligation to forecast problems, the escape velocity idea underscores and even banks upon a sense of science's progress and glory. With this principle, de Grey expands his predictions such that people alive today might live to a thousand years and those born in 2100 even longer.

Interestingly, the history of scientific progress and glory is mobilized against a history of quackery and failure in seeking the fountain of youth. Many gerontologists have cast some of the anti-aging proponents (and specifically the A4M) as profiteering snake oil peddlers, thus highlighting a predation on fear and misuse of science for sales (Butler 2001; Haber 2001; Hayflick 1994; Olshansky and Carnes 2001). The much discussed *Scientific American* position paper on anti-aging medicine signed by the authors and forty-eight endorsers (Olshansky et al. 2002) locates contemporary anti-aging medicine practitioners as "entrepreneurial," "victimizing," and "pseudoscientific." While they state that although "there is every reason to be optimistic that continuing progress in public health and the biomedical sciences will contribute to even longer and healthier lives in the future," they condemn the current practice of anti-aging medicine in clinics *at this moment*. Along with other social scientists (Butler 2000/2001; Haber 2001/2004; Hayflick 2001), the identity of contemporary anti-aging practitioners is linked to a history of snake-oil sales (see also Weil 2007: 41).

In turn, some anti-aging proponents and practitioners classify their own history within the story of the persecuted revolutionary. The identity of a pioneer or revolutionary (Mykytyn 2006a) attempts to invert contemporary thought. Those in the mainstream become linked with quagmire and sheep-like pack mentalities while the revolutionary "outliers" become "cutting-edge." And the popular history of science and medicine is ripe with such stories; Galileo, perhaps, could be the poster-scientist for this history. Anti-aging practitioner Dr. N nicely articulates this point:

> Look at the history of medicine. Look at the history of science. When a new paradigm comes which doesn't make sense initially and it's poo-pooed, the people are persecuted who follow it, and suddenly it's mainstream. (Dr. N., Personal Communication, 16 October 2001)

Invoking Kuhn's theories of paradigm shifts, Dr. N sees anti-aging practitioners as revolutionaries rather than quacks. The pioneering image bequeaths a martyrdom of morality and wisdom read against a conservative and ultimately myopic mainstream (Mykytyn 2006a).

The general disposition of mainstream gerontology toward anti-aging ideas has historically been largely resistant, though it may now be shifting. De Grey proposes that the "central reason [for this resistance] is simple ignorance of the relevant science" which, he argues, is part and parcel of the scientific and intellectual inertia of those who "hold conventional views" (de Grey 2004b). The pain of shifting paradigms, in other words, is a powerful force. Likewise, Richard Miller argues that the discipline of gerontology is characterized by a number of systemic impediments facing the pursuit of anti-aging science (Miller 2004). From difficulties inherent in long-term research in terms of funding, schooling, and tenure issues to difficulties in the very measurement of aging, Miller maintains that the "obstacles blocking the development of the hypothetical discipline of applied gerontology are at this point about 85% political and 15% scientific" (Miller 2004).

In summary, the histories that have been assembled are strategic and significant. They aim to distinguish between what has been proven and what is likely to be proven in the future. This very brief examination of the grounding histories that highlight the rise in average life span, the progress of biotechnology, quackery, and revolutionaries, reveals how predictions plumb history to substantiate not only a particular contextualization of the present, but also the trajectories that may tell us about our future. The history/histories that come to ground a particular vision of the future are far more than an interesting story behind the science; they are political, tactical, contentious, and consequential.

2.5 Moral Obligation as a Basis for Prediction

In addition to grounding predictions in the trends of history and proposing concrete scenarios of how specific predictions might unfold in practice, perceiving a moral basis for making predictions is absolutely critical. With a scarcity of financial and temporal resources and an ethic of practical utility, predictions must seem worthy of the expenditures required for their pursuit. Within the field of anti-aging medicine, the moral obligation to pursue such a future claims a good share of attention (Mykytyn 2009).

The cultural construction of aging as primarily biological and inherently undesirable (Vincent 2006; Mykytyn 2007) plays a tremendous role in the edification of anti-aging predictions. The images of aging as individually painful and societally problematic have deep American/Western roots (Achenbaum 1995a; Cole 1992; Friedan 1993; Gullette 1997; Minois 1987). The growing economic costs of aging, predominantly in the health care sector, compound what is so often regarded as the "problem of aging." While many social scientists advocate for an understanding of aging that transcends its biology, the narrative of aging as a primarily biological process remains stalwart. Coupling the undesirability of aging with the exceptionally powerful ethic of scientific progress (Adams 2004; Franklin 2003; Rabinow 1996; Smith 1999; Toumey 1996) provides a potent and morally charged basis for predicting anti-aging advances.

Proponents predict that an anti-aging intervention will ultimately serve humanity positively, both in terms of mitigating individual suffering as well as alleviating social and financial burdens associated with aging. These predictions rest on critiques of traditional or mainstream models of care. For example, one practitioner I interviewed argued that the current approach to aging constitutes a kind of "victim mentality." In this model "you just accept [that] you have joint pain and all the other symptoms [because] that's just part of aging" (Dr. N., personal communication, 16 October 2001). With all of the recent advances of science, biomedicine is poised for the good fight, but because of its insistence on a fatalistic understanding of aging, laments Dr. N., it never takes a swing. And, as a result, people have come to simply accept suffering.

Anti-aging clinicians often view the current biomedical system, focused as it is on disease, as "wasteful... I mean look at the billions of dollars that are wasted on cardiovascular disease that we can intervene in before it occurs" (Dr. M., personal communication, 1 July 2002). For many practitioners, the current system that concentrates its efforts on cancer, Alzheimer's disease, osteoporosis, etc, acts as an expensive band-aid on an open fracture. Such "malpractice" breeds disdain in these anti-aging practitioners and serves as a major motivation for their migration to anti-aging medicine (Mykytyn 2006a). Attempted anti-aging intervention becomes a way to practice a more preventative, honorable, fiscally prudent, and pain-palliating form of medicine.

Many researchers are posing similar arguments as well. Testifying before the U.S. President's Council on Bioethics in 2002, Jay Olshansky critiqued the "hurdle approach to disease" whereby "whatever is in front of us, we jump over it, only to face another [disease] later." He argues that pushing these hurdles back would mean going "after the aging process itself" (PCBE 2002a). Steven Austad too argued that "slowing aging is really a much more effective approach to preserving health, than is the treatment of individual disease" (PCBE 2002b). Richard Miller's compelling analysis of the obstacles anti-aging science faces not only critiques the disease approach but contextualizes it within a political economy perspective wherein "disease lobbies" and scientists' careers are built upon a disease-structured system (Miller 2004). This kind of critique not only disparages the current biomedical-scientific environments but also offers an alternative whose focus is on "the *preservation* of health" (PCBE 2002b) (emphasis mine).

The personal costs of aging factor significantly in much of this work. Throughout my research I heard of the indignities perpetrated upon individuals by aging: decreased independence, aesthetic changes that can weigh heavily on one's self-image and social mobility, physical pain, loss of vigor, and, especially, intellectual sluggishness and memory loss. While these are widely discussed "quality of life" issues, anti-aging proponents predict a future wherein these are not inevitable, and not often just an oversimplified utopian future free from pain, but one in which the underlying causes of these age associated diseases and situations – aging – becomes the explicit therapeutic entry-point. Opening up this door, proponents argue, could provide opportunities to mitigate such suffering.

The societal costs of *not* pursuing anti-aging medicine and research are discussed and quantified in interesting ways. For example, David Kekich of the Maximum Life Foundation asserts that the elderly are not only a tremendous asset but that American society treats them as throwaways. Thus, based on the valuation of lives from plane crash settlements at one million dollars per life, America shoulders a 1.7 trillion dollar loss per year (Kekich 2004) (see also Appendix B). The prediction of a future based on the status quo, thus, is marked as economically dire. This financial assessment affords a vivid numerical assessment of wisdom and life experience in an attempt to clarify and quantify the anti-aging mandate.

The pursuit of this anti-aging endeavor, now that many researchers can envision it, is sometimes taken explicitly as a moral responsibility. De Grey unambiguously articulates this position: "we risk being responsible for the deaths of *over 100,000 people* **every day** that [Engineered Negligible Senescence] is not developed" (de Grey N.D.) (emphasis original). Responsibility is placed on practitioners of science once a remedy is imagined; once we can envision something we are obligated to pursue it and must be accountable for any ramifications of a lack of the remedy. So, de Grey asserts, the imagination of an anti-aging future demands a discussion of its feasibility, and a positive prediction of its feasibility demands its pursuit.

Anti-aging futures are not always rosy, however. Critics of anti-aging predict much different and often dire outcomes of effective interventions. These predictions, which substantiate critiques of anti-aging and make a case for its hindrance, tend to fall into two main categories (see also Post 2004). In brief, these include the problems that anti-aging medicine tenders for social justice and its threat to natural law.

Social justice concerns (Chapman 2004; Dumas and Turner 2007; Fukuyama 2002; Holstein 2001; Singer 1991) speak to questions of access to care and often predict that those who will benefit from anti-aging medicine will likely be the wealthy. On the other side of this coin, too, lies danger in the likelihood of further medicalization of the aged body and the potential for concomitant ageism (Vincent 2006). Additionally, pursuing anti-aging goals would inherently detract from other research agendas that could help the "underserved". Social justice arguments also invoke the detrimental environmental impact of increased carbon footprints that living longer would inscribe on an already threatened world.

Upon the mantle of natural law, some critics charge that anti-aging interventions would undermine what it means to be human, a humanity that is in part defined by the natural life course that encompasses both aging and mortality (Callahan 1994, 2006; Fukuyama 2002; Kass 2004; Manheimer 2000; Mitchell et al. 2004; Vincent 2006). This could include a detrimental reshuffling of the relationships between the individual, the family, and society (Callahan 1994; PCBE 2003) and the possibility for a lackluster creative drive since, the logic goes, people would have even more time to procrastinate. Even the decline of aging, excruciating and devastating as it may be, presents as a fundamental aspect of human nature; the physiological decline is just one cost of "fully living" and to have fewer limitations would undermine the very fullness of life (PCBE 2003; see also Mykytyn 2009).

Anti-aging proponents decry these oppositions as either ageist, in that critics seek to preserve suffering in older people, or deathist, in that critics actively *choose* mortality. While equity issues may seem the "most serious, and perhaps least tractable" (Horrobin 2005), fears of impending boredom and a potential restructuring of societies, families, and individual experience come across as a "repugnant" position that "den[ies] others a choice to live" (de Grey 2005). In contrast, anti-aging proponents generally invoke a future in which individuals are spared the painful decline and are able to interact and be more fully invested in the world for a much longer time. This longer investment may lead, then, to an increased concern with our ecological and cultural footprint resulting in a more concerned, involved, and devoted citizenry.

Regardless, environmental historian Carolyn Merchant writes that the "is" of science and the "ought" of society are not opposed but contained within each other (Merchant 1983). In anti-aging medicine, this is evident. Scientific practice is charged with moral responsibilities inextricable from innovation and forecasted knowledge. Because anti-aging can be predicted – though debated – and because a future without it can be drawn as foreboding and because technoscientific progress is summarily expectable, the "should" of aging science is fused with the practice of today, and, too, with the understanding of yesterday. Many anti-aging advocates articulate their work within a critique of the "wasteful" biomedical system; anti-aging thus provides a hope for redemption. Predicting an anti-aging future provides one alternative "way out" of such health care problems. Framing anti-aging as fiscally prudent, potentially more clinically effective, and a salve for individual suffering in later life lays bare a moral mandate for its development.

2.6 Conclusion

The questions of whose responsibility it is to predict, what responsibilities are assumed by predicting (or avoiding predictions), what expertise is useful for predicting, what should be laid out for predicting and what cannot be foreseen meet varied answers within science. Nonetheless, some gerontologists argue that predicting equates to scientists' "sometimes unfortunate bent toward hubris" (Austad 1997). Conversely, researchers such as de Grey argue (de Grey 2004) that predicting is a duty of the informed scientist:

> I feel that those with the best information have a duty to state their best-guess timeframe, because that information determines peoples' life choices, whether or not the public's assessment of the reliability of what experts tell them is accurate.

Thus, not only are predictions moored to a moral cause, but they are moral in and of themselves. The duty to predict is linked to the position of imagining what the future may bring. The role of the researcher in predicting is far from institutionalized, but as research monies are competed for and as research findings are publicized predictions circulate.

As anti-aging medicine is increasingly garnering public attention, its predictions are of mounting importance. Rendering clinical aging interventions a feasible prospect has been a particularly tough job in light of so many discussions of hucksterism and the need to combat a deep-seated sense not only that anti-aging is an impossibility, but also that aging represents a proper "natural life course." Nonetheless, clinical attempts to intervene into the aging process began mostly in the early 1990s, and by the late 1990s aging had become increasingly thought of as malleable (Mykytyn 2010). The future of the "new immortality that has come into view only in the last decade or so" (Rose 2004) is based on the morally charged sense of promise, the histories it claims and the maps it offers for its success. Examining a field via its predictions provides a sense of what is socially and scientifically at stake and how the field shaped and is shaped by competing moral and historical claims. The contentiousness within and about this emerging field plays out on the context of these predictions – making strategic and compelling predictions all the more critical to the longevity of anti-aging goals.

References

Achenbaum WA (1995a) Crossing Frontiers: Gerontology Emerges as a Science. Cambridge University Press, Cambridge

Achenbaum WA (1995b) Images of Old Age in America 1790–1970: A Vision and Re-Vision. In Images of Aging. In: Featherstone M, Wernick A (eds) Images of Aging: Cultural Representations of Later Life. Routledge, New York

Adams M (2004) The Quest for Immortality: Visions and Presentiments in Science and Literature. In: Post S, Binstock R (eds) The Fountain of Youth: Scientific, Ethical and Policy Perspectives on a Biomedical Goal. Oxford University Press, Oxford

American Academy of Anti-Aging Medicine (2000a) Validating the Facts and Science of Anti-Aging Medicine. Anti-Aging Medical News Fall

American Academy of Anti-Aging Medicine (2000b) Making the Quantum Leap to Human Immortality in the Year 2029. Anti-Aging Medical News Fall

American Academy of Anti-Aging Medicine (2002a) World Health Organization's Ageing and Health. MCARE 1

American Academy of Anti-Aging Medicine (2002b) Official Position Statement on the Truth about Human Aging Intervention. www.worldhealth.net/p/96,333.html, June, Accessed 10 June 2003

Arking R, Butler R, Chiko B, Fossel M, Gavrilov L, Morley JE, Olshansky SJ, Perls T, Walker RF (2003) Anti-Aging Teleconference: What is Anti-Aging Medicine? J Anti-Aging Med 6(2):91–106

Austad S (1997) Why we age: what science is discovering about the body's journey through life. J. Wiley and Sons, New York

Binstock R (2003) The war on "anti-aging medicine": maintaining legitimacy. The Gerontologist 43:4–14

Blumenfeld, Y. (ed) (1999) Scanning the Future: 20 Eminent Thinkers on the World of Tomorrow. Thames and Hudson, New York

Brown N (2003) Hope against hype – accountability in biopasts, presents and futures. Science Studies 16(2):3–21

Brown N, Michael M (2003) A Sociology of Expectations: Retrospecting Prospects and Prospecting Retrospects. Technology Analysis and Strategic Management, 3–18

Borup M, Brown N, Konrad K, van Lente H (2006) The Sociology of Expectations in Science and Technology. Technology Analysis and Strategic Management, 285–298

Butler R (2000) Editorial: Turning back the clock: Has aging become a 'disease' again – to be prevented, treated, and even cured? Geriatrics 55(7):11

Butler R (2001) Is there an "anti-aging" medicine? Nonscientists seeking to attract consumers to untested remedies. Generations XXV(4):63–65

Callahan D (1994) Manipulating human life: is there no end to it? In: Blank R, Bonnicksen A (eds) Medicine Unbound: The Human Body and Limits of Intervention. Columbia University Press, New York

Callahan D (2006) The Desire for Eternal Life: Scientific Versus Religious Visions. Harvard Divinity Bulletin,www.hds.harvard.edu/news/bulletin/articles/callahan.html Accessed 20 May 2006

Chapman A (2004) The social and justice implications of extending the human life span. In: Post S, Binstock R (eds) The Fountain of Youth: Scientific, Ethical and Policy Perspectives on a Biomedical Goal. Oxford University Press, Oxford

Cole T (1992) Journey of Life: A Cultural History of Aging in America. Cambridge University Press, Cambridge

de Grey A (N.D.) An Engineering Approach to defeating aging: Issues of Mindset and Mobilisation. www.gen.cam.ac.uk/sens/adgbio Accessed 30 May 2003

de Grey A (2004) The Curious Case of the Catatonic Biogerontologists, http://www.longevitymeme.org/articles/viewarticle.cfm?page=2&article_id=19 Accessed 25 August 2005 de Grey A (2005) Life extension, human rights, and the rational refinement of repugnance. J Med Ethics 31(11):659–663

de Grey A, Gavrilov L, Olshansky SJ, Coles LS, Cutler RG, Fossel M, Harman SM (2002) Antiaging technology and pseudoscience. Science 296:656a

de Grey A, Sprott R (2004) SAGE WEBCAST: How Soon Until We Control Aging. www.sagecrossroads.com Accessed 10 December 2004

Dumas A, Turner B (2007) The life extension project: a sociological critique. Health Sociology Review 16(1):5–17

Fischer M (2005) Technoscientific infrastructures and emergent forms of life: a commentary. Am Anthropologist 107(1):55–61

Fortun K, Fortun M (2005) Scientific imaginaries and ethical plateaus in contemporary U.S. toxicology. Am Anthropologist 107(1):43–54

Franklin S (2003) Ethical biocapital: new strategies of cell culture. In: Franklin S, Lock M (eds) Remaking Life & Death: Toward an Anthropology of the Biosciences. School of American Research Advanced Seminar Series, Santa Fe

Franklin S, Lock M (2003) Animation and cessation: the remaking of life and death. In: Franklin S, Lock M (eds) Remaking Life & Death: Toward an Anthropology of the Biosciences. School of American Research Advanced Seminar Series, Santa Fe

Freedonia Group (2005) Freedonia.exnext.com 15 June, Accessed 15 August 2005

Friedan B (1993) The Fountain of Age. Simon and Schuster, New York

Fries J (1980) Aging, natural death, and the compression of morbidity. N Engl J Med 303:13–15

Fujimura J (2003) Future imaginaries: genome scientists as sociocultural entrepreneurs. In: Goodman A, Heath D, Lindee S (eds) Genetic Nature/Culture. University of California Press, Berkeley

Fukuyama F (2002) Our Posthuman Future: Consequences of the Biotechnology Revolution. Picador, New York

Gavrilov L (2004) Pieces of the Puzzle: Aging Research Today and Tomorrow. Longevity Meme, www.longevitymeme.org/articles/viewarticle.cfm?page=1&article_id=12 Accessed 9 August 2004

Geels FW, Smit WA (2000) Failed technology futures: pitfalls and lessons from a historical survey. Futures 32(9):867–885

Giddens A (1998) Risk society: the context of British politics. In: Franklin J (ed) The Politics of Risk Society. Polity Press, Cambridge, UK

Gullette M (1997) Declining to Decline: Cultural Combat and the Politics of the Midlife. University Press of Virginia, Charlottesville

Haber C (2001) Anti-aging: why now? A historical framework for understanding the contemporary enthusiasm. Generations XXV(4):9–14

Haber C (2004) Anti-aging medicine: the history: life extension and history: the continual search for the fountain of youth. J Gerontology: Biol Sci 59A(6):B515–B522

Hayflick L (1994) How and Why We Age. Random House, New York

Hayflick L (2001) Anti-aging medicine: hype, hope, and reality: the science at the root of the question. Generations XXV(4):20–26

Hogle LF (2003) Life/time warranty: rechargeable cells and extendable lives. In: Franklin S, Lock M (eds) Remaking Life & Death: Toward an Anthropology of the Biosciences. School of American Research Advanced Seminar Series, Santa Fe

Holstein M (2001) A feminist perspective on anti-aging medicine: ethical and practical implications. Generations XXV(4):38–43

Horrobin S (2005) The ethics of aging intervention and life extension. In: Rattan S (ed) Aging Interventions and Therapies. World Scientific Publishers, Singapore

Kass L (2004) L'Chaim and its limits: why not immortality? In: Post S, Binstock R (eds) The Fountain of Youth: Scientific, Ethical and Policy Perspectives on a Biomedical Goal. Oxford University Press, Oxford

Kekich D (2004) Introduction to Financing Anti-Aging Research. Paper presented at The Integrative Medical Therapeutics for Anti-Aging Conference and Exposition, Las Vegas, 30 October 2004

Konrad K (2006) The social dynamics of expectations: the interaction of collective and actor-specific expectations on electronic commerce and interactive television. Technol Anal Strategic Manage 18(3/4):429–444

Kurzweil R, Grossman T (2004) Fantastic Voyage: Live Long Enough to Live Forever. Rodale Inc., USA

Manheimer RJ (2000) Aging in the mirror of philosophy. In: Cole TR, Kastenbaum R, Ray RE (eds) Handbook of the Humanities and Aging, 2nd edn. Springer Publishing, New York

Marcus G (ed) (1995) Techno-scientific Imaginaries: Conversations, Profiles, Memoirs. University of Chicago Press, Chicago

Merchant C (1983) The Death of Nature: Women, Ecology and the Scientific Revolution. Harper, San Francisco

Methuselah Foundation (2005) Charity auction in support of the Mprize for longevity research a success. www.mfoundation.org/index.php?pagename=newsdetaildisplay&ID=078, 18 July, Accessed 1 August 2005

Miller R (2004) Extending life: scientific prospects and political obstacles. In: Post S, Binstock R (eds) The Fountain of Youth: Scientific, Ethical and Policy Perspectives on a Biomedical Goal. Oxford University Press, Oxford

Minois G (Transl. Tenison SH) (1987) History of Old Age: From Antiquity to the Renaissance. Chicago University Press, Chicago

Mitchell CB, Orr RD, Salladay SA (2004) Aging, Death, and The Quest For Immortality. Wm. B. Eerdmans Publishing Company

Mitka M (2002) As Americans Age, Geriatricians Go Missing. JAMA 287(14):1792–1793

Mykytyn CE (2006a) Anti-aging medicine: a patient/practitioner movement to redefine aging. Soc Sci Med 62(3):643–653

Mykytyn CE (2006b) Contentious terminology and complicated cartography of anti-aging medicine. Biogerontology 7(4):279–285

Mykytyn CE (2007) Executing Aging: An Ethnography of Process and Event in Anti-Aging Medicine. Dissertation, University of Southern California

Mykytyn CE (2009) Anti-aging medicine is not necessarily anti-death: bioethics and the front lines of practice. Med Studies 1(3):209–228

Mykytyn CE (2010) A history of the future: the emergence of contemporary anti-aging medicine. Soc Health Illness 32(2):181–196

National Institute on Aging (1996) In Search of the Secrets of Aging. Internet Document www.niapublications.org/pubs/secrets-of-aging/index.htm#content Accessed May 2003

Olshansky SJ, Carnes BA (2001) The Quest for Immortality: Science at the Frontiers of Aging. W.W. Norton and Company, New York

Olshansky SJ, Hayflick L, Carnes BA (2002) No truth to the fountain of youth. Scientific Am 286(6):92–95

Post S (2004) Establishing an appropriate ethical framework: the moral conversation around the goal of prolongevity. J Gerontol: Biol Sci 59A(6):B534–B539

Post S, Binstock R (2004) Introduction. In: Post S, Binstock R (eds) The Fountain of Youth: Scientific, Ethical and Policy Perspectives on a Biomedical Goal. Oxford University Press, Oxford

President's Council on Bioethics (2002b) Duration of Life: Is There a Biological Warranty Period? (Transcripts). www.bioethics.gov/transcripts/dec02/session2.html, 12 December, Accessed 15 September 2003

President's Council on Bioethics (2002b) Adding Years to Life: Current Knowledge and Future Prospects (Transcripts). www.bioethics.gov/transcripts/dec02/session1.html, 12 December, Accessed 15 September 2003

President's Council on Bioethics (2003) Beyond Therapy: Biotechnology and the Pursuit of Happiness – A Report of the President's Council on Bioethics. Dana Press, New York

Rabinow P (1996) Making PCR: A Story of Biotechnology. The University of Chicago Press, Chicago

Rescher N (1998) Predicting the Future: An Introduction to the Theory of Forecasting. State University of New York Press, Albany

Rose, M (2004) Biological immortality. In: The Scientific Conquest of Death: Essays on Infinite Lifespans. Libros En Red, Buenos Aires

Singer P. (1991) Research into aging: should it be guided by the interests of present individuals, future individuals, or the species? In: Ludwig F (ed) Life Span Extension: Consequences and Open Questions. Springer Publishing Company, New York

Smith MR (1999) Technology, industrialization, and the idea of progress in America. In: Blumenfeld Y (ed) Scanning the Future: 20 Eminent Thinkers on the World of Tomorrow. Thames and Hudson, New York

Toumey C (1996) Conjuring Science: Scientific Symbols and Cultural Meanings in American Life. Routledge University Press, New Brunswick, NJ

Vincent J (2006) Ageing contested: anti-ageing science and the cultural construction of old age. Sociology 40(4):681–698

Warner H (2004) NIA support for Anti-Aging Research. Paper presented at The Integrative Medical Therapeutics for Anti-Aging Conference and Exposition, Las Vegas, 30 October 2004

Weil A (2007) The Truth about the Fountain of Youth: How to live a Healthy, Vigorous – and Long – Life. AARP Newsletter May/June: 40–41

ID # Chapter 3
Towards Naturalistic Transcendence: The Value of Life and Life Extension to Persons as Conative Processes

Steven Horrobin

Contents

3.1 Introduction – Science and Natural Philosophy 39
3.2 The Value of Life and the Value of Living: The Liberal/Conservative Divide 41
3.3 Persons as Conatively-Driven Processes 44
3.4 Conatus, The Master Value 47
    3.4.1 A Brief History of the Conatus Cluster Concept 47
    3.4.2 Spinoza's Conatus Argument, Physics, and Modern Metaethics 50
    3.4.3 Hobbes, the Conatus, and Evolutionary Theories of Morality 52
    3.4.4 The Conatus and Modern Thermodynamics of Self-Organised Systems . . . 53
3.5 Objectivity, Reductionism, and Emergence: A Note on the Nature
of the Universe, and of the Sciences that Endeavour to Describe It 56
References 59

3.1 Introduction – Science and Natural Philosophy

The inclusion of a philosophical essay in a volume of even speculative applied science, to some may seem out of place. There appear to many to be deep divisions between the disciplines, perhaps rendering commentary between them somewhat irrelevant. However, while there may appear in modern context to be such divisions, these are largely illusory. Just as philosophy must account for the developments of its empirical offspring, Natural Philosophy (as science was known until the mid-nineteenth century) is underpinned by and wedded to philosophical commitments

---

S. Horrobin (✉)
College of Medicine, University of Edinburgh, Edinburgh, Scotland
e-mail: s_horrobin@hotmail.com

that themselves are far from resolved. More will be said about the latter in the final section of the paper, but it is germane at this point to note that the fragmentation of discipline seen especially in the past two centuries has been significantly precipitated by the growing mismatch between human knowledge and human lifespan, rendering the possibility of a true Renaissance individual, a master of all disciplines, a matter of quaint history. Whereas once it was possible to be a Natural Philosopher, and to have a number of other accomplishments to a professional level, it is now not possible in one lifetime to master the whole of a sub-discipline such as biology, let alone "science" as a whole. While lifespans have indeed expanded chronologically (though not in endogenous biological terms) in the same period, measured culturally they have dramatically shortened, and this process and its attendant fragmentation is accelerating (Horrobin 2006).

While some of its component essays are based in pure science, this volume in particular can best be described as largely consisting of highly speculative applied science tending towards a particular aim: that of positively altering human longevity. Any such project requires, at base, some fundamental assumptions as to value: namely that there is indeed a value, amounting to an imperative, to achieve this aim. Such a judgement represents a positive resolution of the question: is it good, or valuable, in the main, to live a longer, rather than a shorter life? But it also represents a resolution of the further question: is it good to live a longer life than indeed has ever been lived by persons thus far? Now it may seem to some readers that a positive answer to this question is a trivial truism, and that the question is not in need of further examination. However, in order to underpin the arguments in favour of life extension, and more pointedly, in order to begin to understand the nature of the motivation represented by such advocacy, it is necessary to attempt to resolve the more fundamental question:

> Other things being equal, so whether or not there are specific reasons such as particular goals or life plans or desires *other than the simple desire for continued existence*, is there anything which might be said to be good about continuing to live, *simpliciter*?

In other words, is there a fundamental motivating principle, or master value, which in some sense objectively underpins the assertion that living longer is better? In still other words, what, if anything, can be said to supply the "goodness", which may be to say the "arrow" of the person's self-projection towards the future, or the "arrow of striving"? If this question remains unaddressed, then the project may suffer from the appearance of foundation upon the ground of personal preference alone, with no explanation of the nature of the preference motivation itself, and as such to be both somewhat whimsical, and lacking in an elucidation of its central motivating principle.

It is important therefore, at least to attempt to trace the structure and origin of this value.

3.2 The Value of Life and the Value of Living: The Liberal/Conservative Divide

When discussing values, it has become common, in the past hundred years or so, to discuss persons. This is because persons are, in the liberal philosophical tradition,[1] by and large thought to be the sole originators of moral value in the universe.[2] Values are not, it is asserted, part of the objective furniture of the universe, such that one could, for example, detect or examine them using fine or powerful scientific instruments (Mackie 1977), but rather are projected upon reality by **our** subjective valuing attitudes, choices and judgements (e.g. Hume 1739; Ayer 1936; Blackburn 2000). So, if I desire a particular object or state of affairs, to *me*, it is valuable, and I make it so. In this tradition, the definition of a person has become synonymous with that of a valuing agent, requiring, at base, the features of self-consciousness, rationality and autonomy (Locke 1690). If neither I nor any other valuing agent either directly or indirectly[3] so desires or values an object or state of affairs, it is held to be valueless, unless and until it comes to be valued by some valuing agent. So values are held, in this tradition, to be essentially subjective and the value of life is then seen in mostly instrumental terms, as the value of *living*,[4] dependent upon some further set of person-projected values. Life is seen as valuable solely by virtue of its usefulness in facilitating other desires, projects, or goals, which are valued in themselves. There is no basic objective motivating value whatsoever.

Conservative views of value have traditionally been at the same time more apparently commonsensical, and more esoteric. Conservatives have by and large held that, while it is trivially true that persons as subjective agents do value things, and that these values are real, they are not the only values, and indeed are wholly secondary to an underlying value or set of values which are in some way objectively written into or at least present in the fabric of the universe, and so are independent of,

---

[1] I use the terms "liberal" and "conservative" in this paper largely as shorthand for what may broadly be described as the two principal camps in modern bioethics, which have their origins in divergent traditions of ethical theory. With regard to the concept of life's value, conservatives generally hold to a content-full canonical concept of the "sanctity of life" often based in person-extrinsic conceptions of life's value as a "gift" of the divine or of "nature" over whose frame, including duration, persons have little or no prerogative. Liberals generally reject such a view, favoring a concept of all value as being generated by the subjects themselves. Among other things this latter view facilitates decisive personal prerogative (of the living person) in matters of life, death, and life extension.

[2] Although this may be thought to be a feature primarily of the liberal tradition, those conservative traditions that accept divine ordinators of value, in which the deity is considered a person, accept a transcendent version of this idea in basic structure, though in these cases *human* persons are held not to be the ordinators of moral value.

[3] Allowing for values to be subsidiary to other valued goods or ends.

[4] In the sense of living persons, since simply being alive, so the simple "value of life", is held not to be sufficient, and indeed not to obtain.

or prior to subjective person-originated values. It may very well be that this latter conviction has in large part been responsible for the uneasiness of the relationship such conservative philosophers have had with the findings and commitments of empirical science, since the latter, given the strongly reductionistic nature of the philosophical assumptions which have underpinned it for much of the past several centuries, has provided little room for the ontology of objective value, and therefore little comfort to such a view. The trouble is that hitherto it has appeared difficult, if not impossible, to account for such values within a naturalistic frame (Mackie 1977). This has led some (rarely strictly philosophers as opposed to theologians) to assert that values are supernatural qualities of divine ordinance, and other, more secular conservatives to argue for a non-natural status that stops short of outright claims of a religious nature (Moore 1903), but whose character, ontology, and epistemological accessibility remains mysterious and largely unexplained.

Liberal philosophers have typically found the concept of objective value dubious and unwarranted, particularly in view of its hitherto apparent lack of plausible compatibility with a reductionistic picture of reality, and so with the findings of a naturalistic, and generally "physicalistic" empirical science, as well as being superfluous to what they regard as the more economical explanation of values as wholly subjectively determined. Conservatives, for their part, have balked at the liberal view, feeling that for such a momentous question, particularly that of the value of life itself, there must be some more objective matter of fact than the mere subjective value of what we may from time to time desire, or else this value disappears into the dangerously undignified marshland of the vagaries of human whimsy. (Holland 2003; Kuhse 1987; Harris 1985; Harris 1999; Macklin 2006; Cohen 2006; Pinker 2008).

Interestingly, in work concerning the ethics of life extension, the traditional lines of the conservative/liberal divide have thus far been rather blurred. Liberal philosophers, despite being habitually sanguine on issues such as abortion, euthanasia and suicide, have often strongly advocated the project of radical life extension (e.g., Stock 2002; Overall 2003; Harris 2004; Caplan 2004), while conservatives, despite professing a commitment to such an absolute and objective value of human life have rather astonishingly been seen to oppose it (e.g. Kass 2001; Fukuyama 2002[5]). As I have argued at greater length elsewhere (Horrobin 2006), I suggest that on the conservative side this represents a failure of true commitment to the traditional conservative elucidations of this purported objective value, which indeed appears vague, esoteric, and often parochial, while on the liberal side it may suggest a greater commitment than is warranted by the assertion of the value of living as being wholly derivative of other, entirely subjective values.

---

[5] Fukuyama's insistence on the value of "human nature" as an objective reality appears ill-at-ease with his assertion that we should not seek to extend life span, for what, in truth, could be more "natural" for us than that we should seek to preserve our lives?

The intuition that there is something more to the value of life's continuance to persons than subjective personal preference that I detect on the part of liberals confronted with the possibility of radical life extension, brings into sharp focus a further problem for the liberal "anti-realist" position on value. Briefly, this problem is that the criticism of the lack of explanation of the place of values in the naturalistic frame applies just as much to the liberal conception, as it does to the conservative. Simply stating that these values are "projected" on reality by persons does not get one to an adequate naturalistic explanation of the nature and origin of values and value motivation *unless one accepts a thoroughgoing dualism with regards to persons*, or else in some other way partly or wholly abandons a strictly reductionist naturalistic metaphysics. Even in the latter case, there remains no real accounting for the place and structure of a semi- or non-reductionistic subjectivist view within a naturalistic schema. If one is committed to naturalism, one must account for persons *themselves*, and all their valuing activity, within the framework of nature, and if one accepts a thoroughgoing reductionism, then one must accept that there must be some available, more fundamental description of values within a natural scientific schema. Values don't, according to the hard reductionist naturalistic schema, which is both causally and *explanatorily*[6] complete in a bottom-up sense, simply appear in a sui-generis manner within the high protectorate of the category of personhood, but rather must be accounted for by appeal to explanation at some more basic level. In other words given a deterministic naturalistic reductionism, persons themselves, at some level, must be subject to determining causal forces, and if this is so, then the values which appear to be person-projected are indeed objective features of the universe, in principle accessible to scientific description, and even measurement and prediction from fundamental physical laws, in very much the manner denied by the classic liberal picture.[7]

In this way it would seem that the liberal schema of value may well suffer much the same kinds of difficulties as does the conservative, vis a vis its relationship to a naturalistic concept of the universe. For to deny this would be something akin to accepting that values somehow magically appear where there are persons, and that they are causally disconnected by their person-predication from more basic causal forces, and this would appear to commit one to an abandonment of either thoroughgoing naturalism or else strict reductionism, if not both, making the liberals vulnerable to precisely the critique which they classically level at the religious and conservative views of value. In this way the liberals might be accused of accepting the prima-facie apparition of values as projections of the person, and consenting to "pay no attention to the little [value] behind the curtain"[8] of personhood itself.

---

[6] Such that one could, for example, in principle look to a function of a unified "Theory of Everything" and be able, solely from that, both to predict, and understand, the value of the smell of cheese, or in the moral case, the respect of privacy. A traditional shorthand for this view is the so-called "demon" of Simon de Laplace.

[7] As referenced in the first paragraph of this section, above.

[8] To paraphrase the "Wizard of Oz".

3.3 Persons as Conatively-Driven Processes

Persons are not objects, but processes. The concept of persons as having some *unchanging* character, such that they are like separated homunculi inhabiting the human body, to whom experiences occur without fundamental effect, and from whose fixed core by the edicts of free will, action initiation and governance through decision by rational process, and of course, therefore, value flows as from a timeless entity, is mistaken. Persons rather are processes, fully embedded in time, which is to say, change.[9]

For what can be said to exist, without the *temporally-extended* interchange of anticipation, desire, hope, plan, perception, experience, decision, sensation, and recollection? It is precisely the interchange of these, and the interchange which is represented by rational processes, which gives us warrant to suppose or assert that there is a person present at all.[10] If one attempts to imagine a being in which no such interchange *ever* occurs, it becomes apparent that one is not imagining a person, and that the actualisation of such a process is therefore requisite for any such an ascription of status. To be a person *is to be such a process*, and a corollary of this is clearly that without the constant motivation of the person-process, there can be no process, and therefore there can be no person. Therefore the motivation of the process itself is a foundational value to persons, and persons cannot be without it.

But what, then, is doing the *motivating* of the process itself at the most fundamental level? It cannot be the person-process *themselves*, since the motivation of that process, at base, is required for the person to exist. Such an idea would involve an unacceptable bootstrapping circularity. Question: what *fundamentally* motivates a person-process? Answer: the person-process themselves. Clearly, this cannot be the *whole* story. Consider rational thought. One's thought moves from one rational object to the next, but what *propels* the process of thought is not contained within the rational objects themselves. Rather they drift or rush past on the river of personhood, carried along by a wholly other current, a current that flows through, informs, constructs, and motivates the objects themselves. Thought is propelled through this rational process by some motivating feature, flowing through, carrying, but not originating *in* the thoughts themselves. The same is true of desire. Any particular desire towards an object *represents a motivation towards that object.*

---

[9] Just as they are embedded in the flux of the natural realm itself.

[10] Sleep and unconsciousness are no counterexample to this, since the process of personhood is not necessarily wholly self-conscious. The processes of the brain are continuous, consciousness aside. Rather the elements that become manifest within, and indeed *as*, self-consciousness require formulation, in the main, in the unconscious. As an experiment to see this in action, next time you speak, and account yourself as speaking and as the speaker, notice how the words appear to be formed wholly unconsciously, and are *observed* to exit the mouth, rather than being consciously hunted-for. More generally, the person-as-process view is far more resilient to the problem of unconsciousness than is the person-as-object view, for it allows analogy of the person-process to a series of overlapping strands of desire, experience, plan, action etc., most of which are unconscious at any particular moment in the conscious life of a person. The unifying continuous process, of course, lies in the continuous process of the life of the brain, the scaffold of consciousness.

But subjectively-selected objects of desire are not themselves the **source** of that motivation, rather they are its **objects**. In this way desires and thoughts do not spring fully-formed from the mind like Athena from Zeus' brow, but rather accumulate their structure and reality at the urging of some more basic psychological motivation.[11] In order to understand the motivating aspects of fully-formed desires, then, one must consider the further question of whether *every* desire whatsoever originates within the *person*, and if not, whether there is some basic desire or motivator, some fundamental driver both of personhood itself and *particularised* desire, or value, consequently, and if so what its nature and origin, and most particularly what its place within a naturalistic account of the world may be. This last question will be addressed further below, but it appears clear from the above analysis of the necessary motivation of person processes to their ontology, that such a motivation is indeed prior to, and generative of the person. But first, it is important to discuss what implications the basic nature of such a motivation, and its necessary *primacy* in the formation of the process of personhood, may have not merely in relation to the subjective values generated secondary to, and downstream of the existence of the person-process, but also for the question *to* persons of the value of their own extension in time.

All desires, all motivations, must extend towards the future. Even the desire to contemplate the past is future-directed. Each and every particularised desire, or selected object of desire, or articulated value, is the object then of this future-directed flow. For each and every particularised desire, there is a prior and more basic future-directed motivation from which it arises, and this *motivation presupposes the value of the continuation of the process which it drives* and which constitutes, at its most sophisticated level (thus far!), a person. Without the urging of this (to borrow a phrase from Bernard Williams) "categorical desire", there would be no person (Williams 1973). As Williams suggests, such "categorical desires" themselves resolve the question of whether it would be "good" or valuable, for the person to continue into the future. Each and every particularised person-predicated subjective desire or value which itself requires the continuation of the person in time represents, in Williams' concept, such a categorical desire. I here suggest that each of these in turn depends upon a fundamental driver that is prior to and necessary for subjective activity itself, and therefore is both fundamental to such *particularised* categorical desires, and is not itself accessible to subjective denial. In this way, if correct, such a desire, or future oriented driver, would constitute a necessary and so objective categorical desire for persons.

Provided that it is accepted that the basic motivating principle that drives and gives rise to personhood itself represents such a "categoric desire" (Horrobin 2009), and that this desire itself represents a value, then it follows that there can be no point

---

[11] The observation that sometimes they *appear* so to do is dealt with by the previous note. The person-process is only partly conscious, and while much of the process of thought structuring occurs at a subconscious level, the assertion holds nonetheless.

at which a person's continuance into the future is not of value to that person. These assertions will be further discussed in the final section of this paper.

So long as a person exists, this basic categorical desire must, *ex hypothesi*, be present as *an intrinsically open-ended, future-directed* driving force. That there may be particular states of affairs, such as the generalised terminal process of the classic biology of aging, or some other more specific terminal illness, or the prospect, say, of being burned alive, which happen to represent the likeliest, or indeed the certain actual future of a particular person, if no intervening suicide occurs, impinges not at all upon this fundamental categoric value. If a person decides to self-euthanise in the face of such prospects, such an action is not in any way a denial of the value of the continuance of their persons *simpliciter*, but rather *solely* represents a choice not to undergo *the particular set of experiences in prospect* which in any case lead to an inevitable end to their person-process.[12]

Persons, therefore, are processes for whom continuance into the future is an absolute and objective value, and the orientation of this value derives from the orientation of the motivating principle of personhood itself. This value is, as stated, not susceptible to subjective denial. If one were to deny it, one's very denial would represent a function of the process of personhood, driven by this "categorical desire", or fundamental motivating value. Consider the following: is it possible for a person to switch off this fundamental motivating drive? Is it possible to cease to desire *simpliciter*? Certainly. One may, for example become utterly demented through degenerative disease, or pick up a gun and blow one's brains out. Equally, it is conceivable that particularised person-predicated desires, plans or projects come to be abandoned or repudiated by persons. But can a person cease to desire, at all, but yet continue to be a *person*? Can a person cease, by this token, to have any future-directed motivating principle, from which all particularised desires both arise, *and in the extended course of the person-process, recur*? It might prima facie seem plausible for this to occur, but to desire to cease to desire is itself a forward directed desire, arising from the fundamental driver of personhood, and to cease to desire *altogether* is to cease to be a person. Therefore a *person* cannot make a desire not to have any further desires *effective*. Imagine your reaction to a person who told you that, by an act of will, at a particular time, say, next Tuesday at 5 O'clock, they will have no further desires, whatever. Such an assertion cannot be believed, and the reasons for this are to be found in the above analysis.

Quite apart from the fundamental metaphysical analysis of the nature of this value as *objective*, such an attempted denial of all desire represents a decision made *for the whole set of possible states of the person-process at once*, and as such treats the person-process as though it were indeed a whole object, present at every moment in the consciousness. But this idea is surely wrong, and is certainly denied, *ex hypothesi*. In this way the suicide always (unless in an act of

---

[12] Indeed it is arguable that such judgements may only be able to be made in context of the concept of the value of some other mode of continued being.

self-euthanasia, which amounts to mercy-killing as described above) commits murder upon both those aspects of themselves of which they are presently forgetful or unaware, and also upon those aspects of themselves which may arise in the future of their person-process as a whole. Perhaps surprisingly, this analysis may further imply that suicides never commit suicide save in reference to some further desire, which is positive, and in view of which they see their present position as contingently trapped in negativity. If they had no desire whatever, they would not be capable of the action of suicide at all. Rather, they would be inert, since they would not, of course, be persons at all, and suicide would indeed be meaningless. A little reflection should bear out the commonsense aspects of this last analysis.

3.4 Conatus, The Master Value

Given an acceptance of a fundamental, non-subjective driver of personhood, and predicator of the value of life's continuance to persons, what can be said about the nature of this objective motivation, this objective categorical desire? If it is objective,[13] and if we are, as I absolutely assert, wholly naturally constituted beings, then such a principle should, in principle, be open to description by natural, even empirical means. Could this be done? What could possibly provide the arrow of the striving towards self-preservation, the striving towards the process of existing, if it is not subjectivity alone? Some things can be said at this point. Firstly, if it is a naturalistic "arrow" we are looking for, then it should be accessible to physics, or at least describable within a language of physical terms. Secondly, if it is correct *ex hypothesi*, that this natural law or principle should, in context of persons, be both the originator, universal driver, and fundamental orientator of value in the universe, perhaps as mediated secondarily by valuing activity in persons, then we come hard against the problem alluded to above: namely, how to reconcile such an idea with a naturalistic account, and in particular with a reductionistic account, which is the predominant model. A full treatment of this subject would amount to a book length treatise, so what follows should be regarded as a kind of speculative introduction.

3.4.1 A Brief History of the Conatus Cluster Concept

There exists a cluster of traditional positions in philosophy that centre on a concept known as the conatus. This constellation of ideas has had various incarnations at various times, with varying degrees of emphasis and scope, but at its core may be stated to be the principle by which things strive to keep themselves in being, or motion. If in the former sense of striving to keep in being, this usually encompasses the thing's striving not only to persist in being, but also towards self-enhancement, or

---

[13] At least insofar as its being irreducible to subjectivity, since *ex hypothesi* the latter arises from it.

self-development. In the most ancient formulations of this concept,[14] it was applied solely to the set of living organisms, but was later extended[15] to include the motions of inanimate as well as animate bodies, so physical dynamics (Sorabji 1988). This more comprehensive view was embraced by the early modern founders of empiricism,[16] who were also, significantly, the founders of reductionist mechanistic views of the natural realm. The comprehensive view of conatus as explaining both the motions of bodies and natural forces such as centripetal and centrifugal "force", *as well as* the striving of living beings towards self-preservation and development reached its height in the writings of such 17th Century rationalists as Descartes, Leibniz, Hobbes and Spinoza. The views of the latter two are the most interesting in the context of this paper. I will deal with these further below, but it is important, at this point, both to untangle the central fault line in this cluster-concept, and at the same time briefly to explore the possible reasons for the general collapse of interest in the conatus hypothesis, subsequent to the 17th Century. I account that this collapse was predicated by three main factors:

1. The conflation of the idea of the conatus as the driver towards self-preservation *and development* of specifically living systems, with that of the (superficially similar but deeply distinct) idea of the driver of the mechanics of motion in systems generally, whether inanimate or animate. In each case these concepts were held to be a *positive force or principle that urges to motion*.
2. The emergence of a strictly reductionistic and mechanistic scientific world view in the course of the 16th to the 18th centuries.
3. The publication of Isaac Newton's *Philosophiæ Naturalis Principia Mathematica* ("Mathematical Principles of Natural Philosophy").

It is perhaps easy to see why the ancient, and I believe still useful, concept of the conatus, as it appertains to the striving towards self-preservation and development of living biological systems became conflated with the quite separate set of attempts to answer the questions concerning the motions of physical bodies generally in the universe. Living bodies are seen to move. It is no great leap, particularly given a reductionistic project and commitment,[17] to consider that the principles of the motion of living versus nonliving bodies may be accounted to be the same. The concept of the conatus was merely borrowed from its old category, and generalised in what appeared a quite logical manner, to apply to and explain the pressing question of what, say, keeps an arrow in flight, *as well as* a duck. That these concepts were only very broadly isomorphic, (the latter for example lacking the crucial element of *self development*, let alone reproduction, in living systems) perhaps did not trouble so much in an age wherein no such concepts were as yet clearly defined. After all,

---

[14] Particularly in the work of Aristotle, and later, the Stoics.

[15] Originally by John Philiponus, in the course of his criticism of Aristotle's theory of motion.

[16] For example by Bernardino Telesio, an important influence on Spinoza.

[17] Evidenced, for example, in Descartes' assertion that all non-human animals were mere automatons.

3 Towards Naturalistic Transcendence

Natural Philosophy in the 17th Century lacked the disciplinary specialisations of physics and biology. Further Descartes, Hobbes and Spinoza, among other major thinkers of this period, held the motion of bodies in general physical dynamics to depend upon some *active* force, which was conceptually isomorphic with the apparent active principle in animal behaviour (Pietarinen 1998).

The publication of Newton's *Principia* on 5 July 1687 caused a revolution in thinking about physical dynamics, and one which, of course, had great consequences for the conatus cluster hypothesis, as it then stood.[18] The idea of an *active* force which kept bodies in motion, so arrows in flight, was demonstrated to be a misunderstanding and instead what caused the continuation of motion in moving bodies was shown to be a *passive* tendency, which Newton dubbed "inertia". According to this, Newton's First Law, arrows stayed in flight because all bodies resisted changes in their states of rest or motion, and would remain in such states until some force was applied to change these same. In the case of arrows, this was of course friction with the air, and the force of gravity (though this only impedes motion, of course, secondarily by occasioning contact with the ground), which meant that in absence of such friction or other impeding force, an arrow would continue to move smoothly and indefinitely, requiring no special "force" to propel it along. There was no need for an *active* principle or force whatever. The same applied to the centrifugal "force", dubbed by Descartes as the *conatus recedendi*, which subsequent to the Principia was recognised to be a "fictitious force" explained by appeal to inertial frames of reference, within which the First and Second Laws are seen to be valid. Simply put, the consequence of the new dynamics of the Principia and the aftermath of its publication was finally to hole below the waterline much of the concept of the conatus as an active principle in physical dynamics.[19] With the collapse of the conatus concept in physical dynamics, and the ascendancy of a reductionist mechanistic view of the universe in large part predicated by the same natural philosophers who had completed the conjunction between the ancient, strictly biological, and the medieaval/early modern physical dynamical aspects of the conatus, the concept cluster as a whole appeared ramshackle and unfit for purpose, and thus largely disappeared from view.

---

[18] That there is some controversy over whether Newton understood, at the time, the full consequences of this himself is neither here nor there, since the *Principia* precipitated the collapse of the conatus hypothesis subsequently in fairly short order, whether Newton himself was directly cognisant of this or not. See Kollerstrom for a critique of the historical implications of Newton's discoveries in this, *qua* Newton (Kollerstrom 1999).

[19] This is of course an oversimplification. However it neatly summarises what did actually happen, whose actual sequence was, as with the development and decline of all ideas in real history, messy and fiendishly complex. For example, while it is true that Descartes saw conatus as an active force ultimately derived from divine power, he considered that this was solely manifest as a primordial impulse from the Divine, the motion proceeding in a smoothly mechanistic manner thereafter (Geroult 1980). A book-length work would be required to tease out all the threads of this transition, but the broad lines of the story are, I believe, correct, and the story as told may be regarded (as with perhaps all history) as a heuristic approximation of the truth.

However, the general collapse of the concept's perceived relevance may well have been premature. For the work done by the initial, ancient concept of biological conatus has yet to be adequately fulfilled and replaced by any modern theory. Taxis, or activity on the part of living organisms (including internal activities, for example maintaining homeostasis), while described as a general phenomenon in biology, are essentially merely assumed, and then described, rather than explained by a general theory, and the striving towards self-preservation which undoubtedly represents an *active* principle made manifest by such taxis is as yet not adequately explained in a hard reductionist, mechanistic manner, such that it is smoothly integrated with the physics of non-biological systems.

Before returning to this claim, however, it is important at this point to outline why some of the particular philosophies of the 17th Century relating to the conatus are of great interest and relevance in context of this paper, and more generally in modern metaethical discourse. The two philosophers who are most relevant in this regard are Hobbes and Spinoza. Each of these described a system of ethics and metaethics whose principal feature was the central role of the conatus of self-preservation and development in human psychology, making this principle in effect the "master value".

3.4.2 *Spinoza's Conatus Argument, Physics, and Modern Metaethics*

Taking Spinoza first, in his magnum opus, Ethics (published posthumously in the Opera Posthuma 1677), he undertakes an astonishingly comprehensive project to elucidate a rational and objective naturalistic explanation for ethics, with the conatus as its lynchpin and guiding principle. From this simple principle he constructed a comprehensive view of metaphysical and metaethical reality. Spinoza echoed Aristotle, Diogenes Laertius, and Cicero in his belief that human affects and cognition had a dependency relation to conatus. This is best expressed in the Scholium to Proposition 9 in Part Three of his magnum opus, Ethics:

> [The conatus], when it is related to the mind alone, is called *will* but when it is related at the same time both to the mind and the body, is called *appetite*,[20] which is therefore nothing but the very essence of man, from the nature of which necessarily follow those things which promote his preservation, and thus he is determined to do those things. Hence there is no difference between appetite and desire, unless in this particular, that desire is generally related to men in so far as they are conscious of their appetites, and it may therefore be defined as appetite of which we are conscious. From what has been said it is plain, therefore,

---

[20] The term appetite or appetitive was often used as a synonym or modifier or the conatus concept, when it was used in context of unconscious or pre-conscious living matter, or the states thereof. Leibniz, for example, primarily calls the conative, the appetitive. Although the latter (of the two following) may be said to be derived from it, the appetitive in the former sense should not be confused with simple appetite, in the sense of hunger for food, but is rather a more general physical principle, cognate with conatus.

that we neither strive for, wish, seek, nor desire anything because we think it good, but, on the contrary, we adjudge a thing to be good because we strive for, wish, seek, or desire it. (Spinoza 2001, f.p. 1677)

In this statement, Spinoza anticipates what has become a central move in modern western analytic moral philosophy, starting with Hume's projectivism (Hume 1739), through Ayer's emotivism (Ayer 1936), to the likes of the quasi-realism of Blackburn (Blackburn 2000). These all rely on the move from the conservative/religious classic position of locating moral psychology in cognition, in which one is cognisant of objective moral *facts* in the world which then influence the affects and result in a conative stimulation or moral motivation, to locating the originator of moral psychology in the affects, such that the affective triggers the conative or motivating aspects (say, to primitive moral exclamations of outrage or approval) and also the cognitive in the construction of more iterated moral concepts and normative theories. However Spinoza goes further than these, stating unequivocally that the conative aspect of mind is the prime moral psychological driver,[21] in this way providing an objective moral "arrow" or moral "master value" that is largely free of the embarrassing difficulties of the classic moral cognitivist picture, which theory, in order to explain the mysterious moral "objects" which are apprehended in cognition, is forced either to postulate supernatural properties or "non-natural" properties (hardly distinguishable, in my view, and equally suspect, requiring a special faculty of "moral intuition", presumably itself natural, which surely begs the question), which cannot be accounted for in a naturalistic frame (Mackie 1977). The non-cognitivist and other expressivist theories, which locate the originator of value in the affects, suffer from the difficulty of explaining motivational internalism. In other words these theories simply don't explain why the affects give rise to motivations, especially moral or valuing motivations. Rather, they simply state that this is the case, or assume the motivational component altogether without explanation. In this way, these theories again suffer from some of the central problems they themselves lay at the door of cognitivism, for if the motivational aspects of moral "facts" or properties require (and lack) naturalistic explanation, the same is true of the affects. Following my analysis of persons necessarily as processes, however, it appears that the affects simply *cannot* be the fundamental motivators, as they are themselves driven into being by some more fundamental motivating principle (Horrobin 2006, 2009). Spinoza's particular concept of the conatus, however, postulates a far more elegant schema than these former, whose primary intended function is precisely that of locating value in a wholly naturalistic frame. As outlined further below, I believe that the conatus at the heart of Spinoza's project of naturalising this psychological and normative *ursprung* may be beginning to be described and located in modern naturalistic scientific terms.

---

[21] Conation, less well recognised generally, is generally the poor cousin in modern psychological theory, but nonetheless represents one of the triumvirate of basic psychological modes conative, affective, cognitive. Though Spinoza's hierarchy is not presently widely accepted in this field, Freud expressed something rather similar to Spinoza's position, and acknowledged this influence in his work.

The relevant propositions of Spinoza's *Ethics* that define the conatus principle in this context are propositions 4 to 9 of Part Three:

Proposition 4: A thing cannot be destroyed except by an external cause.
Proposition 5: In so far as one thing is able to destroy another they are of contrary natures; that is to say, they cannot exist in the same subject.
Proposition 6: Each thing, in so far as it is in itself, endeavours to persevere in its being.
Proposition 7: The effort by which each thing endeavours to persevere in its own being is nothing but the actual essence of the thing itself.
Proposition 8: The effort by which each thing endeavours to persevere in its own being does not involve finite but indefinite time.
Proposition 9: The mind, both in so far as it has clear and distinct ideas, and in so far as it has confused ideas, endeavours to persevere in its being for an indefinite time, and is conscious of this effort. (Spinoza 2001, f.p. 1677)

There is not the space within such a paper adequately to discuss these propositions in detail, but some brief discussion will help clarify the situation, as I think it stands. Spinoza died a decade prior to the publication of Newton's *Principia*, and nearly two centuries prior to Rudolf Clausius' description of the second law of thermodynamics (Clausius 1850), and entropy (Clausius 1865). The latter's significance will be discussed further below, but in the context of the above propositions, it is clear both that this principle in physics constitutes a problem for propositions 4 and 5, but equally one of which Spinoza could not possibly have been aware. With regard to inertia and conatus, Spinoza applied the latter equally to nonliving and living entities, which he referred to as "modes" of the ultimate singular "substance", which term is synonymous with "nature". He made no strict delineation between them as regards the conatus, but I believe that this is precisely because he did not have access to either of the concepts of inertia, or of entropy. These combined provide a reason both to reject talk of conatus in the context of a *general* physical dynamics, but on the other hand in the case of thermodynamics may give us a reason not to reject talk of the conatus as a whole, and indeed to seek to amend his propositions accordingly, or rebuild their like afresh in a new, more comprehensive system. More will be said about the latter below, and I will return to the implications of the other propositions shortly, but first it is important briefly to outline the use of the conatus concept by Hobbes, since it has an equally resonant modern aspect.

3.4.3 Hobbes, the Conatus, and Evolutionary Theories of Morality

Essentially, for Hobbes, conatus was the master value in that each living being strived to preserve its own life, and in context of conscious social beings, and especially self-conscious, rational social beings of the nature of humans, this striving was seen to be best served by entering into contractual arrangements which allowed for peace, and an end to the war of all against all. The war itself was predicated by the requirement for resources to be accumulated by each individual, from food

3 Towards Naturalistic Transcendence

on upwards, in order to preserve themselves in existence.[22] The resource-gathering behaviour was best done, according to a kind of game theory (see e.g. Pietarinen 1998), in localised cooperation, rather than general hostile competition. These contracts may be seen to be either the primitive, prelinguistic social bonds of herds or packs or flocks of animals, or the linguistically iterated, complex, and conceptually abstracted concepts of particular tribes, societies, nations, etc. Crucially, in Hobbes, these larger groupings would then be seen to act in ways that made them appear to be a corporate body, or a single, self-interested person, which for Hobbes was quite literally personified in the body and person of a monarch or dictator (Hobbes 1998, f.p. 1651). These corporate bodies, given finite resources, would of course enter into competition with other such, predicating war on an international scale, but also providing the rationale for an international economics. It doesn't take a great leap of the imagination to see how this Hobbesian model lies in conceptual isomorphism, and considerable accord, with modern evolutionary biological theories of ethics, wherein the "contractarian" impulse, driven by the need for survival, is described as biological "altruism", predicated by "group selection" (Ridley 1997). Indeed considering evolutionary theory as a whole, while it is most certainly correct in its basic premises of evolution by natural selection, it *crucially appears to lack an elucidation of one of its central assumptions:* that living beings do in fact strive to keep themselves in being. This concept is simply *assumed* as a precondition, after which, all else follows according to Darwinian principles, very nicely. However one must be very cautious not to draw any hasty, morally reductionistic conclusions from this nexus. The situation is more complex than it might prima facie appear. I will return to this assertion in the final section of this paper.

3.4.4 The Conatus and Modern Thermodynamics of Self-Organised Systems

So what then could we say in modern scientific terms about this missing principle, about this ancient but possibly magnificently useful concept of the conatus, whose

---

[22] Vitally, this maps very neatly on to the requirement of non-linear open thermodynamic systems, in order to hold themselves far from equilibrium, to feed on "free energy", dumping increased entropy or statistical (Boltzmann) disorder downstream of themselves. Because of the requirement of the second law of thermodynamics that entropy increases in the total system (so, the universe), it can only be decreased within thermodynamically semi-open, but strictly bounded systems (e.g. those bounded within the skin of an organism), or localised groups of systems, even if they are subsystems within a mid-range open thermodynamic system such as a planet, fed with free-energy by a star. Thus such systems, *if driven into being by a conatus-like principle*, will necessarily compete for resources, in order to "swim upstream" of the flow of Gibbs free energy, and thus maintain and develop themselves. The significance of this note will become clearer on reading the remainder of the paper.

conceptual cognate appears to be assumed by the most stable and useful theory in the whole of biology, and possibly the whole of the sciences? Where can we turn for inspiration?

I believe, as an initial port of call, we may turn to an essay by Erwin Schrodinger, originally delivered as a lecture in Dublin in 1943, entitled "What is Life?" (Schrodinger 1944). This essay is especially fascinating because it represents a rare and immensely powerful nexus between the now fragmented disciplines of physics, biology, and philosophy, once whole within Natural Philosophy. The significance to biology of this essay cannot be overstated, since Schrodinger's description within it of the replicating material of living organisms as an aperiodic crystal of a certain size, with the property of replication facilitated by genetic information encoded in a system of covalent chemical bonds accurately gave Watson and Crick their target zone (Watson 1968). Apart from successfully outlining and predicting the "master code" of living, self-replicating systems, another, indeed overriding, aspect of this paper deals with the more metaphysical question of what living matter is, as opposed to non-living matter:

> The large and important and very much discussed question is: How can the events in space and time which take place within the spatial boundary of a living organism be accounted for by physics and chemistry? The preliminary answer which this little book will endeavor to expound and establish can be summarized as follows: The obvious inability of present-day physics and chemistry to account for such events is no reason at all for doubting that they can be accounted for by those sciences. (Schrodinger 1944; Chapter 1, 1st para)

Schrodinger's conclusion is essentially merely to draw a more specific target area, and may very roughly be stated as being that living organisms constitute bounded material systems which prevent themselves falling towards equilibrium, or succumbing to the effects of entropy (as defined by a statistical, so Boltzmann interpretation of the Second Law of Thermodynamics), by feeding on "negative entropy", which is also known as Gibbs free energy available in their environment, and dissipating it through their processes. In more modern language, following the work of the likes of Nobel Laureate physicist Ilya Prigogine (Prigogine 1997), we might describe these same as nonlinear dynamic dissipative open systems that are subsets of an open thermodynamic system in the form of the earth, with the sun providing a constant supply of raw, "free energy". All other "nonliving" systems fall toward equilibrium in a fairly smooth statistically predictable manner, and exhibit no taxis, behaviours, or activities which "swim upstream" of the fall towards equilibrium, and most particularly no "self developing" activities, which exhibit innovation or variation of both strategy and physical form with the common object of maintaining an internally low entropy, or state far from equilibrium, both within the system, and along the genetic line of descendants, as do living systems.

What is on offer here is not, at least not yet, an answer. It is more of an outline, a pregnant absence, like remarkable tracks in the snow, giving some idea that not only is there some unknown or new kind of beast, but what general shape and size it might be, and which direction it appears to run, leading on, perhaps, to where it might be found, studied, and finally fully described. Speaking of this outline,

Schrodinger draws what he calls a "remarkable general conclusion from the model" (Schrodinger 1944):

> ...there is just one general conclusion to be obtained from it and that, I confess, was my only motive for writing this book. From Delbruck's general picture of the hereditary substance it emerges that living matter, while not eluding the 'laws of physics' up to date, is likely to involve 'other laws of physics' hitherto unknown, which, however, once they have been revealed, will form just as integral a part of this science as the former.

This is very much the conclusion that the complex systems researcher and theoretical biologist Stuart Kauffman also draws, and towards the elucidation of which his remarkable book *Investigations* (Kauffman 2003) reads like a training manual for future trackers of this perhaps new thermodynamic principle or law. Whatever final shape this will take, some things are reasonably clear. Most particularly what might be said is that this principle has a *directionality* about it. If the classical thermodynamic principle of entropy (as statistically described by Boltzmann) provides not only a ratchet giving us the arrow of irreversibility in physics, but perhaps even the arrow of temporal directionality itself, as has been suggested (Zeh 2001), then it is not so surprising, perhaps, that a still elusive member of this genus should possess an arrow of its own, perhaps even bestowed by the directionality of the former, classic Second Law: the quasi-opposing arrow of striving towards self-preservation, and self-development in specifically living, biological systems. Yes, I am suggesting that these concepts are not merely isomorphic, but that the paw of the ancient conjectural beast of the conatus of living beings appears to fit precisely into the paw prints of this possible new law.[23]

Returning to Spinoza's propositions relating to the conatus, we may now suggest that, provided one allows for the developments of inertia, separating the conflated cluster-concept into its component parts, and doing away with the general, but not the living-systems case, and provided likewise one allows for a modern understanding of entropy, then while propositions 4 and 5 need at least amendment, propositions 6–9 look remarkably accurate, and appear to be perfectly in line with a putative thermodynamic arrow specific to self-organising systems far (and increasingly far) from equilibrium. Further, they provide a clear route by which, through a process understanding of personhood and subjective value generation, ethics can be joined to modern conceptions in systems biology, and theoretical physics, and thereby provide a coherent route by which norms may be naturalised, and at least one value, the Master Value, may be seen to have a very real, and indeed *objectively* real ontology!

---

[23] It is perhaps no coincidence that Schrodinger chose to preface his essay with a quotation from Spinoza's ethics, specifically Part IV, Proposition 67: Homo liber nulla de re minus quam de morte cogitat; et ejus sapientia non mortis sed vitae meditatio est. (There is nothing over which a free man ponders less than death; his wisdom is, to meditate not on death but on life.)

3.5 Objectivity, Reductionism, and Emergence: A Note on the Nature of the Universe, and of the Sciences that Endeavour to Describe It

Reductionism is true, but very likely it is not the *whole* truth. It is clearly correct that regarding phenomena at each level of the universe, from the cosmic to the macroscopic to the molecular to the atomic to the subatomic, and at every level within this perhaps non-exhaustive list, there is a dependency relation of the "higher" more (*in totam*) energetic interactions and phenomena, to the "lower", less complex, less energetic, and smaller level below, and so on down. In this way, I hold it true that if there were suddenly a universal shift in the value of some atomic or quantum-scale characteristic, it would have a universal effect, and would radically change many or all aspects of the worlds above. This reductionistic holism, and the broad dependency relation it represents, is reasonable, and compelling. At times, however, it has led physical scientists to go a little too far in their claims. Much work in 20th Century physics was directed at the end of finding a unified "Theory of Everything", uniting Quantum Mechanics with Einsteinian Relativity, and all physical forces in one grand equation. String theory was a hoped for candidate, but hope for this appears to be waning, as the number of possible string theories increases towards levels of absurdity (Kauffman 2006). Some, such as the Nobel Prize-winning condensed matter physicist R.B. Laughlin, have gone so far as to suggest that this project may be formally impossible (Laughlin and Pines 2000), arguing for a strong emergentist view of physical reality. Others, such as the Nobel Laureate physicist P.W. Anderson have argued for a weaker version of emergence (Anderson 1972), but one which is, in practice, no less critical of the strong claim to attainability of the reductionist holy grail- the constructionist hypothesis- that given reduction, it will *in fact* be possible to describe all phenomena at each level of the universe by appeal to such a fundamental theory.

Strong emergence essentially is ontological emergence. It states that, as one progresses upwards through levels of complexity, scale, and energetic interaction, one encounters wholly "new" properties or laws that are formally irreducible, meaning formally incalculable from below (so, from a full understanding of properties and laws at a lower, or fundamental level) in a way which would deny the famous conjecture by Simon de Laplace in the introduction of his Essai, commonly known as "Laplace's Demon" (Laplace 1814). Such properties require, in turn, elucidation by means of new laws, which are equally formally incalculable from below. This form of emergence denies the hard reductionist, "theory of everything" model outright, as being a simply incorrect picture of universal ontology.

Weak emergence essentially is epistemological emergence. It states that, as one progresses upwards through such levels, one encounters properties that are in principle calculable, if one were in the mode of Laplace to posit an *infinite* calculating capacity, but that are in fact pragmatically incalculable from within any one universal history, rendering them in practice wholly conjectural. In the latter case, hard reductionism and the theory of everything, with the potential for a *constructivist* account of all possible and actual states from these basic laws and properties, may be formally possible in principle, but is formally *unknowable* in reality.

While a discussion concerning the relative merits of either position might be interesting, it seems clear that for *practical* purposes, the difference between them is irrelevant.

There are some systems biologists who hold out hope for a truly accurate calculability of biological processes, starting with a truly representative in silico model of a cell, which may then dispel talk of emergence altogether, at least at *this* level, revealing it to be merely a contingent present heuristic (Westerhoff and Kell 2007; Snoep and Westerhoff 2005). But this kind of approach seems to ignore the more general lesson that the *apparent* phenomenon of emergence in the apparently ontologically layered universe, with its apparently striking examples of symmetry-breaking and emergent properties, offers us. This lesson, surely, is that whatever we may think we know presently, it is unlikely to be anywhere near the whole story. If, and here are some big "ifs" (but not out of place in this rather speculative volume), the universe is indeed layered as it *prima facie* appears to be, with the quantum giving way to the macroscopic, and the macroscopic, perhaps, giving rise to the psychological realm (so perhaps indeterminism → determinism → freedom of "will" – or at the very least, consciousness, agency, and the reality of conscious valuing agents, or persons), *then what gives us warrant to suppose that the story, the layeredness of the universe with its unfolding spectacular novelty will end at this level?*

Whatever else is true about the nature of the universe, and of our scientific heuristics in attempting to understand it, it is not unreasonable to think that startling novelty of either the ontological or epistemological variety remains an inductively *probable* feature of the universe, such that layers *above* the one to which we have attained, with startling new properties and perhaps attendant new laws await our discovery. Teleology may take the forms of a *final causal* model, following Aristotle, with a fixed, closed destination, but also may take an open-ended form, with the telos an arrow pointing through to worlds and realms beyond, to new layers of possibility meaning and purpose within this, our universe. The conatus is neither a Summum Bonum,[24] nor does it appear that it is a teleology of ends, or final causal teleology, in which there is a fixed end. In the works of Aristotle, this end was that of man, and the highest virtue of man, the Eudaimonic[25] man, the highest good of all. Towards this end, according to Aristotle, did all of nature bend. Rather, agreeing with Spinoza, the

---

[24] A good that may be describable as the ultimate end of all goods, thus the end of all good striving, towards which that striving ought to move. Rather the conatus is better understood to be that which orients the good from below, that from which all other goods may (but do not of necessity) extend, and from which they do of necessity originate rather than that towards which they necessarily intend.

[25] An Aristotelian concept, wherein the goal of life, and indeed the universal goal of Nature, is the attainment of the "good life" for a human person, wherein the person is in a state of perfectly harmonious and virtuous being, or "well being" both with themselves, and with others in their society.

telos of the conatus is intrinsically open-ended, and, given an acceptance of a non-hard-reductionistic, layered natural realm, with emergence a reality,[26] tends towards the perhaps open-ended upward layering of the possible in the natural universe. In the above ways, hard reductionism, with its hubristic assumptions of a nearly complete final discovery of all, may well be a wholly incorrect method of viewing the world, and we may yet find that the bio-teleological arrow of the conatus points us toward an ever distant horizon of the possible, and towards a truly naturalistic transcendence.[27]

Very much more could be said than there is space for here, of course, but I will close with a caution and an exhortation. The caution is that we must be careful not to view an ethical system of naturalistic origin with wholly reductionistic eyes, when considering normative principles. That there may be some emergent law which predicates and governs self-organised dissipative living systems, and that this law has operated in accordance with strictly Darwinian principles hitherto prior to the emergence of self-conscious rational agents, tells us in itself but little of the particular further structures such complex emergent beings have and will continue to project upon the world in terms of value, however driven ultimately by this master value. Our eyes must rather look both to the level we find ourselves at now, that of seemingly free valuing agents capable of culture transcending brute Darwinian principle, whose values are phenomenally real features of the world, to be examined and considered at their own level, and not solely through the lens of the levels which gave them rise, and indeed towards other, more transcendant levels still. In this way, Hobbes' insight, and that of modern evolutionary theories of ethics, while instructive and, I believe, likely correct *insofar as they go*, cannot give us a *fully normative* sense, or better, should not be thought of as *governing* and therefore *limiting* the development of our moral present or indeed *future*.

It may of course be objected that the latter observations mean we may likewise ignore, as previously necessary but now in some sense non-compelling, the imperative of the conatus towards self-preservation *itself*. However I consider this not to be so, and consider that careful reflection on the unique nature of this value will bear this out to the reader. For this is not simply *any* value, emergent as an instrumental expedient towards survival of any one of our ancestors or ancestral groups, in any particular moment of our evolutionary history. Rather, it is the constancy of this arrow at the heart of life, this master value which has enabled the emergence of any values throughout life's history at all. It is the progenitor of all and any value, and while its subsidiaries and offspring will perforce change according to circumstance, this value will not, and perhaps cannot, for without it, all others are nullified- not merely those which are or have been, *but those which have not yet been but might be as well*.

---

[26] Whether ontological or epistemological, but to us, de facto.

[27] For a description of the category of persons as being a transcendent category itself, see Ruiping Fan's paper on the subject (Fan 2000).

This leads, in turn, to the exhortation, which is that we should not consider our own present situation to be in itself the end of the story. The arrow which brought us to this place, which drove us to the emergent reality where we now dwell, while not binding us to the early principles, values and exigencies of our developmental history, indeed does point us towards that unknown horizon, whose possibilities beckon with the subtle promise of transcendence. In the words of Rossetti:

> Think thou and act; to-morrow thou shalt die
> Outstretch'd in the sun's warmth upon the shore,
> Thou say'st: "Man's measur'd path is all gone o'er:
> Up all his years, steeply, with strain and sigh,
> Man clomb until he touch'd the truth; and I,
> Even I, am he whom it was destin'd for."
> How should this be? Art thou then so much more
> Than they who sow'd, that thou shouldst reap thereby?
> Nay, come up hither. From this wave-wash'd mound
> Unto the furthest flood-brim look with me;
> Then reach on with thy thought till it be drown'd.
> Miles and miles distant though the last line be,
> And though thy soul sail leagues and leagues beyond, –
> Still, leagues beyond those leagues, there is more sea.[28]

To which we may perhaps ask in response, and in his own spirit: tomorrow, shalt thou die?[29]

References

Anderson PW (1972) More is Different. Science, New Series, 177(4047):393–396
Ayer AJ (1936) Language, Truth, and Logic. Gollancz, London
Blackburn S (2000) Ruling Passions. Clarendon Press, Oxford
Caplan A (2004) An unnatural process: Why it is not inherently wrong to seek a cure for aging. In: Post SG, Binstock RH (eds) The Fountain of Youth: Cultural, Scientific, and Ethical Perspectives on a Biomedical Goal. Oxford University Press, New York
Clausius R (1850) Über die bewegende Kraft der Wärme. Ann Phys 79:368–397, 500–524
Clausius R (1865) The Mechanical Theory of Heat – with its Applications to the Steam Engine and to Physical Properties of Bodies. John van Voorst, London
Cohen E (2006) Conservative Bioethics & The Search For Wisdom. Hastings Center Report, Vol 36(1), Jan–Feb, pp. 44–56; pp.http://muse.jhu.edu/journals/hastings_center_report/v036/36.1cohen.pdf
Fan R (2000) Can we have a general conception of personhood in bioethics? In Becker GK (ed) The Moral Status of Persons. Perspectives in Bioethics. Rodopi, Amsterdam/Atlanta, GA.
Fukuyama F (2002) Our Posthuman Future: Consequences of the Biotechnology Revolution. Farrar, Straus and Giroux, New York, NY.

---

[28] Dante Gabriel Rossetti, The House of Life: 73, The Choice, III

[29] By this question, I mean to question the inevitability of death in the classic frame of aging, to which Rossetti was undoubtedly referring, as well as to point to unknown possibilities, and not to some confused concept of attaining to supernatural immortality, which can only exist, as I have argued, in the supernatural realm (Horrobin 2006). Of course, while not retreating from this assertion, in the spirit of the paper's exhortation, I may say – who *knows*?

Geroult M (1980) The Metaphysics and Physics of Force in Descartes. In: Gaukroger S (ed) Descartes: Philosophy, Mathematics and Physics. Harvester Press, Sussex

Harris J (1985) The Value of Life: An Introduction to Medical Ethics. Routledge, London

Harris J (1999) The concept of a person and the value of life. Kennedy Institute of Ethics J 9(4):293–308; http://muse.jhu.edu/journals/kennedy_institute_of_ethics_journal/v009/9.4harris.html

Harris J (2004) Immortal Ethics. Ann NY Acad Sci 1019:527–534

Holland S (2003) Bioethics: A Philosophical Introduction. Polity Press, Cambridge.

Horrobin S (2006) Immortality, human nature, the value of life and the value of life extension. Bioethics 20(6):279–292

Horrobin S (2009) The value of life and the value of life's continuance to persons as conatively-driven processes. In: Savulescu J, Ter Meulen R (eds) Enhancing Human Capacities: Ethics, Regulations and European Policy. The Enhance Project

Hobbes T (1998 f.p. 1651) Leviathan. Oxford World's Classics, Oxford University Press, Oxford

Hume D (1990 f.p. 1739) A Treatise of Human Nature. Oxford University Press, Oxford

Kass LR (2001) L'chaim and its limits: why not immortality? First Things 113:17–25

Kauffman S (2003) Investigations. Oxford University Press, New York

Kauffman S (2006) Beyond Reductionism: Reinventing the Sacred. http://www.edge.org/3rd_culture/kauffman06/kauffman06_index.html

Kollerstrom N (1999) How Newton Failed to Discover the Law of Gravity. Ann Sci 331–356; http://www.ucl.ac.uk/sts/nk/newton-gravity.htm

Kuhse H (1987) The Sanctity-of-Life Doctrine in Medicine: A Critique. Clarendon Press, Oxford

Laplace PS (1814) Essai Philosophique sur les Probabilités. Courcier, Paris; http://www.tektonics.org/classics/laplaceprob.pdf

Laughlin RB, Pines D (2000) The Theory of Everything. PNAS 97(1):28–31

Locke J (1997 f.p. 1690) An Essay Concerning Human Understanding. Penguin Classics, London; http://socserv.mcmaster.ca/econ/ugcm/3ll3/locke/Essay.htm

Mackie JL (1990 f.p. 1977) Ethics: Inventing Right and Wrong. Penguin Books, London

Macklin R (2006) The New Conservatives in Bioethics: Who Are They and What do They Seek? Hastings Center Report Vol 36(1), Jan–Feb, pp. 34–43; http://muse.jhu.edu/journals/hastings_center_report/v036/36.1macklin.pdf

Moore GE (2000 f.p. 1903) Principia Ethica. Cambridge University Press, Cambridge

Overall C (2003) Aging, Death, and Human Longevity: A Philosophical Inquiry. University of California Press, California.

Pietarinen J (1998) Hobbes, Conatus and the Prisoner's Dilemma. Paideia Project, Boston University; http://www.bu.edu/wcp/Papers/Mode/ModePiet.htm

Pinker S (2008) The Stupidity of Dignity: Conservative bioethics' latest, most dangerous ploy. The New Republic, May 28; http://www.tnr.com/story_print.html?id=d8731cf4-e87b-4d88-b7e7-f5059cd0bfbd

Prigogine I (1997) The End of Certainty: Time, Chaos, and the New Laws of Nature. The Free Press, New York

Ridley M (1997) The Origin of Virtue. Penguin, London

Schrodinger (1967 f.p. 1944) What Is Life? Mind and Matter. Cambridge University Press, Cambridge; http://home.att.net/~p.caimi/schrodinger.html

Snoep JL, Westerhoff HV. (2005) Silicon cells. In: Alberghina L, Westerhoff HV (eds) Systems Biology. Springer, Berlin

Sorabji R (1988) Matter, Space and Motion: Theories in Antiquity and their Sequel. Duckworth, London

Spinoza B de (2001 f.p. 1677) White WH, Stirling AH (trans.) The Ethics. Wordsworth Classics of World Literature, Hertfordshire

Stock G (2002) Redesigning Humans. Profile Books, London

Watson J (1968) The Double Helix – A Personal Account of the Discovery of the Structure of DNA. Weidenfeld and Nicholson, London

Westerhoff HV, Kell DB (2007) The methodologies of systems biology. In: Boogerd FC, Bruggeman FJ, Hofmeyr J-HS, Westerhoff HV (eds) Systems Biology: Philosophical Foundations. Elsevier, Amsterdam

Williams B (1973) Problems of the Self: Philosophical Papers 1956–1972. Cambridge University Press, Cambridge

Zeh HD (2001) The Physical Basis of The Direction of Time. Springer-Verlag, Berlin and Heidelberg; http://www.time-direction.de/

Chapter 4
The Ethical Basis for Using Human Embryonic Stem Cells in the Treatment of Aging

L. Stephen Coles

Contents

4.1	Introduction: The Ethical Challenge to Stem Cell Therapies	64
4.2	Historical Concepts of Personhood	64
4.3	Potentiality Vs. Actuality: Legally Distinguishable States	66
4.4	The Scientific Facts: Stem Cells, Early Development, and Cultural Responses	67
4.5	Defining Personhood	71
	4.5.1 Fuzzy Definitions and Fuzzy Logic	71
	4.5.2 A Fuzzy Definition of Human Life History	72
	4.5.3 Alternative Views of Personhood and Logical Contradictions	74
	4.5.4 Fuzzy Definitions and the Law	75
4.6	A Fuzzy Definition of Death	75
4.7	Implications of Personhood	76
	4.7.1 Personhood, Sentience, Sapience, Cognition, and Murder	76
	4.7.2 Necessary and Sufficient Conditions for Personhood	78
4.8	The Slippery Slope Argument Against Embryonic Stem-Cell Therapeutics	79
4.9	Technological Approaches to Transcending the Ethical Objections	79
	4.9.1 The Hurlbut Solution	79
	4.9.2 The ACT Solution	80
	4.9.3 The iPS Cell Solution	80
4.10	Conclusion	82
References		83

L.S. Coles (✉)
Institute of Molecular Biology, Department of Chemistry and Biochemistry, Young Hall, University of California, Los Angeles, CA 90095-1569, USA
e-mail: scoles@ucla.edu

G.M. Fahy et al. (eds.), *The Future of Aging*, DOI 10.1007/978-90-481-3999-6_4,
© Springer Science+Business Media B.V. 2010

4.1 Introduction: The Ethical Challenge to Stem Cell Therapies

It has been suggested that histocompatible human embryonic stem cells hold the answer to ameliorating age-related diseases. But in almost all cases to date, there has been a significant problem with developing embryonic stem cells for therapy. In order to extract (or harvest) stem cells from the inner cell mass or *embryoblast* (50–100 cells) of a pre-implantation blastocyst for therapeutic purposes, the blastocyst must be destroyed. Some, such as former Presidential candidate Sen. Sam Brownback (R.-KA), have argued that this intervention would constitute the deliberate taking of a human life (murder in the first degree), and the senator even proposed legislation to criminalize this behavior to the point of prosecuting not just those physicians administering treatment with substantial fines and prison time, but even their patients who were treated off-shore using such cells just as soon as they reentered the US (where the FBI would presumably have jurisdiction to arrest them) (Brownback and Landrieu 2007). Richard Doerflinger, Deputy Director of Pro-Life Activities for the US Conference of Catholic Bishops, has compared embryonic stem-cell research to Nazi-era medical experimentation (Feuerherd 2006) while Dr. James Dobson, Founder of Focus on the Family, has made the same sort of accusations (Hendricks 2006). "One can never justify destroying a life to save a life," President Bush has said (Bush 2003, 2005, 2006, 2008). Opponents of stem-cell research have expressed concern that this work constitutes meddling or tampering with nature ... and might lead to more troubling experiments downstream, such as growing chimera (multi-species embryos or admixed or cytoplasmic embryos, sometimes called *cybrids*) that could be transferred to women's uterus where they might develop (Nardo 2009).

Obviously, there is a wide disparity of opinion about the moral significance of this line of therapeutic research. And although new technologies are coming along that promise to eventually supplant the need for deriving embryonic stem cells from embryos (Section 9), this hasn't happened yet and is not likely to happen in the near future, so the moral issues surrounding the therapeutic use of embryonic stem cells will continue to be pertinent to the future of aging research for some time to come. For these reasons, the purpose of this chapter is to present a mathematical argument, applying Fuzzy Set Theory, that the deliberate destruction of a human pre-embryo is not in the same equivalence class as homicide.

4.2 Historical Concepts of Personhood

Aristotle [384–322 BCE] did not believe that human life begins at conception. To the contrary, as the first embryologist who actually dissected human embryos and made drawings of them c. 324 BCE, Aristotle speculated that "quickening" was the marker for the onset of *personhood*, and that it did not begin until 40–90 days post conception (Wentworth Thompson 1952). (He even coined the term *effluxion* for the spontaneous abortion of a one-week old pre-embryo, which he described as composed of a "flesh-like substance" without distinct parts, so as to distinguish it from a miscarriage or spontaneous abortion when visible parts were present).

Aristotle employed the term *nous* for the rational mind, as distinguished from the physical body. Indeed, he posited three phases of nous – a vegetable, an animal, and ultimately a human nous, which he referred to as *entelechy* (a complete actualization or fulfillment of the human condition). Aristotle's concepts were subsequently adopted by Judaic faith and even by the Roman Catholic Church for many centuries (see the writings of St. Augustine [354–430 AD] (Dods 1952) and St. Thomas Aquinas [1227–1274 AD] (Fathers of the English Dominican Province 1952), who typically described personhood by employing theological terms like *ensoulment* or *hominization* and who continued in the Platonic/Aristotelian tradition).

Another historical aspect of the Roman Catholic Church's conflict with science and personhood involved the Renaissance artist/scientist/engineer Leonardo da Vinci. Leonardo's studies included highly original speculations about the origin of an embryo's cognitive processes (what he termed the "embryo's soul") in his closely-held notebooks. In particular, he included drawings of a fetus in a dissected womb (da Vinci c 1510-1). We were aware that the Catholic Church prohibited Leonardo da Vinci from conducting dissections of human remains in connection with his art and scientific investigations (which he subsequently performed in secret), but we had the wrong idea as to why this was so. Historians have long believed that the dissections themselves were what got him in trouble with the Pope. But Domenico Laurenza has documented that there were no religious or ethical objections to dissections in Italy at that time; the value of an autopsy had not yet been conceived, and the first serious human anatomy book would not be published in Italy for another 30 years (Vesalius 1543). According to Laurenza, it was the clash between Leonardo's Aristotelian view of ensoulment and Pope Leo X's Thomistic view [Aquinas's views regarding the "rational soul" were more neo-Platonic than Aristotelian] that was at the root of the Pope's ban (da Vinci c 1510-1).

Bringing us to more modern times, it was not until 1965 when two Belgian Priests from the University of Louvain (Ernest C. Messenger and Canon Henry de Dorlodot) caused the Catholic Church to begin to reconsider its central dogma (Walter and Shannon 2005), referred to as the theory of *immediate animation*. Arguing from the Aristotelian theory of causality (four causes: material, formal, efficient, and final), if the property of personhood exists in an entity, it must have an efficient cause that preexisted in any prior incarnation of that entity (like arguing that an acorn must contain the DNA of an oak tree). Therefore, if a zygote has the potential of becoming a person, personhood must exist in some fashion in the pre-embryo as well, and therefore the pre-embryo must be accorded all the rights and privileges provided for persons as sacred human beings.

Most recently, the Roman Catholic Church has issued instructions to its clergy regarding these matters (Anonymous 1987). Late last year the church updated its instructions in a 23-page document with 59 scholarly footnotes (Anonymous 2008a). The newest instruction elaborates on the earlier instruction in a manner consistent with its philosophy that research resulting in the destruction of human embryonic stem cells is immoral and even states that the freezing of excess preimplantation embryos for future use ought to be prohibited. Prof. Arthur Caplan, a bioethicist at the University of Pennsylvania, has put forth a compelling mathematical argument explaining the difficulty with adopting so-called "snowflake babies" (Caplan 2003), as has been urged by certain religious groups as well as President Bush.

In more recent discussions, Prof. Robert P. George, a member of former President Bush's President's Council on Bioethics, has written an entire 242 page book with the aim of arguing in favor of the hypothesis that "human personhood begins at conception." Now, when one encounters this argument in op-ed pieces and other media it is easy to ignore, since one merely has to say "consider the source" to note that the argument is based on an evangelical religious argument of the form: "it's true because I say it's true." Sometimes, it's even easier to dismiss the argument since the source is illiterate. However, Prof. George is an exception. He is an articulate spokesman for an anomalous point of view with logical arguments that "connect the dots." Besides being an endowed chair professor of philosophy at a major university (Princeton) he is also a lawyer. He understands the subject matter. The chapter that teaches embryology to the uninitiated is otherwise impeccable, and the book exhaustively covers all counter arguments. Most of his arguments are clear and persuasive (self evident) and don't need such exhaustive exposition. However, when one finds a sentence of the form "$2 + 2 = 5$," where the syntax, punctuation, spelling, etc. are correct, one suspects that this was simply a typographical error. But one wonders about this hypothesis if the presumed typo is repeated in every chapter over and over again. Your assumption that this was a typo must be wrong! It slowly dawns on you that a literate professor of philosophy actually believes something that is incomprehensible to you based on what you know to be the semantics of the ordinary English language. A typical example of such a sentence is: "We did not acquire a rational nature by achieving anything – sentience, sapience, or what have you – but by coming into being [at conception]. If we are persons now, we were persons right from the beginning; we were never nonpersons" (George and Tollefsen 2008; p. 182).

But this assertion is not true. We were fertilized eggs at least a month before we could ever be called a "person" by any sensible definition of the term. Thus, this set of sentences is a linguistic abomination that violates everything that a normal native speaker of English understands by the word "person." We will present our own precise definition of "personhood" later in Section 4.5 of this chapter.

4.3 Potentiality Vs. Actuality: Legally Distinguishable States

It is fine to argue that an acorn is a potential oak tree and ought to be accorded a different degree of moral respect for its potentiality to become a mature tree than say a leaf from such a tree. But what is the legal liability for crushing an acorn as opposed to chopping down a tall oak tree with an ax? Obviously, there are operational consequences under the law that obtain in one case and not the other (especially depending on whose property the mature tree may reside). Our sense of justice dictates that it would be wrong to punish the crushing of an acorn under foot to the same extent that one might be punished for deliberately cutting down a mature tree with a power saw.

Another example would be to imagine a very large pile of identical bricks lying in a heap on the sand. They have the *potential* to be arranged, one-by-one, into an Egyptian-style pyramid, a gothic cathedral, or a commercial shopping center, but until this is done, they are none of these things. As paraphrased from another scientist, if a Home Depot department store hypothetically contains the materials necessary to build three complete houses but then burns down, should the headline in the next morning's newspaper read "Home Depot Burns Down" or "Three houses burn down"? Clearly, a potentiality is not identical to the thing itself.

Is there a time when a potential person, say a blastula, which is a hollow ball of nearly identical cells, should be accorded a different degree of respect (or moral status), than, say, a scrapping of skin from inside one's cheek? Surely, yes. But one should not confuse the respect accorded to a potentiality with the real thing, the *profound* respect that one should have for a new born baby in this case.

While the Roman Catholic Church officially stands against hESC research, a recent poll showed that 72 percent of ordinary American Catholics do support this research. Thus, Catholics can tell the difference between potentiality and actuality even if the clergy might imply there is no difference.

4.4 The Scientific Facts: Stem Cells, Early Development, and Cultural Responses

Before continuing, it's best to establish the biological facts underlying this discussion.

In comparison with routine somatic cells, a *stem cell* has at three special properties:

1. The ability to replicate mitotically for indefinite periods in culture in which the pair of daughter cells are each themselves multipotent stem cells (this can amplify a stem-cell pool as needed);
2. The ability to give rise to a large number of differentiated cells by recursive asymmetric division in which one of the two daughter cells is another stem cell (the tree of divisions is sometimes called a *fish-tail* tree, as opposed to a symmetric *binary* tree). The particular cells that are produced depends exquisitely on the chemokines or differentiating factors present in the extracellular matrix;
3. The ability to give rise to a pair of fully differentiated daughter cells.

By definition, *totipotent* stem cells, when properly stimulated, can give rise to a complete blastocyst including the *trophoblastic* cells that form a placenta.

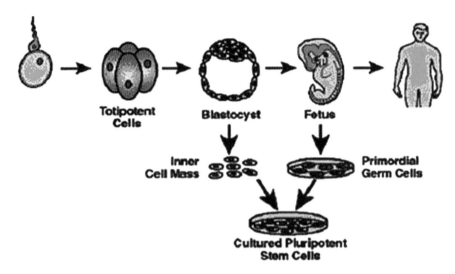

Fig. 4.1 Human embryonic and fetal germ-line stem cells (Reproduced with permission from Prof. James Walter)

Pluripotent stem cells can give rise to any type of cell in the body except for trophoblastic cells; therefore, they can never, by themselves, form a complete (implantable in the uterus) blastocyst representing a potential person. Pluripotent *embryonic* stem cells are found in the Inner Cell Mass, as illustrated in Fig. 4.1 (at the top of the Blastocyst).

Adult stem cells, such as are found in bone marrow, have similar properties to pluripotent stem cells but are normally restricted to an ability to produce cells of determined lineage and are therefore designated as being *multipotent*. Multipotent stem cells can also be found in amniotic fluid during pregnancy, umbilical cord blood and the placenta at birth (some parents are banking these tissues for potential therapeutic use by their child at a future time), and even fluids derived from liposuction.[1]

The process of deriving histocompatible pluripotent embryonic stem cells by somatic cell nuclear transfer (SCNT) is illustrated in Fig. 4.2. This is the same process that was used to create *Dolly* the sheep in Edinburgh ten years ago, and the expectation that therapeutic cloning would provide the foundation for the future of medicine without tissue rejection.

---

[1] **BioEden, Inc.** of Austin, TX is soliciting parents to preserve their children's deciduous teeth on the grounds that the pulp contains potentially valuable stem cells. Another company in Florida (**Cryo-Cell International**) is attempting to sort through menstrual fluid and find valuable stem cells for storage in liquid nitrogen that will be of potential value to these women if they ever come down with an age-related disease in the future.

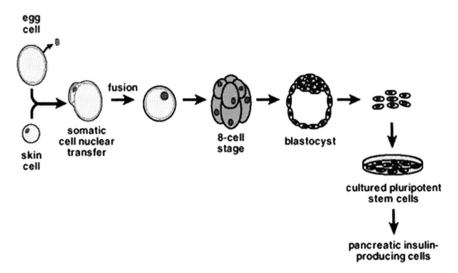

Fig. 4.2 Somatic Cell Nuclear Transfer (SCNT). With this technique, the egg cell nucleus is removed and the egg is fused with a diploid donor cell to create pluripotent stem cells and differentiated tissues such as pancreatic islets genetically identical to the donor (Reproduced with permission from Prof. James Walter)

The six images of Fig. 4.3 provide additional iconic placeholders to further aid our discussion[2]:

Image 4 of Fig. 4.3 shows an early pre-embryo, which by definition includes all stages of development up until implantation; the pre-implantation pre-embryo has no oxygenated blood supply of its own (Wilcox et al. 1999). Although implantation usually takes place on day 9–10, the pre-embryonic period is often considered to last until day 14. This stage includes the blastocyst stage (the blastocyst being a

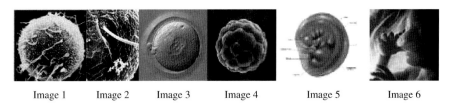

Image 1   Image 2   Image 3   Image 4   Image 5   Image 6

Fig. 4.3 Six critical stages of human development: (1) oocyte surrounded by sperm; (2) single sperm penetration; (3) zygote formation; (4) pre-embryo; (5) embryo; and (6) fetus (Reprinted with permission from: Google Images)

---

[2] Award-winning scientific photographer and biologist David Barlow's *National Geographic Channel* Program "Inside the Living Body" (Barlow 2007) showed for the first time on national television a high-resolution view of an egg ready to erupt from an ovary and begin its journey down the fallopian tube.

hollow ball-shaped structure containing about two hundred cells). A blastocyst starts with a few identical totipotent cells and then begins to differentiate into a sphere with well-defined "North" and "South" poles. *Trophoblastic cells* will go on to form the placenta, while the inner cell mass of *pluripotent cells* will subsequently go on to form the embryo proper. At about day four, the eight- to ten-cell stage *morula* undergoes a process called *compaction* in which the individual blastomeres form tight junctions and gap junctions to facilitate intercellular communications, allowing the genome to further guide the process of embryogenesis.

According to the latest gynecological statistics, somewhere in the range of 40–60 percent of all pre-embryos never even implant in the uterine lining for various reasons (sometimes due to ectopic pregnancies which could be lethal for the mother, but largely due to chromosomal abnormalities, like trisomy or polysomy [disjunctional failure of the spindle apparatus during metaphase of cell-cycle division (Anonymous 2008b)].) Non-implantable pre-embryos are unceremoniously expelled with the next menstrual period two weeks later [metaphorically spoken of as a "disappointed uterus" (Ord 2008).] The potential mother in question was never even aware of the fact that she was "almost pregnant." These abandoned pre-embryos are not mourned, nor are they sought out scrupulously for baptism [nor any other "sacred rite"] to enable them to avoid being permanently exiled to either Hell or "Limbo" (a type of "Purgatory"[3]).

By contrast with the pre-embryo, a human *embryo*, shown in Image 5, is defined as a developing set of structures, post-implantation in the uterine lining that begins with the formation of a placenta possessing a joint blood supply (fetal/maternal circulation). Technically, this is the start of a true *pregnancy* that will go to term in about nine months [~40 weeks], if all goes well. *Gastrulation*, which involves the formation of the three basic tissue types (*endoderm, mesoderm, and ectoderm*), begins in about one more week. *Organogenesis* follows with the formation of a heart and blood vessels, lungs, a nervous system, a digestive tract, and so on. Ultimately, there will be 220 histologically different cell types in the adult human body. This number may be an underestimate when specific markers at the cell-surface antigen-level are further refined.

---

[3] Note that the hypothetical *Limbo of Infants* among Catholics refers to a status of the unbaptized who die in infancy too young to have committed any personal sins, but not having been freed from so-called "original sin." At least from the time of St. Augustine, theologians considered baptism to be essential for salvation and have continued to debate the fate of unbaptized innocents ever since. The Roman Catholic Church has recently rejected the Doctrine of Limbo with the hope that these infants may indeed attain Heaven; however, there are other theologians who hold the opposite opinion, namely that there is no afterlife state intermediate between salvation and damnation and therefore all unbaptized infants are damned to Hell. But in the case of spontaneously aborted pre-embryos how could there be such a baptismal ritual? As is typical, God never saw fit to reveal the subtlety of His ways to men until we acquired a technology sufficient to clearly establish with microscopy what was happening. This is not something that could have been anticipated or guessed beforehand.

Image 6 of Fig. 4.3 shows the transition into what we call a *fetus* after nine weeks [approximately 63 days post conception]. The formation of human morphology (limbs, eyes, ears, etc.) now becomes conspicuous. This is around the time when uterine ultrasound can be used to establish the gender of the future baby (XX female or XY male chromosomes can also diagnose the sex of the baby following amniocentesis or chorionic villus sampling [CVS]).

Opponents of stem-cell research have been known to speak misleadingly of an "*embryo*" being destroyed in the harvesting of human embryonic stem cells from the inner cell mass of a *blastocyst* during SCNT. But, as is made clear by Figs. 4.2 and 4.3, the only tissue being destroyed would be a pre-embryo, and a pre-embryo is **not** an embryo. It is only a stage on the way to becoming an embryo after implantation. One even hears disgraceful comparisons with so-called "partial birth abortions," otherwise known in the medical field as third-trimester "intact dilation and extraction." Nothing could be less biologically accurate.

4.5 Defining Personhood

One of the problems with any natural language such as English as opposed to a technical or mathematical language is the problem of semantic ambiguity. Science is a subset of natural language that strives for clarity and therefore seeks to eliminate ambiguity from the words it employs by providing text-book style definitions as needed. But there is one word in particular that needs a great deal more attention than can be gotten by means of a standard dictionary – that word is the word "*person*".

So exactly what is a human person? We really need a proper definition for the term "personhood." The problem with defining personhood doesn't arise when the person in question is a person in the normal sense. The problem arises at the fringes of life (at the beginning and at the end of life). At the beginning is the problem of "when does a person first become a person?" and at the end of life "how do we know whether a person is still a living person or is in fact a dead person?" These definitions of human life and death have important moral, legal, religious, and medical consequences for scientific researchers.

4.5.1 Fuzzy Definitions and Fuzzy Logic

In the early 1970s, Prof. L. A. Zadeh of the University of California at Berkeley introduced a mathematical notation called Fuzzy Sets (Zadeh 1975; Zadeh and Yage 1987; Meunier et al. 2000) for dealing with uncertainty (but not the same thing as probability theory) that when applied to natural language provided the means to eliminate ambiguity from common English words. For example, Zadeh spoke about the word "tall" not in terms of its dictionary definition but as a graph in which tallness for men was defined by membership in a class of tall men on a scale of

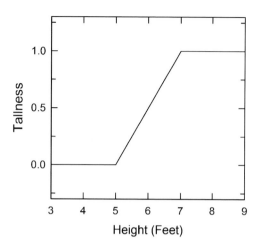

Fig. 4.4 One fuzzy-set definition of the English word "tall"

zero to one as a function of height. The shape of the graph could be a *ramp* or even a *sigmoid* function with short statures below five feet giving us a nearly zero membership in the class of tall men, rising exponentially toward an infection point around 5′ 7″ and then tapering off asymptotically toward 100 percent membership if the man was over 6′ 2″. In particular, for each man in the universe of discourse, we can assign a degree of membership in the fuzzy set "Tall." A very simple way to do this might be with a linear membership function based on the man's height:

Tall(x) = {0,   if Height(x) < 5 feet;
[Height(x) - 5]/2, if 5 ft. <= Height (x) <= 7 feet;
1,   if Height(x) > 7 feet},

which can be graphically represented as in Fig. 4.4.

Zadeh then performed the same sort of computation for male baldness. His initial contribution to the field of fuzzy sets was then to add logical operatives like "and," "or," and "not" (or "very") giving us derived concepts like the class of men who were "bald and not very tall" as a computed function of the independent distributions for baldness and tallness.

This variation gives rise to *Fuzzy Logic*. Applications of *Fuzzy Algorithms* to Control Theory, in which we achieved, for example, smooth braking of subway cars at stations or elevators at upper floors in tall buildings, soon followed.

4.5.2 A Fuzzy Definition of Human Life History

In our case, I would like to apply the methodology of Fuzzy Sets and Fuzzy Logic to define the distribution for "personhood." Based on this formalism, the definition of a person is not something that occurs at a precise time in development that makes a person a person. Instead, it will be **the distribution itself** which becomes the definition of *personhood*.

Certain English words that appear to connote a particular circumscribed point in time like *conception* or *birth* are really not very precise when one is close to the event in time. For example, *conception*, as defined by the penetration of a sperm head into the membranes of an oocyte ending in the mingling of the DNA contents of the sperm and egg to reform a diploid nucleus in the zygote, can take as long as 24 hours. Mothers know that a birth can take anywhere from minutes to hours to days!

For any English term that describes an event, it probably needs to be defined by an interval on a time line rather than by a sharp point. For example, a countable infinity goes on forever in a linear direction, as do the integers, but as we know from the mathematical field of fractal geometry and number theory, there are a countably-infinite number of rational numbers between any two consecutive integers, such as zero and one (Anonymous 2008c); while there are an uncountably-infinite number of irrational numbers between any pair of integers in this same interval (real numbers, like transcendental numbers, as opposed to simply rational numbers). Furthermore, one can look recursively between any pair of such numbers and still find an infinite number of numbers in that interval no matter how close together the original numbers may have been to start with. And so it is with the infinity of linear time (counting time in units of seconds with a real-number clock). Increasing the resolution of the clock to, say, microseconds, can reveal many more biological processes than were previously recognized at lower temporal resolution. All biological processes move continuously along their own developmental time lines without consideration for the English-language terms that we humans place on events that have great significance to us. Microscopic biological time clocks tick relentlessly at their own pace without our permission. And so it is with embryogenesis as a particular case.

On this basis, one possible fuzzy definition of "personhood" is illustrated in Fig. 4.5. It includes the spectrum of prior events that lead up to the birth of a human life and at the end life concludes with death, with time measured in years. Shown as a smaller insert with gestational age measured in weeks, an initial sigmoid (or logistic) function remains at 0 from before conception (Personhood = 0) to a time

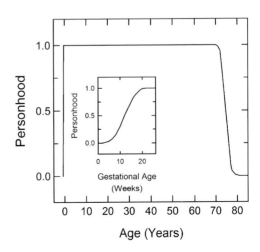

Fig. 4.5 One fuzzy-set definition of "personhood"

when it starts to rise exponentially for the first time at ~2 weeks (post implantation), next going through an inflection point (Personhood = 0.5) during organogenesis at ~12 weeks, and rising with a second exponential that asymptotically converges to become a horizontal line (Personhood = 1.0) at ~22 weeks. (22 weeks is the first time when a fetus is capable of surviving on its own in an incubator; our criterion for full personhood is survival of the neonate outside the womb. The youngest neonate known to survive so far was 21 weeks, 6 days (Anonymous 2007); of course, over the last few decades, this time has decreased inversely with improvements in neonatal technology). The horizontal line then extends out to another symmetrical mirror-imaged sigmoid function we call "death," when an individual finally expires. Thus, our fuzzy notion of "personhood" is the complete distribution of all these events, not just a simple point in time.

4.5.3 Alternative Views of Personhood and Logical Contradictions

The Roman Catholic Church and many other religious faiths arbitrarily state that "human life begins at conception." While it's true that the cells of a living human pre-embryo are human cells (not DNA from a dog or cat or cow or horse) and they're alive (not dead), whether "human life" begins at conception is still a matter of definition. After all, sperm and oocytes are also both alive and they contain human DNA, so therefore "human life" in the literal sense is present before, at, and after conception.

If, however, what we really mean by "human life" is "*personhood*", we need to recognize at least two fallacies associated with the proposition that personhood begins at conception. Firstly, if personhood were to begin at conception then there couldn't be such a thing as identical twins, since the cleavage leading to identical twins occurs long after conception, during or after implantation, and identical twins are obviously different persons (always of the same gender) – and just as obviously they do exist, yet they did not exist at the time of conception. Secondly, although rare, fraternal twins (different "persons") can merge on occasion to form a singleton embryo on implantation (a DNA mosaic), i.e., two allegedly different persons ultimately became one person.

So what is a person? Logicians enjoy pointing out that, among the class of animals, humans can be uniquely characterized as "featherless bipeds." Although this facetious observation may be largely true (chimps and gorillas don't always move on all four limbs), it totally fails to characterize who we are. We need something more substantial, something that has both moral and legal significance. This issue will be addressed more fully below in Section 4.7.

Presumably, the proposition that "human life begins at conception" might be based on the pre-embryo being a potential person with unique DNA. But identical twins have identical DNA yet are clearly different persons. Thus, having unique human DNA is not the essential criterion for personhood.

For the religious conservatives among us, a person is an entity created in the image of God (and is therefore sacred) and therefore must be treated with dignity. Legally, a person is an entity with certain inalienable rights (among which are, ideally, life, liberty, and the pursuit of happiness, as stated in the American Declaration of Independence).

In any case, deliberately taking the life of a "person" with premeditation is punishable as murder in the first degree (a capital offense in most US states). But is this what we want to have happen to medical scientists who harvest stem cells from pre-embryos? Clearly not.

4.5.4 Fuzzy Definitions and the Law

Note, that in the fuzzy distribution, the law provides for an even higher degree of respect for adults who achieve the age of majority that entitles them to certain privileges not accorded to infants or minors, like drinking alcohol, serving in the military, driving a car on a public road, getting married, voting, or even becoming President. At a silly extreme, ambitious parents might imagine that their child should not be considered a person unless he or she attained a graduate degree from a prestigious ivy-league college, like Harvard University. So, we conclude that the definition of personhood and the degree of respect for any such status achieved thus far is really a continuum.

In summary, the serious, significant, or profound respect for the moral and legal status of a human being depends on the stage of life attained *and is therefore a distribution, not a particular point on the curve.* The legal status of a mature fetus (say 22 weeks gestation), which can survive independently of the mother, is different from that of an embryo, which cannot.

4.6 A Fuzzy Definition of Death

Just as with the beginning of life in Fig. 4.5, the notion of death at that other end of the spectrum is described with an inverted-sigmoid fuzzy distribution. In the good old days, if your heart wasn't beating, and you weren't breathing (say, you couldn't "fog a mirror"), you were pronounced dead and that was it. Today, even if your EKG has flat-lined on a monitor (plus 120 seconds to play it safe), that doesn't mean you're actually "dead." The latest definition of death depends on whether you have suffered an irreversible cessation of higher mental functions.

The *biological* definition of death is "the irreversible degradation of the information content of one's brain or CNS (central nervous system)" (Hughes 2007) (equivalently, "the tissue architecture of the brain is entirely obliterated," a turn-of-phrase that neuropathologists fondly teach their residents). But this definition is frequently at odds with the *legal* definition of death. According to the law, a person is officially dead under most circumstances if cardiac arrest has occurred and a licensed physician has declared them deceased and subsequently writes and signs a death certificate citing the primary and secondary causes of death with a time-and-date stamp and submits it to the coroner's office of the local county of the decedent.

One of the curiosities that makes for ambiguity in defining *brain death* is that a few young children have been known to "drown" in a frozen lake for an hour or so and then been subsequently resuscitated. Even finger nails continue to grow for a short time after death. The issue of a patient in a deep coma or in a persistent vegetative state only adds to the fuzziness of the distribution. Recall the infamous case of Mrs. Terry Schiavo, who appeared to be awake but not truly aware of her surroundings for approximately 15 years before her feeding tube was finally removed and she was subsequently declared dead (Eisenberg 2005). An even more curious case was reported in Bagdad in which the cadaver was on the autopsy table, the prosector cut into the body, and discovered that the blood was fresh and the "cadaver" let out a loud scream! So, even experts can be wrong about whether a patient is dead or alive.

Long after a person is officially (and permanently) dead, we do have moral principles in our society and laws that govern our respect for the remains of such a deceased person. In the case where the remains in question are from a fetus >22 weeks gestation (spontaneous or third-trimester therapeutic abortion), which represents the current threshold of viability for neonatologists, then a death certificate must be filed with the county coroner's office before the products of conception can be incinerated as medical waste in the hospital.

In conclusion, pre-embryos, as potential human persons, should be deserving of moral/legal considerations (profound respect) in the same way that donated cadaver parts are. Under current law, they may be exploited for instrumental or teaching purposes so long as they are treated with dignity and disposed of properly.

4.7 Implications of Personhood

For the sake of our proof that harvesting embryonic stem cells from a blastocyst could not constitute the murder of a person, we need to expand on the interval of time in Fig. 4.3 between conception and embryogenesis (organogenesis) to arrive at an adequate level of detail. We need to understand the *necessary* and *sufficient* conditions for personhood, as well as when they could first emerge in the development of an embryo. Scientifically speaking, we don't really know exactly when all these necessary conditions are in place for the first time, but we certainly know when they're not there yet.

4.7.1 Personhood, Sentience, Sapience, Cognition, and Murder

To facilitate our discussion, we need to define certain largely-unfamiliar technical terms.

First, *sentience* is the ability of a multi-cellular organism (or even a single-celled life form) to sense or detect changes in the environment. Humans are said to have five basic senses (sight, hearing, touch, smell, and taste); six if one counts *proprioception* as a separate sense (e.g., correctly knowing the location of one's arm with one's eyes closed).

Next, *sapience* is the ability of a sentient organism or life form to perform calculations on its sensory data with the aim of executing *actions* (gasping, moving, etc.). Biologists also use the term *tropism* for this characteristic ability of life (reflexively withdrawing from noxious environmental stimuli, such as bright light, heat, electrical shocks, etc.).

Cognition (or thinking) can be thought of as a more advanced form of sapience in which the organism possesses an internal model of its external environment, which includes a representation of the organism itself as a distinguished object in that model and subject to autonomous mental manipulation by the organism for the sake of forming various hypotheses). Thus, cognition allows for rudimentary planning, such as how to successfully hide/escape from (or confront) a hungry predator recognized to be higher in the food chain.

If we accept that "personhood" requires a sapient multi-cellular organism, which in turn requires cells devoted to the task of computing changes in state of the environment, we need something that is the equivalent of a nervous system. But as described in Section 4.4 (Fig. 4.3), a nervous system doesn't develop for the first time until *organogenesis*, which in turn could never be expressed until *after* implantation in the uterus. Thus, no procedures performed (surgical or otherwise) on a pre-embryo prior to implantation could be construed as subjecting a "person" to any sort of procedure.

Finally, if one cannot murder an entity that is "not" a person (by legal definition of the range of objects of the verb "to murder") then any procedure applied to a blastocyst (or pre-embryo), including harvesting of cells from the inner cell mass that may destroy the blastocyst as a side effect, cannot possibly be called murder. **QED**.

In fact, it wouldn't even be linguistically meaningful to entertain an English sentence describing such an act. For example, Prof. Noam Chomsky, founder of Transformational Grammar at MIT, formulated the sentence "colorless green ideas sleep furiously" back in 1957 as an example of a syntactically correct English sentence that was semantically meaningless (Chomsky 1969). In the linguistics literature, academic scholars normally place an asterisk after a deliberately anomalous sentence that is being entertained strictly for the sake of argument or to illustrate a point. Therefore, to punctuate our example, the sentence "destroying a blastocyst is equivalent to murder [*]" merits such an asterisk, as it is in the same equivalence class as the set of sentences that includes "furious, colorless, green ideas," i.e., one that is semantically meaningless.

In summary, a pre-embryo is not *human life* in the sense that some would have us believe; certainly, a pre-embryo is human tissue with human DNA, but it is **not** "human life" in the sense of being a *person*. A pre-embryo is only a "potential" person. Of course, it should be entitled to a certain degree of respect as would any human tissue, like the proper disposal of the products of conception (POC) by a hospital or birth clinic in which a woman suffered a miscarriage in the second or third trimester. Such tissues are not haphazardly thrown away in the trash but are routinely incinerated as medical waste.

To demonstrate the foolishness of obfuscating or dissembling about whether all human tissues deserve equal respect, if having human DNA or the presence of living human cells were the only requirement for this exalted status then my finger-nail clippings, my hair follicles, my blood, my saliva, etc. could hypothetically be called "human life." For a fuller discussion of the considerations that must be given to establishing a theory of moral status that extends from animals to humans or even alien beings were they to exist, the last chapter of a new book by bioethicist Ronald Lindsay (Lindsay 2008) should be consulted.

In conclusion, the life of a human person does not "begin at conception," but could only begin three or four weeks later at the earliest, following implantation in the uterus, a time far beyond that at which embryonic stem cells would be collected. Prof. Michael Gazzaniga of the University of California at Santa Barbara, who remains a member of the President's Council on Bioethics, has suggested that the human person comes into being only with the development of a brain (Gazzaniga 2008). Prior to that point, we truly have a human organism, but one lacking the rights of a person.

Therefore, granting full moral respect to a preimplantation embryo makes no sense. But this is not to say that such entities have no moral status whatsoever. They are, in fact, human beings, who under the right circumstances (implantation in a uterus) will potentially become persons (Green 2001).

4.7.2 Necessary and Sufficient Conditions for Personhood

Following in the tradition of philosophical inquiry begun by Immanuel Kant, Georg Hegel, and John Locke (Locke 1690), we now have an adequate basis to describe our main hypothesis that *personhood* (and all the rights and privileges that we normally accord to human beings) requires four jointly necessary and sufficient conditions:

1. Rationality (the ability to anticipate, detect, and avoid logical contradictions in normal speech, and particularly inconsistencies between speech and action);
2. Autonomy (*conation*, independent agency or free will by the entity);
3. Identity (self-awareness or consciousness of self);
4. Instrumentality (a physical embodiment required for computational capacity, thinking, or mind; it is assumed that there cannot be such a thing as a disembodied intelligence. In the case of mammals, such a physical embodiment is obviously the brain).

Focusing on Item 3, identity, we would like to have an operational marker for consciousness or self-awareness. This is difficult when the communication skills of an infant are non-linguistic, yet he or she is certainly self-aware. Infants really are people because they exhibit non-verbal means to demonstrate their self-awareness, and even a live fetus in the womb of the mother shares this skill to a lesser extent.

But we can say with certainty that not having a brain is a conspicuous impediment to self-awareness. Such an entity would have no claim on personhood. Therefore, exploiting such an entity in the laboratory, for any instrumental purpose, regardless of its potentiality, would be neither immoral nor illegal.

4.8 The Slippery Slope Argument Against Embryonic Stem-Cell Therapeutics

A more nuanced argument against stem-cell research was brought by William Hurlbut, a physician and ethicist best known as a member of the President's Council on Bioethics, which we may call "the slippery-slope argument." What Dr. Hurlbut argues is that "until you have a self-conscious being, you don't have moral value. We don't know exactly what consciousness is, but most neurophysiologists don't believe that there's consciousness present before 18 or 20 weeks at the earliest. If that's one's criterion, one could probably justify the instrumental use of human embryos up to maybe 20 weeks. So, without a strong moral principle, you may very well see the argument move from 14 days [what we have been calling pre-embryos] to later stages [slippery slope]. So, at least at the Federal-funding level, we should preserve the principle of the defense of human life from its earliest origins in the one-cell stage" (Fitzgerald 2007).

So, Hurlbut is saying that his way to solve the slippery-slope problem is to go back to the zygote (human life begins at conception) and stop there. But that's only one way (in my view, a silly way) to solve the slippery slope problem. A more precise way would be to exploit pre-embryos (what he calls instrumental use), but not embryos (post implantation) by law. We're still perfectly safe in this 14-day range, as there can be no physical basis for sapience if there is no computational structure that we can identify under the microscope as a nervous system.

4.9 Technological Approaches to Transcending the Ethical Objections

4.9.1 The Hurlbut Solution

Dr. Hurlbut has his own approach to resolving the moral dilemma which he calls *ANT* [for altered nuclear transfer] (Hurlbut 2008) that involves the creation of embryos that are destined to fail to develop into human beings beyond a certain stage due to genetic tinkering with certain critical-path genes. Prof. Rudolph Jaenisch of MIT has even demonstrated a proof-of-concept for this proposal in mice (Meissner and Jaenisch 2005). However, to my knowledge, religious fundamentalists have never seriously embraced this proposal.

4.9.2 The ACT Solution

Another approach is to harvest embryonic stem cells without destroying an embryo, thereby end-running the ethical and moral dilemmas associated with the destruction of an embryo, as proposed by Dr. Robert Lanza, M.D., Chief Scientific Officer of Advanced Cell Technology of Worcester, MA (Klimanskaya et al. 2006). Lanza suggested that it should be possible to develop a cell line from a single cell extracted from an eight-cell *pre-embryo*, allowing the remaining seven cells to develop into a normal embryo. This technique is ethically and procedurally equivalent to a *pre-implantation genetic diagnosis* or *PGD* procedure. According to the Society for Assisted Reproductive Technology (SART), about 5,000 IVF couples opt for PGD every year. Lanza now reports that he has achieved an increase in efficiency of creating stem-cell lines by PGD from only 2 percent in August 2006 to 50 percent today (Chung et al. 2008). However, even though this approach, in principle, does not destroy any embryos, bioethicists and political conservatives still have objected on the grounds that this approach represents what they call the *commodification* of human embryos, an alleged starting point for harvesting raw materials from young human beings.

4.9.3 The iPS Cell Solution

The most recent approach has been based on a new class of cells known as *iPS* (induced pluripotent stem) cells, which are derived from adult rather than embryonic tissues. This cell class was first derived in mice (Nakagawa et al. 2006; Okita et al. 2007; Maherali et al. 2007; Wernig 2007), but within a few short months was reproduced using human cells in tissue culture (Yu et al. 2007; Takahashi et al. 2007). The creation of iPS cells (iPSCs) has been compared by some scientists to "man's first walk on the moon." Four genes (Oct3/4, Sox3, Klf4, and c-Myc) were inserted into fibroblasts using a viral vector to induce the pluripotent state. The c-Myc gene is a known retroviral oncogene and so could not be used in human therapy. However, Yamanka and his coworkers in Kyoto have recently devised a modified protocol that deletes the c-Myc gene, completely avoiding tumorigenesis (Nakagawa et al. 2008). Most recently, James Thomson and his team at the University of Wisconsin have developed a plasmid technique for inserting necessary iPS genes into a cell rather than a viral vector (Yu et al. 2009).

These approaches introduce the problem of contaminating human DNA with ectopic viral genes from the vector used to import one or more reprogramming genes into human cells. These viral genes could conceivably be responsible for triggering cancer at a future time. Recently, an international team of researchers has succeeded in purifying the related protein products of these genes (using recombinant techniques in *E. Coli* bacteria), and these proteins by themselves have been shown to achieve the same reprogramming effect as the original genes did. The proteins were introduced into both mouse and human cells by attaching poly-arginine peptide tails and by using a few other cell-biology tricks (Kim, et al; 2009). Finally, new techniques using only chemokines are being worked on, so that none of these four genes will have to be introduced into the cells to induce pluripotency, and we expect that this result will be reported in the literature soon.

4.9.3.1 Why ECSs Remain Necessary Despite the Existence of iPSCs

It can be stated with confidence that iPSCs could lead to therapeutic interventions. For example, Jaenisch's Lab at MIT has already succeeded in demonstrating that sickle cell anemia can be cured in a mouse model using iPSCs that carry a corrected gene for hemoglobin inserted into the cells using standard genetic-engineering techniques (Hanna et al. 2007). However, scientists at the University of California in Santa Barbara have uncovered approximately 1,000 genes whose upward- or downward-regulation in iPS cells is slightly different from that found in hESCs (Xu et al. 2009), and nobody knows yet what the significance of this difference may be.

Even in the felicitous event that iPS cells prove their promise, embryonic stem cells will still be needed as controls. As carefully explained in a paper published in *Cell Stem Cell* by some of the pioneers in the field ["New Advances in iPS-Cell Research Do Not Obviate the Need for Human Embryonic Stem Cells" (Hyun et al. 2007)], hESCs were an essential step in discovering iPS cells and were not independent of them. The inspiration for iPS cell research came from an earlier stem cell study in which human body cells were reprogrammed by fusion with hESCs. From this earlier study it was hypothesized that hESCs have defined factors that induce pluripotency, thus leading to the first iPS cell breakthrough in 2006. In short, the recent advancements in iPS cell research would not be possible if it were not for the many years of dedicated hESC research that preceded them. Thus, the authors state, "we cannot support the notion that iPS cell research can advance without hESC research."

Many challenges remain before iPSC therapies can become routine in the clinic. These include the poor reprogramming efficiency of iPSCs derived from a human patient skin punch-biopsy (especially those from geriatric patients), the amplification of these cells in culture into millions of cells, the subsequent differentiation of these cells along the specific pathway of the tissue damaged in the patient's body, the route-of-administration (IV or infused into an organ by direct injection), dosage (thousands or millions of cells), frequency-of-administration (over the course of months or years), and the timing of the onset of administration after traumatic injury (curiously, administration too soon after injury turns out to be counterproductive) (E. Marban, 2009, personal communication). Another critical challenge is how to verify that the new iPSCs engraft into the tissue-location of injury and don't just die in a few days by apoptosis or wander aimlessly in the blood stream. Finally, even if the cells wind up in the right location and don't die immediately, will they truly participate in the tissue architecture by fulfilling their proper structure or function (native fate)? There is a risk that the new cells may just happily sit there (taking up space) without carrying out any potentially life-extending function, since the host tissue extra cellular matrix (ECM) has sadly issued contrary instructions, forcing them to remain idle (see Chapter 13 later in this volume). These challenges will probably endure for the next ten years, since they all involve human clinical trials that we know consume significant time, whether done under FDA auspices or not. In the meantime, ESC therapeutics must be allowed to proceed in parallel to ensure that the ultimate solutions utilized therapeutically are the optimal ones.

In a curious twist, the opponents of stem-cell research attempted to take credit for the development of iPSCs on the grounds that their hard-line position created pressure that led to the advent of iPSCs. The reality is that their position has slowed research and the potential for eliminating sickness and disease from the world sooner by drawing a moral line in the sand that has no basis in biology.

4.10 Conclusion

If only a few of the promises of embryonic stem-cell research to cure disease and suffering turn out to come true, the opportunity cost of depriving humanity of this form of therapy for even a few years, could be catastrophic. Yet scientists are sometimes confused about their obligation to provide means to cure disease. Prof. Leonard Hayflick of the University of California in San Francisco has said, "scientists ought not to tamper with longevity, lest we lead ourselves down a slippery slope with unpredictable consequences" (Hayflick 2000).

As J. Craig Venter, scientist, entrepreneur, and genetics pioneer, pointed out, "the revolution in genetics will ... give us the ability to change humanity for its long-term survival. That's not something that has been possible to contemplate before, except in science fiction movies. ... There is probably nothing more important to study about human biology than *stem cells*. The fact that it has been blocked by the Bush Administration on religious grounds is one of the great intellectual tragedies of our century" (Goodell 2007).

Those of us who live in California expect that the successful resolution of various legal challenges to Proposition 71, which finally allowed CIRM to sell its bonds in the Fall of 2007, no longer encumbered by two years of litigation, will permit stem cell research to advance at a rapid pace. Of course, following the 2008 US presidential election, the future could look considerably brighter, now that President Obama has by Executive Order on the morning of January 9, 2009 lifted the ban imposed by the previous Bush Administration back in August of 2001 (Obama 2009). This will allow NIH funding for both embryonic and iPSC research, and such funding could instantly unblock similar bans imposed by individual US states.[4]

---

[4] Not everyone was as pleased with this decision as I was. In particular, 10 of the 18 members of the President's Council on Bioethics created by former President Bush wrote a statement condemning the change in policy on the grounds that potential loopholes in the change could lead to reproductive cloning of human beings in the absence of specific legislation prohibiting such a practice (Meilaender et al. 2009).

So, what can we conclude about the flawed argument that preimplantation embryos deserve our full moral respect and thus cannot be utilized for any therapeutic purpose whatsoever? Our definition of "personhood" above implies that preimplantation embryos do not warrant full moral respect as do post-implantation embryos with a developing nervous system. But this is not to say that preimplantation embryos have no moral status at all. Because of their potential to become human beings, they have a moral status at least as important as the products of conception (POC) as any gynecologist would provide for them in a hospital setting following a therapeutic or spontaneous abortion (miscarriage), which, as we stated earlier, means that these human tissues should be incinerated upon disposal and not just thrown away in the regular trash as one might do with fingernail clippings. Physicians must dispose of human tissues in the manner prescribed by law, and preimplantation embryos should be no exception. All hospitals have such procedures already in place.

Finally, whether one believes our human condition derives from a purposeful act of God or else appeared somehow as a subtle, emergent property of random Darwinian evolution, it is evident that the predicament of human mortality was thrust upon us without our consent. We are, to our knowledge, the only species that has an interest in our own mortality. So how shall we deal with this tragic phenomenon we call "death" in the future, assuming that we will have real choices available for medical intervention?

For the first time in human history, stem cell biology – in conjunction with other technologies described in this book – will allow us to choose which of the following (nearly contemporaneous) quotations is the one that makes sense for us.

> (1) "What a piece of work is man! How noble in reason! How infinite in faculty! In form and moving how express and admirable! In action how like an angel! In apprehension how like a god! The beauty of the world! The paragon of animals!"
> – *Hamlet*, Act II, Scene 2 – William Shakespeare (1601)
> (2) "... the life of man... solitary, poor, nasty, brutish, and short."
> – *Leviathan*, Chapter XIII – Thomas Hobbes (1651)

Hopefully, we will choose the first quotation rather than the second.

References

Anonymous (1987) *Donum Vitae*, "Congregation for the Doctrine of the Faith: Instruction *Donum Vitae* on Respect for Human Life at Its Origins and for the Dignity of Procreation" The Vatican, Rome, Italy

Anonymous (2007) "World's Youngest Baby Born In Miami: Amillia Taylor Born After 21 Weeks, 6 Days" *Miami News*, February 19 (http://www.local10.com/news/11053141/detail.html)

Anonymous (2008a) *Dignitas Personae*, Congregation for the Doctrine of the Faith: Instruction *Dignitas Personae* on Certain Bioethical Questions (The Vatican, Rome, Italy; December 12) http://www.usccb.org/comm/Dignitaspersonae/Dignitas_Personae.pdf

Anonymous (2008b) Prenatal Radiation Exposure: A Fact Sheet for Physicians. CDC and California Public Health (2008) http://bepreparedcalifornia.ca.gov/EPO/Partners/Healthcare Providers/ForHealthcareProviders/PracticeGuidelinesFactSheets/BioterrorismPreparedness/Prenatal+Radiation+Exposure.htm

Anonymous (2008c) Rational Number. Wikipedia http://en.wikipedia.org/wiki/Rational_number
Barlow D (2007) Inside the Human Body, National Geographic Channel TV, Sunday, September 16
Brownback S, Landrieu M (2007) Human Cloning Prohibition Act S.1036 http://brownback.senate.gov/pressapp/record.cfm?id=271575
Bush GW (2003) State of the Union Address http://www.whitehouse.gov/news/releases/2003/01/20030128-20.html
Bush GW (2005) State of the Union Address http://www.whitehouse.gov/news/releases/2005/02/20050202-11.html
Bush GW (2006) State of the Union Address http://www.whitehouse.gov/news/releases/2006/01/20060131-10.html
Bush GW (2008) State of the Union Address http://www.whitehouse.gov/news/releases/2008/01/20080128-13.html
Caplan A (2003) The problem with 'embryo adoption': Why is the government giving money to 'snowflakes'? *MSNBC.COM* http://www.msnbc.msn.com/id/3076556/
Chomsky N (1969) Aspects of the Theory of Syntax. MIT Press, Cambridge, MA
Chung Y, Klimanskaya I, Becker S, Li T, Maserati M, Lu SJ, Zdravkovic T, Genbacev DO, Fisher S, Krtolica A, Lanza R (2008) Human embryonic stem cell lines generated without embryo destruction. Cell Stem Cell 2:113–117
da Vinci, L (c 1510-1) Anatomical Folio Figure E-1, p. 261 and Note 67, p. 285
Dods M (1952) The City of God (translated from Saint Augustine), Great Books of the Western World, No.18; pp. 129–620, University of Chicago Press, Chicago
Eisenberg JB (2005) Using Terri: The Religious Right's Conspiracy To Take Away Our Rights. Harper Collins, San Francisco
Fathers of the English Dominican Province (1952) The Summa Theologica of Saint Thomas Aquinas (translation) Vols. I and II. Great Books of the Western World, Nos. 19 and 20, University of Chicago Press, Chicago, IL
Feuerherd J. (2006), Missouri vote may point to stem cell future. National Catholic Reporter, p. 1 (September 15)
Fitzgerald M (2007) Q&A with William Hurlbut: Embryonic stem cells without embryos. MIT Technol Rev 110:6:22–23.
Gazzaniga MS (2008) Human: The Science Behind What Makes Us Unique Ecco, New York
George RP, Tollefsen C (2008) Embryo: A Defense of Human Life. Doubleday, New York
Goodell J (2007) Interview with Craig Venter, genetics pioneer. Rolling Stone Magazine, 1039:88–90
Green R (2001) The Human Embryo Research Debate: Bioethics in the Vortex of Controversy. Oxford University Press, New York
Hanna J, Wernig M, Markoulaki S, Sun CW, Meissner A, Cassady JP, Beard C, Brambrink T, Wu LC, Townes TM, Jaenisch R (2007) Treatment of sickle cell anemia mouse model with iPS cells generated from autologous skin. Science 318:1920–1923
Hayflick L (2000) Commentary: The future of aging. Nature 408:267–269
Hendricks M (2006) Keyes and friends are way off base. Kansas City Star, p. B1 (September 13)
Hughes J (2007) Cheating death: Dispatches from the war against mortality: Bringing people back from the brink, advances in diagnostics and medicine are yet again changing our definition of death – and may eventually break down the concept of death altogether. New Scientist, 196:44–45
Hurlbut WB (2008) Altered Nuclear Transfer (ANT). http://www.alterednucleartransfer.com.
Hyun I, Hochedlinger K, Jaenisch R, Yamanaka S (2007) New advances in iPS cell research do not obviate the need for human embryonic stem cells. Cell Stem Cell 1:367–368
Kim D, Kim CH, Moon JI, Chung YG, Chang MY, Han BS, Ko S, Yang E, Yul Cha K, Lanza R, Kim, KS (2009) Generation of human induced pluripotent stem cells by direct delivery of reprogramming proteins. Cell Stem Cell 5:472–476
Klimanskaya I, Chung, YS, Becker S, Lu SJ, Lanza R. (2006) Human embryonic stem-cell lines derived from single blastomeres. Nature 444:481–485 [with Addendum on p. 512]

Lindsay RB (2008) Future Bioethics: Overcoming Taboos, Myths, and Dogmas. Prometheus Books, New York

Locke J (1690) An Essay Concerning Human Understanding. In: Britannica Great Books Series, Vol. 35, pp. 93-395, University of Chicago Press, 1952

Maherali N, Sridharan R, Xie W, Utikal J, Eminli S, Arnold K, Stadtfeld M, Yachechko R, Tchieu J, Jaenisch R, Plath K, Hochedlinger K (2007) Directly reprogrammed fibroblasts show global epigenetic remodeling and widespread tissue contribution. Cell Stem Cell 1: 55–70

Meilaender G, McHugh P, Carson B, Eberstadt N, Bethke Elshtain J, Gómez-Lobo A, Hurlbut W, Landry D, Lawler P, Schaub D (2009) Federal funding of embryonic stem cell research. Bioethics Forum (March 25). http://www.thehastingscenter.org/Bioethicsforum/Post.aspx?id =3298

Meissner A, Jaenisch R (2005) Generation of nuclear transfer-derived pluripotent ES cells from cloned Cdx2-deficient blastocysts. Nature 439:212–215. http://www.wi.mit.edu/news/archives/2005/rj_1016.html

Meunier BB, Yager RR, Zadeh LA, Eds. (2000) Uncertainty in Intelligent and Information Systems: Advances in Fuzzy Systems – Applications and Theory, Vol. 20, World Scientific Publishing Company

Nakagawa M, Koyanagi M, Tanabe K, Takahashi K, Ichisaka T, Aoi T, Okita K, Mochiduki Y, Takizawa N, Yamanaka S (2008) Generation of induced pluripotent stem cells without myc from mouse and human fibroblasts. Nat Biotechnol 26: 101–106

Nakagawa M, Koyanagi M, Tanabe K, Takahashi K, Ichisaka T, Takahashi K, Yamanaka S (2006) Induction of pluripotent stem cells from mouse embryonic and adult fibroblast cultures by defined factors. Cell 126:663–676

Nardo D (2009) Cure Quest: The Science of Stem Cell Research. Compass Point Books, Mankato, MN.

Obama B (2009) Removing barriers to responsible scientific research involving human stem cells (March 9). http://www.whitehouse.gov/the_press_office/Removing-Barriers-to-Responsible-Scientific-Research-Involving-Human-Stem-Cells.

Okita K, Ichisaka T, Yamanaka S (2007) Generation of germline-competent induced pluripotent stem cells. Nature 448:313–317

Ord, T (2008) The scourge: Moral implications of natural embryo loss. Am J Bioethics 8: 12–19

Takahashi K, Tanabe K, Ohnuki M, Narita M, Ichisaka, T, Tomoda K, Yamanaka S (2007) Induction of pluripotent stem cells from adult human fibroblasts by defined factors. Cell 131:861–872

Vesalius A (1543) De Humani Corporis Fabrica Librorum Septem and Epitome (Basel, Switzerland)

Walter JJ, Shannon TA (2005) Contemporary Issues in Bioethics: A Catholic Perspective. Roman and Littlefield Publishers, Inc., New York, Footnote 5, Chapter 5, p. 68

Wentworth Thompson D (1952) The History of Animals [Historia Animalium (by Aristotle)] (translation from the Greek) Book VII, Chapter 3, p. 109, Section 583b Great Books of the Western World, No. 9, The Works of Aristotle, Vol. II; University of Chicago Press, Chicago, IL

Wernig M, Meissner A, Foreman R, Brambrink T, Ku M., Hochedlinger K, Bernstein BE, Jaenisch R (2007) In vitro reprogramming of fibroblasts into a pluripotent ES-cell-like state. Nature 448:318–324

Wilcox AJ, Baird DD, Weinberg CR (1999) Time of implantation of the conceptus and loss of pregnancy. The New England J Med 340:1796–1799

Xu N, Papagiannakopoulos T, Pan G, Thomson JA, Kosik KS (2009) MicroRNA-145 regulates OCT4, SOX2, and KLF4 and represses pluripotency in human embryonic stem cells. Cell 137:647–658

Yu J, Vodyanik MA, Smuga-Otto K, Antosiewicz-Bourget J, Frane JL, Tian S, Nie J, Jonsdottir, GA, Ruotti V, Stewart R, Slukvin II, Thomson JA (2007) Induced pluripotent stem-cell lines derived from human somatic cells. Science 318:1917–1920

Yu J, Hu K, Smuga-Otto K, Tian S, Stewart R, Slukvin II, Thomson JA (2009) Human induced pluripotent stem cells free of vector and transgene sequences. Science 324:797–801

Zadeh LA (1975) The calculus of fuzzy restrictions. In: Zadeh LA, Fu K, Tanaka K, Shimura M (eds) Fuzzy Sets and Applications to Cognitive and Decision Making Processes, pp. 1–39. Academic, New York

Zadeh LA, Yage RR (1987) Fuzzy Sets and Applications: Selected Papers by L.A. Zadeh

Chapter 5
Evolutionary Origins of Aging

Joshua Mitteldorf

Contents

5.1	Introduction	88
5.2	Four Perspectives on Evolution of Aging	89
5.3	Theories of Aging Based on Accumulated Damage	90
5.4	Theories of Aging Based on Neglect: Is There Selection Pressure Against Senescence in Nature?	93
5.5	Theories Based on Tradeoffs: Pleiotropy and the Disposable Soma	97
	5.5.1 Classic Pleiotropy: Genetic Tradeoffs	98
	5.5.2 Tradeoffs in Resource Allocation: The Disposable Soma Theory	101
	5.5.3 Adaptive Pleiotropy: An Alternate Theory	103
5.6	Adaptive Theories of Aging	104
	5.6.1 Programmed Death in Semelparous Plants and Animals	106
	5.6.2 Caloric Restriction and Other Forms of Hormesis	107
	5.6.3 How Might Senescence have Evolved as an Adaptation? Two Ideas that Don't Work	108
	5.6.4 How Might Senescence have Evolved as an Adaptation? The Demographic Theory	110
5.7	How could Evolutionary Theory have been So Wrong?	114
5.8	Implications for Medicine	115
	5.8.1 Telomeres	116
	5.8.2 Apoptosis	116
	5.8.3 Inflammation	117
	5.8.4 Caloric Restriction Mimetics	117
	5.8.5 Thymic Involution	118
	5.8.6 Regeneration as a Target of Anti-aging Research	118
5.9	Conclusion	119
References		119

J. Mitteldorf (✉)
Department of Ecology and Evolutionary Biology, University of Arizona, Tucson,
AZ 85720, USA
e-mail: josh@mathforum.org

5.1 Introduction

Centuries of charlatans have bequeathed to anti-aging medicine the suspicion with which it is regarded today. But history is not the only source of skepticism; there is also a pervasive feeling about the benign power of nature, which indeed has been buttressed by evolutionary theory.

Scientists and laymen alike imagine that it must be terribly difficult for technology to do what æons of evolution have failed to accomplish. Natural selection has made us as strong and as smart and as robust as we are, and natural selection has arranged for us the life spans that we know. "Evolution is smarter than you are," according to the Second Law of Leslie Orgel. Who among us has the hubris to try to improve upon natural life spans?

George Williams, along with Peter Medawar (1952) in the 1950s, was responsible for the ideas that underlie the modern evolutionary theory of senescence. In his seminal paper, Williams (1957) theorized that aging derives from a concurrent, multi-factorial deterioration of all bodily systems. If any single system (let alone a single gene) were to substantially depress life span without offering compensatory benefit, the situation would quickly be addressed by natural selection.

> Basic research in gerontology has proceeded with the assumption that the aging process will be ultimately explicated through the discovery of one or a few physiological processes... Any such small number of primary physiological factors is a logical impossibility if the assumptions made in the present study are valid. This conclusion banishes the "fountain of youth" to the limbo of scientific impossibilities where other human aspirations, like the perpetual motion machine and Laplace's "superman" have already been placed by other theoretical considerations. Such conclusions are always disappointing, but they have the desirable consequence of channeling research in directions that are likely to be fruitful.

In the half century since these words were written, a tidal wave of empirical data have indicated to us that life spans are indeed plastic, that human life span is not only extensible in principle, but is now longer by more than a decade (in the USA) than it was in 1957. Dozens of point mutations have been discovered that extend life span in every laboratory model for which they have been sought; and hormetic interventions, especially caloric restriction, have been shown to have a robust and reliable effect.

Evolutionists were quick to embrace Williams's theoretical predictions, but they have been slow to acknowledge the contrapositive: If a single gene can extend life span without profound pleiotropic costs, then the core of the theory is called into question. If a single small molecule can extend life span with side-effects that seem to be primarily beneficial, then we must conclude that natural selection is not in the business of maximizing life span.

The evolutionary implications of biological research on aging, and the medical implications of a revised evolutionary perspective are the subject of this chapter. There is substantial evidence that the pleiotropic theories of aging have failed, and further evidence that natural selection (at the population level) has crafted an aging program, despite its individual cost. I acknowledge that this is not orthodox evolutionary theory; but expositions of orthodox evolutionary theory abound (Stearns 1992; Austad 1999), while the opposition case, based on observation, has only recently been articulated, and so it seems appropriate to summarize here the evidence for genetically programmed aging, and to explore the consequences of such a phenomenon for evolutionary theory.

The bottom line is very promising. When we understand that aging is genetically programmed, the prospect for finding simple interventions with a global effect on aging becomes much brighter. As for "channeling research in directions that are likely to be fruitful," the health impact of life extension technology is poised to demonstrate that life extension research is competitive with the most cost-effective investments that we as a society are making (de Grey 2006a, b).

5.2 Four Perspectives on Evolution of Aging

The evolutionary origin and significance of aging constitute the most fundamental unanswered question in biological science today. There are many ideas on the subject, but they may be broadly categorized in four camps. Most evolutionary theorists are united in a particular camp, which likes to claim the high ground in the debate; in fact, many regard aging as a "solved problem". However, I will argue that the three most popular camps are the least grounded in empirical science. The more we know about the biochemistry and genetics of aging, the more it looks as though a major revision of evolutionary theory will be necessary.

The academic implications of such a revision are potentially profound; but the practical import looms just as large. Fundamental beliefs about the nature of aging affect judgments of our prospects for success in concocting anti-aging therapies. Four different ideas about the meaning of aging lead to contrasting approaches to research and interventions to slow human aging. I have been a proponent of the most optimistic of the four camps, and after reading this chapter, I hope that you, too, might be inspired to believe that anti-aging medicine has exciting near-term prospects.

We will review the history of each of the four camps in turn, examine the logical connections among them, and summarize the evidence for each.

1. *Aging as the accumulation of damage*. The "entropic" theory of aging. This is the oldest idea about what aging is, and remains the most prevalent among scientists and non-scientists alike. Ideas of accumulated damage are so seductive, in fact, that they have tainted the thinking of many a scientist who "ought to know better". It is interesting to note, however, that evolutionary theorists have long recognized the inadequacy of damage-based approaches to explain aging and have sought for an understanding of the significance of aging in the context of natural selection. But for those who *do* believe that aging results from the accumulation of random damage, the only way to slow aging is to prevent or repair that damage, in all its diverse forms.

2. *Aging as irrelevant to evolution.* Natural selection has had no chance to affect the dynamics of senescence, because so few organisms in nature ever attain to an age at which they are affected.
3. *Aging as the result of evolutionary tradeoffs.* "There are no aging genes – only fertility genes that have detrimental side-effects." It is this perspective that has constituted the evolutionary orthodoxy since 1957. If you believe that aging results from evolutionary tradeoffs (most prominently with fertility), then it may be possible to intervene to lengthen life spans, but success is likely to be limited, and to come at a cost to other metabolic functions.
4. *Aging as a genetic program*, evolved independently of side-effects and selected for its own sake. Though there is broad empirical support for such theories, they conflict deeply with the standard framework of population genetics. If programmed life span is your perspective, then you might expect that dramatic extension of life span can result from manipulation of the program at its highest level, effecting a change in the rate of aging without compromising any other metabolic faculties along the way.

5.3 Theories of Aging Based on Accumulated Damage

It is an ancient and enduring misconception that aging derives from a physical necessity. In contemporary analysis, there remain a few advocates for this perspective, most notably Gavrilov and Gavrilova (1978, 2004), who emphasize similarities between the generalized characteristics of mortality curves and failure rates of complex non-living systems.

Through the mid nineteenth century, ideas about wearing out and chemical degradation were sufficiently amorphous that no contradiction was apparent in this position. But with the formulation of thermodynamic laws, it became clear to those who thought most deeply on the subject that there is no physical necessity for aging. It was the German mathematician Rudolf Clausius (1867) who first had the insight that entropy was not just a vague descriptor, but a physical variable that could be quantified precisely. Entropy change finds a precise expression in terms of work, heat, and energy transfer; but entropy also can be understood as a measure of disorder. In a closed physical system, with no energy flowing in or out, entropy must increase monotonically, and thus a closed system can only become less ordered over time.

This indeed is the physical law underlying the degradation of inanimate objects, and the decay of (dead) organic matter. The Second Law of Thermodynamics is a deep cause for the rusting of metal, the melting of ice, and the resistance losses whenever electricity flows through a wire. Closely-related processes on a somewhat larger scale explain why knives get dull, why beaches erode, and why machine parts fail with wear. And even the decay of dead leaves – which is accomplished by insects and bacteria – is closely consonant with the Second Law, because these biological agents of decay have no other source of *free energy*, and can *only* process organic matter in a direction of increased entropy.

But living things are different. It is a defining characteristic of life that it is capable of gathering *free energy* from the environment, creating new order and structure within itself, while dumping its waste entropy back into the surroundings. Green plants take in free energy in the form of sunlight. Animals take in chemical free energy in their food. Both are able to build up from seed to a fully robust soma. This initial feat of anabolism is far more impressive than anything that could be demanded in the way of repair.

We repair an automobile for a time, but at some point in its life, the economic cost and benefit lines cross, and then it is cheaper to replace the entire car than it is to repair it. It is tempting to think that natural selection has followed such a calculus, and that investment in a new offspring becomes at some point preferable to maintaining the soma of the parent. But this analogy is flawed. The reason that deep engine repairs can be so expensive is that the engine must be completely dismantled and re-assembled in order to replace a few dollars' worth of rings and seals at the engine's heart. Somatic repairs are not like this; body parts are repaired from the inside out. Cells can be discarded, recycled and replaced one-by-one, with stem cells serving as the source of fresh, young muscle, bone and nerve tissue. Individual protein molecules can be degraded into amino acids, and new, perfect proteins grown from ribosomal templates. DNA repair is an ongoing process in cell nuclei.

In terms of free energy, these continuous processes are enormously more economical than the creation of a new soma from seed. In addition, there is inevitable attrition, as young animals and seedlings grow into mature adults with very low rates of success. As an energy investment, it is always more efficient to expend resources in repairing damage to a mature, proven parent than to build a new, mature and fertile adult from seed.

This reasoning demonstrates that accumulated damage is not a viable evolutionary explanation for aging. It is not to say that aging has nothing to do with physical and chemical wear. Certainly much of the chemical and physical degradation that characterizes senescence is fairly characterized as wear or stochastic chemical damage. But in principle, there is no reason the body could not repair this damage to its original standard of perfection, and the body should be highly motivated to do so. For example, a muscle that is stressed in exercise with microscopic tears is repaired in such a way that it grows stronger than before it was damaged. A bone that has been fractured will knit back together such that the site of the break is denser and stronger than surrounding bone, and future stresses are unlikely to re-fracture the bone in the same location. The ability to regenerate damaged body parts persists in many lower animals, but has been lost or actually suppressed in birds and mammals (Heber-Katz et al. 2004). It is a compelling evolutionary question: since repair is so efficient and economical, why has natural selection chosen consistently to permit the soma to degrade with age?

In writing theoretically about aging, Hamilton (1966) assumed that repairing a tissue "perfectly" is a standard that can only be approached asymptotically, and which would, theoretically require an infinite amount of energy. The idea of a high energy cost of repair later became the basis of Kirkwood's (1977) Disposable Soma theory for the evolution of aging. But Hamilton's assumption was fallacious. Building the tissue the first time did not require infinite energy, and the body's ongoing repair and recycling systems are far more energy-efficient than its developmental anabolism. Based on the infinite energy assumption, Hamilton claimed to "prove" that somatic degradation at each moment in time is implied by evolutionary theory. Later, Vaupel et al. (2004) exposed this fallacy, introducing a metabolically reasonable model that posited a finite marginal cost of ongoing growth, increasing fertility, and lowered mortality risk. Retracing Hamilton's logic, he "proves" that natural selection on the individual level should favor ever-increasing fertility and ever-decreasing mortality. (He also sites numerous examples in nature of this "negative senescence".) This has not stopped contemporary theorists from modeling senescence based on an assumed high cost of repair (Mangel 2008).

To their credit, most evolutionary theorists of today do not fall into the trap of believing that accumulated damage is a sufficient evolutionary theory of aging. In fact, it was more than a century ago that August Weismann et al. (1891) helped to establish the idea that damage to living things is *not* a viable basis for an evolutionary theory. He originally proposed that compulsory death exists for the purpose of eliminating older individuals, which are inevitably damaged:

> Suppose that an immortal individual could escape all fatal accidents, through infinite time – a supposition which is of course hardly conceivable. The individual would nevertheless be unable to avoid, from time to time, slight injuries to one or other part of its body. The injured parts could not regain their former integrity, and thus the longer the individual lived, the more defective and crippled it would become, and the less perfectly would it fulfill the purpose of its species. Individuals are injured by the operation of external forces and for this reason alone it is necessary that new and perfect individuals should continually arise and take their place, and this necessity would remain even if the individuals possessed the power of living eternally. From this follows, on the one hand the necessity of reproduction and on the other the utility of death. Worn out individuals are not only valueless to the species, but they are even harmful, for they take the place of those which are sound. Hence by the operation of natural selection, the life of our hypothetically immortal individual would be shortened by the amount which was useless to the species.

But later in life, he disavowed his earlier theory, while never explicitly acknowledging that this argument was dependent on the older, flawed ideas about "necessary" wear and tear. The logical flaw was exposed by Peter Medawar (1952) in a much later era: "Weismann assumes that the elders of his race are worn out and decrepit – the very state of affairs whose origin he purports to be inferring – and then proceeds to argue that because these dotard animals are taking the place of the sound ones, so therefore the sound ones must by natural selection dispossess the old."

In addition to these theoretical reasons for believing that accumulated damage to a living thing is not inevitable, there is evidence from the phenomenology of aging that mechanisms of repair are shut down progressively in aged individuals. A crude example: the African elephant is dependent on its teeth to grind immense quantities of savannah grass each day. Teeth wear out over time, and the elephant grows six full sets of teeth in its lifetime; after that, no seventh set is available, and an aged elephant will die of starvation for lack of them. A more common example: Reactive oxygen species (ROS) generated as by-products of the Krebs cycle are quenched by the enzymes catalase, SOD and ubiquinone. Expression of all three have been found to decline with age in humans (Linnane et al. 2002; Tatone et al. 2006). Similarly, expression of anabolic enzymes tends to decline with age, while that of catabolic enzymes trends upward (Epel et al. 2007). There is evidence, too, that DNA damage does not simply accumulate with age, but that repair mechanisms are progressively shut down (Singh et al. 1990; Engels et al. 2007). As a primary mechanism of aging, accumulated oxidative damage has attracted more attention and more support than any other form of damage; yet there is reason to believe that oxidative damage is not the primary source of aging. Naked mole rats live eight times longer than mice of comparable size, though the latter seem to be better protected against oxidative damage (Andziak et al. 2005, 2006). And life spans of mice are generally a few years, while bats live for decades, despite a higher metabolic rate and greater load of mitochondrial ROS (Finch 1990).

It is certainly true that much of the damage we associate with senescence can be traced back to oxidative damage from ROS created as a byproduct of mitochondrial processes; but biochemical protections could be adequate to protect against these hazards with "better than perfect" efficiency. That these protections are rationed, allowing the damage of senescence to accumulate: this is the primary dynamic of aging that cries out for explanation (Sanz et al. 2006). This may be an example of evolution taking advantage of a convenient, pre-existing mechanism, coöpting and regulating its action for a selective purpose.

5.4 Theories of Aging Based on Neglect: Is There Selection Pressure Against Senescence in Nature?

The theory that aging exists because there is no effective selection against it still has currency, despite definitive evidence to the contrary. The idea derives directly from Medawar (1952), who argued purely on theoretical grounds that the force of natural selection must decline with age. Edney and Gill (1968) noted that if selection pressure is truly close to zero, then mutational load alone would be a sufficient explanation for the evolutionary emergence of aging. Two independent areas of study compel the conclusion that this is not the primary explanation for aging: First, demographic surveys indicate that aging does indeed take a substantial bite out of individual fitness in nature (Ricklefs 1998); and second, that the genetic basis for aging has been conserved over æons of evolutionary history indicates that natural selection has not neglected senescence, but on the contrary has shaped aging programs to exacting specifications (Guarente and Kenyon 2000; Kenyon 2001). Evolutionary conservation of aging genetics is an observation particularly damaging to the theory of Edney and Gill, which has come to be known as *Mutation Accumulation*.

Medawar begins from a definition of individual fitness derived from Lotka (1907), via Fisher (1930): a time-weighted average of effective fertility. Fitness consists in leaving more offspring, earlier. Medawar recognized two reasons why a loss of viability late in life would be subject to less selection pressure than a similar loss early in life. First, the Lotka (1907) equation explicitly discounts fertility with time. If natural selection is a race to take over a gene pool with one's progeny, then an offspring produced early contributes more to the exponential rate of increase than would the same offspring born later. Second, there is a background rate of mortality in nature, so that even without senescence, any population cohort would decline exponentially over time. Natural selection is less sensitive to viability of the older individual to the extent that many individuals succumbed to predation, starvation or disease before they can attain to old age.

Following Medawar, this raises the possibility that aging exists in nature because it doesn't matter in nature: that nature is so competitive a realm that few (if any) individuals actually survive to die of old age, so the selection pressure against senescence is close to zero. Early in the history of this idea, Williams (1957) argued against it on general grounds: It may be true that vanishingly few individuals actually die of old age in nature, but that does not mean that senescence takes no toll. The early stages of senescence cause an individual to be slower, or weaker, or more susceptible to disease. Turning Medawar's argument on its head, Williams argued that the state of nature is such a competitive place that small increments in robustness have large consequences for survival. Williams predicted on this basis that the impact in the wild of the early stages of senescence might be substantial.

Thirty years later, the first field surveys proved that Williams had been accurate on this point. Nesse (1988) was the first to report evidence for aging in natural populations, interpreting it as evidence against the Accumulated Mutation theory, and therefore supporting the pleiotropic theory (see section on tradeoffs, below). Promislow (1991) surveyed field data on 56 species of small mammals, concluding that mortality increase with age was readily observable in the wild. Ricklefs (1998) proposed to measure the impact of senescence as the proportion of deaths in the wild that would not have occurred but for senescence. He derived statistical methods for extracting this number from animal demographics, and found a huge range, from less than 2% to more than 78% senescent deaths, with the higher numbers corresponding to long-lived species. Bonduriansky and Brassil (2002) applied a similar methodology to Ricklefs in an intensive study of a single species, the antler fly. They estimated the impact of senescence on the fly's individual fitness as 20%. Thus senescence presents a surprisingly large target for individual selection, despite a short (6 day) life span. Ricklefs (2008) compared causes of bird and mammal mortalities in captivity and in the wild, and concluded "Similar patterns of ageing-related mortality in wild and captive or domesticated populations indicate that most ageing-related death is caused by intrinsic factors, such as tumor and cardiovascular failure, rather than increasing vulnerability to extrinsic causes of mortality." This pulls the legs from under the *Accumulated Mutation* theory.

More surprising yet were results reported by Reznick et al. (2004) concerning guppies in Trinidad river pools. They collected fish from high-predation sites, where extrinsic mortality was great, and nearby sites without predators, where extrinsic mortality was much lower. Intrinsic life spans of the two populations were measured in captivity. The principal finding was that the group subject to predation had *longer* intrinsic life spans, suggesting that senescence in guppies may be adaptive as a population regulatory function (Bronikowski and Promislow 2005). This flies in the face not only of Medawar's hypothesis, but Williams's theory as well. Excessive individual reproductive rates can be a threat to the local population when they cause ecological collapse from crowding. Aging may be a particularly efficient means of stabilizing local population dynamics (Mitteldorf and Pepper 2007). Medawar's hypothesis was that aging evolves because it has no impact in the wild; but Reznick seems to demonstrate that not only does aging show a substantial impact, but under some circumstances natural selection seems to have acted explicitly to enhance that impact. The idea that natural selection could militate so strongly *against* individual fitness is difficult for evolutionary theory to assimilate; but Reznick's finding constitutes direct and affirmative evidence for aging as a genetic program, evolved for its effect on group dynamics.

A second line of evidence against the *Accumulated Mutation* theory is the existence of a conserved genetic basis for aging. Homologous genes associated with aging have been discovered in all laboratory species in which they have been sought, from protists through worms and flies to mammals. During the 1950s–1970s, when modern evolutionary theories of aging had their origin, no one was predicting the existence of these families of genes, which must go back hundreds of millions of years, to the dawn of eukaryotic life. The idea behind the *AM* theory is mutational load, i.e., that selection pressure for the genes that cause aging is so weak that they have not yet been selected out of the genome. It follows that aging genes are expected to be random defects, recently acquired and differing widely from one species to the next. This is the opposite of the observed situation.

A conserved family of aging genes codes for components of the insulin/IGF signaling pathway (Guarente and Kenyon 2000; Kenyon 2001; Barbieri et al. 2003). These hormones affect the timing of growth and development, and also the metabolic response to food. Part of their function signals the soma to age more rapidly in the presence of abundant food. This is true in *C. elegans*, where the link to aging was first characterized, and also in yeast, in fruit flies, and in mice, with details of the implementation varying substantially across taxa.

CLK-1 is a gene originally discovered in worms (Johnson et al. 1993), inactivation of which increases life span via a mechanism independent of the IGF pathway. The homologous gene in mice is MCLK1, and its deletion also leads to enhanced life span (Liu et al. 2005). In fact, the action of CLK-1 is essential for synthesis of ubiquinone, and, as a result, CLK-1 mutants are *less* able to quench the ROS products of mitochondrial metabolism – yet they live longer (LaPointe and Hekimi 2008). The first data were reported for heterozygous CLK-1 +/– worm, and it was thought that the homozygous (–/–) mutant was not viable. More careful experimentation revealed that the –/– worm would develop on a delayed schedule, and subsequently lived to a record ten times the normal *C. elegans* life span (Ayyadevara et al. 2008). This worm has no capacity to synthesize ubiquinone, and its prodigious life span is truly a paradox from the perspective of the damage theories of aging; the only prospect for understanding is provided in the context of *hormesis* (Section 5.6.2).

Other, less well-known conserved genes for aging have been catalogued by Budovsky, Abramovich et al. (2007), who conducted a widely-targeted, computer-based search for networks of genes that regulate aging across disparate taxa.

Apoptosis (Gordeeva et al. 2004) and cellular senescence (Clark 1999) (telomeric aging) are two ancient mechanisms of programmed death at the cellular level. It has been known for decades that these mechanisms remain active in multicellular organisms, including humans; but it has always been assumed that their presence can be explained by an adaptive benefit for the individual. Apoptosis is important for ridding the body of diseased or defective cells, including malignancies. Telomeric aging provides last-ditch protection against tumors, imposing a limit on runaway reproduction. But recent experiments suggest that both mechanisms also play a role in an organismal aging program.

Short telomeres have been associated with increased mortality and shorter life spans in humans (Cawthon et al. 2003; Epel et al. 2006) as well as other mammals (Rudolph et al. 1999; Tomás-Loba et al. 2008) and birds (Pauliny et al. 2006; Haussmann et al. 2005). This is surprising because telomerase is available in the genome to extend telomeres as needed. Thus withholding its expression has the effect of reducing life span. The older, more traditional understanding of the role of telomerase rationing in mammals is that it is a defense against runaway cell proliferation in cancer (Shay 2005). But why, then, would net life span be shortened (including cancer mortality) by the withholding of telomerase? Why should we expect telomerase to increase life span in flatworms, which are not subject to cancer (Joeng et al. 2004)?

Apoptosis has been documented in yeast cells as an altruistic adaptation during periods of food scarcity (Fabrizio et al. 2004), and in phytoplankton that are stressed or overcrowded (Lane 2008; Berman-Frank et al. 2004). There are biochemical parallels between apoptosis at the cellular level and aging at the organismic level, an indication that aging in multicellular organisms has evolved, at least in part, from the ancient program for cell death in protists (Skulachev 2002; Skulachev and Longo 2005). Apoptosis is triggered by mitochondria in a cascade with H_2O_2 as intercellular mediator (Hajnoczky and Hoek 2007). This phenomenon in human neurons may be associated with Alzheimer's disease (Su et al. 1994; Vina et al. 2007). Genetically modified mice in which apoptosis has been downregulated live longer than untreated control mice; this may be an indication that apoptosis remains an active aging agent in mammals (Migliaccio et al. 1999).

For some of these conserved gene families, there are effects on fertility, etc. in addition to longevity, so that a net benefit to individual fitness is realized (Chen et al. 2007). Even so, demonstrating pleiotropic phenomena is not the same as showing that pleiotropy (and therefore aging) is unavoidable. And there remain some genes found in wild specimens, the elimination of which seems to extend life span without cost (Kenyon 2005; Landis and Tower 2005). Other genes play a dual role in development (enhancing individual fitness) and in aging (reducing individual fitness), and it remains a mystery why they are not simply switched off after maturity, for a net fitness benefit (Curran and Ruvkun 2007).

There is one more article of direct evidence contradicting the *MA* theory and all theories of aging based on low selection pressure. When a trait is subject to selection pressure, some competing genes are eliminated, and consequently the variation of that trait within a population decreases. The proper measure of selection pressure (first defined by Fisher (1930)) is called *additive genetic variance*. This quantity has been measured in fruit flies (Promislow et al. 1996; Tatar et al. 1996) with respect to late mortality as the target trait. The measurement is not straightforward, because mortality is a statistical trait that depends on chance factors as well as genetics. Promislow, Tatar et al. had the understanding and the patience to make a clean separation: They established 100 diverse clonal populations in jars of 650 matched flies; they then recorded mortality rates day by day for each jar, and computed the variance in this quantity among the 100 jars. Their finding was that additive genetic variance for mortality is low, and decreasing late in life. This is the signature of an adaptation, and is difficult to reconcile with the idea that aging is caused by random mutations.

Low genetic variance and the existence of families of genes that control aging, conserved over vast evolutionary distances, are both inconsistent with the *Mutation Accumulation* theory of aging. Indeed, they constitute *prima facie* evidence that aging is a genetic program. The only reasons not to accept this conclusion is that it flies in the face of bedrock population genetic theory. This may be interpreted as a fundamental crisis for evolutionary theory. Of course, the fact of Darwinian evolution is not in dispute; but the postulates of the neo-Darwinist school which has dominated evolutionary theory for the last century are called into question.

5.5 Theories Based on Tradeoffs: Pleiotropy and the Disposable Soma

Theories based on evolutionary tradeoffs posit that the deep evolutionary cause of aging is an inescapable tradeoff between preservation of the soma and other tasks essential to fitness, such as metabolism and reproduction. These theories offer a way to explain aging within the context of standard population genetic theory (based on individual fitness). The ideas also have compelling commonsense appeal: selection is not actually crafting senescence, but senescence results as a byproduct of selection going for broke in the quest for maximal fertility.

However, the pivotal word is *inescapable*. Clearly, living things manage simultaneously to do many things well in some areas, while being compelled to compromise in other areas. Only if tradeoffs are truly inescapable are the pleiotropic theories tenable. There is a great deal of empirical support for the existence of pleiotropy that affects aging, but no support for the thesis that the tradeoffs are inescapable. On the contrary, for many aging genes discovered in wild-type animals, no substantial benefit has yet been identified.

If we take the evidence to indicate that tradeoffs involving senescence are *not* inescapable, then there is another natural explanation for pleiotropy, rooted in multilevel selection (MLS) theory (Wilson 1975, 2004; Sober 1998), discussed below.

Pleiotropic theories of aging come in two flavors. The classic flavor, as first formulated by Williams (1957), is based on direct genetic tradeoffs: genes that have benefits for reproduction or robustness in early life, but that cause deterioration leading to death at late ages. The newer flavor, as envisioned by Kirkwood (1977), is based on metabolic tradeoffs. Everything the body does, from metabolism to reproduction to repair, involves the use of food energy; and the energy budget imposes a compromise among these tasks, restricting the body's ability to do a perfect job of repair or regeneration.

5.5.1 Classic Pleiotropy: Genetic Tradeoffs

In proposing this theory in 1957, Williams (1957) imagined individual genes that act antagonistically in different stages of life. His presentation was abstract; even the one example he offered lacked documentation of a mechanism.

> Convincing examples are hard to find... there seems to be little necessity for documenting the existence of the necessary genes. Pleiotropy in some form is universally recognized, and no one has ever suggested that all the effects of a gene need be equally beneficial or harmful, or that they must all be manifest at the same time.

But without explicit examples, Williams assumed that, since aging affects a core component of individual fitness, it must be tied to an equally essential, positive advantage. It was natural to imagine that aging derived from unavoidable side-effects of fertility genes.

In the intervening years, many examples of pleiotropic tradeoffs have been documented. Breeding for longevity often has deleterious consequences for other aspects of fitness. Many genes for aging have been discovered that have a balancing, beneficial effect. But in two respects, Williams's vision has not been borne out: (1) Beneficial side-effects don't seem to be universal, casting doubt on the premise that it is the individual benefits that are responsible for the evolution of aging; and (2) the benefits often seem to be peripheral to the core fitness functions of survival and reproduction.

At the back of Stephen Stearns's textbook (1992) on life history evolution is a table of experiments that were designed to look for evidence of trade-offs between fertility and longevity in diverse animal species. About half the studies find some relationship and half find none. (Taking into account publication bias – that negative results frequently go unreported – the true proportion may be well under half.) This situation has been taken (Partridge et al. 2005) as qualified confirmation of the pleiotropic theory of senescence, given the uncertainty and complexity in all biological experiments. However, the theory depends crucially on the premise that pleiotropy is unavoidable, and that longevity cannot be attained without a sacrifice in fertility, or some other important fitness component. The experiments on their face do not support pleiotropic theory, but indicate instead that trade-offs between fertility and longevity are secondary modifiers of aging genes – not their *raison d'être*.

Across species, it is certainly true that long life span is associated with low fertility. This has often been cited (Partridge et al. 2005) as evidence for the pleiotropic theories (though Williams (1957) himself was astute enough not to suggest this argument). The fact is that this inverse correlation is dictated by physics, ecology and demography, and does not support pleiotropy or any particular theory of aging. Body size is necessarily correlated separately with longevity and low predation risk, both for purely physical reasons (West et al. 1997). Animals that suffer heavy losses to predation must have high reproductive rates in order to maintain their populations. This alone is sufficient to explain an inverse relationship between fertility and longevity across species that have comparable ecological roles. For species that are not ecologically commensurate (e.g., mammals and birds) the predicted relationship between fertility and longevity fails (Holmes et al. 2001).

Similarly, it has been noted that life span extension by caloric restriction (CR) frequently involves a sacrifice in fertility (Finch 2007). Perhaps the mechanism of action is that caloric restriction curtails fertility, and it is the avoided cost of fertility that generates an increase in longevity? But details of the CR data do not support this hypothesis. For example, male mice enjoy as much longevity gain as female mice when subjected to CR, but the fertility cost is much lower (Merry and Holehan 1981). And female reproductive capacity is shut down altogether beyond 40% restriction, but female life span continues to increase linearly right up to the threshold of starvation – around 70% restriction (Ross and Bras 1975; Weindruch et al. 1986). In this regime of extreme caloric restriction where the mouse is no longer expending any resources on reproduction, what reservoir does it draw upon to continue increasing its longevity as calories are withdrawn? This experimental profile is the qualitative signature we might expect if fertility suppression and life extension were independent adaptive responses to food shortage. But it is not at all what we would expect if life extension were achieved only as a secondary result of suppressing reproduction. (Mitteldorf 2004)

A large-scale, direct effort to look for a tradeoff between fertility and longevity was undertaken by the laboratory of Michael Rose in the 1980s and 1990s (Leroi et al. 1994). (As of 2009, their breeding experiment continues.) Fruit flies are bred for longevity by collecting eggs from the longest-lived individuals in each generation. It was Rose's expectation that he would see a decline in fertility as the longevity of the flies was extended by artificial selection. After two years, the experiment reported early results that seemed to confirm this hypothesis, and Rose wrote that

> The scientific significance of these conclusions is that senescence in this population of *Drosophila melanogaster* appears to be due to antagonistic pleiotropy, such that genes which postpone senescence appear to depress early fitness components. Put another way, these results corroborate the hypotheses of a cost of reproduction (Williams 1957, 1966), since prolonged life seems to require reduced early reproductive output. (Rose 1984)

But subsequently, the fertility of the test strain began to increase in comparison with the controls. After twelve years of laboratory evolution, the test strain lived longer (mean 74 vs. 41 days) *and* laid more eggs than the controls (63 vs. 52 eggs in a 24-hour period). In the new data (1991 assays), the fecundity difference was found to be consistent over age, so that the long-lived flies laid more eggs in every day of their lives than did the controls.

This result baldly contradicted the prediction of pleiotropic theory. Rose, however, has remained loyal to the theory in which he was trained, and dismisses his own results as a laboratory artifact (Leroi et al. 1994).

Although the deep, compulsory link between fertility and longevity was not found, the flies bred in this fashion were not super-competitors in the sense that they had other crippling defects that explain why this variety has not evolved in the wild. Other breeding regimens (with the waterflea *Daphnia pulex*) *have* produced such super-competitors: apparently superior in strength, size, survival, longevity, and fertility to the wild type (Spitze 1991). Evolutionary theorists have scrambled to explain why this strain has not appeared in the wild (Reznick et al. 2000), without departing from the broad theoretical framework outlined by Medawar (1952) and Williams (1957).

Much of the experimental work on the genetics of life span has been conducted with the roundworm *C. elegans*. The UCSF laboratory of Cynthia Kenyon, in particular, has conducted an intensive search for examples of mutations that extend life span without identifiable cost to other aspects of the worm's fitness. These are most impressive when they are gene deletions, indicating that the identified gene has evolved in the wild type despite the fact that it has no other effect than to curtail life span. Several examples of such genes have been identified. The *AGE-1* mutant (Johnson et al. 1993), and several of the many *DAF* mutant worms (Kenyon et al. 1993) exhibit normal fertility, and no cost of extended life span has been identified. Similarly, fruit flies that are engineered to overexpress a gene dubbed *dFOXO* seem to have an extended life span without a cost to fertility (Hwangbo et al. 2004). Some mutations in the gene called INDY (for "I'm not dead yet") similarly seem to offer cost-free life extension (Marden et al. 2003).

In one series of experiments, Kenyon and colleagues demonstrated that in *C. elegans* there is a link between reproduction and aging, but it is not a direct metabolic link, but rather is mediated by hormonal signaling (Arantes-Oliveira et al. 2003). The experiment found that germline cells issue a life-shortening signal, while the somatic gonad issues a life-extending signal (Arantes-Oliveira et al. 2002).

There are other experiments as well that indicate that the connection between reproduction and aging in worms is a matter of signaling, rather than a metabolic tradeoff. The difference is crucial in that the *AP* theory only makes sense if these tradeoffs are metabolic preconditions; signaling, by definition, is programmable. Laser ablation of the worm's chemical sensor cells is itself sufficient to generate an increase in life span (despite normal food intake) (Apfeld and Kenyon 1999). Similarly, life span of flies has been found to vary not just with the ingestion of food but with the perception of its smell (Libert et al. 2007). Genetic manipulation of yeast elicits this same behavior (Roux et al. 2009). And experiments with mosaic worms (different genotypes in different tissues) indicate that it is the nervous system that signals life extension associated with mutations of the *DAF* family of genes (Wolkow et al. 2000; Libina et al. 2003). This is direct evidence for the (heretical) notion that fertile life span is subject not to maximization but rather to tight genetic regulation in response to the environment, perhaps for the purpose of enhancing demographic homeostasis (Mitteldorf 2006; Mitteldorf and Pepper 2009).

The actions of the *DAF-2* gene with respect to fertility and longevity are separable according to the timing of the gene's expression. Using RNA interference, it has been demonstrated (Dillin et al. 2002) that if the mutant gene is masked during development, and then expressed in maturity, then it has no effect on fertility, but offers full life extension, by about a factor of two. Differential expression of genes through a lifetime is a universal feature in higher life forms; so why has selection not arranged this *DAF-2* variant to be expressed only after maturity?

Linda Partridge and David Gems are advocates for the pleiotropic theory, but have come to terms squarely with the evidence that some aging genes appear to offer no substantial benefit in trade. They catalog these examples, and put them in the context of other experiments that *do* indicate the presence of tradeoffs. They frankly acknowledge that "Even one genuine exception to the rule that lifespan can be extended only by reducing fecundity implies that there can be no obligate tradeoff between them." (Partridge and Gems 2006) They offer an explanation for these exceptions: they are anomalies deriving from the difference between lab conditions and conditions in the wild. Perhaps the tradeoffs appear only when resources are scarce (as in the wild), and the tradeoffs become unimportant when food is plentiful (as in the lab). Also, the measurement of fertility in laboratory experiments seldom involves all-out, continuous reproductive effort, as we expect to be demanded in the wild. Finally, it could be that the ways in which wild animals adapt when they are bred in a stable, controlled environment reduces their natural life spans, making life extension easier, and, perhaps, less costly.

These are speculations – possible loopholes to avoid the straightforward interpretation of experiment; however, Partridge and Gems justify their construction by appealing to the great body of theory that would be falsified if it turns out that nature has *not* maximized the product of fertility and longevity. I lean the opposite way because there is evidence from other quarters indicating that aging is an adaptation, and because the core of evolutionary theory only *seems* to be well established. More on both these subjects below.

5.5.2 Tradeoffs in Resource Allocation: The Disposable Soma Theory

According to the *Disposable Soma* (*DS*) theory, there are competing metabolic demands for energy: reproduction, immune defense against parasites, locomotion, nervous system, etc. As the body's essential molecules become chemically damaged, repair and maintenance becomes one demand for energy among many. The body is forced to optimize its overall energy use, compromising each of these demands. This is seen as a logical proof that repair and maintenance cannot be perfect, and that damage will inevitably accumulate. It is an irresistibly appealing idea that the phenomena we identify as senescence are the result of this process.

Irresistible, perhaps, but not correct. When Thomas Kirkwood (1977) authored this theory as a young man in 1977, he was unaware of a body of literature – already mature but unknown outside its narrow constituency – on the subject of caloric restriction. Animals that are fed less live longer. This remains the most robust, most effective, and most reliable life extension intervention, broadly applicable across widely separated taxa.

If the fundamental cause of aging were a compromise in the allocation of caloric energy, then it follows that caloric restriction should cause *shortening* of life span (Mitteldorf 2001). When *less* food energy is available, each of the demands on that energy must share the burden, making do with a reduced share of the smaller total. Allocation for repair and maintenance must be smaller, and if the *DS* theory is correct then aging must proceed more rapidly. This is the opposite of what is observed. Reduced caloric intake reliably leads to slower aging and enhanced life span.

For many species, reproduction requires a substantial energy investment. It is a logical possibility that reproduction may be constrained to be either "on" or "off", with no in-between state. Then there will be a point in the caloric restriction curve when reproduction is abruptly shut down, and more energy becomes available on the far side of that line, before the decline in available energy inevitably continues. Shanley and Kirkwood attempted a *DS* model of aging in mice based on this effect, and reported (Shanley and Kirkwood 2000) the limited success of their model in an optimistic light. But in most versions of their model, there was no energy dividend for repair and maintenance, and in the one version showing a dividend, it appeared over a narrow range of caloric restriction, and for lactating female mice only. In contrast, the CR data show that life span is extended linearly as calories are reduced over a broad range, and the experiments are generally performed comparing non-reproducing mice with other non-reproducing mice, male and female (Mitteldorf 2001). Absent the energy of lactation, Shanley's model predicts clearly that the CR mice should have curtailed life spans.

The idea of metabolic tradeoffs is so appealing that other authors (van Noordwijk and de Jong 1986; de Jong and van Noordwijk 1992; de Jong 1993; Cichon 1997) have advocated versions based on allocation of some other resource than energy. Theoretical models of this mechanism work well in the abstract, but authors have lots of wiggle room because they are not constrained by real world data. The primary problem with these theories is that the scarce resource (if it is not energy) remains unspecified. Whatever it is, it must be essential to life, in short supply, and more of it would enable most living things both to live longer and to function better. Until such a resource is specified, the idea remains difficult to evaluate.

A direct and general test applicable to any of the theories of metabolic tradeoff was conducted by Ricklefs and Cadena (2007). They cross-tabulated fertility and longevity for captive birds and mammals in zoos, and found a slight *positive* correlation between fertility and longevity. Similarly, many demographers have sought for evidence of a "cost of reproduction" in humans, and have found no relationship (Le Bourg et al. 1993; Muller et al. 2002) or a small *positive* association (Korpelainen 2000; Lycett et al. 2000) between fertility and longevity. One well-publicized study claimed to discern a cost of reproduction in a historic database of British nobility (Westendorp and Kirkwood 1998), but its methodology was compromised by use of an obscure and inappropriate statistical test (Mitteldorf 2009). Standard linear correlation on the same database reveals a positive correlation (Mitteldorf 2009).

All versions of the *DS* theory rely on the notion that "perfect" repair would be very expensive. This idea was exploded by Vaupel et al. (2004) as related above in Section 5.3.

5.5.3 Adaptive Pleiotropy: An Alternate Theory

Genetic tradeoffs affecting life span appear to be a common feature across taxa, but not a precondition or physical necessity. If we accept at face value evidence that aging itself is an adaptation, then it must be a group-level adaptation, as senescence offers only costs, not benefits to the individual.

The question then arises: how is senescence maintained in the face of individual selection? My own hypothesis is that pleiotropy itself is an adaptation, selected for its effect of suppressing individual competition for longer reproductive life spans. The conservation of aging genetics suggests that aging has been an important target of natural selection over the æons; in this view, pleiotropy is one instrument by which the genetic basis for aging has been protected from the contravening force of individual selection. The conventional view is that antagonistic pleiotropy is an unavoidable physical property of some genes, but empirical data indicate that pleiotropy is neither unavoidable nor universal. The alternative view which I propose is that pleiotropy is a "design feature". Without it, aging would be lost to rapid individual selection, and with it, aging is preserved, creating population-scale benefits to which short-term selection is blind.

Biological evolution operates on competing units at many levels of selection: self-reproducing molecules once competed in a primordial soup. Organelles reproduce independently, competing with each other within cells; cells, whole organisms, communities of individuals, and entire ecosystems compete among themselves. The history of evolution has included transitions from each of these levels to the level above. In order for this to occur, competition at the lower level must be held in check, so that individuals are motivated to cooperate for the common good rather than compete in ways that might subvert it.

For example, before metazoans, evolution was dominated by competition cell-against-cell. But the cells within our bodies no longer compete among themselves for control of resources or for higher rates of reproduction (cancer pathology excepted). The transition to multicellularity was accomplished via a strict segregation of the germ line from the soma. Since DNA in the germ line is identical in each cell of the soma, there is every motivation for somatic cells to perform their functions in a program directed for the benefit of germ line reproduction. We may take this dynamic for granted, but it was no small feat for natural selection to engineer the germ line segregation, effectively lifting the level of competition from cell-against-cell to organism-against-organism.

Another example is sexual reproduction. There is no necessary connection between genetic exchange and reproduction. In protists, for example, genetic exchange is accomplished via conjugation, and reproduction occurs via mitosis. The two processes are separate and unrelated.

Genetic exchange imposes a cost on the individual, and offers benefits for the population: diversity, resistance to epidemics, and adaptability in the face of environmental change. Individual selection would be expected to efficiently eliminate genetic exchange; but instead, we find that genetic exchange is almost universal. In what has been termed the "masterpiece of nature" (Bell 1982), selection has lashed genetic exchange securely to reproduction, so that individuals cannot avoid sharing their genes. The "masterpiece" is the engineering via evolutionary selection that separated the sexes and imposed the imperative (in most multicellular species) against cloning and self-fertilization. Group selection accomplished this feat while struggling uphill against individual selection the whole time.

Consider the organization of the genome into command-and-control genes that call genetic "subroutines" to create organs and implement developmental programs. The genome did not come to be so organized because it was to the advantage of any one gene or any one individual. This is a feature with unknown short-term cost to the individual, but with tremendous benefit to the long-term evolvability of the community.

In these three examples, natural selection at the group level has acted to suppress individual selection, cementing important group-level adaptations in place, despite their cost to the individual.

Perhaps we can understand antagonistic pleiotropy as an adaptation with a similar function. Individual selection tends to do away with senescence, but the group does not permit this to happen, because senescence is important to the diversity and population turnover that keeps the population healthy. Antagonistic pleiotropy is not a physical precondition, a limitation which selection has tried and failed to work around; rather, pleiotropy is a group-level adaptation, evolved for the purpose of preventing individual selection from dislodging senescence. Of necessity, genes that cause aging have had to be tightly integrated with highly desirable traits, or they would have been lost.

5.6 Adaptive Theories of Aging

A broad view of the phenomenology of aging supports the thesis that it is an adaptation in its own right, selected independently. Many details in individual genetic experiments support this conclusion the more strongly. Yet there has been understandable resistance among evolutionary biologists to the idea that aging could be an independent adaptation. The reason for this is that the great body of population genetic theory developed in the 20th century – the "new synthesis" – is founded on concepts of competition at the individual level. It is inconceivable to most population geneticists that group selection could be strong enough to account for aging, in the face of resistance in the form of individual selection. Thus it is not considered sufficient explanation for senescence to note that senescence provides benefits for the population.

To many people, including scientists, the distinction between individual selection and group selection may seem inessential. Without senescence, juveniles of any species would have to compete with full-grown conspecifics at the height of their power, and it would be much more difficult for them to grow to maturity. In order for evolution to operate, novel variations must be tried against real-world conditions. Population turnover is essential for the efficient operation of evolution. It may seem intuitive that senescence might be selected on this basis, as an aid to the efficient operation of evolution itself; however, there is indeed substance to the theoretical view that group selection ought to be weaker than individual selection. We explore the theoretical arguments and some plausible answers below, after outlining some of the affirmative evidence for aging as an independent adaptation.

Several of the phenomena cited above as evidence against one of the three other theories may also be interpreted as affirmative evidence for an adaptive origin of aging:

> *Programmed death in protists*: There are two forms of programmed death in protists. These are apoptosis (Fabrizio et al. 2004) and cellular senescence (via telomeres) (Clark 1999). No other theory has yet been proposed than that these are independent adaptations based on benefits to the microbial community. What is more, both these mechanisms remain active in higher organisms, and have been implicated in human aging (Finch 2007; Clark 2004). If programmed death could evolve half a billion years ago in microbes, this undercuts the theoretical proposition that it is not possible for aging to evolve in more complex life forms.
>
> *Conserved families of aging genes*: Genes that control aging go back half a billion years, and remain closely related in widely separated taxa (Guarente and Kenyon 2000; Kenyon 2001; Budovsky et al. 2007). All other such conserved genes relate to core aspects of cellular metabolism. This constitutes evidence that natural selection has treated aging as a core process.
>
> *Low additive genetic variance*: Additive genetic variance is the technical measure of the variation in natural genotypes upon which selection is acting. High AGV is indicative of random mutations or genetic drift; low AGV is the signature of an adaptation. AGV of late mortality has been found to be low (Promislow et al. 1996; Tatar et al. 1996), declining with advancing age.
>
> *Aging can be accelerated by nervous signaling*: Ablating the chemical sensor of the worm *C. elegans* causes an increase in life span (Apfeld and Kenyon

1999). Fruitfly life span is attenuated not just by abundant food supply, but merely by the odor of food (Libert et al. 2007). And experiments with mosaic worms indicate that genes affecting life span act through the nervous system, rather than through reproductive or endocrine tissue (Libina et al. 2003).

Here are two additional lines of evidence for aging as an adaptation, selected for its own sake:

5.6.1 Programmed Death in Semelparous Plants and Animals

Semelparous organisms are animals and plants whose life histories are organized around a single burst of reproduction. Almost without exception, they die promptly on the completion of reproduction, and they die in ways that appear to be manifestly programmed. Annual plants, the octopus, and Pacific salmon are examples.

Every gardener knows that flowering annuals wither and die shortly after their flowers go to seed. However, if the flowers are removed before they form pods, the plant can be induced to flower repeatedly over an extended time. If it were the burst of reproductive effort that killed the plant, one would not expect the plant to be capable of replacing its flowers so handily. This phenomenon presents itself as a form of programmed death, triggered by the final stages of seeding as a signal.

After tending her eggs, the female octopus stops eating and dies in about ten days (Wodinsky 1977). Lest we doubt that this is an example of programmed death, the animal's behavior can be altered by surgical removal of a pair of endocrine glands, the optic glands, which evidently assert control over a genetic program for maturation, breeding, and dying. With just one optic gland, the octopus lives six weeks more. Without both optic glands, the animal resumes feeding and can survive half a year on average.

The semelparous Pacific salmon is thought to have evolved fairly recently from iteroparous cousins in the Atlantic (Crespi and Teo 2002). The usual hypothesis is that semelparity evolved as a result of the attrition attendant upon the animal's long and difficult migration out to the open ocean and back. But some species of Atlantic salmon undertake comparable migrations and presumably endure a comparable attrition rate for second-time breeding. They have evolved a reproductive burst similar to that of the Pacific species but do not always die afterwards (Crespi and Teo 2002). This suggests that reproductive bursts and accelerated senescence may be separate adaptations.

Many ecologists accept as a matter of course that "genetically programmed, irreversible degeneration subsequent to breeding" is a group-level adaptation (Crespi and Teo 2002). Northcote (1978) has proposed that the carcasses of salmon that have completed their breeding help to fertilize small ponds, sustaining food stocks for the hatchlings. If this explanation is correct, then semelparity in salmon is a case of individual death programmed for communal benefit.

Demonstration of programmed death in semelparous organisms detracts from the plausibility of arguments that programmed senescence in iteroparous organisms is theoretically excluded.

5.6.2 Caloric Restriction and Other Forms of Hormesis

Most animal species that have been studied in laboratory experiments evince a capacity to extend life span when deprived of food (for details, see Chapter 12). This raises the question, why do they not extend life spans when food is plentiful?

The answer from pleiotropy theory is that food deprivation causes a curtailment of fertility, and it is the curtailment of fertility that enables life span to be extended. But much is known about the phenomenology of CR, and the details support a picture in which curtailment of fertility and extension of life span are separable and independent responses.

- CR extends life span comparably in males and females, despite the facts that (1) female investment in reproduction is so much higher to begin with, and (2) males lose less of their fertility than females (Merry and Holehan 1981).
- Moderate caloric restriction (30%) is sufficient to completely shut down reproduction in female mice, yet life span continues to increase linearly as more severe CR is imposed (up to 70%). In the range 30–70%, there is no fertility left to lose, yet longevity continues to increase (Ross and Bras 1975).
- There is direct evidence that *C. elegans* life extension from CR is mediated through the nervous system, rather than through endocrine or reproductive tissues (Apfeld and Kenyon 1999; Libert et al. 2007; Roux et al. 2009). Instead of a direct, metabolic tradeoff between reproduction and repair, this indicates there is a life extension program that can be invoked in response to a signal.

The CR response is but one example of a more general phenomenon known as *hormesis* (Masoro 2007). Modest environmental challenges in many other forms also lead to enhanced longevity. That aerobic exercise increases life span is so ordinary a fact that it is never posed as an evolutionary dilemma. Only because it is less familiar do we find it curious that life span can also be increased by low doses of toxins or ionizing radiation, by electric shocks, by infection, by physical injury, by heat and by cold. Small doses of chloroform have been found to extend the lives of several lab animals (Heywood et al. 1979; Palmer et al. 1979; Roe et al. 1979). A steady, low exposure to gamma radiation extends the life spans of rats (Carlson et al. 1957) and flies (Sacher 1963). Human as well as animal evidence concerning radiation hormesis is reviewed by Luckey (1999). Mice exposed to daily electric shocks enjoy enhanced longevity (Ordy et al. 1967). Immersion in cold water extends the lives of rats (Holloszy and Smith 1986). Heat shock induces life extension in roundworms (Butov et al. 2001; Cypser and Johnson 2002).

Hormetic phenomena suggest that life span is plastic under genetic control, that it can be increased without cost or side-effects in response to a more challenging and competitive environment. Forbes (2000) reviews a wide range of hormetic phenomena, and concludes that fitness hormesis is surprising in the context of evolutionary theory based on individual selection. The essence of the paradox is this: Why is the life extension program not implemented in less challenging times? If genes are available for extending life and, thereby enhancing (individual) fitness, then why is this program ever shut down? Why should animals that have plenty to eat, that are not poisoned or heat-shocked or compelled to exercise vigorously live shorter lives?

5.6.3 How Might Senescence have Evolved as an Adaptation? Two Ideas that Don't Work

Early theories of senescence as an adaptation were based on two ideas that have been discredited for good reason. The first is that damage to the soma inevitably accumulates with age, and that death is programmed in order to rid the population of old, damaged individuals so that they can be replaced with younger, more perfect ones. The second is that aging offers benefits for adaptability in a population, and enhances the rate of evolution.

The idea that somatic damage must inevitably accumulate, and that programmed death was nature's way to rid the population of these imperfect specimens, was Weismann et al.'s (1891) original theory for evolution of aging. He abandoned his theory, as described above (Section 5.3), because he realized there is no necessity for damage to accumulate, and that it is almost always metabolically cheaper to repair somatic damage than to cast aside the soma and create a new one from seed. This objection is sound, but the idea remains appealing, and variations have continued to be published in the modern literature: Kirchner and Roy (1999) present a numerical model in which infections may cause sterility but not death, and aging evolves to rid the community of sterile individuals. Travis and Dytham (Travis 2004; Dytham and Travis 2006) model a mechanism for the evolution of programmed death as an adaptation, given that individual fecundity declines with age. These models escape Weismann's dilemma, but in doing so introduce assumptions that are likely to limit their general applicability.

The idea that aging evolves as an adaptation to accelerate the pace of adaptive change in a population also has a long history. The premise is basically sound, but it is difficult to imagine quantitative models in which the mechanism can work. It is true that if a population is able to evolve more rapidly, it will eventually overtake another population that evolves more slowly, and so enjoys a decisive advantage in the very long run. It is also true that population turnover rate is often limited by the adult death rate, since intraspecific competition between juveniles and full-grown adults has a major impact on the juvenile mortality rate (Cushing and Li 1992). In other words, enhancement of diversity and adaptability are real and substantial adaptive benefits of aging at the population level.

Several variations of this idea have been described qualitatively (Libertini 1988; Bowles 1998; Goldsmith 2003); however, no quantitative model based on this mechanism has ever been published, and I suspect from my own unpublished numerical experiments that none is possible. The problems are first, that the benefits of aging are not well focused on kin of the individuals who display this trait, and second, that the benefits accrue so slowly over time, while the costs are immediate.

Hamilton's Rule says that one of the factors determining whether a group-adaptive trait may be selected is the extent to which communal benefits are focused on relatives of the individual bearing the trait. Classically, parental nurturance of one's own offspring is easy to evolve, and communal food-gathering is greatly facilitated within a hive of sister insects. By this standard, aging (or programmed death) is at the opposite extreme: the only benefit that accrues is to lower competition for resources, creating an opportunity for another individual to exploit them. The one who grasps this opportunity may be a distant relative (perhaps a "cheater" who does not age) or an unrelated individual, or even a member of an entirely different species that shares a common food resource.

The second problem, timing, arises because aging and death impose an immediate cost, in the form of forgone fertility for that part of the life span that is sacrificed, while the benefits accrue very slowly. The benefit of adaptability does not affect fitness but only the rate of change of fitness in a group; thus it is realized very slowly, over evolutionary time. In the language of population genetics, imagine a population divided into competing demes (interbreeding subgroups). Within each deme, natural selection is working *against* aging, because those individuals who die sooner leave fewer offspring. At a higher level, deme against deme, aging is *benefiting* demes in which life spans are shorter, since such demes evolve faster than others. But evolutionary time scales are much longer than individual lifetimes. For the evolutionary advantage to accrue, we must wait long enough for the environment to change, requiring new adaptations; then the requisite adaptation must rise in frequency within the short-lived deme more rapidly than in longer-lived demes, because aging implies a shorter effective generation cycle.[1] Finally, this deme must succeed in competition with other, longer-lived demes. But, of course, the longer-lived demes have been enjoying an enhanced rate of reproduction all along. Why haven't they dominated the short-lived demes long before the advantage in adaptive rate is realized?

There is a related problem concerning "cheaters" – non-aging individuals that might find their way into a deme in which aging predominates. The cheaters would have an advantage in individual fitness, and it is difficult to explain why their progeny would not quickly come to prevail within the deme.

These arguments were well-appreciated by the community of evolutionary theorists (Williams 1966; Maynard Smith 1976) when the subject of group selection was

---

[1] In populations that are at or near their density limit, competition is sufficiently intense that offspring (small and undeveloped) rarely grow to maturity unless an older conspecific dies or migrates, creating a vacancy. It is in this sense that senescence can contribute to a higher population turnover, and a shorter effective generation cycle.

hotly debated in the 1970s; in fact, they were decisive in establishing a consensus that aging could not evolve as an adaptation in its own right.

5.6.4 How Might Senescence have Evolved as an Adaptation? The Demographic Theory

The key to understanding how aging might have evolved as an adaptation is to consider population dynamics. Population genetic thinking focuses on incremental increase in gene frequency within a population close to steady state. But steady state is the exception in nature. Natural populations are often either expanding exponentially or else collapsing toward extinction. Cycles of boom and bust are driven by the tendency of a population that is below carrying capacity to over-exploit resources; the population overshoots its carrying capacity, and starvation drives an exponential decline. The simplest and earliest model of this phenomenon is the Lotka-Volterra equation (Gilpin and Rosenzweig 1972; Maynard Smith and Slatkin 1973), the solution of which involves violent oscillations in population level for most parameter combinations.

The Demographic Theory of Senescence (Mitteldorf 2006) posits that population dynamics leads to frequent extinctions in nature, and thus demographic homeostasis is a major target of natural selection, rivaling the imperative to maximize individual reproductive output in its strength, and tempering individual fitness in its effect. Population cycles can be rapid and lethal, wiping out entire communities in a few generations. (This is in contrast to benefits of diversity or evolvability, which act slowly because they affect the rate of change of fitness, but not fitness itself.) This efficiency is the justification for believing that population dynamics is a force that can compete effectively with the population genetic notion of individual fitness. (The point was first articulated by Gilpin (1975), who brilliantly modeled a group selection process based on population dynamics at a time when computer resources were yet inconvenient and unwieldy.)

Individual selection for increased reproductive success puts populations on a collision course with the unstable regime of Lotka-Volterra dynamics (Lotka 1925; Volterra 1931). A pair of linked equations (which Lotka and Volterra derived independently) is the basis for classical descriptions of predator-prey population dynamics. Reproduction requires food energy, so the predator is driven by individual selection to exploit its prey more and more aggressively as it competes for reproductive success. This process continues, predation and predator fertility evolving ever higher, until exploitation is no longer sustainable; this is the classic, original Tragedy of the Commons (Hardin 1968). Population dynamics then enters a chaotic regime (Mitteldorf 2006). Local extinctions then proceed swiftly, and group selection can operate very efficiently under such conditions. Competition among demes drives population dynamics back from the brink of chaos.

This is a compelling story, but its relationship to aging is not yet clear; for this to be a sound basis for the evolution of aging, it is necessary that senescence should be selected in preference to other solutions for the problem of unstable population dynamics. Senescence certainly tempers individual reproductive output, thus helping to quell the violence of population cycles; but so would restrained predation, or lower fertility. However, senescence offers special advantages for demographic homeostasis, setting it above these other modes of individual restraint.

The problem of volatility may be described as a tendency for a population to grow too rapidly when conditions are favorable, and to collapse too precipitously when resources are scarce. The effects of aging on a population in its growing phase are to limit life span, to diminish the number of offspring per individual lifetime, and thus to suppress the rate of population growth. But when the same population is in collapse, most animals are dying of starvation. Few live to be old enough for old age to contribute to their demise. Thus aging exacts its greatest toll when the population is freely expanding, but does not add appreciably to the rate of decline when the population is collapsing. The CR adaptation further enhances the effectiveness of senescence for modulating unstable dynamics, since it pushes the death rate from senescence even lower under the conditions of starvation that attend the contracting phase of the population cycle (Mitteldorf 2006).

> **Box 5.1 One way in which aging can help to stabilize population dynamics and quench fluctuations that might otherwise lead to extinction**
>
> Without aging, the most vulnerable segment of any population is the very young. This is an inherently unstable situation. Predation, disease, and other hardships are striking the young disproportionately. Adults are reproducing as fast as they can to compensate for the fact that a great proportion of the young are dying before maturity. If the hardships let up, then there is a flood of young attaining maturity, and the population will explode. Conversely, any increase in the severity of the environment might wipe out the entire young population, leading to extinction.
>
> Aging can lessen the tendency of a population to extreme swings. Senescence causes a progressive decline in strength and immune function, so the most vulnerable segment of the population is always the very old, and they are the first victims of both predation and disease. If environmental conditions change, either for the better or the worse, the effect on the population will be much less pronounced, because the coming generation can recover.
>
> This may be illustrated with the ecologists' canonical model of predator-prey dynamics, the Lotka-Volterra equation (Fig. 5.1). Solutions of this equation can be quite unstable. In the first illustration below, you can see how the two populations cycle in response to each other. When the prey population is high, predators prosper and their numbers increase. But then the prey all but disappear, and the predators starve. Only after the predator population collapses does the prey population begin to recover, and the cycle begins anew.

In the case illustrated, the predator and prey populations both range over a factor of 100 in each cycle, and they remain a long time near the minimum level. Most natural populations cannot survive such volatility; at minimum levels, they may fluctuate to zero, and the game is over.

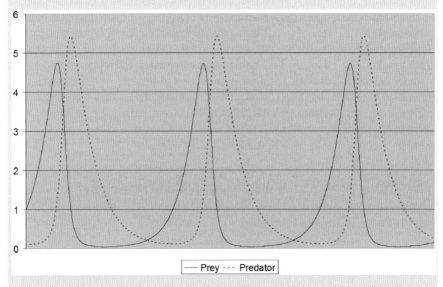

Fig. 5.1 The rise and fall of predator and prey populations according to the classical Lotka-Volterra equation (without maturation or aging)

The classic Lotka-Volterra equation (above) takes no account of age differences. Watch what happens (in the second frame, (Fig. 5.2)) when we allow for two stages of life, in which the young are more vulnerable than the mature. Predators capture the young in preference to the adults. Result: the fluctuations become yet more severe, and each cycle is worse than the last one. If the graph is continued, both predator and prey are driven to extinction.

In the third frame (Fig. 5.3), we add a late stage of life, less agile than the other two. The aged prey are the easiest to catch, followed by the young, and the middle age is most resistant. Result: the dynamic has stabilized considerably, and the fluctuations narrow with each cycle. (If the graph is continued, the wiggles smooth out over time and level off.) Note that the average prey population over time is actually lower than without aging; this reflects the fact that individual fitness has decreased (in both young and old stages of life). Group fitness may be higher, however, because the population is rescued from the risk of extinction at its minimum levels.

It is possible that aging evolved for the purpose of preventing violent swings in population level that can lead to extinction.

5 Evolutionary Origins of Aging

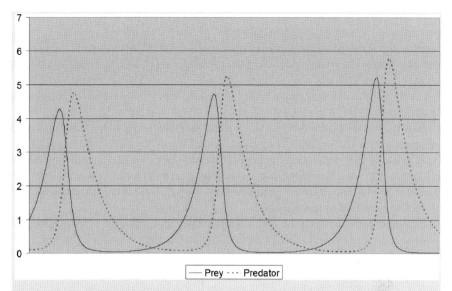

Fig. 5.2 The Lotka-Volterra equation, modified so that young, immature prey have a greater chance of being caught

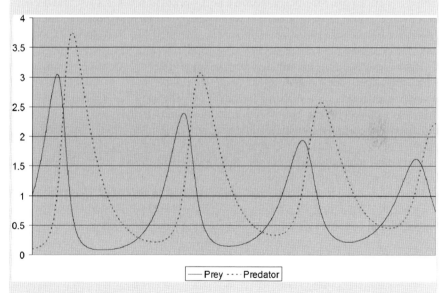

Fig. 5.3 Lotka-Volterra equation modified so that young prey are more vulnerable, but aged prey are even more vulnerable

In Box 5.1, a computational model is described that traces the rise and fall of populations when the youngest members are the most vulnerable to predation, compared to the same population when the oldest, weakened by aging, are even more vulnerable than the youngest. The computation illustrates what common sense tells us: that the old have less effect on population dynamics than the young, so that the weakening of the oldest segment with progressing senescence stabilizes population cycles.

Besides predator-prey dynamics, a second population dynamic effect may also be important in the evolution of aging. According to the Red Queen hypothesis (van Valen 1973), populations maintain diversity as protection against microbial epidemics. This mechanism enjoys widespread credibility as an explanation for the evolution of sex; but there is no reason why the same mechanism cannot also be applied to evolution of senescence.

Like aging, the ubiquity of sexual reproduction is a mystery in the theoretical context of exclusively individual selection. In fact the "cost of males" imposes a full factor of two in fitness compared to hermaphroditism. Currently, one of the best-accepted hypotheses in response to this challenge is the Red Queen mechanism, named for her line in the children's classic by Lewis Caroll (Dodgson 1871), "Now, *here*, you see, it takes all the running *you* can do to keep in the same place. If you want to get somewhere else, you must run at least twice as fast as that!" Because pathogens are so specifically adapted to a host's peculiar biochemistry, and because they evolve so fast, there is powerful evolutionary pressure to change constantly and arbitrarily for the sake of maintaining genetic diversity. The Red Queen hypothesis links the evolution (and maintenance) of sex to the imperative to avoid epidemics.

But senescence also contributes to population diversity (limiting the extent to which any super-robust organism can dominate a gene pool with its offspring), and its individual cost is far less than a factor of two. The same Red Queen mechanism that is so well received by the evolutionary community in the context of sex is also capable of supporting the evolution of aging as an adaptation (Mitteldorf and Pepper 2009).

Starvation and epidemics are the two most lethal threats to an overcrowded, homogeneous population. Population dynamics may provide the toehold that enables senescence to evolve "uphill", in opposition to individual selection. In the long run, the benefits for population diversity and rate of adaptation may play a crucial role in establishing senescence as a core, conserved process.

5.7 How could Evolutionary Theory have been So Wrong?

This question is a matter of primary concern for evolutionary biologists, but a diversion in the context of the present article. The short answer is this: Early evolutionary theorists created a model world in which total populations were static and gene frequency changed differentially in response to small increments in individual reproductive success. A great deal of mathematical theory was developed in the twentieth century based on this premise. Simultaneously, experimental work was done to validate the theory. But evolution in the biosphere is too slow to observe in the course of a human career. Laboratory experiments were performed instead, substituting artificial breeding for natural selection. Artificial selection in these experiments is almost invariably implemented in proportion to the number of offspring produced by each genotype. In other words, the close association of fitness with fertility is built into the experimental protocol. The success of many such experiments in reproducing theoretical predictions has lent the illusion that population genetic theory rests on a firm experimental foundation, when, in fact, the reasoning has been circular. It has been demonstrated that selection for maximal fertility works in the laboratory as theory predicts, but there has never been experimental confirmation that the Malthusian Parameter r is optimized in the wild, as it is in the textbook. Experiments that directly pose the question, "Does natural selection optimize reproductive fitness?", have generated results in the negative (Reznick et al. 2000, 2002).

5.8 Implications for Medicine

It may be that we age simply because we have aging genes: there are time bombs built into our developmental clocks that destroy us on cue. It is possible that lengthening the human life span could be accomplished by interfering with the action of existing biochemical pathways. There is a well-developed science of drugs that block specific biochemical signals, and all this knowledge could be brought rapidly to bear, once target receptors are identified.

There are two ancient mechanisms of programmed cellular death, as old as eukaryotic life:

1. Telomeric aging
2. Apoptosis

For theoretical reasons, it has been assumed that these can play no substantial role in mammalian aging; but recent evidence suggests that both these mechanisms may contribute to diseases of old age.

Four additional, promising targets for intervention are:

3. Inflammation – an infection-fighting mechanism that turns against self in old age;
4. Caloric restriction mimetics – stimulating the metabolism to invoke the CR response without actually reducing consumption;
5. Thymic involution – responsible for much of the decline in immune function that underlies the greater prevalence of infections disease in the elderly;
6. Regeneration – could it be that higher organisms retain a latent ability to rebuild and to heal that has not been lost but only suppressed?

5.8.1 Telomeres

Telomeres are strings of repetitive DNA that serve as replication counters in all eukaryotic cells. They provide the basis for an ancient, ubiquitous mechanism of aging, as old as eukaryotic life (Clark 1999, 2004). Precisely because it is so clearly a "design feature" of evolution, theorists held until recently that telomeric senescence could not play a role in the aging of higher organisms. But recent experiments imply that telomeres do affect the viability of older humans (Cawthon et al. 2003; Epel et al. 2006). The best evidence in humans suggests that long telomeres may lower risk for heart disease (Brouilette et al. 2007; Spyridopoulos and Dimmeler 2007). Frequency of cardiovascular events decreases exponentially with telomere length, with each age-specific standard deviation (620 base pairs) accounting for approximately a factor of 2 (Fitzpatrick et al. 2007).

So strong has been the theoretical preconception that these data have just begun to inspire discussion of telomere extension as a therapeutic target (Fleisig and Wong 2007). Telomere length can be restored by an endogenous enzyme, telomerase, which is encoded in the genome of every eukaryotic cell. There is one experiment with mice engineered to over-express telomerase, in which life extension was observed (Tomás-Loba et al. 2008). This is both puzzling and promising, since wild-type mice already have particularly long telomeres, and it has been demonstrated that mice with no telomerase at all can survive with normal life span for five generations (Rudolph et al. 1999). The prospect that telomerase therapy may enhance life span in humans remains controversial, but at least five companies are already accelerating research directed toward near-term commercial applications (Advanced Cell Technologies, Geron, Phoenix Biomolecular, TA Science, and Sierra Sciences). Much additional evidence linking aging to telomere shortening and a discussion of possibilities for addressing life extension via telomere maintenance or extension are provided in the chapters by Aviv and Bogden and by West elsewhere in this volume.

5.8.2 Apoptosis

Apoptosis, the orderly, programmed dismemberment of a cell, can be triggered by infection, disease, or by external signals. The genetic basis for apoptosis is present in all eukaryotic life, indicating that, like telomeric senescence, it is an ancient, conserved mechanism. Several roles for apoptosis in somatic cells of higher life forms are known: for sculpting the body during development, for limiting infections, and as a barrier against cancer. As with telomeric senescence, theorists have classically recognized the individual organism as the only target of natural selection; hence they have assumed that apoptosis always serves the interest of the individual. But there is some evidence that apoptosis late in life is on a hair trigger; that is, it is destroying healthy cells on a massive scale in aging mammals. In experiments with mice (Migliaccio et al. 1999), knocking out some apoptosis genes has the expected effect of slightly increasing the cancer rate, but the unexpected net effect is substantial extension of the life span. In humans, a role for brain cell apoptosis in dementia is hypothesized (Kruman et al. 1997; Lu et al. 2000). Apoptosis is tightly regulated by a network of promoting and inhibiting signals, so that the optimal therapeutic target is far from apparent; nevertheless, the intriguing possibility exists that simply dialing down the apoptosis response may have a net benefit for life span in humans (Epel et al. 2007).

5.8.3 Inflammation

The primary function of inflammation is to combat infection, but the process has been coöpted in the body's self-destruction program. The relationship between inflammation and arthritis has long been recognized (Zvaifler 1973), but only in the last decade has it become clear that inflammation plays a central role in other diseases of old age: atherosclerosis (Reiss and Glass 2006) and cancer (Coussens and Werb 2002). Aspirin and other NSAIDs are blunt instruments that merely damp the body's inflammation response, yet they lower the mortality risk of heart disease (Eidelman et al. 2003), stroke (Hass et al. 1989) some cancers (Moysich et al. 2005; Flick et al. 2006), and possibly Alzheimer's disease (Sven et al. 2003). "Aspirin may be the world's first anti-aging drug." (D.K. Crawford 2003, Personal communication) See Finch (2007) for extensive background on the role of inflammation in aging.

If crude anti-inflammatories are already saving lives and forestalling morbidity, we may presume that the effect will be much enhanced once there are treatments that are able to down-regulate damage to self without compromising the body's first line defense against infection.

5.8.4 Caloric Restriction Mimetics

The broad anti-aging action of caloric restriction has been recognized since the 1930s, but for many years it was thought to act through a passive alteration of the metabolism; only recently has it become clear that CR sends specific signals that cause the body to invoke a stress response, once again an ancient evolutionary program (Sinclair 2005; Masoro 2007). The simplest intervention inspired by this research is to regulate the body's hormone balance with supplements, emulating the hormone levels characteristic of young or calorically restricted individuals. DHEA is an adrenal steroid and precursor of several sex hormones. Its levels decline steeply with normal aging, and CR tends to delay the decline in laboratory primates (Roth et al. 2002). DHEA levels are inversely correlated with human mortality (Shock et al. 1984), and with the incidence of cardiovascular disease in males (Feldman et al. 2001). This evidence has made DHEA a candidate for an anti-aging supplement (Genazzani et al. 2007), but early experience in standardized clinical trials has been disappointing (Baulieu et al. 2000; Nair et al. 2006).

In the last decade, an earnest search has begun for drugs that may extend life span by invoking the CR response, without the need for reduced dietary intake (Weindruch et al. 2001). Sinclair and coworkers have demonstrated that resveratrol is the active ingredient for the cardioprotective action of red wine (Howitz et al. 2003; Baur and Sinclair 2006), and that it acts in the same gene-silencing pathway as caloric restriction (Bredesen 2004). Treatment with resveratrol extends life span in yeast (Howitz et al. 2003), in worms and flies (Wood et al. 2004; Baur and Sinclair 2006), and in fish (Valenzano et al. 2006). Trials presently underway in mice (Baur et al. 2006) have shown mixed results (Barger et al. 2008; Pearson et al. 2008), with dosage emerging as a key variable (Dudley et al. 2008).

Maintaining insulin sensitivity may be a general strategy for prolonging youthful phenotypes (Brunet 2009). Caloric restriction in mice slows growth of most cancers, but particularly those that rely on insulin signaling (Kalaany and Sabatini 2009). A diet with low glycemic index coupled with periodic fasting may deliver much of the benefit of CR (Goodrick et al. 1990; Anson et al. 2005) in a program that more people can tolerate. 2-deoxyglucose is a sugar analog capable of suppressing insulin response in rats, though its toxic long-term effects preclude consideration as a life extension drug (Roth et al. 2005). Other CR mimetics target the blood lipids and fat deposition (Roth et al. 2005).

5.8.5 Thymic Involution

The thymus is an organ in the upper chest where T-cells are prepared for their immune function. It reaches maximum size and activity just before puberty, and begins its slow, lifelong involution when sex hormones are first released into the blood (Bodey et al. 1997). Loss of thymus function has been linked to the gradual degradation of the immune system with age, exacerbating susceptibility to infectious disease, and, probably, to cancer (Aspinall and Andrew 2000). It may be possible to regrow thymic tissue using stem cell therapy (Gill et al. 2002), or, more simply, to stimulate the body to do so with growth hormones (Fahy 2003). In Chapter 15 of the present volume, Aspinall and Mitchell describe the biology of the thymus and suggest interventions that might be expected to extend its active life.

5.8.6 Regeneration as a Target of Anti-aging Research

Many invertebrates are capable of regenerating damaged body parts. Planaria can regrow in their entirety from a severed head. When starfish are cut into pieces, each piece is capable of generating a new starfish. Among vertebrates, lizards and newts are capable of regenerating severed limbs.

The traditional view of regeneration is that the ability does not exist in mammals and higher organisms because the basis is lacking for rebuilding a complex structure. But it is consistent with the hypothesis of this article that regenerative abilities may simply be "switched off" in mammals. Heber-Katz has garnered intriguing evidence that some regenerative ability remains latent in mice (Heber-Katz et al. 2004), and has speculated on the potential for extending life by re-activating this capability in humans (Heber-Katz et al. 2006).

Wound healing is affected by senescence, and the traditional view is that the problem is systemic; it is both surprising and promising that a single blood factor markedly enhances the healing capacity of aged mice (Beck 1993). It suggests that the capacity to heal and to regenerate both remain latent within older individuals.

5.9 Conclusion

Evolutionary thinking, conscious and unconscious, has limited the scope of what researchers imagine is possible in anti-aging interventions. A broad body of experimental evidence points to an interpretation of aging as an extension of the developmental program, under genetic control. If the body is causing itself to lose faculties with age "by choice", then there may be a finite number of biochemical signals that control all the downstream damage that we associate with senescence. This suggests promising targets for anti-aging intervention.

References

Andziak B et al. (2006) High oxidative damage levels in the longest-living rodent, the naked mole-rat. Aging Cell 5:463–471

Andziak B, O'Connor TP, Buffenstein R (2005) Antioxidants do not explain the disparate longevity between mice and the longest-living rodent, the naked mole-rat. Mech Ageing Dev 126:1206–1212

Anson M, Jones B, de Cabod R (2005) The diet restriction paradigm: a brief review of the effects of every-other-day feeding. AGE 27(1):17–25

Apfeld J, Kenyon C (1999) Regulation of lifespan by sensory perception in *Caenorhabditis elegans*. Nature 402:804–809

Arantes-Oliveira N, Apfeld J, Dillin A, Kenyon C (2002) Regulation of life-span by germ-line stem cells in *Caenorhabditis elegans*. Science 295:502–505

Arantes-Oliveira N, Berman JR, Kenyon C (2003) Healthy animals with extreme longevity. Science 302:611

Aspinall R, Andrew D (2000) Thymic involution in aging. J Clin Immunol 20:250–256

Austad S (1999) Why We Age. Wiley, New York

Ayyadevara S, Alla R et al. (2008) Remarkable longevity and stress resistance of nematode PI3K-null mutants. Aging Cell 7(1):13–22

Barbieri M, Bonafe M, Franceschi C, Paolisso G (2003) Insulin/IGF-I-signaling pathway: an evolutionarily conserved mechanism of longevity from yeast to humans. Am J Physiol Endocrinol Metab 285:E1064–E1071

Barger JL, Kayo T et al. (2008) A low dose of dietary resveratrol partially mimics caloric restriction and retards aging parameters in mice. PLOS One 3(6):e2264

Baulieu EE et al. (2000) Dehydroepiandrosterone (DHEA), DHEA sulfate, and aging: contribution of the DHEAge Study to a sociobiomedical issue. Proc Natl Acad Sci USA 97:4279–4284

Baur JA et al. (2006) Resveratrol improves health and survival of mice on a high-calorie diet. Nature 444:337–342

Baur JA, Sinclair DA (2006) Therapeutic potential of resveratrol: the in vivo evidence. Nat Rev Drug Discov 5:493–506

Beck LS DL, Lee WP, Xu Y, Siegel MW, Amento EP (1993) One systemic administration of transforming growth factor-beta 1 reverses age- or glucocorticoid-impaired wound healing. J Clin Invest 92:2841–2849

Bell G (1982) The Masterpiece of Nature: The Evolution and Genetics of Sexuality. University of California Press, Berkeley

Berman-Frank IH, Bidle KD et al. (2004) The demise of the marine cyanobacterium, Trichodesmium spp., via an autocatalyzed cell death pathway. Limnol Oceanogr 49(4):997–1005

Bodey B, Bodey B Jr, Siegel SE, Kaiser HE (1997) Involution of the mammalian thymus, one of the leading regulators of aging. In Vivo 11:421–440

Bonduriansky R, Brassil CE (2002) Senescence: rapid and costly ageing in wild male flies. Nature 420:377

Bowles JT (1998) The evolution of aging: a new approach to an old problem of biology. Med Hypotheses 51:179–221

Bredesen DE (2004) The non-existent aging program: how does it work? Aging Cell 3:255–259

Bronikowski AM, Promislow DE (2005) Testing evolutionary theories of aging in wild populations. Trends Ecol Evol 20:271–273

Brouilette SW, Moore JS et al. (2007) Telomere length, risk of coronary heart disease, and statin treatment in the West of Scotland Primary Prevention Study: a nested case-control study. Lancet 369(9556):107–114

Brunet A (2009) Cancer: When restriction is good. Nature 458(7239):713–714

Budovsky A, Abramovich A et al. (2007) Longevity network: construction and implications. Mech Ageing Dev 128(1):117–124

Butov A et al. (2001) Hormesis and debilitation effects in stress experiments using the nematode worm *Caenorhabditis elegans*: the model of balance between cell damage and HSP levels. Exp Gerontol 37:57–66

Carlson LD, Scheyer WJ, Jackson BH (1957) The combined effects of ionizing radiation and low temperature on the metabolism, longevity, and soft tissues of the white rat. I. Metabolism and longevity. Radiat Res 7:190–197

Caroll L (pen name of CL Dodgson) (1871) Through the Looking Glass and What Alice Found There. Macmillian & Co, London

Cawthon RM, Smith KR, O'Brien E, Sivatchenko A, Kerber RA (2003) Association between telomere length in blood and mortality in people aged 60 years or older. Lancet 361:393–395

Chen J et al. (2007) A demographic analysis of the fitness cost of extended longevity in *Caenorhabditis elegans*. J Gerontol A Biol Sci Med Sci 62:126–135

Cichon M (1997) Evolution of longevity through optimal resource allocation. Proc Royal Soc B 264:1383–1388

Clark WR (1999) A Means to an End: The Biological Basis of Aging and Death. Oxford University Press, New York, Oxford

Clark WR (2004) Reflections on an unsolved problem of biology: the evolution of senescence and death. Adv Gerontol 14:7–20

Clausius R (1867) On a modified form of the second fundamental theorem in the mechanical theory of heat. In: Hirst TA (ed) The Mechanical Theory of Heat. van Voorst, pp. 111–135

Coussens LM, Werb Z (2002) Inflammation and cancer. Nature 420:860–867

Crespi BJ, Teo R (2002) Comparative phylogenetic analysis of the evolution of semelparity and life history in salmonid fishes. Evol Int J Org Evol 56:1008–1020

Curran SP, Ruvkun G (2007) Lifespan regulation by evolutionarily conserved genes essential for viability. PLoS Genet 3:e56

Cushing JM, Li J (1992) Intra-specific competition and density dependent juvenile growth. Math Biol 54:503–519

Cypser JR, Johnson TE (2002) Multiple stressors in *Caenorhabditis elegans* induce stress hormesis and extended longevity. J Gerontol A Biol Sci Med Sci 57:B109–B114

de Grey AD (2006a) Compression of morbidity: the hype and the reality, part 1. Rejuvenation Res 9:1–2

de Grey AD (2006b) Compression of morbidity: the hype and the reality, part 2. Rejuvenation Res 9:167–168

de Jong G (1993) Covariances between traits deriving from successive allocations of a resource. Funct Ecol 7:75–83

de Jong G, van Noordwijk AJ (1992) Acquisition and allocation of resources: genetic (CO) variances, selection, and life histories. Am Nat 139:749–770

Dillin A, Crawford DK, Kenyon C (2002) Timing requirements for insulin/IGF-1 signaling in *C. elegans*. Science 298:830–834

Dudley J, Das S (2008) Resveratrol, a unique phytoalexin present in red wine, delivers either survival signal or death signal to the ischemic myocardium depending on dose. J Nutr Biochem 20(6):443–452

Dytham C, Travis JM (2006) Evolving dispersal and age at death. Oikos 113:530–538

Edney EB, Gill RW (1968) Evolution of senescence and specific longevity. Nature 220:281–282

Eidelman RS, Hebert PR, Weisman SM, Hennekens CH (2003) An Update on Aspirin in the Primary Prevention of Cardiovascular Disease, pp. 2006–2010

Engels WR, Johnson-Schlitz D, Flores C, White L, Preston CR (2007) A third link connecting aging with double strand break repair. Cell Cycle 6:131–135

Epel ES, Burke, H.M., and Wolkowitz, O.M. (2007) Anabolic and catabolic hormones. In: Aldwin CM (ed) Handbook of Health Psychology and Aging. Guilford, pp. 119–142

Epel ES et al. (2006) Cell aging in relation to stress arousal and cardiovascular disease risk factors. Psychoneuroendocrinology 31:277–287

Fabrizio P et al. (2004) Superoxide is a mediator of an altruistic aging program in Saccharomyces cerevisiae. J Cell Biol 166:1055–1067

Fahy GM (2003) Apparent induction of partial thymic regeneration in a normal human subject: A case report. J Anti-aging Med 6:219–227

Feldman HA, Johannes CB, Araujo AB, Mohr BA, Longcope C, McKinlay JB (2001) Low dehydroepiandrosterone and ischemic heart disease in middle-aged men: prospective results from the Massachusetts Male Aging Study. Am J Epidemiol 153:79–89

Finch CE (1990) Longevity, Senescence and the Genome. University of Chicago Press, Chicago

Finch CE (2007) The Biology of Human Longevity. Elsevier, Burlington, MA

Fisher RA (1930) The Genetical Theory of Natural Selection. The Clarendon Press, Oxford

Fitzpatrick AL, Kronmal RA et al. (2007) Leukocyte telomere length and cardiovascular disease in the cardiovascular health study. Am J Epidemiol 165(1):14–21

Fleisig HB, Wong JMY (2007) Telomerase as a clinical target: current strategies and potential applications. Exp Gerontol 42(1–2):102–112

Flick ED, Chan KA, Bracci PM, Holly EA (2006) Use of nonsteroidal antiinflammatory drugs and non-Hodgkin lymphoma: a population-based case-control study. Am J Epidemiol 164:497–504

Forbes V (2000) Is hormesis an evolutionary expectation? Funct Ecol 14:12–24

Gavrilov LA, Gavrilova NS (2004) The reliability-engineering approach to the problem of biological aging. Ann N Y Acad Sci 1019:509–512

Gavrilov LA, Gavrilova NS et al. (1978) [Basic patterns of aging and death in animals from the standpoint of reliability theory]. Zh Obshch Biol 39(5):734–742

Genazzani AD, Lanzoni C, Genazzani AR (2007) Might DHEA be considered a beneficial replacement therapy in the elderly? Drugs Aging 24:173–185

Gill J, Malin M, Hollander GA, Boyd R (2002) Generation of a complete thymic microenvironment by MTS24(+) thymic epithelial cells. Nat Immunol 3:635–642

Gilpin ME (1975) Group Selection in Predator-Prey Communities. Princeton University Press, Princeton

Gilpin ME, Rosenzweig ML (1972) Enriched predator-prey systems: theoretical stability. Science 177:902–904
Goldsmith T (2003, 2nd edition 2008) The Evolution of Aging. Azinet
Goodrick CL, Ingram DK et al. (1990) Effects of intermittent feeding upon body weight and lifespan in inbred mice: interaction of genotype and age. Mech Ageing Dev 55(1):69–87
Gordeeva AV, Labas YA et al. (2004) Apoptosis in unicellular organisms: mechanisms and evolution. Biochem-Moscow 69(10):1055–1066
Guarente L, Kenyon C (2000) Genetic pathways that regulate ageing in model organisms. Nature 408:255–262
Hajnoczky G, Hoek JB (2007) Cell signaling. Mitochondrial longevity pathways. Science 315:607–609
Hamilton WD (1966) The moulding of senescence by natural selection. J Theor Biol 12(1):12–45
Hardin G (1968) The tragedy of the commons. Science 162:1243–1248
Hass WK et al. (1989) A randomized trial comparing ticlopidine hydrochloride with aspirin for the prevention of stroke in high-risk patients. Ticlopidine Aspirin Stroke Study Group. N Engl J Med 321:501–507
Haussmann MF, Winkler DW et al. (2005) Longer telomeres associated with higher survival in birds. Biol Lett 1(2):212–214
Heber-Katz E, Leferovich J, Bedelbaeva K, Gourevitch D, Clark L (2006) Conjecture: Can continuous regeneration lead to immortality? Studies in the MRL mouse. Rejuvenation Res 9:3–9
Heber-Katz E, Leferovich JM, Bedelbaeva K, Gourevitch D (2004) Spallanzani's mouse: a model of restoration and regeneration. Curr Top Microbiol Immunol 280:165–189
Heywood R et al. (1979) Safety evaluation of toothpaste containing chloroform. III. Long-term study in beagle dogs. J Environ Pathol Toxicol 2:835–851
Holloszy JO, Smith EK (1986) Longevity of cold-exposed rats: a reevaluation of the "rate-of-living theory". J Appl Physiol 61:1656–1660
Holmes DJ, Fluckiger R, Austad SN (2001) Comparative biology of aging in birds: an update. Exp Gerontol 36:869–883
Howitz KT et al. (2003) Small molecule activators of sirtuins extend Saccharomyces cerevisiae lifespan. Nature 425:191–196
Hwangbo DS, Gershman B, Tu MP, Palmer M, Tatar M (2004) Drosophila dFOXO controls lifespan and regulates insulin signalling in brain and fat body. Nature 429:562–566
Joeng KS, Song EJ et al. (2004) Long lifespan in worms with long telomeric DNA. Nat Genet 36(6):607–611
Johnson TE, Tedesco PM, Lithgow GJ (1993) Comparing mutants, selective breeding, and transgenics in the dissection of aging processes of *Caenorhabditis elegans*. Genetica 91:65–77
Kalaany NY, Sabatini DM (2009) Tumours with PI3K activation are resistant to dietary restriction. Nature 458(7239):725–731
Kenyon C (2001) A conserved regulatory system for aging. Cell 105:165–168
Kenyon C (2005) The plasticity of aging: insights from long-lived mutants. Cell 120:449–460
Kenyon C, Chang J, Gensch E, Rudner A, Tabtiang R (1993) A *C. elegans* mutant that lives twice as long as wild type. Nature 366:461–464
Kirchner JW, Roy BA (1999) The evolutionary advantages of dying young: epidemiological implications of longevity in metapopulations. Am Nat 154:140–159
Kirkwood T (1977) Evolution of aging. Nature 270:301–304
Korpelainen H (2000) Fitness, reproduction and longevity among European aristocratic and rural Finnish families in the 1700s and 1800s. Proc Biol Sci 267(1454):1765–1770
Kruman I, Bruce-Keller AJ, Bredesen D, Waeg G, Mattson MP (1997) Evidence that 4-hydroxynonenal mediates oxidative stress-induced neuronal apoptosis. J Neurosci 17:5089–5100
Landis GN, Tower J (2005) Superoxide dismutase evolution and life span regulation. Mech Ageing Dev 126:365–379

Lane N (2008) Origins of death. Nature 453 (29 May 2008):583–585
Lapointe J, Hekimi S (2008) Early mitochondrial dysfunction in long-lived Mclk1+/− mice. J Biol Chem 283(38):26217–26227
Le Bourg E, Thon B et al. (1993) Reproductive life of French-Canadians in the 17–18th centuries: a search for a trade-off between early fecundity and longevity. Exp Gerontol 28(3):217–232
Leroi A, Chippindale, AK, Rose, MR (1994) Long-term evolution of a genetic life-history trade-off in Drosophila: the role of genotype-by-environment interaction. Evolution 48:1244–1257
Libert S, Zwiener J, Chu X, Vanvoorhies W, Roman G, Pletcher SD (2007) Regulation of Drosophila life span by olfaction and food-derived odors. Science 315:1133–1137
Libertini G (1988) An adaptive theory of the increasing mortality with increasing chronological age in populations in the wild. J Theor Biol 132:145–162
Libina N, Berman JR, Kenyon C (2003) Tissue-specific activities of *C. elegans* DAF-16 in the regulation of lifespan. Cell 115:489–502
Linnane AW et al. (2002) Human aging and global function of coenzyme Q10. Ann N Y Acad Sci 959:396–411; discussion 463–395
Liu X, Jiang N, Hughes B, Bigras E, Shoubridge E, Hekimi S (2005) Evolutionary conservation of the clk-1-dependent mechanism of longevity: loss of mclk1 increases cellular fitness and lifespan in mice. Genes Dev 19:2424–2434
Lotka AJ (1907) Relation between birth rates and death rates. Science, N. S. 26:21–22
Lotka AJ (1925) Elements of Physical Biology. Williams and Wilkins, Baltimore
Lu DC et al. (2000) A second cytotoxic proteolytic peptide derived from amyloid beta-protein precursor. Nat Med 6:397–404
Luckey TD (1999) Nurture with ionizing radiation: a provocative hypothesis. Nutr Cancer 34:1–11
Lycett JE, Dunbar RI et al. (2000) Longevity and the costs of reproduction in a historical human population. Proc Biol Sci 267(1438):31–35
Mangel M (2008) Environment, damage and senescence: modelling the life-history consequences of variable stress and caloric intake. Func Ecol 22(3):422–430
Marden JH, Rogina B, Montooth KL, Helfand SL (2003) Conditional tradeoffs between aging and organismal performance of Indy long-lived mutant flies. Proc Natl Acad Sci USA 100: 3369–3373
Masoro EJ (2007) The role of hormesis in life extension by dietary restriction. Interdiscip Top Gerontol 35:1–17
Maynard Smith J (1976) Group selection. Q Rev Biol 51:277–283
Maynard Smith J, Slatkin M (1973) The stability of predator-prey systems. Ecology 54:384–391
Medawar PB (1952) An unsolved problem of biology. Published for the college by H. K. Lewis, London
Merry BJ, Holehan AM (1981) Serum profiles of LH, FSH, testosterone and 5 alpha-DHT from 21 to 1000 days of age in ad libitum fed and dietary restricted rats. Exp Gerontol 16:431–444
Migliaccio E et al. (1999) The p66shc adaptor protein controls oxidative stress response and life span in mammals. Nature 402:309–313
Mitteldorf J (2001) Can experiments on caloric restriction be reconciled with the disposable soma theory for the evolution of senescence? Evol Int J Org Evol 55:1902–1905; discussion 1906
Mitteldorf J (2004) Aging selected for its own sake. Evol Ecol Res 6:1–17
Mitteldorf J (2006) Chaotic population dynamics and the evolution of aging: proposing a demographic theory of senescence. Evol Ecol Res 8:561–574
Mitteldorf J (2009) Female Fertility and Longevity, ArXiv http://arxiv.org/ftp/arxiv/papers/0904/0904.1815.pdf
Mitteldorf J, Pepper J (2007) How can evolutionary theory accommodate recent empirical results on organismal senescence? Theory Biosci 126:3–8
Mitteldorf J, Pepper J (2009) Senescence as an adaptation to limit the spread of disease. J Theor Biol 260:186–195
Moysich KB, Baker JA, Rodabaugh KJ, Villella JA (2005) Regular analgesic use and risk of endometrial cancer. Cancer Epidemiol Biomarkers Prev 14:2923–2928

Muller HG, Chiou JM et al. (2002) Fertility and life span: late children enhance female longevity. J Gerontol A Biol Sci Med Sci 57(5):B202–B206

Nair KS et al. (2006) DHEA in elderly women and DHEA or testosterone in elderly men. N Engl J Med 355:1647–1659

Nesse RM (1988) Life table tests of evolutionary theories of senescence. Exp Gerontol 23:445–453

Northcote TG (1978) Migratory strategies and production in freshwater fishes. In: Gerking SD (ed) Ecology of Freshwater Fish Production. Blackwell Scientific, Oxford, pp. 326–359

Ordy JM, Samorajski T, Zeman W, Curtis HJ (1967) Interaction effects of environmental stress and deuteron irradiation of the brain on mortality and longevity of C57BL/10 mice. Proc Soc Exp Biol Med 126:184–190

Palmer AK, Street AE, Roe FJ, Worden AN, Van Abbe NJ (1979) Safety evaluation of toothpaste containing chloroform. II. Long term studies in rats. J Environ Pathol Toxicol 2:821–833

Partridge L, Gems D (2006) Beyond the evolutionary theory of ageing, from functional genomics to evo-gero. Trends Ecol Evol 21:334–340

Partridge L, Gems D, Withers DJ (2005) Sex and death: what is the connection? Cell 120: 461–472

Pauliny A, Wagner RH et al. (2006). Age-independent telomere length predicts fitness in two bird species. Mol Ecol 15(6):1681–1687

Pearson KJ, Baur JA et al. (2008). Resveratrol delays age-related deterioration and mimics transcriptional aspects of dietary restriction without extending life span. Cell Metab 8(2):157–168

Promislow DE (1991) Senescence in natural populations of mammals: a comparative study. Evolution 45:1869–1887

Promislow DE, Tatar M, Khazaeli AA, Curtsinger JW (1996) Age-specific patterns of genetic variance in *Drosophila melanogaster*. I. Mortality. Genetics 143:839–848

Reiss AB, Glass AD (2006) Atherosclerosis: immune and inflammatory aspects. J Investig Med 54:123–131

Reznick D, Ghalambor C, Nunney L (2002) The evolution of senescence in fish. Mech Ageing Dev 123:773–789

Reznick D, Nunney L, Tessier A (2000) Big houses, big cars, superfleas and the costs of reproduction. Trends Ecol Evol 15:421–425

Reznick DN, Bryant MJ, Roff D, Ghalambor CK, Ghalambor DE (2004) Effect of extrinsic mortality on the evolution of senescence in guppies. Nature 431:1095–1099

Ricklefs R (1998) Evolutionary theories of aging: confirmation of a fundamental prediction, with implications for the genetic basis and evolution of life span. Am Nat 152:24–44

Ricklefs RE (2008) The evolution of senescence from a comparative perspective. Funct Ecol 22(3):379–392

Ricklefs RE, Cadena CD (2007) Lifespan is unrelated to investment in reproduction in populations of mammals and birds in captivity. Ecol Lett 10(10):867–872

Roe FJ, Palmer AK, Worden AN, Van Abbe NJ (1979) Safety evaluation of toothpaste containing chloroform. I. Long-term studies in mice. J Environ Pathol Toxicol 2:799–819

Rose M (1984) Laboratory evolution of postponed senescence in *Drosophila melanogaster*. Evolution 38:1004–1010

Ross MH, Bras G (1975) Food preference and length of life. Science 190:165–167

Roth GS, Lane MA, Ingram DK (2005) Caloric restriction mimetics: the next phase. Ann N Y Acad Sci 1057:365–371

Roth GS et al. (2002) Biomarkers of caloric restriction may predict longevity in humans. Science 297:811

Roux AE, Leroux A et al. (2009) Pro-aging effects of glucose signaling through a G protein-coupled glucose receptor in fission yeast. PLoS Genet 5(3):e1000408

Rudolph KL, Chang S et al. (1999) Longevity, stress response, and cancer in aging telomerase-deficient mice. Cell 96(5):701–712

Sacher GA (1963) The effects of X-rays on the survival of *Drosophila imagoes*. Physiol Zool 36:295–311

Sanz A, Pamplona R, Barja G (2006) Is the mitochondrial free radical theory of aging intact? Antioxid Redox Signal 8:582–599

Shanley DP, Kirkwood TB (2000) Calorie restriction and aging: a life-history analysis. Evolution Int J Org Evolution 54:740–750

Shay JWW (2005) Telomerase and human cancer. In: De Lange T, Lundblad V, blackburn EH (eds) Telomeres, 2nd edn. Cold Spring Harbor Press, Cold Spring Harbor, NY, pp. 80–108

Shock NW et al. (1984) Normal human aging: the Baltimore longitudinal Study of Aging. In: Services USDoHaH (ed) NIH, Baltimore

Sinclair DA (2005) Toward a unified theory of caloric restriction and longevity regulation. Mech Ageing Dev 126:987–1002

Singh NP, Danner DB, Tice RR, Brant L, Schneider EL (1990) DNA damage and repair with age in individual human lymphocytes. Mutat Res 237:123–130

Skulachev VP (2002) Programmed death phenomena: from organelle to organism. Ann N Y Acad Sci 959:214–237

Skulachev VP, Longo VD (2005) Aging as a mitochondria-mediated atavistic program: can aging be switched off? Ann N Y Acad Sci 1057:145–164

Sober EaW, D.S. (1998) Unto Others: The Evolution and Psychology of Unselfish Behavior. Harvard University Press, Cambridge, MA

Spitze K (1991) Chaoborus predation and life history evolution in Daphnia pulex: temporal pattern of population diversity, fitness, and mean life history. Evolution 45:82–92

Spyridopoulos I, Dimmeler S (2007) Can telomere length predict cardiovascular risk? Lancet 369(9556):81–82

Stearns SC (1992) The Evolution of Life Histories. Oxford University Press, Oxford, New York

Su JH, Anderson AJ, Cummings BJ, Cotman CW (1994) Immunohistochemical evidence for apoptosis in Alzheimer's disease. Neuroreport 5:2529–2533

Sven EN et al. (2003) Does aspirin protect against Alzheimer's dementia? A study in a Swedish population-based sample aged =80Â years. Eur J Clin Pharmacol V59:313–319

Tatar M, Promislow DE, Khazaeli AA, Curtsinger JW (1996) Age-specific patterns of genetic variance in *Drosophila melanogaster*. II. Fecundity and its genetic covariance with age-specific mortality. Genetics 143:849–858

Tatone C et al. (2006) Age-dependent changes in the expression of superoxide dismutases and catalase are associated with ultrastructural modifications in human granulosa cells. Mol Hum Reprod 12:655–660

Tomás-Loba AF, Flores I, Fernández-Marcos PJ, Cayuela M et al. (2008) Telomerase reverse transcriptase delays aging in cancer-resistant mice. Cell 135(4):609–622

Travis JM (2004) The evolution of programmed death in a spatially structured population. J Gerontol A Biol Sci Med Sci 59:301–305

Valenzano DR, Terzibasi E, Genade T, Cattaneo A, Domenici L, Cellerino A (2006) Resveratrol prolongs lifespan and retards the onset of age-related markers in a short-lived vertebrate. Curr Biol 16:296–300

van Noordwijk AJ, de Jong G (1986) Acquisition and allocation of resources: their influence on variation in life history tactics. Am Nat 128:137–142

van Valen L (1973) A new evolutionary law. Evol Theor 1:1–30

Vaupel JW, Baudisch A et al. (2004) The case for negative senescence. Theor Popul Biol 65(4):339–351

Vina J et al. (2007) Mitochondrial oxidant signalling in Alzheimer's disease. J Alzheimers Dis 11:175–181

Volterra V (1931) Variations and fluctuations of the number of individuals in animal species living together. In: Chapman RN (ed) Animal Ecology. McGraw-Hill

Weindruch R et al. (2001) Caloric restriction mimetics: metabolic interventions. J Gerontol A Biol Sci Med Sci 56 Spec No 1:20–33

Weindruch R, Walford RL, Fligiel S, Guthrie D (1986) The retardation of aging in mice by dietary restriction: longevity, cancer, immunity and lifetime energy intake. J Nutr 116:641–654

Weismann A, Poulton EB, Schönland S, Shipley AE (1891) Essays Upon Heredity and Kindred Biological Problems, 2d edn. Clarendon press, Oxford

West GB, Brown JH, Enquist BJ (1997) A general model for the origin of allometric scaling laws in biology. Science 276:122–126

Westendorp RG, Kirkwood TB (1998) Human longevity at the cost of reproductive success. Nature 396(6713):743–746

Williams G (1957) Pleiotropy, natural selection, and the evolution of senescence. Evolution 11:398–411

Williams G (1966) Adaptation and Natural Selection. Princeton University Press, Princeton

Wilson DS (1975) A theory of group selection. Proc Natl Acad Sci USA 72:143–146

Wilson DS (2004) What is wrong with absolute individual fitness? Trends Ecol Evol 19:245–248

Wodinsky J (1977) Hormonal inhibition of feeding and death in octopus: control by optic gland secretion. Science 198:948–995

Wolkow CA, Kimura KD, Lee MS, Ruvkun G (2000) Regulation of *C. elegans* life-span by insulin like signaling in the nervous system. Science 290:147–150

Wood JG et al. (2004) Sirtuin activators mimic caloric restriction and delay ageing in metazoans. Nature 430:686–689

Zvaifler NJ (1973) The immunopathology of joint inflammation in rheumatoid arthritis. Adv Immunol 16:265–336

Chapter 6
Precedents for the Biological Control of Aging: Experimental Postponement, Prevention, and Reversal of Aging Processes

Gregory M. Fahy

Contents

6.1	Introduction: The Plasticity of Aging and its Significance	128
6.2	The Phenomenon of Zero Aging	129
	6.2.1 Zero Aging Prior to Sexual Maturation	129
	6.2.2 Zero Aging in Adults After Previous Aging?	134
	6.2.3 Zero Aging in Adults Without Previous Aging	136
	6.2.4 Aging Versus Zero Aging	137
6.3	Life Extension by Blockade of Default Active Life-Shortening Mechanisms in Adults	139
	6.3.1 Semelparous Species	139
	6.3.2 Monocarpic Plants	145
	6.3.3 Iteroparous Species: C. elegans	146
	6.3.4 Insects	153
	6.3.5 Genetically Altered Mice	153
	6.3.6 Pharmacological and Nutritional Interventions in Mammals	159
	6.3.7 Humans	166
	6.3.8 Other Examples	168
	6.3.9 Future Examples: Overcoming Fundamental Design Features of Iteroparous Species	169
6.4	Life Extension by "Negative Reproductive Costs" in Nature	174
	6.4.1 Social Insects	174
	6.4.2 Social Mammals	175
6.5	Segmental Aging Reversal	176
	6.5.1 Reversal of Reduced Transcription	176
	6.5.2 Reversal of Reduced Translation	177
	6.5.3 Reversal of Age-Related Changes in Gene Expression	180
	6.5.4 Reversal of Reduced DNA Repair	181
	6.5.5 Reversal of Impaired Mitochondrial Function	182

G.M. Fahy (✉)
Intervene Biomedical LLC, Box 478, Norco, CA 92860, USA
e-mail: intervene@sbcglobal.net

6.5.6	Reversal of Reduced Regenerative Capacity		187
6.5.7	Prevention and Reversal of Replicative Senescence		188
6.5.8	Reversal of Lipofuscin Accumulation		189
6.5.9	Reversal of Reduced Immune Function		190
6.5.10	Reversal of Thymic Involution		191
6.5.11	Reversal of Lost Reproductive Cycling in Females		194
6.5.12	Reversal of Age-Related Organ Atrophy in Humans and Rodents		195
6.5.13	Reversal of Hair Graying and Balding in Humans		197
6.5.14	Reversal of Multiple Segmental Aging Processes by One Intervention: Thymus Transplantation		198
6.6	Unregulated Aging		198
6.7	Summary and Conclusions		200
References			202

6.1 Introduction: The Plasticity of Aging and its Significance

The task of this chapter is to provide an overview of the general potential of endogenous genetic resources to control aging. Scattered but broad evidence indicates that gene expression patterns may generally control the onset, time course, extent, and nature of age-related changes, including even stochastic and seemingly stochastic changes. If unused genomic life-extending resources exist that can be tapped, or if genetic inducers of aging exist that can be blocked, or both, then substantial pathways to human life extension can be visualized much more readily and may become relevant much sooner than if the genome is fundamentally incapable of controlling aging much beyond what it already accomplishes.

The present survey considers little-known and/or underappreciated classical evidence in combination with selected recent evidence indicating that genomic capabilities for the control of aging and death are surprisingly powerful in widely divergent species. The nature of this evidence in combination with some interesting features of human aging suggests that genetic resources for better control of aging are also likely to reside within the human genome.

Many observations have shown aging in whole or in part to be malleable, and even preventable or reversible, by relatively pedestrian interventions that in almost all cases clearly involve changing gene expression or cellular signaling. Although it seems that in the past many such observations did not fit into either the research framework or the theoretical framework of most investigators and may have therefore been little noted, present paradigms of aging are becoming more compatible with classical observations. Re-examination of some of these older findings and how they come together with recent discoveries provides a hopeful picture of the nature and biological origins of aging.

This chapter is intended to provide a general context for the rest of this book. More circumscribed examination of genetic mechanisms involved in aging and how they may be perturbed to mitigate human aging is provided in other chapters of the present volume. The examples highlighted here are far from comprehensive but are illustrative. They are intended to provide a general perspective on aging, but also

6.2 The Phenomenon of Zero Aging

Caleb Finch is widely and properly credited with establishing the concept and the likely existence of "negligible senescence" in many complex metazoans and plant species (Finch 1990). In this section, we will consider evidence for not just "negligible" senescence, but zero or truly non-existent senescence. That is to say, there are examples in which the measured rate of organismal aging is identical to zero, and we term this phenomenon "zero aging" to emphasize that it is not "negligible" aging (in the sense of demonstrated aging that is small in magnitude) that has been observed, but no aging at all.

The empirical observation of zero aging is of fundamental relevance to the question of whether and how biological systems can control aging. The following examples serve not only as practical demonstrations of the possibility of achieving a prolonged non-aging state in metazoa, but also as clues to the basic nature and timing mechanisms of aging.

6.2.1 Zero Aging Prior to Sexual Maturation

Evidently, aging in many organisms is part of the adult phenotype only and does not normally occur in the juvenile state. The following examples establish this phenomenon in nematodes, mollusks, and beetles, which could hardly be more diverse representatives of the phenomenon, by extending the juvenile period without effect on subsequent adult lifespan. Similar phenomena probably also occur in fish that delay sexual maturation in response to the presence of older fish (Borowsky 1978), and very possibly in mice as well (Biddle et al. 1997), although in the latter two cases insufficient longevity data appear to be available. Zero aging prior to sexual maturation is compatible with evolutionary theory, but is not predicted by and is not obviously compatible with random damage theories of aging.

6.2.1.1 Zero Aging in C. elegans

The larval dauer state of C. elegans is entered into primarily as a result of food deprivation or cues stimulated by crowding and is a condition of complete developmental arrest (Klass and Hirsh 1976). In the experiments depicted in Fig. 6.1, the dauer state was maintained by withholding food for up to six times the length of the normal adult lifespan before refeeding to allow subsequent adult lifespan to be observed. The results showed that the adult lifespan was completely independent of the time previously spent in the dauer state (Klass and Hirsh 1976). This shows that detectable aging does not exist in the dauer state – that the rate of aging in these larvae is equal to zero as far as it relates to subsequent adult lifespan. Further, when the adults developing from prolonged dauer state larvae were allowed to reproduce, they produced as many eggs as control worms, with equal egg viability as for

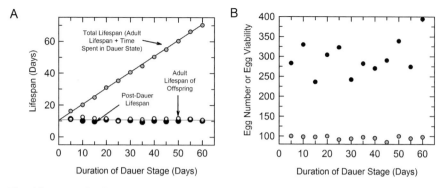

Fig. 6.1 Zero aging in dauer stage C. elegans. (**a**) Under the conditions of these experiments, normal adults live about 10 days, but preventing maturation for up to six times this adult lifespan has no effect on the subsequent adult lifespan. Therefore, the total lifespan (time lived as an adult plus time lived as a larva, *upper line*) is identical to the normal adult lifespan (*black points*) plus the number of days the worms are kept in the dauer condition. Furthermore, the adult lifespans of the offspring of animals previously kept in dauer for up to 60 days (*open points*) was not affected by the time spent in the dauer state. (**b**) Lack of effect of time in dauer on the number of eggs (*black points*) and the percentage viability of eggs (*gray points*) produced after subsequent maturation, indicating that reproduction at a total age up to seven times longer than normal was not adversely affected by accumulated damage. Drawn from data in Klass and Hirsh (1976)

controls, and their offspring had lifespans identical to those of normal adults. This result shows that no heritable life-shortening mutations developed in the dauer larvae even after large multiples of the normal adult lifespan, despite the persistence of movement (Klass and Hirsh 1976) and about 50% of the normal rate of oxygen consumption (Anderson 1978) in the dauer state.

As the authors of this report stated, "If ageing in *C. elegans* is a result of random accumulation of damage (nuclear or cytoplasmic) ... then it is reasonable to presume that prolonging the dauer state would be correlated to an increase in accumulated damage and subsequently a decrease in the post-dauer life span. This result was not found. ... One might expect an increased probability of accumulated damage in the germinal nuclei and cytoplasm and subsequently a reduction in the viability and mean life span of F_1 progeny because the primordial gonad is present throughout the dauer state. This effect was not observed."

These results were replicated and extended by Tom Johnson's lab, but this time using non-dauer larvae. Although the non-dauer larvae could only be starved for 10 days before dying, larvae starved for 10 days and then allowed to mature also resulted in adults with normal, not shortened, lifespans. As the authors concluded, "there is a one day increase in total life-span for each day that the culture is starved. ... In other words, aging is completely blocked and there is no effect on the length of the adult life-span ... and ... under complete starvation larvae continue to move vigorously . . . suggesting that the observed life-span extension is not due to a complete arrest of metabolic processes. Our observations suggest a simple alternative explanation for the extended life-span of arrested larvae and of the dauer: the normal senescent processes follow or are otherwise coupled to the developmental program. By this model, dauers would live longer because the dauer state is distinct from the normal senescent program" (Johnson et al. 1984).

It has recently been proposed that the dauer state is not a non-aging state, but is instead an aging state whose aging is fully reversible upon refeeding (Houthoofd et al. 2002). In my view, the evidence for this interpretation is weak, but an aging state that is fully reversible implies even greater genomic control over aging than zero aging itself.

6.2.1.2 Zero Aging in Mollusks

The idea of zero aging is difficult for many to accept due to the expectation that uncontrollable random damage must always increase over time and can never be entirely prevented or reversed. But the results of Fig. 6.1 are exactly parallel to a similar phenomenon documented in mollusks and illustrated in Fig. 6.2.

The nudibranch, *Phestilla sibogae*, similarly to many other marine invertebrates (Jackson et al. 2002), becomes naturally arrested in both development and aging until it encounters signals in its benthic environment that are permissive of metamorphic induction. In the case of P. sibogae, Miller and Hadfield (Miller and Hadfield 1990) freely fed larvae for 7–28 days before exposing them to a fragment of their coral prey that was permissive of metamorphosis and then followed the effect of delayed maturation on subsequent lifespan. The result was that, just as in the case of the C. elegans dauer larvae, the overall lifespan increased by the same number of days the nudibranchs were held in developmental arrest. In other words, the adult period of life lasted about 45 days whether previous larval development had been arrested for 7 days or for 28 days, meaning that the time spent in the pre-adult state was time spent in a non-aging condition. Even the maximum lifespan increased by 21 days when development was retarded for 21 days (Fig. 6.2, top panel). In addition, the extension of larval life had no effect on subsequent adult egg output (both within the first post-metamorphic week and over the lifespan of the adult; Fig. 6.2, bottom panel).

Unlike the case of the C. elegans dauer larvae, the larval P. sibogae were fully active and were not calorie restricted in any way. The extension of lifespan in this species and probably in many other related species (Miller and Hadfield 1990) by developmental arrest is therefore inconsistent with the "rate of living" theory of aging. This conclusion is further supported by the extreme example of the marine snail, *Fusitriton oregonensis*, which can spend up to 4.6 years in the active larval state with "no sign of senescence associated with long larval duration" and with normal post-metamorphic growth rates, times to reproductive maturation, and reproductive competence, although adult lifespans were not measured (Strathmann and Strathmann 2007).

6.2.1.3 Zero Aging in Insects

An even more dramatic example of the same phenomenon of zero aging in the juvenile state was provided by Beck and Bharadwaj (Beck and Bharadwaj 1972). In a

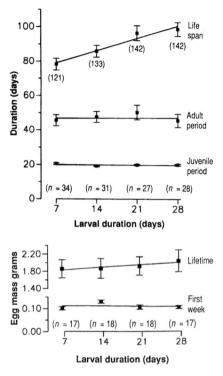

Fig. 6.2 *Top:* Extension of mean and maximum lifespan by extending the larval stage of snails. The time spent in the larval state had no effect on the subsequent time spent in the juvenile period or as an adult. Therefore, the total mean lifespan (larval period + juvenile period + adult period, upper curve) and the maximum lifespan (age of the oldest surviving individual, given by parentheses along the upper curve) rose linearly with time spent in the larval state (for maximum lifespan, note that adding 21 days to the larval period also added 21 days to the maximum lifespan: 121+21=142). Because time spent in the larval state did not shorten either the mean or the maximum lifespan of the adult, the larval state must be a non-aging condition of the organism even though the larvae are fully active metabolically and bear no resemblance to the dauer state of C. elegans. *Bottom:* lack of effect of suspended development on subsequent adult fecundity (egg mass during the first week of reproduction or over the course of the complete adult lifespan). From Miller and Hadfield (1990). Reprinted with permission from AAAS

remarkable experiment, they withdrew food from nearly mature sixth instar larvae of *Trogoderma glabrum* dermestid beetles (carrion beetles) and found that, after 10 days, the larvae began growing biologically younger, as evidenced by an initial "retrogressive ecdysis" followed by subsequent ecdyses at 4-week intervals that led to progressively smaller larvae. The larvae were refed after 20–36 weeks of "retrogression," whereupon they began to regain their previous size and weight through a series of molting cycles. This cycle of retrogression and regrowth could then be repeated.

The normal T. glabrum has a lifespan of 8 weeks from newly laid egg to death of the adult, but in their experiment, Beck and Bharadwaj were able to keep larvae alive for more than two years, which represents a lifespan extension of over 13 fold. This may be the largest artificially induced relative increase in total lifespan achieved for any organism to date.

Furthermore, although these beetles eventually did die, they apparently did not die of any recognizable aging process. Instead, they died while still non-aged juveniles due to a side effect of a developmental peculiarity of the biology of these insects. In control beetles, the fat body cells had a ploidy of three. In insects subjected to 4 twenty-week cycles of retrogression and regrowth (∼91 weeks of total treatment), the fat body cell ploidy rose to 6.8 on average. Similarly, after two cycles of 36-week retrogression and regrowth (∼79 weeks of total treatment), the mean fat body cell ploidy was 8.2. Ploidy increase beyond the diploid condition is a normal phenomenon in these insects, but in the normal course of events, all insects die before this process has any chance to become biologically relevant. In the insects in which aging was precluded by preventing the adult phenotype from arising, however, the authors suggested that the increase in ploidy was able to progress to the point of interfering with metabolic efficiency, leading basically to the death of the animals from starvation. Significantly, the increase in ploidy in this one tissue was the only histological sign of pathology found in these extremely long-lived beetles.

An even more extreme example of life extension by reversed development was reported for *Trogoderma tarsale* in 1917 (Wodsedalek 1917). The author had left some larvae of this species in a drawer without food and forgotten about them, and was very surprised to find all larvae alive 5 months later. He proceeded to starve larvae of different sizes for various times and found that "the last of a large number of specimens lived, without a particle to eat, for the surprisingly long period of five years, one month and twenty-nine days or, to be more specific, . . . a period of 1,884 days." Rather than starving to death, the larvae apparently catabolized their own bodies, growing smaller over time. To quote: "speaking in terms of reduction in size, it is astonishing to note that some of the largest larvae have been reduced to about 1/600 of the maximum larval mass." Regarding retrogression and regrowth as in *Trogoderma glabrum*, it was noted that "when the starved specimens almost reach the smallest size possible and are then given plenty of food, they will again begin growing in size. A number . . . have through alternating periods of 'feasting and fasting' attained that size three times and are now on the way to their fourth 'childhood'." Although adult lifespans were not reported for either control beetles or for previously retrogressed beetles, no pathologies or mortality were noted in beetles that were not starved to death outright. Further, the author noted that these larvae had been subjected to considerable stress (being transported long distances repeatedly, which induced molting responses), but that later larvae, "which have not been so disturbed, show indications of even greater tenacity than is here recorded."

Cicadas provide an interesting natural case of extremely prolonged development with presumably zero aging prior to eclosion, larval forms living for 3, 13, 17 years or more, depending on the species in question, before a 4–6-week adult period whose duration is independent of the prior larval period and amounts to well under 1% of a total 13+ year-long lifespan [for further discussion, see (Finch 1990).]

6.2.1.4 Zero Aging in Mammals?

In 1997, the POSCH-2 mouse was described as a possible model of developmentally postponed reproduction and aging in mammals (Biddle et al. 1997). Female POSCH-2 mice produce their first litters between 25 and 51 weeks of age (mean, 46.8 weeks), in contrast to 9–13 weeks for females of the parental mouse strain, wild-caught *Mus musculus domesticus*, and a mean of 9.5 weeks for C57BL/6J females. In addition, POSCH-2 males do not breed before 6–7 months of age.

The upper age limit for reproduction in this mouse was not determined, but it was noticed that reproduction continues into the third year of life. In contrast, control animals were said to die in their second year of life, whereas deaths and life-shortening pathologies were rarely observed in POSCH-2 mice aged 24 months and beyond.

These results are consistent with the possibility of zero aging in mammals prior to sexual maturation. In this case, the delay in female sexual maturation was on average around 36 weeks, or roughly 9 months, which could easily account for the observed longevity difference between POSCH-2 mice and their strain of origin. Unfortunately, 12 years after the publication of this potentially pivotal paper, the lifespan curve of POSCH-2 mice remains undetermined!

A broadly similar phenomenon has been documented as well in humans (Walker et al. 2009). In this syndrome, termed "disorganized development," normal development beyond the infant stage is blocked and sexual maturation does not take place, the subject remaining infantile in most respects even at the age of 16 as of the time of the report. Development is not totally arrested, however, as changes in dentition and very slow growth, for example, were present. Fascinatingly, telomeres shorten more rapidly in infants than in adults, and as a result the subject's telomere lengths were age-normal or even slightly shorter than what would be expected for a 16 year old, a factor that could become more significant for this case than such an effect in the POSCH-2 mouse unless telomere shortening is corrected. Only 4 such cases seem to be known to date, but they may shed a great deal of light on mechanisms of aging in humans over the next few decades.

6.2.2 Zero Aging in Adults After Previous Aging?

There is some evidence that the induction of diapause in adult Drosophila induces a state of zero or negligible aging for the duration of the time spent in diapause (Tatar et al. 2001b).

A much broader phenomenon, however, is the one described by Carey and colleagues, who reported in 1992 that, contrary to the predictions of Gompertz mortality kinetics, the mortality rate of medflys actually declines at advanced ages: "life expectancy in older individuals increased rather than decreased with age" (Carey et al. 1992). This was not due to confounding factors such as reduced population density-mediated stress with age (Carey et al. 1995; Curtsinger 1995) and appears to apply to many other species (Comfort 1979; Curtsinger 1995; Easton 1997; Vaupel

Fig. 6.3 Human mortality rate curve for all ages, showing the late-life mortality rate plateau for people over the age of 110. Data from many sources (including the Social Security Administration below age 90 and Louis Epstein above age 110), combined and smoothed above age 90 and converted to estimated mortality rates by Donald B. Gennery; figure constructed by Donald B. Gennery. Published for the first time by permission of Donald B. Gennery

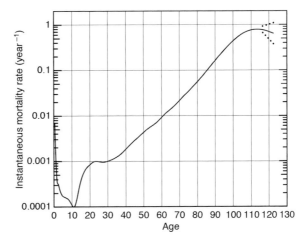

et al. 2008), including humans (Greenwood and Irwin 1939; Gavrilov and Gavrilova 1991; Maier et al. 2010) (Fig. 6.3).

A conventional explanation of this phenomenon is that it is caused by, essentially, the increasingly selective survival of spectacularly robust but still progressively aging subpopulations rather than by the cessation of aging at advanced ages (Partridge and Mangel 1999; de Grey 2003; Vaupel 2010). Others have strongly disagreed with this interpretation and embraced the concept that aging can come to an end beyond a certain age (Rose et al. 2006; Rose et al. 2009; Rose and Mueller 2000), going so far as to declare that the late-life state in which mortality no longer increases is a "phase of life that is as distinct in its fundamental properties as aging is from development" (Rose et al. 2006).

Rose and Mueller (2000) have termed the condition of zero or even negative mortality rate acceleration "immortality" and modeled it as a key phenomenon of life (Mueller and Rose 1996). Biological immortality or the "cessation of aging" (Rose et al. 2006) is generally understood to consist of a zero aging state, such as is observed in many cancer cells that do not die of any senescent processes. Applied to the case of animals, "immortality" in our present state of knowledge is most conveniently equated to a state in which the mortality rate does not increase with time, and certainly does not accelerate with time.

If the biological condition of animals and of people can stop worsening beyond a certain point in time, then there are strong implications about what can and can not be the biological basis of aging. If aging is conceived of as the irreversible accumulation of random damage, then even if the rate of infliction of biological damage stops increasing and becomes constant, the sum of the damage must still increase over time, not stop or reverse itself, as long as damage continues to accumulate at some finite rate. Therefore, only a reduction in the rate of damage accumulation to approximately zero would be expected to result in a mortality rate that stops increasing, and this situation is not predictable by theories that ascribe aging to random stochastic damage accrued during the lifetimes of individuals.

Instead, Mueller and Rose (Mueller and Rose 1996) have modeled the late-life mortality plateau as being the result of the eventual plateauing of the force of natural selection; natural selection imposes a finite burden (Rose et al. 2006) of late-acting life-shortening genes. If aging is induced by antagonistic pleiotropy, it should be limited by the finite number of such pleiotropic genes and the progressive loss of their protective effects in the young (and hence of their deleterious effects in the old) with advancing age (Rose and Mueller 2000; Rose et al. 2006).

If this is the case, a reasonable inference is that selective inhibition of late-acting deleterious genes should extend life, and at much younger ages than those at which mortality rate plateaus are normally observed. If the functions of pleiotropic genes are beneficial only in youth, and harmful only after youth, then inhibiting them after youth should be possible without side effects that are worse than aging itself. The same practical conclusion would also apply on the basis of the theory that natural selection more directly favors life-shortening genes (Bredesen 2004; Mitteldorf 2004, 2009; Longo et al. 2005), but with necessarily waning effectiveness as more and more of the population disappears.

A contrary interpretation of apparently zero aging after clear prior aging could be that aging does not stop but only seems to stop because so much damage has already accumulated that further damage increments simply become too small to affect mortality rates in comparison to the total amount of damage that has previously accrued. However, this interpretation seems unable to explain how mortality rates can actually substantially *decrease* in old age (Carey et al. 1992) nor why the transition from accelerating mortality to constant mortality tends to be so abrupt and clear-cut.

The natural onset of zero or even negative aging in animals that have previously aged still leaves them with aging debilities and a high prevailing risk of death in comparison to young animals. But there are species that can do better, as discussed in the next section.

6.2.3 Zero Aging in Adults Without Previous Aging

As pointed out by Finch (Finch 1990), there are many interesting cases in which aging of normal adult animals or plants is either "negligible" or possibly even non-existent. In the case of animal species, as far as is known, actuarial aging does not occur in, for example, hydras (Brock and Strehler 1963); ant (Donisthorpe 1936; Haskins 1960), bee (Winston 1987), and probably some termite and wasp queens (collectively constituting tens of thousands of species) (Finch 1990); red sea urchins (Ebert 2008); perhaps tubeworms (Lamellibrachia sp., whose numbers and growth rates may decline as a simple exponential function of time and which can conservatively attain at least 170–250 years of age) (Bergquist et al. 2000); sea anemones (Comfort 1979) (estimated [http://www.nhm.ku.edu/inverts/ebooks/ch34.html#survival] to reach ages in excess of 300 years); and certain quahogs (Jones 1983) (recently shown to be able to survive for at least 405 years [http://www.bangor.ac.uk/news/full.php.en?Id=382]). All of these animals, it is to be noted, have nervous systems. In addition, discreet individual living golden corals have been dated as being over 1800 ± 300 years old (Druffel et al. 1995), the oldest known age for any animal species.

The existence of aging in many other highly complex and differentiated animal species is at least open to question. There is little or no evidence for aging in most turtle species (Gibbons 1987). Painted turtles observed in a nature reserve showed no age-related increase in mortality rate and no reduction in reproductive output (and in fact had improved offspring quality) between the ages of 9 and 61 years, the highest age studied (Congdon et al. 2003), and female Blanding's turtles actually have lower mortality and increased reproductive output after the age of 60 compared to younger ages (Congdon et al. 2001). Certain lobsters (Cooper and Uzmann 1977; Finch 1990), and fish (Comfort 1979; Leaman and Beamish 1984; Finch 1990; www.agelessanimals.org) also exhibit no clear aging.

While probably not immortal, it is impressive that the warm-blooded bowhead whale has been shown to be able to sustain the health of its highly developed brain and avoid cataracts to at least the age of about 211 years, to continue to produce sperm to at least the age of about 159 years, and apparently to avoid browning of lens proteins until the age of 135 years and above, and in general shows pathologies only very rarely at any age (George et al. 1999). The oldest documented bowhead, at 211 ± 35 years (estimated age ± 1 estimate SEM), was killed by hunters rather than by old age, and the number of whales killed and therefore made available for study (although far too small for meaningful statistical analysis) was independent of age per 50-year time interval from 101 through 250 years of age (George et al. 1999), suggesting the possibility of no Gompertz mortality rate acceleration after the age of 100.

Last but far from least, the ever-fascinating naked mole rat shows no age-related mortality rate acceleration, has never been observed to develop cancer, and maintains undiminished fecundity throughout life, making it the first mammal to be identified as having achieved negligible senescence (Buffenstein 2008). [Bizarrely, aging in naked mole rats may develop in approximately the last year of a very long (~28–30 year) lifespan, suggesting to Buffenstein a resemblance to the biologically driven onset of aging normally seen in semelparous species (Buffenstein 2008).]

Outside the animal kingdom, the giant fungus, Armillaria bulbosa, can live at least 1,500 years and attain a mass in excess of 100 tons (Smith et al. 1992), and trees such as the Sequoia also fail to demonstrate detectable aging (Harper 1977) and live at least 3,200 years [http://en.wikipedia.org/wiki/Sequoiadendron]), while the Bristlecone pine is known to have reached nearly 5,000 years (LaMarche 1969), and some box huckleberry clones have vintages in excess of 13,000 years (Wherry 1972). The lives of these highly complex species are limited by exogenous events and not by aging.

6.2.4 Aging Versus Zero Aging

The ability of many species to sustain a nominally zero aging state even as adults indicates that aging is not a fundamentally and biologically inescapable process even for a fully differentiated organism. This raises the question of why some complex animal species age and others do not. More generally, if natural selection favors the elimination of aging and if the elimination of aging is possible in organisms as different as bees, turtles, and naked mole rats, why has the elimination of aging not been achieved more often?

It is interesting that C. elegans, P. sibogae, and T. glabrum all have relatively short lifespans, which a priori suggests a relatively limited capacity for self repair compared to longer-lived species, and therefore relatively limited prerequisites for an extended lifespan. Nevertheless, in the examples cited in Section 6.2.1, the mollusks were able to spend a time equivalent to 62% of their adult lifespan in a zero-aging state, the worms were able to spend six times their normal lifespan in a zero-aging state, and the insects were able to spend up to 13 times their normal lifespan in a zero-aging state. This shows that these organisms possess the intrinsic self repair capacity required for them to live dramatically longer than they normally do. The mollusks accomplished major extension of overall lifespan with zero diminution of metabolic activity, and even the worms and beetles could not have had a zero metabolic rate even though they experienced zero aging.

Nevertheless, the adults of all of these species undergo typical aging just as though the life-maintenance resources of their immature forms did not exist. How and why this switch from a non-aging state to an aging state takes place and why zero aging is not the default "choice" of the adults of most species is of direct relevance for better understanding the limits of and the opportunities for intervening into aging in adults.

The traditional explanations such as antagonistic pleiotropy, tradeoffs between self-maintenance and reproduction (see Section 6.4), and the waning force of natural selection after reproduction (Williams 1957; Kirkwood 1977; Rose 1991; Austad and Kirkwood 2008) are not satisfying because they are biochemically vague and are not necessarily compatible with the totality of the data reviewed in this chapter. They don't traditionally (but see Chapter 11) pinpoint the genetic mechanisms that explain why and how some organisms experience an abrupt switch from a non-aging to a rapidly-aging state at the point of sexual maturation while others don't, or how the non-aging state of the juvenile differs biochemically from the aging state of the adult.

Blocking mitochondrial functions in C. elegans only before adulthood extends subsequent adult lifespan, whereas the same intervention done only in the adult has no effect on adult lifespan (Dillin et al. 2002). Similarly, a limited duration of calorie restriction (CR) in three mammalian species (mice, rats, and non-hibernating hamsters) initiated early in life gave different (and more positive) results than either the same duration of CR given later in life or CR given throughout life (Fig. 6.4) (Deyl et al. 1975; Stuchlikova et al. 1975). Such observations point indirectly to a fundamental gerontological difference between the pre-adult and the adult state and indirectly reinforce the notion of zero aging before sexual maturation.

The examples that follow in the remainder of this chapter suggest more specific mechanisms and reasons for the induction of aging in adults.

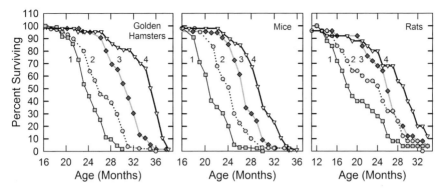

Fig. 6.4 Differing and interacting effects of different developmental states on life extension by calorie restriction (CR) in three animal species. 1: ad libitum feeding throughout life. 2: 50% CR throughout life. 3: 50% CR only during the second 12 months of life. 4: 50% CR only during the first 12 months of life. In every case, restriction during early development extended life more than restriction later in life even though the total time of restriction was the same in both cases. Redrawn from Deyl et al. (1975)

6.3 Life Extension by Blockade of Default Active Life-Shortening Mechanisms in Adults

In most organisms, CR and certain other stresses can induce biochemical responses that protect the organism and extend lifespan. However, as is described at length in this section, the way these life-extending pathways are generally marshaled is by releasing longevity pathways from default inhibition. In other words, the default state of the adult organism, when it is replete with energy [and when it is even actively dissipating energy through physiological mitochondrial uncoupling (McDonald et al. 1996; Speakman et al. 2004)] is to repress longevity pathways. For this reason, experimentally damaging or inhibiting a variety of normal physiological mechanisms that permit this repression can dramatically extend adult lifespan and healthspan.

Semelparous and monocarpic species carry this theme one step further and make what may be subtle in other species quite obvious and undeniable. We begin the discussion with semelparous and monocarpic species to establish the biological precedent of default active life-shortening in adults after sexual maturation.

6.3.1 Semelparous Species

Semelparous species by definition reproduce only once before dying. Some semelparous adults are aphagous (lacking mouthparts, lacking digestive organs, or simply declining to consume food after sexual maturation) and are therefore doomed to die from starvation or malnutrition, a phenomenon that has been called "inarguably, programmed senescence" (Finch 1990). This condition affects nine orders of insects and thousands of individual species, including all 277 genera of mayflies as well as many male spiders, male rotifers, water fleas (Daphnia), marine parasites, nonparasitic and some parasitic lampreys, Pacific salmon, eels, and even the protozoan, Tokophrya (Finch 1990). Other famous examples of semelparous animals include the males of a few species whose females consume them after mating as a result of clearly programmed life-terminating behavior. These cases establish the general principle that evolution can indeed prescribe death for adults immediately after reproduction, regardless of whether theory might suggest this could not evolve due to the possibility of "cheaters" who might emerge without these genetic prescriptions. In addition, they indicate that there are mechanisms other than "reproductive costs" that can closely link death to reproduction (see also Section 6.4).

Significantly, semelparous marsupials and rodents also exist. The death of these relatives of ours is linked to neuroendocrine mechanisms (Section 6.3.1.4) that are similar to those found in diverse non-mammalian semelparous species (Sections 6.3.1.1, 6.3.1.2, and 6.3.1.3) and analogous to life-shortening mechanisms in iteroparous species, including humans (Section 6.3.3).

6.3.1.1 Optic Gland Removal in Octopi

According to Wodinsky (Wodinsky 1977), "the optic glands are the only definitely identified endocrine glands in octopus." They are involved in the control of sexual maturation and they are also responsible for the precipitous death of female octopi about 10 days after hatching of the female's eggs (Wodinsky 1977). Female *Octopus hummelincki* greatly reduce their food intake and change their feeding behavior prior to death. This reduced food intake is not improved after removing one of the two optic glands, but the survival time of the octopi after egg hatching is nevertheless increased from about 10–11 days to about 44–45 days, or about 4-fold (Fig. 6.5). This life extension is apparently limited by starvation, since it is achieved despite the refusal of all food for about the last month of life, with a nearly 50% reduction of body weight by the time of death (vs. a mean weight loss of only about 8% in unoperated controls or 24% in sham-operated controls). This indicates that the death of unoperated females 10 days after egg hatching cannot be due to starvation and must be actively induced.

When both optic glands were removed, the females lived an average of about 143 days after egg hatching, or about 14 times longer than unoperated or sham-operated controls (Fig. 6.5), and began to feed more normally starting at about 2 weeks after surgery. These octopi actually died with weights approximately equal to their maximum weights prior to egg laying. They also stopped incubating their eggs immediately upon recovering from surgery, indicating that the optic glands control not only sexual maturation and death, but also key aspects of reproductive behavior. Evidently the entire life history of the octopus, from birth to reproduction to death, is under endocrine control.

Optic gland removal also increased the maximum survival time of the females after egg hatching from about 19 days in intact controls to ~59 days with the

6 Precedents for the Biological Control of Aging

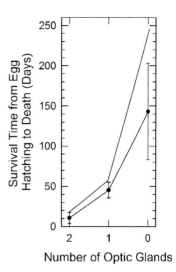

Fig. 6.5 Extension of mean and maximum survival times of *Octopus hummelincki* by removal of one or both optic glands 1–22 days after egg laying (~10–31 days before egg hatching). *Lower line*: mean survival time following egg hatching ± 1 SD; *Upper line*: maximum observed survival time. Drawn from the data of Wodinsky (1977)

removal of one optic gland and 245 days with the removal of both glands (3-fold and 13-fold increases, respectively; Fig. 6.5). Thirteen male octopi were also subjected to bilateral optical gland removal and were found to live longer than normal males and to achieve larger body weights than are observed in unoperated males, but the lifespan extension could not be quantitated because the ages of the operated males were not known prior to surgery.

Wodinsky concluded: "The octopus apparently possesses a specific 'self-destruct' system" and suggested that "the function that this mechanism serves in the female may be to insure survival of the eggs by preventing the predatory female from eating them (although this [still] occurs occasionally). ... in both sexes, this mechanism guarantees the elimination of old, large predatory individuals and constitutes a very effective means of population control." In addition, the debilitation of the female also causes the female to spend more time caring for the eggs before they hatch.

Wodinsky also pointed out that his observations likely pertain to most or all cephalopods including squids and cuttlefish, except the Nautilus, which may live much longer than most other cephalopods. Furthermore, the phenomenon of programmed aging and death may also pertain to "many species (fish, insects, arachnids, and molluscs) [in which] the female ceases to feed, spawns, and dies."

6.3.1.2 Gonadectomy and Hypophysectomy of Lampreys

The death of lampreys is often associated with intestinal and visceral atrophy and attendant starvation that is closely tied to gonadal development (Larson 1980). Hypophysectomy before maturation increased lifespan by as much as 11 months in *Lampetra fluviatilis* (Larsen 1965, 1969), gonadectomy prior to reproduction prevented intestinal atrophy (Larsen 1974; Pickering 1976), testosterone and estrogen administration accelerated intestinal atrophy (Pickering 1976), and both gonadectomy and hypophysectomy slowed body wasting (Larsen 1985).

More dramatic results have been reported for the sea lamprey, *Petromyzon marinus*. In this species, which normally lives for about 7 years, the prevention of sexual maturation can extend lifespan to beyond 20 years (Manion and Smith 1978).

As Finch concluded, "neuroendocrine mechanisms are implicated as regulators of lamprey lifespan" and "these studies show that the proximal pacemaker for senescence in lampreys is gonadal maturation" (Finch 1990).

6.3.1.3 Castration of Salmon

The neuroendocrine suicide of some species of salmon shortly after spawning is a well-known event in adults that is mediated by hyperadrenocorticism secondary to the gonadal secretion of sex hormones (Robertson 1961; McBride and van Overbeeke 1969). Robertson found that gonadectomy at 2.1 years, or about 5 months prior to sexual maturation, extended the maximum lifespans of male salmon by 69% and of female salmon by 74% (Fig. 6.6).

These fish were in captivity, and therefore the results of gonadectomy were not related to reduced physical activity of the castrated fish. In fact, the intact fish may have been more sluggish than the castrated fish, which may have made them

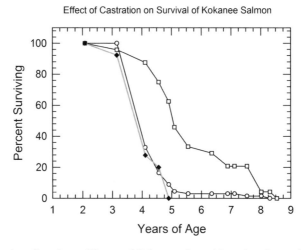

Fig. 6.6 Extension of maximum lifespan of Kokanee salmon (*Oncorhynchus nerka kennerlyi*) by gonadectomy at 2.1 years of age (onset of the curves). *Diamonds, gray line*: intact controls (an unspecified mixture of males and females). *Circles*: castrated males. *Squares*: castrated females. Most castrated males regenerated their gonads rapidly and died on schedule, whereas a small number did not and lived dramatically longer than the normal maximum lifespan. The oldest castrated male and female salmon lived 69 and 74%, respectively, longer than the oldest non-castrated salmon. The original population sizes at the time of castration have been corrected to subtract all salmon subsequently lost to predation or disappearance from the pond and all salmon that were later killed by the investigator. Constructed from the data of Robertson (1961)

differentially susceptible to predation, the intact population being reduced by about 40% by raccoons near the time of sexual maturation, as compared to only about a 15% loss of the castrates (extraneous deaths have been omitted from Fig. 6.6). In addition, the castrated animals weighed slightly more than controls at the time the last controls died, but then continued to grow, reaching approximately 3 times the final weight of controls in the case of males before death, or about six times the final weight of controls in the case of females. Therefore, the greater survival of the castrates is unlikely to be related to calorie restriction or a reduced metabolic rate.

Control salmon did not spawn, presumably because of the captive state of these fish, so the results of Fig. 6.6 cannot be attributed to either the energetic or the physical stresses of spawning. The control females were replete with eggs and the males were replete with milt, but according to Robertson, "the castrated male kokanee with regeneration of only a very small amount of gonadal tissue shows marked hyperplasia of the adrenal cortical tissue and dies with all the typical stigmata of the fish with normally mature testes. It is accordingly clear that the stress of gonad formation is not operating in these fish" (Robertson 1961). In other words, it is clearly the hyperadrenocorticism *per se* that is responsible for the changes that lead to death, not the metabolic demand of growing and maintaining large functioning gonads, and in fact the symptoms of senescence can be induced even in immature rainbow trout by implantation of hydrocortisone alone (Robertson et al. 1963). Therefore, the sum of the available evidence overwhelmingly indicates that the therapeutic results of Fig. 6.6 are due to the interruption of a specific genetic program that ensures that the death of the adult is coupled to sexual maturation and reproduction, just as in the case of other fish, octopi, insects, arachnids, and mollusks.

Intriguingly, gonadectomy in Robertson's experience caused the prolonged retention of juvenile physical characteristics, reminiscent of the phenomenon discussed in Section 6.2 of zero aging prior to sexual maturation followed by aging in the adult phase. Consistent with this analogy, when gonadectomy was delayed until sexual maturation was complete, comparatively little extension of lifespan could be achieved despite the subsequent lack of reproductive cost (Robertson 1961).

Slow regeneration of the gonads was found to occur in almost every male (Robertson 1961), and even regeneration of as little as 2% of the mass of the testes was sufficient to permit the metamorphosis of the salmon from the silvery, streamlined form of the juvenile into the hooked jaw, humped back, and elongated snout morphology of the normal spawning male. Salmon without gonad regeneration were able to retain fully juvenile morphology until at least 5.5 years of age and partially juvenile morphology until the maximum lifespan was reached (Robertson 1961).

It is of interest that sex hormones stimulate adrenal cortical hormone release in mammals as well, including humans (McBride and van Overbeeke 1969). In salmon as in humans, sexual maturation is associated with the onset of significant thymic involution (Robertson and Wexler 1960) and reduced immune function (Maule and Schreck 1987), and in the salmon there is also degeneration of the spleen, liver, intestine, skin, heart, and kidneys (Robertson and Wexler 1960), which is consistent with the killing of various cell types in mammals by excessive glucocorticoid exposure as in Cushing's disease [victims of which appear strikingly old (Baxter and

Tyrrell 1987)]. Furthermore, in the salmon, changes are induced in blood vessels that strongly resemble human coronary artery disease (Robertson et al. 1961). Finally, the life-extending effect of castration in salmon is also qualitatively consistent with the effect of castration in men (Section 6.3.7.1).

6.3.1.4 Castration and Prevention of Mating in Male Marsupial Mice

The males of several "marsupial mouse" (Antechinus) species die within a few weeks after mating and are therefore semelparous (Finch 1990). Like salmon, these terrestrial animals die of effects associated with hypercortisolemia (Diamond 1982; Lee and Cockburn 1985a), including immunosuppression, gastric and duodenal ulcers, and tissue involution including even dramatic testicular involution (though arranged so as to not preclude reproduction) (Finch 1990), and similar effects can be produced by cortisol injections (Barker et al. 1978; Bradley et al. 1980). Apparently in response to gonadal secretions, as in the case of salmon, the adrenal gland hypertrophies (Barnett 1973) and corticosteroids rise approximately 4-fold (Bradley et al. 1980) in preparation for and during a prodigious mating session lasting for up to 12 hours.

Although the death of these animals is often ascribed to the stress of the act of mating, even males that are not allowed to mate show wasting changes during the mating season, becoming sterile in some cases (Woolley 1981), and undergo greatly reduced sensitivity to the feedback suppression of corticosteroid release (Lee and Cockburn 1985b). In addition, elevated testosterone levels (Moore 1974; Bradley et al. 1980) lower the levels of corticosteroid binding globulins (Lee and Cockburn 1985a), which makes the increased corticosteroid levels more lethal. Therefore, although the stress of mating may increase corticosteroid levels, the animals appear primed to be maximally rather than minimally damaged by these elevated hormone levels.

The death of these males after mating dooms them to a lifespan of less than 1 year, whereas the females can live more than 2 years (Cockburn et al. 1983). However, if males are kept separated from females so as to be unable to mate, they can live at least three times longer than normal (Woolley 1971). More definitively, castration can prevent adrenal hyperplasia and hypercortisolemia, thus sparing the animals from major organ degeneration (Moore 1974) and allowing greater survival even under field conditions (Lee and Cockburn 1985a).

The bush rat (*Rattus fuscipes*), which lives in the same parts of Australia as marsupial mice, has evolved a very similar life history to that of Antechinus (Lee and Cockburn 1985a). Another Australian denizen, the smokey mouse (*Pseudomys fumeus*), has conditional semelparity, depending upon the food supply (Lee and Cockburn 1985a). This raises the possibility (Finch 1990) that semelparity of marsupial and mammalian males is related to the periodic population crashes common to small animals like voles and lemmings. The death of males under such circumstances would conserve food for the gestating female and her offspring, consistent with the origin of aging from boom-bust population cycles as proposed in the previous chapter (Mitteldorf 2010) and consistent as well with the observation that the programmed death of monocarpic plants [(Leopold 1961); Section 6.3.2] and salmon may have evolved at least in part to ensure adequate food for growing offspring.

6.3.2 Monocarpic Plants

Genetically-mandated death is clearly not restricted to animals. Although our focus is primarily on animals, it is worth citing a few examples of programmed death in plants to highlight the breadth of this undeniable phenomenon of biology. These examples are relevant also because there may even be some mechanisms of death [particularly involving depletion of mRNA leading to inactivation of protein synthesis (Leopold 1961)] that are broadly similar between monocarpic plants (which, like semelparous animals, by definition reproduce only once and then die) and most mammals (see Sections 6.5.1 and 6.5.2).

While annual plants live only one year before flowering and dying, Mediterranean Agaves can live in good health for 100 years and then suddenly flower and die (Molisch 1938), and the pineapple, *Puya raimondii*, can mature and then immediately die at ages as high as 150 years (Raven et al. 1986). In the case of bamboo, different species are clocked to die at 7, 30, 60, or 120 years of age, and local populations often die at the same time (Janzen 1976; Soderstrom and Calderon 1979). It is obvious from these examples, the existence of non-aging plants, and the clearly inherited nature of monocarpism in species that show an interfertile continuum between monocarpic and polycarpic varieties (Oka 1976) that death and aging in plants are under very strong genetic control.

Furthermore, just as is the case in animals, the rapid death of plants near the time of reproduction is widely acknowledged to be caused by the hormones that are necessary for reproduction (Finch 1990). Monocarpy is also considered to be adaptive and to have independently evolved on countless occasions (Leopold 1961; Finch 1990). Also just as in semelparous animals, rapid death in plants associated with reproduction can be delayed or prevented by removing specific reproductive structures (Molisch 1938; Leopold et al. 1959; Nooden 1988). In both semelparous animals and monocarpic plants, the death of the adult is guaranteed or all but guaranteed as a prerequisite for reproduction.

Given the pervasive occurrence of this situation, the universality of the general mechanisms involved even across kingdoms, and the fact that these mechanisms are induced generally before any possible energetic tradeoff between continued survival and reproduction could take place, the death of semelparous or monocarpic adults presumably evolved to facilitate the survival of the offspring. A final extreme example of this is the tree, *Tachigalia versicolor*, which can grow to a height of up to 40 meters and then flower, produce fruit, and die within a year (Foster 1977), perhaps to open up a gap in the forest canopy that allows the offspring to obtain more sunlight (Finch 1990). At the opposite end of the size spectrum, the altruistic death of yeast (Fabrizio et al. 2004; Longo et al. 2005) [and of bacteria (Sat et al. 2003)] has also been demonstrated.

6.3.3 Iteroparous Species: C. elegans

Might genetic mechanisms also more subtly prescribe slow rather than abrupt death of non-semelparous animals, perhaps for similar reasons? Iteroparous species are by definition species that reproduce more than once. Although the genetic induction of death in semelparous species appears obvious and inarguable, genetic mechanisms that induce death in iteroparous species have traditionally been harder to pinpoint. However, many such mechanisms have recently been elucidated in detail as a result of a mounting ability to nullify them and observe resulting lifespan extensions that can be even more dramatic than the lifespan extensions achievable in semelparous animals. In at least some cases, the particular mechanisms that cause death in iteroparous species seem closely related to the mechanisms that are operative in semelparous species (e.g., Sections 6.3.3.1, 6.3.7, 6.5.9, and 6.5.10). But aging and death seem to be fundamentally built into the physiology of iteroparous species in other ways too, and these mechanisms can theoretically be undone by approaches that have not yet been possible in known longevity mutants, as described in Section 6.3.9.

6.3.3.1 Escape from Default Life Shortening Mediated by Insulin-Like Signaling and Germ Line Signaling

A great many single-gene mutations have been described that extend lifespan in C. elegans [see, e.g.: (Baumeister et al. 2006; Gami and Wolkow 2006; Kenyon 2006)] and other species. Each of these has helped to demonstrate the existence of extensive genetic pathways that shorten lifespan in adult organisms under default conditions and has helped to define the nature of the relevant pathways in increasingly exquisite detail. The extensiveness and degree of integration of the interacting pathways that ensure aging in even such a simple organism as C. elegans is anything but accidental, and implies that these pathways have been highly evolved. One specific example will suffice to illustrate the significance of such observations.

A particularly dramatic and well-known demonstration of active life-shortening by normal physiological mechanisms associated with reproduction has been provided by Cynthia Kenyon's laboratory (Arantes-Oliveira et al. 2003). Using C. elegans, her group showed that combining daf-2 inactivation (induced by combining daf-2 RNA interference with the presence of a weak daf-2 mutation, the e1368 mutation) with laser gonad ablation extended the mean lifespan from 20.7 days for wild type controls to 124.1 days ($p < 0.0001$) (Fig. 6.7). Of the two interventions, gonad ablation produced the greater lifespan extending effect: daf-2 inactivation alone extended maximum lifespan by 180%, whereas the combination of daf-2 inactivation and gonad ablation increased maximum lifespan by 507%. As can be seen in Fig. 6.7, 98% of the gonad-ablated, daf-2-inactivated worms were still alive long after the last wild-type control worm died, and about 85% were still alive when the last daf-2 animal died. Further, at 120 days of age, after half of these

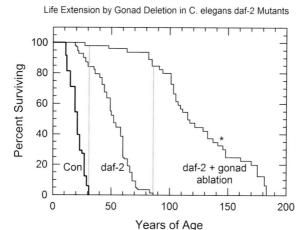

Fig. 6.7 Prolongation of the lifespan of adult C. elegans (Con) by combining a weak daf-2 mutation with daf-2 RNA interference (daf-2) or the same intervention with laser ablation of the gonad (daf-2 + gonad ablation). Note that with the latter treatment, only 2% of the worms die by the time the last control worm has died (*first vertical gray line*), and only 15% of the worms die by the time the last daf-2 worm has died (*second vertical gray line*). *Asterisk* at day 144: surviving worms observed to be as active to blinded observers as 5-day-old control worms (Kenyon 2006). Redrawn from Arantes-Oliveira et al. (2003)

super-worms had died, two-thirds of the survivors were still apparently active and healthy, and movies of even 144-day-old worms indicated that they still retained the vigor and appearance of 5-day-old controls (Arantes-Oliveira et al. 2003; Kenyon 2006). For C. elegans, 120–144 days is comparable to about 500–600 years in humans (Arantes-Oliveira et al. 2003).

Gonad ablation does not extend lifespan in C. elegans merely by reducing the energetic demands of maintaining the gonads or of supporting reproduction as might be conjectured (Kirkwood 1977). In fact, although whole gonad ablation extends life in already long-lived daf-2 mutants, it does not do so in control worms (Kenyon 2006). In control worms, deleting the germ line rather than the whole gonad does extend life, indicating that under normal conditions, a negative effect of germ line cells on lifespan is normally exactly balanced by *positive* effects on lifespan of the rest of the gonad (Hsin and Kenyon 1999; Arantes-Oliveira et al. 2002; Kenyon 2006), despite the energy required for gonadal maintenance.

In *C. elegans*, the germ line normally suppresses kri-1 expression in the intestine and downregulates the production or action of a lipophilic hormone product of DAF-9 that serves as a ligand of DAF-12 (Berman and Kenyon 2006). This combination of actions prevents DAF-16 from moving to the intestinal cell nuclei and maintaining life-extending pathways that were presumably more active prior to maturation of the germ line, thereby shortening lifespan in the adult by up to about 40% (Berman and Kenyon 2006). This germ line-controlled pathway operates in parallel to and is not dependent on signaling through the classical DAF-2 pathway, so it functions specifically to trigger life-shortening when the germ line matures (Berman and Kenyon 2006).

This maturational germ line-induced life-shortening may in part explain why C. elegans seems to transition from a zero-aging state to a normal aging state at the time of sexual maturity. It also seems to be directly analogous to the universal and genetically mandated linkage seen between the onset of reproductive maturity and death in both semelparous animals and monocarpic plants as discussed above (Sections 6.3.1 and 6.3.2, respectively), and the same basic effect has also recently been demonstrated in Drosophila (Flatt et al. 2008).

Because the zero-aging state of dauer larvae has no apparent effect on the lifespan of the adult C. elegans (Section 6.2.1.1 above), it may be possible to combine the six-fold increase in lifespan achieved by prolonging the dauer state with the six-fold increase in lifespan achieved by daf-2 and gonad knock-out in adults and thereby arrive at an overall increase in lifespan approaching 12-fold. In fact, the Shmookler-Reis lab (Ayyadevara et al. 2008) (see also Chapter 20) has achieved something close to this, allowing the creation of sterile but otherwise healthy, vigorous, and metabolically active adult C. elegans that live up to approximately 10-fold longer than wild type worms. Such demonstrations as those of the Kenyon and Shmookler-Reis labs establish beyond doubt that aging is biologically induced and controlled in a complex, post-mitotic, iteroparous organism.

6.3.3.2 A Note on Genetic vs. Stochastic Aging in C. elegans and Other Species

It is often concluded or implied that aging must be driven primarily by stochastic factors rather than by genetic factors because aging and lifespan vary widely even in genetically uniform or near-uniform populations living under closely similar conditions [e.g. (Kirkwood and Finch 2002; Hjelmborg et al. 2006)]. Even the lifespans of monozygotic human twins are only moderately more concordant than the lifespans of fraternal twins or unrelated individuals (Hayakawa et al. 1992; Ljungquist et al. 1998; Hjelmborg et al. 2006). However, the apparent lack of relationship between the age of death and the genotype may in fact be considerably more apparent than real, and we use the opportunity provided by C. elegans to consider how in general genetic control of aging can be compatible with and not excluded by the observation of genotype-independent lifespan variations.

C. elegans exhibits all of the heterogeneity in the onset and progression of aging between individuals as has been seen in other species (Herndon et al. 2002). Nevertheless, as displayed in Fig. 6.7, the population containing the mutation in combination with germ line ablation experienced *almost no deaths and almost no pathology for a time considerably longer than the maximum lifespan of the species.* This clearly shows that *essentially all of the extensive heterogeneity of C. elegans aging over the full course of a normal lifespan can be suppressed completely by genetic influences* (see also Chapter 20 for an even more profound example). The simplest interpretation is that *the stochastic determinants of aging (including random damage) in this species, and most likely in others, are actually subject to, and can be overridden by, genetic control.*

This remarkable effect can be explained in part by the known effects of daf-16 activation in C. elegans (and by the activation of daf-16 counterparts in other species). Longevity mutants are well known to protect against many forms of life-shortening stress, but to focus on just two often-discussed aspects, typical C. elegans longevity mutants have higher expression of antioxidant enzymes (Honda and Honda 1999; Essers et al. 2005; Baumeister et al. 2006) and about 2.5 times more DNA repair capacity (Hyun et al. 2008) than N2 (wild-type) worms. It is therefore reasonable to attribute some of the randomness in wild-type lifespans to random damage caused by default deterministic suppression of the organisms' defenses against random damage.

The rest of the random variation in lifespan between genetically identical individuals may arise because they are not *phenotypically* identical. Genetically identical C. elegans homozygous for all alleles not only differ in gene expression, but show differences in expression that are so stable as to be at least partly *heritable* [(Cypser et al. 2008); Johnson, personal communication]. Further, the expression of a model biomarker of aging in C. elegans, the heat shock protein hsp 16.2, correlates with longevity over a four-fold range of longevity in genetically identical individuals (Rea et al. 2005), strongly indicating that even seemingly "stochastic aging" in this species is at least partly controlled through the phenotype (gene expression pattern) of the adult (Rea et al. 2005; Wu et al. 2006). In 2009, this conclusion was made even stronger by the announcement at the American Aging Association meeting that the variability of expression of the hsp 16.2 biomarker could be reduced by knocking out another gene (Tom Johnson, personal communication), at least suggesting the possibility that genes might actually exist that control the stochasticity of aging.

Like C. elegans, "identical" (monozygotic) human twins can vary substantially in their ages of death (though not to the extent that fraternal twins vary in age of death). However, also like C. elegans, monozygotic twins are not really identical, either genetically or phenotypically. Monozygotic twins do not display identical patterns of gene expression, or even identical numbers of copies of different DNA segments (Bruder et al. 2008), and can vary, for example, in the extent of DNA methylation (Kuratomi et al. 2008), the presence of trisomy 21 (Sethupathy et al. 2007), the inactivation of the paternal vs. the maternal X chromosome in females (Rosa et al. 2007), blood lipid levels (Iliadou et al. 2001), the likelihood of premature ovarian failure (Silber et al. 2008), and even behavioral factors such as risk-taking behavior (Kaminsky et al. 2008). Furthermore, as in the case of C. elegans, variations based on differences in epigenetic control of gene expression are potentially strong enough to be heritable (Holliday 2006; Kaminsky et al. 2009).

Differing expression of the same gene in different individuals can arise from differing influences of RNA regulators of gene expression (Fraga et al. 2005; Mattick and Makunin 2006; Sashital and Butcher 2006; Chuang and Jones 2007; Zaratiegui et al. 2007; Filipowicz et al. 2008), which may explain many aging phenomena (Boehm and Slack 2005, 2006; Syntichaki and Tavernarakis 2006; Wang 2007; Kato and Slack 2008). [In the case of C. elegans, the lin-4 microRNA modifies both developmental event timing and adult lifespan through mechanisms that depend on the insulin-like signaling pathway, daf-16, and HSF-1 (Boehm and Slack 2005), which led the authors to postulate that this microRNA programs adult aging.] Finally, it is conceivable that phenotypic differences might arise as a result of stochastic fluctuations in gene expression on the level of single cells (Elowitz et al. 2002; Fedoroff and Fontana 2002; Raser and O'Shea 2005) or through differing nutrient-genome interactions (Sashital and Butcher 2006), but this has yet to be established. These considerations caution against underestimating the full potential role of gene expression on the basis of surveys that do not distinguish between genotype and phenotype.

A broader point to consider, in evaluating the general significance of genetic versus stochastic factors in normal aging, is that normal genetic controls over aging are not usually apparent until they are broken so as to allow results such as those seen in Figs. 6.5, 6.6, and 6.7 to be observed. Under default conditions, genetic mechanisms driving aging may be *universal or near-universal in wild type populations, so comparing the longevities of sub-populations all of which bear universal aging drivers will not identify the importance of those genetic drivers for aging.*

Nevertheless, indications of a strong genetic influence on mouse (Klebanov et al. 2000) and human aging have begun to emerge, the latter from the study of people who naturally live to extraordinary ages [e.g., (Perls et al. 2002b; Atzmon et al. 2006; Christensen et al. 2006; Martin et al. 2007; Suh et al. 2008)]. For example, the odds of living to 100 are around 17 times higher for siblings of centenarians than for age-matched controls, and these siblings enjoy about a two-fold reduction in mortality rate in comparison to controls over the entire lifespan (Perls et al. 2002b). The offspring of centenarians also have more than a 60% reduction in the risk of dying from all causes, more than a 70% reduction in risk of dying from cancer, and an 85% reduction in the risk of dying from coronary heart disease in comparison to controls whose parents were born in the same year as the centenarians (Terry et al. 2004). A large percentage of individuals who live to greater than 110 years of age remain functionally independent and escape from cardiovascular disease, stroke, hypertension, cancer, diabetes, osteoporosis, and other maladies (Schoenofen et al. 2006). As noted by Perls et al. (Perls et al. 2002a), "centenarians disprove the ageist myth 'the older you get, the sicker you get'; they live 90–95% of their very long lives in excellent health only to experience illnesses in the very last few years of their lives. Thus, it appears that in order to live to 100, one must age relatively slowly and markedly delay and/or escape age-associated diseases." These trends are analogous to the effect seen in Fig. 6.7.

A few specific genes have been found to be associated with the attainment of extraordinary ages in humans [e.g.: (Martin et al. 2007)]. At least one of these (apoC III) is under the control of FOXO 1 (the human analog of daf-16 in C. elegans) (Martin et al. 2007), one is a modification of the human insulin-like growth factor 1 receptor (analogous to the daf-2 mutation in C. elegans) (Suh et al. 2008), while another (CETP) is not present in standard animal models (Martin et al. 2007). It is reasonable to postulate that if the entire human population could somehow be endowed with these genes, or, by pharmacological intervention, at least with the effects of these genes (Perls and Puca 2002; Gami and Wolkow 2006), then some of the seemingly random variation seen in the age of death in younger populations might be suppressed, in analogy to the experiment shown in Fig. 6.7, despite other background differences in genotype and environment.

The lesson from C. elegans is that the ability of genetic influences to not only override but possibly even to control and explain "stochastic" aspects of aging may be far more powerful than is usually considered. Furthermore, counterparts to Fig. 6.7 have been observed in mammals as well (see Section 6.3.5.1 for one example). More broadly, as emphasized by Martin and Finch, if one considers the role of both "environmental" influences (such as calorie restriction) and stochastic influences on aging in rodents in comparison to the role of genetic influences as estimated by comparing the lifespans of rodents to the lifespans of humans, then one must conclude that "the constitutional genome trumps the influence of both stochastic and environmental influences" (Martin and Finch 2008).

If aging and the mortality arising from it are indeed largely controlled by the genome and the epigenome rather than mostly by uncontrollable stochastic factors, the opportunities for genetically-based intervention should be greater. Enforcing protective patterns of gene expression and blunting effects of genes that actively suppress longevity (Bergman et al. 2007) are reasonable goals.

6.3.3.3 Escape from Default Life Shortening Mediated by Protein Synthesis in C. elegans

Although protein synthesis is indispensable for life, life can be extended to a modest extent in C. elegans by inhibiting protein synthesis (Hansen et al. 2007; Pan et al. 2007; Syntichaki et al. 2007), implying that in the default state, lifespan is actually shortened by some consequence of protein synthesis. In C. elegans, blocking protein synthesis by depleting ribosomal proteins increased lifespan by mechanisms that were independent of daf-16 (Hansen et al. 2007). According to a similar study in the same year, depleting the TOR-dependent translation initiation factor eIF4E also extended lifespan by daf-16 independent mechanisms (Syntichaki et al. 2007) whereas the Hansen et al. study found that depleting both eIF4E and other protein synthesis initiation factors extends life through mechanisms that are "completely dependent on the DAF-16/FOXO transcription factor" and simultaneously "appears to block all of the DAF-16-independent lifespan extension that would have been produced by inhibiting translation itself" (Hansen et al. 2007).

Lifespan extension due to inhibition of protein synthesis has been linked to the effect of calorie restriction (Hansen et al. 2007; Kenyon 2010). One suggestion is that reduced protein synthesis may simply conserve energy that is then diverted to increased self-repair (Syntichaki et al. 2007). However, the fact that the effect of inhibiting protein synthesis is mediated by both DAF-16 dependent and DAF-16 independent pathways and that there may be "a switch-like model, in which the activation of a DAF-16-dependent pathway suppresses the operation of the DAF-16-independent pathway" (Hansen et al. 2007) implies that something more subtle and specific than mere energy conservation is involved, and, in fact, inhibiting protein synthesis may extend life in many species by increasing respiration rather than by decreasing ATP consumption [see (Kenyon 2010)]. Inhibiting protein synthesis could also be beneficial due to the suppressed synthesis of proteins whose normal role is to shorten adult lifespan (see next section). In addition, TOR, which promotes protein synthesis, also inhibits autophagy (Hansen et al. 2008). In yeast, life span extension related to inhibition of translation seems to be mediated by induction of Gcn4, a nutrient-sensitive transcription factor (Steffen et al. 2008). Additional studies will be needed to fully define the effects of translation inhibition.

6.3.3.4 Escape from Developmentally Programmed Aging in C. elegans

During the embryonic development of C. elegans, the GATA transcription factors ELT-5 and ELT-6 suppress, and the GATA transcription factor ELT-1 activates, a common target GATA transcription factor known as ELT-3 (Gilleard and McGhee 2001; Koh and Rothman 2001). It was recently discovered that the expression of elt-5 and elt-6 continues and greatly intensifies throughout much of adult life and profoundly suppresses the expression of elt-3, which in turn affects the aging-dependent expression of nearly half (605 out of 1254) of all genes whose expression was found to change throughout the course of normal adult life (Budovskaya et al. 2008). When the repression of elt-3 was partially overcome by knocking out either elt-5 or elt-6 expression, the result was a rightward shift of the mortality curve by a constant 3–4 days or so starting on about day 10 of adulthood.

Thus, the rising influence of ELT-5 and ELT-6 with advancing age creates a default life-shortening effect in the adult and has been interpreted as a "developmental program" because it is believed unlikely to be driven by damage mechanisms (Budovskaya et al. 2008). The authors state: "to our knowledge, this is the first transcriptional circuit accounting for global changes in expression during aging in any organism" [but see also (Boehm and Slack 2005, 2006)]. Still unanswered is the question of what is the specific biochemical reason for the rising expression of elt-5 and elt-6 with age in the adult.

The general explanation suggested for this apparent aging program was "drift of developmental pathways" that has "a neutral effect on fitness in the wild ... as old worms are extremely rare in the wild" (Budovskaya et al. 2008). However, ELT-5 levels rise 50%, ELT-6 levels increase about 30%, and ELT-3 levels in the trunk fall by over 90% (independent of the expression of age-1) between adult day 2 and adult day 5. It is questionable whether such early effects would be truly neutral (without impact on survival or reproduction) in the wild.

It is also worth noting that the effect of elt-3 suppression on over 600 genes amounted to only about a 15% increase in maximum lifespan. Simultaneous knockout of both elt-5 and elt-6 was not reported. By comparison, daf-2 mutants show about a 100% increase in maximum lifespan and the strong suppression of age-1 can achieve about a 900% increase in maximum lifespan. Perhaps comparing gene expression changes in these different cases would give greater insight into the hierarchy of genetic influences over lifespan.

6.3.4 Insects

Many life-extending mutations have been discovered in Drosophila and other insects that appear to interfere with normal life-shortening mechanisms (Harshman 1999; Tatar et al. 2001a; Rogina et al. 2002; Tu et al. 2002; Simon et al. 2003; Hwangbo et al. 2004; Wang et al. 2005; Libert et al. 2007). The aging of insects appears to be controlled to a considerable degree by juvenile hormone and ecdysone through a variety of mechanisms including insulin-like signaling effects (Tatar et al. 2001a, 2003; Simon et al. 2003; Tatar 2004; Amdam et al. 2005; Flatt et al. 2005; Corona et al. 2007; Hodkova 2008). Removal of the influence of juvenile hormone in adult butterflies (Herman and Tatar 2001) and grasshoppers (Pener 1972) by surgical ablation of the corpora allata doubles or triples the adult lifespan, and the induction of diapause in adult Drosophila (which is also controlled by juvenile hormone) appears to induce a state of zero or negligible aging for the duration of the time spent in diapause (Tatar et al. 2001b).

Although reproduction is tightly coupled to aging in insects, this relationship was shown not to be necessary using Drosophila mutants defective for the ecdysone receptor (EcR) (Simon et al. 2003). Males and females heterozygous for EcR mutations very convincingly lived 40–50% longer and were more resistant to stresses regardless of genetic background despite being much more active than non-mutants, and despite having normal body weights (no calorie restriction effect). These flies were not deficient in oogenesis and, in fact, were actually dramatically *more* fertile than wild-type flies in most tests, and never less fertile than non-mutants. (Females with a temperature-sensitive mutation that inhibited ecdysone synthesis had a 42% greater lifespan at the permissive temperature of 29°C compared to the non-permissive temperature of 20°C and were more stress-resistant, but in this case fertility was impaired.)

6.3.5 Genetically Altered Mice

6.3.5.1 Escape from Life Shortening from Insulin-Like Signalling

Aging appears to be controlled in part through insulin-like signaling pathways or their counterparts in yeast, worms, flies, mice, and humans (Longo and Finch 2003; Suh et al. 2008), and a variety of genetic modifications have accordingly been found to extend life in mice by blunting insulin or insulin-like growth factor signaling [e.g.: (Brown-Borg et al. 1996; Flurkey et al. 2001; Bluher et al. 2003; Liang et al. 2003; Richardson et al. 2004; Kopchick et al. 2008)].

The first single gene mutation found to extend life in mammals (Brown-Borg et al. 1996) was the Prophet of Pit-1 (Prop-1) mutation that produces the Ames dwarf mouse phenotype (Andersen et al. 1995). This mouse has a mean lifespan 50–70% longer than controls, and a maximum lifespan 20–50% longer than control mice (Brown-Borg et al. 1996) despite severe physiological derangements, including failure to produce growth hormone, thyroid hormone, and prolactin (Cheng et al. 1983), reduced metabolic rate (Gruneberg 1952) and body temperature (Hunter et al.

1999), and about a three-fold reduction in adult body weight (Liang et al. 2003). As might be expected from the patterns described above for non-mammalian species, dwarf mice also have greatly or completely attenuated fertility (Bartke 1979). An interesting sidelight to the Ames dwarf mouse story is that the Prop-1 mutation, like the daf-2 mutation combined with germ line ablation in C. elegans as discussed in Section 6.3.3.1, almost totally overcomes the stochastic variation in age of death that is observed over the course of normal aging in female control mice (Fig. 6.8).

The Snell dwarf mouse is very similar to the Ames dwarf mouse, having a mutation in the Pit1 gene downstream of the Prop-1 gene (Liang et al. 2003). The fact that both mutants live much longer than controls despite their severe physiological abnormalities is a testament to the relative importance of pathophysiology on the one hand versus normal aging as mediated through default genetic programs on the other. As noted by Flurkey et al., these mutants show that "a single gene can control maximum lifespan and the timing of both cellular and extracellular senescence in a mammal" (Flurkey et al. 2001).

The increased lifespan of dwarf mice is, just as in the case of non-mammalian species, connected to increased antioxidant defenses (Brown-Borg and Rakoczy 2000; Hauck and Bartke 2001; Liang et al. 2003), less oxidation of protein and DNA (Brown-Borg et al. 2001) and increased DNA repair (Huang and Tindall 2007; Salmon et al. 2008). These augmented defenses could also account for the protection against neoplasms (Ikeno et al. 2003) and age-related deficits in ambulation and avoidance learning (Kinney et al. 2001) enjoyed by these mice.

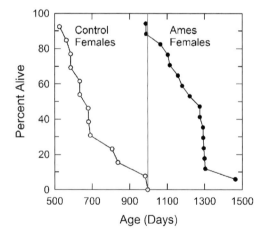

Fig. 6.8 Suppression of stochastic aging in mammals for a time virtually equal to the maximum lifespan of the species (*vertical line near 1000 days*) as a result of the Ames dwarf mouse mutation in the Prophet of Pit-1 (Prop-1) transcription factor gene. At the age of death of the last surviving female control mouse, only two of 17 female Ames dwarf mice ($\sim$10%) had died. In the case of males, the overlap between the controls and the mutant mouse survival curves was more extensive. Redrawn from Brown-Borg et al. (1996)

The "little mouse" is another growth hormone/IGF-1 deficient mutant, in this case the deficiency arising from a mutation in the Ghrhr gene encoding for the growth hormone releasing hormone receptor (Godfrey et al. 1993). It is about twice as large as Snell and Ames dwarf mice, makes thyroid hormone normally, lives less long than the Snell and Ames dwarves, but lives longer when homozygous rather than heterozygous for the mutation (Flurkey et al. 2001), implying a dose-response effect between insulin-like signaling and life shortening. Similarly, growth hormone receptor knockout mice weigh more than Snell/Ames mice but less than little mice, have IGF-1 levels that are about 10% of those of wild-type mice (Coschigano et al. 2000, 2003) as compared to 9–23% in the little mice (Godfrey et al. 1993), and seem to live longer than little mice but not as long as the Snell and Ames dwarves (Coschigano et al. 2000, 2003).

Another mild example of mammalian lifespan extension via reduced insulin-like signaling is that of the heterozygous insulin-like growth factor 1 receptor knockout mouse (Holtzenberger et al. 2003). This mouse has perhaps a 50% reduction in IGF-1 signaling, very little growth stunting (6–8%), undiminished metabolic rate and food intake, intact fertility, variable resistance to paraquat, and a sex-dependent 16–33% increase in mean lifespan.

In a similar model, the insulin receptor was knocked out selectively in adipose tissue, creating the FIRKO (fat insulin receptor knock-out) mouse (Bluher et al. 2002). This mouse experiences a 50–70% decrease in fat mass, a 15–25% decrease in body weight (Bluher et al. 2002), an 18% increase in median lifespan (and about a 14% increase in maximum lifespan) (Bluher et al. 2003), and improved insulin sensitivity with no decrease in food intake (Bluher et al. 2002). The finding that restriction of insulin signaling in adipose tissue alone extends life in mice is reminiscent of observations in flies (Bishop and Guarente 2007; Libert et al. 2007) and worms (Apfeld and Kenyon 1998, 1999; Ailion et al. 1999) demonstrating that organismal aging in these species is orchestrated by central control mechanisms rather than generally originating at the level of individual cells (Boulianne 2001; Mattson et al. 2002; Wolkow 2002).

The homozygous knockout of the $p66^{shc}$ gene in mice results in 50% less intracellular oxidant production (Trinei et al. 2002), less steady state 8-oxo-2-deoxyguanosine in nuclear and mitochondrial DNA in 4 different tissues examined (Trinei et al. 2002), major protection against atherosclerosis (Napoli et al. 2003), oxidizing agents and UV radiation (Migliaccio et al. 1997), and a 30% increase in median lifespan (Migliaccio et al. 1999) with no apparent side effects (Migliaccio et al. 1997, 1999). Very interestingly, the escape from default life shortening caused by the elimination of the product of this gene is probably also mediated in significant part by effectively reduced insulin-like signaling (Nemoto and Finkel 2002; Xi et al. 2008).

Klotho is a protein made by kidneys that (among many other things) inhibits insulin/insulin-like growth factor-1 signaling (Kurosu et al. 2005), thereby activating the FoxO forkhead transcription factor and inducing resistance to oxidative stress, including the expression of manganese superoxide dismutase (Yamamoto et al. 2005), and, when overexpressed two-fold by introduction as a transgene in mice (under the control of the human elongation factor 1-alpha promoter), increasing mean lifespan in mice by 19–31% with no change in body size, food intake, or oxygen consumption (Kurosu et al. 2005). Interestingly, it also reduces fertility (measured as pups per year; pups/litter was not changed) by 38–42%. Klotho polymorphisms in humans have been associated with altered lifespan (Arking et al. 2002, 2005), risk for coronary artery disease (Arking et al. 2003), risk for osteoporosis (Yamada et al. 2005), and risk of stroke (Arking et al. 2005).

The well-known life-extending effects of CR have also been linked to reduced insulin signaling and increased insulin sensitivity and are discussed in detail elsewhere in this volume (Spindler 2010). Bartke and colleagues, however, produced evidence that CR and GH-release mutants act to extend life by at least partly different mechanisms [which is consistent with the fact that Snell dwarf mice actually develop obesity despite their small size (Flurkey et al. 2001)] by finding that CR and the Ames mutation have additive effects on mouse lifespan (Bartke et al. 2001) (Table 6.1). In fact, mild calorie restriction raised median lifespan by identical proportions (about 25%) in wild-type and dwarf mice even though the ad libitum dwarf median lifespan was already 37% longer than the ad libitum control median lifespan (i.e., even greater than CR control mouse median lifespans). The combination of CR and the mutation extended the median lifespan by 69%, to 1226 days (3.4 years). Maximum lifespan may not have been raised as much by CR in dwarf mice as in wild type mice on a percentage basis, but the data of Bartke et al. did not extend to the last mouse to die in the case of the Ames dwarf CR mice, so this point was not adequately defined. Nevertheless, the last mouse recorded in their experiment lived nearly 60% longer than the last surviving control mouse (Table 6.1), which in humans would be the rough equivalent of living to the age of 160.

Table 6.1 Life extension by CR, the Ames mutation, and both[a]

Group or Comparison	Wild-Type	Wild-Type+CR	Ames	Ames+CR
Median LS[a] (Days)	726	915	996	1226
Maximum LS[a] (Days)	885	1185	1304	1396*
% LS increase due to CR (Median LS)		26		23
% LS increase due to CR (Max. LS)		34		7*
% LS increase due to Ames mutation (Median LS)			37	
% LS increase due to Ames mutation (Maximum. LS)			47*	
% LS increase due to Ames mutation plus CR (Median LS)				69
% LS increase due to Ames mutation plus CR (Maximum LS)				58*

[a] *LS* Lifespan, numerical values derived by quantitative inspection of the survival curves of Bartke et al. (2001)
* True maximum lifespan of the CR Ames mice not established

The world record holder for life extension through impaired insulin-like signaling in modified mammals is the Laron dwarf mouse, a transgenic growth hormone receptor knockout mouse that has lived to the age of 1819 days, or 4.98 years, without the addition of CR (Bartke, personal communication).

6.3.5.2 The PEPCK-C (mus) Mouse

In addition to the CR paradigm and the reduced insulin/insulin-like signaling (IIS) paradigm, a new life extension model was recently introduced in the form of the homozygous transgenic cytosolic phosphoenolpyruvate carboxykinase (PEPCK-C) mouse (Hakimi et al. 2007). This mouse has ∼100-fold over-expression of the gluconeogenic and triglyceridogenic PEPCK in the cytoplasm of skeletal muscle cells and about 8% of this level of activity in the heart, where activity is normally undetectable. In response to the effects of the transgene, the livers increase their already-high levels of endogenous (non-transgenic) PEPCK-C by about 60%. Over-expressing the gene in fat leads to obesity, but making the same change in skeletal muscle radically increases the energy output and food intake of these mice, increases the number of skeletal muscle mitochondria (based on mitochondrial DNA content) by 50–100% (Hakimi et al. 2007), and evidently greatly extends lifespan.

PEPCK-C (mus) mice are so hyperactive that they can be identified without genotyping by the age of two weeks (Hanson and Hakimi 2008). At the age of 30 months (the normal control median lifespan), these mice are still able to run about 50% faster than 2–6 month-old (optimal performance age) controls, and there was no decline in performance between 18–24 months and 30 months of age. [The maximal speed was highest at 1–2 months, constant from 2 to 18 months (at which ages the PEPCK-C (mus) mice ran more than twice as fast as controls), and declined approximately 35% at 18–24 months, with no subsequent further decline.]

In the initial study, the authors commented that these mice "survived far longer and looked healthier than controls and that both male and female PEPCK-C^{mus} mice were reproductively active at 21 months of age; one PEPCK-C^{mus} mouse gave birth [to normal sized litters] at 30 months of age" (Hakimi et al. 2007). In a follow-up study it was stated that "they lived almost 2 years longer than the controls and had normal litters of pups at 30–35 months of age (most mice stop being reproductively active at 12–18 months)," (Hanson and Hakimi 2008). These are truly extraordinary effects for a single gene inserted into skeletal muscle, and suggest no tradeoff between reproduction and other energy-requiring processes in these mice. A proper lifespan study for these mice is presently underway.

PEPCK-C mice have a 40% higher VO_{2Max}, post-exercise lactate levels that are up to 84% lower, 7–10 times the cage activity, and up to 30 times the exercise (running) endurance of control mice (Hakimi et al. 2007). These mice eat 60% more but weigh only 57% as much as controls and have only 10% as much total body fat as controls. Significantly, they escape completely from the normal massive age-related increase in body adiposity seen in the controls (3.9-fold increase in visceral fat and 2.2-fold increase in subcutaneous fat) between 6 and 18 months of age. However, their muscles contain anywhere from 40% to 10-fold more triglyceride than control muscles (Hakimi et al. 2007), which is directly proportional to and therefore presumably powers their extraordinary physical performance (Hanson and Hakimi 2008).

They also have 38% lower blood levels of cholesterol, 30% lower levels of free fatty acids, and 39% lower circulating triglyceride levels (Hakimi et al. 2007). Their insulin sensitivity is increased and their insulin levels are "very low," which has been suggested to be the reason for their extraordinary longevity and prolonged fecundity (Hanson and Hakimi 2008), yet their blood glucose levels are increased by 44% compared to control mice and their ketone bodies are elevated by 48% (Hakimi et al. 2007).

In some way, this mouse appears to transcend life-shortening from mitochondrial damage, apoptosis, cancer, and all other factors in the face of a physiologically astronomical rate-of-living. Because these animals appear to shift from glucose metabolism to fat metabolism (Hakimi et al. 2007; Hanson and Hakimi 2008), they may accomplish simultaneous reduction in both insulin signaling and lipotoxicity (Dulloo et al. 2004; Iossa et al. 2004; Sharma et al. 2004) (see also Section 6.3.9), which could be an extremely protective combination that avoids the link between reduced insulin-like signaling in dwarf mouse models and their tendencies to gain fat (Flurkey et al. 2001; Bartke and Brown-Borg 2004).

The mechanisms by which the PEPCK-C transgene transforms so many functions in these mice remain unknown, but several plausible possibilities have been suggested (Hakimi et al. 2007; Hanson and Hakimi 2008). It may increase carbon flow to mitochondria by stimulating succinyl-CoA synthase by providing the GDP needed by this enzyme, and by converting excess mitochondrial oxaloacetic acid, which can't be fully metabolized to CO_2 in mitochondria (and may therefore become a limiting factor in maximum exercise capacity), into cytoplasmic phosphoenolpyruvate and thence into pyruvate. As the key enzyme for glyceroneogenesis, it may allow synthesis of the 3-phosphoglycerol needed to form the triglycerides that are needed to drive skeletal muscle energy output. By increasing both citric acid cycle flux and triglyceride storage, it may upregulate PPARγ, PPARα, and PGC-1α, which ultimately, by acting through NRF-1, NRF-2a, and Tfam, and with the participation of AMP-activated protein kinase (AMPK), increase mitochondrial biogenesis (Puigserver and Spiegelman 2003) and mediate many of the important effects of calorie restriction and sirtuin activation (Weindruch et al. 2008). In support of this proposed mechanism, overexpression of VP16-PPARδ in skeletal muscle enhances mouse running endurance by almost 100% (Wang et al. 2004).

In this connection, it is intriguing that oral administration of resveratrol (Lagouge et al. 2006), and acetylcarnitine [500 mg/kg per day for 4 weeks in C57Bl/6J mice (Narkar et al. 2008)], two non-toxic nutrients already widely consumed by humans, can strongly improve exercise endurance without prior training (Narkar et al. 2008). Acetylcarnitine is an orally active agonist of AMPK (Narkar et al. 2008), a master regulator of metabolism, glucose homeostasis, exercise physiology, and appetite (Hardie 2007; Narkar et al. 2008) and of FOXO3A (Chiacchiera and Simone 2010).

6.3.5.3 Escape from Life Shortening due to Telomere Dysfunction

Consistent with a role for telomere shortening or damage in aging (Chapter 18; Sahin and DePinho 2010), median lifespan was increased 40% in a cancer-resistant mouse by improved telomere maintenance (Tomas-Loba et al. 2008).

6.3.6 Pharmacological and Nutritional Interventions in Mammals

Beyond longevity mutations and calorie restriction, which is dealt with elsewhere in this volume (Spindler 2010), there are a number of other interventions that extend lifespan in mammals, most likely by overriding physiological pro-aging mechanisms. Many of these are pharmacological or "neutraceutical" in nature. Because pharmacological agents generally exert their effects by stimulating or inhibiting endogenous processes, and nutrients or their deficiencies often do the same, pharmacological and nutritional mitigation of aging can provide additional insight into genomic resources and processes for combating aging.

Resveratrol supplementation extends the life spans of fish (Valenzano et al. 2006), flies (Wood et al. 2004), worms (Baur and Sinclair 2006), and yeast (Howitz et al. 2003). In normal mice, resveratrol produces extremely impressive improvements in biochemical and performance characteristics (Baur et al. 2006; Barger et al. 2008; Alcain et al. 2010), yet has not extended the lifespan of normal aging mice (Pearson et al. 2008). The sirtuin/resveratrol story will undoubtedly be a significant one as it continues to unfold, but current controversies (Ledford 2010; Pacholec et al. 2010) must be resolved.

6.3.6.1 CR Mimetics

Some pharmacological agents may have effectiveness as calorie restriction mimetics. It isn't possible to mention all agents in this category here, but metformin is noteworthy as an agent that can recapitulate the gene expression effects of long-term calorie restriction even more effectively than short-term calorie restriction of the same duration when applied to aged long-lived mice (Dhahbi et al. 2005) and that activates AMPK (Shaw et al. 2005). Phenformin is noteworthy for having actually extended lifespan and reduced tumor incidence (Dilman and Anisimov 1980; Anisimov et al. 2003). Similarly, the insulin sensitizing nutrient, chromium picolinate, lowered plasma insulin levels by 60% by the age of 1100 days and raised median lifespan by 25% ($p<0.05$) and maximum lifespan by 26% ($p<0.05$) (the last control died at 1154 days; the first treated animal death took place at 1221 days) in male Long-Evans rats fed 1 microgram of chromium per gram of diet from weaning (Evans and Meyer 1994). Glucose levels were also lowered by 21% by the age of 1000 days and did not rise with age, glycated hemoglobin was reduced by 59% by 1010 days and did not increase with age, and body fat was reduced by 24–28% at death (Evans and Meyer 1994). Other forms of chromium were ineffective and served to provide the control data.

6.3.6.2 Deprenyl

One of the classical features of brain aging is an increase in expression of monoamine oxidase B (Robinson et al. 1972). Deprenyl is a monoamine oxidase B and dopamine reuptake inhibitor that increases dopaminergic and noradrenergic tone (ThyagaRajan et al. 2000; ThyagaRajan and Felten 2002) and has been found to extend the lives of rats (Knoll 1988; Milgram et al. 1990; Norton et al. 1990; Kitani et al. 1993; Kitani et al. 2006), mice (Yen and Knoll 1992), female Syrian hamsters (Stoll et al. 1997) and dogs (Ruehl et al. 1997). In the case of dogs, deprenyl administration to 10–15 year-old beagles for 6 months or more allowed 80% to survive to the end of the study, vs. 39% of controls ($p = 0.017$). Deprenyl also increases brain cell (Seniuk et al. 1994; Semkova et al. 1996) and peripheral (ThyagaRajan and Felten 2002) neural growth factors, the activities of antioxidant enzymes in the brain (Carrillo et al. 1992; Kitani et al. 1999) and in the periphery (Kitani et al. 2002), and immune system functions (ThyagaRajan et al. 1998b; ThyagaRajan and Felten 2002) including the immunological killing of cancer (ThyagaRajan et al. 1998a, 2000; ThyagaRajan and Felten 2002), and reverses critical functional (Kiray et al. 2006) and structural (Zeng et al. 1995; Kiray et al. 2006) age-related changes in the brain. Interestingly, deprenyl also restores IGF-1 levels to youthful values (de la Cruz et al. 1997), reverses age-related cessation of reproductive cycling in rats [(ThyagaRajan et al. 1995); see also Section 6.5.11], and combats beta-amyloid accumulation in the brain (Ono et al. 2006).

6.3.6.3 L-DOPA

The dietary administration of L-DOPA, a free radical generating agent (Golembiowska et al. 2008) that, like deprenyl, also stimulates both brain activity and growth hormone release (Sonntag et al. 1982), has extended median lifespan in male Swiss albino mice after 130 days of age by 50% ($p < 0.001$) and the maximum lifespan (99th percentile survivors) by 12% ($p = 0.014$) (Cotzias et al. 1977) at its maximum tolerated doses of 40 mg/g of diet, despite making the mice clearly sick for long periods as evidenced, in part, by significantly reduced early survival (Cotzias et al. 1974, 1977). L-DOPA was started at 4–5 weeks of age and was continued until death. At a level of 4% of the diet by weight [about 33 times the maximum human dose (Cotzias et al. 1974)], the mice showed, over the first 1–6 months of treatment, hair thinning, an increased death rate, and a blunting of aggressive behavior, but by one year of age hair quality recovered, and the animals looked dramatically more youthful than controls, with dense and smooth hair and lean bodies, and had, at the 13th–14th months of life, drastically improved reproductive performance (1/12 successful matings in the control group, vs. 8/12 with L-DOPA) (Cotzias et al. 1974). Matings over weeks 10–50 of life did not show a difference in male fertility, however (Cotzias et al. 1977). Although weight was reduced in the treated group, food intake was not, at least after the first month of treatment, so the effects of L-DOPA were probably not primarily due to CR (Cotzias et al. 1977).

L-DOPA, as a precursor of dopamine, may act to impede generalized aging by hyperstimulating dopaminergic pathways in the brain whose possible roles in the central regulation of aging (analogous to central regulation of aging in worms and flies) deserve much closer scrutiny. Similar stimulation has been shown to reverse the age-related failure of reproductive cycling in rodent females (Section 6.5.11).

6.3.6.4 Centrophenoxine and DMAE

Centrophenoxine (also known as meclofenoxate), like deprenyl, can inhibit monoamine oxidase B (as well as monoamine oxidase A) (Stancheva and Alova 1988) and increase the expression of antioxidant enzymes (Roy et al. 1983; Bhalla and Nehru 2005). It probably also plays a major role in stimulating turnover of proteins and lipids (autophagy; see also Section 6.5.8) and has a number of other beneficial effects [for a partial listing, see (Zs-Nagy 1989); see also Section 6.5.1]. Based in part on the observation that centrophenoxine extended the mean and maximum lifespans of female *Drosophila melanogaster* by 39% ($p<0.0001$) and 23%, respectively (under conditions in which the treated flies were disadvantaged relative to the controls by the Lansing effect) (Hochschild 1971), Hochschild added centrophenoxine to the drinking water of male Swiss Webster albino mice beginning at the age of 8.6 months (Hochschild 1973b). Survival from the time of treatment onset was on average extended by 27.3%, the effect getting stronger and stronger with advancing age so that the maximum survival time was increased 39.7%; considered over the total lifespan (8.6 months plus survival after the start of the experiment), centrophenoxine increased maximum lifespan by 26.5%.

Centrophenoxine is rapidly hydrolyzed in water, so Hochschild tested one of its hydrolysis products, dimethylaminoethanol (DMAE, as the p-acetamidobenzoate salt), by putting it into the drinking water (28.6 mg of DMAE per liter) of 57 old (604–674 days, or about 21 months) male A/J mice for the rest of their lives (Hochschild 1973a). The mean lifespan of these mice is normally short (490 days, or about 16 months), so treatment was actually begun at an age that was over 30% older than the mean lifespan. The DMAE-treated mice nevertheless lived 49.5% longer on average than the controls, and the maximum survival time was 36.3% longer than for the controls.

DMAE can serve as a precursor to both choline and acetylcholine, but whether this is relevant to the observed increases in lifespan has not been established. Certain forms of dietary choline have been shown to profoundly protect against structural brain aging (Bertoni-Freddari et al. 1985), and centrophenoxine increases choline levels in the brain (Wood and Peloquin 1982), so the connection is suggestive. DMAE alone was only half as effective as centrophenoxine, but it appears that both DMAE and the complete molecule have this effect. It is interesting that centrophenoxine and DMAE have both positive effects on brain function and the ability to extend maximum lifespan about as much as or more than mean lifespan, whereas antioxidants tend to extend mean but not maximum lifespan.

6.3.6.5 Coenzyme Q

The content of coenzyme Q (coQ) in the rat heart declines precipitously at an age coinciding with the onset of strong age-related mortality acceleration in rats (Beyer

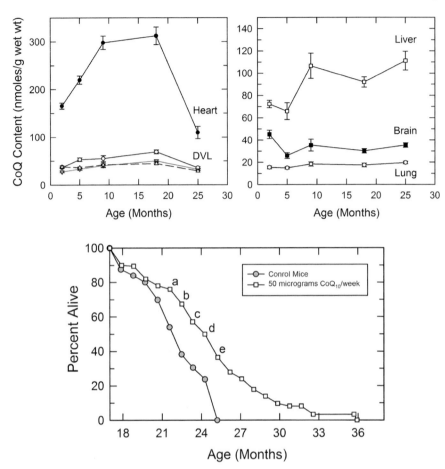

Fig. 6.9 Influence of ubiquinone (coenzyme Q, coQ) on lifespan. *Upper left panel*: abrupt and selective > 50% loss of coQ in male Sprague-Dawley rat hearts coinciding in time with the onset of Gompertz mortality. The same qualitative age dependence can also be seen but to a quantitatively less dramatic extent in other myosystems (the deep vastus lateralis (DVL), soleus (*dashed line*), and gastrocnemius (*gray line*) muscles), but no such phenomenon is apparent in liver, brain, or lung (*upper right panel*, same units of coQ content apply as in the *upper left panel*). Drawn from the data of Beyer et al. (1985). *Lower panel*: extension of survival time in female CF1 mice given 50 μg of emulsified coQ_{10} per mouse once a week by intraperitoneal injection starting from the age of 16–18 months (50 mice in each group). The p value for the survival difference between untreated controls and coQ_{10} treated mice at the ages indicated by a, b, c, d, and e was, respectively, $p < 0.02$, $p < 0.002$, $p < 0.01$, $p < 0.005$, and $p < 0.00005$. Mean survival time was increased by 56% and maximum survival time was increased by 86%. Redrawn from Bliznakov (1981)

et al. 1985) (upper left panel, Fig. 6.9). The potential causal connection between these two events has not been investigated, nor has the cause of this remarkably large and abrupt falloff in coQ. It is mirrored by simultaneous changes in other muscle systems (upper left panel) but not in other investigated organs, whether they are high or low in metabolic rate (e.g., liver, brain, and lung: upper right panel). An outstanding exception may be the thymus, for which up to an 80% deficit in coQ content may develop in female CF-1 mice by the age of 24 months relative to 2.5 month-old controls (Bliznakov 1978).

If coQ synthesis falls in critical organs, it would stand to reason that replacement therapy might have beneficial effects, given the critical role of coQ in oxidative phosphorylation. Interestingly enough, coQ_{10} injections begun at the age of 17 months were found to substantially extend the maximum lifespan of female CF1 mice (Bliznakov 1981) (56% increase in mean survival time and an 86% increase in maximum survival time, Fig. 6.9, lower panel). CoQ_{10} given as 0.1% by weight of the diet from weaning has also increased the mean (but not the maximum) lifespan of long-lived C3H/Sw/Sn x C57Bl0.RIII/Sn hybrid mice by 20% ($p<0.05$) while remarkably increasing their physical activity, improving their appearance, and reducing hind-limb dysfunction (Coles and Harris 1996). On the other hand, coQ_{10} alone or in combination with other agents did not extend lifespan in two more recent studies despite inducing favorable changes in gene expression profiles (Lee et al. 2004; Spindler and Mote 2007). The ability to observe the life-extending effects of coQ replacement may depend on how cancer-prone the chosen strain of mouse is and how susceptible its cancer surveillance mechanisms are to enhancement by coQ; the route of administration (injection vs. dietary), the dose, and the dosage form may also be important.

6.3.6.6 Tryptophan Deprival

It is a strong indication of the physiological origins of aging that a large variety of pathological states brought about by severely disrupting normal physiological functions can extend rather than shorten lifespan and can even extend the period of reproductive competence. In addition to the L-DOPA example cited above, another quintessential example is provided by the effects of tryptophan deficiency. Placing 21-day-old female Long-Evans rats on diets containing only 30 or 40% of the normal dietary content of this essential amino acid resulted in profound early mortality and aging retardation that increased as the magnitude of early mortality increased (Fig. 6.10). A tryptophan deficiency of ~85% (15% of normal dietary tryptophan content) totally blocked the growth of three week old female Long-Evans rats, and maintaining this deficiency state for several months and then re-feeding the rats a normal diet allowed them to reproduce at ages as old as 40 months (vs. <16 months for controls) and to maintain better quality coats and to reduce tumor incidence by the age of 30 months from 36% in controls to 14–18% (Segall and Timiras 1976).

Tryptophan deprival produced profound reductions in brain serotonin levels, particularly early in life (Ooka et al. 1988), and produced tremors in all animals and convulsions in 5 of 27 treated animals (Segall and Timiras 1976). As the authors of this study noted, "delayed aging may originate in the drastic nature of the intervention itself. One possible explanation is that severe nutritional restriction, started early, may delay or prevent programmed events in the CNS . . . . If aging was due to a cascade of events subsequent to maturation, then too, aging would slow" (Ooka et al. 1988).

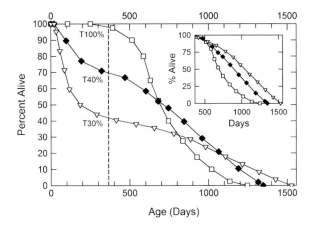

Fig. 6.10 Life extension by tryptophan deprival in rats is increased as the severity of deprivation and its associated early mortality increase. If deaths within the first year (indicated by the vertical dashed line at day 365), which presumably reflect only the side effects of treatment and not aging, are subtracted out and survival is normalized to 100% at 365 days, the life-extending effects of tryptophan deprival become more readily visualized (inset). Redrawn and constructed from Ooka et al. (1988)

Tryptophan is an essential amino acid, and so may be limiting for protein synthesis. It is also an essential precursor of NAD in animals, and NAD is in turn a nutritional activator of sirtuins (see Chapter 10). Tryptophan is also a precursor to serotonin, which in turn is converted into melatonin. The fact that disrupting all of these pathways can so significantly blunt aging is provocative and deserving of further analysis.

6.3.6.7 Methionine Deprival

An 80% reduction in the concentration of methionine in the diet (methionine restriction, or MR) was reported to extend the median lifespan of male Fisher 344 rats by 29% and maximum lifespan by 12% through a mechanism that was largely independent of any effects of altered caloric intake (Orentreich et al. 1993). At this level of methionine deprival, begun at 42 days of age, growth was almost totally arrested, 700 day-old restricted rats weighing at least 75% less than unrestricted rats. Even when normalized to 100 g of body weight, certain key organs (the fat pads, seminal vesicles, and testes) weighed substantially less in the MR rats after 90 days of restriction. Obviously, withholding methionine from the diet, like tryptophan deprival, results in major physiological derangements, and by mechanisms that were not defined in this study, this profound disruption of normal physiological function results in a partial escape from the default life shortening effect of normal physiological control mechanisms despite any adverse health consequences of MR.

A follow-up MR study in the same rat strain and with the same degree of MR found a mean lifespan increase of 42% ($p < 0.0005$), a median lifespan increase of

43%, and a maximum lifespan increase of 44%, with less profound suppression of growth than in the earlier study (just over a 50% weight reduction at the age of 700 days) (Richie et al. 1994). Surprisingly, levels of reduced glutathione (GSH) were conserved in most tissues after long term MR, but blood levels rose remarkably, particularly in the oldest rats, GSH levels falling only in liver and kidney (Richie et al. 1994). In contrast, cysteine levels fell in plasma but not significantly in the liver or the kidney.

Deficiencies of essential amino acids presumably result in major declines in protein synthesis, and it is interesting that experimentally reducing protein synthesis in many different ways extends lifespan in C. elegans (Section 6.3.3.3) and mice (see next section). In addition, methionine is believed to have a special role in regulating transcription, translation, and antioxidant status (Metayer et al. 2008). The mechanisms of lifespan extension by MR remain unknown, but significant mitochondrial protection has been reported (Sanz et al. 2006).

6.3.6.8 Rapamycin

A very recently announced pharmacological extension of mammalian lifespan was achieved by administration of a single dose of rapamycin, a well-known inhibitor of TOR, to genetically heterogeneous mice at three different sites under the auspices of the National Institute on Aging's Interventions Testing Program (Harrison et al. 2009). Given at 600 days of age, the mean survival time beyond 600 days was increased by 28% for males and 38% for females, and the absolute time at which 10% survival was reached was increased on average by 9% for males and 14% for females. Considered over the whole lifespan, the increases in mean lifespan were 9% for males and 13% for females. Unfortunately, when rapamycin was started at 270 days, median lifespans were still only extended by roughly the same amount as were the mean lifespans in the original study started at 600 days, for both males and females.

Also unfortunately, in the two centers reporting the largest increases in male lifespan, male survival prior to initiation of rapamycin treatment at 600 days was already in excess of control male survival, and the diets were different until this age. This may be why, when rapamycin was started at 270 days of age in better controlled conditions, the change in male median lifespan was so low. In addition, mice usually die of cancer, and rapamycin has cancer suppressing effects that might explain the increased longevity in the absence of other anti-aging effects of rapamycin. Furthermore, autopsies revealed no changes in the cause of death in treated vs. control mice. Therefore, although this very high-profile lifespan experiment replicated results previously demonstrated in yeast, worms, and flies [see references in (Harrison et al. 2009)], and although TOR inhibition might be expected to extend lifespan by augmenting autophagy [(Hansen et al. 2008); see also reference in (Kaeberlein and Kennedy 2009)] and by inhibiting protein synthesis (see discussion of tryptophan and methionine restriction above and Section 6.3.3.3), further studies of its effects are necessary.

6.3.7 Humans

In addition to relatively well-known life-extending alleles in humans (see Section 6.3.3.2, "A Note on Genetic vs. stochastic aging in C. elegans and other species"), default life-shortening influences can evidently be blunted in humans through at least two other means, castration and Y-chromosome deletions in men. Although the effects of castration have not been consistent in all studies (Rose 2005; Aspinall and Mitchell 2010), the study described in the next section was careful and thorough enough to be taken seriously and is therefore now reviewed in detail.

6.3.7.1 Castration

In 1969, Hamilton and Mestler reported the first large study of the effect of castration on the mortality of human males and included supplemental data as well on mortality in oophorectomized females (Hamilton and Mestler 1969). The experimental populations were inmates at an institution for the mentally retarded in Kansas who were born between 1871 and 1932. Castrated and intact inmates were matched for exact year of birth and age of admission to the institution and were very similar with respect to degree of mental impairment. The data for this study were assembled over a 33 year period.

In all, 297 Caucasian castrated men were compared to 735 intact men, and 23 oophorectomized women were compared to 309 intact women. The median lifespan of the unoperated men was 55.7 years, and the median lifespan of the castrates was 69.3 years, for a difference of 13.6 years, representing a 24.4% increase in lifespan as the putative result of castration ($p=0.002$). The castrated men actually outlived a population of 883 intact females (median lifespan, 62.6 years) by 6.7 years; this was confirmed in a subset analysis (lower mortality rate, $p=0.002$; death at later ages, $p=0.006$). Women inmates did not benefit from ovariectomy.

The median lifespan within the castrated group for those who survived to at least 40 years of age was strongly dependent upon the age of castration (Fig. 6.11). Those castrated at the age of 8–14 years (generally, before sexual maturation) had estimated median lifespans of 76.3 ± 1.36 years. But even when castration was delayed until age 30–39, castration still extended median lifespan by 4.2 years ($p=0.002$). Overall, lifespan was reduced by 0.28 years for every 1 year delay in instituting castration. The maximum lifespans of the two groups differed by a smaller amount than the median lifespans (86–87 years for the operated subjects, 83–84 years for controls).

The cause of the survival advantage was not a reduced incidence of cardiovascular disease, cancer, or pneumonia. Instead, the death rate from all infections was 38% lower in the castrated males ($p=0.02$), and the death rate from tuberculosis was 61% lower in this group ($p=0.03$). This association has been found in castrated outbred male cats as well, for which castration extended mean and median lifespan ($p<0.001$) and reduced the likelihood of death by infection ($p=0.04$) (Hamilton et al. 1969). In fact, castrated male mental patients had lower rates of infectious and TB-related deaths than did intact females. Orchiectomy also prevented male pattern baldness.

6 Precedents for the Biological Control of Aging

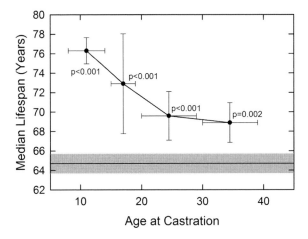

Fig. 6.11 Effect of the age of castration of human mental patients on their estimated median length of life vs. the median length of life in non-castrated inmates; data are for men who survived to at least the age of 40. *Horizontal error* bars = $\frac{1}{2}$ the age range for each age bin. The *horizontal black line* and *shaded area* at the bottom of the figure represent the median estimated lifespan of non-castrated males and its uncertainty (presumably one SEM, but not stated). Drawn from the data of Table 10 of Hamilton and Mestler (1969)

These results document a rare example of strong life extension in humans that is consistent with the effects of removing reproductive structures in semelparous and monocarpic organisms (Sections 6.3.1 and 6.3.2) and in C. elegans (Arantes-Oliveira et al. 2003). Both male and female sex hormones inhibit immune function [see (Fahy 2003) and (Aspinall and Mitchell 2010) for references], so protection from infections is reasonable. Another more mysterious source of protection could involve eliminating default life-shortening effects of the germ line (see Section 6.3.3.1 for a discussion of this complex developmental (maturation-induced) source of default life-shortening). It was surprising that, given that higher testosterone levels are associated with *reduced* mortality from cardiovascular disease, cancer, and all causes in 40–79-year-old men (Khaw et al. 2007), these factors did not overwhelm the protective effects against infection.

6.3.7.2 Y-Chromosome Deletions

Based on a natural experiment in which much genetic information was deleted from the long arm of the human Y chromosome in certain Amish family lineages with an associated large lifespan extension, the Y chromosome may contain genes that induce human mortality. This life-shortening effect of the Y chromosome was reported by Kirby Smith of Johns Hopkins University, Department of Medicine and Biology, at a meeting in 1987 (Holden 1987) but was, unfortunately, never formally published because Smith did not know what to make of it (Smith, personal communication). If correct, however, this insight is too important to remain hidden from view, so it is presented here to get the observations on record and, hopefully, to inspire further investigation.

Smith studied Amish families whose lineage and longevity could be traced back four generations. Some men in one of these families were missing the better part of the long arm of the Y chromosome, and their longevity data were compared to those from non-deleted men in two control Amish families, one living in the same environment and one random control family living in a different environment. The men with the deletion were apparently normal and functional in all respects, so it appears that the deletion did not, in contrast to gonadectomy, reduce reproductive functioning or omit any essential genetic information. On the contrary, men carrying the deletion lived 16–20%, or 12–14 years, longer than those who did not have the deletion, and in fact even outlived the females in all three families, whereas women, as usual, outlived men in both of the control families. As Smith himself put it, "too much Y and you die" (Holden 1987). Smith's exact data appear in Table 6.2.

This observation provides another rare example of major lifespan extension in humans associated with an escape from default physiological controls, and it is broadly consistent with the linkages between mortality and reproduction that have been discussed above for semelparous, monocarpic, and other iteroparous organisms. It also provides an interesting opportunity for identifying human life-shortening genes, because there are very few genes on the long arm of the Y chromosome, and the entire Y chromosome contains only 0.5% of a man's total DNA (Holden 1987).

Table 6.2 Human lifespan extension by a Y chromosome deletion

Family Type	Age at Death: Females (Mean ± 1 SD)	Age at Death: Males (Mean ± 1 SD)
Control Family #1	75.9 ± 6.6	70.7 ± 4.8
Control Family #2	76.4 ± 14.8	68.5 ± 10.0
Y-deleted Family	77.4 ± 10.2	82.3 ± 5.5*

*$p < 0.005$ compared to controls; $n = 14$ males

6.3.8 Other Examples

The above examples provide clear evidence that default life-shortening physiological mechanisms exist in adults and that these mechanisms can be overcome to increase both mean and maximum lifespan. There are other interventions that can extend lifespan either by similar mechanisms or by fortifying the individual with exogenous protective agents or grafts. Although seriously incomplete, a nevertheless helpful database of aging-related genes and interventions is available online for those seeking more information about the genetics and interventions of extended lifespan (Kaeberlein et al. 2002); http://uwaging.org/genesdb/index.php#interventions.

6.3.9 Future Examples: Overcoming Fundamental Design Features of Iteroparous Species

Although, as noted above, longevity mutants have, for very good reasons, led to a great deal of our knowledge of the molecular biology of aging-related genes, there are also some normal genetic mechanisms for the induction of aging and death in higher animals that have been on public display for decades, yet seem to be rarely seen as drivers of aging. Normal developmental events in combination with the insights obtained from longevity mutants seem to show that aging is universally built into the fundamental physiology of iteroparous, non-immortal animals as a basic "ground rule" of life history "design," every bit as much as it is built into the life history plans of semelparous and monocarpic species.

These ideas are illustrated in the schematic of Fig. 6.12. In Fig. 6.12, the portion of the diagram above the main horizontal line illustrating the progression from inception of the individual to successful reproduction is based mainly on the physiology of mammals in general and humans in particular, but the relationships below this line appear to be universal, at least in broad terms (Longo and Finch 2003; Tatar et al. 2003).

For organisms to reproduce, they must develop from birth or hatching into reproductively mature individuals. They are not able to do this without the participation of growth hormone or similar growth factors and insulin/insulin-like growth factors, which are anabolic factors necessary for the growth and maintenance of muscles, visceral organs, and other structures. Organisms must in addition consume calories in order to grow, mature, and sustain life after maturation, and the consumption of calories also stimulates insulin-like signaling. However, eating and insulin-like signaling [the effects of which are not synonymous: e.g. (Bartke et al. 2001; Tsuchiya et al. 2004; Min et al. 2008)] reduce antioxidant defenses (Honda and Honda 1999; Essers et al. 2005; Yamamoto et al. 2005), DNA repair (Licastro et al. 1988; Lipman et al. 1989; Weraarchakul et al. 1989; Guo et al. 1998; Tran et al. 2002; Huang and Tindall 2007; Hyun et al. 2008; Tsai et al. 2008), and other stress resistance

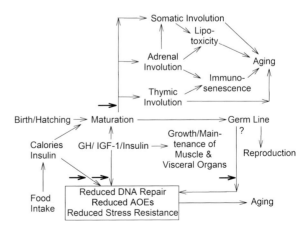

Fig. 6.12 Proposed pathways for the physiological induction of aging in animals. *GH* growth hormone, *IGF-1* insulin-like growth factor 1, *AOEs* antioxidant enzymes. The *four heavy horizontal arrows* indicate where interventions are appropriate to inhibit or prevent physiological points of induction of aging

pathways [summarized in, for example (Bokov et al. 2004)], and increase the rate of production of reactive oxygen species (ROS) (Lambert and Merry 2004). Feeding also reduces protein turnover, allowing damaged proteins and organelles to accumulate with age (Lindell 1982; Lewis et al. 1985; Goto et al. 2001, 2002, 2007; Spindler 2001; Bergamini et al. 2004; Spindler and Dhahbi 2007). In at least these ways, the ground rules of life seem to ensure that the price for growth, reproduction, and anabolism is the inevitable eventual aging that results from eating and insulin-like signaling. Intriguingly, however, as pointed out in Section 6.2.1, the onset of aging can be postponed to about the time of sexual maturation, so the effects of calories, insulin, GH, and IGF-1 may only apply to the adult.

In C. elegans, and most likely in a vast array of other species, and, as we have seen above, conceivably even in man, signals from the germ line may in some way yet to be fully defined participate in the aging of sexually mature organisms. In addition, at least in mammals, and particularly in humans, sexual maturation contributes phenotypic changes (indicated by the upper heavy arrow) that profoundly affect aging (upper part of Fig. 6.12). The three examples illustrated are conceptualized as being thymic involution and, by analogy, adrenal involution and somatic involution.

The production of sex hormones and the peripubertal downregulation of thymic IL-7 contribute to thymic involution (Aspinall and Mitchell 2010), which is universal in all vertebrates (Cardarlli 1989) and leads to immunological (Aspinall and Mitchell 2010) and a great deal of non-immunological [(Fabris et al. 1988); (Section 6.5.14)] aging. Although the relevance of immunological aging is sometimes questioned, many investigators have found that the degree of immunosenescence is predictive of the risk of death in humans (Roberts-Thomson et al. 1974; MacKay et al. 1977; Murasko et al. 1987, 1988; Cawthon et al. 2003; de la Rosa et al. 2006; Hadrup et al. 2006; Wikby et al. 2006), and serial thymus transplants extend the lives of animals despite the surgical trauma involved (Hirokawa and Utsuyama 1984).

The selective loss of DHEA-synthesizing cells in the zona reticularis of the human adrenal gland ("adrenal involution") reverses the peripubertal surge in DHEA synthesis between about 20 and 25 years of age depending on sex (Staton et al. 2004). This event may reduce thermogenesis (Lardy et al. 1995; Lea-Currie et al. 1997; LeBlanc et al. 1998; Ryu et al. 2003; Hampl et al. 2006) and thereby contribute to age-related fat gain and all of its deleterious health consequences ["lipotoxicity," e.g.: (Dulloo et al. 2004; Iossa et al. 2004; Russell 2004; Sharma et al. 2004)]. The loss of DHEA may contribute to other age-related problems as well (Kalimi and Regelson 1990; Bellino et al. 1995), including immunosenescence (Araneo et al. 1995; Straub et al. 1998; Bauer 2005; Corsini et al. 2005; Buford and Willoughby 2008), age-related inflammation (Choi et al. 2008), and somatic involution [(Villareal and Holloszy 2006); see next paragraph]. An unsuspected role for DHEA and/or DHEAS that may be lost with age could be the blunting of the so-called diabetogenic effects of growth hormone (GH) in older humans (Fig. 6.13). GH-induced hyperinsulinemia does not seem to be a problem in young humans who are replete in both GH/IGF-1 and DHEA/DHEAS, but could contribute to type II diabetes if DHEA or DHEAS normally blocks GH-induced hyperinsulinema

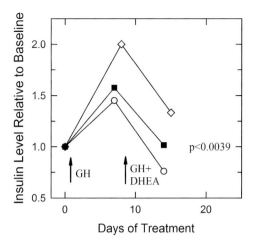

Fig. 6.13 Reversal of growth hormone (GH)-induced "diabetogenic" effect (elevation of fasting insulin levels) by oral DHEA administration. GH regimen begun (*first vertical arrow*) shortly after baseline insulin determination, and simultaneous DHEA administration begun (*second vertical arrow*) shortly after re-documentation of insulin level 7–8 days later while continuing the GH or GH-elevating regimen. *Open circles*: GH release stimulated by daily oral nocturnal arginine intake (15 g) and counteracted by co-administration of 180 mg of pharmaceutical grade DHEA; volunteer age, 46 years. Black squares: subcutaneous injection of 0.018 mg/kg/day of GH every other evening followed by the same regimen plus co-administration of 200 mg DHEA; volunteer age, 46. *White diamonds*: subcutaneous GH ± DHEA as above, but volunteer age = 54. The p value indicated is the result of a paired t-test comparing relative insulin levels after GH elevation before and after DHEA treatment. Assembled from unpublished data of Intervene Biomedical and from data in (Fahy 2001). Published with the permission of Intervene Biomedical

and becomes deficient relative to GH/IGF-1 with age (Fahy 2001). Although the exact and full role of DHEA/DHEA-sulfate in aging has been difficult to establish, and although DHEA does not seem to extend lifespan (Kalimi and Regelson 1990), higher levels of DHEA have been associated with "successful aging" (Roth et al. 2002), and reactivating endogenous DHEA synthesis after 50 years of age was associated with a significant improvement in exercise capacity (Fig. 6.14).

The third consequence of maturation noted in Fig. 6.12 is "somatic involution," by which is meant the gradual atrophy of muscles, skin, and visceral organs, the gradual gain in adiposity, and the progressive loss of stature that occurs in the aged at least in part due to the slow continuing decline of growth hormone and IGF-1 production that follows the very abrupt declines that begin at the age of 13–14 for both boys and girls (MacGillivray 1987). Although the complete etiology of "somatic involution" has not been established, a role for growth hormone and IGF-1 deficiencies is indicated by the ability of exogenous growth hormone administration to reverse visceral (spleen, liver) (Rudman et al. 1991), muscular (Rudman et al. 1991), dermal (Rudman et al. 1990; Rudman et al. 1991), and thymic (Kelley et al. 1986; Goff et al. 1987; Napolitano et al. 2002; Fahy 2003; Aspinall and Mitchell 2010) age-related atrophy as well as to increase lean body mass in general and reduce

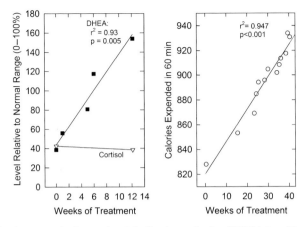

Fig. 6.14 Selective reversal of age-related decline in synthesis of DHEA in a 52-year-old human subject using oral IB007 (*left*) and its effect on aerobic capacity (*right*). *Left*: 0% = the lower limit of the normal clinical age-specific range of serum DHEA levels; 100% = the upper limit of this range. No effects were seen on serum levels of cortisol (*inverted triangles*), total testosterone, or aldosterone (data not shown). *Right*: calorie expenditure as recorded by a commercial exercise machine. No trend in aerobic capacity was noted prior to IB007 treatment (data not shown). Both panels represent unpublished data of Intervene Biomedical; published by permission of Intervene Biomedical. Note: Raising DHEA levels by any means may raise PSA levels in men

adiposity in elderly men (Rudman et al. 1990; Blackman et al. 2002) (see also Sections 6.5.12 and 6.5.14). In addition, there are strong inverse correlations in aging humans between IGF-1 levels and disability and institutionalization (Rudman 1985).

The developmental (genetically programmed) nature of these three phenotypic changes in man seems clear since thymic atrophy, adrenal involution, and the onset of declining growth hormone production are universal and all begin in childhood or at the latest at the beginning of full adulthood and are all precisely clocked and precede any significant signs of aging. Nevertheless, thymic, adrenal, and somatic involution may all increase vulnerability to other life-threatening challenges and ultimately guarantee aging and death if no other cause induces death earlier.

Note that the situation described in Fig. 6.12, in which both eating and growing are coupled to aging and death in part through the dual role of the GH/IGF/insulin system as both an indispensable anabolic system and a system for downregulating life-maintenance enzymes (such as antioxidant defenses and DNA repair systems), tends to make aging inescapable under normal circumstances. To up-regulate this system is, according to present knowledge, presumably to hasten death from mutation (cancer), oxidative damage, and other stresses, but to down-regulate this system is to invite slow deterioration and death from generalized atrophy (Rudman et al. 1991; Huang and Tindall 2007), at least in mammals, perhaps in part due to reduced opposition to FOXO-induced muscle atrophy and generalized apoptotic tendency (Huang and Tindall 2007).

In most or all species studied so far, it has been advantageous for net longevity to down-regulate rather than to maintain insulin-like signaling (for example, Sections 6.3.3.1 and 6.3.5.1). But we could do better by severing the links between insulin-like signaling and reduced self-defense and between maturation and thymic, adrenal, and somatic involution (i.e., by intervening at the sites indicated by the heavy arrows in Fig. 6.12). Would this be biologically feasible?

Viewed globally, the answer should be "yes" because humans are larger than mice yet age and develop cancer more slowly, bowhead whales are larger than humans yet age and apparently develop cancer more slowly still, and aging seems to be avoidable under some circumstances and for some species (Sections 6.2.2 and 6.2.3). More directly, younger animals simultaneously grow and maintain antioxidant (Rao et al. 1990a, 1990b) and DNA [(Busbee et al. 1989); (Fraga et al. 1990) and Section 6.5.4; (Chen et al. 2002) and Section 6.5.5.1] defenses, so it must be biochemically feasible to maintain a similar state after maturation as well (assuming the energy requirements for growth per cell exceed those of maintaining mostly idle reproductive capacity). In addition, the extra energy required to maximize such protective systems as antioxidant defenses and DNA repair appears to be compatible with maintaining wild-type fertility [e.g. (Simon et al. 2003); Section 6.4)] (and maximizing fertility is not a concern in any case for a majority of people).

Viewed on a more biochemical level, the FOXO family of transcription factors functions to augment antioxidant enzymes and DNA repair but also to curtail cell division and to promote muscle fiber atrophy (Huang and Tindall 2007), effects that may contribute to the pathological effects of IGF-1 deficiency. The possibility of removing the latter pathological effects without sacrificing antioxidant defenses and DNA repair is supported by the existence of non-toxic agents that can boost antioxidant defenses (Roy et al. 1983; Carrillo et al. 1992; Kitani et al. 1999) [and possibly DNA repair (Sylvia et al. 1988); see also Section 6.5.4)] while in at least one case simultaneously bringing IGF-1 levels back into the youthful range (de la Cruz et al. 1997) and increasing lifespan (Ruehl et al. 1997; Kitani et al. 2006). Every other day calorie restriction has also been observed to increase lifespan (Goodrick et al. 1990) and to lower insulin levels but not to lower IGF-1 levels in C57Bl/6 mice (Anson et al. 2003), providing a further indication that the links between self-defense and atrophy can be broken. Finally, inhibition of DNA repair by IGF-1 does not seem to be well established in humans.

The strong association between high IGF-1 levels and the risk of cancer [e.g., (Chan et al. 1998); see also discussion in the chapter by Heward in this volume] is the most immediate barrier to growth hormone/IGF1 use and must be recognized as such. However, no clearly increased cancer risk with growth hormone administration in aging humans has been found (Abs et al. 1999; Growth Hormone Research Society 2001; Johannsson and Jørgensen, 2001), and administration of even massive (Everitt 1959) doses of growth hormone to mature rodents does not shorten their lifespans (Everitt 1959; Khansari and Gustad 1991; Kalu et al. 1998), while administration of L-DOPA greatly extends lifespan (Section 6.3.6.3) despite releasing GH (Sonntag et al. 1982). The risk posed by IGF1 depends strongly on levels of IGF1 binding proteins (most prominently IGFBP3), but the effect of growth hormone administration on these binding proteins in elderly GH-deficient patients does not seem to have been thoroughly studied. Hypothetically, a positive effect of growth hormone on cancer surveillance could in theory offset any added cancer risk of conservative GH use.

Fortunately, an enormous amount of work is being done on cancer control, and as this work succeeds, our options with respect to the treatment and prevention of aging will expand. For additional discussion of approaches to minimizing the risk of cancer, see the chapters by Cui and by Steer and Kren, and, for the very long run, the chapters by de Grey and by Freitas, elsewhere in this volume.

Constant use of anabolic hormones may also not be needed for a therapeutic effect. Cycling between anabolism and catabolism so as to maintain lean body mass while still allowing sufficient opportunity for DNA repair and antioxidant defenses might be a possibility. In addition, combining sirtuin activation, which appears capable of reversing or preventing many age-related changes (Baur et al. 2006; Pearson et al. 2008; Alcain et al. 2010), with judicious use of anabolic support of lean body mass might also help to balance the benefits of both.

Thymic involution, adrenal involution, and somatic involution seem to provide no obvious benefits in humans that would outweigh the benefits of their elimination once the hazards associated with such issues as insulin-like signaling can be set aside. Fortunately, methods of eliminating or at least blunting thymic, adrenal, and somatic involution or their effects are already known (Sections 6.5.6, 6.5.10, 6.5.12, Fig. 6.14, and Chapter 15) and will surely be improved in the future.

Another interesting peripubertal phenomenon that may be relevant is pineal involution (Heward 2010), which might be a trigger for the onset of puberty (Macchi and Bruce 2004) and may have deleterious effects later in life (Pierpaoli 2005).

6.4 Life Extension by "Negative Reproductive Costs" in Nature

It is currently believed by many that the virtual slow suicide of the adult by the active withdrawal of life support systems such as DNA repair and immune system function with sexual maturation is necessary in order to pay the "costs of reproduction," but this has not been convincingly shown. The argument that it is energetically necessary for the organism to age specifically in order for the organism to reproduce optimally cannot be universally correct or biologically inevitable because (a) changing gene expression can extend life while maintaining or even increasing fecundity (Simon et al. 2003; Wood et al. 2004; Valenzano et al. 2006), (b) there are animals that both fail to undergo senescence and continue to reproduce until death (Donisthorpe 1936; Congdon et al. 2003; Buffenstein 2008), and (c) there are significant cases in which increased reproduction *extends* life or is directly coupled to extended life rather than shortened life ("negative reproductive costs"). The remainder of this section describes examples of the latter phenomenon.

6.4.1 Social Insects

Dramatic natural examples of aging prevention appear to exist even in adults as a result of modifying the adult pattern of gene expression in connection with increased reproductive output. In most likely tens of thousands of species of ants (Donisthorpe 1936), bees (Winston 1987), termites (Wilson 1971), and other social insects the queen has the same genotype as the workers, but lives much longer.

Rather than by reducing food intake or metabolic rate, honeybee queens are produced as a result of being fed food containing three-fold more fructose and glucose than is received by other workers, ten times as often (Winston 1987). This leads to altered juvenile hormone secretion perhaps secondary to signaling from stretch receptors in the gut (Dietz et al. 1979; Asencot and Lensky 1984), which in turn selects the non-aging queen phenotype. Without reducing food intake or reproductive effort – indeed, while engaging in profligate reproduction and excessive calorie consumption – the queen lives up to 100 times longer (Finch 1990), and ultimately dies not of aging but of assassination when she runs out of stored sperm and becomes incapable of generating further offspring for the hive (Winston 1987).

In *Drosophila* (Tatar et al. 2003) and in *C. elegans* (Murphy et al. 2003), adult death is positively associated with the production of the yolk precursor gene vitellogenin, even though vitellogenin is a powerful antioxidant (Seehuus et al. 2006). In bees, this relationship has been reversed, high levels of vitellogenin promoting the longevity of both queens and favored workers (Seehuus et al. 2006; Corona et al. 2007). The longevity of the queen and workers with higher vitellogenin contents is permitted by a modification of the physiological interrelationships between vitellogenin, juvenile hormone, and insulin/insulin-like signaling (Corona et al. 2007) and may be related at least in part to the antioxidant effects of vitellogenin (Seehuus et al. 2006; Corona et al. 2007).

6.4.2 Social Mammals

Increased reproductive success extends rather than shortens the lifespan of both wild rabbits (von Holst et al. 1999) and Ansell's mole rats (*Cryptomys anselli*) (Table 6.3; (Dammann and Burda 2006a)], the latter being a particularly well-defined example and possibly representative of other eusocial mammals such as *Heterocephalus glaber* and several other *Cryptomys* species. In these mole rats, extraneous factors such as altered social status, changed activity levels, changed food intake, and

Table 6.3 Life extension by reproduction in mole rats (Data derived from Dammann and Burda (2006b)

Measure of longevity	Sex	Non-Breeders	Breeders	p value	Percent gain
Mean Lifespan	Males	4.8 ± 0.5 years	11.1 ± 1.4 years	<0.0001	131
Mean Lifespan	Females	4.3 ± 0.4 years	8.3 ± 0.8 years	<0.0001	93
Maximum Lifespan	Males	7.8 years	19.5 years	–	150
Maximum Lifespan	Females	5.9 years	>14.8 years	–	>153
% Dead at 8 Years	Males	100%	<30%	–	>200
% Dead at 6 years	Females	100%	<10%	–	>800

pre-breeding growth rates (taken as a sign of individual "quality") could be excluded as causal factors, leaving only increased breeding as the cause of the increased longevity.

Remarkably, the *mean* lifespan of breeding males was 42% longer than the *maximum* lifespan of non-breeding males, and breeding female *mean* lifespan was 40.6% longer than the *maximum* lifespan of the non-breeders (Table 6.3). Not only did greater reproductive effort have positive rather than negative effects on the longevity of both sexes, but the ~100% gains in mean lifespans and ~150% gains in maximum lifespans in both sexes due to breeding in this species exceed both the percentage and the absolute longevity gains that are known to be attainable with calorie restriction, and even with the combination of calorie restriction and known longevity mutations, in mammals (cf. Table 6.1). Breeding enabled the males to reach a maximum lifespan of nearly 20 years, vs. less than 8 years without breeding. This constitutes an absolute gain in lifespan that equals the absolute gain in human lifespan achieved by castration or Y-chrormosome deletion (Section 6.3.7), which is impressive considering the ~12-fold shorter maximum lifespan of non-breeding members of this species.

These results and similar but less clear-cut observations in wild rabbits (von Holst et al. 1999) make it clear that there is no inviolable reproductive tradeoff against longevity that arises from a still-undefined and hypothetical "energy budget" limitation in mammals. Instead, central mechanisms may be the true governors of the genetic switch between reproduction and longevity (Tatar et al. 2003).

6.5 Segmental Aging Reversal

So far, we have considered longevity as the main measure of aging, but there are many age-related changes that can be examined in isolation from longevity itself, and the reversal of many such biochemical aspects of aging is a commonplace laboratory observation. Here we can only consider sufficient examples to demonstrate the generality of the reversibility of core aspects of aging on a basic biochemical and physiological level. These examples are referred to as cases of "segmental aging reversal" by analogy to G.M. Martin's concept of "segmental progeroid syndromes" (Martin 1979). The fact that segmental aging reversal is possible suggests the physiological origin of much of the aging process, and the likely reversibility of additional elements of aging in future studies.

6.5.1 Reversal of Reduced Transcription

The chemical constituents of living cells are intrinsically unstable, yet life has persisted for eons on our planet. The way life avoids accumulation of random damage is by continuous synthesis of new, undamaged biomolecules to replace those that have been damaged by oxidation, glycation, chemical reaction, spontaneous chemical decomposition, denaturation, or the like. This process of turnover is essential to life, but turnover apparently slows with aging (Lindell 1982; Goto et al. 2001). Importantly, the global rate of synthesis of messenger RNA slows very substantially with aging, and the global protein synthesis rate slows as well. Slowed rates of synthesis and degradation (turnover) are likely to allow damaged versions of these molecules to accumulate with age (Gershon 1979; Rothstein 1979; Sharma and Rothstein 1980), so reversing this slowing of turnover could have potentially significant implications for all age-related changes that could be secondary consequences of declining RNA and protein turnover with age.

The processes that regulate transcription are complex and numerous, so diffuse damage to the transcriptional apparatus as predicted by random damage theories would be expected a priori to be essentially irreversible. However, as shown in Fig. 6.15, administration of centrophenoxine to rats can completely reverse the large age-related declines in mRNA synthesis observed in their brains and livers (Zs-Nagy and Semsei 1984).

By exposing 22-month-old mouse liver nuclei to varying doses of centrophenoxine in vitro, Webster was able to confirm the same effect in mice and show its dose-dependence, with complete reversal of the decline in transcription past 11 months of age at between 100 and 1000 micromolar centrophenoxine (Webster 1988). When the molecule was allowed to spontaneously hydrolyze, which is a rapid process in water, it lost its ability to reverse the aging of transcription even though its components are still effective antioxidants (Fig. 6.16). The rescue of transcription by centrophenoxine in rats and mice has also been extended to human glial cells in stationary culture by Ludwig-Festl et al. (1983).

The specific mechanism by which centrophenoxine acts to restore transcription has not been investigated. George Webster, who made the observations of Fig. 6.16, noticed that centrophenoxine resembles the dipeptide tyrosyl-valine (Webster 1988) and suggested that it perhaps resembles "a substance which keeps genes switched on in young organisms," such as a peptide hormone regulator of transcription. Centrophenoxine, like deprenyl (which it also resembles), has also been shown to induce antioxidant enzymes in the brain (Roy et al. 1983).

6.5.2 Reversal of Reduced Translation

It is universally agreed that the global rate of protein synthesis declines with aging. In humans, total body protein synthesis has been reported to decline by 37% between youth and old age (Young et al. 1975).

The theoretical importance of decreased protein synthesis in aging was summarized rather well by Webster as follows: "First, continued protein synthesis is essential for the maintenance of the living state. If we stop protein synthesis in an organism by means of a specific inhibitor (cycloheximide, for example), the organism dies. Thus, any large drop in protein synthesis is a potential danger to

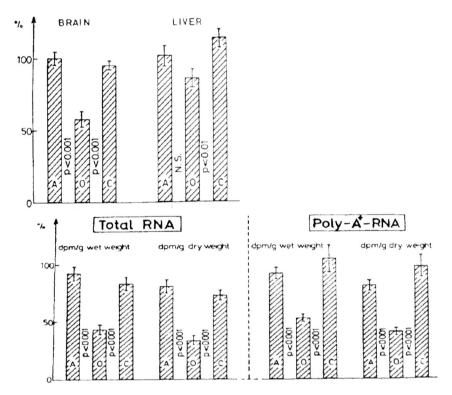

Fig. 6.15 Reversal of age-related deficits in RNA production in 24-month-old female CFY rat cerebral cortex and liver by 2 months of intraperitoneal centrophenoxine (C). *Upper panel*: total RNA. *Bottom panel*: total RNA (*left*) and mRNA (poly-A^+-RNA, *right*) in the cerebral cortex, expressed on either a wet or a dry weight basis (left vs. right data sets for each RNA type). 100% is the value for 1.5 month-old rats in all figures. A=adult (13 months); O=26-month-old control rats (same final age as the treated rats). Reprinted from Experimental Gerontology, Vol 19, I. Zs-Nagy and I. Semsei, "Centrophenoxine increases the rates of total and mRNA synthesis in the brain cortex of old rats: an explanation of its action in terms of the membrane hypothesis of aging", pp. 171–178, Copyright (1984), with permission from Elsevier

the continued functioning of the cell. The second reason … is that the decline could contribute to the other molecular changes [of aging]" (Webster 1985). In other words, if protein synthesis becomes deficient in aging, this deficiency in and of itself could account for many other age-related changes.

In the face of these simple observations, it is paradoxical that inhibiting protein synthesis seems to extend life in many circumstances (see Sections 6.3.3.3 and 6.3.6.8 above). This effect might be mediated by blunting the default expression of life-shortening genes or perhaps in part (ironically) by promoting autophagy (Hansen et al. 2008). Nevertheless, the synthesis of most proteins is presumably more beneficial than harmful, and the ability to maintain or restore global

Fig. 6.16 Dose-dependent reversal of the age-related ~60% decrease in mouse liver nucleus RNA synthesis rate between the ages of 11 and 22 months of age by intact but not by hydrolyzed centrophenoxine. Means ± SD. Modified from a previous version (Fahy 1992) drawn from the data of Webster (1988)

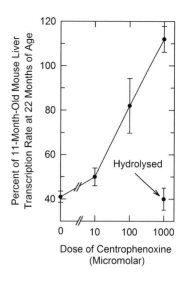

protein synthesis should therefore be generally beneficial, particularly once selective suppression of harmful proteins can be achieved.

The age-related reduction in protein synthesis was once proposed to be driven by a prior drop in transcription of the rate-limiting enzyme for protein synthesis, elongation factor 1a (EF1a), in both mammals and flies by Webster and colleagues (Webster and Webster 1982; Blazejowski and Webster 1983; Webster and Webster 1984; Webster 1985) (Fig. 6.17). However, contradictions later arose concerning both this hypothesis (Sojar and Rothstein 1986; Shikama et al. 1994) and the originally asserted time course of suppression of EF1a expression with age (Vargas and

Fig. 6.17 Coincidence between the age-dependent percentage decline in EF1a activity, EF1a messenger RNA, and total protein synthesis (normalized to the starting age for each curve) in the mouse liver. Collected and redrawn from Blazejowski and Webster (1983) and Webster (1985)

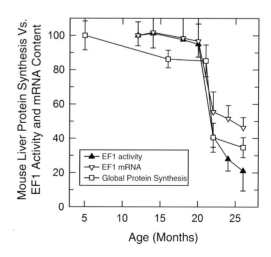

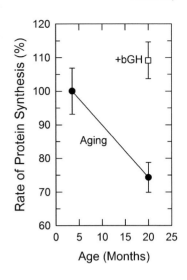

Fig. 6.18 Complete reversal of normal age-related loss of protein synthesis in 20-month-old male Sprague-Dawley rat diaphragm muscle in comparison to 3–4 month-old mouse muscle by administration of bovine growth hormone (2 mg/kg subcutaneously for 8 days). Means ± SEM. Redrawn from Sonntag et al. (1985)

Castaneda 1981; Castaneda et al. 1986), so further analysis is warranted. In the meantime, other controllers of protein synthesis, including TOR, have emerged and deserve further study.

Whatever the specific mechanism for reduced protein synthesis with aging, Sonntag showed that the age-related drop in protein synthesis in rat diaphragm can be completely reversed by the administration of growth hormone (Fig. 6.18) (Sonntag et al. 1985). Administration of a non-optimized dose of L-DOPA, which is a growth hormone releaser (Sonntag et al. 1982), also erased about half of the age-related drop in muscle protein synthesis (Sonntag et al. 1985). Ludwig-Festl et al. were also able to reverse lipofuscin-associated declines in protein synthesis in human glial cells in culture using centrophenoxine (Ludwig-Festl et al. 1983).

Protein synthesis also declines by up to 80% in human mitochondria with age (Hayashi et al. 1994). This decline was completely reversed in the skin cells of a 97-year-old woman by merging these cells with HeLa cells that lacked mitochondria, implying that the defect arises from age-related changes in nuclear gene expression and not from intrinsic deficiencies in the mitochondria themselves (Hayashi et al. 1994). In a more practical intervention, it was shown that mitochondrial protein synthesis can also be at least partly restored by acetylcarnitine administration (Gadaleta et al. 1994).

6.5.3 Reversal of Age-Related Changes in Gene Expression

Spindler's lab has repeatedly shown that short-term, late-life CR can induce the same changes in gene expression as are induced by long-term, earlier-onset CR, meaning that, in essence, late life CR can reverse a panoply of age-related changes in gene expression (Dhahbi et al. 2004, 2005; Spindler 2005; Dhahbi et al. 2006) [see also (Spindler 2010), this volume]. This fundamentally important phenomenon is well-appreciated by many investigators and requires no further elaboration here.

6.5.4 Reversal of Reduced DNA Repair

One central feature of the default pace of aging in the insulin-replete (non-starved) state may be (Niedernhofer et al. 2006) down-regulation of DNA repair [see Sections 6.3.3.2, 6.3.5.1 and 6.3.9; (Licastro et al. 1988; Lipman et al. 1989; Weraarchakul et al. 1989; Guo et al. 1998; Tran et al. 2002; Huang and Tindall 2007; Hyun et al. 2008; Tsai et al. 2008)]. In the example illustrated in Fig. 6.19, DNA damage accumulation is clearly preceded by reduced DNA repair (in this case, measured by excision of 8-hydroxy-2'-deoxyguanosine from DNA and excretion into the urine). Interestingly, this same study showed that this damaged base did not accumulate with age in either the brain or the testis (highly dissimilar tissues) (Fraga et al. 1990).

The normal age-related decline of DNA repair can be reversed by CR. Cabelof et al. found that CR completely reversed age-related declines in base excision repair (BER) capacity in all tissues tested (brain, liver, spleen and testes) and also reversed the age-related decline in the activity, mRNA, and content of a rate-limiting enzyme for nuclear BER (DNA polymerase beta) in all tissues tested. CR also reversed the age-related increase in mutation frequency following exposure to mutagens

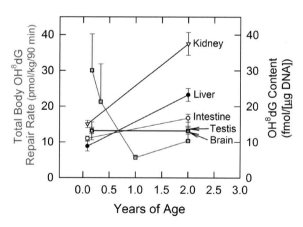

Fig. 6.19 Approximate three to six-fold maturational down-regulation of whole-body excision repair of 8-hydroxy-2'-deoxyguanosine (as determined by the rate of appearance of this damaged base in the urine: *gray boxes*, left axis) after 2 months of age precedes or coincides with the onset of accumulation of 8-hydroxy-2'-deoxyguanosine in the DNA of kidney, liver, and intestinal cells (remaining curves, right axis) in Fischer 344 rats. There is no accumulation of this damaged base in brain (postmitotic tissue) or testis (actively dividing tissue) despite the global reduction in excision repair. Drawn from the data of Fraga et al. (1990)

(Cabelof et al. 2003). Similarly, Guo et al. observed that a 40% overall reduction in UV-stimulated excision repair between the ages of 6 and 24 months in rat hepatocytes was completely reversed by CR (Guo et al. 1998).

Most of the bases added during DNA repair and during normal DNA synthesis are supplied by DNA polymerase α (Fry and Loeb 1986). Intriguingly, there seem to be two forms of DNA polymerase α, a form made by young human cells that has high DNA affinity, high copying fidelity, and high activity (the A2 form) and a form made by older human cells that has low DNA affinity, possibly (Linn et al. 1981; Silber et al. 1985) low copying fidelity, and low activity (the A1 form) (Busbee et al. 1987; Sylvia et al. 1988; Busbee et al. 1989). The A2 share of total human fibroblast DNA polymerase α protein falls from 100% in a 16-week fetus to 78% in a 10-year-old, reaches 65% by the age of 39, and, tellingly, is just 6% by the age of 66 (Busbee et al. 1989).

In mice, CR maintains DNA polymerase α in the liver during aging, in contrast to significant declines in activity in aging ad libitum fed mice (Srivastava et al. 1991), and also maintains DNA repair and improves the fidelity of both DNA polymerase α and DNA polymerase β (Srivastava and Busbee 1992). In terms of reversal, the A2 to A1 transition can be reversed, at least in vitro and at least functionally, by incubation with inositol-1,4-bisphosphate (Sylvia et al. 1988).

6.5.5 Reversal of Impaired Mitochondrial Function

6.5.5.1 Possible Physiological Origins of Mitochondrial Aging

Point mutations in mitochondrial DNA clearly do not cause aging in mice (Vermulst et al. 2007) and do not arise randomly in humans (Wang et al. 2001). Mice with a knockout of the OGG1 glycosylase have 20 times more 8-oxoguanosine in their mitochondrial DNA than normal mice, yet do not exhibit any aging phenotype (Klungland et al. 1999; de Souza-Pinto et al. 2001). Mice with a heterozygous knockout of manganese SOD show accelerated accumulation of 8-oxo-2-deoxyguanosine (8oxodG) in both mitochondrial and nuclear DNA, reduced mitochondrial function, and increased tumor incidence but exhibit no accelerated aging phenotype and no life shortening (mean lifespan = 103% of control, 90th survival percentile = 105% of control, maximum lifespan = 104% of control) (Van Remmen et al. 2003). These results indicate that many specific kinds of mitochondrial oxidative damage do not drive aging. The results described in the next paragraph also indicate that mitochondrial DNA damage and its contributions to aging (Trifunovic et al. 2004) do not necessarily arise as the result of prior oxidative damage. Considering such results, more physiological, as opposed to random, origins for mitochondrial aging should be considered.

Kujoth et al. and Trifunovic et al. reported that age-related mitochondrial DNA damage can be caused by inadequate activity of mitochondrial DNA polymerase γ, a central enzyme for mitochondrial DNA replication and base excision repair (BER), in the absence of oxidative stress (Kujoth et al. 2005; Trifunovic et al. 2005), and

Kujoth et al. showed that the main pathogenic effect of these mutations on the cellular level is the induction of apoptotic proteins and apoptosis itself (Kujoth et al. 2005). Prior to this, Chen et al. demonstrated that brain mitochondrial BER declines precipitously with age, and long before age-related mitochondrial aging seems to appear (Chen et al. 2002).

Combining the information from the latter two studies suggests the hypothesis that mitochondrial DNA damage normally arises as a result of prior physiological (maturational) down-regulation of the enzymes needed for mitochondrial BER (Fig. 6.20), just as BER activity declines with aging more generally (Cabelof et al. 2003; Lu et al. 2004; Wilson and McNeil 2007). DNA polymerase γ is a nuclear-encoded protein (Hudson and Chinnery 2006), which means that its decline with aging is not due to mitochondrial DNA damage.

The nuclear downregulation of mitochondrial BER might be part of a much broader set of events. A pivotal study by Hayashi et al. reviewed in the next section (Hayashi et al. 1994) and studies described in Section 6.5.5.3 indicating that the decline of COX and many other features of mitochondrial aging are fully reversible, implies a general pattern of reversible (physiological) nuclear gene driven mitochondrial aging (presumably representing another cause of active default life shortening).

Mitochondrial aging is also at least partly due to reduced autophagy (Cavallini et al. 2007) and can be at least partly reversed by augmenting autophagy (see Section 6.5.5.3).

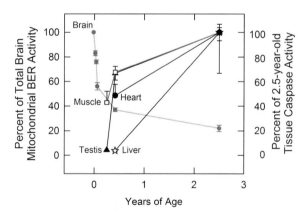

Fig. 6.20 Possible causal sequence of events in mitochondrial aging. *Gray circles*: normal maturational down-regulation of mitochondrial base excision repair (BER) in mouse brain [drawn from the data of Chen et al. (2002)]. *Other symbols*: increased caspase activity resulting from natural aging in 4 different organs [drawn from the data of Kujoth et al. (2005)]. Increased tissue caspase activity with aging was shown to be accelerated in a mouse mutant defective in DNA polymerase gamma, implying that developmental suppression of the same pathway will produce the same result over the course of a normal lifespan

6.5.5.2 Reversal of Mitochondrial Functional Aging in Single Cells

Hayashi et al. fused HeLa cells lacking mitochondria (rho-zero cells) with enucleated fibroblasts taken from a 97 year-old woman (which showed about an 80% decline in COX activity compared to fibroblasts taken from a human fetus), in effect transplanting old mitochondria into HeLa cells. The same experiment was also done using fetal fibroblasts rather than the 97-year-old fibroblasts. The result was that COX activities in the resulting young and old cell derived cybrids were identical (Hayashi et al. 1994), indicating that the mitochondrial defect of aging is caused by nuclear events rather than mitochondrial DNA damage. Furthermore, the decline in COX activity in the old fibroblasts exactly matched an 80% decline in overall mitochondrial protein synthesis in the old versus the fetal cells, and this difference, too, was totally eliminated by fusion with HeLa cells, implying that the aging of at least this feature of human fibroblast mitochondria is entirely secondary to deficiencies in nucleus-directed mitochondrial protein synthesis.

Hayashi et al. next fused rho-zero HeLa cells to the old donor fibroblasts without removing the fibroblast nuclei in advance, in effect transplanting HeLa cell nuclei to fibroblasts rather than transplanting fibroblast mitochondria to HeLa cells. The results were identical (total reversal of reduced COX activity), indicating that the failure of the skin cell nuclei to drive COX synthesis is recessive rather than dominant. It was, however, shown not to be related to the population doubling level of the donor skin cells (Hayashi et al. 1994).

6.5.5.3 Reversal of Mitochondrial Functional Aging in Intact Animals

Reversal of several aspects of mitochondrial aging has been brought about by administration of acetylcarnitine to intact animals. As illustrated in Fig. 6.21, giving acetylcarnitine to old rats can completely reverse aging-induced reductions in myocardial cytochorome c oxidase activity, COX I mRNA levels, ADP transport, and state 3 and state 4 respiration rates (Gadaleta et al. 1994; Paradies et al. 1994). Hoppel's group verified many of these results and extended them by showing that acetylcarnitine given to 24 month-old Fisher 344 rats also increased myocardial cytochrome b and aa_3 content and replaced a 17% drop in complex III activity in interfibrillar mitochondria with a 29% increase in activity compared to the activity of 6-month-old adult controls (Lesnefsky et al. 2006). In the same study, Lesnefsky et al. showed as well that acetylcarnitine restored the stress resistance of the older heart (measured by resistance to an ischemic insult) to that of the young adult heart, but did not change the resistance of young adult hearts, indicating that the effect specifically related to reversing myocardial mitochondrial aging. Significantly, only interfibrillar mitochondria and not sub-sarcolemmal mitochondria were affected by aging.

Acetylcarnitine, like the cell fusion experiments described in the previous section, increases global mitochondrial RNA and protein synthesis (Gadaleta et al. 1994). Within one hour of intraperitoneal administration, acetylcarnitine completely or more than completely reversed major declines in mitochondrial transcription of

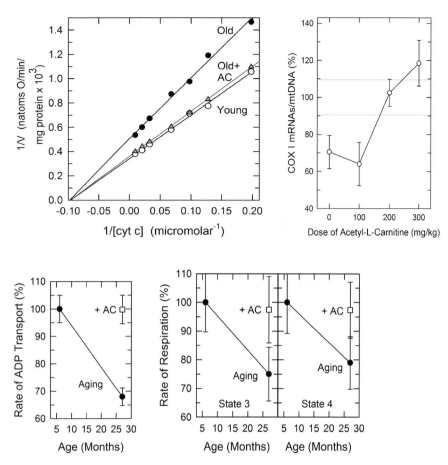

Fig. 6.21 Reversal of mitochondrial aging processes in rat heart mitochondria by acetylcarnitine (AC). *Top Left*: Near-total reversal of the age-related 28% drop in cytochrome c oxidase V_{MAX} three hours after a single injection of 300 mg/kg of AC. Young: 7 mo; Old and Old + AC: 27 mo. Redrawn from Paradies et al. (1994). *Top Right*: Dose-dependent total reversal of the decline in the number of transcripts of the cytochrome c oxidase gene at 27 months of age three hours after a single injection of AC. *Horizontal lines*: upper and lower 7 month old control range (mean ± one SD; data points show ± 1 SD as well). Redrawn from Gadaleta et al. (1994). *Bottom Left*: Total reversal of age-related impairment of ADP transport three hours after a single injection of 300 mg/kg of AC. Redrawn from Paradies et al. (1994). *Bottom Right*: Total reversal of reduced state 3 and state 4 respiration by AC. Drawn from the data of Paradies et al. (1994). ADP:O ratios were not changed by aging or AC

both the 12S rRNA and COX subunit 1 genes observed at 28 months of age in heart and cerebellar and cerebral hemisphere synaptic mitochondria while having no effect on adult (8-month-old) rates of transcription (Gadaleta et al. 1990), and these results are probably representative of the response of mitochondrial transcription in general (Gadaleta et al. 1990, 1994).

In the liver, corresponding age-related deficits of transcription were not observed (Gadaleta et al. 1990). Interestingly, total carnitines decline with age in brain and muscle, but not in liver, and acetylcarnitine in the drinking water prevents this decline and reduces lipofuscin accumulation (Maccari et al. 1990); this suggests that the effect of acetylcarnitine may be partly the result of simply correcting an aging-induced deficiency state, as might be the case as well with coQ replacement (Section 6.3.6.5, "Coenzyme Q").

Orally administered acetylcarnitine is known to activate (Narkar et al. 2008) the master metabolic regulator, AMPK (see also Section 6.3.5.2, "The PEPCK-C (mus) Mouse"), in muscle and to increase expression of UCP3 as well as increasing oxygen consumption, treadmill time (23% longer), and treadmill distance (44% farther) (Narkar et al. 2008). AMPK activates PGC-1α, which increases mitochondrial biogenesis, increases aerobic capacity, induces mitochondrial uncoupling proteins, regulates glucose and fat metabolism, and may be responsible for many of the effects of SIRT1 activation and calorie restriction (Puigserver and Spiegelman 2003; Weindruch et al. 2008). Acetylcarnitine administration was shown to upregulate 32 genes related to oxidative metabolism (Narkar et al. 2008), of which 30 (Narkar et al. 2008) were also upregulated in transgenic mice with a known 100% increase in running endurance due to PPARδ activation (Wang et al. 2004). Thus, the reversal of mitochondrial aging by acetylcarnitine and the prevention of mitochondrial aging in the PEPCK-C (mus) mouse (Hakimi et al. 2007) suggest that mitochondrial aging is associated with reversible age-related downregulation of mitochondrial turnover and related processes. This possibility is further strengthened by the observation that the number of mitochondria in the left ventricle of middle-aged mice could be increased by about 10% by feeding a diet containing added carnitine, inositol, and other ingredients (Lindseth et al. 1984; Fahy 1985).

Mitochondrial BER can be up-regulated in the brain by an ischemic insult (Chen et al. 2003) and in photoreceptor synaptic terminals by intense light treatment (Cortina et al. 2005). The decline in nuclear BER with aging can be reversed by CR (Cabelof et al. 2003). No one seems to have as yet demonstrated the reversal of aging-associated mitochondrial BER declines, but the tools to enable this seem to be in hand.

So far, deliberate reversal of age-related mitochondrial DNA deletions has not been reported, but the possible spontaneous reversal (∼4-fold decline) of one particular lesion was noticed in one individual between the ages of 66 and 82 (Michikawa et al. 1999). Future options for reversing these deletions and other mitochondrial mutations may include repletion of aged cells with young mitochondria by exploiting normal mechanisms for mitochondrial exchange between cells (Spees et al. 2006) and natural mitochondrial DNA complementation mechanisms (Nakada et al. 2002) or the use of enzymatic or transfectional approaches to eliminating corrupted mitochondrial genomes and inserting uncorrupted ones (see Chapter 16).

Such interventions may not be necessary, however, if normal levels of mitochondrial turnover by autophagy can be maintained over the lifespan. Reversal of electron microscopic evidence of hepatic mitochondrial aging, age-related declines in ATP and mitochondrial succinic dehydrogenase, and age-related increases in mitochondrial membrane permeability were all reported after restoring chaperone-mediated autophagy and macroautophagy in 19 month-old mice (Zhang and Cuervo 2008).

6.5.6 Reversal of Reduced Regenerative Capacity

Recently, mechanisms of the profound age-related decline in the ability of skeletal muscle to regenerate following, for example, needle sticks or freezing injury were elucidated and shown to be reversible through a variety of interventions (Conboy et al. 2003, 2005; Brack et al. 2007; Carlson et al. 2009a, b). Muscle regeneration is accomplished by adult muscle stem cells known as satellite cells that reside in muscle and react to injury by proliferating, migrating, and fusing with pre-existing muscle cells to repair or replace them. The satellite cell response is induced by Delta-1, an injury-stimulated ligand of the Notch receptor (Conboy et al. 2003). Delta expression is up-regulated by a diffusible factor produced in response to injury in young but not in older skeletal muscle, and the lack of the Delta-1 response accounts for the aging defect in muscle regeneration (Conboy et al. 2003).

The regenerative defect of aging muscle was first reversed in vivo by injecting old muscle with an antibody to Notch that directly activates Notch and that "markedly improved the regeneration of old muscle, rendering it similar to young muscle" (Conboy et al. 2003). Next, Delta signaling and the regenerative ability of old muscle were restored in vivo by joining the circulatory systems of old animals to those of younger animals (heterochronic parabiosis) (Conboy et al. 2005). This rejuvenating effect was shown not to be due to transfer of replacement satellite cells from the young partner to the old, because the younger parabionts expressed green fluorescent protein, and green satellite cells were not seen in old muscle after heterochronic parabiosis. Instead, it was found that exposure of old satellite cells to young serum in vitro produced the same rejuvenating effect as parabiosis on satellite cell signaling (Conboy et al. 2005).

Old mouse serum was also found to have a negative effect on young mouse muscle stem cells. Brack et al. claimed to be able to largely reverse this effect by adding Wnt inhibitors (sFRP3 or DKK1) to the serum and that heterochronic parabiosis suppresses Wnt signaling in old mouse satellite cells (Brack et al. 2007). However, others found no detectable Wnt in serum at any age in either mice or humans and either no effect of Wnt or sFRP3 on myogenic proliferation or a positive rather than a negative effect of Wnt (Carlson et al. 2009a).

More promisingly, the difference between young and old mouse muscle repair in vivo was successfully eliminated by controlled systemic delivery of a small molecule TGF-β receptor-1 inhibitor that might mimic the effect of an undiscovered endogenous TGF-β1 regulator (Carlson et al. 2009a). Further work showed that satellite cell response to young and old serum and age-related changes in Notch, Delta, and TGF-β1 are extremely similar in mice and humans and are driven at least in humans by downregulation of MAP kinase (Carlson et al. 2009b). Remarkably, both the induction of MAP kinase by the use of fibroblast growth factor 2 and the administration of Delta allowed 70-year-old human satellite cells to respond at least as well as control 20-year-old human satellite cells in vitro (Carlson et al. 2009b).

These findings indicate that muscle stem cell incompetence with aging arises from extrinsic factors rather than by intrinsic defects of the stem cells themselves, and Conboy et al. proposed "that a decline in Notch signaling necessary for cell proliferation and cell fate determination may also occur in progenitor cells and stem cells in other tissues with age and may be a general mechanism underlying the diminished regenerative properties of aged tissues" (Conboy et al. 2003).

Hepatic regenerative ability also declines with age, and this decline was found to be due to impairment of hepatic progenitor cell proliferation caused by the formation of a complex involving cEBP-alpha and the chromatin remodeling factor brahma (Brm) (Conboy et al. 2005). Heterochronic parabiosis dramatically reduced the amount of cEBP-alpha complex found in old livers, making their levels equal to levels seen in the young heterochronic parabionts, and restored youthful regenerative competence to the old livers without transferring competent cells from the young to the old parabiont (Conboy et al. 2005).

For further information, please see Chapter 13 in this volume.

6.5.7 Prevention and Reversal of Replicative Senescence

At one time, the limited capacity of cells to divide in culture ("replicative senescence") as established by Leonard Hayflick was assumed to be caused by cellular damage. In 1998, of course, this prejudice was demolished by the classic experiments of Bodnar et al., who showed that the single and simple intervention of expressing telomerase could completely and permanently rescue four diverse cell types from the formerly mysterious cause of replicative senescence that had presumably been present in all such normal cells throughout the entire history of mammals on earth until their experiment was completed (Bodnar et al. 1998). More recently, extending telomeres in whole cancer-resistant mice was shown, remarkably, to extend median lifespan by over 40% (Tomas-Loba et al. 2008). This observation creates yet another independent mechanism for major life extension in the mouse, and combining this modality with the other modalities described above may have truly impressive effects on the lifespan of a major mammalian model of aging.

Today, the role of telomeres in replicative senescence and the ability of telomerase to overcome the latter [see, e.g., (Shay and Wright 2008; Aviv and Bogden 2010; West 2010)] are too well-known and taken for granted to require extended discussion here. But before the Bodnar et al. experiment, I was told that the Hayflick limit would never be overcome, not even in 1,000 years. This should provide an object lesson about the dangers of automatic pessimism about the supposed inexorability of fundamental aging processes.

6.5.8 Reversal of Lipofuscin Accumulation

Lipofuscin accumulation is a hallmark of the aging of some cells in the body, and may be the result of and a marker for reduced intracellular turnover (Ivy et al. 1984, 1996). Fonseca et al. were able to provide strong evidence that lipofuscin accumulation due to normal aging under natural conditions is reversible by natural physiological mechanisms in vivo. They showed that the lipofuscin content of the left eyestalk of 2–14 year-old crayfish, *Pacifastacus leniusculus*, recaptured from a field pond one year after being released following removal of the right eyestalk for quantitation of its lipofuscin content, was lower than the amount present a year earlier in the right eyestalk, at least when the baseline content of lipofuscin exceeded 1.1% by volume (p value for the linear inverse relationship between initial and final lipofuscin contents was <0.0001) (Fonseca et al. 2005).

In their model (Fonseca et al. 2003), the reversal of the normally expected linear increase in lipofuscin accumulation with age (Belchier et al. 1998) was attributed to compensatory upregulation of repair or other mechanisms (Fonseca et al. 2005), and they presented histological and ultrastructural evidence for a lipofuscin disposal system involving exocytosis of lipofuscin granules in the crayfish, the Norwegian lobster (*Nephrops norvegicus*) and the crab (*Cancer pagurus*) and pointed to apparently overlooked evidence for similar disposal systems in other marine species as well as in frogs (Srebro 1966), rats (Monteiro 1991; Cavanagh et al. 1993), birds (Singh and Mukherjee 1972), and monkeys (Brizzee et al. 1974; El-Ghazzawi and Malaty 1975).

In a mammalian model of spontaneous lipofuscin clearance, lipofuscin-like inclusions in the retinal pigmented epithelium of the eye were induced by injection of leupeptin, which is a protease inhibitor. Over a 12-week follow-up period, the massive over-accumulation of lipofuscin induced by the treatment was completely reversed by spontaneous endogenous mechanisms (Katz et al. 1999).

In 1974, it was reported that 8 weeks of treatment of 24–26 month-old rats with centrophenoxine reduced brain lipofuscin content by 25–42.3% in different brain regions (Riga and Riga 1974). In the case of the pontine reticular formation, the decrease achieved was equal to 64.7% of the amount of lipofuscin accumulated between the age of 8–9 months and 24–26 months, and in cortical cell layer V, the reversal was equal to 53.4% of the total amount otherwise accumulated. Although complete reversal was not achieved, the period of treatment was short compared to the period of aging being treated. No change in brain cell number was seen with aging in this study.

The mobilization and export of lipofuscin from rat spinal ganglion cells in response to centrophenoxine was demonstrated and visualized by electron microscopy of these cells in tissue culture in the same year (Spoerri and Glees 1974), and similar effects were shown as well in the hypothalamus (Spoerri and Glees 1975) and in hearts (Spoerri et al. 1974; Patro et al. 1992). Old guinea pigs treated with centrophenoxine showed dissolution of lipofuscin in the brain at all doses studied (30, 50, and 80 mg/kg, given for 10 weeks), and this lipofuscinolysis continued for several weeks even after discontinuing the lowest concentration of the drug (Glees and Spoerri 1975). Later work confirmed lipofuscinolysis in the rat cerebral cortex, hippocampus, thalamus, basal ganglia, midbrain, medulla oblongata, and spinal cord (Tani and Miyoshi 1977).

In addition to mobilizing lipofuscin, centrophenoxine also reduces lipid peroxidation in the brain (Sharma et al. 1993), presumably as a result of inducing antioxidant enzymes (Roy et al. 1983). The exact mechanism of action of centrophenoxine on lipofuscin mobilization remains a mystery, but both parts of the molecule (p-chlorophenoxyacetate and dimethylaminoethanol) and not just one of these hydrolysis products alone were found to be necessary for dissolving lipofuscin in *Caenorhabditis briggsae* (Zuckerman and Barrett 1978).

Centrophenoxine is not the only drug that can reduce or reverse lipofuscin accumulation. Additional effective agents appear to include deprenyl, piracetam, orotic acid, Geriforte (Riga and Riga 1995), the diethylaminoethyl ester of p-chlorophenoxyacetic acid, kavaine, and other agents (Riga and Riga 1974). In 1984, Kristin Lindseth and colleagues from the Ames Research Center in Moffett Field, California, presented an abstract indicating that left ventricular lipofuscin in middle aged mice could be reduced by about 10% by feeding them a diet enriched in inositol, carnitine, vitamin E (mixed tocopherols), pyridoxine, ascorbic acid, and chromium (Lindseth et al. 1984; Fahy 1985). This result is consistent with a report showing that acetylcarnitine in the drinking water could slow lipofuscin accumulation in male Sprague-Dawley rats (Maccari et al. 1990). Total carnitines decline with age in brain and muscle, so the effects of carnitine and acetylcarnitine may be partly the result of simply correcting an age-related deficiency state.

Autophagy is a major process for turning over both proteinaceous and nonproteinaceous cellular constituents. Recent exciting research in Ana Maria Cuervo's lab has identified the molecular deficiencies in chaperone-mediated autophagy (CMA) that occur in aging mammals and has shown that the correction of these deficiencies can prevent features of liver aging including lipofuscin accumulation, even when this correction is induced late in life (Zhang and Cuervo 2008). These studies may lead to massive advances in aging therapeutics if "cellular constipation" explains much of the pathophysiology of aging, as her laboratory's work (Martinez-Vicente and Cuervo 2007; Ventruti and Cuervo 2007) strongly implies is the case. In addition, the Bergamini lab has reported that CR stimulates autophagy, but that certain drugs can greatly reduce the amount of CR required to achieve this effect (Bergamini et al. 2003) and can have similar effects on their own (Donati et al. 2008).

6.5.9 Reversal of Reduced Immune Function

Very many interventions have been shown to reverse reduced immune function in elderly animals and people. For space reasons, just two are considered here that seem to act at least in part through thymic regeneration, which is a fundamental approach to the reversal of immunosenescence.

The first example is the use of growth hormone to completely reverse the estimated 70–80% reduction in proliferative capacity (Con A and PHA response)

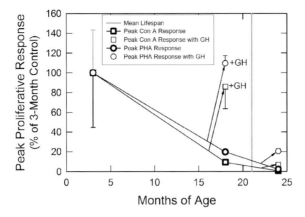

Fig. 6.22 Recovery of immune function in female Wistar-Furth rats treated with growth hormone for 60 days by implantation of GH3 pituitary adenoma cells. GH treatment was instituted at the times indicated by the origins of the arrows. Complete reversal of dramatic impairments is possible within 60 days in animals treated prior to the mean lifespan of this strain of 21 months (indicated by the *vertical gray line*), whereas the gains after 21 months of age were incomplete. The possibility that full restoration of immune function and thymic mass and morphology by 18 months of age could have allowed function to persist through 24 months if GH exposure had been continued was not addressed in this study. The 90–95% decline in immune function observed in control animals just prior to the mean lifespan could be life-shortening and account in part for the length of life of this rat strain. Drawn from the data of Kelley et al. (1986)

that takes place as female Wistar-Furth rats age between 3 and 16 months of life (Fig. 6.22). Similar intervention at a later time (22 months) was much less effective, but longer durations of growth hormone administration at older ages were not attempted.

In humans, thyroid hormone has long been known to stimulate thymic function, as indicated by thymic hyperplasia (Judd and Bueso-Ramos 1990) and increased thymulin production (Fabris et al. 1986). Remarkably, the normal large drop in human thymulin production with age is totally absent in hyperthyroid individuals (Fabris et al. 1986) (Fig. 6.23, left panel), in keeping with a lack of thymic involution in hyperthyroid patients up to the maximum age studied (about 60) (Simpson et al. 1975). Further, age-related attenuation of thymulin levels can be reversed even in initially hypothyroid patients by over-correcting their thyroid deficiency with thyroxine (Fig. 6.23 right panel) (Fabris et al. 1986).

6.5.10 Reversal of Thymic Involution

The involution of the thymus is associated with the onset of puberty in all vertebrates (Cardarlli 1989) but ultimately results in life-threatening impairment of immune system function (Cawthon et al. 2003; Aspinall and Mitchell 2010). However, several dramatic examples of the reversibility of thymic involution have been documented by using growth hormone (Kelley et al. 1986; Goff et al. 1987; Napolitano

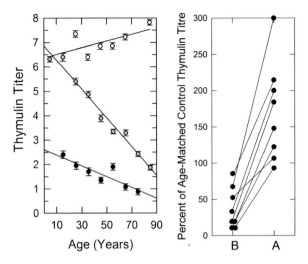

Fig. 6.23 *Left*: Prevention of the normal age-related decline in thymic function (*gray points*), as indicated by serum thymulin levels, by hyperthyroidism (*white points*) and exacerbation of thymulin decline by hypothyroidism (*black points*). Right: reversal of age-related thymulin declines in hypothyroid patients relative to age-matched euthyroid patients (100%) by administration of L-thyroxine: thyroxine restored thymulin levels generally to levels exceeding those of age-matched euthyroid controls (*B* before treatment, *A* after L-thyroxine treatment). Redrawn from Fabris et al. (1986)

et al. 2002; Fahy 2003), drugs that block the synthesis of sex hormones (Greenstein et al. 1987, 1992), and other compensatory stimuli. Figure 6.24 documents the reversal of thymic involution in rats by administration of a male contraceptive (Zoladex) (Greenstein et al. 1987) and in mice by the simple administration of zinc sulfate (Fabris et al. 1990). In the former experiment, no attempt was made to measure immune function, but the treatment was analogous to castration, which increases the rate of graft rejection (Greenstein et al. 1987). Treatment of mice with zinc resulted in increased thymic hormone production, complete recovery of Thy 1.2^+ cell number in the spleen, and partial restoration of PHA responsiveness and cytotoxic natural killer cell activity.

Figure 6.25 illustrates the effectiveness of growth hormone in regenerating the thymus in rats (upper 4 panels). Regeneration was morphologicallhy normal if the treatment was begun soon enough, but less normal when begun at 22 months. The immunological effects of this treatment have been presented in Fig. 6.22.

Dramatic morphological evidence of regeneration of the canine thymus was also reported by (Goff et al. 1987) after giving bovine GH every other day for 30 days (only 14 total treatments) to final ages of 33–55 months, but not when given to older dogs, in general agreement with Kelley and colleagues (Kelley et al. 1986). However, unlike the observations of Kelley et al., dogs were able to respond to growth hormone by making thymulin even when they were too old to actually display thymic hyperplasia within the short treatment time provided.

6 Precedents for the Biological Control of Aging

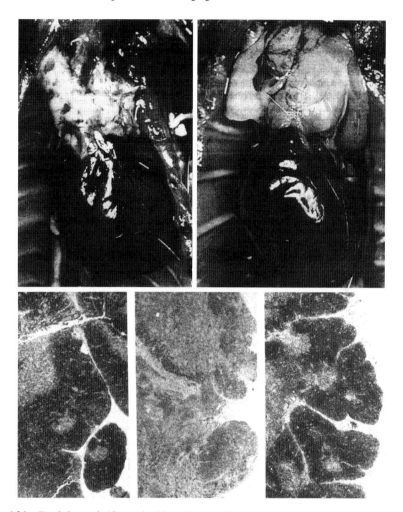

Fig. 6.24 *Top left panel*: 15 month-old rat thymus (flabby white area over the heart). *Top right panel*: 15 month-old rat thymus after 28 days of prior treatment with Zoladex, a testosterone synthesis inhibitor. Reproduced by permission (and with modification); original figure (c) Society for Endocrinology (1987). From (Greenstein et al. 1987). Bottom panels, from left to right: young mouse thymus, 22 month-old mouse thymus, and 22 month-old mouse thymus regenerated by administration of 0.1 mg/mouse/day of zinc sulfate. Modified from (Fabris et al. 1990)

Finally, last but certainly not least, Fig. 6.26 provides similar examples of the regeneration of the human thymus using growth hormone. Unfortunately, measures of the immune responses to these regenerative treatments that would be relevant for the treatment of most older healthy subjects were not successfully completed.

For additional information about possibilities for regenerating, maintaining, or replacing the human thymus, see Chapter 15 (Aspinall and Mitchell 2010) and the summary in (Fabris et al. 1997) or contact Intervene Biomedical.

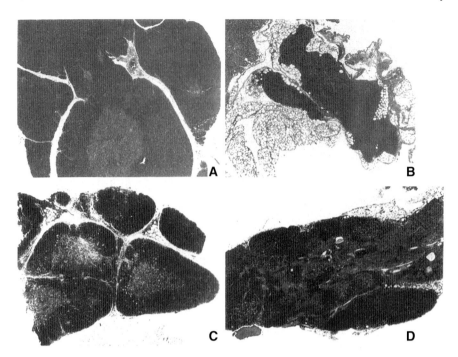

Fig. 6.25 Regeneration of female Wistar-Furth rat (**a–d**) thymus with growth hormone. Image magnifications vary from panel to panel. (**a**) 3 month-old rat thymus (242±20 mg weights). (**b**) 18 month-old rat thymus (difficult to weigh due to replacement with fatty tissue). (**c**) 18 month-old rat thymus following 60 days of growth hormone exposure in vivo from implantation of pituitary GH3 adenoma cells, showing normal structure and mass (208±21 mg). (**d**) 24 month-old rat thymus after treatment with GH3 cells as in **c**; morphology is not normal, but the mass showed substantial improvement (200±36 mg). Note: 24 months is beyond the mean lifespan of this strain. Reproduced with modifications from Kelley et al. (1986)

6.5.11 Reversal of Lost Reproductive Cycling in Females

The loss of reproductive cycling in females is driven partly by ovarian exhaustion and partly by brain mechanisms (Meites 1991; Meites 1993). The latter, at least, can be effectively reversed to allow cycles to resume until ovarian exhaustion becomes the limiting factor and cycles again stop.

The monoamine oxidase inhibitor, deprenyl, was given to non-cycling female Sprague-Dawley rats at 15–16 months of age at either 0.25 or 2.5 mg/kg per day and was found to restore estrous cycling in most of the treated rats while greatly reducing the incidence of pituitary and mammary tumors (ThyagaRajan et al. 1995). A variety of other agents that increase brain catecholamine levels, including L-DOPA and iproniazid, also restore cycling in constant-estrous rats (Huang et al. 1976; Meites 1991; Meites 1993).

Other physiological influences that can restore estrous cycling in old female rats include progesterone and ACTH (Huang et al. 1976). The Meites lab also reported

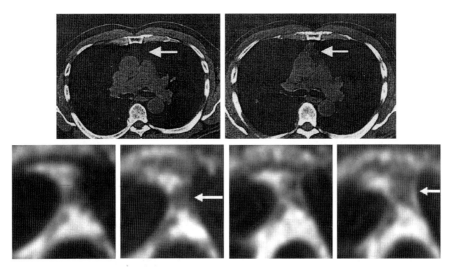

Fig. 6.26 Regeneration of the human thymus. *Top*: Regrowth of the thymus in a 65 year-old AIDS patient after six months of treatment with human growth hormone (1.5–3 mg/day, subcutaneously), as visualized by CT imaging (*Left*: without GH administration; *Right*: after GH administration). Modified from (Napolitano et al. 2002). *Bottom*: two side-by-side before and after comparisons showing early stages of regrowth of thymic functional mass [*gray areas* highlighted by arrows between the aorta (*black* arc at the *bottom*) and the sternum (mottled arc at the very *top*), vs. fatty tissue appearing as white areas] in a 46 year-old normal male subject after 30 days of hGH administration four times a week (0.018 mg/kg), as visualized by magnetic resonance imaging. Image analysis demonstrated a near doubling of *gray* (functional) thymic mass (total functional mass increasing from normal to >3 SD above the established mean for a person of this age). Modified from (Fahy 2003)

that 10 weeks of CR imposed on 15–16 month-old female rats allowed 100% of the restricted rats to resume cycling upon refeeding, whereas prior to CR, only 41% could cycle either regularly or irregularly (Quigley et al. 1987).

6.5.12 Reversal of Age-Related Organ Atrophy in Humans and Rodents

The body can be divided into adipose mass and lean body mass (LBM). Both show major age-related changes [see (Rudman et al. 1991) and references therein]. Between the ages of 40 and 80, men lose lean body mass at the rate of 5% per decade on average, while women lose 2.5% per decade. By the age of 75, men have only half the muscle mass they possessed when they were 30, and between 30 and 70, LBM declines by 30% while adipose mass increases by 50%. Only about 25% of the decrease in LBM can be attributed to disuse atrophy, the rest being due to aging. Most tellingly, the human liver, pancreas, kidney, and brain lose 30% of their age 30 volumes by the age of 75. Obviously, these changes, in addition to being responsible for a great deal of age-related frailty, loss of reserve capacity, and morbidity, will ultimately result in death if they continue long enough. The withering and shrinkage of older people accounts for a great deal of our image of aging.

Because similar effects are also seen in growth hormone deficient children and animals, Rudman postulated in an excellent review of the subject (Rudman 1985) that these changes in aging humans might be due to the well-known declines with aging of human growth hormone and IGF-1. His initial famous (or infamous) experimental investigation of this hypothesis concluded by likening the effects of human growth hormone administration to growth hormone deficient elderly men (>60 years of age) to as much as a 20-year reversal of human aging based on measurements of the reversal of dermal atrophy ($p = 0.07$) and significant ($p < 0.05$) improvements in bone density, overall adipose mass, and LBM (Rudman et al. 1990).

He and his colleagues then went on to publish a little-noted paper of potentially much greater significance that provided specific evidence that the atrophy of visceral organs in growth hormone deficient men over the age of 61 can be stopped and even reversed by growth hormone administration (Rudman et al. 1991) (Fig. 6.27). This

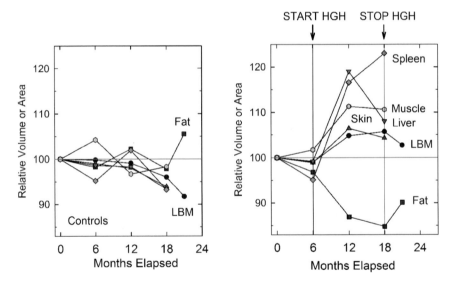

Fig. 6.27 Arrest and reversal of age-related atrophy of visceral organs, muscle, and skin in men over 61 years old by growth hormone administration. *Left*: untreated controls. *Right*: treated with growth hormone for 12 months after a six-month baseline period. Error bars omitted for clarity. In controls, lean body mass (LBM) and skin thickness declined significantly ($p < 0.05$). In treated subjects, LBM, skin thickness, liver size, spleen size, and the sum of 10 muscle areas all increased significantly at both 12 and 18 months ($p < 0.05$), and LBM remained significantly elevated 3 months after the end of treatment, while adipose mass ("fat") significantly declined at 12 and 18 months and remained reduced 3 months after treatment (at 21 months). Drawn from the data of Rudman et al. (1991)

report confirmed, in humans, similar prior observations in old rats indicating that growth hormone could increase the weight of the heart, thymus, liver, spleen, and kidneys (Sonntag and Meites 1988) as well as Rudman's own less detailed prior study. Unfortunately, Rudman did not include measurements of thymic size in his study, but, as noted earlier (Sections 6.5.9 and 6.5.10), restoration of thymic size and function by growth hormone administration to human subjects and animals has since been demonstrated by other laboratories.

6.5.13 Reversal of Hair Graying and Balding in Humans

Another of the most cardinal signs of human aging is the graying of hair. About 50% of individuals have 50% gray hair by the age of 50 (Keogh and Walsh 1965), and hair graying or whitening becomes more and more universal thereafter. The age of onset and the progression of hair color loss (canities) with age are both hereditary (Tobin and Paus 2001), and the latter tends to follow a specific pattern.

Recently, hair graying/whitening was found to be due to the loss of the stem cell precursors to hair follicle melanocytes (Nishimura et al. 2005; Sarin and Artandi 2007), possibly secondary to "premature differentiation or activation of a senescence program" (Nishimura et al. 2005). Presumably, administering exogenous stem cells might be able to correct this defect by re-supplying the missing cells.

In the meantime, existing stem cell melanocyte precursors can apparently be regenerated or re-recruited in at least some individuals by treatment with Imatinib (Gleevec) (Ettienne et al. 2002). In a study on 133 patients with chronic myeloid leukemia treated with Imatinib, about 7% (5 men and 4 women) aged 53–75 years (mean, 63.4 years) were observed to experience "progressive repigmentation of the hair" over a 2 to 14 month period, in one case involving not just repigmentation of hair on the head but also repigmentation of hair on the body. This effect was later postulated to be related to inhibition of c-KIT, a tyrosine kinase receptor whose stimulation leads to the degradation of an activator of the tyrosinase pigmentation gene promoter in melanocytes (Robert et al. 2003). Patients treated with SU11428, which stimulates a similar receptor, experienced reversible depigmentation of all hair starting at around 5–6 weeks and resolving as early as 3 weeks after discontinuing drug treatment.

White hair can also become repigmented as a result of triiodothyronine treatment (Redondo et al. 2007). Repigmentation can also occur spontaneously (Tobin and Paus 2001).

Male pattern baldness is genetic and amazingly common [affecting at least half of the white male population by the age of 50 (Stough et al. 2005)], but can be reversed by benoxaprofen (Fenton et al. 1982), finasteride (Rushton et al. 2002), and minoxidil (Loniten®) (Stough et al. 2005). In women, hair loss tends to have more nutritional origins and can be reversed by dietary supplementation with iron and lysine (Rushton et al. 2002).

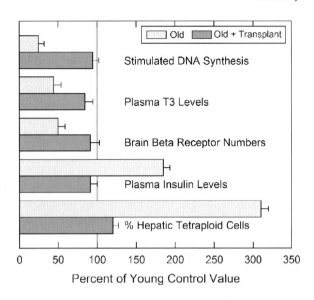

Fig. 6.28 Reversal of major age-related decrements and increments in several widely divergent physiological systems through thymus transplantation in old mice. *From top to bottom*: isoproterenol-stimulated submandibular gland DNA synthesis; plasma levels of triiodothyronine; beta receptor population "density" in the cerebral cortex; plasma insulin concentrations; percent of tetraploid cells in the liver. Modified from Fabris et al. (1988)

6.5.14 Reversal of Multiple Segmental Aging Processes by One Intervention: Thymus Transplantation

Figure 6.28 summarizes the results of several studies of the effects of syngeneic thymus transplantation in old mice (Fabris et al. 1988). The results show the thymus to be a controller not only of immunological aging but also of many aging processes of little seeming relevance to the immune system. Prior to transplantation, major age-related changes were documented in the number of cerebral cortical beta receptors, the ploidy of hepatocytes in the liver, the level of circulating thyroid hormone, the level of insulin, and the ability of the submandibular gland to synthesize DNA in response to isoproterenol stimulation. These are non-trivial aging changes, yet each of these segments of whole-body aging was almost completely reversed by thymus transplantation. Serial thymus transplantation extends lifespan as well despite the surgical trauma of repeated transplantation (Hirokawa and Utsuyama 1984).

6.6 Unregulated Aging

Although not the main subject of this chapter, there are aspects of aging that appear to be beyond the reach of the genome, and a few words about these may be in order.

The first class of problems of this kind is represented by loss of or damage to structures that can't be turned over or replaced by natural mechanisms. Perhaps the most obvious and stark examples are the loss of adult body parts in insects such as wings and legs due to mechanical trauma ["mechanical senescence" (Comfort 1979; Finch 1990)]. In mammals, incontrovertible examples of genomically intractable mechanical senescence are harder to identify and are relatively few in number. It is well recognized that tooth wear is often limiting for lifespan in the wild or in captivity (Finch 1990) [e.g., in kangaroos (Breeden 1967), lions (Schaller 1972), chimpanzees (Zihlman et al. 1989), and elephants (Sikes 1971)]. On the other hand, the mammalian genome is obviously fully capable of tooth replacement, and in fact tooth replacement is continuous throughout life in the entire manatee genus (Domning 1983, 1987; Finch 1990) and in Australian rock wallabies (Domning 1983) and can occur up to six times in elephants (Sikes 1971) and up to two times in humans (triphyodonty). (It is also interesting to contemplate why, if the elephant can replace its teeth so many times and dies largely as a result of not replacing them still more times, evolution has chosen to allow these animals' lifespans to be so precisely limited by such an easily-remedied ailment.)

Cochlear hair cells are probably also lost for mechanical reasons with aging (Johnson and Hawkins 1972). However, they can evidently be regenerated by activating the Atoh-1 gene pathway (Izumikawa et al. 2005).

As far as we know, the lens can't be regenerated and can't be maintained indefinitely (although it can last at least 211 years in Bowhead whales; Section 6.2.3), and the same may be true for a few other structures such as cartilage (Furber 2010), although recent evidence indicates cartilage can be regenerated in humans using autologous adult mesenchymal stem cells (see www.regenexx.com). Fortunately, many structures that can't be naturally maintained can already be replaced artificially (lenses, hips, knees), and most of those that can't be replaced now may be regenerated or replaced in the future (Atala 2010; Ikeda et al. 2009; West 2010). The repair of neural systems may be possible using stem cell and other approaches, and a stem cell trial for spinal cord repair is expected to proceed in 2010.

Harris (personal communication) has speculated that the brain is the ultimate example of a structure that wears out because it can't be regenerated without losing that which gives it value, which is to say, the structures that encode particularly memory but probably other vital functions as well. However, little seems to be known about the turnover of identity-critical structures in the brain, and it is possible that just as cells persist and retain their identities despite turnover of all of their components, memories can persist despite turnover of all of the molecules and connections of which they are composed. Encouragingly, age-related structural changes in the brain can be prevented by certain forms of dietary choline (Bertoni-Freddari et al. 1985) and CR (Moroi-Fetters et al. 1989) and reversed by the use of growth factors such as nerve growth factor (Mervis et al. 1991) and drugs (Giuli et al. 1980; Zeng et al. 1995; Kiray et al. 2006).

The most formidable source of incompletely controllable aging presumably consists of presently irreversible accumulation of mutations and other changes in DNA. Sooner or later, if life continues long enough, nuclear DNA damage should eventually accumulate to the point where it will require correction (Lu et al. 2004; Niedernhofer et al. 2006). Fortunately, the feasibility of a variety of artificial DNA repair technologies, including in vivo block replacement of targeted DNA segments, has already been demonstrated (Kren et al. 1998; Woolf 1998). Block DNA replacement has the significant advantage of being able to correct unknown and diverse genetic defects within predefined critical regions of DNA. Kren et al. were able in 1998 to replace targeted sequences in up to 40% of the rat hepatic factor IX genes present in vivo using a repair construct that was delivered by injection into the tail vein (Kren et al. 1998). A detailed and updated discussion of the prospects for reversing aging-associated DNA damage and gene expression changes is provided by the same authors in this volume (Steer and Kren 2010).

In the human brain, it has been reported that age-related changes in gene expression affecting about 2% of human genes (around 400 genes) are causally linked to promoter damage (Lu et al. 2004) (although this result was obtained using post-mortem brains). In addition to the remedies described in Chapter 21, it will be interesting to watch the development of a method being investigated by George Church at Harvard Medical School (http://www.464.com/downloads/NYT_Mammotharticle.pdf) that might eventually help to enable the repair of 50,000 genetic sites at one time, or 100 times more sites than have been implicated in human brain aging. In the farther future, genetic damage should become correctable using medical nanotechnological approaches if necessary or desirable (Freitas 2010). Finally, if a significant cause of DNA damage is the withholding of DNA repair capacity (Sections 6.5.4 and 6.5.5.1; Figs. 6.19 and 6.20), upregulation of DNA repair should greatly postpone the development of symptomatic DNA damage.

6.7 Summary and Conclusions

This chapter has summarized some of the evidence that the genome contains resources for controlling aging that are not fully deployed under normal circumstances. In fact, such resources generally appear to be actively suppressed, and life-shortening mechanisms appear to be generally activated, with the combined effect of limiting the lifespans and healthspans of most species. This state of affairs seems well established and indicates that aging is ultimately under biological control to a much greater degree than has been classically believed or expected. From a practical point of view, this situation implies that many powerful and relatively feasible opportunities for targeted intervention into human aging are likely to exist.

The examples provided here and elsewhere establish that aging can be modulated by adjusting specific biochemical pathways. However, this does not mean that the effects of presently known interventions will all last indefinitely or will all be free of known or potential side effects. In fact, both fading effectiveness and side effects are known or likely problems of many otherwise effective interventions. However, similar difficulties have not prevented traditional pharmaceuticals from saving lives and alleviating diseases, and evolving experience generally makes it possible to minimize or avoid such problems over time.

In the case of aging interventions, some of the most important side effects may arise from inappropriately balanced actions on different pathways (such as promotion of cell division on the one hand and inhibition of DNA repair on the other), in which case a more balanced interventive approach and/or the development of specific means of activating or blocking particularly targeted pathways independently of one another so as to eliminate side effects while preserving the intended therapeutic effects should provide effective remedies. Tools that can enable specific repression of harmful pathways are already highly advanced and will surely become even more powerful in the relatively near future.

One postulated side effect, reduced fecundity arising from the biological incompatibility that tends to exist between long life and high fecundity, might actually be seen as being mostly an advantage overall for most people considering that overpopulation is the most compelling biological objection to human life extension. But in all likelihood, the link between aging and reproduction in humans can be whatever we wish it to be, thanks in part to the coming advent of stem cell and tissue engineering and regeneration technologies.

Perhaps the most important contraindication to aging interventions of many kinds is the specter of increased cancer risk. Fortunately, fundamental biological remedies (Cui 2010) or engineered remedies based on reversing the DNA damage underlying the life threatening features of cancer (Steer and Kren 2010) may be on the way, and in the meantime, there is a massive anticancer research machine in high gear that, over time, will surely continue to provide more and more satisfactory answers. In the unlikely event that all else fails, more extreme measures can also be envisioned (de Grey 2010; Freitas 2010).

Truly effective intervention into aging is likely to require a multiplicity of interventive strategies, and this will remain a significant challenge for some time. However, the fact that such strategies are likely to be possible in large part by the relatively clear path of inhibiting or augmenting biochemical pathways that already exist, including master pathways that appear to control the global rate and heterogeneity of aging, should be encouraging.

The view of aging described in this chapter, which might be termed a physiological theory of aging, is not a common one. Its central hypothesis, stated in practical and empirical rather than in theoretical terms, is that aging ultimately originates mostly from patterns of gene expression that can be modified so as to postpone, ameliorate, prevent, or reverse aging processes. No single current theory may be able to fully account for the degree of biological control of aging that is indicated by all of the phenomena reviewed here, including the universality of thymic atrophy, for example. However, a combination of developmental theories of aging, which hold that aging in the adult is the evolutionarily inevitable consequence of developmental programs that either continue or stop in the adult, contributing to aging in either case (Clark 2004; de Magalhaes and Church 2005) [see also (Budovskaya et al. 2008) and (Boehm and Slack 2005; Boehm and Slack 2006)] with other divergent evolutionary views of aging (Denckla 1975; Mitteldorf 2010; Rose et al. 2010) has promise for being able to do so.

Biological control over aging is indicated in this review by a large number of specific examples, each of which points to the same final overall conclusion in different ways. The overall conclusion should, therefore, remain valid even if flaws are found in any particular examples. I hope to be forgiven for and look forward to being informed of the latter. In the meantime, I hope that the unusual perspective provided here can be helpful to those who have far more specialized knowledge of the details of biogerontology than I.

Acknowledgements I thank J. Mitteldorf for helpful peer review of this chapter and Don Gennery for providing Fig. 6.3. I thank S. Spindler, S. Mullen, and K. Perrott for gracious assistance in obtaining obscure or elusive source material.

References

Abs R et al. (1999) GH replacement in 1034 growth hormone deficient hypopituitary adults: demographic and clinical characteristics, dosing and safety. Clin Endocrinol (Oxf) 50:703–713
Ailion M, Inoue T, Weaver CI, Holdcraft RW, Thomas JH (1999) Neurosecretory control of aging in *Caenorhabditis elegans*. Proc Nat Acad Sci 96:7394–7397
Alcain FJ, Minor RK, Villalba JM, de Cabo R (2010) Small molecule modulators of sirtuin activity. In: Fahy GM, West M, Coles LS, Harris SB (eds) The Future of Aging: Pathways to Human Life Extension. Springer, Berlin, Heidelberg, New York, pp. 331–356
Amdam GV, Aase AL, Seehuus S-C, Kim-Fondrk M, Norberg K, Hartfelder K (2005) Social reversal of immunosenescence in honey bee workers. Exp Gerontol 40:939–947
Andersen B et al. (1995) The Ames dwarf gene is required for Pit-1 gene activation. Develop Biol 172:495–503
Anderson GL (1978) Response of dauer larvae of *Caenorhabditis elegans* (nematode: Rhabditidae) to thermal stress and oxygen deprivation. Can J Zool 56:1786–1791
Anisimov VN, Semenchenko AV, Yashin AI (2003) Insulin and longevity: antidiabetic biguanides as geroprotectors. Biogerontology 4:297–307
Anson RM et al. (2003) Intermittent fasting dissociates beneficial effects of dietary restriction on glucose metabolism and neuronal resistance to injury from calorie intake. Proc Nat Acad Sci 100:6116–6220
Apfeld J, Kenyon C (1998) Cell nonautonomy of *C. elegans* DAF-2 function in the regulation of diapause and life span. Cell 95:199–210
Apfeld J, Kenyon C (1999) Regulation of lifespan by sensory perception in *Caenorhabditis elegans*. Nature 402:804–809
Araneo B, Dowell T, Woods ML, Daynes R, Judd M, Evans T (1995) DHEAS as an effective vaccine adjuvant in elderly humans. Proof-of-principle studies. Ann NY Acad Sci 774:232–248
Arantes-Oliveira N, Apfeld J, Dillin A, Kenyon C (2002) Regulation of lifespan by germ-line stem cells in *Caenorhabditis elegans*. Science 295:502–505
Arantes-Oliveira N, Berman J, Kenyon C (2003) Healthy animals with extreme longevity. Science 302:611
Arking DE, Atzmon G, Arking A, Barzilai N, Dietz HC (2005) Association between a functional variant of the KLOTHO gene and high-density lipoprotein cholesterol, blood pressure, stroke, and longevity. Circulation Res 96:412–418
Arking DE et al. (2003) KLOTHO allele status and the risk of early-onset occult coronary artery disease. Am J Human Genet 72:1154–1161
Arking DE et al. (2002) Association of human aging with a functional variant of klotho. Proc Nat Acad Sci 99:856–861
Asencot M, Lensky Y (1984) Juvenile hormone induction of "queenliness" on female honey bee (Apis mellifera L.) larvae reared on worker jelly and on stored royal jelly. Comparative Biochem Physiol 78B:109–117
Aspinall R, Mitchell W (2010) Maintenance and restoration of immune system function. In: Fahy GM, West M, Coles LS, Harris SB (eds) The Future of Aging: Pathways to Human Life Extension. Springer Science, Berlin, Heidelberg, New York, pp. 489–520
Atala A (2010) Life extension by tissue and organ replacement. In: Fahy GM, West M, Coles LS, Harris SB (eds) The Future of Aging: Pathways to Human Life Extension. Springer, Berlin, Heidelberg, New York, pp. 543–571

Atzmon G et al. (2006) Lipoprotein genotype and conserved pathway for exceptional longevity in humans. PLoS Biol 4:e113

Austad SN, Kirkwood TBL (2008) Evolutionary theory in aging research. In: Guarente LP, Partridge L, Wallace DC (eds) Molecular Biology of Aging. Cold Spring Harbor Laboratory Press, Cold Spring Harbor, pp. 95–111

Aviv A, Bogden JD (2010) Telomeres and the arithmetic of human longevity. In: Fahy GM, West M, Coles LS, Harris SB (eds) The Future of Aging: Pathways to Human Life Extension. Springer, Berlin, Heidelberg, New York, pp. 573–586

Ayyadevara S, Alla R, Thaden JJ, Shmookler Reis RJ (2008) Remarkable longevity and stress resistance of nematode PI3K-null mutants. Aging Cell 7:13–22

Barger JL, Kayo T, Pugh TD, Prolla TA, Weindruch R (2008) Short-term consumption of a resveratrol-containing nutraceutical mixture mimics gene expression of long-term caloric restriction in mouse heart. Exp Gerontol 43:859–866

Barker K, Beveridge I, Bradley AJ, Lee AK (1978) Observations on spontaneous stress-related mortality among males of the dasyurid marsupial Antechinus stuartii (MacLeay). Aust J Zool 26:435–447

Barnett JL (1973) A stress response in Antechinus stuartii (MacLeay). Aust J Zool 21:501–513

Bartke A (1979) Prolactin-deficient mice. In: Alexander NJ (ed) Animal Models for Research on Contraception and Fertility. Harper & Row, Hagerstown, pp. 360–365

Bartke A, Brown-Borg HM (2004) Life extension in the dwarf mouse. Curr Topics Develop Biol 63:189–225

Bartke A, Wright JC, Mattison JA, Ingram DK, Miller RA, Roth GS (2001) Extending the lifespan of long-lived mice. Nature 414:412

Bauer ME (2005) Stress, glucocorticoids and ageing of the immune system. Stress 8:69–83

Baumeister R, Schaffitzel E, Hertweck M (2006) Endocrine signaling in *Caenorhabditis elegans* controls stress response and longevity. J Endocrinol 190:191–202

Baur JA et al. (2006) Resveratrol improves health and survival of mice on a high-calorie diet. Nature 444:337–342

Baur JA, Sinclair DA (2006) Therapeutic potential of resveratrol: the in vivo evidence. Nat Rev Drug Discov 5:493–506

Baxter JD, Tyrrell JB (1987) The adrenal cortex. In: Felig P, Baxter JD, Boardus AE, Frohman LA (eds) Endocrinology and Metabolism, 2nd ed. McGraw-Hill Book Company, New York, pp. 511–650

Beck SD, Bharadwaj RK (1972) Reversed development and cellular aging in an insect. Science 178:1210–1211

Belchier M, Edsman L, Sheehy MRJ, Shelton PMJ (1998) Estimating age and growth of long-lived temperate freshwater crayfish using lipofuscin. Freshwater Biol 39:439–446

Bellino FL, Daynes RA, Hornsby PJ, Lavrin DH, Nestler JE (eds) (1995) Dehydroepiandrosterone (DHEA) and Aging. The New York Academy of Sciences, New York

Bergamini E, Cavallini G, Donati A, Gori Z (2003) The anti-ageing effects of caloric restriction may involve stimulation of macroautophagy and lysosomal degradation, and can be intensified pharmacologically. Biomed Pharmacother 57:203–208

Bergamini E, Cavallini G, Donati A, Gori Z (2004) The role of macroautophagy in the ageing process, anti-ageing intervention and age-associated diseases. Int J Biochem Cell Biol 36:2392–2404

Bergman A, Atzmon G, Ye K, MacCarthy T, Barzilai N (2007) Buffering mechanisms in aging: a systems approach toward uncovering the genetic component of aging. PLoS Comput Biol 3:e170

Bergquist DC, Williams FM, Fisher CR (2000) Longevity record for deep-sea invertebrate. Nature 403:499–500

Berman JR, Kenyon C (2006) Germ-cell loss extends *C. elegans* life span through regulation of DAF-16 by kri-1 and lipophilic-hormone signaling. Cell 124:1055–1068

Bertoni-Freddari C, Mervis RF, Giuli C, Pieri C (1985) Chronic dietary choline modulates synaptic plasticity in the cerebellar glomeruli of aging mice. Mechanisms of Ageing Develop 30:1–9

Beyer RE et al. (1985) Tissue Coenzyme Q (uniquinone) and protein concentration over the life span of the laboratory rat. Mechanisms of Ageing Develop 32:267–281

Bhalla P, Nehru B (2005) Modulatory effects of centrophenoxine on different regions of ageing rat brain. Exp Gerontol 40:801–806

Biddle FG, Eden SA, Rossler JS, Eales BA (1997) Sex and death in the mouse: genetically delayed reproduction and senescence. Genome 40:229–235

Bishop NA, Guarente L (2007) Two neurons mediate diet-restriction-induced longevity in *C. elegans*. Nature 447:545–549

Blackman MR, Sorkin JD, Munzer T et al. (2002) Growth hormone and sex steroid administration in healthy aged women and men: a randomized controlled trial. J Am Med Assoc 288:2282–2292

Blazejowski CA, Webster GC (1983) Decreased rates of protein synthesis by cell-free preparations from different organs of aging mice. Mechanisms of Ageing Develop 21:345–356

Bliznakov EG (1978) Coenzyme Q deficiency in aged mice. J Med 9:337–346

Bliznakov EG (1981) Coenzyme Q, the immune system and aging. In: Folkers K, Yamamura Y (eds) Biomedical and Clinical Aspects of Coenzyme Q. Elsevier/North-Holland, Amsterdam, pp. 311–323

Bluher M, Kahn BB, Kahn CR (2003) Extended longevity in mice lacking the insulin receptor in adipose tissue. Science 299:572–574

Bluher M et al. (2002) Adipose tissue selective insulin receptor knockout protects against obesity and obesity-related glucose intolerance. Develop Cell 3:25–38

Bodnar AG et al. (1998) Extension of life-span by introduction of telomerase into normal human cells. Science 279:349–352

Boehm M, Slack F (2005) A developmental timing microRNA and its target regulate life span in *C. elegans*. Science 310:1954–1957

Boehm M, Slack FJ (2006) MicroRNA control of lifespan and metabolism. Cell Cycle 5:837–840

Bokov A, Chaudhuri A, Richardson A (2004) The role of oxidative damage and stress in aging. Mechanisms of Ageing Develop 125:811–826

Borowsky R (1978) Social inhibition of maturation in natural populations of *Xiphophorus variatus* (Pices: Poeciliidae). Science 201:933–935

Boulianne GL (2001) Neuronal regulation of lifespan: clues from flies and worms. Mech Ageing Develop 122:883–894

Brack AS et al. (2007) Increased Wnt signaling during aging alters muscle stem cell fate and increases fibrosis. Science 317:807–810

Bradley A, McDonald IR, Lee AK (1980) Stress and mortality in a small marsupial (*Antechinus stuartii*, Macleay). General and Comparative Endocrinol 40:188–200

Bredesen DE (2004) The non-existent aging program: how does it work? Aging Cell 3:255–259

Breeden S (1967) The Life of the Kangaroo. Taplinger, New York

Brizzee KR, Ordy JM, Kaach B (1974) Early appearance and regional differences in intraneuronal and extraneuronal lipofuscin accumulation with age in the brain of a nonhuman primate (*Macaca mullata*). J Gerontol 29:366–381

Brock MA, Strehler BL (1963) Studies on the comparative physiology of aging. 4. Age and mortality of some marine Cnidaria in the laboratory. J Gerontol 18:23–28

Brown-Borg HM, Borg KE, Meliska CJ, Bartke A (1996) Dwarf mice and the aging process. Nature 384:33

Brown-Borg HM, Johnson WT, Rakoczy S, Romanick M (2001) Mitochondrial oxidant generation and oxidative damage in Ames dwarf and GH transgenic mice. J Am Aging Assoc 85:85–100

Brown-Borg HM, Rakoczy SG (2000) Catalase expression in delayed and premature aging mouse models. Exp Gerontol 35:199–212

Bruder CEG et al. (2008) Phenotypically concordant and discordant monozygotic twins display different DNA copy-number-variation profiles. Am J Human Genet 82:763–771

Budovskaya YV, Wu K, Southworth LK, Jiang M, Tedesco P, Johnson T (2008) An elt-3/elt-5/elt-6 GATA transcription circuit guides aging in *C. elegans*. Cell 134:291–303

Buffenstein R (2008) Negligible senescence in the longest-living rodent, the naked mole-rat: insights from a successfully aging species. J Comparative Physiol [B] 178:439–445

Buford TW, Willoughby DS (2008) Impact of DHEA(S) and cortisol on immune function in aging: a brief review. Appl Physiol Nutr Metab 33:429–433

Busbee DL, Sylvia VL, Norman JO et al. (1989) Age-related differences in DNA polymerase alpha specific activity: potential for interaction in DNA repair. In: Wang E, Warner HR (eds) Growth Control During Cell Aging. CRC Press, Boca Raton, pp. 65–87

Busbee DL, Sylvia VL, Stec J, Cernosek Z, Norman JO (1987) Lability of DNA polymerase alpha correlated with decreased DNA synthesis and increased age in human cells. J Natl Cancer Inst 79:1231–1239

Cabelof DC, Yanamadala S, Raffoul JJ, Guo Z, Soofi A, Heydari AR (2003) Caloric restriction promotes genomic stability by induction of base excision repair and reversal of its age-related decline. DNA Repair (Amst) 2:295–307

Cardarlli NF (1989) Involution, age, and puberty. In: Cardarelli NF (ed) The Thymus in Health and Senescence, Volume 1, Thymus and Immunity. CRC Press, Boca Raton, pp. 51–57

Carey JR, Liedo P, Orozco D, Vaupel JW (1992) Slowing of mortality rates at older ages in large medfly cohorts. Science 258:457–461

Carey JR, Liedo P, Vaupel JW (1995) Mortality dynamics of density in the Mediterranean fruit fly. Exp Gerontol 30:605–629

Carlson ME, Conboy MJ, Hsu M, Barchas L, Jeong J, Agrawal A, Mikels AJ, Agrawal S, Schaffer DV, Conboy IM (2009a) Relative roles of TGF-beta1 and Wnt in the systemic regulation and aging of satellite cell responses. Aging Cell 8:676–689

Carlson ME, Suetta C, Conboy MJ, Aagaard P, Mackey A, Kjaer M, Conboy I (2009b) Molecular aging and rejuvenation of human muscle stem cells. EMBO Molecular Med 1:381–391

Carrillo MC, Kanai S, Nokubo M, Ivy GO, Sato Y, Kitani K (1992) (-)Deprenyl increases activities of superoxide dismutase and catalase in striatum but not in hippocampus: the sex and age-related differences in the optimal dose in the rat. Exp Neurol 116:286–294

Castaneda M, Vargas R, Galvan SC (1986) Stagewise decline in the activity of brain protein synthesis factors and relationship between this decline and longevity in two rodent species. Mech Ageing Develop 36:197–210

Cavallini G, Donati A, Taddei M, Bergamini E (2007) Evidence for selective mitochondrial autophagy and failure in aging. Autophagy 3:26–27

Cavanagh JB, Nolan CC, Seville MP, Anderson VER, Leigh PN (1993) Routes of excretion of neuronal lysosomal dense bodies after ventricular infusion of leupeptin in the rat: a study using ubiquitin and PGP 9.5 immunocytochemistry. J Neurocytol 22:779–791

Cawthon RM, Smith KR, O'Brien E, al e (2003) Association between telomere length in blood and mortality in people aged 60 years or older. Lancet 361:393–395

Chan JM et al. (1998) Plasma insulin growth factor-1 and prostate cancer risk: a prospective study. Science 279:563–566

Chen D et al. (2002) Age-dependent decline of DNA repair activity for oxidative lesions in rat brain mitochondria. J Neurochem 81:1273–1284

Chen D, Minami M, Henshall DC, Meller R, Kisby G, Simon RP (2003) Upregulation of mitochondrial base-excision repair capability within rat brain after brief ischemia. J Cereb Blood Flow Metab 23:88–98

Cheng TC, Beamer WG, Phillip JA, Bartke A, Mallonee RL, Dowling C (1983) Etiology of growth hormone deficiency in little, Ames, and Snell dwarf mice. Endocrinology 113:1669–1678

Chiacchiera F, Simone C (2010) The AMPK-FoxO3A axis as a target for cancer treatment. Cell Cycle 9: (in press)

Choi IS, Cui Y, Koh Y-A, Lee H-C, Cho Y-B, Won Y-H (2008) Effects of dehydroepiandrosterone on Th2 cytokine production in peripheral blood mononuclear cells from asthmatics. Korean J Int Med 23:176–181

Christensen K, Johnson T, Vaupel JW (2006) The quest for genetic determinants of human longevity: challenges and insights. Nat Rev Genet 7:436–448

Chuang JC, Jones PA (2007) Epigenetics and microRNAs. Pediatric Res 61(5 Pt 2):24R–29R
Clark WR (2004) Reflections on an unsolved problem of biology: the evolution of senescence and death. Adv Gerontol 14:7–20
Cockburn A, Lee AK, Martin RW (1983) Macrogeographic variation in litter size in Antechinus (*Marsupialia: Dasyuridae*). Evolution 37:86–95
Coles LS, Harris SB (1996) Coenzyme Q-10 and lifespan extension. In: Klatz RM (ed) Advances in Anti-Aging Medicine. Liebert, New York, pp. 205–215
Comfort A (1979) The Biology of Senescence, 3rd ed. Elsevier, New York
Conboy IM, Conboy MJ, Smythe GM, Rando TA (2003) Notch-mediated restoration of regenerative potential to aged muscle. Science 302:1575–1577
Conboy IM, Conboy MJ, Wagers AJ, Girma ER, Weissman IL, Rando TA (2005) Rejuvenation of aged progenitor cells by exposure to a young systemic environment. Nature 433:760–764
Congdon JD, Nagle RD, Kinney OM, van Loben Sells RC, Quinter T, Tinkle DW (2003) Testing hypotheses of aging in long-lived painted turtles (*Chrysemys picta*). Exp Gerontol 38:765–772
Congdon JD, Nagle RD, Kinney OM, Van Loben Sels RC (2001) Hypotheses of aging in a long-lived vertebrate, Blanding's Turtle (*Emydoidea blandingii*). Exp Gerontol 36:813–828
Cooper RA, Uzmann JR (1977) Ecology of juvenile and adult clawed lobsters, Homarus americanus, Homarus gammarus, and Nephrops norvegicus. Circ-CSIRO Div Fish Oceanog 187–208
Corona M et al. (2007) Vitellogenin, juvenile hormone, insulin signaling, and queen honey bee longevity. Proc Nat Acad Sci 104:7128–7133
Corsini E et al. (2005) Age-related decline in RACK-1 expression in human leukocytes is correlated to plasma levels of dehydroepiandrosterone. J Leukocyte Biol 77:247–256
Cortina MS, Gordon WC, Lukiw WJ, Bazan NG (2005) Oxidative stress-induced retinal damage up-regulates DNA polymerase gamma and 8-oxoguanine-DNA-glycosylase in photoreceptor synaptic mitochondria. Exp Eye Res 81:742–750
Coschigano KT, Clemmons D, Bellush LL, Kopchick JJ (2000) Assessment of growth parameters and lifespan of GHR/BP gene-disrupted mice. Endocrinology 141:2608–2613
Coschigano KT, Holland AN, Riders ME, List EO, Flyvbjerg A, Kopchick JJ (2003) Deletion, but not antagonism, of the mouse growth hormone receptor results in severely decreased body weights, insulin, and insulin-like growth factor I levels and increased life span. Endocrinology 144:3799–3810
Cotzias GC, Miller ST, Nicholson AR Jr et al. (1974) Prolongation of the life-span in mice adapted to large amounts of L-Dopa. Proc Nat Acad Sci 71:2466–2469
Cotzias GC, Miller ST, Tang LC et al. (1977) Levodopa, fertility, and longevity. Science 196: 549–550
Cui Z (2010) Near term prospects for broad spectrum amelioration of cancer. In: Fahy GM, West M, Coles LS, Harris SB (eds) The Future of Aging: Pathways to Human Life Extension. Springer, Berlin, Heidelberg, New York, pp. 307–329
Curtsinger JW (1995) Density and age-specific mortality. Genetica 96:179–182
Cypser JR, Tedesco P, Wu D, Park S-K, Johnson T (2008) Secondary phenotypes associated with a GFP reporter that predicts increased survival in *C. elegans*. 37th Annual Meeting of the American Aging Association, Abstract 71
Dammann P, Burda H (2006a) Sexual activity and reproduction delay ageing in a mammal. Curr Biol 16:R117–R118
Dammann P, Burda H (2006b) Sexual activity and reproduction delay ageing in a mammal. Curr Biol 16:R117–R188
de Grey A (2010) WILT: Necessity, feasibility, affordability. In: Fahy GM, West M, Coles LS, Harris SB (eds) The Future of Aging: Pathways to Human Life Extension. Springer, Berlin, Heidelberg, New York, pp. 667–684
de Grey ADNJ (2003) Critique of the demographic evidence for 'late-life non-senescence'. Biochem Soc Trans 31:452–454

de la Cruz CP, Revilla E, Rodriguez-Gomez JA, Vizuete ML, Cano J, Machado A (1997) (-)-Deprenyl treatment restores serum insulin-like growth factor-I (IGF-I) levels in aged rats to young rat level. Eur J Pharmacol 327:215–220

de la Rosa O et al. (2006) Immunological biomarkers of ageing in man: changes in both innate and adaptive immunity are associated with health and longevity. Biogerontology 7:471–481

de Magalhaes JP, Church GM (2005) Genomes optimize reproduction: aging as a consequence of the developmental program. Physiology 20:252–259

de Souza-Pinto NC et al. (2001) Repair of 8-oxodeoxyguanosine lesions in mitochondrial DNA depends on the oxoguanine DNA glycosylase (OGG1) gene and 8-oxoguanine accumulates in the mitochondrial DNA of OGG1-defective mice. Cancer Res 61:5378–5381

Denckla WD (1975) A time to die. Life Science 16:31–44

Deyl Z, Juricova M, Stuchlikova E (1975) The effect of different nutritional regimes upon collagen concentration and survival of rats. Adv Exp Med Biol 53:359–369

Dhahbi JM, Kim HJ, Mote PL, Beaver RJ, Spindler SR (2004) Temporal linkage between the phenotypic and genomic responses to caloric restriction. Proc Nat Acad Sci 101:5524–5529

Dhahbi JM, Mote PL, Fahy GM, Spindler SR (2005) Identification of potential calorie restriction mimetics by microarray profiling. Physiol Genomics 23:343–350

Dhahbi JM, Tsuchiya T, Kim HJ, Mote PL, Spindler SR (2006) Gene expression and physiologic responses of the heart to the initiation and withdrawal of caloric restriction. J Gerontol A Biol Sci Med Sci 61:218–231

Diamond JM (1982) Big-bang reproduction and aging in male marsupial mice. Nature 298: 115–116

Dietz A, Hermann HR, Blum MS (1979) The role of exogenous JH I, JH III, and anti-JH (precocene II) on queen induction of 4.5-day-old worker honey bee larvae. J Insect Physiol 25: 503–512

Dillin A et al. (2002) Rates of behavior and aging specified by mitochondrial function during development. Science 298:2398–2401

Dilman VM, Anisimov VN (1980) Effect of treatment with phenformin, diphenylhydantoin or L-DOPA on life span and tumor incidence in C3H/Sn mice. Gerontology 26:241–246

Domning DP (1983) Marching teeth of the manatee. Nat History 92:8–10

Domning DP (1987) Sea cow family reunion. Nat History 96:64–71

Donati A et al. (2008) In vivo effect of an antilipolytic drug (3,5′-dimethylpyrazole) on autophagic proteolysis and autophagy-related gene expression in rat liver. Biochem Biophys Res Commun 366:786–792

Donisthorpe H (1936) The oldest insect on record. Entomol Record J Variation 48:1–2

Druffel ERM et al. (1995) Gerardia: Bristlecone pine of the deep sea? Geochimica et Cosmochimica Acta 59:5031–5036

Dulloo AG, Gubler M, Montani JP, Seydoux J, Solinas G (2004) Substrate cycling between de novo lipogenesis and lipid oxidation: a thermogenic mechanism against skeletal muscle lipotoxicity and glucolipotoxicity. Int J Obesity-Related Metabolic Disorders 28:S29-S37

Easton DM (1997) Gompertz growth in number dead confirms medflies and nematodes show excess oldster survival. Exp Gerontol 32:719–726

Ebert TA (2008) Longevity and lack of senescence in the red sea urchin Strongylocentrotus franciscanus. Exp Gerontol 43:734–738

El-Ghazzawi E, Malaty HA (1975) Electron microscopic observations on extraneural lipofuscin in the monkey brain. Cell and Tissue Res 161:555–565

Elowitz MB, Levine AJ, Siggia ED, Swain PS (2002) Stochastic gene expression in a single cell. Science 297:1183–1186

Essers MA, de Vries-Smits LM, Barker N, Polderman PE, Burgering BM, Korswagen HC (2005) Functional interaction between beta-catechin and FOXO in oxidative stress signaling. Science 308:1181–1184

Ettienne G, Cony-Makhoul P, Mahon F-X (2002) Imatinib mesylate and gray hair. NE J Med 347:446

Evans GW, Meyer LK (1994) Life span is increased in rats supplemented with a chromium-pyridine 2 carboxylate complex. Adv Sci Res 1:19–23

Everitt A (1959) The effect of pituitary growth hormone on the aging male rat. J Gerontol 14: 415–424

Fabris N, Mocchegiani E, Mariotti M, Pacini F, Pinchera A (1986) Thyroid function modulates thymic endocrine activity. J Clin Endocrinol Metab 62:474–478

Fabris N, Mocchegiani E, Muzzioli M, al e (1988) Neuroendocrine-thymus interactions: perspectives for intervention in aging. Annals NY Acad Sci 521:72–87

Fabris N, Mocchegiani E, Muzzioli M, al. e (1990) Zinc, immunity, and aging. In: Goldstein AL (ed) Biomedical Advances in Aging. Plenum, New York, pp. 271–281

Fabris N, Mocchegiani E, Provinciali M (1997) Plasticity of neuroendocrine-thymus interactions during aging. Exp Gerontol 32:415–429

Fabrizio P et al. (2004) Superoxide is a mediator of an altruistic aging program in Saccharomyces cerevisiae. J Cell Biol 166:1055–1067

Fahy GM (1985) Anti-aging conference report – part I. Anti-Aging News 5:1–5

Fahy GM (2001) Growth hormone therapy and related methods and pharmaceutical compositions. In: US Patent 6,297,212 B1

Fahy GM (2003) Apparent induction of partial thymic regeneration in a normal human subject: a case report. J Anti-Aging Med 6:219–227

Fedoroff N, Fontana W (2002) Small numbers of big molecules. Science 297:1129–1131

Fenton DA, English JS, Wilkinson JD (1982) Reversal of male-pattern baldness, hypertrichosis, and accelerated hair and nail growth in patients receiving benoxaprofen. Br Med J 284: 1228–1229

Filipowicz W, Bhattacharyya SN, Sonenberg N (2008) Mechanisms of post-transcriptional regulation by microRNAs: are the answers in sight? Nat Rev Genet 9:102–114

Finch CE (1990) Longevity, Senescence, and the Genome. The University of Chicago Press, Chicago

Flatt T et al. (2008) Drosophila germ-line modulation of insulin signaling and lifespan. Proc Nat Acad Sci 105:6368–6373

Flatt T, Tu MP, Tatar M (2005) Hormonal pleiotropy and the juvenile hormone regulation of Drosophila development and life history. Bioessays 27:999–1010

Flurkey K, Papaconstantinou J, Miller RA, Harrison DE (2001) Lifespan extension and delayed immune and collagen aging in mutant mice with defects in growth hormone production. Proc Nat Acad Sci 98:6736–6741

Fonseca DB, Sheehy MR, Blackman N, Shelton PM, Prior AE (2005) Reversal of a hallmark of brain ageing: lipofuscin accumulation. Neurobiol Aging 26:69–76

Fonseca DB, Sheehy MRJ, Shelton PMJ (2003) Unilateral eyestalk ablation reduces neurolipofuscin accumulation rate in the contra-lateral eyestalk of a crustacean Pacifastacus leniusculus. J Exp Mar Biol Ecol 289:277–286

Foster RB (1977) Tachigalia versicolor is a suicidal neotropical tree. Nature 268:624–626

Fraga CG, Shigenaga MK, Park J-W, Degan P, Ames BN (1990) Oxidative damage to DNA during aging: 8-hydroxy-2′-deoxyguanosine in rat organ DNA and urine. Proc Nat Acad Sci 87: 4533–4537

Fraga MF et al. (2005) Epigenetic differences arise during the lifetime of monozygotic twins. Proc Nat Acad Sci 102:10604–10609

Freitas R, Jr (2010) Comprehensive nanorobotic control of human morbidity and aging. In: Fahy GM, West M, Coles LS, Harris SB (eds) The Future of Aging: Pathways to Human Life Extension. Springer, Berlin, Heidelberg, New York, pp. 685–805

Fry M, Loeb LA (1986) Animal Cell DNA Polymerases. CRC Press, Boca Raton

Furber JD (2010) Repairing extracellular aging and glycation. In: Fahy GM, West M, Coles LS, Harris SB (eds) The Future of Aging: Pathways to Human Life Extension. Springer, Berlin, Heidelberg, New York, pp. 587–621

Gadaleta MN et al. (1994) Mitochondrial DNA transcription and translation in aged rat. Annals NY Acad Sci 717:150–160

Gadaleta MN, Petruzzella V, Renis M, Fracasso F, Cantatore P (1990) Reduced transcription of mitochondrial DNA in the senescent rat. Tissue dependence and effect of L-carnitine. Eur J Biochem 187:501–506

Gami MS, Wolkow CA (2006) Studies of *Caenorhabditis elegans* DAF-2/insulin signaling reveal targets for pharmacological manipulation of lifespan. Aging Cell 5:31–37

Gavrilov LA, Gavrilova NS (1991) The Biology of Lifespan: A Quantitative Approach. Harwood Academic Publishers, New York

George JC et al. (1999) Age and growth estimates of bowhead whales (Balaena mysticetus) via aspartic acid racemization. Can J Zool 77:571–580

Gershon D (1979) Current status of age altered enzymes: alternative mechanisms. Mech Ageing Dev 9:189–196

Gibbons JW (1987) Why do turtles live so long? BioScience 37:262–269

Gilleard JS, McGhee JD (2001) Activation of hypodermal differentiation in the *Caenorhabditis elegans* embryo by GATA transcription factors ELT-1 and ELT-3. Mol Cell Biol 21:2533–2544

Giuli C, Bertoni-Freddari C, Pieri C (1980) Morphometric studies on synapses of the cerebellar glomerulus: the effect of centrophenoxine treatment in old rats. Mech Ageing Develop 14:265–271

Glees P, Spoerri PE (1975) Centrophenoxin-induced dissolution and removal of lipofuscin. An electron microscopic study. Arzneimittelforschung 25:1543–1548

Godfrey P, Rahal JO, Beamer WG, Copeland NG, Jenkins NA, Mayo KE (1993) GHRH receptor of little mice contains a missense mutation in the extracellular domain that disrupts receptor function. Nat Genet 4:227–232

Goff BL, Roth JA, Arp LH, al. e (1987) Growth hormone treatment stimulates thymulin production in aged dogs. Clin Exp Immunol 68:580–587

Golembiowska K, Dziubina A, Kowalska M, Kaminska K (2008) Paradoxical effects of adenosine receptor ligands on hydroxyl radical generation by L-DOPA in the rat striatum. Pharmacol Rep 60:319–330

Goodrick CL, Ingram DK, Reynolds MA, Freeman JR, Cider N (1990) Effects of intermittent feeding upon body weight and lifespan in inbred mice: interaction of genotype and age. Mech Ageing Develop 55:69–87

Goto S, Takahashi R, Araki S, Nakamoto H (2002) Dietary restriction initiated in late adulthood can reverse age-related alterations of protein and protein metabolism. Annals NY Acad Sci 959:50–56

Goto S et al. (2001) Implications of protein degradation in aging. Annals NY Acad Sci 928:54–64

Goto S, Takahashi R, Radak Z, Sharma R (2007) Beneficial biochemical outcomes of late-onset dietary restriction in rodents. Annals NY Acad Sci 1100:431–441

Greenstein BD, de Bridges EF, Fitzpatrick FTA (1992) Aromatase inhibitors regenerate the thymus in aging male rats. Int J Immunopharmacol 14:541–553

Greenstein BD, Fitzpatrick FT, Kendall MD, Wheeler MJ (1987) Regeneration of the thymus in old male rats treated with a stable analogue of LHRH. J Endocrinol 112:345–350

Greenwood M, Irwin JO (1939) Biostatistics of senility. Human Biol 11:1–23

Growth Hormone Research Society (2001) Critical evaluation of the safety of recombinant human growth hormone administration: Statement from the Growth Hormone Research Society. J Clin Endrocrinol Metab 86:1868–1870

Gruneberg H (1952) The Genetics of the Mouse. Martinus Nijhoff, The Hague

Guo Z, Heydari A, Richardson A (1998) Nucleotide excision repair of actively transcribed versus nontranscribed DNA in rat hepatocytes: effect of age and dietary restriction. Exp Cell Res 245:228–238

Hadrup SR et al. (2006) Longitudinal studies of clonally expanded CD8 T cells reveal a repertoire shrinkage predicting mortality and an increased number of dysfunctional cytomegalovirus-specific T cells in the very elderly. J Immunol 176:2645–2653

Hakimi P et al. (2007) Overexpression of the cytosolic form of phosphoenolpyruvate carboxykinase (GTP) in skeletal muscle repatterns energy metabolism in the mouse. J Biol Chem 282:32844–32855

Hamilton JB, Hamilton RS, Mestler GE (1969) Duration of life in domestic cats: influence of sex, gonadectomy, and inbreeding. J Gerontol 24:427–437

Hamilton JB, Mestler GF (1969) Mortality and survival: comparison of eunuchs with intact men and women in a mentally retarded population. J Gerontol 24:395–411

Hampl R, Starka L, Jansky L (2006) Steroids and thermogenesis. Physiol Res 55:123–131

Hansen M, Chandra A, Mitic LL, Onken B, Driscoll M, Kenyon C (2008) A role for autophagy in the extension of lifespan by dietary restriction in C. elegans. PLoS Genet 4:e24 (21–14)

Hansen M, Taubert S, Crawford D, Libina N, Lee S-J, Kenyon C (2007) Lifespan extension by conditions that inhibit translation in Caenorhabditis elegans. Aging Cell 6:95–110

Hanson RW, Hakimi P (2008) Born to run; the story of the PEPCK-C (mus) mouse. Biochimie 90:838–842

Hardie DG (2007) AMP-activated/SNF1 protein kinases: conserved guardians of cellular energy. Nat Rev Mol Cell Biol 8

Harper JL (1977) Population Biology of Plants. Academic Press, New York

Harrison DE et al. (2009) Rapamycin fed late in life extends lifespan in genetically heterogeneous mice. Nature 460:392–395

Harshman LG (1999) Investigation of the endocrine system in extended longevity lines of Drosophila melanogaster. Exp Gerontol 34:997–1006

Haskins CP (1960) Note on the natural longevity of fertile females of *Aphaenogaster picea*. NY Entomol Soc 58:66–67

Hauck SJ, Bartke A (2001) Effects of growth hormone on hypothalamic catalase and Cu/Zn superoxide dismutase. Free Radicals Biol Med 28:970–978

Hayakawa K et al. (1992) Intrapair differences of physical aging and longevity in identical twins. Acta Genet Med Gemellol (Roma) 41:177–185

Hayashi J-I et al. (1994) Nuclear but not mitochondrial genome involvement in human age-related mitochondrial dysfunction: functional integrity of mitochondrial DNA from aged subjects. J Biol Chem 269:6878–6883

Herman WS, Tatar M (2001) Juvenile hormone regulation of longevity in the migratory monarch butterfly. Proc Royal Soc London B 268:2509–2514

Herndon LA et al. (2002) Stochastic and genetic factors influence tissue-specific decline in ageing C. elegans. Nature 419:808–814

Heward CB (2010) An approach to extending human lifespan today. In: Fahy GM, West M, Coles LS, Harris SB (eds) The Future of Aging: Pathways to Human Life Extension. Springer, Berlin, Heidelberg, New York

Hirokawa K, Utsuyama M (1984) The effect of sequential multiple grafting of syngeneic newborn thymus on the immune functions and life expectancy of aging mice. Mech Ageing Develop 28:111–121

Hjelmborg J, Iachine I, Skytthe A, Vaupel JW, McGue M, al. e (2006) Genetic influence on human lifespan and longevity. Human Genet 119:312–321

Hochschild R (1971) Effect of membrane stabilizing drugs on mortality in Drosophila melanogaster. Exp Gerontol 6:133–151

Hochschild R (1973a) Effect of dimethylaminoethanol on the life span of senile male A-J mice. Exp Gerontol 8:185–191

Hochschild R (1973b) Effect of dimethylaminoethyl p-chlorophenoxyacetate on the life span of male Swiss Webster Albino mice. Exp Gerontol 8:177–183

Hodkova M (2008) Tissue signaling pathways in the regulation of life-span and reproduction in females of the linden bug, *Pyrrhocoris apterus*. J Insect Physiol 54:508–517

Holden C (1987) Why do women live longer than men? Science 238:158–160

Holliday R (2006) Epigenetics: a historical overview. Epigenetics 1:76–80

Holtzenberger M et al. (2003) IGF-1 receptor regulates lifespan and resistance to oxidative stress in mice. Nature 421:182–187

Honda Y, Honda S (1999) The daf-2 gene network for longevity regulates oxidative stress resistance and Mn-superoxide dismutase gene expression in *Caenorhabditis elegans*. FASEB J 13:1385–1393

Houthoofd K et al. (2002) Ageing is reversed, and metabolism is reset to young levels in recovering dauer larvae of *C. elegans*. Exp Gerontol 37:1015–1021

Howitz KT et al. (2003) Small molecule activators of sirtuins extend *Saccharomyces cerevisiae* lifespan. Nature 425:191–196

Hsin H, Kenyon C (1999) Signals from the reproductive system regulate the lifespan of *C. elegans*. Nature 399:308–309

Huang H, Tindall DJ (2007) Dynamic FoxO transcription factors. J Cell Sci 120:2479–2487

Huang HH, Marshall S, Meites J (1976) Induction of estrous cycles in old non-cyclic rats by progesterone, ACTH, ether stress, or L-dopa. Neuroendocrinology 20:21–34

Hudson G, Chinnery PF (2006) Mitochondrial DNA polymerase-gamma and human disease. Hum Mol Genet 15 Spec No 2:R244-R252

Hunter WS, Croson WB, Bartke A, Gentry MV, Meliska CJ (1999) Low body temperature in long-lived Ames dwarf mice at rest and during stress. Physiol Behav 67:433–437

Hwangbo DS, Gersham B, Tu MP, Palmer M, Tatar M (2004) Drosophila dFOXO controls lifespan and regulates insulin signalling in brain and fat body. Nature 429:562–566

Hyun M, Lee J, Lee K, May A, Bohr VA, Ahn B (2008) Longevity and resistance to stress correlate with DNA repair capacity in *Caenorhabditis elegans*. Nucleic Acids Res 36:1380–1389

Ikeda E et al. (2009) Fully functional bioengineered tooth replacement as an organ replacement therapy. PNAS 106:13475–13480

Ikeno Y, Bronson RT, Hubbard GB, Lee S, Bartke A (2003) Delayed occurrence of fatal neoplastic diseases in Ames dwarf mice: correlation to extended longevity. J Gerontol A Biol Sci Med Sci 58:B5–B12

Iliadou A, Lichtenstein P, de Faire U, Pedersen NL (2001) Variation in genetic and environmental influences in serum lipid and apolipoprotein levels across the lifespan in Swedish male and female twins. Am J Med Genet 102:48–58

Iossa S, Mollica MP, Lionetti L, Crescenzo R, Tasso R, Liverini G (2004) A possible link between skeletal muscle mitochondrial efficiency and age-induced insulin resistance. Diabetes 53:2861–2866

Ivy GO, Roopsingh R, Kanai S, Ohta M, Sato Y, Kitani K (1996) Leupeptin causes an accumulation of lipofuscin-like substances and other signs of aging in kidneys of young rats: further evidence for the protease inhibitor model of aging. Annals NY Acad Sci 786:12–23

Ivy GO, Schottler F, Wenzel J, Baudry M, Lynch G (1984) Inhibitors of lysosomal enzymes: accumulation of lipofuscin-like dense bodies in the brain. Science 226:985–987

Izumikawa M et al. (2005) Auditory hair cell replacement and hearing improvement by Atoh1 gene therapy in deaf mammals. Nat Med 11:271–276

Jackson D, Leys SP, Hinman VF, Woods R, Lavin MF, Degnan BM (2002) Ecological regulation of development: induction of marine invertebrate metamorphosis. Int J Develop Biol 46:679–686

Janzen DH (1976) Why bamboos wait so long to flower. Ann Rev Ecol Syst 7:347–391

Johannsson G, Jørgensen J (2001) Safety aspects of growth hormone replacement in adults. Growth Horm IGF Res 11:59–71

Johnson L, Hawkins JE, Jr (1972) Sensory and neuronal degeneration with aging, as seen in microdissections of the human inner ear. Ann Otol 81:178–183

Johnson T, Mitchell D, Kline S, Kemal R, Foy J (1984) Arresting development arrests aging in the nematode *Caenorhabditis elegans*. Mech Ageing Develop 28:23–40

Jones DS (1983) Sclerochronology: Reading the record of the molluscan shell. Am Sci 71:384–391

Judd R, Bueso-Ramos C (1990) Combined true thymic hyperplasia and lymphoid hyperplasia in Grave's disease. Pediatric Pathol 10:829–836

Kaeberlein M, Jegalian B, McVey M (2002) AGEID: a database of aging genes and interventions. Mech Ageing Develop 123:1115–1119

Kaeberlein M, Kennedy BK (2009) A midlife longevity drug? Nature 460:331–332

Kalimi M, Regelson W (eds) (1990) The Biologic Role of Dehydroepiandrosterone (DHEA). Walter de Gruyter, New York

Kalu DN et al. (1998) Aged-rodent models of long-term growth hormone therapy: lack of deleterious effect on longevity. J Gerontol A Biol Sci Med Sci 53:B452-B463

Kaminsky Z et al. (2008) Epigenetics of personality traits: an illustrative study of identical twins discordant for risk-taking behavior. Twin Res Hum Genet 11:1–11

Kaminsky ZA et al. (2009) DNA methylation profiles in monozygotic and dizygotic twins. Nat Genet 41:240–245

Kato M, Slack FJ (2008) microRNAs: small molecules with big roles – *C. elegans* to human cancer. Biol Cell 100:71–81

Katz ML, Rice LM, Gao CL (1999) Reversible accumulation of lipofuscin-like inclusions in the retinal pigment epithelium. Invest Ophthalmol Vis Sci 40:175–181

Kelley KW, Brief S, Westly HJ, al e (1986) GH3 pituitary adenoma cells can reverse thymic aging in rats. Proc Nat Acad Sci 83:5663–5667

Kenyon C (2006) My adventures with genes from the fountain of youth. The Harvey Lectures Series 100:29–70

Kenyon CJ (2010) The genetics of ageing. Nature 464:504–512

Keogh EV, Walsh R (1965) Rate of graying of human hair. Nature 207:877–878

Khansari DN, Gustad T (1991) Effecs of long-term, low-dose growth hormone therapy on immune function and life expectancy of mice. Mech Ageing Develop 57:87–100

Khaw KT et al. (2007) Endogenous testosterone and mortality due to all causes, cardiovascular disease, and cancer in men: European prospective investigation into cancer in Norfolk (EPIC-Norfolk) Prospective Population Study. Circulation 116:2694–2701

Kinney BA, Meliska CJ, Steger RW, Bartke A (2001) Evidence that Ames dwarf mice age differently from their normal siblings in behavioral and learning and memory parameters. Hormones and Behavior 39:277–284

Kiray M et al. (2006) Deprenyl and the relationship between its effects on spatial memory, oxidant stress and hippocampal neurons in aged male rats. Physiol Res 55:205–212

Kirkwood TBL (1977) Evolution of ageing. Nature 270:301–304

Kirkwood TBL, Finch CE (2002) The old worm turns more slowly. Nature 419:794–795

Kitani K, Kanai S, Ivy GO, Carrillo MC (1999) Pharmacological modifications of endogenous antioxidant enzymes with special reference to the effects of deprenyl: a possible antioxidant strategy. Mech Ageing Develop 111:211–221

Kitani K, Kanai S, Miyasaka K, Carrillo MC, Ivy GO (2006) The necessity of having a proper dose of (-)deprenyl (D) to prolong the life spans of rats explains discrepancies among different studies in the past. Annals NY Acad Sci 1067:375–382

Kitani K, Kanai S, Sato Y, Ohta M, Ivy GO, Carrillo MC (1993) Chronic treatment of (-)deprenyl prolongs the life span of male Fischer 344 rats. Further evidence. Life Sci 52:281–288

Kitani K et al. (2002) Why (-)deprenyl prolongs survivals of experimental animals: increase of anti-oxidant enzymes in brain and other body tissues as well as mobilization of various humoral factors may lead to systemic anti-aging effects. Mech Ageing Develop 123:1087–1100

Klass M, Hirsh D (1976) Non-ageing developmental variant of *Caenorhabditis elegans*. Nature 260:523–525

Klebanov S et al. (2000) Heritability of life span in mice and its implication for direct and indirect selection for longevity. Genetica 110:209–218

Klungland A et al. (1999) Accumulation of premutagenic DNA lesions in mice defective in removal of oxidative base damage. Proc Nat Acad Sci 96:13300–13305

Knoll J (1988) The striatal dopamine dependency of life span in male rats. Longevity study with (-)deprenyl. Mech Ageing Develop 46:237–262

Koh K, Rothman JH (2001) ELT-5 and ELT-6 are required continuously to regulate epidermal seam cell differentiation and cell fusion in *C. elegans*. Development 128:2867–2880

Kopchick JJ, Bartke A, Berryman DE (2008) Extended life span in mice with reduction in the GH/IGF-1 axis. In: Guarente LP, Partridge L, Wallace DC (eds) Molecular Biology of Aging. Cold Spring Harbor Laboratory Press, Cold Spring Harbor, pp. 347–369

Kren BT, Bandyopadhyay P, Steer CJ (1998) In vivo site-directed mutagenesis of the factor IX gene by chimeric RNA/DNA oligonucleotides. Nat Med 4:285–290

Kujoth GC et al. (2005) Mitochondrial DNA mutations, oxidative stress, and apoptosis in mammalian aging. Science 309:481–484

Kuratomi G et al. (2008) Aberrant DNA methylation associated with bipolar disorder identified from discordant monozygotic twins. Mol Psychiatr 13:429–441

Kurosu H et al. (2005) Suppression of aging in mice by the hormone klotho. Science 309:1829–1833

Lagouge M et al. (2006) Resveratrol improves mitochondrial function and protects against metabolic disease by activating SIRT1 and PGC-1alpha. Cell 127:1109–1122

LaMarche VC (1969) Environment in relation to age of bristlecome pines. Ecology 50:53–59

Lambert AJ, Merry BJ (2004) Effect of calorie restriction on mitochondrial reactive oxygen species production and bioenergetics: reversal by insulin. Am J Physiol Regulatory Integrative Comparative Physiol 286:R71–R79

Lardy H, Kneer N, Bellei M, Bobyleva V (1995) Induction of thermogenic enzymes by DHEA and its metabolites. Annals NY Acad Sci 774

Larsen LO (1965) Effects of hypophysectomy in the cyclostome, Lampetra fluviatilis (L.) Gray. General Comparative Endocrinol 5:16–30

Larsen LO (1969) Effects of hypophysectomy before and during sexual maturation in the cyclostome, Lampetra fluviatilis L. Gray. General Comparative Endocrinol 12:200–208

Larsen LO (1974) Effects of testosterone and oestradiol on gonadectomized and intact male and female river lampreys (Lampetra fluviatilis L. Gray). General Comparative Endocrinol 24:305–313

Larsen LO (1985) The role of hormones in reproduction and death in lampreys and other species which reproduce once and die. In: Lofts B, Holmes WN (eds) Current Trends in Comparative Endocrinology. Hong Kong University Press, Hong Kong, pp. 613–616

Larson LO (1980) Physiology of adult lampreys, with special regard to natural starvation, reproduction, and death after spawning. Can J Fish Aquatic Sci 37:1762–1779

Lea-Currie YR, Wu SM, McIntosh MK (1997) Effects of acute administration of dehydroepiandrosterone-sulfate on adipose tissue mass and cellularity in male rats. Int J Obesity-Related Metabolic Disorders 21:147–154

Leaman BM, Beamish RJ (1984) Ecological and management implications of longevity in some northeast Pacific ground-fishes. Bull Int N Pac Fish Commn 42:85–97

LeBlanc J, Arvaniti K, Richard D (1998) Effect of dehydroepiandrosterone on brown adipose tissue and energy balance in mice. Hormone and Metabolism Res 30:236–240

Ledford H (2010) Much ado about ageing. Nature 464:480–481

Lee AK, Cockburn A (1985a) Evolutionary Ecology of Marsupials. Cambridge University Press, New York

Lee AK, Cockburn A (1985b) Spring declines in small mammal populations. Acta Zoologica Fennica 173:75–76

Lee CK et al. (2004) The impact of alpha-lipoic acid, coenyzme Q10 and calorie restriction on life span and gene expression patterns in mice. Free Radicals Biol Med 36:1043–1057

Leopold AC (1961) Senescence in plant development: the death of plants or plant parts may be of positive ecological or physiological value. Science 134:1727–1732

Leopold AC, Niedergang-Kamien E, Janick J (1959) Experimental modification of plant senescence. Plant Physiol 34:570–573

Lesnefsky EJ, He D, Moghadas S, Hoppel CL (2006) Reversal of mitochondrial defects before ischemia protects the aged heart. FASEB J 20:E840–E845

Lewis SE, Goldspink DF, Phillips JG, Merry BJ, Holehan AM (1985) The effects of aging and chronic dietary restriction on whole body growth and protein turnover in the rat. Exp Gerontol 20:253–263

Liang H, Masoro EJ, Nelson JF, Strong R, McMahan CA, Richardson A (2003) Genetic mouse models of extended lifespan. Exp Gerontol 38:1353–1364

Libert S, Zwiener J, Chu X, Vanvoorhies W, Roman G, Pletcher SD (2007) Regulation of Drosophila life span by olfaction and food-derived odors. Science 315: 1133–1137

Licastro F, Weindruch R, Davis LJ, Walford RL (1988) Effect of dietary restriction upon the age-associated decline of lymphocyte DNA repair activity in mice. AGE 11:48–52

Lindell TJ (1982) Current concepts: III. Molecular aspects of dietary modulation of transcription and enhanced longevity. Life Sci 31:625–635

Lindseth K, Philpott D, Miquel J (1984) Accelerated aging under simulated hypergravity: protection by dietary enrichment. AGE 7:140

Linn S, Kairis M, Holiday R (1981) Decreased fidelity of DNA polymerase activity isolated from aging human fibroblasts. J Mol Biol 146:2818–2822

Lipman JM, Turturo A, Hart RW (1989) The influence of dietary restriction on DNA repair in rodents: a preliminary study. Mech Ageing Develop 48:135–143

Ljungquist B, Berg S, Lanke J, McClearn GE, Pedersen NL (1998) The effect of genetic factors for longevity: a comparison of identical and fraternal twins in the Swedish Twin Registry. J Gerontol A Biol Sci Med Sci 53:M441-M446

Longo VD, Finch CE (2003) Evolutionary medicine: from dwarf model systems to healthy centenarians? Science 299:1342–1346

Longo VD, Mitteldorf J, Skulachev VP (2005) Programmed and altruistic ageing. Nat Rev Genet 6:866–872

Lu T et al. (2004) Gene regulation and DNA damage in the ageing human brain. Nature 429: 883–891

Ludwig-Festl VM, Grater B, Bayreuther K (1983) Erhohung von Zellstoffwechselleistungen in normalen, diploiden, menschlichen Glia-Zellen in stationaren Zellkulturen induziert durch Meclofenoxat. Arzneim-Forsch 33:495–501

Maccari F, Arseni A, Chiodi P, Ramacci MT, Angelucci L (1990) Levels of carnitines in brain and other tissues of rats of different ages: effect of acetyl-L-carnitine administration. Exp Gerontol 25:127–134

Macchi MM, Bruce JN (2004) Human pineal physiology and functional significance of melatonin. Front Neuroendocrinol 25:177–195

MacGillivray MH (1987) Disorders of growth and development. In: Felig P, Baxter JD, Broadus AE, Frohman LA (eds) Endocrinology and Metabolism, 2nd ed. McGraw-Hill Book Company, New York, pp. 1581–1628

MacKay IR, Whittingham SF, Mathews JD (1977) The immunoepidemiology of aging. In: Makinodan T, Yunis E (eds) Immunology and Aging. Plenum, New York, pp. 35–49

Maier H, Gampe J, Jeune B, Robine J-M, Vaupel JW (2010) Supercentenarians (Springer, in press)

Manion PJ, Smith BR (1978) Biology of larval and metamorphosing sea lampreys, Petromyzon marinus, of the 1960 year class in the Big Garlic River, Michigan: Part 2, 1966–1972. Great Lakes Fisheries Commission

Martin GM (1979) Genetic and evolutionary aspects of aging. Federation Proc 38:1962–1967

Martin GM, Bergman A, Barzilai N (2007) Genetic determinants of human health span and life span: progress and new opportunities. PLOS Genet 3:1121–1130

Martin GM, Finch CE (2008) An overview of the biology of aging: a human perspective. In: Guarente LP, Partridge L, Wallace DC (eds) Molecular Biology of Aging. Cold Spring Harbor Laboratory Press, Cold Spring Harbor, pp. 113–126

Martinez-Vicente M, Cuervo AM (2007) Autophagy and neurodegeneration: when the cleaning crew goes on strike. Lancet Neurol 6:352–361

Mattick JS, Makunin IV (2006) Non-coding RNA. Human Molecular Genetics 15 (Spec No 1):R17–R29

Mattson M, Duan W, Maswood N (2002) How does the brain control lifespan? Ageing Res Rev 1:155–165

Maule AG, Schreck CB (1987) Changes in the immune system of coho salmon (*Oncorhynchus kisutch*) during the parr-to-smolt transformation and after implantation of cortisol. Can J Fish Aquatic Sci 44:161–166

McBride JR, van Overbeeke AP (1969) Hypertrophy of the interrenal tissue in sexually maturing Sockeye salmon (*Oncorhynchus nerka*) and the effect of gonadectomy. J Fish Res Board Canada 26:2975–2985

McDonald RB, Florez-Duquet M, Murtagh-Mark C, Norwitz BA (1996) Relationship between cold-induced thermoregulation and spontaneous rapid body weight loss of aging F344 rats. Am J Physiol 271 (5 Pt 2):R1115–R1122

Meites J (1991) Role of hypothalamic catecholamines in aging processes. Acta Endocrinol (Copenh) 125(Suppl 1):98–103

Meites J (1993) Anti-ageing interventions and their neuroendocrine aspects in mammals. J Reprod Fert Suppl 46:1–9

Mervis RF, Pope D, Lewis R, Dvorak RM, Williams LR (1991) Exogenous nerve growth factor reverses age-related structural changes in neocortical neurons in the aging rat. A quantitative Golgi study. Ann NY Acad Sci 640:95–101

Metayer S et al. (2008) Mechanisms through which sulfur amino acids control protein metabolism and oxidative status. J Nutr Biochem 19:207–215

Michikawa Y, Mazzuchelli F, Bresolin N, Scarlato G, Attardi G (1999) Aging-dependent large accumulation of point mutations in the human mtDNA control region for replication. Science 286:774–779

Migliaccio E et al. (1999) The p66shc adaptor protein controls oxidative stress response and life span in mammals. Nature 402:309–313

Migliaccio E et al. (1997) Opposite effects of the p52shc/p46shc and p66shc splicing isoforms on the EGF receptor-MAP kinase-fos signalling pathway. EMBO J 16:706–716

Milgram NW, Racine RJ, Nellis P, Mendonca A, Ivy GO (1990) Maintenance on L-deprenyl prolongs life in aged male rats. Life Sci 47:415–420

Miller SE, Hadfield MG (1990) Developmental arrest during larval life and life-span extension in a marine mollusc. Science 248:356–358

Min KJ, Yamamoto R, Buch S, Pankratz M, Tatar M (2008) Drosophila lifespan control by dietary restriction independent of insulin-like signaling. Aging Cell 7:199–206

Mitteldorf J (2004) Aging selected for its own sake. Evol Ecol Res 6:1–17

Mitteldorf J (2010) Evolutionary origins of aging. In: Fahy Gm, West M, Coles LS, Harris SB (eds) The Future of Aging: Pathways to Human Life Extension. Springer, Berlin, Heidelberg, New York, pp. 87–126

Molisch H (1938) The Longevity of Plants. Science Press, Lancaster, PA

Monteiro RAF (1991) Genesis, structure and transit of dense bodies in rat neocerebellar corticalcells, namely purkinje neurons – an ultrastructural study. J Hirnforsch 32:593–609

Moore GH (1974) Aetiology of the die-off of male Antechinus stuartii. In: (Ph.D. dissertation). Australian National University, Canberra

Moroi-Fetters SE, Mervis RF, London ED, Ingram DK (1989) Dietary restriction suppresses age related changes in dendritic spines. Neurobiol Aging 10:317–322

Mueller LD, Rose MR (1996) Evolutionary theory predicts late-life mortality plateaus. Proc Nat Acad Sci 93:15249–15253

Murasko DM, Weiner P, Kaye D (1987) Decline in mitogen induced proliferation of lymphocytes with increasing age. Clin Exp Immunol 70:440–448

Murasko DM, Weiner P, Kaye D (1988) Association of lack of mitogen induced lymphocyte proliferation with increased mortality in the elderly. Aging, Immunol Infectious Dis 1:1–6

Murphy CT et al. (2003) Genes that act downstream of DAF-16 to influence the lifespan of *Caenorhabditis elegans*. Nature 424:277–283

Nakada K, Ono T, Hayashi J (2002) A novel defense system of mitochondria in mice and human subjects for preventing expression of mitochondrial dysfunction by pathogenic mutant mtDNAs. Mitochondrion 2:59–70

Napoli C et al. (2003) Deletion of the p66Shc longevity gene reduces systemic and tissue oxidative stress, vascular cell apoptosis, and early atherogenesis in mice fed a high-fat diet. Proc Nat Acad Sci 100:2112–2116

Napolitano LA, Lo JC, Gotway MB et al. (2002) Increased thymic mass and circulating naive CD4 T cells in HIV-1-infected adults treated with growth hormone. AIDS 15:1103–1111

Narkar VA et al. (2008) AMPK and PPARgamma agonists are exercise mimetics. Cell 134: 405–415

Nemoto S, Finkel T (2002) Redox regulation of forkhead proteins through a p66shc-dependent signaling pathway. Science 295:2450–2452

Niedernhofer LJ et al. (2006) A new progeroid syndrome reveals that genotoxic stress suppresses the somatotroph axis. Nature 444:1038–1043

Nishimura EK, Granter SR, Fisher DE (2005) Mechanisms of hair graying: incomplete melanocyte stem cell maintenance in the niche. Science 307:720–724

Nooden LD (1988) Whole plant senescence. In: Nooden LD, Leopold AC (eds) Senescence and Aging in Plants. Academic Press, San Diego

Norton NW, Racine RJ, Nellis P, Mendonca A, Ivy GO (1990) Maintenance on L-deprenyl prolongs life in aged male rats. Life Science 47

Oka HI (1976) Mortality and adaptive mechanisms of Oryza perennis strains. Evolution 30: 380–392

Ono K, Hasegawa K, Naiki H, Yamada M (2006) Anti-Parkinsonian agents have anti-amyloidogenic activity for Alzheimer's beta-amyloid fibrils in vitro. Neurochem Int 48:275–285

Ooka H, Segall PE, Timiras PA (1988) Histology and survival in age-delayed low-tryptophan-fed rats. Mech Ageing Develop 43:79–98

Orentreich N, Matias JR, deFelice A, Zimmerman JA (1993) Low methionine ingestion by rats extends life span. J Nutr 123:269–274

Pacholec M et al. (2010) SRT1720, SRT2183, SRT1460, and resveratrol are not direct activators of SIRT1. J Biol Chem 285:8340–8351

Pan KZ et al. (2007) Inhibition of mRNA translation extends lifespan in *Caenorhabditis elegans*. Aging Cell 6:111–119

Paradies G, Ruggiero FM, Petrosillo G, Gadaleta MN, Quagliariello E (1994) The effect of aging and acetyl-L-carnitine on the function and on the lipid composition of rat heart mitochondria. Annals NY Acad Sci 717:233–243

Partridge L, Mangel M (1999) Messages from mortality: the evolution of death rates in the old. Trends Ecol Evol 14:438–442

Patro N, Sharma SP, Patro IK (1992) Lipofuscin accumulation in ageing myocardium & its removal by meclophenoxate. Indian J Med Res 96:192–198

Pearson KJ et al. (2008) Resveratrol delays age-related deterioration and mimics transcriptional aspects of dietary restriction without extending life span. Cell Metabolism 8:157–168

Pener MP (1972) The corpus allatum in adult acrilids: the inter-relation of its functions and possible correlations with the life cycle. In: Hemming CF, Taylor THC (eds) Proceedings of the International Study Conference on the Current and Future Problems of Acridology. Center for Overseas Pest Research, London, pp. 135–147

Perls T, Levenson R, Regan M, Puca A (2002a) What does it take to live to 100? Mech Ageing Develop 123:231–242

Perls T, Puca A (2002) The genetics of aging – implications for pharmacogenomics. Pharmacogenomics 3:469–484

Perls TT, Wilmoth J, Levenson R, Drinkwater M, M C, al. e (2002b) Life-long sustained mortality advantage of siblings of centenarians. Proc Nat Acad Sci 99:8442–8447

Pickering AD (1976) Stimulation of intestinal degeneration by oestradiol and testosterone implantation in the migrating river lambry, *Lampetra fluviatilis* L. General Comparative Endocrinol 30:340–346

Pierpaoli W (ed) (2005) Reversal of Aging, Resetting the Pineal Clock. New York Academy of Sciences, New York

Puigserver P, Spiegelman BM (2003) Peroxisome proliferator-activated receptor-gamma coactivator 1 alpha (PGC-1alpha): transcriptional coactivator and metabolic regulator. Endrocr Rev 24:78–90

Quigley K, Goya R, Meites J (1987) Rejuvenating effects of 10-week underfeeding period on estrous cycles in young and old rats. Neurobiol Aging 8:225–232

Rao G, Xia E, Nadakavukaren MJ, Richardson A (1990a) Effect of dietary restriction on the age-dependent changes in the expression of antioxidant enzymes in rat liver. J Nutr 120: 602–609

Rao G, Xia E, Richardson A (1990b) Effect of age on the expression of antioxidant enzymes in male Fischer F344 rats. Mech Ageing Dev 31:49–60

Raser JM, O'Shea EK (2005) Noise in gene expression: origins, consequences, and control. Science 309:2010–2013

Raven PH, Evert RF, Eichhorn SE (1986) Biology of Plants, 4th ed. Worth, New York

Rea SL, Wu D, Cypser JR, Vaupel JW, Johnson T (2005) A stress-sensitive reporter predicts longevity in isogenic populations of *Caenorhabditis elegans*. Nat Genet 37:894–898

Redondo P et al. (2007) Repigmentation of gray hair after thyroid hormone treatment. Actas Dermosifiliogr 98:603–610

Richardson A, Liu F, Adamo ML, van Remmen H, Nelson JF (2004) The role of insulin and insulin-like growth factor-1 in mammalian ageing. Best Pract Res Clin Endocrinol Metabol 18:393–406

Richie JPJ, Leutzinger Y, Parthasarathy S, Malloy V, Orentreich N, Zimmerman JA (1994) Methionine restriction increases blood glutathione and longevity in F344 rats. FASEB J 8:1302–1307

Riga D, Riga S (1995) Brain lipofuscinolysis and ceroidolysis – to be or not to be. Gerontology 41(Suppl 2)

Riga S, Riga D (1974) Effects of centrophenoxine on the lipofuscin pigments in the nervous system of old rats. Brain Res 72:265–275

Robert C, Spatz A, Faivre S, Armand J-P, Raymond E (2003) Tyrosine kinase inhibition and grey hair. The Lancet 361:1056

Roberts-Thomson I, Whittingham S, Youngschaiyd U, al e (1974) Aging, immune response and mortality. Lancet 2:368–370

Robertson OH (1961) Prolongation of the life span of Kokanee salmon (*Oncorhynchus nerka kennerlyi*) by castration before beginning of gonad development. Proc NY Acad Sci 47: 609–621

Robertson OH, Hane S, Wexler BC, Rinfret AP (1963) The effect of hydrocortisone on immature rainbow trout (Salmo gairdnerii). General and Comparative Endocrinol 3:422–436

Robertson OH, Wexler BC (1960) Histological changes in the organs and tissues of senile castrated kokanee salmon: Oncorhynchus nerka kennerlyi. General and Comparative Endocrinol 2: 458–472

Robertson OH, Wexler BC, Miller BF (1961) Degenerative changes in the cardiovascular system of the spawning Pacific salmon (*Oncorhynchus tshawytscha*). Circul Res 9:826–834

Robinson DS, Nies A, Davis JN et al. (1972) Ageing, monoamines, and monoamine oxidase levels. Lancet 1:290

Rogina B, Helfand SL, Frankel S (2002) Longevity regulation by Drosophila Rdp3 deacetylase and caloric restriction. Science 298:1745

Rosa A et al. (2007) Differential methylation of the X-chromosome is a possible source of discordance for bipolar disorder female monozygotic twins. Am J Med Genet B Neuropsychiatr Genet October 22nd

Rose MR (1991) The Evolutionary Biology of Aging. Oxford University Press, Oxford

Rose MR (2005) The Long Tomorrow, How Advances in Evolutionary Biology Can Help Us Postpone Aging. Oxford University Press, New York

Rose MR et al. (2010) Evolutionary nutrigenomics. In: Fahy GM, West M, Coles LS, Harris SB (eds) The Future of Aging: Pathways to Human Life Extension. Springer, Berlin, Heidelberg, New York, pp. 357–366

Rose MR, Mueller LD (2000) Ageing and immortality. Philosophical Trans Royal Soc London Series B Biol Sci 355:1657–1662

Rose MR, Rauser CL, Mueller LD, Benford G (2006) A revolution for aging research. Biogerontology 7:269–277

Roth GS et al. (2002) Biomarkers of caloric restriction may predict longevity in humans. Science 297:811

Rothstein M (1979) The formation of altered enzymes in aging animals. Mech Ageing Dev 9: 197–202

Roy D, Pathak DN, Singh R (1983) Effect of centrophenoxine on the antioxidative enzymes in various regions of the aging rat brain. Exp Gerontol 18:185–197

Rudman D (1985) Growth hormone, body composition and aging. J Am Geriatr Soc 33: 800–807

Rudman D, Feller AG, Cohn L, Shetty KR, Rudman IW, Draper MW (1991) effects of human growth hormone on body composition in elderly men. Hormone Res 36(Suppl. 1): 73–81

Rudman D, Feller AG, Nagraj HS et al. (1990) Effect of human growth hormone in men over 60 years old. NE J Med 323:1–6

Ruehl WW, Entriken TL, Muggenburg BA, Bruyette DS, Griffith WC, Hahn FF (1997) Treatment with l-deprenyl prolongs life in elderly dogs. Life Sci 61:1037–1044

Rushton DH, Norris MJ, Dover R, Busuttil N (2002) Causes of hair loss and the developments in hair rejuvenation. Int J Cosmet Sci 24:17–23

Russell AP (2004) Lipotoxicity: the obese and endurance-trained paradox. Int J Obesity-Related Metabol Disorders 28:S66–S71

Ryu JW et al. (2003) DHEA administration increases brown fat uncoupling protein 1 levels in obese OLETF rats. Biochem Biophys Res Commun 303:726–731

Sahin E, DePinho RA (2010) Linking functional decline of telomeres, mitochondria and stem cells during aging. Nature 464:520–528

Salmon AB, Ljungman M, Miller RA (2008) Cells from long-lived mutants exhibit enhanced repair of ultraviolent lesions. J Gerontol A Biol Sci Med Sci 63:219–231

Sanz A, Caro P, Ayala V, Portero-Otin M, Pamplona R, Barja G (2006) Methionine restriction decreases mitochondrial oxygen radical generation and leak as well as oxidative damage to mitochondrial DNA and proteins. FASEB J 20:1064–1073

Sarin KY, Artandi SE (2007) Aging, graying and loss of melanocyte stem cells. Stem Cell Rev 3:212–217

Sashital DG, Butcher SE (2006) Flipping off the riboswitch: RNA structures that control gene expression. ACS Chem Biol 21:341–345

Sat B, Reches M, Engelberg-Kulka H (2003) The *Escherichia coli* mazEF suicide module mediates thymineless death. J Bacteriol 185:1803–1807

Schaller GB (1972) The Serengeti Lion: A Study in Predator-Prey Relations. University of Chicago Press, Chicago

Schoenofen EA et al. (2006) Characteristics of 32 supercentenarians. J Am Geriatr Soc 54: 1237–1240

Seehuus S-C, Norberg K, Gimsa U, Krekling T, Amdam GV (2006) Reproductive protein protects functionally sterile honey bee workers from oxidative stress. Proc Nat Acad Sci 103: 962–967

Segall PE, Timiras PA (1976) Patho-physiologic findings after chronic tryptophan deficiency in rats: a model for delayed growth and aging. Mech Ageing Develop 5:109–124

Semkova I, Wolz P, Schilling M, Krieglstein J (1996) Selegiline enhances NGF synthesis and protects central nervous system neurons from excitotoxic and ischemic damage. Eur J Pharmacol 315:19–30

Seniuk NA, Henderson JT, Tatton WG, Roder JC (1994) Increased CNTF gene expression in process-bearing astrocytes following injury is augmented by R(-)-deprenyl. J Neurosci Res 37:278–286

Sethupathy P et al. (2007) Human microRNA-155 on chromosome 21 differentially interacts with its polymorphic target in the AGTR1 3′ untranslated region: a mechanism for functional single-nucleotide polymorphisms related to phenotypes. Am J Human Genet 81:405–413

Sharma D, Maurya AK, Singh R (1993) Age-related decline in multiple unit action potentials of CA3 region of rat hippocampus: correlation with lipid peroxidation and lipofuscin concentration and the effect of centrophenoxine. Neurobiol Aging 14:319–330

Sharma HK, Rothstein M (1980) Altered enolase in aged Turbatrix aceti results from conformational changes in the enzyme. PNAS 77:5865–5868

Sharma S et al. (2004) Intramyocardial lipid accumulation in the failing human heart resembles the lipotoxic rat heart. FASEB J 18:1692–1700

Shaw RJ et al. (2005) The kinase LKB1 mediates glucose homeostasis in liver and therapeutic effects of metformin. Science 310:1642–1646

Shay JW, Wright WE (2008) Telomeres and telomerase in aging and cancer. In: Guarente LP, Partridge L, Wallace DC (eds) Molecular Biology of Aging. Cold Spring Harbor Laboratory Press, Cold Spring Harbor, pp. 575–597

Shikama N, Ackermann R, Brack C (1994) Protein synthesis elongation factor EF-1alpha expression and longevity in *Drosophila melanogaster*. Proc Nat Acad Sci 91:4199–4203

Sikes SK (1971) The Natural History of the African Elephant. American Elsevier, New York

Silber JR, Fry M, Martin GM et al. (1985) Fidelity of DNA polymerases isolated from regenerating liver chromatin of aging Mus musculus. J Biol Chem 260:1304–1310

Silber SJ et al. (2008) A series of monozygotic twins discordant for ovarian failure: ovary transplantation (cortical versus microvascular) and cryopreservation. Human Reprod 23: 1531–1537

Simon AF, Shih C, Mack A, Benzer S (2003) Steroid control of longevity in *Drosophila melanogaster*. Science 299:1407–1410

Simpson JG, Gray ES, Michie W, Beck JS (1975) The influence of preoperative drug treatment on the extent of hyperplasia of the thymus in primary thyrotoxicosis. Clin Exp Immunol 22: 249–255

Singh R, Mukherjee B (1972) Some observations on the lipofuscin of the avian brain with a review of some rarely considered findings concerning the metabolic and physiologic significance of the neuronal lipofuscin. Acta Ana. (Basel) 83:302–320

Smigrodzki R, Portell FR (2010) Mitochondrial manipulation as a treatment for aging. In: Fahy GM, West M, Coles SL, Harris SB (eds) The Future of Aging: Pathways to Human Life Extension. Springer, Berlin, Heidelberg, New York, pp. 521–541

Smith ML, Bruhn JN, Anderson JB (1992) The fungus *Armillaria bulbosa* is among the largest and oldest living organisms. Nature 356:428–431

Soderstrom TR, Calderon CE (1979) A commentary on the bamboos (*Poaceae: Bambusoideae*). Biotropica 11:161–172

Sojar HT, Rothstein M (1986) Protein synthesis by liver ribosomes from aged rats. Mech Ageing Develop 35:47–57

Sonntag WE, Forman LJ, Miki N, Trapp JM, Gottschall PE, Meites J (1982) L-dopa restores amplitude of growth hormone in old male rats to that observed in young male rats. Neuroendocrinology 34:163–168

Sonntag WE, Hylka VW, Meites J (1985) Growth hormone restores protein synthesis in skeletal muscle of old male rats. J Gerontol 40:689–694

Sonntag WE, Meites J (1988) Decline in GH secretion in aging animals and man. In: Everitt AV, Walton JR (eds) Regulation of Neuroendocrine Aging. Karger, Basel, pp. 111–124

Speakman JR et al. (2004) Uncoupled and surviving: individual mice with high metabolism have greater mitochondrial uncoupling and live longer. Aging Cell 3:87–95

Spees JL, Olson SD, Whitney MJ, Prockop DJ (2006) Mitochondrial transfer between cells can rescue aerobic respiration. Proc Nat Acad Sci 103:1283–1288

Spindler SR (2001) Calorie restriction enhances the expression of key metabolic enzymes associated with protein renewal during aging. Annals NY Acad Sci 928:296–304

Spindler SR (2005) Rapid and reversible induction of the longevity, anticancer and genomic effects of caloric restriction. Mech Ageing Develop 126:960–966

Spindler SR (2010) Biological effects of calorie restriction: implications for modification of human aging. In: Fahy Gm, West M, Coles LS, Harris SB (eds) The Future of Aging: Pathways to Human Life Extension. Springer, Berlin, Heidelberg, New York, pp. 367–438

Spindler SR, Dhahbi JM (2007) Conserved and tissue-specific genic and physiologic responses to caloric restriction and altered IGF1 signaling in mitotic and postmitotic tissues. Annu Rev Nutr 27:193–217

Spindler SR, Mote PL (2007) Screening candidate longevity therapeutics using gene-expression arrays. Gerontology 53:306–321

Spoerri PE, Glees P (1974) The effects of dimethylaminoethyl p-chlorophenoxyacetate on spinal ganglia neurons and satellite cells in culture. Mitochondrial changes in he aging neurons. An electron microscope study. Mech Ageing Develop 3:131–155

Spoerri PE, Glees P (1975) The mode of lipofuscin removal from hypothalamic neurons. Exp Gerontol 10:225–228

Spoerri PE, Glees P, Ghazzawi E (1974) Accumulation of lipofuscin in the myocardium of senile guinea pigs: dissolution and removal of lipofuscin following dimethylaminoethyl p-chlorophenoxyacetate administration. An electron microscopic study. Mech Ageing Develop 3:311–321

Srebro Z (1966) Lipofuscin turnover in the frog brain. Naturwissenschaften 53:590

Srivastava VK, Busbee DL (1992) Decreased fidelity of DNA polymerases and decreased DNA excision repair in aging mice: effects of caloric restriction. Biochem Biophys Res Commun 31

Srivastava VK, Tilley RD, Miller S, Hart R, Busbee D (1991) Effects of aging and dietary restriction on DNA polymerase expression in mice. Exp Gerontol 26:97–112

Stancheva SL, Alova LG (1988) Effect of centrophenoxine, piracetam and aniracetam on the monoamine oxidase activity in different brain structures of rats. Farmakol Toksikol 51: 16–18

Staton BA, Mixon RL, Dharia S, Brissie RM, Parker CR Jr (2004) Is reduced cell size the mechanism for shrinkage of the adrenal zona reticularis in aging? Endocrinol Res 30:529–534

Steer CJ, Kren BT (2010) Reversing age-related DNA damage through engineered DNA repair. In: Fahy GM, West M, Coles LS, Harris SB (eds) The Future of Aging: Pathways to Human Life Extension. Springer, Berlin, Heidelberg, New York, p

Steffen KK et al. (2008) Yeast life span extension by depletion of 60s ribosomal subunits is mediated by Gcn4. Cell 133:292–302

Stoll S, Hafner U, Kranzlin B, Muller WE (1997) Chronic treatment of Syrian hamsters with low-dose selegiline increases life span in females but not males. Neurobiol Aging 18:205–211

Stough D et al. (2005) Psychological effect, pathophysiology, and management of androgenetic alopecia in men. Mayo Clin Proc 80:1316–1322

Strathmann MF, Strathmann RR (2007) An extraordinarily long larval duration of 4.5 years from hatching to metamorphosis for teleplanic veligers of Fusitriton oregonensis. Biol Bull 213: 152–159

Straub RH et al. (1998) Serum dehydroepiandrosterone (DHEA) and DHEA sulfate are negatively correlated with serum interleukin-6 (IL-6), and DHEA inhibits IL-6 secretion from mononuclear cells in man in vitro: possible link between endocrinosenescence and immunosenescence. J Clin Endocrinol Metab 83:2012–2017

Stuchlikova E, Juricova-Horakova M, Deyl Z (1975) New aspects of the dietary effect of life prolongation in rodents. What is the role of obesity in aging? Exp Gerontol 10:141–144

Suh Y et al. (2008) Functionally significant insulin-like growth factor 1 receptor mutations in centenarians. Proc Nat Acad Sci 105:3438–3442

Sylvia V, Curtin G, Norman J, Stec J, Busbee D (1988) Activation of a low specific activity form of DNA polymerase alpha by inositol-1,4-bisphosphate. Cell 54:651–658

Syntichaki P, Tavernarakis N (2006) Signalling pathways regulating protein synthesis during ageing. Exp Gerontol 41:1020–1025

Syntichaki P, Troulinaki K, Tavernarakis N (2007) eIF4E function in somatic cells modulates ageing in *Caenorhabditis elegans*. Nature 445:922–926

Tani F, Miyoshi K (1977) Neuronal lipofuscin in centrophenoxine treated rats. Folia Psychiatr Neurol Jpn 31:104–109

Tatar M (2004) The neuroendocrine regulation of Drosophila aging. Exp Gerontol 39:1745–1750

Tatar M, Bartke A, Antebi A (2003) The endocrine regulation of aging by insulin-like signals. Science 299:1346–1351

Tatar M, Kopelman A, Epstein D, Tu MP, Yin CM, Garofalo RS (2001a) A mutant Drosophila insulin receptor homolog that extends life-span and impairs neuroendocrine function. Science 292:107–110

Tatar M, Priest N, Chien S (2001b) Negligible senescence during reproductive diapause in D. melanogaster. Am Naturalist 158:248–258

Terry D et al. (2004) Lower all-cause, cardiovascular, and cancer mortality in centenarians' offspring. J Am Geriatr Soc 52:2074–2076

ThyagaRajan S, Felten DL (2002) Modulation of neuroendocrine-immune signaling by L-deprenyl and L-desmethyldeprenyl in aging and mammary cancer. Mech Ageing Develop 123:1065–1079

ThyagaRajan S, Felten SY, Felten DL (1998a) Anti-tumor effect of L-deprenyl in rats with carcinogen-induced mammary tumors. Cancer Lett 123:177–183

ThyagaRajan S, Madden KS, Kalvass JC, Dimitrova S, Felten SY, Felten DL (1998b) L-deprenyl-induced increase in IL-2 and NK cell activity accompanies restoration of noradrenergic nerve fibers in the spleens of old F344 rats. J Neuroimmunol 92:9–21

ThyagaRajan S, Madden KS, Stevens SY, Felten DL (2000) Anti-tumor effect of L-deprenyl is associated with enhanced central and peripheral neurotransmission and immune function in rats with carcinogen-induced mammary tumors. J Neuroimmunol 109:95–104

ThyagaRajan S, Meites J, Quadri SK (1995) Deprenyl reinitiates estrous cycles, reduces serum prolactin, and decreases the incidence of mammary and pituitary tumors in old acyclic rats. Endocrinology 136:1103–1110

Tobin DJ, Paus R (2001) Graying: gerontobiology of the hair follicle pigmentary unit. Exp Gerontol 36:29–54

Tomas-Loba A et al. (2008) Telomerase reverse transcriptase delays aging in cancer-resistant mice. Cell 135:609–622

Tran H et al. (2002) DNA repair pathway stimulated by the forkhead transcription factor FOXO3a through the Gadd45 protein. Science 296:530–534

Trifunovic A et al. (2005) Somatic mtDNA mutations cause aging phenotypes without affecting reactive oxygen species production. Proc Nat Acad Sci 102:17993–17998

Trifunovic A et al. (2004) Premature ageing in mice expressing defective mitochondrial DNA polymerase. Nature 429:417–423

Trinei M et al. (2002) A p53-p66Shc signaling pathway controls intracellular redox status, levels of oxidation-damaged DNA and oxidative stress-induced apoptosis. Oncogene 21:3872–3878

Tsai WB, Chung YM, Takahashi Y, Xu Z, Hu MC (2008) Functional interaction between FOXO3a and ATM regulates DNA damage response. Nature Cell Biol 10:460–467

Tsuchiya T, Dhahbi JM, Cui X, Mote PL, Bartke A, Spindler SR (2004) Additive regulation of hepatic gene expression by dwarfism and caloric restriction. Physiol Genomics 17:307–315

Tu MP, Epstein D, Tatar M (2002) The demography of slow aging in male and female Drosophila mutant for the insulin-receptor substrate homologue chico. Aging Cell 1:75–80

Valenzano DR, Terzibasi E, Genade T, Cattaneo A, Domenici L, Cellerino A (2006) Resveratrol prolongs lifespan and retards the onset of age-related markers in a short-lived vertebrate. Curr Biol 16:296–300

Van Remmen H et al. (2003) Life-long reduction in MnSOD activity results in increased DNA damage and higher incidence of cancer but does not accelerate aging. Physiol Genomics 16:29–37

Vargas R, Castaneda M (1981) Role of elongation factor 1 in the translational control of rodent brain protein synthesis. J Neurochem 37:687–694

Vaupel JW et al. (1998) Biodemographic trajectories of longevity. Science 280:855–860

Vaupel JW (2010) Biodemography of human aging. Nature 464:536–542

Ventruti A, Cuervo AM (2007) Autophagy and neurodegeneration. Curr Neurol Neurosci Rep 7:443–451

Vermulst M et al. (2007) Mitochondrial point mutations do not limit the natural lifespan of mice. Nat Genet 39:540–543

Villareal DT, Holloszy JO (2006) DHEA enhances effects of weight training on muscle mass and strength in elderly men and women. Am J Physiol Endocrinol Metabol 291

von Holst D, Hutzelmeyer H, Kaetzke P, Khaschel M, Schonheiter R (1999) Social rank, stress, fitness, and life expectancy in wild rabbits. Naturwissenschaften 86:388–393

Walker RF, Pakula LC, Sutcliffe MJ, Kruk PA, Graakjaer J, Shay JW (2009) A case study of "disorganized development" and its possible relevance to genetic determinants of aging. Mech Ageing Develop 130:350–356

Wang E (2007) MicroRNA, the putative molecular control for mid-life decline. Ageing Res Rev 6:1–11

Wang MC, Bohmann D, Jasper H (2005) JNK extends life span and limits growth by antagonizing cellular and organism-wide responses to insulin signaling. Cell 121:115–125

Wang Y et al. (2001) Muscle-specific mutations accumulate with aging in critical human mtDNA control sites for replication. Proc Nat Acad Sci 98:4022–4027

Wang YX et al. (2004) Regulation of muscle fiber type and running endurance by PPARdelta. PLoS Biol 2:e294

Webster GC (1985) Protein synthesis in aging organisms. In: Sohall RS, Birnbaum LS, Cutler RG (eds) Molecular Biology of Aging: Gene Stability and Gene Expression. Raven Press, New York, pp. 263–289

Webster GC (1988) Protein synthesis. In: Lints FA, Soliman MH (eds) Drosophila as a Model Organism for Ageing Studies. Blackie, London, pp. 119–128

Webster GC, Webster SL (1982) Effects of age on the post-initiation stages of protein synthesis. Mech Ageing Develop 18:369–378

Webster GC, Webster SL (1984) Specific disappearance of translatable messenger RNA for elongation factor one in aging *Drosophila melanogaster*. Mech Ageing Develop 24:335–342

Weindruch R, Colman RJ, Perez V, Richardson AG (2008) How does caloric restriction increase the longevity of mammals? In: Guarente LP, Partridge L, Wallace DC (eds) Molecular Biology of Aging. Cold Spring Harbor Laboratory Press, Cold Spring Harbor, pp. 409–425

Weraarchakul N, Strong R, Wood WG, Richardson A (1989) The effect of aging and dietary restriction on DNA repair. Exp Cell Res 181:197–204

West M (2010) Embryonic stem cells: prospects of regenerative medicine for the treatment of human aging. In: Fahy GM, West M, Coles LS, Harris SB (eds) The Future of Aging: Pathways to Human Life Extension. Springer, Berlin, Heidelberg, New York, pp. 451–487

Wherry ET (1972) Box-huckleberry as the oldest living protoplasm. Castanea 37:94–95

Wikby A et al. (2006) The immune risk phenotype is associated with IL-6 in the terminal decline stage: findings from the Swedish NONA immune longitudinal study of very late life functioning. Mech Ageing Develop 127:695–704

Williams GC (1957) Pleiotropy, natural selection, and the evolution of senescence. Evolution 11:398–411

Wilson DM, III, McNeil DR (2007) Base excision repair and the central nervous system. Neuroscience 145:1187–1200

Wilson EO (1971) The Insect Societies. Belknap Press, Cambridge

Winston ML (1987) The Biology of the Honey Bee. Harvard University Press, Cambridge, MA

Wodinsky J (1977) Hormonal inhibition of feeding and death in octopus: control by optic gland secretion. Science 198:948–951

Wodsedalek JE (1917) Five years of starvation of larvae. Science 46:366–367
Wolkow CA (2002) Life span: getting the signal from the nervous system. Trends Neurosci 25:212–216
Wood JG et al. (2004) Sirtuin activators mimic caloric restriction and delay ageing in metazoans. Nature 430:686–689
Wood PL, Peloquin A (1982) Increases in choline levels in rat brain elicited by meclofenoxate. Neuropharmacology 21:349–354
Woolf TM (1998) Therapeutic repair of mutated nucleic acid sequences. Nature Biotechnology 16:341–344
Woolley P (1971) Differential mortality of Antechinus stuartii (Macleay): nitrogen balance of somatic changes. Aust J Zool 19:347–353
Woolley P (1981) Antechinus bellus, another dasyurid marsupial with post-mating mortality of males. J Mammal 62:381–382
Wu D, Rea SL, Yashin AI, Johnson T (2006) Visualizing hidden heterogeneity in isogenic populations of *C. elegans*. Exp Gerontol 41:261–270
Xi G, Shen X, Clemmons DR (2008) $p66^{shc}$ negatively regulates insulin-like growth factor 1 signal transduction via inhibition of $p52^{shc}$ binding to Src homology 2 domain-containing protein tyrosine phosphatase substrate-1 leading to impaired growth factor receptor-bound protein-2 membrance recruitment. Mol Endocrinol 22:2162–2175
Yamada Y, Ando F, Niino N, Shimokata H (2005) Association of polymorphisms of the androgen receptor and klotho genes with bone mineral density in Japanese women. J Mol Med 83:50–57
Yamamoto M et al. (2005) Regulation of oxidative stress by the anti-aging hormone Klotho. J Biol Chem 280:38029–38034
Yen TT, Knoll J (1992) Extension of lifespan in mice treated with Dinh lang (*Policias fruticosum* L.) and (-)deprenyl. Acta Physiol Hung 79:119–124
Young VR, Steffee WP, Pencharz PB, Winterer JG, Scrimshaw NW (1975) Total human body protein synthesis in relation to protein requirements at various ages. Nature 253:192–194
Zaratiegui M, Irvine DV, Martienssen RA (2007) Noncoding RNAs and gene silencing. Cell 128:763–776
Zeng YC et al. (1995) Effect of long-term treatment with L-deprenyl on the age-dependent microanatomical changes in the rat hippocampus. Mech Ageing Develop 79:169–185
Zhang C, Cuervo AM (2008) Restoration of chaperone-mediated autophagy in aging liver improves cellular maintenance and hepatic function. Nat Med 14:959–965
Zihlman AL, Morbeck ME, Sumner DR (1989) Tales of Gombe chimps as told in their bones. Anthroquest (Leakey Foundation News) No. 40 (Summer):20–22
Zs-Nagy I (1989) On the role of intracellular physicochemistry in quantitative gene expression during aging and the effect of centrophenoxine. A review. Arch Gerontol Geriatr 9:215–229
Zs-Nagy I, Semsei I (1984) Centrophenoxine increases the rates of total and mRNA synthesis in the brain cortex of old rats: an explanation of its action in terms of the membrane hypothesis of aging. Exp Gerontol 19:171–178
Zuckerman BM, Barrett KA (1978) Effects of PCA and DMAE on the namatode [sic] Caenorhabditis briggsae. Exp Aging Res 4:133–139

Part II
The Future of Aging

Chapter 7
An Approach to Extending Human Lifespan Today

Christopher B. Heward

Contents

7.1	Life Extension – The Science of Survival	228
	7.1.1 The Survival Curve	228
	7.1.2 The Human Life-History Curve	229
	7.1.3 Human Life Extension – An Operational Definition	230
7.2	What Is Aging?	231
	7.2.1 Biological Vs. Chronological Aging	231
	7.2.2 Theories of Aging	231
	7.2.3 Symptoms of Aging	232
	7.2.4 Death in America	233
7.3	Aging and the Loss of Homeostasis	233
	7.3.1 Health Maintenance Processes	234
	7.3.2 Oxidative Stress	235
	7.3.3 Dysdifferentiation	235
	7.3.4 The Neuroendocrine System	236
7.4	Maintaining Homeostasis	247
7.5	Some Guidelines for Scientific Preventive Medicine	248
	7.5.1 Step 1: Baseline Data – Patient Starting Point	248
	7.5.2 Step 2: Supplements – Optimum Nutrition	249
	7.5.3 Step 3: Hormones – Maintaining Homeostasis	250
	7.5.4 Step 4: Diet and Exercise	250
	7.5.5 Metabolic Medicine (Nutrition, Exercise and Weight Control)	250
7.6	Summary	270
References		271

C.B. Heward (✉)
Kronos Science Laboratory, Phoenix, AZ 85018, USA

C.B. Heward is deceased

7.1 Life Extension – The Science of Survival

7.1.1 The Survival Curve

The term "life extension" generally refers to increasing the lifespan of an individual or group. One of the most common techniques that scientists have developed to study potential life extension technologies is the survival curve (plotting number or percent of individuals surviving in the population being studied until the last member of the population has died). By generating survival curves for both experimental and control groups and determining the key parameters of "average life expectancy" and "maximum life span", scientists can determine the effectiveness of a life extension protocol. Human survival curves record death from whatever cause, including infant death, accidents, murder, disease, war, and other "natural" causes. When the size of the population under study is large enough, an S-shaped curve usually results. The projected survival curve for people born in the United States in 1950 indicates an average life expectancy of about 59 years and a maximum life span of about 110 years. This is the age at which the oldest members of the group are predicted to die.

7.1.1.1 Squaring the Curve

"Squaring the curve" results when the average life expectancy or health span of a population improves without a concomitant increase in maximum life span. As a result of technological advances, this has been occurring for thousands of years in human beings, but the process seems to be decelerating (Olshansky et al. 1990, 2001). Our average life expectancy is continuing to increase, but maximum lifespan has not changed significantly. Maximum human lifespan seems to be fixed at about 120 years. Note that the survival curve tells us nothing about aging in any single individual.

If we continue to increase average life expectancy without also increasing the maximum life span, we will eventually reach the point where, barring accidents, almost everyone lives a long and healthy lifetime and then dies at or very near the maximum life span for the species. Jeanne Calment, of Arles, France, (born on February 25, 1875) is the current world record holder for longevity. She passed away at the age of 122 (as reported on CNN Headline News) on August 4, 1997. As is common with centenarians, sources close to the Arles city hall, who confirmed Calment's death, gave no precise cause. Calment had spent the past 12 years in a retirement home. She claimed that an occasional glass of Port wine – along with a diet rich in olive oil – were the keys to her longevity. A more likely explanation is that she was lucky and managed to age VERY uniformly.

7.1.1.2 Extending the Curve

What about extending the curve in addition to squaring it? Extending the curve is the expressed goal of many life extensionists and biogerontologists and is what happens when both average life expectancy and maximum life span are increased.

Almost everyone in the population lives longer, and some people live significantly longer, increasing the maximum life span for the population.

Is it *possible* to increase maximum human life span (i.e., extend the curve) and, if so, how can it be achieved? The answer to this question is not yet clear. Indeed, it is a topic of considerable controversy within the gerontology community. Although few biogerontologists, today, take the position that maximum life span in humans will never be significantly extended, there is much disagreement about how and when such maximum life span extension might be achieved. Most scientists do agree, however, that if and when it is achieved, it will require a fundamental change in the rate of aging itself.

Surprisingly, there is also disagreement about the desirability of life extension. This is due in part to the failure of the proponents of life extension and/or anti-aging medicine to clearly define their objectives, and in part to a not-very-well disguised religious perspective on the part of those opposed to lifespan extension. To help focus attention on the specific goals of life extension as defined herein, let's consider the human life-history curve, which depicts the life path of a typical human individual.

7.1.2 The Human Life-History Curve

Although survival curves may be useful tools for scientists studying life extension techniques in target populations, they are of little value in expressing the effects of a clinical intervention program on a specific individual. For this purpose, an individual life-history curve is far more meaningful. Such a theoretical Life-history Curve for a "typical" American charts a single individual's expected level of overall physiological functional capacity during his/her lifetime. For this individual, his/her 80-year life span is divided into four approximately equal periods or phases of life: Development, Vitality, Degeneration, and Senescence. At the end of the final phase is death.

7.1.2.1 Development

The first phase of life is characterized by growth, development and maturation. It begins at birth (actually, at conception) and ends when the individual becomes fully mature. Depending upon the individual, this usually happens between the ages of 15 and 25.

7.1.2.2 Vitality

The second phase of life is young adulthood, roughly the period between the ages of 20 and 40. This phase is characterized by a high degree of sexual activity, excellent health, intellectual and emotional growth, physical strength, and general vigor.

7.1.2.3 Degeneration

The third phase of life is "middle age," the period between approximately 40 and 60 years of age. This phase is characterized by a slow, but progressive, degenerative decline in physiological function. Libido wanes and sexual activity becomes less frequent. At the end of this period women go through menopause. There is a general loss of physical strength due to a loss of muscle mass and an increase in body fat. Flexibility and mobility decline. Wrinkled skin, drooping breasts, graying hair, balding heads, and bifocal glasses declare to the world that "the bloom is off the rose." In addition, insomnia often becomes a problem – minor, at first, but increasing with age. Finally, minor health problems begin to occur more frequently and last longer. General health and vitality are in decline.

7.1.2.4 Morbidity

The final phase of life is characterized by rapid decline and disease. This phase begins roughly at age 60. As the progressive degeneration described above continues and accelerates, symptoms of homeostatic imbalance reach clinical thresholds and one or more diseases are diagnosed. Immune function continues to decline and susceptibility to new infections increases. This phase has been characterized as a constant state of physiological stress, complete with anabolic hormone deficiencies and elevated catabolic hormones (e.g. cortisol) that produce a negative nitrogen balance, and inflammatory factors (e.g. IL-6). Medical intervention, at this point, is usually aimed at relieving the symptoms of disease rather than eliminating their causes.

Perhaps the greatest indictment of the current health care delivery system in the United States today is that it focuses almost exclusively upon treating existing disease, thus prolonging the fourth phase of life. An alternative strategy is to focus upon prolongation of that period of an individual's life span associated with a high level of physiological function and free of age-related disease symptoms. The goal is not merely to postpone death but to enhance life. Such a program aims to improve the quality of human life as well as its quantity. The primary focus of the ideal longevity enhancement program is the prevention of those well-known, age-related degenerative processes that ultimately lead to the many disease states currently associated with old age. It is hoped that this approach may result in extension of vitality and compression of morbidity.

7.1.3 Human Life Extension – An Operational Definition

For the present discussion, I would like to operationally define life extension as the prolongation of that period of an individual's lifespan associated with a high level of physiological function and free of age-related disease symptoms. Based on this definition, life extension is not merely death postponement, but incorporates a measure of the quality of the life being extended. Although this approach quite properly includes some consideration of such things as risk reduction and the appropriate treatment of illness, the primary focus of this chapter will be upon the prevention of well-known age-related degenerative processes that ultimately lead to the many disease states associated with old age.

7.2 What Is Aging?

7.2.1 Biological Vs. Chronological Aging

When asked to define aging, almost everyone will do so in terms of the passage of time. However, biogerontologists, the scientists who study aging in living organisms, now believe that aging may have little or nothing to do with time per se. One only needs to consider the facts that a roundworm is "old" at 7 days, a mouse at 36 months, a dog at 10 years, and a horse at 20 years, whereas humans at these same chronological ages are in their growth and development stage.

Aging is a biochemical phenomenon influenced by both genetics (both between and within species) and the environment. Although there is still much debate about the fundamental biology of aging, most scientists share the view that aging manifests itself as a progressive decline in physiological function associated with an array of degenerative processes characteristic of old age. Within a single species, the symptoms of aging may occur earlier or later in life depending both on genetic factors and environmental factors affecting individuals. Therefore, as a practical matter, aging can be viewed as a measure of biological function rather than chronological age.

Given the above, it is clear that biological aging may occur more quickly or more slowly than chronological aging depending upon a great many interrelated factors that affect the general health status of each individual. It is not being an old person that is undesirable, it is functioning like an old person that is undesirable. Indeed, there are certain progeric diseases (e.g., Down's syndrome, Werner's syndrome, etc.) in which biological aging seems to progress at an accelerated rate. People afflicted with such diseases suffer a broad spectrum of the symptoms normally found in people twice their chronological age. Thus, one might say that the goal of life extension is, "increasing chronological aging by decreasing biological aging."

7.2.2 Theories of Aging

The National Institute on Aging (NIA) was created as the official agency of the United States Government (National Institutes of Health, U.S. Public Health Service) that funds and regulates biomedical aging research in this country. The NIA has officially recognized twelve different theories of aging (see Table 7.1), each attempting to explain the aging process as viewed from the perspective of its author. Others have cataloged over 25 different theories, many of which overlap and all of which describe some aspect of the aging process as determined by the

Table 7.1 Twelve theories of aging

* The Wear-and-tear Theory
* The Rate-of-living Theory
* The Metabolic Theory
* The Endocrine Theory
* The Free-radical Theory
* The Collagen Theory
* The Somatic Mutation Theory
* The Programmed Senescence Theory
* The Error Catastrophe Theory
* The Cross-linking Theory
* The Immunological Theory
* The Redundant Message Theory

scientific community. A complete description of each of these theories is beyond the scope of this chapter. Moreover, the aging process is so complex that it is very likely that a full explanation would include aspects of several of these general theories.

Rather than discussing the relative merits of each theory individually, I have chosen to focus on a more practical view of the aging process. It states, simply, that whatever the ultimate nature of aging may be (genetic, thermodynamic, biochemical, etc.), senescence manifests itself as certain progressive degenerative processes resulting from some combination of biochemical, hormonal and neural imbalances. The advantage of this approach is that many of these imbalances are subject to empirical examination and manipulation. Thus, they may be preventable and/or reversible.

Many characteristics of aging in multicellular organisms are clearly under genetic control at some level in the organism, as deduced from the stable, characteristic patterns of aging and the stability of the maximum life span in most species from generation to generation (Finch and Schneider 1985). However, the extent to which aging changes within the cells of higher organisms are preprogrammed for "intrinsic" senescence remains an open question. If the putative mechanism for the control of aging involves neural and endocrine factors, then the time course of aging should be subject to extensive "extrinsic" manipulation. In other words, if we can understand how biochemical, hormonal, and neural imbalances produce degenerative disease states, then we may be able to successfully intervene.

7.2.3 Symptoms of Aging

Despite uncertainties as to fundamental mechanisms of aging, we are all too familiar with the symptoms and characteristics of old age. The progressive decline in function that occurs as people age includes diminished sensory acuity, wrinkled skin, graying hair, muscle weakness, increased body fat, lack of energy, reduced flexibility, chronic aches and pains, insomnia, failing memory, slower wound healing, and increased susceptibility to infections. No one of these characteristics, however, is necessary or sufficient to define the process.

However, if we look, feel, and act young for our age, then it is likely that we are healthier than average and that senescence is progressing at a slower rate in us than it is in our peers. The converse is also true. People who look, feel and act older than they are likely to fall victim earlier to the diseases of old age.

This approach is not foolproof, however, and, as discussed below, there is no substitute for a complete, objective physical and biochemical examination to estimate one's biological aging rate and overall health status. Even with such an evaluation, the state of the art is such that, as yet, we cannot attach a definite "biological age" to an individual because the validation of the "biomarkers of aging" we use is a work in progress. We still do not have a complete set of measurements that can accurately predict where an individual falls on the spectrum of biological aging. In fact, an observer's estimate of an individual's age is still the best single "biomarker" of aging.

7.2.4 Death in America

As the degenerative processes of biological aging progress, they ultimately lead to the development of one or more of the diseases normally associated with old age. Such diseases occur only rarely in people under the age of thirty, or even forty. Yet, in people over the age of sixty-five, they are very common. Just three of these diseases (i.e., cancer, heart disease and stroke) account for 75% of all deaths in people over the age of 65.

This has led to the question of whether the aging process itself should be viewed as a disease. Gerontologists now agree that maximum life span will not be significantly increased by focusing our medical research efforts on curing the diseases that are now responsible for killing us. In fact, curing the three most deadly diseases in this country (i.e., heart disease, cancer, and stroke) would increase average life expectancy by about 15 years and do virtually nothing to increase maximum life span. Without solving the underlying problem of aging, which causes all of these diseases, the additional 15 years gained would be of questionable quality for most people anyway.

Clearly, a better approach would be to prevent these diseases before they happen rather than trying to cure them after the fact. Only by focusing on the degenerative processes of aging itself are we going to develop the technology to prevent these diseases and, only then will we be able to extend maximum human life span in any significant way.

7.3 Aging and the Loss of Homeostasis

Aging is characterized by the progressive failure of an organism to maintain the constancy of its internal environment (i.e., homeostasis) under conditions of physiological stress. This failure is associated with an increase in the vulnerability and a decrease in the viability of the organism.

It appears that aging may be both a cause and an effect of this diminishing homeostasis. Cumulative molecular damage to nucleic acids, proteins, and lipid membranes (perhaps caused by oxygen and nitrogen free radicals) and/or the random inactivation and activation of genes (epigenetic changes) progressively compromise cellular function. Declining cell function leads to diminished organ function with a concomitant degeneration of all of the major organ systems in the body. At the same time there appear to be shifts in number and function of stem cells. These are the undifferentiated cells held in reserve to repair tissues by replacing cells damaged by "wear and tear." There is evidence that, with aging, stem cells become depleted and/or less responsive to signals that their help is needed.

The loss of cell function impacts the integrity of neuroendocrine and other signaling and control mechanisms, further exacerbating homeostatic imbalances in a wide variety of systems. Thus, as normal cellular repair and defense processes gradually fail, the resulting imbalances manifest themselves as the progressive, degenerative processes associated with the functional decline of "middle age." Eventually, this decline evolves into the disease states commonly associated with old age.

The biochemical characteristics of mammalian aging processes are predicted to be qualitatively similar in different species – even those species having significantly different life spans – because mammalian species share remarkably common biological pathways at the cellular and molecular levels. The lifespans of a mammalian species are therefore determined by how well each species is genetically equipped to cope with the common set of side effects of metabolism and environmental stress causing the damage that results in aging. Thus, the life span of a species is likely not a result of differences in basic design or biochemical makeup, so much as a result of the robustness of its defense and repair mechanisms.

7.3.1 Health Maintenance Processes

The genetic complexity of the processes governing the life span may not be as great as was once believed (Cutler 1972, 1975, 1978). Evidence for this includes the unusually high rate of increase in life span that occurred during primate evolution (Cutler 1975). The mechanism of primate speciation is largely a result of quantitative differences in regulation of a common set of structural genes shared by all species, yet closely related primate species may have a substantial differences in life span (e.g. humans have a 98–98.5% homology compared with chimpanzees, but live twice as long).

These facts suggest that "active" mechanisms called *health maintenance processes* (HMPs) exist and are regulated by a relatively small number of *longevity determinant genes* (LDGs), possibly of the order of 50–100 genes (Cutler 1975, 1978). LDGs would act at the level of gene regulation, governing the activity of numerous HMPs that are similar in all species, independently of what their life span may be. Thus, the extent of global health maintenance or life span of a species would be determined through the regulation of a common set of HMPs shared by all species.

Preliminary results verifying this hypothesis have been achieved with animal models such as *Drosophila* (Orr and Sohal 1994). This conclusion implies that the genetic potential to further increase human health and longevity may already exist in our own genome – simply awaiting proper intervention to alter the activity of a few LDGs regulating the more complex set of HMPs.

7.3.2 Oxidative Stress

In the wild, predators and accidents are the primary causes of mortality and dictate the slopes of survival curves. Those familiar with the essentials of evolutionary biology will intuitively understand why there is no selective pressure that would cause the wild type genome to create additional programming for the benefit of geriatric animals. Simply stated, only genetic variants that promote the likelihood of an animal's reproducing will be selected for in the wild. If, for example, environmental pressures (mainly predation) on a species lead to the death of nearly every individual within a year, then the genome will adapt to provide a program that maximizes reproductive output in the first 12 months of life but not beyond.

Thus it is not surprising that evolutionary and comparative studies done to date seem to indicate that mammalian aging is a result of *passive* rather than *active* biochemical mechanisms. That is, aging is not the result of a "genetic program" evolutionarily selected to deliberately limit good health or shorten life span. Rather, the "side effects" of fundamental metabolic processes found in all mammalian species seem to be at the root of aging. This concept is consistent with much evidence indicating that all biological processes appear to evolve as a result of a complex history of trade-offs with the environment.

Detecting the presence of cellular *oxidative stress* (OS) represents an ideal candidate to test the predicted "passive aging process" hypothesis (Cutler 1972, 1984b; Tolmasoff et al. 1980). OS can potentially be classified as a "primary" aging process. The set of repair and defense mechanisms acting to determine the cells' "oxidative stress state" would represent potential HMPs regulated by a set of LDGs. But how could OS actually cause aging?

7.3.3 Dysdifferentiation

Much of aging can be superficially explained as the result of cells slowly drifting away from their properly differentiated state. This has been called *dysdifferentiation* (Cutler 1982). Adopting this view, it is suggested that OS causes aging not by killing cells directly or through an accumulation of damage making proper cell function impossible, but instead by perturbing cells away from their properly differentiated state (at the level of the DNA) (Cutler 1982). Thus, genes controlling the long-term stability of the differentiated state of cells would represent perhaps the most important class of HMPs controlled by LDGs.

If OS can cause genetic instability leading to dysdifferentiation, then any processes acting to reduce oxidative stress might also act to stabilize the differentiated state of a cell and accordingly would represent part of this important class of HMPs. Antioxidants in the diet, of course, could also play an important role in this process. In fact, there is compelling evidence supporting the prediction that longer-lived species are under less oxidative stress (Cutler 1974, 1984a, 1985, 1991; Tolmasoff et al. 1980). However, some recent data from bats, birds, and naked mole rats, all relatively long-lived life forms, compared with similarly sized but shorter-lived warm-blooded animals, suggest that the ability to repair damage and replace damaged cells may be more important than the level of primary OS damage (Andziak et al. 2006; Buffenstein et al. 2008).

The LDG hypothesis carries with it the far-reaching prediction that despite the vast superficial complexity and variety of aging processes, relatively simple interventions could increase the life span of a species. This prediction can now be tested in experimental animals. Successful experimentation in this area has already led to significant increases in the "health span" in a mammalian model (the mouse).

7.3.4 The Neuroendocrine System

The neuroendocrine regulatory system is an intricate network of communication links connecting the various tissues of the body with each other and with the brain. It is this chemical communication system that functions as the final common pathway through which the body exerts regulatory control and maintains homeostatic balance. A challenge for modern medical science is to determine precisely what causes normal regulatory control to deteriorate as we age. As we begin to understand the underlying mechanisms, we may be able to develop the technology we need to correct the problems at their source. In the meantime, however, it may still be possible to intervene in this aspect of the aging process even before we completely understand it.

Below is a rough summary of the primary regulatory pathways used by the neuroendocrine system of the body to maintain homeostasis. It illustrates a complex system of feedback servos designed to maintain the stability of the internal environment and through the pineal gland, synchronize the activities in the internal environment to match changing conditions in the external environment.

Information from higher centers of the brain (hippocampus, amygdala, etc.) is relayed by neurotransmitters to the hypothalamus, which secretes releasing hormones into the portal vessels, which lead from the hypothalamus to the pituitary gland. These releasing hormones stimulate the release of pituitary trophic hormones, which, in turn, control the secretion of hormones from the peripheral endocrine glands (liver, thyroid, adrenal cortex, gonads, etc.). Hormone secretion by the hypothalamic-pituitary-endocrine axis is regulated by the feedback of peripheral hormones acting on the pituitary, hypothalamus and higher brain centers. The system is further modulated by hormones from the pineal gland acting at each level. Nutrients from the gut and metabolites from all bodily tissues also act on these sites to further modulate hormone secretion. Together, all of the components of this complex system act, in concert, to maintain homeostasis.

There is now considerable evidence to support the hypothesis that the rate of aging in the neuroendocrine system, as in other organs, is modulated by the lifelong action of hormones, nutrients, and metabolites on target tissues (Finch and Schneider 1985). The rate of aging of all tissues studied, including those of the neuroendocrine system, is retarded by dietary restriction and hypophysectomy. Long-term calorie restriction reduces the secretion of most hormones. Hormones may affect aging directly by modifying gene action or indirectly by failing to maintain the homeostatic balance necessary for optimum health. The precise biochemical phenomena that lead to the age-related breakdown in neuroendocrine control mechanisms are as yet unknown, but answers are beginning to be found. In the meantime, any practical approach to retarding the aging process must be limited to efforts at artificially maintaining homeostatic balance using exercise, diet, supplements, drugs, and hormone replacement therapy.

The best evidence that aging phenomena are regulated by the hypothalamic-pituitary-endocrine axis comes from lifetime studies in rodents subjected to dietary restriction, feeding of tryptophan-deficient diets, and/or hypophysectomy (Finch and Schneider 1985). All three of these procedures inhibit growth and maturation and successfully prolong life, retard physiological aging, and delay the onset of age-related pathology (Everitt et al. 1980; Holehan and Merry 1986; Meites et al. 1987; Segall 1979; Segall et al. 1983; Timiras et al. 1982). Since the dietary manipulations modulate hypothalamic-pituitary function, this provides strong evidence that the neuroendocrine system has a regulatory role in development and aging.

The most extensive research on the regulation of aging along the hypothalamic-pituitary-endocrine axis been done in rodents. These studies reveal major inter-species differences in the sites of the principal age-related defects in the neuroendocrine system. Such inter-species differences complicate extrapolation of data from animal studies to humans. Therefore, in the interest of developing a life-extension protocol applicable to humans, the discussion that follows will be limited as much as possible to human data.

7.3.4.1 The Aging Gonad

Age-related ovarian failure leading to menopause in women appears to be due to depletion of ovarian germ cells to the point at which there are insufficient numbers of follicles responding to gonadotropic signaling to sustain a hormonal cycle. Although there is some controversy as to whether the post-natal human ovary can generate new germ cells, it is clear that the number of ovarian follicles decreases progressively from puberty to menopause. Consistent with primary ovarian failure, the menopausal fall in estrogen secretion is accompanied by a compensatory rise in the production of pituitary gonadotropins (FSH and LH).

The potential benefits of postmenopausal estrogen replacement therapy include (a) elimination of vasomotor estrogen deficiency symptoms ("hot flashes"), (b) prevention of atrophy of reproductive organs, (c) greatly reduced rate of osteoporosis, and (d) probably decreased risk of atherosclerosis (Finch and Schneider 1985). The most serious risk of estrogen replacement is a six- to eightfold increase in the incidence of endometrial cancer after five or more years of therapy (Jick et al. 1979; Weiss et al. 1979; Hulka et al. 1980). By cyclically interrupting estrogen therapy with a progestogen the risk of developing endometrial cancer is essentially nullified by eliminating pre-cancerous hyperplastic changes (Whitehead et al. 1981; Lee 1993) and may, in fact, result in reduced risk of endometrial cancer, even compared with untreated women of like age (Gambrell et al. 1979). Breast cancer also appears to occur at a rate 30–50% higher after five years of treatment with some (but not other) regimens of hormone replacement.

In the testis, in contrast to the ovary, there is a constant replenishment of male germ cells from puberty to old age. Testicular failure in men is gradual, and the major age-related defect appears to be in the gonad, since blood gonadotropin levels rise while testosterone levels fall (Hallberg et al. 1976; Horton et al. 1975; Stearns et al., 1974; Nankin 1981). In general, a decline in mean serum testosterone seems to be gradual and continuous from about age 30 onward and to progress into the eighth or ninth decades of life (Vermeulen et al. 1972). It is important to note, however, that individual concentrations display a high degree of variation. Although the majority of men in their 70s to 90s have testosterone well below normal for young men, a significant minority of equally old men have normal or even high normal testosterone concentrations. A few large, well-designed studies of American men have failed to find any age-related decline in serum testosterone levels in exceptionally healthy older men (Harman and Tsitouras 1980; Sparrow et al. 1980). Thus testosterone replacement therapy may be appropriate for some men, but not for others. Because the benefits and risks of testosterone replacement in older men have not been well characterized, caution is strongly advised.

7.3.4.2 The Aging Thyroid

Thyroid aging has been reviewed by several authors [e.g., (Gambert and Tsitouras 1985)]. In healthy subjects there is little or no change in plasma levels of thyroid hormones (T3 and T4) or thyrotropin (TSH). However, thyroxine secretion is halved between ages 20 and 80. This apparent paradox is due to a decrease in the rate of clearance of thyroid hormones from the blood. Thus, although thyroxine is maintained in plasma at constant levels, its synthesis is apparently lowered in response to its decreased uptake and utilization by peripheral tissues. In addition to this change there is evidence of a blunted TSH response to the thyrotropin-releasing hormone. Serum T3 may be lower in the aged due to reduced deiodination of T4 to T3 and an increase in deiodination to the inactive form, reverse T3.

Thyroid hormones (T3 and T4) are predominantly calorigenic in target tissues. They exert their biological effects by raising basal metabolic rate (BMR). It has been known for many years that BMR declines with age. While it was initially felt that this decline was associated with a concomitant change in some fundamental neuroendocrine parameter, when anthropometric changes during aging are adjusted for, BMR is found to remain constant with age (Keys et al. 1973; Tzankoff and Norris

1977). Therefore, except in cases of clinical thyroid deficiency, thyroid hormone replacement is neither necessary nor appropriate for most older people, even when blood levels are on the low range of normal. This is to say, there is no compelling evidence that aging per se results in a thyroid hormone deficiency state.

7.3.4.3 The Aging Adrenal Cortex

Adrenocortical aging has been studied by several workers (Sapolsky et al. 1986; Dilman and Dean 1992; Kalimi and Regelson 1990). In healthy human subjects hypothalamic-pituitary-adrenocortical function is normal in old age, permitting the individual to withstand moderate stress. If anything, the system is over-responsive to stress, and therein lies a potential problem.

Basal plasma cortisol levels do not seem to change in unstressed healthy elderly subjects. However, there is a 25% decrease in the 24-hour resting cortisol secretion compared with younger subjects. The normal plasma cortisol levels are maintained due to the slower metabolism of cortisol in old age. In contrast to the "at rest" observations, there are a number of studies showing that the cortisol levels of older persons go up slightly higher and stay up considerably longer in response to stress, such as trauma, surgery, a burn, or acute infection. This may be a result of the aforementioned characteristic failure of the elderly to restore homeostasis quickly in the face of an environmental challenge, such that the period of physiologically perceived stress is prolonged.

This characteristic tendency for elevated stress-induced cortisol could result in an overall integrated exposure to excess glucocorticoid over time. Even modest increases in integrated cortisol exposure can lead to negative nitrogen balance, muscle and bone loss, fragility of connective tissues and blood vessels, thinning of skin, central obesity, and glucose intolerance. All of these changes are seen both in "normal" aging and in Cushing's syndrome (pathological cortisol excess). Thus there is some suspicion that the rate of aging may be influenced by modest overexposure to cortisol over time. This hypothesis deserves further investigation.

Based on the above, hormone replacement therapy with cortisol is obviously not appropriate.

Dehydroepiandrosterone (DHEA) is a major secretory product of the adrenal cortex, and serum concentrations of its sulfate ester, DHEA sulfate (DHEA-S), are the highest of any steroid in the body. The sulfated form of DHEA represents over 99% of the circulating steroid with less than 1% as free DHEA (Zumoff et al. 1980).

DHEA can be converted to active androgens and estrogens in peripheral tissues, but there is no known biological role of either DHEA or DHEA-S themselves in humans. Plasma concentrations of DHEA-S in humans peak at 20–30 years of age and decline thereafter. On the average, it is estimated that the decrease in DHEA production is approximately 20% for every decade after age 25 and that by 85–90, DHEA production is only 5% of that seen in youth (Orentreich et al. 1984). Like cortisol, DHEA-S concentration exhibits circadian fluctuation but the circadian rhythm is lost in elderly men (Montanini et al. 1988).

The striking linear decrease of DHEA and DHEA-S with age observed in cross-sectional studies has led to the postulation that DHEA and DHEA-S may serve as discriminators of life expectancy and aging. Decreased concentrations of DHEA and DHEA-S have been associated with atherosclerosis, cancer, mortality in general, and death due to cardiovascular disease (Bulbrook and Hayward 1967; Bulbrook et al. 1971; Farewell et al. 1978; Barrett-Connor et al. 1986).

For example, it has been reported that in men, higher plasma DHEA-S concentrations were associated with a 36% reduction in mortality from any cause and a 48% reduction in mortality from cardiovascular disease (Barrett-Connor et al. 1986). Although exogenous administration of DHEA or certain analogues produces an array of beneficial effects on obesity, muscle strength and mass, type II diabetes, cholesterol levels, autoimmunity, cancer initiation and proliferation, osteoporosis, memory, and aging in experimental animals, results in humans have been largely disappointing. No clear benefits of DHEA administration have been observed in several recent controlled human trials (O'Donnell et al. 2006; Igwebuike et al. 2008; Nair et al. 2006).

Nonetheless, there are some reports in humans that exogenous administration of adrenal androgens resulted in favorable changes in cholesterol and lipoprotein levels (Kask 1959; Felt and Stárka 1966). In one study, administration of DHEA to normal men reduced serum low density lipoprotein (LDL) levels and body fat without altering insulin sensitivity (Nestler et al. 1988). DHEA given orally for 28 days at a dose of 1,600 mg produced a 7.5% fall in serum LDL cholesterol, which represents an estimated 14% reduction in risk for the development of cardiovascular disease. The decrease in body fat was without loss of weight, and there was no change in tissue sensitivity to insulin. The change in the lipid profile is in sharp contrast to that seen with other androgens, which typically lower high density lipoprotein (HDL) levels and increase LDL levels. It is more in line with an estrogen effect (Mooradian et al. 1987).

There has been considerable speculation that DHEA replacement may represent a useful component of an effective life-extension protocol, but evidence to date does not favor this idea. The possible side effects of taking DHEA are: (a) alopecia (male pattern baldness), (b) hirsutism, (c) pituitary tumors, (d) liver hypertrophy, and (e) prostate hypertrophy. Although these are potentially serious problems, they have been produced only using very high doses of DHEA. It is not likely that an amount sufficient to re-establish a physiological blood level equivalent to that of a typical 25 year old would cause such problems.

7.3.4.4 The Aging Thymus

The thymus gland consists of two soft pinkish-gray lobes lying in the mediastinum (anterior chest cavity) just above the heart. In humans and other mammals the thymus begins to atrophy shortly after puberty. Removal of the gland in the adult animal usually causes no obvious and immediate harmful effects. On this basis the thymus was considered to be an organ without a function. It is now realized that the thymus is an endocrine and immune system organ that plays a pivotal role in the development of immunological competence (see Chapter 15). It is the master gland of the immune system, and it manufactures a family of substances that regulate the growth and maturation of certain types of white blood cell necessary to generate an immune response.

Immature lymphocytes become competent to participate in the immune response by actual passage through the thymus, where they come into contact with one or more thymic hormones (thymosins) and thymus cells. Those that might damage the body are "weeded out" and others mature and exit to compose the body of circulating T (thymus-derived) lymphocytes. Fully differentiated T cells function as killer cells to kill cells infected with viruses, combat tumor cell development, release substances (lymphokines) that affect macrophage function, and may function as helper cells that promote B cell antibody production or as T-regulatory cells to inhibit immune reaction against self-antigens.

Lymphocytes also respond to thymosins released into the blood by the thymus. Thymosins appear to trigger the maturational progression of several early stages of T-cell development and to augment the capacity of certain mature T cells to respond to antigens. One action of the thymosins may be to act through cGMP to induce the expression of alloantigens on the surface of T cells as they develop their functional capacity for immunological competence. Thymosins have been used for many immunodeficiency states (Goldstein and Goldstein 2009) and are being considered in the field of immunotherapy for the treatment of cancer.

A progressive decline in immune function is one of the cardinal signs of the aging process. Thymosins have been shown to slow or reverse some of the immune deficiencies of aging. The thymus gland shrinks at a very early age, becoming noticeably smaller by puberty. As the thymus shrinks, its hormone output declines. This decline accelerates between 25 and 45 years of age. This decline has led some biogerontologists to speculate that the thymus gland may be a primary pacemaker for aging. Indeed, evidence is accumulating that suggests that shrinkage of the thymus gland, together with diminished output of thymosins, may be partly responsible for the deterioration in immune capacity and the increase in anti-self reactions that characterize aging. Whether thymosins or related compounds will have a role in rejuvenation of the senescent immune system remains to be seen (see Chapter 15).

7.3.4.5 The Aging Pancreas

The pancreas contains islands of endocrine tissue (islets of Langerhans) composed mainly of two functional endocrine cell types – alpha cells, which secrete glucagon, and beta-cells, which secrete insulin. Together, these two pancreatic hormones play a major role in regulating carbohydrate and fat metabolism. Insulin exerts its effect on glucose uptake by stimulating insertion of glucose transport proteins into cell membranes, which allows glucose from the blood to enter muscle and fat cells. The opposing islet hormone, glucagon, stimulates mobilization of glucose into the blood from glycogen stores in liver cells.

The past several years have seen the development of substantial interest in the effect of age on insulin release. Reaven and colleagues (1979, 1980) have demonstrated an age-related decrease in glucose- or leucine- stimulated insulin secretion per beta cell of collagenase-isolated islets. This change is associated with an increase in the number of beta cells per islet in the older Sprague-Dawley rat. The authors suggest that the impaired capacity of the individual beta cell to secrete insulin with advancing age is compensated for by an increase in number of beta cells in older animals in an effort to maintain normal carbohydrate homeostasis.

Numerous studies have evaluated circulating insulin levels in individuals across the adult age range (Reaven and Reaven 1980; Davidson 1979; Muller et al. 1996). Studies employing oral or intravenous glucose challenges have almost uniformly demonstrated that elderly individuals have insulin levels equivalent to, or even slightly greater than, the levels found in their younger counterparts. These studies have often been criticized because the older subjects invariably have higher circulating glucose levels than the younger subjects.

In response to this criticism, Andres and Tobin (1975) developed the hyperglycemic glucose clamp technique designed to maintain steady-state plasma glucose levels. When they measured circulating insulin in subjects of varying ages with the same plasma glucose levels, they found a decrease with age in insulin secretory capacity at all but the highest level of hyperglycemic stimulus. In similar studies employing the hyperglycemic glucose clamp technique, DeFronzo (1979) failed to detect a difference in either the early or the late phases of insulin secretion between healthy young and old subjects. Clearly, the higher insulin levels found in older individuals are not the result of higher levels of glucose.

The next question of interest in this system is the cause of the elevated fasting blood glucose levels that occurs with advancing age. Studies in man have shown no effect of age on basal hepatic glucose production (DeFronzo 1979; Robert et al. 1982; Fink et al. 1983). In response to high physiologic and supraphysiologic levels of insulin, hepatic glucose output is rapidly and almost completely suppressed in both young and old subjects (DeFronzo 1979; Robert et al. 1982). Thus, the age-related increase in fasting blood glucose levels is probably not due to increased output from the liver. However, several studies have shown that reduction in circulating glucose level after bolus injection of insulin is impaired with advancing age (DeFronzo 1979). Applying the insulin clamp technique to individuals over a range of circulating steady-state insulin levels varying from physiologic to supraphysiologic demonstrated that aging, in active non-obese healthy men with normal glucose tolerance on normal diets, is associated with marked increases in insulin resistance (Rowe et al. 1983; Fink et al. 1983).

Consistent with the above, maximal insulin-induced glucose disposal rates were the same in young and old, but the dose-response curve was substantially shifted to the right in the older subjects (Rowe et al. 1983). This was interpreted as indicating an age-associated decline in sensitivity of peripheral tissues to insulin. Studies of the effect of age on forearm glucose uptake (Jackson et al. 1982) demonstrated a marked impairment in glucose disposal with advancing age.

Questions that remain unanswered in this area relate to the mechanisms of insulin insensitivity during aging. At least in the Sprague-Dawley rat, the rate-limiting step may be glucose transport (Goodman et al. 1983). However, these authors found little evidence of deterioration in glucose transport after maturity. The overwhelming bulk of available data indicates that normal aging is not associated with a change in insulin receptor number or affinity in rodents or man. Monocyte (Jackson et al. 1982; Fink et al. 1983) and adipocyte (Fink et al. 1983) insulin receptor binding does not differ between young and old subjects, and insulin receptors on circulating monocytes do not change with age (Rowe et al. 1983). Thus, it is probably safe to conclude that a reduction in insulin receptor number and/or affinity does not occur solely as a function of aging.

The fact that chromium deficiency is probably more common in the elderly may explain, at least in part, the insulin resistance found to be associated with old age. In any case, given the very low toxicity and substantial benefits of supplemental chromium, it should be considered (in the form of chromium picolinate, 200–400 μg/day), among the dietary supplements of choice, in any well-designed life extension program. Still, because chromium picolinate, together with vitamin C, has been shown to produce hydroxyl radicals in vitro, care should be taken to avoid excessive doses. Indeed, it may be argued that one should not use chromium supplementation at all, except in the context of a program in which oxidative stress is carefully monitored on an ongoing basis.

Another possible contributor to insulin resistance is the prevalent imbalance of omega-3 to omega-6 fatty acids in the modern western diet. Recently, it has been shown that omega-3 supplementation significantly improves insulin sensitivity in elderly men and women (Tsitouras et al. 2008).

Several studies have evaluated the influence of aging on the release and response to glucagon in man. In general, fasting levels of glucagon are not influenced by age, nor is the post-glucose-induced suppression of glucagon release (Dudl and Ensick 1977; Berger et al. 1978; Simonson and DeFronzo 1983). In addition, glucagon release kinetics are not impaired with aging (Dudl and Ensick 1977; Simonson and DeFronzo 1983). Apparently, glucagon physiology is spared any significant age-related changes and, therefore, hormone replacement therapy of glucagon is not appropriate.

7.3.4.6 The Aging Pituitary – Growth Hormone

The pituitary gland is often referred to as the master gland. It exerts regulatory control over a wide array of peripheral endocrine functions. Although a great deal of information is known about the normal functioning of the pituitary gland, relatively little is known about pituitary-specific functional changes associated with aging. Furthermore, the majority of pituitary hormones exert their effects by producing changes in the activities of other endocrine tissue. Most of the age-related changes associated with these hormones have been addressed elsewhere in this chapter. The major exception to this statement is growth hormone (GH).

GH is an important anabolic hormone. It acts directly on its target tissues to stimulate lipolysis and to antagonize insulin-induced glucose uptake. In addition, GH stimulates somatomedin (insulin-like growth factor-1, or IGF-1) secretion by the liver and locally in a variety of peripheral tissues including muscle and bone. IGF-1, in turn, causes increased cartilage formation, increased bone growth, increased protein synthesis, and increased lean body mass – including muscle, liver, kidney, spleen, thymus, and red blood cell mass (Finch and Schneider 1985).

GH changes associated with aging are well documented (Finkelstein et al. 1972; Florini et al. 1985), but are not without controversy. Studies in humans on basal GH levels have produced conflicting results, possibly due to the fact that obesity is associated with low basal GH levels (Rabinowitz 1970), and aging is characterized by decreased protein synthesis, lean body mass, and bone formation and increased adiposity (Young et al. 1975). Some studies designed to control for obesity have concluded that obesity is correlated with a decline in basal GH levels with advancing age (Kalk et al. 1973; Rudman et al. 1981). Other, equally well-designed studies, suggest that age-related obesity is not associated with declining basal GH levels (Elahi et al. 1982; Vidalon et al. 1973).

The pulsatile secretion of GH is decreased in elderly human and animal subjects (Ho et al. 1987; Sonntag et al. 1980). Analysis of the total 24-hour secretion pattern indicates that this decrease is almost mainly confined to the first three hours after the onset of sleep. Predictably, the decline in GH secretion is associated with a decline in IGF-1 levels. Since pulsatile secretion of GH may be amplified by estradiol, it is significantly reduced after menopause.

The body composition of a typical 20 year old is approximately 10% bone, 30% muscle, and 20% fat tissue. After about age 50, these ratios have changed significantly. By the age of 75, a typical composition is 8% bone, 15% muscle, and 40% fat. This represents 20% less bone, 50% less muscle and 100% more fat. Since GH enhances protein synthesis, diminished GH secretion in old age may contribute, directly and/or indirectly, to this age-related change in body composition.

Despite the fact that aging presents a partial phenocopy of GH deficiency, studies in which GH has been given to healthy elderly men and women with low to low-normal IGF-1 levels have led to mixed results (Blackman et al. 2002; Giannoulis et al. 2006; Sattler et al. 2009). In general, administration of GH for 3 months up to a year has led to significant increases in lean body mass of 5–10% and decreases in body fat by 12–15%, but no reproducible, significant improvements in muscle strength or physical performance measures. GH given with testosterone to men has resulted in modest but significant improvements in muscle strength and in one study physical performance as well (Sattler et al. 2009). Studies have not shown significant improvements in bone mineral density with GH in aging men or women, but in young adults deficient in GH, bone mass did not show improvement until 18–24 months of treatment, and no studies of similar duration have been published for elderly adults. It is important to note that short-term administration of effective doses of GH has been associated with glucose intolerance, edema, and joint pains in some subjects. Thus, long term GH replacement therapy in the elderly remains a highly controversial issue, since no data are, as yet, available establishing its safety and/or efficacy.

The development of GH secretagogues (GHS, molecules that stimulate the release of GH) is another area of great promise. Although a leading pharmaceutical company in this area terminated its research studies on an orally active potent GH releaser, other secretagogues that act at the ghrelin receptor, thus potentiating the GH response to hypothalamic growth-hormone releasing hormone, are under investigation. Secretagogues have the theoretical advantage of preserving pulsatile GH secretion while amplifying the pulses, with presumed reduction in risk of adverse effects of prolonged GH exposure. In a recent study (White et al. 2009), investigators found that 6 months of treatment with an orally active GHS (capromorelin) in older adults with mild functional limitations produced modest but significant improvements in both muscle strength and physical function.

Another complication of human GH replacement therapy is that females are less responsive to GH than are males. Although, it has been known for some time that female GH-deficient patients do not respond as well as age-matched men to their GH-therapy protocol, the physiological reasons for this sex difference remain obscure. One possibility is that estrogen inhibits hepatic production of IGF-1, resulting in decreased IGF-1 levels while simultaneously increasing growth hormone binding protein (GHBP). This is certainly true for orally administered estrogen, but when estrogen (E) was given transdermally (100 mcg daily), these effects in the liver were not observed. This result has been interpreted to mean that transdermal estrogen avoids "first pass" effects in the liver, but this issue has not been resolved conclusively (Ho 1998).

As a result of the above, many physicians who treat adults with GH have begun switching their female patients to a transdermal estrogen preparation (patch, gels, or creams, etc.) to get the greatest IGF-1 response from GH administration. Some have added an androgenic progestin (Pg) or a natural (neutral androgenic Pg) plus testosterone (T) (T-transdermal) if the women are androgen deficient. The downside of using androgenic Pg's and T is that they can be "unfriendly" to the cardiovascular system in women. Done correctly with appropriate doses, it may be possible to give females the same GH effect as for males, but more research is required. The ideal result of such treatment would be increased fat loss, increased lean mass, increased energy, a greater sense of well being, and all of the other positive parameters seen with correct therapy of adult-onset GH-deficiency diseases.

Since GH stimulates cell division, another common concern regarding GH replacement is its effect on the growth rate of cancer cells. An epidemiological prospective prostate cancer study recently sent a chill through the quarters of responsible clinicians treating GH-deficient patients (Chan et al. 1998). Chan's study noted that there was a strong positive correlation between IGF-1 levels and prostate cancer risk. An IGF-1 level between 99 and 185 (ng/ml) had a relative risk of 1.0. At 186–237 the risk was 1.9. At 238–294 it was 2.8, and at 295–500 the risk was 4.3. Although these data are open to interpretation, it is clear that a meticulous cancer screening program should be part of any GH replacement therapy protocol.

Michael Pollak and colleagues from McGill University (co-collaborator with June M. Chan of the Chan et al. 1998 paper) have found a similar strong association between IGF-1 levels and breast cancer. In a study that examined blood samples from 52 patients with prostate cancer, 52 patients with benign prostatic hyperplasia (BPH), and an equal number of controls, no correlation was found between BPH and IGF-1, but increased IGF-1 levels were associated with increased risk of prostate cancer (Mantzoros et al. 1997). An increase of 60 ng/ml increased the odds ratio by 1.91.

Many senior clinicians, with 10–20 years of experience in treating adults and children with well-defined GH deficiency (failure of GH production from pituitary disease, ablation, and failure to produce GH with two or more standard GH stimulation tests) are skeptical of the above data relating GH to cancer. They argue that persons who are acromegalic have been exposed to supraphysiologic levels of GH and IGF-1 for many years, yet their incidence of bowel and other malignancies is only marginally increased, and there seems to be no increase in mortality from them. Peter H. Sonksen, M.D., President of the GH Research Society and Joint Editor of the Journal Growth Hormone and IGF Research has stated: "Worry about developing malignancies is inevitable, but may not be real."

A counter argument put forward by some non-clinician IGF-1 molecular biologists is that the acromegalics have built-in adaptations, with a different GH and IGF binding protein balance that make them more resistant to the mitogenic-carcinogenic effects of IGF-1. There is little data supporting this view and more research is needed to resolve these important issues. It is also important to consider the differences in cancer risk between child (or young) acromegalics and the elderly. If IGF-1, a powerful growth factor, acts "merely" to accelerate growth of existing small foci of malignancy, the risk that such a vulnerable focus exists is much higher in old than in young persons. Thus, caution should remain the watchword with regard to the issue of human GH effects on cancer risk in older patients.

7.3.4.7 The Aging Hypothalamus

The hypothalamic-pituitary-endocrine axis is a self-regulating system with hypothalamic centers receiving feedback of hormones from the peripheral endocrine glands (thyroid, adrenal cortex, gonads). The wear and tear theory of aging suggest that this system continues to function normally until some components wear out due to long-continued action of hormones or other agents (free radicals, neurotransmitter, food factors, etc.) on its target cells. The programmed senescence theory of aging states that aging in this system is a regulated phenomenon, in which the breakdown involving hormonal action on a target is programmed. Data exist that support both theories.

The fact that the sites of breakdown in this system are species-specific and the fact that the age when lesions develop are more or less fixed and characteristic of each species, lend support to the regulated program theory. The idea of a clock in the hypothalamus timing the rate of aging by controlling the secretion of aging hormones has been questioned by Finch et al., whose studies have shown that there is not a fixed number of estrous cycles in the rodent before the failure of reproduction (Finch et al. 1984). Even if the hypothalamus does not have a direct timing role in aging, it certainly controls the secretion of all hormones by the neuroendocrine system and by the actions of these hormones probably influences the rate of aging in the whole hypothalamic-pituitary-endocrine axis.

7.3.4.8 The Aging Pineal

The pineal gland is a fundamental modulator of the entire neuroendocrine system. It functions as a true "biological clock," secreting its primary neurohormone, melatonin, in a circadian fashion. Melatonin synthesis and release is regulated mainly by the light-dark cycle, as well as by adrenergic regulatory mechanisms (Klein 1985), with peaks during the night. Research with pineal polypeptide extracts supports the premise that the pineal gland controls hypothalamic sensitivity to feedback regulation (Dilman 1970, 1971) and governs the circadian rhythm of the hypothalamo-pituitary system (Quay 1972). Decreased pineal gland activity may be one of the mechanisms responsible for elevating the neuroendocrine threshold to feedback signals.

The striking involution of the pineal gland begins in the early twenties and results in a linear (approximately 1% per year) decrease in melatonin production in men (Nair et al. 1986). In women, pineal melatonin production decreases relatively slowly prior to menopause, more dramatically at menopause, again more slowly after age 60. Pineal involution appears to follow an S-shaped curve with the inflection point centered at the time of menopause.

Pineal involution is characterized by: (1) reduced 24-hour total production of melatonin; (2) reduced magnitude of the night-time melatonin peak; and (3) delayed night-time rise in melatonin levels. Calcification of the pineal gland, which begins in the early twenties, is such a universally known phenomenon that is used as a landmark by radiologists when performing CT scans of the brain.

Melatonin has a number of biological activities, which include its ability to: (1) inhibit the hypothalamo-pituitary system; (2) inhibit secretion of GH (Smythe and Lazarus 1974), insulin and gonadotropins; (3) lower pituitary MSH concentration (Kastin and Schally 1967); (4) increase blood potassium concentration (Chazov and Isachenkov 1974; Cheeseman and Forsham 1974); and (5) suppress the action of somatomedin (IGF-1) (Smythe et al. 1974). Appropriate administration of melatonin leads to improved immunity and may even prolong life span in rodents (Regelson and Pierpaoli 1987; Pierpaoli et al. 1990, 1991), but long-term effects in humans are not well studied.

7.4 Maintaining Homeostasis

If aging is, in part, the result of a loss of homeostasis, then it follows that many of the common symptoms of aging may be preventable by maintaining homeostasis using modern biomedical technology. For example, using a well-designed regimen of antioxidants, it may be possible to significantly reduce the free radical damage that occurs as a normal consequence of oxidative metabolism. In addition, it may soon be possible to "up-regulate" certain genes coding for the body's own protective enzymes (e.g. manganese superoxide dismutase, catalase, glutathione peroxidase, etc.) thus, directly combating a primary cause of aging.

In the meantime, it is currently possible to accurately measure blood, urine, and tissue levels of many neurotransmitters, hormones, nutrients, and metabolites. For most of these substances, the optimum concentrations associated with health and vitality are known or can be reasonably estimated. Thus, it may now be possible to reestablish, maintain, and/or prolong optimum health by artificially modulating those substances to optimum concentrations. This will require a carefully designed and professionally administered therapeutic program. An effective program would probably include a combination of diet, exercise, antioxidants, herbs, vitamin and mineral supplements, drugs, and hormone replacement therapy.

Homeostatic mechanisms in a young adult (i.e. 15–25 years old) are extremely well regulated. The body has an enormous number of tightly controlled feedback systems designed to keep countless physiological and biochemical parameters at optimum levels. Thus, as young adults, our internal environment is rigorously maintained in an ideal, balanced, steady state. This is why the young adult stage of our lives is associated with optimal health and vitality.

Unfortunately, as we grow older (i.e. past the age of about 30 years old), the regulatory mechanisms, which so tightly controlled the homeostatic feedback systems in our youth, begin to fail. This loss of regulatory control becomes progressively worse as we age. In time, the resulting homeostatic imbalances manifest themselves as the non-clinical signs of aging. Eventually, if the imbalances become extreme, the degenerative disease states associated with old age appear. Eventually, one of these disease states will become fatal. This, in a nutshell, is how we grow old and die.

To effectively combat this process, medical science need not understand exactly what causes homeostatic regulatory control to be lost. We need only recognize when such loss has occurred and develop the means to correct the biochemical imbalances that may result from it. Thus, we may forestall the degenerative consequences of such imbalances to reduce or prevent many of the symptoms of aging. The program outlined below is an attempt to develop a balanced regimen to help people maintain homeostasis. A brief description of the elements of this program follows.

7.5 Some Guidelines for Scientific Preventive Medicine

Patients who carefully follow a program that is mindful of the considerations described in the following sections should have an improved chance of enjoying many of the anti-aging and disease preventing benefits of maintaining homeostasis. The program may include (if needed) a combination of antioxidants, herbs, vitamin and mineral supplements, drugs, and hormone replacement therapy, an ongoing monitoring of certain specific biomarkers of health status and aging, and "customized" recommendations for diet and exercise.

7.5.1 Step 1: Baseline Data – Patient Starting Point

The first step of the protocol is to obtain a comprehensive and through understanding of each patient's current physiological status through a three-stage process.

Stage one begins with a review of the patient's complete medical background, including their detailed medical history. This gives the physician a record of any previous medical conditions that may exist. This information is helpful in the development of interventions that address the potential for recurrence of any of these past situations as part of the patient's overall treatment protocol.

Stage two involves obtaining a complete "biochemical picture" of each patient. To accomplish this, a complete battery of more than 200 different chemical blood and urine tests is measured. Tests include various metabolites of oxidative stress, a complete cardiovascular panel, an extensive hormone panel assessing a broad spectrum of endocrine function, and a trace metal panel which measures the most physiologically important trace metals. The complete test results provide an unparalleled view of each patient's current biochemical condition and homeostatic balance. In addition, the tests are designed to help identify specific health risks that may need to be addressed.

Stage three is the measurement of the patient's physiological status as determined through a series of biomarker tests. These tests are designed to assess biological or functional age. They include more than 50 physiological tests measuring such things as reaction time, hearing, vision, static balance, short term memory, flexibility, strength, skin elasticity, bone mineral density, body composition, cardiovascular function, and pulmonary capacity.

Through this three-stage process the treating physician gains greater insight into each patient's functional status. This process will establish each patient's individual baseline data, which is extremely important for two reasons. First, the patient's baseline information is necessary to develop a custom-designed intervention program for that patient. Since no two people are identical, all interventions must be based on each patient's unique physiological and biochemical requirements. Second, the baseline information and data provide a starting point from which to measure each patient's progress over the years. This progress is measured through additional testing, repeated periodically as specified by the patient's custom-designed protocol.

7.5.2 Step 2: Supplements – Optimum Nutrition

Although some scientists maintain that Americans get enough vitamins from a typical "well balanced" diet and that additional supplementation is a waste of money, many scientists do not share this view. There is evidence that the U.S. government's recommended daily allowance (RDA) is inadequate for older adults with respect to many vitamins (e.g., B complex, C, D, E, etc.) and minerals (e.g., zinc, selenium, etc.) and it ignores other important supplements entirely (e.g., co-enzyme Q-10, beta-carotene, etc.). Still it is often difficult to know exactly what supplements an individual should be taking. Therefore, a complete custom tailored regimen of dietary supplements should be developed for every patient based upon his/her initial test results. Kronos physicians design a supplementation regimen that may include vitamins, minerals, herbs, specially derived nutrients, nootropics, and macronutrients. Only in this way will each patient be certain that he/she is receiving exactly the supplementation he or she requires, the correct amount of each supplement (without fillers) and the guaranteed high quality raw materials.

7.5.3 Step 3: Hormones – Maintaining Homeostasis

One promising approach to achieving significant longevity and youthful function involves maintaining optimum homeostatic balance using hormone replacement (supplementation) therapy. In principle, the basic approach is quite simple. The proposed medical protocol determines the hormones in which one is deficient and corrects these deficiencies by careful administration of appropriate amounts of those hormones to offset the deficiency.

Since hormone replacement therapy is a long-term proposition, the physician must take great care to insure that optimum blood levels of the hormones being supplied externally are not exceeded. Excess hormone levels maintained over prolonged periods of time may produce degenerative disease processes in a manner directly analogous to those produced by age-related hormonal imbalances. This problem is avoided by the regular follow-up blood tests of hormone levels and biomarkers. Based upon the results of re-testing, modifications to the patient's hormone replacement protocol can be made.

7.5.4 Step 4: Diet and Exercise

The importance of diet and exercise in achieving optimum health and longevity is indisputable and cannot be overstated. However, entire industries already exist that focus specifically on these important subjects. In the area of diet, it is important that an individual does not eat foods that are not consistent with their customized treatment protocol. The dietary guidelines will include what specific foods to eat and not eat, when and how to eat them, and a simple explanation of calorie counting. The goal is to build an understanding of correct nutrition as a dietary habit of each patient. This is easier than it sounds.

The benefits of exercise in achieving optimum health are well-documented. The program's exercise guidelines should describe the correct way to start an exercise program, what elements of cardiovascular fitness are important, and why resistance training should always be included. A simple training program that can be modified to fit the special needs of each patient is recommended.

7.5.5 Metabolic Medicine (Nutrition, Exercise and Weight Control)

Our modern society may be the first civilization in human history to indulge itself by routinely consuming more calories than it expends. Today, there is relatively little cost and virtually no effort required to access a remarkable glut of delectable food products. Modern technology has made a wide variety of energy-dense foods available all over the developed world and particularly in the United States. Therefore, it is no wonder that obesity is fast becoming the number-one health hazard for our society.

As with all other aspects of the Program, the recommended approach to this subject is science-based and straightforward. In writing this section, I have attempted to cut through the jungle of hype in the lay press and reconcile the disjointed and confusing mountain of contradictory data in the scientific literature. The result is a simple yet effective diet and exercise program. The program is simple because it does not require a lot of complicated food analysis, record keeping, calorie counting or exercise equipment. The program is effective because, like other aspects of the overall program, if followed consistently, it will lead to diminished risk of death from age-related diseases (and obesity-related diseases) and an improved quality of life.

7.5.5.1 Nutrition (Diet)

Food supplies not only the energy we need to function but also the nutrients required to build all tissues and to produce substances used for the chemical processes that take place in our bodies on a continuous basis. There are two broad nutrient categories. Macronutrients (carbohydrates, protein, fats and water) supply energy and structural raw materials for the body. They are needed in large amounts for growth and to maintain and repair body structures. Micronutrients are the vitamins and minerals required in small amounts to help regulate chemical processes. Fiber and certain other substances, although technically not nutrients, are also part of a healthy diet.

Food calories (technically kilocalories) are the measure of the amount of energy in a food. One food calorie represents the amount of heat to raise the temperature of one liter of water by 1 degree Celsius. Carbohydrates and protein contain 4 of these calories per gram; fat contains 9 calories per g; and alcohol contains 7 calories per g (Harper 1971).

There are known to be about 50 essential factors (of which about 45 are nutrients) that must come from the diet (Udo 1993). They include:

Essential Nutrients:
- 20 or 21 minerals;
- 13 vitamins;
- 9 amino acids (10 for children, 11 for premature infants); and
- 2 fatty acids;

Energy Sources:
- starches, sugars and fats;

Other substances:
- water.

Our bodies cannot synthesize these 50 factors, but we must have them to live. These 50 factors must therefore come from outside our bodies, from our surroundings – from our diet.

Deficiencies in these basic nutrients are common. Large, U.S. government-sponsored, surveys have shown that over 60% of the population is deficient in one or more essential nutrients (Mokdad et al. 1999) where a deficiency was an intake less than the U.S. RDA, which is not equivalent to an optimum intake for the best of health or to deal with increased requirements for abnormal or stressful circumstances. The surveys tested for only 13 of the 50 essential nutrients. They found that deficiency ranged from 10 to 80% of the population for 12 of the 13 essential nutrients, with low rates of deficiency for the remaining one. Other surveys and estimates indicate that 10–95% of the population obtains less than the minimum daily requirement of each of another 11 essential nutrients.

Therefore, of about 50 essential nutritional factors, at least 23 are lacking in the foods eaten by a substantial portion of the population. Remember that deficiency leads to progressive degeneration (degenerative diseases), ending in death if adequate quantities of the deficient nutrients are not returned to the diet. When the deficient nutrients are returned, deficiency symptoms are reversed, and the deficiency disease is often cured.

The 1988 Surgeon General's Report on Nutrition and Health concluded that 15 out of every 21 deaths (more that two-thirds) in the U.S. involve dysnutrition (Beasley and Swift 1989). Conditions with malnutrition as at least part of their cause (and the percentage of deaths attributable to each) include: cardiovascular disease (heart disease, stroke and artherosclerosis; 43.8%), cancer (22.4%), diabetes (1.8%), accidents (4.4%), lung disease (3.7%), pneumonia and flu (3.2%), suicides (1.4%), and liver disease (1.1%). All 10 leading causes of death in the U.S. are included in this list. They account for 81.9% of all deaths, which totals just over 2 million people each year in the United States.

Deficiencies, excesses, or imbalances in fats alone are involved in 70% or more of all U.S. deaths. Clearly, optimized nutrition must be a major part of any well-designed optimum health/preventive medicine program.

Micronutrients

Vitamins: Vitamins are organic substances needed to regulate metabolic functions within cells. Vitamins do not supply energy, but one of their functions is to aid in the conversion of macronutrients into energy. A detailed discussion of vitamins is beyond the scope of this chapter. Suffice it to say that, in the absence of strong evidence that one's diet is providing all of the essential vitamins in adequate amounts, it is probably a good idea to supplement one's intake daily with at least the RDA. Actual knowledge of the blood levels of the most important vitamins and cofactors should help the physician to prescribe the correct intake for each patient.

Minerals: Minerals are inorganic substances that serve many functions, including acting as enzyme cofactors, helping to maintain water content, and promoting acid-base balance (pH) in the body. Macrominerals (calcium, phosphorus, chloride, sodium, magnesium, potassium, and sulfur) are present in the body in large amounts. Microminerals, though no less important, are present in smaller amounts. The most important minerals and their functions are discussed below.

Phytochemicals: Although a complete description of phytochemicals is beyond the scope of this discussion, a brief description is provided in Table 7.2. Phytochemicals are naturally occurring chemicals found in fruits, vegetables, beans, and other plant-based food products. These compounds may be responsible for part of the disease-preventing effects of fruits and vegetables.

Phytochemicals have no nutritional value in the traditional sense – that is, they are not "required" for life – but they appear to have highly beneficial effects in maintaining long term health. They have been shown to have a broad spectrum of biological activities, such as inhibiting tumor formation, producing anticoagulant effects, scavenging free radicals, blocking the cancer-promoting effects of certain estrogens, and even lowering cholesterol levels (Giovannucci 1999).

Epidemiological studies have repeatedly reported an inverse relationship between fruit and vegetable intake and the risk of cancer. Yet, except for vitamin E (and perhaps lycopene), such effects have not been associated with the intake of any

Table 7.2 Phytochemicals and their food sources

Phytochemical family	Established beneficial effects	Major food sources
Allylic sulfides	Enhance immune function, anti-tumor effects, facilitate excretion of carcinogens	Onions, garlic, leeks, chives
Carotenoids	Antioxidant activity, anti-cancer effects	Carrots, cooked tomatoes, leafy greens, sweet potatoes, apricots
Flavonoids	Function as antioxidants, Vitamin C sparing effect, inhibit tumor development, prevent oxidation of LDL, help control inflammation	Tea, coffee, citrus fruits, onions, garlic
Indoles, isothiocyanates, and sulforaphane	Stimulate cells to produce potent cancer-fighting enzymes	Cruciferous vegetables (broccoli, cabbage, kale, cauliflower, Brussels sprouts)
Phytoestrogens (isoflavones, coumestans, and lignans)	May help prevent breast cancer or prostate cancer, reduce symptoms of menopause	Soybeans and soy products (tofu, soy milk, etc.)
Phenolic acids (ellagic acid, ferulic acid, etc.)	May prevent DNA damage in cells	Berries, citrus fruits, apples, whole grains, nuts
Polyphenols	Potent anti-cancer effects	Green tea, grapes, wine
Saponins	Anti-bacterial, anti-fungal, and anti-viral effects, may boost immune function,	Beans, legumes, tomatoes, oats, soy products, spinach
Terpenes (perillyl alcohol, limonene, carnosol)	Anti-cancer effects, may help stimulate regression of certain tumors	Cherries, citrus fruit peel, rosemary (spice)

single nutrient. Thus, phytochemicals may best exert their beneficial effects when acting together as group rather than when taken alone as a supplement. Because the number of these substances in plant-based foods is so large (over 170 of them have been counted in oranges and over 1,000 in tomatoes), it may be impossible to develop supplements that will substitute for the foods themselves.

Alas, there is still no substitute for a healthy diet. Clearly, fruits and vegetables should become a major part of the daily diet.

Macronutrients

Carbohydrates: Carbohydrates are starches and sugars obtained from plants. All carbohydrates are broken down in the intestine and converted in the liver into glucose, a sugar that is carried through the bloodstream to the cells, where it is used for energy (Harper 1971). Some glucose is incorporated into glycogen, which is stored in limited amounts in the liver and the muscles for future use. Carbohydrates are also converted into fat (mostly saturated fat) when intake exceeds immediate needs and glycogen storage capacity.

Carbohydrates are the first and most efficient source of energy for vital processes in the body. Cereal grains, potatoes, and rice, the staple foods of most countries, are the principal sources of carbohydrates in the diet. Typically, 50% or more of the daily caloric intake is supplied in the form of carbohydrates. This is about 250–500 g per day in the average diet, but it varies within a wide range. A minimum of 5 g of carbohydrate per 100 calories of the total diet is required to prevent ketosis. This is equivalent to 20% of total calories from carbohydrates.

Contrary to popular belief, carbohydrate intake (not fat) is often the principal variable in gain or loss of weight.

Fiber: Fiber is a form of carbohydrate found in fruits, vegetables, grains, and legumes. Supplying no nutrients or calories, fiber is not digestible, but it is valuable in helping speed foods throughout the digestive system and possibly binding toxins and diluting their concentration in the intestine. Since it is often difficult to get sufficient fiber in a low calorie diet, we recommend supplemental psyllium husks (natural and unsweetened) as a convenient source of added fiber.

High insulin levels have been linked to an increased risk of cardiovascular disease (CVD). Dietary fiber intake reduces insulin secretion by slowing the rate of nutrient absorption following a meal and may help prevent hypertension and hyperlipidemia.

The Coronary Artery Risk Development in Young Adults (CARDIA) study explored the effect of dietary fiber and other dietary components on body weight, insulin levels, and other CVD risk factors in 2,909 healthy young adults, age 18–30, followed over a 10-year period (Ludwig et al. 1999). Body weight, insulin levels, blood lipids, and other CVD risk factors were assessed at baseline and at follow-up examinations at years 2, 5, 7, and 10. Dietary history was assessed at baseline and at year 7. After adjusting for potential confounding factors, dietary fiber was significantly inversely correlated with body weight, waist-to-hip ratio, fasting insulin (after adjusting for body mass index), and 2-hour post-glucose insulin (also adjusting for body mass index) in both whites and blacks. Dietary fiber was inversely correlated with systolic and diastolic blood pressure, triglycerides, LDL-cholesterol, and fibrinogen, and positively related to HDL-cholesterol. After adjusting for fasting insulin, however, only the link with diastolic blood pressure remained significant. Compared to fiber, dietary intake of fat, carbohydrate, and protein had inconsistent or weak associations with all CVD risk factors.

The investigators postulated that dietary fiber may prevent hyperinsulinemia by reducing circulating insulin levels and by preventing obesity associated with insulin resistance. "Fiber may play a greater role in determining CVD risk than total or saturated fat intake," they wrote. "Long-term interventional studies are needed to examine the effects of high-fiber and low glycemic index diets in the prevention of obesity and CVD."

Protein: Inadequate dietary protein intake may be an important cause of sarcopenia (loss of muscle mass and strength) in aging adults. The compensatory response to a long-term deficiency of dietary protein intake is a loss in lean body mass. The protein requirement may be considerably increased by the demands of such things as exercise, growth, increased metabolism, illness (e.g., fever), injury (e.g., burns), and trauma.

Typically, the recommended intake of protein is 10–30% of total calories. Using the currently accepted 1985 WHO nitrogen-balance formula on data from four previous studies, the combined weighted averages yielded an overall protein requirement estimate of 0.91 ± 0.043 g/kg/day. The current U.S. RDA for protein is 0.8 g/kg/day (0.35 g/pound/day) of bodyweight. The RDA is based on data collected, for the most part, on young subjects and may not be appropriate for older adults. For most elderly adults, it is probably better to err on the high side in order to compensate for a common tendency to eat too little protein. More recent data suggest that a safer protein intake for elderly adults is about 1.25 g/kg/day (Campbell et al. 1994). Therefore, on the basis these short-term nitrogen-balance studies, a safe recommended protein intake for older patients (i.e. 40+ years) is 1.0–1.25 g of high quality protein/kg/day. One study found that approximately 50% of 946 healthy free-living men and women, above the age of 60 years, consumed less than the recommended amount of protein (Hartz 1992). A large percentage of homebound older adults consuming their habitual dietary protein intake (0.67 g mixed protein/kg/day) have been shown to be in negative nitrogen balance (Bunker et al. 1987).

The requirement for protein in the diet is, however, not only quantitative; it is also qualitative, since the metabolism of a protein is inextricably connected with that of its constituent amino acids. Certain amino acids are indispensable in the diet because human tissues cannot synthesize them. The human body can manufacture only 13 of the 22 amino acids needed to synthesize human proteins. These 13 are called nonessential amino acids because they do not need to be obtained through the diet. The other 9 are known as essential amino acids, because they must be supplied in food. They are histidine, isoleucine, leucine, lysine, methionine, phenylalanine, threonine, tryptophan, and valine.

In 1991, the Food and Drug Administration (FDA) and the Food and Agricultural Organization/World Health Organization (FAO/WHO) adopted a new method for evaluating protein called the Protein Digestibility Corrected Amino Acid Score

(PDCAAS) (Food and Agriculture Organization 1991). The PDCAAS has been adopted by as the preferred method for the measurement of the protein value in human nutrition. This method compares the pattern of essential amino acids in a protein with the requirements that humans have for essential amino acids and then adjusts for how well that protein is digested. Although the principle of the PDCAAS method has been widely accepted, its relevance for older adults is unclear (Schrafsma 2000).

In addition to the PDCAAS, scientists have used a variety of biological, chemical, and microbiological methods, over the years, to measure protein quality. For humans, the most relevant measure is the Biological Value (BV), which is the amount of body protein (in grams) that can be replaced by 100 g of protein in the adult diet. The higher the protein's BV, the higher the nitrogen retention. Proteins with the highest BV are the most tissue-building (e.g. muscle) and growth promoting. Typical BVs include 49 for beans, 54 for wheat, 74 for soy, 77 for casein, 80 for beef, 91 for cow's milk, 100 for whole eggs, and 104 for whey protein. From such an evaluation, it appears that milk and egg proteins are superior to plant proteins in terms of their bioavailability. Still, well-processed soy protein isolates and concentrates can also serve as a good source of protein when eaten together with other, complementary, protein sources.

Fat: Fats are composed of combinations of fatty acids: saturated, monounsaturated, and polyunsaturated linked to glycerol. Unlike carbohydrates and proteins, they are only slightly modified by the body after ingestion. If they are not burned for energy, then they are stored, virtually intact, in adipose tissue or incorporated, unchanged, into the structure of cell membranes. Thus, the fatty acid ratios within an individual's body (cell membranes and adipose tissue) directly reflect the nature and fatty acid proportions of that person's dietary fat (Simopoulos and Robinson 1999).

Saturated fatty acids (SFAs) are abundant in animal products (meats, cheese, whole milk, and butter), and certain vegetable oils – palm kernel, and coconut – are also saturated. SFAs are the most harmful to health. They raise blood cholesterol levels and possibly contribute to various forms of cancer.

Monounsaturated fatty acids (MUFAs) predominate in foods such as olive oil, canola oil, and avocados. Cholesterol levels drop as MUFAs replace SFAs in the diet.

Polyunsaturated fatty acids (PUFAs) make up the majority of the fat in safflower oil, sunflower oil, corn oil, fish, and some nuts, such as walnuts.

Fatty acids all have the same basic structure, a string of hydrogenated carbon atoms connected by single (saturated) or double (unsaturated) bonds and bounded by a methyl group on one end (omega) of the chain and a carboxylic acid group on the other end (delta). They are named according to the total number of carbon atoms they contain and the distance of the first double bond from the omega end of the molecule. For example, the first double bond in omega-3 fatty acids occurs at the third carbon from the methyl group.

There are two essential fatty acids (Fig. 7.1); one (linolenic acid) is an omega-3 fatty acid and the other (linoleic acid) is an omega-6. Most of the PUFAs in vegetable fats are omega-6 fatty acids, whereas fish and some nuts and fruits contains

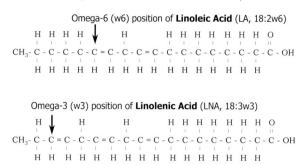

Fig. 7.1 The two essential fatty acids

both omega-3 and omega-6 fatty acids. Americans typically have an abundance of omega-6 in their bodies, but are sorely lacking in omega-3. It has been estimated that 60% of the population gets too much of one essential fatty acid, and that 95% of the population gets too little of the other. It may, therefore, be advisable to supplement with omega-3 (e.g. flax oil) in order to reestablish a proper balance between these two classes of fatty acids in the body. PUFAs, like MUFAs, lower blood cholesterol levels when substituted for SFAs in the diet.

Most of the fats in foods (and in adipose tissue) are in the form of triglycerides. Triglycerides are the major form that fat takes in biological systems. They consist of any combination of three fatty acids attached to a glycerol molecule – for example, two SFAs and one PUFA. As a result, no food contains just one type of fatty acid. Instead, the fat in a particular food is classified as "saturated" or "unsaturated" based on the type of fatty acid that predominates. For example, olive oil is typically thought of as a monounsaturated fat, but it also contains some polyunsaturated and saturated fat: 75% of the oil is MUFA, 14% is SFA and 9% is PUFA. (The percentages do not add up to 100% because other fat-like substances are also present in olive oil.) The effect of fats and oils on health is a topic of much discussion in both the scientific and lay literature.

By now, almost everyone has heard the message that reducing dietary fat can lessen the risk of many chronic diseases. Perhaps the most serious and widespread of these is obesity, and there is no question that excessive fat intake can be a contributing cause. However, fat is not the only cause. While it is true that a high fat intake contributes to obesity, CHD, and some forms of cancer, it is also true that not all types of fat have the same effects on health. SFAs increase blood levels of LDL (bad) cholesterol, while diets low in SFAs reduce LDL levels. Although not everyone responds to the same degree, on average, every 1% reduction in SFA calories reduces total blood cholesterol levels by about 2 mg/dl, mostly from a drop in LDL cholesterol.

Trans fatty acids (TFAs), which raise LDL and lower HDL cholesterol levels, are formed when food manufacturers add hydrogen atoms to unsaturated fats to make them more saturated and thereby improve their shelf life. All foods containing this unnatural harmful ingredient should be strictly avoided. Due to heightened public awareness, these foods are gradually vanishing from the grocery store shelf as manufacturers change their formulations.

Studies suggest that TFAs are even more harmful to health than SFAs because trans fats raise the ratio of LDL to HDL cholesterol more than even saturated fats. An analysis of six studies comparing the effect of TFAs, SFAs, and unsaturated fats determined that for every 1% of calories that comes from TFAs, LDL cholesterol rises by 1.5 mg/dl and HDL cholesterol falls by 0.4 mg/dl (Hu et al. 1999). TFAs also raise blood triglyceride levels – which may increase CHD risk in some people, especially those with diabetes. TFAs may also increase blood levels of Lp(a), a type of lipoprotein that may reduce the ability of certain blood proteins to break down blood clots.

Blood cholesterol levels are affected more by dietary fatty acids than by dietary cholesterol. However, since animal products – the only source of cholesterol – are the leading sources of saturated fat, reducing one's intake of animal fat automatically lowers one's cholesterol intake. A few foods – eggs, lobster, and shrimp – are low in saturated fat but high in cholesterol. New margarines containing plant stanols and sterols can reduce LDL by 7–14% when used daily for a year or more.

Unsaturated fatty acids lower LDL cholesterol levels. One analysis of several studies concluded that when calories supplied by SFAs are reduced by 10% and replaced with unsaturated fats, total cholesterol levels drop by an average 25 mg/dl. Unsaturated fatty acids lower cholesterol partly by stimulating the liver to remove more LDL cholesterol from the bloodstream.

Monounsaturated fatty acids (MUFAs), however, may be preferred over polyunsaturated fatty acids (PUFAs). Replacing saturated fat with PUFAs reduces beneficial HDL cholesterol, while replacing SFAs with MUFAs leaves HDL levels stable. In addition, LDL cholesterol only forms plaques in arterial walls after the fatty acids in LDL are oxidized, and PUFAs are more easily oxidized than MUFAs.

The standard recommendation is to decrease total fat intake to less that 30% of total calories, and to replace those calories with carbohydrates. But diets low in total fat and high in carbohydrates may lower HDL cholesterol and raise triglyceride levels. The rise in triglycerides can be minimized if the switch is made gradually (Katan 1998). The drop in HDL cholesterol may increase CHD risk, especially in women, who appear to be more affected than men. In contrast, when fat intake is not reduced and saturated fats are replaced with MUFAs, HDL cholesterol declines only moderately, and triglyceride levels remain stable.

The decline in HDL cholesterol resulting from replacing saturated fats with carbohydrates may not be as harmful as inherently low HDL cholesterol levels, since LDL cholesterol is reduced as well. Research conducted by Dean Ornish has shown that a very-low-fat diet (about 10% of total calories from fat) combined with smoking cessation, exercise, and stress reduction can reverse the extent of atherosclerotic narrowing of coronary arteries in patients with CHD, despite corresponding reductions of HDL cholesterol levels.

Some studies have suggested that as long as saturated fat intake is low, the total fat content of the diet has no impact on CHD risk. The American Heart Association (AHA) does not recommend diets containing less than 15% of calories from fat.

At this point, following the standard recommendations for fat intake is the prudent course for most people. However, this is true only as long as other factors are kept at optimum levels. For example, saturated fats and "trans" fats must be kept to a minimum. The ratio of omega-6 fatty acids to omega-3 fatty acids should be kept as low as possible. Last, but not lease, total caloric intake must be such that a relatively lean body composition is maintained (less than 20% body fat for males; less than 25% body fat for females; lower is better).

Water

Water can be considered an essential nutrient because it is involved in all body processes. Since an individual's water needs vary with diet, physical activity, environmental temperature, and other factors, it is difficult to pin down an exact water requirement. On average, the National Academy of Sciences (NAS) recommends a daily water intake in the range of 7–11 cups when consuming 2,000 calories per day (Food and Nutrition Board 1989). Although water may be obtained through other liquids, such as milk and juices, it is better to drink water itself to achieve this daily total. Beverages with alcohol and caffeine are not good substitutes for water since both substances have diuretic effects and cause a net loss of fluids from the body. In addition, since our senses of thirst and hunger often become less distinguishable as we grow older, the drinking of anything containing calories is discouraged. We should eat for sustenance and drink to re-hydrate our bodies.

A Nutritional Controversy

The evidence that diet (nutrition) plays an important role in health maintenance is indisputable. Countless scientific studies have been conducted establishing this relationship. Yet, in spite of this fact, there is little consensus among the nutrition experts about what constitutes a proper diet for optimum health. Although everyone seems to agree that obesity is detrimental to health, opinions vary widely regarding how best to combat this problem.

Table 7.3 is a list of ten popular diets (VanTulnen and Adriano 2000). It shows the authors' names and macronutrient composition for each of the diets. Nine of these diets are being or have been promoted in best selling books. The tenth is derived from the dietary guidelines established by the U.S. Department of Agriculture. The numerical values in Table 7.3 were calculated by averaging three to five days of recommended menus from each book.

The USDA's figures were calculated based upon the Food Guide Pyramid, assuming 2,000 calories per day. A quick look at the bottom line of the table clearly reveals that simply quantifying and tabulating the menus in these books

Table 7.3 Nutritional composition of ten popular diets

	Atkins	Eades	Steward	Sears	Sarah	Rolls	Pritikin	Goor	Ornish	RDA
Total fat (g)	110	105	70	30	30	40	15	20	10	66
Percent of total fat (weight)	31.3	31.0	20.0	12.3	8.7	9.0	3.6	3.3	2.5	14.3
Total fat (calories)	990	945	630	270	270	360	135	180	90	594
Percent of total fat (calories)	55.9	56.8	39.6	26.7	19.4	20.2	8.6	7.9	6.0	29.7
Sat. fat (g)	36	34	20	8	10	12	3	5	2	22
Percent of Sat. fat (weight)	10.3	10.0	5.7	3.3	2.9	2.7	0.7	0.8	0.5	4.8
Sat. fat (Calories)	324	306	180	72	90	108	27	45	18	198
Percent of Sat. fat (calories)	18.3	18.4	11.3	7.1	6.5	6.1	1.7	2.0	1.2	9.9
Protein (g)	135	110	100	70	90	95	95	115	60	62
Percent of Protein (weight)	38.5	32.4	28.6	28.8	26.1	21.5	22.7	19.2	14.9	13.5
Protein (calories)	540	440	400	280	360	380	380	460	240	248
Percent of protein (calories)	30.5	26.4	25.2	27.7	25.9	21.3	24.1	20.2	16.1	12.4
Carbs. (g)	60	70	140	115	190	260	265	410	290	290
Percent of carbs. (weight)	17.1	20.6	40.0	47.3	55.1	58.8	63.4	68.3	72.1	63.0
Carbs. (calories)	240	280	560	460	760	1,040	1,060	1,640	1,160	1,160
Percent of carbs. (calories)	13.6	16.8	35.2	45.5	54.7	58.4	67.3	71.9	77.9	57.9
Fiber (g)	10	20	20	20	25	35	40	50	40	20
Percent of fiber (weight)	2.8	5.9	5.7	8.2	7.2	7.9	9.6	8.3	10.0	4.3
Total intake (g)	351	339	350	243	345	442	418	600	402	460
Total caloric intake	1,770	1,665	1,590	1,010	1,390	1,780	1,575	2,280	1,490	2,002

does not result in an "apples to apples" comparison in terms of the absolute quantity of the macronutrients recommended. The diets, as described, are markedly different in caloric intake. "The Zone" by Barry Sears, Ph.D., promotes a carbohydrate:protein:fat ratio of 40:30:30 and contains only 1,000 calories, whereas "Choose to Lose" by Dr. Ron and Nancy Goor, promotes a low fat diet containing almost 2,300 calories (mostly from carbohydrates). Obviously, regardless of the ratio of macronutrients, one will lose more weight on the Sears diet than in the Goor diet. This, however, provides no clue as to the relative merit of the macronutrient ratios bring promoted, and it makes comparison of the diets somewhat difficult.

Even after normalizing their caloric content, the disparity of these diets is astonishing. For a person on a 2,000 calorie diet, Dr. Atkins recommends a daily protein intake of 153 g. while the USDA suggests eating only 69 g of protein. Recommended percent total calories from fat ranges from 56.8% (Eades), to 6.0% (Ornish). Perhaps predictably, carbohydrates show an even wider range of variation among the authors. Dr. Ornish touts the health benefits of a low fat, high carbohydrate (77.9% of calories) diet, while Atkins and Eades point to the dangers of hyperinsulinemia, admonishing us to limit our carbohydrate intake to 13.6% of calories and 16.8% of calories, respectively.

No wonder people are confused and discouraged about controlling their weight. If physicians and scientists cannot agree about what is the best approach, how can a lay person expect to resolve the problem? Fortunately, the situation my not be as hopeless as it seems. Although there is much disagreement about the details of these dietary approaches to weight control, there is also much about which the experts agree. By focusing on fundamental scientific principles and the biochemistry of metabolism about which there is a virtual consensus, it is possible to create a plan that is both scientifically sound and flexible enough to work for everyone.

The guiding principles of this common sense diet are to

- eat foods dense in essential nutrients
- avoid "empty calories" (sugary treats, French fries, etc.)
- minimize total caloric intake to maintain low body weight.

Although caloric restriction to 60–70% of usual or "ad-libitum" intake has been shown to prolong lifespan by as much as 50% in a wide variety of species, including several mammals, it has not yet be shown to do so in humans. However, there is no reason to suppose that such an intervention would not be effective.

Consider the implications of this in the daily life of a typical person who would normally consume (fully fed) 2,400 calories per day to maintain body weight. If this person reduced his caloric intake by 30% (i.e., by 720 calories per day), then he or she might expect to enjoy a 33% reduction in aging rate. This means that the individual would age only 24 hours for every 36 hours of life and (an increase in remaining lifespan of 50%). Since he would gain 12 hours of life every day he was on this 1,680 calorie diet, he would be gaining, essentially, 1 hour of life for each 60 calories of restricted intake. Stated another way, the price he would pay, in terms of lifespan, for each 60 calories in excess of the minimum number required for basic sustenance, is 1 hour of life. Although this calculation is different for everyone because we all have different minimum calorie requirements, it is a reasonable approximation of the cost each of us pays for eating excess calories. It provides a simple "rule of thumb" for explaining, in easy to understand terms, the costs and benefits of restricting one's caloric intake as much as comfortably possible.

Many studies have shown that blood lipid profiles are dramatically improved in patients who follow a "low fat – high carbohydrate" diet. Conversely, other studies clearly demonstrate that "high fat – low carbohydrate diets" can have a similar effect upon blood lipid profiles. Proponents of both types of diet go into great detail explaining why their diet is best. Yet if both types of diet are effective, what is their common characteristic? And the answer is that they both result in a lower than normal dietary intake of calories. This is the single common thread that ties together all the "fad" diets that have worked so well, for so many people, over the years. The only problem with such diets is that people do not stay on them. The sad reality is that caloric restriction, even to a small degree, is difficult, and most people quickly go back to eating the way they did before after they have hit their target weight.

What is required is not really a diet at all, but a mindset, a whole new way of looking at food, for the rest of your life. All food should be eaten consciously, with a full appreciation of its costs and benefits. Weight control is based upon the first law of thermodynamics and the relationship between calories consumed and calories burned. Both factors (diet and exercise) are equally important in achieving optimum weight, fitness, and health. Therefore, a brief discussion of exercise is in order.

7.5.5.2 Exercise

The Relevance of Exercise and Muscle Mass to Aging

The main determinant of energy expenditure in sedentary individuals is fat-free mass (Province et al. 1995), which declines by about 15% between the third and the eighth decade of life, contributing to a lower basal metabolic rate (Cohn et al. 1980) and daily energy expenditure (McGandy et al. 1966) with age. Declining metabolic rate and activity coupled with an energy intake in excess of the declining need for calories contributes to increases in body weight up to the age of 60 (Roberts et al. 1992). In addition to its role in energy metabolism, age-related skeletal muscle alterations may contribute to such age-associated changes as reduction in bone density (Sinaki et al. 1986; Snow-Harter et al. 1990), insulin sensitivity (Kolterman et al. 1980), and aerobic capacity (Fleg and Lakatta 1988).

These data indicate that the preservation of muscle mass and the prevention of sarcopenia can help prevent the decrease in metabolic rate and that strategies for preserving muscle mass with advancing age, as well as for increasing muscle mass and strength in the previously sedentary elderly, may be an important way to increase functional independence and decrease the prevalence of many age-associated chronic diseases. And in fact, participation in regular physical activity (both aerobic and strength training) elicits a number of favorable responses that contribute to healthy aging. These include: (a) cardiovascular fitness; (b) improved strength, muscle mass, and bone density; (c) increased postural stability, flexibility, and prevention of falls; (d) improved mood and cognitive function. Given that, by the year 2030, the number of individuals in the United States 65 years and over will reach 70 million and that persons 85 years and older will be the fastest growing segment of the population, it is imperative to determine the extent to which and mechanisms by which exercise and physical activity can improve health, functional capacity, quality of life, and independence in this population.

One fact is already clear. It is almost never too late to begin exercising. At least one major study found that even patients who began a regular exercise program at age 75 had a lower death rate over the next several years than a similar group of sedentary patients who didn't begin such a program. As a simple rule of thumb, on average, for every hour an individual spends exercising, he will increase his lifespan by about 2 hours (Anonymous 2000).

A more difficult question is not whether to exercise, but how much? Morris Ross found that exercise extended life in ad libitum fed mice but shortened life in mice on CR, essentially equalizing the lifespans of these groups (Ross 1976). Clearly, we have much to learn about the interaction of diet and exercise as determinants of maximum lifespan, but fortunately, the level of exercise likely to produce negative effects is well beyond the capability of the vast majority of persons.

Most human studies agree that the greatest impact of exercise occurs at the lowest level of activity (i.e. the change from a sedentary life-style to moderate daily exercise). However, incremental beneficial (albeit less dramatic) effects are almost always seen, even at the highest level of exercise measured in most studies. For example, The March 2000 issue of *The Harvard Men's Health Watch* reported a linear decrease of cardiac risk factors for runners as the number of miles run per week increased from 0 to 31 miles (statistics based on 8,896 runners) (Anonymous 2000). Both speed and distance turned out to be important. The benefits of exercise continued to increase even at the highest level of activity measured (approximately 5 miles/day, 6 days/week). This is roughly equal to 3,200 calories per week.

But to be sure, more moderate exercise also produces significant health benefits. Compared to an activity index of 0 calories/week, expending 750 calories/week lowered all-cause mortality from about 7.5 deaths/1,000 per year to less than 5.5 deaths/1,000 per year, and going to 2,000 calories/week further lowered all-cause mortality to a minimum of 4 deaths/1,000 per year, no benefit being apparent at higher activity levels up to 3,500 calories/week (Paffenbarger et al. 1993). Two thousand calories/week may seem like a lot, but it's close to the level of exercise deemed optimal by the Harvard Alumni Study.

Table 7.4 shows how long it takes to burn 2,000 calories based on various forms of physical activity. If you don't make time, in an otherwise busy schedule, for what may seem to be a superfluous (non-productive) use of your time, you're actually doing yourself a grave disservice.

Endurance Training

For years, physicians have advised middle-aged patients to get "dynamic" or "aerobic" exercise – physical activity that emphasizes energetic movement of the arms and legs over long periods of time – because of its cardiovascular benefits. Because

Table 7.4 Hours needed to expend 2,000 calories

Activity	Hours
Walking (Strolling)	10
Bowling	8½
Golf	8
Raking Leaves	7
Tennis (Doubles)	6
Walking (Brisk)	5½
Biking (Leisurely)	5½
Tennis (Singles)	4½
Racquetball or Squash	4
Biking (Rigorously)	4
Jogging	4
Skiing (Downhill)	4
Calisthenics (Brisk)	3½
Skiing (Cross Country)	3
Running	3

cardiovascular disease is the major cause of death in older men and women, the effect of endurance exercise training on cardiovascular disease risk factors is of paramount importance.

Despite the obvious benefits of this type of exercise, the cardiovascular response to aerobic exercise in healthy adults decreases 5–15% per decade after the age of 25 years. This is due to decreases in both maximal cardiac output and maximal arteriovenous O_2 difference, which together contribute to an age-associated reduction in VO_2max (Fleg et al. 1995; Ogawa et al. 1992). Much of the age-associated decline in maximal cardiac output is due to the progressive, age-related decrease in maximum heart rate of 6–10 beats per minute per decade.

The cardiovascular responses of older adults to aerobic exercise are qualitatively and, in most cases, quantitatively similar to those of young adults. Heart rate at the same relative work rate (same percent of VO_2max) is lower in older versus younger adults. On the other hand, the heart rate responses of young and older adults are similar at the same absolute work rate (the same walking speed or resistance on a stationary ergometer). Blood pressures are generally higher at both the same absolute and relative work rates in older versus younger adults (Stratton et al. 1994).

Contrary to popular belief, it is now clear that older adults elicit the same 10–30% increases in VO_2max with prolonged endurance exercise training as young adults (Hagberg et al. 1989; Seals et al. 1984a). As with young adults, the magnitude of the increase in VO_2max in older adults is also a function of training intensity, with light-intensity training eliciting minimal or no changes. Furthermore, arterial stiffness is also reported to be lower in older endurance-trained individuals (Vaitkevicius et al. 1993).

Some evidence indicates that high levels of exercise training maintained over a long period of time results in a diminished rate of loss of VO_2max (one of the most reliable biomarkers of aging) with age in older adults. These studies generally report a reduced rate of loss expressed as a percentage of the initial VO_2max value, which could be an artifact of the athletes' initially higher VO_2max. On the other hand, the rate of VO_2max decline for endurance-trained athletes over age 70 appears to be similar to that for sedentary adults (Pollock et al. 1997).

Cross-sectional and intervention studies in older adults consistently indicate that endurance exercise training is associated with lower fasting and glucose-stimulated plasma insulin levels, as well as improved glucose tolerance (if initially impaired) and insulin sensitivity (Hersey et al. 1994; Kirwan et al. 1993; Seals et al. 1984b). Improvements in glucose and insulin metabolism are evident in older adults before changes in body weight or body composition occur, indicating that this is not the result of a reduction in visceral fat.

Endurance exercise training appears to lower blood pressure to the same degree in young and older hypertensive adults, although it is not clear if this benefit increases with increased training intensity. What is clear is that light to moderately intense training (40–70% of VO_2max) is effective in lowering blood pressure in older hypertensive adults.

While it is probably true that older adults improve their plasma lipoprotein and lipid profiles with exercise training, these changes may be secondary to training-induced reductions in body fat stores (Stevenson et al. 1995). The improvements are generally similar to those evident in young adults and include increases in plasma HDL and HDL2 cholesterol levels and reductions in plasma triglyceride levels and the cholesterol:HDL ratio.

Body composition is also improved with endurance exercise training in a similar fashion in older and young adults. The most consistent change is a reduction in the overall percent of body fat with exercise training in older adults, even if body weight is maintained (Seals et al. 1984a). Furthermore, one study reported that intra-abdominal fat decreased by 25% in older men who lost only 2.5 kg of body weight with exercise training (Schwartz et al. 1991). This finding is especially important for older men because intra-abdominal fat is the body fat depot that increases the most with age and is associated with other cardiovascular disease risk factors.

The contraindications to exercise testing and exercise training for older men and women are the same as for young adults (American College of Sports Medicine 1995). The major absolute contraindications precluding exercise testing are recent ECG changes or myocardial infarction, unstable angina, uncontrolled arrhythmias, third degree heart block, and acute congestive heart failure. The major relative contraindications for exercise testing include elevated blood pressures, cardiomyopathies, valvular heart disease, complex ventricular ectopy, and uncontrolled metabolic diseases.

It is of paramount importance to remember that asymptomatic cardiovascular disease and the absolute and relative contraindications precluding exercise testing are much more prevalent in older adults. In addition, there is an increased prevalence of comorbidities in older adults that affect cardiovascular function, including diabetes, hypertension, obesity, and left ventricular dysfunction. Thus, adherence to the general ACSM testing guidelines with respect to the necessity for exercise testing and for medical supervision of such testing is imperative.

Strength Training

Loss of muscle mass (sarcopenia) with age in humans is well documented. The excretion of urinary creatinine, reflecting muscle creatine content and total muscle mass, decreases by nearly 50% between the ages of 20 and 90 years (Lexell et al. 1983). Computed tomography of individual muscles shows that after age 30, there is a decrease in cross-sectional area, decreased muscle density, and increased intramuscular fat. This muscle atrophy may result from a gradual and selective loss of muscle fibers. The number of muscle fibers in the midsection of the *Vastus lateralis* of autopsy specimens is significantly lower in older men (age 70–73 years) compared with younger men (age 19–37 years) (Larsson 1983). The decline is more marked in Type II muscle fibers, which decrease from an average of 60% in sedentary young men to below 30% after the age of 80 years (Jette and Branch 1981), and is directly related to age-related decreases in strength.

A reduction in muscle strength is a major component of normal aging. Data from the Framingham study indicate that 40% of the female population aged 55–64 years, 45% of women aged 65–74 years, and 65% of women aged 75–84 years were unable to lift 4.5 kg (Danneskiold-Samsøe et al. 1984). In addition, similarly high percentages of women in this population reported that they were unable to perform some aspects of normal household work.

In general, approximately 30% of strength is lost between 50 and 70 years. Much of this reduction in strength is due to a selective atrophy of Type II muscle fibers. After the age of 70 years, muscle strength losses may be even more dramatic. Knee extensor strength in a group of healthy 80-yr-old men and women studied in the Copenhagen City Heart Study (Aniansson et al. 1981) was found to be 30% lower than a previous population study (Bassey et al. 1988) of 70 year old men and women. Thus, cross-sectional as well as longitudinal data indicate that muscle strength declines by approximately 15% per decade in the 6th and 7th decade and about 30% thereafter. While other factors may contribute, the overwhelming cause of this loss in strength is the well known age-related decrease in muscle mass.

The decline in muscle strength associated with aging carries with it significant consequences related to functional capacity. A significant correlation between muscle strength and preferred walking speed has been reported for both sexes (Fiatarone et al. 1990). A strong relationship between quadriceps strength and habitual gait speed in frail institutionalized men and women above the age of 86 years supports this concept (Frontera et al. 1990). In older, frail women, leg power was highly correlated with walking speed, accounting for up to 86% of the variance in walking speed. Leg power, which represents a more dynamic measurement of muscle function, may be a useful predictor of functional capacity in the very old.

Strength conditioning is generally defined as training in which the resistance against which a muscle generates force is progressively increased over time. Muscle strength has been shown to increase in response to training between 60 and 100% of the 1 repetition maximum (RM). Strength conditioning results in an increase in muscle size, and this increase in size is largely the result of an increase in contractile protein content.

It is clear that when the intensity of the exercise is low, older subjects achieve only modest increases in strength. A number of studies have demonstrated that, given an adequate training stimulus, older men and women show similar or greater strength gains compared with young individuals as a result of resistance training. Two to threefold increases in muscle strength can be accomplished in a relatively short period of time (3–4 months) (Frontera et al. 1988, 1990). Heavy resistance strength training seems to have profound anabolic effects in older adults. Progressive strength training greatly improves nitrogen retention at all intakes of protein, and for those on marginal protein intakes, this may mean the difference between continued loss or retention of body protein stores (primarily muscle).

Strength training may be an important adjunct to weight loss interventions in the elderly. Significant increases in resting metabolic rate with strength training, in older adults, have been associated with a significant increase in energy intake required to maintain body weight (Campbell et al. 1994). The increased energy expenditure includes increased resting metabolic rate and the energy cost of resistance exercise. Strength training is, therefore, an effective way to increase energy requirements, decrease body fat mass, and maintain metabolically active tissue mass in healthy people. In addition to its effect on energy metabolism, resistance training also improves insulin action in older subjects (Miller et al. 1994).

The effects of a heavy resistance strength training program on bone density in older adults can offset the typical age-associated declines in bone health by maintaining or increasing bone mineral density and total body mineral content (Nelson et al. 1994). However, in addition to its effect on bone, strength training also increases muscle mass and strength, dynamic balance, and overall levels of physical activity. All of these outcomes may result in a reduction in the risk of osteoporotic fractures. In contrast, traditional pharmacological and nutritional approaches to the treatment or prevention of osteoporosis have the capacity to maintain or slow the loss of bone but not the ability to improve balance, strength, muscle mass, or physical activity.

Flexibility Training

Flexibility is a general term, which encompasses the range of motion of single, or multiple joints and the ability to perform specific tasks. The range of motion of a given joint depends primarily on bone, muscle, and connective tissue structure and function, other factors such as pain, and the ability to generate sufficient muscle force. Aging affects the structure of these tissues such that function, in terms of specific range of motion at joints and flexibility in the performance of gross motor tasks, is reduced. The basis for exercise interventions to improve flexibility is that the muscle or connective tissue properties can be improved, joint pain can be reduced, and/or muscle recruitment patterns can be altered.

Soft tissue restraints that may affect flexibility include changes in collagen, which is the primary component of the fibrous connective tissue that forms ligaments and tendons. Aging causes an increase in the crystallinity and cross-linking of the collagen fibers and increases the fibers' diameter, thereby reducing extendibility.

It is evident that flexibility declines with age, with the maximum range of motion occurring in the mid-twenties for men and late-twenties for women (Greey 1955). A study designed to establish population-based normative values indicated losses in the active ranges of motion of the hip and knee that were associated with increasing age (Roach and Miles 1991).

Flexibility training is defined as a planned, deliberate, and regular program of exercises intended to progressively increase the usable range of motion of a joint or set of joints. The effect of a flexibility program can be quantified by changes in joint range of motion and mobility assessment scores. Studies have shown both significant positive effects and no significant effects of exercise on the range of motion of joints in the older adult, depending on the duration of the program, the size of the subject group, the rate of attrition, and the measurement technique.

Few studies have used direct end range of motion exercise (possibly because it would be difficult to maintain subject interest and compliance with such a program). Most studies have used more indirect approaches, such as walking, dance, aerobic exercise, or "general exercise," often mixed with stretching exercises that were hypothesized to have an effect on flexibility. The majority of these studies have demonstrated significant improvements in the range of motion of various joints (neck, shoulder, elbow, wrist, hip, knee, and ankle) in older adults who participated in a program of regular exercise (Morey et al. 1991; Munns 1981).

7.5.5.3 Weight Control

A telephone survey of more than 100,000 people living in the U.S. revealed 64% of men and 78% of women to be engaged in some form of weight control effort (Harper 1971). Nearly 90% of these people said they modified their diet to lose weight. For many of these people, and indeed, it seems, for the entire nation, the focus of their efforts was to cut back on the consumption of fatty foods rather than total calories. The fallacy of this approach to weight control is illustrated by observations of Allred showing that while consumption of dietary fat declined significantly and at an accelerating rate between 1955 and 1990, the incidence of obesity increased exponentially over the same period (Allred 1995). Clearly, relying exclusively upon reductions in fat intake to control body weight does not work.

At the heart of this dilemma is a simple mathematical equation based upon the first law of thermodynamics: *Calories Consumed − Calories Burned = Net Calories Stored As Fat*. This equation may be the one aspect of nutrition and weight control about which there is almost universal agreement. Thus, good physical health and weight management requires a long-term balance between energy consumed and energy expended.

Fat cells evolved to provide a portable storage facility for precious (scarce) calories in times of famine. This energy efficient adaptation has served our species well over most of the last 140 thousand years. Only in recent times has a surplus of calories caused a problem. There is little doubt that the abundance of high-calorie foods plus little need to exercise has meant bigger waistlines and new health risks for modern man.

Given the dramatic impact that even a slight imbalance in the above equation can have on body weight over time, it is a wonder that our current availability of surplus calories has not produced an even greater obesity problem than it has. Consider the situation of a typical, moderately active, American who requires 2,500 calories to maintain his body weight. A positive imbalance of as little as 10% (250 calories) can result in alarming weight gain over time. Stored as fat, 250 calories per day is equal to about 24 pounds of extra body fat every year. Just two 12 oz. cans of beer or soda can contain more than 300 calories. It is easy to see why obesity is such a problem in the United States today. Over the course of one year, the simple act of quenching one's thirst by drinking a soft drink or beer twice a day can result in a weight gain in excess of 25 pounds. Such a seemingly minor imbalance can become a serious health hazard if maintained for a prolonged period.

Fortunately, via our hypothalamic satiety center, the body normally does an excellent job of regulating our caloric intake to within a very close approximation to that required for perfect balance. All it needs is a little help from the higher brain centers and a state of optimum health and weight control can be achieved. This help comes in the form of the herein recommended approach to nutrition, exercise and weight control. The combination of eliminating empty calories and a moderate exercise program are all that is necessary for most people to maintain optimum weight control and fitness.

Body mass index (BMI) is a convenient (but not perfect) method for estimating whether one is overweight or within the optimal range for good health. BMI is the weight in kilograms divided by the square of the height in meters. However, for certain body types, BMI can be misleading. For example, a muscular body building athlete and his "couch potato" friend are both six feet tall and weigh the same (228 pounds). Thus, they both have a BMI of 31 (see Table 7.5, which provides BMI values calculated in relation to weight in pounds and height in feet and inches). However, the body builder is lean and fit (% body fat by DEXA is 15%) whereas the couch potato is dangerously overweight (% body fat by DEXA is 35%). Clearly, this illustrates the importance of good judgment in determining when intervention (of any type) is appropriate.

Shaper et al. showed that all-cause mortality is minimal and independent of BMI between BMI values of 20 and 28; at higher and lower values, all-cause mortality was found to rise. However, the increase in all-cause mortality when BMI was less than 20 was due largely to cancer deaths of smokers who, as a group, tended to fall into this category. When the data are corrected for this, protection from mortality related to cardiovascular disease continued to decline below a BMI of twenty (Shaper et al. 1997). Despite the lack of effect of BMI between 20 and 28, people with a BMI of more than 25 should probably increase exercise and reduce caloric intake, and those whose BMI exceeds 27 may be candidates for drug therapy, particularly if they also have any of the following co-morbidities: hypertension, diabetes, coronary artery disease, hyperlipidemia, or stroke.

Table 7.5 BMI calculation from height (feet/inches) and weight (pounds)

BMI	23	24	25	26	27	28	29	30	31	32	33	34
Ht	Weight (lbs.)											
4'10"	109	114	119	124	129	134	138	143	149	153	158	163
4'11"	115	120	124	128	133	138	143	148	154	158	164	169
5'0"	118	123	128	133	138	143	148	153	159	164	169	175
5'1"	122	127	132	137	143	148	153	158	165	169	175	180
5'2"	124	130	136	142	147	153	158	164	170	175	181	186
5'3"	131	136	141	146	152	158	163	169	175	181	187	192
5'4"	133	139	145	151	157	163	169	174	181	187	193	199
5'5"	138	144	150	156	162	168	174	180	187	193	199	205
5'6"	143	149	155	161	167	173	179	186	192	199	205	211
5'7"	145	152	159	166	172	178	185	191	198	205	211	218
5'8"	150	157	164	171	177	184	190	197	204	211	218	224
5'9"	155	162	169	176	182	189	196	203	210	217	224	231
5'10"	160	167	174	181	188	195	202	207	216	223	230	237
5'11"	165	172	179	186	193	200	208	215	222	230	237	244
6'0"	170	177	184	191	199	206	213	221	228	236	244	251
6'1"	173	181	189	197	204	212	219	227	236	243	251	258
6'2"	178	186	194	202	210	218	225	233	241	250	258	265
6'3"	184	192	200	208	216	224	232	240	248	256	264	272

7.6 Summary

The currently recommended approach to prolongation of a healthy lifespan involves the careful assessment of individual health vulnerabilities and a search for deficiencies of vitamins, other nutrients, and hormones that can be ameliorated. Lifestyle interventions include maintenance of low normal BMI in the 24–26 range by a diet containing reduced calories but enriched fiber and essential nutrients, especially those found in fruits and vegetables. In addition, regular, vigorous aerobic and resistance exercise as well as attention to flexibility are required. Beyond lifestyle, appropriate medications should be employed as needed to reduce risk factors such as high blood pressure or high levels of LDL cholesterol. Finally, a careful and judicious approach to hormone replacement is recommended for those individuals in whom a clear cut hormone deficiency state can be documented. The promise of such "radical interventions" as stem cell replacement and regenerative medicine, extension of telomeres using telomerase inducers, and specific regimens of antioxidant or other anti-aging factors targeting specific cellular pathways remains to be determined by scientific research, but such interventions will undoubtedly further contribute to an integrated approach to life-extension as time goes on.

Acknowledgements Dr. Heward agreed to write this chapter approximately 1 month before being diagnosed as having Stage IV esophageal cancer, which claimed his life less than three months later. During the extremely short time available, Dr. Heward wrote almost all of the text of this chapter and assembled its supporting information, but was not able to put the chapter into its final form. To accomplish the latter, his colleague, Dr. S. Mitchell Harman, spent many painstaking hours consolidating, reorganizing, and polishing the material into essentially the form that appears here. The Editor-in-Chief performed final editing and assembled the bibliography from several raw reference lists and reference fragments provided by Dr. Heward. Unfortunately, under the circumstances, not every reference intended for this chapter by Dr. Heward could be located and incorporated by press time, and figures had to be replaced by descriptions in most cases, but what remains is believed to fully represent what Dr. Heward intended for this chapter. The Editors deeply appreciate Dr. Heward's extraordinary dedication to this chapter and Dr. Harman's extraordinary assistance with its presentation, and are extremely saddened by the fact that Dr. Heward was never able to see his chapter published in its final form.

References

Allred JB (1995) Too much of a good thing? An overemphasis on eating low-fat foods may be contributing to the alarming increase in overweight among US adults. J Am Diet Assoc 95:417–418

American College of Sports Medicine (1995) Guidelines for Exercise Testing and Prescription, 5th Ed. Williams and Wilkins, Baltimore, pp. 1–373

Andres R, Tobin JD (1975) Aging and the disposition of glucose. Adv Exp Biol Med 61:239–249

Andziak B, O'Connor TP, Qi W, DeWaal EM, Pierce A, Chaudhuri AR, Van Remmen H, Buffenstein R (2006) High oxidative damage levels in the longest-living rodent, the naked mole-rat. Aging Cell 5:463–471

Aniansson A, Grimby G, Hedberg M, Krotkiewski M (1981) Muscle morphology, enzyme activity and muscle strength in elderly men and women. Clin Physiol 1:73–86

Anonymous (2000) The healthiness of the long-distance runner. Harvard Men's Health Watch March: 6–8

Barrett-Connor E, Shaw KT, Yen SS (1986) A prospective study of dehydroepiandrosterone sulfate, mortality, and cardiovascular disease. New Engl J Med 315:1519–1524

Bassey EJ, Bendall MJ, Pearson M (1988) Muscle strength in the triceps surae and objectively measured customary walking activity in men and women over 65 years of age. Clin Sci 74: 85–89

Beasley JD, Swift JJ (1989) The Kellogg Report: The Impact of Nutrition, Environment, & Lifestyle on the Health of Americans. Annandale-on Hudson, NY. The Institute of Health Policy and Practice, The Bard College Center

Berger D, Crowther R, Floyd JC, Pelz S, Fajans SS (1978) Effects of age on fasting levels of pancreatic hormones in man. J Clin Endocrinol Metab 47:1183–1189

Blackman MR, Sorkin JD, Münzer T, Bellantoni MF, Busby-Whitehead J, Stevens TE, Jayme J, O'Connor KG, Christmas C, Tobin JD, Stewart KJ, Cottrell E, St Clair C, Pabst KM, Harman SM (2002) Growth hormone and sex steroid administration in healthy aged women and men: a randomized controlled trial. JAMA 288:2282–2292

Bulbrook RD, Hayward JL (1967) Abnormal urinary steroid excretion and subsequent breast cancer. A prospective study in the Island of Guernsey. Lancet 1:519–522

Bulbrook RD, Hayward JL, Spicer CC (1971) Relation between urinary androgen and corticoid excretion and subsequent breast cancer. Lancet 2:395–398

Buffenstein R, Edrey YH, Yang T, Mele J (2008) The oxidative stress theory of aging: embattled or invincible? Insights from non-traditional model organisms. Age (Dordr) 30:99–109

Bunker VM, Lawson MS, Stansfield MF, Clayton BE (1987) Nitrogen balance studies in apparently healthy elderly people and those who are housebound. Br J Nutr 57:211–221

Campbell WW, Crim MC, Dallal GE, Young VR, Evans WJ (1994) Increased protein requirements in the elderly: new data and retrospective reassessments. Am J Clin Nutr 60:167–175

Chan JM, Stampfer MJ, Giovannucci E, Gann PH, Ma J, Wilkinson P, Hennekens CH, Pollak M (1998) Plasma insulin growth factor-1 and prostate cancer risk: a prospective study. Science 279:563–566

Chazov EI, Isachenkov VA (1974) The Pineal Gland. Nauka, Moscow
Cheeseman DN, Forsham PH (1974) Inhibition of induced ovulation by a highly purified extract of the bovine pineal gland. Proc Soc Biol Med 146:722–724
Cohn SH, Vartsky D, Yasumura S, Sawitsky A, Zanzi I, Vaswani A, Ellis KJ (1980) Compartmental body composition based on total-body nitrogen, potassium, and calcium. Am J Physiol 239:E524–E530
Cutler RG (1972) Transcription of reiterated DNA sequence classes throughout the life span of the mouse. In: Strehler BL (ed): Advances in Gerontological Research Vol. 4, Academic Press, New York, pp. 219–321
Cutler RG (1974) Redundancy of information content in the genome of mammalian cells as a protective mechanism determining aging rate. Mech Ageing Devel 2:381–408
Cutler RG (1975) Evolution of human longevity and the genetic complexity governing aging rate. Proc Nat Acad Sci USA 72:4664–4668
Cutler RG (1978) Evolutionary biology of senescence. In: Behnke JA, Finch CE, Moment GB (eds): The Biology of Aging, Plenum Press, New York, pp. 311–360
Cutler RG (1982) The dysdifferentiative hypothesis of mammalian aging and longevity. In: Giacobini E, Filogamo G, Vernadakis A (eds): The Aging Brain Cellular and Molecular Mechanisms of Aging in the Nervous System, Aging, Vol. 20, Raven Press, New York, pp. 1–19
Cutler RG (1984a) Carotenoids and retinol: Their possible importance in determining longevity of mammalian species. Proc Natl Acad Sci USA 81:7627–7631
Cutler RG (1984b) Antioxidants, aging and longevity. In: Pryor W (ed): Free Radicals in Biology, Vol. VI, Academic Press, New York, pp. 371–428
Cutler RG (1985) Peroxide-producing potential of tissues: correlation with the longevity of mammalian species. Proc Natl Acad Sci USA 82:4798–4802
Cutler RG (1991) Antioxidants and aging. Amer L Clin Nutr 53:373s–379s
Danneskiold-Samsøe B, Kofod V, Munter J, Grimby G, Schnohr P, Jensen G (1984) Muscle strength and functional capacity in 78–81-year-old men and women. Eur J Appl Physiol Occup Physiol 52:310–314
Davidson MB (1979) The effect of aging on carbohydrate metabolism: a review of the English literature and a practical approach to the diagnosis of diabetes mellitus in the elderly. Metabolism 28:688–705
Defronzo RA (1979) Glucose intolerance and aging: evidence for tissue insensitivity to insulin. Diabetes 28:1095–1101
Dilman VM (1970) Elevating mechanism of aging and cancer. Need for prophylactic normalization of disturbances in energetic and reproductive homeostasis caused by age increase of resistance of the hypothalamus to inhibition. Vopr Onkol 16:45–53
Dilman VM (1971) Age-associated elevation of hypothalamic threshold to feedback control and its role in development, aging and disease. Lancet 1:1211–1219
Dilman VM, Dean W (1992) The Neuroendocrine Theory of Aging and Degenerative Disease. The Center for Bio-Gerontology, Pensacola, FL
Dudl RJ, Ensick JW (1977) Insulin and glucose relationships during aging in man. Metabolism 26:33–41
Elahi D, Muller DC, Tzankoff SP, Andres R, Tobin JD (1982) Effect of age and obesity on fasting levels of glucose, insulin, glucagon, and growth hormone in man. J Gerontol 37:385–391
Everitt AV, Seedsman NJ, Jones F (1980) The effects of hypophysectomy and continuous food restriction, begun at ages 70 and 400 days, on collagen aging, proteinuria, incidence of pathology and longevity in the male rat. Mech Ageing Dev 12:161–72
Farewell VT, Bulbrook RD, Hayward JL (1978) Early Diagnosis of Breast Cancer: Methods & Results. Gustaf-Fischer Verlag, Stuttgart, p. 43
Felt V, Stárka L (1966) Metabolic effects of dehydroepiandrosterone and Atromid in patients with hyperlipaemia. Cor Vasa 8:40–48
Fiatarone MA, Marks EC, Ryan ND, Meredith CN, Lipsitz LA, Evans WJ (1990) High-intensity strength training in nonagenarians. Effects on skeletal muscle. JAMA 263:3029–3034

Finch CE, Felicio LS, Mobbs CV, Nelson JF (1984) Ovarian and steroidal influences on neuroendocrine aging processes in female rodents. Endocr Rev 5:467–497

Finch CE, Schneider EL (1985) Handbook of the Biology of Aging, 2nd Ed. Van Nostrand Reinhold, New York

Fink RI, Kolterman OG, Griffin J, Olefsky JM (1983) Mechanisms of insulin resistance in aging. J Clin Invest 71:1523–1535

Finkelstein JW, Roffwarg HP, Boyar RM, Kream J, Hellman L (1972) Age-related change in the twenty-four-hour spontaneous secretion of growth hormone. J Clin Endocrinol Metab 35:665–670

Fleg JL, Lakatta EG (1988) Role of muscle loss in the age-associated reduction in VO_2 max. J Appl Physiol 65:1147–1151

Fleg JL, O'Connor F, Gerstenblith G, Becker LC, Clulow J, Schulman SP, Lakatta EG (1995) Impact of age on the cardiovascular response to dynamic upright exercise in healthy men and women. J Appl Physiol 78:890–900

Florini JR, Prinz PN, Vitiello MV, Hintz RL (1985) Somatomedin-C levels in healthy young and old men: relationship to peak and 24-hour integrated levels of growth hormone. J Gerontol 40:2–7

Food and Agriculture Organization (1991) Protein quality evaluation; report of the joint FAO/WHO expert consultation. FAO Food and Nutrition Paper 51, Rome, Italy

Food and Nutrition Board (1989) Recommended Dietary Allowances, 10th Ed. National Academy Press, Washington, DC

Frontera WR, Meredith CN, O'Reilly KP, Knuttgen HG, Evans WJ (1988) Strength conditioning in older men: skeletal muscle hypertrophy and improved function. J Appl Physiol 64:1038–1044

Frontera WR, Meredith CN, O'Reilly KP, Evans WJ (1990) Strength training and determinants of VO_2max in older men. J Appl Physiol 68:329–333

Gambert SR, Tsitouras PD (1985) Effect of age on thyroid hormone physiology and function. J Am Geriat Soc 33:360–365

Gambrell RD Jr, Massey FM, Castaneda TA, Ugenas AJ, Ricci CA (1979) Reduced incidence of endometrial cancer among postmenopausal women treated with proestogens. J Am Geriatr Soc 27:389–394

Giannoulis MG, Sonksen PH, Umpleby M, Breen L, Pentecost C, Whyte M, McMillan CV, Bradley C, Martin FC (2006) The effects of growth hormone and/or testosterone in healthy elderly men: a randomized controlled trial. J Clin Endocrinol Metab 91:477–484

Giovannucci E (1999) Tomatoes, tomato-based products, lycopene, and cancer. J Natl Cancer Inst 91: 74–79

Goldstein AL, Goldstein AL (2009) From lab to bedside: emerging clinical applications of thymosin alpha1. Expert Opin Biol Ther 9:593–608

Goodman MN, Dluz SM, McElaney MA, Belur E, Ruderman NB (1983) Glucose uptake and insulin sensitivity in rat muscle: changes during 3–96 weeks of age. Am J Physiol 244:E93–E100

Greey CW (1955) A Study of Flexibility in Selected Joints of Adult Males Ages 18–72. Doctoral dissertation, University of Michigan

Hagberg JM, Graves JE, Limacher M, Woods DR, Leggett SH, Cononie C, Gruber JJ, Pollock ML (1989) Cardiovascular responses of 70–79 year old men and women to exercise training. J Appl Physiol 66:2589–2594

Hallberg MC, Wieland RG, Zorn EM, Furst BH, Wieland JM (1976) Impaired Leydig cell reserve and altered serum androgen binding in the aging male. Fertil Steril 27:812–814

Harman SM, Tsitouras PD (1980) Reproductive hormones in aging men. I. Measurement of sex steroids, basal luteinizing hormone, and Leydig cell response to human chorionic gonadotropin. J Clin Endocrinol Metab 51:35–40

Harper HA (1971) Review of Physiological Chemistry. Lange Medical Publications, Los Altos

Hartz SC, Russell RM, Rosenberg IH (eds) (1992) Nutrition in the Elderly: The Boston Nutritional Status Survey. Smith-Gordon, London, pp. 1–287

Hersey WC 3rd, Graves JE, Pollock ML, Gingerich R, Shireman RB, Heath GW, Spierto F, McCole SD, Hagberg JM (1994) Endurance exercise training improves body composition and plasma insulin responses in 70- to 79-year-old men and women. Metabolism 43:847–854

Ho KY, Evans WS, Blizzard RM, Veldhuis JD, Merriam GR, Samojlik E, Furlanetto R, Rogol AD, Kaiser DL, Thorner MO (1987) Effects of sex and age on the 24-hour profile of growth hormone secretion in man: importance of endogenous estradiol concentrations. J Clin Endocrinol Metab 64:51–58

Ho KY (1998) Progesterone modulates the effects of estrogen on IGF-1 in postmenopausal women. Growth Hormone IGF Res 8:316 (abstract)

Holehan AM, Merry BJ (1986) The experimental manipulation of ageing by diet. Biol Rev Camb Philos Soc 61:329–368

Horton R, Hsieh P, Barberia J, Pages L, Cosgrove M (1975) Altered blood androgens in elderly men with prostate hyperplasia. J Clin Endocrinol Metab 41:793–796

Hu FB, Stampfer MJ, Manson JE, Rimm EB, Wolk A, Colditz GA, Hennekens CH, Willett WC (1999) Dietary intake of alpha-linolenic acid and risk of fatal ischemic heart disease among women. Am J Clin Nutr 69:890–897

Hulka BS, Fowler WC Jr, Kaufman DG, Grimson RC, Greenberg BG, Hogue CJ, Berger GS, Pulliam CC (1980) Estrogen and endometrial cancer: cases and two control groups from North Carolina. Am J Obstet Gynecol 137:92–101

Igwebuike A, Irving BA, Bigelow ML, Short KR, McConnell JP, Nair KS (2008) Lack of dehydroepiandrosterone effect on a combined endurance and resistance exercise program in postmenopausal women. J Clin Endocrinol Metab 93:534–538

Jackson RA, Blix PM, Matthews JA, Hamling JB, Din BM, Brown DC, Belin J, Rubenstein AH, Nabarro JD (1982) Influence of aging on glucose homeostasis. J Clin Endocrinol Metab 55:840–848

Jette AM, Branch LG (1981) The Framingham Disability Study: II. Physical disability among the aging. Am J Public Health 71:1211–1216

Jick H, Watkins RN, Hunter JR, Dinan BJ, Madsen S, Rothman KJ, Walker AM (1979) Replacement estrogens and endometrial cancer. N Engl Med 300:218–222

Kalimi M, Regelson W (1990) The Biological Role of Dehydroepiandrosterone (DHEA). de Gruyter, New York

Kalk WJ, Vinik AI, Pimstone BL, Jackson PU (1973) Growth hormone response to insulin hypoglycemia in the elderly. J Gerontol 28:431–433

Kask E (1959) 17-Ketosteroids and arteriosclerosis. Angiology 10:358–368

Kastin AJ, Schally AV (1967) Autoregulation of release of melanocyte stimulating hormone from the rat pituitary. Nature 213:1238–1240

Katan MB (1998) Effect of low-fat diets on plasma high density lipoprotein concentrations. Am J Clin Nutr 67(suppl):573 S–576S

Keys A, Taylor HL, Grande F (1973) Basal metabolism and age of adult man. Metabolism 22:579–587

Kirwan JP, Kohrt WM, Wojta DM, Bourey RE, Holloszy (1993) Endurance exercise training reduces glucose-stimulated insulin levels in 60- to 70-year-old men and women. J Gerontol 48:M84–M90

Klein D (1985) Photoneural regulation of the mammalian pineal gland. In: Photoperiodism, Melatonin and the Pineal. Pitman, London (Ciba Foundations Symp 117), pp. 38–56

Kolterman OG, Insel J, Saekow M, Olefsky JM (1980) Mechanisms of insulin resistance in human obesity: evidence for receptor and postreceptor defects. J Clin Invest 65:1272–1284

Larsson L (1983) Histochemical characteristics of human skeletal muscle during aging. Acta Physiol Scand 117:469–471

Lee JR (1993) Optimal Health Guidelines, 2nd Ed. BLL Publishing, Sebastopol, California

Lexell J, Henriksson-Larsen K, Wimblod B, Sjostrom M (1983) Distribution of different fiber types in human skeletal muscles: effects of aging studied in whole muscle cross sections. Muscle Nerve 6:588–595

Ludwig DS, Pereira MA, Kroenke CH, Hilner JE, Van Horn L, Slattery ML, Jacobs DR Jr (1999) Dietary fiber, weight gain, and cardiovascular disease risk factors in young adults. JAMA 282: 1539–1546

Mantzoros CS, Tzonou A, Signorello LB, Stampfer M, Trichopoulos D, Adami HO (1997) Insulin-like growth factor 1 in relation to prostate cancer and benign prostatic hyperplasia. Br J Cancer 76:1115–1118

McGandy RB, Barrows CH Jr, Spanias A, Meredith A, Stone JL, Norris AH (1966) Nutrient intakes and energy expenditure in men of different ages. J. Gerontol. 21:581–587

Meites J, Goya R, Takahashi S (1987) Why the neuroendocrine system is important in aging processes. Exp Gerontol 22:1–15

Miller JP, Pratley RE, Goldberg AP, Gordon P, Rubin M, Treuth MS, Ryan AS, Hurley BF (1994). Strength training increases insulin action in healthy 50- to 65-yr-old men. J Appl Physiol 77:1122–1127

Mokdad AH, Serdula MK, Dietz WH, Bowman BA, Marks JS, Koplan JP (1999) The spread of the obesity epidemic in the United States, 1991–1998. JAMA 282:1519–1522

Montanini V, Simoni M, Chiossi G, Baraghini GF, Velardo A, Baraldi E, Marrama P (1988) Age-related changes in plasma dehydroepiandrosterone sulphate, cortisol, testosterone and free testosterone circadian rhythms in adult men. Hormone Res 29:1–6

Mooradian AD, Morley JE, Korenman SG (1987) Biological actions of androgens. Endocr Rev 8:1–28

Morey MC, Cowper PA, Feussner JR, DiPasquale RC, Crowley GM, Sullivan RJ Jr (1991) Two-year trends in physical performance following supervised exercise among community-dwelling old veterans. J Am Geriatr Soc 39:549–554

Muller DC, Elahi D, Tobin JD, Andres R (1996) Insulin response during the oral glucose tolerance test: the role of age, sex, body fat and the pattern of fat distribution. Aging (Milano) 8:13–21

Munns K (1981) Effects of exercise on the range of joint motion in elderly subjects. In: Smith EL, Serfass RC (Eds): Exercise and Aging: The Scientific Basis. Enslow, Short Hills, NJ, pp. 1–191

Nair NP, Hariharasubramanian N, Pilapil C, Isaac I, Thavundayil JX (1986) Plasma melatonin – An index of brain aging in humans? Biol Psychiatry 21:141–150

Nair KS, Rizza RA, O'Brien P, Dhatariya K, Short KR, Nehra A, Vittone JL, Klee GG, Basu A, Basu R, Cobelli C, Toffolo G, Dalla Man C, Tindall DJ, Melton LJ 3rd, Smith GE, Khosla S, Jensen MD (2006) DHEA in elderly women and DHEA or testosterone in elderly men. N Engl J Med 355:1647–1659

Nankin HR (1981) Leydig cell function in men. J S C Med Assoc 77:531–535

Nelson ME, Fiatarone MA, Morganti CM, Trice I, Greenberg RA, Evans WJ (1994) Effects of high-intensity strength training on multiple risk factors for osteoporotic fractures. JAMA 272:1909–1914

Nestler JE, Barlascini CO, Clore JN, Blackard WG (1988) Dehydroepiandrosterone reduces serum low density lipoprotein levels and body fat but does not alter insulin sensitivity in normal men. J Clin Endocrinol Metab 66:57–61

O'Donnell AB, Travison TG, Harris SS, Tenover JL, McKinlay JB (2006) Testosterone, dehydroepiandrosterone, and physical performance in older men: results from the Massachusetts Male Aging Study. J Clin Endocrinol Metab 91:425–431

Ogawa T, Spina RJ, Martin WH 3rd, Kohrt WM, Schechtman KB, Holloszy JO, Ehsani AA (1992) Effects of aging, sex and physical training on cardiovascular responses to exercise. Circulation 86:494–503

Olshansky SJ, Carnes BA, Cassel C (1990) In search of Methuselah: Estimating the upper limits to human longevity. Science 250:634–640

Olshansky SJ, Carnes BA, Desesquelles A (2001) Prospects for human longevity. Science 291:1491–1492

Orr, WC, Sohal RS (1994) Extension of life span by overexpression of superoxide dismutase and catalase in Drosophila melanogaster. Science 263:1128–1130

Orentreich N, Brind JL, Rizer RL, Vogelman JH (1984) Age changes and sex differences in serum dehydroepiandrosterone sulfate concentrations throughout adulthood. J Clin Endocrinol Metab 59:551–555

Paffenbarger RS Jr, Hyde RT, Wing AL, Lee IM, Jung DL, Kampert JB (1993) The association of changes in physical-activity level and other lifestyle characteristics with mortality among men. N Engl J Med 328:538–545

Pierpaoli W, Yi CX, Dall'Ara A (1990) Aging-postponing effects of circadian melatonin: experimental evidence, significance and possible mechanisms. Int J Neurosci 51:339–340

Pierpaoli W, Dall'Ara A, Pedrinis E, Regelson W (1991) The pineal control of aging. The effects of melatonin and pineal grafting on the survival of older mice. Ann NY Acad Sci 621: 291–313

Pollock ML, Mengelkoch LJ, Graves JE, Lowenthal DT, Limacher MC, Foster C, Wilmore JH (1997) Twenty-year follow-up of aerobic power and body composition of older track athletes. J Appl Physiol 82:1508–1516

Province MA, Hadley EC, Hornbrook MC, Lipsitz LA, Miller JP, Mulrow CD, Ory MG, Sattin RW, Tinetti ME, Wolf SL (1995) The effects of exercise on falls in elderly patients. A preplanned meta-analysis of the FICSIT Trials. Frailty and Injuries: Cooperative Studies of Intervention Techniques. JAMA 273:1341–1347

Quay WB (1972) Pineal homeostatic regulation of shifts in the circadian activity rhythm during maturation and aging. Ann NY Acad Sci 34:239–254

Rabinowitz P (1970) Some endocrine and metabolic aspects of obesity. Ann Rev Med 21:241–258

Reaven EP, Gold G, Reaven GM (1979) Effect of age on glucose-stimulated insulin release by the beta-cell of the rat. J Clin Invest 64:591–599

Reaven EP, Gold G, Reaven G (1980) Effect of age on leucine-induced insulin secreton by the beta-cell. J Gerontol 35:324–328

Reaven GM, Reaven EP (1980) Effects of age on various aspects of glucose and insulin metabolism. Mol Cell Biochem 31:37–47

Regelson W, Pierpaoli W (1987) Melatonin: a rediscovered antitumor hormone? Its relation to surface receptors; sex steroid metabolism, immunologic response, and chronobiologic factors in tumor growth and therapy. Cancer Invest 5: 379–385

Roach KE, Miles TP (1991) Normal hip and knee active range of motion: the relationship to age. Phys Ther 70:656–665

Robert JJ, Cummins JC, Wolfe RR, Durkot M, Matthews DE, Zhao XH, Bier DM, Young VR (1982) Quantitative aspects of glucose production and metabolism in healthy elderly subjects. Diabetes 31: 203–211

Roberts SB, Young VR, Fuss P, Heyman MB, Fiatarone M, Dallal GE, Cortiella J, Evans WJ (1992) What are the dietary energy needs of adults? Int J Obes Relat Metab Disord 16:969–976

Ross MH (1976) Nutrition and longevity in experimental animals. Curr Concepts Nutr 4:43–57

Rowe JW, Minaker KL, Pallotta JA, Flier JS (1983) Characterization of the insulin resistance of aging. J Clin Invest 71:1581–1587

Rudman D, Kutner MH, Rogers CM, Lubin MF, Fleming GA, Bain RP (1981) Impaired growth hormone secretion in the adult population: relation to age and adiposity. J Clin Invest 67:1361–1369

Sapolsky RM, Krey LC, McEwen BS (1986) The neuroendocrinology of stress and aging: the glucocorticoid cascade hypothesis. Endocr Rev 7:284–301

Sattler FR, Castaneda-Sceppa C, Binder EF, Schroeder ET, Wang Y, Bhasin S, Kawakubo M, Stewart Y, Yarasheski KE, Ulloor J, Colletti P, Roubenoff R, Azen SP (2009) Testosterone and growth hormone improve body composition and muscle performance in older men. J Clin Endocrinol Metab 94:1991–2001

Schrafsma G (2000) The protein digestibility-corrected amino acid score. J Nutr 30:1865S-1867S

Schwartz RS, Shuman WP, Larson V, Cain KC, Fellingham GW, Beard JC, Kahn SE, Stratton JR, Cerqueira MD, Abrass IB (1991) Effect of intensive endurance exercise training on body fat distribution in young and older men. Metabolism 40:545–551

Seals DR, Hagberg JM, Hurley BF, Ehsani AA, Holloszy JO (1984a) Endurance training in older men and women. I. Cardiovascular responses to exercise. J Appl Physiol 57:1024–1029

Seals DR, Hagberg JM, Hurley BF, Ehsani AA, Holloszy JO (1984b) Effects of endurance training on glucose tolerance and plasma lipid levels in older men and women. JAMA 252:645–649

Segall PE (1979) Interrelations of dietary and hormonal effects in aging. Mech Ageing Dev 9:515–525

Segall PE, Timiras PS, Walton JR (1983) Low tryptophan diets delay reproductive aging. Mech Ageing Dev 23:245–252

Shaper AG, Wannamethee SG, Walker M (1997) Body weight: implications for the prevention of coronary heart disease, stroke, and diabetes mellitus in a cohort study of middle aged men. Brit Med J 314:1311–1317

Simonson DC, DeFronzo RA (1983) Glucagon physiology and aging: evidence for enhanced hepatic sensitivity. Diabetolgia 25:1–7

Simopoulos AP, Robinson J (1999) The Omega Diet: Lifesaving Nutritional Program Based on the Diet of the Island of Crete. Harper, New York

Sinaki M, McPhee MC, Hodgson SF, Merritt JM, Offord KP (1986) Relationship between bone mineral density of spine and strength of back extensors in healthy postmenopausal women. Mayo Clin Proc 61:116–122

Smythe GA, Lazarus L (1974) Suppression of human growth hormone secretion by melatonin and cyproheptadine. J Clin Invest 54:116–121

Smythe GA, Stuart MC, Lazarus L (1974) Stimulation and suppression of somatomedin activity by serotonin and melatonin. Experientia 30:1356–1357

Snow-Harter C, Bouxsein M, Lewis B, Charette S, Weinstein P, Marcus R (1990) Muscle strength as a predictor of bone mineral density in young women. J Bone Miner Res 5:589–595

Sonntag WE, Steger RW, Forman LJ, Meites J (1980) Decreased pulsatile release of growth hormone in old male rats. Endocrinology 107:1875–1879

Sparrow D, Bosse R, Rowe JW (1980) The influence of age, alcohol consumption, and body build on gonadal function in men. J Clin Endocrinol Metab 51:508–512

Stearns EL, MacDonnell JA, Kaufman BJ, Padua R, Lucman TS, Winter JS, Faiman C (1974) Declining testicular function with age. Hormonal and clinical correlates. Am J Med 57:761–766

Stevenson E, Davy K, Seals D (1995) Hemostatic, metabolic, and androgenic risk factors for coronary heart disease in physically active and less active postmenopausal women. Arterioscler Thromb 15:669–677

Stratton JR, Levy WC, Cerqueira MD, Schwartz RS, Abrass IB (1994) Cardiovascular responses to exercise. Effects of aging and exercise training in healthy men. Circulation 89:1648–1655

Timiras PS, Choy VJ, Hudson DB (1982) Neuroendocrine pacemaker for growth, development and ageing. Age Ageing 11:73–88

Tolmasoff JM, Ono T, Cutler RG (1980) Superoxide dismutase: correlation with life span and specific metabolic rate in primate species. Proc Natl Acad Sci USA 77:2777–2781

Tsitouras PD, Gucciardo F, Salbe AD, Heward C, Harman SM (2008) High omega-3 fat intake improves insulin sensitivity and reduces CRP and IL6, but does not affect other endocrine axes in healthy older adults. Horm Metab Res 40:199–205

Tzankoff SP, Norris AH (1977) Effect of muscle mass decrease on age-related BMR changes. J Appl Physiol 43:1001–1006

Udo E (1993) Fats That Heal – Fats That Kill. Alive Books, Burnaby

Vaitkevicius PV, Fleg JL, Engel JH, O'Connor FC, Wright JG, Lakatta LE, Yin FC, Lakatta EG (1993) Effects of age and aerobic capacity on arterial stiffness in healthy adults. Circulation 88:1456–1462

VanTulnen I, Adriano J (2000) Rating the Diet Books. Nutrition Action Health Letter: May

Vermeulen A, Rubens R, Verdonck L (1972) Testosterone secretion and metabolism in male senescence. J Clin Endocrinol Metab 34:730–735

Vidalon C, Khurana RC, Chae S, Gegick CG, Stephan T, Nolan S, Danowski TS (1973) Age-related changes in growth hormone in non-diabetic women. J Am Ger Soc 21:253–255

Weiss NS, Szekely DR, English DR, Schweid AI (1979) Endometrial cancer in relation to patterns of menopausal estrogen use. JAMA 242:261–264

White HK, Petrie CD, Landschulz W, MacLean D, Taylor A, Lyles K, Wei JY, Hoffman AR, Salvatori R, Ettinger MP, Morey MC, Blackman MR, Merriam GR, Capromorelin Study Group (2009) Effects of an oral growth hormone secretagogue in older adults. J Clin Endocrinol Metab 94:1198–1206

Whitehead MI, Townsend PT, Pryse-Davies J, Ryder TA, King RJ (1981) Effects of estrogens and progestins on the biochemistry and morphology of the postmenopausal endometrium. N Engl J Med 305:1599–1605

Young VR, Steffee WP, Pencharz PB, Winterer JC, Scrimshaw NS (1975) Total human body protein synthesis in relation to protein requirements at various ages. Nature 253:192–194

Zumoff B, Rosenfeld RS, Strain GW, Levin J, Fukushima DK (1980) Sex differences in the twenty-four-hour mean plasma concentrations of dehydroisoandrosterone (DHA) and dehydroisoandrosterone sulfate (DHAS) and the DHA to DHAS ratio in normal adults. J Clin Endocrinol Metab 51:330–333

Chapter 8
Near Term Prospects for Ameliorating Cardiovascular Aging

Roger Yu, Kaveh Navab, and Mohamad Navab

Abbreviations

ABCA1	ATP-binding cassette transporter A1
AHA	American Heart Association
apo	apolipoprotein
CCA	cell co-culture assay
CETP	cholesteryl ester transfer protein
CFA	cell-free assay
CHD	coronary heart disease
CHF	congestive heart failure
CVD	cardiovascular disease
DM	diabetes mellitus
eNOS	endothelial NO synthase
EPC	endothelial progenitor cell
HDL	high-density lipoprotein
HDL-C	HDL cholesterol level
HO-1	hemeoxygenase-1
HTN	hypertension
ICAM-1	intercellular adhesion molecule-1
IL	interleukin
JNC 7	The Seventh Report of the Joint National Committee
LCAT	lecithin cholesterol acyltransferase
LDL	low-density lipoprotein
LDL-C	LDL cholesterol level
LDLR	LDL receptor
LOOH	lipid hydroperoxide
LV	left ventricular
MCA	monocyte chemotactic activity

R. Yu (✉)
Hospital Medicine Program, Beth Israel Deaconess Medical Center, 330 Brookline Ave, Span 2, Boston, MA 02215, USA
e-mail: RogerYu.MD@gmail.com

MCP-1	monocyte chemoattractant protein-1
MIP-1α	macrophage inflammatory protein-1α
MM-LDL	minimally-modified LDL
MWM	Morris Water Maze
NCEP ATP III	National Cholesterol Education Panel, Adult Treatment Panel III
NO	nitric oxide
OX-LDL	oxidized LDL
PAF-AH	platelet activating factor acetylhydrolase
PLTP	phospholipid transfer protein
PON	paraoxonase
RA	rheumatoid arthritis
SAA	serum amyloid A
SLE	systemic lupus erythematosus
sPLA2	secretory phospholipase A2
SSc	scleroderma, or systemic sclerosis
T-CAT	T-maze continuous alternation task
Tsk-/+	tight skin
VCAM-1	vascular cell adhesion molecule-1

Contents

8.1	Introduction	281
8.2	Cardiovascular Aging	282
	8.2.1 Aging of the Heart	282
	8.2.2 Aging of the Vascular Network	283
	8.2.3 Atherosclerosis and Cardiovascular Aging	283
	8.2.4 Traditional CVD Risk Factors and Cardiovascular Aging	283
	8.2.5 Modification of CVD Risk Factors	284
	8.2.6 Novel Risk Factors and Biomarkers for CHD	285
8.3	Oxidized Lipids, Inflammation, and Atherogenesis	285
	8.3.1 HDL Composition and Function	286
	8.3.2 How Good is the HDL?	288
	8.3.3 Dysfunctional HDL	290
	8.3.4 How HDL Becomes Dysfunctional	292
	8.3.5 Making Bad HDL Better	293
8.4	Apolipoprotein A-I Mimetic Peptides	293
	8.4.1 Action of ApoA-I Mimetic Peptides	294
	8.4.2 ApoA-I Mimetic Peptides in Animal Models of Atherosclerosis	295
	8.4.3 D-4F and Vasodilation, Vessel Wall Thickness, and Endothelial Health	296
	8.4.4 D-4F and Improvement in Brain Arteriole Inflammation	297
	8.4.5 D-4F in Diseases with Cardiovascular Complications	297
	8.4.6 D-4F in Non-Atherosclerotic Diseases	298

8.4.7 HDL-Directed Therapy 298
8.4.8 Other Peptides with Lipid-Binding Activity 299
8.4.9 Beyond Cardiovascular Inflammation: HDL and LDL Particle Size 299
8.5 Summary and Conclusions 300
References 302

8.1 Introduction

Cardiovascular disease (CVD) is the leading cause of death in the United States, and its prevalence increases with age. The American Heart Association (AHA) estimates that about 80 million American adults (about 1 in 3) had one or more types of CVD (Lloyd-Jones et al. 2009). The types with the highest rates of mortality are coronary heart disease (CHD), stroke, congestive heart failure (CHF), and hypertension (HTN) (Chart 8.1).

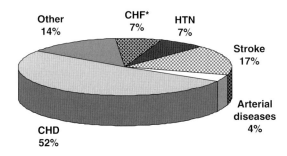

Chart 8.1 Percentage breakdown of deaths due to cardiovascular disease (United States: 2006, preliminary). Redrawn from Lloyd-Jones et al. (2009). *Not a true underlying cause. *CHD* coronary heart disease, *CHF* congestive heart failure, *HTN* hypertension

The gradual accumulation of insults to the cardiovascular system inevitably leads to a large burden of disease in the aging population and an increase in hospital admissions and deaths attributable to acute myocardial infarction, CHD, CHF, arrhythmias, and stroke. The indication for undergoing invasive procedures such as coronary angiography, percutaneous coronary intervention, coronary artery bypass surgery, and pacemaker implantation increases as well. The AHA estimated financial burden of CVD for 2009 to be $475.3 billion (Lloyd-Jones et al. 2009).

Current guidelines for primary and secondary prevention of CVD and management of its associated risk factors are well described in statements by AHA and American College of Cardiology (Smith et al. 2006). Constantly evolving, preventative measures have reduced the prevalence of most of the modifiable risk factors over the last four decades and have been implicated as one of the major reasons for decreased mortality from CVD (Gregg et al. 2005; Ergin et al. 2004).

Despite the noted decrease in mortality, the impact on human lives continues to be striking—in 2005, nearly 2400 Americans died each day from CVD (Lloyd-Jones et al. 2009). Though epidemiological studies have identified HTN, diabetes mellitus (DM), dyslipidemia, smoking tobacco, sedentary lifestyles, and genetic factors as CVD risk factors, advancing age unequivocally confers the major risk (Lakatta and Levy 2003a).

The fact that age is the most important predictor of CVD has several possible explanations – increased exposure to known cardiovascular risk factors, increased time for the disease to occur, or perhaps, over time, the structure and function of the heart and the body's network of blood vessels undergo changes through mechanisms that have not yet been fully described. In otherwise healthy persons without clinically apparent disease, there are changes in the cardiovascular system that were once considered to be normal aging, but studies reveal that they are independent predictors of disease (Lakatta and Levy 2003a). This may explain why some people with only minimal risk factors, often those who are older, go on to develop worse than predicted disease. Cardiovascular aging results from a combination of processes that, over time, leads to compromise in normal function of the cardiovascular system. Like the traditional risk factors for CVD, these age-related changes can be considered to be targets for preventative therapy.

While there is no way to address the vast spectrum of aging through one therapy, much progress has been made in combating some of the major components of cardiovascular aging. Whether attributed to aging itself, modifiable risk factors, or other conditions with cardiovascular involvement, CVD is an end result and remains the leading cause of mortality in the United States. As such, some of the most promising prospects for combating cardiovascular aging address the various contributing factors to CVD, such as the interplay between low-density lipoproteins (LDL) and high-density lipoproteins (HDL) and its effects on the vascular system. One class of compounds that show promise are the apolipoprotein (apo)A-I mimetic peptides, which have been found to have significant effect on various aspects of cardiovascular pathology including atherosclerosis as well as several non-atherosclerotic diseases. ApoA-I mimetic peptides may be effective in ameliorating cardiovascular aging and prove to be a beneficial adjunct to the current strategy of battling CVD, which increasingly burdens the healthcare system as the American population ages.

8.2 Cardiovascular Aging

8.2.1 Aging of the Heart

Cardiovascular aging results from a combination of processes that, over time, leads to compromise in the normal function of the heart and the body's network of blood vessels. Age-related changes to the heart include the following: left ventricular (LV) wall thickening, cardiac myocyte enlargement, impaired LV diastolic relaxation, depressed LV ejection fraction, lower maximum heart rate, and altered heart rhythm

(Lakatta and Levy 2003b). The ability of regulatory mechanisms to increase cardiac output is decreased because of deficiencies in the sympathetic modulation of heart rate, afterload, and myocardial contractility (Lakatta and Levy 2003b). Mitochondrial damage and lipofuscin deposition are hallmarks of aged cardiac myocytes and are a cumulative result of oxidative modification of lipids and proteins over a person's life (Terman et al. 2004). While these changes per se do not necessarily result in clinical heart disease, they do lower the cardiac reserve and the threshold for which signs and symptoms may develop for various CVD such as CHF and arrhythmias.

8.2.2 Aging of the Vascular Network

As with other organ systems, the vascular system undergoes modification from various processes that, over time, leads to changes in vessel structure and function. Large elastic arteries become thicker and more dilated with age. This thickening consists mostly of intimal media thickening, which is not necessarily from atherosclerosis. Endothelial dysfunction, arterial stiffening, and widening arterial pulse pressure are also observed with increasing age (Lakatta and Levy 2003a). Studies in aging rats have revealed that nitric oxide (NO)-dependent endothelial vasodilation is attenuated in older rats likely through the reduced activity of endothelial NO synthase (eNOS) and NO (Lakatta 2003). When combined with the traditional CVD risk factors, these age-related changes are further exacerbated and provide an even more suitable milieu for atherosclerosis to develop (Fig. 8.1).

8.2.3 Atherosclerosis and Cardiovascular Aging

Atherosclerosis is an important etiology of acute dissection and aneurysm formation in the aorta (Hagan et al. 2000; Isselbacher 2005), which are important sequelae of cumulative insults to large vessels. The process of cardiovascular aging leads to an increased prevalence of aneurysms and aortic disease, along with valvular and carotid disease. The prevalence of valve disease is less than 20% for patients in the age groups 20–29 and 30–39, 40% for those aged 60–69, and greater than 60% in those over 80 years old (Croft et al. 2004). For calcific aortic valve disease, the pathogenesis appears to be similar to the process of atherosclerosis (Otto 2002). Acute aortic dissection occurs primarily in older patients, the mean age being about 63 years old (Hagan et al. 2000). Age is the primary risk factor for carotid artery stenosis (Zimarino et al. 2001). Though age confers the major risk, many other factors are important in the progression of CVD.

8.2.4 Traditional CVD Risk Factors and Cardiovascular Aging

Lifestyle choices, environmental exposure, and diseases may stress and impair the cardiovascular system. A high saturated fat diet has been shown to decrease flow-mediated vasodilation that is not correlated with increased LDL-cholesterol level

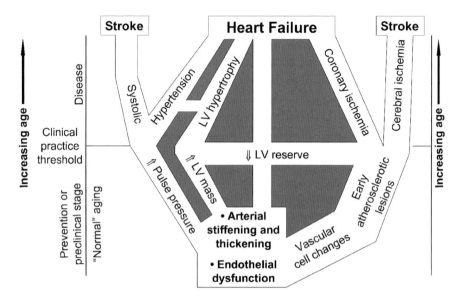

Fig. 8.1 Changes in the vasculature through arterial stiffening and thickening and endothelial dysfunction may be the root of some major cardiovascular diseases. *LV* left ventricular. Redrawn from Lakatta and Levy (2003b)

(LDL-C) (Keogh et al. 2005). Cigarette smoking exerts its damage on the vascular system through the disruption of peripheral and coronary vasodilation (Puranik and Celermajer 2003). Endothelial sloughing has been found to occur in diabetic patients as evidenced by the presence of circulating endothelial cells (McClung et al. 2005). Several other conditions have also been associated with significant cardiovascular complications. This is especially true for many of the rheumatologic diseases that create a chronic inflammatory response in the body (McMahon et al. 2006; Weihrauch et al. 2007). Interestingly, sickle cell disease indirectly causes vascular dysfunction. Episodic ischemic events caused by sickling cells causes injured liver cells to release xanthine oxidase, which can interfere with NO-dependent vasodilation (Ou et al. 2003). With increasing understanding of the mechanisms that increase the aging of the cardiovascular system, a more targeted approach may be employed in trying to halt or even reverse the damage done to the cardiovascular system.

8.2.5 Modification of CVD Risk Factors

Guidelines for primary and secondary prevention of CVD and management of its associated risk factors are well described in statements by AHA and American College of Cardiology (Smith et al. 2006; Grundy et al. 1998). In brief, recommended lifestyle modifications include smoking cessation, regular aerobic exercise and an increase in routine daily physical activities, and weight loss if overweight (i.e., body mass index >25 kg/m^2). DM management should achieve a HgbA1c of <7%, blood pressure control is as recommended in The Seventh Report of the Joint National Committee (JNC 7), while lipid management is as described in NCEP ATP III with the exception of the option of tighter control of LDL-C for those with known CHD or other atherosclerotic diseases (Chobanian et al. 2003; Expert Panel 2001). In patients with high risk-scores for CHD, left ventricular dysfunction, or having a history of coronary event, the addition of aspirin, ACE inhibitors, angiotensin receptor blockers, or beta-blockers should be added as indicated by the guidelines. All patients with known CVD should regularly be vaccinated against influenza.

8.2.6 Novel Risk Factors and Biomarkers for CHD

Despite aggressive modification of risk factors through lifestyle modifications and the wide array of pharmacotherapeutics, CVD still remains the foremost cause of death in the United States and of those deaths more than 50% are from CHD (Lloyd-Jones et al. 2009). Non-invasive angiography and serum biomarkers that correlate with presence of disease or a higher likelihood of developing disease have been aggressively sought after. Cardiac computed tomography has been studied for the last 20 years and has been shown to be a reliable tool in assessing coronary artery calcified plaque and noncalcified plaque. The total amount of coronary calcium can quantify atherosclerotic burden and predicts CAD events beyond standard risk factors (Budoff et al. 2006).

In atherosclerosis, the role of inflammation has become well established (Ross 1999). Therefore various systemic inflammatory markers are being investigated as potential markers for CVD. A non-exhaustive list include oxidized LDL (OX-LDL), pro-inflammatory cytokines such as interleukin(IL)-1, IL-6, and tumor necrosis factor alpha, adhesion molecules such as intercellular adhesion molecule-1 (ICAM-1), vascular cell adhesion molecule-1 (VCAM-1), selectins such as E-selectin and P-selectin, and other acute-phase reactants such as serum amyloid A (SAA), c-reactive protein, fibrinogen, and leukocyte count (Pearson et al. 2003). Other molecules that have been investigated include homocysteine and lipoprotein (a) (Ridker et al. 2001). The relationship between inflammation and lipids is increasingly being studied, particularly with regard to HDL.

8.3 Oxidized Lipids, Inflammation, and Atherogenesis

Coronary artery atherosclerosis is the principle process responsible for CHD. Oxidation and inflammation are central to the initiation and propagation of atherogenesis (Navab et al. 2004). Atherosclerotic plaques begin forming through the accumulation of lipid-laden macrophages (known as foam cells) in artery walls. Oxidation of the components of LDL plays a key role in foam cell development (Navab et al. 2004). LDL that enters into the subendothelial space of artery walls

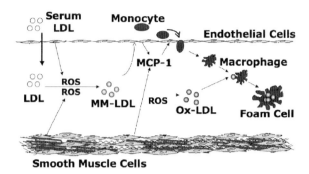

Fig. 8.2 Foam cell formation. Serum low-density lipoprotein (LDL) enter the subendothelial space. Mild oxidation of LDL by reactive oxygen species (ROS) change LDL into minimally modified LDL (MM-LDL), which induces chemoattractants such as monocyte chemoattractant protein-1 (MCP-1). Further modification of LDL by ROS converts it into oxidized LDL (ox-LDL) that gets engulfed by macrophages. Lipid-laden macrophages become foam cells (Navab et al. 2004)

becomes altered by reactive oxygen species (Navab et al. 2004). When phospholipids in LDL become oxidized, the LDL becomes minimally modified LDL (MM-LDL), a potent inducer of monocyte adhesion molecules and chemoattractants (Navab et al. 2004). MM-LDL is highly inflammatory; through the actions of chemoattractants such as monocyte chemoattractant protein-1 (MCP-1) and IL-8, MM-LDL increases recruitment of monocytes into the subendothelial space where they differentiate into macrophages (Navab et al. 2004). When further oxidation causes alteration of apoB, LDL is then characterized as OX-LDL, which scavenger receptors on macrophages recognize (Navab et al. 2004) (Fig. 8.2).

Lipid oxidation contributes to vascular inflammation and is more likely in the presence of systemic inflammation. Leukocytes from patients with systemic lupus erythematosus (SLE) and rheumatoid arthritis (RA) show increased lipid peroxidation and LDL oxidation (Ansell et al. 2005). Systemic infection and metabolic syndrome also have increased LDL oxidation and lipid peroxidation (Ansell et al. 2005). This is likely why patients with metabolic syndrome and inflammatory conditions such as SLE and RA have higher prevalence of CVD (Bernatsky et al. 2006; Mutru et al. 1985). Lipid hydroperoxides (LOOH) can also cause impaired endothelial function by decreasing nitric oxide production and vasomotor response of the artery wall (Ansell et al. 2005).

In short, a large burden of oxidative stress impairs endothelial function, promotes foam cell formation, and propagates the atherosclerotic process. Healthy HDL is able to oppose this process and protect artery walls from atherogenesis.

8.3.1 HDL Composition and Function

The life cycle of HDL is directly linked to one of its major atheroprotective properties—the ability to participate in reverse cholesterol transport (Fig. 8.3). The

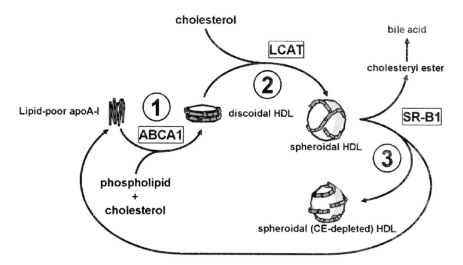

Fig. 8.3 The formation of high-density lipoprotein (HDL). Lipid-poor apoA-I through interactions with ATP-binding cassette A1 (ABCA1) receives phospholipid and unesterified cholesterol and becomes pre-β HDL, or discoidal HDL (Step 1). Esterification of cholesterol through lecithin cholesterol acyltransferase leads to the formation mature HDL, or spheroidal HDL (Step 2). Cholesteryl esters are excreted to the liver through binding of scavenger receptor B1 (SR-B1) and lipid-poor apoA-I is regenerated (Step 3). Reprinted from Catte et al. (2006) with permission from Biophysical Society

assembly of HDL begins with lipid-poor apoA-I receiving phospholipids and unesterified cholesterol through interaction with ATP-binding cassette transporter A1 (ABCA1). This leads to the formation of pre-β HDL that has a discoidal shape. The enzyme lecithin cholesterol acyltransferase (LCAT) esterifies cholesterol turning pre-β HDL into mature HDL with a spherical shape. This cholesteryl ester-rich HDL interacts with receptors, such as the scavenger receptor B1, to unload cholesterol to the liver for excretion into bile, regenerating lipid-poor HDL (Catte et al. 2006).

Mature HDL is comprised of an outer amphipathic layer of free cholesterol, phospholipids, and several types of apolipoprotein (A-I, AII, C, E, AIV, J, and D) and an inner core consisting of triglycerides and cholesterol esters. ApoA-I is the principle protein in HDL, and is generally believed to be responsible for the efflux of cholesterol from cells. Other HDL-associated enzymes are paraoxonase (PON), platelet activating factor acetylhydrolase (PAF-AH), LCAT, and cholesteryl ester transfer protein (CETP) (Ansell et al. 2005).

It is well established that high HDL cholesterol levels (HDL-C) are predictive of lower risk for events related to atherosclerosis. What makes HDL protective against CHD? The best-known antiatherogenic property is its ability to promote cholesterol efflux from cells (Ansell et al. 2005). HDL also has antithrombotic, anti-inflammatory, and antioxidant properties. The anti-inflammatory functions include limiting lipid peroxidation, influencing expression of cytokines, modulating the recruitment and adhesion of monocytes, and altering other aspects of endothelial function (Ansell et al. 2005). Recently, however, it has been recognized that HDL can also moderate and enhance inflammation in atherogenesis (Ansell et al. 2005). In the setting of acute phase or chronic systemic inflammatory response, HDL can lose its antiatherogenic and anti-inflammatory properties and become pro-inflammatory and proatherogenic (Ansell et al. 2005).

8.3.2 How Good is the HDL?

Although epidemiological data shows a highly significant association between low HDL-C and clinical events caused by coronary artery disease, having normal HDL-C is not highly specific for predicting against CHD (Ansell et al. 2003). Extrapolating data from the Framingham study reveals the limitation of using HDL-C as a predictor. Of all of the patients with cardiac events, 44% of men had HDL-C $\geq$40 mg/dL and 43% of women had HDL-C $\geq$50 mg/dL (Ansell et al. 2003) (Fig. 8.4). Since a significant number of people who have normal HDL-C still had events, researchers have been searching for other qualitative measures of HDL (Navab et al. 2004).

Ansell and colleagues performed a study to see if HDL isolated from patients with known CHD were different in quality when compared with samples obtained from healthy controls. Rather than measuring HDL-C quantitatively, the inflammatory status of HDL was evaluated and given a value termed the HDL inflammatory index. Two assays were developed to measure the HDL inflammatory index. In the artery wall cell co-culture assay (CCA), baseline migration of monocytes into the subendothelial space (also known as monocyte chemotactic activity, or MCA) was given a value of 1.0. If HDL added to that assay increased MCA, the index value would be more than 1.0 and the HDL would be considered pro-inflammatory; if MCA decreased, then the index value would be less than 1.0 and the HDL would be considered anti-inflammatory. The cell-free assay (CFA) is a system where an oxidized phospholipid is combined with a fluorescein compound that generates a fluorescent signal when oxidized. Like the artery wall cell co-culture assay, the baseline fluorescent signal was given a value of 1.0. If HDL added to the CFA

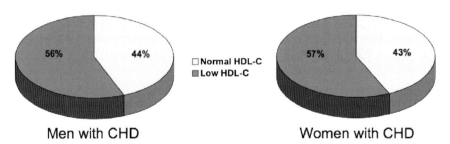

Fig. 8.4 Data from the Framingham study show that 44% of men and 43% of women who had a cardiac event also had a normal high-density lipoprotein cholesterol level (HDL-C), *CHD* coronary heart disease (Ansell et al. 2003)

increased the fluorescent signal, then the index value would be more than 1.0 and the HDL would be considered pro-inflammatory; if the fluorescent signal decreased, then the index value would be less than 1.0 and the HDL would be considered anti-inflammatory (Ansell et al. 2003).

HDL inflammatory index was used when comparing HDL quality in two groups of statin-naïve patients with known CHD or CHD risk equivalent by NCEP ATP III criteria against age- and gender-matched controls (Ansell et al. 2003) (Fig. 8.5). Group 1 consisted of 26 patients, three that had abnormally low HDL. Using the CCA, HDL from 20 out of the 26 patients were pro-inflammatory and had inflammatory index ≥ 1.0; all 26 patients had inflammatory index >0.6. HDL isolated from all 26 controls were anti-inflammatory and had an inflammatory index <1.0; 24 out of the 26 controls had inflammatory index <0.6 (Ansell et al. 2003).

The patients in Group 1 were then given six weeks of simvastatin therapy and the inflammatory index was again measured. There was a highly significant reduction in the inflammatory index in Group 1 patients, but HDL was still significantly more inflammatory than HDL from controls (Ansell et al. 2003) (Fig. 8.6).

Group 2 patients consisted of 20 patients with CHD and high HDL-C, none were diabetic, and none were on a statin. Similar to Group 1, 18 of the 20 patients had

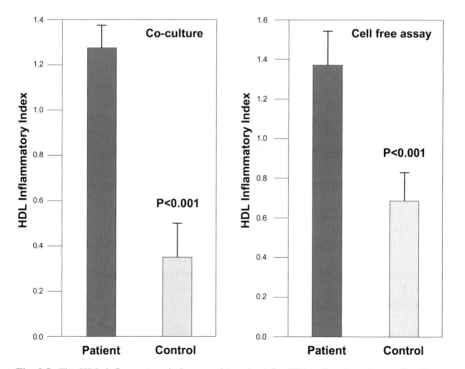

Fig. 8.5 The HDL inflammatory index was determined for HDL using the artery wall cell coculture and the cell-free assays. In this study, the patients all had known coronary heart disease (CHD) or CHD risk equivalent. The data shown are for patients and their age- and gender-matched healthy controls. Values shown are means ± SD, *HDL* high-density lipoprotein. Redrawn from Navab et al. (2004)

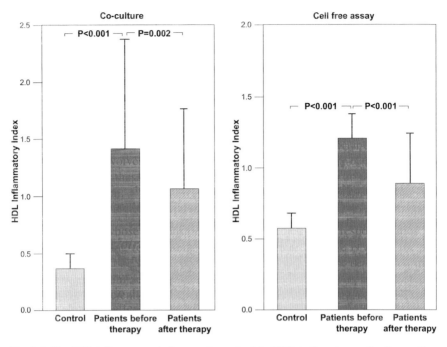

Fig. 8.6 The HDL inflammatory index was determined for HDL in the artery wall cell co-culture and cell-free assays. The HDL for patients with coronary heart disease was pro-inflammatory initially. The data shown are for patients before and after 6 weeks of simvastatin treatment (40 mg/day) and for healthy age- and gender-matched controls. Values shown are means $\pm$ SD, *HDL* high-density lipoprotein. Redrawn from Navab et al. (2004)

inflammatory index ≥ 1.0 using the CCA and only one had inflammatory index <0.6. All 20 of the controls had inflammatory index <0.6. Using the CFA, 19 of the 20 patients had inflammatory index >1.0. Eighteen of the 20 controls had inflammatory index <1.0 (Ansell et al. 2003).

The experiment conducted by Ansell and colleagues demonstrates that patients with CHD or CHD risk equivalent were better separated from controls using the HDL inflammatory index rather than HDL-C. Both CCA and CFA were similar in their abilities to measure inflammatory status of HDL (Ansell et al. 2003). Based on this study, statins appear able to reduce the inflammatory status of HDL. In addition, the HDL inflammatory index is a viable new biomarker for CHD and one that is modifiable with pharmacotherapy.

8.3.3 Dysfunctional HDL

HDL is altered by inflammatory states, whether acute or chronic—higher levels of pro-inflammatory HDL have been found in patients with chronic inflammatory disease as well as during phases of acute stress (Ansell et al. 2007). Pro-inflammatory

HDL is dysfunctional and loses much of the atheroprotective character found in anti-inflammatory HDL (Ansell et al. 2007).

Several conditions cause a chronic inflammatory environment and increased oxidative stress (Fig. 8.7). Patients with CHD, DM, metabolic syndrome, rheumatologic conditions, chronic kidney disease, and obstructive sleep apnea all have pro-inflammatory HDL (Ansell et al. 2007). HDL from patients with poorly controlled DM has been found to have decreased capacity for cholesterol efflux from macrophages (Gowri et al. 1999). Patients with metabolic syndrome have been shown to have pro-inflammatory HDL and higher levels of LOOH when compared to controls (Ansell et al. 2007). HDL from patients with metabolic syndrome also has impaired anti-oxidative properties (Hansel et al. 2004). Patients with rheumatologic diseases, such as SLE and RA, are known to have a very significant increased

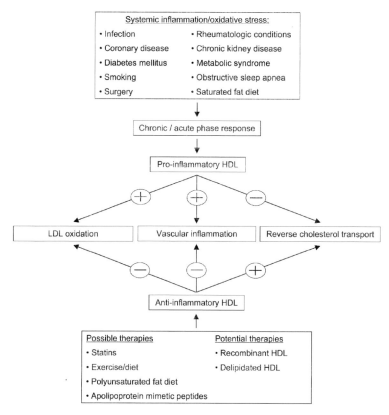

Fig. 8.7 Shown are the effects of systemic inflammation/oxidative stress from a number of conditions that can have an adverse impact on the anti-inflammatory effects of high-density lipoprotein (HDL), including inhibition of low-density lipoprotein (LDL) oxidation, reduction in vascular inflammation, and promotion of cholesterol efflux. Emerging data support the potential of statins, apolipoprotein mimetic peptides, and lifestyle changes to improve these functions of HDL. Redrawn from Ansell et al. (2007)

risk for developing CHD. The presence of pro-inflammatory HDL has also been reported in 44.7% of patients with SLE, 20.1% of patients with RA, and only in 4.1% of healthy controls (McMahon et al. 2006). Weihrauch and colleagues demonstrated that patients with systemic sclerosis (SSc), also known as scleroderma, had higher levels of pro-inflammatory HDL compared to controls (Weihrauch et al. 2007). HDL in patients with chronic kidney disease have notable differences in their content; triglyceride content is higher while apoA-I and apolipoprotein A-II are reduced and LCAT activity is decreased (Ansell et al. 2007).

When the body experiences a transient period of systemic inflammation, pro-inflammatory HDL is also found. In elective surgery, HDL that was anti-inflammatory prior to surgery becomes pro-inflammatory post-op and then reverts to an anti-inflammatory type. Influenza virus infection also produces a transient pro-inflammatory HDL state that may explain the increased MI risk during flu epidemics (Ansell et al. 2007). People eating a diet high in saturated fats from coconut oil also have impaired function when compared to those eating a diet with fats being predominantly polyunsaturated from safflower oil (Nicholls et al. 2006). HDL isolated from the group eating high saturated fat content induced endothelial cells to express higher levels of the adhesion molecules, ICAM-1 and VCAM-1, compared to controls who were fasting. In contrast, HDL isolated from the polyunsaturated fat group had a reduction in endothelial cell adhesion molecule expression (Nicholls et al. 2006).

8.3.4 How HDL Becomes Dysfunctional

Multiple mechanisms are responsible for pro-inflammatory characteristics of HDL. Glycation is one such mechanism of impairing HDL function and may be especially significant in diabetics. The activity of PON, an antioxidant enzyme in HDL, is reduced by 35% after incubation in a glucose solution for one week (Hedrick et al. 2000). In addition, incubation of HDL in glucose led to an inability to inhibit monocyte chemotaxis (Hedrick et al. 2000). Other studies show that glycation of LDL and HDL increases the susceptibility of LDL to oxidation and decreases the antioxidant qualities in HDL (Ansell et al. 2007).

Changes to the protein and lipid composition of HDL are observed when HDL is in a pro-inflammatory state. SAA, ceruloplasmin, apoJ, and secretory phospholipase A2 (sPLA2) are found in pro-inflammatory HDL at higher concentrations while other normal components found in anti-inflammatory HDL, such as apoA-I, apoA-II, PON, PAF-AH, CETP, LCAT, and phospholipid transfer protein (PLTP), are reduced (Ansell et al. 2005). Phospholipid composition in HDL can influence the ability to inhibit endothelial adhesion molecule expression (Baker et al. 2000).

Enzymatic and structural changes are other factors that make HDL more dysfunctional. ApoA-I is subject to oxidation. Myeloperoxidase, a product of white blood cell activation, can cause modification of apoA-I through nitrotyrosination and chlorotyrosination at specific amino acid sites within apoA-I that also confer pro-inflammatory characteristics (Ansell et al. 2007). HDL can be hydrolyzed by sPLA2 in such a manner that its capacity to promote efflux of cholesterol from cells is reduced. The mechanism is associated, though indirectly, with downregulation of ABCA1 on macrophages. SPLA2 activity can also lead to the production of pro-inflammatory fatty acids.

In a given individual, the composition of HDL falls along the continuum between anti-inflammatory and pro-inflammatory and is influenced by the net oxidative stress present. The key to reducing CHD may lie in modifying the inflammatory state of HDL to revert it back to a protective form.

8.3.5 Making Bad HDL Better

The question remains—can dysfunctional HDL regain its ability to be atheroprotective? As with many of the other risk factors for CVD, lifestyle modifications in the form of dietary improvements and exercise can improve the quality of HDL. A group of patients with metabolic syndrome were put on a 21-day lifestyle intervention in overweight patients with metabolic syndrome (Roberts et al. 2006). The diet these patients were given consisted of high-fiber grains, vegetables, fruits, reduced saturated fat, and proteins from plants and fish. They were also supervised on a walk of up to one hour. BMI fell from 33.1 to 32.1, LDL-C fell from 126 to 94, and HDL-C decreased from 44 to 39. Despite the fall in HDL-C, the HDL inflammatory index fell from 1.14 to 0.94 and PAF-AH increased.

Another group of patients were fed meals containing predominantly polyunsaturated fat. Levels of ICAM-1 and VCAM-1 diminished. This was seen as early as 3 hours after the meal and was sustained until 6 hours after. Saturated fat diets resulted in pro-inflammatory HDL by 6 hours leading to increased levels of both adhesion molecules (Nicholls et al. 2006).

Simvastatin, and likely the other statins in the drug class, has demonstrated anti-inflammatory characteristics. The Ansell study described previously showed that six weeks of simvastatin therapy was able to improve the quality of HDL as evidenced by the decrease in the HDL inflammatory index (Ansell et al. 2003). See Fig. 8.6.

Experimentally, apoA-I has also been found to decrease the inflammatory response in artery wall co-cultures. When LOOH levels reach a certain threshold in LDL to transform it into MM-LDL, MCP-1 is induced, thus increasing MCA. Human apoA-I added to the artery wall co-culture prior to LDL addition inhibits oxidation of LDL and results in a marked reduction in LDL-induced MCA (Navab et al. 2005b).

8.4 Apolipoprotein A-I Mimetic Peptides

The importance of apoA-I in atherosclerosis was established by testing in animal models (Plump et al. 1994) and human studies (Nissen et al. 2003). ApoA-I comprises 243-amino acids and therefore requires intravenous administration.

Manufacture of such a large protein is complex and cost prohibitive. As a result, research has been directed toward developing smaller peptides that produce similar results to apoA-I, but that are simpler to manufacture and administer.

The search for peptides smaller than apoA-I, but with similar lipid binding activity, led to the synthesis of 18-amino acid peptides called apoA-I mimetic peptides (Navab et al. 2005b). These peptides do not possess sequence homology with apoA-I; they instead emulate the class A amphipathic helices of apoA-I that are thought to be responsible for the lipid binding activity of apoA-I. These peptides were named based on the number of phenylalanine residues (F being the one letter name for phenylalanine) they contain, for example 2F, 3F, 4F, 5F, 6F, and 7F. The position and number of phenylalanine residues determines the ability of the peptides to sequester inflammatory lipids. Of the variants, 4F and 5F were found to have significant biologic activity in reducing atherosclerotic lesions (see Section 8.4.2 ApoA-I Mimetic Peptides in Animal Models of Atherosclerosis) (Garber et al. 2001; Navab et al. 2002). Most studies on apoA-I mimetic peptides focus on 4F because of its superior ability to inhibit LDL-induced MCA (Datta et al. 2001). The amino acid sequence for 4F is Ac-D-W-F-K-A-F-Y-D-K-V-A-E-K-F-K-E-A-F-NH2 (Navab et al. 2005b).

A peptide that could be orally administered led to the synthesis of D-4F. Enzymatic digestion in mammals recognizes proteins with L-amino acids. 4F synthesized from L-amino acids is rapidly degraded when given orally to mice. D-4F is synthesized from D-amino acids and is found intact in the circulation after oral administration. Oral administration of D-4F rendered HDL from LDL receptor (LDLR)-null mice on a Western diet and HDL from apoE-null mice on a chow diet anti-inflammatory (Navab et al. 2002).

8.4.1 Action of ApoA-I Mimetic Peptides

Atherosclerosis and other vascular pathology are mediated in part by oxidized lipids and other lipid oxidation products. ApoA-I mimetic peptides such as D-4F have the ability to sequester pro-inflammatory lipid oxidation products and this seems to be the major mechanism that they employ. Actually, D-4F is superior to apoA-I in inhibiting LDL-induced MCA through that mechanism. When apoA-I binds LOOH, the LOOH are still able to participate in forming pro-inflammatory oxidized phospholipids that can induce MCA. In contrast, D-4F is able to effectively sequester LOOH in a way that it cannot participate in oxidizing phospholipids and therefore inhibits LDL-induced MCA (Navab et al. 2005b).

Using fast protein liquid chromatography, the plasma distribution of D-4F was studied. One hour after oral administration of fluorescent D-4F, the peptide elutes at the same time as pre-β HDL and mature HDL. At two hours it elutes only with the latter particles. The peptide was cleared from the plasma by 8 hours (Navab et al. 2005b).

Mature HDL particles are constantly changing because apoA-I particles are always leaving the large HDL particle to generate smaller lipid-poor apoA-I particles. ApoA-I mimetic peptides such as D-4F dramatically accelerate this process (Navab et al. 2005b). Oral D-4F causes the movement of LOOH from very low-density lipoprotein, intermediate density lipoprotein, and mature HDL to pre-β HDL. Despite the increase in LOOH content, the pre-β HDL fractions were still anti-inflammatory according to the HDL inflammatory index. Oral D-4F also increases PON activity and formation of pre-β HDL in monkeys. In humans, nanomolar amounts of D-4F causes the formation of pre-β HDL, reduces lipoprotein LOOH content, increases PON activity and converts pro-inflammatory HDL to anti-inflammatory (Navab et al. 2005b).

In vitro, D-4F decreases HDL LOOH content and increases PON activity. PON is reversibly inhibited by LOOH and when D-4F peptide interacts with HDL, LOOH is sequestered and can no longer affect PON activity. This creates a positive feedback loop where the peptide helps to increase PON activity and the increased PON activity further decreases LOOH (Navab et al. 2005b).

Overall, the major effects of D-4F are decreased oxidative stress, reduced systemic inflammation, and improvement in the atheroprotective quality of HDL. These characteristics are what make apoA-I mimetic peptides efficacious in models of atherosclerosis and other diseases with a significant inflammatory response.

8.4.2 ApoA-I Mimetic Peptides in Animal Models of Atherosclerosis

The first evidence that an apoA-I mimetic peptide could inhibit atherosclerosis in vivo was demonstrated in an experiment conducted by Garber and colleagues (Garber et al. 2001). 5F was injected into mice fed an atherogenic diet. Control mice received mouse apoA-I or phosphate buffer solution. The only significant difference in the lipid profiles was that HDL-C was lower in the mice receiving 5F or mouse apoA-I as a percent of the total cholesterol. HDL from mice given the buffer solution did not decrease LDL-induced MCA. HDL from mice given mouse apoA-I was weakly anti-inflammatory and HDL from mice given 5F was significantly anti-inflammatory. Atherosclerosis lesion area was significantly reduced in the 5F-treated mice but not in mice receiving mouse apoA-I (Garber et al. 2001).

D-4F administered orally to LDLR-/- mice on Western diet, atherosclerotic lesions decreased by 79%. When added to the drinking water of apoE-null mice, D-4F decreased lesions by approximately 75% at the lowest dose tested. This occurred independent of changes in total plasma or HDL-C (Navab et al. 2002).

D-4F has been shown to dramatically synergize with statins to reduce atherosclerotic lesions and increase intestinal synthesis of apoA-I. D-4F and pravastatin, when given in combination at oral doses that were ineffective when given as single agents, rendered HDL anti-inflammatory in mice and monkeys and prevented atherosclerosis in their young, as well as causing regression of established lesions in older apoE-null mice (Navab et al. 2005a).

The observation that apoA-I mimetic peptides have a remarkable effect in curbing atherogenesis spawned interest in testing the peptide in various aspects of aging such as endothelial health and cognitive impairment. Being an anti-inflammatory agent, apoA-I has also been tested in other inflammatory diseases and conditions.

8.4.3 D-4F and Vasodilation, Vessel Wall Thickness, and Endothelial Health

Endothelial cells play an important role in maintaining vascular health. With vascular aging, vessels become thicker and stiffer. Insults to the vascular system can cause dysfunction of the NO-dependent endothelial vasodilation. One of the earliest physiological and most sensitive changes in hypercholesterolemia is the loss of endothelium- and eNOS-dependent vasodilation, which appears to occur before structural changes in the vessel wall (Ou et al. 2005). Native LDL and OX-LDL can both dramatically decrease eNOS function. It is thought that D-4F protects vascular endothelial cells function by binding it and removing oxidized lipids (Ou et al. 2005).

Short-term D-4F treatments reduced wall thickness in vessels from hypercholesteremic LDLR-null mice before vasodilation was fully restored. Improvements in vessel architecture preceded improvements in vascular physiology. D-4F decreased wall thickness even in vessels with preexisting disease in mice with HDL that contains apoA-I, but not in mice whose HDL was apoA-I deficient. D-4F still improved vasodilation in mice lacking apoA-I (Ou et al. 2005).

Type 1 DM has been shown to reduce both the number and function of bone marrow-derived endothelial progenitor cells (EPC). This could potentially contribute to the formation of atherosclerotic disease. Growing evidence suggest that proper vascular function relies not only on mature endothelial cells but also on EPC. EPC have been shown to contribute to vascular remodeling in atherosclerosis and other CVD. HDL has been shown to provide vascular protection by increasing EPC in apoE-null mice (Peterson et al. 2007). Hemeoxygenase-1 (HO-1) is strongly induced by its substrate heme and by oxidant stress. HO-1 has a robust ability to protect against oxidative insult in CVD. Endothelial cell dysfunction, demonstrated by the reduced expression of $CD31^+$ and/or thrombomodulin, has been reported within atherosclerotic blood vessels. The antioxidant effects of HO-1 arise from its capacity to degrade the heme moiety from destabilized heme proteins. D-4F, by increasing HO-1 and eNOS and decreasing circulating oxidants, protects EPC function. D-4F prevents vascular damage, rendering endothelial cells more resistant to oxidants in DM (Peterson et al. 2007).

Endothelial cell dysfunction, demonstrated by the reduced expression of $CD31^+$ and/or thrombomodulin, has been reported within atherosclerotic blood vessels.

The number of circulating endothelial cells in peripheral blood was significantly elevated in diabetic rats compared with control animals. D-4F attenuated endothelial cell sloughing in diabetic rats, but not control rats (Peterson et al. 2007). D-4F increased $CD31^+$ and thrombomodulin (Peterson et al. 2007). D-4F also increased EPC function, eNOS expression, and HO-1 protein expression increased (Peterson et al. 2007). Statins also increase HO-1 and reduce adhesion molecule expression, which may explain the mechanism of statins to exert an anti-oxidant effect. D-4F modulates the EPC phenotype, as reflected by the increases in HO-1 and eNOS function that may influence $CD31^+$ and thrombomodulin levels. This all promotes endothelial cell survival and vascular protection (Peterson et al. 2007).

8.4.4 D-4F and Improvement in Brain Arteriole Inflammation

Cognitive impairment is notable in the presence of brain arteriole inflammation. LDLR-null mice on a Western diet have inflammation in large arteries and endothelial dysfunction in small arteries that are improved with D-4F. Western diet caused a marked increase in the percent of brain arterioles with associated macrophages (microglia) that was reduced by oral D-4F but not by a placebo peptide. D-4F reduced the percent of brain arterioles associated with CCL3/macrophage inflammatory protein-1α (MIP-1α) and CCL2/MCP-1. Cognitive performance in the T-maze continuous alternation task (T-CAT) and in the Morris Water Maze (MWM) was impaired by a Western diet and was significantly improved with D-4F. Plasma lipid levels, blood pressure, and arteriole lumen size was unchanged (Buga et al. 2006).

D-4F decreases hyperlipidemia-induced brain arteriole inflammation (microglia, MIP-1α, MCP-1), brain arteriole wall thickness, and brain arteriole smooth muscle α-actin. D-4F improves hyperlipidemia-induced cognitive impairment in T-CAT and MWM. D-4F did not alter blood pressure or plasma lipids. D-4F reduces MCP-1 and MIP-1α expression in nonvascular brain cells (Buga et al. 2006).

8.4.5 D-4F in Diseases with Cardiovascular Complications

Weihrauch and colleagues used tight skin (Tsk-/+) mice as a model for systemic sclerosis (SSc). Tsk-/+ mice have a defect in fibrillin-1, resulting in replication of many of the myocardial and vascular features seen in humans with SSc. SSc is an autoimmune disorder that causes marked increases in fibrosis of the skin and internal organs. Vascular features are impaired vasodilation, endothelial cell senescence, and impaired angiogenesis. Histologic evaluation of hearts from people who died of complications of SSc reveals varying degrees of myocardial inflammation (Weihrauch et al. 2007).

D-4F improved acetylcholine-induced vasodilation in Tsk-/+ mice to approximately half of that achieved by control mice. D-4F reduced pro-inflammatory HDL and angiostatin levels in Tsk-/+ mice as well as patients with SSc. Angiostatin and autoantibodies against oxidized phosphatidylcholine are markedly increased in the interstitium of hearts from Tsk-/+ mice compared to controls. D-4F treatments essentially prevent the increases in myocardial inflammation and generation of angiostatic factors. Tsk-/+ mice have impaired angiogenic responses to vascular endothelial growth factor stimulation. D-4F is sufficient to restore angiogenic responses (Weihrauch et al. 2007).

Why does D-4F improve vascular function in Tsk-/+ mice that are a model for SSc? SSc is a disease that puts the body in a chronic state of oxidative stress. This is evident in the pro-inflammatory nature of HDL in Tsk-/+ mice and SSc patients. The loss of atheroprotective function of HDL is a common pathway to CVD in SSc patients. D-4F improves HDL quality and decreases angiostatin in Tsk-/+ mice demonstrating the importance of the atheroprotective trait of normal HDL in protecting against vascular disease (Weihrauch et al. 2007).

8.4.6 D-4F in Non-Atherosclerotic Diseases

Influenza A infection leads to increased IL-6 production, increased activity of caspases and interferons α/β, and induction of MCP-1 production through increasing oxidized phospholipids. When infected cells were incubated with D-4F, production and release of pro-inflammatory oxidized phospholipids were inhibited. Cellular viral titers also dramatically dropped by 50% after D-4F treatment. D-4F inhibits the viral induction of interferons. D-4F treatment inhibits caspase activation and cytokine production after viral infection, and additionally, inhibits viral replication (Van Lenten et al. 2004).

8.4.7 HDL-Directed Therapy

ApoA-I mimetic peptides such as D-4F have great potential to improve cardiovascular function and reversing the effects of cardiovascular aging. Atherosclerosis is the hallmark of CHD and is the process responsible for the majority of CVD. Much research confirms that the large burden of oxidative stress, whether from the environment or disease, can overwhelm the body's cardiovascular system. Environmental stressors, such as cigarette smoking or eating a meal consisting of high levels of saturated fats, increases inflammation in the body as evidenced by the increase in pro-inflammatory HDL and the impairment of vessel function. Acute phase response and chronic inflammatory conditions all create a milieu in the body that promotes increased oxidative stress and the development of vascular inflammation. ApoA-I mimetic peptides can curb the damage done by these conditions by sequestering pro-inflammatory lipid oxidation products such as oxidized phospholipids and LOOH. In this way, anti-inflammatory function of HDL returns with the formation of more pre-β HDL and improvement in HDL-associated anti-oxidant enzyme activity in PON, LCAT, PAF-AH, and HO-1.

8.4.8 Other Peptides with Lipid-Binding Activity

Other peptides have also had effect in lipid binding and decreasing inflammation in the body. In the last several years, several studies have been performed involving apoA-I mimetic peptides. A more minor apolipoprotein in HDL, apoJ, has also been found to be able to bind inflammatory lipids like apoA-I (Navab et al. 2005c). An apoJ mimetic peptide has been found to effectively bind and sequester the LOOH necessary for oxidation of the LDL phospholipids, which when oxidized cause the artery wall cells to produce MCP-1. Unlike D-4F, apoJ does not increase pre-β HDL. Like D-4F, D-apoJ improved inflammatory properties of HDL apoE-null mice. Oral D-apoJ dramatically reduced atherosclerosis in apoE-null mice and was associated with a significant increase in HDL-C and an increase in HDL-PON activity. The magnitude of the reduction in atherosclerosis in the apoE-null mice approaches that reported for D-4F (Navab et al. 2005c). D-apoJ reduced LOOH in Cynomolgus monkeys and improved lipoprotein inflammatory properties. Common mechanism for the beneficial effect of D-4F and D-apoJ may relate to the ability of both peptides to bind and sequester oxidized lipids and activate antioxidant enzymes such as PON. LOOH also inhibits LCAT. Altered activities of LCAT, PON, PAF-AH in atherosclerosis susceptible mice correlate with plasma levels of oxidized lipids (Forte et al. 2002). Binding and sequestering of oxidized lipids that inhibit a series of antioxidant enzymes may release those enzymes to further destroy lipid oxidation products (Navab et al. 2005c).

Another set of peptides that are too small to form class A amphipathic helices demonstrate ability to reduce lipoprotein LOOH, increase PON activity, increase HDL-C, and render HDL anti-inflammatory, and reduce atherosclerosis in apoE null mice. KRES and FREL, tetrapeptides that are named after their respective amino acid sequences, have been found to have lipid-binding activity (Navab et al. 2005d). KERS was not biologically active. KRES and FREL were found associated with HDL whereas KERS was not. This shows that the ability of peptides to interact with lipids, remove LOOH, and activate antioxidant enzymes associated with HDL determines their anti-inflammatory and anti-atherogenic properties regardless of their ability to form amphipathic helices (Navab et al. 2005d).

8.4.9 Beyond Cardiovascular Inflammation: HDL and LDL Particle Size

The targeting of pro-inflammatory HDL with apoA-I mimetic peptides, statins, and lifestyle modifications will likely increase the ability of people to age well, but other characteristics of lipoproteins may also be important. Evidence supporting the association of HDL and LDL particle size with exceptional longevity has been reported (Barzilai et al. 2003; Heijmans et al. 2006). Barzilai and associates demonstrated in a case-controlled study that Ashkenazi Jews with exceptional longevity and good health (i.e., living independently at the age of 95) and their first-degree offspring had a notably larger HDL and LDL particle size than controls. While offspring of centenarians had strikingly larger HDL and LDL particle sizes than age-matched controls, the centenarians themselves had even larger lipoprotein sizes (Barzilai et al. 2003). There were further associations of larger HDL and LDL particle sizes with lower prevalence of hypertension, metabolic syndrome, and CVD (Barzilai et al. 2003).

Heijmans and colleagues conducted a study in an outbred European population in Leiden to attempt to confirm those observations made by Barzilai. Though larger HDL particle sizes were found in long-lived Ashkenazi families, this finding was not replicated in the study in Leiden (Heijmans et al. 2006). They did, however, report evidence that a low concentration of small LDL particles was associated with human longevity (Heijmans et al. 2006). These results indicate that the association of lipoprotein size and longevity may not be exclusive to individuals with exceptional familial longevity, but rather may apply to the population at large.

8.5 Summary and Conclusions

D-4F is a novel apoA-I mimetic peptide that is in Phase 1 clinical trials (Bloedon et al. 2006, 2008). There may be a significant role for D-4F in addressing CVD. As described above, apoA-I mimetic peptides have an extraordinary capacity to treat the damage done by many different processes from poor diet, smoking, and the many diseases that have a negative impact on the cardiovascular system (Tables 8.1 and 8.2).

Substantial data from animal studies demonstrate the efficacy of apoA-I mimetic peptides in ameliorating many aspects of cardiovascular aging. The beneficial effects of apoA-I mimetic peptides may be through one common pathway—the

Table 8.1 In vitro and ex vivo effects of apolipoprotein A-I mimetic peptide D-4F. After Navab et al. (2006)

In vitro
Incubated mouse, monkey, or human plasma
- Forms pre-β HDL
- Reduces lipid hydroperoxide concentrations in plasma HDL and LDL
- Increases the activity of the antioxidant enzyme paraoxonase in HDL
- Converts pro-inflammatory HDL to anti-inflammatory HDL
- Reduces LDL sensitivity to oxidation by artery wall cells in culture
- Renders artery wall cells less capable of oxidizing LDL in culture
- Reduces superoxide and increases nitric oxide production in cultured endothelial cells
- Promotes cellular cholesterol efflux from cholesterol-loaded human macrophages
- Reduces caspase-3, caspase-8, and caspase-9 activity and interleukin-6 levels in A-549 cell-cultured human type II pneumocytes after infection with influenza A virus stimulation
- Reduces interleukin-6 levels in human type II pneumocytes in culture

Ex vivo
Atherosclerosis and sickle-cell disease mouse models
- Improves vasoreactivity in arteries even in the absence of apolipoprotein A-I

Table 8.2 In vivo effects of apolipoprotein A-I mimetic peptide D-4F. After Navab et al. (2006)

Oral administration to normal mice or monkeys
- Increases pre-β HDL levels in the circulation
- Reduces lipid hydroperoxide concentrations in plasma, HDL, and LDL
- Increases the activity of the antioxidant enzyme paraoxonase in HDL
- Converts pro-inflammatory HDL to anti-inflammatory HDL
- Reduces LDL sensitivity to oxidation

Oral administration to apolipoprotein A-I knockout mice
- Restores the vasoreactive function of HDL

Oral administration to apolipoprotein E knockout mice
- Increases cholesterol efflux from injected macrophages into the plasma, liver, and feces (i.e., reverse cholesterol efflux)
- Reduces association of myeloperoxidase with and reduces oxidative and functional damage to apolipoprotein A-I, thus preserving the ability of apolipoprotein A-I to cause cholesterol efflux through ABCA1
- Reduces artery wall thickness in LDL-receptor knockout mice fed a high-fat high-cholesterol diet

Oral administration to apolipoprotein E and LDL-receptor knockout mice
- Reduces atherosclerotic lesions

Oral administration together with pravastatin
Monkeys
- Renders HDL anti-inflammatory

Apolipoprotein E knockout mice
- Increases intestinal apolipoprotein A-I synthesis and plasma HDL levels
- Causes the regression of already established atherosclerotic lesions
- Reduces macrophage content of atherosclerotic lesions

In a rat model of diabetes
- Restores the activities of the vascular antioxidant enzymes hemoxygenase-1 and extracellular superoxide dismutase
- Prevents the accumulation of superoxide anion and loss of vasoreactivity
- Inhibits arterial endothelial sloughing and reduces the number of circulating endothelial cell particles

restoration of the normal anti-atherogenic functions of HDL. In each of the various conditions where compromise of the cardiovascular system was found (i.e., impaired endothelial health, vessel wall thickening, atherosclerosis, myocardial inflammation) the HDL was pro-inflammatory as evaluated by the HDL inflammatory index. Various animal models of disease and in vitro studies of human HDL show that D-4F is effective in not only converting HDL from pro-inflammatory to anti-inflammatory, but that the improvement in HDL quality is also associated with a concomitant improvement in the health of the cardiovascular system.

The future clinical laboratory will have an assay to measure the HDL inflammatory index to determine if HDL is anti-inflammatory or pro-inflammatory. The inflammatory status of HDL is modifiable; D-4F and other compounds with similar lipid-binding activity are able to improve the quality of HDL without necessarily affecting the quantity. Further clinical trials must now be conducted to determine if improvement of HDL quality will lead to improvement of the structure and function of the cardiovascular system, and ultimately, if that will translate into a decrease in CVD morbidity and mortality. ApoA-I mimetic peptides, such as D-4F, may become part of a prospective new treatment strategy termed HDL-directed therapy, and, in turn, HDL-directed therapy may be the next arm of treatment added to the battle against CVD.

Lipoprotein metabolism is intricately involved in the pathogenesis of many diseases as well as in the maintenance of good health and longevity. Further research in HDL and LDL metabolism may find a link between particle size and the inflammatory properties of these lipoproteins, determine if apoA-I mimetic peptides and related molecules can affect the particle size of HDL and LDL, and if any of these effects will have any influence on aging.

References

Ansell BJ, Fonarow GC, Fogelman AM (2007) The paradox of dysfunctional high-density lipoprotein. Curr Opin Lipidol 18:427–434

Ansell BJ, Navab M, Hama S, Kamranpour N, Fonarow G, Hough G, Rahmani S, Mottahedeh R, Dave R, Reddy ST, Fogelman AM (2003) Inflammatory/antiinflammatory properties of high-density lipoprotein distinguish patients from control subjects better than high-density lipoprotein cholesterol levels and are favorably affected by simvastatin treatment. Circulation 108:2751–2756

Ansell BJ, Watson KE, Fogelman AM, Navab M, Fonarow GC (2005) High-density lipoprotein function: recent advances. J Am Coll Cardiol 46:1792–1798

Baker PW, Rye KA, Gamble JR, Vadas MA, Barter PJ (2000) Phospholipid composition of reconstituted high density lipoproteins influences their ability to inhibit endothelial adhesion molecule expression. J Lipid Res 41:1261–1267

Barzilai N, Atzmon G, Schechter C, Schaefer EJ, Cupples AL, Lipton R, Cheng S, Shuldiner AR (2003) Unique lipoprotein phenotype and genotype associated with exceptional longevity. JAMA 290.2030–2040

Bernatsky S, Boivin JF, Joseph L, Manzi S, Ginzler E, Gladman DD, Urowitz M, Fortin PR, Petri M, Barr S, Gordon C, Bae SC, Isenberg D, Zoma A, Aranow C, Dooley MA, Nived O, Sturfelt G, Steinsson K, Alarcón G, Senécal JL, Zummer M, Hanly J, Ensworth S, Pope J, Edworthy S, Rahman A, Sibley J, El-Gabalawy H, McCarthy T, St Pierre Y, Clarke A, Ramsey-Goldman R (2006) Mortality in systemic lupus erythematosus. Arthritis Rheum 54:2550–2557

Bloedon LT, Dunbar RL, Duffy D, Pinell-Salles P, Navab M, Fogelman A, Radar DJ (2006) Abstract 1496: Oral administration of the apolipoprotein A-I mimetic peptide D-4F in humans with CHD improves HDL anti-inflammatory function after a single dose. Circulation 114:II 288-II 289

Bloedon LD, Dunbar R, Duffy D, Pinell-Salles P, Norris R, Degroot BJ, Movva R, Navab M, Fogelman AM, Rader DJ (2008) Safety, pharmacokinetics, and pharmacodynamics of oral apoA-I mimetic peptide D-4F in high-risk cardiovascular patients. J Lipid Res 49:1344–1352

Budoff MJ, Achenback S, Blumenthal RS, Carr JJ, Goldin JG, Greenland P, Guerci AD, Lima JAC, Rader DJ, Rubin GD, Shaw LJ, Wiegers SE (2006) Assessment of coronary artery disease by cardiac computed tomography: a scientific statement from the American Heart Association Committee on Cardiovascular Imaging and Intervention, Council on Cardiovascular Radiology and Intervention, and Committee on Cardiac Imaging, Council on Clinical Cardiology. Circulation 114:1761–1791

Buga GM, Frank JS, Mottino GA, Hendizadeh M, Hakhamian A, Tillisch JH, Reddy ST, Navab M, Anantharamaiah GM, Ignarro LJ, Fogelman AM (2006) D-4F decreases brain arteriole inflammation and improves cognitive performance in LDL receptor-null mice on a Western diet. J Lipid Res 47:2148–2160

Catte A, Patterson JC, Jones MK, Jerome WG, Bashtovyy D, Su Z, Gu F, Chen J, Aliste MP, Harvey SC, Li L, Weinstein G, Segrest JP (2006) Novel changes in discoidal high density lipoprotein morphology: a molecular dynamics study. Biophys J 90:4345–4360

Chobanian AV, Bakris GL, Black HR, Cushman WC, Green LA, Izzo JL Jr, Jones DW, Materson BJ, Oparil S, Wright JT Jr, Roccella EJ (2003) The Seventh Report of the Joint National Committee on Prevention, Detection, Evaluation, and Treatment of High Blood Pressure: the JNC 7 report. JAMA 289:2560–2572

Croft LB, Donnino R, Shapiro R, Indes J, Fayngersh A, Squire A, Goldman ME (2004) Age-related prevalence of cardiac valvular abnormalities warranting infectious endocarditis prophylaxis. Am J Cardiol 94:386–389

Datta G, Chaddha M, Hama S, Navab M, Fogelman AM, Garber DW, Mishra VK, Epand RM, Epand RF, Lund-Katz S, Phillips MC, Segrest JP, Anantharamaiah GM (2001) Effects of increasing hydrophobicity on the physical-chemical and biological properties of a class A amphipathic helical peptide. J Lipid Res 42:1096–1104

Ergin A, Muntner P, Sherwin R, He J (2004) Secular trends in cardiovascular disease mortality, incidence, and case fatality rates in adults in the United States. Am J Med 117:219–227

Expert Panel on Detection, Evaluation, and Treatment of High Blood Cholesterol in Adults (2001) Executive Summary of the Third Report of the National Cholesterol Education Program (NCEP) Expert Panel on Detection, Evaluation, and Treatment of High Blood Cholesterol in Adults (Adult Treatment Panel III). JAMA 285:2486–2497

Forte TM, Subbanagounder G, Berliner JA, Blanche PJ, Clermont AO, Jia Z, Oda MN, Krauss RM, Bielicki JK (2002) Altered activities of anti-atherogenic enzymes lecithin: cholesterol acyltransferase, paraoxonase, and platelet-activating factor acetylhydrolase in atherosclerosis susceptible mice. J Lipid Res 43:477–485

Garber DW, Datta G, Chaddha M, Palgunachari MN, Hama SY, Navab M, Fogelman AM, Segrest JP, Anantharamaiah GM (2001) A new synthetic class A amphipathic peptide analog protects mice from diet-induced atherosclerosis. J Lipid Res 42:545–552

Gowri MS, Van der Westhuyzen DR, Bridges SR, Anderson JW (1999) Decreased protection by HDL from poorly controlled type 2 diabetic subjects against LDL oxidation may be due to abnormal composition of HDL. Arterioscler Thromb Vasc Biol 19:2226–2233

Gregg EW, Cheng YJ, Cadwell BL, Imperatore G, Williams DE, Flegal KM, Narayan KMV, Williamson DF (2005) Secular trends in cardiovascular disease risk factors according to body mass index in US adults. JAMA 293:1868–1874

Grundy SM, Balady GJ, Criqui MH, Fletcher G, Greenland P, Hiratzka LF, Houston-Miller N, Kris-Etherton P, Krumholz HM, LaRosa J, Ockene IS, Pearson TA, Reed J, Washington HM, Smith SC Jr (1998) Primary prevention of coronary heart disease: guidance from Framingham: a statement for healthcare professionals from the AHA Task Force on Risk Reduction. Circulation 97:1876–1887

Hagan PG, Nienaber CA, Isselbacher EM, Bruckman D, Karavite DJ, Russman PL, Evangelista A, Fattori R, Suzuki T, Oh JK, Moore AG, Malouf JF, Pape LA, Gaca C, Sechtem U, Lenferink S, Deutsch HJ, Diedrichs H, Marcos y Robles J, Llovet A, Gilon D, Das SK, Armstrong WF, Deeb GM, Eagle KA (2000) The International Registry of Acute Aortic Dissection (IRAD): new insights into an old disease. JAMA 283:897–903

Hansel B, Giral P, Nobecourt E, Chantepie S, Bruckert E, Chapman MJ, Kontush A (2004) Metabolic syndrome is associated with elevated oxidative stress and dysfunctional dense high-density lipoprotein particles displaying impaired antioxidative activity. J Clin Endocrinol Metab 89:4963–4971

Hedrick CC, Thorpe SR, Fu MX, Harper CM, Yoo J, Kim SM, Wong H, Peters AL (2000) Glycation impairs high-density lipoprotein function. Diabetologia 43:312–320

Heijmans BT, Beekman M, Houwing-Duistermaat JJ, Cobain MR, Powell J, Blauw GJ, Van Der Ouderaa F, Westendorp RGJ, Slagboom PE (2006) Lipoprotein particle profiles mark familial and sporadic human longevity. PLoS Medicine 3:e495

Isselbacher EM (2005) Thoracic and abdominal aortic aneurysms. Circulation 111:816–828

Keogh JB, Grieger JA, Noakes M, Clifton PM (2005) Flow-mediated dilatation is impaired by a high-saturated fat diet but not by a high-carbohydrate diet. Arterioscler Thromb Vasc Biol 25:1274–1279

Lakatta EG (2003) Arterial and cardiac aging: major shareholders in cardiovascular disease enterprises: part III: cellular and molecular clues to heart and arterial aging. Circulation 107:490–497

Lakatta EG, Levy D (2003a) Arterial and cardiac aging: major shareholders in cardiovascular disease enterprises: part I: aging arteries: a "set up" for vascular disease. Circulation 107:139–146

Lakatta EG, Levy D (2003b) Arterial and cardiac aging: major shareholders in cardiovascular disease enterprises: part II: the aging heart in health: links to heart disease. Circulation 107:346–354

Lloyd-Jones D, Adams R, Carnethon M, De Simone G, Ferguson TB, Flegal K, Ford E, Furie K, Go A, Greenlund K, Haase N, Hailpern S, Ho M, Howard V, Kissela B, Kittner S, Lackland D, Lisabeth L, Marelli A, McDermott M, Meigs J, Mozaffarian D, Nichol G, O'Donnell C, Roger V, Rosamund W, Sacco R, Sorlie P, Stafford R, Steinberger J, Thom T, Wasserthiel-Smoller S, Wong N, Wylie-Rosett J, Hong Y, American Heart Association Statistics Committee and Stroke Statistics Subcommittee (2009) Heart disease and stroke statistics-2009 update: a report from the American Heart Association Statistics Committee and Stroke Statistics Subcommittee. Circulation 119: e21–e181

McClung JA, Naseer N, Saleem M, Rossi GP, Weiss MB, Abraham NG, Kappas A (2005) Circulating endothelial cells are elevated in patients with type 2 diabetes mellitus independently of HbA1c. Diabetologia 48:345–350

McMahon M, Grossman J, FitzGerald J, Dahlin-Lee E, Wallace DJ, Thong BY, Badsha H, Kalunian K, Charles C, Navab M, Fogelman AM, Hahn BH (2006) Pro-inflammatory high-density lipoprotein as a biomarker for atherosclerosis in patients with systemic lupus erythematosus and rheumatoid arthritis. Arthritis Rheum 54:2541–2549

Mutru O, Laakso M, Isomaki H, Koota K (1985) Ten year mortality and causes of death in patients with rheumatoid arthritis. Br Med J 290:1797–1799

Navab, M, Anantharamaiah GM, Hama S, Garber DW, Chaddha M, Hough G, Lallone R, Fogelman AM (2002) Oral administration of an apo A-I mimetic peptide synthesized from D-amino acids dramatically reduces atherosclerosis in mice independent of plasma cholesterol. Circulation 105:290–292

Navab M, Anantharamaiah GM, Reddy ST, Van Lenten BJ, Ansell BJ, Fonarow GC, Vahabzadeh K, Hama S, Hough G, Kamranpour N, Berliner JA, Lusis AJ, Fogelman AM (2004) The oxidation hypothesis of atherogenesis: the role of oxidized phospholipids and HDL. J Lipid Res 45:993–1007

Navab M, Anantharamaiah GM, Hama S, Hough G, Reddy ST, Frank J, Garber DW, Handattu S, Fogelman AM (2005a) D-4F and statins synergize to render HDL antiinflammatory in mice and monkeys and cause lesion regression in old apolipoprotein E-null mice. Arterioscler Thromb Vasc Biol 25:1426–1432

Navab M, Anantharamaiah GM, Reddy ST, Hama S, Hough G, Grijalva CR, Yu N, Ansell BJ, Datta G, Garber DW, Fogelman AM (2005b) Apolipoprotein A-I mimetic peptides. Arterioscler Thromb Vasc Biol 25:1325–1331

Navab M, Anantharamaiah GM, Reddy ST, Van Lenten BJ, Wagner AC, Hama S, Hough G, Bachini E, Garber DW, Mishra VK, Palgunachari MK, Fogelman AM (2005c) An oral apoJ peptide renders HDL antiinflammatory in mice and monkeys and dramatically reduces atherosclerosis in apolipoprotein E-null mice. Arterioscler Thromb Vasc Biol 25:1932–1937

Navab M, Anantharamaiah GM, Reddy ST, Hama S, Hough G, Frank JS, Grijalva VR, Ganesh VK, Mishra VK, Palgunachari MN, Fogelman AM (2005d) Oral small peptides render HDL antiinflammatory in mice and monkeys and reduce atherosclerosis in apoE null mice. Circ Res 97:524–532

Navab M, Anantharamaiah G, Reddy S, Fogelman A (2006) Apolipoprotein A-I mimetic peptides and their role in atherosclerosis prevention. Nat Clin Prac Cardiovasc Med 3:540–547

Nicholls SJ, Lundman P, Harmer JA, Cutri B, Griffiths KA, Rye KA, Barter PJ, Celermajer DS (2006) Consumption of saturated fat impairs the anti-inflammatory properties of high-density lipoproteins and endothelial function. J Am Coll Cardiol 48:715–720

Nissen SE, Tsunoda T, Tuzcu EM, Schoenhagen P, Cooper CJ, Yasin M, Eaton GM, Lauer MA, Sheldon WS, Grines CL, Halpern S, Crowe T, Blankenship JC, Kerensky R (2003) Effect of recombinant apoA-I Milano on coronary atherosclerosis in patients with acute coronary syndromes: a randomized controlled trail. JAMA 290:2292–2300

Otto CM (2002) Calcification of bicuspid aortic valves. Heart 88:321–322

Ou J, Ou Z, Jones DW, Holzhauer S, Hatoum OA, Ackerman AW, Weihrauch DW, Gutterman DD, Guice K, Oldham KT, Hillery CA, Pritchard KA Jr (2003) L-4F, an apolipoprotein A-1 mimetic, dramatically improves vasodilation in hypercholesterolemia and sickle cell disease. Circulation 107, 2337–2341

Ou J, Wang J, Xu H, Ou Z, Sorci-Thomas MG, Jones DW, Signorino P, Densmore JC, Kaul S, Oldham KT, Pritchard KA Jr (2005) Effects on D-4F on vasodilation and vessel wall thickness in hypercholesterolemic LDL receptor-null and LDL receptor/apolipoprotein A-I double knockout mice on Western diet. Circ Res 97:1190–1197

Pearson TA, Mensah GA, Alexander RW, Anderson JL, Cannon RO, Criqui M, Fadl YY, Fortmann SP, Hong Y, Myers GL, Rifai N, Smith SC Jr, Taubert K, Tracy RP, Vinicor F (2003) Markers of inflammation and cardiovascular disease: application to clinical and public health practice: a statement for healthcare professionals from the Centers for Disease Control and Prevention and the American Heart Association. Circulation 107:499–511

Peterson SJ, Husney D, Kruger AL, Olszanecki R, Ricci F, Rodella LF, Stacchiotti A, Rezzani R, McClung JA, Aronow WS, Ikehara S, Abraham NG (2007) Long-term treatment with the apolipoprotein A1 mimetic peptide increases antioxidants and vascular repair in type 1 diabetic rats. J Pharmacol Exp Ther 322:514–520

Plump AS, Scott CJ, Breslow JL (1994) Human apolipoprotein A-I gene expression increase high density lipoprotein and suppresses atherosclerosis in the apolipoprotein E-deficient mouse. Proc Natl Acad Sci 91:9607–9611

Puranik R, Celermajer DS (2003) Smoking and endothelial function. Prog Cardiovasc Dis 45:443–458

Ridker PM, Stampfer MJ, Rifai N (2001) Novel risk factors for systemic atherosclerosis: a comparison of C-reactive protein, fibrinogen, homocysteine, lipoprotein(a), and standard cholesterol screening as predictors of peripheral arterial disease. JAMA 285:2481–2485

Roberts CK, Ng C, Hama S, Eliseo AJ, Barnard RJ (2006) Effect of a short-term diet and exercise intervention on inflammatory/anti-inflammatory properties of HDL in overweight/obese men with cardiovascular risk factors. J Appl Physiol 101:1727–1732

Ross R (1999) Atherosclerosis-an inflammatory disease. N Engl J Med 340:115–126

Smith SC Jr, Allen J, Blair SN, Bonow RO, Brass LM, Fonarow GC, Grundy SM, Hiratzka L, Jones D, Krumholz HM, Mosca L, Pasternak RC, Pearson T, Pfeffer MA, Taubert KA (2006) AHA/ACC guidelines for secondary prevention for patients with coronary and other atherosclerotic vascular disease: 2006 update: endorsed by the National Heart, Lung, and Blood Institute. Circulation 113:2363–2372

Terman A, Dalen H, Eaton JW, Neuzil J, Brunk UT (2004) Aging of cardiac myocytes in culture: oxidative stress, lipofuscin accumulation, and mitochondrial turnover. Ann NY Acad Sci 1019:70–77

Van Lenten BJ, Wagner AC, Navab M, Anantharamaiah GM, Hui EKW, Nayak DP, Fogelman AM (2004) D-4F, an apolipoprotein A-I mimetic peptide, inhibits the inflammatory response induced by influenza A infection of human type II pneumocytes. Circulation 110:3252–3258

Weihrauch D, Xu H, Shi Y, Wang J, Brien J, Jones DW, Kaul S, Komorowski RA, Csuka ME, Oldham KT, Pritchard KA (2007) Effects of D-4F on vasodilation, oxidative stress, angiostatin, myocardial inflammation, and angiogenic potential in tight-skin mice. Am J Physiol Heart Circ Physiol 293:h1432–h1441

Zimarino M, Cappelletti L, Venarucci V, Gallina S, Scarpignato M, Acciai N, Calafiore AM, Barsotti A, De Caterina R (2001) Age-dependence of risk factors for carotid stenosis: an observational study among candidates for coronary arteriography. Atherosclerosis 159:165–173

Chapter 9
Near Term Prospects for Broad Spectrum Amelioration of Cancer

Zheng Cui

Contents

9.1	Introduction: Aging and Cancer	308
9.2	Immunosurveillance for Cancer	309
9.3	The Cancer Resistant (SR/CR) Mouse	310
	9.3.1 The Phenotype of the Mice	310
	9.3.2 The Involvement of Neutrophils	313
	9.3.3 The Genetics and Genomics of Cancer Resistance	314
9.4	Cancer-Resistant Humans	315
9.5	The Cancer Killing Activity (CKA)	316
	9.5.1 In SR/CR Mice	316
	9.5.2 In Humans	316
9.6	Granulocyte Effector Mechanism and Targets	318
	9.6.1 Known Granulocyte Roles and Components	318
	9.6.2 Normal Granulocyte Functions: Chemotaxis and Surface Contact Formation	319
9.7	Cancer Cells as Targets of Granulocytes	320
	9.7.1 Bacterial Infection and Cancer Regression in Humans	320
	9.7.2 Charge-Mediated Granulocyte Recognition of Cancer Cells?	321
9.8	Problems with Conventional Immunotherapies	322
9.9	Human Granulocytes and Their Applications in Therapeutic Settings	322
9.10	Can Granulocytes with High Levels of CKA be Effective in Treating Cancer?	323
	9.10.1 In Mice	323
	9.10.2 In Human Cancer Patients	324
9.11	Conclusions	325
References		326

Z. Cui (✉)
Section of Tumor Biology, Department of Pathology, Wake Forest University School of Medicine, Winston-Salem, NC 27157, USA
e-mail: zheng.cui@gmail.com

9.1 Introduction: Aging and Cancer

Cancer is currently the most feared disease not only because of the disease's painful and debilitating symptoms and often-fatal outcome but also because of the excruciating side-effects associated with many conventional treatments. It is often said in some countries that 1/3 of cancer patients were scared to death, 1/3 were treated to death and only 1/3 actually died of the disease. This may not be entirely precise, but nevertheless reflects a sentiment from the perspectives of both patients and physicians in confronting this horrible disease.

We now have 12.4 million cancer patients in the U.S. There are 1.4 million new cancer cases, and about 600 thousand cancer deaths in this country, each year. The difference between new cancer cases and cancer deaths is alarming. It implies that the cancer patient population in this country (counting both patients with active disease and disease in remission) will increase by 800 thousand people per year. At this rate, we will have 20 million cancer patients 10 years from now in the U.S. alone. Some of these will be long term remission survivors, but many others will be living with active disease, or with disease that will return. If we strive for life extension in humans, it will be nearly impossible without solving the biggest challenge that shortens the human lifespan most inexorably today. In other words, if we don't solve the cancer problem, it will be very difficult to think about healthy life extension in meaningful terms.

Cancer is an aging-related disease. There are very different views on how aging impacts on cancer development in humans. A dominant view believes that somatic cells accumulate mutations during aging until the point where mutations cause cells to be changed into cancer cells that then clonally expand into populations of cancer cells. The premise of this theory is that cancer begins when the first cancer cell is formed and undergoes uncontrolled clonal expansion.

Another view believes that having cancer cells in the body is not necessarily a problem. This theory holds that cancer cells are formed continuously in our bodies on a daily basis but that their presence in our body can reach a dynamic balance with our body's pre-existing ability to remove them, after they formed. Such a hypothetical natural ability to remove cancer cells was termed a cancer "surveillance system" about 100 years ago by Paul Ehrlich (Ehrlich 1909). As long as this balance is maintained, the presence of cancer cells would not pose any health problem. Clinically significant malignancies can form only when such a dynamic balance is tilted in the direction of having more cancer cells, less surveillance against them, or both. Factors that can impact any side of this dynamic balance can influence the risks for developing clinically significant cancers.

But neither view can deny that aging is undoubtedly the most obvious risk factor for cancer. Cancer is the biggest killer for people between the ages of 45 and 85 (Heron 2007; Jemal et al. 2008).

9.2 Immunosurveillance for Cancer

The potential presence of a cancer surveillance system is a very exciting proposition since understanding of such a surveillance system may lead to its use for therapeutic purposes. The question is what the basis of such a surveillance system might be. In recent decades, it has been believed that such a surveillance system is based on the immune system (Thomas 1959; Burnet 1970, 1971; Dunn et al. 2004).

Thus, a philosophy of boosting cancer patients' own immune system to cure cancer has led to the development of many immunotherapies for cancer, such as therapies for stimulating patients' own T lymphocytes, B lymphocytes or natural killer cells. However, after decades of effort, it has become slowly evident that simply stimulating the immune system that was presumed to have failed in the first place to cause cancer, doesn't give clear and consistent clinical benefits to cancer patients. Is it possible that the failed immune systems in cancer patients are simply not correctable? Or that correction of an initial defect is not enough, once a large enough cancer burden is present?

Or, is it just possible that some other part of immune system responsible for cancer surveillance might have escaped our investigative attention?

In the 1950s, one of the most interesting observations suggesting that humans have such a cancer surveillance system was made by Chester Southem (Southam et al. 1957; Moore et al. 1957; Southam 1958; Southem and Moore 1958, unknown author for TIME magazine 1957). Although this set of studies became highly controversial in many respects (Preminger 2002; Lerner 2004), the scientific message of these studies may deserve a revisit, especially now, when our understanding of human biology has become much better than it was 50 years ago.

Southem was trying to see whether a protective specific system against cancer was absent from cancer patients but was present in young healthy humans. So, he transplanted 2 groups of human subjects with live cancer cells. When HeLa cells, a strain of human cervical cancer cells, were transplanted into the group of young healthy human subjects, no tumors developed. On the other hand, large malignant tumors developed from the transplanted HeLa cells in the group of cancer patients already with advanced malignancies.

Southem suggested that there might be a specific cancer rejection mechanism in the young healthy humans to eliminate the transplanted cancer cells, and that such a mechanism had been lost in cancer patients. If this idea was correct, it might have been the first observation that a profound cancer surveillance mechanism is present in young healthy humans to remove established cancer cells, that that this system is absent from cancer patients.

Due to obvious ethical and legal concerns, this line of study has been discontinued for a long time. Years later, serious doubts were also cast over the interpretation of these findings. Southem believed that the growth of cancer cells in cancer patients could be caused by a loss of specific cancer surveillance function while other immune aspects of the cancer patients remained intact, such as anti-viral functions via antibody production. But the idea of healthy young humans having a natural ability to specifically reject cancer cells was eventually abandoned, and replaced by the idea that there might be a broad general suppression of immune function in cancer patients. Thus, the growth of transplanted cancer cells in cancer patients was not believed later by others to be due to a loss of specific cancer rejection, but rather due to a loss of heterotransplant rejection ability via general immune suppression.

However, there is a problem with this view. As we have now come to know, heterotransplant rejection remains largely intact in most patients with solid malignancies, even at very advanced stages. The question therefore is why, if the heterotransplant rejection remains intact in most cancer patients, did the transplanted cancer cells still grow in them? Is it possible that failure of tumor growth in healthy young humans was due to the presence of a specific innate anti-cancer surveillance system? What could be the basis of such a cancer rejection system?

These historic debates have also had direct clinical consequences. Most immunotherapies are based on the philosophy that effective anticancer activity of the immune system can only be engineered and could not be present naturally. On the other hand, if there is indeed a naturally existing anticancer activity in the immune system of healthy young humans, then a different philosophy for cancer treatment should be employed. That is, instead of trying to engineer a new therapeutic activity in cancer patients, we should try to find this existing activity elsewhere and simply transfer it into cancer patients to replenish its loss for therapeutic purposes.

9.3 The Cancer Resistant (SR/CR) Mouse

9.3.1 The Phenotype of the Mice

In 1999, about 10 years ago, my laboratory was using aggressive mouse cancer cell line sarcoma 180 (S180) to induce ascites formation, simply for the purpose of generating more body fluid in mice to collect more antibodies for other research interests. S180 is notoriously known to be highly malignant and extremely aggressive when transplanted anywhere in the body of the mouse, in all mouse strains (Alfaro et al. 1992). Even when just a few hundred S180 cells were transplanted, the recipient mice would uniformly develop malignancies and metastasis and die in just a few weeks after transplantation. When S180 cells were transplanted into the peritoneal cavity, they would plug up the lymphatic drainage system. As a result, body fluid would accumulate in the peritoneal cavity as ascites fluid in 10–14 days.

Intraperitoneal injection of S180 to induce ascites-associated cancer in mice is a highly reliable model for cancer induction and cancer identification. There had never been a report of resistance in mice that failed to develop cancer with S180. We had therefore never confronted before in my lab any mouse that did not develop cancer after S180 challenge using this procedure, after treating several hundred or even perhaps over a thousand mice. So we were very surprised when one BALB/c mouse failed to develop ascites after S180 challenge, while other injected mice in the same experiment all developed ascites and died (Cui et al. 2003; Cui 2003).

Initially we thought we must have forgotten to inject this mouse with cancer cells. But after repeated challenges of S180 cells, we realized that we might have a unique mouse in our hands. This mouse was healthy, active and apparently highly resistant to S180. Our initial thought was that if S180 could induce the worst cancer pathology and rapid lethality in other mice, resistance to S180 in this unique mouse could mean resistance to other cancers as well. We then challenged the mouse with another aggressive lethal mouse leukemia cell line, L5178Y, and found that the mouse survived this as well, while other control mice all died (Cui et al. 2003). This cancer-resistant phenotype was later confirmed independently (Koch et al. 2008).

After realizing the unique cancer resistance of this mouse, a difficult decision was made to breed this male mouse with other normal female mice (which themselves were later proven to lack cancer-resistance). The decision was difficult because it required re-allocation of considerable resources from other projects before we even knew whether the cancer resistance trait could be passed on to this unique mouse's offspring. But when we screened the resulting pups using S180 cells, over 40% of the pups showed the same cancer-resistant phenotype of being able to survive otherwise-lethal challenges with cancer cells. Such an inheritance pattern suggested that this phenotype is a dominant trait.

In last 10 years, this cancer-resistant phenotype has been passed on from this founder mouse to several thousand offspring in more than 40 generations. This kind of productivity may have made this founder mouse one of the most productive mice in the modern history of lab mice. The founder mouse and his offspring were named SR/CR mice. For a detailed description of the discovery of SR/CR mice, a reflective article published in 2003 in Cancer Immunity (an online open source publication) is available (Cui 2003).

Perhaps the most amazing event in cancer suppression in SR/CR mice was our observation that some mice injected with cancer cells would initially seem to be non-resistant, but would then show spontaneous regression of the cancer. In some offspring of SR/CR mice, the S180 cells induced large amounts of ascites, but then the ascites would suddenly disappear, and the mice were thereafter healthy and resistant.

Initially, we were completely baffled by the seemingly miraculous disappearance overnight of large ascites. The phenomenon happened unpredictably at first, but after a series of fortuitous events, a trend began to emerge. Most ascites regressions occurred when the pups were screened with S180 for the first time at a much older age, i.e., at 5–6 months old instead of 5–6 weeks old as we usually did. When exposed to S180 at the young age of 5–6 weeks, the mice that carry the genetic factor mount a rapid anticancer response to eliminate cancer cells within days or even hours of exposure. Ascites never develops. However, at the adult age of 5–6 months, the resistance seemed to have disappeared at first exposure to cancer cells, the mice reacting like wild type (WT) mice. But after a few days of ascites growth, the resistance mechanism seemed to suddenly kick in, and began to rapidly destroy the cancer cells. This delayed onset of resistance allowed the ascites to develop first and then regress at a later point. This was the first time that spontaneous regression of a well-established cancer had been demonstrated reproducibly in a well-controlled laboratory environment.

Spontaneous regression of advanced malignancy is a truly amazing event, and has baffled several generations of scientists. There have been many hypotheses for how it can happen (Cole 1974, 1976, 1981; Challis and Stam 1990; Stoll 1992; Papac 1996, 1998; Bodey 2002). Our observations indicated that aging seems to play a critical role in determining how fast cancer resistance can be deployed, and that this in turn determines if the resistant phenotype is displayed immediately, in the form of complete resistance in young hosts, or in a delayed manner, in the form of spontaneous regression in older hosts.

It was later found that if SR/CR mice were left unchallenged for over a year, they could lose the resistance phenotype entirely. However, if SR/CR mice were challenged frequently with live cancer cells, the resistance could be kept up for their entire lifespan of more than 2 years, in most cases.

SR/CR mice also resisted a wide spectrum of different cancer cells, including EL4, B16, MethA and LL2, much better than WT mice, but with a wide range of maximum tolerated doses (MTDs). We also found that the resistance level to highly aggressive cancer cells with low MTDs could be greatly enhanced by co-injection with S180 cells, which induces infiltration of granulocytes and macrophages in this strain (Z. Cui et al. unpublished results).

It turned out that the co-injection effect was due to the ability of S180 cells to produce a large number of small peptides, some of which are highly chemo-attractive. Cancer cells with low MTDs in SR/CR mice don't produce chemo-attractants as S180 does. Co-injection with S180 cells that produce chemo-attractants would be expected to cause host leukocytes to infiltrate the site of the cancer cells. Once the host leukocytes are in place, both S180 cells and the co-injected cancer cells with low MTDs can apparently be recognized and destroyed. This finding suggests that chemotaxis and subsequent killing of cancer cells by immune cells of SR/CR mice result from sensing two different sets of signals provided by cancer cells.

There are several take-home messages from these mouse studies. First, the genetic background seems to play a critical role in whether an individual can be cancer-resistant or not, or at what levels. Second, the aging process can diminish the activity of cancer-resistance, even when the genetic factor is present at high levels in youth. No wonder cancer is an age-related disease. Third, frequent exposures to live cancer cells, not debris of dead cancer cells, was a key for retaining cancer resistance throughout life in our cancer-resistant mice, although exposure to live cancer cells is lethal when the genetic factor is not present. These behaviors are consistent with the general behaviors of the mammalian immune system, including that of humans: the system gets better with exercise.

These lessons may have profound implications in cancer therapy and cancer prevention settings.

9.3.2 The Involvement of Neutrophils

One of the most obvious questions was that of what actually happened to the cancer cells after they were transplanted into SR/CR mice. The clues came as soon as we began to look into the peritoneal cavity after the transplantation of cancer cells. When we washed out the contents of the peritoneal cavity at desired time points after cancer cell transplantation, it was apparent that cancer cells were either killed already or were being surrounded by some smaller host cells to form something that immunologists generally refer to as "rosettes", indicating that cellular immune reactions might be responsible for the death of cancer cells. Further examination of these rosettes, however, revealed another big surprise. The cancer cells were surrounded and apparently attacked by cells that were easily recognizable by their very unique cellular morphology: most of them were neutrophils with segmented nuclei. Many neutrophils were seen making very tight surface contacts with cancer cells (Cui et al. 2003).

Initially, we thought that these neutrophils, usually considered to be phagocytes, were there only to clean up the battle field littered with cellular debris, after the cancer cells had been killed by something else. The field of cancer immunology, then as now, has been dominated by the idea that cancer cells can only be killed by macrophages or by lymphocytes such as cytotoxic T lymphocytes and natural killer cells. Neutrophil granulocytes are normally considered as "anti-bacterial cells" and were rarely known for killing cancer cells. However, the involvement of neutrophils in killing cancer cells in the cancer-resistant mice was overwhelming and undeniable, and, years later, after intensive studies of the role of neutrophils in SR/CR mice, we are convinced that neutrophils in the cancer-resistant mice were the major functional components for eliminating cancer cells.

This conviction is based on 3 lines of evidence. First, we found that the resistant phenotype could be abolished by specific depletion of neutrophils from SR/CR mice. Second, highly purified neutrophils from SR/CR mice could independently kill cancer cells in in vitro cancer cell killing experiments. Third, highly purified neutrophils from SR/CR mice could be transplanted to confer a cancer-resistant phenotype in genetically non-cancer-resistant mice.

Neutrophils have been shown to infiltrate the site of cancer cells and form tight surface contact zones with cancer cells (Hicks et al. 2006). Upon formation of cell rosettes with infiltrating leukocytes, cancer cells were killed via both cytolysis and apoptosis, a dual-killing event that had not been reported before. Direct contact between cancer cells and neutrophils is required for cancer cell killing, since separating cancer cells from neutrophils by a membrane that allows only the exchange of small diffusible molecules between the two cell populations completely abolished the killing event.

Although laboratory mice are similar to humans in many aspects, the most profound difference is in their immune systems. The leukocyte (white cell) system in WT mice is predominantly lymphocytic (or largely composed of lymphocytes) whereas it is predominantly myelocytic in humans, where the largest fraction of peripheral cells consists of neutrophils. However, in many SR/CR mice, the leukocyte system has been switched to a human-like and predominantly myelocytic pattern, which is consistent with the trait as an innate immune response. Adaptive immunity seems to play an indirect but critical role in the resistance, possibly by supporting the functional development of neutrophils and macrophages as primary effector cells. Thus, we found that the resistant phenotype can be bred into the T cell maturation-deficient nude mouse background (Cui et al. 2003), but the level of resistance is much weaker in the athymic background than in a thymic background, suggesting that the functions of innate and adaptive immunities are inseparable from one another.

SR/CR mice are the first model system for a cancer surveillance system that fits perfectly with the theory proposed by Ehrlich in which such a cancer surveillance system must be innate and pre-existing (Ehrlich 1909). However, the surprise is that such a system is mediated primarily by effector cells such as neutrophils that have been largely overlooked. Is it possible that the reason we have such a large number of granulocytes even in the absence of infections is because granulocytes have to perform other important roles in our body? Being the major component of a critical cancer surveillance system that has to work continuously to remove cancer cells being generated continuously seems to justify the sheer numbers of granulocytes present in our bodies.

9.3.3 The Genetics and Genomics of Cancer Resistance

Cancer resistance in our experience is a dominant trait and is defined by a germ line-transmissible non-sex-linked genetic element (Cui et al. 2003). After 5 rounds of high-resolution mapping, no fixed chromosomal linkage could be found (Cui et al., unpublished results), suggesting it is not a conventional Mendelian mutation but rather something that may be associated with transposons or paramutations. Full genome sequencing could eventually reveal the nature of the genetic component responsible for this phenotype.

Microarray studies of quiescent effector cells from SR/CR mice showed that expression of most cell surface receptors related to leukocyte activation and of cytoskeletal genes related to cell migration are significantly elevated. This is consistent with the cellular actions of finding and killing cancer cells (Cui et al. unpublished results). The most interesting activation of the SR/CR effector cells is the specific elevation of the interferon (INF) signaling pathway. Activation of the INF pathway is usually related to responses to RNA virus infection (Barber 2001; Matsumoto et al. 2004; Uematsu and Akira 2007). Given the information obtained from the genomic studies indicating that a transposon-like element might be responsible for the resistance trait, we hypothesize that the SR/CR genome may harbor a transposable retroviral-like element that can constitutively and endogenously activate neutrophils and macrophages via the INF pathway.

Integration of retroviral components into mammalian genomes is not a rare event. In fact, it is considered by some as the major driving force for evolution, by increasing the sizes and complexity of mammalian genomes. Up to 40% of the mouse and human genomes is composed of retroviral remnants (Gifford and Tristem 2003; Mikkelsen and Pedersen 2000; Silva et al. 2004; Sverdlov 1998; Waterston et al. 2002; Doerfler 1991, , 2006; Maksakova et al. 2006, 2008). Some of these remnants are expressed only during gestation and are silent after birth. It is highly conceivable that such a retroviral component is integrated behind a myelocyte-specific promoter and that its expression after gestation could become a constitutive and endogenous stimulator for neutrophils and macrophages, leading to innate granulocyte response to cancer cells. However, the eventual validation of this hypothesis may have to wait for the availability of full genome sequencing at a reasonable cost.

9.4 Cancer-Resistant Humans

Although the studies of Southem cited above provide direct and compelling evidence of a cancer resistance phenotype in humans, there is also more indirect evidence in support of the same idea. The truth is that most people never develop cancer for several decades, despite hundreds of thousands of DNA damage events per cell per day accumulated in over 100 trillion cells. As an even more extreme case, human centenarians go even farther, out-living the average life expectancy by several decades without having cancer. If the CKA concept is correct, it is possible that their granulocytes may still have a high level of CKA that may have been lost in other human populations in old age. Fortunately, this is a testable hypothesis.

Another example is provided by cigarette smoking. Cigarette smoking is perhaps one of the best examples of a group of humans intentionally, willingly, and reliably exposing themselves to known carcinogens for a long time. This behavior has dramatically increased the incidence of lung cancer, for example, about 60-fold from 20 per 100,000 in males who have never smoked to about 1,260 per 100,000 in male smokers (2 packs/day) (Freedman et al. 2008). But it is equally intriguing that over 98% of the smokers escape from cancer based on the same statistics. Even when integrated over the entire lifespan, the lifetime lung cancer incidence is only about 17% for male smokers and 11% for females (Villeneuve and Mao 1994). So far, there is no scientific explanation for why the vast majority of heavy smokers don't develop lung cancer. Most of these lucky smokers believe that they must have "good genes". Our studies in SR/CR mice and healthy humans with good CKA scores (see Section 9.5.2) may provide an explanation for why some humans are cancer-resistant. It is conceivable that the newly discovered CKA in granulocytes can play a profound role in surveying the entire body and seeking out cancer cells for destruction.

9.5 The Cancer Killing Activity (CKA)

9.5.1 In SR/CR Mice

The resistance trait in SR/CR mice can also be reflected by a newly developed in vitro assay for quantifying the ability of effector cells to kill cancer cells at fixed ratios in cell cultures. This activity is termed cancer-killing activity, or CKA. The CKA assay is longer than and different from conventional cytotoxicity assays based on radioactivity release from labeled target cells. The major modification involves use of a different detection concept, particularly morphology monitoring, which requires a time frame of days instead of a few hours, since the major cell killing event using the SR/CR effector cells occurs after 12 hours of incubation. This turned out to be a highly predictive assay since high CKA level (the percentage of cancer cells killed in the assay within a given time) in the extracted leukocytes is always associated with the resistance phenotype in the mice (Hicks et al. 2006), and low levels of CKA are associated with WT mice.

Recently, this assay has been adapted into a 96-well platform with electronic readout of target cell detachment from the culture surface after being killed. The ability to use this in vitro assay to measure in vivo cancer resistance (Hicks et al. 2006) gave us the opportunity to explore cancer resistance in humans.

9.5.2 In Humans

If such a highly effective activity for directly killing cancer cells without harming normal cells is similarly present in human white blood cells, is it possible that a similar transfer of desired leukocytes can also introduce new CKA in the human recipients for cancer prevention or even treatment? To begin to answer this question, we tested several populations of humans using the CKA assay. Here is a summary of what we found from these studies.

CKA for in vitro killing of known human cancer cells, such as HeLa cells and MCF7 cells, etc., was detected in human leukocytes (Fig. 9.1). The levels of CKA vary from individual to individual. Some people have specific CKA levels that were much higher even than SR/CR mice whereas others were low, like that of WT mice. Most people were somewhere in between. Leukocyte fractionation studies indicated that over 85% of total CKA in circulation was present in the granulocyte fraction, and the rest was in the monocyte (macrophage) fraction. CKA in human populations was also highly dynamic. CKA was higher in younger populations than in older populations, indicating its decline during aging. This is consistent with the rising cancer rate in older populations. CKA was also lower in cancer patients than in healthy populations, consistent with CKA as a valid index of cancer surveillance activity, and consistent as well with the original theory proposed by Ehrlich (Ehrlich 1909).

It is known that stress hormones can profoundly suppress immune functions (Epel 2009; Leonard 2000; Pruett 2001; Stefanski 2001; Dhabhar 2000; Gleeson and Bishop 2000; McEwen 2000; Cohen et al. 2001), and, consistent with this, we observed preliminary evidence that CKA is sensitive to emotional stresses. Under

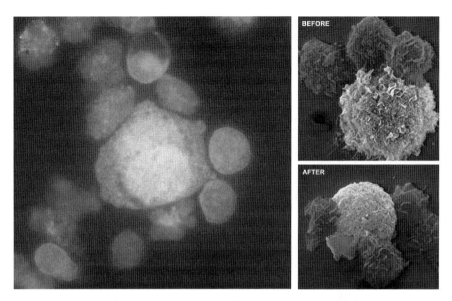

Fig. 9.1 *Left*: A cancer cell (the large central cell with the pale nucleus) is being attacked by a group of mouse granulocytes from cancer-resistant donors. *Right*: Scanning electron microscopy of neutrophils attaching to and killing a mouse cancer cell

normal conditions, CKA was relatively stable in a given person week over week. One day, we found that a particular person who showed relatively stable CKA in our routine monitoring procedure suddenly had no CKA at all. After carefully evaluating what may have happened to this person in the days leading up to this low CKA point, the only thing that stood out from the routine was that the day before the low CKA measurement, he was engaged in a heated argument and was apparently highly stressed as a result. Three days later, tests showed that his CKA had returned to its normal level. In another case, one of my students was sent to a conference and had to make an oral presentation for the first time in his life under this new condition. It appeared that he was nervous days or even weeks before the presentation. Upon return from this conference, his white cell count was so low we couldn't get enough cells to run a single assay. A week later, his CKA was still low but gradually returned to his normal level after several weeks. It appeared that the longer the exposure to stress, the longer it takes to recover from its effect. These anecdotal findings of the relationship between CKA and stress need to be confirmed in better–designed and more extensive studies but were, at least, consistent with knowledge that stress hormones, and cortisol in particular, are strong immune suppressors.

In the early years of these studies, we had to confront what we called at the time "technical problems" with the assay. The assay could suddenly stop working, and we would begin to change things until it started to work again a few months later. This kind of problem and trouble-shooting continued for several years until we realized that all of these "technical problems" seemed to occur when winter came. In the fall of 2006, we decided to test the theory that the "technical problems" could actually be an effect of "seasonality". We continued to perform the CKA assays without changing anything, even when the assays "stopped" working. Sure enough, CKA returned the next spring in all the individuals we monitored, and thereafter continued to follow much the same seasonal pattern. We have since verified the exact same seasonality phenomenon of CKA over additional years.

It appears that winters have a profound impact on CKA. But we needed some kind of additional validation for this idea. If the winter seasons have such a profound impact on the major cancer surveillance activity, is it possible that cancer incidence rates in the geographical regions where there are no major seasonal changes could be much lower than those of the regions where there are clear-cut seasons? By examining the available databases of WHO for worldwide cancer incidence rates, it turns out that the age-adjusted cancer rates in equatorial countries are 4 times lower than those in North America, Europe, and Australia (web link: http://www-dep.iarc.fr/). Although this apparent connection between winters and lowered cancer immunity needs further validation, it suggests the interesting possibility that the immune system is suppressed by winters and that this in turn can lead to more winter-related health problems, such as flu, cancer, or even death. And in fact, the general mortality rate is always higher in winters than in summers, a pattern that has been consistent ever since the data were first collected decades ago (Hajek et al. 1984).

Cancer development usually takes at least 2–3 years, a time frame suggested from studies of nuclear accidents or traumatic carcinogenic events. Therefore, CKA seasonality can't be expected to map onto any comparable annual pattern of cancer death rates. Nevertheless, the significantly different cancer rates at different geological locations with and without "winters" give interesting clues about the impact of cold seasons on immunity and possibly on cancer etiology.

The effector functions of granulocytes are mediated by highly coordinated, multi-component cellular processes (see next section). Therefore, anything that is capable of compromising any component of these processes, directly or indirectly interfering with the general health of granulocytes, can compromise the effector functionality or CKA. The dynamic property of CKA in human granulocytes, which varies with individual genetic backgrounds, prior exposures that strengthen general functionality, aging, stress hormone exposure, and seasonality, may reflect the final functional output of the effector mechanisms of the granulocytes.

9.6 Granulocyte Effector Mechanism and Targets

9.6.1 Known Granulocyte Roles and Components

Granulocytes are the first line defense of the human immune system. They are the first to arrive at the sites of infection, wounds, or malignant neoplasia. Granulocytes have a large repertoire of effector mechanisms to fight against pathogenic or abnormal cells. They are known as phagocytes to engulf small pathogens before killing them internally (Nathan 2006; Witko-Sarsat et al. 2000). The other effector mechanism of granulocytes is degranulation to release the cytotoxic contents of granules, such as lactoferrin, cathelicidin, myeloperoxidase, bactericidal/permeability increasing protein (BPI), defensins (also known as human neutrophil peptides), serine proteases, neutrophil elastase, cathepsin G, and gelatinase (Lehrer et al. 1993; Lichtenstein et al. 1988; Zanetti et al. 1995; Dale et al. 2008).

Like hemoglobin in red blood cells, human neutrophil peptides (HNPs) are the most abundant proteins in human neutrophils. HNPs constitute 5% of total neutrophil proteins and 50% of azurophil granule proteins and are the major antimicrobial effectors (Lehrer et al. 1993; De Smet and Contreras 2005; Lehrer 2004; Rice et al. 1987; Ganz et al. 1985). HNPs are a group of small peptides with 20–40 amino acids that are rich in arginine, lysine and cysteine (Selsted and Harwig 1989). HNPs are also called "cation peptides" since they are highly positively charged due to their arginine and lysine residues.

HNPs have an amphipathic folded rod-like structure with one side being hydrophobic and the other side being hydrophilic and positively charged (Hill et al. 1991; White et al. 1995). The primary targets of HNPs are negatively charged lipid bilayer membranes (Lehrer et al. 1993). The activation mechanism of HNPs is the cleavage of negatively charged leader peptides that neutralize the positive charges in the propeptide stage (Lehrer et al. 1993).

After degranulation, the mature amphipathic HNPs exert two major effector actions on the target cells. First, HNPs polymerize to form barrel-like pores on the negatively charged plasma membranes of target cells (Lehrer et al. 1993). The hydrophobic side of the monomeric HNP lines up against the hydrophobic portion of the target bilayer membranes whereas the hydrophilic side of the HNP forms a hydrophilic pore with other similarly lined-up HNPs. As a result, the permeability of the attacked plasma membrane increases dramatically, causing colloid cell swelling and ultimately target cell rupture (cytolysis). Second, HNPs, once they enter the cytoplasm of target cells that contain mitochondria, can bind to and neutralize negatively-charged outer mitochondrial membranes, leading to dissipation of the mitochondrial transmembrane potential, a well-known mechanism for triggering rapid apoptosis (Green and Reed 1998).

9.6.2 Normal Granulocyte Functions: Chemotaxis and Surface Contact Formation

The effector mechanism of granulocytes also requires 2 additional stages prior to degranulation. In the first stage, granulocytes must migrate, or infiltrate, to the site of action via a process of chemotaxis in response to a gradient of chemotactic molecules formed around the site of the lesion. Chemoattractants can be small peptides, such as f-MLP, or larger cytokines (Demaurex et al. 1994; Ellis and Beaman 2004; Funaro et al. 2004; McLoughlin et al. 2004).

In the second stage, granulocytes upon arrival must bind to target cells. This also takes place in two steps, the first being charge-mediated attraction and the second being the formation of tight surface-to-surface interactions before degranulation can take place.

9.6.2.1 Charge-Mediated Attraction

Metabolically active bacteria, for example, tend to be negatively charged on their surface due to their intrinsic metabolic patterns (Lehrer et al. 1988). Although bacteria have the ability to carry out a rudimentary form of oxidative phosphorylation that can result in a negatively charged outer membrane, this pathway is used sparingly or not at all in the human body, where glucose is plentiful. Instead, the major source for generating ATP is via glycolysis, which utilizes a large amount of six-carbon carbohydrates, such as glucose, to produce a large amount of three-carbon by-products such as lactic acid. It is evidently the latter process that causes bacterial surfaces to become highly acidic and negatively charged. The unique negative charge of microbes sets them apart from normal cells, which are predominantly charge-neutral on their surfaces because most normal cells in the human body have functional mitochondria, and the lactates produced from glycolysis are immediately funneled into oxidative phosphorylation for more efficient ATP production. Therefore, granulocytes with highly charged cation peptides can specifically recognize metabolically active microbes directly based on their surface charges.

9.6.2.2 Formation of Surface Contacts

Direct injection of HNPs into the blood stream has very little therapeutic effect on bacteria, since HNPs can be immediately neutralized by serum proteins, such as albumin and α2-microglobulin (Lehrer et al. 1993). Therefore, the release of HNPs from neutrophils occurs in a sealed enclosure called the "contact zone" formed between the plasma membranes of neutrophils and target cells. Degranulation by neutrophils can allow the local concentration of HNPs in the sealed contact zone to reach 10 mg/ml to facilitate pore formation in the target plasma membranes and coating of the outer membranes of bacteria.

The membrane enclosure at the sites of degranulation or the contact zones is essential for the effector actions of HNPs. This property is also a highly effective "turn-off" mechanism to ensure that the released HNPs, after the target cells are ruptured, won't be able to randomly harm surrounding cells. Once the target cells are ruptured, HNPs are immediately exposed to extracellular proteins and neutralized.

9.7 Cancer Cells as Targets of Granulocytes

9.7.1 Bacterial Infection and Cancer Regression in Humans

In the mid to late 1800s and before the advent of antibiotics, bacterial infections were often untreatable diseases. However, clinicians at the time noticed a profound association between bacterial infections and miraculous regressions of incurable malignancies (Busch 1867). Coley, shortly before 1900, was intrigued and decided to pursue this phenomenon in a new way. He began to culture bacterial cells and then injected live bacteria directly into cancer lesions (Coley 1893). However, his practice was not without risks despite strikingly good results: he was able to induce tumor regression in as many as 40% of his treated patients, but about 30% of these patients succumbed to sepsis (Coley 1893). These observations support the possibility that granulocytes, which are elicited by and which attack bacteria, can also profoundly cross-react against human cancer cells, or, in other words, that human cancer cells can be targets of granulocytes in vivo.

A critical transition made by Coley to confront the problem of sepsis was to avoid using live bacteria and instead to use heat-killed bacteria and later bacterial extracts known also as Coley's toxins (Coley 1908). Since this critical transition, however, the clinical results were never as good as when he used live bacteria. The major component of Coley's toxin was eventually found to be lipopolysaccharide (LPS). LPS was eventually dismissed as a cancer treatment after disappointing results were obtained in several large clinical trials (Starnes 1992; Wiemann and Starnes 1994; Richardson et al. 1999; Bickels et al. 2002; Zacharski and Sukhatme 2005; Tsung and Norton 2006).

We now know that a common theme of human responses to all bacterial infections is a massive increase of human granulocytes entering the circulation within days or even hours. Conversely, the granulocyte increase in the circulation has long been a reliable diagnostic parameter of bacterial infections in clinical settings. Responses to LPS, on the other hand, don't involve such a granulocyte increase. Is it possible that the poor clinical results after Coley made the transition from using live bacteria to using Coley's toxins were related to the difference in induction of granulocytes that can also kill cancer cells in addition to bacteria?

9.7.2 Charge-Mediated Granulocyte Recognition of Cancer Cells?

Cancer cells tend to have dysfunctional mitochondria, and thus tend to rely strongly on anaerobic glycolysis (Bartrons and Caro 2007; Nickelsen 2007; Warburg 1920, 1956a, Hockel and Vaupel 2001; Langman and Burr 1947, 1949). It has long been known that cancer cells use a large amount of glucose, a phenomenon widely used for modern cancer detection technologies, such as fluorodeoxyglucose PET imaging, and produce a large amount of lactic acid. As early as the 1930s, Otto Warburg first noticed that, unlike normal cells, cancer cells used very little oxygen and produced very little ATP from oxidative phosphorylation in mitochondria (Warburg 1956b), requiring additional glycolysis to compensate, a phenomenon later called "the Warburg effect" (Bartrons and Caro 2007).

Aggressive cancer cells are always metabolically active cells. To accommodate the ATP requirement for rapid proliferation, aggressive cancer cells must utilize glycolysis at much higher levels with high consumption of glucose and high production of lactate. Perhaps for this reason, it has been found that metabolically active cancer cells, like bacteria, produce negative charges on their surfaces (Langman and Burr, 1947, 1949), setting themselves apart from normal cells that have functional mitochondria. Such negative charges make cancer cells, like bacteria, perfect targets for neutrophils and HNPs but only when granulocyte effector functions are not compromised.

9.8 Problems with Conventional Immunotherapies

The initial proposal of a cancer surveillance system by Ehrlich and the striking experimental evidence provided by Southem suggested that such a surveillance system must be pre-existing, or innate, since one doesn't have to get cancer in order to develop cancer immunity. However, most cancer immunotherapies in recent years have been centered on the themes of how to identify cancer-specific antigens and how to develop therapies to target cancer cells that express cancer-specific antigens by enhancing the activities of cancer patients' own adaptive immune system. But most immunotherapies based on targeting specific antigens, such as cancer vaccines and T cell manipulations, have been met with unsatisfactory clinical results.

In retrospect, target immunotherapies faced a difficult dilemma. Because of their unstable genomes that are changing all the time, cancer cells can be divided into two different groups: positive for a specific antigen and negative for the same specific antigen. There has rarely been a case in which a specific cancer surface antigen is expressed on all cancer cells in a patient. Therefore, targeted therapies would not have had any effect on the antigen-negative cells. On the other hand, the presence of cancer cells positive for a specific antigen indicates that these cells have already managed to escape host detection, probably due to some unknown defects in the immune system. Without knowing anything about the immune defects, simply trying to enhance the already failed immunity, such as by vaccination or stimulation of T cell function, has not been very successful.

There is another obstacle facing target immunotherapies: induction of adaptive immunity requires several weeks or even months. Although vaccines and T cell therapies have shown success in many preventive settings, they have rarely worked well in therapeutic settings after the diseases have already fully developed. This may be both because bacteria and cancer cells are more likely to become problematic in individuals who have pre-existing defects in their immune functions and because pathogen loads are too high in the therapeutic settings for the immune system to handle in the time available.

9.9 Human Granulocytes and Their Applications in Therapeutic Settings

In humans, granulocytes are the most abundant leukocytes in the circulation, comprising up to 70% of circulating white cells. The name "granulocyte" came from their unique morphology of containing cytotoxic granules in their cytoplasm. Over 95% of granulocytes are neutrophils, whose granules are neutral pink, whereas basophil granules are dark blue, and eosinophil granules are bright red, in H&E staining. Neutrophils are also called polymorphonuclear (PMN) cells or "polys" for short, since their nuclei are characteristically divided into 2–5 lobes, making them the most identifiable white cells in blood smears. In a normal human adult, there are approximately $2-3 \times 10^{11}$ granulocytes. Over 90% of granulocytes reside in tissues and bone marrow and 10% are in the circulation.

The in vitro lifespan of granulocytes is short. In the standard storage bags at room temperature after apheresis collection, the half-life of collected granulocytes is 24 hours. The in vivo lifespan of granulocytes is largely unknown. Some earlier studies measured the disappearing rates of radioactively labeled, newly infused granulocytes from the circulation and estimated the half life (Brubaker et al. 1977). The major problem of this design is that when granulocytes disappear from the circulation, it doesn't mean that they are dead. Since 90% of granulocytes are present in tissues, disappearance of the labeled granulocytes from the circulation may simply mean that they are marginated into tissues. A later study showed that the lifespan of a neutrophil was approximately 14–16 days (Gordon 1994). In our experiments using highly purified granulocytes from SR/CR mice that were transferred into WT recipients, the cancer resistant phenotype derived from the functionality of transferred granulocytes could be detected for up to several weeks after transfer, before it eventually failed.

Granulocytes are made in the bone marrow and can be demarginated into circulation by administration of mobilization agents, such as G-CSF or pathological conditions, such as bacterial infections, wounds, or cancer lesions.

Neutrophils are the most identified cancer-infiltrating white cells. However, there is no linkage of their infiltration to cancer prognosis. Human bodies, especially when young, have massive capacities to regenerate and to demarginate granulocytes in response to various conditions. For example, bacterial infections can cause granulocyte levels in the circulation to increase by up to a hundred-fold. After apheresis, even with mobilization, granulocyte counts in the circulation of donors can be restored in 1–2 hours. This massive capacity of replenishment and surge can be profoundly diminished by aging, by cytotoxic chemotherapies, and in cancer patients.

The guidelines for granulocyte donation set by the FDA, after reviewing a large volume of human study results, allows a healthy person to donate up to 24 times per year or twice a month, which is much more frequently than for red cell donation (once every 3 months). In fact, donor granulocyte infusion has been a standard clinical practice for over 40 years with excellent safety data (Schiffer et al. 1977; Schiffer 1979, 1983; Hester et al. 1995). Donor granulocyte collection via apheresis for allogeneic uses is a standard service at many facilities of the American Red Cross across the nation.

9.10 Can Granulocytes with High Levels of CKA be Effective in Treating Cancer?

9.10.1 In Mice

To test whether infusion of unmanipulated leukocytes from selected cancer-resistant donors with high levels of CKA could be therapeutic against cancer cells

("Leukocyte Infusion Therapy" or LIFT treatment), we went back to the mouse cancer models. Two models have been studied.

The first mouse cancer model was the lethal subcutaneous tumors induced by S180 cells. If not treated, the mice died in 4–5 weeks after S180 transplantation. When treated with donor leukocytes from SR/CR mice, the tumors reached to a volume of 20% of body weight before a rapid regression in all treated mice in a few days (Hicks et al. 2006). The treated mice then lived a normal or perhaps even a longer lifespan, the latter conceivably due to acquisition of potent immunity via leukocyte infusion.

The second mouse model we treated with LIFT was a prostate-specific knockout of the PTEN gene, a genetic manipulation that generates lethal malignancy and death in 100% of the genetically affected mice (Wang et al. 2006). The average lifespan was about 7 months. Leukocytes from SR/CR donor mice were infused at the age of 5 weeks. All the treated mice lived much longer, to an average of 14 months. Upon their natural deaths, the treated mice were necropsied, and prostates were examined histologically. Gross examinations showed that the treated prostates were just as enlarged as the untreated control prostate cancers. However, histology showed that the entire prostates in the treated mice had become scar tissue, without any residual normal structure of prostates (Cui et al, unpublished results). These data indicated that the injected donor leukocytes didn't begin to attack prostatic cancer until it became fully advanced into late-stage malignancy. This result was consistent with the possibility that neutrophils recognize cancer cells when they present the fully developed signature of the unique metabolic pattern of cancer, which only occurs when malignancy is highly advanced. Early stages of tumor growth, such as hyperplasia, may not have sufficient metabolic signatures for recognition by effector cells. These studies may have been the first time that LIFT was able to treat mouse malignancies that could not be treated by any other therapy known to date.

9.10.2 In Human Cancer Patients

Based on the above-mentioned information, we hypothesize that granulocytes with high levels of CKA collected from young healthy donors can be used as highly effective, low toxicity treatments for human cancers. LIFT treatment is a unique new entry into the arena of immunotherapies for cancer treatment. If Paul Ehrlich was correct in that cancer is caused, at least in some cases, primarily by a damaged cancer surveillance system that now turns out to be very likely the CKA in the granulocyte, LIFT is now based on a new philosophy of replacing a damaged immune system with a new younger system that has a desired and validated therapeutic functionality.

The ultimate test of the LIFT idea is not our interpretation of the mechanisms of granulocyte actions, but rather whether or not such a new idea actually works in a real clinical oncology setting. There has been significant debate on whether it is premature to take an idea from the lab bench into clinical testing in such a rapid manner before we fully understand and prove the mechanism of action. However, we feel that there will be plenty of time for us to work out the mechanistic aspects after we know whether or not LIFT is therapeutically effective. If the LIFT idea is correct, many lives might be saved, or deaths delayed, without the severe side effects of some conventional cancer therapies. Granulocyte transfer, as a treatment for infections in neutropenic patients, is already well tested and established as safe. Thus, the dangers are small, and the potential benefit is high. And we are encouraged by early indications that LIFT could work in treating various types of human malignancies without any adverse side effect.

Only time will tell, but it may not be long before answers begin to become available. A clinical trial of LIFT for treating human malignancies is just now beginning. Once donors are qualified, granulocytes will be collected via apheresis after these donors are mobilized to increase the number of granulocytes in the circulation. The collected fresh granulocytes will be immediately infused into cancer patient recipients as a treatment. Between the time of infusion and the time of rejection as allografts by the new host's immune system, the donor granulocytes can have a therapeutic window of up to several weeks in the new hosts. Such a therapeutic window should give the donor granulocytes plenty of opportunity to find and kill cancer cells as indicated by the preclinical study results. If the donor granulocytes have not shown the expected therapeutic function in the first two weeks, for example, they probably would never be functional. On the other hand, additional cancer killing, if needed, can presumably be provided by additional fresh infusions of granulocytes, with no limits that are apparent at present.

9.11 Conclusions

If the concept of CKA turns out to be correct, there could be many applications in anti-aging strategies for human life extension.

First of all, CKA can be used to guide LIFT for cancer treatment. With CKA screening, LIFT can have better or more effective donors. Second, the CKA assay might be used in cancer prevention and risk management. For example, a low level of CKA might allow cancer prevention by restoring CKA before cancer actually develops, and this test might also inspire more frequent testing for early cancer detection. Thus, CKA could serve in a way analogous to measurements of cholesterol levels for prevention of cardiovascular diseases. Third, CKA can be used to screen for compounds or dietary components that can boost its level. Fourth, CKA might be used to guide the establishment and efficacy of methods for granulocyte banking. If the best therapeutic or preventative agent to restore lost CKA is one's own granulocytes, then storing granulocytes away while young for use 20, 30, or 40 years later would allow older people to have immunity younger than the rest of their bodies.

The only limitations for our futures will be our own imaginations and our determination to make improvements in how we manage our aging process.

References

Alfaro G, Lomeli C, Ocadiz R, Ortega V, Barrera R, Ramirez M, Nava G (1992) Immunologic and genetic characterization of S180, a cell line of murine origin capable of growing in different inbred strains of mice. Vet Immunol Immunopathol 30:385–398

Barber GN (2001) Host defense, viruses and apoptosis. Cell Death Differ, 8: 113–126

Bartrons, R. and J. Caro (2007) Hypoxia, glucose metabolism and the Warburg's effect. J Bioenerg Biomembr 39(3):223–229

Bodey B (2002) Spontaneous regression of neoplasms: new possibilities for immunotherapy. Expert Opin Biol Ther 2:459–476

Bickels J et al. (2002) Coley's toxin: historical perspective. Isr Med Assoc J 4:471–472

Brubaker LH, Essig LJ, Menge CE (1977) Neutrophil life span in paroxysmal nocturnal hemoglobinuria. Blood 50:657–662

Busch W (1867) Aus der sitzung de medicinischen section vom. Berl Klin Wochenschr 5:137–145

Burnet FM (1970) The concept of immunological surveillance. Prog Exp Tumor Res 13:1–27

Burnet FM (1971) Immunological surveillance in neoplasia. Transplant Rev 7:3–25

Challis GB, Stam HJ (1990) The spontaneous regression of cancer. A review of cases from 1900 to 1987. Acta Oncol 29:545–550

Cohen S, Miller GE, Rabin BS (2001) Psychological stress and antibody response to mmunization: a critical review of the human literature. Psychosom Med 63:7–18

Cole WH (1974) Spontaneous regression of cancer: the metabolic triumph of the host? Ann NY Acad Sci 230:111–141

Cole WH (1976) Spontaneous regression of cancer and the importance of finding its cause. Natl Cancer Inst Monogr 44:5–9

Cole WH (1981) Efforts to explain spontaneous regression of cancer. J Surg Oncol 17:201–209

Coley WB (1893) The treatment of malignant tumors by repeated inoculations of erysipelas. With a report of ten original cases. Clin Orthop Relat Res 1991:3–11

Coley WB (1908) The treatment of sarcoma with the mixed toxins of erysipelas and bacillus prodigious. Boston Med Surg J 158:175–182

Cui Z (2003) The winding road to the discovery of the SR/CR mice. Cancer Immun 3:14–23

Cui Z et al. (2003) Spontaneous regression of advanced cancer: identification of a unique genetically determined, age-dependent trait in mice. Proc Natl Acad Sci USA 100:6682–6687

Dale DC, Boxer L, Liles WC (2008) The phagocytes: neutrophils and monocytes. Blood 112:935–945

Demaurex N et al. (1994) Characterization of receptor-mediated and store-regulated Ca2+ influx in human neutrophils. Biochem J 297:595–601

De Smet, K. and R. Contreras, (2005) Human antimicrobial peptides: defensins, cathelicidins and histatins. Biotechnol Lett 27(18):1337–1347

Dhabhar FS (2000) Acute stress enhances while chronic stress suppresses skin immunity. The role of stress hormones and leukocyte trafficking. Ann NY Acad Sci 917:876–893

Doerfler W (1991) Patterns of DNA methylation–evolutionary vestiges of foreign DNA inactivation as a host defense mechanism. A proposal. Biol Chem Hoppe Seyler 372:557–564

Doerfler W (2005) On the biological significance of DNA methylation. Biochemistry (Mosc) 70:505–524

Doerfler W (2006) De novo methylation, long-term promoter silencing, methylation patterns in the human genome, and consequences of foreign DNA insertion. Curr Top Microbiol Immunol 301:125–175

Dunn GP, Old LJ, Schreiber RD (2004) The immunobiology of cancer immunosurveillance and immunoediting. Immunity 21:137–48

Ellis TN, Beaman BL (2004) Interferon-gamma activation of polymorphonuclear neutrophil function. Immunology 112(1):2–12

Ehrlich P (1909) Ueber den jetzigen Stand der Karzinomforschung. Ned. Tijdschr Geneeskd 5:273–290

Epel ES (2009) Psychological and metabolic stress: a recipe for accelerated cellular aging? Hormones (Athens) 8:7–22
Freedman ND, Leitzmann MF, Hollenbeck AR, Schatzkin A, Abnet CC (2008) Cigarette smoking and subsequent risk of lung cancer in men and women: analysis of a prospective cohort study. Lancet Oncol 9:649–656
Funaro A et al. (2004) CD157 is an important mediator of neutrophil adhesion and migration. Blood 104:4269–4278
Ganz T et al. (1985) Defensins. Natural peptide antibiotics of human neutrophils. J Clin Invest 76:1427–1435
Gifford R, Tristem M (2003) The evolution, distribution and diversity of endogenous retroviruses. Virus Genes 26:291–315
Gleeson M, Bishop NC (2000) Special feature for the Olympics: effects of exercise on the immune system: modification of immune responses to exercise by carbohydrate, glutamine and antioxidant supplements. Immunol Cell Biol 78:554–561
Gordon MY (1994) Origin and development of neutrophils. In: Hellewell PG, Williams TJ (eds) Immunopharmacology of Neutrophils. Academic Press Limited, London, pp. 1–26
Green DR, Reed JC (1998) Mitochondria and apoptosis. Science 281:1309–1312
Hajek ER, Gutiérrez JR, Espinosa dGA (1984) Seasonality of mortality in human populations of Chile as related to a climatic gradient. Intl J Biometeorol 28:29–38
Hester JP et al. (1995) Collection and transfusion of granulocyte concentrates from donors primed with granulocyte stimulating factor and response of myelosuppressed patients with established infection. J Clin Apher 10:188–1893
Heron M (2007) Deaths: leading causes for 2004. Natl Vital Stat Rep 56:1–95
Hicks AM et al. (2006) Transferable anticancer innate immunity in spontaneous regression/complete resistance mice. Proc Natl Acad Sci USA 103:7753–7758
Hill CP et al. (1991) Crystal structure of defensin HNP-3, an amphiphilic dimer: mechanisms of membrane permeabilization. Science 251:1481–1485
Hockel M, Vaupel P (2001) Tumor hypoxia: definitions and current clinical, biologic, and molecular aspects. J Natl Cancer Inst 93:266–2676
Jemal A, Thun MJ, Ries LA, Howe HL, Weir HK, Center MM, Ward E, Wu XC, Eheman C, Anderson R, Ajani UA, Kohler B, Edwards BK (2008) Annual report to the nation on the status of cancer, 1975–2005, featuring trends in lung cancer, tobacco use, and tobacco control. J Natl Cancer Inst 100:1672–1694
Koch J et al. (2008) Frequency of the cancer-resistant phenotype in SR/CR mice and the effect of litter seriation. In Vivo 22:565–569
Langman L, Burr HS (1947) Electrometric studies in women with malignancy of cervix uteri. Science 105:209–210
Langman L, Burr HS (1949) A technique to aid in the detection of malignancy of the female genital tract. Am J Obstet Gynecol 57:274–281
Lehrer RI et al. (1988) Modulation of the in vitro candidacidal activity of human neutrophil defensins by target cell metabolism and divalent cations. J Clin Invest 81:1829–1835
Lehrer RI, Lichtenstein AK, Ganz T (1993) Defensins: antimicrobial and cytotoxic peptides of mammalian cells. Annu Rev Immunol 11:105–128
Lehrer RI (2004) Primate defensins. Nat Rev Microbiol 2(9)727–738
Leonard B (2000) Stress, depression and the activation of the immune system. World J Biol Psychiatr 1:17–25
Lerner BH (2004) Sins of omission-Cancer research without informed consent. N Engl J Med 351:628–630
Lichtenstein AK et al. (1988) Synergistic cytolysis mediated by hydrogen peroxide combined with peptide defensins. Cell Immunol 114:104–116
Maksakova IA et al. (2006) Retroviral elements and their hosts: insertional mutagenesis in the mouse germ line. PLoS Genet 2:e2

Maksakova IA, Mager DL, Reiss D (2008) Keeping active endogenous retroviral-like elements in check: the epigenetic perspective. Cell Mol Life Sci 65:3329–3347

Matsumoto M et al. (2004) Toll-like receptor 3: a link between toll-like receptor, interferon and viruses. Microbiol Immunol 48:147–154

McEwen BS (2000) The neurobiology of stress: from serendipity to clinical relevance. Brain Res 886:172–189

McLoughlin RM et al. (2004) Differential regulation of neutrophil-activating chemokines by IL-6 and its soluble receptor isoforms. J Immunol 172:5676–83

Mikkelsen JG, Pedersen FS (2000) Genetic reassortment and patch repair by recombination in retroviruses. J Biomed Sci 7:77–99

Moore AE, Rhoads CP, Southam CM (1957) Homotransplantation of human cell lines. Science 125:158–160

Nathan C (2006) Neutrophils and immunity: challenges and opportunities. Nat Rev Immunol 6:173–182

Nickelsen K (2007) Otto Warburg's first approach to photosynthesis. Photosynth Res 92: 109–120

Papac RJ (1996) Spontaneous regression of cancer. Cancer Treat Rev 22:395–423

Papac RJ (1998) Spontaneous regression of cancer: possible mechanisms. In Vivo 12:571–578

Preminger BA (2002) The case of Chester M. Southam: research ethics and the limits of professional responsibility. Pharos Alpha Omega Alpha Honor Med Soc 65(2):4–9

Pruett SB (2001) Quantitative aspects of stress-induced immunomodulation. Int Immunopharmacol 1:507–520

Richardson MA et al. (1999) Coley toxins immunotherapy: a retrospective review. Altern Ther Health Med 5:42–47

Rice WG et al. (1987) Defensin-rich dense granules of human neutrophils. Blood 70:757–765

Schiffer CA, Aisner J, Wiernik PH (1977) Current status of granulocyte transfusion therapy. Prog Clin Biol Res 13:281–292

Schiffer CA (1979) Filtration leukapheresis: summary and perspectives. Exp Hematol 7(4 Suppl): 42–47

Schiffer CA (1983) Granulocyte transfusion therapy. Cancer Treat Rep 67:113–119

Selsted ME, Harwig SS (1989) Determination of the disulfide array in the human defensin HNP-2. A covalently cyclized peptide. J Biol Chem 264:4003–4007

Silva JC, Loreto EL, Clark JB (2004) Factors that affect the horizontal transfer of transposable elements. Curr Issues Mol Biol 6:57–71

Southam CM, Moore AE, Rhoads CP (1957) Homotransplantation of human cell lines. Science 125:158–160

Southam CM (1958) Homotransplantation of human cell lines. Bull N Y Acad Med 34:416–423

Southem CM, Moore AE (1958) Induced immunity to cancer cell homografts in man. Ann N Y Acad Sci 73:635–653

Starnes CO (1992) Coley's toxins in perspective. Nature 357:11–12

Stefanski V (2001) Social stress in laboratory rats: behavior, immune function, and tumor metastasis. Physiol Behav 73:385–391

Stoll BA (1992) Spontaneous regression of cancer: new insights. Biotherapy 4:23–30

Sverdlov ED (1998) Perpetually mobile footprints of ancient infections in human genome. FEBS Lett 428:1–6

Thomas L (1959) Discussion of Medawar reactions to homologous tissue antigens in relation to hypersensitivity. In: Lawrence H (ed) Cellular and Humoral Aspects of the Hypersensitivity States. Paul Hoeberg, New York

Tsung K, Norton JA (2006) Lessons from Coley's Toxin. Surg Oncol 15:25–28

Uematsu S, Akira S (2007) Toll-like receptors and Type I interferons. J Biol Chem 282:15319–15323

Unknown author (1957) Cancer volunteers. TIME (http://www.time.com/time/magazine/article/0,9171,936841,00.html)

Villeneuve PJ, Mao Y (1994) Lifetime probability of developing lung cancer, by smoking status, Canada. Can J Pub Health 85:385–388

Wang S, Garcia AJ, Wu M, Lawson DA, Witte ON, Wu H (2006) Pten deletion leads to the expansion of a prostatic stem/progenitor cell subpopulation and tumor initiation. Proc Natl Acad Sci USA 103:1480–1485

Warburg O (1920) Uber die Geschwindigkeit der photochemischen Kohlensaurezersetzung in lebenden Zellen. Biochem Z 103:188–217

Warburg O (1956a) On the origin of cancer cells. Science 123:309–314

Warburg O (1956b) On respiratory impairment in cancer cells. Science 124:269–270

Waterston RH et al. (2002) Initial sequencing and comparative analysis of the mouse genome. Nature 420:520–562

White SH, Wimley WC, Selsted ME (1995) Structure, function, and membrane integration of defensins. Curr Opin Struct Biol 5:521–527

Wiemann B, Starnes CO (1994) Coley's toxins, tumor necrosis factor and cancer research: a historical perspective. Pharmacol Ther 64:529–564

Witko-Sarsat V et al. (2000) Neutrophils: molecules, functions and pathophysiological aspects. Lab Invest 80:617–653

Zacharski LR, Sukhatme VP (2005) Coley's toxin revisited: immunotherapy or plasminogen activator therapy of cancer? J Thromb Haemost 3:424–427

Zanetti M, Gennaro R, Romeo D (1995) Cathelicidins: a novel protein family with a common proregion and a variable C-terminal antimicrobial domain. FEBS Lett 374: 1–5

Chapter 10
Small Molecule Modulators of Sirtuin Activity

Francisco J. Alcaín, Robin K. Minor, José M. Villalba, and Rafael de Cabo

Contents

10.1	Introduction	331
10.2	Members of the Mammalian Sirtuin Family, Subcellular Localizations, Chromosome Mapping and Regulation of Expression	333
10.3	Biological Functions and Physiological Roles	335
	10.3.1 Nuclear Sirtuins	335
	10.3.2 Cytoplasmic Sirtuin	337
	10.3.3 Mitochondrial Sirtuins	338
10.4	Structural Basis for the Mechanism of Action of Sirtuin Modulators	339
10.5	Sirtuin Modulator Compounds: Inhibitors	340
	10.5.1 Sirtuin Inhibitors in Cancer Therapy	343
	10.5.2 Sirtuin Inhibitors in Other Diseases	345
10.6	Sirtuin Modulator Compounds: Activators	345
	10.6.1 Cardioprotective Effect	348
	10.6.2 Neuroprotective Effect	348
	10.6.3 Metabolic Diseases	349
10.7	Conclusion	351
References		351

10.1 Introduction

Histone deacetylases (HDACs) play a central role in the epigenetic regulation of gene expression. Acetylation in the ε-amino group of lysine residues of chromatin-interacting proteins regulates genome architecture and gene expression. This

---

J.M. Villalba (✉)
Departamento de Biología Celular, Fisiología e Inmunología, Facultad de Ciencias, Universidad de Córdoba 14014- Córdoba, Spain
e-mail: bc1vimoj@uco.es

process also results in a gain-of-function or loss-of-function for many proteins. In addition to histones, many non-histone proteins are also subject to post-translational modification by acetylation/deacetylation, such as transcription factors and proteins that are involved in DNA repair and replication, metabolism, cytoskeletal dynamics, apoptosis, protein folding and cellular signaling. N-ε-acetylation is reversible and is dynamically controlled by opposing actions of lysine acetyltransferases (histone acetyltransferases, or HATs) and lysine deacetylases (HDACs) (Yang and Seto 2008).

The first human HDAC was isolated by Taunton et al. on the basis of its high affinity with the yeast histone deacetylase inhibitor trapoxin (Taunton et al. 1996). To date, HDACs comprise a family of 18 genes that are subdivided into four classes (Haberland et al. 2009). Classes I, II, and IV are referred to as "classical HDACs". Class I enzymes are ubiquitously expressed, located primarily in the nucleus, and have high homology to the yeast RPD3 gene. Class II deacetylases show more tissue specificity and can be located both in the nucleus and the cytoplasm, presenting homology to HDA1 histone deacetylases existing in yeast (Gray and Ekström 2001). HDA IV is structurally related to both class I and class II HDACs, but to date, very little information about its expression or function is available (Witt et al. 2009). These classes of deacetylases use zinc-dependent mechanisms.

The third class of histone deacetylases are also known as sirtuins, as they share homology with the yeast Silent Information Regulator (SIR 2) gene (Gray and Ekström 2001). The most studied effect of sirtuins is their potential to promote longevity and healthy aging in a variety of species, potentially delaying the onset of age-related neurodegenerative disorders (Michan and Sinclair 2007). In this respect sirtuins are generally thought of as prolongevity factors because they are related to an increase of life span in yeast, worms, and flies. However, whether the mammalian Sir2 ortholog, Sirt1, has a similar pro-longevity role remains unclear (Kaeberlein 2008). Aside from their effects on longevity, the sirtuin family has also been linked to gene repression, the control of metabolic processes, apoptosis and cell survival, DNA repair, development, and inflammation (Michan and Sinclair 2007).

In view of the many functions of sirtuins in cells, researchers have devoted considerable energy to harnessing the activities of the various sirtuins because of their potential to address a broad range of diseases including obesity, diabetes, cancer, inflammation, and cardiovascular, neuronal and age-related diseases. In this review we begin by introducing the mammalian sirtuins and give an overview of their known activities in the context of their subcellular localizations. Next we review the structural characteristics of the sirtuins and discuss how their structure influences their sensitivity to chemical modulators. We follow this with a review of compounds currently known to activate or inhibit sirtuins, and finally, we discuss the degree to which existing data support the use of sirtuin-based therapies for the treatment of human diseases.

10.2 Members of the Mammalian Sirtuin Family, Subcellular Localizations, Chromosome Mapping and Regulation of Expression

Seven proteins constitute the mammalian sirtuin family, and although deacetylase activity has not been reported for all members, all situins contain a conserved catalytic core domain of 275 amino acids and have a stoichiometric requirement for the cofactor nicotinamide adenine dinucleotide (NAD^+) to deacetylate substrates from histones to transcriptional regulators (Brittain et al. 2007). Based on sequence similarities, eukaryotic sirtuins have been divided into four broad phylogenetic groups. Class I sirtuins include SIRT1 (the ortholog of yeast Sir2), SIRT2 and SIRT3. Class II only includes SIRT4; SIRT5 is the only member of sirtuin Class III; SIRT6 and SIRT7 belong to Class IV. It is important to note that there is no obvious correlation between this classification system and the specific biological functions of the sirtuins.

The genomic sequence of human sirtuins is available for SIRT1, SIRT2, SIRT3, SIRT4, and SIRT6 (Frye 2000), and Mahlknecht's group has characterized and mapped human SIRT1, SIRT2, SIRT5, SIRT6 and SIRT7 (Voelter-Mahlknecht et al. 2005, 2006a, b; Voelter-Mahlknecht and Mahlknecht 2006). The SIRT1, SIRT6 and SIRT7 genes encode a single protein, whereas the SIRT2 and SIRT5 genes encode two isoforms each. SIRT3 has been identified in chromosome 11 (11p15) in a region known for silencing specific parental alleles (Onyango et al. 2002).

Mammalian sirtuins are found in a variety of subcellular locations. SIRT1, SRT6 and SIRT7 are nuclear proteins (Voelter-Mahlknecht et al. 2006b; Vaziri et al. 2001; Ford et al. 2006) with distinct subnuclear localizations. For example, SIRT1 is detected in the nuclei but not the nucleoli, whereas SIRT7 is a widely-expressed nucleolar protein associated with active rRNA. SIRT2, while predominantly a cytoplasmic protein, can also be located in the nucleus (North et al. 2003) . SIRT3, SIRT4 and SIRT5 are mitochondrial proteins, although SIRT3 has also been identified to move from the nucleus to mitochondria during cellular stress (Michan and Sinclair 2007). Interestingly, SIRT3 is expressed in two forms, a 44-kDa long form and a 28-kDa short form. In this case, form determines localization, as the long form is localized in the mitochondria, nucleus, and cytoplasm while the short form is localized exclusively in the mitochondria (Scher et al. 2007; Sundaresan et al. 2008).

Expression of sirtuin mRNA is reportedly ubiquitous, but differs quantitatively between tissues (Michishita et al. 2005). Table 10.1 summarizes both the subcellular and the tissue-specific localization of the different sirtuins.

Environmental, dietary and pathological conditions are just some of the variables that are known to modulate the expression of sirtuins. SIRT1, SIRT2, SIRT3 and SIRT6 have been reported to be up-regulated by exogenous factors such as calorie restriction (CR, a low-calorie dietary regimen without micronutrient malnutrition), external cold temperatures, and oxidative stress (Cohen et al. 2004;

Table 10.1 Localization, substrates and mechanism of action of different sirtuins

Sirtuin	Subcellular localization	High tissue expression	Substrates	Mechanism of action
SIRT1	Nuclear	Brain, skeletal muscle, heart, kidney and uterus	Histones H1, H3 and H4, transcription factors, p53, FOXO family, Ku70, p300, NF-κB, PGC-1α, PPAR-γ, UCP2, Acetyl-CoA synthetase 1	NAD-dependent deacetylase
SIRT2	Cytosolic	Brain	α-tubulin	NAD-dependent deacetylase
SIRT3	Mitochondrial Nuclear Cytoplasmic	Brain, heart, liver, kidney and brown adipose tissue	Acetyl-CoA synthetase 2 isocitrate dehydrogenase 2 Ku70	NAD-dependent deacetylase
SIRT4	Mitochondria	Kidney, uterus, lung, testis, brain and eye	Glutamate dehydrogenase	ADP-ribosyl transferase
SIRT5	Mitochondrial	Brain, testis, heart muscle and lymphoblast	Cytochrome c	NAD-dependent deacetylase
SIRT6	Nuclear (associated with heterochromatin)	Muscle, brain, heart, ovary and bone cells	Histone H3	NAD-dependent deacetylase, ADP-ribosyl transferase
SIRT7	Nucleolar	Blood cells	RNA polymerase I	NAD-dependent deacetylase

Crujeiras et al. 2008; Schwer and Verdin 2008; Kanfi et al. 2008). However, the general assumption that SIRT1 is up-regulated by caloric restriction is currently questioned, because there may be differences between acute starvation and long-term CR in terms of enzymatic activity and expression levels of sirtuins as well as tissue-specific differences in the SIRT1 response (Chen et al. 2008; Bordone et al. 2006).

Sirtuin activity is also modulated by endogenous conditions. Yeast Sir2 and mammalian SIRT1 activities are up-regulated by increasing the $NAD^+/NADH$ ratio and down-regulated by increasing levels of nicotinamide (Lin et al. 2004). Furthermore, an endogenous nuclear protein has been recognized in mammalian cells, the so-called active regulator of SIRT1 (AROS). AROS directly regulates SIRT1 function by physical interaction with the catalytic domain of SIRT1, and AROS binding is required to stimulate SIRT1 to deacetylate p53 and suppress p53 transcriptional activity (Kim et al. 2007a).

Interestingly, while cancer cells are known to over-express SIRT1 (Lim 2006; Ashrafn et al. 2006; Stünkel et al. 2007), AROS has shown a remarkably similar expression profile in multiple cancer cell lines, suggesting that AROS may be a tumor cell survival factor and a target for tumor therapy (Kim et al. 2007a). Another chemotherapy approach involving SIRT1 may be the downregulation of SIRT1 by using antisense oligonucleotides. This has been shown to increase acetylation of the tumor suppressor p53 as well as to increase Bax expression in A549 lung cancer cells and to decrease proliferation (Sun et al. 2007). SIRT2, on the other hand, which is located at 19q13.2, a region known to be frequently deleted in human gliomas, has been shown to be down-regulated in gliomas, suggesting that SIRT2 deletion may permit tumor development and that its repletion may act as a tumor suppressive therapy (Hiratsuka et al. 2003).

10.3 Biological Functions and Physiological Roles

10.3.1 Nuclear Sirtuins

10.3.1.1 SIRT1

Sirtuins show different subcellular localizations, and consequently display different substrate specificity. SIRT1 is the closest to Sir2 in terms of sequence and enzymatic activity, and is also the most extensively studied mammalian sirtuin.

SIRT1 is a nuclear protein expressed predominantly in brain, skeletal muscle, heart, kidney and uterus, that deacetylates lysine residues in histones H3-H4 and mediates heterochromatin formation by recruitment and deacetylation of H1 (Vaquero et al. 2004; Michishita et al. 2005). In addition to its classical role as a histone deacetylase, SIRT1 has been associated with an increase in lifespan and cell survival associated with stress resistance. SIRT1 deacetylates a large number of key substrates and is involved, for example, in the DNA damage response by binding to the transcription factor p53 protein in vivo and deacetylating p53, resulting in negative regulation of p53-mediated transcriptional activity (Vaziri et al. 2001; Smith 2002). SIRT1 also increases FOXO3's ability to induce cell cycle arrest and resistance to oxidative stress but inhibits FOXO3's ability to induce cell death (Brunet et al. 2004) by sequestering the proapoptotic factor Bax away from mitochondria after DNA repair factor Ku70 deacetylation and, thereby, inhibiting stress-induced apoptotic cell death. SIRT1 also deacetylates the transcriptional integrator p300 that serves to integrate diverse signaling pathways involved in metabolism and cellular differentiation (Bouras et al. 2005).

Cancer cell lines, as well as tissue samples obtained from cancer patients, reveal higher endogenous levels of SIRT1 expression compared with normal cells (Ashrafn et al. 2006; Kim et al. 2007a; Stünkel et al. 2007; Sun et al. 2007). Based on the elevated levels of SIRT1 in cancers, it was hypothesized that SIRT1 serves as a tumor promoter (Lim 2006). SIRT1 deacetylates the tumor suppressor p53 to inhibit its transcriptional activity, resulting in reduced apoptosis in response to various genotoxic stimuli. As p53 activity is down-regulated by SIRT1, this raises the possibility that inhibition of SIRT1 may suppress cancer cell proliferation. However, CR is well known to both extend lifespan and increase SIRT1 expression in mammals, and it is well established that CR exerts strong tumor-suppressive effects (Weindruch and Walford 1982). Further, ectopic induction of SIRT1 in the APCmin/+ mouse model of colon cancer significantly reduces tumor formation, proliferation, and animal morbidity in the absence of CR (Firestein et al. 2008), and a significant inverse correlation between the presence of nuclear SIRT1 and the oncogenic form of β-catenin has been found in 81 different human colon tumor specimens (Firestein et al. 2008). It has also been reported that SIRT1 activity augments apoptosis in response to TNFα by deacetylating and inhibiting a subunit of NF-κB (Yeung et al. 2004).

Aside from its tumor suppressive effects, p53 is also strongly implicated in neuronal apoptosis, and increasing evidence suggests that activation of p53 contributes to neurodegeneration in Alzheimer's disease (AD) (Qin et al. 2006; Gan 2007). Thus, SIRT1-mediated p53 inhibition may play a neuroprotective role in some neurodegenerative diseases, such as AD or Huntington's disease (HD) and Wallerian neurodegeneration (Michan and Sinclair 2007; Outeiro et al. 2008).

A CR diet increases sirtuin activities in mammals and regulates many physiological processes that are beneficial in age-related disorders. Thus, sirtuins are a link between nutrient availability and energy metabolism.

SIRT1 protein binds to and represses genes controlled by the fat regulator peroxisome proliferator-activated receptor-γ (PPAR-γ). Activation of Sirt1 by resveratrol decreases fat accumulation in differentiated adipocytes, and RNA interference of SIRT1 enhances it (Picard et al. 2004). SIRT1 also promotes fat mobilization in white adipocytes by repressing PPAR-γ (Picard et al. 2004; Nemoto et al. 2005).

SIRT1 forms a stable complex with the mitochondrial biogenesis coactivator PGC-1α and induces gluconeogenic genes and hepatic glucose output through PGC-1α deacetylation. Six months of CR in healthy overweight but non-obese humans caused an increase in the expression of genes involved in mitochondrial biogenesis in skeletal muscle (Civitarese et al. 2007). Moreover, it has been reported that SIRT1 functions as a positive regulator of insulin secretion in pancreatic β-cells in rodents by directly repressing the uncoupling protein (UCP) gene UCP2, which uncouples mitochondrial respiration from ATP production and reduces the proton gradient across the mitochondrial membrane (Bordone et al. 2006).

These results link SIRT1 to the control of metabolic processes such as metabolic syndrome and insulin resistance. Also, the improvement in mitochondrial functions could be promising in the treatment of many neurodegenerative diseases, because some of their common biochemical processes include mitochondrial dysfunction and reactive oxygen species generation (Outeiro et al. 2008; Westphal et al. 2007).

10.3.1.2 SIRT6

Mouse SIRT6 is a broadly expressed, predominantly nuclear protein, with the highest levels in muscle, brain, and heart. Human SIRT6 appears to be most predominantly expressed in bone cells and in the ovaries, while it is practically absent in the bone marrow (Liszt et al. 2005; Voelter-Mahlknecht et al. 2006b). Although no NAD^+-dependent deacetylase activity was detected in earlier studies (Liszt et al. 2005), it has been recently reported that SIRT6 associates specifically with telomeres and deacetylates specifically the histone H3K9 that modulates chromatin structure at telomeres (Michishita et al. 2008). Deacetylation of H3K9 represses NF-κB transcriptional activity, and SIRT6 deficiency leads to the activation of transcriptional programs observed in aged tissues, including genes controlled by NF-κB (Kawahara et al., 2009).

SIRT6 plays a key role in DNA repair and maintenance of genomic stability in mammalian cells through the DNA base excision repair pathway. SIRT6 knockout animals are born at a normal Mendelian ratio, and, though somewhat smaller than wild-type littermates, they appear relatively normal until approximately two weeks of age, when they develop profound metabolic defects and ultimately die by four weeks of age (Mostoslavsky et al. 2006).

In vitro, SIRT6 is increased under nutrient deprivation in cultured cells, in mice after fasting, and in rats fed a CR diet. The increase in SIRT6 levels is due to stabilization of SIRT6 protein and not because of an increase in SIRT6 transcription (Kanfi et al. 2008).

p53 positively regulates SIRT6 protein levels under standard growth conditions but has no role in the nutrient-dependent regulation of SIRT6 (Kanfi et al., 2008). These observations imply that at least two sirtuins are involved in regulation of lifespan by nutrient availability.

10.3.1.3 SIRT7

SIRT7 is a widely expressed nucleolar protein that is associated with active rRNA genes, regulates transcription, and is required for cell viability in mammals (Ford et al. 2006). SIRT7 is inactivated by phosphorylation via the CDK1-cyclin B pathway during mitosis, then dephosphorylated by a phosphatase at the exit from mitosis before onset of rDNA transcription (Grob et al. 2009). Depletion of SIRT7 stops cell proliferation and triggers apoptosis (Ford et al. 2006). SIRT7-deficient mice have reduced mean and maximum lifespans and develop inflammatory cardiomyopathy (Vakhrusheva et al. 2008). SIRT7 interacts with p53 and efficiently deacetylates p53 in vitro, conferring resistance to cytotoxic and oxidative stress in neonatal primary cardiomyocytes (Vakhrusheva et al. 2008).

SIRT7 appears to be most predominantly expressed in the blood and in $CD33^+$ myeloid bone marrow precursor cells, while the lowest levels are found in the ovaries and skeletal muscle (Voelter-Mahlknecht et al. 2006d). Levels of SIRT7 expression were significantly increased in breast cancer (Ashrafn et al. 2006).

10.3.2 Cytoplasmic Sirtuin

Human SIRT2 is a predominantly cytoplasmic protein that colocalizes with microtubules. Knockdown of SIRT2 via siRNA results in tubulin hyperacetylation (North et al. 2003). α-Tubulin is the main cytoplasmic substrate of SIRT2 (Li et al. 2007), and inhibition of SIRT2-mediated tubulin deacetylation has been involved in resistance to axonal degeneration in a mutant mouse called slow Wallerian degeneration because tubulin becomes more stable as it is acetylated (Suzuki and Koike 2007).

Human SIRT2 is known to be most predominantly expressed in the brain. Double immunofluorescence established that SIRT2 expression in the central nervous system is primarily within oligodendrocytes, and localized to the outer and juxtanodal loops in the myelin sheath (Li et al. 2007). SIRT2 requires a proteolipid protein (PLP)/DM20 for its transport into the myelin compartment (Werner et al. 2007).

Inhibition of SIRT2 rescued α-synuclein toxicity and protected against dopaminergic cell death both in vitro and in a Drosophila model of Parkinson's disease (PD) (Outeiro et al. 2007). PD is characterized by a loss of dopaminergic neurons due to the development of Lewy bodies containing α-synuclein in the substantia nigra, so compounds that ameliorate α-synuclein fibril formation may be therapeutic in this kind of neurodegenerative disease.

Like SIRT1, SIRT2 binds and deacetylates FOXO3a, thus elevating the expression of FOXO target genes such as superoxide dismutase and reducing cellular levels of reactive oxygen species. However, another target gene activated by SIRT2 is the pro-apoptotic factor Bim, thus promoting cell death when cells are under severe stress (Wang et al. 2007).

Human SIRT2 is severely reduced in a large number of human brain tumour cell lines, so the absence of SIRT2, a potential tumour suppressor, could play a key role in the regulation of the cell-cycle within a multistep pathway that leads to full cellular transformation and, finally, the development of cellular malignancy (Voelter-Mahlknecht et al. 2005). Indeed, ectopic expression of SIRT2 in these cell lines suppressed colony formation and modified the microtubule network (North and Verdin 2004).

10.3.3 Mitochondrial Sirtuins

10.3.3.1 SIRT3

Among the seven sirtuins found in humans, SIRT3 is the only sirtuin shown to correlate with an extended human lifespan: a variant of the SIRT3 gene has been found to be virtually absent in males older than 90 years (Bellizzi et al. 2005; Rose et al. 2003). SIRT3 is localized to the mitochondrial inner membrane and, as would be expected, is particularly high in tissues rich in mitochondria like brain, heart, liver, kidney and brown adipose tissue (Onyango et al. 2002).

SIRT3 exhibits high catalytic efficiency against acetylated mitochondrial Acetyl-CoA synthetase (AceCS), and SIRT3-mediated activation of AceCS plays an important role in adaptive thermogenesis in brown adipose tissue (Hallows et al. 2006). CR up-regulates SIRT3 expression in both white and brown adipose, as does cold exposure (Shi et al. 2005; Hallows et al. 2006). Furthermore, SIRT3 is down-regulated in the brown adipose tissue of several genetically obese mice (Shi et al. 2005). In addition, SIRT3 deacetylates and activates isocitrate dehydrogenase 2, an enzyme that catalyzes a key regulation point of the Krebs cycle and recycles cellular antioxidants (Schlicker et al. 2008). However, although SIRT3 activates some mitochondrial functions and plays an important role in adaptive thermogenesis in brown adipose, SIRT3-deficient mice show normal overall metabolism under standard laboratory conditions, as well as normal adaptive thermogenesis (Lombard et al. 2007).

10.3.3.2 SIRT4

SIRT4 is a mitochondrial sirtuin that lacks detectable NAD^+-dependent deacetylase activity. Rather, SIRT4 uses NAD^+ for ADP-ribosylation of glutamate dehydrogenase, repressing its enzymatic activity. Through this mechanism, SIRT4 plays a role in regulating the ability of β-cells to secrete insulin in response to amino acids by inhibiting glutamate dehydrogenase, an effect that can be reversed by CR (Haigis et al. 2006). Knockdown of SIRT4 via siRNA enhances insulin secretion in response to glucose in the insulin-producing cell line INS-1E, and SIRT4 overexpression suppresses insulin secretion in β-cells (Ahuja et al., 2007). Recently, single nucleotide polymorphisms in the SIRT4 gene have been associated with type II diabetes mellitus (Reiling et al. 2009). These observations imply an important role for mitochondrial SIRT4 in the regulation of insulin secretion.

10.3.3.3 SIRT5

The third mitochondrial sirtuin, SIRT5, is a soluble protein mainly localized to the intermembrane space, although this protein has been also detected in the matrix (Schlicker et al. 2008; Nakamura et al. 2008). SIRT5 can deacetylate cytochrome c, a protein of the mitochondrial electron transport chain with central functions in both oxidative metabolism and initiation of apoptosis (Schlicker et al. 2008).

SIRT5-deficient mice are born at a Mendelian ratio, are fertile, and appear healthy until at least 18 months of age (Lombard et al. 2007). The human SIRT5 gene is most predominantly expressed in brain, testis, heart muscle cells, and in lymphoblasts (Voelter-Mahlknecht et al. 2006a; Michishita et al. 2005).

10.4 Structural Basis for the Mechanism of Action of Sirtuin Modulators

Crystal structures of many of the sirtuins have been elucidated. Bacterial sirtuins have much shorter N and C termini and are more conserved as a group than human sirtuins (Smith et al. 2002). Human SIRT1 consists of a large domain having a Rossmann fold where the NAD^+-binding domain resides, and a small domain containing a three-stranded zinc ribbon motif. Both domains are connected by several conserved loops. A conserved large groove at the interface of the two domains is the likely site of catalysis based on mutagenesis (Finnin et al. 2001; Min et al. 2001).

Sirtuins catalyse a deacetyletion reaction that is proposed to require the formation of a ternary enzyme-NAD^+-acetyl-lysine substrate complex prior to catalysis and in which one NAD^+ molecule is consumed by each acetyl-lysine hydrolyzed. This hydrolysis yields the deacetylated, O-acetyl-ADP-ribose substrate and nicotinamide, and both products are known to modulate Sir2 activity (reviewed in Denu 2005). However, nicotinamide levels in rat tissues are very low because this molecule is rapidly used in the synthesis of NAD^+ and other pyridine nucleotides, meaning it is very unlikely that nicotinamide is a major inhibitor of sirtuins in vivo (Adams and Klaidman 2007). That said, nicotinamide concentrations as high as 100 mM have been reported in cultured mouse embryonic stem cells, providing evidence that nicotinamide concentrations could be a factor regulating sirtuin activities in some situations (Yang and Sauve 2006). A nicotinamide analog, isonicotinamide, which competes with nicotinamide binding but does not react appreciably with the enzyme intermediate, increases Sir2 activity (Sauve et al. 2005).

Another modulator of sirtuin activity is resveratrol (RSV), a polyphenol found in red wine. RSV acts by lowering the Michaelis constant of SIRT1 for both the acetylated substrate and NAD^+ (Howitz et al. 2003). A model has also been proposed where RSV binds to the N-terminus of SIRT1, thereby inducing a conformational change and lowering the Km for the substrate. Mutations to Sirt1 have demonstraed that amino acids 183–225 are important for defining the compound binding site, and specific mutations to SIRT1-E230K and Sir2-D223K prevent activation by RSV without affecting basal activity (Yang et al. 2007). RSV and other novel small-molecule activators like SRT1720 share this single allosteric site despite their structural dissimilarities (Milne and Denu 2008).

10.5 Sirtuin Modulator Compounds: Inhibitors

Because sirtuins play an important role in gene activation and silencing in organisms from prokaryotes to humans, their modulation can be beneficial for a wide variety of diseases. In addition to cancer therapy, sirtuin inhibitors have also been proposed in the treatment of Parkinson's disease (Outeiro et al. 2007), leishmaniosis (Vergnes et al. 2005), and human immunodeficiency virus infection (Pagans et al. 2005), among others. This has made the characterization of small molecules that modify sirtuin activity a hot topic. In addition to nicotinamide, several specific inhibitors of Sir2 have been described, including splitomicin and its analogues (Bedalov et al. 2001; Hirao et al. 2003, Posakony et al. 2004), sirtinol (Grozinger et al. 2001), AGK2 (Outeiro et al. 2007), tenovin (Lain et al. 2008), suramin and its analogues (Schuetz et al. 2007; Trapp et al. 2007), cambinol (Heltweg et al. 2006) and salermide (Lara et al. 2009).

In a cell-based screen for inhibitors of yeast Sir2p, Bedalov et al. (2001) showed that splitomicin (Fig. 10.1) selectively inhibits Sir2 by blocking its active site with an IC_{50} of 60 μM. In an in vitro Sir2 inhibition screen, the most potent compounds had activities similar to that of splitomicin (Posakony et al. 2004). The orientation of

Fig. 10.1 Splitomicin

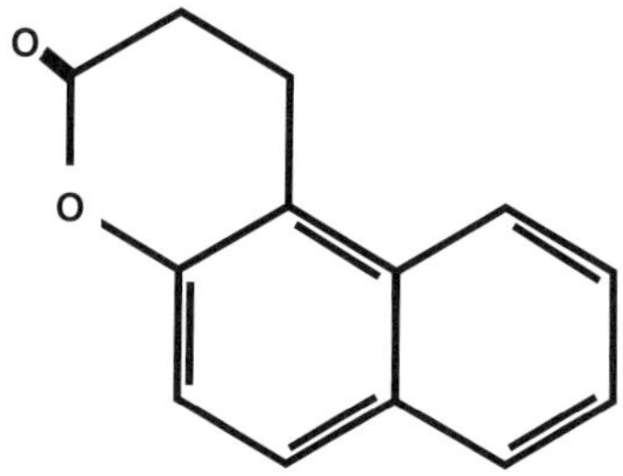

the β-phenyl group is important for inhibition of sirtuin activity, and a link between inhibition and anticancer activity has been established (Neugebauer et al. 2008).

Using similar methodology, Grozinger et al. (2001) concurrently discovered another sirtuin inhibitor, sirtinol (2-[(2-hydroxy-naphthalen-1-ylmethylene)-amino]-N-(1-phenyl-ethyl)-benzamide) (Fig. 10.2). Sirtinol is able to inhibit yeast Sir2p transcriptional silencing activity in vivo, and yeast Sir2p and human SIRT1 deacetylase activity in vitro with IC_{50} values of 70 and 45 µM, respectively. Two sirtinol analogs have also been developed (m- and p-sirtinol) and they are 2- to 10-fold more potent than sirtinol against human SIRT1 and SIRT2 enzymes (Pagans et al. 2005).

Bedalov's group has also identified and characterized a chemically stable compound that shares a β-naphtol pharmacophore with sirtinol and splitomicin. Cambinol (Fig. 10.3) inhibits human SIRT1 and SIRT2 deacetylase activity in vitro with IC_{50} values of 56 and 59 µM, respectively, and to a lesser extent inhibits SIRT5 (Heltweg et al. 2006). Modifications to the core structure of cambinol, in particular by incorporation of substitutents at the N1-position, leads to an increase in potency against SIRT2 in vitro that is not seen for SIRT1 (Medda et al. 2009).

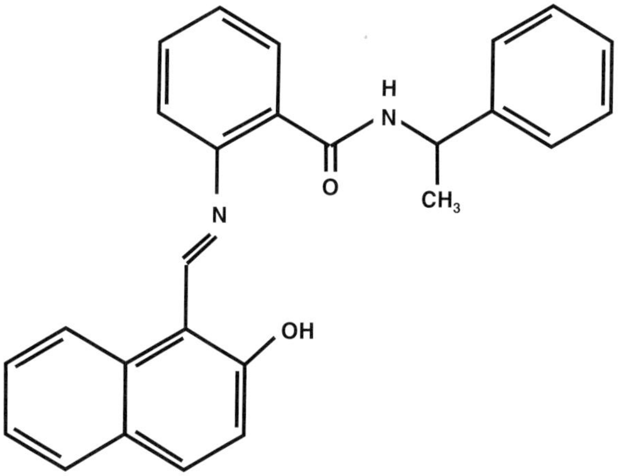

Fig. 10.2 Sirtinol

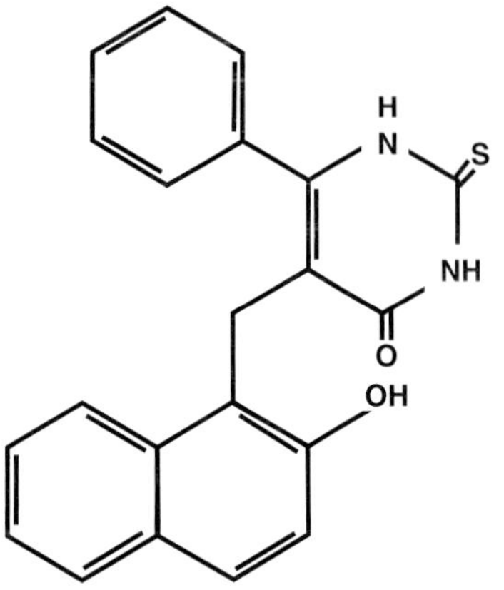

Fig. 10.3 Cambinol

Suramin (Fig. 10.4), developed in 1916 and used for the treatment of sleeping sickness and onchocerciasis, is an inhibitor of mammalian SIRT1 at an IC_{50} value of at 297 nM, SIRT2 at 1.15 μM and SIRT5 at 22 μM. Suramin inhibits SIRT5 activity by mimicking the contacts of the nicotinamide ribose of the cofactor in the B-pocket, the reaction product nicotinamide in the C-pocket, and the peptide in the substrate-binding site. Suramin binding sites in SIRT2 and SIRT5 are similar. A diverse set of suramin analogues has been designed, and some of them showed selective inhibition of SIRT1 and SIRT2 in the two-digit nanomolar range (Schuetz et al. 2007; Trapp et al. 2007).

Finally, Lara et al. (2009) synthesized salermide (N-{3-[(2-hydroxy-naphthalen-1-ylmethylene)-amino]-phenyl}-2-phenyl-propionamide) (Fig. 10.5), which inhibits 80% of SIRT1 and SIRT2 activities in an in vitro assay at 90 and 25 μM, respectively.

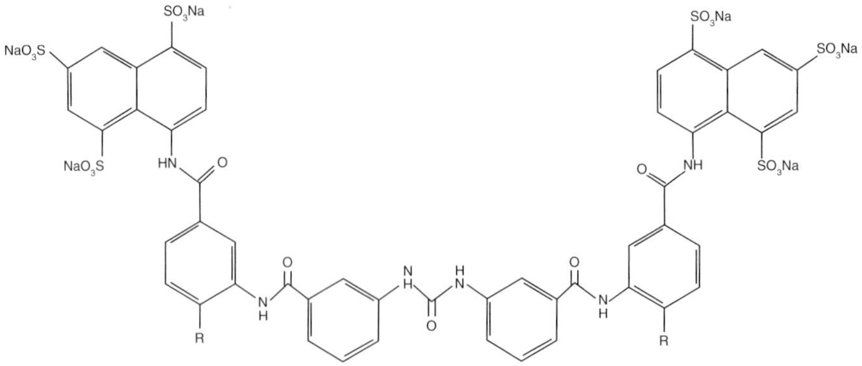

Fig. 10.4 Suramin

Fig. 10.5 Salerrmide

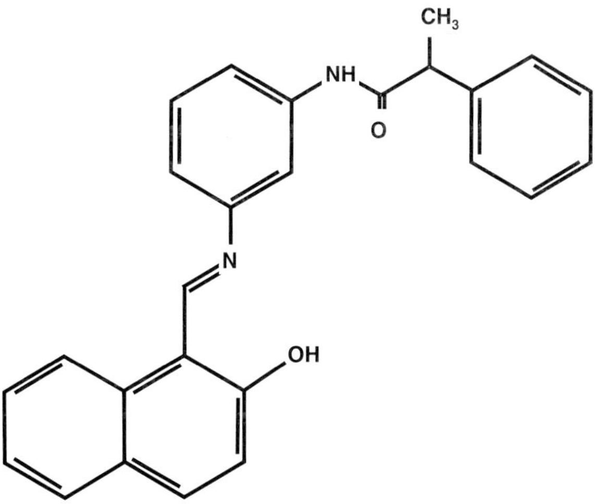

10.5.1 Sirtuin Inhibitors in Cancer Therapy

That p53 plays a central role in preventing tumor development is clear; more than 50% of adult human tumors are characterized by inactivating mutations or deletions of the p53 gene. SIRT1 is known to regulate p53 activity by deacetylating this protein, resulting in reduced apoptosis in response to various genotoxic stimuli (Vaziri et al. 2001). Furthermore, based on the observations that SIRT1 is up-regulated in tumor cells, SIRT1 expression may play an important role in promoting cell growth and thus, inhibitors of SIRT1 such as sirtinol may have anticancer potential.

SIRT2, on the other hand, regulates the dynamics of microtubules and tubulin-associated cellular events by controlling the levels of acetylated α-tubulin (North et al. 2003). In addition, SIRT2 was reported to block the entry to chromosome condensation in glioma cell lines through nucleo-cytoplasmic shuttling in response to mitotic stress (Inoue et al. 2007). Thus, SIRT2 inhibitors may have chemotherapeutic properties for selective cytotoxicity.

Sirtinol induced senescence-like growth arrest in human breast cancer cells (MCF-7) and lung cancer cells (H1299), and this growth arrest was accompanied by attenuated responses to growth factors (Ota et al. 2006). Splitomicin also inhibited proliferation, but at higher concentrations. Furthermore, treatment of the human prostate cancer cell line PC3 with sirtinol inhibited cell growth and increased sensitivity to camptothecin and cisplatin, two well-known anti-cancer drugs (Mai et al. 2005).

Cambinol has also shown potent anti-cancer activity against Burkitt lymphoma cell lines by a mechanism involving BCL6 acetylation (Heltweg et al. 2006). Importantly, it also displayed less toxicity towards primary cells, as cambinol doses of 100 mg/kg were well tolerated in Balb-c mice and reduced tumor growth without inducing obvious toxicity to animals (Heltweg et al. 2006).

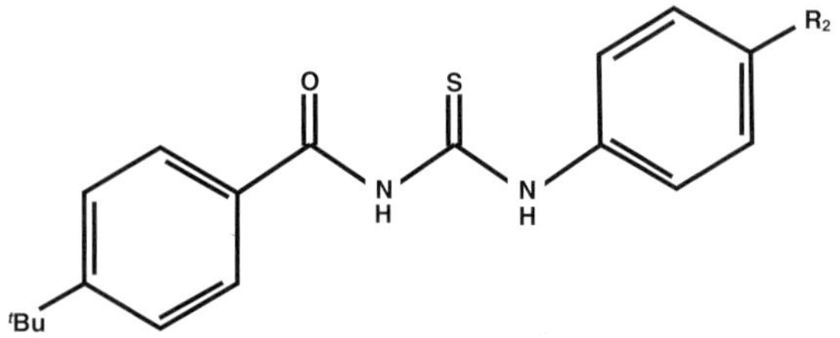

Fig. 10.6 Tenovins

	R_2
Tenovin-1	NHCOCH$_3$
Tenovin-6	NHCO(CH$_2$)$_4$NMe$_2$ · HCL

One sirtuin inhibitor traces its very discovery to its anti-cancer utility. Using a cell-based screen for small molecules able to activate p53 and decrease tumor growth, Lain et al. (2008) found two SIRT1 inhibitors, tenovin-1 and its more water-soluble analog tenovin-6 (Fig. 10.6). These both decrease tumor growth in vitro at one-digit micromolar concentrations and delay tumor growth in vivo without significant general toxicity (Lain et al. 2008).

Salermide was well tolerated by mice at concentrations up to 100 μM and induced p53-independent apoptosis in cancer cells but not normal cells (Lara et al. 2009). Instead, salermide reactivated proapoptotic genes epigenetically repressed exclusively in cancer cells by SIRT1-mediated deacetylation of H4K16, providing further clarification of one molecular mechanism by which SIRT1 may exert its oncogenic effects in cancer. The reduction in the number of cells was dependent on the cell type and was not mediated by SIRT2.

AC-93253 (Fig. 10.7), a new selective SIRT2 inhibitor with an IC$_{50}$ value of 6.0 μM, exhibited potent selective cytotoxicity with IC$_{50}$ values ranging from 10 to 100 nM in four different cancer cell lines derived from prostate (DU145), pancreas (MiaPaCa2), and lung (A549 and NCI-H460). AC-93253 was dramatically less toxic in normal human endothelial and epithelial cell lines. AC-93253 presents a chemical scaffold distinct from small molecules identified to date, and significantly enhanced acetylation of tubulin, p53, and histone H4 (Zhang et al. 2009).

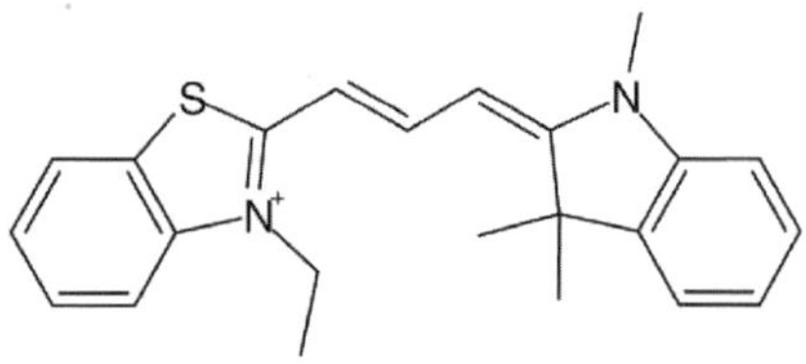

Fig. 10.7 AC-93253

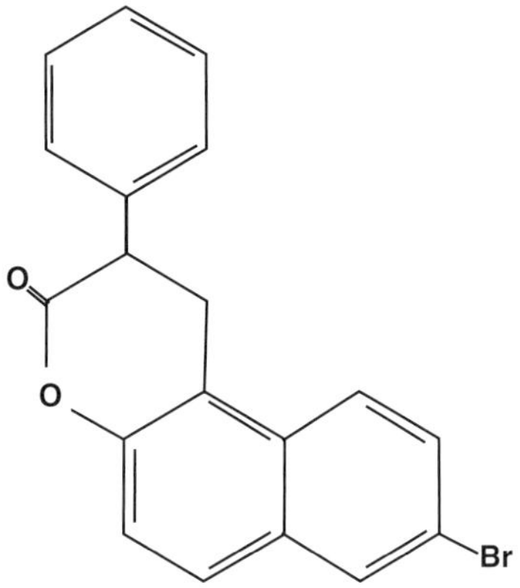

Fig. 10.8 HR73

Finally, suramin posseses antitumor effects and is also a potent reverse transcriptase inhibitor (Perabo and Müller 2005).

10.5.2 Sirtuin Inhibitors in Other Diseases

Pagans et al. (2005) have identified a splitomicin derivative, HR73 (Fig. 10.8), that inhibits SIRT1 activity in vitro with an IC_{50} of less than 5 µM. HR73 treatment results in the inhibition of Tat, an essential protein for the transcriptional activation of the integrated human immunodeficiency (HIV) provirus. These results implicate SIRT1 as a novel therapeutic target for HIV infection, and methods of inhibiting SIRT1 Tat deacetylase activity could be useful for treating human immunodeficiency virus infections.

Splitomicin-mediated inhibition of SIRT1 activity attenuates gene silencing in Fragile X mental retardation syndrome by silencing of FMR1 gene in the cells of Fragile X patients acting downstream of DNA methylation (Biacsi et al. 2008).

Sirtinol administration has been shown to attenuate hepatic injury following trauma-hemorrhage (Liu et al. 2008a) and to protect nematode and cell models of muscular dystrophy (Catoire et al. 2008). Vergnes et al. (2005) have also demonstrated that sirtinol induced apoptosis in *Leishmania infantum*, inhibiting significantly the in vitro proliferation of this axenic amastigote, suggesting a therapy for leishmaniosis that may be especially beneficial to immunodepressed patients.

Finally, AGK2, a potent and selective SIRT2 inhibitor with a calculated IC_{50} of 3.5 µM, reduced α-synuclein-mediated toxicity in a dose-dependent manner in α-synuclein-transfected H4 neuroglioma cells (Outeiro et al., 2007).

10.6 Sirtuin Modulator Compounds: Activators

On the basis of the positive health effects of CR in mammals and the associated increases in SIRT1, the discovery and development of SIRT1-activating drugs holds considerable interest because of the potential to address a broad range of diseases

Fig. 10.9 Butein and piceatannol

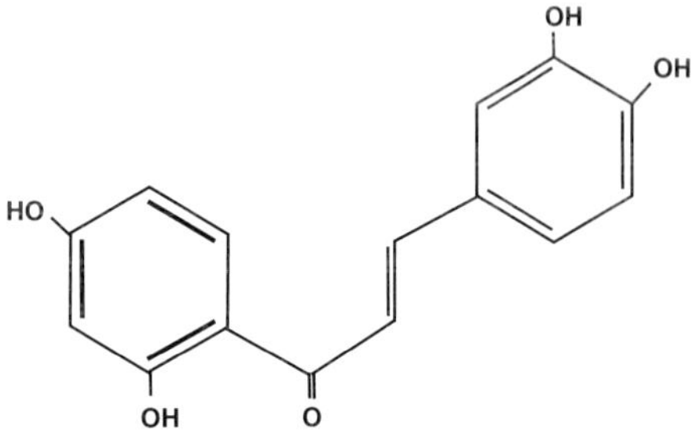

Butein (3,4,2',4'-tetrahydroxychalcone)

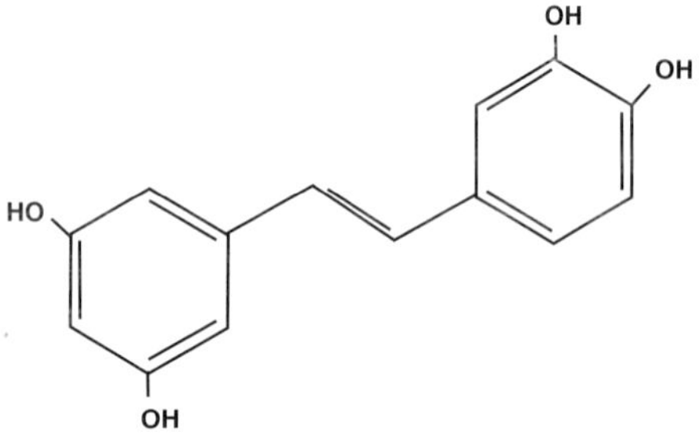

Piceatannol (3,5,3',4'-tetrahydroxy-*trans*-stilbene)

(Elliott and Jirousek 2008). Howitz et al. (2003) screened small molecules for in vitro SIRT1 activation and identified several polyphenols able to stimulate SIRT1 acitivity many fold. Of them, two structurally-similar compounds, quercetin and piceatannol (Fig. 10.9), stimulated SIRT1 activity five- and eight-fold, respectively.

Another polyphenol, resveratrol (RSV) (Fig. 10.10), containing two phenol rings connected by a 2-carbon methylene bridge, increased SIRT1 activity more than 13-fold. RSV has since been shown to stimulate the activity of native sirtuins in vivo and was the first molecule shown to mimic CR through sirtuin activation (Howitz et al. 2003; Wood et al. 2004). One of the most abundant natural sources of RSV is *Vitis vinifera*, commonly used to make wine, and the skin holds the highest concentration (Celotti et al. 1996). RSV has been hypothesized to bind the non-catalytic N-terminus of SIRT1 to induce a conformational change that lowers the Michaelis constant for both the acetylated substrate and NAD^+. In yeast, resveratrol mimics CR by increasing DNA stability and extending lifespan by 70% (Howitz et al. 2003). Among mammalian sirtuins, only SIRT1 is responsive to RSV (Borra et al. 2005).

The nature of RSV's action on SIRT1 in vivo remains controversial. RSV needs to reach concentrations of 10–200 μM to activate SIRT1, but in humans oral RSV shows low bioavailability such that peak plasma levels of RSV and metabolites only reach 491±90 ng/ml (about 2 μM) after an oral dose of 25 mg, with a plasma half-life of 9.2±0.6 h (Walle et al. 2004). Even though RSV is reported to be safe and well-tolerated even at doses up to 5 g/day for 7 consecutive days (Elliott and

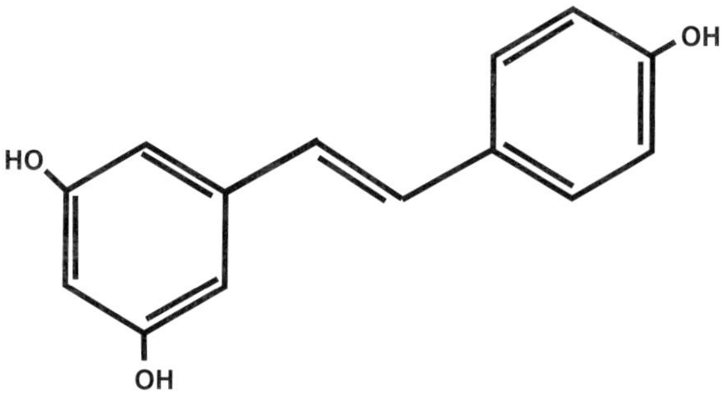

Fig. 10.10 Resveratrol (3,5, 4′-trihydroxy-*trans*-stilbene)

Jirousek 2008), the consumption of high doses of resveratrol might still be insufficient to reach systemic levels that are able to activate SIRT1 for extended periods. Thus, the development of new molecules that activate sirtuins at low concentrations, with great specificity and bioavailability, is a promising field of medical chemistry.

Nayagam et al. (2006) have screened 147,000 compounds and have found 3 quinoxaline-based compounds that are as potent as RSV at activating SIRT1. These compounds were potent lipolytic agents and also showed anti-inflammatory properties in vitro, but their utility in animal models has been not reported yet (Nayagam et al. 2006).

Milne et al. (2007) have also identified activators of human SIRT1 using a high-throughput screen that are many times more effective than RSV. SRT1720 (Fig. 10.11) is the most potent, with an $EC_{1.5}=0.16$ μM and 781% maximal SIRT1

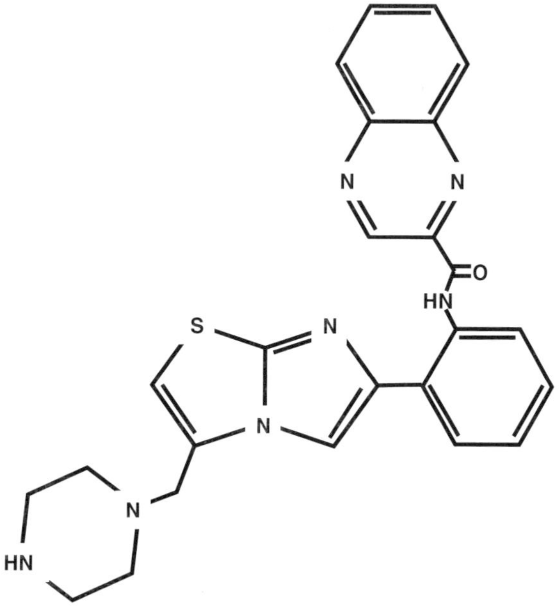

Fig. 10.11 SRT1720

activation. SRT1720 binds to the SIRT1 enzyme–peptide substrate complex at an allosteric site aminoterminal to the catalytic domain, lowering the Michaelis constant for acetylated substrates and activating SIRT1 at the same molecular site as RSV (Milne et al. 2007). SRT1720 exhibited a pharmacokinetic profile suitable for in vivo evaluation in mouse models (bioavailability = 50%, terminal $t_{1/2}$ ~ 5 h, AUC = 7,892 $ngh^{-1}ml^{-1}$) (Milne et al. 2008). Other compounds, like SRT1460 and SRT2183, had similar but less pronounced effects.

10.6.1 Cardioprotective Effect

The CR mimetic effects of SIRT1 could explain the so-called French Paradox, which is that the wine-drinking French have a low incidence of coronary heart disease despite low levels of exercise and a high fat diet (Das and Das 2007). Additionally, expression of SIRT1 has been shown to be reduced in diabetic obese *db/db* mouse heart, correlating with contractile dysfunction (Dong and Ren 2007). Thus, activators of SIRT1 may mediate a powerful cardioprotective effect on organisms.

The most studied sirtuin activator is RSV. In a rat model of cigarette smoking, artery stiffness, increased reactive oxygen species production and upregulated inflammatory markers were abrogated by both RSV treatment and SIRT1 overexpression (Csiszar et al. 2008). RSV also induced mitochondrial biogenesis in cultured human coronary arterial endothelial cells as well as in the aortas of diabetic mice, and knockdown of SIRT1 prevented the RSV-induced mitochondrial biogenesis (Csiszar et al. 2009).

One future target for cardioprotective sirtuin therapies is SIRT3. SIRT3 is a stress-responsive factor in cardiomyocytes, and its overexpression protects cells against stress-mediated cell death through promoting Ku70/Bax interaction by deacetylating Ku70 and thereby making cells resistant to Bax-mediated cell damage (Sundaresan et al., 2008).

10.6.2 Neuroprotective Effect

Age is a major risk factor for a variety of neurodegenerative disorders. Diseases like Alzheimer's disease (AD), Parkinson's disease (PD), Huntington's disease (HD) and amyotrophic lateral sclerosis (ALS) are increasingly prevalent in aging societies and have become major medical, economical and social challenges to modern societies. Although neurodegenerative disorders are cell type specific, many of their underlying pathogenic processes are similar. These include protein misfolding, oligomerization and aggregation, proteolysis, mitochondrial dysfunction, abnormal energy metabolism, stress responses, inflammation, and pro-apoptotic responses, among others (Outeiro et al. 2008).

In animal models of AD, PD and HD and stroke, CR can protect neurons against degeneration. CR has also been shown to enhance synaptic plasticity and restore function following injury (Mattson et al. 2003). In the Tg2576 transgenic mouse strain, a model that shows AD-type amyloid neuropathology, CR increased SIRT1 expression levels in the brains of 10-month-old mice, and SIRT1 promoted α-secretase activity and attenuated Aβ peptide generation in primary Tg2576 neuron cultures. Overexpression of SIRT1 or pharmacological activation of SIRT1 by NAD^+ also promotes α-secretase activity and attenuates the generation of αβ peptides in embryonic Tg2576 mouse neurons in vitro (Zhang et al. 2002; Qin et al. 2006). Treatment with RSV prevented neurodegeneration and improved associative learning in transgenic mouse models of AD and ALS (Kim et al. 2007b). Novel therapeutic strategies directed to increase the supply of NAD^+ or directly influence SIRT1 activation may be effective for treatment of diseases characterized by axonopathy and neurodegeneration (Araki et al. 2004; Qin et al. 2006).

10.6.3 Metabolic Diseases

Obesity has reached epidemic proportions globally and is a major contributor to chronic disease and disability. Metabolic syndrome comprises a cluster of risk factors for cardiovascular disease and insulin resistance associated with obesity, and is clearly a contributing factor to type 2 diabetes mellitus disease progression. Both congenital and environmental factors, such as exercise and eating habits, contribute to the disease. Many individuals in developed countries consume a hypercaloric, micronutrient- and phytonutrient-sparse diet that contributes to their risk for metabolic disorders. On the other hand, moderate wine consumption is associated with the prevention of metabolic syndrome in numerous epidemiological studies on diverse ethnic groups, and may be in part due to the presence of RSV in red wine (Liu et al. 2008b).

Diseases of metabolism can often be framed in a context of general mitochondrial malfunction and increased reactive oxygen species production. Deacetylation of the master regulator of mitochondrial biogenesis, PGC-1α, by SIRT1 leads to increases in mitochondrial size and number and induces hepatic glucose output, providing new and promising therapies for metabolic diseases (Rodgers et al. 2005). While obesity and high saturated fatty acid intakes decrease PGC-1α and mitochondrial gene expression (Crunkhorn et al. 2007), RSV has been shown to increase both PGC-1α activity and mitochondrial number in mice fed a high fat diet (Baur et al. 2006).

Regarding glucose homeostasis, induction of SIRT1 activity with SRT1720 or RSV has been shown to reduce acetylated CREB-regulated transcription coactivator 2 (CRC2), lowering circulating blood glucose concentrations in part through inhibition of hepatic gluconeogenesis (Liu et al. 2008c). SIRT1 activity also represses PPAR-γ, thereby promoting fat mobilization, increasing insulin sensitivity, positively regulating insulin secretion from pancreatic β cells, and elevating glucose uptake in muscle cells by AMP-activated protein kinase (AMPK) activation, thus further linking SIRT1 to the control of metabolic processes such as metabolic syndrome and insulin resistance (Picard et al. 2004; Bordone et al. 2006; Breen et al. 2008). As CR reverses insulin resistance and decreases visceral fat (Barzilai et al. 1998), activation of SIRT1 by RSV or other compounds that mimic the transcriptional effects of CR could attenuate some pathological consequences of high-calorie diets, thereby improving the overall health of the population.

The most important therapeutic intervention in the metabolic syndrome is lifestyle change, with a focus on weight reduction and regular physical activity. Another strategy to overcome the metabolic syndrome is the development of dietary supplements that mimic the transcriptional effects of CR to attenuate at least some of the pathological consequences of high-calorie diets.

Mice fed a diet that induces obesity but that was supplemented with high doses of RSV (200 or 400 mg/kg/day) tended to gain less weight, showed improved insulin sensitivity, had increased muscle aerobic capacity, and had improved motor function compared to controls. The plasma level of RSV was dose-related and ranged from 10–120 ng/ml (Lagouge et al. 2006).

In a long-term study, mice treated at middle-age to the end-of-life with RSV showed an overall health improvement, including improved bone health, a reduction in cataracts and cardiovascular dysfunction, and improved balance and motor coordination. RSV induced gene expression patterns in multiple tissues that paralleled those induced by CR, supporting the idea that RSV can mimic many effects of CR in vivo. Interestingly, the study also found a significant increase in lifespan in both the resveratrol-treated group on a high-calorie diet and the resveratrol-treated group on a calorie restriction diet, but the treatments did not extend lifespan of mice on a standard diet when started at one year of age (Pearson et al. 2008).

Mice treated with SRT1720, a specific and potent synthetic activator of SIRT1, were totally spared from diet-induced obesity by inhibition of fat accumulation rather than because of altered feeding behavior or increased locomotor activity (Feige et al. 2008). SRT1720 also improved their glucose tolerance and insulin sensitivity and enhanced endurance (Feige et al. 2008). RSV supplemented in a commercial formulation called SRT501 and SRT1720 recapitulated a molecular signature that overlaps with that of CR, such as enhanced mitochondrial biogenesis, improved metabolic signaling pathways, and blunted pro-inflammatory pathways in mice fed a high calorie diet (Smith et al. 2009).

Despite the reports of low RSV bioavailability (see above), Prolla and colleagues found that dietary RSV was able to activate SIRT1 in a sustainable way, even at a much lower dose than suspected (4.9 mg/kg/day), mimicking the effects of CR in insulin-mediated glucose uptake in muscle and increasing the expression of genes that protect the heart from aging processes (Barger et al. 2008a, b). Moreover, the expression of genes related to glucose, lipid metabolism, and oxidative phosphorylation was shifted to a pattern similar to long-term CR (Barger et al. 2008a, b). Since RSV undergoes extensive biotransformation during transepithelial transport and only trace amounts of unchanged RSV are detected in plasma (Walle et al. 2004) an open question is if RSV itself can accumulate to bioactive levels in target organs or whether its effects are mediated by metabolites. In support of this, it has been shown that RSV derivatives extend lifespan in yeast (Yang et al. 2007).

10.7 Conclusion

The mammalian sirtuins, because they hold sway over such a wide range of cellular targets and locations, are a family of targets that could potentially be harnessed to regulate a wide range of metabolic and age-associated diseases. While further research is needed to explore the potential of the different enzymes, a growing body of research on SIRT1 suggests that its modulation can lead to multiple improvements in animal models, including reduced cancer proliferation, enhanced glucose utilization, increased mitochondrial biogenesis, and potentially even increased lifespan. For that, RSV and the new sirtuin-modulating compounds have been proposed for treating and/or preventing a wide variety of diseases and disorders including diseases or disorders related to aging or stress, diabetes, obesity, neurodegenerative diseases, cardiovascular disease, blood clotting disorders, inflammation, and cancer, as well as diseases or disorders that would benefit from increased mitochondrial activity. Future research should be directed to better understand the relevance of sirtuin activities to human conditions, and whether modulation of sirtuin activity (whether repressed or enhanced) is safe and effective against human disorders and age related diseases.

References

Adams JD Jr, Klaidman LK (2007) Sirtuins, nicotinamide and aging: a critical review. Lett Drug Design Disc 4:44–48

Ahuja N, Schwer B, Carobbio S et al. (2007) Regulation of insulin secretion by SIRT4, a mitochondrial ADP-ribosyltransferase. J Biol Chem 282:33583–33592

Araki T, Sasaki Y, Milbrandt J (2004) Increased nuclear NAD biosynthesis and SIRT1 activation prevent axonal degeneration. Science 305:1010–1013

Ashrafn N, Zino S, MacIntyre A et al. (2006) Altered sirtuin expression is associated with node-positive breast cancer. Br J Cancer 95:1056–1061

Barger JL, Kayo T, Pugh TD et al. (2008a) Short-term consumption of a resveratrol-containing nutraceutical mixture mimics gene expression of long-term caloric restriction in mouse heart. Exp Gerontol 43:859–866

Barger JL, Kayo T, Vann JM et al. (2008b) A low dose of dietary resveratrol partially mimics caloric restriction and retards aging parameters in mice. PLoS ONE 3:e2264

Barzilai N, Banerjee S, Hawkins M et al. (1998) Caloric restriction reverses hepatic insulin resistance in aging rats by decreasing visceral fat. J Clin Invest 101:1353–1361

Baur JA, Pearson KJ, Price NL et al. (2006) Resveratrol improves health and survival of mice on a high calorie diet. Nature 444:337–342

Bedalov A, Gatbonton T, Irvine WP et al. (2001) Identification of a small molecule inhibitor of Sir2p. Proc Natl Acad Sci USA 98:15113–15118

Bellizzi D, Rose G, Cavalcante P et al. (2005) A novel VNTR enhancer within the SIRT3 gene, a human homologue of SIR2, is associated with survival at oldest ages. Genomics 85:258–263

Biacsi R, Kumari D, Usdin K (2008) SIRT1 inhibition alleviates gene silencing in Fragile X mental retardation syndrome. PLoS Genet 4:e1000017

Bordone L, Motta MC, Picard F et al. (2006) Sirt1 regulates insulin secretion by repressing UCP2 in pancreatic β-cells. PLoS Biol. 4:210–220

Borra, MT, Smith BC, Denu JM (2005) Mechanism of human SIRT1 activation by resveratrol. J Biol Chem 280:17187–17195

Bouras, T, Fu, M, Sauve, AA et al. (2005) SIRT1 deacetylation and repression of p300 involves lysine residues 1020/1024 within the cell cycle regulatory domain 1. J Biol Chem 280: 10264–10276

Breen DM, Sanli T, Giacca A, Tsiani E (2008) Stimulation of muscle cell glucose uptake by resveratrol through sirtuins and AMPK. Biochem Biophys Res Commun 374:117–122

Brittain D, Weinmann H, Ottow E (2007) Recent advances in the medicinal chemistry of hystone deacetylase inhibitors. Ann Rep Med Chem 42:337–348

Brunet A, Sweeney LB, Sturgill JF et al. (2004) Stress-dependent regulation of FOXO transcription factors by the SIRT1 deacetylase. Science 303:2011–2015

Catoire H, Pasco MY, Abu-Baker A et al. (2008) Sirtuin inhibition protects from the polyalanine muscular dystrophy protein PABPN1. Hum Mol Genet 17:2108–2117

Celotti E, Ferrarini R, Zironi R, Conte, LS (1996) Resveratrol content of some wines obtained from dried Valpolicella grapes: Recioto and Amarone. J Chromatogr A 730:47–52

Chen D, Bruno J, Easlon E et al. (2008) Tissue-specific regulation of SIRT1 by calorie restriction. Genes Dev 22:1753–1757

Civitarese AE, Carling S, Heilbronn LK et al. (2007) Calorie restriction increases muscle mitochondrial biogenesis in healthy humans. PLoS Med 4:484–494

Cohen HY, Miller C, Bitterman KJ et al. (2004) Calorie restriction promotes mammalian cell survival by inducing the SIRT1 deacetylase. Science 305:390–392

Crujeiras AB Parra D, Goyenechea E, Martínez JA (2008) Sirtuin gene expression in human mononuclearcells is modulated by caloric restriction. Eur J Clin Invest 38: 672–678

Crunkhorn S, Dearie F, Mantzoros C et al. (2007) Peroxisome proliferator activator receptor gamma coactivator-1 expression is reduced in obesity: potential pathogenic role of saturated fatty acids and p38 mitogen-activated protein kinase activation. J Biol Chem 282:15439–15450

Csiszar A, Labinskyy N, Pinto JT et al. (2009) Resveratrol induces mitochondrial biogenesis in endothelial cells. Am J Physiol Heart Circ Physiol 297:H13–20

Csiszar A, Labinskyy N, Podlutsky A et al. (2008) Vasoprotective effects of resveratrol and SIRT1: attenuation of cigarette smoke-induced oxidative stress and proinflammatory phenotypic alterations. Am J Physiol Heart Circ Physiol 294:H2721-H2735

Das S, Das DK (2007) Resveratrol: a therapeutic promise for cardiovascular diseases. Recent Patents Cardiovasc Drug Discov 2:133–138

Denu JM (2005) Vitamin B3 and sirtuin function. Trends Biochem Sci 30:479–483

Dong F, Ren J (2007) Fidarestat improves cardiomyocyte contractile function in db/db diabetic obese mice through a histone deacetylase Sir2-dependent mechanism. J Hypertens 25: 2138–2147

Elliott PJ, Jirousek M (2008) Sirtuins: Novel targets for metabolic disease. Curr Opin Inves Drugs 9:371–378

Feige JN, Lagouge M, Canto C et al. (2008) Specific SIRT1 activation mimics low energy levels and protects against diet-induced metabolic disorders by enhancing fat oxidation. Cell Metab 8:347–358

Finnin MS, Donigian JR, Pavletich NP (2001) Structure of the histone deacetylase SIRT2. Nat Struct Biol 8:621–625

Firestein R, Blander G, Michan S et al. (2008) The SIRT1 deacetylase suppresses intestinal tumorigenesis and colon cancer growth. PLoS ONE 3:e2020

Ford E, Voit R, Liszt G et al. (2006) Mammalian Sir2 homolog SIRT7 is an activator of RNA polymerase I transcription. Genes Dev 20:1075–1080

Frye RA (2000) Phylogenetic classification of prokaryotic and eukaryotic Sir2-like proteins. Biochphys Res Comunn 273:793–798

Gan L (2007) Therapeutic potencial of sirtuin-activating compounds in alzheimer's disease. Drug News Perspect 20:233–239

Gray SG, Ekström TJ (2001) The human histone deacetylase family. Exp Cell Res 262:75–83

Grob A, Roussel P, Wright JE et al. (2009) Involvement of SIRT7 in resumption of rDNA transcription at the exit from mitosis. J Cell Sci 122:489–498

Grozinger CM, Chao ED, Blackwell HE et al. (2001) Identification of a class of small molecule inhibitors of the sirtuin family of NAD-dependent deacetylases by phenotypic screening. J Biol Chem 276:38837–38843

Haberland M, Montgomery RL, Olson EN (2009) The many roles of histone deacetylases in development and physiology: implications for disease and therapy. Nat Rev Genet 10 32–42

Haigis MC, Mostoslavsky R, Haigis KM et al. (2006) SIRT4 inhibits glutamate dehydrogenase and opposes the effects of calorie restriction in pancreatic beta cells. Cell 126:941–954

Hallows WC, Lee S, Denu JM (2006) Sirtuins deacetylate and activate mammalian acetyl-CoA synthetases. Proc Natl Acad Sci USA 103:10230–10235

Heltweg B, Gatbonton T, Schuler AD et al. (2006) Antitumor activity of a small-molecule inhibitor of human silent information regulator 2 enzymes. Cancer Res 66:4368–4377

Hirao M, Posakony J, Nelson M et al. (2003) Identification of selective inhibitors of NAD+-dependent deacetylases using phenotypic screens in yeast. J Biol Chem 278:52773–52782

Hiratsuka M, Inoue T, Toda T et al. (2003) Proteomics-based identification of differentially expressed genes in human gliomas: down-regulation of SIRT2 gene. Biochem Biophys Res Commun 309:558–566

Howitz KT, Bitterman KJ, Cohen HY et al. (2003) Small molecule activators of sirtuins extend Saccharomyces cerevisiae lifespan. Nature 425:191–196

Inoue T, Hiratsuka M, Osaki M et al. (2007) SIRT2, a tubulin deacetylase, acts to block the entry to chromosome condensation in response to mitotic stress. Oncogene 26:945–957

Kaeberlein M (2008) The ongoing saga of sirtuins and aging. Cell Metab 8:4–5

Kanfi Y, Shalman R, Peshti V et al. (2008) Regulation of SIRT6 protein levels by nutrient availability. FEBS Lett 582:543–548

Kawahara TL, Michishita E, Adler AS et al. (2009) SIRT6 links histone H3 lysine 9 deacetylation to NF-κB dependent gene expression and organismal lifespan. Cell 136:62–74

Kim EJ, Kho JH, Kang MR, Um SJ (2007a) Active regulator of SIRT1 cooperates with SIRT1 and facilitates suppression of p53 activity. Mol Cell 2:277–290

Kim D, Nguyen MD, Dobbin MM et al. (2007b) SIRT1 deacetylase protects against neurodegeneration in models for Alzheimer's disease and amyotrophic lateral sclerosis. EMBO J 26:3169–3179

Lagouge M, Argmann C, Gerhart-Hines Z et al. (2006) Resveratrol improves mitochondrial function and protects against metabolic disease by activating SIRT1 and PGC-1α. Cell 127:1109–1122

Lain S, Hollick JJ, Campbell J et al. (2008) Discovery, in vivo activity, and mechanism of action of a small-molecule p53 activator. Cancer Cell 13:454–463

Lara E, Calvanese V, Altucci L et al. (2009) Salermide, a Sirtuin inhibitor with a strong cancer-specific proapoptotic effect. Oncogene 28:781–791

Li W, Zhang B, Tang J et al. (2007) Sirtuin 2, a mammalian homolog of yeast silent information regulator-2 longevity regulator, is an oligodendroglial protein that decelerates cell differentiation through deacetylating α-tubulin. J Neurosci 27:2606–2616

Lim CS. (2006) SIRT1: tumor promoter or tumor suppressor? Med Hypotheses 67: 341–344

Lin SJ, Ford E, Haigis M et al. (2004) Calorie restriction extends yeast life span by lowering the level of NADH. Genes Dev 18:12–16

Liszt G, Ford E, Kurtev M, Guarente L. (2005) Mouse Sir2 homolog SIRT6 is a nuclear ADP-ribosyltransferase. J Biol Chem 280:21313–21320

Liu FC, Day YJ, Liou JT et al. (2008a) Sirtinol attenuates hepatic injury and pro-inflammatory cytokine production following trauma-hemorrhage in male Sprague-Dawley rats. Acta Anaesthesiol Scand 52:635–640

Liu L, Wang Y, Lam KS, Xu A (2008b) Moderate wine consumption in the prevention of metabolic syndrome and its related medical complications. Endocr Metab Immune Disord Drug Targets 8:89–98

Liu Y, Dentin R, Chen D et al. (2008c) A fasting inducible switch modulates gluconeogenesis via activator/coactivator exchange. Nature 456:269–273

Lombard DB, Alt FW, Cheng HL et al. (2007) Mammalian Sir2 Homolog SIRT3 Regulates Global Mitochondrial Lysine Acetylation. Mol Cell Biol 27:8807–8814

Mai A, Massa S, Lavu S, et al. (2005) Design, synthesis, and biological evaluation of sirtinol analogues as class III histone/protein deacetylase (Sirtuin) inhibitors. *J Med Chem* 48: 7789–7795

Mattson MP, Duan W, Guo Z (2003) Meal size and frequency affect neuronal plasticity and vulnerability to disease: cellular and molecular mechanisms. J Neurochem 84:417–431

Medda F, Russell RJ, Higgins M et al. (2009) Novel cambinol analogs as sirtuin inhibitors: synthesis, biological evaluation, and rationalization of activity. J Med Chem 52:2673–2682

Michan S, Sinclair D (2007) Sirtuins in mammals: insights into their biological function. Biochem J 404:1–13

Michishita E, McCord RA, Berber E (2008) SIRT6 is a histone H3 lysine 9 deacetylase that modulates telomeric chromatin. Nature 452:492–496

Michishita E, Park JY, Burneskis JM et al. (2005) Evolutionarily conserved and nonconserved cellular localizations and functions of human SIRT proteins. Mol Biol Cell 16:4623–4635

Milne JC, Lambert PD, Schenk S et al. (2007) Small molecule activators of SIRT1 as therapeutics for the treatment of type 2 diabetes. Nature 50:712–716

Milne JC, Denu JM (2008) The Sirtuin family: therapeutic targets to treat diseases of aging. Curr Opin Chem Biol 12:11–17

Milne JC, Lambert PD, Smith JJ et al. (2008) Activation of the protein deacetylase SIRT1 blunts pro-inflammatory pathways in vivo. In: Abstracts of the American Diabetes Association's 68th Annual Scientific Sessions, San Francisco, CA, June 2008

Min J, Landry J, Sternglanz R, Xu RM (2001) Crystal structure of a SIR2 homolog-NAD complex. Cell 105:269–279

Mostoslavsky R, Chua KF, Lombard DB et al. (2006) Genomic instability and aging-like phenotype in the absence of mammalian SIRT6. Cell 124:315–329

Nakamura Y, Ogura M, Tanaka D, Inagaki N (2008) Localization of mouse mitochondrial SIRT proteins: shift of SIRT3 to nucleus by co-expression with SIRT5. Biochem Biophys Res Commun 366:174–179

Nayagam VM, Wang X, Tan YC et al. (2006) SIRT1 modulating compounds from high-throughput screening as anti-inflammatory and insulin-sensitizing agents. J Biomol Screen 11:959–967

Nemoto S, Fergusson MM, Finkel T (2005) SIRT1 functionally interacts with the metabolic regulator and transcriptional coactivator PGC-1 α. J Biol Chem 280:16456–16460

Neugebauer RC, Uchiechowska U, Meier R et al. (2008) Structure-activity studies on splitomicin derivatives as sirtuin inhibitors and computational prediction of binding mode. J Med Chem 51:1203–1213

North BJ, Marshall BL, Borra MT et al. (2003) The human Sir2 ortholog, SIRT2, is an NAD^+-dependent tubulin deacetylase. Mol Cell 11:437–444

North BJ, Verdin E (2004) Sirtuins: Sir2-related NAD-dependent protein deacetylases. Genome Biol. 5:224

Onyango, P. Celic, I McCaffery, JM et al. (2002) SIRT3, a human SIR2 homologue, is an NAD-dependent deacetylase localized to mitochondria. Proc Natl Acad Sci USA 99: 13653–13658

Ota H, Tokunaga E, Chang K et al. (2006) Sirt1 inhibitor, Sirtinol, induces senescence-like growth arrest with attenuated Ras-MAPK signaling in human cancer cells. Oncogene 25:176–185

Outeiro TF, Kontopoulos E, Altmann SM et al. (2007) Sirtuin 2 inhibitors rescue alpha-synuclein-mediated toxicity in models of Parkinson's disease. Science 317:516–519

Outeiro TF, Marques O, Kazantsev A (2008) Therapeutic role of sirtuins in neurodegenerative disease. Biochim Biophys Acta 1782:363–369

Pagans S, Pedal A, North BJ et al. (2005) SIRT1 regulates HIV transcription via Tat deacetylation. PLoS Biol 3:e41

Pearson KJ, Baur JA, Lewis KN et al. (2008) Resveratrol delays age-related deterioration and mimics transcriptional aspects of dietary restriction without extending life span. Cell Metab 8:157–168

Perabo FG, Müller SC (2005) New agents in intravesical chemotherapy of superficial bladder cancer. Scand J Urol Nephrol 39:108–116

Picard F, Kurtev M, Chung N et al. (2004) Sirt1 promotes fat mobilization in white adipocytes by repressing PPAR-gamma. Nature 429:771–776

Posakony J, Hirao M, Stevens S et al. (2004) Inhibitors of Sir2: evaluation of splitomicin analogues. J Med Chem 47:2635–2644

Qin W, Yang T, Ho L et al. (2006) Neuronal SIRT1 activation as a novel mechanism underlying the prevention of Alzheimer disease amyloid neuropathology by calorie restriction. J Biol Chem 281:21745–21754

Reiling E, van Vliet-Ostaptchouk JV, van't Riet E et al. (2009) Genetic association analysis of 13 nuclear-encoded mitochondrial candidate genes with type II diabetes mellitus: the DAMAGE study. Eur J Hum Genet doi: 10.1038/ejhg.2009.4

Rodgers JT, Lerin C, Haas WG. et al. (2005) Nutrient control of glucose homeostasis through a complex of PGC-1α and SIRT1. Nature 434:113–118

Rose G, Dato S, Altomare K et al. (2003) Variability of the SIRT3 gene, human silent information regulator Sir2 homologue, and survivorship in the elderly. Exp Gerontol 38:1065–1070

Sauve AA, Moir RD, Schramm VL, Willis IM (2005) Chemical activation of Sir2-dependent silencing by relief of nicotinamide inhibition. Mol Cell 17:595–601

Scher MB, Vaquero A, Reinberg D (2007) SirT3 is a nuclear NAD^+-dependent histone deacetylase that translocates to the mitochondria upon cellular stress. Genes Dev 21:920–928

Schlicker C, Gertz M, Papatheodorou P et al. (2008) Substrates and regulation mechanisms for the human mitochondrial sirtuins sirt3 and sirt5. J Mol Biol 382:790–801

Schuetz A, Min J, Antoshenko T et al. (2007) Structural basis of inhibition of the human NAD^+-dependent deacetylase SIRT5 by suramin. Structure 15:377–389

Schwer B, Verdin E (2008) Conserved metabolic regulatory functions of sirtuins. Cell Metab 7:104–112

Shi T, Wang F Stieren E, Tong Q (2005) SIRT3, a mitochondrial sirtuin deacetylase, regulates mitochondrial function and thermogenesis in brown adipocytes. J Chem Biol 280:13560–13567

Smith JS (2002) Human Sir2 and the "silencing" p53 activity. Trends Cell Biol 12:404–406

Smith JS, Avalos J, Celic I et al. (2002) SIR2 family of NAD^+-dependent. Methods Enzymol 353:282–300

Smith JJ, Kenney RD, Gagne DJ (2009) Small molecule activators of SIRT1 replicate signaling pathways triggered by calorie restriction in vivo. BMC Syst Biol 10:3–31

Stünkel W, Peh BK, Tan YC et al. (2007) Function of the SIRT1 protein deacetylase in cancer. Biotechnol J 2:1360–1368

Sun Y, Sun D, Li F et al. (2007) Downregulation of Sirt1 by antisense oligonucleotides induces apoptosis and enhances radiation sensitization in A549 lung cancer cells. Lung Cancer 58:21–29

Sundaresan NR, Samant SA, Pillai VB et al. (2008) SIRT3 is a stress responsive deacetylase in cardiomyocytes that protects cells from stress-mediated cell death by deacetylation of Ku-70. Mol Cell Biol 28:6384–6401

Suzuki K, Koike T (2007) Mammalian Sir2-related protein (SIRT) 2-mediated modulation of resistance to axonal degeneration in slow Wallerian degeneration mice: a crucial role of tubulin deacetylation. Neuroscience 147:599–612

Taunton J, Hassig CA, Schreiber SL (1996) A mammalian histone deacetylase related to the yeast transcriptional regulator Rpd3p. Science 272:408–411

Trapp J, Meier R, Hongwiset D et al. (2007) Structure-activity studies on suramin analogues as inhibitors of NAD+-dependent histone deacetylases (sirtuins). ChemMedChem 2:1419–1431

Vakhrusheva O, Smolka C, Gajawada P et al. (2008) Sirt7 increases stress resistance of cardiomyocytes and prevents apoptosis and inflammatory cardiomyopathy in mice. Circ Res 102:703–10

Vaquero A, Scher M, Lee D et al. (2004) Human SirT1 interacts with histone H1 and promotes formation of facultative heterochromatin. Mol Cell 16:93–105

Vaziri H, Dessain SK, Eaton EN et al. (2001) hSIR2SIRT1 Functions as an NAD-Dependent p53 Deacetylase. Cell 107:149–159

Vergnes B, Vanhille L, Ouaissi A, Sereno D (2005) Stage-specific antileishmanial activity of an inhibitor of SIR2 histone deacetylase. Acta Trop 94:107–115

Voelter-Mahlknecht S, Ho, AD, Mahlknecht, U (2005) FISH-mapping and genomic organization of the NAD-dependent histone deacetylase gene, Sirtuin 2 (Sirt2). Int J Oncol 27:1187–1196

Voelter-Mahlknecht S, Ho AD, Letzel S, Mahlknecht U (2006a) Assignment of the NAD-dependent deacetylase sirtuin 5 gene (SIRT5) to human chromosome band 6p23 by in situ hybridization. Cytogenet Genome Res 112:208–212

Voelter-Mahlknecht S, Ho AD, Mahlknecht, U (2006b) Chromosomal organization and fluorescence in situ hybridization of the human Sirtuin 6 gene. Int J Oncol 28:447–456

Voelter-Mahlknecht S, Mahlknecht, U (2006) Cloning, chromosomal characterization and mapping of the NAD-dependent histone deacetylases gene sirtuin 1. Int J Mol Med 17:59–67

Voelter-Mahlknecht S, Ho AD, Letzel S, Mahlknecht U (2006d) Fluorescence in situ hybridization and chromosomalorganization of the human Sirtuin 7 gene. Int J Oncol 28:899–908

Westphal CH, Dipp MA, Guarente L (2007) A therapeutic role for sirtuins indiseases of aging? Trends Biochem Sci 32:555–560

Walle T, Hsieh F, DeLegge MH et al. (2004) High absorption but very low bioavailability of oral resveratrol in humans. Drug Metab Dispos 32:1377–1382

Wang F, Nguyen M, Qin FX, Tong Q (2007) SIRT2 deacetylates FOXO3a in response to oxidative stress and caloric restriction. Aging Cell 6:505–514

Weindruch R, Walford RL (1982) Dietary restriction in mice beginning at 1 year of age: effect on life-span and spontaneous cancer incidence. Science 215:1415–1418

Werner HB, Kuhlmann K, Shen S et al. (2007) Proteolipid protein is required for transport of sirtuin 2 into CNS myelin. J Neurosci 27:7717–7730

Witt O, Deubzer HE, Milde T, Oehme I (2009) HDAC family: What are the cancer relevant targets? Cancer Lett 277:8–21

Wood JG, Rogina B, Lavu S et al. (2004). Sirtuin activators mimic caloric restriction and delay ageing in metazoans. Nature 430:686–689

Yang H, Baur JA, Chen A et al. (2007) Design and synthesis of compounds that extend yeast replicative lifespan. Aging Cell 6:35–43

Yang T, Sauve AA (2006) NAD metabolism and sirtuins: metabolic regulation of protein deacetylation in stress and toxicity. AAPS J 8:E632–643

Yang XJ, Seto E (2008) The Rpd3/Hda1 family of lysine deacetylases from bacteria and yeast to mice and men. Nat Rev 9:206–218

Yeung F, Hoberg JE, Ramsey CS et al. (2004) Modulation of NF-kappaB-dependent transcription and cell survival by the SIRT1 deacetylase. EMBO J 23:2369–2380

Zhang Y, McLaughlin R, Goodyer C, LeBlanc A (2002) Selective cytotoxicity of intracellular amyloid beta peptide1–42 through p53 and Bax in cultured primary human neurons. J Cell Biol 156: 519–522

Zhang Y, Au Q, Zhang M et al. (2009) Identification of a small molecule SIRT2 inhibitor with selective tumor cytotoxicity. Biochem Biophys Res Commun 386:729–733

Chapter 11
Evolutionary Nutrigenomics

Michael R. Rose, Anthony D. Long, Laurence D. Mueller, Cristina L. Rizza, Kennedy C. Matsagas, Lee F. Greer, and Bryant Villeponteau

Contents

11.1	Introduction: Aging Arises from a Failure of Adaptation, Not Cumulative Damage	357
11.2	Making SENSE: Strategies for Engineering Negligible Senescence Evolutionarily	358
11.3	Using Double-Screen Genomics to Identify Targets for Intervention	359
11.4	Pharmaceutical Versus Nutritional Intervention Strategies	360
11.5	Genescient Uses Two Key Accelerators to Identify Evolutionary Nutrigenomic Agents	362
11.6	Genescient Progress Report	363
11.7	Conclusions: Prospects for Evolutionary Nutrigenomics	364
References		365

11.1 Introduction: Aging Arises from a Failure of Adaptation, Not Cumulative Damage

The common assumption among gerontologists is that aging is a process of inexorably accumulating damage or disharmony. This assumption has held sway since Aristotle, the chief source of variation in the articulation of this assumption being the historically prevalent fashions in biological thought, from the four Greek elements of air, fire, water, and earth to contemporaneous notions about oxidation, free radicals, and the like (Rose 2007). The falsity of this assumption is revealed by three obdurate biological facts:

 (i) there are organisms like fissile sea anemones and Hydra which show no detectable aging;
 (ii) species are sustained by unbroken cell lineages that are hundreds of millions of years old, whether those lineages engage in sex or not; and

M.R. Rose (✉)
Genescient, LLC, Irvine, CA, USA
e-mail: mrose@genescient.com

(iii) in some laboratory cohorts of sufficient size, actuarial aging comes to a halt at late adult ages (Rose 2008).

None of these now well-established features of aging are compatible with the view that aging is simply and solely a result of inexorably accumulating damage, disharmony, or the like.

Instead, aging is due to sustained declines in Hamilton's Forces of Natural Selection (Hamilton 1966; Charlesworth 1980; Rose 1991; Rose et al. 2007). Natural selection is what produces adaptation, the term "adaptation" referring to attributes useful for survival and reproduction. As the power of natural selection declines, a decline that Hamilton's Forces quantify explicitly and from first principles, adaptation is expected to decline. This is how evolutionary biologists explain aging.

Cases in which aging does not occur at all, such as strictly and symmetrically fissile species or evolving germ lines, are instances where Hamilton's Forces do not decline at any point. Notably, it has recently been shown that the apparent cessation of aging late in adult life is also explicable in terms of Hamilton's Forces (Mueller and Rose 1996; Rose et al. 2002; Rauser et al. 2006). That is, there is a direct correspondence between situations in which aging is not observed and circumstances in which Hamilton's Forces do not decline. This is one of many types of empirical evidence that support the Hamiltonian explanation of aging (Rose 1991; Rose et al. 2007). Significantly for Popperian scientists, there are no well-attested refutations of the Hamiltonian theory of aging. This is a significant advantage of the Hamiltonian theory of aging for evolutionary geneticists, physicists, and other scientists who practice "strong inference" (vid. Platt 1964).

Naturally enough, evolutionary biologists have been able to readily and substantially postpone aging by manipulating Hamilton's Forces (Rose and Charlesworth 1980; Luckinbill et al. 1984; Rose 1991), a track record that is unmatched by attempts to manipulate aging based on non-evolutionary gerontological theories. This is not surprising, because most mainstream gerontological theories are variants of Aristotle's original error about aging. Our conclusion is that the Hamiltonian gerontology provides the best scientific foundation for properly thought-out attempts to substantially intervene in the process of aging (Rose 1991, 2008).

11.2 Making SENSE: Strategies for Engineering Negligible Senescence Evolutionarily

In this article, we principally address this question of how to use Hamilton's Forces to ameliorate human aging. This is a question that we have long pursued (e.g. Rose 1984, 2005, 2008), generally without making any material headway. There was a singular reason for this past failure: evolutionary biologists had not been given the resources to pursue any of their proposals as to how we might ameliorate aging in humans, despite their notable successes both at explaining aging scientifically and at slowing aging in laboratory populations. We will not comment here on why this regrettable situation was allowed to subsist, and turn instead to recent developments that have proven surprisingly positive.

We have recently summarized alternative strategies for engineering negligible senescence based on the scientific foundations supplied by evolutionary biology ("SENSE"), rather than the typical foundations supplied by current fashions in cell and molecular gerontology (Rose 2008). These SENSE strategies are based on using Hamilton's Forces to produce model organisms with slowed aging and then reverse-engineering the biology of those organisms to discover interventions that can be used to ameliorate human aging.

The chief issue within Hamiltonian gerontology has been the best type of organism to use in this project. In the 1980s, it was supposed that only a mammalian species would yield Hamiltonian results that could be reliably reverse-engineered (e.g. Rose 1984), because of the close evolutionary relationship among mammalian species and the then-considerable difficulty of discerning genetic commonalities between humans and less-related species, such as insects and nematodes. But with the advent of powerful genomic technologies circa the year 2000, it became apparent that it might be possible to use fruit flies that had been forced to evolve slower aging using Hamiltonian methods (Rose et al. 2004), the so-called "Methuselah Flies," as an alternative to mice, so long as "SENSE Methuselah Mice" were not available (Rose 2008). Thus the most immediate prospect for SENSE is the use of Hamiltonian, or SENSE, Methuselah Flies as a source of genomic information with which to develop useful ways to ameliorate human aging. It is this prospect which is our chief concern here.

11.3 Using Double-Screen Genomics to Identify Targets for Intervention

With the whole-genome sequencing of both humans and fruit flies circa 2000, as well as the concomitant advent of whole-genome tools for measuring gene expression, there is now an "information superhighway" connecting fruit flies and humans. It is now trivial to identify corresponding genes ("orthologs") between these species. Furthermore, it is easy to compare genetic and gene-expression differences between Methuselah Flies produced using Hamiltonian methods with their matched controls.

In 2006, Genescient LLC took advantage of these technologies to compare whole-genome gene-expression patterns in Methuselah Flies with their matched controls. We found about 1,000 genes showing statistically significant differences in expression. These genes are thus presumptive candidates for the genetic changes that underlie the substantially ameliorated aging achieved using Hamiltonian methods in fruit flies. Even more exciting for the purpose of reverse-engineering interventions, in 2007 we found that more than 700 of these genes had matching "orthologous" loci in the human genome.

Fig. 11.1 Genescient identified human aging genes. Compared low and high risk patients with chronic diseases to wild-type and long-lived flies to obtain orthologs related to aging

Genescient performed Genome-wide screens to identify genomic changes (Δ's) that are required for *Drosophila* health and longevity

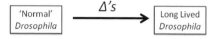

Genescient used Genome-wide screens to identify genomic changes (Δ's) for at-risk human patients with low disease risks versus high disease risks

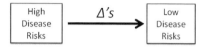

Fig. 11.2 Genetic overlap identifies shared aging pathways. Overlap of Drosophila longevity genes and human disease genes generating greater health risks

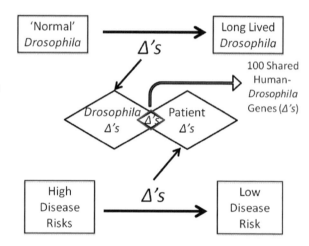

The second phase of our work was to use extant human genome-wide association studies (GWAS) to ascertain whether any of these orthologous loci were statistically associated with reduced risks of aging-associated disease. So far, we have more than 100 human genes showing such statistical associations with risks of contracting such diseases. These genes thus became Genescient's targets for intervention.

Figures 11.1 and 11.2 give a crude graphical outline of the two basic genomic screens that we have performed, searching for genomic changes that are associated with increased lifespan in Drosophila as well as better chronic disease outcomes in human subjects.

11.4 Pharmaceutical Versus Nutritional Intervention Strategies

So long as one supposes that aging is due to just a few "master regulatory" genes (e.g. Guarente and Kenyon 2000) or a handful of fully delimited types of accumulating damage (e.g. de Grey and Rae 2007), then it is reasonable to suppose that massively effective "anti-aging" pharmaceuticals might be discovered. In other words, so long as aging is NOT conceived in Hamiltonian terms, then anti-aging pharmaceuticals should be possible, despite the failure of all attempts to produce any such agents throughout the lengthy history of the many attempts to do exactly that.

But on the Hamiltonian view of aging, and given Genescient's own genomic results described above, it is only to be expected that aging involves a failure of natural selection across hundreds of genes, with many physiological mechanisms of aging produced by these failures of adaptation. The prospects for developing FDA-approvable pharmaceuticals for such a "Many-Headed Monster" as aging (cf. Rose and Long 2002) are extremely doubtful, if not hopeless. But does this mean that there are no prospects for ameliorating human aging? We don't think so.

Understanding how natural selection creates adaptations is the key to understanding Genescient's evolutionary nutrigenomic strategy. There are a few cases (such as those involving short-term selection for resistance to antibiotics, heavy metals, or pesticides) where natural selection produces adaptations based on single genetic changes. But when the genetic basis of typical adaptations is studied, adaptations such as body size or resistance to cancer, it is commonly found that many genes underlie these adaptations. Thus it is no surprise to evolutionary geneticists that the genetic basis of the several-fold extension of Methuselah Fly lifespans involves hundreds of genes. To re-shape aging, we will need to make appropriate adjustments involving many biochemical pathways.

At this point, a reader of the gerontological literature might point to the "longevity mutants" that have been such an obsession in recent gerontological research (reviewed in Guarente and Kenyon 2000, as well as Arking 2006). Such "longevity" mutations have been known for more than fifty years (e.g. Maynard Smith 1958). When they are studied with greater and greater care for their side-effects, which has not always been standard practice within the gerontological research community, they are characteristically found to show debilitating effects on fitness, reproduction, and related functions (Van Voorhies et al. 2006). The pathways that are "knocked-out" by these "longevity mutants" typically produce extended lifespan in a manner that is achieved at great physiological cost, such as sterility, dwarfism, or metabolic hibernation. Such side-effects make these "longevity genes" unlikely targets for the extension of "human healthspan," although they might be useful targets for preserving the lives of hospitalized patients under extreme medical conditions, in which reproductive incapacitation or impaired cognition might not be issues.

To return to the Hamiltonian perspective, the technological problem is evidently one of re-tuning hundreds of genetically-defined mechanisms of aging. This problem is readily solved by natural selection, as the Hamiltonian Methuselah Flies directly demonstrate. Furthermore, we know from detailed studies of individual loci in these flies that these manifold re-tunings do not involve "knocking-out" or otherwise destroying normal genetic mechanisms (Rose et al. 2004; Teotonio et al. 2009). Instead, as Genescient's own genomic findings show in great detail, the evolutionary changes that lead to greatly slowed aging involve relatively subtle changes in gene frequency and gene expression, in most cases. Such subtle changes are not effects that pharmaceuticals notably emulate.

Instead, the best strategy for emulating the effects of natural selection in extending lifespan several-fold is nutritional supplementation. This does NOT mean ingesting large quantities of hundreds of supplements that one or another molecular biologist supposes might be beneficial based on results obtained using in vitro cell cultures. Indeed, the wholesale failure of such research to produce extended human healthspans suggests the greatest caution in using guidelines derived from such work.

The Hamiltonian perspective suggests instead using nutritional supplements in the same manner as evolution uses genetic variants of small effect. Genetic changes that do not have massively disruptive effects, unlike the "longevity mutants," are likely to alter a number of physiological mechanisms a small amount. Similarly, nutritional supplements that do not have drastic effects are likely to have moderate effects on a number of physiological mechanisms. This does NOT, however, mean that these effects are on the whole benign. Just that they are moderate.

In Hamiltonian experiments in which we evolve postponed aging in model animal species, natural selection "screens" such moderate, often diffuse, genetic effects, favoring those variants that have benign effects accumulated over the entire spectrum of physiological functions. Likewise, we have to screen candidate "nutrigenomic agents" for their benefits, just as natural selection would. Just because a nutritional supplement seems like it should be beneficial based on our genomic findings is no guarantee that it will in fact be useful in the amelioration of human aging.

11.5 Genescient Uses Two Key Accelerators to Identify Evolutionary Nutrigenomic Agents

Genescient's evolutionary nutrigenomic approach is based on emulating natural selection, using nutritional supplements in lieu of genetic variation, with two major "accelerators," as follows:

Accelerator 1: We choose candidate substances based on biochemical associations between the effects of candidate substances and the pathways that Genescient has identified genomically. Mutation supplies "blind variation" for natural selection to act on. While we are very far from supposing that contemporary biochemistry is infallible, the physiological mechanisms and gene products disclosed by Genescient's Hamiltonian genomics provide valuable clues that can be combined with the published literature and small-molecule databases to direct us toward some nutritional supplement choices over others. Thus, our first accelerator lets us *do even better than natural selection*, by using our proprietary genomic insights combined with extant biochemical information.

Accelerator 2: We first test our candidate nutrigenomic agents using fruit fly healthspan assays, followed by human functional tests. Since we operate within the Hamiltonian paradigm, we are not interested in substances that might increase longevity at great functional cost. Instead, we are interested in supplements that will enhance longevity, fertility, cognitive function, physical performance, *et cetera*, all at the same time. This means that we seek substances that can be shown to have both long-term and short-term benefits.

In principle, all our developmental research could be performed on human subjects. But to seek measurable benefits for human healthspan, when we expect such benefits to be of small magnitude for any single supplement, is commercially hopeless. We would never get access to the funding required to test dozens, not to say eventually hundreds, of candidate nutritional supplements over large test groups of human subjects over decades, prior to marketing the compounded nutrigenomic agents as commercial products.

This is where the evolutionary foundations of our nutrigenomic strategy pay particularly large dividends. All the genetic mechanisms targeted by our nutrigenomic candidate substances have been implicated in *both* fruit flies and humans. If we have an excellent nutritional supplement based appropriately on genes that are associated with healthspan in both fruit flies and humans, it should benefit both fruit flies *and* human subjects. We can readily screen for lifelong benefits in fruit flies. Genescient scientists have decades of expertise in accurately and efficiently characterizing longevity, mortality rates, fecundity, male mating success, and related lifelong indicators of healthspan in fruit flies. We can also readily screen for useful, short-term, functional benefits in humans, because there is no other organism for which we have better metrics for short-term function. Naturally enough, we start with fruit fly tests, passing candidate nutrigenomic agents on to human testing only once we have cleared them for lifelong benefits in fruit flies. Figure 11.3 summarizes this R&D strategy.

11.6 Genescient Progress Report

It might be useful to summarize Genescient's progress to date in broad terms. [We are in the process of writing up and submitting detailed "data papers" for publication in the scientific literature.] That way the reader will have a more concrete idea

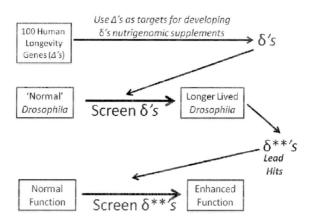

Fig. 11.3 Shared aging genes are used to develop nutrigenomic treatments. Substances (δ's) are identified that act on human disease pathways and then are tested in fly aging assays. Lead hits that extend fly healthspan are screened in humans for enhanced function

of what Genescient has accomplished, by way of materially embodying its R&D strategy.

1. We have already performed an extensive genomic inventory of the gene-expression changes that underlie the several-fold extension of lifespan in Hamiltonian Methuselah Flies. We have found about 1,000 genes for which there are statistically significant and consistent changes in gene expression that result from selection for increased adaptation at later ages, including survival to, and function at, much later ages than is normal for laboratory fruit flies.
2. We have used GWAS databases to identify more than 100 genetic loci that are associated with *both* increased fly lifespan *and* decreased risk of chronic human diseases, such as cardiovascular and metabolic disorders.
3. We have used the key loci identified in step 2 to choose nutritional supplements that we regard as candidates for nutrigenomic agents that might give enhanced healthspan in both humans and fruit flies.
4. One significant result from this initial screen is that high doses were often less effective than low or moderate doses. Therefore, more of a longevity compound is not necessarily better.

We are now performing trials to test for the effects of combinations of nutrigenomic agents that have passed through our R&D program. We also plan human trials to test for short-term functional effects of our candidate nutrigenomic agents. Following the completion of these tests, we would like to produce nutrigenomic products for sale in the marketplace.

11.7 Conclusions: Prospects for Evolutionary Nutrigenomics

Genescient's evolutionary nutrigenomic R&D strategy is expandable in many directions, and on a very large scale. Here are several ways we can build on what we have accomplished to this point.

1. The genomic work we have done to this point with the Hamiltonian Methuselah Flies is just a start. New genomic technologies are being released rapidly, from tiling arrays to large-scale rapid re-sequencing. The use of any and all of these technologies will reveal still more detail concerning the genomic foundations of the Hamiltonian prolongation of healthspan.
2. Likewise, human population genomics is a rapidly burgeoning field. As more human genomic data become available, we will find still more genes that are key for aging in both fruit flies and humans.
3. Given enough resources, we can select on mice using Hamiltonian strategies, as originally proposed 25 years ago. Such mice would provide still more genomic information concerning the genetic controls on human aging.

4. Small-molecule databases are rapidly improving, thanks to the application of high-throughput methods for detecting interactions between individual gene-products and candidate small molecules. These databases will furnish more candidates for testing as potential nutrigenomic agents.
5. We look forward to the development of better, and more widely-accepted, protocols for testing human functions, whether cognitive, athletic, or metabolic.

There is nothing easy or magical about the Hamiltonian approach to the amelioration of human aging. But it may well be the best strategy for radical extension of human healthspan that is both scientifically well-founded and experimentally supported. We regard the alternatives as more challenging, however widely accepted within conventional gerontology or geriatrics. As we have said before (Rose 2008), the difficulty of materially extending useful healthspan helps reveal which of the contending scientific and technological approaches to aging are well-founded, and which are not.

References

Arking R (2006) The Biology of Aging: Observations and Principles, 3rd edn. Oxford University Press, New York
Charlesworth B (1980) Evolution in Age-Structured Populations. Cambridge University Press, London
De Grey A, Rae M (2007) Ending Aging, The Rejuvenation Breakthroughs that could Reverse Human Aging in our Lifetime. Methuselah Foundation
Guarente L, Kenyon C (2000) Genetic pathways that regulate ageing in model organisms. Nature 408:255–262
Hamilton WD (1966) The moulding of senescence by natural selection. J Theoret Biol 12:12–45
Luckinbill LS, Arking R, Clare MJ, Cirocco WC (1984) Selection for delayed senescence in *Drosophila melanogaster*. Evolution 38:996–1003
Maynard Smith J (1958) The effects of temperature and of egg-laying on the longevity of *Drosophila subobscura*. J Exp Biol 35:832–842
Mueller LD, Rose MR (1996) Evolutionary theory predicts late-life mortality plateaus. Proc Natl Acad Sci USA 93:15249–15253
Platt JR (1964) Strong inference. Science 146:347–353
Rauser CL, Mueller LD, Rose MR (2006) The evolution of late life. Aging Res Rev 5:14–32
Rose MR (1984) The evolutionary route to Methuselah. New Scientist 103:15–18
Rose MR (1991) Evolutionary Biology of Aging. Oxford University Press, New York
Rose MR (2005) The Long Tomorrow, How Evolutionary Biology can help us Postpone Aging. Oxford University Press, New York
Rose MR (2007) End of the line. Quart Rev Biol 82:395–400
Rose MR (2008) Making SENSE: strategies for engineering negligible senescence evolutionarily. Rejuvenation Res 11:527–534
Rose M, Charlesworth B (1980) A test of evolutionary theories of senescence. Nature 287:141–142
Rose MR, Long AD (2002) Ageing: The many-headed monster. Curr Biol 12:R311–R312
Rose MR, Drapeau MD, Yazdi PG, Shah KH, Moise DB, Thakar RR, Rauser CL, Mueller LD (2002) Evolution of late-life mortality in *Drosophila melanogaster*. Evolution 56:1982–1991
Rose MR, Passananti HB, Matos M (eds) (2004) Methuselah Flies, A Case Study in the Evolution of Aging. World Scientific, Singapore

Rose MR, Rauser CL, Benford G, Matos M, Mueller LD (2007) Hamilton's forces of natural selection after forty years. Evolution 61:1265–1276

Teotonio H, Chelo IM, Bradic M, Rose MR, Long AD (2009) Experimental evolution reveals natural selection on standing genetic variation. Nat Genet doi:10.1038/ng.289

Van Voorhies W, Curtsinger JW, Rose MR (2006) Do longevity mutants always show trade-offs? Exp Gerontol 41:1055–1058

Chapter 12
Biological Effects of Calorie Restriction: Implications for Modification of Human Aging

Stephen R. Spindler

Contents

12.1	Introduction	368
12.2	Lifespan Effects of Caloric Restriction (CR)	369
	12.2.1 Phylogenetic Conservation of the CR Response	369
	12.2.2 CR in Nonhuman Primates	371
	12.2.3 CR in Humans	371
	12.2.4 Age and CR Responsiveness	374
	12.2.5 CR Intensity and the Lifespan Response	375
	12.2.6 CR Duration and the Lifespan Response	375
	12.2.7 Rapidity and Reversibility of the Shift to the "CR State"	376
	12.2.8 Does CR Induce a Physiological "Memory"?	377
12.3	Dietary Composition, Meal Frequency, and Lifespan	377
	12.3.1 Protein Restriction	378
	12.3.2 Tryptophan Restriction	379
	12.3.3 Methionine Restriction	379
	12.3.4 Mechanisms of Lifespan Extension by Protein, Methionine and Tryptophan Restriction	380
	12.3.5 Meal Frequency and Intermittent Fasting	381
	12.3.6 Specific Micronutrients and Lifespan	381
12.4	Evolutionary Origin of the Health Benefits of CR	383
12.5	Longevity Mutations and the Molecular Mechanisms of CR	383
12.6	Reproductive Effects of CR in Rodents	385
12.7	Reproductive Effects of CR in Humans	387
12.8	Anticancer Effects of CR in Rodents	387
	12.8.1 Apoptosis and the Anticancer Effects of CR	388
	12.8.2 IGFI, Insulin Receptor Signaling, and Cancer	389
	12.8.3 Can Life History Account for the Anticancer Effects of CR?	390

S.R. Spindler (✉)
Department of Biochemistry, University of California, Riverside, CA 92521, USA
e-mail: spindler@ucr.edu

G.M. Fahy et al. (eds.), *The Future of Aging*, DOI 10.1007/978-90-481-3999-6_12,
© Springer Science+Business Media B.V. 2010

12.9	Metabolic Effects of Aging and CR	391
12.10	CR and Autophagy	392
12.11	CR and Glucocorticoids	393
12.12	Cardiovascular Effects of CR	394
12.13	Immunological Effects of CR	395
12.14	Neurological Effects of CR	396
12.15	Genic Effects of CR	397
12.16	CR and Specific Transcription Factors	397
	12.16.1 SirT1	398
	12.16.2 PGC-1α	399
	12.16.3 AMPK	400
12.17	Hypothermia and CR	403
12.18	Exercise and CR	404
12.19	Adiposity and the Health Effects of CR	405
12.20	Human BMI, Morbidity, and Mortality	406
12.21	The Search for Longevity Therapeutics	408
12.22	Conclusions	410
References		411

12.1 Introduction

During the 1994 Miss USA contest, Miss Alabama purportedly was asked, "If you could live forever, would you and why?" She is said to have replied, "I would not live forever, because we should not live forever, because, if we were supposed to live forever, then we would live forever, but we cannot live forever, which is why I would not live forever". Although this is beautifully circular logic, she is not alone in this view. Dr. Leon Kass, who was President Bush's Chair of the President's Council on Bioethics from 2001 to 2005, wrote, "Confronted with the growing moral challenges posed by biomedical technology, let us resist the siren song of the conquest of aging and death" (Kass 2001).

Despite the opinions expressed above, humans have been linearly extending their average lifespans since about 1840, thereby making steady progress in the "conquest of aging and death" (Oeppen and Vaupel 2002). By 1900, average lifespans in Europe and the USA had increased from about 22 to 35 years of age to around 60. Today, the average lifespan world wide has risen to roughly 63 years, and in Japan, the world leader in longevity, the life expectancy for women is almost 85 years (Oeppen et al. 2002).

Because of these gains, by the 1960s, the shape of the human survival curve in developed countries had begun to resemble that of research animals in a vivarium. Maximum lifespan is often defined as the lifespan of the longest lived 10% of a cohort born at about the same time. Because it is mostly the old who die, people born today have a better chance of reaching extreme old age than do the current "old". For example, the expectancy of living to 100 years for boys and girls born today in the

United Kingdom is 18.1 and 23.5%, respectively (Anonymous 2007). In contrast, a 40-year-old man and woman have only an 8 and 11.7% chance of reaching 100 years of age, respectively. As of May 26, 2008, the Los Angeles Gerontology Research Group had verified only 74 living supercentenarians (people living to 110 years of age or above) [64 Females and 10 Males; (http://www.grg.org/Adams/Tables.htm)]. Only one human, Mme. Jeanne Calment, is credibly documented to have survived to 122 years of age.

Demographers tell us that the shape of our lifespan curve today suggests we are approaching the theoretical limits of the lifespan of our species (Olshansky et al. 2001). It has been estimated that if we successfully conquer all the diseases which currently kill us, including cancer and cardiovascular disease, which are the major killers in industrialized societies, we will extend our average lifespan by only about 15 years (Olshansky et al. 2001). Thus, even if we are lucky enough to escape all the diseases currently killing us, we will still die when we encounter the "wall" looming at approximately 110 years of age. This is part and parcel of the health care dilemma facing the developed world. Most of the spending for health-related research and care by individuals, governments, private companies and foundations, doctors, hospitals and other health care providers are focused on capturing that final 15 years of life.

12.2 Lifespan Effects of Caloric Restriction (CR)

Scientists have known since the 1930's that diets which reduce calories below the level required for maximum fertility and fecundity, while avoiding malnutrition, can extend mean and maximum lifespan (the lifespan of the longest lived 10% of a cohort) of laboratory rats by 40% or more (McCay et al. 1935). Such dietary regimens are often termed "caloric restriction" (CR) or "dietary restriction". Early reports showed that CR reduces the incidence and severity of many of the diseases that limit the lifespan of rats (McCay et al. 1935). Subsequent investigations confirmed that CR, begun at weaning and continued throughout life, extends survival, reduces the incidence of tumors and other diseases, and shifts these diseases to later ages in laboratory rats and mice (Cheney et al. 1980; Weindruch and Walford 1988; Masoro 2006).

12.2.1 Phylogenetic Conservation of the CR Response

The longevity and health effects of CR appear to be widespread phylogenetically. As summarized by Masoro (Masoro 2006), they include species from three Kingdoms (Animalia, Fungi and Protoctista), and four phyla (Rotifera, Nematoda, Arthropoda and Chordata) within Animalia. Responsive species include dog (Labrador Retriever) (Lawler et al. 2008), rodents (rats, mice, hamsters) (Weindruch and Walford 1988), nematode (Caenorhabditis elegans) (Klass 1977), rotifer (Asplanchna brightwelli and Philodena acuticornis) (Fanestil and Barrows 1965; Verdone-Smith and Enesco 1982), spider (Frontinella pyramitela; a.k.a. bowl and doily spider) (Austad 1989), fruit fly (Drosophila melanogaster) (Min et al. 2007), the Mediterranean fruit fly (Ceratitis capitata; a.k.a. medfly) (Carey et al. 2005), guppies (Lebistes reticulatus, Peters), and zebrafish (Danio rerio) (Comfort 1963; Keller et al. 2006). Nutritional insufficiency also appears to extend the chronological and replicative lifespan of baker's yeast (Saccharomyces cerevisiae), although the meaning of these observations and their relationship to the lifespan of multicellular eukaryotes remains unclear (Kaeberlein and Powers 2007; Fabrizio and Longo 2007; Michan and Sinclair 2007).

Recently, gerontologists have become more sensitive to the idea that the "universality" of the CR response could be the result of reporting bias. The health and/or longevity effects of CR may not be universal, even within species. For example, while CR begun at 4 months of age increases median and maximum lifespan of C57BL/6 and B6D2F1 mice, it fails to alter the lifespan of DBA/2 mice, at least using the methods of these studies (Forster et al. 2003). A CR regimen has not been described which is capable of extending the lifespan of the housefly (Musca domestica) (Cooper et al. 2004). In addition, some age-related diseases, such as certain mouse models of Parkinson's disease, also may be unresponsive to CR (Armentero et al. 2008).

These data may not indicate that CR cannot extend the lifespan of these species or strains. Responsiveness to CR can depend on subtleties of the treatment protocol. CR was long thought incapable of extending the lifespan of middle-aged and older mice. Only when CR was introduced in a stepwise fashion was lifespan extension obtained in mice 12 months of age or older (Weindruch and Walford 1982; Dhahbi et al. 2004). Harrison and Archer found a shortening of lifespan when CR was introduced to C57BL/6J mice immediately after weaning (Harrison and Archer 1987). However, other laboratories obtained robust lifespan extension with this mouse strain when CR was initiated in middle age (e.g. Pugh et al. 1999). Initial reports indicated that medflies were unresponsive to CR (Carey et al. 2002). However, clear evidence of a CR response was found when protocols similar to those used for Drosophila were employed (Davies et al. 2005; Carey et al. 2005).

Triggering the CR response in some strains or species may require specific husbandry techniques and/or dietary regimen. Differences in the husbandry of some strains of dwarf mice can determine whether shortened or lengthened lifespan is observed (Bartke 2008b). One interpretation of such results is that certain stresses [e.g. CR; reduced insulin and insulin-like growth factor-I (IGFI) signaling] induce a repair and survival-related physiological response. For lifespan extension to be observed, the response must obviate the negative effects of the inducer and redirect molecular priorities to pathways designed for stress resistance, repair, and survival (Schumacher et al. 2008). Whether a given stress produces extended longevity in a specific strain or species will depend on the severity of the stress, the degree to which it induces the response pathway, and the potency of the response pathway. These are all polygenic traits. Thus, it is not surprising that genetic background can have significant influences on the response.

12.2.2 CR in Nonhuman Primates

Studies from two colonies of rhesus macaques suggest that the effects of long-term CR (LTCR) in nonhuman primates recapitulate many of the physiological, hematological, hormonal, immunological and biochemical effects produced in rodents (Kemnitz et al. 1993; Mattison et al. 2003; Maswood et al. 2004; Roth et al. 2004; Mattison et al. 2005; Anderson et al. 2009). Approximately 30% LTCR (a 30% reduction in calories) initiated in macaques of various ages, decreases body weight and adiposity (Colman et al. 1998; Lane et al. 1999); improves glucoregulatory functions and increases insulin sensitivity (Lane et al. 1995, 1999; Gresl et al. 2003); produces favorable changes in blood triglyceride and lipid profiles (Edwards et al. 1998); reduces serum levels of C-reactive protein (Edwards et al. 1998); delays male skeletal and sexual maturation; delays the age-associated decline in serum dehydroepiandrosterone and melatonin normally found in ad libitum fed controls (Lane et al. 1999; Mattison et al. 2003); reduces oxidative damage in skeletal muscle (Zainal et al. 2000); and attenuates the development of sarcopenia (Colman et al. 2008). LTCR also produces a trend toward reduced cardiovascular disease, diabetes, neoplastic disease and liver failure as causes of mortality (Lane et al. 1999; Roth et al. 1999). CR monkeys develop diabetes later in life and with a lower incidence than ad libitum fed controls (Bodkin et al. 1995). CR initiated during adulthood may delay T-cell aging and preserve naïve CD8 and CD4 T cells into advanced age, although the timing and method of introduction of CR appears to be crucial to this effect (Messaoudi et al. 2008). LTCR also reduces the production of local inflammatory mediators and the risk of inflammatory periodontal disease among male macaques (Reynolds et al. 2009). Recently, investigators at the Wisconsin National Primate Research Center have published an analysis of their survival data which found a statistically significant increase in the lifespan of CR rhesus macaques (Colman et al. 2009). However, their posthoc analysis excludes deaths deemed not "age related". Further, the analysis did not use a multiple testing procedure to compensate for the posthoc design. Thus, their conclusions must be regarded as provisional.

Results from a third colony of rhesus macaque have been interpreted as evidence that CR extends the lifespan of these primates (Bodkin et al. 2003). However, statistical and methodological concerns make these conclusions equivocal (Lane et al. 2004). Further survival data from the two ongoing studies discussed above will be needed to conclusively determine whether CR is effective at extending the lifespan of nonhuman primates.

12.2.3 CR in Humans

Some have argued that the inevitable increase in entropy makes it unlikely that CR will prolong human health- and lifespan (Hayflick 2004). Phelan and Rose (2005), Rose and Demetrius (2005) and De Grey (2005) have used somewhat different lines of reasoning to argue that the life-history of humans presents little selective pressure for a robust CR response (Phelan et al. 2005; Demetrius 2005; De Grey 2005). However, as with the nonhuman primates discussed above, it is unlikely that we will have an unequivocal resolution of this debate in the near future.

The limited evidence available for humans suggests that CR produces physiological effects that are similar to those found in rodents and monkeys (Verdery and Walford 1998; Walford et al. 1999, 2002; Weyer et al. 2000). However, there are few studies in the literature of human health and longevity using nutritious, low calorie diets. Kagawa reported that in the 1970s the death rates from cerebral vascular disease, malignancy and heart disease on Okinawa Island were 59, 69 and 59% of those found in the rest of Japan (Kagawa 1978). The mortality rate for 60–64 year olds living on Okinawa was half of that found elsewhere in Japan. The incidence of centenarians on the island was two to forty times greater than that of other Japanese communities. He suggested this good fortune resulted from CR. Okinawan school children consumed only 62% of the calories of other Japanese school children during the early 1960s (Hokama et al. 1967). Kagawa reported in 1978 that Okinawan adults consumed 20% fewer calories relative to the national average in Japan (Kagawa 1978). Okinawans who emigrated to the USA and began to consume a more typical Western diet had mortality and morbidity rates similar to others in the USA (Kagawa 1978). The results of this study are supported by subsequent studies using six decades of archived population data on elderly Okinawans (aged 65 years and over) regarding diet composition, energy intake, energy expenditure, anthropometry, plasma DHEA, mortality from age-related diseases, and current survival patterns (Willcox et al. 2007). Willcox et al. found low caloric intake and negative energy balance, little weight gain with age, life-long low BMI, higher plasma DHEA levels during aging, low risk of mortality from age related diseases, and survival patterns consistent with extended mean and maximum lifespan in this cohort (Willcox et al. 2007).

A role of CR in human health is also supported by a study of 60 healthy seniors (average age of 72 years at the start of the study) who received a dietary regimen averaging 1500 kcal/day for 3 years, versus an equal number of controls consuming 2300 kcal/day (Vallejo 1957). The CR group consumed 2300 kcal/day every-other-day, and one liter of milk and 500 g of fruit on the alternate day. Analysis of these data indicates that CR significantly lowered the rates of hospital admissions (123 versus 219 days; $p<0.001$) and numerically lowered deaths (6 versus 13; Stunkard 1976).

A longitudinal CR study conducted by Walford and colleagues on eight healthy nonobese humans eating a low-calorie nutrient-dense diet for 2 years in Biosphere 2 (1750–2100 kcal/day), found 50 CR-related changes in physiologic, hematologic, hormonal, and biochemical parameters which resemble those of CR rodents and monkeys (Walford et al. 1999, 2002; Weyer et al. 2000).

More recently, a number of studies have been performed on groups of volunteers subjected to CR for 6 months or one year. Studies of a group of individuals consuming a nutrient-rich, low-calorie diet for an average of 6 years (BMI 19.6 +/− 1.9) and age-matched healthy controls eating typical American diets (BMI

25.9 +/− 3.2) found that the CR group had markedly reduced serum total cholesterol, low-density lipoprotein cholesterol, ratio of total cholesterol to high-density lipoprotein cholesterol, triglycerides, fasting glucose, fasting insulin, C-reactive protein, platelet-derived growth factor AB, and systolic and diastolic blood pressure (Fontana et al. 2004). Carotid artery intima-media thickness was ~40% lower in the CR group. High-density lipoprotein cholesterol was higher with CR. Another study of healthy, middle aged subjects practicing CR, this time for an average of 6.5 years, found lower heart chamber viscoelasticity and stiffness, lower blood pressure, and lower serum C-reactive Protein, Tumor Necrosis Factor-α, and Transforming Growth Factor-β1 levels than were found in age- and gender-matched controls consuming Western diets (Meyer et al. 2006). Together, these studies suggest that longer-term CR initiated in older humans reduces blood pressure, systemic inflammation, myocardial fibrosis, and other risk factors for cardiovascular disease.

The cardiovascular effects of CR and exercise were investigated in a one year controlled study of sedentary, nonobese middle-aged men and women randomly assigned to a 20% CR diet, an exercise regimen that increased energy expenditure by 20% with no increase in energy intake, and controls simply given guidelines for a healthy lifestyle. The study found that both CR and exercise produced similar reductions in coronary heart disease risk factors, including body fat mass, plasma LDL-cholesterol, total cholesterol/HDL ratio, homeostasis model assessment of insulin resistance index, and serum C-reactive protein levels (Fontana et al. 2007). A similar 6 month study using other human volunteers reported very similar results (Lefevre et al. 2009).

Human CR appears to have effects on sex steroids, insulin, and IGFI which are similar to those found in rhesus monkeys and rodents. A cross-sectional comparison of individuals eating a low protein CR diet for an average of 4.4 years (mean BMI 21.3), endurance runners (mean BMI 21.6), and age- and sex-matched sedentary controls eating Western diets (mean BMI 26.5) found that especially the low-protein and CR group, but also the runners, had significantly lower plasma levels of insulin, free sex hormones, leptin, and C-reactive protein, and higher sex hormone-binding globulin than the control group (Fontana et al. 2006a). A one year controlled study of sedentary, nonobese middle-aged men and women randomly assigned one year of exercise training, CR, or a healthy lifestyle control group, found that weight loss induced by either means produced improvement in glucose tolerance and insulin action (Weiss et al. 2006). A low-protein CR diet also reduced plasma IGFI and the ratio of IGFI to IGF binding protein-3 more than endurance running (Fontana et al. 2006a). Thus, a low-protein, CR diet reduces risk factors for atherosclerosis, and growth factors and hormones linked to an increased risk of cancer. Interestingly, CR alone in humans was not found to reduce serum IGFI levels (see below).

In a weight loss study, a group of overweight male and female volunteers (mean BMI of 27.5) were subjected to 25% CR diets and a combined diet and exercise regimen for 6 months (ending BMI of 24.8). The CR diets and combined diet reduced fasting insulin levels, but not DHEAS or glucose levels. Core body temperature and sedentary 24-hour energy expenditure decreased in all the CR groups (Heilbronn et al. 2006). This decrease may be due to the CR-related decrease in serum T_3 concentrations found by Fontana and Holloszy in lean and weight-stable healthy humans consuming a nutrient dense, CR diet for 3–15 years (Fontana et al. 2006b). The decrease is also similar to that found in CR rodents. A decrease in DNA damage and an increase in muscle mitochondrial DNA and the expression of genes encoding proteins involved in mitochondrial function has also been reported in muscle biopsies of humans subjected to 6 months of CR or CR and exercise (Heilbronn et al. 2006; Civitarese et al. 2007).

Clear downsides have been identified for CR in humans in addition to the decrease in free sex hormones and increase in sex hormone-binding globulin mentioned above. Twelve months of CR or exercise initiated in healthy middle aged men and women (BMI 23.5–29.9) led to a loss in lean body mass in both groups, but loss of more thigh muscle volume and composite knee flexion strength in the CR group relative to the exercisers (Weiss et al. 2007). VO_2 max also decreased in the CR group, but increased in the exercisers. One year of CR in humans also decreased bone mineral density at the total hip, intertrochanter and spine of the CR group but not for an age and sex matched group of the exercisers (Villareal et al. 2006). Thus, CR-induced weight loss reduces muscle mass and physical work capacity, and bone mineral density at clinically important fracture sites.

Despite the results suggesting that CR can lead to lifespan extension in humans, this conclusion is not supported by prospective cohort studies of the relationship between BMI and longevity in humans (see below). These studies and their possible meanings are discussed later in this review.

12.2.4 Age and CR Responsiveness

Older mice can respond to CR when it is incrementally introduced (Dhahbi et al. 2004). Ad libitum fed, male, C3B6F1 mice, are poised to begin the accelerated mortality phase of their lifespan at 19 months of age. If they are shifted to a CR diet at that time, they rapidly assume a new mortality trajectory which is characterized by fewer cancer-related deaths, and an increase in both mean and maximum lifespan (Spindler 2005). In these mice, the decrease in mortality results almost entirely from reduced rates of tumor-associated deaths. The shift in lifespan is accompanied by a concomitant change in the global patterns of gene expression in the mice, especially in the liver, to a pattern which recapitulates most key features of life-long caloric restriction (Dhahbi et al. 2004; Spindler 2005). Likewise, shifting mice from a life-long CR diet to a control diet at an older age rapidly reverts most CR-specific gene expression in the heart and liver back to the expression levels found in control fed mice (Dhahbi et al. 2004, 2006). Thus, some strains of mice appear capable of rapidly shifting to a CR or control physiological state, even at advanced ages. Whether such shifts occur at advanced ages for all strains of mice, or for mice older than 19 months of age is not known.

Rats may be less responsive to CR initiated at advanced ages. Restriction of old Long Evans (18 months old) and F344 × BN F1 hybrid rats (18 and 26 month old) to approximately one third fewer calories than their control fed littermates produced no increase in longevity (Lipman et al. 1995, 1998). Interpretation of these studies is confounded by the absence of a positive control. No animals in these studies achieved lifespan extension. Thus, it remains possible that the rats were simply refractory to CR under the conditions used.

Short-term CR (STCR) in middle age or older mice and rats improves protein turnover, upregulates proteasome activity in liver and skeletal muscle, decreases protein carbonyls in liver mitochondria and skeletal muscle, reduces oxidative DNA damage, and increases carbonyl modifications in histones to levels found in young animals (Goto et al. 2007). These results suggest that STCR in middle aged and old rodents produces a wide spectrum of positive effects on protein metabolism.

12.2.5 CR Intensity and the Lifespan Response

The effects of CR on lifespan appear to be dose-responsive. In a compilation of 24 published survival studies, the increase in survival appears to be proportional to the decrease in calories (Merry 2002). The gene-expression effects of CR also are dose responsive. Caloric restriction at 20 or 50% of ad libitum intake proportionately increases insulin receptor (IR) mRNA and decreases GRP78 (BiP) and GRP94 mRNA in young, middle-aged and old mice (Spindler et al. 1990, 1991).

A frequent confound to such studies is the issue of food consumption by the control group. Ad libitum feeding is often used for the control group. However, ad libitum intake can be strongly influenced by the caloric density, palatability, and physical form of the food. We have observed that mice fed Purina laboratory chow ad libitum tend to be leaner, probably from eating fewer calories, than mice fed hard packed AIN-93 M diet (unpublished observations). Unfortunately, the number of calories consumed by the ad libitum group is seldom measured or reported.

12.2.6 CR Duration and the Lifespan Response

Walford and colleagues reported that the effects of CR on lifespan are proportional to the length of time on CR, whether the CR period was before weaning, after weaning, early in life, or after 15 months of age (Cheney et al. 1983). Data compiled from 21 independent studies confirms that the duration of CR is directly proportional to the increase in longevity in rodents, irrespective of when CR is begun (Merry 2002). To a first approximation, this relationship appears to be linear irrespective of the period of life during which CR is administered (however, see "Does CR induce a physiological *memory*", below and "Age and CR responsiveness", above) (Weindruch and Walford 1982; Cheney et al. 1983; Yu et al. 1985; Beauchene et al. 1986; Merry 2002; Dhahbi et al. 2004).

12.2.7 Rapidity and Reversibility of the Shift to the "CR State"

The most widely accepted evolutionary explanation for the existence of the CR response holds that it evolved early in metazoans as an adaptation to boom and bust cycles in the food supply (see below). A key element of this theory is the idea that animals should be capable of rapidly switching between the CR and the ad libitum fed state. However, there is relatively little evidence that this switching occurs. We addressed this question by shifting a cohort of 19 month old, male, B6C3F1 mice from life-long control feeding to CR (Dhahbi et al. 2004; Spindler 2005). The accelerated mortality phase of the lifespan curve begins soon after 19 months of age with this mouse strain. Linear-regression and breakpoint analysis could be used to estimate the length of time required to initiate lifespan extension (Spindler 2005). The lifespan effects of CR appeared to begin within 8 weeks of its initiation. The extension of lifespan appeared to be almost entirely due to reduced tumor-related mortality. Because hepatocellular carcinoma is the major cause of mortality in this mouse strain, later-life CR appears to strongly reduce the growth rate of these tumors. In unpublished studies, we have found that preexisting liver tumors shrink to approximately half their size by 8 weeks after the introduction of CR. This reduction in tumor size does not appear to involve a reduction in the number of dividing cells (unpublished results). Thus, late-life CR may increase rates of apoptosis in preexisting liver tumors.

Importantly, the rate of response to CR appears to be organ-specific. In 8 weeks, heart gene expression does not shift as completely from the control to the CR state. Eight weeks after shifting control mice to CR, relatively few of the changes in cardiac gene expression found in LTCR were produced (Dhahbi et al. 2006). These results are similar to those found in the white adipose tissue of mice (Higami et al. 2004). Only a few LTCR-responsive transcripts were differentially expressed 23 days after a shift to CR. Thus, in heart and adipose, CR-related gene expression is induced more slowly and/or less completely after the commencement of CR. Changes in cardiac physiology, such as reduction of perivascular extracellular matrix, are only slowly reversible after the introduction of CR (Dhahbi et al. 2006). Further, age-related cardiomyocyte loss and hypertrophy in the left ventricle probably cannot be reversed by the introduction of CR in older mice (Dhahbi et al. 2006). Therefore, not all of the responses to CR are as rapid as its effects on tumor growth.

Using gene-expression as a surrogate for the CR-related physiological state, we attempted to quantify how rapidly the physiological effects of CR are reversed after the cessation of CR. Shifting old LTCR mice to control feeding was accompanied by an almost complete shift of the LTCR-responsive genes in the liver and heart to control gene-expression levels within 8 weeks (Cao et al. 2001; Dhahbi et al. 2004; Spindler 2005). These results suggest that mice can shift rapidly from the CR to the control state.

As indicated above, the rapidity and reversibility of the CR state is consistent with the prevailing theory for the adaptive value of CR. It is also consistent with studies performed in Drosophila showing that the short-term rate of mortality (which determines lifespan in this species) is rapidly responsive to dietary calories

(Mair et al. 2003). Shifting flies from a control to a CR diet decelerated their short-term risk of death within 2 days, while shifting from CR to control had the reverse effect, also within 2 days. The data from flies and mice suggest that the effects of CR are either phylogenetically conserved or result from convergent evolution (Mair et al. 2005). In either case, the rapidity and reversibility of the CR response is adaptive and integral to its physiological role.

12.2.8 Does CR Induce a Physiological "Memory"?

The results cited above suggest that an interval of CR in young rodents can produce a "memory" which persists during subsequent control feeding to extend lifespan (reviewed by Klebanov 2007). In one such study, CD rats restricted in caloric intake between 21 and 70 days of age had increased life expectancy over rats fed ad libitum throughout their life (Ross 1972). In another study, Fisher 344 rats fed a 40% CR diet from 1.5 to 6 months of age followed by ad libitum feeding for the remainder of their life had modestly extended lifespan (Yu et al. 1985; Maeda et al. 1985). Numerically similar, but not statistically significant, results have been reported for other rat strains (Nolen 1972; Ross et al. 1993; Merry et al. 2008). In mice, Walford and colleagues found that preweaning CR, achieved by reducing the opportunity for suckling, followed by lifelong control feeding, consistently produced a numerically longer mean and/or maximum lifespan in both B10C3F1 and C57BL/6 J mice (Weindruch et al. 1979; Cheney et al. 1980, 1983).

An important caveat to the interpretation of the studies cited above is that the early CR results in a smaller rodent, which consumes less food throughout the remainder of its lifespan, thereby inducing mild CR, and a subsequent increase in lifespan (e.g. Cheney et al. 1980). Walford and his colleagues noted that in their studies and those of Widdowson, a preweaning period of CR led to both a slight increase in lifespan and a decrease in average body weight throughout the remainder of life (Widdowson 1964; Cheney et al. 1980). This effect may lead to results such as those reported by Grasl-Kraupp, who found that three months of 40% CR early in life resulted in resistance to nafenopin-induced tumorigenesis during a subsequent 9 month period of ad libitum feeding (Grasl-Kraupp et al. 1994). Neither food intake nor weights were reported. Thus, it is highly likely that this memory effect involves an ongoing low level of CR, and not a more mysterious and persistent physiological change.

12.3 Dietary Composition, Meal Frequency, and Lifespan

The gross composition of otherwise balanced diets with altered proportions of fat, protein, carbohydrate, or minerals does not alter the lifespan of rats. Longevity is extended only if the number of calories consumed is reduced (Ross and Bras 1973; Iwasaki et al. 1988; Masoro 1990). In general, dietary composition appears to be of secondary importance to caloric intake, within a wide range of protein to carbohydrate ratios. However, the ratio of protein to carbohydrate does differentially affect the distribution of tumors among the tissues of the ad libitum fed and CR rats (Ross and Bras 1973). This ratio can also affect the health and longevity of Drosophila (Mair et al. 2005; Skorupa et al. 2008; Lee et al. 1008).

12.3.1 Protein Restriction

While some have reported that the restriction of dietary protein in the absence of CR has a negative effect on survival (Davis et al. 1983), most studies in rodents report that protein restriction enhances longevity irrespective of caloric intake (Pamplona and Barja 2006). Pamplona and Barja compiled eleven published studies of the relationship between protein restriction and lifespan in rats and mice (Pamplona et al. 2006). Ten of the eleven studies in their survey found that reductions in dietary protein increase maximum lifespan by an overall average of ~20%. They point out that this extension appears to be about half that frequently reported for LTCR.

Protein restriction shares many of the highly pleiotropic health and physiological effects of CR in addition to its extension of lifespan. These effects include delayed puberty; decreased growth; transiently decreased metabolic rate; preserved cell-mediated immunity; reduced oxidative damage to proteins (Youngman et al. 1992); reduced glomerulosclerosis in mice (Doi et al. 2001); reduced nephropathy and cardiomyopathy in rats (Maeda et al. 1985); enhanced hepatic resistance to toxic and oncogenic insult (Rodrigues et al. 1991); and decreased preneoplastic lesions and tumors.

In Drosophila, protein restriction (achieved by a reduction in the concentration of yeast in the diet) has a greater effect on lifespan than the restriction of dietary carbohydrate (achieved by a reduction in the concentration of carbohydrate) (Mair et al. 2005). However, such studies in Drosophila are confounded by uncertainties about the relationship between food dilution and caloric intake (Carvalho et al. 2005; Mair et al. 2005; Min et al. 2007). Further, in Drosophila, olfaction appears to be a key method of sensing nutrient availability (Libert et al. 2007). The odor of yeast can partially reverse lifespan extension by food dilution-related CR (Libert et al. 2007). Thus, food dilution may involve determinants of lifespan distinct from an actual reduction in calories consumed.

Serum IGFI levels are reduced in rodents by either CR or protein restriction. However, in humans, 1–6 years of CR without protein restriction has no effect on total or free serum IGFI levels or the IGFI to IGFI binding protein-3 (IGFBP-3) ratios (Fontana et al. 2008). In contrast, CR with accompanying mild protein restriction does reduce serum IGFI levels and the IGFI to IGFBP-3 ratio (Fontana et al. 2008). Thus, reduced protein intake is required for CR-mediated lowering of IGFI levels in humans. Since reduced IGFI signaling appears to be a key to the longevity and anticancer effects of CR in rodents and lower eukaryotes, reduced protein intake may be required for a similar longevity response in humans. In this regard, long-term

CR in adult humans also decreases serum T_3, also a powerful mitogen for some cell types (Fontana et al. 2006c). This may be one mechanism by which CR reduces basal metabolic rate and core body temperature in humans, and reduces cancer incidence and tumor growth.

12.3.2 Tryptophan Restriction

In rodents, restriction of specific amino acids can extend longevity. Thirty and 40% reductions in dietary tryptophan produce elevated initial mortality, but delayed later-life mortality in the survivors, leading to an increase in maximum lifespan (from 30.5 months in control rats to 36.3 months in tryptophan restricted rats) (Segall and Timiras 1976; Segall 1979; Timiras et al. 1984). This does not appear to be a simple CR effect, since the tryptophan restricted rats and mice consume slightly more food than controls (De Marte and Enesco 1986a).

Low tryptophan diets appear to decrease multiple biomarkers of aging in rats and mice. These diets delay reproductive senescence (Segall and Timiras 1976; Segall et al. 1983), senescent-related deterioration of the coat in female rats (Segall and Timiras 1976), age-related tumor onset in rats and mice (Segall and Timiras 1976; De Marte and Enesco 1986b), and senescence-associated loss of temperature homeostasis in rats (Segall and Timiras 1975). For example, female rats fed a low tryptophan diet from 3 weeks of age and returned to a control diet at older ages were capable of producing offspring at up to 28 months of age (Segall and Timiras 1976). No control fed rats were capable of having litters after 17 months of age. Tryptophan restricted mice also enjoy a slight (~10%) numerical increase in lifespan at most ages, although it is unclear whether this increase is statistically significant (De Marte and Enesco 1986a).

The initial increase in mortality after the introduction of low tryptophan diets is reminiscent of the early mortality reported for dwarf mice (Fabris et al. 1971; Shire 1973) and older mice abruptly shifted to CR in old age (Weindruch and Walford 1988). In these cases, elevated mortality appears to result from animal husbandry. Caging of dwarf mice with retired breeders led to extended, rather than shortened lifespans, and a more gradual introduction of CR in older mice led to extended lifespans for them as well (Weindruch and Walford 1982; and see below).

12.3.3 Methionine Restriction

Restriction of dietary methionine by 80% also reproduces many of the physiological effects of CR. The effects of methionine restriction are robust. It has extended lifespan in four strains of rats and one mouse strain (Orentreich et al. 1993; Miller et al. 2005; Malloy et al. 2006; Komninou et al. 2006; Pamplona et al. 2006). Like CR, methionine restriction produces highly pleiotropic effects on the health and physiology of rodents. Methionine restriction preserves insulin sensitivity; decreases preneoplastic, aberrant crypt foci induced by azoxymethane treatment; prevents age-associated increases in serum lipids; reduces visceral fat mass, IGFI, basal insulin, glucose, and leptin levels; and increases serum adiponectin levels (Malloy et al. 2006). Methionine restriction produces these effects without an apparent reduction in dietary calorie intake (Malloy et al. 2006). Disrupted IGFI signaling produces many of these same effects on physiology and lifespan (Bartke 2005).

12.3.4 Mechanisms of Lifespan Extension by Protein, Methionine and Tryptophan Restriction

Relatively little is known about the molecular mechanisms of the physiological effects of protein or specific amino acid restriction. Methionine restriction, CR, and mutations that disrupt GH signaling reduce the blood levels of insulin, glucose (Masoro et al. 1992; Dhahbi et al. 2001), IGFI (Oster et al. 1995; Sonntag et al. 1999), and thyroid hormones (Weindruch and Walford 1988) relative to the appropriate control groups (Miller et al. 2005; Bartke et al. 2008). Decreased plasma glucose levels do not appear to be an important factor in the antiaging action of CR (McCarter et al. 2007). However, their similar regulation of several endocrine systems suggests that they may share this mechanism of action. Protein and amino acid restriction may extend lifespan by directly altering the rate or accuracy of translation or protein processing (Reviewed in Kaeberlein and Kennedy 2008; Tavernarakis 2008), patterns of DNA methylation, glutathione levels (Richie et al. 1994), stress resistance (Miller et al. 2005), or hormesis (Gems and Partridge 2008).

The rate of translational initiation or elongation appears to be intimately involved in lifespan regulation (Tavernarakis 2008). Tryptophan and methionine are both essential amino acids, and methionine is the initiating amino acid for translation. Further, protein restriction may limit the amount of essential amino acids available for protein synthesis. Downregulation of protein translation in response to reduced nutrient availability is a highly conserved longevity pathway for invertebrates (Kaeberlein and Kennedy 2008; Tavernarakis 2008). Deceleration of protein translation robustly extends the lifespan of C. elegans (Hansen et al. 2007; Pan et al. 2007; Syntichaki et al. 2007). Eight of 25 longevity-related genes conserved between yeast and C. elegans modulate protein translation (including TOR, S6 kinase, large subunit ribosomal proteins, and translation initiation factors) (Smith et al. 2008). Thus, a single nutrient-responsive longevity pathway may be conserved downstream of TOR, a nutrient-responsive protein kinase (Kaeberlein and Kennedy 2008).

Downregulation of protein synthesis may redirect dietary energy away from reproduction and toward repair; thereby reducing protein aggregation and proteotoxicity through decreased demands on the chaperone, repair and degradation pathways (Kaeberlein and Kennedy 2008). It may also enhance the relative rate of turnover of whole body protein through enhanced proteasomal degradation, autophagy, and (in mitotic tissues) apoptosis (Spindler and Dhahbi 2007; Cuervo 2008; Kaeberlein and

Kennedy 2008). CR increases the rate of protein and lipid turnover in mitotic and postmitotic tissues, and cell turnover in mitotically competent tissues (reviewed in Spindler and Dhahbi 2007).

12.3.5 Meal Frequency and Intermittent Fasting

The effects of CR are sufficiently robust that meal frequency and composition are not crucially important to its actions. Lifespan is extended by CR in rodents whether food is presented three times per week (Cheney et al. 1983), as a single daily meal (Nelson and Halberg 1986), or as 2 (Masoro et al. 1995) or 6 meals per day (Nelson and Halberg 1986). A dietary paradigm sometimes termed "every-other-day feeding" (EOD) or "intermittent fasting" has been reported to extend rodent lifespan since the mid 1940s (Carlson and Hoelzel 1946; Goodrick et al. 1983). Although it has been studied as a dietary paradigm distinct from CR, more recent studies measuring food intake have shown that EOD induces mild (~20%) CR (e.g. Caro et al. 2008; Spindler and Mote, unpublished observations). Not surprisingly, the effects of EOD feeding are very similar to those of mild CR. They include a lower incidence of diabetes, lower fasting blood glucose and insulin concentrations, and other effects comparable to those of CR (reviewed in Varady and Hellerstein 2007).

In humans, reduced meal frequency may be detrimental to health. A randomized, crossover study in normal weight humans used two 8-week periods during which subjects consumed sufficient calories to maintain their weight as either one or as three meals per day (Stote et al. 2007). Subjects consuming one meal per day had significant increases in blood pressure and total LDL- and HDL-cholesterol. No significant effects were found on heart rate, body temperature, or most other blood variables. Thus, reduced meal frequency in normal weight humans may produce adverse effects on serum LDL-cholesterol and blood pressure.

12.3.6 Specific Micronutrients and Lifespan

A few specific nutrients and nutrient combinations have been reported to extend the lifespan of metazoans. Many such studies have significant confounds associated with them, and these studies are not further considered here. They include studies in which caloric intake or subject weights were not measured (thus confounding the potential treatment effects with those of CR); studies which are underpowered statistically and thus of questionable significance; studies using enfeebled, transgenic, or short-lived mouse strains; studies using treatments with pro-oxidants, toxins, carcinogens or radiation to induce enfeeblement; and studies using transplanted or induced tumors. Such studies may have value in another context, but not in the context of this review.

Variations in a number of nutrients have been reported to lengthen lifespan. Replacing sucrose with starch extends the mean lifespan of rats by about 9%, independent of the total calories consumed (Murtagh-Mark et al. 1995). These effects do not appear related to body weight or energy absorption. Two combinations of vitamins, minerals and other dietary/cellular components reportedly extend the lifespan of male C57BL/6 mice (Bezlepkin et al. 1996; Lemon et al. 2005). One combination, fed isocalorically to male C57BL/6 mice, reportedly extended their lifespan by 9.5 and 16% (mean and maximum, respectively) (Bezlepkin et al. 1996). Another combination fed ad libitum to C57BL/6JxSJL hybrid mice of both sexes, extended mean longevity by 11% (Lemon et al. 2005). The supplemented and unsupplemented groups were not significantly different in weight, suggesting the effects may have been independent of the calories consumed. However, our experience using daily feeding of diets containing the latter of these supplement combinations to male B6C3F1 mice is that the supplemented mice eat approximately 5% fewer calories than controls (S.R. Spindler and P.L. Mote, unpublished observations). In other studies, 1% creatine supplemented diets fed isocalorically beginning at 1 year of age reportedly increases the mean (9%) and maximum (3.5%) lifespan of C57BL/6J female mice (Bender et al. 2007). Male C57BL/6 mice fed diets containing either green tea polyphenols (11.7%) or tetrahydrocurcumin (6.5%) from the age of 13 months have significantly longer average lifespans (Kitani et al. 2007). The tetrahydrocurcumin effects may have been CR-related since the body weight of the supplemented mice were significantly below that of the control mice for the first year after beginning the supplementation. In contrast to the above results, high fat and high carbohydrate diets can shorten the lifespan of rodents (as well as humans, see below) (French et al. 1953).

It is common for natural products cited as extending the longevity of laboratory animals to actually induce voluntary CR in the ad libitum fed test mice and rats. The case of tetrahydrocurcumin cited above is one example. Another is oral treatment of mice with a 15% kombucha extract, which was reported to extend the lifespan of C57BL/6 mice by 5% (males) and 2% (females) (Hartmann et al. 2000). Kombucha extract has an anoretic effect, reducing the weight of the female mice from an average of 25 to 21 g and male mice from 30 to 22 g at autopsy. In another example, data have been published suggesting that treatment of Fisher 344 rats with Ginkgo biloba extract extended their lifespan (Winter 1998). However, these studies were underpowered statistically and the article presents insufficient information for the reader to exclude an anoretic CR-related effect.

Thus, enhanced consumption of a limited number of natural dietary components has been shown to extend the lifespan of rodents. However, many, even most, such studies are unintepretable either because effects of the tested agents on caloric consumption cannot be excluded or because of other methodological confounds.

A few nutrients and plant components have been reported to extend fruit fly longevity. These include boron (Massie 1994); green tea catechin extract (Li et al. 2007), α-lipoic acid (Bauer et al. 2004); resveratrol (Bauer et al. 2004; Wood et al. 2004), although these results could not be repeated in similar studies by others (Bass et al. 2007); Rhodiola rosea root extract (Jafari et al. 2007); Rosa damascena in Drosophila (Jafari et al. 2008); and Vitamin E in Zaprionus paravittiger (Kakkar et al. 1996). A confound associated with most of these studies again involves the potential effects of the supplements on food consumption, which is rarely measured. Compound screening has also been performed using C. elegans, with similar potential confounds.

12.4 Evolutionary Origin of the Health Benefits of CR

A study performed using bowl and doily spiders provides a striking example of the evolutionary conservation of the response to CR (Austad 1989). Austad compared the longevity of the spiders in the field to that of captive spiders fed different numbers of fruit flies per week. Adult female spiders lived about 8 days in the field. However, in the protected laboratory environment the mean adult lifespan was 42.3, 63.9, or 81.3 days when they were allowed to consume five, three, or one Drosophila per day. The longevity and health responses to CR in species from three Kingdoms and four phyla within Animalia suggests that it is either an ancient trait which has been selectively retained after arising very early in evolution, or that it has arisen multiple times by convergent evolution.

The most widely accepted hypothesis for this plasticity of lifespan and the similarities in the CR response among species is that it has been selected to allow individuals to shift energy usage away from growth and reproduction and towards maintenance and stress resistance during times of nutritional stress (Holliday 1989; Masoro and Austad 1996; Kirkwood and Austad 2000). This shift presumably allows individuals to expend energy on metabolic processes, behaviors, and metabolism that enhance survival. CR delays reproductive maturation in juvenile animals, and by reducing reproductive hormones, behavior and fertility in adults, limits the number of offspring (see below). When the food supply is restored (in the Spring for example), refeeding induces maturation in prepubertal animals, and returns adult animals to fertility (see below).

12.5 Longevity Mutations and the Molecular Mechanisms of CR

The search for the mechanisms of action of CR received a major boost when it became clear that components of the insulin and IGFI signaling pathways and their homologues were key regulators of longevity in C. elegans, Drosophila and mice (Kenyon 2005; Russell and Kahn 2007; Bartke 2008a). For example, Ames dwarf mice, which are homozygous for a loss-of-function mutation at the *Prop1* locus, exhibit a 40–70% increase in mean and maximum lifespan compared with their phenotypically normal siblings (Brown-Borg et al. 1996). The Ames mutation ablates cell lineages of the anterior pituitary, leading to a combination of endocrine abnormalities that includes low levels of circulating growth hormone (GH), IGFI, thyroid-stimulating hormone, thyroid hormones, and prolactin (Sornson et al. 1996). These same hormone systems are down regulated by CR in rodents, monkeys and humans (see references elsewhere in this review). In Ames and Snell mice, the absence of these hormones postpones many degenerative age-related changes in physiology, including the development of neoplastic diseases, immune system decline, and collagen cross-linking (Flurkey et al. 2001; Ikeno et al. 2003). Snell dwarf mice, which are homozygous for a lose-of-function mutation in the *Pit1* locus, are deficient in the same anterior pituitary cell types and hormones as the Ames mice, and have an essentially identical longevity and physiological phenotype (Flurkey et al. 2001). These phenotypes are very similar to those found in CR mice. Genome wide microarray studies suggest that CR and the Ames mutation affect many of the same genes, pathways and processes (Tsuchiya et al. 2004). Ames mice are responsive to CR (Bartke et al. 2001). Such data are sometimes mistakenly thought to indicate that CR and the Ames mutation act on distinct signaling and genic pathways. However, on both theoretical (Gems et al. 2002) and experimental (Tsuchiya et al. 2004) grounds, it is more likely that CR and insulin/IGFI signaling use many of the same molecular pathways to extend health and lifespan in mammals.

Among the affected endocrine signaling systems, the IGFI system appears important for the mammalian CR response. CR strongly downregulates GH and IGFI levels in mammals. Growth hormone receptor knockout (GHRKO) mice have robustly extended lifespan and reduced IGFI levels (Coschigano et al. 2003). Mice homozygous for a loss-of-function mutation in the GH releasing hormone receptor gene (Lin et al. 1993), termed "little mice", have reduced GH levels and a ~25% increase in lifespan compared to congenic controls (Flurkey et al. 2001). However, type 1 insulin-like growth factor-I receptor (IGFR) knockout mice do not have as robust a longevity phenotype. Only female mice with heterozygous deletion of the IGFI receptor have extended lifespan (~25% extension) compared to their phenotypically normal siblings (Holzenberger et al. 2003). Males have no increase in lifespan. These results suggest that in addition to the IGFI signaling pathway, other signaling systems perturbed in the Ames, Snell and GHRKO mice are also involved in the CR response.

The importance of insulin and IGFI signaling is reinforced by studies of the lifespan effects of the *Klotho* gene. Mice that overexpress *Klotho* live 31% (males) and 19% (females) longer than congenic controls (Kurosu et al. 2005). An alternatively spliced mRNA produced by the *Klotho* gene encodes a circulating protein which blocks autophosphorylation of the occupied IGFI receptor and promotes dephosphorylation of activated insulin and IGFI receptors, thereby inhibiting their downstream signaling (Kurosu et al. 2005).

The involvement of IGFI signaling in longevity is further supported by studies in which disruption of downstream mediators of this pathway increased the lifespan of C. elegans, Drosophila, and mice. Female mice carrying a homozygous knockout of the insulin receptor substrate (IRS)-1 gene live 32% longer than wild type controls (Selman et al. 2008). Deletion of IRS-2 did not extend lifespan. However, the short lifespan of the control and mutant mice in these studies also is a cause for concern (Barger et al. 2003; Liang et al. 2003). In other studies, knockout mice with reduced IRS-2 levels either through-out the body or just in the brain have an ~18% extension of lifespan (Taguchi et al. 2007). These mice have no change in body weight or food consumption, and have reduced fasting levels of insulin and glucose, and improved insulin and glucose tolerance. In Drosophila, mutational inactivation of the *chico* gene, which encodes a fly homologue of the IRS 1–4 genes, reduces insulin/IGFI signaling and results in a dwarf phenotype with extended lifespan (~50% extension) (Clancy et al. 2001). Thus, reduced signaling through the insulin/IGFI signaling cascade is one likely mediator of the CR response. However, other pathways disrupted in Ames, Snell and GHRKO mice also appear to be involved.

GH receptor signaling may be important in the longevity effects of CR. As discussed above, GHRKO mice have low levels of IGFI and GH receptor signaling and significantly extended lifespans (Coschigano et al. 2003; Holzenberger et al. 2003). They are also unresponsive to CR (Bonkowski et al. 2006), suggesting that the GH receptor, or a gene it regulates, is required for the CR longevity response. But the GHRKO longevity study has the weakness that both the control and mutant mice have short lifespans relative to most wild-type mice (Barger et al. 2003; Liang et al. 2003). These short lifespans raise the possibility that the knockouts may have simply restored a more normal lifespan to enfeebled "control" mouse strains (Barger et al. 2003; Liang et al. 2003).

12.6 Reproductive Effects of CR in Rodents

Studies of Drosophila, rats and mice often find that control females outlive control males, and the same is true of their CR counterparts (e.g. Cheney et al. 1983; Turturro et al. 1999; Sheldon et al. 1995; Magwere et al. 2004; Partridge et al. 2005). Female medflies live significantly longer than male flies at all food levels (Davies et al. 2005). However, in mice females are not always longer lived (Blackwell et al. 1995). While the molecular basis for these sexual dimorphisms is not known, they appear to extend to the effects of CR on disease states. Male and female G93A mice (which overexpress the mutant human Cu/Zn-SOD gene, making them an animal model of amyotrophic lateral sclerosis) respond differentially to STCR (Hamadeh and Tarnopolsky 2006). STCR promotes the onset of disease and shortens male, but not female lifespan. These effects are very likely the results of the differential effects of the sex steroids on processes such as cancer growth (see below).

CR and mutations that reduce IGFI signaling attenuate, delay, or abolish fertility and fecundity in Drosophila, C. elegans, and female rats and mice (Klass 1977; Holehan and Merry 1985a; Weindruch and Walford 1988; Kenyon et al. 1993; Chippindale et al. 1993; Chapman and Partridge 1996; Gems et al. 1998; Tissenbaum and Ruvkun 1998; Bruning et al. 2000; Burks et al. 2000; Tatar et al. 2001; Clancy et al. 2001; Chandrashekar and Bartke 2003; Giannakou et al. 2004). CR, begun after weaning, delays sexual maturation and the loss of estrous cyclicity in aging female mice, perhaps by retarding the rate of follicular depletion (Merry and Holehan 1979; Holehan and Merry 1985b; Nelson et al. 1985; Gonzales et al. 2004). Oocyte numbers are higher in CR females when compared to age-matched ad libitum controls (Lintern-Moore and Everitt 1978; Nelson et al. 1985). Female mice and rats maintained on CR from an early age and returned to ad libitum feeding later in life are restored to reproductive cycling and reproductive competence at chronological ages where ad libitum animals have long ceased these activities (Osborne et al. 1917; Ball et al. 1947; Visscher et al. 1952; Holehan and Merry 1985a; Nelson et al. 1985).

CR later in life causes female rats and mice to cease reproductive cycling and reproduction (Ball et al. 1947; Visscher et al. 1952; Nelson et al. 1985). Even moderate levels of CR initiated in rodents during adulthood can sustain the function of the female reproductive axis into advanced chronological age (Selesniemi et al. 2008). These findings suggest that the beneficial effects of CR on female reproductive function are at least partly mediated via maintenance of the ovarian follicular reserve. However, CR affects the secretion patterns of hormones produced by the hypothalamus and pituitary, which participate in the control of ovarian function (Martin et al. 2008). Hence, the effects of CR on reproductive performance probably also involves the hypothalamic–pituitary–gonadal axis.

LTCR reduces male reproductive activity. LTCR, male CFY Sprague-Dawley rats sire significantly fewer litters at most ages (Merry and Holehan 1981). They also have a delayed and reduced pubertal peak of serum testosterone, delayed puberty, and reduced testosterone levels (Merry and Holehan 1981). Males subjected to 25% CR were less able to attract females, and those on 50% CR were both less able to attract and to sexually arouse females (Govic et al. 2008). Thirty days of CR in adult male Wistar rats reduces serum and testicular testosterone concentrations (Santos et al. 2004). One study found that LTCR increases serum testosterone levels at all ages in Lobund-Wistar rats (Snyder et al. 1988). However, control fed Lobund-Wistar rats already have high testosterone levels. Thus, together the results suggest that CR delays and suppresses sexual maturation, reproductive hormone levels, and reproduction in both male and female rats, extends reproductive lifespan, and delays reproductive senescence when older female rodents are shifted from CR to an ad libitum diet.

The response to CR is also sexually dimorphic in Drosophila (Magwere et al. 2004). Here, the peak extension of longevity occurs at lower food concentrations for males than for females, perhaps because females devote energy to egg-laying (Magwere et al. 2004). In addition, CR increases the longevity of females 60%, while for males the increase is only 30%. Male and female Drosophila also exhibit sex-specific differences in the sensitivity of their lifespan to insulin/IGFI signaling, nutrient/energy demand, and the allocation and utilization of energy (Clancy et al. 2001, 2002).

Sex-related differences have been reported in the energy balance of young CR Wistar rats. CR females deactivate facultative thermogenesis to a greater degree than males, conserving metabolically active organ mass and decreasing adipose depots to a greater extent than restricted males. This ability likely has survival advantages for females when food is limiting (Valle et al. 2005). Thus, both the physiology and hormonal responsiveness of males and females appear to lead to sex specific differences in their responsiveness to CR.

12.7 Reproductive Effects of CR in Humans

In humans, fasting and extremely low calorie diets suppress the pituitary-gonadal axis, and free testosterone levels in males (Veldhuis et al. 1993; and references therein). However, neither the Biosphere study nor the CALERIE studies have reported a decrease in sex steroid levels in adult humans subjected to a CR diet for up to two years (Walford et al. 2002; Holloszy and Fontana 2007; Redman et al. 2009). The reasons for the dissimilarity between these results and those found in rodents and fasting humans is unclear. Obvious possibilities are that the Biosphere or CALERIE diets were not of sufficient severity or length to produce suppressive effects on sex steroid levels. This seems unlikely based on the decrease in BMI and the length of time involved in these studies. More likely, humans may not be as readily susceptible to suppression of the hypothalamic-pituitary-gonadal axis as rodents. Differences in the life-histories of primates and rodents may select for variable degrees of reproductive sensitivity to nutrient deprivation. The Biosphere and CALERIE studies did not detect a decrease in serum IGFI or the IGFI to IGFBP-3 ratio in humans (Walford et al. 2002; Holloszy and Fontana 2007; Redman et al. 2009). However, moderate protein-restriction can rapidly downregulate serum IGFI levels in CR humans (Fontana et al. 2006a, 2008). Thus, the nutrient composition of the diets used in the human CR studies cited above may have a pivotal effect on whether IGFI, sex steroids, and reproductive activity are affected by CR.

12.8 Anticancer Effects of CR in Rodents

The anticancer effects of CR were recognized a century ago (Moreschi 1909), long before its effects on longevity were published (McCay et al. 1935; Reviewed in Weindruch and Walford 1988). A widely accepted model of carcinogenesis proposes that it involves 3 stages: *initiation*, an initial mutation leading to cell division; *promotion*, accrual of additional mutations in cell growth or proliferation-related genes, accompanied by some clonal expansion; and *progression*, significant clonal expansion and genetic damage resulting in a gain of tumor mass either through increased rates of cell proliferation or reduced rates of apoptosis (Hursting et al. 2003). The relative rates of proliferation and apoptosis during the promotion and progression stages of tumorigenesis are the major determinants of the rates of tumor onset and growth (Grasl-Kraupp et al. 1997; Hursting et al. 2003; Patel et al. 2004). LTCR can reduce carcinogenesis at each stage of this multistage model (Hursting et al. 2003; Patel et al. 2004). Fasting, STCR and LTCR reduce cellular proliferation and increase apoptosis in a wide variety of organs and tissues, including liver (Merry and Holehan 1985; Grasl-Kraupp et al. 1994; James and Muskhelishvili 1994; Muskhelishvili et al. 1996; Hikita et al. 1999; Higami et al. 2000), bladder (Lok et al. 1988; Dunn et al. 1997), skin (Merry and Holehan 1985; Lok et al. 1988), kidney (Merry and Holehan 1985), heart (Merry and Holehan 1985), mammary gland, esophagus, jejunum (Lok et al. 1988), and colorectum (Lok et al. 1988;

Premoselli et al. 1997). LTCR increases the age of onset and decreases the rate of growth and number of metastases produced by many model tumors, including hepatomas, mammary carcinomas, and prostatic tumors (Hursting et al. 2003). CR suppresses the carcinogenic action of several classes of chemicals, inhibits several forms of radiation-induced cancer, and inhibits neoplasia in early-tumor-onset knockout and transgenic mouse models (Patel et al. 2004).

Even when begun late in life (at the beginning of the accelerated mortality phase of their lifespan) CR decreases tumor-associated mortality by 3.1-fold within 8 weeks, and extends both mean and maximum lifespan in C3B6F1 mice (Dhahbi et al. 2004; Spindler 2005). These effects appear to be due to a rapid, CR-related reduction in the rate of tumor growth (Dhahbi et al. 2004; Spindler 2005). Because CR does not reduce the apparent rate of tumor-cell division under these circumstances, its anti-cancer effects probably result from an increase in the rate of tumor cell death (S.R. Spindler, Y. Higami and I. Shimokawa, unpublished results).

Despite the potent anticancer effects of CR, it is important to recognize that there are reports of tumor-induction regimens and/or tumor types against which CR may be ineffective. For example, reducing mean body weight by up to 25% through chronic dietary restriction did not reduce prostate carcinogenesis induced in male Wistar-Unilever rats by a sequential regimen of cyproterone acetate, testosterone propionate, and N-methyl-N-nitrosourea followed by testosterone implants (McCormick et al. 2007). In contrast, others have reported that male Wistar rats subjected to CR after a similar prostate cancer induction protocol had longer prostate cancer-free survival and lower cancer rates than control fed rats (Boileau et al. 2003). Thus, the model system used can affect the results of the studies. In addition, there are no data indicating that such models of carcinogenesis recapitulate the spontaneous tumors against which CR is effective in vivo.

12.8.1 Apoptosis and the Anticancer Effects of CR

Preneoplastic and neoplastic cells are more sensitive to apoptotic cell death than normal cells, and are selectively eliminated by CR. In one example in mice, 40% food restriction for 3 months significantly reduced the volume of liver occupied by preneoplastic foci (Grasl-Kraupp et al. 1994). The relative rates of proliferation and apoptosis during the promotion and progression stages of carcinogenesis are the major determinants of the rates of tumor onset and growth (Hursting et al. 2003; Patel et al. 2004). There is compelling evidence that LTCR, STCR, and many of the longevity-enhancing mutations in mice can dramatically delay tumor-associated mortality and increase apoptosis in at least some mitotic tissues, including the liver and lung (Kennedy et al. 2003; Massaro et al. 2004, 2007; reviewed in Spindler and Dhahbi 2007). LTCR increases the age of onset and decreases the rate of growth and number of metastases produced by model tumors, including hepatomas, mammary carcinomas, and prostatic tumors (Weindruch and Walford 1988; Hursting et al. 2003). LTCR suppresses the carcinogenic action of several classes of chemicals, inhibits several forms of radiation-induced cancer, and inhibits neoplasia in early-tumor-onset knockout and transgenic mouse models (Hursting et al. 2003; Patel et al. 2004).

The rate of apoptosis in the liver of mice is 3-times higher in hepatocytes from CR mice at all ages (James and Muskhelishvili 1994; Muskhelishvili et al. 1995; James et al. 1998). Increased hepatocyte apoptosis is associated with a significantly lower incidence of spontaneous hepatomas throughout the life of LTCR mice, although the mechanism remains unknown. Even brief periods of CR enhance apoptosis and reduce tumor incidence. For example, 1 to 3 months of food restriction can significantly increase the latency and reduce the incidence of spontaneous cancer over the entire lifespan of a mouse (Klebanov 2007). Just one week of CR induces apoptosis in preneoplastic liver cells of old mice (Muskhelishvili et al. 1996). Forty-percent food restriction for 3 months eliminates 20–30% of liver cells through apoptosis, and reduces the number and volume of chemically induced preneoplastic foci by 85% (Grasl-Kraupp et al. 1994). CR also enhances apoptosis in mitotically competent cells of other organs, including jejunum, colon, bladder, and dexamethasone treated lymph node and spleen lymphocytes of MRL/lpr mice (Luan et al. 1995; Holt et al. 1998).

12.8.2 IGFI, Insulin Receptor Signaling, and Cancer

Oncogenesis may involve activation of oncogenes, overexpression of growth factors, inappropriate growth factor signaling, and/or inactivation of tumor suppressor genes (Alexia et al. 2004; Pandini et al. 2004; Boissan et al. 2005; Huether et al. 2005a, b; Alexia et al. 2006; Hopfner et al. 2006). LTCR, STCR, and many of the longevity-enhancing mutations in mice dramatically reduce insulin and IGFI serum levels and post-receptor signaling, delay tumor-associated mortality, and increase apoptosis in mitotic tissues (Dhahbi et al. 2001; Dominici et al. 2002; Kennedy et al. 2003; Koubova and Guarente 2003; Bartke 2005, 2006). IGFR plays a major role in mitogenesis, transformation and protection from apoptosis. In many tumors, the level of IGFR expression correlates with disease stage, survival, development of metastases, and tumor de-differentiation (Scharf and Braulke 2003; Sedlaczek et al. 2003; Yao et al. 2003). Reduction of IGFR signaling in mice by mutation, genetic manipulation, or Klotho overexpression enhances the rate of tumor cell apoptosis, reduces the number and the volume of preneoplastic lesions, delays tumor-associated mortality, and extends lifespan (Bartke and Brown-Borg 2004; Bartke 2006). A part of the antineoplastic effects of CR is probably a direct result of reduced IGFR signaling (Dunn et al. 1997; Hopfner et al. 2006). Cells that do not express IGFR are resistant to transformation by any means (Baserga et al. 2003).

LTCR also increases insulin sensitivity (glucose uptake and utilization) about 4-fold in rats and mice, and reduces blood insulin levels to about half those found in control rats and mice (Masoro et al. 1992; Dhahbi et al. 2001). These lower insulin levels may also be important in the anticancer effects of CR. According to the classical view, the IR predominantly mediates anabolic effects, while the IGFR predominantly mediates antiapoptotic, mitogenic, and transforming effects (Baserga et al. 2003; Biddinger and Kahn 2006). However, there are two IRs. Insulin receptor B (IRB) is the long-recognized IR isoform found in insulin-responsive tissues. IRB mediates insulin-responsive glucose uptake, glucose metabolism, protein synthesis and cell growth in muscle and adipose. IRA is a splice variant derived from the same primary transcript as IRB. IRA has a mitogenic and transforming role in oncogenesis. Both the IRA and IRB are expressed in most insulin responsive tissues (Kosaki and Webster 1993). Insulin, IGFI and IGFII are high affinity activators of IRA. Activation of IRA can promote growth and protect malignant cells from apoptosis (Kalli et al. 2002; Sciacca et al. 2002; Vella et al. 2002). Binding of IGFII or insulin to IRA induces a different, but partially overlapping, gene expression profile (Pandini et al. 2003, 2004). In cells expressing IRA, IGFII stimulates the Shc/ERK branch of the insulin/growth factor signaling pathway, inducing mitogenesis and migration more potently than insulin (Frasca et al. 1999). In contrast, insulin is more potent than IGFI in stimulating signaling through the IRS/AKT pathway, leading to enhanced glucose metabolism, protein synthesis, and cell growth (Frasca et al. 1999).

Thus, the CR mediated reductions in serum levels of IGFI and insulin may be keys to its anticancer activity. However, the potential for effects of CR on other growth factors (especially IGFII), and their receptors is unexplored. Other autocrine and paracrine mediators of tumor cell growth and division may be altered by CR, and be important in its anticancer effects. We still do not know whether the effects of CR on insulin and IGFI levels are the primary mechanism for reduced carcinogenesis in mammals.

12.8.3 Can Life History Account for the Anticancer Effects of CR?

Relatively few cancers arise from postmitotic cells (Wright and Deshmukh 2006). Some or all cancers probably arise from mitotically competent cells, perhaps stem cells (Chiang and Massague 2008). Cell division is required to genetically fix oncogenic mutations. Selection can act only on pre-reproductive or reproductively active members of a population, and in many species few animals live long enough to die of cancer in the wild (e.g. Phelan and Austad 1989). Highly effective molecular mechanisms have evolved to suppress the development of cancer until after reproduction ceases (Sharpless 2005). These include mechanisms for high fidelity DNA replication and repair, and tumor suppressor genes such as p53, Rb, p16^{INK4a} and ARF (Janzen et al. 2006; Krishnamurthy et al. 2006; Molofsky et al. 2006).

The anticancer effects of CR may have co-evolved with another phenotype, which is subject to natural selection. We have argued that this trait is the metabolic role of tissues as reservoirs of metabolic energy (Spindler and Dhahbi 2007). CR animals alternate between balanced waves of turnover and resynthesis (Spindler and Dhahbi 2007). The minimal energy stores in CR animals demand that the immediate postprandial period for them is characterized by intense biosynthetic activity, including cell division, under the powerfully anabolic influence of insulin (Spindler and Dhahbi 2007). During the later postprandial period, as insulin levels decline and glucagon and other catabolic hormones rise, high relative rates of protein, lipid, organelle and cellular turnover take place (Spindler and Dhahbi 2007). Thus, CR drives repeated cycles of cell and tissue turnover in mitotically competent tissues and tissue turnover and repair in postmitotic tissues. These effects seem to be major contributors to the multifaceted health and longevity effects of CR.

12.9 Metabolic Effects of Aging and CR

Aging produces a decline in the autophagic and apoptotic turnover of cells, organelles, membranes, carbohydrates, and proteins (Bergamini 2006; Donati 2006). It also decreases the expression and capacity of enzymes required for mobilizing protein for the production of metabolic energy (Dhahbi et al. 1999, 2001; Spindler 2001; Dhahbi and Spindler 2003; Hagopian et al. 2003a, b; Spindler et al. 2003, 2007). During the postabsorptive state, when blood insulin and glucose levels wane, and glucagon, glucocorticoids and catecholamines increase, most tissues begin to utilize amino acids derived from protein turnover to generate energy via the tricarboxylic acid (Krebs) cycle. This drives the degradation of glycogen and the autophagic degradation of proteins, organelles, and membranes. The glucagon/insulin ratio is a major regulator of autophagy in vivo (Meijer and Codogno. 2006). Insulin represses autophagy. During the postabsorptive period, the high glucagon/insulin ratio activates autophagy (Meijer et al. 2006). Amino acid catabolism leads to the transport of carbon and nitrogen to the liver, mostly as glutamine, for synthesis into glucose and for nitrogen disposal as urea. These changes are consistent with the decrease in whole body protein turnover found with age in rodents (Goodman et al. 1980; Lewis et al. 1985; el Haj et al. 1986; Goldspink et al. 1987; Merry et al. 1987, 1991a, 1992). Decreased macromolecular turnover may underlie the age-related accumulation of oxidatively and otherwise damaged protein. This decrease may exacerbate the effects of the oft reported age-related increase in oxidant production by isolated muscle mitochondria (e.g. Bejma and Ji 1999; Mansouri et al. 2006).

Broadly speaking, CR appears to lead to increased catabolism of protein and lipid derived from proteolysis, autophagy, and the autophagic and apoptotic death of mitotically competent cells to generate substrates for energy. For example, CR stimulates autophagy in rat liver (Cavallini et al. 2001; Donati et al. 2001; Bergamini et al. 2007). This is apparent in the genic and enzymatic shift toward increased protein, lipid and organelle degradation and disposal (turnover) (Feuers et al. 1989; Dhahbi et al. 1999, 2001; Tillman et al. 1996; Cao et al. 2001; Spindler 2001; Dhahbi and Spindler 2003; Hagopian et al. 2003a, b; Spindler et al. 2003, 2007; Tsuchiya et al. 2004). CR increases the enzymatic capacity of the muscle for the mobilization of carbon for glucose synthesis, and the disposal of nitrogen as urea

(Tillman et al. 1996; Dhahbi et al. 1999, 2001; Cao et al. 2001; Spindler 2001; Dhahbi and Spindler 2003; Spindler et al. 2003). Together, these changes enhance energy generation from protein in the tissues, and gluconeogenesis in the liver to supply glucose for energy production by the brain and other organs, and disposal of the nitrogen produced from protein turnover as urea. We found very similar changes in metabolic enzyme expression in Ames dwarf mice, which are also long-lived (Tsuchiya et al. 2004).

These metabolic responses to CR are apparent in their effect on organ physiology. For example, CR decreases the rate of DNA and protein synthesis, and decreases the RNA content of the liver, kidney, heart and small intestine in rats (Lewis et al. 1985; Merry and Holehan 1985; el Haj et al. 1986; Goldspink et al. 1987; Merry et al. 1987, 1991a, 2002; Merry and Holehan 1991b). After the initiation of either CR or fasting, mitotically competent tissues such as liver and lung undergo a profound, rapid, and reversible loss of cells (via necrosis, apoptosis and/or autophagic cell death), proteins, and lipids (Kouda et al. 2004; Massaro et al. 2004). Intracellular turnover in metazoans involves degradation of cytoplasmic and nuclear proteins by cellular calpains (Sorimachi et al. 1997) and the proteosome (Myung et al. 2001), degradation of mitochondria by mitochondrial proteases (Bakala et al. 2003), autophagic degradation of membranes, mitochondria, ribosomes, ER, and peroxisomes (Bergamini 2006; Donati 2006; Kadowaki et al. 2006), and cellular turnover initiated by apoptotic, necrotic and autophagic cell death (Jin et al. 2005). Nutritional stress increases the rate of these processes (Finn and Dice 2006). For example, fasting a rat for 48 hours reduces liver weight by half and liver proliferative index by 85%, while increasing its apoptotic index by 2.5-fold (Kouda et al. 2004). The number of lung alveoli in mice is reduced by 35% after 72 hours of 33% CR (Massaro et al. 2004). Fifteen days of CR reduces alveolar number by 45% (Massaro et al. 2004). Refeeding for 72 hours fully restores alveolar number (Massaro et al. 2004). Thus, CR appears to tip the regulatory balance in some mitotic tissues toward apoptosis and the degradation of cellular carbohydrate, protein, and lipid. Despite this increase in macromolecular degradation and decreased rates of macromolecular synthesis and cell division, LTCR animals are able to maintain their organ mass by balancing these effects to produce higher rates of protein turnover and smaller visceral organ sizes (Weindruch and Sohal 1997). LTCR (30%) reduces the weight of the heart, liver, kidney, spleen, prostate, and skeletal muscle of rats by 25 to 50% (Weindruch and Sohal 1997). This turnover may drive many of the beneficial effects of CR on health and lifespan.

12.10 CR and Autophagy

The enhancement of autophagy by CR may be necessary, but not sufficient for its effects on lifespan. Autophagy is induced by nutrient deprivation (Mizushima and Klionsky 2007). It thereby provides amino acids and substrates for energy production when nutrients are scarce. The rate of autophagy declines with age in essentially all tissues (Reznick and Gershon 1979; Cuervo et al. 2005). CR enhances autophagy (Cavallini et al. 2001; Donati et al. 2001). Further, in wild-type and long-lived worm mutants, mutational reduction of autophagy shortens lifespan (Melendez et al. 2003; Hars et al. 2007). In contrast, enhanced expression of the autophagy-related 8a gene, a rate limiting autophagy gene, in older fly brains, extends average adult lifespan by 56% and promotes resistance to oxidative stress (Simonsen et al. 2008). In C. elegans, mutational inactivation of essential autophagy-related genes abolishes the extended longevity phenotype of eat-2 mutants, which are a model of CR (Morck and Pilon 2006; Jia and Levine 2007; Hansen et al. 2008; Toth et al. 2008). These results suggest that autophagy is required for the CR-related longevity response. However, enhanced autophagy does not appear to be sufficient for lifespan extension. In C. elegans, inhibiting genes required for autophagy blocks the longevity effects of CR (Hansen et al. 2008). But, lifespan extension by Daf-2 (IR/IGFR) longevity mutations requires both autophagy and DAF-16/FOXO. Mutation of daf-16 leaves autophagy active, but blocks lifespan extension by the Daf-2 longevity mutations (Hansen et al. 2007).

The role of autophagy in the effects of CR on longevity in mammals remains unclear at present. However, the importance of turnover and renewal, its decline with age, and its enhancement with CR strongly suggests that autophagy also is central to the longevity effects of CR in mammals. Pharmacological enhancement of autophagy may prove therapeutic in protein-aggregation related neurodegenerative disorders such as Huntington's disease and some forms of Parkinson's disease (see Sarkar et al. 2009 and reference therein).

12.11 CR and Glucocorticoids

Glucocorticoids also may contribute to the anticancer effects of CR (Nelson et al. 1995; Masoro 1995). CR transiently increases the diurnal level of free plasma corticosterone in mice and rats (Sabatino et al. 1991; Klebanov et al. 1995; Han et al. 2001). Further, at least some of the anticancer effects of CR may require glucocorticoids. Either 20 or 40% CR significantly decreases the incidence and multiplicity of skin papillomas initiated with 7,12-dimethylbenzanthracene (DMBA) and promoted with 12-O-tetradecanoylphorbol- 13-acetate (TPA) in female, SENCAR mice (Stewart et al. 2005). Adrenalectomy of these mice partially reverses the inhibition of papilloma multiplicity and incidence by CR. Corticosterone supplementation to the CR-adrenalectomized mice restores much of the ability of CR to reduce papilloma incidence and multiplicity. Glucocorticoid replacement restores the effectiveness of CR at inhibiting DMBA- and TPA-induced skin, and DMBA-induced pulmonary carcinogenesis in adrenalectomized mice (Pashko and Schwartz 1992, 1996). Therefore, at least some of the anticancer effects of CR may require glucocorticoids. The reason for this dependence is not known. However, glucocorticoids do suppress cellular proliferation and enhance apoptosis in some cell types, including osteoblasts, lymphocytes and keratinocytes (Weinstein 2001; Budunova et al.

2003; O'Brien et al. 2004; Herold et al. 2006). Thus, elevation of glucocorticoids in combination with suppression of sex steroids, insulin and IGFI may cooperate to produce a part of the anticarcinogenic activity of CR.

12.12 Cardiovascular Effects of CR

Aging impairs cardiovascular capacity, contractility, and diastolic and systolic function (McGuire et al. 2001). In rodents and humans, three major age-associated changes markedly affect myocardial performance. First, the development of myocardial fibrosis, a hallmark of cardiac aging in both humans and rats, is initiated by cellular necrosis and apoptosis (Eghbali et al. 1989; Anversa et al. 1990). Cell death induces reparative interstitial and perivascular collagen deposition, which plays a key role in the development of fibrosis in aged human and rodent hearts (De Souza 2002). Fibrosis decreases cardiac distensibility and increases diastolic pressure, impairing coronary hemodynamics and lowering coronary reserve (Janicki 1992; Varagic et al. 2001). Second, the decline in the number of cardiomyocytes with age due to necrosis and apoptosis is followed by compensatory myocyte hypertrophy (Colucci 1997; Lushnikova et al. 2001). This remodeling and the weakening and bulging of the ventricular wall is the most common cardiac manifestation of aging (Lakatta 2000). Remodeling necessitates increased atrial and ventricular filling pressure. These general age-related changes appear to underlie cardiac arrhythmias, dysfunction, and failure. Third, an age-related impairment in mitochondrial bioenergetics appears to contribute to myocardial stiffness, apoptosis, atrophy and compensatory hypertrophy (Moreau et al. 2004; van Raalte et al. 2004).

CR is known to reduce the incidence and increase the age of onset of cardiovascular diseases (Wagh and Stone 2004). Approximately 40% of male C57BL/6 mice develop cardiomyopathy by 1000 days of age (Turturro et al. 2002). Aging produces extensive changes in cardiac gene expression in mice, including changes consistent with a metabolic shift from fatty acid to carbohydrate metabolism, increased expression of extracellular matrix genes, and reduced protein synthesis (Lee et al. 2002). LTCR produces changes in gene expression in the heart consistent with preserved fatty acid metabolism, reduced DNA damage, decreased innate immune activity, and cytoskeletal reorganization (Lee et al. 2002). We found that the initiation of CR in older mice rapidly reduces the expression of genes associated with extracellular matrix and cytoskeletal structure and dynamics, cell motility, and inflammation (Dhahbi et al. 2006). The initiation of CR also increases the expression of genes associated with PPARα signal transduction and fatty acid metabolism for energy production (Dhahbi et al. 2006). Beginning CR in older mice reduces natriuretic peptide precursor type B expression along with the expression of a number of forms of collagen, and reduces perivascular collagen deposition in the heart (Dhahbi et al. 2006). LTCR also preserves smaller cardiomyocytes in the left ventricle of the heart of old-LTCR mice, suggesting reduced age-related cell death and hypertrophy (Dhahbi et al. 2006; Spindler and Dhahbi 2007). Together these changes are consistent with a rapid CR-related reduction in the forces leading to the development of myocardial fibrosis, tissue remodeling and hemodynamic stress.

12.13 Immunological Effects of CR

The immune system declines with age, impairing the response to vaccinations and infections (McElhaney 2005; Effros 2007). Elderly people are particularly susceptible to influenza, with 80–90% of mortalities from infection with influenza virus occurring in individuals aged 65 years and older (Trzonkowski et al. 2009). Elderly individuals also suffer more frequently from autoimmunity (Yung and Julius 2008).

The data regarding the effects of age on innate immune function are contradictory. Some find an age-related decline in the function of neutrophils, macrophages and natural killer cells, while other studies detect no such changes (Gomez et al. 2008). However, studies in mice and humans have shown that the adaptive immune system is deregulated by aging (reviewed by Dorshkind et al. 2009). Thymic involution reduces the number of naïve T-cells produced with age (Jamieson et al. 1999). Also, B-cell production in the bone marrow is reduced with age (Jamieson et al. 1999; Johnson et al. 2002; Miller and Allman 2003; Van der et al. 2003). Lymphocytes from older humans have impaired ability to undergo immunoglobulin class-switching recombination (Frasca et al. 2008). There also is an age-related decline in CD4+ T helper cell function which may contribute to this impairment (Haynes and Swain 2006). CD4+ T cells, but more so CD8+ T cells, undergo clonal expansion in the elderly, limiting the repertoire of antigens to which the resident T cells can respond and blocking the niches that would otherwise be colonized by newly formed immune cells (Haynes and Swain 2006).

CR can inhibit the age-related decline in immune function in mice, non-human primates and humans. While aging causes a shift from naïve to memory cells, CR increases the number of naïve T cells and the diversity of the T-cell repertoire (Miller 1996; Pawelec et al. 1999; Nikolich-Zugich and Messaoudi 2005; Messaoudi et al. 2006). CR also retards the age-related decline in antigen presentation, T-cell proliferation, and antibody production in response to influenza vaccination (Effros et al. 1991; Pahlavani 2000). However, CR also can increase mortality in mice in response to influenza virus infection, probably due to lack of energy reserves (Ritz and Gardner 2006). Further, in contrast to the benefits of CR initiated during early adulthood, CR initiated in young male rhesus monkeys accelerates the loss of naïve T cells, decreases T-cell repertoire diversity, and increases the frequency of T-cell clonal expansions (Messaoudi et al. 2008), predisposing them to a reduced immune response after infection (Messaoudi et al. 2004). CR initiated in old rhesus monkeys produces lymphopenia and decreases T-cell proliferative capacity, suggesting that CR initiated at older ages can have adverse affects in primates (Messaoudi et al. 2008). Thus, little is known about how CR will affect the immune system of humans, especially elderly humans.

12.14 Neurological Effects of CR

Neuronal loss promotes the cognitive, sensory, and motor impairments found with age (Kuhn et al. 1996). CR appears to reduce neuronal damage in response to toxins and other stressors, protect against neurodegenerative diseases, and increase neurogenesis and synaptic plasticity, even in old animals (Gillette-Guyonnet and Vellas 2008). Aging is accompanied by a reduction in synaptic contacts, synaptic strength, neural plasticity, brain neurogenesis (Burke and Barnes 2006; Kuhn et al. 1996), and spinal cord densities and neurogenesis (Burke and Barnes 2006; Segovia et al. 2006; Fontan-Lozano et al. 2008). Aging alters the expression of the neurotransmission related genes N-methyl- D-aspartate receptor, brain derived neurotrophic factor, Trk-B, and α-synuclein (Mattson et al. 2001). For some of the genes, CR counteracts these changes (Mattson et al. 2001).

The cells of the aging brain experience increased levels of oxidative stress (Serrano and Klann 2004; Zecca et al. 2004; Martin et al. 2006), damaged protein (Gray et al. 2003; Trojanowski and Mattson 2003; Martin et al. 2006), and damaged DNA (Lu et al. 2004; Kyng and Bohr 2005). Aging also leads to changes in the normal brain that are similar to, but less severe than, the changes found in some age-related neurological diseases. These changes include amyloid-β accumulation (Trojanowski and Mattson 2003), as found in Alzheimer's disease; α-synuclein accumulation and dopamine depletion in substantia nigra neurons, as found in Parkinson's disease (Grondin et al. 2002); Cu/Zn-SOD accumulation in motor neurons, as found in amyotrophic lateral sclerosis (Rakhit et al. 2004); and reduced BDNF levels, as found in Alzheimer's disease and Huntington's disease (Zuccato et al. 2003; Mattson et al. 2004).

LTCR either reduces the rate of damage, or reverses some of the age-related degenerative changes described above (Idrobo et al. 1987; Ingram et al. 1987; Stewart et al. 1989; Pitsikas et al. 1991; Pitsikas and Algeri 1992; Eckles et al. 1997; Eckles-Smith et al. 2000). LTCR increases neurotrophic factor levels and attenuates neurochemical and behavioral deficits in a primate model of Parkinson's disease (Maswood et al. 2004). CR protects neurons and improves the functional outcome in rodent and monkey models of stroke, Alzheimer's, Parkinson's and Huntington's diseases (Mattson et al. 2004, and references therein). CR attenuates Alzheimer's disease type brain amyloidosis in squirrel monkeys (Saimiri sciureus) (Qin et al. 2006). CR also increases the levels of brain-derived neurotrophic factor and heat-shock proteins in neurons and stimulates neurogenesis in the hippocampus [(Mattson et al. 2004) and references therein]. Neurons of LTCR rats and mice are more resistant to oxidative, metabolic and excitotoxic stressors than those of controls (reviewed in Martin et al. 2006). CR induces analgesia in acute and chronic models of pain (los Santos-Arteaga et al. 2003; Hargraves and Hentall 2005).

Late-onset, STCR can bring about many of the beneficial neurological effects found in animals subjected to LTCR. CR initiated late in life can reverse the age-related downregulation of neural cell adhesion molecule and polysialylated-neural cell adhesion molecule in various brain regions, and reverse the age-related upregulation of astrocytic glial fibrillary acidic protein in old rats (Kaur et al. 2008). These results suggest that late-onset STCR can have beneficial effects on neuroplasticity. Related results have been found in humans. High caloric intake induced obesity increases the risk of age-related cognitive decline (Knecht et al. 2008). Just 3 months of 20% CR in elderly humans significantly improves verbal memory scores, in concert with a decrease in fasting plasma levels of insulin and high sensitive C-reactive protein (Witte et al. 2009).

12.15 Genic Effects of CR

Here we focus on results of large scale gene-expression studies in mammals. Genetic studies in lower eukaryotes have been reviewed elsewhere (Mair and Dillin 2008). Large-scale microarray studies have allowed a number of important broad conclusions to be drawn regarding aging and the effects of CR in mammals. First, aging appears to shift tissues and cells toward patterns of gene expression consistent with enhanced inflammation, stress, and the age-related pathologies of each tissue type studied (e.g. Lee et al. 1999, 2000, 2002; Cao et al. 2001; Kayo et al. 2001; Dhahbi et al. 2004, 2006; Tsuchiya et al. 2004; Spindler and Dhahbi 2007). Second, the majority of the genic responses of tissues to CR appear to be tissue- and cell-specific, and tailored to resisting or reversing the characteristic age-related dysfunctions of specific tissue or cell-types (e.g. Lee et al. 1999, 2000, 2002; Cao et al. 2001; Kayo et al. 2001; Dhahbi et al. 2004, 2006; Tsuchiya et al. 2004; Spindler and Dhahbi 2007).

Finally, there appears to be a broad difference in the way in which mitotically competent and postmitotic tissues respond to CR. Consistent with evidence cited elsewhere in this article, the genic responses of heart and liver to CR suggest that mitotically-competent tissues shift toward increased susceptibility to apoptotic cell death, while postmitotic tissues shift toward increased cellular repair and stress resistance (Spindler and Dhahbi 2007). For example, in heart we and others have found that CR produces gene expression and histochemical changes associated with reduced fibrosis, tissue remodeling, apoptosis and blood pressure, and alterations in signal transduction associated with enhanced lipid catabolism for energy production and contractility (Edwards et al. 2003; Gonzalez et al. 2004; Dhahbi et al. 2006). In contrast, in liver, CR produces changes in gene expression and enzymology associated with the differentiated functions of the liver, including induced gluconeogenesis, enhanced detoxification, enhanced apoptosis of damaged cells, and reduced glycolysis and fatty acid biosynthesis (reviewed in Spindler and Dhahbi 2007).

12.16 CR and Specific Transcription Factors

A number of transcription factors and transcriptional co-activators have been implicated in the health and longevity effects of CR. Among these, perhaps none have received more attention to date than SirT1 (silent mating type information regulation 2 homolog S. cerevisiae), PGC-1α (peroxisome proliferator activated receptor γ co-activator-1-α), and AMPK (AMP-activated protein kinase). We consider the evidence for this involvement below.

12.16.1 SirT1

Orthologues of the Sir gene encode a family of NAD-dependent protein deacetylases that are termed "sirtuins" (Imai et al. 2000; Landry et al. 2000). Under some conditions, sirtuins appear to be required for lifespan extension by CR in yeast (Lin et al. 2000), C. elegans (Wang and Tissenbaum 2006), and Drosophila (Rogina and Helfand 2004). Further, an additional copy of the orthologous gene can increase lifespan of yeast (Kaeberlein et al. 1999), Caenorhabditis elegans (Tissenbaum and Guarente 2001), and Drosophila (Rogina and Helfand 2004) by 18–50%. However, the relationship between lifespan and these sirtuin orthologues is complex and still poorly defined, even in yeast. For example, at 0.5% glucose, which is defined as CR in some yeast studies, lifespan extension requires Sir2 (the yeast orthologue of SirT1) and nicotinamide adenine dinucleotide (NAD) (Lin et al. 2000). However, at 0.05% glucose, replicative lifespan is extended in yeast by a different mechanism that is independent of Sir2 (Kaeberlein et al. 2004, 2005).

The mammalian SirT1 sirtuin gene is the closest human ortholog to the yeast Sir2 gene. It encodes a sirtuin that deacetylates transcription factors and coactivators with key roles in metabolism and stress resistance, including p53 (Vaziri et al. 2001), FOXO proteins (Motta et al. 2004; Brunet et al. 2004), PGC-1α (Rodgers et al. 2005), and NF-κB (nuclear factor-κB) (Yeung et al. 2004). For example, PGC-1α is activated when it is deacetylated by SirT1. The potential role of SirT1 in aging and CR is even less defined than that of Sir2 in yeast. CR does not extend the lifespan of SirT1 null mice (Chen et al. 2005). However, the meaning of this observation is unclear. SirT1-null mice have multiple, severe abnormalities (McBurney et al. 2003). They are small, with developmental defects of the retina and heart. On an inbred background, they usually do not survive postnatally (Cheng et al. 2003). On an outbred genetic background, most do survive to adulthood (McBurney et al. 2003), but these mice are small and sterile, with craniofacial abnormalities, and an eyelid inflammatory condition. They are hypermetabolic, utilize ingested food inefficiently, have inefficient liver mitochondria, and have elevated rates of lipid oxidation (Boily et al. 2008). The absence of lifespan extension in these mice has been interpreted as evidence that SirT1 is mechanistically required for the CR response (Haigis and Guarente 2006). It seems equally possible that the molecular pathways engaged by CR remain functional in SirT1-null mice, but are already fully induced by the stress of the gene knockout (Schumacher et al. 2008). Mice carrying knockins of human DNA repair-deficient progeroid syndrome genes die prematurely. Despite these short lifespans, there are significant similarities between the genome-wide gene-expression patterns of the progeroid mice and those of long-lived CR mice, Ames and Snell dwarf mice, CR-Ames dwarf mice, and growth hormone receptor knockout mice (Schumacher et al. 2008). The endocrine, metabolic, and gene expression changes induced by longevity-associated manipulations such as CR and dwarfism appear to induce a "survival"-related genetic program which is also strongly induced by the progeroid mutations. Similarly, the SirT1 knockout mice may already have fully induced this survival response, and be unable to respond further to CR.

12.16.2 PGC-1α

The mitochondrial capacity for oxidative phosphorylation (ATP production) declines with advancing age in human skeletal muscle (Short et al. 2005). This decrease is accompanied by an age-related decline in muscle mitochondrial number and function (Short et al. 2005). The decline also involves a decrease in the levels of mitochondrial DNA, mRNA, and protein, and a decrease in the tissue levels of citrate synthase activity. Because mitochondria participate in glucose and lipid catabolism, mitochondrial dysfunction reduces muscle carbohydrate and lipid uptake and catabolism, thereby increasing systemic dyslipidemia and hyperglycemia (Petersen et al. 2007; Kim et al. 2008). These steps appear to be key to the age-related development of insulin resistance and metabolic syndrome (Eckel et al. 2005; Petersen et al. 2007; Kim et al. 2008). Metabolic syndrome aggravates age-related inflammation, hypertension and cardiovascular disease (Kim et al. 2008).

PGC-1α is a transcriptional co-activator and central regulator of mitochondrial biogenesis, oxidative phosphorylation, hepatic gluconeogenesis, fatty acid oxidation and muscle fiber type (Puigserver et al. 1998; Lopez-Lluch et al. 2008; Scarpulla 2008; Ventura-Clapier et al. 2008; Canto and Auwerx 2009). It integrates the activity of a diverse set of transcription factors, including PPARα, NRF-1, NRF-2, ERRx and mtTFA (Puigserver et al. 1998; Puigserver et al. 1999; Schreiber et al. 2004; Gleyzer et al. 2005). An increase in PGC-1α level or activity can upregulate the expression of genes associated with mitochondrial biogenesis, fatty acid oxidation, and the switching of muscle fiber types to type I fibers, which have a higher mitochondrial content and oxidative rate than type IIb fibers (Puigserver et al. 1998; Wu et al. 1999; Vega et al. 2000; Canto and Auwerx 2009). The induction of PGC-1α expression can also have a salutary effect on the survival of at least some cell types. For example, lentiviral overexpression of PGC-1α in the striatum protects neurons from atrophy in the R6/2 mouse model of Huntington's disease (Cui et al. 2006).

The activity of the PGC-1α gene is regulated by multiple factors. One of these is TORC 1 (transducer of regulated cAMP response element-binding protein), the coactivator of the cAMP response element-binding protein (CREB). TORC 1 and the other two members of the TORC family, TORC 2 and 3, strongly induce PGC-1α expression in skeletal muscle cells (Wu et al. 2006). PGC-1α induction leads to upregulation of the mitochondrial respiratory chain and TCA cycle genes, leading to increased cellular respiration and fatty acid oxidation (Wu et al. 2006).

While mitochondrial number and function decline with age, CR maintains PGC-1α levels in muscle during aging, thereby preserving mitochondrial function and biogenesis (Baker et al. 2006; Hepple et al. 2006). Maintenance of muscle PGC-1α levels may involve signals originating in the hypothalamus. CR and the longevity-related dwarf mutations enhance fatty acid oxidation and inhibit fatty acid synthesis (Tsuchiya et al. 2004; Spindler and Dhahbi 2007). Pharmacologic inhibition of fatty acid synthase has been shown to increase the number of mitochondria in white and red (soleus) skeletal muscle by increasing the levels of malonyl-CoA (the substrate of fatty acid synthase) in the hypothalamus (Cha et al. 2006). Malonyl-CoA accumulation increases signaling through the sympathetic nervous system to the skeletal muscle. In the muscle, this signaling induces the expression of the β-adrenergic signaling molecules norepinephrine, β3-adrenergic receptor, and cAMP, thereby inducing PGC-1α and estrogen receptor-related receptor-α (ERRα) expression and mitochondrial biogenesis (Cha et al. 2006, and references therein). Exercise induces mitochondrial biogenesis in muscle through a signaling cascade beginning with elevated intracellular free calcium and culminating in enhanced PGC-1α activity (Wu et al. 2002).

12.16.3 AMPK

The mammalian AMPK is a heterotrimer of a catalytic α subunit, and two regulatory subunits, β and γ (Hardie et al. 2003; Carling 2004). There are multiple genes for each of these subunits (α1, α2, β1, β2, γ1, γ2, γ3). The use of alternative promoters and alternative splice sites further increases the structural complexity of this kinase. In mammals, AMPK is allosterically activated by AMP (Hardie et al. 2003; Carling 2004). Thus, it is readily activated by an increase in the intracellular ratio of AMP to ATP. While this response makes AMPK sensitive to intracellular energy charge, it also responds to and regulates food intake and systemic energy expenditure through its responsiveness to hormonal signals in the central nervous system and peripheral tissues (Kahn et al. 2005). AMPK is also activated by exercise (Zhou et al. 2001).

Elevated intracellular AMP binds to AMPK, which increases the interaction of the kinase with LKB1, which is itself a kinase. This interaction leads to phosphorylation and activation of AMPK. This activation initiates a signaling cascade which stimulates glucose uptake and fatty acid oxidation, and downregulates protein synthesis in cultured skeletal muscle cells (Alessi et al. 2006). The phosphorylation and activation of TSC2 (tuberous sclerosis 2 protein) by activated AMPK protects cultured HEK293 (human embryonic kidney 293) cells (an adenovirus-DNA transformed cell line) from energy deprivation-induced apoptosis, suggesting that AMPK also can enhance cell survival (Inoki et al. 2003).

AMPK activity is induced by the systemically active cytokines leptin and adiponectin, which are also termed adipokines since they are secreted by adipose. The AMPK pathway is required for the metabolic- and insulin-sensitizing actions of these adipokines. Leptin selectively stimulates threonine 172 (Thr-172)-phosphorylation of the AMPK α2-catalytic subunit of the kinase in skeletal muscle, activating it (Minokoshi et al. 2002). Stimulation involves an early, direct effect of leptin on muscle, as well as an indirect, more sustained activation involving the action of the hypothalamic-sympathetic nervous system on α adrenergic receptors leading to fatty acid oxidation in muscle (Minokoshi et al. 2002). Adiponectin stimulates AMPK phosphorylation and activity in muscle and liver in vivo and in vitro (Yamauchi et al. 2002). In muscle, AMPK activation is required for adiponectin responsive stimulation of fatty acid oxidation and glucose transport (Yamauchi et al. 2002). In liver, adiponectin mediated inhibition of PEPCK and glucose 6 phosphatase expression and hepatic glucose production requires AMPK activation (Yamauchi et al. 2002). AMPK also regulates food intake by responding to hormonal and nutrient signaling in the hypothalamus (Reviewed in Kahn et al. 2005).

Reduced skeletal and heart muscle AMPK activity may be a key factor responsible for reduced mitochondrial biogenesis, insulin resistance and impaired lipid metabolism in older animals and humans (Reznick et al. 2007; Qiang et al. 2007). Age-related reductions in mitochondrial number and function may contribute to the dysregulated intracellular lipid metabolism, leading to the increased insulin resistance found in older animals and humans (Petersen et al. 2003). In skeletal muscle cells, activated AMPK enhances fatty-acid oxidation by directly phosphorylating and activating PGC-1α and acetyl-CoA carboxylase 2 (ACC2) (Merrill et al. 1997; Winder et al. 2006; Jager et al. 2007).

Enhanced AMPK signaling is capable of extending lifespan in C. elegans, Drosophila and yeast (Tschape et al. 2002; Apfeld et al. 2004; Harkness et al. 2004). However, its molecular linkage to the lifespan effects of CR in these organisms remains uncertain. For example, the C. elegans ortholog of the AMPKα subunit aak-2 it is not required for the lifespan effects of eat-2 mutants, which are thought to mimic CR in C. elegans (Curtis et al. 2006).

The role of AMPK in the longevity effects of CR in mammals also is unclear at present. Fasting does appear to be a potent inducer of AMPK activity in mammals. Just 6 hours of food deprivation in rats increases the level of the Thr-172-phosphorylated, AMPK α-subunit in gastrocnemius muscle by approximately 4-fold (de Lange et al. 2006). In these studies, fasting also increased the level of nuclear PGC-1α and PPARγ protein (de Lange et al. 2006).

However, CR does not appear to produce effects that recapitulate those of fasting. In mice, LTCR did not change AMPKα1 and -α2 activities measured in cell-free extracts of heart or skeletal muscle (Gonzalez et al. 2004). In contrast, CR did increase liver AMPK activity in cell-free extracts by approximately 20% (Gonzalez et al. 2004). In another study, STCR increased the myocardial level of Thr172-phosphorylated AMPKα in both young and old male Fisher 344 rats (Shinmura et al. 2005). However, others report that LTCR produces little effect on the level of Thr-172-phosphorylated AMPKα in the quadriceps femoris muscle of Wistar rats, and decreases its level in the liver (To et al. 2007).

In contrast, AMPK clearly is activated in mammals by metformin treatment and by exercise (Zhou et al. 2001; Hadad et al. 2008). Metformin is a small-molecule biguanide which increases peripheral insulin sensitivity (Widen et al. 1992), increases glucose uptake in skeletal muscle (Galuska et al. 1991; Hundal et al. 1992), decreases hepatic gluconeogenesis (Johnson et al. 1993; Hundal et al.

2000), and inhibits complex I of the mitochondrial respiratory chain (El Mir et al. 2000; Owen et al. 2000; Detaille et al. 2002; Zou et al. 2004). Metformin therapy reduces blood glucose levels in type 2 diabetics, and inhibits the development of metabolic syndrome (Orchard et al. 2005). Metabolic syndrome is associated with increased cardiovascular- and diabetes-associated morbidity and mortality. At least some of the physiological pathways engaged by CR also appear to be targeted by metformin. Eight weeks of oral metformin treatment of old mice reproduces approximately 75 and 74% of the gene expression changes found using Affymetrix microarrays that were produced by LTCR or by 8 weeks of CR, respectively (Dhahbi et al. 2005).

Metformin usage may be associated with a reduced risk of cancer in humans and animals. A small-scale, population-based case-control study found that diabetics who take metformin are 23% less likely to get cancer compared with those taking sulfonylurea (Evans et al. 2005). The risk reduction was 40% for people who took the drug for a longer time. A second population based cohort study found that new users of metformin had a significantly reduced risk of cancer mortality relative to those using sulfonylureas or exogenously administered insulin (Bowker et al. 2006). A confound in these studies is that they do not distinguish the potential pro-cancer effects of elevated insulin exposure in the sulfonylurea and insulin treated groups from the putative protective effects of metformin on the development of cancer.

In female, transgenic HER-2/neu mice, a breast-cancer model, chronic metformin (100 mg/kg body weight) administration increased mean and maximum life spans by 8 and 13%, respectively (Anisimov et al. 2005a; Anisimov et al. 2005b). However, the metformin treated mice decreased their food consumption at 4 and 6 mo of age, suggesting that "voluntary CR" may be the source of this anticancer response. Phenformin, a biguanide that is structurally and functionally closely related to metformin, extended the lifespan of C3H mice by ~23% and reduced tumor incidence by 80% (Dilman and Anisimov 1980; Anisimov et al. 2003). Recent studies have shown that metformin inhibits breast cancer cell growth, colony formation and induces cell cycle arrest in vitro (Alimova et al. 2009).

Metformin treated SHR mice, a normal but short-lived strain, experience nearly a 38 and 21% increase in mean and maximum lifespan (Anisimov et al. 2008). However, metformin decreases food consumption in these mice. The magnitude of the increase in lifespan seems greater than that expected of the modest decrease in calorie consumption. Thus, metformin may extend the lifespan of this mouse strain directly. In another study, a low dose of metformin increased the survival of male R6/2 mice by approximately 20% (Ma et al. 2007). R6/2 mice are a transgenic model of Huntington's disease.

Together, these results suggest that metformin may extend the lifespan of short-lived or enfeebled mouse strains by reducing cancer-related mortality. However, where measured, compounds which extend the lifespan of short-lived, enfeebled mice are not similarly effective in long-lived strains (Spindler 2010).

AMPK signaling pathways cross-talk with other signaling systems thought to regulate lifespan. AMPK activation limits energy consuming processes such as translation initiation and protein synthesis directly by phosphorylating and inhibiting the activity of key factors (Bolster et al. 2002; Inoki et al. 2003). As discussed in Section 12.3.4, the rate of translational initiation or elongation can strongly influence lifespan in lower eukaryotes. AMPK phosphorylates and inhibits the activity of TSC2, which in turn inhibits Rheb activity (Ras homolog enriched in brain), and finally mTOR (mammalian target of rapamycin) activity (Bolster et al. 2002; Inoki et al. 2003). Phosphorylation of TSC2 by AMPK is required for translation regulation and cell size control in response to energy deprivation. Furthermore, TSC2 and its phosphorylation by AMPK protect cells from energy deprivation-induced apoptosis.

Another example of cross-talk involving AMPK is its direct phosphorylation and activation of PGC-1α during nutrient deprivation (Jager et al. 2007). In primary mouse muscle cells, activation of PGC-1α by AMPK results in feed-forward, auto-induction of the PGC-1α promoter, enhancing its transcription and that of glucose transporter 4 and a number of key mitochondrial oxidative metabolism genes (Jager et al. 2007).

The induction of AMPK activity also inhibits the activity of mTOR (Gwinn et al. 2008). mTOR is a nutrient-activated, signal-transduction factor activated by PGC-1α in skeletal muscle (Cunningham et al. 2007). The inhibition of mTOR by rapamycin decreases the expression of PGC-1α, ERRα, and a number of key, nuclear respiratory factors, resulting in decreased mitochondrial gene expression and reduced oxygen consumption (Cunningham et al. 2007). Thus, a negative feedback loop may exist in which AMPK both induces PGC-1α activity and inhibits mTOR activation by PGC-1α, thereby limiting the feed-forward response of PGC-α to nutrient deprivation.

12.17 Hypothermia and CR

CR lowers the mean body temperature of homeotherms such as mice, rats, monkeys and humans (Weindruch and Walford 1988; Walford and Spindler 1997; Walford et al. 2002). This decrease in temperature is likely due in part to lowering of tri-iodothyronine (T_3) levels. Long-lived mutant mouse strains have reduced serum T_3 levels and core body temperatures, including the Ames, Snell and the growth hormone receptor/binding protein knockout dwarf mice (Hauck et al. 2001; Bartke and Brown-Borg 2004). However, thermal regulation is more complex than just thyroid hormone levels (Swoap 2008). The decrease in body temperature in response to 40% CR can vary from 1.5 to 5.8°C, depending on mouse strain, indicative of its multigenic character (Rikke et al. 2003).

Lowering the body temperature of poikilothermic vertebrates, such as annual fish, by 3 to 5°C can substantially prolong their lifespans (Liu and Walford 1966, 1970, 1972). Further, the lifespans of ectothermic animals such as C. elegans (Klass 1977) and Drosophila (Lamb 1968) are inversely-related to environmental temperature. This seems to strongly suggest that the decrease in body temperature and the longevity effects of CR are causally related, at least in poikilotherms. However, in these animals, CR and temperature actually may use different pathways to regulate longevity (Miquel et al. 1976; Mair et al. 2003).

Body temperature appears to contribute to the health and longevity benefits of CR in mice (Turturro and Hart 1991; Walford and Spindler 1997). Support for this idea has come from studies of transgenic mice overexpressing uncoupling protein 2 only in hypocretin neurons (Conti et al. 2006). These mice have a 0.3–0.5°C reduction of the core body temperature, and a 12% increase in male and a 20% increase in female median lifespan, independent of caloric consumption (Conti et al. 2006). Another line of evidence supporting a link between body temperature and lifespan comes from a study of C57BL/6 mice housed either at the normal housing temperature of 20–23°C, or at 30°C (Koizumi et al. 1996). At the normal temperature, 40% CR produces a 1°C decrease in core body temperature, while this decrease is prevented at 30°C (Koizumi et al. 1993; Jin and Koizumi 1994). Housing at 30°C reverses the life-extending effects of CR in C57BL/6 mice, largely by decreasing its anti-lymphoma action (Koizumi et al. 1996). This reduction may involve loss of the general inhibitory action of CR on cellular proliferation. Housing LTCR Fischer 344 rats at 30 °C largely cancels the normal CR-related reduction in cellular proliferation found in the jejunum, epidermis, and lung of these animals (Jin and Koizumi 1994). Maintaining C57BL/6 female mice at 30°C also antagonizes the suppressive effect of LTCR on white blood cell counts (Koizumi et al. 1993). These counts are considered indicative of the cellular proliferation rates in the bone marrow. Thus, decreasing core body temperature may be both sufficient to extend lifespan and required for the antiproliferative and anticancer effects of CR.

12.18 Exercise and CR

The relationship between exercise and biomarkers of health in humans is discussed above. The question of whether exercise will produce effects similar to CR has been addressed in rodents by Holloszy and his colleagues (Holloszy and Smith 1987; Holloszy and Schechtman 1991; Holloszy 1992, 1997). Exercise reproduces a number of the same physiological effects as LTCR in rats. Exercised rats are smaller and have less fat than sedentary animals fed the same number of calories. Young rats, about 4 months of age, will run voluntarily almost 8 kilometers per day if a running wheel is placed in their cages. As they age, their running distances decline. But, if their food intake is reduced by 8% of ad libitum consumption, their running behavior declines much more slowly, and will continue at reduced levels even into extreme old age (Holloszy et al. 1985). The runners are much leaner and smaller overall than sedentary control rats fed the same number of calories as the runners. However, while running significantly increased the mean lifespan of the rats by about 10%, it did not increase their maximum lifespan (Holloszy and Schechtman 1991;

Holloszy 1997). CR is capable of increasing both the mean and maximum lifespan of rats.

The meaning of these results remains unclear. Gerontologists often aver that the extension of average and maximum lifespan must be mechanistically different, because only CR is known to extend maximum lifespan. They often conjecture that extension of maximum lifespan means that CR slows something termed *the underlying rate of aging*. However, CR can extend the maximum and average lifespan of mice by slowing the rate of tumor growth (Dhahbi et al. 2004; Spindler 2005). It seems unlikely that very many gerontologists would regard a regimen of cancer chemotherapy as *slowing the underlying rate of aging*.

12.19 Adiposity and the Health Effects of CR

There is a direct relationship between adiposity, especially visceral fat mass, and mortality in humans (Takata et al. 2007; Fair and Montgomery 2009). CR clearly reduces body fat, especially visceral fat, in mice, rats, rhesus and cynomolgus monkeys, and humans (reviewed in Masoro 2005). Adipose mass appears to be inversely related to longevity, and therefore may be important in the longevity effects of CR. Initially, several studies seemed to suggest that reduced fat mass is not required for the longevity effects of CR (Bertrand et al. 1980; Harrison et al. 1984). The strongest piece of evidence came from a study in which ob/ob mice, which are leptin-deficient, hyperphagic, and obese, were compared to congenic lean mice lacking the mutation. CR ob/ob mice live twice as long as ad libitum-fed lean mice, even though the ob/ob mice have over twice the adipose weight of the congenic lean mice (Harrison et al. 1984). These results argue that adipose fat mass is not a key determinant of longevity in CR mice. This conclusion is consistent with a previous longitudinal study that found body weight throughout life not to be correlated with the lifespan of ad libitum fed rats (Bertrand et al. 1980).

More recent studies using the fat-specific insulin-receptor knockout (FIRKO) mouse, in which the IR has been specifically knocked out in adipose tissue, suggests that adipose may have an important effect on lifespan, perhaps through the action of a secreted factor. FIRKO mice have increased median (18%) and maximum (14%) lifespan compared with phenotypically normal littermate controls (Bluher et al. 2003). The knockout renders adipose cells insulin resistant, but the mice are insulin sensitive and lean despite food intakes equivalent to those of controls. The mice also have increased systemic insulin sensitivity and resistance to diabetes, even when fed a high-fat diet. The authors suggest that these effects may be due to the reduction in fat mass. Thus, a systemically-active secretory product from adipose tissue may be responsible. As discussed by Russell and Kahn (2007), leptin, which is secreted by adipose, is unlikely to be this factor, since leptin levels in FIRKO mice are similar to those in controls. Adiponectin, which is secreted by adipose, is a possibility, since its level increases in CR rats and humans (Zhu et al. 2004; Weiss et al.

2006), and it increases in FIRKO mice (Bluher et al. 2002). Other possible factors include steroid hormones (Russell and Kahn 2007). Adipose both synthesizes and metabolizes sex steroids (Belanger et al. 2002). The diffusible factor also could be an unidentified adipokine or small molecule.

12.20 Human BMI, Morbidity, and Mortality

CR men and women undergo many of the same metabolic adaptations that occur in CR rodents and monkeys, including decreased metabolic, hormonal, and inflammatory risk factors for diabetes, cardiovascular disease, and cancer (see above for references). Indeed, most published studies focus on the relationship between BMI and disease-related mortality because caloric intake is difficult to measure in population based studies. Most of these studies find a positive association between elevated BMI (obesity) and risk of mortality from all causes, including cancer and cardiovascular disease, and reduced BMI with a lower risk of mortality from cardiovascular disease and many types of cancer (Reviewed in Takata et al. 2007; Fair and Montgomery 2009). For example, a prospective cohort study of Swedish women diagnosed and treated for anorexia prior to age 40 found that they had about half the risk of breast cancer relative to age-matched controls (Michels and Ekbom 2004). A prospective cohort study of Canadian women who consumed in excess of 2,406 kcal/ day had a significantly higher risk of breast cancer than women who consumed less than 1,630 kcal/day (Silvera et al. 2006). Data from the Shanghai Breast Cancer Study showed that women with higher BMIs or higher calorie intakes and lower levels of physical activity were at increased risk of breast cancer (Malin et al. 2005). In both case-control and cohort studies, obesity has been consistently associated with higher risk of colorectal cancer for both men and women (Slattery et al. 1997; Fair and Montgomery 2009). Colon cancer risk in men with a BMI > 30 is increased by up to 80% (McMillan et al. 2006). A meta-analysis of 14 studies was used to estimate that excess weight accounts for 39% of cancers of the endometrium and approximately 25% of cancers of the kidney and gallbladder in both sexes (Bergstrom et al. 2001). An analysis of data from the first National Health and Nutrition Examination Survey (NHANES) I, and the NHANES Epidemiologic Follow-up Study encompassing 14,407 participants in the age group of 25–74 years found that all neoplasms in women decreased with decreasing BMI from approximately a BMI of 29 to a BMI of 17 (Sunder 2005). A similar decrease was not found for men. Together, the results suggest that lower body weight, and presumably lower caloric intake is associated with reduced mortality from all causes, including cancer and cardiovascular disease.

Conversely, a number of such studies show that there is an association between low BMI and increased mortality from all causes, including cardiovascular disease, and cancer in the middle aged and elderly. For example, a community-based longitudinal study of the association between BMI and all-cause mortality in 80–84 year olds found that total mortality in the overweight group was 52% less than that in the normal weight group (Takata et al. 2007). In this study, underweight individuals had all-cause mortality 3.9 times higher, and mortality from cancer 17.7 times higher than overweight individuals. Further, mortality from cardiovascular disease was 4.6 times higher in the underweight compared to the normal-weight subjects. These results are in agreement with studies on individuals aged 60 and older (Janssen et al. 2005), on adults and the elderly (Pitner 2005; Shahar et al. 2005), on those 65–84 years of age (Sergi et al. 2005), on those averaging 71 years of age (Wassertheil-Smoller et al. 2000), on those averaging 73 years of age (Corrada et al. 2006), and on those of middle-age (Flegal et al. 2005; Hayashi et al. 2005; Gu et al. 2006). Others also have reported that low BMI is associated with greater mortality and/or morbidity from cardiovascular disease and cancer (Wassertheil-Smoller et al. 2000; Cui et al. 2005; Hu et al. 2005; Chen et al. 2006).

A number of investigators have found "U shaped" survival versus BMI curves, suggesting that being either underweight or overweight increases the risk of mortality. For example, a plot of BMI versus estimated hazards ratio for individuals of 50–75 years of age had a minimum hazards ratio from a BMI of 25.7 to one of 28.9 for men and from 24.5 to 29.0 for women (Sunder 2005). This range is higher that that typically considered to be healthy. A population study of Chinese adults found that a BMI from 23.0 to 23.9 had a relative risk of all cause mortality of 1.00, and BMIs that were higher or lower had an increased risk of mortality from all causes (Gu et al. 2006). One possible interpretation of these results is that low BMI among adults and the elderly leads to reduced resistance to the stress of disease, or predisposes to mortality from disease. This interpretation is consistent with the results cited above showing that by decreasing energy reserves, CR can increase the mortality of influenza infected mice (Ritz and Gardner 2006).

The studies above challenge the idea that human CR will lead to extended longevity. However, there are a number of possible confounds to the interpretation of these studies. Normal weight or overweight people who survive into old age may be less susceptible to obesity related diseases. The quality of the diets consumed by the low BMI individuals are difficult to assess, and may lack nutrients important to longevity. The lower weight individuals in the studies are not CR because their caloric intake reflects their individual ad libitum set-points, and not a reduction from that set-point. Still, the absence of a verifiable 160 year old lean human is a challenge to the view that human CR will lead to robust lifespan extension.

The lack of a readily detectable effect of CR on human lifespan may be due to the effects of diet composition on IGFI levels. There is overwhelming evidence that IGFI and IGFR signaling are important to lifespan in metazoans. Therefore, the effects of CR on human IGFI levels are likely to be important. Neither the Biosphere nor the CALERIE studies reported a decrease in serum IGFI or the IGFI to IGFBP-3 ratio in humans after CR diets lasting from six months to two years (Walford et al. 2002; Holloszy and Fontana 2007; Redman and Ravussin 2009). In contrast, Fontana et al. found that total and free IGFI concentrations are significantly lower in moderately protein-restricted individuals (Fontana et al. 2006a, 2008). In a recent study, reducing protein intake from an average of 1.67 g/kg body weight to 0.95 g/kg body weight per day for 3 weeks in six volunteers already practicing CR reduced serum IGFI from 194 to 152 ng/ml (Fontana et al. 2008). Thus, unlike rodents, humans may need to reduce both protein and calorie intake to lower their serum IGFI levels and their IGFI to IGFBP-3 ratios. Diet composition may be especially important for humans to enjoy the longevity benefits of a CR lifestyle.

12.21 The Search for Longevity Therapeutics

As discussed above, there are many published epidemiological and clinical studies suggesting that reduced calorie diets decrease the incidence of cardiovascular disease, type 2 diabetes, and cancer in humans. As reviewed above, there is much we do not know about how low calorie diets and low BMI affect human longevity, especially in the elderly. It is unlikely that many will embrace a CR lifestyle, thereby exposing themselves to these uncertainties. Although sales of weight reduction products produce billions in profits for companies every year, almost none of these products are effective in the long run. The rate of recidivism after weight loss is nearly 100%. Even in mice, hunger induced by a CR diet does not diminish over time (Hambly et al. 2007). There are powerful genetic influences on weight. Family studies demonstrate that tendencies toward obesity or thinness are largely inherited (Maes et al. 1997a). Twin and adoption studies also indicate that weight is largely genetically determined (Stunkard et al. 1986, 1990; Sorensen et al. 1999).

The problems associated with maintaining a healthy weight are well-illustrated by the tragic story of Dr. Stuart M. Berger (Adler 1994; Sullivan 1994). He transformed himself from a lonely, obese child in Brooklyn into a trim, rich, celebrity nutrition doctor. He wrote the *Southampton Diet, Dr. Berger's Immune Power Diet*, and *Forever Young*, advocating among other things the preventative power of steamed broccoli. He died in his sleep in his Manhattan apartment at the age of 40 at a weight of 365 pounds. Dr. Berger knew what to do. But, in the end he succumbed to the physiological prompting to eat. That prompting tells us that we are "starving to death," even if the opposite is true. Thus, it is not very likely that we will soon avail ourselves of the benefits of being even the "ideal weight" specified by the Metropolitan Height and Weight Tables for Men and Women (Anonymous 1999). Because humans are not experimental animals maintained in clean, low stress environments, it is debatable whether the potential downsides of CR in humans will be greater than the potential benefits.

Despite these limitations, the existence of multiple dietary interventions and a number of natural mutations, gene knockout, and transgene overexpressing mice with extended longevities suggests that druggable therapeutic targets exist in mammals (Brown-Borg et al. 1996; Zhou et al. 1997; Flurkey et al. 2001; Holzenberger et al. 2003; Coschigano et al. 2003; Kurosu et al. 2005; Taguchi et al. 2007; Selman et al. 2008). However, the identification of these targets continues to be challenging. There have been numerous attempts to use surrogate assays for identifying compounds that can reproduce the longevity and health effects of CR and the mammalian longevity mutations.

One strategy has been to use rodent-lifespan assays. Unfortunately the literature produced by this effort is largely confounded and unreliable (Spindler 2010). Repeated key word searches of the online PubMed database found 86 lifespan studies performed using healthy, long-lived rodents (Spindler 2010). One of these reports is a summary of 21 lifespan studies performed with melatonin, which report contradictory results (Anisimov et al. 2006). Of the studies remaining, 18 report no effect, or a shortening of lifespan. Of the 48 remaining studies, 5 would be difficult to repeat because the preparation and composition of the treatment agents are published in difficult to obtain journals (Kohn 1971; Emanuel and Obukhova 1978; Anisimov et al. 1989, 1997; Ghanta et al. 1990). Twenty nine of the remaining studies are likely artifacts of induced "voluntary" CR, or this possibility cannot be excluded by the data given or discussed in the publication (Spindler 2010). In most of these studies, neither food intake nor body weight was measured or reported. In a few cases the authors write that there was no change in body weight or food intake. However, no data are shown, no indication is given of when or how many times during the study the measurements were taken, the means and standard deviations of the measurements are omitted, and the statistics applied to the data are not described. Of the remaining studies, all but 8 suffer from other confounds (Spindler 2010).

Only 2 of the 8 unconfounded studies report both food consumption and lifespan extension (Yen and Knoll 1992; Caldeira da Silva et al. 2008). Deprenyl and Dinh lang root extract reportedly extended the lifespan of mice (Yen and Knoll 1992), and dinitrophenol slightly extended the median and mean lifespan of a normal mouse strain with a short lifespan (Caldeira da Silva et al. 2008). These are the only unconfounded studies in the literature showing an increase in longevity by a chemical agent. The 6 remaining studies found lifespan extension but used body weight as a surrogate measure of food consumption. These 6 studies are: 2-mercaptoethanol administered orally in food to male BC3F1 mice (Heidrick et al. 1984); cornstarch vs. sucrose-containing diets fed to Fischer 344 rats (Murtagh-Mark et al. 1995); deprenyl fed to Syrian hamsters (Stoll et al. 1994, 1997); ginkgo biloba extract administered orally to male Fisher rats (Winter 1998); green tea polyphenols administered in drinking water to male C57BL/6 mice (Kitani et al. 2007); PBN fed to C57BL/6 J male mice (Saito et al. 1998); and piperoxane administered by injection to Fisher 344 rats (Compton et al. 1995). However, body weight can be an unreliable surrogate measure of caloric intake. Supplements that alter water retention, oxygen utilization (e.g. uncouple oxidative phosphorylation), calorie adsorption in the gastrointestinal tract, or body fat mass independent of caloric intake could change the number of calories consumed without changing body weight. Thus, the only way to insure that food consumption is unchanged by a treatment is to feed a known amount of food to the animals, and monitor its consumption (or lack of consumption).

Review of the studies outlined above led us to design and execute an ongoing lifespan study involving 58 treatment groups of mice (Spindler 2010). These studies observed a number of design parameters that are essential for obtaining meaningful rodent lifespan data (Spindler 2010). These are: (1) A long-lived, healthy rodent strain should be used, preferably an F1 or further outcrossed strain. (2) The diets should be fed in measured amounts. The presence of any uneaten food pellets can be noted, and the subsequent day's allotment of food adjusted accordingly. The amount of food fed to the entire study group can be reduced in stages if food remains in the treatment groups. (3) Only chemically defined diets, such as the AIN diets, should be used for any study. (4) The animals should be fed daily. This ensures that they are dosed daily with an effective concentration of the test compounds. (5) A positive control, such as a CR group, must be used. Most lifespan studies are published without a positive control. However, negative results are difficult to interpret in the absence of a positive control.

Another approach we and others have explored is screening compounds for their ability to reproduce a surrogate marker of extended health and longevity. We chose to use the genome wide, gene expression signature of LTCR as the surrogate marker (Spindler and Mote 2007). Gene expression patterns appear to be reliable indicators of biological status. For example, the gene expression patterns of primary tumor cells appear to be useful for predicting clinical outcomes such as chemosensitivity, metastases, and survival (Rosenwald et al. 2002; Vasselli et al. 2003). In mice, both the anticancer and the gene-expression response of the liver to CR is rapid (Dhahbi et al. 2004). A similarly rapid, but less extensive response also was found in heart (Dhahbi et al. 2006). We found that 8 weeks of metformin treatment of old mice was superior to even 8 weeks of CR at reproducing the global gene expression signature of LTCR (Dhahbi et al. 2005). As discussed in Section 12.16 above, metformin increases insulin sensitivity, decreases hepatic gluconeogenesis, inhibits the development of metabolic syndrome, may reduce cancer risk in humans and animals, and extends the lifespan of short-lived or enfeebled mice. These results indicate that gene-expression profiling may be useful for the identification of potential longevity therapeutics.

Thus, results of our work, and those of others (Corton et al. 2004; Ingram et al. 2006; Miller et al. 2007; Caldeira da Silva et al. 2008; Pearson et al. 2008; Strong et al. 2008; Anisimov et al. 2008), suggest that longevity therapeutics exist, that we are developing procedures to identify them, and that someday they will be developed sufficiently to be used to extend the healthy human lifespan. When this goal is successfully attained, we will have fulfilled one of the most ancient of human goals (Walford 1998).

12.22 Conclusions

We know that it is possible to extend the mean and/or maximum lifespan of mammals by reducing dietary calories, protein, methionine, or tryptophan; or by reducing insulin and/or IGFI signaling. These interventions also delay the onset of deleterious age-related physiological changes and diseases. At least in mice and Drosophila, lifespan extension can be reversibly induced by CR, even in "middle age". The longevity and health effects of CR are phylogenetically widespread. The CR effect appears to be an adaptation to boom and bust cycles in the food supply. Whether human lifespan can be similarly extended by CR or any means remains open to debate. However, studies in humans indicate that CR produces many of the same physiologic, hematologic, hormonal, and biochemical changes produced in species which experience lifespan extension. In humans, non-human primates and rodents, CR provides protection from type 2 diabetes, cardiovascular and cerebral vascular disease, age-related immunological decline, malignancy, hepatotoxicity, liver fibrosis and failure, sarcopenia, systemic inflammation, and DNA damage. It produces intense organelle, lipid and protein turnover and renewal, enhances muscle mitochondrial biogenesis, affords neuroprotection, and extends mean and in some cases maximum lifespan. CR also may increase apoptosis rates in tissues capable of cell replacement and in tumors. The lifespan effects of CR are dose and duration responsive in rodents. Meal frequency appears to have little if any effect on the lifespan response. In mice, CR rapidly induces powerful antineoplastic effects in the liver and lung, and salutary effects on the physiology of the cardiovascular system. CR induces transient increases in serum glucocorticoid levels, and glucocorticoids appear to be required for at least some of the anticancer effects of CR in rodents. In dogs, CR mitigates the onset of arthritis. CR, amino acid restriction and disrupted insulin and IGF-I signaling also down-regulate thyroid hormones and reproductive hormones and activity. CR lowers the mean body temperature of homeotherms such as mice, rats, monkeys and humans, and this appears to contribute to its health and longevity effects in mice. Hypothermia also appears to contribute to the enhanced longevity of long-lived mutant mouse strains. In addition to its effects on the insulin and IGF-I signaling systems, CR may also induce lifespan extension by altering the activity of transcription factors SirT1, PGC-1α AMPK. Paradoxically, low body weight in middle aged and elderly humans is associated with increased mortality. Thus, further enhancement of human longevity may require pharmaceutical interventions. A number of surrogate assays are currently being applied in the search for authentic longevity pharmaceuticals. The literature claiming to have identified drugs or nutrients that extend the lifespan of rodents is largely confounded by design flaws, most commonly by induced "voluntary" CR. We suggest a number of design parameters which will avoid such confounds in future studies. When we are successful in identifying authentic longevity therapeutics, we will have fulfilled one of the most ancient of human goals.

Acknowledgements The author would like to thank Patricia L. Mote for critical reading this manuscript and helpful comments and suggestions.

References

Adler J (1994) The World's Biggest Diet Doctor: Stuart Berger thought he knew how to stay young forever. He was wrong. Newsweek Magazine issue dated March 21, 1994. http://www.newsweek.com/id/116461

Alessi DR, Sakamoto K, Bayascas JR (2006) Lkb1-dependent signaling pathways. Annu Rev Biochem 75:137–163

Alexia C, Fourmatgeat P, Delautier D, Groyer A (2006) Insulin-like growth factor-I stimulates H(4)II rat hepatoma cell proliferation: Dominant role of PI-3'K/Akt signaling. Exp Cell Res 312:1142–1152

Alexia C, Lasfer M, Groyer A (2004) Role of constitutively activated and insulin-like growth factor-stimulated ERK1/2 signaling in human hepatoma cell proliferation and apoptosis: evidence for heterogeneity of tumor cell lines. Ann NY Acad Sci 1030:219–229

Alimova IN, Liu B, Fan Z, Edgerton SM, Dillon T, Lind SE, Thor AD (2009) Metformin inhibits breast cancer cell growth, colony formation and induces cell cycle arrest in vitro. Cell Cycle 8:909–915

Anderson RM, Shanmuganayagan D, Weindruch R (2009) Caloric Restriction and Aging: Studies in Mice and Monkeys. Toxicol Pathol 37:47–51

Anisimov VN, Berstein LM, Egormin PA, Piskunova TS, Popovich IG, Zabezhinski MA, Kovalenko IG, Poroshina TE, Semenchenko AV, Provinciali M, Re F, Franceschi C (2005a) Effect of metformin on life span and on the development of spontaneous mammary tumors in HER-2/neu transgenic mice. Exp Gerontol 40:685–693

Anisimov VN, Berstein LM, Egormin PA, Piskunova TS, Popovich IG, Zabezhinski MA, Tyndyk ML, Yurova MV, Kovalenko IG, Poroshina TE, Semenchenko AV (2008) Metformin slows down aging and extends life span of female SHR mice. Cell Cycle 7:2769–2773

Anisimov VN, Egormin PA, Bershtein LM, Zabezhinskii MA, Piskunova TS, Popovich IG, Semenchenko AV (2005b) Metformin decelerates aging and development of mammary tumors in HER-2/neu transgenic mice. Bull Exp Biol Med 139:721–723

Anisimov VN, Loktionov AS, Khavinson VK, Morozov VG (1989) Effect of low-molecular-weight factors of thymus and pineal gland on life span and spontaneous tumour development in female mice of different age. Mech Ageing Dev 49:245–257

Anisimov VN, Mylnikov SV, Oparina TI, Khavinson VK (1997) Effect of melatonin and pineal peptide preparation epithalamin on life span and free radical oxidation in *Drosophila melanogaster*. Mech Ageing Dev 97:81–91

Anisimov VN, Popovich IG, Zabezhinski MA, Anisimov SV, Vesnushkin GM, Vinogradova IA (2006) Melatonin as antioxidant, geroprotector and anticarcinogen. Biochim Biophys Acta. 1757:573–589

Anisimov VN, Semenchenko AV, Yashin AI (2003) Insulin and longevity: antidiabetic biguanides as geroprotectors. Biogerontology 4:297–307

Anonymous (1999) 1996, 1999 Metropolitan Life Insurance Company Build Study. Society of Actuaries and Association of Life Insurance Medical Directors of America, 1980. http://www.bcbst com/MPManual/HW htm

Anonymous (2007) WHO, WHAT, WHY?: what are my chances of living to 100? BBC News Magazine Tuesday, 24 April 2007

Anversa P, Palackal T, Sonnenblick EH, Olivetti G, Meggs LG, Capasso JM (1990) Myocyte cell loss and myocyte cellular hyperplasia in the hypertrophied aging rat heart. Circ Res 67:871–885

Apfeld J, O'Connor G, McDonagh T, Distefano PS, Curtis R (2004) The AMP-activated protein kinase AAK-2 links energy levels and insulin-like signals to lifespan in *C. elegans*. Genes Dev 18:3004–3009

Armentero MT, Levandis G, Bramanti P, Nappi G, Blandini F (2008) Dietary restriction does not prevent nigrostriatal degeneration in the 6-hydroxydopamine model of Parkinson's disease. Exp Neurol 212:548–551

Austad SN (1989) Life extension by dietary restriction in the bowl and doily spider, Frontinella pyramitela. Exp Gerontol 24:83–92

Bakala H, Delaval E, Hamelin M, Bismuth J, Borot-Laloi C, Corman B, Friguet B (2003) Changes in rat liver mitochondria with aging. Lon protease-like reactivity and N(epsilon)-carboxymethyllysine accumulation in the matrix. Eur J Biochem 270:2295–2302

Baker DJ, Betik AC, Krause DJ, Hepple RT (2006) No decline in skeletal muscle oxidative capacity with aging in long-term calorically restricted rats: effects are independent of mitochondrial DNA integrity. J Gerontol A Biol Sci Med Sci 61:675–684

Ball ZB, Barnes RH, Visscher MB (1947) The effects of dietary caloric restriction on maturity and senescence, with particular reference to fertility and longevity. Am J Physiol 150: 511–519

Barger JL, Walford RL, Weindruch R (2003) The retardation of aging by caloric restriction: its significance in the transgenic era. Exp Gerontol 38:1343–1351

Bartke A (2005) Minireview: Role of the growth hormone/insulin-like growth factor system in Mammalian aging. Endocrinology 146:3718–3723

Bartke A (2006) Long-lived Klotho mice: new insights into the roles of IGF-1 and insulin in aging. Trends Endocrinol Metab 17:33–35

Bartke A (2008a) Impact of reduced insulin-like growth factor-1/insulin signaling on aging in mammals: novel findings. Aging Cell 7:285–290

Bartke A (2008b) Insulin and aging. Cell Cycle 7:3338–3343

Bartke A, Bonkowski M, Masternak M (2008) How diet interacts with longevity genes. Hormones. (Athens) 7:17–23

Bartke A, Brown-Borg H (2004) Life extension in the dwarf mouse. Curr Top Dev Biol 63:189–225

Bartke A, Wright JC, Mattison JA, Ingram DK, Miller RA, Roth GS (2001) Extending the lifespan of long-lived mice. Nature 414:412

Baserga R, Peruzzi F, Reiss K (2003) The IGF-1 receptor in cancer biology. Int J Cancer 107:873–877

Bass TM, Weinkove D, Houthoofd K, Gems D, Partridge L (2007) Effects of resveratrol on lifespan in *Drosophila melanogaster* and *Caenorhabditis elegans*. Mech Ageing Dev 128:546–552

Bauer JH, Goupil S, Garber GB, Helfand SL (2004) An accelerated assay for the identification of lifespan-extending interventions in *Drosophila melanogaster*. Proc Natl Acad Sci USA 101:12980–12985

Beauchene RE, Bales CW, Bragg CS, Hawkins ST, Mason RL (1986) Effect of age of initiation of feed restriction on growth, body composition, and longevity of rats. J Gerontol 41:13–19

Bejma J, Ji LL (1999) Aging and acute exercise enhance free radical generation in rat skeletal muscle. J Appl Physiol 87:465–470

Belanger C, Luu-The V, Dupont P, Tchernof A (2002) Adipose tissue intracrinology: potential importance of local androgen/estrogen metabolism in the regulation of adiposity. Horm Metab Res 34:737–745

Bender A, Beckers J, Schneider I, Holter SM, Haack T, Ruthsatz T, Vogt-Weisenhorn DM, Becker L, Genius J, Rujescu D, Irmler M, Mijalski T, Mader M, Quintanilla-Martinez L, Fuchs H, Gailus-Durner V, de Angelis MH, Wurst W, Schmidt J, Klopstock T (2007) Creatine improves health and survival of mice. Neurobiol Aging 29: 1404–1411

Bergamini E (2006) Autophagy: A cell repair mechanism that retards ageing and age-associated diseases and can be intensified pharmacologically. Mol Aspects Med 27:403–410

Bergamini E, Cavallini G, Donati A, Gori Z (2007) The role of autophagy in aging: its essential part in the anti-aging mechanism of caloric restriction. Ann NY Acad Sci 1114:69–78

Bergstrom A, Pisani P, Tenet V, Wolk A, Adami HO (2001) Overweight as an avoidable cause of cancer in Europe. Int J Cancer 91:421–430

Bertrand HA, Lynd FT, Masoro EJ, Yu BP (1980) Changes in adipose mass and cellularity through the adult life of rats fed ad libitum or a life-prolonging restricted diet. J Gerontol 35: 827–835

Bezlepkin VG, Sirota NP, Gaziev AI (1996) The prolongation of survival in mice by dietary antioxidants depends on their age by the start of feeding this diet. Mech Ageing Dev 92:227–234

Biddinger SB, Kahn CR (2006) From Mice to Men: Insights into the Insulin Resistance Syndromes. Annu Rev Physiol. 68:123–158

Blackwell BN, Bucci TJ, Hart RW, Turturro A (1995) Longevity, body weight, and neoplasia in ad libitum-fed and diet- restricted C57BL6 mice fed NIH-31 open formula diet. Toxicol Pathol 23:570–582

Bluher M, Kahn BB, Kahn CR (2003) Extended longevity in mice lacking the insulin receptor in adipose tissue. Science 299:572–574

Bluher M, Michael MD, Peroni OD, Ueki K, Carter N, Kahn BB, Kahn CR (2002) Adipose tissue selective insulin receptor knockout protects against obesity and obesity-related glucose intolerance. Dev Cell 3:25–38

Bodkin NL, Alexander TM, Ortmeyer HK, Johnson E, Hansen BC (2003) Mortality and morbidity in laboratory-maintained Rhesus monkeys and effects of long-term dietary restriction. J Gerontol A Biol Sci Med Sci 58:212–219

Bodkin NL, Ortmeyer HK, Hansen BC (1995) Long-term dietary restriction in older-aged rhesus monkeys: effects on insulin resistance. J Gerontol A Biol Sci Med Sci 50:B142-B147

Boileau TW, Liao Z, Kim S, Lemeshow S, Erdman JW, Jr., Clinton SK (2003) Prostate carcinogenesis in N-methyl-N-nitrosourea (NMU)-testosterone-treated rats fed tomato powder, lycopene, or energy-restricted diets. J Natl Cancer Inst 95:1578–1586

Boily G, Seifert EL, Bevilacqua L, He XH, Sabourin G, Estey C, Moffat C, Crawford S, Saliba S, Jardine K, Xuan J, Evans M, Harper ME, McBurney MW (2008) SirT1 regulates energy metabolism and response to caloric restriction in mice. PLoS ONE 3:e1759

Boissan M, Beurel E, Wendum D, Rey C, Lecluse Y, Housset C, Lacombe ML, Desbois-Mouthon C (2005) Overexpression of insulin receptor substrate-2 in human and murine hepatocellular carcinoma. Am J Pathol 167:869–877

Bolster DR, Crozier SJ, Kimball SR, Jefferson LS (2002) AMP-activated protein kinase suppresses protein synthesis in rat skeletal muscle through down-regulated mammalian target of rapamycin (mTOR) signaling. J Biol Chem 277:23977–23980

Bonkowski MS, Rocha JS, Masternak MM, Al Regaiey KA, Bartke A (2006) Targeted disruption of growth hormone receptor interferes with the beneficial actions of calorie restriction. Proc Natl Acad Sci USA 103:7901–7905

Bowker SL, Majumdar SR, Veugelers P, Johnson JA (2006) Increased cancer-related mortality for patients with type 2 diabetes who use sulfonylureas or insulin. Diabetes Care 29:254–258

Brown-Borg HM, Borg KE, Meliska CJ, Bartke A (1996) Dwarf mice and the ageing process. Nature 384:33

Brunet A, Sweeney LB, Sturgill JF, Chua KF, Greer PL, Lin Y, Tran H, Ross SE, Mostoslavsky R, Cohen HY, Hu LS, Cheng HL, Jedrychowski MP, Gygi SP, Sinclair DA, Alt FW, Greenberg ME (2004) Stress-dependent regulation of FOXO transcription factors by the SIRT1 deacetylase. Science 303:2011–2015

Bruning JC, Gautam D, Burks DJ, Gillette J, Schubert M, Orban PC, Klein R, Krone W, Muller-Wieland D, Kahn CR (2000) Role of brain insulin receptor in control of body weight and reproduction. Science 289:2122–2125

Budunova IV, Kowalczyk D, Perez P, Yao YJ, Jorcano JL, Slaga TJ (2003) Glucocorticoid receptor functions as a potent suppressor of mouse skin carcinogenesis. Oncogene 22:3279–3287

Burke SN, Barnes CA (2006) Neural plasticity in the ageing brain. Nat Rev Neurosci 7:30–40

Burks DJ, Font dM, Schubert M, Withers DJ, Myers MG, Towery HH, Altamuro SL, Flint CL, White MF (2000) IRS-2 pathways integrate female reproduction and energy homeostasis. Nature 407:377–382

Caldeira da Silva CC, Cerqueira FM, Barbosa LF, Medeiros MH, Kowaltowski AJ (2008) Mild Mitochondrial Uncoupling in Mice Affects Energy Metabolism, Redox Balance and Longevity. Aging Cell 7:552–560

Canto C, Auwerx J (2009) PGC-1alpha, SIRT1 and AMPK, an energy sensing network that controls energy expenditure. Curr Opin Lipidol 20:98–105

Cao SX, Dhahbi JM, Mote PL, Spindler SR (2001) Genomic profiling of short- and long-term caloric restriction effects in the liver of aging mice. Proc Natl Acad Sci USA 98:10630–10635

Carey JR, Liedo P, Harshman L, Zhang Y, Muller HG, Partridge L, Wang JL (2002) Life history response of Mediterranean fruit flies to dietary restriction. Aging Cell 1:140–148

Carey JR, Liedo P, Muller HG, Wang JL, Zhang Y, Harshman L (2005) Stochastic dietary restriction using a Markov-chain feeding protocol elicits complex, life history response in medflies. Aging Cell 4:31–39

Carling D (2004) The AMP-activated protein kinase cascade--a unifying system for energy control. Trends Biochem Sci 29:18–24

Carlson AJ, Hoelzel F (1946) Apparent prolongation of the life-span of rats by intermittent fasting. J Nutr 31:363–375

Caro P, Gomez J, Lopez-Torres M, Sanchez I, Naudi A, Portero-Otin M, Pamplona R, Barja G (2008) Effect of every other day feeding on mitochondrial free radical production and oxidative stress in mouse liver. Rejuvenation Res 11:621–629

Carvalho GB, Kapahi P, Benzer S (2005) Compensatory ingestion upon dietary restriction in *Drosophila melanogaster*. Nat Methods 2:813–815

Cavallini G, Donati A, Gori Z, Pollera M, Bergamini E (2001) The protection of rat liver autophagic proteolysis from the age-related decline co-varies with the duration of anti-ageing food restriction. Exp Gerontol 36:497–506

Cha SH, Rodgers JT, Puigserver P, Chohnan S, Lane MD (2006) Hypothalamic malonyl-CoA triggers mitochondrial biogenesis and oxidative gene expression in skeletal muscle: Role of PGC-1alpha. Proc Natl Acad Sci USA 103:15410–15415

Chandrashekar V, Bartke A (2003) The role of insulin-like growth factor-I in neuroendocrine function and the consequent effects on sexual maturation: inferences from animal models. Reprod Biol 3:7–28

Chapman T, Partridge L (1996) Female fitness in *Drosophila melanogaster*: an interaction between the effect of nutrition and of encounter rate with males. Proc Biol Sci 263:755–759

Chen D, Steele AD, Lindquist S, Guarente L (2005) Increase in activity during calorie restriction requires Sirt1. Science 310:1641

Chen Z, Yang G, Zhou M, Smith M, Offer A, Ma J, Wang L, Pan H, Whitlock G, Collins R, Niu S, Peto R (2006) Body mass index and mortality from ischaemic heart disease in a lean population: 10 year prospective study of 220,000 adult men. Int J Epidemiol 35:141–150

Cheng HL, Mostoslavsky R, Saito S, Manis JP, Gu Y, Patel P, Bronson R, Appella E, Alt FW, Chua KF (2003) Developmental defects and p53 hyperacetylation in Sir2 homolog (SIRT1)-deficient mice. Proc Natl Acad Sci USA 100:10794–10799

Cheney KE, Liu RK, Smith GS, Leung RE, Mickey MR, Walford RL (1980) Survival and disease patterns in C57BL/6 J mice subjected to undernutrition. Exp Gerontol 15:237–258

Cheney KE, Liu RK, Smith GS, Meredith PJ, Mickey MR, Walford RL (1983) The effect of dietary restriction of varying duration on survival, tumor patterns, immune function, and body temperature in B10C3F1 female mice. J Gerontol 38:420–430

Chiang AC, Massague J (2008) Molecular basis of metastasis. N Engl J Med 359:2814–2823

Chippindale AK, Leroi A, Kim SB, Rose MR (1993) Phenotypic plasticity and selection in Drosophila life history evolution. 1. Nutrition and the cost of reproduction. J Evol Biol. 6:171–193

Civitarese AE, Carling S, Heilbronn LK, Hulver MH, Ukropcova B, Deutsch WA, Smith SR, Ravussin E (2007) Calorie restriction increases muscle mitochondrial biogenesis in healthy humans. PLoS Med 4:e76

Clancy DJ, Gems D, Hafen E, Leevers SJ, Partridge L (2002) Dietary restriction in long-lived dwarf flies. Science 296:319

Clancy DJ, Gems D, Harshman LG, Oldham S, Stocker H, Hafen E, Leevers SJ, Partridge L (2001) Extension of life-span by loss of CHICO, a Drosophila insulin receptor substrate protein. Science 292:104–106

Colman RJ, Anderson RM, Johnson SC, Kastman EK, Kosmatka KJ, Beasley TM, Allison DB, Cruzen C, Simmons HA, Kemnitz JW, Weindruch R (2009) Caloric Restriction Delays Disease Onset and Mortality in Rhesus Monkeys. Science 325:201–204

Colman RJ, Beasley TM, Allison DB, Weindruch R (2008) Attenuation of sarcopenia by dietary restriction in rhesus monkeys. J Gerontol A Biol Sci Med Sci 63:556–559

Colman RJ, Roecker EB, Ramsey JJ, Kemnitz JW (1998) The effect of dietary restriction on body composition in adult male and female rhesus macaques. Aging (Milano) 10:83–92

Colucci WS (1997) Molecular and cellular mechanisms of myocardial failure. Am J Cardiol 80:15L–25L

Comfort A (1963) Effect of delayed and resumed growth on the longevity of a fish (Lebistes reticulatus, Peters) in captivity. Gerontologia 49:150–155

Compton DM, Dietrich KL, Smith JS (1995) Influence of the alpha 2 noradrenergic antagonist piperoxane on longevity in the Fischer-344 rat: a preliminary report. Psychol. Rep 77:139–142

Conti B, Sanchez-Alavez M, Winsky-Sommerer R, Morale MC, Lucero J, Brownell S, Fabre V, Huitron-Resendiz S, Henriksen S, Zorrilla EP, de Lecea L, Bartfai T (2006) Transgenic mice with a reduced core body temperature have an increased life span. Science 314:825–828

Cooper TM, Mockett RJ, Sohal BH, Sohal RS, Orr WC (2004) Effect of caloric restriction on life span of the housefly, Musca domestica. FASEB J 18:1591–1593

Corrada MM, Kawas CH, Mozaffar F, Paganini-Hill A (2006) Association of body mass index and weight change with all-cause mortality in the elderly. Am J Epidemiol 163:938–949

Corton JC, Apte U, Anderson SP, Limaye P, Yoon L, Latendresse J, Dunn C, Everitt JI, Voss KA, Swanson C, Kimbrough C, Wong JS, Gill SS, Chandraratna RA, Kwak MK, Kensler TW, Stulnig TM, Steffensen KR, Gustafsson JA, Mehendale HM (2004) Mimetics of caloric restriction include agonists of lipid-activated nuclear receptors. J Biol Chem 279: 46204–46212

Coschigano KT, Holland AN, Riders ME, List EO, Flyvbjerg A, Kopchick JJ (2003) Deletion, but not antagonism, of the mouse growth hormone receptor results in severely decreased body weights, insulin, and insulin-like growth factor I levels and increased life span. Endocrinology 144:3799–3810

Cuervo AM (2008) Calorie restriction and aging: the ultimate "cleansing diet". J Gerontol A Biol Sci Med.Sci 63:547–549

Cuervo AM, Bergamini E, Brunk UT, Droge W, Ffrench M, Terman A (2005) Autophagy and aging: the importance of maintaining "clean" cells. Autophagy 1:131–140

Cui L, Jeong H, Borovecki F, Parkhurst CN, Tanese N, Krainc D (2006) Transcriptional repression of PGC-1alpha by mutant huntingtin leads to mitochondrial dysfunction and neurodegeneration. Cell 127:59–69

Cui R, Iso H, Toyoshima H, Date C, Yamamoto A, Kikuchi S, Kondo T, Watanabe Y, Koizumi A, Wada Y, Inaba Y, Tamakoshi A (2005) Body mass index and mortality from cardiovascular disease among Japanese men and women: the JACC study. Stroke 36:1377–1382

Cunningham JT, Rodgers JT, Arlow DH, Vazquez F, Mootha VK, Puigserver P (2007) mTOR controls mitochondrial oxidative function through a YY1-PGC-1alpha transcriptional complex. Nature 450:736–740

Curtis R, O'Connor G, Distefano PS (2006) Aging networks in *Caenorhabditis elegans*: AMP-activated protein kinase (aak-2) links multiple aging and metabolism pathways. Aging Cell 5:119–126

Davies S, Kattel R, Bhatia B, Petherwick A, Chapman T (2005) The effect of diet, sex and mating status on longevity in Mediterranean fruit flies (Ceratitis capitata), Diptera: Tephritidae Exp Gerontol 40:784–792

Davis TA, Bales CW, Beauchene RE (1983) Differential effects of dietary caloric and protein restriction in the aging rat. Exp Gerontol 18:427–435

De Grey AD (2005) The unfortunate influence of the weather on the rate of ageing: why human caloric restriction or its emulation may only extend life expectancy by 2–3 years. Gerontology 51:73–82

de Lange P, Farina P, Moreno M, Ragni M, Lombardi A, Silvestri E, Burrone L, Lanni A, Goglia F (2006) Sequential changes in the signal transduction responses of skeletal muscle following food deprivation. FASEB J 20:2579–2581

De Marte ML, Enesco HE (1986a) Influence of low tryptophan diet on survival and organ growth in mice. Mech Ageing Dev 36:161–171

De Marte ML, Enesco HE (1986b) Influence of low tryptophan diet on survival and organ growth in mice. Mech Ageing Dev 36:161–171

De Souza RR (2002) Aging of myocardial collagen. Biogerontology 3:325–335

Demetrius L (2005) Of mice and men. When it comes to studying ageing and the means to slow it down, mice are not just small humans. EMBO Rep 6:S39–S44

Detaille D, Guigas B, Leverve X, Wiernsperger N, Devos P (2002) Obligatory role of membrane events in the regulatory effect of metformin on the respiratory chain function. Biochem Pharmacol 63:1259–1272

Dhahbi JM, Kim HJ, Mote PL, Beaver RJ, Spindler SR (2004) Temporal linkage between the phenotypic and genomic responses to caloric restriction. Proc Natl Acad Sci USA 101: 5524–5529

Dhahbi JM, Mote PL, Fahy GM, Spindler SR (2005) Identification of potential caloric restriction mimetics by microarray profiling. Physiol Genomics 23:343–350

Dhahbi JM, Mote PL, Wingo J, Rowley BC, Cao SX, Walford R, Spindler SR (2001) Caloric restriction alters the feeding response of key metabolic enzyme genes. Mech Ageing Dev 122:35–50

Dhahbi JM, Mote PL, Wingo J, Tillman JB, Walford RL, Spindler SR (1999) Calories and aging alter gene expression for gluconeogenic, glycolytic, and nitrogen-metabolizing enzymes. Am J Physiol 277:E352–E360

Dhahbi JM, Spindler SR (2003) Aging of the liver. In: Aspinall R (ed) Aging of the Organs and Systems. Kluwer Academic Publisher, Dordrecht, The Netherlands, pp. 271–291

Dhahbi JM, Tsuchiya T, Kim HJ, Mote PL, Spindler SR (2006) Gene expression and physiologic responses of the heart to the initiation and withdrawal of caloric restriction. J Gerontol A Biol Sci Med Sci 61:218–231

Dilman VM, Anisimov VN (1980) Effect of treatment with phenformin, diphenylhydantoin or L-dopa on life span and tumour incidence in C3H/Sn mice. Gerontology 26:241–246

Doi SQ, Rasaiah S, Tack I, Mysore J, Kopchick JJ, Moore J, Hirszel P, Striker LJ, Striker GE (2001) Low-protein diet suppresses serum insulin-like growth factor-1 and decelerates the progression of growth hormone-induced glomerulosclerosis. Am J Nephrol 21:331–339

Dominici FP, Hauck S, Argentino DP, Bartke A, Turyn D (2002) Increased insulin sensitivity and upregulation of insulin receptor, insulin receptor substrate (IRS)-1 and IRS-2 in liver of Ames dwarf mice. J Endocrinol 173:81–94

Donati A (2006) The involvement of macroautophagy in aging and anti-aging interventions. Mol Aspects Med 27:455–470

Donati A, Cavallini G, Paradiso C, Vittorini S, Pollera M, Gori Z, Bergamini E (2001) Age-related changes in the autophagic proteolysis of rat isolated liver cells: effects of antiaging dietary restrictions. J Gerontol A Biol Sci Med Sci 56:B375–B383

Dorshkind K, Montecino-Rodriguez E, Signer RA (2009) The ageing immune system: is it ever too old to become young again? Nat Rev Immunol 9:57–62

Dunn SE, Kari FW, French J, Leininger JR, Travlos G, Wilson R, Barrett JC (1997) Dietary restriction reduces insulin-like growth factor I levels, which modulates apoptosis, cell proliferation, and tumor progression in p53-deficient mice. Cancer Res 57:4667–4672

Eckles KE, Dudek EM, Bickford PC, Browning MD (1997) Amelioration of age-related deficits in the stimulation of synapsin phosphorylation. Neurobiol Aging 18:213–217

Eckel RH, Grundy SM, Zimmet PZ (2005) The metabolic syndrome. Lancet 365:1415–1428

Eckles-Smith K, Clayton D, Bickford P, Browning MD (2000) Caloric restriction prevents age-related deficits in LTP and in NMDA receptor expression. Brain Res Mol Brain Res 78: 154–162

Edwards IJ, Rudel LL, Terry JG, Kemnitz JW, Weindruch R, Cefalu WT (1998) Caloric restriction in rhesus monkeys reduces low density lipoprotein interaction with arterial proteoglycans. J Gerontol A Biol Sci Med Sci 53:B443–B448

Edwards MG, Sarkar D, Klopp R, Morrow JD, Weindruch R, Prolla TA (2003) Age-related impairment of the transcriptional responses to oxidative stress in the mouse heart. Physiol Genomics 13:119–127

Effros RB (2007) Role of T lymphocyte replicative senescence in vaccine efficacy. Vaccine 25:599–604

Effros RB, Walford RL, Weindruch R, Mitcheltree C (1991) Influences of dietary restriction on immunity to influenza in aged mice. J Gerontol 46:B142–B147

Eghbali M, Eghbali M, Robinson TF, Seifter S, Blumenfeld OO (1989) Collagen accumulation in heart ventricles as a function of growth and aging. Cardiovasc Res 23:723–729 el Haj AJ, Lewis SE, Goldspink DF, Merry BJ, Holehan AM (1986) The effect of chronic and acute dietary restriction on the growth and protein turnover of fast and slow types of rat skeletal muscle. Comp Biochem Physiol A 85:281–287

El Mir MY, Nogueira V, Fontaine E, Averet N, Rigoulet M, Leverve X (2000) Dimethylbiguanide inhibits cell respiration via an indirect effect targeted on the respiratory chain complex I. J Biol Chem 275:223–228

Emanuel NM, Obukhova LK (1978) Types of experimental delay in aging patterns. Exp Gerontol 13:25–29

Evans JMM, Donnelly LA, Emslie-Smith AM, Alessi DR, Morris AD (2005) Metformin and reduced risk of cancer in diabetic patients. BMJ 330:1304–1305

Fabris N, Pierpaoli W, Sorkin E (1971) Hormones and the immunological capacity. 3. The immunodeficiency disease of the hypopituitary Snell-Bagg dwarf mouse. Clin Exp Immunol 9:209–225

Fabrizio P, Longo VD (2007) The chronological life span of Saccharomyces cerevisiae. Methods Mol Biol 371:89–95

Fair AM, Montgomery K (2009) Energy balance, physical activity, and cancer risk. Methods Mol Biol 472:57–88

Fanestil DD, Barrows CH Jr (1965) Aging in the rotifer. J Gerontol 20:462–469

Ferder L, Inserra F, Romano L, Ercole L, Pszenny V (1993) Effects of angiotensin-converting enzyme inhibition on mitochondrial number in the aging mouse. Am J Physiol 265:C15–C18

Feuers RJ, Duffy PH, Leakey JA, Turturro A, Mittelstaedt RA, Hart RW (1989) Effect of chronic caloric restriction on hepatic enzymes of intermediary metabolism in the male Fischer 344 rat. Mech Ageing Dev 48:179–189

Finn PF, Dice JF (2006) Proteolytic and lipolytic responses to starvation. Nutrition 22:830–844

Flegal KM, Graubard BI, Williamson DF, Gail MH (2005) Excess deaths associated with underweight, overweight, and obesity. JAMA 293:1861–1867

Flurkey K, Papaconstantinou J, Miller RA, Harrison DE (2001) Lifespan extension and delayed immune and collagen aging in mutant mice with defects in growth hormone production. Proc Natl Acad Sci USA 98:6736–6741

Fontan-Lozano A, Lopez-Lluch G, Delgado-Garcia JM, Navas P, Carrion AM (2008) Molecular bases of caloric restriction regulation of neuronal synaptic plasticity. Mol Neurobiol 38:167–177

Fontana L, Klein S, Holloszy JO (2006a) Long-term low-protein, low-calorie diet and endurance exercise modulate metabolic factors associated with cancer risk. Am J Clin Nutr 84:1456–1462

Fontana L, Klein S, Holloszy JO, Premachandra BN (2006b) Effect of long-term calorie restriction with adequate protein and micronutrients on thyroid hormones. J Clin Endocrinol Metab 91:3232–3235

Fontana L, Klein S, Holloszy JO, Premachandra BN (2006c) Effect of long-term calorie restriction with adequate protein and micronutrients on thyroid hormones. J Clin Endocrinol Metab 91:3232–3235

Fontana L, Meyer TE, Klein S, Holloszy JO (2004) Long-term calorie restriction is highly effective in reducing the risk for atherosclerosis in humans. Proc Natl Acad Sci USA 101:6659–6663

Fontana L, Villareal DT, Weiss EP, Racette SB, Steger-May K, Klein S, Holloszy JO (2007) Calorie restriction or exercise: effects on coronary heart disease risk factors. A randomized, controlled trial. Am J Physiol Endocrinol Metab 293:E197–E202

Fontana L, Weiss EP, Villareal DT, Klein S, Holloszy JO (2008) Long-term effects of calorie or protein restriction on serum IGF-1 and IGFBP-3 concentration in humans. Aging Cell 7:681–687

Forster MJ, Morris P, Sohal RS (2003) Genotype and age influence the effect of caloric intake on mortality in mice. FASEB J 17:690–692

Frasca D, Landin AM, Lechner SC, Ryan JG, Schwartz R, Riley RL, Blomberg BB (2008) Aging down-regulates the transcription factor E2A, activation-induced cytidine deaminase, and Ig class switch in human B cells. J Immunol 180:5283–5290

Frasca F, Pandini G, Scalia P, Sciacca L, Mineo R, Costantino A, Goldfine ID, Belfiore A, Vigneri R (1999) Insulin receptor isoform A, a newly recognized, high-affinity insulin-like growth factor II receptor in fetal and cancer cells. Mol Cell Biol 19:3278–3288

French CE, Ingram RH, Uram JA, Barron GP, Swift RW (1953) The influence of dietary fat and carbohydrate on growth and longevity in rats. J Nutr 51:329–339

Galuska D, Zierath J, Thorne A, Sonnenfeld T, Wallberg-Henriksson H (1991) Metformin increases insulin-stimulated glucose transport in insulin-resistant human skeletal muscle. Diabete Metab 17:159–163

Gems D, Partridge L (2008) Stress-response hormesis and aging: "that which does not kill us makes us stronger". Cell Metab 7:200–203

Gems D, Pletcher S, Partridge L (2002) Interpreting interactions between treatments that slow aging. Aging Cell 1:1–9

Gems D, Sutton AJ, Sundermeyer ML, Albert PS, King KV, Edgley ML, Larsen PL, Riddle DL (1998) Two pleiotropic classes of daf-2 mutation affect larval arrest, adult behavior, reproduction and longevity in *Caenorhabditis elegans*. Genetics 150:129–155

Ghanta VK, Hiramoto NS, Hiramoto RN (1990) Thymic peptides as anti-aging drugs: effect of thymic hormones on immunity and life span. Int J Gerontol 51:371–372

Giannakou ME, Goss M, Junger MA, Hafen E, Leevers SJ, Partridge L (2004) Long-lived Drosophila with overexpressed dFOXO in adult fat body. Science 305:361

Gillette-Guyonnet S, Vellas B (2008) Caloric restriction and brain function. Curr Opin Clin Nutr Metab Care 11:686–692

Gleyzer N, Vercauteren K, Scarpulla RC (2005) Control of mitochondrial transcription specificity factors (TFB1M and TFB2M) by nuclear respiratory factors (NRF-1 and NRF-2) and PGC-1 family coactivators. Mol Cell Biol 25:1354–1366

Goldspink DF, el Haj AJ, Lewis SE, Merry BJ, Holehan AM (1987) The influence of chronic dietary intervention on protein turnover and growth of the diaphragm and extensor digitorum longus muscles of the rat. Exp Gerontol 22:67–78

Gomez CR, Nomellini V, Faunce DE, Kovacs EJ (2008) Innate immunity and aging. Exp Gerontol 43:718–728

Gonzales C, Voirol MJ, Giacomini M, Gaillard RC, Pedrazzini T, Pralong FP (2004) The neuropeptide Y Y1 receptor mediates NPY-induced inhibition of the gonadotrope axis under poor metabolic conditions. FASEB J 18:137–139

Gonzalez AA, Kumar R, Mulligan JD, Davis AJ, Weindruch R, Saupe KW (2004) Metabolic adaptations to fasting and chronic caloric restriction in heart, muscle, and liver do not include changes in AMPK activity. Am J Physiol Endocrinol Metab 287:E1032–E1037

Goodman MN, Larsen PR, Kaplan MM, Aoki TT, Young VR, Ruderman NB (1980) Starvation in the rat. II. Effect of age and obesity on protein sparing and fuel metabolism. Am J Physiol 239:E277–E286

Goodrick CL, Ingram DK, Reynolds MA, Freeman JR, Cider NL (1983) Effects of intermittent feeding upon growth, activity, and lifespan in rats allowed voluntary exercise. Exp Aging Res 9:203–209

Goto S, Takahashi R, Radak Z, Sharma R (2007) Beneficial biochemical outcomes of late-onset dietary restriction in rodents. Ann NY Acad Sci 1100:431–41

Govic A, Levay EA, Hazi A, Penman J, Kent S, Paolini AG (2008) Alterations in male sexual behaviour, attractiveness and testosterone levels induced by an adult-onset calorie restriction regimen. Behav Brain Res 190:140–146

Grasl-Kraupp B, Bursch W, Ruttkay-Nedecky B, Wagner A, Lauer B, Schulte-Hermann R (1994) Food restriction eliminates preneoplastic cells through apoptosis and antagonizes carcinogenesis in rat liver. Proc Natl Acad Sci USA 91:9995–9999

Grasl-Kraupp B, Ruttkay-Nedecky B, Mullauer L, Taper H, Huber W, Bursch W, Schulte-Hermann R (1997) Inherent increase of apoptosis in liver tumors: implications for carcinogenesis and tumor regression. Hepatology 25:906–912

Gray DA, Tsirigotis M, Woulfe J (2003) Ubiquitin, proteasomes, and the aging brain. Sci.Aging Knowledge.Environ. 44:pe30

Gresl TA, Colman RJ, Havighurst TC, Byerley LO, Allison DB, Schoeller DA, Kemnitz JW (2003) Insulin sensitivity and glucose effectiveness from three minimal models: effects of energy restriction and body fat in adult male rhesus monkeys. Am J Physiol Regul Integr Comp Physiol 285:R1340–R1354

Grondin R, Zhang Z, Yi A, Cass WA, Maswood N, Andersen AH, Elsberry DD, Klein MC, Gerhardt GA, Gash DM (2002) Chronic, controlled GDNF infusion promotes structural and functional recovery in advanced parkinsonian monkeys. Brain 125:2191–2201

Gu D, He J, Duan X, Reynolds K, Wu X, Chen J, Huang G, Chen CS, Whelton PK (2006) Body weight and mortality among men and women in China. JAMA 295:776–783

Gwinn DM, Shackelford DB, Egan DF, Mihaylova MM, Mery A, Vasquez DS, Turk BE, Shaw RJ (2008) AMPK phosphorylation of raptor mediates a metabolic checkpoint. Mol Cell 30:214–226

Hadad SM, Fleming S, Thompson AM (2008) Targeting AMPK: a new therapeutic opportunity in breast cancer. Crit Rev Oncol Hematol 67:1–7

Hagopian K, Ramsey JJ, Weindruch R (2003a) Caloric restriction increases gluconeogenic and transaminase enzyme activities in mouse liver. Exp Gerontol 38:267–278

Hagopian K, Ramsey JJ, Weindruch R (2003b) Influence of age and caloric restriction on liver glycolytic enzyme activities and metabolite concentrations in mice. Exp Gerontol 38: 253–266

Haigis MC, Guarente LP (2006) Mammalian sirtuins—emerging roles in physiology, aging, and calorie restriction. Genes Develop 20:2913–2921

Hamadeh MJ, Tarnopolsky MA (2006) Transient caloric restriction in early adulthood hastens disease endpoint in male, but not female, Cu/Zn-SOD mutant G93A mice. Muscle Nerve 34:709–719

Hambly C, Mercer JG, Speakman JR (2007) Hunger does not diminish over time in mice under protracted caloric restriction. Rejuvenation Res 10:533–542

Han ES, Evans TR, Shu JH, Lee S, Nelson JF (2001) Food restriction enhances endogenous and corticotropin-induced plasma elevations of free but not total corticosterone throughout life in rats. J Gerontol A Biol Sci Med Sci 56:B391–B397

Hansen M, Chandra A, Mitic LL, Onken B, Driscoll M, Kenyon C (2008) A role for autophagy in the extension of lifespan by dietary restriction in *C. elegans*. PLoS Genet 4:e24

Hansen M, Taubert S, Crawford D, Libina N, Lee SJ, Kenyon C (2007) Lifespan extension by conditions that inhibit translation in *Caenorhabditis elegans*. Aging Cell 6:95–110

Hardie DG, Scott JW, Pan DA, Hudson ER (2003) Management of cellular energy by the AMP-activated protein kinase system. FEBS Lett 546:113–120

Hargraves WA, Hentall ID (2005) Analgesic effects of dietary caloric restriction in adult mice. Pain 114:455–461

Harkness TA, Shea KA, Legrand C, Brahmania M, Davies GF (2004) A functional analysis reveals dependence on the anaphase-promoting complex for prolonged life span in yeast. Genetics 168:759–774

Harrison DE, Archer JR (1987) Genetic differences in effects of food restriction on aging in mice. J Nutr 117:376–382

Harrison DE, Archer JR, Astle CM (1984) Effects of food restriction on aging: Separation of food intake and adiposity. Proc Natl Acad Sci USA 81:1835–1838

Hars ES, Qi H, Ryazanov AG, Jin S, Cai L, Hu C, Liu LF (2007) Autophagy regulates ageing in *C. elegans*. Autophagy 3:93–95

Hartmann AM, Burleson LE, Holmes AK, Geist CR (2000) Effects of chronic kombucha ingestion on open-field behaviors, longevity, appetitive behaviors, and organs in C57BL/6 mice: a pilot study. Nutrition 16:755–761

Hauck SJ, Hunter WS, Danilovich N, Kopchick JJ, Bartke A (2001) Reduced levels of thyroid hormones, insulin, and glucose, and lower body core temperature in the growth hormone receptor/binding protein knockout mouse. Exp Biol Med (Maywood) 226:552–558

Hayashi R, Iwasaki M, Otani T, Wang N, Miyazaki H, Yoshiaki S, Aoki S, Koyama H, Suzuki S (2005) Body mass index and mortality in a middle-aged Japanese cohort. J Epidemiol 15:70–77

Hayflick L (2004) "Anti-aging" is an oxymoron. J Gerontol A Biol Sci Med Sci 59:B573–B578
Haynes L, Swain SL (2006) Why aging T cells fail: implications for vaccination. Immunity 24:663–666
Heidrick ML, Hendricks LC, Cook DE (1984) Effect of dietary 2-mercaptoethanol on the life span, immune system, tumor incidence and lipid peroxidation damage in spleen lymphocytes of aging BC3F1 mice. Mech Ageing Dev 27:341–358
Heilbronn LK, de Jonge L, Frisard MI, DeLany JP, Larson-Meyer DE, Rood J, Nguyen T, Martin CK, Volaufova J, Most MM, Greenway FL, Smith SR, Deutsch WA, Williamson DA, Ravussin E, for the Pennington CALERIE Team (2006) Effect of 6-month calorie restriction on biomarkers of longevity, metabolic adaptation, and oxidative stress in overweight individuals: a randomized controlled trial. JAMA 295:1539–1548
Hepple RT, Baker DJ, McConkey M, Murynka T, Norris R (2006) Caloric restriction protects mitochondrial function with aging in skeletal and cardiac muscles. Rejuvenation Res. 9:219–222
Herold MJ, McPherson KG, Reichardt HM (2006) Glucocorticoids in T cell apoptosis and function. Cell Mol Life Sci 63:60–72
Higami Y, Pugh TD, Page GP, Allison DB, Prolla TA, Weindruch R (2004) Adipose tissue energy metabolism: altered gene expression profile of mice subjected to long-term caloric restriction. FASEB J 18:415–417
Higami Y, Shimokawa I, Ando K, Tanaka K, Tsuchiya T (2000) Dietary restriction reduces hepatocyte proliferation and enhances p53 expression but does not increase apoptosis in normal rats during development. Cell Tissue Res 299:363–369
Hikita H, Vaughan J, Babcock K, Pitot HC (1999) Short-term fasting and the reversal of the stage of promotion in rat hepatocarcinogenesis: role of cell replication, apoptosis, and gene expression. Toxicol Sci 52:17–23
Hokama T, Argaki H, Sho H et al. (1967) Nutrition survey of school children in Okinawa. Sci Bull Coll Agr Univ Ryukyus 14:1–15
Holehan AM, Merry BJ (1985a) Lifetime breeding studies in fully fed and dietary restricted female CFY Sprague-Dawley rats. 1. Effect of age, housing conditions and diet on fecundity. Mech Ageing Dev 33:19–28
Holehan AM, Merry BJ (1985b) The control of puberty in the dietary restricted female rat. Mech Ageing Dev 32:179–191
Holliday R (1989) Food, reproduction and longevity: Is the extended lifespan of calorie-restricted animals an evolutionary adaptation? Bioessays 10:125–127
Holloszy JO (1992) Exercise and food restriction in rats. J Nutr 122:774–777
Holloszy JO (1997) Mortality rate and longevity of food-restricted exercising male rats: a reevaluation. J Appl Physiol 82:399–403
Holloszy JO, Fontana L (2007) Caloric restriction in humans. Exp Gerontol 42:709–712
Holloszy JO, Schechtman KB (1991) Interaction between exercise and food restriction: effects on longevity of male rats. J Appl Physiol 70:1529–1535
Holloszy JO, Smith EK (1987) Effects of exercise on longevity of rats. Fed Proc 46:1850–1853
Holloszy JO, Smith EK, Vining M, Adams S (1985) Effect of voluntary exercise on longevity of rats. J Appl Physiol 59:826–831
Holt PR, Moss SF, Heydari AR, Richardson A (1998) Diet restriction increases apoptosis in the gut of aging rats. J Gerontol: Biol Sci 53A:B168-B172
Holzenberger M, Dupont J, Ducos B, Leneuve P, Geloen A, Even PC, Cervera P, Le Bouc Y (2003) IGF-1 receptor regulates lifespan and resistance to oxidative stress in mice. Nature 421:182–187
Hopfner M, Huether A, Sutter AP, Baradari V, Schuppan D, Scherubl H (2006) Blockade of IGF-1 receptor tyrosine kinase has antineoplastic effects in hepatocellular carcinoma cells. Biochem Pharmacol 71:1435–1448
Hu G, Tuomilehto J, Silventoinen K, Barengo NC, Peltonen M, Jousilahti P (2005) The effects of physical activity and body mass index on cardiovascular, cancer and all-cause mortality among 47 212 middle-aged Finnish men and women. Int J Obes (Lond) 29:894–902

Huether A, Hopfner M, Baradari V, Schuppan D, Scherubl H (2005a) EGFR blockade by cetuximab alone or as combination therapy for growth control of hepatocellular cancer. Biochem Pharmacol 70:1568–1578

Huether A, Hopfner M, Sutter AP, Schuppan D, Scherubl H (2005b) Erlotinib induces cell cycle arrest and apoptosis in hepatocellular cancer cells and enhances chemosensitivity towards cytostatics. J Hepatol 43:661–669

Hundal HS, Ramlal T, Reyes R, Leiter LA, Klip A (1992) Cellular mechanism of metformin action involves glucose transporter translocation from an intracellular pool to the plasma membrane in L6 muscle cells. Endocrinology 131:1165–1173

Hundal RS, Krssak M, Dufour S, Laurent D, Lebon V, Chandramouli V, Inzucchi SE, Schumann WC, Petersen KF, Landau BR, Shulman GI (2000) Mechanism by which metformin reduces glucose production in type 2 diabetes. Diabetes 49:2063–2069

Hursting SD, Lavigne JA, Berrigan D, Perkins SN, Barrett JC (2003) Calorie restriction, aging, and cancer prevention: mechanisms of action and applicability to humans. Annu Rev Med 54:131–152

Idrobo F, Nandy K, Mostofsky DI, Blatt L, Nandy L (1987) Dietary restriction: effects on radial maze learning and lipofuscin pigment deposition in the hippocampus and frontal cortex. Arch Gerontol Geriatr 6:355–362

Ikeno Y, Bronson RT, Hubbard GB, Lee S, Bartke A (2003) Delayed occurrence of fatal neoplastic diseases in Ames dwarf mice: correlation to extended longevity. J Gerontol A Biol Sci Med Sci 58:291–296

Imai S, Armstrong CM, Kaeberlein M, Guarente L (2000) Transcriptional silencing and longevity protein Sir2 is an NAD-dependent histone deacetylase. Nature 403:795–800

Ingram DK, Weindruch R, Spangler EL, Freeman JR, Walford RL (1987) Dietary restriction benefits learning and motor performance of aged mice. J Gerontol 42:78–81

Ingram DK, Zhu M, Mamczarz J, Zou S, Lane MA, Roth GS, deCabo R (2006) Calorie restriction mimetics: an emerging research field. Aging Cell 5:97–108

Inoki K, Zhu T, Guan KL (2003) TSC2 mediates cellular energy response to control cell growth and survival. Cell 115:577–590

Iwasaki K, Gleiser CA, Masoro EJ, McMahan CA, Seo EJ, Yu BP (1988) Influence of the restriction of individual dietary components on longevity and age-related disease of Fischer rats: the fat component and the mineral component. J Gerontol 43:B13–B21

Jafari M, Felgner JS, Bussel II, Hutchili T, Khodayari B, Rose MR, Vince-Cruz C, Mueller LD (2007) Rhodiola: a promising anti-aging Chinese herb. Rejuvenation Res 10:587–602

Jafari M, Zarban A, Pham S, Wang T (2008) Rosa damascena decreased mortality in adult Drosophila. J Med Food 11:9–13

Jager S, Handschin C, St Pierre J, Spiegelman BM (2007) AMP-activated protein kinase (AMPK) action in skeletal muscle via direct phosphorylation of PGC-1alpha. Proc Natl Acad Sci USA 104:12017–12022

James SJ, Muskhelishvili L (1994) Rates of apoptosis and proliferation vary with caloric intake and may influence incidence of spontaneous hepatoma in C57BL/6 × C3H F1 mice. Cancer Res 54:5508–5510

James SJ, Muskhelishvili L, Gaylor DW, Turturro A, Hart R (1998) Upregulation of apoptosis with dietary restriction: Implications for carcinogenesis and aging. Environ Health Perspectives 106:307–312

Jamieson BD, Douek DC, Killian S, Hultin LE, Scripture-Adams DD, Giorgi JV, Marelli D, Koup RA, Zack JA (1999) Generation of functional thymocytes in the human adult. Immunity 10:569–575

Janicki JS (1992) Myocardial collagen remodeling and left ventricular diastolic function. Braz J Med Biol Res 25:975–982

Janssen I, Katzmarzyk PT, Ross R (2005) Body mass index is inversely related to mortality in older people after adjustment for waist circumference. J Am Geriatr Soc 53:2112–2118

Janzen V, Forkert R, Fleming HE, Saito Y, Waring MT, Dombkowski DM, Cheng T, DePinho RA, Sharpless NE, Scadden DT (2006) Stem-cell ageing modified by the cyclin-dependent kinase inhibitor p16INK4a. Nature 443:421–426

Jia K, Levine B (2007) Autophagy is required for dietary restriction-mediated life span extension in *C. elegans*. Autophagy 3:597–599

Jin Y, Koizumi A (1994) Decreased cellular proliferation by energy restriction is recovered by increaseing housing temperature in rats. Mech Ageing Dev 75:59–67

Jin Z, El Deiry WS (2005) Overview of cell death signaling pathways. Cancer Biol Ther 4:139–163

Johnson AB, Webster JM, Sum CF, Heseltine L, Argyraki M, Cooper BG, Taylor R (1993) The impact of metformin therapy on hepatic glucose production and skeletal muscle glycogen synthase activity in overweight type II diabetic patients. Metabolism 42:1217–1222

Johnson KM, Owen K, Witte PL (2002) Aging and developmental transitions in the B cell lineage. Int Immunol 14:1313–1323

Kadowaki M, Karim MR, Carpi A, Miotto G (2006) Nutrient control of macroautophagy in mammalian cells. Mol Aspects Med 27:426–443

Kaeberlein M, Hu D, Kerr EO, Tsuchiya M, Westman EA, Dang N, Fields S, Kennedy BK (2005) Increased life span due to calorie restriction in respiratory-deficient yeast. PLoS Genet 1:e69

Kaeberlein M, Kennedy BK (2008) Protein translation, 2008. Aging Cell 7:777–782

Kaeberlein M, Kirkland KT, Fields S, Kennedy BK (2004) Sir2-independent life span extension by calorie restriction in yeast. PLoS Biol 2:E296

Kaeberlein M, McVey M, Guarente L (1999) The SIR2/3/4 complex and SIR2 alone promote longevity in Saccharomyces cerevisiae by two different mechanisms. Genes Dev 13:2570–2580

Kaeberlein M, Powers RW III (2007) Sir2 and calorie restriction in yeast: a skeptical perspective. Ageing Res Rev 6:128–140

Kagawa Y (1978) Impact of Westernization on the nutrition of Japanese: changes in physique, cancer, longevity and centenarians. Prev Med 7:205–217

Kahn BB, Alquier T, Carling D, Hardie DG (2005) AMP-activated protein kinase: ancient energy gauge provides clues to modern understanding of metabolism. Cell Metab 1:15–25

Kakkar R, Bains JS, Sharma SP (1996) Effect of vitamin E on life span, malondialdehyde content and antioxidant enzymes in aging *Zaprionus paravittiger*. Gerontology 42:312–321

Kalli KR, Falowo OI, Bale LK, Zschunke MA, Roche PC, Conover CA (2002) Functional insulin receptors on human epithelial ovarian carcinoma cells: implications for IGF-II mitogenic signaling. Endocrinology 143:3259–3267

Kass, LR (2001) L'Chaim and Its Limits: Why Not Immortality? First Things (May 2001) http://www.firstthings.com/article.php3?id_article=2188

Kaur M, Sharma S, Kaur G (2008) Age-related impairments in neuronal plasticity markers and astrocytic GFAP and their reversal by late-onset short term dietary restriction. Biogerontology 9:441–454

Kayo T, Allison DB, Weindruch R, Prolla TA (2001) Influences of aging and caloric restriction on the transcriptional profile of skeletal muscle from rhesus monkeys. Proc Natl Acad Sci USA 98:5093–5098

Keller ET, Keller JM, Fillespie G (2006) The use of mature zebrafish (*Danio rerio*) as a model for human aging. In: Conn PM (ed) Handbook of Models for Human Aging. Elsevier, Amsterdam, The Netherlands, pp. 299–314

Kemnitz JW, Weindruch R, Roecker EB, Crawford K, Kaufman PL, Ershler WB (1993) Dietary restriction of adult male rhesus monkeys: design, methodology, and preliminary findings from the first year of study. J Gerontol 48:B17-B26

Kennedy MA, Rakoczy SG, Brown-Borg HM (2003) Long-living Ames dwarf mouse hepatocytes readily undergo apoptosis. Exp Gerontol 38:997–1008

Kenyon C (2005) The plasticity of aging: insights from long-lived mutants. Cell 120:449–460

Kenyon C, Chang J, Gensch E, Rudner A, Tabtiang R (1993) A *C. elegans* mutant that lives twice as long as wild type. Nature 366:461–464

Kim JA, Wei Y, Sowers JR (2008) Role of mitochondrial dysfunction in insulin resistance. Circ Res 102:401–414

Kirkwood TB, Austad SN (2000) Why do we age? Nature 408:233–238

Kitani K, Osawa T, Yokozawa T (2007) The effects of tetrahydrocurcumin and green tea polyphenol on the survival of male C57BL/6 mice. Biogerontology 8:567–573

Klass MR (1977) Aging in the nematode *Caenorhabditis elegans*: major biological and environmental factors influencing life span. Mech Ageing Dev 6:413–429

Klebanov S (2007) Can short-term dietary restriction and fasting have a long-term anticarcinogenic effect? Interdiscip Top Gerontol 35:176–92

Klebanov S, Diais S, Stavinoha WB, Suh Y, Nelson JF (1995) Hyperadrenocorticism, attenuated inflammation, and the life- prolonging action of food restriction in mice. J Gerontol A Biol Sci Med Sci 50:B79–B82

Knecht S, Ellger T, Levine JA (2008) Obesity in neurobiology. Prog Neurobiol 84:85–103

Kohn RR (1971) Effect of antioxidants on life-span of C57BL mice. J Gerontol 26:378–380

Kohn RR, Leash AM (1967) Long-term lathyrogen administration to rats, with special reference to aging. Exp Mol Pathol 7:354–361

Koizumi A, Roy NS, Tsukada M, Wada Y (1993) Increase in housing temperature can alleviate decrease in white blood cell counts after energy restriction in C57BL/6 female mice. Mech Ageing Dev 71:97–102

Koizumi A, Wada Y, Tuskada M, Kayo T, Naruse M, Horiuchi K, Mogi T, Yoshioka M, Sasaki M, Miyamaura Y, Abe T, Ohtomo K, Walford RL (1996) A tumor preventive effect of dietary restriction is antagonized by a high housing temperature through deprivation of torpor. Mech Ageing Dev 92:67–82

Komninou D, Leutzinger Y, Reddy BS, Richie JP Jr (2006) Methionine restriction inhibits colon carcinogenesis. Nutr Cancer 54:202–208

Kosaki A, Webster NJ (1993) Effect of dexamethasone on the alternative splicing of the insulin receptor mRNA and insulin action in HepG2 hepatoma cells. J Biol Chem 268: 21990–21996

Koubova J, Guarente L (2003) How does calorie restriction work? Genes Dev 17:313–321

Kouda K, Nakamura H, Kohno H, Ha-Kawa SK, Tokunaga R, Sawada S (2004) Dietary restriction: effects of short-term fasting on protein uptake and cell death/proliferation in the rat liver. Mech Ageing Dev 125:375–380

Krishnamurthy J, Ramsey MR, Ligon KL, Torrice C, Koh A, Bonner-Weir S, Sharpless NE (2006) p16INK4a induces an age-dependent decline in islet regenerative potential. Nature 443:453–457

Kuhn HG, Dickinson-Anson H, Gage FH (1996) Neurogenesis in the dentate gyrus of the adult rat: age-related decrease of neuronal progenitor proliferation. J Gerontol 16:2027–2033

Kurosu H, Yamamoto M, Clark JD, Pastor JV, Nandi A, Gurnani P, McGuinness OP, Chikuda H, Yamaguchi M, Kawaguchi H, Shimomura I, Takayama Y, Herz J, Kahn CR, Rosenblatt KP, Kuro-o M (2005) Suppression of aging in mice by the hormone Klotho. Science 309:1829–1833

Kyng KJ, Bohr VA (2005) Gene expression and DNA repair in progeroid syndromes and human aging. Ageing Res Rev 4:579–602

Lakatta EG (2000) Cardiovascular aging in health. Clin Geriatr Med 16:419–444

Lamb MJ (1968) Temperature and lifespan in Drosophila. Nature 220:808–809

Landry J, Sutton A, Tafrov ST, Heller RC, Stebbins J, Pillus L, Sternglanz R (2000) The silencing protein SIR2 and its homologs are NAD-dependent protein deacetylases. Proc Natl Acad Sci USA 97:5807–5811

Lane MA, Ball SS, Ingram DK, Cutler RG, Engel J, Read V, Roth GS (1995) Diet restriction in rhesus monkeys lowers fasting and glucose- stimulated glucoregulatory end points. Am J Physiol 268:E941–E948

Lane MA, Ingram DK, Roth GS (1999) Calorie restriction in nonhuman primates: effects on diabetes and cardiovascular disease risk. Toxicol Sci 52:41–48

Lane MA, Mattison JA, Roth GS, Brant LJ, Ingram DK (2004) Effects of long-term diet restriction on aging and longevity in primates remain uncertain. J Gerontol A Biol Sci Med Sci 59:405–407

Lawler DF, Larson BT, Ballam JM, Smith GK, Biery DN, Evans RH, Greeley EH, Segre M, Stowe HD, Kealy RD (2008) Diet restriction and ageing in the dog: major observations over two decades. Br J Nutr 99:793–805

Lee CK, Allison DB, Brand J, Weindruch R, Prolla TA (2002) Transcriptional profiles associated with aging and middle age-onset caloric restriction in mouse hearts. Proc Natl Acad Sci USA 99:14988–14993

Lee CK, Klopp RG, Weindruch R, Prolla TA (1999) Gene expression profile of aging and its retardation by caloric restriction. Science 285:1390–1393

Lee CK, Weindruch R, Prolla TA (2000) Gene-expression profile of the ageing brain in mice. Nat Genet 25:294–297

Lefevre M, Redman LM, Heilbronn LK, Smith JV, Martin CK, Rood JC, Greenway FL, Williamson DA, Smith SR, Ravussin E (2009) Caloric restriction alone and with exercise improves CVD risk in healthy non-obese individuals. Atherosclerosis 203:206–213

Lemon JA, Boreham DR, Rollo CD (2005) A complex dietary supplement extends longevity of mice. J Gerontol A Biol Sci Med Sci 60:275–279

Lewis SE, Goldspink DF, Phillips JG, Merry BJ, Holehan AM (1985) The effects of aging and chronic dietary restriction on whole body growth and protein turnover in the rat. Exp Gerontol 20:253–263

Li YM, Chan HY, Huang Y, Chen ZY (2007) Green tea catechins upregulate superoxide dismutase and catalase in fruit flies. Mol Nutr Food Res 51:546–554

Liang H, Masoro EJ, Nelson JF, Strong R, McMahan CA, Richardson A (2003) Genetic mouse models of extended lifespan. Exp Gerontol 38:1353–1364

Libert S, Zwiener J, Chu X, Vanvoorhies W, Roman G, Pletcher SD (2007) Regulation of Drosophila life span by olfaction and food-derived odors. Science 315:1133–1137

Lin SC, Lin CR, Gukovsky I, Lusis AJ, Sawchenko PE, Rosenfeld MG (1993) Molecular basis of the little mouse phenotype and implications for cell type-specific growth. Nature 364:208–213

Lin SJ, Defossez PA, Guarente L (2000) Requirement of NAD and SIR2 for life-span extension by calorie restriction in Saccharomyces cerevisiae [see comments]. Science 289:2126–2128

Lintern-Moore S, Everitt AV (1978) The effect of restricted food intake on the size and composition of the ovarian follicle population in the Wistar rat. Biol Reprod 19:688–691

Lipman RD, Smith DE, Blumberg JB, Bronson RT (1998) Effects of caloric restriction or augmentation in adult rats: longevity and lesion biomarkers of aging. Aging (Milano) 10:463–470

Lipman RD, Smith DE, Bronson RT, Blumberg J (1995) Is late-life caloric restriction beneficial? Aging (Milano) 7:136–139

Liu RK, Walford RL (1966) Increased Growth and Life-span with Lowered Ambient Temperature in the Annual Fish, Cynolebias adloffi. Nature 212:1277–1278

Liu RK, Walford RL (1970) Observations on the lifespans of several species of annual fishes and of the world's smallest fishes. Exp Gerontol 5:241–246

Liu RK, Walford RL (1972) The effect of lowered body temperature on lifespan and immune and non-immune processes. Gerontologia 18:363–388

Lok E, Nera EA, Iverson F, Scott F, So Y, Clayson DB (1988) Dietary restriction, cell proliferation and carcinogenesis: a preliminary study. Cancer Lett 38:249–255

Lopez-Lluch G, Irusta PM, Navas P, de Cabo R (2008) Mitochondrial biogenesis and healthy aging. Exp Gerontol 43:813–819

los Santos-Arteaga M, Sierra-Dominguez SA, Fontanella GH, Delgado-Garcia JM, Carrion AM (2003) Analgesia induced by dietary restriction is mediated by the kappa-opioid system. J Gerontol. 23:11120–11126

Lu T, Pan Y, Kao SY, Li C, Kohane I, Chan J, Yankner BA (2004) Gene regulation and DNA damage in the ageing human brain. Nature 429:883–891

Luan X, Zhao W, Chandrasekar B, Fernandes G (1995) Calorie restriction modulates lymphocyte subset phenotype and increases apoptosis in MRL/lpr mice. Immunol Lett 47:181–186

Lushnikova EL, Nepomnyashchikh LM, Klinnikova MG (2001) Morphological characteristics of myocardial remodeling during compensatory hypertrophy in aging Wistar rats. Bull Exp Biol Med 132:1201–1206

Ma TC, Buescher JL, Oatis B, Funk JA, Nash AJ, Carrier RL, Hoyt KR (2007) Metformin therapy in a transgenic mouse model of Huntington's disease. Neurosci Lett 411:98–103

Maeda H, Gleiser CA, Masoro EJ, Murates I, McMahan CA, Yu BP (1985) Nutritional influences on aging of Fischer 344 rats: II. Pathology. J Gerontol 40:671–688

Maes HH, Neale MC, Eaves LJ (1997a) Genetic and environmental factors in relative body weight and human adiposity. Behav Genet 27:325–351

Maes HH, Neale MC, Eaves LJ (1997b) Genetic and environmental factors in relative body weight and human adiposity. Behav Genet 27:325–351

Magwere T, Chapman T, Partridge L (2004) Sex differences in the effect of dietary restriction on life span and mortality rates in female and male *Drosophila melanogaster*. J Gerontol A Biol Sci Med Sci 59:3–9

Mair W, Dillin A (2008) Aging and survival: the genetics of life span extension by dietary restriction. Annu Rev Biochem 77:727–754

Mair W, Goymer P, Pletcher SD, Partridge L (2003) Demography of dietary restriction and death in Drosophila. Science 301:1731–1733

Mair W, Piper MD, Partridge L (2005) Calories do not explain extension of life span by dietary restriction in Drosophila. PLoS Biol 3:e223

Malin A, Matthews CE, Shu XO, Cai H, Dai Q, Jin F, Gao YT, Zheng W (2005) Energy balance and breast cancer risk. Cancer Epidemiol Biomarkers Prev 14:1496–1501

Malloy VL, Krajcik RA, Bailey SJ, Hristopoulos G, Plummer JD, Orentreich N (2006) Methionine restriction decreases visceral fat mass and preserves insulin action in aging male Fischer 344 rats independent of energy restriction. Aging Cell 5:305–314

Mansouri A, Muller FL, Liu Y, Ng R, Faulkner J, Hamilton M, Richardson A, Huang TT, Epstein CJ, Van Remmen H (2006) Alterations in mitochondrial function, hydrogen peroxide release and oxidative damage in mouse hind-limb skeletal muscle during aging. Mech Ageing Dev 127:298–306

Martin B, Golden E, Carlson OD, Egan JM, Mattson MP, Maudsley S (2008) Caloric restriction: impact upon pituitary function and reproduction. Ageing Res Rev 7:209–224

Martin B, Mattson MP, Maudsley S (2006) Caloric restriction and intermittent fasting: two potential diets for successful brain aging. Ageing Res Rev 5:332–353

Masoro EJ (1990) Assessment of nutritional components in prolongation of life and health by diet. Proc Soc Exp Biol Med 193:31–34

Masoro EJ (1995) Glucocorticoids and aging. Aging (Milano) 7:407–413

Masoro EJ (2005) Overview of caloric restriction and ageing. Mech Ageing Dev 126:913–922

Masoro EJ (2006) Dietary restriction-induced life extension: a broadly based biological phenomenon. Biogerontology 7:153–155

Masoro EJ, Austad SN (1996) The evolution of the antiaging action of dietary restriction: A hypothesis. J Gerontol: Biol Sci 51A:B387–B391

Masoro EJ, McCarter RJ, Katz MS, McMahan CA (1992) Dietary restriction alters characteristics of glucose fuel use. J Gerontol 47:B202–B208

Masoro EJ, Shimokawa I, Higami Y, McMahan CA, Yu BP (1995) Temporal pattern of food intake not a factor in the retardation of aging processes by dietary restriction. J Gerontol 50A:B48–B53

Massaro D, DeCarlo MG, Baras A, Hoffman EP, Clerch LB (2004) Calorie-related rapid onset of alveolar loss, regeneration, and changes in mouse lung gene expression. Am J Physiol Lung Cell Mol Physiol 286:L896–L906

Massaro DJ, Alexander E, Reiland K, Hoffman EP, Massaro GD, Clerch LB (2007) Rapid onset of gene expression in lung, supportive of formation of alveolar septa, induced by refeeding mice after calorie restriction. Am J Physiol Lung Cell Mol Physiol 292:L1313–1326

Massie HR (1994) Effect of dietary boron on the aging process. Environ Health Perspect 102 (Suppl 7):45–48

Maswood N, Young J, Tilmont E, Zhang Z, Gash DM, Gerhardt GA, Grondin R, Roth GS, Mattison J, Lane MA, Carson RE, Cohen RM, Mouton PR, Quigley C, Mattson MP, Ingram DK (2004) Caloric restriction increases neurotrophic factor levels and attenuates neurochemical and behavioral deficits in a primate model of Parkinson's disease. Proc Natl Acad Sci USA 101:18171–18176

Mattison JA, Black A, Huck J, Moscrip T, Handy A, Tilmont E, Roth GS, Lane MA, Ingram DK (2005) Age-related decline in caloric intake and motivation for food in rhesus monkeys. Neurobiol Aging 26:1117–1127

Mattison JA, Lane MA, Roth GS, Ingram DK (2003) Calorie restriction in rhesus monkeys. Exp Gerontol 38:35–46

Mattson MP, Duan W, Lee J, Guo Z (2001) Suppression of brain aging and neurodegenerative disorders by dietary restriction and environmental enrichment: molecular mechanisms. Mech Ageing Dev 122:757–778

Mattson MP, Duan W, Wan R, Guo Z (2004) Prophylactic activation of neuroprotective stress response pathways by dietary and behavioral manipulations. NeuroRx 1:111–116

McBurney MW, Yang X, Jardine K, Hixon M, Boekelheide K, Webb JR, Lansdorp PM, Lemieux M (2003) The mammalian SIR2alpha protein has a role in embryogenesis and gametogenesis. Mol Cell Biol 23:38–54

McCarter R, Mejia W, Ikeno Y, Monnier V, Kewitt K, Gibbs M, McMahan A, Strong R (2007) Plasma glucose and the action of calorie restriction on aging. J Gerontol A Biol Sci Med Sci 62:1059–1070

McCay CM, Crowell MF, Maynard LA (1935) The effect of retarded growth upon the length of the life span and upon the ultimate body size. J Nutr 10:63–79

McCormick DL, Johnson WD, Haryu TM, Bosland MC, Lubet RA, Steele VE (2007) Null effect of dietary restriction on prostate carcinogenesis in the Wistar-Unilever rat. Nutr Cancer 57:194–200

McElhaney JE (2005) The unmet need in the elderly: designing new influenza vaccines for older adults. Vaccine 23(Suppl 1):S10-S25

McGuire DK, Levine BD, Williamson JW, Snell PG, Blomqvist CG, Saltin B, Mitchell JH (2001) A 30-year follow-up of the Dallas Bedrest and Training Study: I. Effect of age on the cardiovascular response to exercise. Circulation 104:1350–1357

McMillan DC, Sattar N, McArdle CS (2006) ABC of obesity. Obesity and cancer. BMJ 333:1109–1111

Meijer AJ, Codogno P (2006) Signalling and autophagy regulation in health, aging and disease. Mol Aspects Med 27:411–425

Melendez A, Talloczy Z, Seaman M, Eskelinen EL, Hall DH, Levine B (2003) Autophagy genes are essential for dauer development and life-span extension in *C. elegans*. Science 301:1387–1391

Merrill GF, Kurth EJ, Hardie DG, Winder WW (1997) AICA riboside increases AMP-activated protein kinase, fatty acid oxidation, and glucose uptake in rat muscle. Am J Physiol 273:E1107–E1112

Merry BJ (2002) Molecular mechanisms linking calorie restriction and longevity. Int J Biochem Cell Biol 34:1340–1354

Merry BJ, Goldspink DF, Lewis SE (1991a) The effects of age and chronic restricted feeding on protein synthesis and growth of the large intestine of the rat. Comp Biochem Physiol A 98:559–562

Merry BJ, Holehan AM (1979) Onset of puberty and duration of fertility in rats fed a restricted diet. J Reprod Fertil 57:253–259

Merry BJ, Holehan AM (1981) Serum profiles of LH, FSH, testosterone and 5 alpha-DHT from 21 to 1000 days of age in ad libitum fed and dietary restricted rats. Exp Gerontol 16:431–444

Merry BJ, Holehan AM (1985) In vivo DNA synthesis in the dietary restricted long-lived rat. Exp Gerontol 20:15–28

Merry BJ, Holehan AM (1991b) Effect of age and restricted feeding on polypeptide chain assembly kinetics in liver protein synthesis in vivo. Mech Ageing Dev 58:139–150

Merry BJ, Holehan AM, Lewis SE, Goldspink DF (1987) The effects of ageing and chronic dietary restriction on in vivo hepatic protein synthesis in the rat. Mech Ageing Dev 39:189–199

Merry BJ, Kirk AJ, Goyns MH (2008) Dietary lipoic acid supplementation can mimic or block the effect of dietary restriction on life span. Mech Ageing Dev 129:341–348

Merry BJ, Lewis SE, Goldspink DF (1992) The influence of age and chronic restricted feeding on protein synthesis in the small intestine of the rat. Exp Gerontol 27:191–200

Messaoudi I, Fischer M, Warner J, Park B, Mattison J, Ingram DK, Totonchy T, Mori M, Nikolich-Zugich J (2008) Optimal window of caloric restriction onset limits its beneficial impact on T-cell senescence in primates. Aging Cell 7:908–919

Messaoudi I, Lemaoult J, Guevara-Patino JA, Metzner BM, Nikolich-Zugich J (2004) Age-related CD8 T cell clonal expansions constrict CD8 T cell repertoire and have the potential to impair immune defense. J Exp Med 200:1347–1358

Messaoudi I, Warner J, Fischer M, Park B, Hill B, Mattison J, Lane MA, Roth GS, Ingram DK, Picker LJ, Douek DC, Mori M, Nikolich-Zugich J (2006) Delay of T cell senescence by caloric restriction in aged long-lived nonhuman primates. Proc Natl Acad Sci USA 103:19448–19453

Meyer TE, Kovacs SJ, Ehsani AA, Klein S, Holloszy JO, Fontana L (2006) Long-term caloric restriction ameliorates the decline in diastolic function in humans. J Am Coll Cardiol 47:398–402

Michan S, Sinclair D (2007) Sirtuins in mammals: insights into their biological function. Biochem J 404:1–13

Michels KB, Ekbom A (2004) Caloric restriction and incidence of breast cancer. JAMA 291:1226–1230

Miller JP, Allman D (2003) The decline in B lymphopoiesis in aged mice reflects loss of very early B-lineage precursors. J Immunol 171:2326–2330

Miller RA (1996) The aging immune system: primer and prospectus. Science 273:70–74

Miller RA, Buehner G, Chang Y, Harper JM, Sigler R, Smith-Wheelock M (2005) Methionine-deficient diet extends mouse lifespan, slows immune and lens aging, alters glucose, T4, IGF-I and insulin levels, and increases hepatocyte MIF levels and stress resistance. Aging Cell 4:119–125

Miller RA, Harrison DE, Astle CM, Floyd RA, Flurkey K, Hensley KL, Javors MA, Leeuwenburgh C, Nelson JF, Ongini E, Nadon NL, Warner HR, Strong R (2007) An aging Interventions Testing Program: study design and interim report. Aging Cell 6:565–575

Min KJ, Flatt T, Kulaots I, Tatar M (2007) Counting calories in Drosophila diet restriction. Exp Gerontol 42:247–251

Minokoshi Y, Kim YB, Peroni OD, Fryer LG, Muller C, Carling D, Kahn BB (2002) Leptin stimulates fatty-acid oxidation by activating AMP-activated protein kinase. Nature 415: 339–343

Miquel J, Lundgren PR, Bensch KG, Atlan H (1976) Effects of temperature on the life span, vitality and fine structure of *Drosophila melanogaster*. Mech Ageing Dev 5:347–370

Mizushima N, Klionsky DJ (2007) Protein turnover via autophagy: implications for metabolism. Annu Rev Nutr 27:19–40

Molofsky AV, Slutsky SG, Joseph NM, He S, Pardal R, Krishnamurthy J, Sharpless NE, Morrison SJ (2006) Increasing p16INK4a expression decreases forebrain progenitors and neurogenesis during ageing. Nature 443:448–452

Morck C, Pilon M (2006) *C. elegans* feeding defective mutants have shorter body lengths and increased autophagy. BMC Dev Biol 6:39

Moreau R, Heath SH, Doneanu CE, Harris RA, Hagen TM (2004) Age-related compensatory activation of pyruvate dehydrogenase complex in rat heart. Biochem Biophys Res Commun 325:48–58

Moreschi C (1909) Beziehungen zwischen ernahrung und tumorwachstum [Relationship between nutrition and tumor growth]. Zeitschrift f Immunitätsforschung 2:651–675

Motta MC, Divecha N, Lemieux M, Kamel C, Chen D, Gu W, Bultsma Y, McBurney M, Guarente L (2004) Mammalian SIRT1 represses forkhead transcription factors. Cell 116:551–563

Murtagh-Mark CM, Reiser KM, Harris R, Jr., McDonald RB (1995) Source of dietary carbohydrate affects life span of Fischer 344 rats independent of caloric restriction. J Gerontol A Biol Sci Med Sci 50:B148–B154

Muskhelishvili L, Hart RW, Turturro A, James SJ (1995) Age-related changes in the intrinsic rate of apoptosis in livers of diet-restricted and ad libitum-fed B6C3F1 mice. Am J Pathol 147:20–24

Muskhelishvili L, Turturro A, Hart RW, James SJ (1996) Pi-class glutathione-S-transferase-positive hepatocytes in aging B6C3F1 mice undergo apoptosis induced by dietary restriction. Am J Pathol 149:1585–1591

Myung J, Kim KB, Crews CM (2001) The ubiquitin-proteasome pathway and proteasome inhibitors. Med Res Rev 21:245–273

Nelson JF, Gosden RG, Felicio, LS (1985) Effect of dietary restriction on estrous cyclicity and follicular reserves in aging C57BL/6 J mice. Biol Reprod 32:515–522

Nelson JF, Karelus K, Bergman MD, Felicio LS (1995) Neuroendocrine involvement in aging: evidence from studies of reproductive aging and caloric restriction. Neurobiol Aging 16:837–843

Nelson W, Halberg F (1986) Meal-timing, circadian rhythms and life span of mice. J Nutr 116:2244–2253

Nikolich-Zugich J, Messaoudi I (2005) Mice and flies and monkeys too: caloric restriction rejuvenates the aging immune system of non-human primates. Exp Gerontol 40:884–893

Nolen GA (1972) Effect of various restricted dietary regimens on the growth, health and longevity of albino rats. J Nutr 102:1477–1493

O'Brien CA, Jia D, Plotkin LI, Bellido T, Powers CC, Stewart SA, Manolagas SC, Weinstein RS (2004) Glucocorticoids act directly on osteoblasts and osteocytes to induce their apoptosis and reduce bone formation and strength. Endocrinology 145:1835–1841

Oeppen J, Vaupel JW (2002) Demography. Broken limits to life expectancy. Science 296:1029–1031

Olshansky SJ, Carnes BA, Desesquelles A (2001) Demography. Prospects for human longevity. Science 291:1491–1492

Orchard TJ, Temprosa M, Goldberg R, Haffner S, Ratner R, Marcovina S, Fowler S (2005) The effect of metformin and intensive lifestyle intervention on the metabolic syndrome: the Diabetes Prevention Program randomized trial. Ann Intern Med 142:611–619

Orentreich N, Matias JR, DeFelice A, Zimmerman JA (1993) Low methionine ingestion by rats extends life span. J Nutr 123:269–274

Osborne TB, Mendel LB, Ferry EL (1917) The effect of retardation of growth upon the breeding period and duration of life of rats. Science 45:294–295

Oster MH, Fielder PJ, Levin N, Cronin MJ (1995) Adaptation of the growth hormone and insulin-like growth factor-I axis to chronic and severe calorie or protein malnutrition. J Clin Invest 95:2258–2265

Owen MR, Doran E, Halestrap AP (2000) Evidence that metformin exerts its anti-diabetic effects through inhibition of complex 1 of the mitochondrial respiratory chain. Biochem J 348(Pt 3):607–614

Pahlavani MA (2000) Caloric restriction and immunosenescence: a current perspective. Front Biosci 5:D580–D587

Pamplona R, Barja G (2006) Mitochondrial oxidative stress, aging and caloric restriction: the protein and methionine connection. Biochim Biophys Acta 1757:496–508

Pan KZ, Palter JE, Rogers AN, Olsen A, Chen D, Lithgow GJ, Kapahi P (2007) Inhibition of mRNA translation extends lifespan in Caenorhabditis elegans. Aging Cell 6:111–119

Pandini G, Conte E, Medico E, Sciacca L, Vigneri R, Belfiore A (2004) IGF-II binding to insulin receptor isoform A induces a partially different gene expression profile from insulin binding. Ann N Y Acad Sci 1028:450–456

Pandini G, Medico E, Conte E, Sciacca L, Vigneri R, Belfiore A (2003) Differential gene expression induced by insulin and insulin-like growth factor-II through the insulin receptor isoform A. J Biol Chem 278:42178–42189

Partridge L, Gems D, Withers DJ (2005) Sex and death: what is the connection? Cell 120: 461–472

Pashko LL, Schwartz AG (1992) Reversal of food restriction-induced inhibition of mouse skin tumor promotion by adrenalectomy. Carcinogenesis 13:1925–1928

Pashko LL, Schwartz, AG (1996) Inhibition of 7,12-dimethylbenz[a]anthracene-induced lung tumorigenesis in A/J mice by food restriction is reversed by adrenalectomy. Carcinogenesis 17:209–212

Patel AC, Nunez NP, Perkins SN, Barrett JC, Hursting SD (2004) Effects of energy balance on cancer in genetically altered mice. J Nutr 134:3394S–3398S

Pearson KJ, Baur JA, Lewis KN, Peshkin L, Price NL, Labinskyy N, Swindell WR, Kamara D, Minor RK, Perez E, Jamieson HA, Zhang Y, Dunn SR, Sharma K, Pleshko N, Woollett LA, Csiszar A, Ikeno Y, Le Couteur D, Elliott PJ, Becker KG, Navas P, Ingram DK, Wolf NS, Ungvari Z, Sinclair DA, de Cabo R (2008) Resveratrol Delays Age-Related Deterioration and Mimics Transcriptional Aspects of Dietary Restriction without Extending Life Span. Cell Metab 8:157–168

Petersen KF, Befroy D, Dufour S, Dziura J, Ariyan C, Rothman DL, DiPietro L, Cline GW, Shulman GI (2003) Mitochondrial dysfunction in the elderly: possible role in insulin resistance. Science 300:1140–1142

Petersen KF, Dufour S, Savage DB, Bilz S, Solomon G, Yonemitsu S, Cline GW, Befroy D, Zemany L, Kahn BB, Papademetris X, Rothman DL, Shulman GI (2007) The role of skeletal muscle insulin resistance in the pathogenesis of the metabolic syndrome. Proc Natl Acad Sci USA 104:12587–12594

Phelan JP, Austad SN (1989) Natural selection, dietary restriction, and extended longevity. Growth Dev Aging 53:4–6

Phelan JP, Rose MR (2005) Why dietary restriction substantially increases longevity in animal models but won't in humans. Ageing Res Rev 4:339–350

Pitner JK (2005) Obesity in the elderly. Consult Pharm 20:498–513

Pitsikas N, Algeri S (1992) Deterioration of spatial and nonspatial reference and working memory in aged rats: protective effect of life-long calorie restriction. Neurobiol Aging 13:369–373

Pitsikas N, Garofalo P, Manfridi A, Zanotti A, Algeri S (1991) Effect of lifelong hypocaloric diet on discrete memory of the senescent rat. Aging (Milano) 3:147–152

Popovich IG, Voitenkov BO, Anisimov VN, Ivanov VT, Mikhaleva II, Zabezhinski MA, Alimova IN, Baturin DA, Zavarzina NY, Rosenfeld SV, Semenchenko AV, Yashin AI (2003) Effect of delta-sleep inducing peptide-containing preparation Deltaran on biomarkers of aging, life span and spontaneous tumor incidence in female SHR mice. Mech Ageing Dev 124:721–731

Premoselli F, Sesca E, Binasco V, Franchino C, Tessitore L, (1997) Cell death and cell proliferation contribute to the enhanced growth of foci by fasting in rat medial colon. Boll Soc Ital Biol Sper 73:71–76

Pugh TD, Oberley TD, Weindruch R (1999) Dietary intervention at middle age: caloric restriction but not dehydroepiandrosterone sulfate increases lifespan and lifetime cancer incidence in mice. Cancer Res 59:1642–1648

Puigserver P, Adelmant G, Wu Z, Fan M, Xu J, O'Malley B, Spiegelman BM (1999) Activation of PPARgamma coactivator-1 through transcription factor docking. Science 286:1368–1371

Puigserver P, Wu Z, Park CW, Graves R, Wright M, Spiegelman BM (1998) A cold-inducible coactivator of nuclear receptors linked to adaptive thermogenesis. Cell 92:829–839

Qiang W, Weiqiang K, Qing Z, Pengju Z, Yi L (2007) Aging impairs insulin-stimulated glucose uptake in rat skeletal muscle via suppressing AMPKalpha. Exp Mol Med 39:535–543

Qin W, Chachich M, Lane M, Roth G, Bryant M, de Cabo R, Ottinger MA, Mattison J, Ingram D, Gandy S, Pasinetti GM (2006) Calorie restriction attenuates Alzheimer's disease type brain amyloidosis in Squirrel monkeys (Saimiri sciureus). J Alzheimers Dis 10:417–422

Rakhit R, Crow JP, Lepock JR, Kondejewski LH, Cashman NR, Chakrabartty A (2004) Monomeric Cu,Zn-superoxide dismutase is a common misfolding intermediate in the oxidation models of sporadic and familial amyotrophic lateral sclerosis. J Biol Chem 279:15499–15504

Redman LM, Ravussin E (2009) Endocrine alterations in response to calorie restriction in humans. Mol Cell Endocrinol 299:129–136

Reynolds MA, Dawson DR, Novak KF, Ebersole JL, Gunsolley JC, Branch-Mays GL, Holt SC, Mattison JA, Ingram DK, Novak MJ (2009) Effects of caloric restriction on inflammatory periodontal disease. Nutrition 25:88–97

Reznick AZ, Gershon D (1979) The effect of age on the protein degradation system in the nematode Turbatrix aceti. Mech Ageing Dev 11:403–415

Reznick RM, Zong H, Li J, Morino K, Moore IK, Yu HJ, Liu ZX, Dong J, Mustard KJ, Hawley SA, Befroy D, Pypaert M, Hardie DG, Young LH, Shulman GI (2007) Aging-associated reductions in AMP-activated protein kinase activity and mitochondrial biogenesis. Cell Metab 5:151–156

Richie JP, Jr., Leutzinger Y, Parthasarathy S, Malloy V, Orentreich N, Zimmerman JA (1994) Methionine restriction increases blood glutathione and longevity in F344 rats. FASEB J 8:1302–1307

Rikke BA, Yerg JE, III, Battaglia ME, Nagy TR, Allison DB, Johnson TE (2003) Strain variation in the response of body temperature to dietary restriction. Mech Ageing Dev 124:663–678

Ritz BW, Gardner EM (2006) Malnutrition and energy restriction differentially affect viral immunity. J Nutr 136:1141–1144

Rodgers JT, Lerin C, Haas W, Gygi SP, Spiegelman BM, Puigserver P (2005) Nutrient control of glucose homeostasis through a complex of PGC-1alpha and SIRT1. Nature 434: 113–118

Rodrigues MA, Sanchez-Negrette M, Mantovani MS, Sant'ana LS, Angeleli AY, Montenegro MR, de Camargo JL (1991) Liver response to low-hexachlorobenzene exposure in protein- or energy-restricted rats. Food Chem Toxicol 29:757–764

Rogina B, Helfand SL (2004) Sir2 mediates longevity in the fly through a pathway related to calorie restriction. Proc Natl Acad Sci USA 101:15998–16003

Rosenwald A, Wright G, Chan WC, Connors JM, Campo E, Fisher RI, Gascoyne RD, Muller-Hermelink HK, Smeland EB, Giltnane JM, Hurt EM, Zhao H, Averett L, Yang L, Wilson WH, Jaffe ES, Simon R, Klausner RD, Powell J, Duffey PL, Longo DL, Greiner TC, Weisenburger DD, Sanger WG, Dave BJ, Lynch JC, Vose J, Armitage JO, Montserrat E, Lopez-Guillermo A, Grogan TM, Miller TP, LeBlanc M, Ott G, Kvaloy S, Delabie J, Holte H, Krajci P, Stokke T, Staudt LM (2002) The use of molecular profiling to predict survival after chemotherapy for diffuse large-B-cell lymphoma. N Engl J Med 346:1937–1947

Ross MH (1972) Length of life and caloric intake. Am J Clin Nutr 25:834–838

Ross MH, Bras G (1973) Influence of protein under- and overnutrition on spontaneous tumor prevalence in the rat. J Nutr 103:944–963

Ross TK, Darwish HM, Moss VE, DeLuca HF (1993) Vitamin D-influenced gene expression via a ligand-independent, receptor-DNA complex intermediate. Proc Natl Acad Sci USA 90:9257–9260

Roth GS, Ingram DK, Lane MA (1999) Calorie restriction in primates: will it work and how will we know? J Am Geriatr Soc 47:896–903

Roth GS, Mattison JA, Ottinger MA, Chachich ME, Lane MA, Ingram DK (2004) Aging in rhesus monkeys: relevance to human health interventions. Science 305:1423–1426

Russell SJ, Kahn CR (2007) Endocrine regulation of ageing. Nat Rev Mol Cell Biol 8:681–691

Sabatino F, Masoro EJ, McMahan CA, Kuhn RW (1991) Assessment of the role of the glucocorticoid system in aging processes and in the action of food restriction. J Gerontol 46:B171–B179

Saito K, Yoshioka H, Cutler RG (1998) A spin trap, N-tert-butyl-alpha-phenylnitrone extends the life span of mice. Biosci Biotechnol Biochem 62:792–794

Santos AM, Ferraz MR, Teixeira CV, Sampaio FJ, da Fonte RC (2004) Effects of undernutrition on serum and testicular testosterone levels and sexual function in adult rats. Horm Metab Res 36:27–33

Sarkar S, Ravikumar B, Rubinsztein DC (2009) Autophagic clearance of aggregate-prone proteins associated with neurodegeneration. Methods Enzymol 453:83–110

Scarpulla RC (2008) Transcriptional paradigms in mammalian mitochondrial biogenesis and function. Physiol Rev 88:611–638

Schreiber SN, Emter R, Hock MB, Knutti D, Cardenas J, Podvinec M, Oakeley EJ, Kralli A (2004) The estrogen-related receptor alpha (ERRalpha) functions in PPARgamma coactivator 1alpha (PGC-1alpha)-induced mitochondrial biogenesis. Proc Natl Acad Sci USA 101:6472–6477

Scharf JG, Braulke T (2003) The role of the IGF axis in hepatocarcinogenesis. Horm Metab Res 35:685–693

Schumacher B, van dP, I, Moorhouse MJ, Kosteas T, Robinson AR, Suh Y, Breit TM, van Steeg H, Niedernhofer LJ, van Ijcken W, Bartke A, Spindler SR, Hoeijmakers JH, van der Horst GT, Garinis GA (2008) Delayed and accelerated aging share common longevity assurance mechanisms. PLoS Genet 4:e1000161

Sciacca L, Mineo R, Pandini G, Murabito A, Vigneri R, Belfiore A (2002) In IGF-I receptor-deficient leiomyosarcoma cells autocrine IGF-II induces cell invasion and protection from apoptosis via the insulin receptor isoform A. ncogene. 21:8240–8250

Sedlaczek N, Hasilik A, Neuhaus P, Schuppan D, Herbst H (2003) Focal overexpression of insulin-like growth factor 2 by hepatocytes and cholangiocytes in viral liver cirrhosis. Br J Cancer 88:733–739

Segall PE (1979) Interrelations of dietary and hormonal effects in aging. Mech Ageing Dev 9:515–525

Segall PE, Timiras PS (1975) Age-related changes in the thermoregulatory capacity of tryptophan-deficient rats. Fed Proc 34:83–85

Segall PE, Timiras PS (1976) Patho-physiologic findings after chronic tryptophan deficiency in rats: a model for delayed growth and aging. Mech Ageing Dev 5:109–124

Segall PE, Timiras PS, Walton JR (1983) Low tryptophan diets delay reproductive aging. Mech Ageing Dev 23:245–252

Segovia G, Yague AG, Garcia-Verdugo JM, Mora F (2006) Environmental enrichment promotes neurogenesis and changes the extracellular concentrations of glutamate and GABA in the hippocampus of aged rats. Brain Res Bull 70:8–14

Selesniemi K, Lee HJ, Tilly JL (2008) Moderate caloric restriction initiated in rodents during adulthood sustains function of the female reproductive axis into advanced chronological age. Aging Cell 7:622–629

Selman C, Lingard S, Choudhury AI, Batterham RL, Claret M, Clements M, Ramadani F, Okkenhaug K, Schuster E, Blanc E, Piper MD, Al Qassab H, Speakman JR, Carmignac D, Robinson IC, Thornton JM, Gems D, Partridge L, Withers DJ (2008) Evidence for lifespan extension and delayed age-related biomarkers in insulin receptor substrate 1 null mice. FASEB J 22:807–818

Sergi G, Perissinotto E, Pisent C, Buja A, Maggi S, Coin A, Grigoletto F, Enzi G (2005) An adequate threshold for body mass index to detect underweight condition in elderly persons: the Italian Longitudinal Study on Aging (ILSA) J Gerontol A Biol Sci Med Sci 60:866–871

Serrano F, Klann E (2004) Reactive oxygen species and synaptic plasticity in the aging hippocampus. Ageing Res Rev 3:431–443

Shahar A, Shahar D, Kahar Y, Nitzan-Kalusky D (2005) [Low-weight and weight loss as predictors of morbidity and mortality in old age]. Harefuah 144:443–452

Sharpless NE (2005) INK4a/ARF: a multifunctional tumor suppressor locus. Mutat Res 576:22–38

Sheldon WG, Bucci TJ, Hart RW, Turturro A (1995) Age-related neoplasia in a lifetime study of ad libitum-fed and food-restricted B6C3F1 mice. Toxicol Pathol 23:458–476

Shinmura K, Tamaki K, Bolli R (2005) Short-term caloric restriction improves ischemic tolerance independent of opening of ATP-sensitive K+ channels in both young and aged hearts. J Mol Cell Cardiol 39:285–296

Shire JG (1973) Growth hormone and premature ageing. Nature 245:215–216

Short KR, Bigelow ML, Kahl J, Singh R, Coenen-Schimke J, Raghavakaimal S, Nair KS (2005) Decline in skeletal muscle mitochondrial function with aging in humans. Proc Natl Acad Sci USA 102:5618–5623

Silvera SA, Jain M, Howe GR, Miller AB, Rohan TE (2006) Energy balance and breast cancer risk: a prospective cohort study. Breast Cancer ResTreat 97:97–106

Simonsen A, Cumming RC, Brech A, Isakson P, Schubert DR, Finley KD (2008) Promoting basal levels of autophagy in the nervous system enhances longevity and oxidant resistance in adult Drosophila. Autophagy 4:176–184

Skorupa DA, Dervisefendic A, Zwiener J, Pletcher SD (2008) Dietary composition specifies consumption, obesity, and lifespan in *Drosophila melanogaster*. Aging Cell 7:478–490

Slattery ML, Potter J, Caan B, Edwards S, Coates A, Ma KN, Berry TD (1997) Energy balance and colon cancer--beyond physical activity. Cancer Res 57:75–80

Smith ED, Tsuchiya M, Fox LA, Dang N, Hu D, Kerr EO, Johnston ED, Tchao BN, Pak DN, Welton KL, Promislow DE, Thomas JH, Kaeberlein M, Kennedy BK (2008) Quantitative evidence for conserved longevity pathways between divergent eukaryotic species. Genome Res 18:564–570

Snyder DL, Wostmann BS, Pollard M (1988) Serum hormones in diet-restricted gnotobiotic and conventioal Lobund-Wistar rats. J Gerontol 43:168–173

Sonntag WE, Lynch CD, Cefalu WT, Ingram RL, Bennett SA, Thornton PL, Khan AS (1999) Pleiotropic effects of growth hormone and insulin-like growth factor (IGF)-1 on biological aging: inferences from moderate caloric-restricted animals. J Gerontol A Biol Sci MedSci 54:B521–B538

Sorensen TI, Price RA, Stunkard AJ, Schulsinger F (1989) Genetics of obesity in adult adoptees and their biological siblings. BMJ 298:87–90

Sorimachi H, Ishiura S, Suzuki K (1997) Structure and physiological function of calpains. Biochem J 328:721–732

Sornson MW, Wu W, Dasen JS, Flynn SE, Norman DJ, O'Connell SM, Gukovsky I, Carriere C, Ryan AK, Miller AP, Zuo L, Gleiberman AS, Andersen B, Beamer WG, Rosenfeld MG (1996) Pituitary lineage determination by the Prophet of Pit-1 homeodomain factor defective in Ames dwarfism. Nature 384:327–333

Spindler SR (2001) Caloric restriction enhances the expression of key metabolic enzymes associated with protein renewal during aging. Ann NY Acad Sci 928:296–304

Spindler SR (2005) Rapid and reversible induction of the longevity, anticancer and genomic effects of caloric restriction. Mech Ageing Dev 126:960–966

Spindler SR (2010) Design of rodent lifespan studies for the identification of potential longevity therapeutics. Submitted for publication

Spindler SR, Crew MD, Mote PL, Grizzle JM, Walford RL (1990) Dietary energy restriction in mice reduces hepatic expression of glucose-regulated protein 78 (BiP) and 94 mRNA. J Nutr 120:1412–1417

Spindler SR, Dhahbi JM (2007) Conserved and Tissue-Specific Genic and Physiologic Responses to Caloric Restriction and Altered IGFI Signaling in Mitotic and Postmitotic Tissues. Annu Rev Nutr 27:193–217

Spindler SR, Dhahbi JM, Mote PL (2003) Protein turnover, energy metabolism and aging. In: Mattson MP (ed) Energy Metabolism and Lifespan Determination: Advances in Cell Aging and Gerontology, Vol. 14. Elsevier, Amsterdam, The Netherlands, pp. 69–86

Spindler SR, Grizzle JM, Walford RL, Mote PL (1991) Aging and restriction of dietary calories increases insulin receptor mRNA, and aging increases glucocorticoid receptor mRNA in the liver of female C3B10RF1 mice. J Gerontol 46:B233–B237

Spindler SR, Mote PL (2007) Screening candidate longevity therapeutics using gene-expression arrays. Gerontology 53:306–321

Stewart J, Mitchell J, Kalant N (1989) The effects of life-long food restriction on spatial memory in young and aged Fischer 344 rats measured in the eight-arm radial and the Morris water mazes. Neurobiol Aging 10:669–675

Stewart JW, Koehler K, Jackson W, Hawley J, Wang W, Au A, Myers R, Birt DF (2005) Prevention of mouse skin tumor promotion by dietary energy restriction requires an intact adrenal gland and glucocorticoid supplementation restores inhibition. Carcinogenesis 26: 1077–1084

Stoll S, Hafner U, Kranzlin B, Muller WE (1997) Chronic treatment of Syrian hamsters with low-dose selegiline increases life span in females but not males. Neurobiol Aging 18: 205–211

Stoll S, Hafner U, Pohl O, Muller WE (1994) Age-related memory decline and longevity under treatment with selegiline. Life Sci 55:2155–2163

Strong R, Miller RA, Astle CM, Floyd RA, Flurkey K, Hensley KL, Javors MA, Leeuwenburgh C, Nelson JF, Ongini E, Nadon NL, Warner HR, Harrison DE (2008) Nordihydroguaiaretic acid and aspirin increase lifespan of genetically heterogeneous male mice. Aging Cell 7: 641–650

Stote KS, Baer DJ, Spears K, Paul DR, Harris GK, Rumpler WV, Strycula P, Najjar SS, Ferrucci L, Ingram DK, Longo DL, Mattson MP (2007) A controlled trial of reduced meal frequency without caloric restriction in healthy, normal-weight, middle-aged adults. Am J Clin Nutr 85:981–988

Stunkard AJ (1976) Nutrition, aging and obesity. In: Rockstein M, Sussman ML (eds) Nutrition, Longevity, and Aging: Proceedings of a Symposium on Nutrition, Longevity, and Aging, held in Miami, Florida, February 26–27, 1976. Academic Press, New York, pp. 253–284

Stunkard AJ, Harris JR, Pedersen NL, McClearn GE (1990) The body-mass index of twins who have been reared apart. N Engl J Med 322:1483–1487

Stunkard AJ, Sorensen TI, Hanis C, Teasdale TW, Chakraborty R, Schull WJ, Schulsinger F (1986) An adoption study of human obesity. N Engl J Med 314:193–198

Sullivan R (1994) Obituaries: Dr. Stuart M. Berger, Health Advice Columnist, 40. New York Times Issue published March 1, 1994. http://query.nytimes.com/gst/fullpage.html?res= 9E03E6D7143AF932A35750C0A962958260

Sunder M (2005) Toward generation XL: anthropometrics of longevity in late 20th-century United States. Econ Hum Biol 3:271–295

Swoap SJ (2008) The pharmacology and molecular mechanisms underlying temperature regulation and torpor. Biochem Pharmacol 76:817–24

Syntichaki P, Troulinaki K, Tavernarakis N (2007) eIF4E function in somatic cells modulates ageing in *Caenorhabditis elegans*. Nature 445:922–926

Taguchi A, Wartschow LM, White MF (2007) Brain IRS2 signaling coordinates life span and nutrient homeostasis. Science 317:369–372

Takata Y, Ansai T, Soh I, Akifusa S, Sonoki K, Fujisawa K, Awano S, Kagiyama S, Hamasaki T, Nakamichi I, Yoshida A, Takehara T (2007) Association between body mass index and mortality in an 80-year-old population. J Am Geriatr Soc 55:913–917

Tatar M, Kopelman A, Epstein D, Tu MP, Yin CM, Garofalo RS (2001) A mutant Drosophila insulin receptor homolog that extends life-span and impairs neuroendocrine function. Science 292:107–110

Tavernarakis N (2008) Ageing and the regulation of protein synthesis: a balancing act? Trends Cell Biol 18:228–235

Tillman JB, Dhahbi JM, Mote PL, Walford RL, Spindler SR (1996) Dietary calorie restriction in mice induces carbamyl phosphate synthetase I gene transcription tissue specifically. J Biol Chem 271:3500–3506

Timiras PS, Hudson DB, Segall PE (1984) Lifetime brain serotonin: regional effects of age and precursor availability. Neurobiol Aging 5:235–242

Tissenbaum HA, Guarente L (2001) Increased dosage of a sir-2 gene extends lifespan in *Caenorhabditis elegans*. Nature 410:227–230

Tissenbaum HA, Ruvkun G (1998) An insulin-like signaling pathway affects both longevity and reproduction in *Caenorhabditis elegans*. Genetics 148:703–717

To K, Yamaza H, Komatsu T, Hayashida T, Hayashi H, Toyama H, Chiba T, Higami Y, Shimokawa I (2007) Down-regulation of AMP-activated protein kinase by calorie restriction in rat liver. Exp Gerontol 42:1063–1071

Toth ML, Sigmond T, Borsos E, Barna J, Erdelyi P, Takacs-Vellai K, Orosz L, Kovacs AL, Csikos G, Sass M, Vellai T (2008) Longevity pathways converge on autophagy genes to regulate life span in *Caenorhabditis elegans*. Autophagy 4:330–338

Trojanowski JQ, Mattson MP (2003) Overview of protein aggregation in single, double, and triple neurodegenerative brain amyloidoses. Neuromol Med 4:1–6

Trzonkowski P, Mysliwska J, Pawelec G, Mysliwski A (2009) From bench to bedside and back: the SENIEUR Protocol and the efficacy of influenza vaccination in the elderly. Biogerontology 10:83–94

Tschape JA, Hammerschmied C, Muhlig-Versen M, Athenstaedt K, Daum G, Kretzschmar D (2002) The neurodegeneration mutant lochrig interferes with cholesterol homeostasis and Appl processing. EMBO J. 21:6367–6376

Tsuchiya T, Dhahbi JM, Cui X, Mote PL, Bartke A, Spindler SR (2004) Additive regulation of hepatic gene expression by dwarfism and caloric restriction. Physiol Genomics 17:307–315

Turturro A, Duffy P, Hass B, Kodell R, Hart R (2002) Survival characteristics and age-adjusted disease incidences in C57BL/6 mice fed a commonly used cereal-based diet modulated by dietary restriction. J Gerontol A Biol Sci Med Sci 57:B379–B389

Turturro A, Hart RW (1991) Longevity-assurance mechanisms and caloric restriction. Ann NY Acad Sci 621:363–72

Turturro A, Witt WW, Lewis S, Hass BS, Lipman RD, Hart RW (1999) Growth curves and survival characteristics of the animals used in the Biomarkers of Aging Program. J Gerontol A Biol Sci Med Sci 54:B492–B501

Valle A, Catala-Niell A, Colom B, Garcia-Palmer FJ, Oliver J, Roca P (2005) Sex-related differences in energy balance in response to caloric restriction. Am J Physiol Endocrinol Metab 289:E15–E22

Vallejo EA (1957) [Hunger diet on alternate days in the nutrition of the aged]. Prensa Med Argent 44:119–120

Van der PE, Sherwood EM, Blomberg BB, Riley RL (2003) Aged mice exhibit distinct B cell precursor phenotypes differing in activation, proliferation and apoptosis. Exp Gerontol 38:1137–1147

van Raalte DH, Li M, Pritchard PH, Wasan KM (2004) Peroxisome proliferator-activated receptor (PPAR)-alpha: a pharmacological target with a promising future. Pharm Res 21:1531–1538

Varady KA, Hellerstein MK (2007) Alternate-day fasting and chronic disease prevention: a review of human and animal trials. Am J Clin Nutr 86:7–13

Varagic J, Susic D, Frohlich E (2001) Heart, aging, and hypertension. Curr Opin Cardiol 16:336–341

Vasselli JR, Shih JH, Iyengar SR, Maranchie J, Riss J, Worrell R, Torres-Cabala C, Tabios R, Mariotti A, Stearman R, Merino M, Walther MM, Simon R, Klausner RD, Linehan WM (2003) Predicting survival in patients with metastatic kidney cancer by gene-expression profiling in the primary tumor. Proc Natl Acad Sci USA 100:6958–6963

Vaziri H, Dessain SK, Ng EE, Imai SI, Frye RA, Pandita TK, Guarente L, Weinberg RA (2001) hSIR2(SIRT1) functions as an NAD-dependent p53 deacetylase. Cell 107:149–159

Vega RB, Huss JM, Kelly DP (2000) The coactivator PGC-1 cooperates with peroxisome proliferator-activated receptor alpha in transcriptional control of nuclear genes encoding mitochondrial fatty acid oxidation enzymes. Mol Cell Biol 20:1868–1876

Veldhuis JD, Iranmanesh A, Evans WS, Lizarralde G, Thorner MO, Vance ML (1993) Amplitude suppression of the pulsatile mode of immunoradiometric luteinizing hormone release in fasting-induced hypoandrogenemia in normal men. J Clin Endocrinol Metab 76:587–593

Vella V, Pandini G, Sciacca L, Mineo R, Vigneri R, Pezzino V, Belfiore A (2002) A novel autocrine loop involving IGF-II and the insulin receptor isoform-A stimulates growth of thyroid cancer. J Clin Endocrinol Metab 87:245–254

Ventura-Clapier R, Garnier A, Veksler V (2008) Transcriptional control of mitochondrial biogenesis: the central role of PGC-1alpha. Cardiovasc Res 79:208–217

Verdery RB, Walford RL (1998) Changes in plasma lipids and lipoproteins in humans during a 2-year period of dietary restriction in Biosphere 2. Arch Intern Med 158:900–906

Verdone-Smith C, Enesco HE (1982) The effect of temperature and of dietary restriction on lifespan and reproduction in the rotifer Asplanchna brightwelli. Exp Gerontol 17:255–262

Villareal DT, Fontana L, Weiss EP, Racette SB, Steger-May K, Schechtman KB, Klein S, Holloszy JO (2006) Bone mineral density response to caloric restriction-induced weight loss or exercise-induced weight loss: a randomized controlled trial. Arch Intern Med 166:2502–2510

Visscher MB, King JT, Lee YC (1952) Further studies on influence of age and diet upon reproductive senescence in strain A female mice. Am J Physiol 170:72–76

Wagh A, Stone NJ (2004) Treatment of metabolic syndrome. Expert Rev Cardiovasc Ther 2:213–228

Walford RL (1998) Children of the elderberry bush (for Alex Comfort). Exp Gerontol 33:189–190

Walford RL, Mock D, MacCallum T, Laseter JL (1999) Physiologic changes in humans subjected to severe, selective calorie restriction for two years in Biosphere 2: health, aging, and toxicological perspectives. Toxicol Sci 52:61–65

Walford RL, Mock D, Verdery R, MacCallum T (2002) Calorie Restriction in Biosphere 2: Alterations in Physiologic, Hematologic, Hormonal, and Biochemical Parameters in Humans Restricted for a 2-Year Period. J Gerontol A Biol Sci Med Sci 57:B211–B224

Walford RL, Spindler SR (1997) The response to calorie restriction in mammals shows features also common to hibernation: A cross-adaptation hypothesis. J Gerontol: Biol Sci 52A: B179–B183

Wang Y, Tissenbaum HA (2006) Overlapping and distinct functions for a *Caenorhabditis elegans* SIR2 and DAF-16/FOXO. Mech Ageing Dev 127:48–56

Wassertheil-Smoller S, Fann C, Allman RM, Black HR, Camel GH, Davis B, Masaki K, Pressel S, Prineas RJ, Stamler J, Vogt TM (2000) Relation of low body mass to death and stroke in the systolic hypertension in the elderly program. The SHEP Cooperative Research Group. Arch Intern Med 160:494–500

Weindruch R, Sohal RS (1997) Seminars in medicine of the beth Israel deaconess medical center. Caloric intake and aging. N Engl J Med 337:986–994

Weindruch R, Walford RL (1982) Dietary restriction in mice beginning at 1 year of age: effect on life-span and spontaneous cancer incidence. Science 215:1415–1418

Weindruch R, Walford RL (1988) The Retardation of Aging and Disease by Dietary Restriction. Charles C. Thomas, Springfield, IL

Weindruch RH, Kristie JA, Cheney KE, Walford RL (1979) Influence of controlled dietary restriction on immunologic function and aging. Fed Proc 38:2007–2016

Weinstein RS (2001) Glucocorticoid-induced osteoporosis. Rev Endocr Metab Disord 2:65–73

Weiss EP, Racette SB, Villareal DT, Fontana L, Steger-May K, Schechtman KB, Klein S, Ehsani AA, Holloszy JO (2007) Lower extremity muscle size and strength and aerobic capacity decrease with caloric restriction but not with exercise-induced weight loss. J Appl Physiol 102:634–640

Weiss EP, Racette SB, Villareal DT, Fontana L, Steger-May K, Schechtman KB, Klein S, Holloszy JO (2006) Improvements in glucose tolerance and insulin action induced by increasing energy expenditure or decreasing energy intake: a randomized controlled trial. Am J Clin Nutr 84:1033–1042

Weyer C, Walford RL, Harper IT, Milner M, MacCallum T, Tataranni PA, Ravussin E (2000) Energy metabolism after 2 y of energy restriction: the Biosphere 2 experiment. Am J Clin Nutr 72:946–953

Widdowson EM (1964) Diet and Bodily Constitution. Ciba Foundation Study Group No 17. Little and Brown, Boston

Widen EI, Eriksson JG, Groop LC (1992) Metformin normalizes nonoxidative glucose metabolism in insulin-resistant normoglycemic first-degree relatives of patients with NIDDM. Diabetes 41:354–358

Willcox BJ, Willcox DC, Todoriki H, Fujiyoshi A, Yano K, He Q, Curb JD, Suzuki M (2007) Caloric restriction, the traditional Okinawan diet, and healthy aging: the diet of the world's longest-lived people and its potential impact on morbidity and life span. Ann NY Acad Sci 1114:434–455

Winder WW, Taylor EB, Thomson DM (2006) Role of AMP-activated protein kinase in the molecular adaptation to endurance exercise. Med Sci Sports Exerc 38:1945–1949

Winter JC (1998) The effects of an extract of Ginkgo biloba, EGb 761, on cognitive behavior and longevity in the rat. Physiol Behav 63:425–433

Witte AV, Fobker M, Gellner R, Knecht S, Floel A (2009) Caloric restriction improves memory in elderly humans. Proc Natl Acad Sci USA 106:1255–1260

Wood JG, Rogina B, Lavu S, Howitz K, Helfand SL, Tatar M, Sinclair D (2004) Sirtuin activators mimic caloric restriction and delay ageing in metazoans. Nature 430:686–689

Wright KM, Deshmukh M (2006) Restricting apoptosis for postmitotic cell survival and its relevance to cancer. Cell Cycle. 5:1616–1620

Wu H, Kanatous SB, Thurmond FA, Gallardo T, Isotani E, Bassel-Duby R, Williams RS (2002) Regulation of mitochondrial biogenesis in skeletal muscle by CaMK. Science 296: 349–352

Wu Z, Huang X, Feng Y, Handschin C, Feng Y, Gullicksen PS, Bare O, Labow M, Spiegelman B, Stevenson SC (2006) Transducer of regulated CREB-binding proteins (TORCs) induce PGC-1alpha transcription and mitochondrial biogenesis in muscle cells. Proc Natl Acad Sci USA 103:14379–14384

Wu Z, Puigserver P, Andersson U, Zhang C, Adelmant G, Mootha V, Troy A, Cinti S, Lowell B, Scarpulla RC, Spiegelman BM (1999) Mechanisms controlling mitochondrial biogenesis and respiration through the thermogenic coactivator PGC-1. Cell 98:115–124

Yao X, Hu JF, Daniels M, Yien H, Lu H, Sharan H, Zhou X, Zeng Z, Li T, Yang Y, Hoffman AR (2003) A novel orthotopic tumor model to study growth factors and oncogenes in hepatocarcinogenesis. Clin Cancer Res 9:2719–2726

Yamauchi T, Kamon J, Minokoshi Y, Ito Y, Waki H, Uchida S, Yamashita S, Noda M, Kita S, Ueki K, Eto K, Akanuma Y, Froguel P, Foufelle F, Ferre P, Carling D, Kimura S, Nagai R, Kahn BB, Kadowaki T (2002) Adiponectin stimulates glucose utilization and fatty-acid oxidation by activating AMP-activated protein kinase. Nat Med 8:1288–1295

Yen TT, Knoll J (1992) Extension of lifespan in mice treated with Dinh lang (Policias fruticosum L.) and (-)deprenyl. Acta Physiol Hung 79:119–124

Yeung F, Hoberg JE, Ramsey CS, Keller MD, Jones DR, Frye RA, Mayo MW (2004) Modulation of NF-kappaB-dependent transcription and cell survival by the SIRT1 deacetylase. EMBO J 23:2369–2380

Youngman LD, Park J-YK, Ames BN (1992) Protein oxidation associated with aging is reduced by dietary restriction of protein or calories. Proc Natl Acad Sci USA 89: 9112–9116

Yu BP, Masoro EJ, McMahan CA (1985) Nutritional influences on aging of Fischer 344 rats: I. Physical, metabolic, and longevity characteristics. J Gerontol 40:657–670

Yung RL, Julius A (2008) Epigenetics, aging, and autoimmunity. Autoimmunity 41:329–335

Zainal TA, Oberley TD, Allison DB, Szweda LI, Weindruch R (2000) Caloric restriction of rhesus monkeys lowers oxidative damage in skeletal muscle. FASEB J 14:1825–1836

Zecca L, Youdim MB, Riederer P, Connor JR, Crichton RR (2004) Iron, brain ageing and neurodegenerative disorders. Nat Rev Neurosci 5:863–873

Zhou G, Myers R, Li Y, Chen Y, Shen X, Fenyk-Melody J, Wu M, Ventre J, Doebber T, Fujii N, Musi N, Hirshman MF, Goodyear LJ, Moller DE (2001) Role of AMP-activated protein kinase in mechanism of metformin action. J Clin Invest 108:1167–1174

Zhou Y, Xu BC, Maheshwari HG, He L, Reed M, Lozykowski M, Okada S, Cataldo L, Coschigamo K, Wagner TE, Baumann G, Kopchick JJ (1997) A mammalian model for Laron syndrome produced by targeted disruption of the mouse growth hormone receptor/binding protein gene (the Laron mouse). Proc Natl Acad Sci U S A 94:13215–13220

Zhu M, Miura J, Lu LX, Bernier M, deCabo R, Lane MA, Roth GS, Ingram DK (2004) Circulating adiponectin levels increase in rats on caloric restriction: the potential for insulin sensitization. Exp Gerontol 39:1049–1059

Zou MH, Kirkpatrick SS, Davis BJ, Nelson JS, Wiles WG, Schlattner U, Neumann D, Brownlee M, Freeman MB, Goldman MH (2004) Activation of the AMP-activated protein kinase by the antidiabetic drug metformin in vivo. Role of mitochondrial reactive nitrogen species. J Biol Chem 279:43940–43951

Zuccato C, Tartari M, Crotti A, Goffredo D, Valenza M, Conti L, Cataudella T, Leavitt BR, Hayden MR, Timmusk T, Rigamonti D, Cattaneo E (2003) Huntingtin interacts with REST/NRSF to modulate the transcription of NRSE-controlled neuronal genes. Nat Genet 35:76–83

Chapter 13
Calibrating Notch/TGF-β Signaling for Youthful, Healthy Tissue Maintenance and Repair

Morgan Carlson and Irina M. Conboy

Contents

13.1	Introduction: Aging and the Loss of Tissue Regenerative Potential	439
13.2	Cellular and Molecular Mechanisms of Muscle Repair	440
	13.2.1 Aging of the Local Organ Environment and Regeneration	441
13.3	Aging of the Systemic Environment and Its Influence on Organ Stem Cell Regenerative Potential	443
13.4	Conclusion: TGF-β/Notch Signal Integration and Tissue Restoration	444
References		447

13.1 Introduction: Aging and the Loss of Tissue Regenerative Potential

A central feature of vertebrate aging is the profound decline and impairment of productive tissue repair. With advancing age, the body's capacity to regenerate new tissue can no longer compete with an accelerating rate of tissue dysfunction. Such amassing imbalance encompasses multiple organ systems, and is ultimately characteristic of numerous degenerative disorders (e.g. Parkinson's, Alzheimer's, muscle atrophy, etc.) (Grounds 1998; Thomas 2001). It is clear that the repair of injured tissues, and maintenance of healthy tissues, greatly relies on the activity of adult organ stem and progenitor cells (Thomas 2001). Unfortunately, with old age, the properties of these cells become altered in ways that lead to their poor performance during the process of tissue regeneration/repair. Moreover, this age-associated regenerative decline is not exclusive to specific tissues, as it has been identified within multiple distinct subsets of adult stem cells, including blood, intestinal epithelia, skeletal

---

I.M. Conboy (✉)
Irina Conboy, B108B Stanley Hall, Department of Bioengineering, UC Berkeley, Berkeley, CA 94720-1762, USA
e-mail: iconboy@berkeley.edu

muscle, cardiac muscle, and brain (Kuhn et al. 1996; Morrison et al. 1996; Martin et al. 1998; Conboy et al. 2003; Capogrossi 2004).

Adult skeletal muscle represents a perfect example of a tissue that robustly regenerates throughout adult life, but fails to do so in old age (Grounds 1998; Renault et al. 2002). The precise reason for the age-related regenerative decline of this organ system remains ill-defined, and the relative roles of muscle cell intrinsic changes versus alterations in the aged environment are under continued examination. Curiously, recent reports demonstrate that old muscle stem cells, despite normally failing to engage in productive tissue repair, retain their intrinsic ability to regenerate. This occurs upon exposure to young niches themselves, or through 'youthful' molecular recalibration (to be discussed in greater detail below) (Conboy et al. 2005; Carlson and Conboy 2007; Carlson et al. 2008).

The observation that old stem cells retain their intrinsic capacity for youthfulness sets forth a very interesting question – can the onset of aging be delayed by compensating tissue deterioration/aging with unremitting tissue repair, thus rendering the body into a state of youthful homeostasis? If so, we can logically conclude that practical therapies for the treatment of age-related illnesses, and of aging itself, will emerge when this regenerative/degenerative process balance is properly understood and, moreover, deliberately regulated to promote a desired phenotype. The rest of this chapter explores these concepts further, dedicating particular attention to relevant progress and discoveries made in the field of muscle stem cells and regeneration.

13.2 Cellular and Molecular Mechanisms of Muscle Repair

Skeletal muscle is maintained and repaired by endogenous stem cells, called satellite cells, which largely constitute the regenerative potential in this organ (Zammit et al. 2002; Sherwood et al. 2004). Satellite cells reside in direct contact with the differentiated, multinucleated muscle cells (myofibers) and are positioned beneath their basal lamina. Further, satellite cells comprise roughly 2–5% of all adult muscle nuclei, and express specific genetic markers, such as M-cadherin, CD34, Myf5, Pax7, and syndecan-4 (Morgan and Partridge 2003). It is important to note, however, that this population is heterogeneous, as no single marker identifies all of the cells, although 90–95% of satellite cells are reported to co-express CD34 and M-cadherin (Beauchamp et al. 2000; Conboy et al. 2003). Additionally, CXCR4 and β1-integrin have been characterized as markers of a highly myogenic satellite cell subset (Sherwood et al. 2004).

In the resting adult muscle, the vast majority of satellite cells remain mitotically quiescent. However, in response to injury, quiescence is broken and these stem cells engage in a program of coordinated cell-expansion followed by differentiation, first into early myogenic progenitor cells. This initial pool of early progenitors gives rise to fusion-competent myoblasts, capable of continued short-term division and ultimately fusion, in order to repair damaged myofibers, or to generate de-novo myofibers (Morgan and Partridge 2003).

13.2.1 Aging of the Local Organ Environment and Regeneration

13.2.1.1 Notch: A Key Molecular Determinant of Successful Skeletal Muscle Repair

At the molecular level, muscle repair is tightly regulated by the evolutionarily-conserved Notch signaling pathway (well characterized in the regulation of embryonic organogenesis in many evolutionarily distinct species) (rtavanis-Tsakonas et al. 1999; Conboy and Rando 2002; Hansson et al. 2004). Inactive Notch is expressed as a cell-surface receptor, which becomes cleaved upon the binding of its specific ligands (e.g. Delta, Jagged or Serrate), that can be expressed on the surface of a neighboring cell or the same cell as Notch (rtavanis-Tsakonas et al. 1999). The cleaved cytoplasmic component of Notch translocates to the nucleus, where it binds to DNA in a complex with the RbpJ/SuH proteins to activate Notch-specific gene expression, such as the bHLH transcription factors, Hes and Hey (rtavanis-Tsakonas et al. 1999; Hansson et al. 2004). Typically, Notch promotes cell proliferation and inhibits terminal tissue specification in embryonic development (rtavanis-Tsakonas et al. 1999; Hansson et al. 2004).

During regeneration of injured skeletal muscle, the activity of Notch pathway is conserved (Conboy and Rando 2002). Specifically, Notch activates satellite cell proliferation and is critically important for the productive expansion of myogenic progenitor cells, and for the prevention of their precocious terminal differentiation by antagonizing the Wnt pathway (Conboy and Rando 2002; Brack et al. 2008). Conversely, forced inhibition of Notch activation dramatically impairs muscle regeneration, thereby further strengthening its role as a key molecular determinant of repair (Conboy et al. 2003; Carlson et al. 2008).

Among the first biochemical changes associated with the response of muscle to injury is the up-regulation of the Notch ligand, Delta, in both the satellite cells and myofibers adjacent to the injury site (Conboy and Rando 2002; Conboy et al. 2003). This provides both a positional cue for Notch activation and promotes cell expansion near the injury site, which is important for efficient tissue repair. Dividing myogenic progenitor cells up-regulate the Notch antagonist Numb, which is localized asymmetrically in these cells, reminiscent of the situation in embryonic organogenesis (Conboy and Rando 2002). The inhibition of Notch by Numb attenuates the proliferation of progenitor cells and promotes their terminal myogenic differentiation (Conboy and Rando 2002). Thus, consistent with its role in embryogenesis, Notch induces the expansion of the myogenic progenitor cells, while Numb attenuates Notch signaling to promote the differentiation of these cells into fusion competent myoblasts and, ultimately, multinucleated myofibers.

With age, however, few myoblasts are produced in response to injury, and thus not enough cells are available to form new myofibers (Schultz and Lipton 1982; Bockhold et al. 1998; Conboy et al. 2003). Interestingly, such decline in the generation of myoblasts in aged muscle is not primarily caused by the physical loss of satellite cells (Conboy et al. 2003). Rather, the ability of the individual aged satellite cells to break quiescence and proliferate in response to injury is severely diminished (Schultz and Lipton 1982; Conboy et al. 2003). At the molecular level, expression of the Notch receptor is not changed with age. However, Delta expression is not induced in injured, aged muscle, thus resulting in poor Notch activation and impaired muscle tissue repair (Conboy et al. 2003). Importantly, forced activation of Notch signaling by a "pseudo-ligand" restores the regenerative potential to aged muscle, namely activation of aged satellite cells, cell expansion and effective tissue repair (Conboy et al. 2003). Thus, satellite cell regenerative potential is not irreversibly lost with age, but rather it is not triggered in old muscle.

13.2.1.2 TGF-β Signaling and the Inhibition of Muscle Stem Cell Responses

The TGF-β superfamily includes roughly 35 proteins, which fall into smaller subfamilies such as the TGF-βs themselves, bone morphogenic proteins (BMPs), activins and nodals (Derynck and Zhang 2003; de Caestecker 2004; Feng and Derynck 2005). At the molecular level, TGF-β proteins are soluble, secreted ligands that bind to and stabilize the heteromeric complexes of their specific receptors: TGF RI, RII and RIII (Derynck and Zhang 2003; ten Dijke and Hill 2004; Feng and Derynck 2005). There are also multiple accessory transmembrane proteins, such as betaglycan, and endoglin, which modulate the affinity of TGF-β proteins for their specific receptors (ten Dijke and Hill 2004). In conventional TGF-β signaling, receptor-ligand binding induces the activation of TGF R1, which then triggers intracellular Smad protein phosphorylation. These receptor-activated Smads (R-Smads) form heteromeric molecular complexes (which include the common Smad4), and translocate to the nucleus where they drive TGF-β specific gene expression (Derynck and Zhang 2003; Feng and Derynck 2005).

Importantly, changes in TGF-β levels and/or responsiveness have been associated with aging in many tissues, including: heart, bone, brain, prostate, mammary gland and skeletal muscle (McCaffrey 2000; de Souza 2002; Moerman et al. 2004; Masliah et al. 2001; Huang and Lee 2003; Muraoka-Cook et al. 2004, Javelaud and Mauviel 2004; Roberts et al. 2003, Carlson et al. 2008; Carlson and Conboy 2009). The precise regulation of TGF-β signaling levels is critically important, not only for organ formation but also for organ function. For example, excess TGF-β1 leads to scarring in many tissues, and causes Alzheimer's-like microvascular degeneration in the brain (Javelaud 2004; Roberts 2003; Wyss-Coray et al. 2000). In comparison, TGF-β1 null mice die at 3-4 weeks of age from multifocal inflammatory disease and have profound neuronal cell-death (Clark and Coker 1998; Brionne et al. 2003). In this regard, TGF-β1 has been shown to be one of the main factors inducing cellular senescence in response to oxidative stress (Frippiat et al. 2001); TGF-β signaling has also been linked to carcinogenic transformations, a prominent phenotype of aging (Akhurst and Derynck 2001). However, the mechanisms by which such age-related changes influence the regenerative potential of organ stem cells have not been well defined.

In regard to skeletal muscle, the muscle-specific TGF-β class member, myostatin (GDF-8,) has been shown to induce the CDK inhibitors p15(Ink4B) and p21, and thus inhibits cell proliferation (Moustakas et al. 2002; Thomas et al. 2000).

In vivo, myostatin mRNA was shown to progressively accumulate during muscle repair (Armand et al. 2003), while the mRNA levels of its inhibitor, folistatin, were shown to be present in mono-nucleated muscle cells located near the injury site, and in newly-formed myofibers (Armand et al. 2003). Indeed, these studies strongly suggest that myostatin is a physiologic regulator of muscle repair in adult. However, recent work demonstrates that myostatin levels do not change with age, while the ubiquitous TGF-β1 does (Carlson et al. 2008; Carlson and Conboy 2009). In this respect, physiological aging of the differentiated muscle niche has been defined by a signaling imbalance, arising from excessive TGF-β1 production, coupled with insufficient Notch signaling activity. Such TGF-β1 overproduction interferes with old muscle stem cell regenerative responses by inducing heightened TGF-β/pSmad3 signaling intensity, thereby driving the up-regulation of CDK inhibitor levels (e.g. p15, p16, p21 and p27) and thwarting endogenous stem cell proliferation (Carlson et al. 2008).

While the activity of TGF-β proteins is likely to be associated with the aging process, the cellular and molecular mechanisms of such associations are not well understood. Understanding how the TGF-β family controls cell behavior is complicated by the fact that these proteins are highly pleiotropic cytokines, capable of orchestrating a diversity of responses, such as proliferation, differentiation, migration and apoptosis (Roberts 1998; Massague et al. 2000; Moustakas 2002; Javelaud 2004). It is also critical to note that TGF-β proteins act as morphogens, with the particular effect on gene expression and cell behavior not only determined by what specific TGF-β cytokine is present, but also by the intensity and duration of the signal (ten Dijke and Hill, 2004). Accordingly, in contrast to inhibitory aged levels, "young" levels of bio-active TGF-β do not interfere with regeneration-specific molecular signaling and muscle repair (Carlson et al. 2008; Carlson and Conboy 2009; Carlson et al. 2009). In this regard, experimental attenuation of Smad3 signaling strength in aged muscle stem cells restores muscle repair, and is coincident with a youthful molecular profile of the Notch/TGF-β signaling balance. Therefore, in order to progress towards therapeutic applications, it is imperative to determine when TGF-β/pSmad signaling becomes "negative" to cellular homeostasis and cell-fate determination, within actively regenerating tissues.

13.3 Aging of the Systemic Environment and Its Influence on Organ Stem Cell Regenerative Potential

Early heterochronic transplantation studies suggested that the intrinsic potential of aged satellite cells to repair muscle remains largely intact (Zacks and Sheff 1982; Carlson and Faulkner 1989). In these studies, minced muscle from young or old donors (rats or mice) was transplanted into either young or old recipients. The ability of transplanted muscle to regenerate was subsequently evaluated. From these experiments, it was determined that the age of the host environment determined the regenerative outcome, as both young and old muscle effectively regenerated in young but not in old animals (Zacks and Sheff 1982; Carlson and Faulkner 1989).

These early studies were particularly informative, but had not excluded the possibility that host cells may have contributed to the effective regeneration seen in young animals. In addition, they could not clearly distinguish among the many possible host factor contributions to the muscle stem cell regenerative potential (e.g. local tissue and immunologic, paracrine, neural, and angiogenic).

In order to determine whether the regenerative potential of satellite cells can be influenced solely by the age of the systemic environment, a more recent study examined muscle regeneration in parabiotically-paired young and old mice (Conboy et al. 2005). In this system, two mice are connected surgically and develop a shared circulatory system, thus exposing the tissues of each mouse to the systemic milieu of the different age. These experiments revealed that aged muscle regeneration was greatly improved in old mice paired with young mice ("heterochronic" parabiosis), as compared to with another old mouse ("isochronic" parabiosis). This enhanced regeneration was paralleled by an increase in satellite cell activation and the generation of myoblasts following injury. Furthermore, the progenitor cell response had the youthful molecular signature of increased activation of Delta-Notch signaling (Conboy et al. 2005). The improved regeneration of aged muscle was unequivocally due to the enhanced activation of the aged, resident satellite cells and not to any contribution from young cells present in common circulation (Conboy et al. 2005).

Interestingly, it was recently reported that aged muscle and blood sera contain soluble inhibitors of muscle stem cell regenerative capacity (Carlson and Conboy 2007), thus suggesting that (1) heterochronic milieu from parabiosed animals had a rejuvenating effect on old organ stem cells, largely because of an attenuation of aged systemic factors and, (2) the aged circulation has a pronounced inhibitory influence on young stem and progenitor cells repair capacity (Conboy et al. 2005; Carlson and Conboy 2007). Along these lines, it was also reported that, unlike their adult stem cell counterparts, human embryonic stem cells produce "pro-regenerative" factors that positively regulate postnatal tissue repair, in vivo and in vitro, and have the unique ability to rescue stem cell function within the old niche (Carlson and Conboy 2007). Establishing the molecular mechanisms of these phenomena should thus be of high priority. The basis of such findings is relevant and applicable for the future of stem-cell based therapeutic applications of age-related disease.

It is reasonable to conclude that systemic aging is not exclusive to muscle aging, but likely determines the efficiency of stem cell regenerative potential in many tissues (Conboy et al. 2005; Carlson and Conboy 2007).

13.4 Conclusion: TGF-β/Notch Signal Integration and Tissue Restoration

As outlined above, the deterioration of organ regenerative capacity is largely attributed to age-related inhibitory changes within the extrinsic cues that govern stem cell regenerative responses. These events, as they broadly apply to various stem cell subsets, are summarized and depicted in Fig. 13.1. Muscle stem cell responses

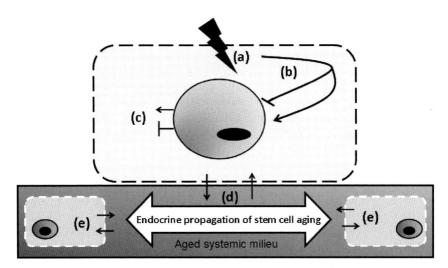

Fig. 13.1 Self-reinforcing age-specific loop of stem cell inhibition by local and circulatory niches The effect of the local organ and systemic environments on stem cell aging is summarized. (**a**) An age-specific transition within the stem cell niche is believed to coincide with the changed expression levels of natural agonists and antagonists of signal transduction pathways, which eventually inhibits stem cell function (**b**). Altered levels of signaling diminish the responsiveness of the endogenous organ stem cells, and contribute to deregulation of their signaling feedback to the local organ environment (**c**). Combined with endocrine factors within the aged systemic environment, which also act locally on organ stem cell populations (**d**), these secreted factors, and those deposited within the immediate extracellular compartments likely create biochemical imbalances that affect the regenerative responses of endogenous stem cells within multiple organ systems (**e**). Understanding the specific molecular changes responsible for the regenerative failure of aged stem cells and tissues will reveal key points of therapeutic intervention for diseased and age-induced degenerative states

are not triggered in the old due to a decline in Notch activation, and can be rejuvenated by forced local activation of Notch, for example, by exposure to young circulation, or by factors produced by human embryonic stem cells (Conboy et al. 2005; Carlson and Conboy 2007). This evidence strongly suggests that stem cells endogenous to old organs have a continuing capacity for maintenance and repair, but are inhibited by the aged differentiated niche. Additionally, physiological aging of the differentiated muscle niche is defined by excessive TGF-β production, which interferes with old muscle stem cell regenerative responses by inducing heightened TGF-β/pSmad signaling and CDK inhibitor levels (Carlson et al. 2008). In this regard, muscle repair success is determined by an antagonistic balance between TGF-β/pSmad signaling levels and adequate Notch pathway activation. However, in contrast to inhibitory aged levels, "young" levels of bio-active TGF-β do not interfere with regeneration-specific molecular signaling and muscle repair.

Promisingly, aged muscle stem cells are rejuvenated by a "youthful" recalibration of Notch and TGF-β (Conboy et al. 2005; Carlson and Conboy 2007; Carlson et al. 2008), which based on our preliminary data might be conserved between mice and humans. Such findings emphasize that mammalian organ stem cells remain intrinsically young and suggest strategies for healthy tissue maintenance in aged individuals. Equally importantly, young muscle stem cells become instantly "aged" biochemically and fail in their regenerative responses in the context of old niches, or when signal transduction intensity is experimentally altered to the levels typically of old organs (Carlson and Conboy 2007; Carlson et al. 2008). Therefore, transplantation therapies in aged individuals might not be successful until the molecules responsible for blocking regenerative responses in the old are identified and functionally neutralized.

Let's assume that organ stem cell responses are boosted in older individuals and the regenerative capacity is at the "young" level. Will this strategy yield healthy tissue homeostasis and rejuvenated organs? This particular question arises because it was postulated that a progressive decline in regenerative responses of old organs serves to protect against cancers, such that the abandoned replacement of damaged body components by adult stem cells, while causing accumulation of somatic damage, counteracts the accumulation of DNA damage and minimizes genomic instability (Beausejour and Campisi 2006; Vijg and Campisi 2008). However, our published and current work demonstrates that in the old, muscle stem cells remain numerous, avoid or repair DNA damage, and exhibit normal telomerase regulation (O'Connor et al. 2009). Therefore, a "youthful" regenerative capacity is likely to yield healthy tissue with good genomic stability (Carlson and Conboy 2009; Carlson et al. 2009). Quite interestingly, such age-specific flares of cancer might be unrelated to DNA damage or deregulation of telomerase in organ stem cells. Nevertheless, preserving the balance between enhanced stem cell responses and tight regulation of stem cell proliferation (typical of young mammals), is very important for any therapy aimed to boost tissue regenerative capacity.

In order to progress towards therapeutic applications, we believe it will be critical to determine when TGF-β/pSmad signaling becomes "negative", or imbalanced to stem cell function. In the future, one such approach could be through the engineering of self-attenuating synthetic circuitries, designed for maintaining TGF-β/pSmad and Notch signaling homeostasis (i.e. youthful recalibration). Such constructs would also facilitate the study of recalibration of variable TGF-β/Notch signaling intensities and their effects on stem cell regenerative potential and tissue-specific repair. From the gene therapy perspective, viral methods would be suitable for such approaches, using our current animal models, but would likely not be particularly amenable, nor practical, for human use. Thus, it would be important to develop alternatives that combine the safety of viral-free DNA delivery with prolonged and efficient expression of exogenous genes.

In this regard, plasmid-based gene delivery is often inefficient due to the fact that transient expression is typically turned off after just a few days. Moreover, stable integration of plasmid DNA into the mammalian genome is highly infrequent and also subject to gene silencing (Colosimo et al. 2000; Kushibiki and Tabata 2004). In contrast to such less-effective and traditional means, the use of de-novo, bio-engineered systems may be required to generate therapeutic approaches that will be applicable for humans. For example, nanoparticle gene therapy systems have been developed that allow sustained delivery of plasmid DNA by exploiting polymer degradation to control plasmid release (Capan et al. 1999; Jeong and Park 2002; Shea et al. 1999). Such alternatives that combine the safety of viral-free DNA delivery, coupled with sustained gene expression, would be desirable.

These methods could be further improved through their integration with molecularly-defined, tissue-specific biomatrices. In this manner, such bioengineered constructs could be designed to have controlled release of specific TGF-β signaling attenuators (e.g. dominant negative TGF-β receptor, or shRNA to Smad3) (Carlson et al. 2008). Combined, these methods would hopefully promote the safe and effective rejuvenation of local tissue repair throughout organismal aging, and would certainly be applicable for attenuating newly-discovered signaling network imbalances within various stem cell/organ niches.

References

Akhurst RJ, Derynck R. (2001) TGF-beta signaling in cancer – a double-edged sword. Trends Cell Biol Nov;11(11):S44–S51

Armand AS, Della GB, Launay T, Charbonnier F, Gallien CL, Chanoine C (2003) Expression and neural control of follistatin versus myostatin genes during regeneration of mouse soleus. Dev Dyn 227:256–265

Beauchamp JR et al. (2000) Expression of CD34 and Myf5 defines the majority of quiescent adult skeletal muscle satellite cells. J Cell Biol 151:1221–1234

Beausejour CM, Campisi J (2006) Ageing: balancing regeneration and cancer. Nature 443:404–405

Bockhold KJ, Rosenblatt JD, Partridge TA (1998) Aging normal and dystrophic mouse muscle: analysis of myogenicity in cultures of living single fibers. Muscle Nerve 21:173–183

Brack AS, Conboy IM, Conboy MJ, Shen J, Rando TA (2008) A temporal switch from notch to Wnt signaling in muscle stem cells is necessary for normal adult myogenesis. Cell Stem Cell 2:50–59

Brionne TC, Tesseur I, Masliah E, Wyss-Coray T (2003) Loss of TGF-beta 1 leads to increased neuronal cell death and microgliosis in mouse brain. Neuron 40:1133–1145

Capan Y, Woo BH, Gebrekidan S, Ahmed S, DeLuca PP (1999) Preparation and characterization of poly (D,L-lactide-co-glycolide) microspheres for controlled release of poly(L-lysine) complexed plasmid DNA. Pharm Res 16:509–513

Capogrossi MC (2004) Cardiac stem cells fail with aging: a new mechanism for the age-dependent decline in cardiac function. Circ Res 94:411–413

Carlson BM, Faulkner JA (1989) Muscle transplantation between young and old rats: age of host determines recovery. Am.J.Physiol 256:C1262–C1266

Carlson ME, Conboy IM (2007) Loss of stem cell regenerative capacity within aged niches. Aging Cell 3:371–382

Carlson ME, Conboy MJ, Hsu M, Barchas L, Jeong J, Agrawal A, Mikels AJ, Agrawal S, Schaffer DV, Conboy IM. (2009) Relative roles of TGF-beta 1 and Wnt in the systemic regulation and aging of satellite cell responses. Aging Cell 8(6):617–623

Carlson ME, Hsu M, Conboy IM (2008) Imbalance between pSmad3 and Notch induces CDK inhibitors in old muscle stem cells. Nature 454:528–532

Carlson C, Suetta MJ, Conboy P, Aagaard A, Mackey M, Kjaer, Conboy IM (2009) Molecular aging and rejuvenation of human muscle stem cells. EMBO Mol Med 1(8–9):381–391

Clark DA, Coker R (1998) Transforming growth factor-beta (TGF-beta). Int J Biochem Cell Biol 30:293–298

Colosimo A et al. (2000) Transfer and expression of foreign genes in mammalian cells. Biotechniques 29:314–318, 320–312, 324 passim

Conboy IM, Conboy MJ, Smythe GM, Rando TA (2003) Notch-mediated restoration of regenerative potential to aged muscle. Science 302:1575–1577

Conboy IM, Conboy MJ, Wagers AJ, Girma ER, Weissman IL, Rando TA (2005) Rejuvenation of aged progenitor cells by exposure to a young systemic environment. Nature 433:760–764

Conboy IM, Rando TA (2002) The regulation of Notch signaling controls satellite cell activation and cell fate determination in postnatal myogenesis. Dev Cell 3:397–409

de Caestecker M (2004) The transforming growth factor-beta superfamily of receptors. Cytokine Growth Factor Rev 15:1–11

de Souza RR (2002) Aging of myocardial collagen. Biogerontology 3(6):325–335

Derynck R, Zhang YE (2003) Smad-dependent and Smad-independent pathways in TGF-beta family signalling. Nature 425:577–584

Feng XH, Derynck R (2005) Specificity and versatility in tgf-beta signaling through Smads. Annu Rev Cell Dev Biol 21:659–693

Frippiat C, Chen QM, Zdanov S, Magalhaes JP, Remacle J, Toussaint O (2001) Subcytotoxic H_2O_2 stress triggers a release of transforming growth factor-beta 1, which induces biomarkers of cellular senescence of human diploid fibroblasts. J Biol Chem 276:2531–2537

Grounds MD (1998) Age-associated changes in the response of skeletal muscle cells to exercise and regeneration. Ann NY Acad Sci 854:78–91

Hansson EM, Lendahl U, Chapman G (2004) Notch signaling in development and disease. Semin Cancer Biol 14:320–328

Huang X, Lee C (2003 Sep) Regulation of stromal proliferation, growth arrest, differentiation and apoptosis in benign prostatic hyperplasia by TGF-beta. Front Biosci 8:s740–749

Javelaud D, Mauviel A (2004 Jul) Mammalian transforming growth factor-betas: Smad signaling and physio-pathological roles. Int J Biochem Cell Biol 36(7):1161–1165

Jeong JH, Park TG (2002) Poly(L-lysine)-g-poly(D,L-lactic-co-glycolic acid) micelles for low cytotoxic biodegradable gene delivery carriers. J Control Release 82:159–166

Kuhn HG, Dickinson-Anson H, Gage FH (1996) Neurogenesis in the dentate gyrus of the adult rat: age-related decrease of neuronal progenitor proliferation. J Neurosci 16:2027–2033

Kushibiki T, Tabata Y (2004) A new gene delivery system based on controlled release technology. Curr Drug Deliv 1:153–163

Martin K, Potten CS, Roberts SA, Kirkwood TB (1998) Altered stem cell regeneration in irradiated intestinal crypts of senescent mice. J Cell Sci 111(Pt 16):2297–2303

Masliah E, Ho G, Wyss-Coray T (2001, Nov–Dec) Functional role of TGF beta in Alzheimer's disease microvascular injury: lessons from transgenic mice. Neurochem Int 39(5–6):393–400

Massague J, Blain SW, Lo RS (2000, Oct 13) TGFbeta signaling in growth control, cancer, and heritable disorders. Cell 103(2):295–309

McCaffrey TA (2000, Mar–Jun) TGF-betas and TGF-beta receptors in atherosclerosis. Cytokine Growth Factor Rev 11(1–2):103–114

Moerman EJ, Teng K, Lipschitz DA, Lecka-Czernik B (2004, Dec) Aging activates adipogenic and suppresses osteogenic programs in mesenchymal marrow stroma/stem cells: the role of PPAR-gamma2 transcription factor and TGF-beta/BMP signaling pathways. Aging Cell 3(6):379–389

Morgan JE, Partridge TA (2003) Muscle satellite cells. Int J Biochem Cell Biol 35:1151–1156

Morrison SJ, Wandycz AM, Akashi K, Globerson A, Weissman IL (1996) The aging of hematopoietic stem cells. Nat Med 2:1011–1016

Moustakas A, Pardali K, Gaal A, Heldin CH (2002, Jun 3) Mechanisms of TGF-beta signaling in regulation of cell growth and differentiation. Immunol Lett 82(1–2):85–91

Muraoka-Cook RS, et al. (2004, Dec 15) Conditional overexpression of active transforming growth factor beta1 in vivo accelerates metastases of transgenic mammary tumors. Cancer Res 64(24):9002–9011

O'Connor MS, Carlson ME, Conboy IM (2009) Differentiation rather than aging of muscle stem cells abolishes their telomerase activity. Biotechnol Prog 25(4): 1130–1137

Renault V, Thornell LE, Eriksson PO, Butler-Browne G, Mouly V (2002) Regenerative potential of human skeletal muscle during aging. Aging Cell 1:132–139

Roberts AB (1998) Molecular and cell biology of TGF-beta. Miner Electrolyte Metab 24 (2–3):111–119

Roberts AB, Russo A, Felici A, Flanders KC (2003, May) Smad3: a key player in pathogenetic mechanisms dependent on TGF-beta. Ann N Y Acad Sci 995:1–10

rtavanis-Tsakonas S, Rand MD, Lake RJ (1999) Notch signaling: cell fate control and signal integration in development. Science 284:770–776

Schultz E, Lipton BH (1982) Skeletal muscle satellite cells: changes in proliferation potential as a function of age. Mech Ageing Dev 20:377–383

Shea LD, Smiley E, Bonadio J, Mooney DJ (1999). DNA delivery from polymer matrices for tissue engineering. Nat Biotechnol 17:551–554

Sherwood RI et al. (2004) Isolation of adult mouse myogenic progenitors: functional heterogeneity of cells within and engrafting skeletal muscle. Cell 119:543–554

ten Dijke P, Hill CS (2004) New insights into TGF-beta-Smad signalling. Trends Biochem Sci 29(5):265–273

Thomas DR (2001) Age-related changes in wound healing. Drugs Aging 18:607–620

Thomas M, et al. (2000, Dec 22) Myostatin, a negative regulator of muscle growth, functions by inhibiting myoblast proliferation. J Biol Chem 275(51):40235–40243

Vijg J, Campisi J (2008) Puzzles, promises and a cure for ageing. Nature 454:1065–1071

Wyss-Coray T, Lin C, von Euw D, Masliah E, Mucke L, Lacombe P (2000, Apr) Alzheimer's disease-like cerebrovascular pathology in transforming growth factor-beta 1 transgenic mice and functional metabolic correlates. Ann N Y Acad Sci 903:317–323

Zacks SI, Sheff MF (1982) Age-related impeded regeneration of mouse minced anterior tibial muscle. Muscle Nerve 5:152–161

Zammit PS et al. (2002) Kinetics of myoblast proliferation show that resident satellite cells are competent to fully regenerate skeletal muscle fibers. Exp Cell Res 281:39–49

Chapter 14
Embryonic Stem Cells: Prospects of Regenerative Medicine for the Treatment of Human Aging

Michael D. West

Contents

14.1	Introduction	452
14.2	Implications of the Germ-Line/Soma Dichotomy	453
	14.2.1 Origins of the Cellular Theory of Aging	453
	14.2.2 Telomere Dynamics in Cellular Aging and Immortalization	454
	14.2.3 Cellular Aging and Age-Related Disorders	461
	14.2.4 Cellular Aging In Vivo	463
	14.2.5 Strategies for Telomerase-Based Therapy	465
14.3	The Pluripotency of ES Cells	466
	14.3.1 Differentiation of Murine ES Cells	466
	14.3.2 Differentiation of Human ES Cells	466
14.4	The Immortality of Cultured Germ-Line Cells	467
	14.4.1 Telomerase Regulation in Somatic Vs. Germ-Line Stem Cells	467
	14.4.2 Genetic Drift in Cultured ES Cells	470
14.5	Reprogramming and the Reversal of Cellular Aging	470
	14.5.1 Somatic Cell Nuclear Transfer	470
	14.5.2 Telomere Dynamics During SCNT	470
	14.5.3 Telomere Dynamics During Induced Pluripotency	472
14.6	Prospects for Heterochronic Tissue Regeneration	473
	14.6.1 Macular Degeneration	474
	14.6.2 Neurodegenerative Diseases	475
	14.6.3 Vascular, Hematopoietic, Immunological, and Other Disorders	475
14.7	Conclusion	477
References		477

M.D. West (✉)
CEO, BioTime, Inc. and Embryome Sciences, Inc., Alameda, CA, USA
e-mail: mwest@biotimemail.com

14.1 Introduction

Despite the scope and gravity of the problem of human aging for the individual and society (Butler 2008), the field of gerontology has largely remained a backwater of medical research. There is, for example, no consensus within the scientific community as to the central mechanisms of aging itself. This demission likely reflects less the degree of curiosity of members of the research community than the persistent skepticism that a phenomenon so seemingly inextricably linked to the normal life cycle could be altered to any substantial degree. In addition, many researchers quickly ascribe the aging process to entropy and hence conclude that reversing aging is for all practical purposes, a physical impossibility.

These precipitate conclusions regarding the mechanisms of aging ignore the obvious reality that the germ-line lineage of cells, unlike their somatic counterparts, replicates in an immortal continuum of mitotic and meiotic events (McLaren 1992, 2001). In the case of some vertebrate parthenotes, this immortal succession of cells succeeds despite the absence of sexual reproduction (Halliday and Adler 1986). Therefore, despite the widespread assumption that aging is due to inevitable and irreparable wear-and-tear, germ-line cells must possess sufficient repair mechanisms to allow a sufficient number of the cells the potential for immortality and the perpetuation of the species. This chapter addresses the postulate that germ-line cells cultured in vitro, like their in vivo counterparts, possesses a similar repair capacity allowing immortal growth and an unending capacity to generate all human somatic cell lineages. We suggest that fundamental insights relating to the biology of aging can be garnered from the comparative study of the known molecular mechanisms distinguishing immortal germ-line cells from their mortal counterparts, and that it is possible to draw inferences regarding the feasibility of applying the emerging tools of regenerative medicine to interventional gerontology.

The term *regenerative medicine* refers to a constellation of novel technologies surrounding human embryonic stem (hES) cells including the means to manufacture, differentiate, and implement the cells in therapy. While murine ES (mES) cells have been utilized for years to produce mice with germ-line genetic modifications (Robertson et al. 1986), until the first isolation and stable culture of hES cells from the inner cell mass of blastocyst-stage preimplantation embryos (Thomson et al. 1998) and human embryonic germ (hEG) cells from the primordial germ cells of the embryonic gonadal ridge (Shamblott et al. 1998), very little research was performed on ES cell differentiation in vitro and little consideration was given to the use of these cells as a manufacturing modality for human cell therapy. The topics discussed in this chapter will be focused on hES cells, though many of the comments could apply to both hES and hEG cell types.

Human ES cells appear to differ from most fetal and adult-derived stem cells in regard to their twin properties of replicative immortality (while maintaining germ-line telomere length) and totipotency. Since human aging presents as widespread somatic cell dysfunction, human ES cells may offer a novel methodology to produce embryonic precursors useful in regenerating aged tissue. While the fundamental etiology of age-related cell dysfunction is currently a matter of debate, the fact that many cell types are lost or display alterations in phenotype with age, and that this cell dysfunction plays a role in many age-related pathologies is beyond reasonable dispute. Some salient examples include the loss of dopaminergic neurons leading to Parkinson's disease (Toulouse and Sullivan 2008), the dysfunction of the endothelium of the coronaries leading to atherosclerosis (Minamino and Komuro 2008a, b), the loss of functional myocardium leading to heart failure (Genovese et al. 2007), the loss of articular chondrocytes in the weight-bearing joints during osteoarthritis (da Silva et al. 2008), and the dysfunction of the retinal pigment epithelium of the retina implicated in the pathogenesis of age-related macular degeneration (McLeod et al. 2002).

A comprehensive list of such disorders is beyond the scope of this chapter, since virtually every organ system in the human body shows age-related pathology. In each of the above-mentioned conditions as well as many others, the opportunities to utilize new human embryonic progenitor cell types in regenerative medicine offer important new potential therapeutic strategies. Since the twin properties of replicative immortality and totipotency distinguish these new cell lines, we will review the basic cellular and molecular biology of germ-line cells.

14.2 Implications of the Germ-Line/Soma Dichotomy

14.2.1 Origins of the Cellular Theory of Aging

Modern cellular gerontology traces its origins to the 19th century and the advent of cytological studies of development and aging. These early studies led the German naturalist August Weismann to postulate that the substance of heredity is transmitted via an immortal lineage of cells called the *germ-line*. His proposal contrasted with Darwin's theory of *pangenesis* that postulated the existence of subsets of the genetic material packaged in vesicles called *gemmules* that were transported from the somatic cells to the reproductive cells where they were assembled for the purpose of procreation. Darwin's model of heredity and species immortality was a complex interplay of somatic and reproductive components. Weismann's simpler hypothesis predicted a lineage of germ-line cells that possessed the impressive capacity to replicate indefinitely, periodically generating new somatic cell lineages, with the hereditary information being transmitted solely through the germ-line (Weismann 1891). A corollary of Weismann's model of the germ-line/soma dichotomy was that the role of somatic cell lineages was to build a body to facilitate the successful sexual transmission and recombination of the germ-line cells. Following the successful rearing of reproductively-competent progeny, the fate of somatic cell lineages had little selective value to the germ-line cells, and indeed, the viability of surplus somatic cells may compete for scarce nutritional resources.

This line of reasoning apparently led Weismann to postulate the first theory of cellular aging; namely, that the first appearance of aging and programmed death coincided with the divergence of the germ-line and somatic cells. He also proposed that the repression of replicative immortality in the somatic cell lineages might have survival value for the germ-line cells and might, in fact, be a central mechanism underlying human aging. He wrote, "Death takes place because a worn-out [somatic] tissue cannot for ever renew itself, and because a capacity for increase by means of cell-division is not everlasting, but finite" (Weismann 1891). Weismann was prescient enough to recognize that not all cells in a particular tissue would have to reach the end of their replicative lifespan in order for age-related pathology to emerge. As he put it, his theory, "does not, however, imply that the immediate cause of death lies in the imperfect renewal of cells, for death would in all cases occur long before the reproductive power of the cells had been completely exhausted. Functional disturbances will appear as soon as the rate at which the worn-out cells are renewed becomes slow and insufficient" (Weismann 1891).

The test of the Weismann hypothesis of cellular aging awaited the advent of in vitro cell culture in the early 20th century. The first long-term culture of chick somatic cells led Alexis Carrel to the erroneous conclusion that cells freed from the influences of the intact organism display an intrinsic immortality (Carrel 1912). This then led to early advances in organ culture with the intent of maintaining tissues such as the heart in an *ex vivo* senescence-free state to facilitate the extension of human lifespan (Friedman 2007). It wasn't until the 1960s that Leonard Hayflick finally challenged the prevailing consensus by forcefully arguing that human somatic cells display an intrinsic finite replicative capacity (now known as the "Hayflick limit") similar to that predicted by Weismann, unless they are derived from malignant tumors or otherwise transformed in vitro (Hayflick and Moorhead 1961; Hayflick 1965, 1992).

In the early 1970s, Robert Dell'Orco reported that the timing of the in vitro senescence of cultured fibroblasts appeared to reflect the number of times the cells replicated rather than the amount of time the cells had been cultured (Dell'Orco et al. 1973). That is, the clock of cellular aging appeared to be a *replicometer* (measuring cell cycle events) as opposed to a chronometer (measuring metabolic time). In 1975, Woodring Wright demonstrated that the transplant of senescent cell nuclei into young cytoplasts led to the induction of senescence while the transplant of young cell nuclei into senescent cytoplasts led to young cells (Wright and Hayflick 1975). Taken together, these observations suggested that the nucleus contained a replicometer recording cell doublings in mortal somatic cells and, of course, a clock that could be reset indefinitely in germ-line cells. The question for theoretical biologists then became: "What molecular structure could be utilized by a cell to 'count' and 'remember' up to 100 cell division events?"

14.2.2 Telomere Dynamics in Cellular Aging and Immortalization

In a spectacular display of foresight, on simply learning that human somatic cells showed a finite replicative lifespan in vitro, Alexey Olovnikov authored the theory of "marginotomy" now known as the "telomere hypothesis of cellular aging."

Olovnikov theorized that cell aging was due to the lack of a terminal DNA polymerase (now designated *telomerase*) maintaining chromosome termini (*telomeres*) in somatic cells, leading to progressive telomere shortening and ultimately cell cycle arrest (Olovnikov 1971). He also proposed that immortal cells (such as germ-line and neoplastic cells) express this terminal transferase, resulting in telomere maintenance and hence replicative immortality. A year later James Watson wrote a similar paper on the end replication problem without, however, suggesting a linkage to cell aging and immortalization (Watson 1972). When asked whether Watson knew of Olovnikov's earlier paper when writing his own, Watson replied "No. I have had at least two good ideas in my life and the end replication problem was one of them" (J.D. Watson 1994, personal communication).

More than a decade later, telomere terminal transferase activity was reported for the first time in extracts of the immortal protozoan *Tetrahymena* (Greider and Blackburn 1985). Later, the enzyme was shown to require an RNA component that encoded the template for a six-base pair G-rich repetitive sequence of which the telomeres are composed (Greider and Blackburn 1987, 1989). Even before the human telomere sequence had been determined, Howard Cooke authored the first report that shortening occurred during in vivo aging in the telomeres adjacent to the pseudoautosomal region of the sex chromosomes in human somatic cells but that they remained constant in length in the germ-line over time (Cooke and Smith 1986). Cooke speculated that germ-line cells maintain telomeres and immortality through telomere terminal transferase (telomerase).

The sequencing of human telomeric DNA (Moyzis et al. 1988) and the demonstration that vertebrate telomeres are capped with several kilobases of the tandem repetitive sequence $5'$-TTAGGG/$3'$-AATCCC, opened the door to the facile measurement of mean telomere length. This was initially performed by digesting genomic DNA with restriction enzymes having 4-base pair recognition sites followed by measurement of mean telomere length on Southern blots using (TTAGGG)$_4$ telomeric repeat probe. Because telomeres generally do not contain palindromes, the digested products generally contain intact telomere repeat arrays and when hybridized to telomeric repeat probe allow a quantification of mean telomere restriction fragment (TRF) length. It should be recognized, however, that TRFs also contain degenerate subtelomeric sequences designated the "X" region, of approximately 3,000 bp. Therefore, TRF length is both a measurement of mean length (not the length of the shortest telomere), and TRF length includes the X region (Fig. 14.1). Therefore, cells with a mean TRF length of 5,000 bp likely have a mean length of TTAGGG repeats of approximately 2,000 bp, and consequently, one of 92 telomeres may have lost telomeric repeats entirely, triggering a DNA damage checkpoint arrest if one assumes an appropriately wide length distribution of telomere lengths.

Early TRF analysis showed markedly shorter telomere lengths in blood compared to sperm DNA (Allshire et al. 1988, 1989) consistent with Cooke's prior observation. In addition, a progressive shortening of telomeres was later reported in serially passaged human somatic cells (Harley et al. 1990; Harley et al. 1994). A series of additional evidence in support of the telomere hypothesis came with the

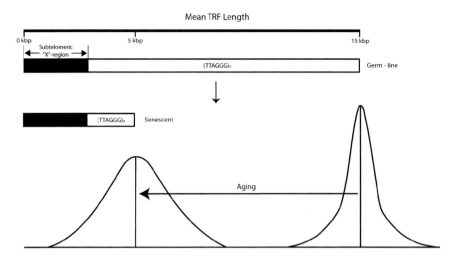

Fig. 14.1 The structure of the human telomere. Germ-line cells maintain a mean TRF length of approximately 15 kbp through the activity of telomerase and associated proteins. In the absence of telomerase activity, the average length of telomeres shorten in somatic cells during each replication until the mean TRF length is approximately 5 kbp. While there is genetic diversity in the subtelomeric "X-region", a mean TRF length of 5 kbp could mean that one telomere has lost terminal sequences of TTAGGG repeats, triggering cell cycle arrest

reports that yeast mutants displaying the unusual phenotype of *ever shorter telomeres* (EST) transformed yeast from an immortal organism to a mortal one (Lundblad and Szostak 1989). In addition, virtually all immortal human tumor cell lines but not normal human somatic cells expressed telomerase and had stable telomeres (Morin 1989; Hastie 1990; Kim et al. 1994) and a 3′ overhang was observed at chromosome termini suggestive of a lack of complete replication of the lagging strand (Henderson and Blackburn 1989).

The precise mechanism of telomere shortening in human mortal somatic cell replicative senescence is not known with certainty since multiple factors may play a role. These include the unidirectional 5′-3′ activity of DNA polymerase and the involvement of labile RNA primers on the lagging end of the replicated strand. As shown in Fig. 14.2, assuming the RNA primer were laid down randomly on the overhang, there could be a loss of DNA on the lagging strand. The presence of a 3′ overhang suggests that there may be a 5′ exonuclease and, indeed, several such nucleases have been identified, including the *WRN* helicase/exonuclease gene that is mutated in the premature aging disorder known as Werner syndrome (Eller et al. 2006). In addition, there is a leading strand replication defect as well, since the 3′ overhang cannot be replicated without a template.

Regardless of the precise mechanisms, TRF length is generally observed to shorten by (25–200) bp per cell doubling in cultured mortal cell types from an initial germ-line (sperm) TRF length of approximately 15 kbp to approximately 5–7 kbp at

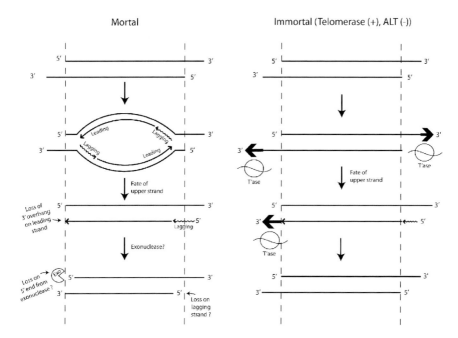

Fig. 14.2 Proposed mechanisms for telomere shortening. In the Olovnikov-Watson model, during semiconservative DNA replication, an RNA primer initiates DNA polymerization in the 5′ to 3′ direction. DNA polymerization in the region of the removed RNA primer can be initiated from the adjacent 3′ DNA. However, because there is no DNA 5' to the RNA primer on the terminal lagging strand, a small segment of telomeric DNA may be lost in telomerase (–) mortal cells depending on where that primer hybridizes on the 3′ overhang. In addition, leading strand synthesis will not be able to re-create the 3′ overhang and a 5′ exonuclease may recreate an overhang at the expense of further 5′ telomeric nucleotides. In immortal cells, telomerase primes the extension of the 3′ end off of a mobile RNA template

the Hayflick limit. This led to the concept that cells that possess sufficient reserves of telomere repeats are reproductively competent. Cells that have exhausted their telomeric reserve (as measured by one telomere having no terminal repeats) are at their replicative limit (sometimes also referred to as M1, for first type of replicative mortality.)

In recent years, considerable detail has been added to our understanding of the structure of telomeres. It is now accepted that telomeres are often capped in a D-loop wherein the telomere circles back and the 3′ overhang is inserted within the DNA helix. The structures called *t loops* may play a role in stabilizing telomere ends, but also may reflect processes in the germ-line or even in somatic cells that allow gene conversion or recombination to repair or maintain telomeres (Griffith et al. 1999). It would be anticipated that telomeric repeats, like other tandem repeats, pose particular problems due to the chance of hairpin loops and other inappropriate structures. Assuming that there was evolutionary pressure to repress telomerase expression in the soma, there may have been selection for mechanisms that block the simple process of the formation of a hairpin loop or t loop that allows the extension of the 3′end.

The proteins that bind the telomere repeats are collectively called the *telosome* or alternatively *shelterin* (Liu et al. 2004; de Lange 2005). Some of these, like the telomere-specific telomeric binding proteins *TERF1*, *TERF2*, and helicases such as *BLM, WRN*, and *LSH*, show markedly differential expression in germ-line vs somatic cells (unpublished data). Many of these telosome components likely play an important role in maintaining the integrity of the telomeric DNA such as regulating illegitimate homologous recombination within the repeated sequences in somatic cells to prevent runaway telomere extension, and repairing telomere strand breaks. As of today there is only a rudimentary understanding of how these components cooperate in regulating telomere maintenance and recombination.

It is important to recognize that the repression of telomerase activity in somatic cells may have consequences other than consigning cells simply to a mortal phenotype. The alternate recombination mechanisms to maintain telomeres in somatic cells, as well as the repression of telomerase activity except, perhaps, to cap double-strand breaks (DSBs), may leave somatic cells particularly vulnerable to irreparable telomeric DSBs. Consistent with this hypothesis, genes that begin to be expressed at about the same time as with the repression of telomerase reverse transcriptase (*TERT*), such as *LMNA* and *XPA*, (Constantinescu et al. 2006) have been shown to play a role in DSB DNA damage repair (DDR) in Hutchinson-Gilford progeria (Liu et al. 2008).

Regardless of whether the loss of telomeric repeats is due to progressive telomere shortening or stochastic fragmentation, the resulting cells could then be detected by DDR pathways as DSBs. These lesions have serious consequences for a cell not only because of the potential loss of information, but also because of the risk of genetic instability and cancer (Khanna and Jackson 2001). Therefore, DSBs induce a DDR response that includes the triggering of DNA damage-specific gene expression and protein modifications (D′Adda Di Fagagna et al. 2003) through, in part, the aggregation of proteins like gamma-*H2AX* at the break through the activity of *ATM* (Meier et al. 2007). Therefore, human cell senescence is generally viewed as the phenotype of a cell responding to serious DNA damage such as DSBs, or telomere loss.

It has long been recognized that infection of human cells with oncogenic viruses such as the simian virus SV40, papillomavirus, adenovirus, or Epstein-Barr virus can lead to the immortalization of cells in vitro (Counter et al. 1992, 1994; Shiga et al. 1997; Gallimore et al. 1997). The spontaneous immortalization of normal (uninfected) human cells cultured in vitro has rarely been reported, and when it has, the cells are generally stem cells such as basal keratinocytes that already express low levels of *TERT* (Rea et al. 2006). The long-term culture of cells with low levels of telomerase expression may lead to the artifactual selection of subsets of cells with higher levels of telomerase expression until eventually immortal clones prevail.

In the majority of human cells where no telomerase is expressed, infection with SV40 virus typically extends the replicative lifespan by approximately 30%, with

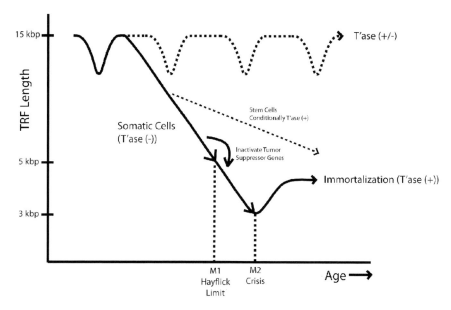

Fig. 14.3 Telomere Dynamics from Embryogenesis to M1, M2, and Immortalization. Telomeres are maintained in the germ-line by telomerase activity, though this activity and telomere length appears to vary with development. Early in differentiation telomerase is repressed in somatic cell types leading eventually to Hayflick senescence (M1). The bypass of M1 can be accomplished through the repression of tumor suppression mechanisms by viral transforming proteins such as SV40 T-antigen and cells show an extended lifespan until they senesce in crisis (M2). Rare foci of immortalized cells may then appear at a low frequency that appear to maintain telomeres by reactivated telomerase expression or the ALT pathway

the cells then entering a second type of senescence called *crisis* or *M2* (Mortality 2) characterized by marked heterogeneity and cell death followed by the relatively rare emergence of immortal foci (Girardi et al. 1965; Walen and Arnstein 1986). Careful quantitation of the frequency of the emergence of immortal foci showed that approximately 1 in 10^6 human cells in SV40-induced crisis will emerge as immortal cell lines (Shay and Wright 1989).

SV40 large T-antigen likely exerts its effects by binding to and inactivating the tumor suppressor genes p53 and Rb (Green 1989; DeCaprio et al. 1988; Scheidtmann and Haber 1990) that mediate the DNA damage checkpoint arrest signaled by at least one telomere being critically shortened (Fig. 14.3). This is consistent with the high frequency of dicentric chromosomes observed in cells during crisis (Ray et al. 1992) and their premalignant counterparts in vivo (Morgan et al. 1986).

The mechanism responsible for escaping M2 appears to be the reactivation of telomerase activity or the "alternative lengthening of telomeres" (ALT) pathway that appears to extend telomeres through abnormal regulation in t-loops leading to homologous recombination and rolling circle DNA replication and the appearance of extrachromosomal circles of telomeric DNA (Muntoni et al. 2008; Wang et al. 2004). The down-regulation of T-antigen in cells that have inactivated the M2 mechanisms (immortalized cells) can lead to cells becoming reversibly senescent in M1 (Shay et al. 1992), suggesting that the initial senescence observed in normal cells approaching the Hayflick limit, designated mortality 1 (M1), is distinct from the mortality horizon during crisis, called mortality 2 (M2) (Wright et al. 1989; Shay et al. 1993).

Despite the mounting evidence that telomeres were the clocking mechanism of cell senescence, a definitive case required the cloning of the telomerase RNA component *TERC* (Feng et al. 1995) and the catalytic telomerase reverse transcriptase *TERT* (Nakamura et al. 1997). Expression studies indicated that the RNA template was expressed in both mortal and immortal cells, while the catalytic component was generally expressed only in immortal cells. Therefore, the transfection of the differentially-expressed catalytic component *TERT* into somatic cells allowed the first direct test of the telomere hypothesis. The results demonstrated that telomerase is sufficient to impart the immortal phenotype in most cultured human cells without conferring the other characteristics of a transformed cell, such as growth in soft agar (Bodnar et al. 1998).

Subsequent studies showed that another type of senescence in some human cell strains such as WI-38 cells and in cultured mouse cells appears to be unrelated to telomeric attrition, is not rescued by the exogenous expression of *TERT*, and instead appears to be related to oxygen toxicity (Forsyth et al. 2003). DSBs are observed in such cells in both telomeric and internal regions of the chromosome, and culture at lowered oxygen tension is reported to largely eliminate the premature senescence (Di Micco et al. 2008). In addition to some cell types being sensitive to ambient oxygen tension and internal DSBs, there is considerable evidence that such high oxygen concentrations can cause the abrupt fragmentation of telomeric DNA in some cell types leading to a stress induced premature senescence (SIPS) (Toussaint O. et al. 2000) due to the sensitivity of triple G's, such as those resident in the telomere, to oxidative damage (von Zglinicki 2002; Oikawa and Kawanishi 1999).

Another important distinction is that human and mouse cells differ significantly in telomere biology. In contrast to humans, cells from inbred mouse strains typically show very long and complex telomeres, and labile expression of *TERT* (Di Micco et al. 2008; Prowse and Greider CW 1995). As a result, mouse cells often spontaneously immortalize in vitro. In addition, the senescence of murine fibroblasts likely reflects oxygen toxicity rather than telomere shortening since it occurs relatively early in their lifespan (despite long telomeres), and transgenic mouse cells overexpressing *TERT* still senesce. The profound differences in telomere biology between species is important to consider when interpreting data related to aging such as the reports of correlations of maximum proliferative capacity of cultured cells from animals with varied lifespans. For instance, if murine fibroblast senescence occurs at approximately 20 doublings in vitro due to oxygen toxicity, and human dermal fibroblast senescence occurs at approximately 50 doublings due to telomere shortening, then studies correlating relative species lifespan with cell culture lifespan based on these data would not generate a meaningful result.

14.2.3 Cellular Aging and Age-Related Disorders

If aging on a tissue level is triggered by a dominant deleterious effect brought about by the presence of senescent cells, rather than the loss of a beneficial effect of young cells, then age-related changes could begin to occur with the accumulation of a minority of cells reaching their replicative lifespan as Weismann proposed. There appear to be candidate dominant and deleterious genes identified in some cell types during senescence.

14.2.3.1 Senstatic Activation

Generally speaking, senescent cells often express an inflammatory-like state of gene expression called *senstatic activation*. Young replication-competent cells such as dermal fibroblasts replicate only when necessary to maintain or repair tissue. They are generally in a state of quiescence and maintain the extracellular matrix (ECM) of the dermis through the secretion of a balance of ECM components and proteases that turn over those components at a relatively slow rate. In the case of damage to the dermis resulting from trauma or inflammation, these quiescent cells re-enter the cell cycle and simultaneously up-regulate proteolytic enzymes that facilitate the migration of immune cells into the tissue and allow the repair of the lesion through increased proteolysis and turnover. Once repair is complete, young cells can re-enter quiescence in order to maintain the tissue.

The definition of a senescent cell is a cell that can no longer enter the cell cycle in response to growth signals such as serum. However, paradoxically, senescent cells often do not display a pattern of ECM gene expression of a quiescent cell. Despite the fact that they are not entering the cell cycle, senescent cells often display an ECM pattern of gene expression similar to a highly activated cell. In the case of dermal fibroblasts for example, senescent cells express constitutively high expression of interstitial collagenase and plasminogen activators (West et al. 1989, 1996) implicated in the age-related thinning and friability of the dermis (West 1994). Significantly, even when senescent cells are cultured at high density or low serum (conditions that would induce quiescent gene expression in a young cell) senescent cells constitutively express genes characteristic of a highly activated cell. Depending on the differentiated state of the cell, the inappropriate senstatic activation of the cells may involve diverse genes, but in general, these genes are similar to inflammatory patterns of gene expression and often can be correlated with the age-related pathology in that particular tissue.

14.2.3.2 Telomere Damage

Further evidence supporting a role for telomere maintenance in age-related disease comes from the study of premature aging syndromes. While not ideal models of human aging, several of these disorders such as Werner syndrome (WS) and Hutchinson-Gilford syndrome (progeria) show accelerated onset of a wide array of age-related pathologies including skin aging, osteoporosis, cataracts, type II diabetes, and coronary disease (Ding and Shen 2008; Kudlow et al. 2007).

The affected gene for WS is the *WRN* helicase (Yu et al. 1996), a protein now recognized to protect the integrity of telomeres by interacting with DSBs and fragmented telomeres, and when mutated in Werner syndrome, to cause an accelerated onset of critically-short telomeres on sister chromatids (Hasenmaile et al. 2003; Bai and Murnane 2003; Crabbe et al. 2004). Similar *WRN*-mediated disruption of D-loops in vitro has been reported (Orren et al. 2002).

The transfection of *TERT* into WS fibroblasts rescues the premature senescence of the cells in vitro (Wyllie et al. 2000; Crabbe et al. 2007). Evidence of the causes of telomere damage in WS come from studies showing that *WRN* associates with the amino terminus of *TRF2*, a telomere-specific binding protein that plays a role in telomere maintenance and telomere length regulation (Machwe et al. 2004). The transfection of the mutant form of *TRF2* designated *TRF2DeltaB* lacking the amino-terminal basic domain into normal fibroblasts causes rapid telomere loss, extrachromosomal circles of telomeric DNA, and senescence (Wang et al. 2004). When the mutant form of *TRF2* was transfected into *TERT* immortalized WS fibroblasts, no rapid telomere fragmentation was observed (Li et al. 2008). However, wild type *WRN* transfection did induce telomere loss, extrachromosomal circles of telomeric DNA, and senescence. In *TERT*-expressing WS cells, fragmented t loops existed as extrachromosomal circles similar to ALT cells (Wang et al. 2004). The addition of wild type *WRN* to the *TERT*-expressing WS cells decreased the extrachromosomal telomere circles, consistent with the telomere-protective activity of *WRN* (Li et al. 2008).

Improved mouse models of human aging and stem cell biology were produced when the *Wrn* and *Terc* knockouts were combined in mice. While mice null for *Wrn* appear normal (Lombard et al. 2000), the breeding of *Wrn* null with *Terc* null mice (the latter having been previously bred until they had relatively short telomeres) resulted in offspring displaying a constellation of age-related pathologies known to be common between aging and WS in humans, including: alopecia, graying of hair, osteoporosis, type II diabetes, and cataracts, disorders not commonly observed in wild type mice (Chang et al. 2004). Cells from the mice also showed gamma-H2AX DNA damage foci.

These data suggest that telomere damage resulting from either replicative senescence or spontaneous fragmentation not repairable in somatic cells may be a common feature of WS and normal aging. The data also suggest that means to reactivate telomerase and restore telomere length may be useful in providing a therapeutic effect in those particular age-related diseases that overlap in *WRN* and normal aging since telomerase would be expected to cap such DSBs with telomeric repeats as it did in the WS cells. Such diseases could include atherosclerosis, cataracts, osteoporosis, dermal atrophy, and type II diabetes (Epstein et al. 1966).

14.2.3.3 Cancer

The murine system has been utilized to test the effects of altered telomerase expression on cancer incidence. Knockout of telomerase activity leads to the shortening of telomeres and an increased resistance to cancer in mice (Gonzales-Suarez et al. 2000). More directed studies utilizing the fact that constitutive expression of *TERT* driven by the p56lck promoter limits expression to thymocytes and T cells resulted in evidence of an increase in the frequency of lymphoma (Canela et al. 2004). Similar directed expression in transgenic mice expressing *TERT* under the control of the keratin 5 promoter leads to an increased frequency of epidermal tumors but also improved wound healing (Gonzales-Suarez et al. 2001).

Experiments designed to overexpress *TERT* in transgenic mice are reported to lead to increased telomerase activity in at least a subset of tissues such as the mammary gland, and a significantly increased frequency of invasive intra-epithelial mammary carcinoma (Artandi et al. 2002). To test whether increased telomerase activity would increase mouse lifespan in a genetic background of increased tumor suppression, transgenic mice harboring a 50% increased gene dosage of *p53*, or a 50% increased gene dosage of *Arf* and *p16* (that displayed decreased incidence of cancer (Garcia-Cao et al. 2002; Matheu et al. 2004, 2007)) were crossed with mice overexpressing *TERT*. The latter mice showed up to a 50% extension of median lifespan when factoring out cancer-related death (Tomas-Loba et al. 2008). Since aging mice often die of cancer and other pathologies unrelated to those observed in normal human aging, it would be useful to measure the extension of median lifespan in mice engineered to have short telomeres, tissue-restricted expression of *TERT* using the human promoter elements (Horikawa et al. 2005), and pathologies more closely related to human aging.

It seems reasonable to speculate that tight control of telomerase expression, even in stem cell compartments that occasionally utilize telomerase, such as lymphocytes and keratinocytes, occurred coincident with the evolution of longer lifespans, perhaps as a result of selecting for an increased resistance to cancer incidence. The obvious tradeoff is that human somatic cells are virtually always mortal with a replicative lifespan sufficient only to insure fitness of the individual long enough to reproduce and rear young. In other words, it seems reasonable to propose that aging was selected for in evolutionary history at least in part as a tumor suppressive mechanism, an example of the evolutionary theory known as *antagonistic pleiotropy* (Ungewitter and Scrable 2008).

14.2.4 Cellular Aging In Vivo

Even with August Weismann's first proposal that human aging reflects an intrinsic cellular clock limiting the replicative lifespan of somatic cells, the question arose as to whether aging on a tissue level resulted in mitotic arrest per se, or some secondary event associated with senescence. Weismann theorized that age-related pathology could result prior to the global senescence of all cells in a particular tissue (Weismann 1891). If this is the case, then the simple test of determining whether the remaining replicative lifespan of mass cultures of cells aged in vivo correlated inversely with donor lifespan could be fatally flawed. The maximum achievable lifespan of a cell culture would then reflect the replicative lifespan of the youngest cell in the explanted culture rather than reflecting an accumulating number of senescent cells present.

Initial attempts to determine whether cellular aging correlated with in vivo age suggested that the replicative lifespan of cultures was inversely proportional to donor age (Schneider and Mitsui 1976). Later studies could not reproduce this result (Cristofalo et al. 1998), but the primary culture of cells from aged donors did appear to contain an increased percentage of cells with a senescent morphology (Hayflick 1968) not typically seen in explants from young donors. The use of a senescence-associated beta galactosidase assay at pH 6 (SA-beta-gal) (Cristofalo and Kabakjian 1975; Itahana et al. 2007) showed evidence of accumulating senescent cells in numerous tissues as a function of age (Le Maitre et al. 2007; Bandyopadhyay et al. 2005). However, the marker is not necessarily senescence-specific (Cristofalo 2005). The alternative analysis of colony size distribution has the potential to quantify the replicative capacity of individual clones, and the use of this technique has shown an age-dependent loss of proliferative lifespan (Smith et al. 2002), though even this technique might not measure stochastic fragmentation of telomeres where a minority of cells cannot proliferate at all.

Measurements of TRF length in peripheral blood cells have shown shortening during in vivo age, but such results were performed by the analysis of TRF length on Southern blots and therefore are complicated by a genetic diversity in lengths likely due to variability in the subtelomeric X region (Cawthon et al. 2003). Nevertheless, studies focused on centenarians show that the cohort without any of the symptoms of hypertension, heart failure, acute MI, peripheral vascular disease, dementia, cancer, stroke, COPD, or diabetes had statistically significantly longer peripheral blood lymphocyte TRF lengths compared to the cohort with at least two of these disorders (Terry et al. 2008).

Most of these studies suffer from a lack of definition of the particular differentiated, progenitor or hematopoietic stem cells being assayed. Nevertheless, they do suggest that significant telomere attrition occurs in blood cells during aging in vivo. Improved high-throughput quantitative fluorescence in situ hybridization (Q-FISH) protocols using interphase cells now allow a more precise analysis of the effects of gender, age, and disease on telomere length (Canela et al. 2007).

Another assay of senescence, though less specific than TRF measurement, utilizes the detection of DSBs. As mentioned earlier, the lack of telomerase expression in somatic cells can lead to both replicative senescence and an impaired ability to cap internal DSBs with telomeric repeats, and therefore the cell may become terminally incapable of being repaired. The assay for senescence-associated DDR foci (SDFs) or alternatively telomere dysfunction-induced foci (TIF) detects DSB-associated ATM or gamma-*H2AX*, a variant form of histone H2A that binds to such lesions (Takai et al. 2003). The marker is inclusive for shortened telomeres in replicative senescence (Nakamura et al. 2008), the premature senescence of Werner syndrome and Hutchinson-Gilford cells (Nakamura et al. 2008; Liu et al. 2008), and the senescence of mouse cells due to oxygen toxicity (Di Micco et al. 2008). Though not specific to telomere fragmentation, the SDF assay is perhaps more relevant than the SA-beta-gal assay as an aging marker, since the SA-beta-gal assay also frequently identifies quiescent cells without any significant DNA damage (Cristofalo 2005). The assay of SDF during human aging in vivo has shown increased DSBs in peripheral lymphocytes (Nakamura et al. 2008) and an age-dependent increase of up to 15% of dermal fibroblasts in baboons (Jeyapalan 2007).

Human aging is a syndrome similar in many respects to Werner syndrome in that affected cells often include continuous or discontinuously replicating cells. In initial surveys of telomerase activity in various adult tissues, activity was confined to sites of known stem cells, such as basal keratinocytes, intestinal crypt, and hematopoietic cell progenitor expansion, such as the germinal centers of lymph nodes (Wright et al. 1996). Indeed, it was from a cDNA library prepared from an inflamed tonsil that the first sequence tag of *TERT* was identified (West, 2003).

14.2.5 Strategies for Telomerase-Based Therapy

With the mounting evidence of the importance of telomere dysfunction in age-related disease, it is relevant to ask what novel therapeutic opportunities are apparent to restore telomere length or repair fragmented telomeres. At least three strategies appear feasible in the intact organism. First, there may exist small molecule drug candidates that could effectively and safely reactivate telomerase activity and extend telomere length in vivo. For example, screens of natural product extracts for compounds capable of activating telomerase activity have demonstrated that the root of the Chinese herb *Astragalus* contains a molecule designated TAT2 capable of inducing telomerase activity in several cultured cell types. While not yet reported to profoundly elevate telomerase activity or telomere length in cultured cells, the compound is reported to significantly improve immune function in stressed peripheral lymphocytes from HIV+ donors (Fauce et al. 2008), cells that have previously been shown to display accelerated telomeric attrition (Effros et al. 1996).

An attractive alternative to small molecule activators of *TERT* is targeted telomerase gene therapy. This approach has demonstrated some feasibility in animal models (Mogford et al. 2006), though its application in many tissues may suffer from the inefficiency or immunogenicity of current gene therapy vectors. This targeted application would reduce the risk of the complication of facilitating the progression of premalignant lesions throughout the body and there is now a significant body of literature detailing which cell types respond to exogenous *TERT* by markedly extending telomere length.

Third, it will likely be possible to utilize telomerase positive totipotent germ-line cells to produce any particular cell type that would be expected to repress telomerase upon differentiation and then possess a long proliferative lifespan. The impressive feature of the use of germ-line cells in rebuilding the aged human body is that these cells have the intrinsic capacity to generate young somatic lineages indefinitely (McLaren 1992, 2001) as opposed to many strategies for intervention in aging that likely can only delay the onset of age-related sequelae.

14.3 The Pluripotency of ES Cells

14.3.1 Differentiation of Murine ES Cells

The field of embryonic stem cells has a colorful history, beginning with the research of Leroy Stevens. While researching the origins of cancer, Stevens observed a strain of mice that spontaneously generated teratocarcinomas, a tumor generating a host of complex tissues such as hair, teeth, and indeed, representatives of all three embryonic germ layers (Stevens 1980). He inbred the mice to yield the well-known mouse strain designated 129. When the inner cell masses from blastocysts of the 129 strain were subsequently cultured in appropriate conditions, a developmental stasis occured such that the cells retained pluripotency (Evans and Kaufman 1981; Martin 1981). In fact, since these cells also retain the capacity to differentiate into germ cells, the cells could perhaps optimally be designated *totipotent*. Stevens called these cells *embryonic stem (ES) cells*, though they are only a stem cell in an in vitro sense. That is, the term stem cell is usually applied to cells that have the capacity to self-renew (in a mortal or an immortal sense) or to differentiate into another type of cell. ES cells are not believed to self-renew in *vivo*, but rather are thought to be an evanescent cell type, rapidly differentiating into embryonic progenitors and subsequently into terminally differentiated germ-line and somatic cells.

The gold standard test of pluripotency in the mouse system is the transplantation of candidate cells into the blastocyst of an allogeneic mouse and the subsequent generation of a chimera to determine whether any of the injected cells engrafted in the animal. Providing that the candidate cells contribute to all three germ layers as well as the germ-line, they are considered ES cells (Nagy et al. 2003). The unique capacity for murine ES cells to propagate endlessly in vitro, be genetically modified, and then be used to obtain germ-line transmission into animals, allows the creation of transgenic mice (Robertson et al. 1986). However, candidate ES cells from other species such as rat, rabbit, cattle, and pigs have generally been "ES-like" cells that are capable of contributing to all somatic cell lineages but not to the germ-line (Brevini et al. 2008). This may reflect ontogeny recapitulating phylogeny, i.e. as Weismann pointed out, in evolutionary history the first differentiation step may have been the division of germ-line and soma. Careful studies of mES cell cultures demonstrated that the cells could be shifted back and forth between a germ-line competent state with the typical ES cell clumped morphology (perhaps corresponding to the inner cell mass of the blastocyst) and a state where they were no longer capable of differentiating into germ cells (perhaps being primitive ectoderm-like (EPL) cells) (Rathjen et al. 1999).

14.3.2 Differentiation of Human ES Cells

The transplantation of hES cells into a recipient allogeneic embryo to generate a chimera is widely accepted to be an unethical test of pluripotency (Lanza et al. 2001). Therefore, there is no clear definition whether hES cells are more similar to

ES or EPL cells. Instead, the commonly-used test of pluripotency for hES cells is the transplantation of the cells into an immunocompromised mouse and the generation of a mass of somatic tissues (a benign tumor designated a teratoma). If such tissue includes derivatives of all three germ-layers, such as, for instance, intestinal epithelium (endodermal), skeletal muscle (mesodermal), and neurons (ectodermal), then the candidate cells are considered pluripotent hES cells. Given the relatively flattened morphology of hES cells (more similar to murine EPL than murine ES cells), and the fact that most mammalian species other than mice are ES-like, it is reasonable to conclude that hES cells would be best described as hEPL in regard to their differentiation potential.

Current hES cell directed differentiation protocols generally utilize information from the mouse developmental biology literature relating to inductive factors that appear to play a role in early embryonic development. The presence of cell types of interest are then identified utilizing well-established techniques such as immunocytochemistry. Examples of such approaches would be the studies to differentiate pancreatic beta cells utilizing activin (D'Amour et al. 2006). Extrapolating from these studies, the corresponding manufacturing protocols would then be an initial scaling of hES cells in large quantities, followed by specific differentiation protocols, and finally cell purification. The difficulties facing the industrial manufacture of cell types using this modality are based on the staggering pluripotency of hES cells. There are literally thousands of distinct cell types arising from hES cells in vitro and there is currently a paucity of data on the molecular markers that allow assay development and purification to the degree required for clinical grade therapeutics (West and Mason 2007).

One solution to simplify the manufacture of purified product from hES cells utilizes the long initial telomere length of hES cells and the robust proliferative potential of embryonic progenitor cells (the intermediate cell types between hES cells and fully differentiated cells). Utilizing a combinatorial cloning protocol (Fig. 14.4), wherein a screen is employed to expose hES cells to an array of diverse differentiation and propagation conditions where the final purification step utilizes clonal propagation, we have demonstrated that it is possible to isolate over 140 diverse embryonic progenitor cell lines that are scalable in vitro and show temporal and site-specific homeobox gene expression (West et al. 2008).

14.4 The Immortality of Cultured Germ-Line Cells

14.4.1 Telomerase Regulation in Somatic Vs. Germ-Line Stem Cells

As mentioned earlier, many mouse cells show labile telomerase expression, unlike their human counterparts where telomerase activity appears to be tightly regulated. Therefore, it is problematic to interpret the considerable literature on replicative capacity in murine ES and somatic stem cells such as the hematopoietic stem cells.

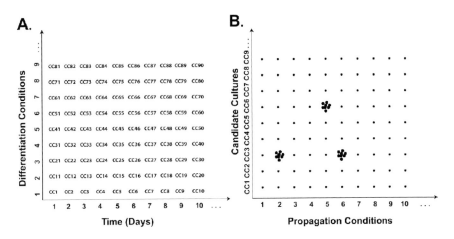

Fig. 14.4 Combinatorial Cloning of Embryonic Progenitors. The ACTCellerate protocol offers a novel modality for manufacturing embryonic progenitors from hES cells in a highly purified state. (**a**) Cultures of hES cells are first exposed to an array of differentiation protocols involving diverse growth factors for varying times to generate an array of candidate cultures (CCs) enriched in diverse cell types. (**b**) Each candidate culture is then plated in another array of diverse culture conditions at clonal densities to allow the isolation of novel purified and scalable embryonic progenitor cell types

Nevertheless, murine ES cells would appear to be intrinsically immortal since they have been passaged for hundreds of doublings in vitro since their initial isolation in 1981, and provided that clones are periodically analyzed to remove aneuploid cells, they are reported to produce mice with the same efficiencies over time and such mice show no evidence of premature aging as a result of such extensive propagation (Suda et al. 1987). Consistent with the necessity of telomerase in maintaining this immortality, murine *Terc* knockout ES cells lacked telomerase activity and senesced after approximately 450 doublings, consistent with the long telomeres in the mouse germ-line (Niida et al. 1998).

In regard to the proliferative potential of murine somatic stem cells, early studies of the proliferative potential of hematopoietic stem cells (HSCs) in mice suggested that the cells lost little if any clonal proliferative capacity during one mouse lifespan, but did show defects after serial transplantations (Harrison et al. 1978). However, such studies are complicated by the fact that, as mentioned earlier, mouse cells have unusually long telomeres, frequently spontaneously immortalize, and serial engraftment may pose problems other than reflecting the proliferative lifespan of the cells. Experiments with the telomerase knockout mouse showed that disruption of *Terc* resulted in no phenotype in the first generation, but after six generations, telomere shortening led to stem cell defects, such as reproductive impairment, and blood cell abnormalities (Lee et al. 1998). It should be recognized that the knock-out of telomerase in an animal would not necessarily be expected to mirror human aging, not only because of the long telomeres in mice, but more importantly because the knockout would be expected to adversely affect telomerase-dependent stem cell populations, while in normal animals, such cells may be relatively protected in aging due to telomerase activity.

It is commonly stated that in humans, telomerase expression is restricted to the germ-line, embryonic somatic cells, and in the adult, restricted to tightly regulated expression in certain stem cells (Wright et al. 1996). As mentioned previously, telomerase activity is reported in human ES cells, hematopoietic cells, basal keratinocytes, and intestinal crypt cells (Sharma et al. 1995; Chiu et al. 1996; Yasumoto et al. 1996). Consistent with the above, mutations affecting telomerase, including *DKC1*, a nucleolar protein that is associated with *TERC*, can lead to dyskeratosis congenita, an inherited bone marrow failure syndrome displaying segmental premature aging, disorders in nail and hair growth, and markedly shortened telomeres in peripheral blood cells (Vulliamy et al. 2001a). Mutations in the *TERC* gene have also been implicated in the same disease (Vulliamy et al. 2001b). Consistent with an important role in telomerase (+) stem cells, *DKC1* is relatively highly expressed in both CD34(+), CD133(+), and hES cells compared to telomerase (–) somatic cells (unpublished).

Despite being a mutipotent stem cell that generates many billions of downstream cells daily, the human HSC is believed to be generally quiescent, dividing on average only once every 1–2 years (Vickers et al. 2000). However, in the lymph node germinal center, B and T lymphocytes may be required to rapidly clonally expand in response to antigen to fight life-threatening infection. Therefore, it seems reasonable to conclude that *TERT* may be expressed at low levels in the HSCs and their progeny, but actual extension of telomeres may be tightly regulated to proceed only in the context of the germinal center. Indeed, as predicted, such lymphocytes show markedly extended telomeres (Norrback et al. 2001). Some other stem cells such as the intestinal crypt cell may be similar in regard to rarely dividing and accomplishing the replication of large numbers of downstream cells by the use of downstream proliferating progenitors. The basal keratinocyte may divide only once every 1–2 years in the human (West 1994).

Measurements of TRF length in human blood cells are consistent with the above model. Mixed peripheral blood leukocytes of monozygotic and dizygotic twin pairs aged 2–95 decreased by approximately 31 bp/year, consistent with the rate of turnover of HSCs and rates of telomere loss in mortal cells (Slagboom et al. 1994). In another study (Rufer et al. 1999), TRF lengths were measured in more than 500 individuals 0–90 years of age, including 36 monozygotic and dizygotic twins, and showed average rates of loss in granulocytes of 39 bp/yr and in lymphocytes of 59 bp/yr.

Interestingly, while the germ-line clearly maintains a replicative immortality in the combination of mitotic and meiotic events, telomeres may vary in length over the reproductive cycle, and the cells show varying levels of telomerase expression depending on the stage of the reproductive cycle. For instance, the highly proliferative human and avian primordial germ cells cultured in vitro (EG cells) appear to be negative for telomerase activity (unpublished). In addition, mature oocytes and sperm appear to express low or undetectable telomerase activity that increased with the activation of gene expression at the blastocyst stage of development (Liu et al.

2007). Also relevant is that studies suggest that telomeres are extended in the early cleavage stages where telomerase is expressed at low levels, and are lengthened at this particular embryonic stage even in telomerase null embryos (Liu et al. 2007). This suggests that telomeric chromatin is permissive for recombination mechanisms perhaps similar to the ALT pathway, perhaps playing an important role in setting the initial germ-line telomere length.

14.4.2 Genetic Drift in Cultured ES Cells

While hES cells maintain telomeres and appear to possess replicative immortality, as seen in the murine system (Suda et al. 1987), hES cells would be expected to drift into aneuploidy during extended culture. The reason for this drift is simply that unlike normal embryogenesis, where ES cells are present for only a few days, the long term propagation and subculturing of hES cells would put the culture under conditions to continuously select for the most rapidly replicating cells, which may be cells with an abnormal genotype. Initial studies of the karyotype of long-term cultures of hES cells indeed demonstrate that the cells have a propensity for specific alterations that need to be part of assay development in the manufacture of therapeutic products (Draper et al. 2004).

14.5 Reprogramming and the Reversal of Cellular Aging

14.5.1 Somatic Cell Nuclear Transfer

One remaining issue to resolve is the problem of histocompatibility between the hES-derived cells and the patient. The demonstration of the feasibility of somatic cell nuclear transfer (SCNT) in animal cloning (Wilmut et al. 1997) leads to the question of the feasibility of the use of SCNT to produce autologous hES cells (Lanza et al. 1999a, b). It is worth noting that most SCNT-derived "clones" are not strictly clones in that their mitochondrial genome is generally derived from the oocyte donor rather than the somatic cell (Evans et al. 1999). However, SCNT-derived tissues transplanted in large animal models show no evidence of transplant rejection despite the presence of allogeneic mitochondrial DNA (Lanza et al., 2002).

14.5.2 Telomere Dynamics During SCNT

The initial report that animals derived by SCNT displayed relatively short telomeres (Shiels et al. 1999) led to the initial conclusion that oocyte-mediated reprogramming cytoplasm could reverse differentiation but not cellular aging. However, subsequent studies using more carefully defined cells for donor chromatin and improved assays of telomere length demonstrated for the first time that SCNT had the potential to

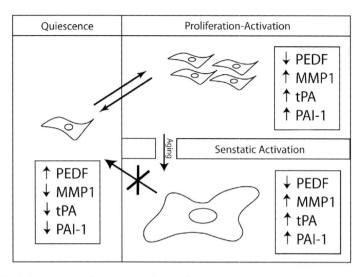

Fig. 14.5 Quiescence vs. Senescence. Young discontinuously replicating cell types generally reside in a quiescent state wherein they maintain extracellular proteins. However, in the case of trauma or inflammation, these cells may reversible re-enter the cell cycle, proliferate, and express a pattern of gene expression useful in the degradation and remodeling of extracellular proteins. When finished repairing the lesion, the cells may then re-enter quiescence. Senescent cells not only become unresponsive to growth stimuli for proliferation, they become terminally arrested in an activated state that may play an important role in age-related pathology

reverse both development and cellular aging in bovine species (Lanza et al. 2000). Shortly thereafter, this observation was extended to other species and other conditions of SCNT (Wakayama et al. 2000; Clark et al. 2003). Interestingly, the use of senescent donor nuclei resulted not only in reprogrammed cells with longer TRF length than the original population, but the cells displayed an ability to induce quiescent gene expression, such as *PEDF* as shown in Fig. 14.5, at levels greater than normal young cells. One explanation for this would be that the removal of telomeric repeats by passaging the cells to M1 and then the recreation of new telomeres in the nuclear transfer embryo led to more uniform tracts of telomeric repeats in the cloned vs. wild type cells. This could lead to decreased incidence of intermittent DDR in the cells and therefore a more uniform entrance into quiescence.

In the course of nuclear transfer experiments, it is common to include controls to assay oocyte activation frequency in the absence of enucleation or nuclear transfer (Cibelli et al. 2002a). An offshoot of human and nonhuman primate cloning experiments was the demonstration that it is possible to derive diploid ES cell lines from activated oocytes in the absence of nuclear transfer, a process called *parthenogenesis* (Cibelli et al. 2002b). This offers a means of creating ES cells containing only DNA from a reproductively-competent female patient that may have important utility in preventing transplant rejection. Potentially complicating the use of these cells, however, is the relative lack of paternal imprinting and large segments of homozygosity within the genome that on the one hand may allow the production of HLA-homozygous reduced complexity libraries (RCL) that would simplify the matching of tissue antigens, and on the other hand may cause difficulties due to homozygous deleterious alleles. Like hES cells from biparental embryos, such cells appear to maintain germ-line telomere length and telomerase activity in the undifferentiated state and to rapidly repress telomerase expression during differentiation (unpublished results).

14.5.3 Telomere Dynamics During Induced Pluripotency

While SCNT offers great potential as a means of reversing the aging of human cells to produce young histocompatible cell and tissue grafts, the practicalities of obtaining human egg cells and performing nuclear transfer on a large scale are daunting. In addition, initial experiments demonstrated that while it was possible to remodel human somatic cell nuclei into pseudo pronuclei, embryonic development generally ceases before blastocyst formation (Cibelli et al. 2001). The efficiencies of human SCNT would likely improve markedly as it has in other mammalian species with access to adequate oocytes to optimize the parameters of oocyte maturation, enucleation, nuclear transfer, activation, and embryo culture. It would, of course, also be informative to study telomere dynamics and telosome components in embryos reconstructed from senescent cells. However, the relative difficulty of obtaining human oocytes has made such research difficult.

An important alternative to SCNT is exemplified by newer technologies designed to analyze the key components requisite in cellular reprogramming, a field called *analytical reprogramming technology* (ART). Recent studies suggest that the exogenous expression of specific factors such as *MYC, KLF4, OCT4*, and *SOX2* (Takahashi et al. 2007) or *LIN28, NANOG, OCT4*, and *SOX2* (Yu et al. 2007) are capable of transforming differentiated cells into cells similar in phenotype to hES cells. Since these cells did not have their immediate origin with a human embryo, they have been designated induced pluripotent stem (iPS) cells. The iPS protocol has the advantage of being more practical to implement on a high-throughput platform, and since they do not utilize human embryos, they alleviate the ethical concerns of some segments of our society.

While demonstrating the feasibility of reprogramming cells using defined factors, many current iPS protocols present serious challenges for clinical implementation. The use of retroviral transduction of transcription factors results in iPS cells retaining the genetic modifications, and even if such modifications were targeted for removal by recombinases, even a small number of residual genes could perhaps lead to the risk of malignancy. Alternative approaches to reprogramming employ the genetic modification of surrogate cells and the use of those cells, or extracts from those cells, as a replacement of an oocyte and the nuclear transfer protocol. This family of reprogramming technologies is summarized in Fig. 14.6.

In the case of reprogramming by permeabilization in the presence of cytoplasm of undifferentiated cells such as embryonal carcinoma (EC) cells (Taranger et al.

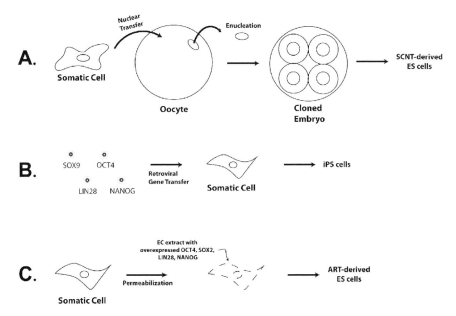

Fig. 14.6 Methods of Generating Patient-Specific Embryonic Stem Cells. Autologous hES cells may be produced by SCNT by enucleation of a mature MII oocyte, followed by the transfer of the donor cell and activation to generate a reconstituted cloned embryo useful in to production of hES cells. Alternatively, analytical reprogramming technolgies offer the possibility of transferring critical factors useful in reprogramming through the use of retroviral vectors to make induced pluripotent stem (iPS) cells, or the preparation of extracts from transformed EC cell lines that overexpress such transcription factors and the introduction of these factors by permeabilization

2005), differentiated cells have been transformed into cells showing undifferentiated markers including *OCT4* and *TERT*. Such reprogramming has been shown to include the reprogramming of DNA methylation and histone modifications in the regulatory regions of *OCT4* and *NANOG* (Freberg et al. 2007). A comparison of iPS and ART protocols will clarify which protocol provides the most reliable modality to reset germ-line telomere length and hES-like gene expression. The characterization of the fundamental differences in telosome composition in the germ-line and soma, such as altered expression of *TERF1, BLM,* and *LMNA*, as well as other components may be useful in improving the restoration of embryonic telomere structure and global gene expression to optimize embryonic progenitors for human therapeutic use (Lillard-Wetherell et al. 2004).

14.6 Prospects for Heterochronic Tissue Regeneration

Assuming it will be practical to generate clinical-grade human embryonic progenitors of diverse types genetically identical to aged patients, it is important to understand whether the engraftment of such cells into aged and diseased tissues would be safe and efficacious.

Early studies in experimental embryology where relatively large segments of embryonic tissue were transplanted from one embryo to another, demonstrated an impressive plasticity in tissue transplantation not observed in the fetus or adult. For example, regions of the hindbrain from the quail could be successfully removed and engrafted in the chick, leading to a living chimera useful in tracking the fate of populations of cells such as the neural crest (Le Douarin 1984). However, the transplantation of embryonic cells or tissue anlagen into various sites of an aged animal with fully differentiated or even diseased tissue (heterochronic transplantation) has received far less attention.

Nevertheless, several age-related degenerative diseases appear to be well suited for such therapy. Studies are currently underway to evaluate the safety and efficacy of the transplantation of hES cell-derived retinal pigment epithelium (RPE) into the subretinal space to ameliorate the progression of age-related macular degeneration into the advanced form of the disease (Klimanskaya et al. 2004). More provocatively, hemangioblasts and myocardial progenitors show promise in the ability to restore contractility to the heart following myocardial dysfunction (Lanza et al. 2004; Wu et al. 2006). The ability to generate neurons such as dopaminergic neurons may prove beneficial in the treatment of age-related neurodegenerative diseases including Parkinson's disease (Roy et al. 2006). Lastly, evidence of cellular senescence in circulating blood cells (Cawthon et al. 2003), suggests that the engraftment of young autologous HSCs may provide benefit to patients at risk of infections or anemia. We will briefly discuss these opportunities in turn.

14.6.1 Macular Degeneration

Age-related macular degeneration (AMD) is the leading cause of blindness in the >60 year-old cohort, and afflicts greater than 30 million people worldwide (Friedman et al. 2004). The juxtaposition of the retinal pigment epithelium to the pathology in both geographic atrophy (dry form) and neovascularization (wet form) of the disease led early researchers to speculate that an age-dependent dysfunction of the RPE cells was an important triggering event in the pathogenesis of the disease (Matsunaga et al. 1999). Histology supports this conclusion with large variant or lost RPE cells localized near the pathology.

The transplantation of young fetal-derived RPE cells into sites of retinal lesions in the rat Royal College of Surgeons (RCS) model of retinal degeneration has demonstrated the potential for the cells to preserve photoreceptors and slow loss of visual function (Coffey et al. 2002; Lund et al. 2001). This led to later efforts to transplant RPE cells into the subretinal space of humans to treat retinal degeneration (da Cruz et al. 2007; Binder et al. 2004). When human ES cell-derived RPE cells were isolated (Klimanskaya et al. 2004) and subsequently tested in the RCS model, similar if not better results were reported (Lund et al. 2006). Work remains, however, in optimizing purification strategies and providing a formulation that promotes the reestablishment of an RPE monolayer in the subretinal space (Haruta 2005). In addition, while the intraocular space is relatively immunoprivileged, the trauma of disease or of transplantation may be enough of a stimulus to trigger a rejection response (Zhang and Bok, 1998), necessitating the use of reprogramming or RCL banking strategies. Lastly, while the RCS rat carries a mutation in the *Mertk* gene leading to dysfunctional RPE cells (D'Cruz et al. 2000) and is therefore a useful model of RPE-based retinal degeneration, it is not macular degeneration per se, and the relative benefits of cell-based therapy as opposed to current pharmacological modalities, will likely only be obvious once human clinical data are in hand.

14.6.2 Neurodegenerative Diseases

Neurodegenerative diseases such as Parkinson's disease and stroke can have devastating consequences to the individual and society. The apparent ability of primitive vertebrates such as the salamander and newt to not only regenerate limbs but also large segments of the brain, eye, and midbrain (Tsonis 2008), suggest that human embryonic progenitors may recapitulate these earlier phylogenetic traits and allow the isolation of cells capable of regenerating even regions of the brain. While the transplantation of fetal-derived neurons has not yet proven an effective means of therapy (McKay and Kittappa 2008), efforts persist to find an in vitro-derived preparation of dopaminergic cells that may prove an effective means of therapy (Barberi et al. 2003). Simpler strategies, such as the transplantation of hES cell-derived myelinating cells into acute spinal cord injuries to improve clinical outcome are already approaching human clinical trials (Coutts and Keirstead 2008).

14.6.3 Vascular, Hematopoietic, Immunological, and Other Disorders

It has long been recognized that the aging of the vascular system is one of the most universal manifestations of aging and carries the highest risk for mortality in the United States (American Heart Association 2008). The vascular endothelium serves as an important regulator of vascular integrity and maintenance, and therefore the aging of the endothelium and strategies to regenerate youthful function is of strategic import. Age-related vascular disorders are characterized by alterations in endothelial maintenance that include an impairment of vascular tone due in part to decreased NO production (Amrani et al. 1996) and increased production of endothelin and angiotensins (Komatsumoto and Nara 1995; Stauffer et al. 2008), likely linked to what is referred to as *endothelial dysfunction* (Minamino and Komuro 2008b; Yavuz et al. 2008). Age-related disease is also characterized by alterations in thrombosis due in part to increased *PAI-1* expression (Yamamoto et al. 2005) and decreased prostacylin (Nakajima et al. 1997; Tokunaga et al. 1991). Interestingly, the senescence of endothelial cells in vitro leads to a striking up-regulation of *PAI-1* (West et al. 1996) and endothelin-1 (Sato et al. 1993) and decrease in prostacyclin (Sato et al. 1993; Neubert et al. 1997). Therefore, there is considerable interest in the role cellular aging may be playing in the aging of the vascular system.

The demonstration that much of the repair of injured endothelium comes not only from the replication of adjacent endothelial cells, but also from bone marrow-derived circulating endothelial progenitor cells (CEPCs), led to two questions. First, do telomeres shorten in the CEPCs with age, and second, can young CEPCs incorporate into the vasculature simply by IV infusion in an old animal?

While there is no consensus as to the molecular markers of CEPCs, commonly used antigens for their isolation and characterization include CD34 and CD133. Age and hypertension were reported to be independent risk factors for low numbers of CD34(+)/CD133(+)/CD45(low) CEPCs (Umemura et al. 2008) and CD34+/CD133+/CD45(low) cells are reported to be reduced in number and colony-forming ability in patients with endothelial dysfunction compared to normals (Boilson et al. 2008).

The inverse relationship between the number and function of CEPCs with coronary disease severity suggests that such assays could have diagnostic value (Balbarini et al. 2007; Spyridopoulos and Dimmeler 2007). Shortened telomeres in peripheral blood leukocytes correlates with a 3.18-fold higher risk of mortality from coronary disease and an 8.54-fold higher risk of death from infectious disease (Cawthon et al. 2003). Furthermore, the generation of hemangioblasts capable of circulating in the blood and participating in either hematopoiesis or vascular differentiation (Lu et al. 2007), offers a novel therapeutic strategy for atherosclerosis and potentially other manifestations of aging such as stroke, peripheral vascular disease (Adams et al. 2007), skin aging (Chung and Eun 2007), kidney disease (Long et al. 2005), liver dysfunction (Le Couteur et al. 2007), and vascular disorders associated with ALS (Zhong et al. 2008).

Initial experiments to test the engraftment of young SCNT-derived hemangioblasts in an old animal showed successful incorporation of the cells into the bone marrow and differentiation into circulating blood cells and incorporation of the cells into aortic endothelium (Lanza et al. 2005). Therefore, hES cell-derived hemangioblasts show promise for delivery via IV infusion, with the potential to target regions of the vascular tree afflicted with age-related vascular dysfunction (Lanza et al. 2005).

Human HSCs produced by reprogramming may offer an attractive vehicle to restore the aged hematopoietic system with young stem cells. These therapies would obviously have usefulness not only in the treatment of age-related immune dysfunction and aplastic anemia, but also for patients in need of a histocompatible HSC graft following chemotherapy. They, like CEPCs, may have the advantage of being deliverable by simple IV infusion. While it is unlikely hematopoietic stem cells are the only cells capable of homing to their natural niche, most other cell types will likely require tissue-specific formulations including, in some applications, the attachment of the cells to matrices and surgical implantation (Chapter 17).

14.7 Conclusion

The continuing elaboration of all of the potential embryonic progenitors that can be propagated in vitro from hES cells and formulated to make effective grafts may provide novel therapies for a wide range of age-related degenerative disease.

The project to isolate human embryonic stem cells began with the anticipation that such primitive cells might display both a totipotency and replicative immortality in vitro. Since the original publication of the propagation of these cells (Thomson et al. 1998), their totipotency has been well-documented, though their replicative immortality must be inferred based on the abundant expression of telomerase and apparent maintenance of 15 kbp TRF length over long-term passaging. As was the case with mES cells, the cultivation of hES cells makes it feasible to perform genetic modifications (including gene targeting), and subsequently generate virtually all somatic cell types early in their developmental and replicative lifespan, on a scale practical for use in transplantation.

Current studies of the molecular biology of the inherited genetic diseases modeling premature human aging, Werner syndrome and Hutchinson-Gilford syndrome, both point to the pathogenesis of a wide array of age-related pathologies such as atherosclerosis, skin aging, osteoporosis, type II diabetes, as being ultimately rooted in telomere dysfunction. Since telomere extension via exogenous telomerase expression rescues these cells in vitro, and extends the lifespan of mice, there is a reasonable basis to conclude that the resetting of telomere length during reprogramming may effectively repair age-related telomere damage, as appears to be the case in animals cloned from old donors.

Reprogramming technologies, such as iPS and ART, allow the facile resetting of pluripotency and youthful telomere length in aged somatic cells. In the future this could allow health care professionals to create young embryonic progenitors that are genetically identical to the patient and potentially useful in regenerating a host of cells and tissues afflicted with age-related degenerative disease. While such reprogramming would not remove somatic mutations, affordable ultrahigh throughput sequencing of the human genome is now a reality, and therefore it is easy to imagine that cells with a healthy genome could be identified from an aged individual and could be utilized in therapy. This approach would mirror the manner in which natural selection perpetuates the human species by favoring those gametes with healthy genomes.

In summary, by following nature's lessons relating to the molecular basis of the immortal perpetuation of the human species, useful insights into the design of new therapies for the greatest medical challenge of our time may one day become a reality.

References

Adams B, Xiao Q, Xu Q (2007) Stem cell therapy for vascular diseases. Trends Cardiovasc Med 17:246–251

Allshire RC, Gosden JR, Cross SH, Cranston G, Rout D, Sugawara N, Szostak JW, Fantes PA, Hastie ND (1988) Telomeric repeat from T thermophila cross hybridizes with human telomeres. Nature 332:656–659

Allshire RC, Dempster M, Hastie ND (1989) Human telomeres contain at least three types of G-rich repeat distributed non-randomly. Nucleic Acids Res 17:4611–4627

American Heart Association (2008) Heart disease and stroke statistics–2006 update A report from the American Heart Association Statistics Committee and Stroke Statistics Subcommittee. Circulation 113: e85–e151

Amrani M, Goodwin AT, Gray CC, Yacoub MH (1996) Ageing is associated with reduced basal and stimulated release of nitric oxide by the coronary endothelium Acta Physiol Scand 157:79–84

Artandi SE Alson S, Teitze MK, Sharpless NE, Ye S, Greenberg RA, Castrillon DH, Horner JW, Weiler SR, Carrasco RD, DePinho RA (2002) Constitutive telomerase expression promotes mammary carcinomas in aging mice. Proc Natl Acad Sci USA 99:8191–8196

Bai Y, Murnane JP (2003) Telomere instability in a human tumor cell line expressing a dominant-negative *WRN* protein. Hum Genet 113:337–347

Balbarini A, Barsotti MC, Di Stefano R, Leone A, Santoni T (2007) Circulating endothelial progenitor cells characterization, function, and relationship with cardiovascular risk factors. Curr Pharm Des 13:1699–1713

Bandyopadhyay D, Gatza C, Donehower LA, Medrano EE (2005) Analysis of cellular senescence in culture in vivo: the senescence-associated beta-galactosidase assay. Curr Protoc Cell Biol Chapter 18 :Unit 189

Barberi T, Klivenyi P, Calingasan NY, Lee H, Kawamata H, Loonam K, Perrier AL, Bruses J, Rubio ME, Topf N, Tabar V, Harrison NL, Beal MF, Moore MA, Studer L (2003) Neural subtype specification of fertilization and nuclear transfer embryonic stem cells and application in parkinsonian mice. Nat Biotechnol 21:1200–1207

Binder S, Krebs I, Hilgers RD, Abri A, Stolba U, Assadoulina A, Kellner L, Stanzel BV, Jahn C, Feichtinger H (2004) Outcome of transplantation of autologous retinal pigment epithelium in age-related macular degeneration: a prospective trial. Invest Ophthalmol Vis Sci 45:4151–4160

Bodnar AG, Ouellette M, Frolkis M, Holt SE, Chiu C-P, Morin GB, Harley CB, Shay JW, Lichtsteiner S, Wright WE (1998) Extension of cell life-span by introduction of telomerase into normal human cells. Science 279:349–352

Boilson BA, Kiernan TJ, Harbuzariu A, Nelson RE, Lerman A, Simari RD (2008) Circulating CD34+ cell subsets in patients with coronary endothelial dysfunction. Nat Clin Pract Cardiovasc Med 5:489–496

Brevini TA, Antonini S, Pennarossa G, Gandolfi F (2008) Recent progress in embryonic stem cell research and its application in domestic species. Reprod Domest Anim 43(Suppl 2):193–199

Butler, RN (2008) The longevity revolution: the benefits and challenges of living a long life. Publlic Affairs, New York

Canela A, Martin-Caballero J, Flores JM, Blasco MA (2004) Constitutive expression of *Tert* in thymocytes leads to increased incidence and dissemination of T-cell lymphoma in *Lck-Tert* Mice. Mol Cell Biol 24:4275–4293

Canela A, Vera E, Klatt P, Blasco MA (2007) High-thoughput telomere length quantification by FISH and its application to human population studies. Proc Natl Acad Sci USA 104:5300–5305

Carrel A (1912) On the permanent life of tissues outside of the organism. J Exp Med 15:516–527

Cawthon RM, Smith KR, O'Brien E, Sivatchenko A, Kerber RA (2003) Association between telomere length in blood and mortality in people aged 60 years or older. Lancet 361:393–395

Chang S, Multani AS, Cabrera NG, Naylor ML, Laud P, Lombard D, Pathak S, Guarente L, DePinho RA (2004) Essential role of limiting telomeres in the pathogenesis of Werner syndrome. Nat Genet 36:877–882

Chiu C-P, Dragowska W, Kim NW, Vaziri H, Yui J, Thomas TE, Harley CB, Lansdorp, PM (1996) Differential expression of telomerase activity in hematopoietic progenitors from adult human bone marrow. Stem Cells 14:239–248

Chung JH and Eun HC (2007) Angiogenesis in skin aging and photoaging. J Dermatol 34:593–600

Cibelli JB, Kiessling AA, Cunniff K, Richards C, Lanza RP, West MD (2001) Somatic cell nuclear transfer in humans: Pronuclear and early embryonic development. e-biomed: J Regen Med 2:25–31

Cibelli JB, Lanza RP, Campbell KHS, West MD (2002a) Principles of Cloning. Academic Press, New York

Cibelli JB, Grant KA, Chapman KB, Cunniff K, Worst T, Green HL, Walker SJ, Gutin PH, Vilner L, Tabar V, Dominko T, Kane J, Wettstein PJ, Lanza RP, Studer L, Vrana KE, West MD (2002b) Parthenogenetic stem cells in nonhuman primates. Science 295:819

Clark AJ, Ferrier P, Aslam S, Burl S, Denning C, Wylie D, Ross A, de Sousa P, Wilmut I, Cui W (2003) Proliferative lifespan is conserved after nuclear transfer. Nat Cell Biol 5(6):535–538

Coffey PJ, Girman S, Wang SM, Hetherington L, Keegan DJ, Adamson P, Greenwood J, Lund RD (2002) Long-term preservation of cortically dependent visual function in RCS rats by transplantation. Nat Neurosci 5:53–56

Constantinescu D, Gray HL, Sammak PJ, Schatten GP, Csoka AB (2006) Lamin A/C expression is a marker of mouse and human embryonic stem cell differentiation. Stem Cells 24(1):177–185

Cooke HJ, Smith BA (1986) Variability at the telomeres of the human X/Y pseudoautosomal region. Cold Spring Harb Symp Quant Biol 51 Pt1:213–219

Counter CM, Avilion AA, LeFeuvre CE, Stewart NG, Greider CW, Harley CB, Bacchetti S (1992) Telomere shortening associated with chromosome instability is arrested in immortal cells which express telomerase activity. EMBO J 11:1921–1929

Counter CM, Botelho FM, Wang P, Harley CB, Bacchetti S (1994) Stabilization of short telomeres and telomerase activity accompany immortalization of Epstein-Barr virus-transformed human B lymphocytes. J Virol 68:3410–3414

Coutts M, Keirstead HS. (2008) Stem cells for the treatment of spinal cord injury. Exp Neurol 209(2):368–77

Crabbe L, Verdun RE, Haggblom CI, Karlseder J (2004) Defective telomere lagging strand synthesis in cells lacking *WRN* helicase activity. Science 306:1951–1953

Crabbe L, Jauch A, Naeger CM, Holtgreve-Grez H, Karlseder J (2007) Telomere dysfunction as a cause of genomic instability in Werner syndrome. Proc Natl Acad Sci USA 104:2205–2210

Cristofalo VJ (2005) SA beta Gal staining: biomarker or delusion. Exp Gerontol 40:836–838

Cristofalo VJ, Kabakjian J (1975) Lysosomal enzymes and aging in vitro: subcellular enzyme distribution and effect of hydrocortisone on cell life-span. Mech Ageing Dev 4:19–28

Cristofalo VJ, Allen RG, Pignolo RJ, Martin BG, Beck JC (1998) Relationship between donor age and the replicative lifespan of human cells in culture: a reevaluation. Proc Natl Acad Sci USA 95:10614–10619

D'Adda Di Fagagna F, Reaper PM, Clay-Farrace L, Fiegler H, Carr P, Von Zglinicki T, Saretzki G, Carter NP, Jackson SP (2003) A DNA damage checkpoint response in telomere-initiated senescence. Nature 426:194–198

D'Amour KA, Bang AG, Eliazer S, Kelly OG, Agulnick AD, Smart NG, Moorman MA, Kroon E, Carpenter MK, Baetge EE (2006) Production of pancreatic hormone-expressing endocrine cells from human embryonic stem cells. Nat Biotech 24:1392–1401

da Cruz L, Chen FK, Ahmado A, Greenwood J, Coffey P (2007) RPE transplantation and its role in retinal disease. Prog Retin Eye Res 26:598–635

D'Cruz PM, Yasumura D, Weir J, Matthes MT, Abderrahim H, LaVail MM, Vollrath D (2000) Mutation of the receptor tyrosine kinase gene *Mertk* in the retinal dystrophic RCS rat. Hum Mol Genet 9:645–651

Da Silva MA, Yamada N, Clarke NM, Roach HI (2008) Cellular and epigenetic features of a young healthy and a young osteoarthritic cartilage compared with aged control and OA cartilage. J Orthop Res Nov 4 [Epub ahead of print]

DeCaprio JA, Ludlow JW, Figge J, Shew J-Y, Huang C-M, Lee W-H, Marsilio E, Paucha E, Livingston DM (1988) SV40 large tumor antigen forms a specific complex with the product of the retinoblastoma susceptibility gene. Cell 54:275–283 de Lange T (2005) Shelterin: the protein complex that shapes and safeguards human telomeres. Genes Dev 19:2100–2110

Dell'Orco RT, Mertens JG, Kruse PF (1973) Doubling potential, calendar time, and senescence of human diploid cells in culture. Exp Cell Res 77(1):36–360

Di Micco R, Cicalese A, Fumagalli M, Dobreva M, Verrecchia A, Pelicci PG, di Fagagna F (2008) DNA damage response activation in mouse embryonic fibroblasts undergoing replicative senescence and following spontaneous immortalization. Cell Cycle 7:3601–3606

Ding SL, Shen CY (2008) Model of human aging: recent findings on Werner's and Hutchinson-Gilford progeria syndromes. Clin Interv Aging 3:431–444

Draper JS, Smith K, Gokhale P, Moore HD, Maltby E, Johnson J, Meisner L, Zwaka TP, Thomson JA, Andrews PW (2004) Recurrent gain of chromosomes 17q and 12 in cultured human embryonic stem cells. Nat Biotechnol 22:53–54

Effros RB, Allsopp R, Chiu C-P, Hausner MA, Hirji K, Wang L, Harley CB, Villeponteau B, West MD, Giorgi JV (1996) Shortened telomeres in the expanded CD28- CD8+ cell subset in HIV disease implicate replicative senescence in HIV pathogenesis. AIDS 10:F17–F22

Eller MS, Liao X, Liu S, Hanna K, Bäckvall H, Opresko PL, Bohr VA, Gilchrest BA (2006) A role for *WRN* in telomere-based DNA damage responses. Proc Natl Acad Sci USA 103:15073–15078

Epstein CJ, Martin GM, Schultz AI, Motulsky AG (1966) Werner's syndrome: a review of its symptomatology, natural history, pathologic features, genetics and relationship to the natural aging process. Medicine (Baltimore) 45:177–221

Evans MJ, Kaufman MH (1981) Establishment in culture of pluripotential cells from mouse embryos. Nature 292:154–156

Evans MJ, Gurer C, Loike JD, Wilmut I, Schnieke AE, Schon EA (1999) Mitochondrial DNA genotypes in nuclear transfer-derived cloned sheep. Nat Genet 23:90–93

Fauce SR, Jamieson BD, Chin AC, Mitsuyasu RT, Parish ST, Ng HL, Kitchen CM, Yang OO, Harley CB, Effros RB (2008) Telomerase-based pharmacologic enhancement of antiviral function of human CD8+ T lymphocytes. J Immunol 181:7400–7406

Feng J, Funk WD, Wang S-S, Weinrich SL, Avilion AA, Chiu C-P, Adams R, Chang E, Allsopp RC, Siyuan Le J-Y, West MD, Harley CB, Andrews WH, Greider CW, Villeponteau BV (1995) The RNA Component of Human Telomerase. Science 269:1236–1241

Forsyth NR, Evans AP, Shay JW, Wright WE (2003) Developmental differences in the immortalization of lung fibroblasts by telomerase. Aging Cell 2:235–243

Freberg CT, Dahl JA, Timoskainen S, Collas P (2007) Epigenetic reprogramming of *OCT4* and *NANOG* regulatory regions by embryonal carcinoma cell extract. Mol Biol Cell 18:1543–1553

Friedman DM (2007) The Immortalists. HarperCollins Publishers, New York

Friedman DS, O'Colmain BJ, Munoz B, Tomany SC, McCarty C, de Jong PT, Nemesure B, Mitchell P, Kempen J (2004) Prevalence of age-related macular degeneration in the United States. Arch Ophthalmol 122:564–572

Gallimore PH, Lecane PS, Roberts S, Rookes SM, Grand RJ, Parkhill J (1997) Adenovirus type 12 early region 1 B 54 K protein significantly extends the life span of normal mammalian cells in culture. J Virol 71:6629–6640

Garcia-Cao I, Garcia-Cao M, Martin-Caballero J, Criado LM, Klatt P, Flores JM, Weill JC, Blasco MA, Serrano M (2002) "Super p53" mice exhibit enhanced DNA damage response, are tumor resistant and age normally. EMBO J 21:6225–6235

Genovese J, Cortes-Morichetti M, Chachques E, Frati G, Patel A, Chachques JC (2007) Cell based approaches for myocardial regeneration and artificial myocardium. Curr Stem Cell Res Ther 2:121–127

Girardi AJ, Jensen FC, Koprowski H (1965) SV40-induced transformation of human diploid cells: Crisis and recovery. J Cell Physiol 65:69–83

Gonzales-Suarez E, Samper E, Flores JM, Blasco MA (2000) Telomerase-deficient mice with short telomeres are resistant to skin tumorigenesis. Nat Genet 26:114–117

Gonzales-Suarez E, Samper E, Ramirez A, Flores JM, Martin-Caballero J, Jorcano JL, Blasco MA (2001) Increased epidermal tumors and increased skin wound healing in transgenic mice overexpressing the catalytic subunit of telomerase, *mTERT*, in basal keratinocytes. EMBO J 20:2619–2630

Green MR (1989) When the products of oncogenes and anti-oncogenes meet. Cell 56:1–3

Greider CW, Blackburn EH (1985) Identification of a specific telomere terminal transferase activity in Tetrahymena extracts. Cell 43:405–413

Greider CW, Blackburn EH (1987) The telomere terminal transferase of Tetrahymena is a ribonucleoprotein enzyme with two kinds of primer specificity. Cell 51:887–898

Greider CW, Blackburn EH (1989) A telomeric sequence in the RNA of Tetrahymena telomerase required for telomere repeat synthesis. Nature 337:331–337

Griffith JD, Comeau L, Rosenfield S, Stansel RM, Bianchi A, Moss H, de Lange T (1999) Mammalian telomeres end in a large duplex loop. Cell 97:503–514

Halliday TR, Adler K (eds) (1986) Reptiles & Amphibians. Torstar Books, p. 101

Harley CB, Kim NW, Prowse KR, Weinrich SL, Hirsch KS, West MD, Bacchetti S, Hirte HW, Counter CM, Greider CW, Piatyszek MA, Wright WE, Shay JW (1994) Telomerase, Cell Immortality, and Cancer. Cold Spring Harbor Symposium on Quantitative Biology LIX:307–315 Cold Spring Harbor Press, Cold Spring Harbor, NY

Harley, CB, Futcher, AB, Greider, CW (1990) Telomeres shorten during ageing of human fibroblasts. Nature 345:458–460

Harrison DE, Astle CM, Delaittre JA (1978) Loss of proliferative capacity in immunohemopoietic stem cells caused by serial transplantation rather than aging. J Exp Med 147:1526–1531

Haruta M (2005) Embryonic stem cells: potential source for ocular repair. Semin Ophthalmol 20:17–23

Hasenmaile S, Pawelec G, Wagner W (2003) A lack of telomeric non-reciprocal recombination (TENOR) may account for the premature proliferation blockade of Werner's syndrome fibroblasts. Biogerontology 4:253–273

Hastie ND, Dempster M, Dunlop MG, Thompson AM, Green DK, Allshire RC (1990) Telomere reduction in human colorectal carcinoma and with ageing. Nature 346:866–868

Hayflick L (1965) The limited in vitro lifetime of human diploid cell strains. Exp Cell Res 37:614–636

Hayflick L (1968) Human cells and aging. Sci Am 218:32–37

Hayflick L (1992) Aging, longevity, and immortality in vitro. Exp Gerontol 27:363–368

Hayflick L, Moorhead PS (1961) The serial cultivation of human diploid cell strains. Exp Cell Res 25:585–621

Henderson ER, Blackburn EH (1989) An overhanging 3' terminus is a conserved feature of telomeres. Mol Cell Biol 9:345–348

Horikawa I, Chiang YJ, Patterson T, Feigenbaum L, Leem S-H, Michishita E, Larionov V, Hodes RJ, Barrett JC (2005) Differential cis-regulation of human versus mouse *TERT* gene expression in vivo: Identification of a human-specific repressive element. Proc Natl Acad Sci USA 102:18437–18442

Itahana K, Campisi J, Dimri, GP (2007) Methods to detect biomarkers of cellular senescence: the senescence-associated beta-galactosidase assay. Methods Mol Biol 371:21–31

Jeyapalan JC, Ferreira M Sedivy JM Herbig U (2007) Accumulation of senescent cells in mitotic tissue of aging primates. Mech Ageing Dev 128:36–44

Khanna KK, Jackson SP (2001) DNA double-strand breaks: signaling, repair and the cancer connection Nat Genet 27:247–254

Kim NW, Piatyszek MA, Prowse KR, Harley CB, West MD, Ho PLC, Coviello GM, Wright WE, Weinrich SL, Shay JW (1994) Specific association of human telomerase activity with immortal cells and cancer. Science 266:2011–2014

Klimanskaya I, Hipp J, Rezai KA, West M, Atala A, Lanza R (2004) Derivation and comparative assessment of retinal pigment epithelium from human embryonic stem cells using transcriptomics. Cloning Stem Cells 6(3):217–245

Komatsumoto S, and Nara M (1995) Changes in the level of endothelin-1 with aging. Nippon Ronen Igakkai Zasshi 32:664–669

Kudlow BA, Kennedy BK, Monnat RJ Jr (2007) Werner and Hutchinson-Gilford progeria syndromes: mechanistic basis of human progeroid diseases. Nat Rev Mol Cell Biol 8:394–404

Lanza RP, Cibelli JB, West MD (1999a) Human therapeutic cloning. Nat Med 5(9):975–977

Lanza RP, Cibelli JB, West MD (1999b) Prospects for the use of nuclear transfer in human transplantation. Nat Biotech 17(12):1171–1174

Lanza RP, Cibelli JB, Blackwell C, Cristofalo VJ, Francis MK, Baerlocher GM, Mak J, Schertzer M, Chavez EE, Sawyer N, Lansdorp PM, West MD (2000) Extension of cell life-span and telomere length in animals cloned from senescent somatic cells. Science 288:665–669

Lanza RP, Cibelli JB, West MD, Dorff E, Tauer C, Green RM (2001) The ethical reasons for stem cell research. Science 292(5520):1299b

Lanza RP, Chung HY, Yoo JJ, Wettstein PJ, Blackwell C, Borson N, Hofmeister E, Schuch G, Soker S, Moraes CT, West MD, Atala A (2002) Generation of histocompatible tissues using nuclear transplantation. Nat Biotech 20(7):689–696

Lanza R, Moore MA, Wakayama T, Perry AC, Shieh J-H, Hendrikx J, Leri A, Chimenti S, Monsen A, Nurzynska D, West MD, Kajstura J, Anversa P (2004) Regeneration of infarcted heart with stem cells derived by nuclear transplantation. Circ Res 94(6):820–827

Lanza R, Shieh J-H, Wettstein PJ, Sweeney RW, Wu K, Weisz A, Borson N, Henderson B, West MD, Moore MAS (2005) Long-term bovine hematopoietic engraftment with clone-derived stem cells. Cloning Stem Cells 7(2):95–106

Le Couteur DG, Cogger VC, McCuskey RS, DE Cabo R, Smedsrød B, Sorensen KK, Warren A, Fraser R (2007) Age-related changes in the liver sinusoidal endothelium: a mechanism for dyslipidemia. Ann NY Acad Sci 1114:79–87

Le Douarin NM (1984) Ontogeny of the peripheral nervous system from the neural crest and the placodes: a developmental model studied on the basis of the quail-chick chimaera system. Harvey Lect 1984–1985;80:137–186

Le Maitre CL, Freemont AJ, Hoyland JA (2007) Accelerated cellular senescence in degenerate intervertebral discs: a possible role in the pathogenesis of intervertebral disc degeneration. Arthritis Res Ther 9:R45

Lee HW, Blasco MA, Gottlieb GJ, Horner JW 2nd, Greider CW, DePinho RA (1998) Essential role of mouse telomerase in highly proliferative organs. Nature 392:569–574

Li B, Jog SP, Reddy S, Comai L (2008) *WRN* controls formation of extrachromosomal telomeric circles and is required for *TRF2DeltaB*-mediated telomere shortening. Mol Cell Biol 28:1892–1904

Lillard-Wetherell K, Machwe A, Langland GT, Combs KA, Behbehani GK, Schonberg SA, German J, Turchi JJ, Orren DK, Groden J (2004) Association and regulation of the *BLM* helicase by the telomeric proteins TRF1 and TRF2. Hum Mol Genet 13:1919–1932

Liu D, O'Connor MS, Qin J, Songyang, Z (2004) Telosome, a mammalian telomere-associated complex formed by multiple telomeric proteins. J Biol Chem 279:51338–51342

Liu L, Bailey SM, Okuka M, Muñoz P, Li C, Zhou L, Wu C, Czerwiec E, Sandler L, Seyfang A, Blasco MA, Keefe DL (2007) Telomere lengthening early in development. Nat Cell Biol 9:1436–1441

Liu Y, Wang Y, Rusinol AE, Sinensky MS, Liu J, Shell SM, Zou Y (2008) Involvement of xeroderma pigmentosum group A *(XPA)* in progeria arising from defective maturation or prelamin A. FASEB J 22:603–611

Lombard DB, Beard C, Johnson B, Marciniak RA, Dausman J, Bronson R, Buhlmann JE, Lipman R, Curry R, Sharpe A, Jaenisch R, Guarente L (2000) Mutations in the *WRN* gene in mice accelerate mortality in a p53-null background. Mol Cell Biol 20:3286–3291

Long DA, Mu W, Price KL, Johnson RJ (2005) Blood vessels and the aging kidney. Nephron Exp Nephrol 101:e95–e99

Lu SJ, Feng Q, Caballero S, Chen Y, Moore MA, Grant MB, Lanza, R (2007) Generation of functional hemangioblasts from human embryonic stem cells. Nat Methods 4:501–509

Lund RD, Adamson P, Sauve Y, Keegan DJ, Girman SV, Wang S, Winton H, Kanuga N, Kwan AS, Beauchene L, Zerbib A, Hetherington L, Couraud PO, Coffey P, Greenwood J (2001) Sub-retinal transplantation of genetically modified human cell lines attenuates loss of visual function in dystrophic rats. Proc Natl Acad Sci USA 98:9942–9947

Lund RD, Wang S, Klimanskaya I, Holmes T Ramos-Kelsey R, Lu B, Girman S, Bischoff N, Sauve Y, Lanza R (2006) Human embryonic stem cell-derived cells rescue visual function in dystrophic RCS rats. Cloning Stem Cells 8:189–199

Lundblad V, Szostak JW (1989) A mutant with a defect in telomere elongation leads to senescence in yeast. Cell 57:633–643

Machwe A, Xiao L, Orren DK (2004) *TRF2* recruits the Werner syndrome (*WRN*) exonuclease for processing of telomeric DNA. Oncogene 23:149–156

Martin GM (1981) Isolation of pluripotent cell line from early mouse embryos cultured in medium conditioned by teratocarcinoma stem cells. Proc Natl Acad Sci USA 78:7634–7638

Matheu A, Pantoja C, Efeyan A, Criado LM, Martin-Caballero J, Flores JM, Klatt P, Serrano M (2004) Increased gene dosage of *Ink4a/Arf* results in cancer resistance and normal aging. Genes Dev 18:2736–2746

Matheu A, Maraver A, Klatt P, Flores I, Garcia-Cao I, Borras C, Flores JM, Viña J, Blasco MA, Serrano M (2007) Delayed aging through damage protection by the *Arf/p53* pathway. Nature 448:375–379

Matsunaga H, Handa JT, Aotaki-Keen A, Sherwood SW, West MD, Hjelmeland LM (1999) Beta-galactosidase histostochemistry and telomere loss in senescent retinal pigment epithelial cells. Invest Ophthalmol Vis Sci 40:197–202

McKay R, Kittappa R (2008) Will stem cell biology generate new therapies for Parkinson's disease? Neuron 58:659–661

McLaren A (1992) Embryology. The quest for immortality. Nature 359:482–483

McLaren A (2001) Mammalian germ cells: birth, sex, and immortality. Cell Struct Funct 26:119–122

McLeod DS, Taomoto M, Otsuji T, Green WR, Sunness JS, Lutty GA (2002) Quantifying changes in RPE and choroidal vasulature in eyes with age-related macular degeneration. Invest Ophthalmol Vis Sci 43:1986–1993

Meier A, Fiegler H, Muñoz P, Ellis P, Rigler D, Langford C, Blasco MA, Carter N, Jackson SP (2007) Spreading of mammalian DNA-damage response factors studied by ChIP-chip at damaged telomeres. EMBO J 26:2707–2718

Minamino T, and Komuro I (2008a) Role of telomeres in vascular senescence. Front Biosci 13:2971–2979

Minamino T, Komuro I (2008b) Vascular aging: insights from studies on cellular senescence, stem cell aging, and progeroid syndromes. Nat Clin Pract Cardiovasc Med 5:637–648

Mogford JE, Liu WR, Reid R, Chiu C-P, Said H, Chen SJ, Harley CB, Mustoe TA (2006) Adenoviral human telomerase reverse transcriptase dramatically improves ischemic wound healing without detrimental immune response in an aged rabbit model. Hum Gene Ther 17:651–660

Morgan R, Jarzabek V, Jaffe JP, Hecht BK, Hecht F, Sandberg AA (1986) Telomeric fusion in pre-T-cell acute lymphoblastic leukemia. Hum Genet 73:260–263

Morin GB (1989) The human telomere terminal transferase enzyme is a ribonucleoprotein that synthesizes TTAGGG repeats. Cell 59:521–529

Moyzis RK, Buckingham JM, Cram LS, Dani M, Deaven LL, Jones MD, Meyne J, Ratliff RL, Wu JR (1988) A highly conserved repetitive DNA sequence, (TTAGGG)n, present at the telomeres of human chromosomes. Proc Natl Acad Sci USA 85:6622–6626

Muntoni A, Neumann AA, Hills M, Reddel RR (2008) Telomere elongation involves intra-molecular DNA replication in cells utilizing Alternative Lengthening of Telomeres. Hum Mol Genet 2008 Dec 18 [Epub ahead of print]

Nagy A, Gertsenstein M, Vintersten K, Behringer R (eds) (2003) Manipulating the mouse embryo: a laboratory manual. Cold Spring Harbor Laboratory Press, Cold Spring Harbor, NY

Nakajima M, Hashimoto M, Wang F, Yamanaga K, Nakamura N, Uchida T, Yamanouchi K (1997) Aging decreases the production of PGI2 in rat aortic endothelial cells. Exp Gerontol 32:685–693

Nakamura AJ, Chiang YJ, Hathcock KS, Horikawa I, Sedelnikova OA, Hodes RJ, Bonner WM (2008) Both telomeric and non-telomeric DNA damage are determinants of mammalian cellular senescence. Epigenetics Chromatin 1:6

Nakamura TM, Morin GB, Chapman KB, Weinrich SL, Andrews WH, Lingner J, Harley CB, Cech TR (1997) Telomerase catalytic subunit homologs from fission yeast and human. Science 277:955–959

Neubert K, Haberland A, Kruse I, Wirth M, Schimke I (1997) The ratio of formation of prostacyclin/thromboxane A2 in HUVEC decreased in each subsequent passage. Prostaglandins 54:447–462

Niida H, Matsumoto T, Satoh H, Shiwa M, Tokutake Y, Furuichi Y, Shinkai, Y (1998) Severe growth defect in mouse cells lacking the telomerase RNA component. Nat Genet 19: 203–206

Norrback KF, Hultdin M, Dahlenborg K, Osterman P, Carlsson R, Roos G (2001) Telomerase regulation and telomere dynamics in germinal centers. Eur J Haematol 67:309–317

Oikawa S, Kawanishi S (1999) Site-specific DNA damage at GGG sequence by oxidative stress may accelerate telomere shortening. FEBS Lett 453:365–368

Olovnikov AM (1971) Principles of marginotomy in template synthesis of polynucleotides. Doklady Akad Nauk SSSR 201:1496–1499

Orren DK, Theodore S, Machwe A (2002) The Werner syndrome helicase/exonuclease (*WRN*) disrupts and degrades D-loops in vitro. Biochemistry 41:13483–13488

Prowse KR, Greider CW (1995) Developmental and tissue-specific regulation of mouse telomerase and telomere length. Proc Natl Acad Sci USA 92(11):4818–4822

Rathjen J, Lake JA, Bettess MD, Washington JM, Chapman G, Rathjen PD (1999) Formation of a primitive ectoderm like cell population, EPL cells, from ES cells in response to biologically derived factors. J Cell Sci 112(Pt 5):601–612

Ray FA, Meyne J, Kraemer PM (1992) SV40 T antigen induced chromosomal changes reflect a process that is both clastogenic and aneuploidogenic and is ongoing throughout neoplastic progression of human fibroblasts. Mutat Res 284:265–273

Rea MA, Zhou L, Qin Q, Barrandon Y, Easley KW, Gungner SF, Phillips MA, Holland WS, Gumerlock PH, Rocke DM, Rice RH (2006) Spontaneous immortalization of human epidermal cells with naturally elevated telomerase. J Invest Dermatol 126:2507–2515

Robertson E, Bradley A, Kuehn M, Evans M. (1986) Germ-line transmission of genes into cultured pluripotent cells by retroviral vector. Nature 323(6087):445–448

Roy NS, Cleren C, Singh SK, Yang L, Beal MF, Goldman SA (2006) Functional emgraftment of human ES cell-derived dopaminergic neurons enriched by coculture with telomerase-immortalized midbrain astrocytes. Nat Med 12:1259–1268

Rufer N, Brummendorf TH, Kolvraa S, Bischoff C, Christensen K, Wadsworth L, Schulzer M, Lansdorp PM (1999) Telomere fluorescence measurements in granulocytes and T lymphocyte subsets point to a high turnover of hematopoietic stem cells and memory T cells in early childhood. J Exp Med 190:157–167

Sato I, Kaji K, Morita I, Nagao M, Murota S (1993) Augmentation of endothelin-1, prostacyclin and thromboxane A2 secretion associated with in vitro ageing in cultured human embilical vein endothelial cells. Mech Ageing Dev 71:73–84

Scheidtmann KH, Haber A (1990) Simian virus 40 T antigen induces or activates a protein kinase which phosphorylates the transformation-associated protein p53. J Virol 64:672–679

Schneider EL, Mitsui Y (1976) The relationship between in vitro cellular aging and in vivo human age. Proc Natl Acad Sci USA 73:3584–3588

Shamblott MJ, Axelman J, Wang S, Bugg EM, Littlefield JW, Donovan PJ, Blumenthal PD, Huggins GR, Gearhart JD (1998) Derivation of pluripotent stem cells from cultured human primordial germ cells. Differentiation of immortal cells inhibits telomerase activity. Proc Natl Acad Sci USA 95:13726–13731

Sharma HW, Sokoloski JA, Perez JR, Maltese JY, Sartorelli AC, Stein CA, Nichols G, Khaled Z, Telang NT, Narayanan R (1995) Differentiation of immortal cells inhibits telomerase activity. Proc Natl Acad Sci USA 92:12343–12346

Shay JW, Wright WE (1989) Quantitation of the frequency of immortalization of normal human diploid fibroblasts by SV40 large T-antigen. Exp Cell Res 184:109–118

Shay JW, West MD, Wright WE (1992) Re-expression of senescent markers in de-induced reversibly immortalized cells. Exp Gerontol 27:477–492

Shay JW, Wright WE, Werbin H (1993) Toward a molecular understanding of human breast cancer: a hypothesis. Breast Cancer Res Treat 25:83–94

Shiels PG, Kind AJ, Campbell KH, Waddington D, Wilmut I, Colman A, Schnieke AE (1999) Analysis of telomere lengths in cloned sheep. Nature 399(6734):316–317

Shiga T, Shirasawa H, Shimizu K, Dezawa M, Masuda Y, Simizu B (1997) Normal human fibroblasts immortalized by introduction of human papillomavirus type 16 (HPV-16) E6-E7 genes. Microbiol Immunol 41:313–319

Slagboom PE, Droog S, Boomsma DI (1994) Genetic determination of telomere size in humans: A twin study of three age groups. Am J Hum Genet 55:876–882

Smith JR, Venable S, Roberts TW, Metter EJ, Monticone R, Schneider EL (2002) Relationship between in vivo age and in vitro aging: assessment of 669 cell cultures derived from members of the Baltimore Longitudinal Study of Aging. J Gerontol A Biol Sci Med Sci 57:B239–246

Spyridopoulos I, Dimmeler S (2007) Can telomere length predict cardiovascular risk? Lancet 365:81–82

Stauffer BL, Westby CM, DeSouza CA (2008) Endothelin-1, aging, and hypertension. Curr Opin Cardiol 23:350–355

Stevens LC. (1980) Teratocarcinogenesis and spontaneous parthenogenesis in mice. Results Probl Cell Differ 11:265–274.

Suda Y, Suzuki M, Ikawa Y, Aizawa S (1987) Mouse embryonic stem cells exhibit indefinite proliferative potential. J Cell Physiol 133:197–201

Takahashi K, Tanabe K, Ohnuki M, Narita M, Ichisaka T, Tomoda K, Yamanaka S (2007) Induction of pluripotent stem cells from adult human fibroblasts by defined factors. Cell 131:861–872

Takai H, Smogorzewska A, de Lange T (2003) DNA damage foci at dysfunctional telomeres. Curr Biol 13:1549–1556

Taranger CK, Noer A, Sørensen AL, Håkelien AM, Boquest AC, Collas P (2005) Induction of dedifferentiation, genomewide transcriptional programming, and epigenetic reprogramming by extracts of carcinoma and embryonic stem cells. Mol Biol Cell 16:5719–5735

Terry DF, Nolan VG, Andersen SL, Perls TT, Cawthon R (2008) Association of longer telomeres with better health in centenarians. J Gerontol A Biol Sci Med Sci 63:809–812

Thomson JA, Itskovitx-Eldor J, Shapiro SS, Waknitz MA, Swiergierl JJ, Marshall VS, Jones JM (1998) Embryonic stem cell lines derived from human blastocysts. Science 282:1145–1147

Tokunaga O, Yamada T, Fan JL, Watanabe T (1991) Age-related decline in prostacyclin synthesis by human aortic endothelial cells: Qualitative and quantitative analysis. Am J Pathol 138:941–949

Tomas-Loba A, Flores I Fernandez-Marcos PJ, Cayuela ML, Maraver A, Tejera A, Borras C, Matheu A, Klatt P, Flores JM, Vina J, Serrano M, Blasco MA (2008) Telomerase reverse transcriptase delays aging in cancer-resistant mice. Cell 135:609–622

Toulouse A, Sullivan AM (2008) Progress in parkinson's disease – where do we stand? Prog Neurobiol Jun 5 [Epub ahead of print]

Toussaint O, Medrano EE, von Zglinicki T (2000) Cellular and molecular mechanisms of stress-induced premature senescence (SIPS) of human diploid fibroblasts and melanocytes. Exp Gerontol 35:927–945

Tsonis PA. (2008) Stem cells and blastema cells. Curr Stem Cell Res Ther 3(1):53–54

Umemura T, Soga J, Hidaka T, Takemoto H, Nakamura S, Jitsuiki D, Nishioka K, Goto C, Teragawa H, Yoshizumi M, Chayama K, Higashi Y (2008) Aging and hypertension are independent risk factors for reduced number of circulating endothelial progenitor cells. Am J Hypertens 21:1203–1209

Ungewitter E and Scrable H (2008) Antagonistic pleiotropy and p53. Mech Ageing Dev Jul 1 [Epub ahead of print]

Vickers M, Brown GC, Cologne JB Kyoizumi S (2000) Modelling haemopoietic stem cell division by analysis of mutant red cells. Br J Haematol 110:54–62

von Zglinicki T (2002) Oxidative stress shortens telomeres. Trends Biochem Sci 27:339–344

Vulliamy TJ, Knight SW, Mason PJ, Dokal I (2001a) Very short telomeres in the peripheral blood of patients with X-linked and autosomal dyskeratosis congenital. Blood Cells Mol Dis 27:353–357

Vulliamy T, Marrone A, Goldman F, Dearlove A, Bessler M, Mason PJ, Dokal I (2001b) The RNA component of telomerase is mutated in autosomal dominant dyskeratosis congenital. Nature 413:432–435

Wakayama T, Shinkai Y, Tamashiro KL, Niida H, Blanshard DC, Ogura A, Tanemura K, Tachibana M, Perry AC, Colgan DF, Mombaerts P, Yanagimachi R (2000) Cloning of mice to six generations. Nature 407(6802):318–319

Walen KH, Arnstein P (1986) Induction of tumorigenesis and chromosomal abnormalities in human amniocytes infected with simian virus 40 and Kirsten sarcoma virus. In Vitro Cell Dev Biol 22:57–65

Wang RC, Smogorzewska A, de Lange T (2004) Homologous recombination generates T-loop-sized deletions at human telomeres. Cell 119:355–368

Watson JD (1972) Origin of concatemeric T7 DNA. Nat New Biol 239:197–201

Weismann A (1891) Essays upon heredity and kindred biological problems Vol I. Clarendon Press

West MD (1994) The cellular and molecular biology of skin aging. Arch Dermatol 130:87–95

West MD (2003) The immortal cell. Doubleday, New York

West MD, Pereira-Smith OM, Smith JR (1989) Replicative senescence of human skin fibroblasts correlates with a loss of regulation and overexpression of collagenase activity. Exp Cell Res 184:138–147

West MD, Shay JW, Wright WE, Linskens MHK (1996) Altered expression of plasminogen activator and plasminogen activator inhibitor during cellular senescence. Exp Gerontol 31:175–193

West MD, Sargent RG, Long J, Brown C, Chu J-S, Kessler S, Derugin N, Sampathkumar J, Burrows C, Vaziri H, Williams R, Chapman KB, Larocca D, Loring JF, Murai J (2008) The ACTCellerate Initiative: large-scale combinatorial cloning of novel human embryonic stem cell derivatives. Reg Med 3(3):287–308

West MD, Mason C (2007) Mapping the human embryome:1 to 10e13 and all the cells in between. Reg Med 2(4):329–333

Wilmut I, Schnieke AE, McWhir J, Kind AJ, Campbell KH (1997) Viable offspring derived from fetal and adult mammalian cells. Nature 385(6619):810–813

Wright WE, Hayflick L (1975) Nuclear control of cellular aging demonstrated by hybridization of anucleate and whole cultured normal human fibroblasts. Exp Cell Res 96: 133–121

Wright WE, Pereira-Smith OM, Shay JW (1989) Reversible cellular senescence: Implications for immortalization of normal human diploid fibroblasts. Mol Cell Biol 9:3088–3092

Wright WE, Piatyszek MA, Rainey WE, Byrd W, Shay JW (1996) Telomerase activity in human germline and embryonic tissues and cells. Dev Genet 18:173–179

Wu SM, Fujiwara Y, Cibulsky SM, Clapham DE, Lien CL, Schultheiss TM, Orkin SH (2006) Developmental origin of a bipotential myocardial and smooth muscle cell precursor in the mammalian heart Cell 127:1137–1150

Wyllie FS, Jones CJ, Skinner JW, Haughton MF, Wallis C, Wynford-Thomas D, Faragher RG, Kipling D (2000) Telomerase prevents the accelerated cell aging of Werner syndrome fibroblasts. Nat Genet 24:16–17

Yamamoto K, Takeshita K, Kojima T, Takamatsu J, Saito H (2005) Aging and plasminogen activator inhibitor-1 (PAI-1) regulation: implication in the pathogenesis of thrombotic disorders in the elderly. Cardiovasc Res 66:276–85

Yasumoto S, Kunimura C, Kikuchi K, Tahara H, Ohji H, Yamamoto H, Ide T, Utakoji T (1996) Telomerase activity in normal human epithelial cells Oncogene 13:433–439

Yavuz BB, Yavuz B, Sener DD, Cankurtaran M, Halil M, Ulger Z, Nazli N, Kabakci G, Aytemir K, Tokgozoglu L, Oto A, Ariogul S (2008) Advanced age is associated with endothelial dysfunction in healthy elderly subjects. Gerontology 54:153–156

Yu CE, Oshima J, Fu YH, Wijsman EM, Hisama F, Alisch R, Matthews S, Nakura J, Miki T, Ouais S, Martin GM, Mulligan J, Schellenberg GD (1996) Positional cloning of the Werner's syndrome gene. Science 272:258–262

Yu J, Vodyanik MA, Smuga-Otto K, Antosiewicz-Bourget J, Frane JL, Tian S, Nie J, Jonsdottir GA, Ruotti V, Stewart R, Slukvin II, Thomson JA (2007) Induced pluripotent stem cell lines derived from human somatic cells. Science 318:1917–1920

Zhang X, Bok D (1998) Transplantation of retinal pigment epithelial cells and immune response in the subretinal space. Invest Ophthalmol Vis Sci 39:1021–1027

Zhong Z, Deane R, Ali Z, Parisi M, Shapovalov Y, O'Banion MK, Stojanovic K, Sagare A, Boillee S, Cleveland DW, Zlokovic BV (2008) ALS-causing *SOD1* mutants generate vascular changes prior to motor neuron degeneration. Nat Neurosci 11:420–422

Chapter 15
Maintenance and Restoration of Immune System Function

Richard Aspinall* and Wayne A. Mitchell

Contents

15.1	The Pursuit of Immortality	489
15.2	The Formation and Aging of the Immune System	494
	15.2.1 T Cell Formation in the Thymus	494
	15.2.2 Immune Function and the Effects of Age	497
	15.2.3 Infection Shaping the Repertoire	502
15.3	Thymic Rejuvenation	504
	15.3.1 Compounds Associated with Growth Hormone Production	504
	15.3.2 Castration	505
	15.3.3 Interleukin 7	505
	15.3.4 Keratinocyte Growth Hormone	508
	15.3.5 Potential Drawbacks of Thymic Rejuvenation	508
15.4	Complete Rebuilding of the Thymus In Vivo	510
15.5	Complete Rebuilding of the Thymus In Vitro	510
15.6	Designer Restoration	511
15.7	Specific T Cells Immortalised with Telomerase	512
15.8	Concluding Remarks	513
References		514

15.1 The Pursuit of Immortality

The pursuit of immortality has fascinated the human race throughout its history. Many individuals have the desire to live forever, and an early example of this was Gilgamesh, the King of Uruk. He wanted to be immortal and sought the help of Utnapishtim who with his wife had survived the Great Flood, for which they had

---

R. Aspinall (✉)
Cranfield Health, Vincent Building, Cranfield University, Cranfield, Bedfordshire, MK43 0AL, UK
e-mail: r.aspinall@cranfield.ac.uk

*Chair of Translational Medicine.

Fig. 15.1 The flood tablet at the British Museum

been granted immortality by the Gods. According to the deciphered cuneiform text on the clay "flood tablet" seen in the British Museum (Fig. 15.1), Utnapishtim offered Gilgamesh immortality if he could stay awake for six days and seven nights. During the task Gilgamesh fell asleep, and when woken he was distraught. It was only through the intervention of Utnapishtim's wife that Gilgamesh is given a second chance. This time he is told of a secret plant at the bottom of the sea which will make him young again. So he binds rocks to his feet and sinks to the bottom of the sea, finds the plant and takes it even though its thorns prick his hands. He then cuts the rocks from his feet and the sea delivers him up to the shore. But Gilgamesh is wary and rather than try it immediately he decides to take the plant back to Uruk and test it out on an old man first to make sure it works. Whilst on his journey home he stops to bathe in cool water, and a snake, sensing the plant, glides up, takes it, and eats it (which is why snakes shed their skin) then crawls away. Gilgamesh returns to find the plant gone; he falls to his knees and weeps and accepts his eventual fate.

There are reports that some individuals, unlike Gilgamesh, have succeeded in attaining immortality. One of these is, or was, the Comte de St Germain (Fig. 15.2). The provenance of the Comte is vague but he moved in aristocratic circles and came to prominence in the 18th century. Although he seldom ate in public, he was often "at table," where his conversational skills, his ability to speak many languages, and the casual way in which he dispensed diamonds brought him an appreciative audience. All this would be relatively commonplace were it not for his assertion that

15 Maintenance and Restoration of Immune System Function

Fig. 15.2 The Comte de St Germain

he was at least 300 years old and that he knew the secret formula for the Elixir of Life. This boast was apparently substantiated by records suggesting he was about 40 in 1710, and again in 1760 he is reportedly recorded to be about 40 by an elderly witness who had met him previously in 1710.

The Comte appears in the writings of other individuals, including Horace Walpole, who in 1745 thought that the Comte travelling in Edinburgh and later London might be a spy. Voltaire thought he was the "man who knows everything and who never dies," but Giacomo Casanova thought he was a celebrated and learned impostor.

The Comte was a well travelled man, moving from France through Holland and then appearing in Russia at the time when Catherine the Great gained the throne. After this he was seen in Belgium where he turns iron into gold and later appears in Bavaria. There is a report of a M. Saint-Germain being Governor of Chengalaput in India, along with reports from several theosophists who claim to have met the Comte in the late 18th and early 19th century. Apparently, in 1972 a man appeared on French television claiming to be the Comte de Saint Germain and to prove his claim he turned lead into gold on a camp stove in front of the cameras.

Attractive as it is to believe in the existence of such individuals, there are a few problems. One of the most difficult ones to conjure with is their ability to survive illness following infection.

If immortal individuals exist, then they must be subject to the same risks of infection as those individuals surrounding them. Immortality does not provide them with a shield from infection, so any immortal must be subject to the same degree of bombardment with potential pathogens every day as the individuals they walk amongst. Since pathogenic organisms are constantly changing their characteristics, an immortal individual must have an immune system which can cope with all potential variants of all possible pathogens that they could meet.

To put this into perspective, we need to determine the degree of risk of infection. In making this calculation, I want to consider an average individual (John), alive today and working on the 3rd floor of a building in a major city such as London. John travels to work using the metro and after a short walk takes the lift to his desk. During this journey a number of individuals enter John's zone of infection or "social distance," an area around an individual in which they may become infected via the airborne route (http://www.socialdistancing.org/determining-your-social-distance-group-plan/). In his daily journey to work, approximately 30 individuals will enter John's zone. With ten journeys per week this makes 300 possible interactions or 13,800 per year (I have allowed him 6 weeks holiday per year). And this is infection by the airborne route only. Transmission of infection is also possible through touching handrails, buttons or other inanimate surfaces, especially since potentially infectious agents can persist on these surfaces for some weeks (Kramer et al. 2006). Moreover, this takes into account only John's working life, and further contacts could occur through his social life.

The chances of John contacting a potential pathogen are therefore quite high, but his immune system should cope with the majority of these organisms. Consider a pathogen entering John's throat and growing there. Figure 15.3 shows the cellular basis of the response. Components of the pathogen may cross the epithelial barrier (1) into the tissues and be picked up by wandering dendritic cells (2) and taken to the local lymph node (3), where they would be presented to T cells (4). Output from the lymph node is reduced, there is extensive proliferation of lymphocytes that have the correct receptor for the pathogen, and this leads to enlargement of the lymph node (5), producing the classic appearance of swollen glands.

From these beginnings, there is proliferation of antigen specific cells (clonal expansion) and the production of effector lymphocytes [both B (antibody producing cells) and T (cytotoxic) cells], both of which in addition to components of the innate immune system such as the complement system direct the full force of the immune

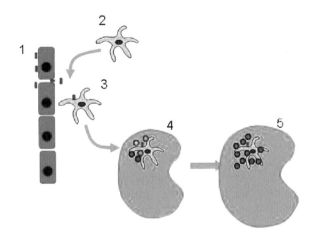

Fig. 15.3 Development of swollen glands

system against the organism. If the immune response is successful, then there is a reduction in the amount of the potential pathogen and its elimination.

From this point on, the immune system no longer requires large numbers of lymphocytes specific for that potential pathogen, and so many of these cells die. Some survive in the body and can remain for considerable periods, through their continued division. These cells and their daughters are termed memory cells. Clinically, memory has been shown to persist for up to 50 years after antigen exposure (Michie et al. 1992). The function of memory cells is to provide an immune response of greater magnitude at a pace which is faster than the primary response if the same pathogen is encountered again.

When the Comte de St. Germain was present in the UK and Europe, the prevalence of transmissible diseases such as tuberculosis, small-pox, cholera, and diphtheria, would have ensured that with his extensive travels, he would have met individuals with these diseases or other potential pathogens. He may have been immune to a broader spectrum of pathogens than many individuals or may have supplemented his diet with potential immunostimulants such as oysters, as did Casanova, or derived some benefit from his elixir (a tea of elder flowers, fennel, and senna pods, soaked in spirits of wine), but even with these nostrums, sooner or later I believe he would have succumbed to disease through an infectious agent. More than likely this would have been transmitted either through direct contact or through food or drink (levels of hygiene especially in food preparation were much lower than they are currently). Because medical treatment regimens were not as advanced as they are now, my contention is that at some point the Comte de St Germain would have died from an infection related illness, either because his memory cells for a specific pathogen had reached their limit of replication (of which, more later) and not been replaced, or because he had been exposed to a pathogen later in life that his immune system had not seen before. Although some would propose that the cells of the Comte's immune system were immortal, this would more likely lead to his death through developing leukaemia or lymphoma (see chapter by Aviv and Bogden in this volume). Even if we make the fairly safe assumption that this was not the case and also make the assumption that the Comte's immune system did not show any signs of aging, then my contention would still be that he would die from an infection related illness. My basis for the latter is that potential pathogenic organisms have a doubling time which is very short, and through mutation they may produce novel variants able to evade our immune response. A good example of this is a disease known as the "English Sweat" (sudor anglicus), which swept through the UK in four epidemics in 1508, 1517, 1528 and 1551, producing an illness characterised by headaches, myalgia, fever, profuse sweating, dyspnoea and usually death. After the last outbreak, the disease abruptly disappeared (Thwaites et al. 1997), and attempts to identify the pathogen with any degree of certainty have proved difficult.

Even with a competent immune system, it is possible that not all organisms will be identified by the antigen receptors on the surface of our lymphocytes, i.e. there are "holes" in the immune repertoire. The risk therefore of meeting a pathogen for which we do not possess a specific receptor will be a factor of both time and geography. The Comte was at risk because of travelling and meeting many individuals and living a long time, suggesting that sooner or later he would be infected by a pathogen to which he couldn't make a response.

On the other hand, we who live in the early 21st century do not face the same challenges as the Comte. We are fortunate enough to enjoy the many advantages of improved sanitation, deadly disease eradication, public health, extensive disease control measures, and the possibility of assisting our immune systems with vaccines, adoptive immunotherapy, exogenous antibodies, and exogenous immune cells targeted to antigens our own immune systems may not adequately recognize, all of which now have demonstrated efficacy in patients. The benefits of having an immune system that never wears out today may therefore be far greater than those the Comte could have enjoyed. To explore how immune system preservation might be attained and what it might mean in our present era, let us now turn to an examination of how the immune system is formed, maintained, and may be improved upon.

15.2 The Formation and Aging of the Immune System

The components that produce an immune response directed against a specific pathogen are the T cells, B cells and dendritic cells. The prefix T stands for thymus, and the B stands for Bone Marrow.

15.2.1 T Cell Formation in the Thymus

The thymus is derived from cells originating in very different parts of the embryo. Careful dissection of the cellular components of the thymus reveals epithelial cells derived from precursors in the third pharyngeal pouch (Zamisch et al 2005; Blackburn and Manley 2004), fibroblasts, and cells of the haematopoietic lineage including neutrophils, mast cells, dendritic cells, macrophages, and thymocytes, which represent all stages of the T cell development pathway. During development the epithelial components from the pouch on both sides of the presumptive pharyngeal region take distinctive shape and move towards the heart, collecting a contribution from the neual crest (Bockman and Kirby 1984) as they go. During this migration, they attract multipotential stem cells first from the fetal liver and then from the bone marrow, which proliferate and begin the process of differentiation which will eventually result in the production of new T cells. Both of these migrating masses eventually meet and come together to form a two lobed thymus situated at the base of the great vessels overlaying the heart.

The majority of T cells in the body have an antigen receptor on their surface which is made up of 2 protein chains called α and β. These cells are produced in the thymus from a small number of multipotential progenitors derived from the bone marrow in adults. These progenitor cells migrate to the thymus, where

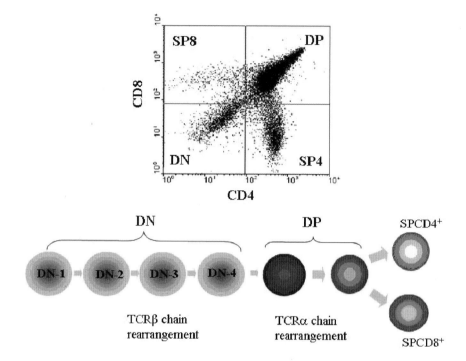

Fig. 15.4 Scheme of αβ+T cell development and FACS profile of thymocytes

they initially express neither the CD4 nor the CD8 molecules on their surface that characterise T cells. They are thus termed CD4⁻CD8⁻ [double negative (DN)] cells (see Fig. 15.4).

The DN-1 population contains a variety of cells, only a small number of which have true potential to colonise the thymus. The earliest T lineage progenitors (ETP) are Lineage⁻cKit^{high} Sca-1^{high} and IL-7Rα$^{neg/low}$ and can not be maintained by IL-7 in vitro (Allman et al. 2003). Up-regulation and expression of the IL-7Rα can be found within the DN-1 to DN-4 populations.

Cells within the DN-1 population are multipotential, whilst those at DN-2 have lost the capacity to form B cells, but can still produce either T cells or dendritic cells (Shortman and Wu 1996; Wu et al. 1996). By the time the cells are within the DN-3 population, they are committed to becoming T cells and have undergone extensive rearrangement of the TCRβ chain genes (see Fig. 15.5) (Capone et al. 1998). Expression of the TCRβ chain at the thymocyte surface requires a TCRα chain equivalent (Fehling and von Boehmer 1997) (the pre-TCRα). These cells then undergo expansion and differentiation so that they become CD4⁺CD8⁺ (double positive or DP) thymocytes. These immature thymocytes are the largest subpopulation in the thymus and are located in the densely packed cortical region of each thymic

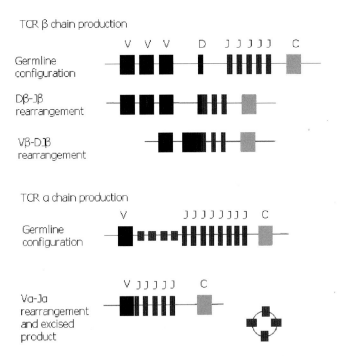

Fig. 15.5 Genetic rearrangement for the production of the αβ T cell receptor

lobule. It is in the double positive stage when the TCRα chain undergoes rearrangement (see Fig. 15.5) (Petrie et al. 1993), after which there is TCRαβ-dependent selection.

The antigen binding regions of both the TCRα chain and the TCRβ chain are extremely variable. This variability is generated by constructing each chain from a large number of variable region genes and by gene rearrangement when joining the genetic elements that contribute to the final chain as shown in Fig. 15.5. Calculations based on the number of variable region genes and the variation generated during rearrangement suggest that there are somewhere in the region of 10^{18} potential TCRαβ antigen receptors that can be generated in humans (Janeway et al. 2004).

Having generated such diversity in the double positive population, care has to be taken to ensure that no self-reactive cells emerge into the peripheral T cell pool, and so during selection many of these double positive cells are prevented from maturing further. This leaves a small percentage that can develop into mature thymocytes expressing either CD4 or CD8 alone, and these are located in the medullary region of each thymic lobe. These cells show a higher level of expression of the TCRαβ than the double positive cells, and these cells undergo negative selection. In this process, those cells whose antigen receptor interacts strongly with stromal cells of the thymus and with dendritic cells that have returned from the periphery are eliminated. The removal of these strong responders ensures that any potentially self-reactive cells do not enter the peripheral T cell pool.

Only a fraction of the single positive cells in the medulla are eventually exported to the periphery to join the peripheral T cell pool. These are specifically the naïve T lymphocyte pool. As discussed above, these cells circulate around the body until they meet the antigen that fits their receptor, at which time they are driven into division.

15.2.2 Immune Function and the Effects of Age

Because of the daily bombardment of potential pathogens that we receive, our immune systems must be continually active. Long term survival of this onslaught requires that our immune responses are successful and that many naïve T cells recognise many potential pathogens and many memory T cells are generated. To reach a ripe old age we must have combated many pathogens and should have built up a prodigious bank of memory T cells which could provide a rapid and protective response to a wide variety of potential pathogens. Theoretically then, the older we get, the more likely it is that we will have already met a specific pathogen and should have the memory cells ready to combat the infection. In other words, our immune system should get better as we get older. But it doesn't. Epidemiological surveys, clinical observations, and laboratory tests all reveal that the immune system declines with age.

Epidemiological evidence reveals that older individuals are often the first to be affected by new or emerging pathogens. In the first outbreak of West Nile Virus in the USA in 1999, the median age of the 59 patients was 71 years, with 73% aged 60 years or greater (Nash et al. 2001). The cost of this outbreak in New York alone was estimated by the UK's Department of Health (Department of Health 2002) to be almost $100 million.

Clinicians recognise that in addition to this susceptibility to new pathogens, older individuals often have difficulties in dealing with pathogens they have previously overcome. For example, there are persistent problems associated with the yearly return of influenza and respiratory syncytial virus (RSV). In England and Wales in 2000, there were 56,838 deaths from pneumonia and influenza, of which 53,833 were in the population over 75. On average, over 400,000 general practitioner consultations annually are attributable to influenza and influenza-like-illness in England and Wales, and during such epidemics of influenza, in excess of over 11,000 elderly respiratory hospital admissions in England cost the UK health service more than £22 million. In the USA from 1990 to 1999, influenza and RSV accounted for 51,203 and 17,358 deaths annually, respectively (Thompson et al. 2003).

Common problems also include reactivation of herpes zoster virus (Schmader 2001) or the response to cytomegalovirus (Pawelec et al. 2004) and an increase in the incidence and severity of some infections. For example, urinary tract infections, lower respiratory tract infections, skin and soft tissue infections, intra-abdominal infections, infective endocarditis, bacterial meningitis, tuberculosis, and herpes zoster have a higher incidence in the elderly, and show a higher mortality rate than younger adult patients with the same disease (Yoshikawa 1997, 2000).

Vaccination alone is not the answer. The ability of current influenza vaccine to induce protection is age related, with an efficacy of between 70 and 90% in those under 65. For those over 65 this efficacy is reduced to 30–40% (Hannoun et al. 2004). And this problem with providing a good response to vaccines is not just associated with influenza. For example, a recent study analysing responses to hepatitis B showed that efficacy was compromised by the age of the individual. In this study, 45 healthy elderly (average age 74) and 37 healthy young controls (average age 28) were vaccinated with hepatitis B, and whilst all of the young individuals developed a protective titre, only 42% of the elderly cohort reached this level (Looney et al. 2001).

Much work has been carried out recently on age-related changes in the immune system. Most of these studies agree that alterations in the characteristics of T cells are at the centre of this reduced immunity. Age-associated immune dysfunction is preceded by atrophy of the thymus (θ in Fig. 15.6), and current evidence suggests a causal link between these two events.

As outlined above, cells in the thymus undergo periods of expansion, selection and cell death. In the thymus of older individuals, we can safely assume that these processes continue, probably at the same rate but on populations that are much

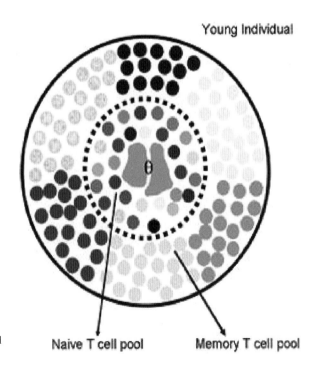

Fig. 15.6 The thymus (θ) and the peripheral T cell pool in a young individual

reduced in size. One point on which all of those working on age-associated thymic atrophy agree is that there is massive loss of total thymocyte numbers with age, and it is this loss that must contribute substantially to the reduction seen in the overall size of the thymus.

For the human thymus, there are studies that indicate no significant variation in the distribution of the main thymic subsets according to the age of the donor (Bertho et al. 1997), and studies in the mouse provide a similar story. Between the ages of 3 and 20 months in the mouse, the number of CD4$^+$CD8$^+$ thymocytes is reduced by more than 10^8, but at both 3 and 20 months of age this major thymic subset is about 85% of the total population of thymocytes (Aspinall 1997). Changes in number but not percentages are also seen for the single positive thymocyte populations. This would seem to indicate that there is a bottleneck in T cell development early in the pathway that leads to this massive reduction in downstream populations.

The pertinent early stage of T cell development, i.e. before expression of CD4 and CD8, was subject to longitudinal analysis in male C57Bl/10 mice. The results revealed that all subpopulations defined by their expression of CD44 and CD25 declined with age with the exception of the CD44$^+$CD25$^-$ population (the DN-1 subset), whose numbers remained unchanged throughout life, but whose progeny, the CD44$^+$CD25$^+$ population (the DN-2 subset), decline in number with age (Aspinall 1997). No change in the number of DN-1 cells with age, but declines in all downstream populations suggests either that these cells had an intrinsic deficiency preventing their differentiation or that the thymic microenvironment was insufficient to support their progeny. At the moment, all available evidence suggests that the defect lies in the thymic environment.

What are the effects of the reduction in thymic output on the peripheral T cell pool? A schematic outline of the changes is seen by comparing Figs. 15.6 and 15.7.

In all individuals, T cells produced by the thymus enter the naïve T cell pool where they move round the body, circulating from the blood to the secondary lymphatic tissue (spleen, lymph nodes) and back to the blood again, until they meet their specific antigen. Analysis of the blood of large numbers of young individuals aged 20–30 showed that about 50% of their blood T cells had a phenotype which would place them in the naïve T cell pool (Hulstaert et al. 1994). As mentioned above, a successful response by a naïve T cell leads to activation, proliferation and eventual entry into the memory T cell pool. So as we age and meet more potential pathogens, our memory T cell pool should increase in size, which appears to be the case (Cossarizza et al. 1996).

Previous estimates of the number of T cells in the body have been worked out based on the notion that about 2% of the total body T cells are circulating in the blood at any one time (Trepel 1974). A normal 70 kg individual has an estimated 5 litres of blood, and each microlitre of blood contains approximately 1.5×10^3 T cells. Thus the blood contains 7.5×10^9 T cells, which constitutes about 2% of the total T cell pool. This suggests that the peripheral T cell pool in a normal individual is somewhere in the region of 3.75×10^{11} cells. As we mentioned above, there are somewhere in the region of 10^{18} potential TCR$\alpha\beta$ receptors, so at the most extreme

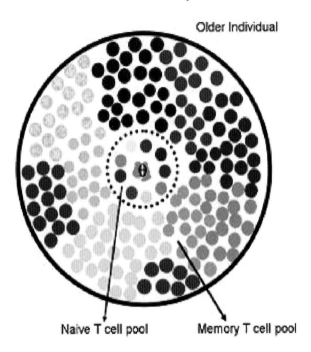

Fig. 15.7 The thymus (θ) and the peripheral T cell pool in an older individual

estimate only 0.0000375% of the total T cell repertoire will be in the pool at any one time.

This is an extreme estimate because it assumes there are no clones of T cells in the peripheral pool, but only individual cells expressing individual receptors. However, we know that this is not the case. We know that receptors are shared by clones of cells, and a recent paper tried to estimate the number of clones in healthy individuals and how these change with age. Using TCR genes as clonal markers, the study found between 44,000 and 110,000 clones in healthy young (20–30 years old) adults. In 4 healthy active older (>75 years old) adults, clone numbers were generally lower (3600 5500, 14,000 and 97,000 clones), and the clones were usually larger in size (Dare et al. 2006).

Perhaps the most astonishing observation is that despite the constant state of flux of the immune system, with its proliferation and cell death, the total number of cells in the peripheral T cell pool remains relatively constant, constrained between upper and lower limits throughout the life of a healthy individual. Such close maintenance of numbers across a lifespan is unusual on several counts. First, the immune system although anatomically defined is not physically constrained. There are no external forces preventing the unrestricted proliferation of lymphocytes. Second, the organs of the immune system are extremely flexible, capable of accommodating lymphocyte expansions. And third, since there are no absolutely sterile environments, each of us exists in a sea of antigens capable of inducing more and more memory cells.

So how are the numbers of lymphocytes maintained within defined limits throughout life? One idea is that for each lymphocyte there is an ecological niche which supplies all of the resources (cytokines, chemokines, MHC molecules, antigens, co-stimulatory ligands, adhesion molecules, etc.) for the survival of the cell (Almeida et al. 2005; Freitas and Rocha 2000). The number of niches remains constant, and so there is competition for each niche, and this controls the size of the peripheral T cell pool. As we age and the naïve T cell pool becomes considerably reduced in size (Fig. 15.7), more ecological niches become available, and these need to be filled in order to maintain the total number of T cells in the body. The obvious source of more T cells is through proliferation of residents of the peripheral T cell pool, which in the main will be memory T cells.

But proliferation in the memory T cell pool, driven initially by antigen and later by cytokines, is not without consequence. T cells, like most other cells in the body, have a finite replicative capacity, and their continued proliferation may be associated with the accumulation of DNA damage and the loss of telomere repeat sequences (Pawelec et al. 2002).

This cell division also affects burst size. As seen in Fig. 15.8, at one point along the replicative pathway (at point A), a cell will have the capacity to produce a large burst of cells because it is at an early phase of the replicative life-span. However, a cell at position B further down the replicative lifespan pathway would produce a considerably reduced burst size. Since the burst size may be critical in driving the immune response, a reduced burst size may lead to poorer and less than effective immunity. In addition, proliferation may not be undertaken at the same rate in all memory cells.

There is evidence of enlarged clones of T cells in many older individuals (Posnett et al. 1994). Eventually, this repeated proliferation leads to increasing numbers of

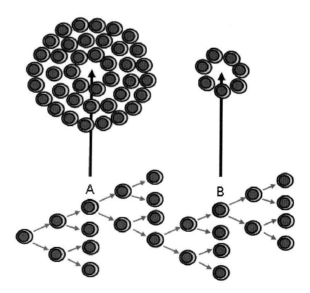

Fig. 15.8 Different burst size capabilities along the replicative lifespan pathway

peripheral T cells that have lost the capacity to respond successfully to normal mitogenic signals. This contributes to the state of immunosenesence in vivo that has been equated to the senescence seen in vitro following repeated replication (Effros and Pawelec 1997). In both mice and humans, aging appears to be associated with increasing numbers of senescent T cells (Voehringer et al. 2001, 2002), whose presence may not be benign. Two studies suggest that high numbers of senescent CD8 T cells correlate with poorer responses to influenza vaccine (Saurwein-Teissl et al. 2002; Goronzy et al. 2001).

15.2.3 Infection Shaping the Repertoire

The chance of exposure to potentially pathogenic organisms is a product of both time and location, and so increasing age may be associated with increasing antigen burden. If we were able to analyse the peripheral T cell pool of an individual, it would reveal a detailed history of his or her immunological provenance, showing which infections the person had conquered, plus those infections against which the individual was currently making responses. So the peripheral pool of T cells is shaped by previous and ongoing infections, and a major driving force in this process may be silent sub-clinical infections.

Of the viruses that can cause silent or latent infections, probably the most important group are the herpes viruses. There are more than 100 herpes viruses, of which 8 routinely infect humans. The human herpes viruses (HHV) have recently been classified and numbered (Table 15.1). Because of the nature of the routes of infection and the less than overt nature of the symptoms following infections, many individuals may be silent carriers of HHV-4 (Nishiwaki et al. 2006), HHV-5 (Nishiwaki

Table 15.1 Human herpes viruses

Type	Other name	Disease association
HHV-1	Herpes simplex-1	Mainly oral herpes
HHV-2	Herpes simplex-2	Mainly genital herpes
HHV-3	Varicella-Zoster virus	Chickenpox and shingles
HHV-4	Epstein Barr virus	Infectious mononucleosis (glandular fever)
HHV-5	Cytomegalovirus,	Few symptoms initially in healthy individuals, associated with immunodeficiency later in life
HHV-6	Roseolovirus	HHV-6A and B subtypes are recognised. HHV-6 causes roseola and occasional seizures and encephalitis in infants, and two thirds of patients with Mesial Temporal Lobe Epilepsy (MTLE) may have chronic HHV-6B infection
HHV-7	Roseolovirus	HHV-7 has also been associated with roseola in infants.
HHV-8	Kaposi sarcoma associated herpesvirus	Kaposi's sarcoma, primary effusion lymphoma

et al. 2006), and HHV-6 & 7 (Leong et al. 2007; Ward 2005). Problems with these infections come as we age and fail to control them adequately.

For HHV-4, a recent study suggests a link between this virus and the onset of specific cancer in the elderly (Shimoyama et al. 2006). HHV-4 has often been associated with diseases of the lymphoid system in the immunocompromised, because during primary infection, HHV-4 establishes lifelong latency in the memory B cell compartment (Gottschalk et al. 2005). Analysis of a series of elderly patients with HHV-4 associated B cell lymphoproliferative disease in the absence of any overt immunodeficiency (Shimoyama et al. 2006) revealed individuals with a neoplasia of B cell origin where the median age of onset was 71 years of age. The authors suggested that this disease may be related to the age-associated deterioration seen in the immune system (Shimoyama et al. 2006).

Recently, there has been a great deal of work carried out to establish the changes to the immune system wrought by the HHV-5 virus. Individuals are usually infected silently, probably during the early years, and the virus remains latent, hiding in cells of the myeloid lineage, and is kept in check by a healthy immune system. It does, however, have an impact on the naïve T cell pool, as shown by a study that revealed that individuals chronically infected with HHV-5 have lower numbers of naïve T cells than individuals who were HHV-5 negative (Almanzar et al. 2005). It is not known whether this reduction was due to an alteration in thymopoiesis, producing fewer thymic emigrants, or whether naïve T cells were leaving the naïve pool more quickly either through death or transformation into memory cells.

In addition, as we age, the immune system appears to become increasingly obsessed with this virus. The control of HHV-5 within the body is mainly attributed to $CD8^+$ T cells, and in older individuals up to 50% of $CD8^+$ T cells may be specific for HHV-5, as judged by tetramer staining (Khan et al. 2002).

It is possible that this increased response to HHV-5 may lead to impaired immune responses (Hadrup et al. 2006) to other viruses (Khan et al. 2004) and other potential pathogens, and may be an explanation for the findings linking infection with HHV-5 to potential mortality in the elderly carried out by Dr Anders Wikby and his group in Sweden. They carried out a longitudinal immunological study in order to establish predictive factors for longevity, first working on a group of normal octogenarians in the OCTA study in Sweden, which later became the NONA study as these individuals entered their nineties. Within this group, they were able to describe the "immune risk phenotype," which was characterised by a CD4:CD8 ratio of <1, poor in vitro T cell proliferation, an increase in the number of $CD8^+CD28^-$ cells (or $CD8^+CD28^-CD27^-$ in the very old), low numbers of B cells, and the presence of CD8 T cells which were HHV-5 tetramer positive. Within both the OCTO and NONA groups, they were able to show that the immune risk phenotype was predictive for mortality, which was even more apparent when cognitive impairment was included in the calculations (Wikby et al. 2005).

Less work has been carried out on HHV-6 and HHV-7, probably because these have been characterised more recently, but current work suggests that infection with these viruses is widespread. A recent study in blood donors in London (UK) showed 0.8% had high viral copy numbers of HHV-6 (all samples were HHV-6 variant B)

(Leong et al. 2007), but the values for HHV-7 were much higher, with another study showing that HHV-7 DNA could be detected in 78% of HIV-seronegative healthy blood donors (Boutolleau et al. 2005).

15.3 Thymic Rejuvenation

As we get older and the thymus has undergone considerable atrophy, every pathogen that is defeated provides a "Pyrrhic victory" for our immune response, even if it is a pathogen we have defeated previously. This is because the T cells making the response cannot be renewed indefinitely. Without thymic input, the "average age" of all of the T cells in the peripheral T cell pool, measured in terms of post-thymic divisions, will increase. As we age, more and more T cells will reach their replicative limit. How, then, to replenish this diminishing resource of T cells?

Probably the best way, but not the only way, is to reverse the atrophy seen in the thymus and renew thymopoiesis, thereby increasing thymic output. This should provide new T cells with a diverse range of receptors to rejuvenate the peripheral T cell pool. There are several ways in which thymopoiesis can be restored, including using growth hormone (French et al. 2002; Napolitano et al. 2002; Fahy 2003), growth hormone sectretagogues, castration, IL-7, and keratinocyte growth factor.

15.3.1 Compounds Associated with Growth Hormone Production

In recent years, a class of synthetic nonpeptidyl compounds known as Growth Hormone Secretagogues (GHS) has been discovered. GHS have the ability to synergize with natural GH-releasing factor and induce calcium flux within rat pituitary cells to cause growth hormone release. The full description of the discovery process and the development of GHS can be found elsewhere (Smith 2005; Smith et al. 2005). Old (20–24 month) BALB/c or B6 mice given doses of GHS showed a significant increase in the cellularity of their thymuses and also an increased resistance to tumour metastasis (Koo et al. 2001).

More recent data has provided compelling evidence that the peptide hormone ghrelin plays a significant role in promoting T cell output from the aging thymus (Dixit et al. 2007). Ghrelin binds specifically to the growth hormone secretagogue receptor (GHS-R), which has been shown to undergo a decline in expression in the thymus with advancing age. Dixit et al., in their studies on mice, demonstrated that by 12 months of age there was a dramatic reduction in ghrelin expression, which corresponded to infiltration of adipocytes.

The authors suggest that the loss of ghrelin and its receptor with advancing age might contribute to the involution process. To examine this possibility, they infused ghrelin into mice of ages ranging from 2 to 24 months. Treated mice showed a significant increase in thymic mass, but interestingly, this effect was only observed in the old mice, not in the young mice. Furthermore, the architecture of both the cortical and medullary components of the thymic epithelial cells were significantly restored in mice where ghrelin had been infused. Thymic output as determined by TREC measurement (see below) and the diversity of the TCR repertoire were also improved (Dixit et al. 2007).

Taken together, these results suggest a significant role for ghrelin and other agents that initiate the GHS function for the rejuvenation of the T cell immune response.

15.3.2 Castration

Experiments in rodents have shown that castration of old animals results in the reversal of thymic atrophy and the production of new T cells (Kendall et al. 1990; Heng et al. 2005; Sutherland et al. 2005). This does seem to have an effect on survival, although studies on castration and lifespan suggest that castrated animals have a slightly longer lifespan (Drori and Folman 1976) than intact animals.

Unfortunately, the data from human studies is more mixed. Analysis of the eunuchs of the Chinese and Ottoman courts, of the Skoptzy (a religious sect who believed that after expulsion from the Garden of Eden, Adam and Eve had the forbidden fruit implanted on their bodies, forming the testicles and breasts; removal of these sexual organs would therefore restore them to the pristine state before the fall), and of the castrato singers reveals that castration did not lead to an increased lifespan (Jenkins 1998). However, contrary results were reported from a mental institute in Kansas that regularly castrated patients with "behavioral difficulties:" per 1000 cases analysed, 300 of whom were castrated, the records showed that the castrated men lived on average 14 years longer than the intact individuals (Hamilton and Mestler 1969) (see also Chapter 6).

Currently, castration can be achieved by chemical means rather than by drastic surgery. Total androgen blockade using luteinising hormone-releasing hormone analogues alone or in combination with other anti-androgens is a treatment option for patients with prostate cancer (Msaouel et al. 2007), permitting some analysis of the clinical potential of sex steroid ablation. The results have been encouraging: elderly males undergoing sex steroid ablation therapy for prostatic carcinoma provided some indication of an increase in the percentage of recent thymic emigrants after treatment as measured using the sjTREC assay (Sutherland et al. 2005). Later work with mice seems to show that the cytokine IL-7 may have a role in this process (Goldberg et al. 2007).

15.3.3 Interleukin 7

Interleukin 7 (IL-7) is a pleiotropic cytokine with a central role in the development and maintenance of T cells (Fry and Mackall 2002a). The human IL-7 gene consists of 6 exons on chromosome 8q12–13, which together code for a protein whose molecular weight is 25kD after glycosylation. IL-7 is classified as a type 1 short-chain cytokine of the hematopoietin family. The protein can be detected in many stromal tissues including epithelial cells in the thymus, bone marrow and gut, keratinocytes, fetal liver, adult liver, dendritic cells, and follicular dendritic cells (Fry and Mackall 2002a) and also in astrocytes (Michaelson et al. 1996). However, within these tissues the availability of IL-7 may be altered because of its propensity to bind to extracellular matrix-associated glycosaminoglycan, heparan sulfate, and fibronectin (Clarke et al. 1995; Kimura et al. 1991) as well as to its appropriate receptor.

The IL-7 receptor is composed of 2 chains, α (CD127) and γ (CD132). Successful interaction between IL-7 and its receptor leading to signal transduction begins with IL-7 first interacting with the IL-7Rα chain and forming a stable 1:1 intermediate. This intermediate then interacts with the γ chain, forming an active ternary complex. The merger of these two chains by IL-7 orients their cytoplasmic domains so that their respective kinases (Janus and phophatidylinositol 3-kinases) can phosphorylate sequence elements on the cytoplasmic domains. Transcriptional activators bind to the phosphorylated sequence elements of these chains and become phophorylated, after which they dissociate from the chains, oligomerise, and move to the nucleus, where they elevate transcription (Jiang et al. 2005).

Experiments in the mouse have made it abundantly clear that interaction of the IL-7 ligand with the IL-7Rα chain is essential for development of the immune system. IL-7Rα knockout mice reveal the crucial role that ligand-receptor interaction have in T cell development at early (CD4-CD8-) stages (Peschon et al. 1994), at post-selection expansion stages (Hare et al. 2000), and in the survival and proliferation of cells in the periphery (Maraskovsky et al. 1996). Deletion of the γ chain of the receptor, a 65–70 kD glycoprotein chain used as part of the receptor for IL-2, IL-4, IL-7, IL-9, IL-15 and IL-21 (Habib et al. 2003), results in limited thymopoiesis and lymphopoenia (Di Santo et al. 1999).

Within the intrathymic T cell developmental pathway, IL-7 may act both as an essential co-factor in the rearrangement of the TCR β chain genes (Muegge et al. 1993) and also be required for the survival of cells at both early stages in the pathway (Kim et al. 1998; Oosterwegel et al. 1997) as well as being important in post selection expansion at later downstream stages (Hare et al. 2000) and in the survival and proliferation of cells in the periphery (Maraskovsky et al. 1996). Recent studies indicate that long lived memory cells, which are the most crucial cells to generate in order to make an effective and long-lasting vaccine response, all express the IL-7 receptor, indicating the importance of IL-7 for responsiveness to vaccines (Kaech et al. 2003). The importance of IL-7 in vaccine responsiveness was also indicated by recent studies showing that supplementation with IL-7 can augment immune responses to an otherwise weak immunogen and act as a potent vaccine adjuvant (Melchionda et al. 2005).

15.3.3.1 Therapeutic Usage of IL-7

The ability of IL-7 to induce T cell production either through increased thymic output or through proliferation of peripheral T cells (Fry et al. 2001) coupled with the close linkage between the levels of IL-7 and survival of both mature and immature T lymphocytes lead to its championing as a therapeutic agent (Fry and Mackall 2002a, b). Previous work showed that a central element in the process of age-associated thymic atrophy was a reduction in the levels of IL-7 in the intrathymic microenvironment (Andrew and Aspinall 2002). This in turn leads to a significant reduction in the number of thymocytes and lower thymic output (Andrew and Aspinall 2001).

Treatment of old mice with recombinant IL-7 reverses thymic atrophy, increases thymic output, and improves subsequent immune responses (Pido-Lopez et al. 2002). However, one of the problems with this approach is that the drug is introduced into the body at a location far away from its site of action, where it is required at a specific concentration in order to be effective. This means that the drug has to be injected at a high concentration so that when it diffuses through the tissues to reach its target organ it can be present there at the effective concentration. There are therefore chances of the drug producing unwanted side-effects.

To solve this problem, a novel compound was developed which associates a therapeutic agent with a targeting moiety in a single molecule. The approach made use of the knowledge that the chemokine CCL25 is produced in the thymus (Corral et al. 2000) and binds to the chemokine receptor CCR9 for which it is the only known ligand. A fusion protein was created between the extracellular portion of CCR9 and IL-7 that, when used as a therapeutic agent in old animals, results in the accumulation of the fusion protein in the thymus, the reversal of age-associated thymic atrophy, a significant increase in the production of new T cells, and a significant improvement in anti-viral responses in old animals (Henson et al. 2005).

Before extending this work further, it was essential to carry out a proof of principle study in old female rhesus macaques (aged 18.5–23.9 years). There is the potential for IL-7 to produce several side effects, including neutropaenia, anaemia (Storek et al. 2003), and bone loss (Toraldo et al. 2003), so animals were carefully followed after they were treated with recombinant simian IL-7 according to a two phase regimen (Aspinall et al. 2007).

Treated animals showed an increase in the number of $CD3^+T$ cells in their blood in addition to increases in TREC levels per T cell following both phases of treatment. TREC (T cell receptor excision circles) are small circles of DNA excised from the genome during the production of $\alpha\beta^+$ T cells. They do not replicate with the cell, so as the T cell divides, the TREC are passed to only one of the progeny cells. TREC are therefore at their highest in a population of cells that contains many recent thymic emigrants.

Following vaccination with influenza (strain A/PR/8/34), IL-7 treated animals showed more than a four-fold rise in geometric mean haemagglutination inhibition titres, which was not apparent in saline treated animals. Further analysis of these studies showed that animals with the highest titres and the best proliferation against influenza antigen were amongst those with the highest TREC per T cell levels after the second phase of treatment. Moreover, such treatment was not associated with bone loss, nor with anaemia nor neutropaenia (Aspinall et al. 2007).

15.3.4 Keratinocyte Growth Hormone

Human keratinocyte growth factor, forms of which are also known as palifermin or Kepivance®, have been used clinically to reduce the incidence and duration of mucositis (inflammation and ulceration of the cells lining the mouth and gastrointestinal tract) in patients with cancer who receive high doses of chemotherapy and radiation therapy (http://www.fda.gov/cder/drug/infopage/palifermin/paliferminQA.htm), usually followed by stem cell replacement. In the clinic, the drug is normally given by intravenous injection, and treatment with KGF is not without its drawbacks. Side effects of the drug include skin rash, pruritis, erythema and taste alteration.

Keratinocyte growth factor is a member of the fibroblast growth factor family (which has at least 22 members, with roles in controlling cell migration, proliferation and differentiation), is also known as fibroblast growth factor-7, and is normally produced by cells of mesenchymal origin. It acts to induce epithelial cell proliferation through a subset of FGF receptor isoforms (FGFR2b) expressed predominantly by epithelial cells (Finch and Rubin 2006).

Keratinocyte growth factor has been used recently in old mice, where it has been shown to increase thymopoiesis and improve T cell dependent antibody responses (Min et al. 2007). The authors also noted that there was replacement of thymic epithelial cells in addition to a correction of the structural deformities which occur in old thymuses, such as the loss of the cortico-medullary junction. Keratinocyte growth factor has also been used in primates, to evaluate its effectiveness in restoring thymic architecture and thymic ouput after irradiation and subsequent stem cell replacement. In this study, the animals treated with KGF showed a well-preserved thymic architecture at 1 year after transplantation, whilst most of the untreated animals showed thymic atrophy. In addition, thymic output was improved following treatment, as was the ability to make T cell dependent antibody responses, and the majority of KGF-treated animals showed a broader TCR repertoire compared with the control animals 1 year after transplantation (Seggewiss et al. 2007).

The receptor which binds KGF is present on thymic epithelial cells, which produce IL-7. Since treatment with KGF increases IL-7 transcripts in the thymus, the most likely mechanism of action is through the presence of more epithelial cells expressing more IL-7.

15.3.5 Potential Drawbacks of Thymic Rejuvenation

Evolutionary theory suggests that genes that benefit us early in life are favoured by natural selection over those whose benefits arrive later (Charlesworth 2000). In humans, thymic atrophy begins as early as one year of age and proceeds at a rate of 3% per year (George and Ritter 1996), and it appears to be a common feature of mammals. The widespread nature of this phenomenon has led evolutionary theorists to believe that thymic atrophy must have considerable biological benefits, so its reversion might be associated with significant problems.

Some suggestions as to why the thymus atrophies with age (George and Ritter 1996) include the possibility that continued thymic output throughout life would lead to an increase in the incidence of autoimmune disease or that thymic atrophy is a means of optimising energy investment. Production of T cells in the thymus requires considerable energy, and those individuals who divert this energy to the production of progeny or to their continued survival might be favoured and selected.

If we can overcome the energy requirement problem, then we still have to deal with the possibility that the reversion of thymic atrophy and the increase in the production of new T cells may be associated with autoimmune disease. As discussed in Section 15.2.1, a current view is that the first line of defence against self-reactive T cells is clonal deletion in the thymus. Selection within the developing thymocyte population by thymic epithelial cells ensures that self reactive cells are deleted from the peripheral T cell repertoire (Caton 2003). Any which slip through this selection process are held in check by mechanisms in the periphery, including regulatory T cells, which prevent their activation (Jordan et al. 2001). If such peripheral mechanisms fail, then T cells recognise self and drive a chronic persistent autoimmune response. This implies that the fundamental breakdown that can lead to immune disease lies in the escape of forbidden clones within the recent thymic emigrants.

Analysis of the range of autoimmune diseases (see Table 15.2) shows two striking points. The first is that most sufferers are women, and the second is that for most cases, the peak time of onset is in the region of 30–60 years of age.

The gender-related skewing of the incidence of autoimmunity has been noted previously and has been put down to a difference in the environment provided for the differentiation of T cells towards a Th1 phenotype in women (Whitacre et al. 1999), but this does not account for the presence of forbidden clones in the peripheral T cell pool. Recent work showed that in humans, the output from the thymus is gender dependent, with females showing a higher thymic output for longer in their lifespan than males (Pido-Lopez et al. 2001). If thymic selection changes with age, then greater thymic output for longer in females compared with males could lead to more forbidden clones within the recent thymic emigrant pool.

Table 15.2 Autoimmune disease and age

Autoimmune disease	Ratio ♀:♂	age of onset
Primary Biliary Cirrhosis	9:1	40–50
Chronic Active Hepatitis	6:1	40–60
Pernicious Anemia	3:2	>60
Rheumatoid Arthritis	3:1	30–40
Sjorgens syndrome	9:1	~50
Progressive sytemic sclerosis	2:1	~30
Autoimmune hypothyroidism	10:1	40–60
Graves disease	7:1	~30
Addisons disease	1.8:1	~30
Multiple sclerosis	2:1	~30

One current hypothesis proposes there are changes in the thymic epithelial cells with age which alter their ability to perform both positive and negative selection at the same level at which this selection process occurs in the young. Thus, at middle age, more potential self-reactive T cells may leave the thymus in females than in males. Increases in the numbers of self reactive T cells means that eventually the conditions exist for them to evade the normal control systems, and they become activated and undergo clonal expansion and cause overt disease. The increased incidence of autoimmune disease in females compared with males would therefore be a by-product of increased thymic output and poorer selection processes. Any rejuvenation of the thymus in the elderly may benefit from ensuring that the MHC Class II$^+$ epithelial cells also undergo some rejuvenation.

It is interesting to note that women live longer than men. This gender related skewing in survival appears to be independent of many environmental or extrinsic factors (diet, lifestyle choices) because sampling from several different geographical locations at different times reveals that in all of these different places and at different times, women live longer than men (Eskes and Haanen 2007). It is tempting to speculate that improved immunity in women following the higher thymic output which occurs for longer in their lifespan than for men, may in part be responsible for this increased longevity (Aspinall 2000).

15.4 Complete Rebuilding of the Thymus In Vivo

Although there are undoubtedly problems with some of the early lymphocyte precursors in older individuals, most workers in the field would agree that the major problem is with the thymic environment. The recent characterisation of the stem cell that can form the epithelial component of the thymus (Bennett et al. 2002) and its ability to generate a fully functional thymus microenvironment when transplanted into mice (Gill et al. 2002) has lead to the idea that this might be the way forward to rejuvenate the immune system. If the human equivalent could be identified, then it could be possible to use these thymic stem cells to rebuild a thymus in vivo.

15.5 Complete Rebuilding of the Thymus In Vitro

The idea here is to generate populations of T cells that have undergone positive and negative selection in vitro from autologous stem cells. This population would have a diverse repertoire and should be able to be transferred into an immunodeficient recipient.

Initial studies have already shown that T cells can be generated from bone marrow stem cells in vitro using a thymus generated in vitro from skin fibroblasts and keratinocytes (Clark et al. 2005). These cells are grown separately as a monolayer and then harvested and combined and grown on a tantalum coated matrix (Fig. 15.9) until they produce a thymus microenvironment equivalent. When seeded with stem

Fig. 15.9 Tanatalum coated matrix

cells of bone marrow origin, the skin fibroblasts and keratinocytes cause the stem cells to undergo differentiation and commitment to the T cell lineage. There is rearrangement of the TCRβ and TCRα genes and the production of mature cells with a diverse antigen receptor repertoire.

This is an exciting discovery since it lays the foundation for the possibility that we may be able to generate new T cells in vitro from our own skin and stem cells. When these cells are injected back into the host, there should be no graft versus host disease because all of the tissues are of host origin. In addition, responses to antigen should also be self-restricted.

There are a number of technical problems to overcome with this system. For example, these cultures were carried out using fetal calf serum, and a clinically applicable system must use either autologous serum or be serum-free.

15.6 Designer Restoration

One of the design strengths of the immune system is its ability to cope with most forms of infection. As we have discussed previously, there may only be a select number of organisms that are problematic in the elderly. Rather than rejuvenate the whole of the immune system, a smarter option may be to consider rejuvenating the immune system with cells that are limited in their specificity.

The link between some aspects of immunosenesence and HHV-5 has lead some to wonder whether the state of immunosenesence was infectious (Pawelec et al.

2004). Vaccination would be useful, but to be fully effective would have to be carried out in young children prior to infection. For those older individuals currently infected by HHV-5, another option would be treatment with anti-viral agents such as ganciclovir, but the potential side effects including diarrhoea, nausea, headache, dizziness, and confusion would argue against its widespread use.

A third option has become apparent from a series of recent experiments. The first was a report on treatment of immunosuppressed patients in the post-transplant period, for whom reactivation of HHV-5 remains a significant cause of morbidity and mortality. These patients received an infusion of HHV-5-specific CD8 T cells selected from healthy donors using HLA-peptide tetramers. These patients' HHV-5 viraemia was reduced in every case, with eight patients out of the nine treated managing to clear the infection, including one patient who had a prolonged history of HHV-5 infection that was refractory to antiviral therapy (Cobbold et al. 2005). This indicates the central role of effective CD8 T cells in the control of the virus and its consequences.

The second report was a method for generating HHV-5-specific T cells in vitro from a relatively small (500 ml) amount of blood drawn from recipients. Both $CD4^+$ and $CD8^+$ cells were generated in this process, and these cells were capable of lysing HHV-5 infected fibroblasts (Rauser et al. 2004). From these two reports it would seem feasible in the near future to generate HHV-5-specific T cells in vitro and transfer them to HLA compatible older individuals for whom HHV-5 infection poses a threat to their immune status.

15.7 Specific T Cells Immortalised with Telomerase

The above protocol requires that healthy young individuals donate blood from which to generate HHV-5-specific T cells for use in HLA matched individuals on the grounds that these cells will be at an earlier stage of their replicative lifespan than any derived from older individuals carrying the infection. The problems associated with this approach are mainly related to not getting an exact HLA match and so inducing graft versus host disease.

Several recent papers have reported the generation of T cells with defined specificity that have been transformed with htert (human telomerase reverse transcriptase) (Verra et al. 2004; Dagarag et al. 2004; Rufer et al. 2001), which prolongs their lifespan. This makes it possible to obtain antigen-specific functional lymphocytes from older individuals, transform them as these studies have described, and then transfer them back into the donor in order to boost a specific immune response.

The only drawback with this is the question of the mortality of these cells and the possibility that they may be transformed in vivo into cells with leukaemic potential. These questions may be answered fairly soon, since a group has generated hTERT-transformed CD8 T cells from rhesus macaques that are specific for an immunodominant SIV gag epitope, and these cells have similar characteristics to non-transformed cells. It is likely that these transformed cells will be used to test the feasibility of such an adoptive transfer approach (Andersen et al. 2007).

15.8 Concluding Remarks

At the beginning of this chapter, we put forward the notion that past claims for immortality were unlikely to be true, especially for those who come into regular contact with a wide range of individuals because of the likely risk of infection. The principle reason is that our immune system winds down with age, and so we become increasingly more susceptible to infectious disease as we grow older. The primary event which precedes this immune decline is the age-associated atrophy of the thymus. Production and export of new T cells declines, and this leaves us with a peripheral T cell pool shaped by previous encounters. Reduced thymic output coupled with the powerful homeostatic force that maintains T cell numbers creates an immune system that contains fewer naïve T cells and more cells closer to their replicative endpoint as we get older. This would seem to provide an explanation for the increased vulnerability to infection with aging, especially for pathogens we have not met previously.

The risk of meeting such a pathogen would seem to be more likely today because of the ease with which we can now travel between countries. Such a change in population movement would provide the ideal opportunity for the dissemination of different infectious agents that were previously geographically restricted. A recent study has reported the emergence for the first time of 335 infectious diseases between 1940 and 2004 that have had a significant impact on global health (Jones et al. 2008). In addition, we are also vulnerable to infections we have conquered in the past because our immune response, which depends on cell proliferation for its effectiveness, contains more and more cells that have reached their replicative limit.

These are not intractable problems. They instead provide a base on which we can build a series of potential therapies to rejuvenate the immune system, the foundation of which has to be the rejuvenation of the thymus. Some immediate thoughts that arise from this are whether thymic rejuvenation has to be a continuous process or whether sufficient improvements in immunity can be generated by a short term therapy that could be repeated every few years or whenever necessary. The difficult question is how to judge when competent immunity is restored without compromising the safety of the individual. In vitro testing could be carried out against a number of potential pathogens, but what about the variants of these organisms? Also, how would we detect a potential hole in the repertoire that could lead to infection and predispose you to another disease?

More importantly, it may be of limited benefit to rejuvenate one's own immune system whilst all those around you fail to improve their own and act as reservoirs of infection. It is unlikely that any rejuvenation therapy would be without side effects, so in the next few years we need to define biomarkers that can be used to identify individuals whose immune systems are on the point of decline or who have specific defects which leave them open to infection. These individuals would be candidates for treatment, and any therapy used in this way must be capable of being easily deployed in the community in order to have good compliance.

This makes the injection route as a delivery system a less viable option and favours therapies that could be delivered by other means. This is not a new problem and has been considered by others, for example in the treatment of insulin-dependent diabetes. In this example, blood sugar levels are controlled by the regular injection of insulin, but more recently, approval has been given to have a therapy based around an inhaled form of insulin (Black et al. 2007). If we could deliver an agent to reverse thymic atrophy, for example, interleukin 7, keratinocyte growth factor, or any of the other thymic rejuvenating factors by this route, we could provide a more amenable form of treatment that could be used in the community. Provided the side effects were low, there should be greater compliance, and one could begin to imagine an older population whose immune responses were not in decline.

With this in mind, perhaps we could imagine what changes there would be in the future if these therapies were available. Earlier in this chapter we discussed the costs as well as the number of deaths directly attributable to pneumonia and influenza in the elderly. Because of age-related physiological changes, diseases such as influenza may present atypically, without the usual fever and so may be missed initially. This can lead to longer periods in hospital and an associated loss of physical condition. Older individuals may not have the reserves to achieve complete recovery, and so significant and permanent disability may follow a bout of acute influenza (McElhaney 2005). Trying to provide protection through vaccination may prove difficult because of the inadequacy of the immune system in older individuals. This is a rather bleak and costly picture of what is in store for some of us, but if a means of reversing age-related depression of immunity becomes available and is in widespread use, then improving immune function in the elderly could significantly reduce the mortality and morbidity associated not only with influenza but also with a broad range of bacterial and viral infections. Moreover, an improved immune system would also enable older individuals to make better responses to vaccines. Whilst this may reduce the spread of infections, it may also prove beneficial in other diseases. For example, vaccines developed for prostate cancer (Sonpavde et al. 2007) may require boosting of the immune system to be fully effective.

A strategy to rejuvenate the immune response in older individuals, to provide them with an immune system that would make an adequate response both to vaccines and to potentially pathogenic agents, would have a significant impact on healthcare provided there were good compliance. It's difficult to be precise on the actual amount of money that would be saved, but one could believe that such a treatment would lead to a considerable extension of healthspan and a substantial reduction in healthcare costs.

References

Allman D, Sambandam A, Kim S, Miller JP, Pagan A, Well D, Meraz A, Bhandoola A (2003) Thymopoiesis independent of common lymphoid progenitors. Nat Immunol 4:168–174

Almanzar G, Schwaiger S, Jenewein B, Keller M, Herndler-Brandstetter D, Wurzner R, Schonitzer D, Grubeck-Loebenstein B (2005) Long-term cytomegalovirus infection leads to significant changes in the composition of the CD8+ T-cell repertoire, which may be the basis for an imbalance in the cytokine production profile in elderly persons. J Virol 79:3675–3683

Almeida AR, Rocha B, Freitas AA, Tanchot C (2005) Homeostasis of T cell numbers: from thymus production to peripheral compartmentalization and the indexation of regulatory T cells. Semin Immunol 17:239–249

Andersen H, Barsov EV, Trivett MT, Trubey CM, Giavedoni LD, Lifson JD, Ott DE, Ohlen C (2007) Transduction with human telomerase reverse transcriptase immortalizes a rhesus macaque CD8+ T cell clone with maintenance of surface marker phenotype and function. AIDS Res Hum Retroviruses 23:456–465

Andrew D, Aspinall R (2001) Il-7 and not stem cell factor reverses both the increase in apoptosis and the decline in thymopoiesis seen in aged mice. J Immunol 166:1524–1530

Andrew D, Aspinall R (2002) Age-associated thymic atrophy is linked to a decline in IL-7 production. Exp Gerontol 37:455–463

Aspinall R (1997) Age-associated thymic atrophy in the mouse is due to a deficiency affecting rearrangement of the TCR during intrathymic T cell development. J Immunol 158:3037–3045

Aspinall R (2000) Longevity and the immune response. Biogerontology 1:273–278

Aspinall R, Pido-Lopez J, Imami N, Henson SM, Ngom PT, Morre M, Niphuis H, Remarque E, Rosenwirth B, Heeney JL (2007) Old rhesus macaques treated with interleukin-7 show increased trec levels and respond well to influenza vaccination. Rejuvenation Res 10:5–18

Bennett AR, Farley A, Blair NF, Gordon J, Sharp L, Blackburn CC (2002) Identification and characterization of thymic epithelial progenitor cells. Immunity 16:803–814

Bertho JM, Demarquay C, Moulian N, Van De Meeren A, Berrih-Akhnin S, Gourmelon P (1997) Phenotypic and immunohistological analyses of the human adult thymus: evidence for an active thymus during adult life. Cell Immunol 179:30–40

Black C, Cummins E, Royle P, Philip S, Waugh N (2007) The clinical effectiveness and cost-effectiveness of inhaled insulin in diabetes mellitus: a systematic review and economic evaluation. Health Technol Assess 11:1–126

Blackburn CC, Manley NR (2004) Developing a new paradigm for thymus organogenesis. Nat Rev Immunol 4:278–289

Bockman DE, Kirby ML (1984) Dependence of thymus development on derivatives of the neural crest. Science 223:498–500

Boutolleau D, Bonduelle O, Sabard A, Devers L, Agut H, Gautheret-Dejean A (2005) Detection of human herpesvirus 7 DNA in peripheral blood reflects mainly CD4+ cell count in patients infected with HIV. J Med Virol 76:223–228

Capone M, Hockett RD Jr, Zlotnik A (1998) Kinetics of T cell receptor beta, gamma, and delta rearrangements during adult thymic development: T cell receptor rearrangements are present in CD44(+)CD25(+) Pro-T thymocytes. Proc Natl Acad Sci USA 95:12522–12527

Caton AJ (2003) Mechanisms, manifestations, and failures of self-tolerance. Immunol Res 27:161–168

Charlesworth B (2000) Fisher, Medawar, Hamilton and the evolution of aging. Genetics 156:927–931

Clark RA, Yamanaka K, Bai M, Dowgiert R, Kupper TS (2005) Human skin cells support thymus-independent T cell development. J Clin Invest 115:3239–3249

Clarke D, Katoh O, Gibbs RV, Griffiths SD, Gordon MY (1995) Interaction of interleukin 7 (IL-7) with glycosaminoglycans and its biological relevance. Cytokine 7:325–330

Cobbold M, Khan N, Pourgheysari B, Tauro S, McDonald D, Osman H, Assenmacher M, Billingham L, Steward C, Crawley C, Olavarria E, Goldman J, Chakraverty R, Mahendra P, Craddock C, Moss PA (2005) Adoptive transfer of cytomegalovirus-specific CTL to stem cell transplant patients after selection by HLA-peptide tetramers. J Exp Med 202:379–386

Corral L, Hanke T, Vance RE, Cado D, Raulet DH (2000) NK cell expression of the killer cell lectin-like receptor G1 (KLRG1), the mouse homolog of MAFA, is modulated by MHC class I molecules. Eur J Immunol 30:920–930

Cossarizza A, Ortolani C, Paganelli R, Barbieri D, Monti D, Sansoni P, Fagiolo U, Castellani G, Bersani F, Londei M, Franceschi C (1996) CD45 isoforms expression on CD4+ and CD8+ T cells throughout life, from newborns to centenarians: implications for T cell memory. Mech Ageing Dev 86:173–195

Dagarag M, Evazyan T, Rao N, Effros RB (2004) Genetic manipulation of telomerase in HIV-specific CD8+ T cells: enhanced antiviral functions accompany the increased proliferative potential and telomere length stabilization. J Immunol 173:6303–6311

Dare R, Sykes PJ, Morley AA, Brisco MJ (2006) Effect of age on the repertoire of cytotoxic memory (CD8+CD45RO+) T cells in peripheral blood: the use of rearranged T cell receptor gamma genes as clonal markers. J Immunol Methods 308:1–12

Department of Health. Getting ahead of the curve. A strategy fior combatting infectious disease. 1-10-2002. Ref Type: Generic

Di Santo JP, Aifantis I, Rosmaraki E, Garcia C, Feinberg J, Fehling HJ, Fischer A, von Boehmer H, Rocha B (1999) The common cytokine receptor gamma chain and the pre-T cell receptor provide independent but critically overlapping signals in early alpha/beta T cell development. J Exp Med 189:563–574

Dixit VD, Yang H, Sun Y, Weeraratna AT, Youm YH, Smith RG, Taub DD (2007) Ghrelin promotes thymopoiesis during aging. J Clin Invest 117:2778–2790

Drori D, Folman Y (1976) Environmental effects on longevity in the male rat: exercise, mating, castration and restricted feeding. Exp Gerontol 11:25–32

Effros RB, Pawelec G (1997) Replicative senescence of T cells: does the Hayflick limit lead to immune exhaustion? Immunol Today 18:450–454

Eskes T, Haanen C (2007) Why do women live longer than men? Eur J Obstet Gynecol Reprod Biol 133:126–133

Fahy GM (2003) Apparent induction of partial thymic regeneration in a normal human subject: a case report. J Anti Aging Med 6:219–227

Fehling HJ, von Boehmer H (1997) Early alpha beta T cell development in the thymus of normal and genetically altered mice. Curr Opin Immunol 9:263–275

Finch PW, Rubin JS (2006) Keratinocyte growth factor expression and activity in cancer: implications for use in patients with solid tumors. J Natl Cancer Inst 98:812–824

Freitas AA, Rocha B (2000) Population biology of lymphocytes: the flight for survival. Annu Rev Immunol 18:83–111

French RA, Broussard SR, Meier WA, Minshall C, Arkins S, Zachary JF, Dantzer R, Kelley KW (2002) Age-associated loss of bone marrow hematopoietic cells is reversed by GH and accompanies thymic reconstitution. Endocrinology 143:690–699

Fry TJ, Connick E, Falloon J, Lederman MM, Liewehr DJ, Spritzler J, Steinberg SM, Wood LV, Yarchoan R, Zuckerman J, Landay A, Mackall CL (2001) A potential role for interleukin-7 in T-cell homeostasis. Blood 97:2983–2990

Fry TJ, Mackall CL (2002b) Interleukin-7 and immunorestoration in HIV: beyond the thymus. J Hematother Stem Cell Res 11:803–807

Fry TJ, Mackall CL (2002a) Interleukin-7: from bench to clinic. Blood 99:3892–3904

George AJ, Ritter MA (1996) Thymic involution with ageing: obsolescence or good housekeeping? [see comments]. Immunol Today 17:267–272

Gill J, Malin M, Hollander GA, Boyd R (2002) Generation of a complete thymic microenvironment by MTS24(+) thymic epithelial cells. Nat Immunol 3:635–642

Goldberg GL, Alpdogan O, Muriglan SJ, Hammett MV, Milton MK, Eng JM, Hubbard VM, Kochman A, Willis LM, Greenberg AS, Tjoe KH, Sutherland JS, Chidgey A, Van Den Brink MR, Boyd RL (2007) Enhanced immune reconstitution by sex steroid ablation following allogeneic hemopoietic stem cell transplantation. J Immunol 178:7473–7484

Goronzy JJ, Fulbright JW, Crowson CS, Poland GA, O'Fallon WM, Weyand CM (2001) Value of immunological markers in predicting responsiveness to influenza vaccination in elderly individuals. J Virol 75:12182–12187

Gottschalk S, Heslop HE, Rooney CM (2005) Adoptive immunotherapy for EBV-associated malignancies. Leuk Lymphoma 46:1–10

Habib T, Nelson A, Kaushansky K (2003) IL-21: a novel IL-2-family lymphokine that modulates B, T, and natural killer cell responses. J Allergy Clin Immunol 112:1033–1045

Hadrup SR, Strindhall J, Kollgaard T, Seremet T, Johansson B, Pawelec G, thor SP, Wikby A (2006) Longitudinal studies of clonally expanded CD8 T cells reveal a repertoire shrinkage predicting mortality and an increased number of dysfunctional cytomegalovirus-specific T cells in the very elderly. J Immunol 176:2645–2653

Hamilton JB, Mestler GE (1969) Mortality and survival: comparison of eunuchs with intact men and women in a mentally retarded population. J Gerontol 24:395–411

Hannoun C, Megas F, Piercy J (2004) Immunogenicity and protective efficacy of influenza vaccination. Virus Res 103:133–138

Hare KJ, Jenkinson EJ, Anderson G (2000) An essential role for the IL-7 receptor during intrathymic expansion of the positively selected neonatal T cell repertoire. J Immunol 165:2410–2414

Heng TS, Goldberg GL, Gray DH, Sutherland JS, Chidgey AP, Boyd RL (2005) Effects of castration on thymocyte development in two different models of thymic involution. J Immunol 175:2982–2993

Henson SM, Snelgrove R, Hussell T, Wells DJ, Aspinall R (2005) An IL-7 Fusion Protein That Shows Increased Thymopoietic Ability. J Immunol 175:4112–4118

Hulstaert F, Hannet I, Deneys V, Munhyeshuli V, Reichert T, De Bruyere M, Strauss K (1994) Age-related changes in human blood lymphocyte subpopulations. II. Varying kinetics of percentage and absolute count measurements Age-related changes in human blood lymphocyte subpopulations. II. Varying kinetics of percentage and absolute count measurements. Clin Immunol Immunopathol 70:152–158

Janeway CA, Travers P, Walport M, Schlomchik MJ (2004) Immunobiology: The immune system in health and disease, 6th edn. Taylor and Fracis Inc

Jenkins JS (1998) The voice of the castrato. Lancet 351:1877–1880

Jiang Q, Li WQ, Aiello FB, Mazzucchelli R, Asefa B, Khaled AR, Durum SK (2005) Cell biology of IL-7, a key lymphotrophin. Cytokine Growth Factor Rev 16:513–533

Jones KE, Patel NG, Levy MA, Storeygard A, Balk D, Gittleman JL, Daszak P (2008) Global trends in emerging infectious diseases. Nature 451:990–993

Jordan MS, Boesteanu A, Reed AJ, Petrone AL, Holenbeck AE, Lerman MA, Naji A, Caton AJ (2001) Thymic selection of CD4+CD25+ regulatory T cells induced by an agonist self-peptide. Nat Immunol 2:301–306

Kaech SM, Tan JT, Wherry EJ, Konieczny BT, Surh CD, Ahmed R (2003) Selective expression of the interleukin 7 receptor identifies effector CD8 T cells that give rise to long-lived memory cells. Nat Immunol 4:1191–1198

Kendall MD, Fitzpatrick FT, Greenstein BD, Khoylou F, Safieh B, Hamblin A (1990) Reversal of ageing changes in the thymus of rats by chemical or surgical castration. Cell Tissue Res 261:555–564

Khan N, Hislop A, Gudgeon N, Cobbold M, Khanna R, Nayak L, Rickinson AB, Moss PA (2004) Herpesvirus-specific CD8 T cell immunity in old age: cytomegalovirus impairs the response to a coresident EBV infection. J Immunol 173:7481–7489

Khan N, Shariff N, Cobbold M, Bruton R, Ainsworth JA, Sinclair AJ, Nayak L, Moss PA (2002) Cytomegalovirus seropositivity drives the CD8 T cell repertoire toward greater clonality in healthy elderly individuals. J Immunol 169:1984–1992

Kim K, Lee CK, Sayers TJ, Muegge K, Durum SK (1998) The trophic action of IL-7 on pro-T cells: inhibition of apoptosis of pro-T1, -T2, and -T3 cells correlates with Bcl-2 and Bax levels and is independent of Fas and p53 pathways. J Immunol 160:5735–5741

Kimura K, Matsubara H, Sogoh S, Kita Y, Sakata T, Nishitani Y, Watanabe S, Hamaoka T, Fujiwara H (1991) Role of glycosaminoglycans in the regulation of T cell proliferation induced by thymic stroma-derived T cell growth factor. J Immunol 146:2618–2624

Koo GC, Huang C, Camacho R, Trainor C, Blake JT, Sirotina-Meisher A, Schleim KD, Wu TJ, Cheng K, Nargund R, McKissick G (2001) Immune enhancing effect of a growth hormone secretagogue. J Immunol 166:4195–4201

Kramer A, Schwebke I, Kampf G (2006) How long do nosocomial pathogens persist on inanimate surfaces? A systematic review. BMC Infect Dis 6:130

Leong HN, Tuke PW, Tedder RS, Khanom AB, Eglin RP, Atkinson CE, Ward KN, Griffiths PD, Clark DA (2007) The prevalence of chromosomally integrated human herpesvirus 6 genomes in the blood of UK blood donors. J Med Virol 79:45–51

Looney RJ, Hasan MS, Coffin D, Campbell D, Falsey AR, Kolassa J, Agosti JM, Abraham GN, Evans TG (2001) Hepatitis B immunization of healthy elderly adults: relationship between naive CD4+ T cells and primary immune response and evaluation of GM-CSF as an adjuvant. J Clin Immunol 21:30–36

Maraskovsky E, Teepe M, Morrissey PJ, Braddy S, Miller RE, Lynch DH, Peschon JJ (1996) Impaired survival and proliferation in IL-7 receptor-deficient peripheral T cells. J Immunol 157:5315–5323

McElhaney JE (2005) The unmet need in the elderly: designing new influenza vaccines for older adults. Vaccine 23(Suppl 1):S10–S25

Melchionda F, Fry TJ, Milliron MJ, McKirdy MA, Tagaya Y, Mackall CL (2005) Adjuvant IL-7 or IL-15 overcomes immunodominance and improves survival of the CD8(+) memory cell pool. J Clin Invest 115:1177–1187

Michaelson MD, Mehler MF, Xu H, Gross RE, Kessler JA (1996) Interleukin-7 is trophic for embryonic neurons and is expressed in developing brain. Dev Biol 179:251–263

Michie CA, McLean A, Alcock C, Beverley PC (1992) Lifespan of human lymphocyte subsets defined by CD45 isoforms. Nature 360:264–265

Min D, Panoskaltsis-Mortari A, Kuro O, Hollander GA, Blazar BR, Weinberg KI (2007) Sustained thymopoiesis and improvement in functional immunity induced by exogenous KGF administration in murine models of aging. Blood 109:2529–2537

Msaouel P, Diamanti E, Tzanela M, Koutsilieris M (2007) Luteinising hormone-releasing hormone antagonists in prostate cancer therapy. Expert Opin Emerg Drugs 12:285–299

Muegge K, Vila MP, Durum SK (1993) Interleukin-7: a cofactor for V(D)J rearrangement of the T cell receptor beta gene. Science 261:93–95

Napolitano LA, Lo JC, Gotway MB, Mulligan K, Barbour JD, Schmidt D, Grant RM, Halvorsen RA, Schambelan M, McCune JM (2002) Increased thymic mass and circulating naive CD4 T cells in HIV-1-infected adults treated with growth hormone. AIDS 16:1103–1111

Nash D, Mostashari F, Fine A, Miller J, O'Leary D, Murray K, Huang A, Rosenberg A, Greenberg A, Sherman M, Wong S, Layton M (2001) The outbreak of West Nile virus infection in the New York City area in 1999. N Engl J Med 344:1807–1814

Nishiwaki M, Fujimuro M, Teishikata Y, Inoue H, Sasajima H, Nakaso K, Nakashima K, Sadanari H, Yamamoto T, Fujiwara Y, Ogawa N, Yokosawa H (2006) Epidemiology of Epstein-Barr virus, cytomegalovirus, and Kaposi's sarcoma-associated herpesvirus infections in peripheral blood leukocytes revealed by a multiplex PCR assay. J Med Virol 78:1635–1642

Oosterwegel MA, Haks MC, Jeffry U, Murray R, Kruisbeek AM (1997) Induction of TCR gene rearrangements in uncommitted stem cells by a subset of IL-7 producing, MHC class-II-expressing thymic stromal cells. Immunity 6:351–360

Pawelec G, Akbar A, Caruso C, Effros R, Grubeck-Loebenstein B, Wikby A (2004) Is immunosenescence infectious? Trends Immunol 25:406–410

Pawelec G, Barnett Y, Forsey R, Frasca D, Globerson A, McLeod J, Caruso C, Franceschi C, Fulop T, Gupta S, Mariani E, Mocchegiani E, Solana R (2002) T cells and aging, January 2002 update. Front Biosci 7:d1056–d1183

Peschon JJ, Morrissey PJ, Grabstein KH, Ramsdell FJ, Maraskovsky E, Gliniak BC, Park LS, Ziegler SF, Williams DE, Ware CB, et al. (1994) Early lymphocyte expansion is severely impaired in interleukin 7 receptor-deficient mice. J Exp Med 180:1955–1960

Petrie HT, Livak F, Schatz DG, Strasser A, Crispe IN, Shortman K (1993) Multiple rearrangements in T cell receptor alpha chain genes maximize the production of useful thymocytes. J Exp Med 178:615–622

Pido-Lopez J, Imami N, Andrew D, Aspinall R (2002) Molecular quantitation of thymic output in mice and the effect of IL-7. Eur J Immunol 32:2827–2836

Pido-Lopez J, Imami N, Aspinall R (2001) Both age and gender affect thymic output: more recent thymic migrants in females than males as they age. Clin Exp Immunol 125:409–413

Posnett DN, Sinha R, Kabak S, Russo C (1994) Clonal populations of T cells in normal elderly humans: the T cell equivalent to "benign monoclonal gammapathy". J Exp Med 179:609–618

Rauser G, Einsele H, Sinzger C, Wernet D, Kuntz G, Assenmacher M, Campbell JD, Topp MS (2004) Rapid generation of combined CMV-specific CD4+ and CD8+ T-cell lines for adoptive transfer into recipients of allogeneic stem cell transplants. Blood 103:3565–3572

Rufer N, Migliaccio M, Antonchuk J, Humphries RK, Roosnek E, Lansdorp PM (2001) Transfer of the human telomerase reverse transcriptase (TERT) gene into T lymphocytes results in extension of replicative potential. Blood 98:597–603

Saurwein-Teissl M, Lung TL, Marx F, Gschosser C, Asch E, Blasko I, Parson W, Bock G, Schonitzer D, Trannoy E, Grubeck-Loebenstein B (2002) Lack of antibody production following immunization in old age: association with CD8(+)CD28(-) T cell clonal expansions and an imbalance in the production of Th1 and Th2 cytokines. J Immunol 168:5893–5899

Schmader K (2001) Herpes zoster in older adults. Clin Infect Dis 32:1481–1486

Seggewiss R, Lore K, Guenaga FJ, Pittaluga S, Mattapallil J, Chow CK, Koup RA, Camphausen K, Nason MC, Meier-Schellersheim M, Donahue RE, Blazar BR, Dunbar CE, Douek DC (2007) Keratinocyte growth factor augments immune reconstitution after autologous hematopoietic progenitor cell transplantation in rhesus macaques. Blood

Shimoyama Y, Oyama T, Asano N, Oshiro A, Suzuki R, Kagami Y, Morishima Y, Nakamura S (2006) Senile Epstein-Barr virus-associated B-cell lymphoproliferative disorders: a mini review. J Clin Exp Hematop 46:1–4

Shortman K, Wu L (1996) Early T lymphocyte progenitors. Annu Rev Immunol 14:29–47

Smith RG (2005) Development of growth hormone secretagogues. Endocr Rev 26:346–360

Smith RG, Jiang H, Sun Y (2005) Developments in ghrelin biology and potential clinical relevance. Trends Endocrinol Metab 16:436–442

Sonpavde G, Spencer DM, Slawin KM (2007) Vaccine therapy for prostate cancer. Urol Oncol 25:451–459

Storek J, Gillespy T, III, Lu H, Joseph A, Dawson MA, Gough M, Morris J, Hackman RC, Horn PA, Sale GE, Andrews RG, Maloney DG, Kiem HP (2003) Interleukin-7 improves CD4 T-cell reconstitution after autologous CD34 cell transplantation in monkeys. Blood 101:4209–4218

Sutherland JS, Goldberg GL, Hammett MV, Uldrich AP, Berzins SP, Heng TS, Blazar BR, Millar JL, Malin MA, Chidgey AP, Boyd RL (2005) Activation of thymic regeneration in mice and humans following androgen blockade. J Immunol 175:2741–2753

Thompson WW, Shay DK, Weintraub E, Brammer L, Cox N, Anderson LJ, Fukuda K (2003) Mortality associated with influenza and respiratory syncytial virus in the United States. JAMA 289:179–186

Thwaites G, Taviner M, Gant V (1997) The English sweating sickness, 1485 to 1551. N Engl J Med 336:580–582

Toraldo G, Roggia C, Qian WP, Pacifici R, Weitzmann MN (2003) IL-7 induces bone loss in vivo by induction of receptor activator of nuclear factor kappa B ligand and tumor necrosis factor alpha from T cells. Proc Natl Acad Sci USA 100:125–130

Trepel F (1974) Number and distribution of lymphocytes in man. A critical analysis. Klin Wochenschr 52:511–515

Verra NC, Jorritsma A, Weijer K, Ruizendaal JJ, Voordouw A, Weder P, Hooijberg E, Schumacher TN, Haanen JB, Spits H, Luiten RM (2004) Human telomerase reverse transcriptase-transduced human cytotoxic T cells suppress the growth of human melanoma in immunodeficient mice. Cancer Res 64:2153–2161

Voehringer D, Blaser C, Brawand P, Raulet DH, Hanke T, Pircher H (2001) Viral infections induce abundant numbers of senescent CD8 T cells. J Immunol 167:4838–4843

Voehringer D, Koschella M, Pircher H (2002) Lack of proliferative capacity of human effector and memory T cells expressing killer cell lectinlike receptor G1 (KLRG1). Blood 100:3698–3702

Ward KN (2005) The natural history and laboratory diagnosis of human herpesviruses-6 and -7 infections in the immunocompetent. J Clin Virol 32:183–193

Whitacre CA, Reingold SC, O'Looney PA, Task Force (1999) A Gender Gap in Autoimmunity. Science 283:1277–1278
Wikby A, Ferguson F, Forsey R, Thompson J, Strindhall J, Lofgren S, Nilsson BO, Ernerudh J, Pawelec G, Johansson B (2005) An immune risk phenotype, cognitive impairment, and survival in very late life: impact of allostatic load in Swedish octogenarian and nonagenarian humans. J Gerontol A Biol Sci Med Sci 60:556–565
Wu L, Li CL, Shortman K (1996) Thymic dendritic cell precursors: relationship to the T lymphocyte lineage and phenotype of the dendritic cell progeny. J Exp Med 184:903–911
Yoshikawa TT (1997) Perspective: aging and infectious diseases: past, present, and future. J Infect Dis 176:1053–1057
Yoshikawa TT (2000) Epidemiology and unique aspects of aging and infectious diseases. Clin Infect Dis 30:931–933
Zamisch M, Moore-Scott B, Su DM, Lucas PJ, Manley N, Richie ER (2005) Ontogeny and regulation of IL-7-expressing thymic epithelial cells. J Immunol 174:60–67

Chapter 16
Mitochondrial Manipulation as a Treatment for Aging

Rafal Smigrodzki and Francisco R. Portell

Contents

16.1	Introduction to the Mitochondrial Theory of Aging	521
	16.1.1 Mitochondrial Genome Deterioration with Aging	521
	16.1.2 Biochemical Aspects of Age-Related Mitochondrial Dysfunction	524
	16.1.3 Animal Models of Mitochondrial Aging	527
	16.1.4 Age-Related Conditions with Known Mitochondrial Contributions	529
	16.1.5 Mitochondrial Mutations as Causal Vs. Epiphenomenal Aspects of Aging	530
16.2	Methods for Mitochondrial Genome Manipulation	530
	16.2.1 Current State of the Art	530
	16.2.2 Technical Prerequisites for Mitochondrial Gene Therapy – Gene Delivery Vector	532
	16.2.3 Technical Prerequisites for Mitochondrial Gene Therapy – Human mtDNA Source	532
	16.2.4 Strategies for Mitochondrial DNA Gene Therapy – Supplementation	532
	16.2.5 Strategies for Mitochondrial DNA Gene Therapy – Replacement	533
16.3	Conclusion – Prospects for Mitochondrial Gene Therapy in Disease and Aging	534
	16.3.1 Initial Targets for Mitochondrial Gene Therapy	534
	16.3.2 Application of Mitochondrial Gene Therapy in Aging-Related Conditions	535
References		535

16.1 Introduction to the Mitochondrial Theory of Aging

16.1.1 Mitochondrial Genome Deterioration with Aging

Mitochondria are cellular organelles involved in the generation of energy by oxidative phosphorylation. They are also crucial for the process of controlled cellular

---

R. Smigrodzki (✉)
Gencia, Inc., Charlottesville, VA, USA
e-mail: rafal@genciabiotech.com

death or apoptosis, as well as metabolism of lipids, nucleotides and some amino acids. Mitochondria are unique among cytoplasmic components in that they contain their own genome, a circular DNA molecule of 16, 569 bp (in humans), coding for some components of the electron transport chain (ETC) – a series of five macromolecular enzymatic complexes responsible for oxidative phosphorylation. There are many copies of the mitochondrial genome (mtDNA) in each human cell, ranging from 2 to 3 copies in platelets to about 200,000 in the ovum, with an average of 4,900 per cell (Bhat and Epelboym 2004).

Mutations are frequently observed in mtDNA and have been proven to cause a number of pathological conditions, including both early onset, deadly diseases such as MELAS (Mitochondrial Encephalopathy, Lactic Acidosis, Stroke-like episodes) (Schmiedel et al. 2003), and milder conditions of middle age, such as MIDD (Maternally inherited Diabetes and Deafness) (Barrett 2001). The presence of a mutation in all mtDNA copies in a cell or organism is termed homoplasmy. Mutations can also affect only a fraction of mtDNA copies – this is called heteroplasmy. The percentage of affected genomes has an impact on the phenotypic effects of a mutation. For example, the most common MELAS mutation, the A3243G transition, may be present at up to 5% heteroplasmy (Cervin et al. 2004) without obvious pathological symptoms, may cause MIDD at 5–20% heteroplasmy, and finally may result in the full MELAS syndrome at heteroplasmy levels 70% and above (Kato et al. 2002). While most of the known pathogenic mutations have a relatively high threshold for causing symptoms (from 70 to as high as 95%), there are mutations capable of causing severe multisystem disorders at heteroplasmy levels as low as 13%, with detectable biochemical dysfunction in cell culture at 4% (Yechoor et al. 2004).

There is extensive evidence for the accumulation of mitochondrial mutations with aging. For the purpose of a discussion it is useful to divide age related mtDNA mutations into two classes based on their abundance: microheteroplasmy, and macroheteroplasmy.

Microheteroplasmy is the presence of single-base substitutions and frameshifts which appear to be randomly distributed along the genome (with the exception of the control region which has a higher frequency of such mutations) and are found in all tissues that have been so far examined for their presence (Smigrodzki and Khan 2005). Each of these mutations in an individual may be found at less than 1% heteroplasmy but thousands of such mutations are present in each cell of an aged individual, adding to a considerable burden of mutations, comparable with mutation burdens in classical mitochondrial disorders. There is a continuous, gradual increase in microheteroplasmic mutation frequency during aging, from less than 50 per million basepairs in the neonate, to over 160 per million in aged adults (Lin et al. 2002; Simon et al. 2004). This translates into 1.1–1.3 missense (aminoacid-altering) mutations per mitochondrial genome in the elderly. An additional form of random mutational damage found in the elderly are mtDNA deletions, which are known to accumulate in muscle and spread clonally, accounting for as much as 70% for all mitochondrial genomes in 80-year old humans (Chabi et al. 2005).

Age-related macroheteroplasmy are mutations that accumulate at relatively high levels, up to 90–100%. These mutations have a non-random distribution – they are found predominantly in the mitochondrial control region, the D-loop, and accumulate only in some tissues, such as peripheral blood (Zhang et al. 2003). Certain specific long deletions such as the "common deletion" may also accumulate to very high levels in some tissues (Lai et al. 2003). A significant accumulation of deletions of varying lengths has also been observed in such postmitotic tissues as muscle and brain tissue (e.g. the substantia nigra) (Kraytsberg et al. 2006; Bender et al. 2006). A correlation between the presence of deletions in individual sarcomers and mitochondrial dysfunction in these sarcomers has been found, supporting the notion that deletions may have a causative role in age-associated muscle dysfunction (Pak and Aiken 2004; Gokey et al. 2004).

The exact mechanism of mtDNA mutation accumulation in aging is not entirely clear. Two contributing processes are most likely present: de novo generation of mutations and clonal expansion of mutations present from birth. Mitochondrial DNA is replicated by DNA polymerase-gamma which is transcribed from a nuclear-encoded gene (PolG). This enzyme has a higher error rate than nuclear DNA polymerases, especially under the in vivo conditions of unequal precursor (deoxynucleotide) concentrations (Song et al. 2005). Since mtDNA is undergoing continuous recycling in all tissues, with a half-life of as little as 2 weeks (Menzies and Gold 1971), replication errors are likely to accumulate inexorably over the lifetime of the individual. The mutation spectrum due to pol-gamma errors is consistent with microheteroplasmy, that is, randomly distributed, ubiquitously present substitutions and small frameshifts. Another source of mutations is damage to mtDNA due to various chemical and physical insults, such as reactive oxygen species (ROS) (Lee and Wei 2007) and other free radicals, or UV radiation (Birket and Birch-Machin 2007) and to a lesser extent (Prithivirajsingh et al. 2004), ionizing radiation. Since DNA repair mechanisms in mitochondria are less active than in the nucleus (Ledoux et al. 2007), random mutations due to these mechanisms will accumulate, especially in tissues with high levels of ROS generation, such as muscle or substantia nigra.

Clonal expansion is a process where the frequency of mutations in a cell or tissue changes due to either random drift or specific replication advantage. Simulations indicate that random drift is capable of causing rapid and profound changes in mutation frequency in single cells (Chinnery et al. 2002), however, some form of replicative advantage is most likely needed to explain tissue-wide macroheteroplasmic accumulation of mutations. Replicative advantage may be present either at the level of genomes within each cell, or at the level of individual cells within a tissue. Independently of the mechanism, clonal expansion is likely to be the mechanism responsible for macroheteroplasmy.

While this is not the focus of the present review, the mechanism of resetting the "mitochondrial aging clock" with each generation has received considerable attention. Since there is some accumulation of mutations between conception and sexual maturity, without some way of removing the accumulated mutations from the germline, each subsequent generation would be born with an ever larger fraction of mutant genomes. During ontogeny, about 400,000 oocytes are generated and spend many years in a quiescent state, with very low levels of mitochondrial oxidative phosphorylation, presumably accumulating mutations at a much lower rate than other tissues (Cetica et al. 2002). According to current concepts, oocytes undergo a stringent selection after a switch to oxidative metabolism during their further maturation (Fan et al. 2008). Cells exhibiting biochemical symptoms of mitochondrial dysfunction (increased ROS production, impaired ATP generation) undergo apoptosis and only oocytes with still intact mitochondrial genomes mature into ova. Given large numbers of starting cells, it is statistically likely that some of them will remain free of mutations for many years – yet when the last intact oocytes are depleted, menopause ensues.

16.1.2 Biochemical Aspects of Age-Related Mitochondrial Dysfunction

The mitochondrial genome codes for 13 protein subunits of the ETC, as well as 22 transfer RNAs and two ribosomal RNAs which are necessary for the translation of the mitochondrial protein subunits. The ETC accepts electrons from oxidative phosphorylation substrates, such as NADPH and succinate, and transfers them to oxygen in a series of reactions coupled to the pumping of protons into the mitochondrial matrix, producing an electrochemical gradient, $\Delta\Psi$, across the mitochondrial inner membrane. Movement of protons out of the matrix through the ATP synthase, the last complex of the ETC, is coupled to the synthesis of ATP, the primary intracellular chemical energy carrier. Oxidative phosphorylation allows complete oxidation of many nutrients, producing significantly more energy than other processes, such as glycolysis. In most mammalian cells, oxidative phosphorylation is the main source of energy and its impairment therefore is likely to be associated with an inability to meet metabolic requirements.

Mitochondrially encoded subunits are present in complexes I, III, IV and V but the vast majority of subunits are encoded in the nucleus, transcribed in the cytoplasm and subsequently imported into mitochondria. Thus the phenotypic effects of mtDNA mutations are at least initially mediated by perturbations in either the quantity or structure of only mitochondrially encoded ETC subunits.

It is useful to divide phenotypic manifestations of mutations into primary and secondary. Primary manifestations are directly related to abnormal amount or function of proteins coded by the mutated genes, while secondary manifestations represent adaptive or maladaptive responses of cells to the primary manifestations.

The most common primary biochemical phenotypic effects of mitochondrial mutations are presumed to be declines in ETC activity, and increased ROS production.

Mitochondrially encoded ETC subunits are believed to be important for the initiation of ETC complex assembly, thus mutations in these subunits reduce the number of active ETC subunits (Lazarou et al. 2007). Similar effects may occur with mutations in mitochondrial tRNA and rRNA genes, which may impair globally the translation of all mitochondrial ETC subunits. Severely diminished activity of one or more elements of the ETC (complex I, III, IV, and V) leads to impaired generation of ATP and frequently cellular death due to loss of energy needed for basic processes, such as maintenance of membrane potentials, protein and other macromolecule production and replacement. Diminished oxidative phosphorylation is seen in MELAS, NARP and other conditions. Marked impairment of ETC activity is also found in aging (Greco et al. 2003). It is however important to realize that reduced ETC activity may also be due to other causes, including nuclear gene mutations, and adaptive responses to stress (see discussion of secondary manifestations of mutations below).

ROS, such as superoxide, hydrogen peroxide and hydroxyl radicals are toxic byproducts of metabolism, generated in the mitochondria, and to a lesser extent in other cellular locations such as the plasma membrane (Bedard and Krause 2007). ROS are capable of reacting with a wide range of cellular targets, leading to inactivation of enzyme prosthetic groups, crosslinking of proteins, and widespread loss of metabolic function. The primary mitochondrial sources are complexes I and III, although complex II and non-ETC sources such as α-ketoglutarate dehydrogenase complex (Stuart and Brown 2006), glycerol-3-phosphate dehydrogenase and others, are also involved. ROS production is enhanced by ETC complexes present in the reduced state (i.e. after accepting an electron from a reducer, or electron donor). The electron may be transferred to e.g. an oxygen molecule, producing the highly reactive superoxide anion, O_2^-. ETC complexes stay in the reduced state if there is a blockage downstream in the transfer chain or if there is relative overloading of the upstream parts of the chain. Thus many mitochondrial poisons, such as rotenone, azide, or MPTP, cause high ROS production. Overloading of the ETC is known to occur with caloric overfeeding, and in some genetic contexts this results in increased ROS generation (Vidal-Puig et al. 2000). High intracellular levels of ROS are believed to accelerate the accumulation of mutations, especially in the mitochondrial genome which is located very close to the main sources of ROS, although there is no absolute correlation between ROS levels and mtDNA mutation accumulation (Trifunovic and Larsson 2008). Many classical mitochondrial diseases are associated with high levels of ROS. For example, mutations causative of LHON, or Leber's Hereditary Optic Neuropathy (Lenaz et al. 2004) tend to affect the binding of coenzyme Q to complex I genes, resulting in elevated ROS production and increased sensitivity to exogenous ROS (such as hydrogen peroxide). Similar effects are also observed in MELAS, PEO (Progressive External Ophthalmoplegia), and in cytochrome b mutations. Increased ROS production is well substantiated in human aging. Caloric restriction diminishes the energy load on the ETC, leading to reductions in ROS production and is known to increase longevity, which is consistent with ROS being an important part of the mechanism of aging.

In response to the primary effects of mutations, the cell mobilizes an array of secondary, compensatory mechanisms. These include the modulation of mitochondrial biogenesis, varying the number of mitochondria and ATP output. Turnover of mitochondria, removing components damaged by ROS, may be stimulated. Antioxidant defenses may be upregulated. Mitochondrial activity may also be downregulated – this occurs for example in Alzheimer's disease where the amyloid beta protein

(A-beta) has a direct inhibitory effect on the mitochondrial ATPase (complex V), occurring within 30 minutes of exposure (Mark et al. 1995). Exposure to A-beta also is known to depress mitochondrial function by reducing the expression of cytochrome c (a mitochondrial gene) and TFAM (mitochondrial transcription factor A), the product of a nuclear gene (Hong et al. 2007). This process may represent an adaptive response to increased production of ROS from the ETC complexes transcribed from mutated mtDNA copies accumulating during aging. However, the exact connection between amyloid and ROS is further complicated by the fact that some formulations of amyloid, especially in the presence of metal ions (copper, iron, zinc) (Smith et al. 2007), may directly increase the production of ROS, predominantly from non-mitochondrial sources (Abramov et al. 2005). An interaction of A-beta with A-beta alcohol dehydrogenase (ABAD) is also known to occur in mitochondria (Takuma et al. 2005) and increases ROS production. Very high levels of amyloid induce apoptosis and cell death. Thus a variety of both adaptive and maladaptive mitochondrial processes may be related to amyloid present in various concentrations. The biochemical outcome of a mitochondrial mutation is at present quite difficult to predict, since it depends on the levels of the mutation, as well as on the nuclear genetic background, the type of cells affected, and the availability of oxidative phosphorylation substrates. Thus, mice genetically engineered to accumulate high levels of random mtDNA mutations do not show increased levels of ROS (Mott et al. 2001; Trifunovic et al. 2005), or changed levels of antioxidants (Trifunovic et al. 2005), and only in some situations have diminished ATP production (Trifunovic et al. 2004) – yet they still do exhibit significant tissue dysfunction (Mott et al. 1999; Trifunovic et al. 2004; Kujoth et al. 2005). This sounds a note of caution, indicating that the mechanisms responsible for the mitochondrial contribution to aging are not well understood and extensive compensatory processes are likely to complicate further research efforts.

It is very important to realize that diminished mitochondrial activity may be either seen as a primary effect of an mtDNA mutation, a secondary response to e.g. increased ROS production from other mtDNA mutations, or a regulatory response to non-mitochondrial abnormalities. Detailed analysis of the state of the cell, including levels of mitochondrial biogenesis factors such as PGC1alpha, and cybrid experiments are necessary to differentiate between the three possibilities (Indo et al. 2007). Secondary mitochondrial activity depression in response to elevated endogenous ROS production from mitochondria may be very important for long term viability of cells, by reducing the rate of further mtDNA accumulation. It is therefore crucial to make the distinction between primary and secondary ETC activity reduction in any decisions regarding therapy – attempts to rev up downregulated mitochondria in e.g. AD may have a deleterious effect by removing a protective feedback loop responding to already damaged mtDNA. In this situation only the removal or inactivation of damaged mtDNA is likely to provide long-term improvement of function.

16.1.3 Animal Models of Mitochondrial Aging

The mitochondrial hypothesis of aging has received considerable support recently in the form of mitochondrial mutator mice. MtDNA is replicated by a specific mitochondrial DNA polymerase, pol γ (Fridlender et al. 1972). Certain mutations in the PolG gene coding for pol gamma diminish the fidelity of replication and are known to cause a severe early onset disease in humans, Alper's syndrome (Naviaux et al. 1999). Interestingly, a much milder mitochondrial disease, chronic progressive external ophthalmoplegia (CPEO), is also caused by polG mutations leading to the accumulation of mtDNA deletions (Van et al. 2002). A few strains of mice with PolG mutations have been produced, showing pathologies of different severity. The first one expresses the abnormal polymerase gamma in the heart in the first few weeks after birth, and develops a severe dilated cardiomyopathy, leading to death within a few months (Mott et al. 1999). Two other models (Trifunovic et al. 2004; Kujoth et al. 2005) express mutated forms of polymerase gamma ubiquitously and develop sarcopenia, anemia, kyphosis, reduced life span, hair graying and alopecia, cardiac enlargement with functional alterations, reduced fertility, accelerated thymic involution, osteoporosis, and age-related hearing loss (presbycusis) (Niu et al. 2007), thus partially reprising the aging phenotype at a younger age.

As noted previously, the polG mutator mice do not exhibit increased levels of ROS, or changes in endogenous antioxidant levels (Mott et al. 2001; Trifunovic et al. 2005). They do however have decreases in ETC (Trifunovic et al. 2004) activity levels and decreased mitochondrial ATP production, as well as increased activation of pro-apoptotic factors such as caspase 3 (Zhang et al. 2005). The mutation levels are between 3 and 2,500 times higher than observed in wild-type mice, depending on the methods used for estimation. In addition to the point mutations the mice also accumulate mtDNA molecules with a specific deletion affecting the region between the origins of replication of the light and heavy strands, and a reduction in mtDNA copy number by about 40%. A very important observation is that heterozygous mutator mice which have mutation frequencies as much as 500 times as high as wild-type mice, do not exhibit an abnormal phenotype (Vermulst et al. 2007). This finding was interpreted as evidence that mitochondrial mutations do not play a role in the aging of wild-type mice.

The combination of dramatically different mutation levels and lack of increased ROS production therefore set these models apart from natural aging. In addition, the time of onset of mutation accumulation is different: the mutator mice seem to start accumulating large amounts of mutations at a very early, even embryonic age, while wild-type mice seem to have a very low mutation burden at birth followed by an exponential accumulation. These findings raise the possibility that the enormous accumulation of mutations in the mutator mice is due in large part to generation of mutations in progenitor cells (both embryonic and tissue-specific stem cells), followed by selection for cells carrying relatively benign, non-ROS generating mutations, while stem cells burdened with higher ROS levels are eliminated very early. Under this interpretation the mutations in mutator mice would be present ubiquitously, have predominantly an impact on ATP production, would not generate

ROS, and early pathological phenotypes would be noted also in some tissues highly dependent on continued stem cell activity, such as blood (leading to anemia seen in mutator mice). MtDNA mutations in aging would in contrast accumulate preferentially (but not exclusively) in terminally differentiated tissues consisting of very long-lived cells, such as muscle and brain, and would be on average more toxic, causing both increased ROS production and ATP reduction.

The concept of age-related mtDNA mutations as different from mutations observed in polG mutator mice is supported by the following observations:

1) Age-related mitochondrial dysfunction and mutations are indeed observed extensively in the long-lived cells of the nervous system (Kraytsberg et al. 2006; Lin et al. 2002; Simon et al. 2004) and in muscle (Chabi et al. 2005). Very slow or absent cell turnover in these tissues may limit the ability to remove highly mutated cells by apoptosis and to replace them with less damaged cells derived from stem cells, a process that is active in epithelia and the immune system.

2) There is a wide range of toxicity among mitochondrial mutations, from compatible with survival at homoplasmic levels, to causing severe dysfunction at the 13% level (Yechoor et al. 2004). It is known that the most toxic mutations are preferentially removed from the germ line, presumably by ovum and zygote apoptosis, yielding an almost "blank slate" in terms of mutation load (Fan et al. 2008). It is also known that stem cells indeed undergo apoptosis at much lower ROS levels than differentiated cells (Oguro and Iwama 2007). Since the selection mechanism in germ line and stem cells is triggered by ROS, it would still allow a large accumulation of less toxic mutations in mutator mice while preferentially removing ROS-generating mutations. However, the apoptotic way of removing ROS-generating mutations is only possible in tissues with high turnover (as long as a viable stem cell pool exists) but it cannot freely proceed in low-turnover tissues, which could lead to a higher percentage of ROS-generators among mutations in terminally differentiated cells.

At present there is insufficient information on the mutation spectra differences between mutator and wild-type mice to confirm or exclude this interpretation. A detailed, nucleotide-by-nucleotide comparison of mutations in these two groups would need to be performed to yield a catalogue of mutations with differing toxic potential.

Overall, mutator mice are a valuable resource in research on the mechanisms of mitochondrial mutation accumulation but, just as progerias, they are only an imperfect model of aging. A crucial experiment determining the contribution of mtDNA mutations to aging might transfer the complete spectrum of mutations between e.g. young vs. old or mutator vs. wild-type. Further developments in mtDNA manipulation technology will hopefully help in answering the mtDNA and aging question in animals.

16.1.4 Age-Related Conditions with Known Mitochondrial Contributions

Abnormalities in mitochondrial function are well-documented in most of the common age-related pathological conditions, including cancer (Chen et al. 2007), the metabolic syndrome and diabetes (Maassen et al. 2007; Davidson and Yannicelli 2006), Parkinson's disease (Schapira 2008), Alzheimer's disease (Swerdlow and Khan 2004), hypertension (Puddu et al. 2007), and macular degeneration (Decanini et al. 2007).

A detailed analysis of the evidence in favor of mtDNA dysfunction in these conditions is beyond the scope of this review; however, we can describe some of the common biochemical themes related to this subject.

There is a significant degree of maternal inheritance of some of these conditions, especially hypertension (Yang et al. 2007), diabetes (Sun et al. 2003), and to a lesser degree, Alzheimer's (Edland et al. 1996). This is consistent with either inheritance of mitochondrial polymorphisms or mitochondrial microheteroplasmy.

There are decreases in mitochondrial ETC activity, either global or specifically affecting some of the ETC complexes. This is evident in Parkinson's disease (complex I loss) (Parker et al. 1989), Alzheimer's disease (complex IV loss) (Parker et al. 1990) and metabolic syndrome/diabetes (Kelley et al. 2002). In cancer there is a much more profound loss of ETC than in other conditions, to the point where neoplastic cells rely primarily on glycolysis for their energetic needs (Godinot et al. 2007).

There are increases in ROS generation, evident in diabetes/metabolic syndrome (Nicolson 2007), Alzheimer's disease (Lovell and Markesbery 2007), hypertension (Oliveira-Sales et al. 2008), cancer (Franco et al. 2008) and Parkinson's disease (Sayre et al. 2005).

In most situations these indicators of mitochondrial dysfunction can be transferred with mitochondria in cytoplasmic hybrid (cybrid) experiments, consistent with mitochondrial rather than nuclear genome dysfunction. This is well-documented in PD (Cassarino et al. 1997), AD (Swerdlow et al. 1997; Khan et al. 2000), cancer (Bonora et al. 2006; Lee et al. 2007a) and diabetes (de Andrade et al. 2006; Tawata et al. 2000).

Research on the familial, early-onset forms of these conditions reveals either well-defined mtDNA mutations (in MIDD) (Murphy et al. 2008), or a plethora of nuclear genes acting in or on mitochondrial targets in PD (Schapira 2008) and AD.

Overall, the research results on age-related conditions are consistent with a permissive role of accumulated mtDNA mutations, modified by multiple nuclear factors, and in some cases further influenced by inherited and early-acquired mutations (mutations accumulating in ova before fertilization).

16.1.5 Mitochondrial Mutations as Causal Vs. Epiphenomenal Aspects of Aging

Based on the foregoing discussion we can arrive at the following conclusions:

1. There is convincing evidence that mitochondrial mutations accumulate with aging
2. There is convincing evidence that significant mitochondrial bioenergetic dysfunction develops during human aging
3. There is suggestive but not conclusive evidence that age-related mitochondrial mutations are at least in part the cause of age-related mitochondrial dysfunction
4. The mechanism by which age-related mitochondrial mutations may contribute to aging is not sufficiently understood and will require a substantial amount of further research.

16.2 Methods for Mitochondrial Genome Manipulation

16.2.1 Current State of the Art

The mitochondrial genome, despite its small size and relative simplicity, has been found to be very difficult to clone and manipulate. There are no published reports on full-length cloning of human mtDNA, although murine mtDNA has been cloned in a low copy number vector (Yoon and Koob 2003). Yoon and Koob confirmed in this study that mtDNA contains transcription initiation sites active in bacteria, leading to expression of mitochondrial polycistronic RNA which is then processed to yield mitochondrial tRNA molecules that interfere with protein synthesis and thus may be highly toxic to the host cell.

Introduction of mitochondrial genomes into target cells has been accomplished by fusion with donor cells containing their own mitochondria (Trounce and Pinkert 2007). The target cells may be first depleted of their own mtDNA by incubation with ethidium bromide or dideoxycytidine (these are referred to as rho-zero cells), so as to promote take-over by donor mitochondria (King and Attardi 1989). Donor cells may be enucleated or be naturally anuclear, such as platelets. The resulting cytoplasmic hybrid cells, or cybrids, thus contain mtDNA from one source against a background of nuclear DNA from another source. While such cells are a valuable tool in mitochondrial research, the cybrid technology does not have a significant therapeutic potential in aging.

Full-length mtDNA introduction into isolated mitochondria has been achieved (Yoon and Koob 2003) but delivery in situ, inside cells or tissues, has been described only for very short DNA fragments, up to about 300 bp. Delivery of DNA into mitochondria in situ, inside cells or tissues, has been achieved with short oligonucleotides conjugated to a mitochondrial leader sequence (MLS) from a protein which is normally imported into mitochondria (Vestweber and Schatz 1989).

Peptide-nucleic acids (PNAs) have been delivered to mitochondria. Delivery to the surface of mitochondria has been achieved using polyethylenimine (Lee et al. 2007b), mitochondriotropic dequalinium compounds (D'souza et al. 2003), and liposome-like particles conjugated to an MLS (Boddapati et al. 2005), but no information on intramitochondrial delivery and expression of DNA with these methods is available. The protected location of mtDNA, surrounded by the inner and outer mitochondrial membranes as well as the plasma membrane, has no doubt contributed to difficulties in delivery. It is possible to deliver longer DNA molecules to mitochondria by biolistic transformation (the "gene gun") where tungsten microparticles coated with DNA are shot at high speeds into cells, but this method suffers from obvious limitations as a form of therapy (Bonnefoy and Fox 2007). Similar limitations pertain to the electroporation of isolated mitochondria which has been used to successfully transfect DNA molecules up to a few kilobases in length (Yoon and Koob 2003).

Given the problems in direct manipulation of mtDNA, a radical idea has been proposed to circumvent the need to deliver DNA to mitochondria, and instead express mitochondrial genes in the nucleus – so called allotopic expression (Grasso et al. 1991). It has been observed that during evolution there was a progressive transfer of genes from the mitochondrial genome to the nucleus, reducing the number of mitochondrial genes from many hundreds in some unicellular eukaryotes to only 13 in multicellular animals. The allotopic expression proposal would complete the gene transfer process and the 13 mitochondrial subunits of the ETC would be produced in the cytoplasm for subsequent import into mitochondria just as thousands of other proteins. So far only the ATP6 and ND4 genes have been successfully produced allotopically and used to correct a pathological phenotype in vitro (Bonnet et al. 2007; Pineau et al. 2005). This approach however suffers from the general limitations of nuclear gene therapy – low transformation efficiency, expression silencing, and risk of mutagenesis and oncogenesis. Additionally, the problem of controlling the expression levels of allotopic genes to match the local needs of mitochondria under various metabolic conditions would need to be solved. According to the colocalization for redox regulation hypothesis of John Allen, the close spatial relationship between the individual mitochondrial genomes and the mitochondrial membranes may be needed to assure fine tuning of local ETC complex assembly and activity in response to local (i.e. subcellular level) influences (Allen 1993). Full transfer of genetic control to the nucleus would significantly slow down this regulation process. Furthermore, since the mutated mtDNA is still present in the cell and likely to produce abnormal proteins, a simple introduction of allotopic protein may be insufficient to reverse the phenotype in general. The enthusiasm for allotopic expression as a therapeutic avenue is also tempered by the observation that the nuclear genome already contains a large number of integrated copies of the mitochondrial genome and its fragments, so-called mitochondrial pseudogenes (Perna and Kocher 1996). It appears that the process of transferring genetic control from mitochondria to the nucleus has stopped after reaching the current number of 13 genes, many hundreds of millions of years ago, despite numerous instances of the remaining genes themselves being introduced into the nucleus (Turner et al. 2003). This could indicate that much more than simple transfer and expression of genes may be needed to make the allotopic approach viable.

16.2.2 Technical Prerequisites for Mitochondrial Gene Therapy – Gene Delivery Vector

Clinically successful mitochondrial gene therapy would require the ability to deliver full-length mtDNA, both wild-type and engineered, to a wide range of cells in vivo. So far there are no published, peer-reviewed reports of such technical capability, but we believe such vectors may be announced soon.

The ideal vector would be non-viral, produced in bacterial rather than eukaryotic cells, non-immunogenic and would be able to transduce mtDNA into a wide range of tissue targets.

16.2.3 Technical Prerequisites for Mitochondrial Gene Therapy – Human mtDNA Source

Therapeutic applications of mitochondrial gene therapy would require a source of full-length mitochondrial DNA. It is possible to produce full-length human mtDNA by PCR, followed by ligation to circularize the molecule. PCR introduces a significant amount of random mutations in addition to the mutations already present in the PCR template. Using PCR-generated DNA to research and possibly treat age-related accumulation of random mutations might actually worsen the situation. Clearly, a mutation-free source of mtDNA is needed.

Relatively mutation-free mtDNA could be isolated from fetal or neonatal tissues but this would pose significant difficulties in procurement.

The best source of mtDNA for therapy is cloning. Cloning allows for the selection of a single mtDNA molecule, and by screening a sufficient number of clones it is possible to find mutation-free clones even if the starting material (tissue-isolated mtDNA) does contain a significant mutation load. Once such a mutation-free clone is found, it can be grown in essentially identical batches in an industrial process for years. Such cloned DNA contains minor amounts of mutations introduced during the fermentation process and is the preferred material for gene therapy (Przybylowski et al. 2007). However, full-length cloning of human mitochondrial DNA has never been reported. It is possible to create full-length mtDNA by ligation of shorter cloned fragments but this procedure is likely to suffer from low efficiency, greatly increasing cost.

16.2.4 Strategies for Mitochondrial DNA Gene Therapy – Supplementation

Each human cell contains on average 4,900 copies of mtDNA and the phenotype caused by a mitochondrial mutation is very dependent on the proportion of mutated genomes in the cell, the heteroplasmic ratio. This raises the possibility that phenotypes could be manipulated by shifting the heteroplasmic ratios towards wild-type, without actually removing the mutated genomes. Cybrid experiments where varying ratios of pathological mitochondria were fused with rho-zero cells indicate that indeed this is a viable strategy (Schoeler et al. 2005; Liu et al. 2004).

For many mutations there appear to be thresholds below which the mutations have very few if any deleterious effects (Rossignol et al. 2003). Shifting of the heteroplasmic ratio just below the threshold might therefore be sufficient in some cases to achieve significant effects.

However, this approach may have limitations in the treatment of age-related mtDNA mutations. It has been hypothesized that mitochondria containing mutated mtDNA may have a replicative advantage leading to progressive accumulation of mutations over time (de Grey 1997). It is known that the mitochondrial DNA polymerase does have a significant error rate under physiological conditions. Whether both or only the latter of these mechanisms are active, it is clear that the results of mtDNA supplementation are likely to be transient, due to continued accumulation of mutations. This would necessitate repeated mitochondrial gene therapy treatments.

Another limitation is the need to deliver a large number of mtDNA copies per cell to achieve meaningful heteroplasmic ratio shifts. Assuming no selective pressures are operating, diluting a MELAS mutated genome from 90 to 50% would require the delivery of 3,900 non-mutated mtDNA copies per cell and this may present significant technical difficulties even for a highly efficacious transfection method.

16.2.5 Strategies for Mitochondrial DNA Gene Therapy – Replacement

Three groups have reported the application of restriction enzymes to shift heteroplasmic ratios in vitro and in vivo (Srivastava and Moraes 2001; Tanaka et al. 2002; Bayona-Bafaluy et al. 2005). If a mitochondrial DNA mutation produces a restriction site absent from wild-type mtDNA, then the mutated genomes should be in principle capable of being selectively degraded by the restriction enzyme, while leaving wild-type mtDNA intact. This has been confirmed, showing a robust and persistent heteroplasmic shift after delivering a recombinant restriction enzyme produced in the cytoplasm. The shift occurs very quickly, within about 6 hours, and there are no signs of toxicity which could be expected if the enzyme was damaging nuclear or wild-type mitochondrial DNA.

If a robust mitochondrial gene therapy were to allow the expression of exogenous genes in mitochondria, the restriction enzyme could be produced in situ and could be used to destroy endogenous mtDNA while allowing the delivered mtDNA constructs to take over. This strategy is likely to be superior to simple supplementation, should allow complete replacement of endogenous mtDNA and lends itself to modifications assuring only transient restriction enzyme expression. In principle only one engineered mtDNA construct per cell would be needed to achieve permanent, full genome replacement.

16.3 Conclusion – Prospects for Mitochondrial Gene Therapy in Disease and Aging

16.3.1 Initial Targets for Mitochondrial Gene Therapy

At least 500,000 Americans suffer from classical mitochondrial conditions, that is conditions with a well-defined maternally inherited mitochondrial mutation shown to be responsible for symptoms (Schaefer et al. 2004). The paradigmatic classical mitochondrial disease, MELAS, usually manifests in neonates, or in young children, and has an average survival of about 6 years after diagnosis (Debray et al. 2007). There are no treatments proven to improve survival, although treatment with dichloroacetate is used to mitigate the severity of episodes of lactic acidosis (Mori et al. 2004). For the most common form of classical mitochondrial disease, MIDD, there is symptomatic treatment with insulin but patients still develop significant complications (blindness, kidney dysfunction, circulatory problems) and have a much higher mortality than the general population (Suzuki et al. 2003).

Classical mitochondrial conditions, and especially the most severe ones, are the most appropriate early targets for mitochondrial genome replacement therapy. There is a combination of severe symptoms, early mortality, a clearly defined cause directly addressable by mitochondrial genome replacement, and lack of efficacious treatment alternatives that improves the odds of obtaining regulatory approval for this completely novel form of therapy. It must be noted that all previous attempts at gene therapy have been directed at the nuclear genome and the large amount of experience gained with these therapies is not directly applicable to mitochondria.

The differences between the nuclear and mitochondrial approaches may play in favor of mitochondrial gene therapy: For example, in the mitochondria there is no risk of insertional mutagenesis that could result in activation of oncogenes, which caused leukemia in 2 out of 14 patients treated for SCID with nuclear gene therapy (Hacein-Bey-Abina et al. 2003). The ability to transduce mitochondria might therefore be useful in non-mitochondrial disease: Mitochondria could be used to produce the mitochondrial proteins normally translated in the cytoplasm, such as frataxin, whose mutations are responsible for Friedreich's ataxia (Campuzano et al. 1997), and possibly treat this condition without the usual risks associated with conventional nuclear gene therapy. Even conditions unrelated to mitochondria, such as SCID, could be treated if the therapeutic protein could be exported from mitochondria into the cytoplasm or other cellular compartments. This idea is more plausible given the observation that some mitochondrial-translated proteins are exported to the cell membrane (Gingrich et al. 2004).

16.3.2 Application of Mitochondrial Gene Therapy in Aging-Related Conditions

While therapeutic applications of mitochondrial gene therapy in classical mitochondrial conditions could become viable in less than a decade, applications in aging and age-related conditions are likely to require a longer development period.

Obtaining regulatory approval for clinical studies in aging will be hampered by the following: The causative nature of mitochondrial mutations in aging-related conditions has not been established with sufficient confidence. We are absolutely confident that the MELAS mutation causes MELAS, and therefore replacing the mutated genome may benefit the patient – we lack this certitude in Parkinson's, or sporadic type II diabetes. The symptoms of many aging-related conditions can be managed to some extent and their nature is chronic – it is therefore more difficult to justify a genetic intervention using a completely new technology that may have unforeseen consequences.

Nevertheless, we do expect that mitochondrial gene therapy will be eventually used in the treatment of aging-related conditions, initially in Parkinson's, then perhaps in diabetes, Alzheimer's and cardiomyopathy. The body of evidence in favor of a mitochondrial contribution to aging is steadily growing. A successful treatment of a classical mitochondrial disease by mitochondrial gene therapy will make it much easier to obtain regulatory approval both from the efficacy and safety standpoints.

If we may give free rein to our imagination, we expect that in 15–20 years everybody older than 45–50 years would be advised by their physician to have regular mitochondrial health checkups – a complete sequencing of their mitochondrial genomes, with analysis of mutation burden. Patients exceeding a certain threshold would be referred to the hospital infusion center, where a standard preparation of mutation-free, recombinant mtDNA constructs complexed to a non-viral vector with mitochondrial specificity would be infused. Within a few hours, the newly delivered constructs would produce the restriction enzyme needed for removal of endogenous mtDNA, replicate to correct levels and then shut down the production of restriction enzyme. A quick check would be performed to confirm successful replacement and the patient would be sent home, protected from many aspects of aging for perhaps 10–15 years, until the next round of mtDNA replication would be needed.

The technologies needed to realize this vision already exist as proof of concept, and continued development work will hopefully transform them into effective therapeutics.

References

Abramov AY, Jacobson J, Wientjes F, Hothersall J, Canevari L, Duchen MR (2005) Expression and modulation of an NADPH oxidase in mammalian astrocytes. J Neurosci 25:9176–9184

Allen JF (1993) Control of gene expression by redox potential and the requirement for chloroplast and mitochondrial genomes. J Theor Biol 165:609–631

Barrett TG (2001) Mitochondrial diabetes, DIDMOAD and other inherited diabetes syndromes. Best Pract Res Clin Endocrinol Metab 15:325–343

Bayona-Bafaluy MP, Blits B, Battersby BJ, Shoubridge EA, Moraes CT (2005) Rapid directional shift of mitochondrial DNA heteroplasmy in animal tissues by a mitochondrially targeted restriction endonuclease. Proc Natl Acad Sci USA 102:14392–14397

Bedard K, Krause KH (2007) The NOX family of ROS-generating NADPH oxidases: physiology and pathophysiology. Physiol Rev 87:245–313

Bender A, Krishnan KJ, Morris CM, Taylor GA, Reeve AK, Perry RH, Jaros E, Hersheson JS, Betts J, Klopstock T, Taylor RW, Turnbull DM (2006) High levels of mitochondrial DNA deletions in substantia nigra neurons in aging and Parkinson disease. Nat Genet 38:515–517

Bhat HK, Epelboym I (2004) Quantitative analysis of total mitochondrial DNA: competitive polymerase chain reaction versus real-time polymerase chain reaction. J Biochem Mol Toxicol 18:180–186

Birket MJ, Birch-Machin MA (2007) Ultraviolet radiation exposure accelerates the accumulation of the aging-dependent T414G mitochondrial DNA mutation in human skin. Aging Cell 6:557–564

Boddapati SV, Tongcharoensirikul P, Hanson RN, D'souza GG, Torchilin VP, Weissig V (2005) Mitochondriotropic liposomes. J Liposome Res 15:49–58

Bonnefoy N, Fox TD (2007) Directed alteration of Saccharomyces cerevisiae mitochondrial DNA by biolistic transformation and homologous recombination. Methods Mol Biol 372:153–166

Bonnet C, Kaltimbacher V, Ellouze S, Augustin S, Benit P, Forster V, Rustin P, Sahel JA, Corral-Debrinski M (2007) Allotopic mRNA localization to the mitochondrial surface rescues respiratory chain defects in fibroblasts harboring mitochondrial DNA mutations affecting complex I or v subunits. Rejuvenation Res 10:127–144

Bonora E, Porcelli AM, Gasparre G, Biondi A, Ghelli A, Carelli V, Baracca A, Tallini G, Martinuzzi A, Lenaz G, Rugolo M, Romeo G (2006) Defective oxidative phosphorylation in thyroid oncocytic carcinoma is associated with pathogenic mitochondrial DNA mutations affecting complexes I and III. Cancer Res 66:6087–6096

Campuzano V, Montermini L, Lutz Y, Cova L, Hindelang C, Jiralerspong S, Trottier Y, Kish SJ, Faucheux B, Trouillas P, Authier FJ, Durr A, Mandel JL, Vescovi A, Pandolfo M, Koenig M (1997) Frataxin is reduced in Friedreich ataxia patients and is associated with mitochondrial membranes. Hum Mol Genet 6:1771–1780

Cassarino DS, Fall CP, Swerdlow RH, Smith TS, Halvorsen EM, Miller SW, Parks JP, Parker WD Jr, Bennett JP Jr (1997) Elevated reactive oxygen species and antioxidant enzyme activities in animal and cellular models of Parkinson's disease. Biochim Biophys Acta 1362:77–86

Cervin C, Liljestrom B, Tuomi T, Heikkinen S, Tapanainen JS, Groop L, Cilio CM (2004) Cosegregation of MIDD and MODY in a pedigree: functional and clinical consequences. Diabetes 53:1894–1899

Cetica P, Pintos L, Dalvit G, Beconi M (2002) Activity of key enzymes involved in glucose and triglyceride catabolism during bovine oocyte maturation in vitro. Reproduction 124:675–681

Chabi B, de Camaret BM, Chevrollier A, Boisgard S, Stepien G (2005) Random mtDNA deletions and functional consequence in aged human skeletal muscle. Biochem Biophys Res Commun 332:542–549

Chen Z, Lu W, Garcia-Prieto C, Huang P (2007) The Warburg effect and its cancer therapeutic implications. J Bioenerg Biomembr 39:267–274

Chinnery PF, Samuels DC, Elson J, Turnbull DM (2002) Accumulation of mitochondrial DNA mutations in ageing, cancer, and mitochondrial disease: is there a common mechanism? Lancet 360:1323–1325

D'souza GG, Rammohan R, Cheng SM, Torchilin VP, Weissig V (2003) DQAsome-mediated delivery of plasmid DNA toward mitochondria in living cells. J Control Release 19;92:189–197

Davidson MH, Yannicelli HD (2006) New concepts in dyslipidemia in the metabolic syndrome and diabetes. Metab Syndr Relat Disord 4:299–314

de Andrade PB, Rubi B, Frigerio F, van den Ouweland JM, Maassen JA, Maechler P (2006) Diabetes-associated mitochondrial DNA mutation A3243G impairs cellular metabolic pathways necessary for beta cell function. Diabetologia 49:1816–1826 de Grey AD (1997) A proposed refinement of the mitochondrial free radical theory of aging. Bioessays 19:161–166

Debray FG, Lambert M, Chevalier I, Robitaille Y, Decarie JC, Shoubridge EA, Robinson BH, Mitchell GA (2007) Long-term outcome and clinical spectrum of 73 pediatric patients with mitochondrial diseases. Pediatrics 119:722–733

Decanini A, Nordgaard CL, Feng X, Ferrington DA, Olsen TW (2007) Changes in select redox proteins of the retinal pigment epithelium in age-related macular degeneration. Am J Ophthalmol 143:607–615

Edland SD, Silverman JM, Peskind ER, Tsuang D, Wijsman E, Morris JC (1996) Increased risk of dementia in mothers of Alzheimer's disease cases: evidence for maternal inheritance. Neurology 47:254–256

Fan W, Waymire KG, Narula N, Li P, Rocher C, Coskun PE, Vannan MA, Narula J, MacGregor GR, Wallace DC (2008) A mouse model of mitochondrial disease reveals germline selection against severe mtDNA mutations. Science 319:958–962

Franco R, Schoneveld O, Georgakilas AG, Panayiotidis MI (2008) Oxidative stress, DNA methylation and carcinogenesis. Cancer Lett 266:6–11

Fridlender B, Fry M, Bolden A, Weissbach A (1972) A new synthetic RNA-dependent DNA polymerase from human tissue culture cells (HeLa-fibroblast-synthetic oligonucleotides-template-purified enzymes). Proc Natl Acad Sci USA 69:452–455

Gingrich JR, Pelkey KA, Fam SR, Huang Y, Petralia RS, Wenthold RJ, Salter MW (2004) Unique domain anchoring of Src to synaptic NMDA receptors via the mitochondrial protein NADH dehydrogenase subunit 2. Proc Natl Acad Sci USA 101:6237–6242

Godinot C, de LE, Hervouet E, Simonnet H (2007) Actuality of Warburg's views in our understanding of renal cancer metabolism. J Bioenerg Biomembr 39:235–241

Gokey NG, Cao Z, Pak JW, Lee D, McKiernan SH, McKenzie D, Weindruch R, Aiken JM (2004) Molecular analyses of mtDNA deletion mutations in microdissected skeletal muscle fibers from aged rhesus monkeys. Aging Cell 3:319–326

Grasso DG, Nero D, Law RH, Devenish RJ, Nagley P (1991) The C-terminal positively charged region of subunit 8 of yeast mitochondrial ATP synthase is required for efficient assembly of this subunit into the membrane F0 sector. Eur J Biochem 199:203–209

Greco M, Villani G, Mazzucchelli F, Bresolin N, Papa S, Attardi G (2003) Marked aging-related decline in efficiency of oxidative phosphorylation in human skin fibroblasts. FASEB J 17:1706–1708

Hacein-Bey-Abina S, Von Kalle C, Schmidt M, McCormack MP, Wulffraat N, Leboulch P, Lim A, Osborne CS, Pawliuk R, Morillon E, Sorensen R, Forster A, Fraser P, Cohen JI, de Saint BG, Alexander I, Wintergerst U, Frebourg T, Aurias A, Stoppa-Lyonnet D, Romana S, Radford-Weiss I, Gross F, Valensi F, Delabesse E, Macintyre E, Sigaux F, Soulier J, Leiva LE, Wissler M, Prinz C, Rabbitts TH, Le Deist F, Fischer A, Cavazzana-Calvo M (2003) LMO2-associated clonal T cell proliferation in two patients after gene therapy for SCID-X1. Science 302: 415–419

Hong WK, Han EH, Kim DG, Ahn JY, Park JS, Han BG (2007) Amyloid-beta-peptide reduces the expression level of mitochondrial cytochrome oxidase subunits. Neurochem Res 32: 1483–1488

Indo HP, Davidson M, Yen HC, Suenaga S, Tomita K, Nishii T, Higuchi M, Koga Y, Ozawa T, Majima HJ (2007) Evidence of ROS generation by mitochondria in cells with impaired electron transport chain and mitochondrial DNA damage. Mitochondrion 7:106–118

Kato Y, Miura Y, Inagaki A, Itatsu T, Oiso Y (2002) Age of onset possibly associated with the degree of heteroplasmy in two male siblings with diabetes mellitus having an A to G transition at 3243 of mitochondrial DNA. Diabet Med 19:784–786

Kelley DE, He J, Menshikova EV, Ritov VB (2002) Dysfunction of mitochondria in human skeletal muscle in type 2 diabetes. Diabetes 51:2944–2950

Khan SM, Cassarino DS, Abramova NN, Keeney PM, Borland MK, Trimmer PA, Krebs CT, Bennett JC, Parks JK, Swerdlow RH, Parker WD Jr, Bennett JP Jr (2000) Alzheimer's disease cybrids replicate beta-amyloid abnormalities through cell death pathways. Ann Neurol 48:148–155

King MP, Attardi G (1989) Human cells lacking mtDNA: repopulation with exogenous mitochondria by complementation. Science 246:500–503

Kraytsberg Y, Kudryavtseva E, McKee AC, Geula C, Kowall NW, Khrapko K (2006) Mitochondrial DNA deletions are abundant and cause functional impairment in aged human substantia nigra neurons. Nat Genet 38:518–520

Kujoth GC, Hiona A, Pugh TD, Someya S, Panzer K, Wohlgemuth SE, Hofer T, Seo AY, Sullivan R, Jobling WA, Morrow JD, Van Remmen H, Sedivy JM, Yamasoba T, Tanokura M, Weindruch R, Leeuwenburgh C, Prolla TA (2005) Mitochondrial DNA mutations, oxidative stress, and apoptosis in mammalian aging. Science 309:481–484

Lai LP, Tsai CC, Su MJ, Lin JL, Chen YS, Tseng YZ, Huang SK (2003) Atrial fibrillation is associated with accumulation of aging-related common type mitochondrial DNA deletion mutation in human atrial tissue. Chest 123:539–544

Lazarou M, McKenzie M, Ohtake A, Thorburn DR, Ryan MT (2007) Analysis of the assembly profiles for mitochondrial- and nuclear-DNA-encoded subunits into complex I. Mol Cell Biol 27:4228–4237

Ledoux SP, Druzhyna NM, Hollensworth SB, Harrison JF, Wilson GL (2007) Mitochondrial DNA repair: a critical player in the response of cells of the CNS to genotoxic insults. Neuroscience 145:1249–1259

Lee HC, Hsu LS, Yin PH, Lee LM, Chi CW (2007a) Heteroplasmic mutation of mitochondrial DNA D-loop and 4977-bp deletion in human cancer cells during mitochondrial DNA depletion. Mitochondrion 7:157–163

Lee HC, Wei YH (2007) Oxidative stress, mitochondrial DNA mutation, and apoptosis in aging. Exp Biol Med (Maywood) 232:592–606

Lee M, Choi JS, Choi MJ, Pak YK, Rhee BD, Ko KS (2007b) DNA delivery to the mitochondria sites using mitochondrial leader peptide conjugated polyethylenimine. J Drug Target 15: 115–122

Lenaz G, Baracca A, Carelli V, D'Aurelio M, Sgarbi G, Solaini G (2004) Bioenergetics of mitochondrial diseases associated with mtDNA mutations. Biochim Biophys Acta 1658: 89–94

Lin MT, Simon DK, Ahn CH, Kim LM, Beal MF (2002) High aggregate burden of somatic mtDNA point mutations in aging and Alzheimer's disease brain. Hum Mol Genet 11: 133–145

Liu CY, Lee CF, Hong CH, Wei YH (2004) Mitochondrial DNA mutation and depletion increase the susceptibility of human cells to apoptosis. Ann NY Acad Sci 1011:133–45.:133–145

Lovell MA, Markesbery WR (2007) Oxidative DNA damage in mild cognitive impairment and late-stage Alzheimer's disease. Nucleic Acids Res 35:7497–7504

Maassen JA, 'T Hart LM, Ouwens DM (2007) Lessons that can be learned from patients with diabetogenic mutations in mitochondrial DNA: implications for common type 2 diabetes. Curr Opin Clin Nutr Metab Care 10:693–697

Mark RJ, Hensley K, Butterfield DA, Mattson MP (1995) Amyloid beta-peptide impairs ion-motive ATPase activities: evidence for a role in loss of neuronal Ca^{2+} homeostasis and cell death. J Neurosci 15:6239–6249

Menzies RA, Gold PH (1971) The turnover of mitochondria in a variety of tissues of young adult and aged rats. J Biol Chem 246:2425–2429

Mori M, Yamagata T, Goto T, Saito S, Momoi MY (2004) Dichloroacetate treatment for mitochondrial cytopathy: long-term effects in MELAS. Brain Dev 26:453–458

Mott JL, Zhang D, Farrar PL, Chang SW, Zassenhaus HP (1999) Low frequencies of mitochondrial DNA mutations cause cardiac disease in the mouse. Ann NY Acad Sci 893:353–357

Mott JL, Zhang D, Stevens M, Chang S, Denniger G, Zassenhaus HP (2001) Oxidative stress is not an obligate mediator of disease provoked by mitochondrial DNA mutations. Mutat Res 474:35–45

Murphy R, Turnbull DM, Walker M, Hattersley AT (2008) Clinical features, diagnosis and management of maternally inherited diabetes and deafness (MIDD) associated with the 3243A>G mitochondrial point mutation. Diabet Med 25:383–399

Naviaux RK, Nyhan WL, Barshop BA, Poulton J, Markusic D, Karpinski NC, Haas RH (1999) Mitochondrial DNA polymerase gamma deficiency and mtDNA depletion in a child with Alpers' syndrome. Ann Neurol 45:54–58

Nicolson GL (2007) Metabolic syndrome and mitochondrial function: molecular replacement and antioxidant supplements to prevent membrane peroxidation and restore mitochondrial function. J Cell Biochem 100:1352–1369

Niu X, Trifunovic A, Larsson NG, Canlon B (2007) Somatic mtDNA mutations cause progressive hearing loss in the mouse. Exp Cell Res 313:3924–3934

Oguro H, Iwama A (2007) Life and death in hematopoietic stem cells. Curr Opin Immunol 19: 503–509

Oliveira-Sales EB, Dugaich AP, Carillo BA, Abreu NP, Boim MA, Martins PJ, D'Almeida V, Dolnikoff MS, Bergamaschi CT, Campos RR (2008) Oxidative stress contributes to renovascular hypertension. Am J Hypertens 21:98–104

Pak JW, Aiken JM (2004) Low levels of mtDNA deletion mutations in ETS normal fibers from aged rats. Ann NY Acad Sci 1019:289–293

Parker WD Jr, Boyson SJ, Parks JK (1989) Abnormalities of the electron transport chain in idiopathic Parkinson's disease. Ann Neurol 26:719–723

Parker WD Jr, Filley CM, Parks JK (1990) Cytochrome oxidase deficiency in Alzheimer's disease. Neurology 40:1302–1303

Perna NT, Kocher TD (1996) Mitochondrial DNA: molecular fossils in the nucleus. Curr Biol 6:128–129

Pineau B, Mathieu C, Gerard-Hirne C, De Paepe R, Chetrit P (2005) Targeting the NAD7 subunit to mitochondria restores a functional complex I and a wild type phenotype in the Nicotiana sylvestris CMSII mutant lacking nad7. J Biol Chem 280:25994–26001

Prithivirajsingh S, Story MD, Bergh SA, Geara FB, Ang KK, Ismail SM, Stevens CW, Buchholz TA, Brock WA (2004) Accumulation of the common mitochondrial DNA deletion induced by ionizing radiation. FEBS Lett 571:227–232

Przybylowski M, Bartido S, Borquez-Ojeda O, Sadelain M, Riviere I (2007) Production of clinical-grade plasmid DNA for human Phase I clinical trials and large animal clinical studies. Vaccine 25:5013–5024

Puddu P, Puddu GM, Cravero E, De PS, Muscari A (2007) The putative role of mitochondrial dysfunction in hypertension. Clin Exp Hypertens 29:427–434

Rossignol R, Faustin B, Rocher C, Malgat M, Mazat JP, Letellier T (2003) Mitochondrial threshold effects. Biochem J 370:751–762

Sayre LM, Moreira PI, Smith MA, Perry G (2005) Metal ions and oxidative protein modification in neurological disease. Ann Ist Super Sanita 41:143–164

Schaefer AM, Taylor RW, Turnbull DM, Chinnery PF (2004) The epidemiology of mitochondrial disorders–past, present and future. Biochim Biophys Acta 1659:115–120

Schapira AH (2008) Mitochondria in the aetiology and pathogenesis of Parkinson's disease. Lancet Neurol 7:97–109

Schmiedel J, Jackson S, Schafer J, Reichmann H (2003) Mitochondrial cytopathies. J Neurol 250:267–277

Schoeler S, Szibor R, Gellerich FN, Wartmann T, Mawrin C, Dietzmann K, Kirches E (2005) Mitochondrial DNA deletions sensitize cells to apoptosis at low heteroplasmy levels. Biochem Biophys Res Commun 332:43–49

Simon DK, Lin MT, Zheng L, Liu GJ, Ahn CH, Kim LM, Mauck WM, Twu F, Beal MF, Johns DR (2004) Somatic mitochondrial DNA mutations in cortex and substantia nigra in aging and Parkinson's disease. Neurobiol Aging 25:71–81

Smigrodzki RM, Khan SM (2005) Mitochondrial microheteroplasmy and a theory of aging and age-related disease. Rejuvenation Res 8:172–198

Smith DG, Cappai R, Barnham KJ (2007) The redox chemistry of the Alzheimer's disease amyloid beta peptide. Biochim Biophys Acta 1768:1976–1990

Song S, Pursell ZF, Copeland WC, Longley MJ, Kunkel TA, Mathews CK (2005) DNA precursor asymmetries in mammalian tissue mitochondria and possible contribution to mutagenesis through reduced replication fidelity. Proc Natl Acad Sci USA 102:4990–4995

Srivastava S, Moraes CT (2001) Manipulating mitochondrial DNA heteroplasmy by a mitochondrially targeted restriction endonuclease. Hum Mol Genet 10:3093–3099

Stuart JA, Brown MF (2006) Mitochondrial DNA maintenance and bioenergetics. Biochim Biophys Acta 1757:79–89

Sun F, Cui J, Gavras H, Schwartz F (2003) A novel class of tests for the detection of mitochondrial DNA-mutation involvement in diseases. Am J Hum Genet 72:1515–1526

Suzuki S, Oka Y, Kadowaki T, Kanatsuka A, Kuzuya T, Kobayashi M, Sanke T, Seino Y, Nanjo K (2003) Clinical features of diabetes mellitus with the mitochondrial DNA 3243 (A-G) mutation in Japanese: maternal inheritance and mitochondria-related complications. Diabetes Res Clin Pract 59:207–217

Swerdlow RH, Khan SM (2004) A "mitochondrial cascade hypothesis" for sporadic Alzheimer's disease. Med Hypotheses 63:8–20

Swerdlow RH, Parks JK, Cassarino DS, Maguire DJ, Maguire RS, Bennett JP Jr, Davis RE, Parker WD Jr (1997) Cybrids in Alzheimer's disease: a cellular model of the disease? Neurology 49:918–925

Takuma K, Yao J, Huang J, Xu H, Chen X, Luddy J, Trillat AC, Stern DM, Arancio O, Yan SS (2005) ABAD enhances Abeta-induced cell stress via mitochondrial dysfunction. FASEB J 19:597–598

Tanaka M, Borgeld HJ, Zhang J, Muramatsu S, Gong JS, Yoneda M, Maruyama W, Naoi M, Ibi T, Sahashi K, Shamoto M, Fuku N, Kurata M, Yamada Y, Nishizawa K, Akao Y, Ohishi N, Miyabayashi S, Umemoto H, Muramatsu T, Furukawa K, Kikuchi A, Nakano I, Ozawa K, Yagi K (2002) Gene therapy for mitochondrial disease by delivering restriction endonuclease SmaI into mitochondria. J Biomed Sci 9:534–541

Tawata M, Hayashi JI, Isobe K, Ohkubo E, Ohtaka M, Chen J, Aida K, Onaya T (2000) A new mitochondrial DNA mutation at 14577 T/C is probably a major pathogenic mutation for maternally inherited type 2 diabetes. Diabetes 49:1269–1272

Trifunovic A, Hansson A, Wredenberg A, Rovio AT, Dufour E, Khvorostov I, Spelbrink JN, Wibom R, Jacobs HT, Larsson NG (2005) Somatic mtDNA mutations cause aging phenotypes without affecting reactive oxygen species production. Proc Natl Acad Sci USA 102:17993–17998

Trifunovic A, Larsson NG (2008) Mitochondrial dysfunction as a cause of ageing. J Intern Med 263:167–178

Trifunovic A, Wredenberg A, Falkenberg M, Spelbrink JN, Rovio AT, Bruder CE, Bohlooly Y, Gidlof S, Oldfors A, Wibom R, Tornell J, Jacobs HT, Larsson NG (2004) Premature ageing in mice expressing defective mitochondrial DNA polymerase. Nature 429:417–423

Trounce IA, Pinkert CA (2007) Cybrid models of mtDNA disease and transmission, from cells to mice. Curr Top Dev Biol 77:157–183

Turner C, Killoran C, Thomas NS, Rosenberg M, Chuzhanova NA, Johnston J, Kemel Y, Cooper DN, Biesecker LG (2003) Human genetic disease caused by de novo mitochondrial-nuclear DNA transfer. Hum Genet 112:303–309

Van GG, Martin JJ, Van BC (2002) Progressive external ophthalmoplegia and multiple mitochondrial DNA deletions. Acta Neurol Belg 102:39–42

Vermulst M, Bielas JH, Kujoth GC, Ladiges WC, Rabinovitch PS, Prolla TA, Loeb LA (2007) Mitochondrial point mutations do not limit the natural lifespan of mice. Nat Genet 39: 540–543

Vestweber D, Schatz G (1989) DNA-protein conjugates can enter mitochondria via the protein import pathway. Nature 338:170–172

Vidal-Puig AJ, Grujic D, Zhang CY, Hagen T, Boss O, Ido Y, Szczepanik A, Wade J, Mootha V, Cortright R, Muoio DM, Lowell BB (2000) Energy metabolism in uncoupling protein 3 gene knockout mice. J Biol Chem 275:16258–16266

Yang Q, Kim SK, Sun F, Cui J, Larson MG, Vasan RS, Levy D, Schwartz F (2007) Maternal influence on blood pressure suggests involvement of mitochondrial DNA in the pathogenesis of hypertension: the Framingham Heart Study. J Hypertens 25:2067–2073

Yechoor VK, Patti ME, Ueki K, Laustsen PG, Saccone R, Rauniyar R, Kahn CR (2004) Distinct pathways of insulin-regulated versus diabetes-regulated gene expression: An in vivo analysis in MIRKO mice. Proc Natl Acad Sci USA 101:16525–16530

Yoon YG, Koob MD (2003) Efficient cloning and engineering of entire mitochondrial genomes in Escherichia coli and transfer into transcriptionally active mitochondria. Nucleic Acids Res 31:1407–1415

Zhang D, Mott JL, Chang SW, Stevens M, Mikolajczak P, Zassenhaus HP (2005) Mitochondrial DNA mutations activate programmed cell survival in the mouse heart. Am J Physiol Heart Circ Physiol 288:H2476–H2483

Zhang J, Asin-Cayuela J, Fish J, Michikawa Y, Bonafe M, Olivieri F, Passarino G, De Benedictis G, Franceschi C, Attardi G (2003) Strikingly higher frequency in centenarians and twins of mtDNA mutation causing remodeling of replication origin in leukocytes. Proc Natl Acad Sci USA 100:1116–1121

Chapter 17
Life Extension by Tissue and Organ Replacement

Anthony Atala

Contents

17.1 Introduction 544
17.2 Native Cells 545
17.3 Biomaterials 546
    17.3.1 Naturally Derived Materials 547
    17.3.2 Acellular Tissue Matrices 547
    17.3.3 Synthetic Polymers 547
17.4 Alternate Cell Sources: Stem Cells and Nuclear Transfer 548
17.5 Tissue Engineering of Specific Structures 552
    17.5.1 Urethra 552
    17.5.2 Bladder 554
    17.5.3 Male and Female Reproductive Organs 554
    17.5.4 Kidney 556
    17.5.5 Blood Vessels 558
    17.5.6 Heart 559
    17.5.7 Liver 560
    17.5.8 Articular Cartilage and Trachea 560
17.6 Cellular Therapies 561
    17.6.1 Bulking Agents 561
    17.6.2 Injectable Muscle Cells 563
    17.6.3 Endocrine Replacement 563
    17.6.4 Angiogenic Agents 564
    17.6.5 Anti-Angiogenic Agents 564
17.7 Summary and Conclusion 565
References 565

A. Atala (✉)
Wake Forest Institute for Regenerative Medicine, Wake Forest University School of Medicine, Medical Center Boulevard, Winston-Salem, NC 27157, USA
e-mail: aatala@wfubmc.edu

17.1 Introduction

Patients suffering from diseased and injured organs may be treated with transplanted organs. However, there is a severe shortage of donor organs that is worsening yearly. As modern medicine increases the human lifespan, the aging population grows, and the need for organs grows with it, as aging organs are generally more prone to failure. Scientists in the field of regenerative medicine and tissue engineering apply the principles of cell transplantation, material science, and bioengineering to construct biological substitutes that can prolong life by reducing mortality from these diseases. Therapeutic cloning, where the nucleus from a donor cell is transferred into an enucleated oocyte in order to extract pluripotent embryonic stem cells, offers a potentially limitless source of cells for tissue engineering applications. The stem cell field is also advancing rapidly, opening new options for therapy. This chapter reviews recent advances that have occurred in regenerative medicine and describes applications of these new technologies that offer promise of a longer life.

The field of regenerative medicine encompasses various areas of technology, such as tissue engineering, stem cells, and cloning. Tissue engineering, one of the major components of regenerative medicine, follows the principles of cell transplantation, materials science, and engineering towards the development of biological substitutes that can restore and maintain normal function. Tissue engineering strategies generally fall into two categories: the use of acellular matrices, which depend on the body's natural ability to regenerate for proper orientation and direction of new tissue growth, and the use of matrices with cells. Acellular tissue matrices are usually prepared by manufacturing artificial scaffolds, or by removing cellular components from tissues via mechanical and chemical manipulation to produce collagen-rich matrices (Dahms et al. 1998; Piechota et al. 1998; Yoo et al. 1998; Chen et al. 1999). These matrices tend to slowly degrade on implantation and are generally replaced by the extracellular matrix (ECM) proteins that are secreted by the ingrowing cells. Cells can also be used for therapy via injection, either with carriers such as hydrogels or alone.

When cells are used for tissue engineering, a small piece of donor tissue is dissociated into individual cells. These cells are either implanted directly into the host, or are expanded in culture, attached to a support matrix, and then reimplanted into the host after expansion. The source of donor tissue can be heterologous (such as bovine), allogeneic (same species, different individual), or autologous. Ideally, both structural and functional tissue replacement will occur with minimal complications. The preferred cells to use are autologous cells, where a biopsy of tissue is obtained from the host, the cells are dissociated and expanded in culture, and the expanded cells are implanted back into the same host (Cilento et al. 1994; Yoo et al. 1998, 1999; Amiel and Atala 1999; Atala 1998, 1999, 2001; Oberpenning et al. 1999; Amiel et al. 2006). The use of autologous cells, although it may cause an inflammatory response, avoids rejection and thus the deleterious side effects of immunosuppressive medications can be avoided.

Most current strategies for tissue engineering depend upon a sample of autologous cells from the diseased organ of the host. Aging itself is not a barrier to successful applications of engineered tissues, as cells can be expanded from young and old patients alike. However, for many patients with extensive end-stage organ failure, a tissue biopsy may not yield enough normal cells for expansion and transplantation. In other instances, primary autologous human cells cannot be expanded from a particular organ, such as the pancreas. In these situations, stem cells are envisioned as being an alternative source of cells from which the desired tissue can be derived. Stem cells can be derived from discarded human embryos (human embryonic stem cells), from fetal tissue, or from adult sources (bone marrow, fat, skin).

Therapeutic cloning has also played a role in the development of the field of regenerative medicine. This type of cloning, which has also been called nuclear transplantation and nuclear transfer, involves the introduction of a nucleus from a donor cell into an enucleated oocyte to generate an embryo with a genetic makeup identical to that of the donor. Stem cells can be derived from this source, which may have the future potential to be used therapeutically.

17.2 Native Cells

One of the limitations of applying cell-based regenerative medicine techniques to organ replacement has been the inherent difficulty of growing specific cell types in large quantities. Even when some organs, such as the liver, have a high regenerative capacity in vivo, cell growth and expansion in vitro may be difficult. By studying the privileged sites for committed precursor cells in specific organs, as well as exploring the conditions that promote differentiation, one may be able to overcome the obstacles that limit cell expansion in vitro. For example, urothelial cells could be grown in the laboratory setting in the past, but only with limited expansion. Several protocols were developed over the past two decades that identified the undifferentiated cells, and kept them undifferentiated during their growth phase (Cilento et al. 1994; Scriven et al. 1997) (Liebert et al. 1991, 1997; Puthenveettil et al. 1999). Using these methods of cell culture, it is now possible to expand a urothelial strain from a single specimen that initially covered a surface area of 1 cm^2 to one covering a surface area of 4202 m^2 (the equivalent of one football field) within 8 weeks (Cilento et al. 1994). These studies indicated that it should be possible to collect autologous bladder cells from human patients, expand them in culture, and return them to the donor in sufficient quantities for reconstructive purposes (Cilento et al. 1994) (Freeman et al. 1997; Liebert et al. 1997; Puthenveettil et al. 1999) (Liebert et al. 1991; Harriss 1995; Nguyen et al. 1999). Major advances have been achieved within the past decade on the possible expansion of a variety of primary human cells, with specific techniques that make the use of autologous cells possible for clinical application.

17.3 Biomaterials

For cell-based tissue engineering, the expanded cells are seeded onto a scaffold synthesized with the appropriate biomaterial. In tissue engineering, biomaterials replicate the biologic and mechanical function of the native ECM found in tissues in the body by serving as an artificial ECM. Biomaterials provide a three-dimensional space for the cells to form into new tissues with appropriate structure and function, and also can allow for the delivery of cells and appropriate bioactive factors (e.g. cell adhesion peptides, growth factors), to desired sites in the body (Kim and Mooney 1998). As the majority of mammalian cell types are anchorage-dependent and will die if no cell-adhesion substrate is available, biomaterials provide a cell-adhesion substrate that can deliver cells to specific sites in the body with high loading efficiency. Biomaterials can also provide mechanical support against in vivo forces such that the predefined three-dimensional structure is maintained during tissue development. Furthermore, bioactive signals, such as cell-adhesion peptides and growth factors, can be loaded along with cells to help regulate cellular function.

The ideal biomaterial should be biodegradable and bioresorbable to support the replacement of normal tissue without inflammation. Incompatible materials are destined for an inflammatory or foreign-body response that eventually leads to rejection and/or necrosis. Degradation products, if produced, should be removed from the body via metabolic pathways at an adequate rate that keeps the concentration of these degradation products in the tissues at a tolerable level (Bergsma et al. 1995). The biomaterial should also provide an environment in which appropriate regulation of cell behavior (adhesion, proliferation, migration, and differentiation) can occur such that functional tissue can form. Cell behavior in the newly formed tissue has been shown to be regulated by multiple interactions of the cells with their microenvironment, including interactions with cell-adhesion ligands (Hynes 1992) and with soluble growth factors. Since biomaterials provide temporary mechanical support while the cells undergo spatial reorganization into tissue, the properly chosen biomaterial should allow the engineered tissue to maintain sufficient mechanical integrity to support itself in early development, while in late development, it should have begun degradation such that it does not hinder further tissue growth (Kim and Mooney 1998).

Generally, three classes of biomaterials have been utilized for engineering tissues: naturally derived materials (e.g., collagen and alginate), acellular tissue matrices (e.g., bladder submucosa and small intestinal submucosa), and synthetic polymers such as polyglycolic acid (PGA), polylactic acid (PLA), and poly(lactic-co-glycolic acid) PLGA). These classes of biomaterials have been tested in respect to their biocompatibility (Pariente et al. 2001, 2002). Naturally derived materials and acellular tissue matrices have the potential advantage of biological recognition. However, synthetic polymers can be produced reproducibly on a large scale with controlled properties such as strength, degradation rate, and microstructure.

17.3.1 Naturally Derived Materials

Collagen is the most abundant and ubiquitous structural protein in the body, and may be readily purified from both animal and human tissues with an enzyme treatment and salt/acid extraction (Li 1995). Collagen implants, under normal conditions, degrade through a process involving phagocytosis of collagen fibrils by fibroblasts (Arora et al. 2000). This is followed by sequential attack by lysosomal enzymes including cathepsins B1 and D. Under inflammatory conditions, the implants can be rapidly degraded largely by matrix metalloproteins (MMPs) and collagenases (Arora et al. 2000). However, the in vivo resorption rate of a collagen implant can be regulated by controlling the density of the implant and the extent of intermolecular crosslinking. The lower the density, the greater the space between collagen fibers and the larger the pores for cell infiltration, leading to a higher rate of implant degradation. Collagen contains cell adhesion domain sequences (e.g., RGD) that may assist to retain the phenotype and activity of many types of cells, including fibroblasts (Silver and Pins 1992) and chondrocytes (Sams and Nixon 1995).

Alginate, a polysaccharide isolated from seaweed, has been used as an injectable cell delivery vehicle (Smidsrod and Skjak-Braek 1990) and a cell immobilization matrix (Lim and Sun 1980) owing to its gentle gelling properties in the presence of divalent ions such as calcium. Alginate is relatively biocompatible and approved by the Food and Drug Administration (FDA) for human use as wound dressing material. Alginate is a family of copolymers of D-mannuronate and L-guluronate. The physical and mechanical properties of alginate gel are strongly correlated with the proportion and length of polygluronic block in the alginate chains (Smidsrod and Skjak-Braek 1990).

17.3.2 Acellular Tissue Matrices

Acellular tissue matrices are collagen-rich matrices prepared by removing cellular components from tissues. The matrices are often prepared by mechanical and chemical manipulation of a segment of tissue (Dahms et al. 1998; Piechota et al. 1998; Chen et al. 1999; Yoo et al. 1998). The matrices slowly degrade upon implantation, and are replaced and remodeled by ECM proteins synthesized and secreted by transplanted or ingrowing cells.

17.3.3 Synthetic Polymers

Polyesters of naturally occurring a-hydroxy acids, including PGA, PLA, and PLGA, are widely used in tissue engineering. These polymers have gained FDA approval for human use in a variety of applications, including sutures (Gilding 1981). The ester bonds in these polymers are hydrolytically labile, and these polymers degrade by nonenzymatic hydrolysis. The degradation products of PGA, PLA, and PLGA are nontoxic natural metabolites and are eventually eliminated from the body in the form of carbon dioxide and water (Gilding 1981). The degradation rate of these polymers can be tailored from several weeks to several years by altering crystallinity, initial molecular weight, and the copolymer ratio of lactic to glycolic acid. Since these polymers are thermoplastics, they can be easily formed into a three dimensional scaffold with a desired microstructure, gross shape, and dimension by various techniques, including molding, extrusion, solvent casting (Mikos et al. 1994), phase separation techniques, and gas foaming techniques (Harris et al. 1998). Many applications in tissue engineering often require a scaffold with high porosity and ratio of surface area to volume. Other biodegradable synthetic polymers, including poly(anhydrides) and poly(ortho-esters), can also be used to fabricate scaffolds for tissue engineering with controlled properties (Peppas and Langer 1994).

17.4 Alternate Cell Sources: Stem Cells and Nuclear Transfer

Human embryonic stem cells exhibit two remarkable properties: the ability to proliferate in an undifferentiated but pluripotent state (self-renewal), and the ability to differentiate into many specialized cell types (Brivanlou et al. 2003). They can be isolated by aspirating the inner cell mass from the embryo during the blastocyst stage (5 days post-fertilization), and are usually grown on feeder layers consisting of mouse embryonic fibroblasts or human feeder cells (Richards et al. 2002). More recent reports have shown that these cells can be grown without the use of a feeder layer (Amit et al. 2004) and thus avoid the exposure of these human cells to mouse viruses and proteins. These cells have demonstrated longevity in culture by maintaining their undifferentiated state for at least 80 passages when grown using current published protocols (Thomson et al. 1998; Reubinoff et al. 2000).

Human embryonic stem cells have been shown to differentiate into cells from all three embryonic germ layers in vitro. Skin and neurons have been formed, indicating ectodermal differentiation (Schuldiner et al. 2000, 2001; Reubinoff et al. 2001; Zhang et al. 2001). Blood, cardiac cells, cartilage, endothelial cells, and muscle have been formed, indicating mesodermal differentiation (Kaufman et al. 2001; Kehat et al. 2001; Levenberg et al. 2002). Pancreatic cells have been formed, indicating endodermal differentiation (Assady et al. 2001). In addition, as further evidence of their pluripotency, embryonic stem cells can form embryoid bodies, which are cell aggregations that contain all three embryonic germ layers while in culture, and can form teratomas in vivo (Itskovitz-Eldor et al. 2000).

An alternate source of stem cells is the amniotic fluid and placenta. Amniotic fluid and the placenta are known to contain multiple partially differentiated cell types derived from the developing fetus. We isolated stem cell populations from these sources, called amniotic fluid and placental stem cells (AFPSC) that express embryonic and adult stem cell markers (De Coppi et al. 2007). The undifferentiated stem cells expand extensively without feeders and double every 36 hours. Unlike human embryonic stem cells, AFPSC do not form tumors in vivo. Lines maintained for over 250 population doublings retained long telomeres and a normal karyotype. AFS cells are broadly multipotent. Clonal human lines verified by retroviral marking can be induced to differentiate into cell types representing each embryonic germ layer, including cells of adipogenic, osteogenic, myogenic, endothelial, neuronal and hepatic lineages. In this respect, they meet a commonly accepted criterion for pluripotent stem cells, without implying that they can generate every adult tissue. Examples of differentiated cells derived from AFS cells and displaying specialized functions include neuronal lineage secreting the neurotransmitter L-glutamate or expressing G-protein-gated inwardly rectifying potassium (GIRK) channels, hepatic lineage cells producing urea, and osteogenic lineage cells forming tissue engineered bone. The cells could be obtained either from amniocentesis or chorionic villous sampling in the developing fetus, or from the placenta at the time of birth. The cells could be preserved for self use, and used without rejection, or they could be banked. A bank of 100,000 specimens could potentially supply 99% of the US population with a perfect genetic match for transplantation. Such a bank may be easier to create than with other cell sources, since there are approximately 4.5 million births per year in the USA (De Coppi et al. 2007).

In addition, stem cells for tissue engineering could be generated through cloning procedures. There has been tremendous interest in the field of nuclear cloning since the birth of the cloned sheep Dolly in 1997, but frogs were the first successfully cloned vertebrates derived from nuclear transfer (Gurdon et al. 1958; Gurdon 1962a, b, c) but the nuclei were derived from non-adult sources. In the past fifteen years, tremendous advances in nuclear cloning technology have been reported, indicating the relative immaturity of the field. Dolly was not the first cloned mammal to be produced via nuclear transfer; in fact, live lambs were produced in 1996 using nuclear transfer and differentiated epithelial cells derived from embryonic discs (Campbell et al. 1996). The significance of Dolly was that she was the first mammal to be derived from an adult somatic cell using nuclear transfer (Wilmut et al. 1997). Since then, animals from several species have been grown using nuclear transfer technology, including cattle (Cibelli et al. 1998), goats (Baguisi et al. 1999), mice (Wakayama et al. 1998), and pigs (Betthauser et al. 2000; De Sousa et al. 2002).

Two types of nuclear cloning, reproductive cloning and therapeutic cloning, have been described, and a better understanding of the differences between the two types may help to alleviate some of the controversy that surrounds these technologies (Colman and Kind 2000; Vogelstein et al. 2002). Banned in most countries for human applications, reproductive cloning is used to generate an embryo that has the identical genetic material as its cell source. This embryo can then be implanted into the uterus of a female to give rise to an infant that is a clone of the donor. On the other hand, therapeutic cloning is used to generate early stage embryos that are explanted in culture to produce embryonic stem cell lines whose genetic material is identical to that of its source. These autologous stem cells have the potential to become almost any type of cell in the adult body, and thus would be useful in tissue and organ replacement applications (Hochedlinger et al. 2004). Therefore, therapeutic cloning, which has also been called somatic cell nuclear transfer, may provide an alternative source of transplantable cells. Figure 17.1 shows the strategy

Therapeutic Cloning Strategies

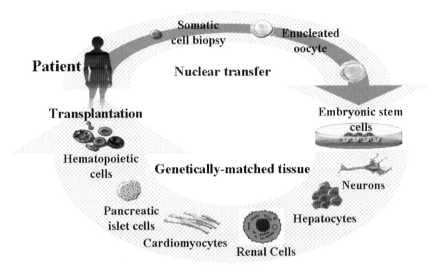

Fig. 17.1 Strategy for therapeutic cloning and tissue engineering

of combining therapeutic cloning with tissue engineering to develop tissues and organs. According to data from the Centers for Disease Control (CDC), an estimated 3000 Americans die every day of diseases that could have been treated with stem cell derived tissues (Lanza et al. 1999, 2001) With current allogeneic tissue transplantation protocols, rejection is a frequent complication because of immunologic incompatibility, and immunosuppressive drugs are usually necessary (Hochedlinger et al. 2004). The use of transplantable tissue and organs derived from therapeutic cloning may potentially lead to the avoidance of immune responses that typically are associated with transplantation of non-autologous tissues (Lanza et al. 1999).

While promising, somatic cell nuclear transfer technology has certain limitations that require further improvements before therapeutic cloning can be applied widely in replacement therapy. Currently, the efficiency of the overall cloning process is low. The majority of embryos derived from animal cloning do not survive after implantation (Solter 2000; Rideout et al. 2001; Hochedlinger and Jaenisch 2002). To improve cloning efficiency, further improvements are required in the multiple complex steps of nuclear transfer, such as enucleation and reconstruction, activation of oocytes, and cell cycle synchronization between donor cells and recipient oocytes (Dinnyes et al. 2002).

Recently, exciting reports of the successful transformation of adult cells into pluripotent stem cells through a type of genetic "reprogramming" have been published. Reprogramming is a technique that involves de-differentiation of adult somatic cells to produce patient-specific pluripotent stem cells, without the use of embryos. Cells generated by reprogramming would be genetically identical to the somatic cells (and thus, the patient who donated these cells) and would not be rejected. Yamanaka was the first to discover that mouse embryonic fibroblasts (MEFs) and adult mouse fibroblasts could be reprogrammed into an "induced pluripotent state (iPS)" (Takahashi and Yamanaka 2006). This group used mouse embryonic fibroblasts (MEFs) engineered to express a neomycin resistance gene from the *Fbx15* locus, a gene expressed only in ES cells. They examined 24 genes that were thought to be important for embryonic stem cells and identified 4 key genes that, when introduced into the reporter fibroblasts, resulted in drug-resistant cells. These were Oct3/4, Sox2, c-Myc, and Klf4. This experiment indicated that expression of the four genes in these transgenic MEFs led to expression of a gene specific for ES cells. Mouse embryonic fibroblasts and adult fibroblasts were co-transduced with retroviral vectors, each carrying one of the four genes, and transduced cells were selected via drug resistance. The resultant iPS cells possessed the immortal growth characteristics of self-renewing ES cells, expressed genes specific for ES cells, and generated embryoid bodies in vitro and teratomas in vivo. When iPS cells were injected into mouse blastocysts, they contributed to a variety of cell types. However, although iPS cells selected in this way were pluripotent, they were not identical to ES cells. Unlike ES cells, chimeras made from iPS cells did not result in full-term pregnancies. Gene expression profiles of the iPS cells showed that they possessed a distinct gene expression signature that was different from that of ES cells. In addition, the epigenetic state of the iPS cells was somewhere between that found in somatic cells and that found in ES cells, suggesting that the reprogramming was incomplete.

These results were improved significantly by Wernig and Jaenisch in July 2007 (Wernig et al. 2007). Fibroblasts were infected with retroviral vectors and selected for the activation of endogenous *Oct4* or *Nanog* genes. Results from this study showed that DNA methylation, gene expression profiles, and the chromatin state of the reprogrammed cells were similar to those of ES cells. Teratomas induced by these cells contained differentiated cell types representing all three embryonic germ layers. Most importantly, the reprogrammed cells from this experiment were able to form viable chimeras and contribute to the germ line like ES cells, suggesting that these iPS cells were completely reprogrammed. Wernig et al. observed that the number of reprogrammed colonies increased when drug selection was initiated later (day 20 rather than day 3 post-transduction). This suggests that reprogramming is a slow and gradual process and may explain why previous attempts resulted in incomplete reprogramming.

It has recently been shown that reprogramming of human cells is possible (Takahashi et al. 2007; Yu et al. 2007). Yamanaka showed that retrovirus-mediated transfection of *OCT3/4, SOX2, KLF4,* and *c-MYC* generates human iPS cells that are similar to hES cells in terms of morphology, proliferation, gene expression, surface markers, and teratoma formation. Thompson's group showed that retroviral transduction of *OCT4, SOX2, NANOG, and LIN28* could generate pluripotent stem cells without introducing any oncogenes (c-MYC). Both studies showed that human iPS were similar but not identical to hES cells.

Another concern is that these iPS cells contain three to six retroviral integrations (one for each factor) which may increase the risk of tumorigenesis. Yamanaka et al. studied the tumor formation in chimeric mice generated from Nanog-iPS cells and found 20% of the offspring developed tumors due to the retroviral expression of c-myc (Okita et al. 2007). An alternative approach would be to use a transient expression method, such as adenovirus-mediated system, since both Jaenisch and Yamanaka showed strong silencing of the viral-controlled transcripts in iPS cells (Meissner et al. 2007; Okita et al. 2007). This indicates that these viral genes are only required for the induction, not the maintenance, of pluripotency.

Another concern is the use of transgenic donor cells for reprogrammed cells in the mouse studies. Both studies used donor cells from transgenic mice harboring a drug resistance gene driven by *Fbx15, Oct3/4,* or *Nanog* promoters so that, if these ES cell-specific genes were activated, the resulting cells could be easily selected using neomycin. However, the use of genetically modified donors hinders clinical applicability for humans. To assess whether iPS cells can be derived from nontransgenic donor cells, wild type MEF and adult skin cells were retrovirally transduced with Oct3/4, Sox2, c-Myc, and Klf4 and ES-like colonies were isolated by morphology alone, without the use of drug selection for *Oct4* or *Nanog* (Meissner et al. 2007). IPS cells from wild type donor cells formed teratomas and generated live chimeras. This study suggests that transgenic donor cells are not necessary to generate IPS cells.

Although this is an exciting phenomenon, it is unclear why reprogramming adult fibroblasts and mesenchymal stromal cells have similar efficiencies (Takahashi and Yamanaka 2006). It would seem that cells that are already multipotent could be reprogrammed with greater efficiency, since the more undifferentiated donor nucleus the better SCNT performs (Blelloch et al. 2006). This further emphasizes our limited understanding of the mechanism of reprogramming, yet the potential for this area of study is exciting.

17.5 Tissue Engineering of Specific Structures

Investigators around the world, including our laboratory, have been working towards the development of several cell types and tissues and organs for clinical application.

17.5.1 Urethra

Various biomaterials without cells, such as PGA and acellular collagen-based matrices from small intestine and bladder, have been used experimentally (in animal models) for the regeneration of urethral tissue (Atala et al. 1992; Chen et al. 1999, 2000) (Olsen et al. 1992; Kropp et al. 1998; Sievert et al. 2000). Some of these biomaterials, like acellular collagen matrices derived from bladder submucosa, have also been seeded with autologous cells for urethral reconstruction. Our laboratory has been able to replace tubularized urethral segments with cell-seeded collagen matrices (Atala 2002; De Filippo et al. 2002).

Acellular collagen matrices derived from bladder submucosa by our laboratory have been used experimentally and clinically. In animal studies, segments of the urethra were resected and replaced with acellular matrix grafts in an onlay fashion. Histological examination showed complete epithelialization and progressive vessel and muscle infiltration, and the animals were able to void through the neo-urethras (Chen et al. 1999). These results were confirmed in a clinical study of patients with hypospadias and urethral stricture disease (El-Kassaby et al. 2003). Decellularized cadaveric bladder submucosa was used as an onlay matrix for urethral repair in patients with stricture disease and hypospadias (Fig. 17.2). Patent, functional neo-urethras were noted in these patients with up to a 7-year follow-up. The use of an off-the-shelf matrix appears to be beneficial for patients with abnormal urethral conditions and obviates the need for obtaining autologous grafts, thus decreasing operative time and eliminating donor site morbidity.

Unfortunately, the above techniques are not applicable for tubularized urethral repairs. The collagen matrices are able to replace urethral segments only when used in an onlay fashion. However, if a tubularized repair is needed, the collagen

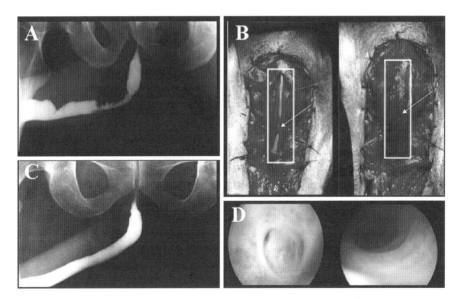

Fig. 17.2 Tissue engineering of the urethra using a collagen matrix. (**a**) Representative case of a patient with a bulbar stricture. (**b**) During the urethral repair surgery, strictured tissue is excised, preserving the urethral plate on the *left side*, and matrix is anastamosed to the urethral plate in an onlay fashion on the *right*. The *boxes* in both photos indicate the area of interest, including the urethra, which appears white in the *left photograph*. In the *left photograph*, the arrow indicates the area of stricture in the urethra. On the *right*, the *arrow* indicates the repaired stricture. Note that the engineered tissue now obscures the native white urethral tissue in an onlay fashion in the *right photograph*. (**c**) Urethrogram 6 months after repair. (**d**) Cystoscopic view of urethra before surgery on the *left side*, and 4 months after repair on the *right side*

matrices should be seeded with autologous cells to avoid the risk of stricture formation and poor tissue development (De Filippo et al. 2002). Therefore, tubularized collagen matrices seeded with autologous cells can be used successfully for total penile urethra replacement.

17.5.2 Bladder

Currently, gastrointestinal segments are commonly used as tissues for bladder replacement or repair. However, gastrointestinal tissues are designed to absorb specific solutes, whereas bladder tissue is designed for the excretion of solutes. Due to the problems encountered with the use of gastrointestinal segments, numerous investigators have attempted alternative materials and tissues for bladder replacement or repair.

The success of cell transplantation strategies for bladder reconstruction depends on the ability to use donor tissue efficiently and to provide the right conditions for long term survival, differentiation, and growth. Urothelial and muscle cells can be expanded in vitro, seeded onto polymer scaffolds, and allowed to attach and form sheets of cells (Atala et al. 1993b). These principles were applied in the creation of tissue engineered bladders in an animal model that required a subtotal cystectomy with subsequent replacement with a tissue engineered organ in beagle dogs (Oberpenning et al. 1999). Urothelial and muscle cells were separately expanded from an autologous bladder biopsy, and seeded onto a bladder-shaped biodegradable polymer scaffold. The results from this study showed that it is possible to tissue engineer bladders that are anatomically and functionally normal.

A clinical experience involving engineered bladder tissue for cystoplasty reconstruction was conducted starting in 1999. A small pilot study of seven patients was reported, using a collagen scaffold seeded with cells either with or without omentum coverage, or a combined PGA-collagen scaffold seeded with cells and omental coverage (Fig. 17.3). The patients reconstructed with the engineered bladder tissue created with the PGA-collagen cell-seeded scaffolds showed increased compliance, decreased end-filling pressures, increased capacities and longer dry periods (Atala et al. 2006) . Although the experience is promising in terms of showing that engineered tissues can be implanted safely, it is just a start in terms of accomplishing the goal of engineering fully functional bladders. Further experimental and clinical work is being conducted.

17.5.3 Male and Female Reproductive Organs

Reconstructive surgery is required for a wide variety of pathologic penile conditions, such as penile carcinoma, trauma, severe erectile dysfunction, and congenital conditions such as ambiguous genitalia, hypospadias, and epispadias. One of the

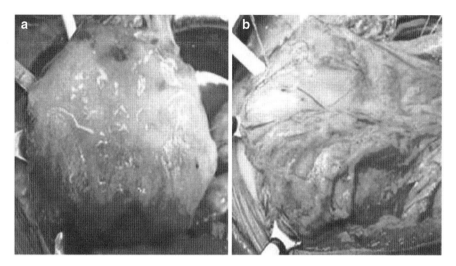

Fig. 17.3 Construction of engineered bladder. (**a**) Engineered bladder anastamosed to native bladder with running 4-0 polyglycolic sutures. (**b**) Implant covered with fibrin glue and omentum

major limitations of phallic reconstructive surgery is the scarcity of sufficient autologous tissue.

The major components of the phallus are corporal smooth muscle and endothelial cells. The creation of autologous functional and structural corporal tissue *de novo* would be beneficial. Autologous cavernosal smooth muscle and endothelial cells were harvested, expanded, and seeded on acellular collagen matrices and implanted in a rabbit model (Kershen et al. 2002; Kwon et al. 2002). Histologic examination confirmed the appropriate organization of penile tissue phenotypes, and structural and functional studies, including cavernosography, cavernosometry, and mating studies, demonstrated that it is possible to engineer autologous functional penile tissue. Our laboratory is currently working on increasing the size of the engineered constructs.

Congenital malformations of the uterus may have profound implications clinically. Patients with cloacal exstrophy and intersex disorders may not have sufficient uterine tissue present for future reproduction. We investigated the possibility of engineering functional uterine tissue using autologous cells (Wang et al. 2003). Autologous rabbit uterine smooth muscle and epithelial cells were harvested, then grown and expanded in culture. These cells were seeded onto preconfigured uterine-shaped biodegradable polymer scaffolds, which were then used for subtotal uterine tissue replacement in the corresponding autologous animals. Upon retrieval 6 months after implantation, histological, immunocytochemical, and Western blot analyses confirmed the presence of normal uterine tissue components. Biomechanical analyses and organ bath studies showed that the functional characteristics of these tissues were similar to those of normal uterine tissue. Breeding studies using these engineered uteri are currently being performed.

Similarly, several pathologic conditions, including congenital malformations and malignancy, can adversely affect normal vaginal development or anatomy. Vaginal reconstruction has traditionally been challenging due to the paucity of available native tissue. The feasibility of engineering vaginal tissue in vivo was investigated (De Filippo et al. 2003). Vaginal epithelial and smooth muscle cells of female rabbits were harvested, grown, and expanded in culture. These cells were seeded onto biodegradable polymer scaffolds, and the cell-seeded constructs were then implanted into nude mice for up to 6 weeks. Immunocytochemical, histological, and Western blot analyses confirmed the presence of vaginal tissue phenotypes. Electrical field stimulation studies in the tissue-engineered constructs showed similar functional properties to those of normal vaginal tissue. When these constructs were used for autologous total vaginal replacement, patent vaginal structures were noted in the tissue-engineered specimens, while the non-cell-seeded structures were noted to be stenotic (De Filippo et al. 2003).

17.5.4 Kidney

We applied the principles of both tissue engineering and therapeutic cloning in an effort to produce genetically identical renal tissue in a large animal model, the cow (*Bos taurus*) (Lanza et al. 2002). Bovine skin fibroblasts from adult Holstein steers were obtained by ear notch, and single donor cells were isolated and microinjected into the perivitelline space of donor enucleated oocytes (nuclear transfer). The resulting blastocysts were implanted into progestin-synchronized recipients to allow for further in vivo growth. After 12 weeks, cloned renal cells were harvested, expanded in vitro, then seeded onto biodegradable scaffolds. The constructs, which consisted of the cells and the scaffolds, were then implanted into the subcutaneous space of the same steer from which the cells were cloned to allow for tissue growth.

The kidney is a complex organ with multiple cell types and a complex functional anatomy that renders it one of the most difficult to reconstruct (Amiel and Atala 1999; Auchincloss and Bonventre 2002). Previous efforts in tissue engineering of the kidney have been directed toward the development of extracorporeal renal support systems made of biological and synthetic components (Aebischer et al. 1987a, b; Ip et al. 1988; Lanza et al. 1996; MacKay et al. 1998; Humes et al. 1999a, b; Amiel et al. 2000; Joki et al. 2001) and ex vivo renal replacement devices are known to be life-sustaining. However, there would be obvious benefits for patients with end-stage kidney disease if these devices could be implanted long term without the need for an extracorporeal perfusion circuit or immunosuppressive drugs.

Cloned renal cells were seeded on scaffolds consisting of three collagen-coated cylindrical silastic catheters (Fig. 17.4a). The ends of the three membranes of each scaffold were connected to catheters that terminated into a collecting reservoir. This created a renal neoorgan with a mechanism for collecting the excreted urinary fluid (Fig. 17.4b). These scaffolds with the collecting devices were transplanted

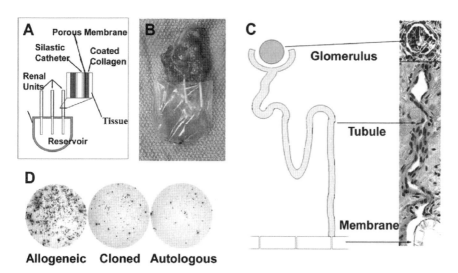

Fig. 17.4 Combining therapeutic cloning and tissue engineering to produce kidney tissue. (**a**) Illustration of the tissue-engineered renal unit. (**b**) Renal unit seeded with cloned cells, three months after implantation, showing the accumulation of urine-like fluid. (**c**) Clear unidirectional continuity between the mature glomeruli, their tubules, and silastic catheter. (**d**) Elispot analyses of the frequencies of T cells that secrete IFNγ after stimulation with allogeneic renal cells, cloned renal cells, or nuclear donor fibroblasts. Cloned renal cells produce fewer IFNγ spots than the allogeneic cells, indicating that the rejection response to cloned cells is diminished. The presented wells are single representatives of duplicate wells

subcutaneously into the same steer from which the genetic material originated, and then retrieved 12 weeks after implantation.

Chemical analysis of the collected urine-like fluid, including urea nitrogen and creatinine levels, electrolyte levels, specific gravity, and glucose concentration, revealed that the implanted renal cells possessed filtration, reabsorption, and secretory capabilities.

Histological examination of the retrieved implants revealed extensive vascularization and self-organization of the cells into glomeruli and tubule-like structures. A clear continuity between the glomeruli, the tubules, and the silastic catheter was noted that allowed the passage of urine into the collecting reservoir (Fig. 17.4c). Immunohistochemical analysis with renal-specific antibodies revealed the presence of renal proteins, RT-PCR analysis confirmed the transcription of renal specific RNA in the cloned specimens, and Western blot analysis confirmed the presence of elevated renal-specific protein levels.

Since previous studies have shown that bovine clones harbor the oocyte mitochondrial DNA (Evans et al. 1999; Hiendleder et al. 1999; Steinborn et al. 2000), the donor egg's mitochondrial DNA (mtDNA) was thought to be a potential source of immunologic incompatibility. Differences in mtDNA-encoded proteins expressed by cloned cells could stimulate a T-cell response specific for mtDNA-encoded minor histocompatibility antigens when the cloned cells are implanted back into the original nuclear donor (Fischer Lindahl et al. 1991). We used nucleotide sequencing of the mtDNA genomes of the clone and fibroblast nuclear donor to identify potential antigens in the muscle constructs. Only two amino acid substitutions were noted to distinguish the clone and the nuclear donor and, as a result, a maximum of two minor histocompatibility antigens could be defined. Given the lack of knowledge regarding peptide-binding motifs for bovine MHC class I molecules, there is no reliable method to predict the impact of these amino acid substitutions on bovine histocompatibility.

Oocyte-derived mtDNA was also thought to be a potential source of immunologic incompatibility in the cloned renal cells. Maternally transmitted minor histocompatibility antigens in mice have been shown to stimulate both skin allograft rejection in vivo and cytotoxic T lymphocytes expansion in vitro (Fischer Lindahl et al. 1991) that could prevent the use of these cloned constructs in patients with chronic rejection of major histocompatibility matched human renal transplants (Hadley et al. 1992; Yard et al. 1993). We tested for a possible T-cell response to the cloned renal devices using delayed-type hypersensitivity testing in vivo and Elispot analysis of interferon-gamma secreting T-cells in vitro. Both analyses revealed that the cloned renal cells showed no evidence of a T-cell response, suggesting that rejection will not necessarily occur in the presence of oocyte-derived mtDNA (Fig. 17.4d). This finding may represent a step forward in overcoming the histocompatibility problem of stem cell therapy (Yard et al. 1993).

These studies demonstrated that cells derived from nuclear transfer can be successfully harvested, expanded in culture, and transplanted in vivo with the use of biodegradable scaffolds on which the single suspended cells can organize into tissue structures that are genetically identical to that of the host. These studies were the first demonstration of the use of therapeutic cloning for regeneration of tissues in vivo.

17.5.5 Blood Vessels

Xenogenic or synthetic materials have been used as replacement blood vessels for complex cardiovascular lesions. However, these materials typically lack growth potential, and may place the recipient at risk for complications such as stenosis, thromboembolization, or infection (Matsumura et al. 2003). Tissue-engineered vascular grafts have been constructed using autologous cells and biodegradable scaffolds and have been applied in dog and lamb models (Shinoka et al. 1995, 1997, 1998; Watanabe et al. 2001). The key advantage of using these autografts is that they degrade in vivo, and thus allow the new tissue to form without the long term presence of foreign material (Matsumura et al. 2003). Application of these techniques from the laboratory to the clinical setting has begun, with autologous vascular cells harvested, expanded, and seeded onto a biodegradable scaffold (Shin'oka et al. 2001). The resultant autologous construct was used to replace a stenosed pulmonary artery that had been previously repaired. Seven months after implantation, no evidence of graft occlusion or aneurysmal changes was noted in the recipient.

17.5.6 Heart

In the United States, over 5 million people currently live with some form of heart disease, and many more are diagnosed each year. While many medications have been developed to assist the ailing heart, the treatment for end-stage heart failure still remains transplantation. Unfortunately, as with other organs, donor hearts are in short supply, and even when a transplant can be performed, the patient must endure the side effects created by lifelong immunosuppression. Thus, alternatives are desperately needed, and the development of novel methods to regenerate or replace damaged heart muscle using tissue engineering and regenerative medicine techniques is an attractive option.

Cell therapy for infracted areas of the heart is attractive, as these methods involve a rather simple injection into the damaged area of a patient's heart, rather than a rigorous surgical procedure, to complete. Various types of stem cells have been investigated for their potential to regenerate damaged or dead heart tissue in this manner. Skeletal muscle cells, bone marrow stem cells (both mesenchymal and hematopoietic), amniotic fluid stem cells, and embryonic stem cells have been used for this purpose. In this technique, cells are suspended in a biocompatible matrix that can range from simple normal saline to complex yet biocompatible hydrogels depending on the type of injection to be performed. The cells are either injected into the damaged area of the heart itself, or they are injected into the coronary circulation with the hope that they will home to the damaged area, take up residence there, and begin to repair the tissue. However, injectable therapies have been shown to be relatively inefficient, and cell loss is quite substantial. Newer methods of tissue engineering include the development of engineered "patches," which are comprised of cells adhered to a biomaterial, that can theoretically be used to replace the damaged area of the heart. These techniques have promise, but require further research into the optimal cell types and biomaterials for this purpose before they can be used extensively in the clinic (see (Jawad et al. 2007) for an excellent review of these methods).

However, the methods described above could only be used in cases where a relatively small section of heart muscle was damaged. In cases where a large area or even the whole heart has become nonfunctional, a more radical approach may be required. In these situations, the use of a bioartificial heart would be ideal, as rejection would be avoided and the problems associated with a mechanical heart (such as thromboembolus formation) would be abolished. To this end, Ott et al. recently developed a novel heart construct in vitro using decellularized cadaveric hearts. By reseeding the tissue scaffold that remained after a specialized decellularization process with various types of cells that make up a heart (cardiomyocytes, smooth muscle cells, endothelial cells, and fibrocytes) and culturing the resulting construct in a bioreactor system designed to mimic physiologic conditions, this group was able to produce a construct that could generate pump function on its own (Ott et al. 2008). This study suggests that production of bioartificial hearts may one day be possible.

17.5.7 Liver

The liver can sustain a variety of insults, including viral infection, alcohol abuse, surgical resection of tumors, and acute drug-induced hepatic failure. The current therapy for liver failure is liver transplantation. However, this therapy is limited by the shortage of donors and the need for lifelong immunosuppressive therapy. Cell transplantation has been proposed as a potential solution for liver failure. This is based on the fact that the liver has enormous regenerative potential in vivo suggests that in the right environment, it may be possible to expand liver cells in vitro in sufficient quantities for tissue engineering (Bhandari et al. 2001). Many approaches have been tried, including development of specialized media, co-culture with other cell types, identification of growth factors that have proliferative effects on these cells, and culture on three-dimensional scaffolds within bioreactors (Bhandari et al. 2001).

Extracorporeal bioartificial liver devices that use porcine hepatocytes have been designed and applied. These devices are designed to filter and purify the patient's blood as would the patient's own liver, and the blood is returned to the patient in a manner similar to kidney dialysis. Another cell-based approach is the injection of liver cell suspensions. This has been performed in animal models. Intraportal hepatocyte injection has also been used in patients with Crigler-Najjar Syndrome Type 1 (Fox et al. 1998); however, complications such as portal vein thrombosis and pulmonary embolism are major concerns, especially when large cell numbers are used (Nieto et al. 1989). Finally, cells including stem cells, oval progenitor cells, and mature hepatocytes have been seeded onto liver shaped biocompatible matrices to engineer artificial, implantable livers. These have been tested in various animal models (Gilbert et al. 1993; Kaufmann et al. 1999), however the transplantation efficiency as well as the functionality of these constructs must be improved substantially before the technology can be moved into the clinic.

17.5.8 Articular Cartilage and Trachea

Full-thickness articular cartilage lesions have limited healing capacity and thus represent a difficult management issue for the clinicians who treat adult patients with damaged articular cartilage (Hunter 1995; O'Driscoll 1998). Large defects can be associated with mechanical instability and may lead to degenerative joint disease if left untreated (Buckwalter and Lohmander 1994; Buckwalter and Mankin 1998). Chondrocytes were expanded and cultured onto biodegradable scaffolds to create engineered cartilage for use in large osteochondral defects in rabbits (Schaefer et al.

2002). When sutured to a subchondral support, the engineered cartilage was able to withstand physiologic loading and underwent orderly remodeling of the large osteochondral defects in adult rabbits, providing a biomechanically functional template that was able to undergo orderly remodeling when subjected to quantitative structural and functional analyses.

Few treatment options are currently available for patients who suffer from severe congenital tracheal pathology, such as stenosis, atresia, and agenesis, due to the limited availability of autologous transplantable tissue in the neonatal period. Tissue engineering in the fetal period may be a viable alternative for the surgical treatment of these prenatally diagnosed congenital anomalies, as cells could be harvested and grown into transplantable tissue in parallel with the remainder of gestation. Chondrocytes from both elastic and hyaline cartilage specimens have been harvested from fetal lambs, expanded in vitro, then dynamically seeded onto biodegradable scaffolds (Fuchs et al. 2002). The constructs were then implanted as replacement tracheal tissue in fetal lambs. The resultant tissue-engineered cartilage was noted to undergo engraftment and epithelialization, while maintaining its structural support and patency. Furthermore, if native tracheal tissue is unavailable, engineered cartilage may be derived from bone marrow-derived mesenchymal progenitor cells as well (Fuchs et al. 2003).

17.6 Cellular Therapies

17.6.1 Bulking Agents

Injectable bulking agents can be endoscopically used in the treatment of both urinary incontinence and vesicoureteral reflux. The advantages in treating urinary incontinence and vesicoureteral reflux with this minimally invasive approach include the simplicity of this quick outpatient procedure and the low morbidity associated with it. Several investigators are seeking alternative implant materials that would be safe for human use (Kershen and Atala 1999).

The ideal substance for the endoscopic treatment of reflux and incontinence should be injectable, nonantigenic, nonmigratory, volume stable, and safe for human use. Toward this goal long term studies were conducted to determine the effect of injectable chondrocytes in vivo (Atala et al. 1993a). It was initially determined that alginate, a liquid solution of gluronic and mannuronic acid, embedded with chondrocytes, could serve as a synthetic substrate for the injectable delivery and maintenance of cartilage architecture in vivo. Alginate undergoes hydrolytic biodegradation and its degradation time can be varied depending on the concentration of each of the polysaccharides. The use of autologous cartilage for the treatment of vesicoureteral reflux in humans would satisfy all the requirements for an ideal injectable substance.

Chondrocytes derived from an ear biopsy can be readily grown and expanded in culture. Neocartilage formation can be achieved in vitro and in vivo using chondrocytes cultured on synthetic biodegradable polymers. In these experiments, the cartilage matrix replaced the alginate as the polysaccharide polymer underwent biodegradation. This system was adapted for the treatment of vesicoureteral reflux in a porcine model (Atala et al. 1994). These studies showed that chondrocytes can be easily harvested and combined with alginate in vitro, the suspension can be easily injected cystoscopically, and the elastic cartilage tissue formed is able to correct vesicoureteral reflux without any evidence of obstruction.

Two multicenter clinical trials were conducted using this engineered chondrocyte technology. Patients with vesicoureteral reflux were treated at ten centers throughout the US. The patients had a similar success rate as with other injectable substances in terms of cure (Fig. 17.5). Chondrocyte formation was not noted in patients who had treatment failure. It is supposed that the patients who were cured have a biocompatible region of engineered autologous tissue present, rather than a foreign material (Diamond and Caldamone 1999). Patients with urinary incontinence were also treated endoscopically with injected chondrocytes at three different medical centers. Phase 1 trials showed an approximate success rate of 80% at follow-up 3 and 12 months postoperatively (Bent et al. 2001). Several of the clinical trials involving bioengineered products have been placed on hold because of the costs involved with the specific technology. With a bioengineered product, costs are usually high because of the biological nature of the therapies involved. As with any therapy, the

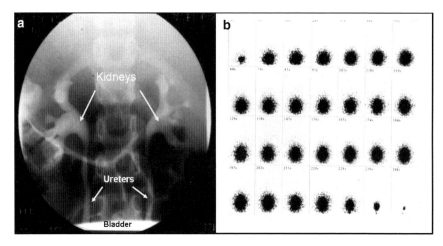

Fig. 17.5 Autologous chondrocytes for the treatment of vesicoureteral reflux. (**a**) Preoperative voiding cystourethrogram of a patient with bilateral reflux. A catheter was inserted into the bladder via the urethra, and contrast material was instilled intravesically. Here, contrast material can be seen within both ureters and within the kidneys, indicating reflux is present. (**b**) Postoperative radionuclide cystogram of the same patient 6 months after injection of autologous chondrocytes. A catheter was inserted into the bladder via the urethra, and a radioactive solution was inserted into the bladder. The bladder was scanned during filling and emptying phases. This panel includes sequential images of the bladder as it was filled and emptied. This shows a normal, round bladder that fills and empties properly. If reflux had been present, the ureters would have been visible in the scan above the round bladder

cost that the medical health care system can allow for a specific technology is limited. Therefore, the costs of bioengineered products have to be lowered for them to have an impact clinically. This is currently being addressed for multiple tissue-engineered technologies. As the technologies advance over time, and the volume of the application is considered, costs will naturally decrease.

17.6.2 Injectable Muscle Cells

The potential use of injectable cultured myoblasts for the treatment of stress urinary incontinence has been investigated (Chancellor et al. 2000; Yokoyama et al. 2000). Myoblasts were labeled with fluorescent latex microspheres (FLM) in order to track them after injection. Labeled myoblasts were directly injected into the proximal urethra and lateral bladder walls of nude mice with a micro-syringe in an open surgical procedure. Tissue harvested up to 35 days post-injection contained the labeled myoblasts, as well as evidence of differentiation of the labeled myoblasts into regenerative myofibers. The authors reported that a significant portion of the injected myoblast population persisted in vivo. Similar techniques of sphincteric derived muscle cells have been used for the treatment of urinary incontinence in a pig model (Strasser et al. 2004). The fact that myoblasts can be labeled and survive after injection and begin the process of myogenic differentiation further supports the feasibility of using cultured cells of muscular origin as an injectable bioimplant.

The use of injectable muscle precursor cells has also been investigated for use in the treatment of urinary incontinence due to irreversible urethral sphincter injury or maldevelopment. Muscle precursor cells are the quiescent satellite cells found in each myofiber that proliferate to form myoblasts and eventually myotubes and new muscle tissue. Intrinsic muscle precursor cells have previously been shown to play an active role in the regeneration of injured striated urethral sphincter (Yiou et al. 2003a). In a subsequent study, autologous muscle precursor cells were injected into a rat model of urethral sphincter injury, and both replacement of mature myotubes as well as restoration of functional motor units was noted in the regenerating sphincteric muscle tissue (Yiou et al. 2003b). This is the first demonstration of the replacement of both sphincter muscle tissue and its innervation by the injection of muscle precursor cells. As a result, muscle precursor cells may be a minimally invasive solution for urinary incontinence in patients with irreversible urinary sphincter muscle insufficiency.

17.6.3 Endocrine Replacement

Patients with testicular dysfunction and hypogonadal disorders are dependent on androgen replacement therapy to restore and maintain physiological levels of serum testosterone and its metabolites, dihydrotestosterone and estradiol (Machluf et al. 2000, 2003). Currently available androgen replacement modalities, such as testosterone tablets and capsules, Depo-Provera injections, and skin patches may be associated with fluctuating serum levels and complications such as fluid and nitrogen retention, erythropoiesis, hypertension, and bone density changes (Santen and Swerdloff 1990). Since Leydig cells of the testes are the major source of testosterone in men, implantation of heterologous Leydig cells or gonadal tissue fragments have previously been proposed as a method for chronic testosterone replacement (Tai et al. 1989; van Dam et al. 1989). These approaches, however, were limited by the failure of the tissues and cells to produce testosterone.

Encapsulation of cells in biocompatible and semipermeable polymeric membranes has been an effective method to protect against a host immune response as well as to maintain viability of the cells while allowing the secretion of desired therapeutic agents (Tai and Sun 1993; De Vos et al. 1997). Alginate poly-L-lysine-encapsulated Leydig cell microspheres were used as a novel method for testosterone delivery in vivo (Machluf et al. 2003). Elevated stable serum testosterone levels were noted in castrated adult rats over the course of the study, suggesting that microencapsulated Leydig cells may be a potential therapeutic modality for testosterone supplementation.

17.6.4 Angiogenic Agents

The engineering of large organs will require a vascular network of arteries, veins, and capillaries to deliver nutrients to each cell. One possible method of vascularization is through the use of gene delivery of angiogenic agents such as vascular endothelial growth factor (VEGF) with the implantation of vascular endothelial cells (EC) in order to enhance neovascularization of engineered tissues. Skeletal myoblasts from adult mice were cultured and transfected with an adenovirus encoding VEGF and combined with human vascular endothelial cells (Nomi et al. 2002). The mixtures of cells were injected subcutaneously in nude mice, and the engineered tissues were retrieved up to 8 weeks after implantation. The transfected cells were noted to form muscle with neovascularization by histology and immunohistochemical probing with maintenance of their muscle volume, while engineered muscle of nontransfected cells had a significantly smaller mass of cells with loss of muscle volume over time, less neovascularization, and no surviving endothelial cells. These results indicate that a combination of VEGF and endothelial cells may be useful for inducing neovascularization and volume preservation in engineered tissue.

17.6.5 Anti-Angiogenic Agents

The delivery of anti-angiogenic agents may help to slow tumor growth for a variety of neoplasms. Encapsulated hamster kidney cells transfected with the angiogenesis inhibitor endostatin were used for local delivery on human glioma cell line xenografts (Joki et al. 2001). The release of biologically active endostatin led to significant inhibition of endothelial cell proliferation and substantial reduction in tumor weight. Continuous local delivery of endostatin via encapsulated endostatin-secreting cells may be an effective therapeutic option for a variety of tumor types.

17.7 Summary and Conclusion

Regenerative medicine efforts are currently underway experimentally for virtually every type of tissue and organ within the human body. As regenerative medicine incorporates the fields of tissue engineering, cell biology, nuclear transfer, and materials science, personnel who have mastered the techniques of cell harvest, culture, expansion, transplantation, as well as polymer design are essential for the successful application of these technologies to extend human life. Various tissues are at different stages of development, with some already being used clinically, a few in preclinical trials, and some in the discovery stage. Recent progress suggests that engineered tissues may have an expanded clinical applicability in the future and may represent a viable therapeutic option for those who would benefit from the life-extending benefits of tissue replacement or repair.

Acknowledgment The author wishes to thank Jennifer L. Olson, Ph.D. for editorial assistance with this manuscript.

References

Aebischer P, Ip TK, Miracoli L, Galletti PM (1987a) Renal epithelial cells grown on semipermeable hollow fibers as a potential ultrafiltrate processor. ASAIO Trans 33:96–102

Aebischer P, Ip TK, Panol G, Galletti PM (1987b) The bioartificial kidney: progress towards an ultrafiltration device with renal epithelial cells processing. Life Support Syst 5:159–168

Amiel GE, Atala A (1999) Current and future modalities for functional renal replacement. Urologic Clinics N Am 26:235–246

Amiel GE, Yoo JJ, Atala A (2000) Renal therapy using tissue-engineered constructs and gene delivery. World J Urol 18:71–79

Amiel GE et al. (2006) Engineering of blood vessels from acellular collagen matrices coated with human endothelial cells. Tissue Eng 12:2355–2365

Amit M, Shariki C, Margulets V, Itskovitz-Eldor J (2004) Feeder layer- and serum-free culture of human embryonic stem cells. Biol Reproduction 70:837–845

Arora PD, Manolson MF, Downey GP, Sodek J, McCulloch CAG (2000) A novel model system for characterization of phagosomal maturation, acidification, and intracellular collagen degradation in fibroblasts. J Biol Chem 275:35432–35441

Assady S, Maor G, Amit M, Itskovitz-Eldor J, Skorecki KL, Tzukerman M (2001) Insulin production by human embryonic stem cells. Diabetes 50:1691–1697

Atala A (1998) Autologous cell transplantation for urologic reconstruction. J Urol 159:2–3

Atala A (1999) Creation of bladder tissue in vitro and in vivo. A system for organ replacement. Adv Exp Med Biol 462:31–42

Atala A (2001) Bladder regeneration by tissue engineering. [see comment]. BJU Int 88:765–770

Atala A (2002) Experimental and clinical experience with tissue engineering techniques for urethral reconstruction. Urologic Clinics of N Am 29:485–492

Atala A, Bauer SB, Soker S, Yoo JJ, Retik AB (2006) Tissue-engineered autologous bladders for patients needing cystoplasty. Lancet 367:1241–1246

Atala A et al. (1993a) Injectable alginate seeded with chondrocytes as a potential treatment for vesicoureteral reflux. J Urol 150:745–747

Atala A, Freeman MR, Vacanti JP, Shepard J, Retik AB (1993b) Implantation in vivo and retrieval of artificial structures consisting of rabbit and human urothelium and human bladder muscle. J Urol 150:608–612

Atala A, Kim W, Paige KT, Vacanti CA, Retik AB (1994) Endoscopic treatment of vesicoureteral reflux with a chondrocyte-alginate suspension. J Urol 152:641–643; discussion 644

Atala A, Vacanti JP, Peters CA, Mandell J, Retik AB, Freeman MR (1992) Formation of urothelial structures in vivo from dissociated cells attached to biodegradable polymer scaffolds in vitro. J Urol 148:658–662

Auchincloss H, Bonventre JV (2002) Transplanting cloned cells into therapeutic promise. [comment]. Nat Biotechnol 20:665–666

Baguisi A et al. (1999) Production of goats by somatic cell nuclear transfer. Nat Biotechnol 17:456–461

Bent AE, Tutrone RT, McLennan MT, Lloyd LK, Kennelly MJ, Badlani G (2001) Treatment of intrinsic sphincter deficiency using autologous ear chondrocytes as a bulking agent. Neurourol Urodynamics 20:157–165

Bergsma JE, Rozema FR, Bos RR, Boering G, de Bruijn WC, Pennings AJ (1995) In vivo degradation and biocompatibility study of in vitro pre-degraded as-polymerized polyactide particles. [see comment]. Biomaterials 16:267–274

Betthauser J et al. (2000) Production of cloned pigs from in vitro systems. [see comment]. Nat Biotechnol 18:1055–1059

Bhandari RN et al. (2001) Liver tissue engineering: a role for co-culture systems in modifying hepatocyte function and viability. Tissue Eng 7:345–357

Blelloch R, Wang Z, Meissner A, Pollard S, Smith A, Jaenisch R (2006) Reprogramming efficiency following somatic cell nuclear transfer is influenced by the differentiation and methylation state of the donor nucleus 1. Stem Cells 24:2007–2013

Brivanlou AH, Gage FH, Jaenisch R, Jessell T, Melton D, Rossant J (2003) Stem cells. Setting standards for human embryonic stem cells. [see comment]. Science 300:913–916

Buckwalter JA, Lohmander S (1994) Operative treatment of osteoarthrosis. Current practice and future development. J Bone Joint Surg – Am 76:1405–1418

Buckwalter JA, Mankin HJ (1998) Articular cartilage repair and transplantation. Arthritis Rheumatism 41:1331–1342

Campbell KH, McWhir J, Ritchie WA, Wilmut I (1996) Sheep cloned by nuclear transfer from a cultured cell line. [see comment]. Nature 380:64–66

Chancellor MB et al. (2000) Preliminary results of myoblast injection into the urethra and bladder wall: a possible method for the treatment of stress urinary incontinence and impaired detrusor contractility. Neurourol Urodynamics 19:279–287

Chen F, Yoo JJ, Atala A (1999) Acellular collagen matrix as a possible "off the shelf" biomaterial for urethral repair. Urology 54:407–410

Chen F, Yoo JJ, Atala A (2000) Experimental and clinical experience using tissue regeneration for urethral reconstruction. World J Urol 18:67–70

Cibelli JB et al. (1998) Cloned transgenic calves produced from nonquiescent fetal fibroblasts. [see comment]. Science 280:1256–1258

Cilento BG, Freeman MR, Schneck FX, Retik AB, Atala A (1994) Phenotypic and cytogenetic characterization of human bladder urothelia expanded in vitro. J Urol 152:665–670

Colman A, Kind A (2000) Therapeutic cloning: concepts and practicalities. Trends Biotechnol 18:192–196

Dahms SE, Piechota HJ, Dahiya R, Lue TF, Tanagho EA (1998) Composition and biomechanical properties of the bladder acellular matrix graft: comparative analysis in rat, pig and human. Br J Urol 82:411–419

De Coppi P et al. (2007) Isolation of amniotic stem cell lines with potential for therapy. [see comment]. Nat Biotechnol 25:100–106

De Filippo RE, Yoo JJ, Atala A (2002) Urethral replacement using cell seeded tubularized collagen matrices. J Urol 168:1789–1792; discussion 1792–1783

De Filippo RE, Yoo JJ, Atala A (2003) Engineering of vaginal tissue in vivo. Tissue Eng 9:301–306

De Sousa PA et al. (2002) Somatic cell nuclear transfer in the pig: control of pronuclear formation and integration with improved methods for activation and maintenance of pregnancy. Biol Reproduction 66:642–650

De Vos P, De Haan B, Van Schilfgaarde R (1997) Effect of the alginate composition on the biocompatibility of alginate-polylysine microcapsules. Biomaterials 18:273–278

Diamond DA, Caldamone AA (1999) Endoscopic correction of vesicoureteral reflux in children using autologous chondrocytes: preliminary results. J Urol 162:1185–1188

Dinnyes A, De Sousa P, King T, Wilmut I (2002) Somatic cell nuclear transfer: recent progress and challenges. Cloning Stem Cells 4:81–90

El-Kassaby AW, Retik AB, Yoo JJ, Atala A (2003) Urethral stricture repair with an off-the-shelf collagen matrix. J Urol 169:170–173; discussion 173

Evans MJ, Gurer C, Loike JD, Wilmut I, Schnieke AE, Schon EA (1999) Mitochondrial DNA genotypes in nuclear transfer-derived cloned sheep. Nat Genet 23:90–93

Fischer Lindahl K, Hermel E, Loveland BE, Wang CR (1991) Maternally transmitted antigen of mice: a model transplantation antigen. Ann Rev Immunol 9:351–372

Fox IJ et al. (1998) Treatment of the Crigler-Najjar syndrome type I with hepatocyte transplantation. [see comment]. N Engl J Med 338:1422–1426

Freeman MR et al. (1997) Heparin-binding EGF-like growth factor is an autocrine growth factor for human urothelial cells and is synthesized by epithelial and smooth muscle cells in the human bladder. J Clin Invest 99:1028–1036

Fuchs JR, Hannouche D, Terada S, Vacanti JP, Fauza DO (2003) Fetal tracheal augmentation with cartilage engineered from bone marrow-derived mesenchymal progenitor cells. J Pediatr Surg 38:984–987

Fuchs JR, Terada S, Ochoa ER, Vacanti JP, Fauza DO (2002) Fetal tissue engineering: in utero tracheal augmentation in an ovine model. J Pediatr Surg 37:1000–1006; discussion 1000–1006

Gilbert JC, Takada T, Stein JE, Langer R, Vacanti JP (1993) Cell transplantation of genetically altered cells on biodegradable polymer scaffolds in syngeneic rats. Transplantation 56:423–427

Gilding D (1981) Biodegradable Polymers. In: Williams D (ed) Biocompatibility of clinical implant materials. CRC Press, Boca Raton, FL, pp. 209–232

Gurdon JB (1962a) The developmental capacity of nuclei taken from intestinal epithelium cells of feeding tadpoles. J Embryol Exp Morphol 10:622–640

Gurdon JB (1962b) Multiple genetically identical frogs. J Heredity 53:5–9

Gurdon JB (1962c) The transplantation of nuclei between two species of Xenopus. Develop Biol 5:68–83

Gurdon JB, Elsdale TR, Fischberg M (1958) Sexually mature individuals of Xenopus laevis from the transplantation of single somatic nuclei. Nature 182:64–65

Hadley GA, Linders B, Mohanakumar T (1992) Immunogenicity of MHC class I alloantigens expressed on parenchymal cells in the human kidney. Transplantation 54:537–542

Harris LD, Kim BS, Mooney DJ (1998) Open pore biodegradable matrices formed with gas foaming. J Biomed Mater Res 42:396–402

Harriss DR (1995) Smooth muscle cell culture: a new approach to the study of human detrusor physiology and pathophysiology. Br J Urol 75(Suppl 1):18–26

Hiendleder S, Schmutz SM, Erhardt G, Green RD, Plante Y (1999) Transmitochondrial differences and varying levels of heteroplasmy in nuclear transfer cloned cattle. Mol Reproduction Develop 54:24–31

Hochedlinger K, Jaenisch R (2002) Nuclear transplantation: lessons from frogs and mice. Curr Opin Cell Biol 14:741–748

Hochedlinger K, Rideout WM, Kyba M, Daley GQ, Blelloch R, Jaenisch R (2004) Nuclear transplantation, embryonic stem cells and the potential for cell therapy. Hematol J 5(Suppl 3):S114–S117

Humes HD, Buffington DA, MacKay SM, Funke AJ, Weitzel WF (1999a) Replacement of renal function in uremic animals with a tissue-engineered kidney. [see comment]. Nat Biotechnol 17:451–455

Humes HD, MacKay SM, Funke AJ, Buffington DA (1999b) Tissue engineering of a bioartificial renal tubule assist device: in vitro transport and metabolic characteristics. Kidney Int 55: 2502–2514

Hunter W (1995) Of the structure and disease of articulating cartilages. 1743. Clin Orthopaedics Related Res:3–6

Hynes RO (1992) Integrins: versatility, modulation, and signaling in cell adhesion. Cell 69: 11–25

Ip TK, Aebischer P, Galletti PM (1988) Cellular control of membrane permeability. Implications for a bioartificial renal tubule. ASAIO Trans 34:351–355

Itskovitz-Eldor J et al. (2000) Differentiation of human embryonic stem cells into embryoid bodies compromising the three embryonic germ layers. Mol Med 6:88–95

Jawad H, Ali NN, Lyon AR, Chen QZ, Harding SE, Boccaccini AR (2007) Myocardial tissue engineering: a review. J Tissue Eng Regenerative Med 1:327–342

Joki T et al. (2001) Continuous release of endostatin from microencapsulated engineered cells for tumor therapy. [see comment]. Nat Biotechnol 19:35–39

Kaufman DS, Hanson ET, Lewis RL, Auerbach R, Thomson JA (2001) Hematopoietic colony-forming cells derived from human embryonic stem cells. Proc Natl Acad Sci USA 98: 10716–10721

Kaufmann PM, Kneser U, Fiegel HC, Kluth D, Herbst H, Rogiers X (1999) Long-term hepatocyte transplantation using three-dimensional matrices. Transplantation Proc 31:1928–1929

Kehat I et al. (2001) Human embryonic stem cells can differentiate into myocytes with structural and functional properties of cardiomyocytes. [see comment]. J Clin Invest 108: 407–414

Kershen RT, Atala A (1999) New advances in injectable therapies for the treatment of incontinence and vesicoureteral reflux. Urologic Clin N Am 26:81–94

Kershen RT, Yoo JJ, Moreland RB, Krane RJ, Atala A (2002) Reconstitution of human corpus cavernosum smooth muscle in vitro and in vivo. Tissue Eng 8:515–524

Kim BS, Mooney DJ (1998) Development of biocompatible synthetic extracellular matrices for tissue engineering. Trends Biotechnol 16:224–230

Kropp BP et al. (1998) Rabbit urethral regeneration using small intestinal submucosa onlay grafts. Urology 52:138–142

Kwon TG, Yoo JJ, Atala A (2002) Autologous penile corpora cavernosa replacement using tissue engineering techniques. J Urol 168:1754–1758

Lanza RP et al. (2002) Generation of histocompatible tissues using nuclear transplantation. [see comment]. Nat Biotechnol 20:689–696

Lanza RP, Cibelli JB, West MD (1999) Prospects for the use of nuclear transfer in human transplantation. Nat Biotechnol 17:1171–1174

Lanza RP, Cibelli JB, West MD, Dorff E, Tauer C, Green RM (2001) The ethical reasons for stem cell research. [comment]. Science 292:1299

Lanza RP, Hayes JL, Chick WL (1996) Encapsulated cell technology. Nat Biotechnol 14: 1107–1111

Levenberg S, Golub JS, Amit M, Itskovitz-Eldor J, Langer R (2002) Endothelial cells derived from human embryonic stem cells. Proc Natl Acad Sci USA 99:4391–4396

Li ST (1995) Biologic biomaterials: tissue derived biomaterials (collagen). In: JD B (ed) The biomedical engineering handbook. CRS Press, Boca Raton, FL, pp. 627–647

Liebert M et al. (1991) Stimulated urothelial cells produce cytokines and express an activated cell surface antigenic phenotype. Seminars Urol 9:124–130

Liebert M et al. (1997) Expression of mal is associated with urothelial differentiation in vitro: identification by differential display reverse-transcriptase polymerase chain reaction. Differentiation 61:177–185

Lim F, Sun AM (1980) Microencapsulated islets as bioartificial endocrine pancreas. Science 210:908–910

Machluf M, Orsola A, Atala A (2000) Controlled release of therapeutic agents: slow delivery and cell encapsulation. World J Urol 18:80–83

Machluf M, Orsola A, Boorjian S, Kershen R, Atala A (2003) Microencapsulation of Leydig cells: a system for testosterone supplementation. Endocrinology 144:4975–4979

MacKay SM, Funke AJ, Buffington DA, Humes HD (1998) Tissue engineering of a bioartificial renal tubule. ASAIO J 44:179–183

Matsumura G, Miyagawa-Tomita S, Shin'oka T, Ikada Y, Kurosawa H (2003) First evidence that bone marrow cells contribute to the construction of tissue-engineered vascular autografts in vivo. Circulation 108:1729–1734

Meissner A, Wernig M, Jaenisch R (2007) Direct reprogramming of genetically unmodified fibroblasts into pluripotent stem cells 1. Nat Biotechnol 25:1177–1181

Mikos AG, Lyman MD, Freed LE, Langer R (1994) Wetting of poly(L-lactic acid) and poly(DL-lactic-co-glycolic acid) foams for tissue culture. Biomaterials 15:55–58

Nguyen HT et al. (1999) Cell-specific activation of the HB-EGF and ErbB1 genes by stretch in primary human bladder cells. In Vitro Cell Develop Biol – Animal 35:371–375

Nieto JA, Escandon J, Betancor C, Ramos J, Canton T, Cuervas-Mons V (1989) Evidence that temporary complete occlusion of splenic vessels prevents massive embolization and sudden death associated with intrasplenic hepatocellular transplantation. Transplantation 47: 449–450

Nomi M, Atala A, Coppi PD, Soker S (2002) Principals of neovascularization for tissue engineering. Mol Aspects Med 23:463–483

O'Driscoll SW (1998) The healing and regeneration of articular cartilage. J Bone Joint Surg – Am 80:1795–1812

Oberpenning F, Meng J, Yoo JJ, Atala A (1999) De novo reconstitution of a functional mammalian urinary bladder by tissue engineering. [see comment]. Nat Biotechnol 17:149–155

Okita K, Ichisaka T, Yamanaka S (2007) Generation of germline-competent induced pluripotent stem cells 1. Nature 448:313–317

Olsen L, Bowald S, Busch C, Carlsten J, Eriksson I (1992) Urethral reconstruction with a new synthetic absorbable device. An experimental study. Scandinavian J Urol Nephrol 26: 323–326

Ott HC et al. (2008) Perfusion-decellularized matrix: using nature's platform to engineer a bioartificial heart. Nature Med 14:213–221

Pariente JL, Kim BS, Atala A (2001) In vitro biocompatibility assessment of naturally derived and synthetic biomaterials using normal human urothelial cells. J Biomedical Mater Res 55:33–39

Pariente JL, Kim BS, Atala A (2002) In vitro biocompatibility evaluation of naturally derived and synthetic biomaterials using normal human bladder smooth muscle cells. J Urol 167:1867–1871

Peppas NA, Langer R (1994) New challenges in biomaterials. [see comment]. Science 263: 1715–1720

Piechota HJ, Dahms SE, Nunes LS, Dahiya R, Lue TF, Tanagho EA (1998) In vitro functional properties of the rat bladder regenerated by the bladder acellular matrix graft. J Urol 159: 1717–1724

Puthenveettil JA, Burger MS, Reznikoff CA (1999) Replicative senescence in human uroepithelial cells. Adv Exp Med Biol 462:83–91

Reubinoff BE et al. (2001) Neural progenitors from human embryonic stem cells. [see comment]. Nat Biotechnol 19:1134–1140

Reubinoff BE, Pera MF, Fong CY, Trounson A, Bongso A (2000) Embryonic stem cell lines from human blastocysts: somatic differentiation in vitro. [see comment] [erratum appears in Nat Biotechnol 2000 May;18(5):559]. Nat Biotechnol 18:399–404

Richards M, Fong CY, Chan WK, Wong PC, Bongso A (2002) Human feeders support prolonged undifferentiated growth of human inner cell masses and embryonic stem cells. [see comment]. Nat Biotechnol 20:933–936

Rideout WM 3rd, Eggan K, Jaenisch R (2001) Nuclear cloning and epigenetic reprogramming of the genome. Science 293:1093–1098

Sams AE, Nixon AJ (1995) Chondrocyte-laden collagen scaffolds for resurfacing extensive articular cartilage defects. Osteoarthritis Cartilage 3:47–59

Santen R, Swerdloff R (1990) Clinical aspects of androgen therapy. In: Workshop Conference on Androgen Therapy: Biologic and Clinical Consequences. Pennsylvania State, Philadelphia, PA

Schaefer D et al. (2002) Tissue-engineered composites for the repair of large osteochondral defects. Arthritis Rheumatism 46:2524–2534

Schuldiner M et al. (2001) Induced neuronal differentiation of human embryonic stem cells. Brain Res 913:201–205

Schuldiner M, Yanuka O, Itskovitz-Eldor J, Melton DA, Benvenisty N (2000) Effects of eight growth factors on the differentiation of cells derived from human embryonic stem cells. Proc Natl Acad Sci USA 97:11307–11312

Scriven SD, Booth C, Thomas DF, Trejdosiewicz LK, Southgate J (1997) Reconstitution of human urothelium from monolayer cultures. J Urol 158:1147–1152

Shin'oka T, Imai Y, Ikada Y (2001) Transplantation of a tissue-engineered pulmonary artery. N Engl J Med 344:532–533

Shinoka T et al. (1995) Tissue engineering heart valves: valve leaflet replacement study in a lamb model. Annals Thoracic Surg 60:S513–516

Shinoka T et al. (1997) Tissue-engineered heart valve leaflets: does cell origin affect outcome? Circulation 96:II-102–107

Shinoka T et al. (1998) Creation of viable pulmonary artery autografts through tissue engineering. J Thoracic Cardiovasc Surg 115:536–545; discussion 545–536

Sievert KD, Bakircioglu ME, Nunes L, Tu R, Dahiya R, Tanagho EA (2000) Homologous acellular matrix graft for urethral reconstruction in the rabbit: histological and functional evaluation. J Urol 163:1958–1965

Silver FH, Pins G (1992) Cell growth on collagen: a review of tissue engineering using scaffolds containing extracellular matrix. J Long-Term Effects Medical Implants 2:67–80

Smidsrod O, Skjak-Braek G (1990) Alginate as immobilization matrix for cells. Trends Biotechnol 8:71–78

Solter D (2000) Mammalian cloning: advances and limitations. Nat Rev Genet 1:199–207

Steinborn R et al. (2000) Mitochondrial DNA heteroplasmy in cloned cattle produced by fetal and adult cell cloning. Nat Genet 25:255–257

Strasser H et al. (2004) Stem cell therapy for urinary stress incontinence. Exp Gerontol 39:1259–1265

Tai IT, Sun AM (1993) Microencapsulation of recombinant cells: a new delivery system for gene therapy. FASEB J 7:1061–1069

Tai J, Johnson HW, Tze WJ (1989) Successful transplantation of Leydig cells in castrated inbred rats. Transplantation 47:1087–1089

Takahashi K, Yamanaka S (2006) Induction of pluripotent stem cells from mouse embryonic and adult fibroblast cultures by defined factors 2. Cell 126:663–676

Takahashi K et al. (2007) Induction of pluripotent stem cells from adult human fibroblasts by defined factors. Cell 131:861–872

Thomson JA et al. (1998) Embryonic stem cell lines derived from human blastocysts. [see comment] [erratum appears in Science 1998 Dec 4;282(5395):1827]. Science 282:1145–1147

van Dam JH, Teerds KJ, Rommerts FF (1989) Transplantation and subsequent recovery of small amounts of isolated Leydig cells. Archiv Androl 22:123–129

Vogelstein B, Alberts B, Shine K (2002) Genetics. Please don't call it cloning! [see comment]. Science 295:1237

Wakayama T, Perry AC, Zuccotti M, Johnson KR, Yanagimachi R (1998) Full-term development of mice from enucleated oocytes injected with cumulus cell nuclei. [see comment]. Nature 394:369–374

Wang T, Koh C, Yoo JJ (2003) Creation of an engineered uterus for surgical reconstruction. In: American Academy of Pediatrics Section on Urology, New Orleans, LA

Watanabe M et al. (2001) Tissue-engineered vascular autograft: inferior vena cava replacement in a dog model. Tissue Eng 7:429–439

Wernig M et al. (2007) In vitro reprogramming of fibroblasts into a pluripotent ES-cell-like state 1. Nature 448:318–324

Wilmut I, Schnieke AE, McWhir J, Kind AJ, Campbell KH (1997) Viable offspring derived from fetal and adult mammalian cells. [see comment] [erratum appears in Nature 1997 Mar 13;386(6621):200]. Nature 385:810–813

Yard BA et al. (1993) Analysis of T cell lines from rejecting renal allografts. Kidney Int – Suppl 39:S133–S138

Yiou R, Lefaucheur JP, Atala A (2003a) The regeneration process of the striated urethral sphincter involves activation of intrinsic satellite cells. Anatomy Embryol 206:429–435

Yiou R, Yoo JJ, Atala A (2003b) Restoration of functional motor units in a rat model of sphincter injury by muscle precursor cell autografts. Transplantation 76:1053–1060

Yokoyama T, Huard J, Chancellor MB (2000) Myoblast therapy for stress urinary incontinence and bladder dysfunction. World J Urol 18:56–61

Yoo JJ, Meng J, Oberpenning F, Atala A (1998) Bladder augmentation using allogenic bladder submucosa seeded with cells. Urology 51:221–225

Yoo JJ, Park HJ, Lee I, Atala A (1999) Autologous engineered cartilage rods for penile reconstruction. J Urol 162:1119–1121

Yu J et al. (2007) Induced pluripotent stem cell lines derived from human somatic cells. Science:1151526

Zhang SC, Wernig M, Duncan ID, Brustle O, Thomson JA (2001) In vitro differentiation of transplantable neural precursors from human embryonic stem cells. [see comment]. Nat Biotechnol 19:1129–1133

Chapter 18
Telomeres and the Arithmetic of Human Longevity

Abraham Aviv and John D. Bogden

Contents

18.1	Introduction	573
18.2	Life Expectancy Vs. Lifespan	574
18.3	Exceptional Longevity	575
18.4	The Biology Vs. Chronology of Aging	576
18.5	Telomere Dynamics and Replicative Senescence	576
	18.5.1 Telomere Dynamics In Vivo	577
18.6	Slowing Down the Rate of Age-Dependent Telomere Shortening in Proliferative Tissues by Therapeutic Intervention	580
18.7	Conclusions	580
References		581

18.1 Introduction

The objective of this review is to explore the relationship between telomere length and the human lifespan. To achieve this objective we will discuss several ideas, including the current gap between basic and clinical research on aging and longevity, life expectancy versus lifespan, exceptional longevity, telomere dynamics and replicative senescence, the role of genetic and environmental factors, and therapeutic interventions.

The scenario below is probably familiar to most researchers who seek to understand aging. A study is published in a medical journal. It reports an association between factor X and aging-related disease Y. Criticisms of the study get under way to a familiar libretto that can be summed up as follows: This association does not

A. Aviv (✉)
The Center of Human Development and Aging, University of Medicine and Dentistry of New Jersey, New Jersey Medical School, Newark, NJ, USA
e-mail: avivab@umdnj.edu

show causality and as such it tells us little about the mechanisms that cause the disease. Another study is published in *Nature*. Through elegant genetic manipulation, its authors unequivocally show that gene S is a player in the lifespan of organism Z, usually a nematode or a fly, though occasionally a mouse. To the accolades of the scientific community, the authors of the work declare that their findings have considerable potential for increasing human longevity and they proceed to establish a biotechnological company, which will use drugs that target the products of gene S to improve the health of humans, if not lengthen their lives. Sound familiar?

This brings us to the term "arithmetic" in the title of this chapter. The "arithmetic" to which we refer denotes the sum of factors built into an equation. The biological equation which is the focus of this chapter is that of human longevity, and not the lifespan of worms and flies in the laboratory. There is no doubt that the fundamentals of evolution and biology are shared by all organisms, but the arithmetic of human longevity is likely to differ substantially from that of any model organism.

Findings in model organisms are indeed exciting and important and have already opened up new lines of research that improve our understanding of the biology of aging, human aging included. However, it is premature to even contemplate lengthening the human lifespan or healthspan based on findings on organisms far removed from humans. Though aging, a major determinant of longevity, is not an immutable process, it is the most complex of all complex human traits. And thus far, our understanding of complex human traits is rudimentary at best. Given that we cannot manipulate the biology of humans, as we do with model organisms, we must resort to epidemiological studies to gain a better understanding of factors that define aging and hence longevity.

As published articles on the associations between telomere length in white blood cells (WBC) and aging-related diseases keep piling up, it is clear that WBC telomere length might be a factor in the arithmetic of human longevity. The key question, however, is whether WBC telomere length is an active player which contributes to the arithmetic of longevity or merely describes it.

18.2 Life Expectancy Vs. Lifespan

Life expectancy and lifespan epitomize the difference between what might be and what is. Lifespan is a precise trait, defined by the age at death of the individual. (Here lifespan is defined in its narrow sense as the age at death from biological causes.) In contrast, life expectancy is a demographic entity that estimates, based on historical data, the projected mean lifespan of a population. This distinction between life expectancy and lifespan is at the core of the controversy about whether in America and elsewhere life expectancy — which has increased by roughly three decades in the last 150 years — will keep rising during the 21st century (Tuljapurkar et al. 2000; Olshansky et al. 2001, 2005a; Oeppen and Vaupel 2002; Tuljapurkar 2005).

The ability to accurately predict life expectancy depends in large measure on understanding the biological constraints imposed by evolution on the lifespan of humans. These constraints define the individual's maximal lifespan (IMLS), namely, the oldest age that might be attained under the most favorable environmental circumstances by the individual. The stakes are high in finding out what factors determine the IMLS in the general population, as their elucidation will reveal by how much and perhaps for how long the rise in life expectancy will continue.

There is a world of difference between the causes of increased life expectancy in the distant past and recent years. The rapid rise in life expectancy starting in the second half of the 19th century resulted primarily from reduced mortality among the young. Improved sanitation, better nutrition, immunization and the treatment of childhood maladies assured that most children survived to adulthood. Safer workplace conditions and the introduction of antibiotics and therapeutic modalities to treat parasitic infestations further extended lifespan. But more recently, a further increase in life expectancy was due mainly to the prevention and treatment of aging-related diseases (Olshansky et al. 2005b).

This is not to say that there will be no additional increase in the human lifespan. However, the future life expectancy trajectory might be constrained by the IMLS, or what Carnes et al. (2003) refer to as the 'biological warranty period.' As the trajectory of life expectancy slowly converges to the 'biological warranty period,' the innate limits of the human lifespan may ultimately dominate future changes in the arithmetic of longevity. The question then is: are there specific biological parameters that define the IMLS? These parameters are likely to be deciphered by studying humans who live a generation or so longer than other mortals, namely, exceptionally old humans.

18.3 Exceptional Longevity

In its broader context, aging is an expression of the temporal, progressive and collective outcome of multiple malfunctions that nibble away at the well-being of the individual. But the features that forecast the outmost boundary of lifespan may arise from biological systems and perhaps cells, reaching the point that they are no longer able to sustain all of the vital tasks that assure survival of the individual. Evidently, this endpoint arises at different ages in individuals who have withstood or avoided age-related diseases to attain an old age. In some, the IMLS may be reached at 70 or 75 years, while in preciously few others after they have surpassed the century mark.

The exceptionally old tend to exhibit compressed morbidity (Fries 1980) in that many of them experience aging-related diseases and infirmity only shortly before death (Hitt et al. 1999; Anderson et al. 2005; Perls 2004). Such findings reinforce the proposition that by being relatively impervious to aging-related diseases in their younger years (Cutler 1975), the life history of the exceptionally old is that of successful aging. What's more, the trait of survival to very old age appears to cluster within families (Perls et al. 2002, 1998; Kerber et al. 2001; Gudmundsson et al. 2000; Atzmon et al. 2004). Accordingly, a number of genes have been implicated as factors that contribute to exceptional longevity and several if not many others are candidate genes for the trait (Schachter et al. 1994; Puca et al. 2001; Barzilai and

Shuldiner 2001; Geesaman et al. 2003; Barzilai et al. 2003; Franceschi et al. 2005; Atzmon et al. 2006; Suh et al. 2008). Thus, like aging itself, exceptional longevity is a complex trait, meaning that a host of genetic and environmental factors ultimately determine lifespan, which is uniquely fashioned in each individual.

18.4 The Biology Vs. Chronology of Aging

As odd as it may seem, at present the calendar is the most objective criterion to gauge the progression of aging. From the demographic standpoint, an individual is as old as the number of days that have elapsed since her birth date. But aging, like lifespan, is not a demographic entity; it is an intrinsic biological quality. This premise is at the heart of the thesis that the pace of aging and its phenotypic manifestations differ not only among species but also among members of any species. In theory, the pace of aging might be attenuated, resulting in an increase in the IMLS.

The concept of biological age supports the distinction between the aging of cells and tissues within the organism and the overall, i.e., aging of the entire organism. Various cells and tissues may age at different rates within and among species, and stochastic factors are probably central to this process (Herndon et al. 2002; Kirkwood and Austad 2000). In some persons the kidneys, for instance, might be biologically older than the heart, while in others the heart might be biologically older than the kidneys. In addition, the underlying mechanisms of aging may vary in different cell types and tissues so that they age in various fashions and at different rates. Tissues that are primarily post-mitotic, including neurons and skeletal muscle, probably age differently than those dominated by proliferating cells, such as the hematopoietic system. Replicative aging of proliferating cells relates in part to the fact that after a finite number of divisions these cells reach the Hayflick limit, at which they stop dividing (Hayflick and Moorhead 1961).

18.5 Telomere Dynamics and Replicative Senescence

If replicative aging is a determinant in the overall biology of human aging, telomere dynamics (telomere length and attrition rate) may partially chronicle this process. There are several routes that lead to replicative senescence, and telomere attrition is one of them. As cells replicate in culture, their telomeres (the TTAGGG tandem repeats at chromosomal ends) undergo progressive attrition, until critically short telomeres relay a signal that switches on the DNA damage response, which includes cell cycle checkpoint activation (Wright and Shay 2002). Although oxidative stress can activate the DNA damage response (Chen and Ames 1994; Severino et al. 2000), it also accelerates telomere erosion per cell division because the telomeric G triplets are highly sensitive to oxidative damage (Oikawa and Kawanishi 1999; von Zglinicki 2003).

While cell replication and oxidative stress promote telomeric erosion, a specialized ribonucleoprotein reverse transcriptase, telomerase, is capable of adding telomeric repeats onto the ends of chromosomes (Blackburn 2005). However, most somatic cells, including non-embryonic somatic stem cells, express insufficient activity of the enzyme to arrest telomere erosion with cell replication. Thus, telomere length of proliferating cells is in large measure a record of their replicative history, cumulative oxidative stress burden, and future replicative potential.

18.5.1 Telomere Dynamics In Vivo

It has been proposed that humans, the longest living land mammals, are endowed with relatively short telomeres as a nearly fail-safe system against tumor formation (Wright and Shay 2002). As such, short telomeres may be a trade-off that balances resistance to cancer against curtailment of the human lifespan through replicative senescence. But unfortunately, the concept that telomere dynamics is a determinant in replicative senescence in cultured cells has quickly taken on a life of its own, stimulating widespread conjecture about the role of telomeres in human aging. The ultimate question, then, is whether the dynamics of human telomeres in vivo mirror their behavior in vitro (Aviv 2004). This question has been explored by studying telomeres in WBCs (Aviv 2008).

Mean telomere length in WBCs is the same in newborn boys and girls (Okuda et al. 2002; Akkad et al. 2006) but longer in women than men (Jeanclos et al. 2000; Nawrot et al. 2004; Vasan et al. 2008; Bekaert et al. 2007; Benetos et al. 2001; Hunt et al. 2008). Age-adjusted WBC telomere length is relatively short in persons who display aging-related diseases and those who have diseases, environmental exposures or lifestyle habits known to diminish the human lifespan. These include cardiovascular disease (Brouilette et al. 2003, 2007; Benetos et al. 2004; van der Harst et al. 2007), obesity (Vasan et al. 2005, 2008; Hunt et al. 2008; Gardner et al. 2005), sedentary lifestyle (Cherkas et al. 2008), dementia (Panossian et al. 2003; Honig et al. 2006) cigarette smoking (Nawrot et al. 2004; Vasan et al. 2008; Valdes et al. 2005), low socio-economic status (Cherkas et al. 2006) and unhealthy environmental circumstances (Bekaert et al. 2007). It is noteworthy that WBC telomere length is largely a surrogate marker of telomere length in hematopoietic stem cells (HSCs) (Aviv 2008). Thus, shortened telomeres in WBCs mirror the same phenomenon in HSCs.

Telomere length is highly variable among newborns (Okuda et al. 2002; Akkad et al. 2006). It follows that accelerated WBC telomere attrition during extra-uterine life is not the sole explanation for the relatively short WBC telomeres during adult life. In fact, several scenarios might characterize WBC telomere dynamics within the human lifespan: (a) starting with short WBC telomere length (at birth) and undergoing slow attrition; (b) starting with long WBC telomeres and undergoing fast attrition; (c) starting with long WBC telomeres and undergoing slow attrition; and (d) starting with short WBC telomeres and undergoing fast attrition. Furthermore, the rate of WBC telomere attrition is faster during early childhood than in adult life, and a substantial fraction ($\sim 30\%$) of telomere shortening occurs before the age of 20 years. (Rufer et al. 1999; Zeichner et al. 1999; Frenck et al. 1998).

18.5.1.1 The Potential Link Between WBC Telomere Biology and Aging-Related Diseases and Lifespan in Humans

The key issue in telomere epidemiology is whether in itself, WBC telomere dynamics in vivo only registers the cumulative effect of other determinants of aging or whether, through yet poorly understood mechanisms, it is an active participant in the process. At present, there is no compelling evidence that replicative senescence in any cell type is a common phenomenon that shapes human aging in the general population. That said, recent studies have shown that reduced numbers and proliferative potential of epithelial progenitor cells (EPCs) are associated with atherosclerosis and vascular dysfunction (Fadini et al. 2006a; Hill et al. 2003; Kunz et al. 2006; Ni et al. 2006; Schmidt-Lucke et al. 2005; Werner et al. 2005; Werner and Nickening 2007). EPCs are derived from the HSC pool and they possess the ability of homing to vascular sites with denuded endothelium, where they are integrated into the vascular wall, retarding atherosclerosis. The numbers of circulating EPCs, their proliferative potential, or both are diminished in older persons (Fadini et al. 2006b; Heiss et al. 2005; Hoetzer et al. 2007; Kunz et al. 2006; Xiao et al. 2007), in individuals with insulin resistance (Cubbon et al. 2007; Fadini et al. 2006b; Murphy et al. 2007), in those with a sedentary lifestyle (Hoetzer et al. 2007; Van Craenenbroeck et al. 2008; Steiner et al. 2005) and in smokers (Michaud et al. 2006). Though the number of EPCs might be lower in postmenopausal women than men (Xiao et al. 2007), their function, expressed in the replicative potential (colony forming units) and migratory activity, is considerably better in women (Hoetzer et al. 2007). In principle, diminished EPC reserves may mirror telomere shortening in HSCs (Imanishi et al. 2005; Oeseburg et al. 2007; Satoh et al. 2007), thereby providing a mechanistic explanation for the associations between WBC telomere length and atherosclerosis.

But might circulating cells other than EPCs, vital for survival, run out of telomeres and actually senesce during the human lifespan? There is no clear answer to this question, though there is evidence of a build-up of WBC subsets with extremely short telomeres in individuals older than 90 years (Kimura et al. 2007). If WBC telomere length were a determinant of the IMLS, it should forecast mortality in the elderly, but until recently the evidence from various studies had not achieved a consensus (Honig et al. 2006; Cawthon et al. 2003; Martin-Ruiz et al. 2005; Bischoff et al. 2006; Harris et al. 2006). Utilizing the powerful same-sex twin model, more recent studies clearly show that in elderly twins, the co-twin with the shorter WBC telomere length is more likely to die first (Kimura et al. 2008a; Bakaysa et al. 2007).

Assuming that telomere shortening does play a role in human longevity, the type of WBCs that may be the best arbiter of the IMLS is worth knowing. Based on their lineage, WBCs are traditionally stratified into two major categories: myeloid and lymphoid. Myeloid-derived WBCs are terminally differentiated, so that they do not divide in the peripheral circulation and their biological half-life is short, perhaps days or even hours. In contrast, T lymphocytes and B lymphocytes do differentiate and divide outside the bone marrow and their biological half-life is long (months and even many years in the case of memory cells). In vivo, T and B-lymphocytes, as well as granulocytes, demonstrate age-dependent telomere erosion (Effros et al. 2005; Weng 2001). Of these two lineages, T lymphocytes, particularly $CD8^+$ cells, have been the focus of intensive research because their status is linked to the decline in immune functions with aging in humans (Effros et al. 2005; Weng 2001; Vallejo 2005).

After multiple rounds of replication, brought about by repeated antigenic stimulation, cultured $CD8^+$ cells are no longer able to enter mitosis. At this stage, they exhibit the loss of the CD28 antigen, a major immunoglobulin molecule expressed on the membrane of T lymphocytes, where it serves to amplify the T-cell receptor signal (Vallejo 2005; Effros et al. 1994). In addition, the loss of CD28 expression in vitro is associated with resistance to apoptosis (Spaulding et al. 1999; Vallejo et al. 2000; Gunturi et al. 2003).

$CD28^-$ T cells are rare in newborn blood (Brzezinska et al. 2003), but their number progressively increases with age. The loss of CD28 expression is more apparent in $CD8^+$ than in $CD4^+$ cells (Vallejo 2005)—a phenomenon that may indicate differences in turnover rates and antigenic stimulation of lymphocyte subsets. In addition, $CD8^+$ $CD28^-$ T cells are probably the descendents of naïve $CD8^+$ $CD28^+$ T cells that have evolved into memory cells (Effros et al. 2005; Vallejo 2005). With the passage of time and successive rounds of cell division and antigenic stimulation, such cells become terminally differentiated, ultimately losing their replicative potential (Effros et al. 2005; Hsu et al. 2005; Vallejo et al. 2004), perhaps due to telomere attrition (Monteiro et al. 1996; Batliwalla et al. 2000). In this context, memory cells exhibit shorter telomeres than naïve ones (Rufer et al. 1999; Weng et al. 1995). All told, the loss of CD28 expression in vivo may signal a state of torpor, akin to replicative senescence and resistance to apoptosis in vitro (Effros et al. 2005; Vallejo 2005; Spaulding et al. 1999; Vallejo et al. 2000; Gunturi et al. 2003; Hsu et al. 2005; Valenzuela and Effros 2002). The rapid accumulation of $CD28^-$ cells in the elderly may hence be a key factor in weakening the immune system and thereby forecasting a shorter lifespan.

18.5.1.2 Genetic Vs. Environmental Factors as Potential Determinants Linking Telomere Dynamics to Aging-Related Diseases and Life Span

WBC telomere length is heritable (Jeanclos et al. 2000; Nawrot et al. 2004; Hunt et al. 2008; Slagboom et al. 1994; Vasa-Nicotera et al. 2005; Andrew et al. 2006) and, as noted above, modified by environmental factors. Moreover, WBC telomere length in offspring is positively associated with paternal age at the time of birth of the offspring (Unryn et al. 2005; De Meyer et al. 2007; Kimura et al. 2008), a phenomenon that might be driven by the emergence of a subset of sperm with longer telomeres in older men (Kimura et al. 2008). These findings suggest a role of genetic as well as epigenetic factors in WBC telomere biology. Whether or not telomere dynamics in HSCs or in WBCs are directly involved in aging-related diseases and longevity is still an open question. If telomere shortening does cause a decline in the production of EPCs and leads to the accumulation of $CD8^+$ $CD28^-$ T cells, telomeres might play an active role in diseases of aging and in longevity.

As both telomere length at birth and its rate of shortening afterwards are highly variable, the ultimate connection of telomere length with aging and longevity might depend on factors that are in control of telomere dynamics. We know little about the factors that determine telomere length at birth. However, inflammation and oxidative stress, two players in the biology of aging and longevity (Finch and Crimmins 2004; Beckman and Ames 1998), may exert some of their effects by accelerating the rate of telomere shortening in HSCs and WBCs. Accordingly, the odds of surviving to the IMLS are better in persons endowed with longer telomeres at birth and a slower rate of age-dependent WBC telomere erosion. In some measure, this rate might be attenuated by a healthy lifestyle, such as avoiding smoking and obesity and being physically active.

18.6 Slowing Down the Rate of Age-Dependent Telomere Shortening in Proliferative Tissues by Therapeutic Intervention

Attempts to ameliorate and perhaps cure monogenetic lethal diseases, the etiologies of which appear to relate to dysfunctions in telomerase or its regulatory proteins, such as dyskeratosis congenita or pulmonary fibrosis (Tsakiri et al. 2007; Vulliomy and Dokal, 2006) by medical means, e.g., bone marrow transplantation, seem reasonable. However, attempting to use drugs or other modalities to attenuate age-dependent telomere shortening or elongate telomeres in the general population is not justified at present. Just because shorter WBC telomere length is associated with aging-related diseases does not mean that elongating telomeres in these or other cells would be medically beneficial. Could the trade-off for elongating telomeres be cancer and perhaps other disorders that relate to aberrant growth and proliferation? We really don't know. Accordingly, large-scale clinical trials of compounds that might putatively prolong life, enhance well-being, or both by telomere elongation are needed before such compounds are marketed. At present the future utility of telomere elongation for the prevention of cancer (dePinho 2000) and the prevention and treatment of other disorders of aging is promising but requires rigorous testing for proof of efficacy without significant adverse effects.

18.7 Conclusions

The 125th anniversary issue of *Science* focused on the future by posing 25 "Big Questions" that face the scientific community today. One of these questions is: How much can human lifespan be extended (Couzin 2005)? The same topic was featured in *Nature* (Abbott 2004). Although it is difficult to answer this question, part of the answer is tied up in the simpler question: "how long should telomeres be?"

Unfortunately, we don't yet know the answer to this question, and elongation of telomeres by pharmaceutical or other means might be a risky endeavor if evolution endowed humans with shorter telomeres as an anti-cancer feature or as protection against other serious diseases. In their article entitled *The Future of Aging Therapies*, EC Hadley et al. (2005) state: "Few interventions have unmitigated benefits; it is unlikely that aging therapies will be an exception. Though human studies suggest that factors contributing to exceptional longevity also contribute to exceptional health over the lifespan, potential adverse effects of any putative human intervention need to be explored thoroughly."

What is needed to enhance the quality of life of the elderly is development of safe modalities to avoid or delay major afflictions of aging without prolonging the travails of frailty and decrepitude of old age. The future utility of telomere elongation for the prevention and treatment of disorders of aging is promising but can only be evaluated by future research.

References

Abbott A (2004) Growing old gracefully. Nature 428:116–118
Akkad A, Hastings R, Konje JC, Bell SC, Thurston H, Williams B (2006) Telomere length in small-for-gestational-age babies. Int J Obstet Gynecol 113:318–323
Anderson SL, Terry DF, Wilcox MA, Babineau T, Malek K, Perls TT (2005) Cancer in the oldest old. Mech Aging Dev 126:263–267
Andrew T, Aviv A, Falchi M, Surdulescu GL, Gardner JP, Lu X, Kimura M, Kato BS, Valdes AM, Spector TD (2006) Mapping genetic loci that determine leukocyte telomere length in a large sample of unselected, female sib-pairs. Am J Hum Genet 78:480–486
Atzmon G, Rincon M, Schecter CB, Shldiner AR, Lipton RB, Bergman A, Barzilai N (2006) Lipoprotein genotype and conserved pathway for exceptional longevity in humans. PloS Biol e113, Epub, Apr 4
Atzmon G, Schechter C, Greiner W, Davidson D, Rennert G, Barzilai N (2004) Clinical phenotype of families with longevity. J Am Geriat Soc 52:274–277
Aviv A (2004) Telomeres and human aging: facts and fibs. Sci Aging Knowledge Environ 51:43
Aviv A (2008) The epidemiology of human telomeres: faults and promises. J Gerontol. A. Biol Sci Med Sci 63:979–983
Bakaysa SL, Mucci LA, Slagboom PE, Boomsma DI, McClean GE, Pederson NL (2007) Telomere length predicts survival independent of genetic influences. Aging Cell 6:769–774
Barzilai N, Atzmon G, Schechter C, Schaefer EJ, Cupples AL, Lipton R, Cheng S, Shuldiner AR (2003) Unique lipoprotein phenotype and genotype associated with exceptional longevity. J Am Med Assoc 290:2030–2040
Barzilai N, Shuldiner AR (2001) Searching for human longevity genes: the future history of gerontology in the post-genomic era. J Gerontol A Biol Sci Med Sci 56:M83–M87
Batliwalla FM, Rufer N, Lansdorp PM, Gregersen PK (2000) Oligoclonal expansions in the $CD8^+CD28^-$ T cells largely explain the shorter telomeres detected in this subset. Hum Immunol 61:951–958
Beckman KB, Ames BN (1998) The free radical theory of aging matures. Physiol Rev 78: 547–581
Bekaert S, De Meyer T, Reitzschel ER, De Buyzere ML, De Bacquer D, Langlois M, Segers P, Cooman L, Van Damme P, Cassiman P, Van Criekinge W, Verdonck P, De Backer GG, Gillebert

TC, Van Oostveldt P; Asklepios investigators (2007) Telomere length and cardiovascular risk factors in middle-aged population free of overt cardiovascular disease. Aging Cell 6:639–647

Benetos A, Gardner JP, Zureik M, Labat C, Xiaobin L, Adamopoulos C, Temmar M, Bean KE, Thomas F, Aviv A (2004) Short telomeres are associated with increased carotid atherosclerosis in hypertensive subjects. Hypertension 43:182–185

Benetos A, Okuda K, Lajemi M, Kimura M, Thomas F, Skurnick J, Labat C, Bean K, Aviv A (2001) Telomere length as an indicator of biological aging: the gender effect and relation with pulse pressure and pulse wave velocity. Hypertension 37 (part 2):381–385

Bischoff C, Peterson HC, Graakjaer J, Andersen-Ranberg K, Vaupel JW, Bohr VA, Kolvraa S, Christensen K (2006) No association between telomere length and survival among elderly and oldest old. Epidemiology 17:190–194

Blackburn EH (2005) Telomeres and telomerase: their mechanismas and the effects of altering their functions. FEBS Lett 579:859–862

Brouilette S, Singh RK, Thompson JR, Goodall AH, Samani NJ (2003) White cell telomere length and risk of premature myocardial infarction. Arterioscler Thromb Vasc Biol 23:842–846

Brouilette SW, Moore JS, McMahon AD, Thompson JR, Ford I, Shepherd J, Packard CJ, Samani NJ (2007) Telomere length, risk of coronary heart disease, and statin treatment in the West of Scotland Primary Prevention Study: a nested case-control study. Lancet 369:107–114

Brzezinska A, Magalska A, Sikora E (2003) Proliferation of $CD8^+$ in culture of human T cells derived from peripheral blood of adult donors and cord blood of newborns. Mech Ageing Dev 124:379–387

Carnes BA, Olshansky SJ, Grahn D (2003) Biological evidence for limits to the duration of life. Biogerontology 4:31–45

Cawthon RM, Smith KR, O'Brien E, Sivatchenko A, Kerber RA (2003) Association between telomere length in blood and mortality in people aged 60 years or older. Lancet 361:393–395

Chen Q, Ames BN (1994) Senescence-like growth arrest induced by hydrogen peroxide in human diploid fibroblasts F65 cells. Proc Natl Acad Sci USA 91:4130–4134

Cherkas LF, Aviv A, Valdes AM, Hunkin JL, Gardner JP, Surdulescu GL, Kimura M, Spector TD (2006) The effects of social status on biological ageing as measured by white-blood-cell telomere length. Aging Cell 5:361–365

Cherkas LF, Hunkin JL, Kato BS, Gardner JP, Surdulescu GL, Kimura M, Spector TD, Aviv A (2008) Physical activity is associated with longer telomeres in leukocytes. Arch Int Med 168:154–158

Cubbon RM, Rajwani A, Whearcroft SB (2007) The impact of insulin resistance on endothelial function, progenitor cells and repair. Diabetes Vasc Dis Res 4:103–111

Couzin J (2005) How much can human life span be extended? Science 309:83

Cutler RG (1975) Evolution of human longevity and the genetic complexity governing aging rate. Proc Natl Acad Sci USA 72:4664–4668

De Meyer T, Rietzschel ER, De Buyzere ML, De Bacquer D, Van Criekinge W, De Backer GG, Gillebert TC, Van Oostveldt P, Bekaert S; on behalf of the Asklepios investigators (2007) Paternal Age at Birth is an Important Determinant of Offspring Telomere Length. Hum Mol Genet 16:3097–102

dePinho RA (2000) The age of cancer. Nature 408:248–254

Effros RB, Boucher N, Porter V, Zhu X, Spaulding C, Walford RL, Kronenberg M, Cohen D, Schachter F (1994) Decline in $CD28^+$ T cells in centenarians and in long-term T cell cultures: a possible cause for both in vivo and in vitro immunosenesence. Exp Gerontol 29:601–609

Effros RB, Dagarag M Spaulding C, Man J (2005) The role of $CD8^+$ T-cell replicative senescence in human aging. Immunol Rev 205:147–157

Fadini GP, de Kreutzenberg SV, Coracina A et al. (2006a) Circulating $CD34^+$ cells, metabolic syndrome, and cardiovascular risk. Eur Heart J. 27:2247–2255

Fadini GP, Coracina A, Baesso I, Agostini C, Tiengo A, Avogaro A, de Kreutzenberg SV (2006b) Peripheral blood $CD34^+KDR^+$ endothelial progenitor cells are determinants of subclinical atherosclerosis in a middle-aged general population. Stroke 37:2277–2282

Murphy C, Kanganayagam GS, Jiang B et al. (2007) Vascular dysfunction in healthy Indian Asians is associated with insulin resistance and reduced endothelial progenitor cells. Arterioscler Thromb Vasc Biol 27:936–942

Finch CE, Crimmins EM (2004) Inflammatory exposure and historical changes in human life-spans. Science 305:1736–1739

Franceschi C, Olivieri F, Marchegiani F, Cardelli M, Cavallone L, Capri M, Salvioli S, Valensin S, De Benedictis G, Di Iorio A, Caruso C, Paolisso G, Monti D (2005) Genes involved in immune response/inflammation, IGF1/insulin pathway and response to oxidative stress play a major role in the genetics of human longevity: the lesson of centenarians. Mech Ageing Dev 126:351–361

Frenck RW Jr, Blackburn EH, Shannon KM (1998) The rate of telomere sequence loss in human leukocytes varies with age. Proc Natl Acad Sci USA 95(12):5607–5610

Fries JF (1980) Aging, natural death, and the compression of morbidity. New Engl J Med 303: 130–135

Gardner JP, Li S, Srinivasan SR, Chen W, Kimura M, Lu X, Berenson GS, Aviv A (2005) Rise in insulin resistance is associated with escalated telomere attrition. Circulation 111:2171–2177

Geesaman BJ, Benson E, Brewster SJ, Kunkel LM, Blanche H, Thomas G, Perls TT, Daly MJ, Puca AA (2003) Haplotype-based identification of a microsomal transfer protein marker associated with the human lifespan. Proc Natl Acad Sci USA 100:14115–14120

Gudmundsson H, Frigge M, Gulcher JR, Stefansson K (2000) Inheritance of human longevity in Iceland. Europ J Hum Genet 8:743–749

Gunturi A, Berg RE, Forman J (2003) Preferential survival of CD8 T and NK cells expressing high levels of CD94. J Immunol 170:1737–1745

Hadley EC, Lakatta EG, Morrison-Bogorad M, Warner HR, Hodes RJ (2005) The future of aging therapies. Cell 120:557–567

Harris SE, Deary I, MacIntyre A, Lamb KJ, Radhakrishnan K, Starr JM, Whalley LJ, Shiels PG (2006) The association between telomere length, physical health, cognitive ageing, and mortality in non-demented older people. Neurosci Lett 406:260–264

Hayflick L, Moorhead PS (1961) The serial cultivation of human diploid cell strains. Exp Cell Res 25:585–621

Herndon LA, Schmeissner PJ, Dudaronke JM, Brown PA, Listner KM, Sakano Y, Paupard MC, Hall DH, Driscoll M (2002) Stochastic and genetic factors influence tissue-specific decline in ageing *C. elegans*. Nature 419:808–814

Heiss C, Keymel S, Niesler U, Ziemann J, Kelm M Kalka C (2005) Impaired progenitor cell activity in age-related endothelial dysfunction. J Am Coll Cardiol 45:1441–1448

Hill JM, Zalos G, halcox JPJ, Schenke WH, Waclawiw MA, Quyymi AA, Finkel T (2003) Circulating endothelial progenitor cells, vascular function, and cardiovascular risk. N Eng J Med 348:593–600

Hitt R, Young-Xu Y, Silver M, Perls T (1999) Centenarians: the older you get, the healthier you have been. Lancet 354:652:

Hoetzer GL, MacEneaney OJ, Irmiger HM, Keith R, Van Guilder GP, Stauffer BL, DeSouza CA (2007) Gender differences in circulating endothelial progenitor cell colony-forming capacity and migratory activity in middle-aged adults. Am J Cardiol 99:46–48

Honig LS, Schupf N, Lee JH, Tang MX, Mayeux R (2006) Shorter telomeres are associated with mortality in those with *APOE* ϵ4 and dementia. Ann Neurol 60:181–187

Hsu HC, Scott DK, Mountz JD (2005) Impaired apoptosis and immune senescence—cause or effect? Immunol Rev 205:130–146

Hunt SC, Chen W, Gardner JP, Kimura M, Srinivasan SR Berenson GR, Aviv A (2008) Leukocyte telomeres are longer in African Americans than in Whites: The NHLBI Family Heart Study and the Bogalusa Heart Study. Aging Cell. 7:451–458

Imanishi T, Hano T, Nishio I. Angiotensin II accelerates endothelial progenitor cell senescence through induction of oxidative stress (2005) J Hypertens 23:97–104

Jeanclos E, Schork NJ, Kyvik KO, Kimura M, Skurnick JH, Aviv A (2000) Telomere length inversely correlates with pulse pressure and is highly familial. Hypertension 36:195–200

Kerber RA, O'Brien E, Smith KR, Cawthon RM (2001) Familial excess longevity in Utah Geneologies. J Gerentol A: Biol Sci Med Sci 56: B130-B139

Kimura M, Barbieri M, Gardner JP, Skurnick J, Cao X, van Riel N, Rizzo MR, Paoliso G, Aviv A (2007) Leukocytes of exceptionally old persons display ultra-short telomeres. Am J Physiol Regul Integr Comp Physiol 293: R2210–2217

Kimura M, Hjelmborg JVB, Gardner JP, Bathum L, Brimacombe M, Lu X, Christiansen L, Vaupel LW, Aviv A (2008a) Short leukocyte telomeres forecast mortality: a study in elderly Danish twins. Am J Epidemiol 167:799–806

Kimura, M, Cherkas LF, Kato BS, Demissie S, Hjelmborg JB, Brimacombe, M, Cupples LA, Hunkin JL, Gardner JP, Lu X, Cao X, Sastrasinh M, Province M, Hunt SC, Christensen K, Levy D, Spector TD, and Aviv A (2008b) Offspring's leukocyte telomere length, paternal age, and telomere elongation in sperm. PLoS Genet Feb 15;4(2):e37 [Epub ahead of print]

Kirkwood TB, Austad SN (2000) Why do we age? Nature 408:233–238

Kunz GA, Liang G, Cuculi F, Gregg D, Vata KC. Shaw LK, Goldschmidt-Clermont PJ, Dong C, Taylor DA, Peterson ED (2006) Circulating endothelial progenitor cells predict coronary artery severity. Am Heart J 152:190–195

Martin-Ruiz CM, van Heemst GJ, von Zglinicki T, Westendorp RG (2005) Telomere length in white blood cells is not associated with morbidity in the oldest old: a population-based study. Aging Cell 4:287–290

Michaud SE, Dussault S, Haddad P, Groleau J, Rivard A (2006) Circulating endothelial progenitor cells from healthy smokers exhibit impaired functional activities. Atherosclerosis 187:423–432

Monteiro J, Batliwalla F, Ostrer H, Gregersen PK (1996) Shortened telomeres in clonally expanded $CD28^-$ $CD8^+$ T cells imply a replicative history that is distinct from their $CD28^+$ $CD8^+$ counterparts. J Immunol 156:3587–3590

Nawrot TS, Staessen JA, Gardner JP, Aviv A (2004) Telomere length and possible link to X chromosome. Lancet 363:507–510

Ni GP, Coracina A, Baesso I, Agostini C, Tiengo A, Avogaro A, de Kreutzenberg SV (2006) Peripheral blood $CD34^+KDR^+$ endothelial progenitor cells are determinants of subclinical atherosclerosis in a middle-aged general population. Stroke 37:2277–2282

Oeppen J, Vaupel JW (2002) Broken limits to life expectancy. Science 296:1029–1031

Oeseburg H, Westenbrink BD, de Boer RA, van Gilst WH, van der Harst P (2007) Can critically short telomeres cause functional exhaustion of progenitor cells in postinfarction heart failure? J Am Coll Cardiol 50:1909–1913

Oikawa S, Kawanishi S (1999) Site-specific DNA damage at GGG sequence by oxidative stress may accelerate telomere shortening FEBS Lett 453:365–368

Okuda K, Bardeguez A, Gardner JP, Rodriguez P, Ganesh V, Kimura M, Skurnick J, Awad G, Aviv A (2002) Telomere length in the newborn. Pediat Res 52:377–381

Olshansky SJ, Carnes BA, Desesquelles A (2001) Prospects for human longevity. Science 291:1491–1492

Olshansky SJ, Passaro DJ, Hershow RC, Layden J, Carnes BA, Brody J, Hayflick L, Butler NB, Allison DB, Ludwig DS (2005a) A potential decline in life expectancy in the United States in the 21st century. New Engl J Med 352:1138–1145

Olshansky SJ, Grant M, Brody J, Carnes BA (2005b) Biodemographic perspectives for epidemiologists. Emerg Themes Epidemiol 2:10

Panossian LA, Porter VR, Valenzuela HF, Zhu X, Reback E, Masterman D, Cummings JL, Effros RB (2003) Telomere shortening in T cells correlates with Alzheimer's disease status. Neurobiol Aging 24:77–84

Perls T (2004) Centenarians who avoid dementia. Trends Neurosci 27:633–636

Perls TT, Bubrick E, Wager CG, Vijg J, Kruglyak L (1998) Siblings of centenarian live longer. Lancet 351:1560

Perls TT, Wilmoth J, Levenson R, Drinkwater M, Cohen M, Bogan H, Joyce E, Brewster S, Kunkel L, Puca A (2002) Life-long sustained mortality advantage of siblings of centenarians. Proc Natl Acad Sci USA 99:8442–8447

Puca AA, Daly MJ, Brewster SJ, Matise TC, Barrett J, Shea-drinkwater M, Kang S, Joce E, Nicoli J, Benson E, Kunkel LM, Perls T (2001) A genome-wide scan for linkage to human exceptional longevity identifies a locus on chromosome 4. Proc Natl Acad Sci USA 98: 10505–10508

Rufer N, Brummendorf TH, Kolvraa S, Bischoff C, Christensen K, Wadsworth L, Schulzer M, Lansdorp PM (1999) Telomere fluorescence measurements in granulocytes and T lymphocytes subsets point to a high turnover of hematopoieticstem cells and memory T cells in early childhood. J Exp Med 190:157–167

Satoh M, Ishikawa Y, Takahashi Y, Itoh T, Minami Y, Nakamura M (2007) Association between oxidative DNA damage and telomere shortening in circulating endothelial progenitor cells obtained from metabolic syndrome patients with coronary artery disease. Atherosclerosis [Epub ahead of print]

Schachter F, Faure-Delanef L, Guenot F, Rouger H, Froguel P, Lesueur-Ginot L, Cohen D (1994) Genetic associations with human longevity at the APOE and ACE loci. Nat Genet 6:29–32

Schmidt-Lucke C, Rossing L, Fichtlscherer S, Vasa M, Britten M, Kamper U, Dimmeler S, Zeiher AM (2005) Reduced number of circulating endothelial progenitor cells predicts future cardiovascular events: proof of concept for the clinical importance of endogenous vascular repair. Circulation 111:2981–2987

Severino J, Allen RG, Balin S, Balin A, Cristofalo VJ (2000) Is β-galactosidase staining a marker of senescence in vitro and in vivo? Exp Cell Res 257:162–171

Slagboom PE, Droog S, Boomsma DI (1994) Genetic determination of telomere size in humans: A twin study of three age groups. Am J Hum Genet 55:876–882

Spaulding C, Guo W, Effros RB (1999) Resistance to apoptosis in human $CD8^+$ T cells that reach replicative senescence after multiple rounds of antigen-specific proliferation. Exp Gerontol 34:633–644

Suh Y, Atzmon G, Cho MO, Hwang D, Liu B, Leahy DJ, Barzilai N, Cohen P (2008) Functionally significant insulin-like growth factor receptor mutations in centenarians. Proc Natl Acad Sci USA March 3 [Epub a head of print]

Tsakiri KD, Cronkite JT, Kuan PJ, Xing C, Raghu G, Weissler JC, Rosenthal RL, Shay JW, Garcia CK (2007) Adult-onset pulmonary fibrosis caused by mutation in telomerase. Proc Natl Acad Sci USA 104:7552–7557

Tuljapurkar S, Li N, Boe C (2000) A universal pattern of mortality decline in G7 countries. Nature 405:789–792

Tuljapurkar S (2005) Future mortality: a bumpy road to Shangri-La? Sci Aging Knowledge Environ Apr 6, 2005 (14):pe9

Unryn BM, Cook LS, Riabowol KT (2005) Paternal age is positively linked to telomere length of children. Aging Cell 4:97–101

Valdes A, Andrew T, Gardner JP, Kimura M, Oelsner E, Cherkas LF, Aviv A, Spector TD (2005) Obesity, cigarette smoking, and telomere length in women. Lancet 366:662–664

Steiner S, Niessner A, Ziegler S, Richter B, Seidinger D, Pleiner J, Penka M, Wolzt M, Huber K, Wojta J, Minar E, Kopp CW (2005) Endurance training increases the number of endothelial progenitor cells in patients with cardiovascular risk and coronary artery disease. Atherosclerosis 181:305–310

Valenzuela HF, Effros RB (2002) Divergent telomerase and CD28 expression patterns in human CD4 and CD8 T cells following repeated encounters with the same antigenic stimulus. Clin Immunol 105:117–125

Vallejo AN, Schirmer M, Weyand CM, Goronzy JJ (2000) Clonality and longevity of $CD4^+CD28^{null}$Tcells are associated with defects in apoptotic pathways. J Immunol 165:6301–6307

Vallejo AN, Weyand CM, Gorozny JJ (2004) T-cell senescence: a culprit of immune abnormalities in chronic inflammation and persistent infection. Trends Mol Med 10:119–124

Vallejo AN (2005) CD28 extinction in human T cells: altered functions and the program of T-cell senescence. Immunol Rev 205:158–169 van der Harst P, van der Steege G, de Boer RA, Voors AA, Hall AS, Mulder MJ, Van Gilst WH, van Veldhuisen DJ (2007) Telomere length of circulating leukocytes is decreased in patients with chronic heart failure. J Am Coll Cardiol 49:1459–1464

Vasan RS, Demissie S, Kimura M, Cupples LA, Rifai N, White C, Wang TJ, Gardner JP, Cao X, Benjamin EJ, Levy D, Aviv A (2008) Association of leukocyte telomere length with circulating biomarkers of the renin-angiotensin-aldosterone system: the Framingham Heart Study. Circulation. 117:1138–1144

Vasa-Nicotera M, Brouilette S, Mangino M, Thompson JR, Braund P, Clemitson JR, Mason A, Bodycote CL, Raleigh SM, Louis E, Samani NJ (2005) Mapping of a major locus that determines telomere length in humans. Am J Hum Genet 76:147–151

Van Craenenbroeck EM, Vrints CJ, Haine SE, Vermeulen K, Goovaerts I, Van Tendeloo VF, Hoymans VY, Conraads VM (2008) A maximal exercise bout increases the number of circulating CD34+/KDR+ endothelial progenitor cells in healthy subjects. Relation with lipid profile. J Appl Physiol. Jan 24 [Epub ahead of print]

Vulliomy T, Dokal I (2006) Dyskeratosis congenita. Semin Hematol 43:157–166

von Zglinicki T (2003) Replicative senescence and the art of counting. Exp Gerontol 38:1259–1264

Weng N (2001) Interplay between telomere length and telomerase in human leukocyte differentiation and aging. J Leukoc Biol 70:861–867

Weng NP, Levine BL, June CH, Hodes RJ (1995) Human naïve and memory T lymphocytes differ in telomeric length and replicative potential. Proc Natl Acad Sci USA 92:11091–11094

Werner N, Nickening G (2007) Endothelial progenitor cells in health and atherosclerotic disease. Ann of Med 39:82–90

Werner N, Kosiol S, Schiegl T, Ahlers P, Walenta K, Link A, Bohm M, Nickening G (2005) Circulating endothelial progenitor cells and cardiovascular outcomes. N Engl J Med 353:999–1007

Wright WE, Shay JW (2002) Historical claims and current interpretations of replicative aging. Nature Biothechnol 20:682–688

Xiao Q, Klechi S, Patel S, Oberhollenzer F, Weger S, Mayr A, Meltzer B, Reindi M, Hu Y, Willeit J, Xu Q (2007) Endothelial progenitor cells, cardiovascular risk factors, cytokine levels and atherosclerosis– results from a large population-based study. PloS Oneissue 10:e975

Zeichner SL, Paulombo P, Feng Y, Xiao X, Gee D, Sleeasman J, Goodenow M, Biggar R, Dimitrov D (1999) Rapid telomere shortening in children. Blood 93:2824–2830

Chapter 19
Repairing Extracellular Aging and Glycation

John D. Furber

Contents

19.1	Introduction: Pathologies Caused by Aging Extracellular Proteins	588
19.2	Normal Functions of the ECM	589
19.3	Maintenance and Turnover of the ECM	590
19.4	Age-Related Deterioration of the ECM: Anatomy, Chemistry, Structures, and Mechanisms of ECM Pathologies	592
	19.4.1 Glycation Pathways	592
	19.4.2 Lipoxidation Pathways	594
	19.4.3 Amino Acid Isomerization, Deamidation, and Oxidation	595
	19.4.4 ECM Protein Strand Breakage	595
	19.4.5 Mechanical Consequences of Protein Alterations	597
	19.4.6 Altered Cell-Matrix Integrin Binding	598
	19.4.7 Cell-Matrix Interactions: Receptors, Signaling, and Inflammation . . .	598
	19.4.8 Extracellular Amyloidosis	600
	19.4.9 β-Amyloid Plaques in the Brain	601
19.5	Present and Possible Future Therapeutic Approaches for Better Maintenance and Repair of the ECM	601
	19.5.1 Diet, Fasting, and Calorie Restriction	601
	19.5.2 Exercise	602
	19.5.3 Inhibitors of Glycation, Lipoxidation, and AGE Formation	602
	19.5.4 Deglycators and Crosslink Breakers	606
	19.5.5 Tuned Electromagnetic Energy	610
	19.5.6 Removing β-Amyloid Plaques	610
	19.5.7 Enhancing Turnover of ECM by FLCs	610
	19.5.8 General Therapy Design Considerations	612
	19.5.9 Therapy Usage and Frequency	613
19.6	Summary and Conclusions	614
References	. . . .	614

J.D. Furber (✉)
Legendary Pharmaceuticals, PO Box 14200, Gainesville, FL 32604, USA
e-mail: johnfurber@LegendaryPharma.com

19.1 Introduction: Pathologies Caused by Aging Extracellular Proteins

Stiffness, arthritis, and cataracts have long been associated with aging humans and other mammals. In recent decades, important biochemical bases of these, and other, progressive age-associated pathologies have been identified. They are caused, at least in part, by accumulating chemical modifications to long-lasting structural proteins in the extracellular matrix (ECM) (Kohn 1978; Cerami et al. 1987; Vasan et al. 2001; Verzijl et al. 2003; DeGroot et al. 2004). Over time, chemical and mechanical changes accumulate in long-lasting extracellular structural proteins (LESPs), profoundly affecting the growth, development, and death of cells, as well as the mechanical operation of bodily systems. The LESPs stay in place for a very long time. Molecular modifications can remain unrepaired, and accumulate with age. It is now apparent that several types of accumulating chemical modifications are especially damaging to human physiological functioning. Extracellular aging is a major player in the interrelated processes of human aging (Cerami et al. 1987; Robert et al. 2008).

Chemical reactions, importantly glycation, lipoxidation, oxidation, nitration, amino acid isomerization, each change the LESPs in the ECM, as do protein strand breaks, wound healing, scar (cicatrix) formation, photoaging of the skin, and the actions of macrophages, infections, and inflammation. Important consequences include:

- Changed mechanical properties of tissues
- Changed environmental niches for cells, which affect their health and development
- Vicious cycles of progressively increasing damage.

Three processes are especially significant causes of pathogenic LESP modifications: glycation, lipoxidation, and strand breaks (Cerami et al. 1987; Januszewski et al. 2003; Robert et al. 2008). Glycation, formerly called "nonenzymatic glycosylation", is the spontaneous covalent bonding of a sugar to a macromolecule, such as a protein (Eble et al. 1983; Bucala and Cerami 1992). Lipoxidation occurs when oxidation of lipids produces reactive lipid fragments that covalently bond to proteins (Miyata et al. 1999). The chemical group attached to the protein is referred to as an "adduct". The phrases, "advanced glycation endproducts" (AGEs) and "advanced lipoxidation endproducts" (ALEs) have been used to describe the wide array of chemical species that eventually result from glycation and lipoxidation reactions (Cefalu et al. 1995; Januszewski et al. 2003). Glycation and AGEs have been studied for many years in connection with diabetic complications and physiological senescence. More recently, the Baynes lab pointed out that lipoxidation pathways also create some of the same damaging endproducts (Januszewski et al. 2003; Miyata et al. 1999).

AGEs and ALEs have been established as strong contributors to many progressive diseases of aging: vascular diseases (such as atherosclerosis, systolic hypertension, pulmonary hypertension, and poor capillary circulation) (Cerami et al. 1987; Bucala and Cerami 1992; Vaitkevicius et al. 2001;Vlassara and Palace 2003), erectile dysfunction (Usta et al. 2004, 2006); kidney disease (Vasan et al. 2001; Vlassara and Palace 2003), stiffness of joints and skin, osteoarthritis and rheumatoid arthritis (deGroot et al. 2004; Drinda et al. 2005; Steenvoorden et al. 2007; Verzijl et al. 2003), cataracts, retinopathy (Vasan et al. 2001), peripheral neuropathy (Bucala and Cerami 1992), Alzheimer's Dementia (Ulrich and Cerami 2001; Perry and Smith 2001), impaired wound healing, urinary incontinence, complications of diabetes, cardiomyopathies (such as diastolic dysfunction, left ventricular hypertrophy, and congestive heart failure) (Bucala and Cerami 1992), and solid cancers and metastasis (Taguchi et al. 2000).

In nondiabetic people, LESP aging occurs very slowly. This is consistent with our understanding that these age-associated diseases occur late in life because passage of time is required for sufficient damage to accumulate on LESPs. It is noteworthy that these same diseases emerge at an earlier age in diabetic individuals, whose average blood sugar and lipid concentrations are higher than normal, thus driving the deleterious reactions faster (Cerami et al. 1987; Bucala and Cerami 1992; Januszewski et al. 2003).

19.2 Normal Functions of the ECM

Our bodies are constructed of cells and extracellular materials. "The **extracellular matrix** consists of macromolecules secreted by cells into their immediate environment. These macromolecules form a region of noncellular material in the interstices between the cells" (Gilbert 2000). Some authors also refer to soluble extracellular materials as the "aqueous phase of the matrix" (Fawcett 1986). The structural molecules of the ECM include proteins, glycoproteins, and proteoglycans. The ECM holds cells together and co-creates the microenvironments in which they live (Spencer et al. 2007). It includes noncellular portions of bones, cartilage, tendons, and ligaments, as well as epithelial basement membranes, the renal glomerular basement membrane, and the fibrous meshworks that give strength to blood vessels, skin, tissues, and organs.

The most abundant protein in extracellular matrix is collagen. It is found in several variants throughout the body, principally as strong, straight structural fibers, providing strength to bones, cartilage, and tissues. Type IV collagen is a flat sheet that forms basement membranes.

Another important extracellular protein is elastin, whose wrinkled meshwork provides elastic properties to tissues. Elastic fibers are assembled extracellularly from elastin and several glycoproteins (Shifren and Mecham 2006). Elastic fibers form a shock-absorber to the hemodynamic pulses of the cardiovascular system. The resilience of lung tissue, arteries, and skin are due to elastic fibers (Wagenseil and Mecham 2007).

Laminin, vitronectin, and fibronectin are extracellular proteins that are important in cell adhesion, differentiation, and migration over the ECM. Integrin receptors on cells attach to a conserved sequence of amino acids: arginine-glycine-aspartate (RGD sequence), which is part of these proteins (Gilbert 2000). The composition of the ECM influences gene expression and differentiation state in resident cells. Signals are sent to cell nuclei through receptor pathways and via cytoskeletal contacts (Boudreau and Bissell 1998; Spencer et al. 2007).

19.3 Maintenance and Turnover of the ECM

Natural cellular processes slowly replace the aging collagen (Bucala and Cerami 1992). The natural turnover and remodeling of ECM proteins occurs at differing rates in various tissues during aging. The average turnover time of collagen is characteristically different in each different human tissue (Sell et al. 2005). Turnover can remove aged ECM and replace it with new, undamaged ECM. However, in many human tissues, the rate of turnover is slower than the rate of AGE accumulation. Furthermore, elastic fiber repair or replacement is imperfect; there is clearly an accumulation of damaged elastin with age (Robert et al. 2008; Wagenseil and Mecham 2007; Shifren and Mecham 2006).

Turnover requires removal of old molecules and replacement by new molecules *in the proper arrangement*. To the extent that old ECM is digested, removed, and replaced, some of the chemical modifications or damage, such as glycation or isomerization, would be removed and digested or excreted to the urine (Bucala and Cerami 1992; Ahmed and Thornalley 2003; Vlassara and Palace 2003). The complex details of ECM degradation are reviewed elsewhere (Robert et al. 2008; Murphy and Reynolds 2002; Everts et al. 1996). Cells of the fibroblast lineage (FLCs), in the connective tissue, degrade and replace LESPs. FLCs include fibroblasts, chondrocytes, osteoblasts, adipocytes, smooth muscle cells, macrophages, and mesenchymal stem cells (MSCs) (Alberts et al. 2002). FLCs can secrete digestive enzymes that cleave the collagen strands so that the resulting fragments may be phagocytosed and digested further within lysosomes (Everts et al. 1996; Murphy and Reynolds 2002). Additionally, vascular endothelial cells and renal mesangial cells may participate in AGE elimination by endocytosis (Vlassara and Palace 2003). After phagocytosis and intracellular digestion, some low-molecular-weight glycated molecules may be released to the circulatory system and cleared through the kidney (Vlassara and Palace 2003).

New collagen molecules are synthesized inside the fibroblasts, as three peptide chains which twist together, like a rope, into a triple helix, stabilized by hydrogen bonds and disulfide bonds (Lodish et al. 2000; Alberts et al. 2002; Piez 2002). These rodlike procollagen molecules are secreted, by exocytosis from Golgi vesicles, into the extracellular space, where their ends are trimmed off. Fibroblasts pull and arrange them into place as intermolecular electrostatic and hydrophobic interactions guide the assembly of collagen fibrils, which can aggregate into larger collagen fibers. After assembly, collagen molecules and fibrils are stabilized and strengthened by dilysine crosslinks (Lodish et al. 2000). These beneficial crosslinks are formed under regulated enzymatic control, and result in the mature collagen fibers. Similarly, elastin molecules are held together by beneficial di-, tri-, and tetralysine crosslinks, which are enzymatically formed after elastin strands are extruded into the ECM (Shifren and Mecham 2006; Mathews and van Holde 1990). Later, over the years, very slow processes of non-enzymatic glycation form additional crosslinks and adducts, which are pathogenic, and which accumulate over the lifetime of the collagen and elastin fibers.

Data indicate that the rate of formation of new AGEs and crosslinks per gram of collagen is the same among all of the human tissues studied. Therefore, differences in accumulation of glycated residues are apparently due to differences in collagen turnover rates of the different tissues (Verzijl et al. 2000; Sell et al. 2005). Consequently, it has been possible to use glycation accumulation to estimate turnover times for collagen in various other tissues. The results correlate well with turnover times calculated by measuring racemization of aspartate residues in collagen (Verzijl et al. 2000, 2001). Sell and colleagues reviewed collagen turnover rates in discussing their own measurements of glycation crosslinks (Sell et al. 2005). Kidney glomerular basement membrane (GBM) appears to turn over fairly quickly compared with skin, which has collagen molecules more than 15 years old. Collagen in articular cartilage reportedly has a turnover half-life of between 60 and 500 years (Verzijl et al. 2000). Sivan, et al, report a turnover half-life of cartilage in human intervertebral disks of 95 years in young adults, but turnover slows to 215 years in older adults (Sivan et al. 2008).

As the number of glycation crosslinks increases over time, the collagen fibrils are held more tightly together, making the ECM stiffer and perhaps less accessible to fibroblasts, macrophages, and enzymes that might attempt to digest and turn it over (DeGroot et al. 2001c). Furthermore, some AGEs, such as the abundant adduct, N-ε-carboxymethyllysine (CML), trigger apoptotic signals in the fibroblasts (Alikhani et al. 2005). The fibroblast population declines in number over the years, and many fibroblasts become "senescent." Senescent fibroblasts do not turn over ECM properly. Not only do they synthesize less ECM proteins but, they secrete excessive amounts of inflammatory cytokines and matrix metalloproteinases (MMPs), which digest ECM proteins without replacing them properly (Campisi 2005; Benanti et al. 2002). Similarly, articular chondrocytes decline in number and slow their production of proteoglycans, contributing to osteoarthritis and deterioration of articular cartilage (Taniguchi et al. 2009; DeGroot et al. 1999).

These events reduce ECM turnover rate, which extends turnover time, thus allowing more time for more AGEs and crosslinks to form (Vater et al. 1979; DeGroot et al. 2001a, b). These factors appear to create a vicious cycle of slowing the turnover rate (DeGroot et al. 2001c). Observations show an exponential increase in crosslinking with age in human skin (Sell et al. 1993, 2005 1993), cartilage (Verzijl et al. 2000), and lens (Cheng et al. 2004). In contrast, crosslinking increases very gradually with age in kidney GBM because the LESP turnover rate there is rapid enough to avoid a vicious cycle (Sell et al. 1993, 2005).

The rate of collagen turnover in human tendons and skeletal muscles is increased by physical exercise, as described in Section 19.5.2 (Kjær et al. 2006). Orthodontists have long noted that fibrous joints and bone undergo increased remodeling in response to mechanical stress (Murphy and Reynolds 2002).

Inflammation induces a less desirable form of ECM remodeling. FLCs secrete additional digestive enzymes, including MMPs, to rapidly open up the ECM (Everts et al. 1996). Their purpose is to allow immune cells to move through the tissue, to search for pathogens. This rapid, inflammatory digestion of ECM is not restored as perfectly as during normal turnover and remodeling.

Scar formation is a form of ECM remodeling occurring during mammalian wound healing. It has evolved to be rapid, to mend tissues and stop fluid loss, but the resulting collagen cicatrix patch is not a perfect match to the surrounding tissue.

19.4 Age-Related Deterioration of the ECM: Anatomy, Chemistry, Structures, and Mechanisms of ECM Pathologies

A variety of processes change the LESPs during aging. Sugar, lipids, and oxygen react with ECM proteins to produce adducts and crosslinks, which we refer to as AGEs/ALEs. These reactions are variously referred to as glycation, glycoxidation, glyco-oxidation, nonezymatic glycosylation, and lipoxidation. Receptor molecules on cell surfaces react to AGEs/ALEs, triggering harmful inflammatory responses. During aging, some cells inappropriately attack the ECM by secreting extracellular proteases. LESP turnover also slows because the FLCs senesce and decline in number. Meanwhile, excess fibronectin molecules accumulate, at least in mouse skin (Labat-Robert 2004). Basement membranes thicken (Kohn 1978). Slow chemical reactions convert several protein residues to other amino acids, which may affect local shape and charge of the protein. Various serum proteins aggregate to form extracellular (EC) protein deposits referred to as amyloid. In some regions of the aging brain, protein fragments of the amyloid precursor protein (APP) aggregate extracellularly to form EC deposits, called "β-amyloid plaques", which are often associated with Alzheimer's disease.

19.4.1 Glycation Pathways

Glycation is the spontaneous covalent attachment of a sugar to a macromolecule, such as protein, phospholipid, or DNA. Occurring without the need for enzymatic facilitation, glycation is quite distinct from the beneficial, enzymatically controlled, glycosylation of proteins, glycoproteins, and proteoglycans. Interstitial fluid allows reactive sugars from the blood to diffuse to protein strands of the ECM, where a complex network of spontaneous reactions takes place, as reviewed in many references (Monnier et al. 2003; Ulrich and Cerami 2001; Rahbar and Figarola 2003; Kikuchi et al. 2003; Metz et al. 2003; Furber 2006).

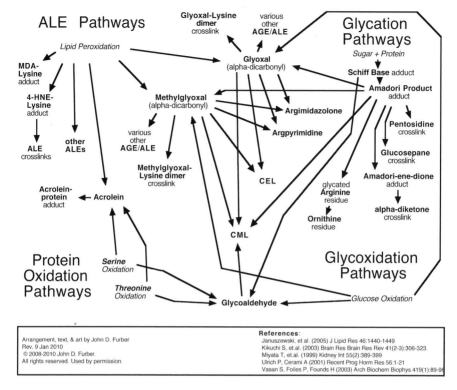

Fig. 19.1 Chemistry of ECM protein aging

The initial reaction is frequently a covalent bonding between glucose and a side chain of lysine in the protein strand (Eble et al. 1983). (See Fig. 19.1) The open-chain form of glucose has a reactive aldehyde group which attacks the reactive ε-amino group of the lysine side chain. These two groups join to form a Schiff base (Cerami et al. 1987), causing loss of lysine's positive charge.

$$\text{glucose} + (\text{lysine in protein}) \Longrightarrow \text{Schiff Base} \Longrightarrow \text{Amadori product} \Longrightarrow$$
$$\Longrightarrow \text{various intermediates and endproducts}$$

The initial Schiff base is unstable and reversible, so often the glucose detaches, leaving the protein unchanged. But sometimes, the Schiff base rearranges its bonds, resulting in various structures called Amadori products. The Amadori products are also unstable, and so many revert back to the Schiff base. The rest undergo further reactions and rearrangements over time to form various stable end products, called AGEs (Cerami et al. 1987). Some of the intermediate products are quite reactive. The conversion of Amadori products to final, stable AGEs sometimes proceeds by bonding with other reactive species. The Amadori adduct on a glycated protein will sometimes bond to a reactive side group of a nearby protein chain. In this case, the former sugar becomes a permanent covalent crosslink between adjacent protein chains or between domains of a folded protein. Several pathways are illustrated in Fig. 19.1.

Glucose is not the most reactive sugar (Ulrich and Cerami 2001), but it is by far the most abundant sugar in the blood (Cerami et al. 1987). Collagen is the most abundant ECM protein. A variety of different AGEs and AGE-crosslinks are formed in tissues via a complex brew of interacting reactions. Oxidation is involved in some of these reactions. Sometimes, glycated arginine decomposes to become ornithine (Sell and Monnier 2004).

Transition metal ions, such as copper and iron, increase the rate of glycation, probably by producing hydrogen peroxide and free radicals (Sajithlal et al. 1999; Xiao et al. 2007). Many glycation intermediates and end products, such as CML and N-ε-carboxyethyllysine (CEL) (Fig. 19.1) bind transition metals, generate free radicals, oxidize proteins and lipids, and accelerate additional glycoxidation reactions (Saxena et al. 1999; Requena and Stadtman 1999).

A variety of crosslink structures have been produced in vitro from glycated proteins and amino acids. Many of them have been found in vivo, as well . Chemically identifying crosslink structures has been difficult because some analytical procedures can destroy most AGEs before they can be characterized (Bucala and Cerami 1992; Biemel et al. 2002). At our present state of knowledge, almost all of the pathogenic extracellular glycation crosslink structures that accumulate in humans during aging appear to be one of two kinds: α-diketone crosslinks (Ulrich and Cerami 2001; Ulrich and Zhang 1997), or glucosepane (Biemel et al. 2002; Sell et al. 2005). The proposed reaction pathways forming these crosslinks are illustrated in Fig. 19.1.

The α-diketone crosslink is believed to form after a sugar adduct transforms into an Amadori ene-dione, which can attack the side chain of a lysine, cysteine, or histidine residue on a nearby protein chain. The crosslink contains two adjacent carbonyl carbons, forming an α-dicarbonyl structure called an α-diketone crosslink (Ulrich and Cerami 2001).

Glucosepane is an AGE crosslink formed between a glycated lysine residue in one protein chain and an arginine residue in a nearby chain. The side chain of arginine has a reactive δ-guanidino group, which can react with oxoaldehydes and other electrophiles. Glucosepane forms after a sugar adduct transforms into the dicarbonyl glycation adduct, dideoxyosone, which cyclizes and is attacked by the reactive guanidino group of a nearby arginine side chain. These covalently bond, forming the crosslink, glucosepane (Biemel et al. 2001).

In the condensation reactions of glycation and crosslinking, the positive charges on the lysine and the arginine are lost.

19.4.2 Lipoxidation Pathways

Oxidation and fragmentation of lipids can result in several reactive small molecules that can covalently bond to protein residue side chains. The Baynes lab has pointed out that lipoxidation reactions have some common intermediate species with the glycation pathways, and can also result in some of the same endproducts (Januszewski et al. 2003). Important reactive intermediates common to glycation and lipoxidation are glyoxal and methylglyoxal, as illustrated in Fig. 19.1. CML and CEL adducts are common to both the AGE and ALE pathway. In contrast, other ALE protein adducts are produced by lipoxidation, but not by glycation, such as 4-hydroxynonenal-lysine (HNE-Lys) and malondialdehyde-lysine (MDA-Lys) (Miyata et al. 1999; Januszewski et al. 2005).

19.4.3 Amino Acid Isomerization, Deamidation, and Oxidation

Asparagine (L-Asn), an uncharged residue, can deamidate, via a series of reactions, to become negatively charged aspartate (L-Asp or D-Asp) or isoaspartate (L-IsoAsp or D-IsoAsp) (Clarke 2003; Shimizu et al. 2005). The change in charge or shape might have some effect on the properties of LESPs, but this has not been reported. By similar pathways, L-Asp can isomerize to D-Asp or to L-IsoAsp or D-IsoAsp (Clarke 2003; Shimizu et al. 2005). This can affect integrin binding, discussed in Section 19.4.6. Other pathological consequences of these changes have been proposed (Ritz-Timme and Collins 2002). Shimizu has observed that amyloid-β peptides in Alzheimer brains contain high levels of IsoAsp in place of Asp, and suggests that this might result in abnormal folding and deposition of β-amyloid in plaques and vascular amyloids (Shimizu et al. 2005).

Racemization of aspartate residues has been used to estimate LESP turnover rate in various tissues and at various ages, as was noted in Section 19.3 (Verzijl et al. 2000). Over time, increasing amounts of D-Asp can be detected in collagen and elastin protein chains (Ritz-Timme and Collins 2002; Sell and Monnier 2004). Although humans have an endogenous intracellular enzyme, PCMT1 or PIMT, which can reverse some of these conversions in intracellular proteins (DeVry et al. 1996; Clarke 2003), it is largely unable to access and repair ECM isomerization. Small amounts of PIMT are released into the ECM at sites of injury, but it cannot travel far into the matrix and does not reach most isomerized residues (Weber and McFadden 1997).

Proteins can be oxidized to create AGE/ALE adducts without the presence of sugar or lipids. During inflammation, macrophages produce EC hypochlorous acid in their immediate vicinity, which can oxidize nearby serine and threonine residues, resulting in acrolein, glycoaldehyde, and CML, as shown in Fig. 19.1 (Anderson et al. 1999; Miyata et al. 1999).

19.4.4 ECM Protein Strand Breakage

Over time, attacks by EC proteases, as well as simple mechanical stresses, create breaks in the protein chains of the ECM, including collagen, elastin, and fibronectin (Li et al. 1999; Wang and Lakatta 2002; Wang et al. 2003; Labat-Robert 2004;

Robert et al. 2008). In some situations, EC proteases such as MMPs are secreted by "senescent" dermal fibroblasts and other FLCs (Parrinello et al. 2005). Furthermore, senescent dermal fibroblasts downregulate TIMP-1, thus restricting normal regulation of MMP activity (Labat-Robert 2004). In other situations, proteases are secreted as part of inflammatory responses to signals from cell surface receptors, when they are activated by AGEs or by fragments of elastin or fibronectin (see Section 19.4.7 and Fig. 19.2). Skin fibroblast secretion of proteases also increases in response to sunburn (Labat-Robert 2004). Protein strand breaks can cause weakening of collagen, fragmentation of elastin and fibronectin, and loss of tissue elasticity.

It is worth remembering that ECM strand lysis and digestion are not always harmful; they are sometimes part of a controlled process of ECM turnover, remodeling, or regeneration, as described in Section 19.3. However, aging and inflammatory processes can result in excessive degradation of ECM that does not get regenerated and leads to tissues becoming thinner, weaker, or stiffer.

As it ages, elastin is degraded via a multi-step process described by Robert, weakening tissue and reducing elasticity (Robert et al. 2008). Its elastic properties arise because its hydrophobic residues gather together in puckers, when not under tensile stress, shrinking the structure. Like a spring, as stress increases, the puckers pull apart, allowing the strands to extend. When tensile force is less, they can pull together again. Over time, calcium ions and lipids bind to these

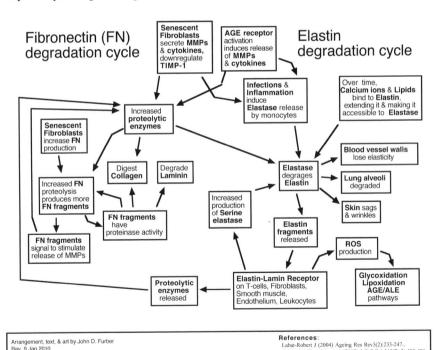

Fig. 19.2 Elastin and fibronectin degradation cycles of ECM protein strand lysis during aging

hydrophobic residues, reducing their mutual hydrophobic attractions for each other. This reduces elasticity because it is easier for the strands to stay in their extended state. Furthermore, the calcium and lipid-bound, extended elastin strands expose vulnerable sites for cutting by extracellular proteases. The lysed chains are no longer elastic, and they release protein fragments that activate inflammatory responses when they bind to the elastin-laminin receptor on cells (Robert et al. 2008), as described in Section 19.4.7 and Fig. 19.2. Like an old rubber band, the tissue loses elasticity and strength. Apparently, the elastic fibers are not readily replaced; perhaps they are never correctly replaced in arterial walls or lung alveoli (Robert et al. 2008; Wagenseil and Mecham 2007; Shifren and Mecham 2006; Finch 2007).

MMPs also lyse fibronectin strands, creating fibronectin fragments. Some fibronectin fragments are themselves proteolytic, having the ability to lyse collagen, laminin, and fibronectin. This produces a vicious cycle of LESP degradation shown in Fig. 19.2. Furthermore, some fibronectin fragments expose cryptic binding sites not available on intact fibronectin. Binding to cell surface receptors triggers a variety of deleterious cell responses (Labat-Robert 2004) described in Section 19.4.7.

19.4.5 Mechanical Consequences of Protein Alterations

Glycation adducts and crosslinks interfere directly with the mechanical properties of LESPs. Changes in charge, and the spaces occupied by adducts, can affect the conformation and behavior of proteins. Glycation adducts occupy space, and so may alter folding, shape, and function of proteins. Electrostatic charge distribution also affects folding and function. At physiological pH, the ε-amino side chain of lysine is positively charged. The guanidino side chain of arginine is also positively charged. Glycation or crosslinking converts these positively charged sites to neutral sites. Crosslinks bind together adjacent protein strands, reducing flexibility and elasticity of the tissue.

Elasticity is very important to cardiovascular function. Systolic blood pressure increases when the shock-absorbing elasticity of the artery walls is reduced (Vasan et al. 2001). High systolic blood pressure increases the risk for hemorrhagic stroke in the brain. It also increases back-pressure to the heart. The heart responds by increasing muscle mass, thickening its wall. A thicker, stiffer heart is less efficient at refilling after each contraction, resulting in diastolic heart failure (DHF). Reduced elasticity in the capillary walls restricts circulation to peripheral tissues. Mechanical elasticity of arteries is also important to maintaining healthy endothelial function, because nitric oxide (NO) signaling is reduced when stretching is limited (Zieman et al. 2007).

Glycation crosslinking of the corpus cavernosum contributes to erectile dysfunction (Usta et al. 2004, 2006). Crosslinking of the urinary bladder decreases its extensibility and capacity, resulting in the need for more frequent urination.

Glycation crosslinking is also believed to attach soluble plasma proteins to LESPs and to proteins on the surfaces of endothelial cells. This could contribute to inflammatory immune responses, to the development of atherosclerosis, and to the thickening of basement membranes, which can impair kidney function (Ulrich and Cerami 2001; Vasan et al. 2001).

As discussed in Section 19.3, glycation crosslinks and adducts could be mechanically restricting the ability of FLCs to turn over ECM, resulting in a vicious cycle.

19.4.6 Altered Cell-Matrix Integrin Binding

Cells bind to the ECM through cell surface integrin molecules. These integrins are also essential to cell migration over and through the ECM. The integrins recognize and bind to specific peptide motifs in the EC structural proteins or glycoproteins, importantly DGEA in collagen and RGD in fibronectin, vitronectin, and laminin (Lanthier and Desrosiers 2004; Gilbert 2000). When arginine (R) or aspartate (D) in a binding motif undergoes a chemical change that alters its shape or charge, the binding strength of cells to that EC protein is reduced because their integrin receptors no longer have that RGD or DGEA sequence to bind to (Lanthier and Desrosiers 2004; Sell and Monnier 2004). As noted earlier, arginine can lose its positive charge in several ways by attachment of glycation adducts or formation of crosslinks. It can also decompose to ornithine. Aspartate can isomerize. Loss of attachment to the ECM can affect a cell's gene expression profile and differentiation state, and may increase the propensity of cells to become cancerous (Boudreau and Bissell 1998; Sell and Monnier 2004; Spencer et al. 2007). In some cases, cells die as a result. "The chondrocytes that produce the cartilage of our vertebrae and limbs can survive and differentiate only if they are surrounded by an extracellular matrix and are joined to that matrix through their integrins (Hirsch et al. 1997). If chondrocytes from the developing chick sternum are incubated with antibodies that block the binding of integrins to the extracellular matrix, they shrivel up and die" (Gilbert 2000).

19.4.7 Cell-Matrix Interactions: Receptors, Signaling, and Inflammation

Several distinct cell-surface receptors are activated by AGEs (Kass 2003). Other receptors are activated by fragments of lysed fibronectin or elastin.

Historically, some of the AGE receptors have had different names. Vlassara and Palace review several AGE receptors, which are found on the surfaces of various cell types (Vlassara and Palace 2003). One specific AGE receptor complex is composed of three subunits: R1, R2, and R3. Ohgami describes several other AGE receptors: RAGE, galectin-3, 80 K-H, OST-48, CD-36, SR-A-I and SR-A-II. SR-A are multiligand macrophage scavenger receptors (MSR) of the class A family. CD-36 is a multiligand scavenger receptor of the class B family (SR-B). CD-36 is expressed on macrophages and smooth muscle cells (Ohgami et al. 2001).

One specific receptor was named RAGE (Receptor of AGEs) by Stern's group (Stern et al. 2002). Stern's review of RAGE notes the complexity of the RAGE signaling system (Stern et al. 2002). RAGE is a member of the immunoglobulin superfamily of cell surface receptors. It is found on a variety of cell types, including macrophages and endothelial cells. It binds and is activated by various ligands, including amyloid fibrils, amphoterin, S100/calgranulins, CML and probably other AGEs. Upon binding a ligand, RAGE induces multiple signaling pathways within the cell (Stern et al. 2002). RAGE signaling activates inflammatory pathways, and inflammation is known to contribute to several processes important in aging (Vlassara and Palace 2003; Finch 2007). RAGE signaling also induces transdifferentiation of kidney epithelial cells to become myofibroblasts, thus impairing kidney function (Jerums et al. 2003). RAGE and CD-36 activation by AGEs/ALEs appear to contribute to the development of foam cells during atherogenesis (Vlassara and Palace 2003; Ohgami et al. 2001). RAGE activation stimulates oxidant stress and upregulates cell surface adhesion molecules and cytokines, stimulating vascular inflammation, remodeling, and atherogenesis (Zieman et al. 2007). Confusingly, some authors refer to *all* AGE receptors as "RAGE".

Not only do AGE receptors initiate signaling in response to AGE binding, but also the presence of AGE causes increased expression of the RAGE and R3 receptors (Candido et al. 2003).

The macrophage scavenger receptor (MSR, probably SR-A and CD-36), and other closely related receptors, appear to trigger an attack on AGE-modified proteins by macrophages (Araki et al. 1995). Glycation adducts on the surface of articular cartilage and in synovial fluid are major factors in the development of osteoarthritis and rheumatoid arthritis probably through inflammatory mechanisms (deGroot et al. 2004; Drinda et al. 2005; Steenvoorden et al. 2007; Verzijl et al. 2003). Glycated peripheral nerve myelin is attacked by macrophages, contributing to peripheral neuropathy (Cerami et al. 1987). Glycation can crosslink immunoglobulins to kidney glomerular basement membrane; this then initiates complement-mediated damage (Bucala and Cerami 1992).

Although the consequences of AGE receptor activation by AGEs/ALEs are generally deleterious, these inflammation pathways are probably an inappropriate immune activity that could, on occasion, be protective against infections. AGE receptor signaling may also help to activate removal of AGE-damaged proteins by phagocytosis (Bucala and Cerami 1992).

Some glycation intermediates and end products generate free radicals, causing additional damage by oxidation and inflammation. CML generates free radicals, and is considered to be the major signaling ligand implicated in causing inflammatory diseases and cancers (Taguchi et al. 2000; Kislinger et al. 1999; Monnier et al. 2003). The complex associations between inflammation and age-related pathologies have been reviewed in Finch's recent book, *The Biology of Human Longevity* (Finch 2007), which covers in-vivo glycooxidation, ingestion of AGEs in cooked and processed foods, and more details on the role of AGE receptors in inflammation.

AGEs also contribute to endothelial dysfunction by degrading endothelial nitric oxide synthase (eNOS), which results in decreased NO concentrations (Bucala et al.

1991; Bucala and Cerami 1992; Dong et al. 2008). NO signaling causes vasodilation, so low NO contributes to high blood pressure. (Huang et al. 1995; Zieman et al. 2007) Decreased NO also contributes to erectile dysfunction (Haimes 2005).

As described in Section 19.4.4, elastin and fibronectin become fragmented during aging. Protein fragments from degraded elastin act as agonists binding to the elastin-laminin receptor. This upregulates the release of elastase endopeptidases, and the production of reactive oxygen species (ROS), which can cause a vicious cycle of further damage to elastin fibers, shown in Fig. 19.2 (Robert et al. 2008; Labat-Robert 2004).

Protein fragments from degraded fibronectin (Section 19.4.4 and Fig. 19.2) bind to receptors on cell surfaces, generating signals that result in inflammation, tissue degradation, and tumor progression (Labat-Robert 2004). Kume and colleagues found that AGEs in cell culture inhibited the proliferation of human MSCs, induced apoptosis, and inhibited differentiation into adipose tissue, cartilage, and bone (Kume et al. 2005).

19.4.8 Extracellular Amyloidosis

"Amyloidosis is a clinical disorder caused by extracellular deposition of insoluble abnormal fibrils, derived from aggregation of misfolded, normally soluble, protein. About 23 different unrelated proteins are known to form amyloid fibrils in vivo" (Pepys 2006). Pepys further notes that these extracellular deposits interfere with the proper functioning of the surrounding tissues, resulting in pathologies that can become fatal. Although amyloidosis is rarely cited as a cause of human death, one type, transthyretin (TTR-amyloid) is frequently found at autopsy in the hearts, kidneys, and lungs of people aged over 80 (Pepys 2006). The first population-based autopsy study found TTR-amyloidosis in 25% of humans aged 85 or more from southern Finland (Tanskanen et al. 2008). Pepys and Lachmann propose that amyloidosis may contribute to several diseases of the elderly. Furthermore, they suggest that if not addressed, TTR-amyloidosis might become a more serious problem at transcentenarian ages if human lifespan is increased by successful treatment of other age-associated diseases (Pepys 2006; Lachmann and Hawkins 2006).

Amyloid deposits resist attack by phagocytosis and most enzymes. Apparently the SAP protein, normally found in blood, binds to amyloid deposits and protects them. An experimental therapy, directed at SAP, is currently in human trials. The drug crosslinks soluble SAP, thus preventing it from binding to amyloid deposits. If the therapy is successful, the body's natural scavengers would then clear up the amyloid deposits (Pepys 2006; Lachmann and Hawkins 2006).

Nattokinase is a bacterial serine protease enzyme found in the fermented Japanese soybean food called, "*natto*". Preliminary experiments have shown that this enzyme can degrade several kinds of amyloid molecules in vitro (Hsu et al. 2009). It is interesting that it remains active in the bloodstream after oral assimilation, and that it is part of a traditional human food. Further research is needed to determine whether it can clear up TTR-amyloid or other deposits in older people. Even if not, its structure might inform future rational drug design efforts (Section 19.5.8).

19.4.9 β-Amyloid Plaques in the Brain

Extracellular deposits (β-amyloid plaques) of amyloid-β protein (A-β) accumulate in some brains as they age. A significant constituent is a 42 amino acid fragment of APP, "amyloid-$β_{1-42}$" or "A-$β_{42}$". Although often associated with Alzheimer's disease, there is considerable debate regarding whether these β-amyloid plaques are very harmful (Castellani et al. 2007). However, it is generally agreed that *in solution*, A-$β_{42}$ produces reactive oxygen species (ROS), which can damage nearby neurons. *Soluble* A-$β_{42}$ also activates RAGE, which contributes to neurotoxicity (Sturchler et al. 2008). Importantly, the plaques are in dissociable equilibrium with the soluble A-$β_{42}$, and thus can serve as a reservoir of the toxic species (Adlard et al. 2008).

19.5 Present and Possible Future Therapeutic Approaches for Better Maintenance and Repair of the ECM

This section examines prospects for therapies to slow AGE formation or to repair EC damage. The importance of glycation in diabetes and aging has led to searches for therapies that inhibit the glycation reactions or safely remove the products of glycation. Glycoxidation moieties, AGEs, and crosslinks might be chemically removed from ECM by drugs or bioengineered enzymes. Enhancement of natural ECM turnover and replacement could regenerate damaged tissues.

19.5.1 Diet, Fasting, and Calorie Restriction

As a non-enzymatic chemical reaction, we would expect glycation rate to increase with greater blood sugar concentration (Eble et al. 1983). Where glycation rate exceeds turnover rate, we expect to see accumulation of AGEs. In fact, the glycation rate does change with glucose concentration as expected. AGE/ALE accumulation rate is higher in diabetics, who have higher average blood sugar and lipid levels. Glycation rate decreases with calorie restriction, which lowers average blood sugar level. Rats fed calorie-restricted diets have less glycation crosslinking than rats that consume more calories (Lingelbach et al. 2000; Cefalu et al. 1995). Furthermore, Snell dwarf mice produce no growth hormone, and consequently have lower average blood sugar levels than control animals. Collagen glycation rates increase more slowly with age in Snell dwarves; they have much lower rates of cancer in old age; and they live longer (Flurkey et al. 2001; Alderman et al. 2009).

Thinking about therapeutic regimens, although it would be impossible to reduce the blood sugar and lipid concentrations to zero, average levels could be lowered by exercise, periodic fasting, or by constant or intermittent calorie restriction. Consequently, any of these alone, or in combination, should slow the rate of glycation.

High-temperature cooking produces AGEs/ALEs which, if ingested, would contribute to the body's AGE burden. The greatest quantity are created by frying or broiling foods containing fats or meats. Few are found in boiled or raw vegetarian foods (Goldberg et al. 2004). High levels of heat-stable glycation adduct residues, CML and CEL, were found in pasteurized and sterilized milk (Ahmed et al. 2005).

Inflammatory markers in the blood of diabetic humans and animals increased substantially after a few weeks on a high-AGE diet (Vlassara et al. 2002). This indicates that AGEs/ALEs do enter the systemic circulation from food digestion and increase inflammation. Similarly, although CR often improves the health and extends the lifespan of laboratory mice, when nondiabetic mice are maintained on a CR diet that is cooked to increase dietary AGEs, they have higher serum AGEs, oxidative stress, inflammatory markers, organ damage, and shorter lifespans than matched CR controls that received the same total calories, but not cooked food (Cai et al. 2008). A cautious person with an interest in optimizing health and lifespan might choose diets that minimize ingested AGEs and ALEs.

19.5.2 Exercise

Exercise increases the rate of turnover of collagen in human tendons and skeletal muscles, resulting in improved strength and flexibility (Kjær et al. 2006). As Kjær and colleagues observe, tendons contain fibroblasts. Weight-bearing exercise induces the surrounding tissue to release growth factors (IGF-1, IGF-1 binding proteins, TGF-β, and IL-6), which induce fibroblasts to remodel the collagen of the ECM. Collagen degradation is increased during the first day after exercise. However, new collagen synthesis is upregulated in tendon and in skeletal muscle for the first three days following intense exercise. Thus, they caution, to prevent overuse injury, it is important to space out exercise sessions. "If training sessions are too close to one another, an athlete may not gain maximum benefit from the stimulated collagen synthesis, but is instead likely to be in a net state of collagen catabolism." (Kjær et al. 2006).

There appears to be a synergistic benefit to combining exercise and crosslink breaker therapy (described in Section 19.5.4).

19.5.3 Inhibitors of Glycation, Lipoxidation, and AGE Formation

Many of the studies of potential glycation inhibitors do not look at LESP glycation in normally aging humans. They look instead at levels of soluble AGEs and reactive glycation intermediates in the blood of diabetic humans and rats. These blood levels do not accumulate over time, so they are not useful as biomarkers of aging. To the extent that a glycation inhibitor could reduce these blood levels in nondiabetic humans, then it *might* slow the rate of accumulation of glycation adducts and crosslinks in the extracellular matrix. That, however, is speculative at this time.

There are several intervention points in the cascade of events leading to production of AGEs/ALEs (including AGE crosslinks). Table 19.1 lists several dozen compounds that inhibit AGE production. Some are lipid membrane soluble, while others are hydrophilic. Many of these inhibitory compounds exhibit multiple modes of action, such as: trapping reactive carbonyls, interacting with dicarbonyls, quenching ROS, preventing autoxidation, chelating metals such as copper, inhibiting nitric oxide synthase (NOS), combatting inflammation, binding to glucose, inhibiting crosslinking of proteins, inhibiting early Amadori reactions, or inhibiting post-Amadori reactions. A few of these inhibitors are also able to break AGE crosslinks after they have formed. Crosslink breakers are examined in more detail in the next section.

Table 19.1 Inhibitors of glycation

Inhibitor	Notes	Ref
ALT-946 = N-(2-acetamidoethyl) hydrazinecarboximidamide HCl		V1,T,J,R3
ALT-462 = triazine derivative		V1
ALT-486 = benzoic acid derivative		V1
aminoguanidine = pimagedine	DCI, EAi, TRC, NOSI, MC, SAi	V1,T,J,R3
ascorbate = vitamin C	AO	R3
aspirin	AOp, AO, AI	R3
benfotiamine	LS	R3
benzoic acid	AOp, AO	R3
carnosine = β-alanylhistidine	AO, MC, TRC	R3
carotenoids	AO	R3
cinnamon, aqueous extract	AO, TRC	P
curcumin	AO, AI, Ci	R3
cysteine	TG	F, S
desferoxamine		R3
diaminophenazine = 2,3 DAP	DCI, MC	R3
Diclofenac = Voltran	AI	R3
EGCG = epigallocatechin gallate	Ci	W
fasting	BGL	F
garlic		A
glutathione	TG	F, S
histidine	MC, TRC	
Ibuprofen	AI	R3
Indomethacin	AI	R3
Inositol	AO, Gb	R3
LR-9 = 4-(2-naphtylcarboxamido) phenoxyisobutyric acid	MC, TRC	R3
LR-series # 20, 102	MC, PAi, CB	R3
LR-23	CB	R3

Table 19.1 (continued)

Inhibitor	Notes	Ref
LR-90	MC, PAi, TRC	R3
luteolin	AO, EAi, PAi, Ci	W
metformin = Glucophage = dimethylbiguanide	DCI, EAi, PAi, CB	R0, R3
MEAG = morpholino-ethyl aminoguanidine		V1
OPB-9195 = (±)-2-isopropylidenhydrazono-4-oxo-thiazolidin-5-ylacetalinide	DCI, MC	V1, R3
PABA	AOp, AO	R3
D-penicillamine		R3
pentoxyfylline		R0, R3
Pioglitazone	DCI, MC	R0, R3
Probucol	AO	R3
Pyridoxamine	PAi, LEi, DCI, MC	Me, V1, R3
quercetin	AO, EAi, Ci	W
resveratrol = 3,4,5-trihydroxystilbene		R3
rutin	EAi, Ci	W
salicylic acid	AO, AOp	R3
Tenilsetam = (+)-3-(-2-thienyl)-2-piperazine	Ci	R3
thiamine pyrophosphate = Vitamin B1	PAi	V1, R3
thyme		Mo
Tocopherol = vitamin E	AO	R3

Abbreviations: *AO* Antioxidant, *AI* Antiinflammatory, *BGL* Lowers blood glucose, *CB* Cross link breaker, *Ci* Inhibits cross link formation, *DAOi* Diamine oxidase inhibitor, *DCI* Interacts with dicarbonyls, *EAi* Early Amadori stage inhibitor, *Gb* Binds to glucose, *LS* Lipid soluble, *LEi* Lipoxidation endproduct inhibitor, *MC* Metal chelator, *AOp* Prevents autoxidation, *PAi* Post Amadori inhibition, *SAi* Inhibits semicarbazide-sensitive amine oxidase, *TG* Transglycation, *TRC* Traps reactive carbonyls
References: A = Ahmad et al. 2007; F = Furber 2006; J = Jerums et al. 2003; Me = Metz et al. 2003; Mo = Morimitsu et al. 1995; P = Peng et al. 2008; R0 = Rahbar et al. 2000; R3 = Rahbar and Figarola 2003; S = Szwergold 2005; T = Thornalley 2003; V = Vasan et al. 2001; W = Wu and Yen 2005

Some well-known antioxidant or anti-inflammatory substances appear to inhibit AGE formation: aspirin (Bucala and Cerami 1992), ibuprofen, inositol, probucol, vitamins C and E, carotenoids, salicylic acid, PABA, and benzoic acid. Rahbar and Figarola conclude that because not all antioxidants inhibit AGE formation, those that do are employing another mechanism of action. They note that in clinical trials of diabetic patients, treatments with antioxidants that don't inhibit AGE formation do not improve their condition (Rahbar and Figarola 2003). Aspirin acetylates specific primary amino groups, thereby blocking their glycation (Bucala and Cerami 1992).

Aminoguanidine (AG or pimagedine) has been well studied in clinical trials of diabetic patients. It is a nucleophilic compound that traps reactive carbonyl groups (Ulrich and Cerami 2001). In addition to inhibiting AGE formation, it also inhibits NOS (Jerums et al. 2003). However, there have been safety concerns and apparently low clinical efficacy (Thornalley 2003). Human side effects included pernicious anemia and anti-nuclear antibodies. In rat studies, pancreas and kidney tumors developed (Rahbar and Figarola 2003).

Pyridoxamine (PM) is the 4-aminomethyl form of vitamin B6. PM inhibits formation of AGEs and ALEs, apparently by reacting with dicarbonyl intermediates. In diabetic rats, oral PM stayed in the blood longer, and had greater therapeutic benefit than similar doses of AG (Metz 2003). The Baynes lab has showed that PM breaks dicarbonyl compounds in vitro (Yang et al. 2003). Although they were unable to show in vivo breaking of AGEs, this might be worthy of further study by other labs.

Some radical trapping compounds alter branchpoints in the AGE formation reaction network, inhibiting the formation of some AGEs, while increasing the formation of others. For example, 6-dimethylaminopyridoxamine (dmaPM) and Trolox each inhibit the formation of glucosepane crosslinks in vitro, but increase the production of other glycation products (Culbertson et al. 2003). This is especially interesting because, as discussed in Section 19.5.4.4, no breaker for glucosepane crosslinks has yet been identified.

Metformin (N,N-dimethylimidodicarbonimidic diamide mono-hydrochloride) (glucophage) (pKa = 12.4) is a drug prescribed to improve glucose tolerance in type-2 diabetes. It has also been shown to inhibit glycation in vitro (Rahbar et al. 2000), to bind dicarbonyl glycation intermediates, inactivating them (Beisswenger and Ruggiero-Lopez 2003), and to break glycation crosslinks in vitro (Rahbar and Figarola 2003).

Benfotiamine is a lipid soluble analog of thiamine (vitamin B1). In diabetic rats, it effectively reversed neuropathy and reduced accumulation of glycation intermediates (Stracke et al. 2001). Its effect on normally aging humans has not been reported. However, its mode of action seems to control pathways that are induced by diabetic hyperglycemia (Hammes et al. 2003). It would therefore not be helpful in nondiabetic situations, such as normal aging.

Carnosine (β-alanyl-L-histidine) is a dipeptide that is heavily marketed as a nutritional supplement. Its putative ability to inhibit protein glycation or crosslinking in humans is still under investigation. Hipkiss, who has been studying carnosine for years, notes that "carnosine *may* be an effective anti-glycating agent, *at least in model systems*" (emphasis added) (Hipkiss 2005).

Glutathione and cysteine may have anti-glycating ability. The glucose-lysine Schiff base can spontaneously donate its sugar moiety to nucleophiles such as cysteine and glutathione, restoring the protein to its original, unglycated condition. This has been observed in vitro without any enzymes present. The sugar binds to the sulfur atom of the cysteine. Szwergold et al. propose that this reaction also occurs spontaneously within cells, and that the glycated glutathione or cysteine is then pumped out of the cell. In support of their model is the observation that glycated cysteine is found in human urine, and that levels are higher in diabetic urine (Szwergold et al. 2005). They did not comment on the possibility of transglycation taking sugar from extracellular collagen. Cysteine and even glutathione may be small enough to go wherever glucose goes among the collagen molecules. Thus, there may be possible benefits to therapeutic use of oral N-acetylcysteine (NAC) or parenteral glutathione to increase concentrations of these nucleophiles in the extracellular fluid that bathes collagen. NAC is commonly available as a nutritional supplement. Some clinics offer intravenous glutathione injections. Note however, that this reaction deglycates only the earliest step in the glycation pathway. After the glycation has proceeded to form Amadori products, AGEs, or crosslinks, transglycation does not occur. Nonetheless, even partial inhibition of glycation may be beneficial.

In general, AGE inhibitors are tested in vitro and in vivo. In diabetic models, they slow down the rates of physiological deterioration to some extent. However, for long-lasting benefits and rejuvenation, we must look for therapies that actually reverse or repair accumulated LESP damage, which has already occurred, including crosslinks, glycation, fragmentation, and lipoxidation.

19.5.4 Deglycators and Crosslink Breakers

Within mammalian cells, endogenous mechanisms exist for reversing glycation (Section 19.5.4.1). Outside cells, in the ECM, glycation is destroyed wherever the ECM is turned over. Several approaches are being explored to design therapies to break crosslinks or remove glycation adducts on ECM proteins. Some are based on small molecule drug designs. Others are based on adapting strategies from intracellular enzymes or fungal enzymes. A significant consideration is that much of the collagen matrix is densely packed so that glycation crosslinks may not be accessible to large enzyme molecules. If a large enzyme cannot travel to its target crosslink, it cannot break it. Perhaps this problem might be circumvented if small molecule crosslink breakers could loosen up the ECM enough for larger enzymes to get in and finish the job.

19.5.4.1 Intracellular Enzymatic Deglycation

Enzymes have been found in some cells that are able to remove Amadori adducts from intracellular proteins. In mammals, fructosamine 3-kinases (FN3Ks) have been found to act as Amadoriases. They phosphorylate Amadori products, which then spontaneously deglycate, leaving the original proteins good as new (Szwergold et al. 2001). However, Amadoriases do not work on AGEs or crosslinks, because their chemical structure is changed from the early Amadori structure. Furthermore, Amadoriases are inside the cell and they require ATP. This presents problems because crosslinked collagen is outside the cell, and a source of extracellular ATP is not available. So FN3Ks are not useful for repairing ECM (Monnier et al. 2003). However, they might serve as a starting point for future development of useful drugs or designer enzymes.

19.5.4.2 Fungal Amadoriase Enzymes

Enzymes that are able to deglycate small Amadori products, such as glycated amino acids, have been isolated from fungi. However, the enzymes discovered so far do not deglycate proteins. This is apparently due to both steric hindrance and electostatic interactions (Monnier et al. 2003). Their mechanism is to oxidize the fructosylamino Amadori product, releasing the original unglycated amine (such as lysine), along with hydrogen peroxide and oxidized sugar (such as glucosone). Thus, they are also called *"fructosyl amine oxidases"*. An advantage of this reaction is that it does not require ATP, so it could take place outside of cells. A disadvantage is that both hydrogen peroxide and glucosone are reactive, and could cause further oxidative damage. Although these enzymes do not deglycate collagen, they have been sequenced, and the structure has been determined (Collard et al. 2008). They might suggest strategies for development of new agents.

19.5.4.3 Thiazolium Salts and Other Small Molecules

Several small molecules have been reported to have the ability to chemically cleave some of the glycation crosslinks or adducts in LESPs. Torrent Pharmaceuticals was granted several patents covering crosslink-breaking by pyridinium structures, and later published promising results with diabetic rats treated with compound "TRC4149" (Pathak et al. 2008). Rahbar, at City of Hope, was granted patents for the crosslink-breaking ability of several other structures, including metformin (Rahbar and Figarola 2003). However, his recent publications have focused on their glycation-inhibition rather than crosslink-breaking (Rahbar 2007; Figarola et al. 2008). The crosslink-breaker furthest along in human clinical trials is a thiazolium salt discovered by Cerami and colleagues.

In the early-1990s, Ulrich and Cerami were examining thiazolium compounds for their ability to interact with α-dicarbonyl structures in advanced Amadori products (Ulrich and Zhang 1997; Ulrich and Cerami 2001). These thiazolium compounds contain a nucleophilic catalytic carbon (position #2) analogous to thiamine (vitamin B-1) and a second nucleophilic carbon, attached to the nitrogen, nearby. These two carbons could interact with the two carbonyls of α-dicarbonyl structures (Vasan et al. 1996). They were surprised to discover that these compounds not only inhibited the progression of Amadori products to crosslinks, but they were also able to break model crosslinks in vitro (Ulrich and Cerami 2001). Many similar thiazolium compounds were tested and found to have crosslink-breaking activity. Patent rights were assigned to Alteon Pharmaceuticals (later renamed Synvista Therapeutics). Animal testing showed promising results in reversing collagen crosslinking, and improved functioning of kidneys, penile erections, heart, arteries, and other organ systems in aged or diabetic animals (Asif et al. 2000; Vaitkevicius et al. 2001; Usta et al. 2004, 2006). Similar beneficial results have been reported by Cheng and colleagues at the Beijing Institute of Pharmacology and Toxicology, who have been testing a structurally similar thiazolium compound, "C36" (Cheng et al. 2007).

Alteon chose alagebrium, *3-(2-phenyl-2-oxoethyl)-4,5-dimethylthiazolium chloride*, to use in their clinical trials. Early papers refer to this compound and its close relatives as *"ALT-711"*. Some of the early testing was done with bromide analogs (PTB), with or without the methyl groups. PTB was abandoned by Alteon in favor of the dimethyl chloride, alagebrium, because PTB is less active and unstable (Ulrich and Cerami 2001). PTB degrades rapidly in aqueous solution. Furthermore, bromides may have undesirable side effects (Thornalley and Minhas 1999; Vasan et al. 2001, 2003).

Alagebrium is now the crosslink breaker furthest in clinical development for human oral therapeutic use. Alagebrium appears to be effective at partially reversing some human pathologies, probably by breaking α-diketone crosslinks in collagen and elastin (Vasan et al. 1996). Possibly, it also reacts with other α-dicarbonyl glycation intermediates or endproducts, such as methylglyoxal (MGO) (Yang et al. 2003; Haimes 2007).

In 2003, the Baynes lab published a report suggesting that thiazolium bromides "do not break Maillard crosslinks in skin and tail collagen from diabetic rats" (Yang et al. 2003). This is a controversial claim, contradicting a large number of studies, which show evidence that thiazolium salts do break crosslinks in tail tendon collagen from diabetic rats (Vasan et al. 1996, 2001, 2003; Ulrich and Cerami 1997; Wolffenbuttel et al. 1998; Cheng et al. 2007). The situation is confounded because different techniques were used by different labs, so we cannot say, with certainty, why their results differ. Note, however, that the Baynes report did not use the stable alagebrium chloride, but rather, the less active, unstable bromide salts (Yang et al. 2003).

Interestingly, the Baynes group did acknowledge that the thiazolium halides produce beneficial clinical physiological results in vivo. However, they proposed different mechanisms of action. They suggested that alagebrium might be inhibiting the production of new crosslinks, as well as inhibiting glycoxidation reactions. Then, over a period of time, they reasoned, natural turnover of collagen would result in a reduction in the number of crosslinks, creating the appearance of crosslinks being broken (Yang et al. 2003). However, the Baynes hypothesis appears to be inconsistent with the multiyear long collagen turnover times calculated by independent labs (Sell et al. 2005), and the rapid in vivo benefits observed with alagebrium (Asif et al. 2000; Kass et al. 2001; Vaitkevicius et al. 2001).

Jerums and colleagues report that alagebrium treatment reduced kidney damage (Jerums et al. 2003). There are also reports that alagebrium treatment reverses the AGE-stimulated progression of several pathologic markers in the hearts of diabetic rats, including collagen solubility and expression of the AGE receptors RAGE and R3 (Candido et al. 2003; Kass 2003; Tikellis et al. 2008).

Phase 2 clinical trials of alagebrium began in 1998 (Vasan et al. 2003). As of mid-2009, several phase 2 trials had been completed, but Synvista had stopped further trials citing lack of funds. By 2007, about 1000 people had taken alagebrium in various phase 2b clinical trials (Haimes 2007). So far, the safety profile of the drug appears to be excellent in human subjects. Concerns arose in December 2004 regarding liver cell irregularities in male Sprague-Dawley rats that had been given alagebrium throughout their whole lives. After investigating, FDA allowed continuation of clinical trials. Apparently, Sprague-Dawley rats have exhibited similar changes in response to other approved drugs, such as statins. It appears that this breed of lab rat is not a reliable model for long-term human drug safety tests, although it was long been used because it is easy to handle (Creel 1980).

Alagebrium treatments have produced improvements in DHF patients, for whom ventricular hypertrophy was reduced and heart function was improved (Little et al. 2005). Other patients with systolic hypertension showed improvement in arterial pulse pressure and arterial compliance (Kass et al. 2001; Bakris et al. 2004). Endothelial function was also improved, probably because removal of AGE crosslinks allowed better stretch-mediated release of NO (Zieman et al. 2007).

Preliminary results indicate that alagebrium is able to repair erectile dysfunction, probably due to improved vascular compliance, NO signaling, and endothelial function (Coughlan et al. 2007). This was first reported in studies of diabetic rats (Usta et al. 2004, 2006). This author has heard firsthand reports from several men remarking on their improvement after several weeks or months of oral alagebrium (100–300 mg per day).

There appears to be a synergistic benefit of combining exercise (see Section 19.5.2) and alagebrium therapy. To the extent that alagebrium breaks LESP crosslinks and improves flexibility, exercise would be easier and tissue remodeling would be facilitated (Haimes 2007). This author has heard firsthand reports from several people remarking on their improved exercise tolerance after several weeks of oral alagebrium (100–300 mg per day). Two people noted that reduced arthritis allowed them to hike longer in the hills.

In June 2005, Alteon announced that it had granted a nonexclusive worldwide license to Avon Products, Inc. for the use of 2-amino-4,5-dimethylthiazole HBr to improve skin wrinkles and elasticity. Very soon after, Avon brought out its "Age Intensive" skin cream, containing this substance as a minor ingredient. The product is popular, although clinical comparisons with common moisturizers have not been published.

Anecdotally, several longtime users of alagebrium have told the author that they noticed improvements in bladder capacity, peripheral neuropathy, erectile function, kidney function, angina pectoris, or joint pain after several months of usage. Each was taking 100–400 mg per day, orally.

Several people have been giving alagebrium to their elderly dogs (ages 10–16 years), mixed with food or water. They told the author that their dogs had previously been exhibiting arthritis, low energy, and restricted movement. After about a month on alagebrium, their dogs were running and jumping as though they were several years younger. Their subjective assessment was that the alagebrium treatments had given their pets two additional years of quality life. Dosage was approximately 1–2 mg/kg per day.

19.5.4.4 Glucosepane Crosslink Breakers

So far, no small molecule has been identified that breaks glucosepane crosslinks. However, because an assay has not yet been implemented to test for glucosepane breakers, it is possible that some of the small-molecule breakers described in Section 19.5.4.3 might actually break glucosepane, yet we would not know it.

A drug discovery effort targeted at breaking glucosepane crosslinks might yield therapeutic leads. The isoimidizole structure at its core may be unique enough that a chemical agent could cleave it while not harming other essential extracellular structures.

Besides small molecule drugs, it is also possible that enzymes might be discovered or designed that could break glucosepane. However, there is not much space within the tightly packed collagen matrix where the crosslink is located, so enzymes might not fit. Nevertheless, we cannot rule out the possibility that a small enzyme might slip in, first breaking the most exposed crosslinks, and thereby opening the collagen matrix to access the more cryptic crosslinks. Perhaps in combination with alagebrium, other small molecules, and exercise, glucosepane-breaking enzymes might be even more effective.

As noted in Section 19.5.3, a couple of compounds have been found to inhibit glucosepane formation in vitro. Development of a drug to inhibit glucosepane formation in vivo could be beneficial until a therapy to remove glucosepane is developed.

19.5.5 Tuned Electromagnetic Energy

It is attractive to speculate that laser frequencies might exist that would safely penetrate tissues, while coupling energetically enough with crosslink structures to break them. Experiments with tunable lasers could explore frequencies in search of effective ones. There is no assurance of success. Even if cleaved, the crosslinks might quickly reform by the reverse reaction. Nevertheless, I predict that the costs of preliminary experiments on pieces of meat could be low and the potential payoff high. A physics lab that has a tunable laser, in collaboration with a biochemist who can assay crosslinks in animal tissue, could yield answers in a very short time.

19.5.6 Removing β-Amyloid Plaques

Considerable work is underway to find treatments for Alzheimer's disease. A promising approach is directed at solubilising and flushing out the extracellular β-amyloid plaques, by removing the metals around which they aggregate. An 8-hydroxyquinoline agent, PBT2, in clinical trials sponsored by Prana Biotechnology, is showing early success (Adlard et al. 2008).

19.5.7 Enhancing Turnover of ECM by FLCs

Human FLCs have the means to digest LESPs, and to replace them with newly synthesized fibers (Bucala and Cerami 1992; Murphy and Reynolds 2002). Unlike crosslink-breaking enzymes, which might be unable squeeze between collagen fibrils to reach crosslinks, enzymes secreted by FLCs to digest ECM start at the outside of the collagen fiber and chew their way in, so steric hindrance is not a problem.

Even cartilage and bone can be remodeled by appropriate cell types. Future developments might stimulate or reprogram FLCs to more quickly digest and replace age-damaged ECM in a controlled fashion. We might speculate that future bioengineers could integrate AGE receptors into signaling systems in FLCs to target these activated FLCs to turn over glycated ECM.

An important challenge will be to ensure that the turnover is well regulated, to prevent either thinning and loss of ECM or excess, disorganized fibrosis and cicatrix formation. Obviously, inducing widespread scar formation would not be a desirable fix for AGE accumulation. Ideally, working fiber-by-fiber, even the strands reinforcing blood vessels might be replaced without catastrophic system failure.

With advancing age, the population of FLCs declines and becomes less active at turning over LESPs (Campisi 2005). It is reasonable to foresee that a successful therapy would expand the numbers of FLCs, and also stimulate their activity of turning over LESPs. For example, platelet-derived growth factor (PDGF) and insulin-like growth factor-1 (IGF-1) have long been known to promote growth and mitosis of mesenchymal/fibroblast lineage cells (Bucala and Cerami 1992). Recent work at the University of Glasgow has shown that inserting an extra copy of the TERT gene into chondrocytes from articular cartilage results in longer telomeres and increased replicative lifespan, without neoplastic transformation. So far, the Glasgow results have been reported only for cell cultures of chondrocytes from young dogs (Nicholson et al. 2007). More work is needed to reveal whether altered integrin binding in old cartilage (Section 19.4.6) would harm the transgenic chondrocytes, or whether the activated FLCs could turn over the old ECM before it could harm them. Careful work could refine the optimal dosage, timing, and combinations of factors to expand cell numbers and induce differentiation into cell types best able to turn over ECM.

FLC stimulation might be done either in the body or in cell culture. In the body, biological response modifiers such as signaling molecules could be administered or gene therapy vectors might be injected. These agents might be designed to act directly on FLCs or they might work indirectly through other cells, which would signal to the FLCs. However, dosing of the target cells could not be uniform or precise, or responsively tailored to observed progress on the differentiation path. Furthermore, it might be difficult to prevent unintended cell populations from proliferating in response to systemically administered therapies. These issues might not be problematical if the treatment could be something like restoring youthful levels of hormones and other signals. There is still much to be learned.

An alternative method would be to extract and treat FLCs in culture. Fibroblasts, bone marrow stem cells, or MSCs could be treated *ex vivo* to increase their numbers. Then they could be monitored while differentiation agents are used to enhance their activity. Finally, the activated autologous cells would be injected into the patient to increase regeneration of the ECM (see also Chapter 14).

As noted in Section 19.3, exercise and mechanical force can increase the rate of collagen turnover and ECM remodeling by fibroblasts in various human tissues. Close examination of the signaling pathways and cytoskeletal responses to exercise and force could reveal clues to developing more general ECM rejuvenation therapies.

Useful lessons about enhancing human ECM turnover may also be learned by studying the regeneration of amphibians, such as the axolotl (*Ambystoma mexicanum*). Some amphibians and invertebrates are able to replace whole body parts after amputation. As Muneoka and colleagues note in their review, axolotls repair wounds and amputations perfectly, without scar formation. For example, axolotl limb regeneration results in a perfectly formed new limb, with new bone, new joints, new ECM, and new cells, all in exactly the correct pattern (Muneoka et al. 2008). Importantly, in the early phase of regeneration, the ECM at the wound site is extensively remodeled by migrating dermal fibroblasts, which have positional information to correctly rebuild the regenerating structure (Rinn et al. 2006). Collagen in the stump is first digested and then new collagen is created as the wound site is remodeled. Subsequently, additional ECM is built and populated by cells to rebuild the entire limb (Gardiner 2005).

It is encouraging that in humans, repair of oral mucosa wounds inside the mouth does not involve scar formation; it somewhat resembles amphibian regeneration (Schrementi et al. 2008). Furthermore, Muneoka, Han, and Gardiner point out, "wounds in [human] fetal skin heal without forming scars–yielding perfect skin regeneration and indicating that the switch to a fibrotic [scar-forming] response arises with the developmental maturation of the skin." This suggests that the human genome still possesses the ancient genes needed to accomplish regeneration (Muneoka et al. 2008). An important challenge will be to learn how to activate those inherent abilities, in a controlled manner, to remodel ECM that has become aged and glycated. Furthermore, of course, activation presumably would need to occur without prior wounding, in order to safely remodel critical structures, such as arterial walls and lung alveoli. Scheid and colleagues have observed that transforming growth factor β3 (TGFβ3) is expressed in regenerating fetal wounds, and that it promotes epithelial and mesenchymal cell migrations and cell-ECM interactions (Tredget and Ding 2009; Scheid et al. 2002). Subsequently Ferguson and colleagues demonstrated reduced scar formation during adult human wound healing treated with TGFβ3 (Ferguson et al. 2009). This suggests that factors might be found to induce adult FLCs to regenerate and repair age-damaged tissues.

19.5.8 General Therapy Design Considerations

"Rational drug design" (RDD) looks at a target structure (crosslink or adduct) to figure out what sort of molecule would effectively break it or remove it. Interactive molecular models in silico (in computers) are very helpful in these studies. Designers must bring the active sites of the agent and the target molecules close enough to interact. If the agent is not properly shaped, steric hindrance can prevent active site contact. Large molecules such as proteins may have particular problems squeezing among collagen fibrils to reach crosslinks or adducts. Electrostatic interactions can also affect apposition of active sites. Furthermore, reactions must be energetically favored. Local chemistry predicts whether the reaction will move forward. If the target bonds are not sufficiently energetic to be catalytically broken, then the agent, or nearby reactants such as oxygen, must provide some of the energy to move the reaction forward. We would also like some small products to move away quickly, to decrease the reverse reaction rate. There is some evidence that crosslinks broken by alagebrium might relink within a few weeks. This would suggest that alagebrium leaves reactive pieces in place, which can reassemble.

"High throughput screening" (HTS) creates a standardized chemical version of the target structure inside thousands of tiny reaction vessels. With a standardized assay, thousands of compounds are tested for any that show effectiveness. When promising lead compounds are discovered, variations on the structure are tested to find those with the best performance.

The best leads from RDD and HTS are used as starting points for creating families of similar structures, which are extensively tested in vitro. Compounds that look promising in vitro are next tested in animals for efficacy, side effects, and toxicity, as well as for the pharmacokinetics of absorption, distribution, metabolism, and excretion (ADME). RDD modeling can also be helpful in predicting whether problems such as collateral molecular damage might be caused by candidate breakers, and in determining whether such damage might be reparable. The structures of biomolecules can be compared with glucosepane to determine whether they share any structural motifs that might be damaged by the candidate agent.

Perhaps, in the distant future, engineers will compete with biologists to see if they can repair aging ECM better with tiny, nonliving *nanobot* machines (see Chapter 23).

19.5.9 Therapy Usage and Frequency

If the therapeutic agent is a large molecule, such as a protein or enzyme, it might be injected or implemented through gene therapy because proteins get digested when taken orally, and they are not well absorbed from the GI tract. Small molecule agents can often be made in an orally bioavailable form. (See Section 19.5.7 for a discussion of FLC therapy administration.)

An effective therapy might repair the ECM so well that it need be repeated only at multiyear intervals. Less effective therapies might leave reactive residues or require more frequent re-treatments, perhaps even daily. If glycation inhibitors are used instead of repair therapies, continual use would be required for maximum effect, and even then glycation could probably not be completely halted. Perhaps, some combination of therapies will prove to be the best available treatment strategy.

Large-molecule therapies might stimulate dangerous antigenic responses, especially if they are administered repeatedly. However, in the future, techniques might be developed to control antigenic responses to large molecule therapies. That problem is under intense study by many labs that are developing protein therapies for a variety of conditions.

19.6 Summary and Conclusions

Damage to extracellular proteins, including strand breaks, crosslinks, and AGE/ALE adducts impair the structure and function of the ECM, causing or contributing to many diseases of aging. Furthermore, with increasing age, the rate of turnover and repair of the damaged ECM declines, and damage accumulates faster. Good diet and glycation inhibitors can slow the accumulation of damage. Weight-bearing exercise stimulates natural turnover and remodeling of ECM in tendons and skeletal muscles. Thiazolium compounds can repair a portion of the AGE crosslinks, and provide clinical improvements of several age-associated pathologies. Perhaps a series of future drug discoveries will remove the entire menagerie of pathogenic crosslinks and adducts. Alternatively, a straightforward, complete therapy for extracellular aging might involve stimulating fibroblast lineage cells to more rapidly replace and regenerate the damaged ECM with newly synthesized ECM, as they move through it.

Acknowledgements I would like to thank George M. Martin, David A. Spiegel, Steven G. Clarke, Ulf T. Brunk, Duncan MacLaren, and especially my editor, Gregory M. Fahy, for their careful reading of earlier drafts or portions of this chapter, and for their helpful comments.

References

Adlard PA, Cherny RA, Finkelstein DI, Gautier E, Robb E, Cortes M, Volitakis I, Liu X, Smith JP, Perez K, Laughton K, Li QX, Charman SA, Nicolazzo JA, Wilkins S, Deleva K, Lynch T, Kok G, Ritchie CW, Tanzi RE, Cappai R, Masters CL, Barnham KJ, Bush AI (2008) Rapid srestoration of cognition in Alzheimer's transgenic mice with 8-hydroxy quinoline analogs is associated with decreased interstitial aβ. Neuron 59(1):43–55

Ahmad MS, Pischetsrieder M, Ahmed N (2007) Aged garlic extract and S-allyl cysteine prevent formation of advanced glycation endproducts. Eur J Pharmacol 561(1–3):32–38

Ahmed N, Thornalley PJ (2003) Quantitative screening of protein biomarkers of early glycation, advanced glycation, oxidation and nitrosation in cellular and extracellular proteins by tandem mass spectrometry multiple reaction monitoring. Biochem Soc Trans 31(Pt 6):1417–1422

Ahmed N, Mirshekar-Syahkal B, Kennish L, Karachalias N, Babaei-Jadidi R, Thornalley PJ (2005) Assay of advanced glycation endproducts in selected beverages and food by liquid chromatography with tandem mass spectrometric detection. Mol Nutr Food Res 49(7): 691–699

Alberts B, Johnson A, Lewis J, Raff M, Roberts K, Walter P (2002) Molecular Biology of the Cell. Garland, New York

Alderman JM, Flurkey K, Brooks NL, Naik SB, Gutierrez JM, Srinivas U, Ziara KB, Jing L, Boysen G, Bronson R, Klebanov S, Chen X, Swenberg JA, Stridsberg M, Parker CE, Harrison DE, Combs TP (2009) Neuroendocrine inhibition of glucose production and resistance to cancer in dwarf mice. Exp Gerontol 44(1–2):26–33

Alikhani Z, Alikhani M, Boyd CM, Nagao K, Trackman PC, Graves DT (2005) Advanced glycation end products enhance expression of pro-apoptotic genes and stimulate fibroblast apoptosis through cytoplasmic and mitochondrial pathways. J Biol Chem 280(13):12087–12095

Anderson MM, Requena JR, Crowley JR, Thorpe SR, Heinecke JW (1999) The myeloperoxidase system of human phagocytes generates N-ε-(carboxymethyl)lysine on proteins: a mechanism for producing advanced glycation end products at sites of inflammation. J Clin Invest 104(1):103–113

Araki N, Higashi T, Mori T, Shibayama R, Kawabe Y, Kodama T, Takahashi K, Shichiri M, Horiuchi S (1995) Macrophage scavenger receptor mediates the endocytic uptake and degradation of advanced glycation end products of the Maillard reaction. Eur J Biochem 1;230(2):408–415

Asif M, Egan J, Vasan S, Jyothirmayi GN, Masurekar MR, Lopez S, Williams C, Torres RL, Wagle D, Ulrich P, Cerami A, Brines M, Regan TJ (2000) An advanced glycation endproduct cross-link breaker can reverse age-related increases in myocardial stiffness. Proc Natl Acad Sci USA 97(6):2809–2813

Bakris GL, Bank AJ, Kass DA, Neutel JM, Preston RA, Oparil S (2004) Advanced glycation endproduct cross-link breakers: A novel approach to cardiovascular pathologies related to the aging process. Am J Hypertens 17(12 Pt 2):23S–30S

Beisswenger P, Ruggiero-Lopez D (2003) Metformin inhibition of glycation processes. Diabetes Metab (4 Pt 2):6S95–103

Benanti JA, Williams DK, Robinson KL, Ozer HL, Galloway DA (2002) Induction of extracellular matrix-remodeling genes by the senescence-associated protein APA-1. Mol Cell Biol 22(21):7385–7397

Biemel KM, Reihl O, Conrad J, Lederer MO (2001) Formation pathways for lysine-arginine cross-links derived from hexoses and pentoses by maillard processes: unraveling the structure of a pentosidine precursor. J Biol Chem 276(26):23405–12

Biemel KM, Friedl DA, Lederer MO (2002) Identification and quantification of major maillard cross-links in human serum albumin and lens protein: evidence for glucosepane as the dominant compound. J Biol Chem 277(28):24907–24915

Boudreau N, Bissell MJ (1998) Extracellular matrix signaling: integration of form and function in normal and malignant cells. Curr Opin Cell Biol 10(5):640–646

Bucala R, Tracey KJ, Cerami A (1991) Advanced glycosylation products quench nitric oxide and mediate defective endothelium-dependent vasodilatation in experimental diabetes. J Clin Invest. 87(2):432–438

Bucala R, Cerami A (1992) Advanced glycosylation: chemistry, biology, and implications for diabetes and aging. Adv Pharmacol 23:1–34

Cai W, He JC, Zhu L, Chen X, Zheng F, Striker GE, Vlassara H (2008) Oral glycotoxins determine the effects of calorie restriction on oxidant stress, age-related diseases, and lifespan. Am J Pathol 173(2):327–336

Campisi J (2005) Aging, tumor suppression and cancer: high wire-act!. Mech Ageing Dev 126(1):51–58

Candido R, Forbes JM, Thomas MC, Thallas V, Dean RG, Burns WC, Tikellis C, Ritchie RH, Twigg SM, Cooper ME, Burrell LM (2003) A breaker of advanced glycation end products attenuates diabetes-induced myocardial structural changes. Circ Res 92(7): 785–792

Castellani RJ, Zhu X, Lee HG, Moreira PI, Perry G, Smith MA (2007) Neuropathology and treatment of Alzheimer disease: did we lose the forest for the trees? Expert Rev Neurother 7(5):473–85

Cefalu WT, Bell-Farrow AD, Wang ZQ, Sonntag WE, Fu MX, Baynes JW, Thorpe SR (1995) Caloric restriction decreases age-dependent accumulation of the glycoxidation products, N-ε-(carboxymethyl)lysine and pentosidine, in rat skin collagen. J Gerontol A Biol Sci Med Sci 50(6):B337–341

Cerami A, Vlassara H, Brownlee M (1987) Glucose and aging. Sci Am 256(5):90–96

Cheng R, Feng Q, Argirov OK, Ortwerth BJ (2004) Structure elucidation of a novel yellow chromophore from human lens protein. J Biol Chem 279(44):45441–9

Cheng G, Wang LL, Long L, Liu HY, Cui H, Qu WS, Li S (2007) Beneficial effects of C36, a novel breaker of advanced glycation endproducts cross-links, on the cardiovascular system of diabetic rats. Br J Pharmacol 152(8):1196–1206

Clarke S (2003) Aging as war between chemical and biochemical processes: protein methylation and the recognition of age-damaged proteins for repair. Ageing Res Rev 2(3): 263–285

Collard F, Zhang J, Nemet I, Qanungo KR, Monnier VM, Yee VC (2008) Crystal structure of the deglycating enzyme fructosamine oxidase (Amadoriase II). J Biol Chem 283(40):27007–27016

Coughlan MT, Forbes JM, Cooper ME (2007) Role of the AGE crosslink breaker, alagebrium, as a renoprotective agent in diabetes. Kidney Int Suppl (106):S54–S60

Creel D (1980) Inappropriate use of albino animals as models in research. Pharmacol Biochem Behav 12(6):969–977

Culbertson SM, Vassilenko EI, Morrison LD, Ingold KU (2003) Paradoxical impact of antioxidants on post-Amadori glycoxidation: counterintuitive increase in the yields of pentosidine and N-ε-carboxymethyllysine using a novel multifunctional pyridoxamine derivative. J Biol Chem 278(40):38384–38394

DeGroot J, Verzijl N, Bank RA, Lafeber FP, Bijlsma JW, TeKoppele JM (1999) Age-related decrease in proteoglycan synthesis of human articular chondrocytes: the role of nonenzymatic glycation. Arthritis Rheum 42(5):1003–1009

DeGroot J, Verzijl N, Budde M, Bijlsma JW, Lafeber FP, TeKoppele JM (2001a) Accumulation of advanced glycation end products decreases collagen turnover by bovine chondrocytes. Exp Cell Res 266(2):303–310

DeGroot J, Verzijl N, Jacobs KM, Budde M, Bank RA, Bijlsma JW, TeKoppele JM, Lafeber FP (2001b) Accumulation of advanced glycation endproducts reduces chondrocyte-mediated extracellular matrix turnover in human articular cartilage. Osteoarthritis Cartilage 9(8):720–726

DeGroot J, Verzijl N, Wenting-Van Wijk MJ, Bank RA, Lafeber FP, Bijlsma JW, TeKoppele JM (2001c) Age-related decrease in susceptibility of human articular cartilage to matrix metalloproteinase-mediated degradation: the role of advanced glycation end products. Arthritis Rheum 44(11):2562–2571

DeGroot J, Verzijl N, Wenting-van Wijk MJ, Jacobs KM, Van El B, Van Roermund PM, Bank RA, Bijlsma JW, TeKoppele JM, Lafeber FP (2004) Accumulation of advanced glycation end products as a molecular mechanism for aging as a risk factor in osteoarthritis. Arthritis Rheum 50(4):1207–1215

DeVry CG, Tsai W, Clarke S (1996) Structure of the human gene encoding the protein repair L-isoaspartyl (D-aspartyl) O-methyltransferase. Arch Biochem Biophys 335(2):321–32

Dong Y, Wu Y, Wu M, Wang S, Zhang J, Xie Z, Xu J, Song P, Wilson K, Zhao Z, Lyons T, Zou MH (2008) Activation of Protease Calpain by Oxidized and Glycated LDL Increases the Degradation of Endothelial Nitric Oxide Synthase. J Cell Mol Med 13(9a):2899–2910

Drinda S, Franke S, Rüster M, Petrow P, Pullig O, Stein G, Hein G (2005) Identification of the receptor for advanced glycation end products in synovial tissue of patients with rheumatoid arthritis. Rheumatol Int 25(6):411–413

Eble AS, Thorpe SR, Baynes JW (1983) Nonenzymatic glucosylation and glucose-dependent cross-linking of protein. J Biol Chem 258(15):9406–9412

Everts V, Van der Zee E, Creemers L, Beertsen W (1996) Phagocytosis and intracellular digestion of collagen, its role in turnover and remodeling. Histochem J 28, 229–245

Fawcett DW (1986) A Textbook of Histology, 11th edn. Saunders, Philadelphia

Ferguson MW, Duncan J, Bond J, Bush J, Durani P, So K, Taylor L, Chantrey J, Mason T, James G, Laverty H, Occleston NL, Sattar A, Ludlow A, O'Kane S (2009) Prophylactic administration of avotermin for improvement of skin scarring: three double-blind, placebo-controlled, phase I/II studies. Lancet 373(9671):1264–1274

Figarola JL, Loera S, Weng Y, Shanmugam N, Natarajan R, Rahbar S (2008) LR-90 prevents dyslipidaemia and diabetic nephropathy in the Zucker diabetic fatty rat. Diabetologia 51(5):882–891

Finch CE (2007) The Biology of Human Longevity: Inflammation, Nutrition, and Aging in the Evolution of Life Spans. Academic, Amsterdam

Flurkey K, Papaconstantinou J, Miller RA, Harrison DE (2001) Lifespan extension and delayed immune and collagen aging in mutant mice with defects in growth hormone production. Proc Natl Acad Sci USA 98(12):6736–6741

Furber JD (2006) Extracellular glycation crosslinks: prospects for removal. Rejuvenation Res 9(2):274–278

Gardiner DM (2005) Ontogenetic decline of regenerative ability and the stimulation of human regeneration. Rejuvenation Res 8(3):141–153

Gilbert SF (2000) Developmental Biology, 6th edn. Sinauer, Sunderland, MA

Goldberg T, Cai W, Peppa M, Dardaine V, Baliga BS, Uribarri J, Vlassara H (2004) Advanced glycoxidation end products in commonly consumed foods. J Am Diet Assoc 104(8):1287–1291

Haimes HB (2005) Alagebrium: Intervention on the A.G.E. pathway modifies deficits caused by aging and diabetes. Paper presented at strategies for engineered negligible senescence (SENS), 2nd conference, Cambridge, England, 7–11 September 2005

Haimes HB (2007) Hardening of the arteries: breaking the ties that bind. Paper presented at Edmonton aging symposium, Edmonton, Canada, 30–31 March 2007

Hammes HP, Du X, Edelstein D, Taguchi T, Matsumura T, Ju Q, Lin J, Bierhaus A, Nawroth P, Hannak D, Neumaier M, Bergfeld R, Giardino I, Brownlee M (2003) Benfotiamine blocks three major pathways of hyperglycemic damage and prevents experimental diabetic retinopathy. Nat Med 9(3):294–299

Hipkiss AR (2005) Glycation, ageing and carnosine: Are carnivorous diets beneficial? Mech Ageing Dev 126(10):1034–1039

Hirsch MS, Lunsford LE, Trinkaus-Randall V, Svoboda KK (1997) Chondrocyte survival and differentiation in situ are integrin mediated. Dev Dyn 210(3):249–263

Hsu RL, Lee KT, Wang JH, Lee LY, Chen RP (2009) Amyloid-degrading ability of nattokinase from Bacillus subtilis natto. J Agric Food Chem 57(2):503–508

Huang PL, Huang Z, Mashimo H, Bloch KD, Moskowitz MA, Bevan JA, Fishman MC (1995) Hypertension in mice lacking the gene for endothelial nitric oxide synthase. Nature 377(6546):239–242

Januszewski AS, Alderson NL, Metz TO, Thorpe SR, Baynes JW (2003) Role of lipids in chemical modification of proteins and development of complications in diabetes. Biochem Soc Trans 31(Pt 6):1413–1416

Januszewski AS, Alderson NL, Jenkins AJ, Thorpe SR, Baynes JW (2005) Chemical modification of proteins during peroxidation of phospholipids. J Lipid Res 46:1440–1449

Jerums G, Panagiotopoulos S, Forbes J, Osicka T, Cooper M (2003) Evolving concepts in advanced glycation, diabetic nephropathy, and diabetic vascular disease. Arch Biochem Biophys 419(1):55–62

Kass DA, Shapiro EP, Kawaguchi M, Capriotti AR, Scuteri A, deGroof RC, Lakatta EG (2001) Improved arterial compliance by a novel advanced glycation end-product crosslink breaker. Circulation 104(13):1464–1470

Kass DA (2003) Getting better without AGE: new insights into the diabetic heart. Circ Res 92(7):704–706

Kikuchi S, Shinpo K, Takeuchi M, Yamagishi S, Makita Z, Sasaki N, Tashiro K (2003) Glycation – a sweet tempter for neuronal death. Brain Res Brain Res Rev 41(2–3):306–323

Kislinger T, Fu C., Huber B, Qu W, Taguchi A, Du Yan S, Hoffmann M, Yan SF, Pischetsrieder M, Stern D, Schmidt AM (1999) N-ε-(carboxymethyl)lysine adducts of proteins are ligands for receptors for advanced glycation end products that activate cell signaling pathways and modulate gene expression. J Biol Chem 274:31740 - 31749

Kjær M, Magnusson P, Krogsgaard M, Møller JB, Olesen J, Heinemeier K, Hansen M, Haraldsson B, Koskinen S, Esmarck B, Langberg H (2006) Extracellular matrix adaptation of tendon and skeletal muscle to exercise. J Anat 208:445–450

Kohn RR (1978) Principles of Mammalian Aging, 2nd edn. Prentice-Hall, Englewood Cliffs, NJ

Kume S, Kato S, Yamagishi S, Inagaki Y, Ueda S, Arima N, Okawa T, Kojiro M, Nagata K (2005) Advanced glycation end-products attenuate human mesenchymal stem cells and prevent cognate differentiation into adipose tissue, cartilage, and bone. J Bone Miner Res 20(9):1647–1658

Labat-Robert J (2004) Cell-matrix interactions in aging: role of receptors and matricryptins. Ageing Res Rev 3(2):233–247

Lachmann HJ, Hawkins PN (2006) Systemic amyloidosis. Curr Opin Pharmacol (2):214–20

Lanthier J, Desrosiers RR (2004) Protein L-isoaspartyl methyltransferase repairs abnormal aspartyl residues accumulated in vivo in type-I collagen and restores cell migration. Exp Cell Res 293(1):96–105

Li Z, Froehlich J, Galis ZS, Lakatta EG (1999) Increased expression of matrix metalloproteinase-2 in the thickened intima of aged rats. Hypertension 33(1):116–23

Lingelbach LB, Mitchell AE, Rucker RB, McDonald RB (2000) Accumulation of advanced glycation endproducts in aging male Fischer 344 rats during long-term feeding of various dietary carbohydrates. J Nutr 130(5):1247–1255

Little WC, Zile MR, Kitzman DW, Hundley WG, O'Brien TX, Degroof RC (2005) The effect of alagebrium chloride (ALT-711), a novel glucose cross-link breaker, in the treatment of elderly patients with diastolic heart failure. J Card Fail 11(3):191–195

Lodish H, Berk A, Zipursky SL, Matsudaira P, Baltimore D, Darnell J (2000) Molecular Cell Biology. W.H. Freeman, New York

Mathews CK, van Holde KE (1990) Biochemistry. Benjamin/Cummings, Redwood City

Metz TO, Alderson NL, Thorpe SR, Baynes JW (2003) Pyridoxamine, an inhibitor of advanced glycation and lipoxidation reactions: a novel therapy for treatment of diabetic complications. Arch Biochem Biophys 419(1):41–49

Miyata T, van Ypersele de Strihou C, Kurokawa K, Baynes JW (1999) Alterations in nonenzymatic biochemistry in uremia: origin and significance of "carbonyl stress" in long-term uremic complications. Kidney Int 55(2):389–399

Monnier VM, Sell DR, Saxena A, Saxena P, Subramaniam R, Tessier F, Weiss MF (2003) Glycoxidative and carbonyl stress in aging and age-related diseases. In: Cutler RG, Rodriguez H (eds) Critical Reviews of Oxidative Stress and Aging: Advances in Basic Science, Diagnostics and Intervention, vol 1. World Scientific, Singapore, pp. 413–433

Morimitsu Y, Yoshida K, Esaki S, Hirota A (1995) Protein glycation inhibitors from thyme (Thymus Vulgaris). Biosci Biotechnol Biochem 59(11):2018–2021

Muneoka K, Han M, Gardiner DM (2008) Regrowing human limbs. Sci Am 298(4):56–63

Murphy G, Reynolds JJ (2002) Extracellular matrix degradation. In: Royce PM, Steinmann B (eds) Connective Tissue and its Heritable Disorders. Wiley-Liss, Wilmington, pp. 343–384

Nicholson IP, Gault EA, Foote CG, Nasir L, Bennett D (2007) Human telomerase reverse transcriptase (hTERT) extends the lifespan of canine chondrocytes in vitro without inducing neoplastic transformation. Vet J 174(3):570–576

Ohgami N, Nagai R, Ikemoto M, Arai H, Kuniyasu A, Horiuchi S, Nakayama H (2001) Cd36, a member of the class b scavenger receptor family, as a receptor for advanced glycation end products. J Biol Chem 276(5):3195–3202

Parrinello S, Coppe JP, Krtolica A, Campisi J (2005) Stromal-epithelial interactions in aging and cancer: senescent fibroblasts alter epithelial cell differentiation. J Cell Sci 118(Pt 3):485–496

Pathak P, Gupta R, Chaudhari A, Shiwalkar A, Dubey A, Mandhare AB, Gupta RC, Joshi D, Chauthaiwale V (2008) TRC4149 a novel advanced glycation end product breaker improves hemodynamic status in diabetic spontaneously hypertensive rats. Eur J Med Res 13(8):388–398

Peng X, Cheng KW, Ma J, Chen B, Ho CT, Lo C, Chen F, Wang M (2008) Cinnamon bark proanthocyanidins as reactive carbonyl scavengers to prevent the formation of advanced glycation endproducts. J Agric Food Chem 56(6):1907–1911

Pepys MB (2006) Amyloidosis. Annu Rev Med 57:223–241

Perry G, Smith MA (2001) Active glycation in neurofibrillary pathology of Alzheimer's disease: N-(Carboxymethyl) lysine and hexitol-lysine. Free Radic Biol Med 31(2):175–180

Piez KA (2002) Research on collagen in the author's laboratory, 1952–1982. In: Royce PM, Steinmann B (eds) Connective tissue and its heritable disorders. Wiley-Liss, Wilmington, pp. 1–11

Rahbar S, Natarajan R, Yerneni K, Scott S, Gonzales N, Nadler JL (2000) Evidence that pioglitazone, metformin and pentoxifylline are inhibitors of glycation. Clin Chim Acta 301(1–2):65–77

Rahbar S, Figarola JL (2003) Novel inhibitors of advanced glycation endproducts. Arch Biochem Biophys 419:63–79

Rahbar S (2007) Novel inhibitors of glycation and AGE formation. Cell Biochem Biophys 48(2–3):147–157

Requena JR, Stadtman ER (1999) Conversion of lysine to N(epsilon)-(carboxymethyl)lysine increases susceptibility of proteins to metal-catalyzed oxidation. Biochem Biophys Res Commun 264(1):207–211

Scheid A, Wenger RH, Schäffer L, Camenisch I, Distler O, Ferenc A, Cristina H, Ryan HE, Johnson RS, Wagner KF, Stauffer UG, Bauer C, Gassmann M, Meuli M (2002) Physiologically low oxygen concentrations in fetal skin regulate hypoxia-inducible factor 1 and transforming growth factor-beta3. FASEB J 16(3):411–413

Rinn JL, Bondre C, Gladstone HB, Brown PO, Chang HY (2006) Anatomic demarcation by positional variation in fibroblast gene expression programs. PLoS Genet 2(7):e119

Ritz-Timme S, Collins MJ (2002) Racemization of aspartic acid in human proteins. Ageing Res Rev 1(1):43–59

Robert L, Robert AM, Fülöp T (2008) Rapid increase in human life expectancy: will it soon be limited by the aging of elastin? Biogerontology 9(2):119–133

Sajithlal GB, Chithra P, Chandrakasan G (1999) An in vitro study on the role of metal catalyzed oxidation in glycation and crosslinking of collagen. Mol Cell Biochem 194(1–2):257–263

Saxena AK, Saxena P, Wu X, Obrenovich M, Weiss MF, Monnier VM (1999) Protein aging by carboxymethylation of lysines generates sites for divalent metal and redox active copper binding: relevance to diseases of glycoxidative stress. Biochem Biophys Res Commun 260(2):332–338

Schrementi ME, Ferreira AM, Zender C, DiPietro LA (2008) Site-specific production of TGF-β in oral mucosal and cutaneous wounds. Wound Repair Regen 16(1):80–86

Sell DR, Carlson EC, Monnier VM (1993) Differential effects of type 2 (non-insulin-dependent) diabetes mellitus on pentosidine formation in skin and glomerular basement membrane. Diabetologia 36(10):936–941

Sell DR, Monnier VM (2004) Conversion of arginine into ornithine by advanced glycation in senescent human collagen and lens crystallins. J Biol Chem 279(52):54173–54184

Sell DR, Biemel KM, Reihl O, Lederer MO, Strauch CM, Monnier VM (2005) Glucosepane is a major protein cross-link of the senescent human extracellular matrix: relationship with diabetes. J Biol Chem 280(13):12310–12315

Shifren A, Mecham RP (2006) The stumbling block in lung repair of emphysema: elastic fiber assembly. Proc Am Thorac Soc 3(5):428–433

Shimizu T, Matsuoka Y, Shirasawa T (2005) Biological significance of isoaspartate and its repair system. Biol Pharm Bull 28(9):1590–1596

Sivan SS, Wachtel E, Tsitron E, Sakkee N, van der Ham F, Degroot J, Roberts S, Maroudas A (2008) Collagen turnover in normal and degenerate human intervertebral discs as determined by the racemization of aspartic acid. J Biol Chem 283(14):8796–801

Spencer VA, Xu R, Bissell MJ (2007) Extracellular matrix, nuclear and chromatin structure, and gene expression in normal tissues and malignant tumors: a work in progress. Adv Cancer Res 97:275–294

Steenvoorden MM, Toes RE, Ronday HK, Huizinga TW, Degroot J (2007) RAGE activation induces invasiveness of RA fibroblast-like synoviocytes in vitro. Clin Exp Rheumatol. 25(5):740–742

Stern DM, Yan SD, Yan SF, Schmidt AM (2002) Receptor for advanced glycation endproducts (RAGE) and the complications of diabetes. Ageing Res Rev 1(1):1–15.

Stracke H, Hammes HP, Werkmann D, Mavrakis K, Bitsch I, Netzel M, Geyer J, Kopcke W, Sauerland C, Bretzel RG, Federlin KF (2001) Efficacy of benfotiamine versus thiamine on function and glycation products of peripheral nerves in diabetic rats. Exp Clin Endocrinol Diabetes 109(6):330–336

Sturchler E, Galichet A, Weibel M, Leclerc E, Heizmann CW (2008) Site-specific blockade of RAGE-Vd prevents amyloid-β oligomer neurotoxicity. J Neurosci 28(20):5149–5158

Szwergold BS, Howell SK, Beisswenger PJ (2001) Human fructosamine-3-kinase (FN3K): purification, sequencing, substrate specificity and evidence of activity in vivo. Diabetes 50:2139–2147

Szwergold BS, Howell SK, Beisswenger PJ (2005) Transglycation - a potential new mechanism for deglycation of Schiff's bases. Ann NY Acad Sci 1043:845–864

Taguchi A, Blood DC, del Toro G, Canet A, Lee DC, Qu W, Tanji N, Lu Y, Lalla E, Fu C, Hofmann MA, Kislinger T, Ingram M, Lu A, Tanaka H, Hori O, Ogawa S, Stern DM, Schmidt AM (2000) Blockade of RAGE-amphoterin signalling suppresses tumour growth and metastases. Nature 405:354–360

Taniguchi N, Caramés B, Ronfani L, Ulmer U, Komiya S, Bianchi ME, Lotz M (2009) Aging-related loss of the chromatin protein HMGB2 in articular cartilage is linked to reduced cellularity and osteoarthritis. Proc Natl Acad Sci USA 106(4):1181–1186

Tanskanen M, Peuralinna T, Polvikoski T, Notkola IL, Sulkava R, Hardy J, Singleton A, Kiuru-Enari S, Paetau A, Tienari PJ, Myllykangas L (2008) Senile systemic amyloidosis affects 25% of the very aged and associates with genetic variation in alpha2-macroglobulin and tau: a population-based autopsy study. Ann Med 40(3):232–239

Thornalley PJ, Minhas HS (1999) Rapid hydrolysis and slow alpha,β-dicarbonyl cleavage of an agent proposed to cleave glucose-derived protein cross-links. Biochem Pharmacol 57:303–307

Thornalley PJ (2003) Use of aminoguanidine (Pimagedine) to prevent the formation of advanced glycation endproducts. Arch Biochem Biophys 419(1):31–40

Tikellis C, Thomas MC, Harcourt BE, Coughlan MT, Pete J, Bialkowski K, Tan A, Bierhaus A, Cooper ME, Forbes JM (2008) Cardiac inflammation associated with a Western diet is mediated via activation of RAGE by AGEs. Am J Physiol Endocrinol Metab 295(2):E323–E330

Tredget EE, Ding J (2009) Wound healing: from embryos to adults and back again. Lancet 373(9671):1226–1228

Ulrich P, Zhang X (1997) Pharmacological reversal of advanced glycation end-product-mediated protein crosslinking. Diabetologia 40:S157–S159

Ulrich P, Cerami A (2001) Protein glycation, diabetes, and aging. Recent Prog Horm Res 56:1–21

Usta MF, Kendirci M, Bivalacqua TJ, Gur S, Hellstrom WJG, Foxwell NA, Cellek S (2004) Delayed administration of ALT-711, but not of aminoguanidine, improves erectile function in streptozotocin diabetic rats: curative versus preventive medicine. Paper presented at the 11th World Congress of the International Society for Sexual and Impotence Research, Buenos Aires, October 2004

Usta MF, Kendirci M, Gur S, Foxwell NA, Bivalacqua TJ, Cellek S, Hellstrom WJ (2006) The breakdown of preformed advanced glycation end products reverses erectile dysfunction in streptozotocin-induced diabetic rats: preventive versus curative treatment. J Sex Med 3(2):242–250

Vaitkevicius PV, Lane M, Spurgeon H, Ingram DK, Roth GS, Egan JJ, Vasan S, Wagle DR, Ulrich P, Brines M, Wuerth JP, Cerami A, Lakatta EG (2001) A cross-link breaker has sustained effects on arterial and ventricular properties in older rhesus monkeys. Proc Natl Acad Sci USA 98(3):1171–1175

Vasan S, Zhang X, Zhang X, Kapurniotu A, Bernhagen J, Teichberg S, Basgen J, Wagle D, Shih D, Terlecky I, Bucala R, Cerami A, Egan J, Ulrich P (1996) An agent cleaving glucose- derived protein crosslinks in vitro and in vivo. Nature 382(6588):275–278

Vasan S, Foiles PG, Founds HW (2001) Therapeutic potential of AGE inhibitors and breakers of AGE protein cross-links. Expert Opin Investig Drugs 10(11):1977–1987

Vasan S, Foiles P, Founds H (2003) Therapeutic potential of breakers of advanced glycation end product-protein crosslinks. Arch Biochem Biophys 419(1):89–96

Vater CA, Harris ED Jr, Siegel RC (1979) Native cross-links in collagen fibrils induce resistance to human synovial collagenase. Biochem J 181(3):639–645

Verzijl N, DeGroot J, Thorpe SR, Bank RA, Shaw JN, Lyons TJ, Bijlsma JW, Lafeber FP, Baynes JW, TeKoppele JM (2000) Effect of collagen turnover on the accumulation of advanced glycation end products. J Biol Chem 275(50):39027–39031

Verzijl N, DeGroot J, Bank RA, Bayliss MT, Bijlsma JW, Lafeber FP, Maroudas A, TeKoppele JM (2001) Age-related accumulation of the advanced glycation endproduct pentosidine in human articular cartilage aggrecan: the use of pentosidine levels as a quantitative measure of protein turnover. Matrix Biol 20(7):409–417

Verzijl N, Bank RA, TeKoppele JM, DeGroot J (2003) AGEing and osteoarthritis: a different perspective. Curr Opin Rheumatol (5):616–622

Vlassara H, Cai W, Crandall J, Goldberg T, Oberstein R, Dardaine V, Peppa M, Rayfield EJ (2002) Inflammatory mediators are induced by dietary glycotoxins, a major risk factor for diabetic angiopathy. Proc Natl Acad Sci USA 99(24):15596–15601

Vlassara H, Palace MR (2003) Glycoxidation: the menace of diabetes and aging. Mt Sinai J Med 70(4):232–241

Wagenseil JE, Mecham RP (2007) New insights into elastic fiber assembly. Birth Defects Res C Embryo Today 81(4):229–240

Wang M, Lakatta EG (2002) Altered regulation of matrix metalloproteinase-2 in aortic remodeling during aging. Hypertension 39(4):865–873

Wang M, Takagi G, Asai K, Resuello RG, Natividad FF, Vatner DE, Vatner SF, Lakatta EG (2003) Aging increases aortic MMP-2 activity and angiotensin II in nonhuman primates. Hypertension 41(6):1308–1316

Weber DJ, McFadden PN (1997) Injury-induced enzymatic methylation of aging collagen in the extracellular matrix of blood vessels. J Protein Chem 16(4):269–281

Wolffenbuttel BH, Boulanger CM, Crijns FR, Huijberts MS, Poitevin P, Swennen GN, Vasan S, Egan JJ, Ulrich P, Cerami A, Lévy BI (1998) Breakers of advanced glycation end products restore large artery properties in experimental diabetes. Proc Natl Acad Sci USA 95(8):4630–4634

Wu CH, Yen GC (2005) Inhibitory effect of naturally occurring flavonoids on the formation of advanced glycation endproducts. J Agric Food Chem 53(8):3167–3173

Xiao H, Cai G, Liu M (2007) Fe^{2+}-catalyzed non-enzymatic glycosylation alters collagen conformation during AGE-collagen formation in vitro. Arch Biochem Biophys 468(2):183–192

Yang S, Litchfield JE, Baynes JW (2003) AGE-breakers cleave model compounds, but do not break Maillard crosslinks in skin and tail collagen from diabetic rats. Arch Biochem Biophys 412(1):42–46

Zieman SJ, Melenovsky V, Clattenburg L, Corretti MC, Capriotti A, Gerstenblith G, Kass DA (2007) Advanced glycation endproduct crosslink breaker (alagebrium) improves endothelial function in patients with isolated systolic hypertension. J Hypertens 25(3):577–583

Chapter 20
Methuselah's DNA: Defining Genes That Can Extend Longevity

Robert J. Shmookler Reis and Joan E. McEwen

Contents

20.1	Overview	623
20.2	Extreme Longevity of *C. elegans*	624
	20.2.1 The Normal Function of Many Genes Limits Longevity	624
	20.2.2 Ten-Fold Extension of *C. elegans* Lifespan by Single Mutations in the *age-1* Gene	625
20.3	Extreme Longevity of *S. cerevisiae*	628
	20.3.1 Measurement of Yeast Lifespan	628
	20.3.2 Ten-Fold Extension of Chronological Lifespan	629
20.4	Intersections Between Pathways, in Diverse Species, Allow Integrated Responses .	631
	20.4.1 Implications for Life Extension	632
20.5	Regulating Metabolism and Lifespan: Tissue Specificity in Worms, Flies, and Mice	632
20.6	What Molecular Features do Long-Lived Yeast and Nematodes Have in Common?	633
20.7	Prospects for Translation to Mammals	634
References	. . . .	635

20.1 Overview

Genetic mutations capable of extending metazoan lifespan were first discovered in the nematode *C. elegans* (Apfeld and Kenyon 1999; Cypser and Johnson 1999; Friedman and Johnson 1988; Hansen et al. 2005; Kenyon et al. 1993; Klass 1983; Lin et al. 1997, 2001). Mutations with substantial effects on lifespan were later discovered in Baker's yeast (*S. cerevisiae*), Drosophila, mice, and rats, either by targeted quests for mutations in the corresponding pathway of these species

---

R.J. Shmookler Reis (✉)
Department of Geriatrics, University of Arkansas for Medical Sciences, and Research Service, Central Arkansas Veterans Healthcare System, Little Rock, AR 72205, USA
e-mail: rjsr@uams.edu

(Bonkowski et al. 2006; Clancy et al. 2001; Holzenberger et al. 2003; Tatar et al. 2001) or by independent discovery (Bartke et al. 2001; Kurosu et al. 2005; Lin et al. 1998; Marden et al. 2003; Migliaccio et al. 1999).

The observation that random mutations (expected to impair function) frequently increase lifespan implies that *the primary function of those genes is something other than ensuring survival*. Other genes, when mutated, limit longevity. Before concluding that these mutations identify "longevity assurance genes" (D'Mello et al. 1994; Ayyadevara et al. 2005), it is necessary to consider the trivial explanation that they affect an essential function or pathway of the organism. Through studies of single gene mutations and/or overexpression, several signal-transduction pathways regulating longevity have been identified, and it is becoming increasingly clear that these pathways intersect with one another, perhaps defining a "network" of regulatory and protective genes that impact survival in many circumstances. In addition, environmental conditions, including nutrient availability, can profoundly affect lifespan and act, at least in part, through the signal transduction pathways that have been identified genetically to influence lifespan.

In this chapter, we will discuss two examples of extreme longevity, in which the lifespans of model organisms, *C. elegans* and *S. cerevisiae*, have been extended by about 10-fold. We will attempt to interpret these effects with reference to the principle longevity pathways affected in each system.

20.2 Extreme Longevity of *C. elegans*

20.2.1 The Normal Function of Many Genes Limits Longevity

Reduction-of-function mutations, identified in mutagenesis screens (usually for traits other than lifespan), extend *C. elegans* longevity by factors ranging from 1.1 to almost 3-fold. The first mutation to be identified as long-lived, an allele of *age-1*, emerged from a sib-screen for long-lived mutants arising from chemical mutagenesis (Klass 1983). The mean longevity of these *age-1(hx546)* homozygotes was increased by 40–65%, depending on the incubation temperature (Friedman and Johnson 1988). The *daf-2* mutation, first identified as a temperature-sensitive mutant that forms dauer larvae constitutively at $\geq 25°C$ (Riddle et al. 1981), was discovered over a decade later to have a two-fold extended adult lifespan (Kenyon et al. 1993). Fifteen independent *daf-2* alleles tested in an isogenic background conferred lifespan extensions of 1.1- to 2.5-fold at either $15°C$ or $22.5°C$ (Gems et al. 1998).

Daf-2 encodes a cell membrane receptor for insulin-like ligands, while *age-1* encodes the catalytic (p110) subunit of a class-I PI3 kinase. These are two components of the insulin/IGF-1 (IIS) response pathway of nematodes, regulating dauer formation, fecundity, stress resistance, and longevity. Addition of a second mutation inactivating the *daf-16* gene, which encodes a FOXO transcription factor that is phosphorylated via the DAF-2/AGE-1/PDK-1/AKT kinase cascade, abrogates all previously reported phenotypes of both *age-1* and *daf-2* mutations (Guarente and Kenyon 2000; Larsen et al. 1995; Tissenbaum and Ruvkun 1998).

Although a remarkable variety of life-extending mutations has been discovered in the nematode *C. elegans*, none surpassed the 2- to 3-fold increases reported for genetic disruptions of the insulin-Igf1 signaling (IIS) pathway (Gems et al. 1998; Larsen et al. 1995; Lin et al. 1997; Tissenbaum and Ruvkun 1998). Other species, mutated in corresponding pathways, showed more modest life extensions. In Drosophila, no single mutations have increased lifespan by more than 2-fold (Rogina et al. 2000; Tatar et al. 2001), and in mice the record is roughly 1.5-fold (Bartke et al. 2001; Bluher et al. 2003; Clancy et al. 2001; Holzenberger et al. 2003; Kurosu et al. 2005; Tatar et al. 2001; Tirosh et al. 2004). Greater life extensions have been achieved through the additive effects of several distinct interventions; e.g., specific allelic combinations of mutations to *daf-2* and *daf-12* (encoding a nuclear hormone receptor) together boost worm longevity by 3.5-fold (Larsen et al. 1995) or 4.4-fold (Gems et al. 1998) at 25.5°C. Combining interventions, *e.g.* augmenting *daf-2* mutation with germ-cell ablation or dietary restriction, can extend *C. elegans* lifespan by 4- to 6-fold (Arantes-Oliveira et al. 2003; Houthoofd et al. 2003). Combining the effects of a mutation impairing pituitary development with caloric restriction produced a record 1.8-fold life extension in mice (Bartke et al. 2001). Long-lived mutations often confer resistance to stresses (Lithgow and Walker 2002; Lin et al. 1998; Murakami 2006); however, several natural genetic variants for lifespan, quantitative trait alleles, vary widely in resistance to diverse stresses (Shmookler Reis et al. 2006).

20.2.2 Ten-Fold Extension of C. elegans Lifespan by Single Mutations in the age-1 Gene

20.2.2.1 Conditions that Lead to Extreme Longevity of Strains with Strong *age-1* Mutations

C. elegans strains bearing nonsense mutations in the *age-1* gene cannot be propagated as "pure" strains but instead are maintained as heterozygotes, with one copy of the mutated gene and one wild-type copy on a marked "balancer" chromosome. These worms, almost all hermaphrodites, self-fertilize to produce about one-fourth homozygotes with both *age-1* copies mutated. If we call those homozygous *age-1* mutants the "F1" generation, their progeny ("F2's") are all homozygous for the mutation, and from the start they are clearly quite different from their parents – despite being genetically identical. F2 mutants develop slowly at 15° or 20°C, and never complete development at 25°C. Those that develop into adults are several fold more resistant to stresses, either oxidative or electrophilic, and 10-fold longer-lived than wild-type worms [(Ayyadevara et al. 2008) and Fig. 20.1]. Their entire lifespans are prolonged, including maximum lifespan (see right panel of Fig. 20.1), and at any age they appear as active and robust as normal worms one-tenth their age. In contrast, the F1 homozygous mutants are intermediate in stress resistance (Ayyadevara et al. 2008) and only about twofold longer lived than wild-type worms (Morris et al. 1996).

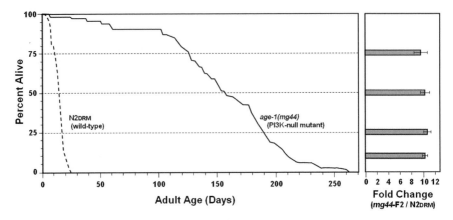

Fig. 20.1 Survivals of *age-1(mg44)* F2 homozygotes vs. N2DRM controls. Survivals are plotted for *age-1(mg44)* F2 (*solid line*) and N2DRM (*dashed*) [redrawn from Ayyadevara et al. (2008)]. The *right panel* shows *mg44*-F2 / N2DRM ratios, ±SEM, for *adult* survival to the 25th, 50th, 75th and 90th percentiles

Why there is such a dramatic difference between these two generations, despite identical genes, is a topic of active research in our laboratory. It is almost certain to be a consequence of something in the egg cytoplasm, which in the case of the F1 worms was made by the heterozygous *age-1/+* parent – but we do not yet know whether that "something" is normal *age-1* messenger RNA, the PI3 Kinase enzyme it encodes, the PIP_3 product of that enzyme's activity, or something else downstream of those components.

20.2.2.2 Phosphatidyl Inositol 3-Kinase (PI3K)

The *age-1* gene of *C. elegans* encodes a phosphatidyl inositol 3-kinase catalytic (p110) subunit. PI3Ks belong to an enzyme family capable of catalyzing the donation of a phosphate group from ATP to the 3 position of the inositol ring in a phosphatidylinositide. Specific PI3K enzymes transform phosphatidylinositol (PtdIns or PI) substrate to PtdIns(3)P, PtdIns(4)P to PtdIns(3,4)P_2, and PtdIns(4,5)P_2 (abbreviated PIP_2) to PtdIns(3,4,5)P_3 (abbreviated PIP_3) (Fig. 20.2).

The mammalian tumor-suppressor gene *Pten* encodes a phosphatase able to remove the 3-phosphate from PIP_3, and thus opposes PI3K action. *Pten* of mammals

Fig. 20.2 Structure of PIP_3. R_1 and R_2 are fatty-acid chains that vary among these molecules

and nematodes converts PtdIns(3,4,5)P_3 to PtdIns(4,5)P_2, and PtdIns(3,4)P_2 to PtdIns(4)P (Hawkins et al. 2006; Ogg and Ruvkun 1998; Vanhaesebroeck et al. 2001). The SHIP1 and SHIP2 phosphatidylinositol 5-phosphatases also deplete PIP$_3$ by removing the 5-phosphate to form PtdIns(3,4)P_2. The 3 classes of mammalian PI3Ks (I, II, and III) differ in structure, substrates, and activators.

From its sequence and structure, the *C. elegans* AGE-1 protein is a class-I PI3K catalytic subunit, able to convert PI(4,5)P_2 to PI(3,4,5)P_3. Class I PI3Ks are heterodimers of a catalytic (p110) subunit, of which mammals have four varieties (Hawkins et al. 2006) and *C. elegans*, one (WormBase), and a regulatory subunit, of which mammals have seven known varieties, and worms have at least one (AAP-1) (WormBase). AGE-1 most closely resembles p110-α and -β catalytic subunits (Morris et al. 1996), refining its identity as class IA. The nematode has fewer PI3K varieties than are found in mammals, so it is possible that AGE-1 also shares properties of mammalian class-IB PI3Ks. F39B1.1, another *C. elegans* p110 protein, is more closely related to class-II PI3K homologs. The third PI3K p110 gene of *C. elegans*, *vps-34*, is orthologous to yeast *vps34*, a class-III PI3K. Enzymes of this class are involved in vesicular trafficking, endocytosis, autophagy, and secretion (Vieira et al. 2001; Backer 2008). Consistent with this, VPS34 regulates yeast endocytosis, and *C. elegans vps-34* is implicated in endocytic trafficking between cell compartments, including trafficking of low-density lipoprotein (LDL) receptors (Roggo et al. 2002).

In vitro, class II and III PI3Ks have high affinity for their normal PIP substrates, PI and PI(4)P, but also phosphorylate the 3 position of PI(4,5)P_2 at very low efficiency (Hawkins et al. 2006). If this were to occur in vivo, it could ameliorate the phenotype of *age-1*-null mutants.

20.2.2.3 Targets of Phosphatidyl Inositol 3,4,5-Phosphate (PIP$_3$) and Role in Insulin/IGF1-Like Signaling

PIP$_3$ is normally present at roughly 0.1% of the levels of its precursor, PtdIns(4,5)P_2, which is itself a rather low-abundance molecule. PIP$_3$, however, can increase up to 100-fold in activated cells (Pettitt et al. 2006). The scarcity of this molecular species, coupled with the fact that PTEN (which removes the 3-position phosphate of PIP$_3$) is a tumor suppressor in mammals, suggests that PIP$_3$ is far from innocuous and must be held in tight check to prevent excessive cell proliferation.

In the IIS cascade, the action of PI3K is thought to be entirely effected through the generation of PIP$_3$ (Hawkins et al. 2006; Vanhaesebroeck et al. 2001), although the p110 catalytic subunit is also a protein kinase able to phosphorylate its own regulatory subunit (Gami et al. 2006; Remenyi et al. 2006). PIP$_3$ binds to a pleckstrin-homology (PH) domain in PDK-1 (phosphoinositide-dependent kinase1), activating it to phosphorylate a Thr residue of AKT-1/PKB, which (in a heterodimer with AKT-2) then phosphorylates the transcription factor DAF-16/FOXO to restrain it to the cytoplasm where it is inactive (Berdichevsky et al. 2006; Tissenbaum and Ruvkun 1998).

PIP_3 plays two roles in AKT-1 activation: it binds the AKT-1 PH domain, tethering it to the membrane, and it also allosterically exposes a phosphorylation site to PDK-1 upon binding a second, PH domain site (Gami et al. 2006; Remenyi et al. 2006). Thus, although it is required stoichiometrically for anchoring many signal-transduction molecules to membranes, where they would thereby be brought into proximity, PIP_3 may act catalytically to transiently alter the conformation of AKT-1 (and perhaps other targets) to enable their phosphorylation.

We postulate that PIP_3 can dissociate from AKT-1 and circulate to other target molecules, in which case a single molecule of PIP_3 may be sufficient to activate all of a cell's AKT-1. Thus, removal from a cell of the last traces of PIP_3 might have profound consequences for all AKT-mediated signaling, as well as any other pathways with a comparable dependence on PIP_3. This may be why the F2 homozygous *age-1(mg44)* mutants live so much longer than their genetically identical parents, which still retain some traces of PIP_3 (via carry-over from their heterozygous parents). If true, then the key to extreme longevity would lie in extinguishing all PIP_3-dependent signaling – something that has thus far been accomplished only in *C. elegans* and which will be challenging to achieve in higher organisms which are believed to require such signaling in specific cell types and during development.

20.3 Extreme Longevity of *S. cerevisiae*

20.3.1 Measurement of Yeast Lifespan

Studies of yeast aging have chiefly employed two experimental paradigms – replicative lifespan and chronological aging (Kaeberlein et al. 2001; Piper 2006). Replicative lifespan is defined as the number of times a yeast mother cell is able to produce a bud. The method involves microscopic examination of yeast cells on agar medium, microdissection of newly formed buds away from mother cells, with a tally maintained of the number of buds produced by each mother cell until there are no further cell divisions (D'Mello et al. 1994). This system is considered an appropriate model for aging of mitotically active cells. Chronological aging involves prolonged aerobic incubation of yeast cells in water or spent culture medium after they have achieved stationary phase growth in a minimal complete medium containing glucose as a carbon source (Fabrizio and Longo 2003). After initial inoculation into this medium, the cells grow exponentially by mostly fermentative metabolism, converting glucose to ethanol. When glucose is exhausted, the cells undergo a major reprogramming of gene expression and metabolism in order to utilize ethanol as a carbon and energy source. As such, the glycolytic pathway is downregulated and respiration and gluconeogenesis pathways are upregulated (Haurie et al. 2001). This phase is termed the *diauxic transition* or shift. After exhaustion of ethanol, the cells enter stationary phase and maintain viability for days.

The yeast chronological aging model has been likened to chronological aging of metazoan post-mitotic tissues because the yeast stationary-phase cells are non-dividing yet viable for a long period of time (Fabrizio and Longo 2003). This initial analogy of yeast stationary phase and/or starved cells to metazoan post-mitotic cells may have been simplistic, but the fact that key longevity pathways involved in metazoan longevity have been implicated as well in longevity of yeast chronologically aged cells (see Sections 20.3.2 and 20.6), validates the yeast model.

20.3.2 Ten-Fold Extension of Chronological Lifespan

Ablation of two genes, combined with extreme dietary restriction, led to a 10-fold extension of yeast lifespan (Wei et al. 2008). A *ras2*Δ*sch9*Δ double mutant exhibited 5-fold extension of chronological lifespan. When subjected to extreme dietary restriction (incubation in water, with no added nutrients), the lifespan was further doubled, for a final lifespan extension of ten-fold.

Both the *RAS2* and *SCH9* nutrient-sensing pathways inhibit another protein kinase, *RIM15*, which integrates signals from the *RAS2*, *SCH9* and TOR (target of rapamycin) nutrient sensing pathways in order to promote nuclear localization of transcription factors *MSN2*, *MSN4* and *GIS1* during conditions of nutrient scarcity. The latter transcription factors positively regulate expression of numerous longevity assurance proteins required for survival of environmental stresses.

Yeast don't encode a DAF-16/FOXO transcription factor, but the MSN2, MSN4 and GIS1 factors may be functional equivalents, as indicated in part by their positive regulation of stress-response genes, for example SOD (superoxide dismutase) genes (Samuelson et al. 2007; Longo 2003). Another parallel between DAF-16/FOXO in nematodes, and the MSN2/MSN4 and/or GIS1 functions in yeast, can be seen in their similar role in achieving or maintaining the non-dividing G_0 state of the cell cycle (Burgering and Medema 2003; Cameroni et al. 2004).

20.3.2.1 Role of *RAS2* in Yeast Longevity

The *RAS2* protein, along with the closely related *RAS1* protein, governs the choice between stress-response and cell-cycling pathways in accord with environmental status, thus enhancing replicative longevity in the presence of stress (Shama et al. 1998). *RAS1* expression favors replicative longevity under conditions of transient stress and limits lifespan otherwise (Jazwinski 1999).

This effect of *RAS1* may be due to its role in regulating inositol phospholipid turnover in response to glucose. The inositol phospholipid turnover products diacylglycerol and inositol triphosphate then act in a parallel manner to cAMP to signal the cell cycle to transition past the G_0/G_1 stage (Kaibuchi et al. 1986). *RAS2*, through regulation of adenyl cyclase activity, has a positive effect on replicative lifespan after sub-lethal stress (Jazwinski 1999). In contrast, adenyl cyclase inhibits chronological lifespan, due to activation of Protein Kinase A, which directly inhibits the RIM15 and the MSN2/MSN4 stress-response transcription factors (Longo 2003; Wei et al. 2008).

20.3.2.2 Role of *SCH*9 in Yeast Longevity

Deletion of *SCH*9 leads to a 3-fold extension in mean chronological lifespan, apparently through increased resistance to oxidative and thermal stress (Longo 2003). *SCH*9 is homologous to AKT kinases, which modulate longevity of *C. elegans* and Drosophila via the IGF-1 signaling pathway (Kenyon 2001). The *SCH*9 kinase is also the functional ortholog of the mammalian S6 Kinase 1, which phosphorylates ribosomal protein S6 in response to nutrient signaling through the target of rapamycin complex 1 (TORC1) (Urban et al. 2007), and it is this activity that appears to be relevant to longevity regulation in yeast (Urban et al. 2007; Wanke et al. 2005).

The hypothesis that TOR inhibition of protein translation is key to TOR regulation of longevity is supported by a recent study in yeast in which depletion of 60S ribosomal subunits by inactivation of various ribosomal proteins led to increased lifespan (Steffen et al. 2008). Similarly, in *C. elegans*, it was shown that mean and maximum lifespan are increased after inhibition of protein translation (Hansen et al. 2007).

20.3.2.3 Effect of Dietary Restriction on Yeast Longevity

A distinction has been noted between the effects and potential mechanisms of life extension conferred by moderate and severe dietary restriction (DR) in yeast (Bishop and Guarente 2007). In moderate dietary restriction, yeast are provided with a low level of glucose, while in severe restriction, yeast are incubated in the absence of glucose or other nutrients that could be used as carbon and energy sources.

The proposal of parallel pathways was, in part, an effort to explain the apparently contradictory roles of sirtuins (e.g., *SIR*2) in life extension by moderate vs. severe DR. *SIR*2 promotes longevity under moderate DR, but reduces it during severe restriction. Other differences in mechanisms of life extension are implied by the involvement of the SCH9 and TOR proteins only in severe DR, and the involvement of elevated mitochondrial respiration in life extension by moderate but not severe DR (see (Bishop and Guarente 2007) for additional details).

In the case of ten-fold extension of lifespan reported recently (Wei et al. 2008), severe DR conferred a further doubling of the 5-fold lifespan extension due to a $ras2\Delta sch9\Delta$ double mutant. The *RIM*15 gene product was required for some, but not all, of the severe DR effect, indicating the existence of one or more nutrient-sensing mediators in addition to *RIM*15. Candidates are the energy-sensing AMP-activated protein kinase, known to regulate longevity in several organisms (Ashrafi et al. 2000; Curtis et al. 2006; Greer et al. 2007a), *GCN*4, a transcription factor implicated in TOR regulation of protein translation (Steffen et al. 2008), and perhaps the LAT1 gene product, the E2 subunit of pyruvate dehydrogenase, whose role in the regulation of yeast lifespan is complex but may include setting the rate of mitochondrial respiration, regulating NAD+/NADH ratios, and/or signaling the status of carbon flux from glycolysis (Easlon et al. 2007).

20.4 Intersections Between Pathways, in Diverse Species, Allow Integrated Responses

Signal transduction pathways do not function in isolation, but cross-talk with other pathways (Gami et al. 2006; Berdichevsky et al. 2006; Kondo et al. 2005; Troemel et al. 2006; Matsumoto et al. 2006). Interactions are especially well documented between the IIS and the JNK and p38/MAPK stress- and cytokine-response pathways in *C. elegans* (Kondo et al. 2005; Troemel et al. 2006), invoking interactions with AMP-activated kinase (AMPK) and TOR complexes (TORC 1 and 2). In addition, insulin secretion is regulated via Wnt/β-catenin signaling (Rulifson et al. 2007), and AKTs can be activated by many other inputs, including DNA-dependent protein kinase (DNA-PK) (Ashrafi et al. 2000; Dragoi et al. 2005; Sester et al. 2006).

Multiple pathways converge through their common activation of AKTs, with effects mediated by many routes in addition to DAF-16/FOXO. It is known in mammals (and presumed to be true of other taxa) that activation of Thr phosphorylation in AKT-1 by PDK-1 requires a prior Ser phosphorylation, usually by TORC2 (target of rapamycin, complex 2) (Hawkins et al. 2006). AKTs directly *activate* at least 14 targets, including IRS-1 (providing feedback reinforcement of its own activation), RAF-1, eNOS, NFkB, and several cell-cycling and anti-apoptotic genes. They indirectly stimulate synthesis of both glycogen (via GSK-3) and proteins (via TSCs and TOR). AKTs exert *inhibitory* effects on at least 11 targets, including DAF-16/FOXO, androgen receptors (in mammals), and pro-apoptotic gene products.

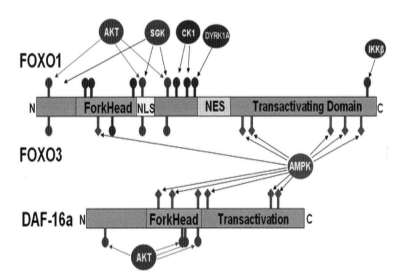

Fig. 20.3 Phosphorylation sites and the implicated kinases for each in FOXO1a, FOXO3, and DAF-16a. Key: *ForkHead*, winged-helix DNA-binding domain; *NLS*, nuclear localization signal; *NES*, nuclear export signal; *SGK*, serum/glucocorticoid stimulated kinase; *CK1*, casein kinase 1; *AMPK*, AMP-activated protein kinase. *DYRK1A*, dual-specificity Tyr-phosphorylated and regulated kinase 1A; *IKKβ*, IkB (NFkB inhibitor) kinase β. AKT/SGK sites have been defined in FOXO1, FOXO3, and DAF-16a (Lin et al. 2001). AMPK sites have been defined in FOXO3 and DAF-16 (Greer et al. 2007a, b)

Increasingly, DAF-16/FOXO is implicated as a convergence point for multiple pathways. Its nuclear entry, and hence transcriptional activity, depend on phosphorylation of at least 3 sites by AKT complex or SGK-1 (see Fig. 20.3), without which it is inactive. An additional 6 or more other sites on DAF-16 are phosphorylated by AMP-activated protein kinase (AMPK), activating a specific subset of transcriptional targets that include a number of signal-transduction and oxidative-stress-resistance genes (Greer and Brunet 2008; Greer et al. 2007a, b). Crosstalk between IIS and nutrient sensing thus takes place within DAF-16 (or FOXO in mammals).

20.4.1 Implications for Life Extension

Crosstalk among pathways may reflect in part the relatively small repertoire of cellular responses, elicited by a diverse spectrum of stimuli. *It allows an integrated response to signals that arrive in unpredictable patterns and combinations.* The interplay of signaling cascades may be analogous to a neural network or genetic algorithm for machine learning, in that the association between inputs and outputs cannot be expressed by an equation or a set of rules. We propose that the impressive gains in survival and stress resistance, conferred by a single mutated gene in *age-1(mg44)* F2 worms, arise from a novel orchestration of signaling pathways that invoke mechanisms beyond those known to be affected by IIS.

In *C. elegans*, to a far greater extent than in mammals, IIS has been represented as a linear pathway, augmented by crosstalk to Sir2, MAPK, TOR, and AMPK. In view of the potential for complex kinase interactions, and a central role of PIP_3 depletion, we argue that these interactions form an inherently nonlinear pathway. "Positive-feedback circuitry" via insulinlike peptides (Murphy et al. 2007), and also via transcriptional repression of multiple kinases by DAF-16/FOXO (Tazearslan et al. 2009) adds further complexity, and creates the genetic equivalent of a switch with two preferred states. In an extreme mutant, it even has the potential to lock the system into a single permitted state in which IIS and several other signaling pathways are chronically silenced and unresponsive (Tazearslan et al. 2009). Whereas the normal interactions between pathways are commonly referred to as "cross-talk", the interactions we observed in this very stress-resistant and long-lived mutant would be better described as "cross-silencing".

20.5 Regulating Metabolism and Lifespan: Tissue Specificity in Worms, Flies, and Mice

Worms. Neurons are pivotal to IIS control of lifespan in *C. elegans*. Hypomorphic mutations of *daf-2* (encoding the insulin/IGF-1 receptor homolog) or *age-1* (encoding PI3K) extend lifespan (Ayyadevara et al. 2008; Dorman et al. 1995; Friedman and Johnson 1988; Johnson 1990; Kenyon et al. 1993; Kimura et al. 1997). Expression of wild-type *age-1* or *daf-2* transgenes only in neurons (but not in muscle or gut) "rescues" the corresponding long-lived mutants, restoring normal longevity (Wolkow et al. 2000).

Nematode longevity is thus limited by neuronal metabolic activity or signaling. The key role of neurons in determining *C. elegans* lifespan is further implied by life-extension effects of mutations impairing sensory-neuron function (Apfeld and Kenyon 1999). Nerve cells and especially the intestine are major sources of INS-7, the only insulin-like peptide shown to restrict lifespan (Murphy et al. 2007). In contrast, *metabolic* effects of *age-1* and *daf-2* mutants, affecting lipid utilization, are rescued primarily by expression in muscle (Wolkow et al. 2000).

Flies. Cu/Zn-SOD transgenes also extend *D. melanogaster* lifespan when targeted to motor neurons, but not when targeted to other tissues or expressed globally (Parkes et al. 1998). This implies that oxidative damage to the nervous system limits longevity, and that countervailing effects apply in other tissues, to explain lack of protection when expression is global.

Mice. Mouse lifespan is extended by global attenuation of *Igf-1* receptor signaling without changes in body size, activity, and metabolism (Holzenberger et al. 2003), implying that Igf-1 signaling plays a similar role in worms and mammals. Mice heterozygous for *Irs2* knockout, whether the knockout is global or restricted to brain, live 17–18% longer than controls; only the brain knockout results in obesity and hyperactivity (Taguchi et al. 2007).

Clearly this life extension hinges on brain effects of IIS, but it is far from clear whether the signaling occurs through receptors for insulin or IGF-1, or hybrid receptors (Blakesley et al. 1996; Kulkarni 2002). *Irs-2* was assumed to respond only to insulin receptor (Taguchi et al. 2007), but Igf-1R also modifies Irs-1 and Irs-2 (Blakesley et al. 1996; Byron et al. 2006). This could be seen as either Igf-1 signaling via IRSs, which is well established (Blakesley et al. 1996; Kulkarni 2002), or crosstalk between these pathways comparable to IIS crosstalk with JNK, AMPK, and S6K through IRS phosphorylation (Hansen et al. 2007; Troemel et al. 2006; Wang et al. 2005). Deletion of PI3K p110γ fails to extend life, perhaps because it is normally expressed in only a few tissues, including macrophages, where its absence impairs innate immunity and inflammation (Chang et al. 2007).

20.6 What Molecular Features do Long-Lived Yeast and Nematodes Have in Common?

We have performed extensive comparisons of the very long-lived *age-1*-null F2 adult worms and weaker *age-1* mutants, or wild-type adults (Tazearslan et al. 2009; Ayyadevara et al. 2009), to ask what RNA transcripts and proteins differ between them. We found that the nematode *let-60* gene, encoding a small RAS membrane-associated GTPase, is transcriptionally downregulated in F2 mutant worms. Like the yeast RAS2 GTPase, *C. elegans* LET-60/RAS governs the ERK/MAPK signaling pathway, and all five tested genes lying downstream of *let-60* in that pathway showed nearly as much (3- to 5-fold) downregulation as *let-60* (5.5-fold) (Tazearslan et al. 2009). We also examined the nematode *SCH9* homologs, encoding AKT-1,

AKT-2 and S6 kinase. Although none showed diminished transcripts, strong transcriptional silencing was seen for 4 out of 5 tested components of the TOR pathway responsible for sensing nutrient availability, and which regulates the S6 kinase gene (Tazearslan et al. 2009).

There is thus circumstantial evidence that a combination of three perturbations in yeast (*ras2*Δ, *sch9*Δ, and extreme dietary restriction) might be able to "phenocopy" a strong *age-1* mutation in *C. elegans* to achieve an equivalent life extension by synthetically reproducing several of its key downstream effects. It is striking that both 10x long-lived mutants (yeast and worm) in some ways mimic the quiescent G_0 state of stem cells, by increasing expression of G_0 cell cycle-associated functions such as resistance to oxidative and other environmental stresses (Cameroni et al. 2004; Tothova et al. 2007; Yamazaki et al. 2006).

If this "combination of ingredients" indeed underlies extraordinary life extension in two model systems as distant as yeast and nematodes, it may prove (like the insulin/IGF-1 pathway) to coordinate a hyper-longevity state that is conserved even in mammals.

20.7 Prospects for Translation to Mammals

The use of animal models provides powerful exploratory tools for the discovery of new genetic pathways, or (as has been more often the case for longevity) unanticipated effects of altering previously known pathways and genes. The facile genetics of simpler animals enhances the potential for such discoveries, and the brevity of their lifespans has also facilitated advances in aging research. However, the very act of employing such models initially required a level of optimism, with regard to conservation of mechanisms, that has only quite recently been vindicated.

Despite its late arrival, vindication has been compelling for several shared pathways limiting or extending the duration of life. The *age-1* mutant alleles that multiply longevity tenfold in the nematode lie at the center of the insulin/IGF-1 signaling pathway, already known to be remarkably conserved, and to extend life when attenuated in taxa as distantly divergent as nematodes, insects and mammals. It thus seems very likely that whatever mechanisms underlie the far greater effects of extreme disruption of this pathway in the worm, will also have counterparts in mammals.

It will, however, require extensive research to reveal the details of those mechanisms (presumed here, but not yet known, to be multiple), and technical advances to permit the selective application of the same mechanisms (or their downstream effectors) in cell types and/or at times that avoid deleterious effects known to be lethal to dividing mammalian cells, although innocuous in postmitotic nematode adults. Mutations that confer both pro-survival and anti-proliferative effects would not be tolerated in yeast, unicellular organisms that require cell division for propagation. Thus, comparative studies of the downstream components of survival pathways, in yeast and multicellular animals, may be quite informative for translation to mammals because mutations with anti-proliferative consequences would have been weeded out by natural selection acting on yeast.

Extrapolation from worms (or yeast) to mammals is thus not far-fetched, but neither is it fully predictable through a process as simple as multiplication. That is, a two-fold or ten-fold increase in nematode lifespan may not translate into a proportional enhancement of human survival. Gains in longevity from caloric restriction, or gene mutations studied thus far, exert smaller effects in Drosophila than in nematodes, and smaller still in mammals.

This is not a surprising outcome, since mammals have already evolved the means to live many times the span of a worm's life. From an evolutionary perspective, worms would derive no obvious benefit from longer life, whereas mammals and their predecessors may have already been subjected to 100–200 million years of natural selection favoring traits such as greater size and complexity of development, parental protection of offspring, additional immune mechanisms, and increased learning capacity, that indirectly contributed to greater longevity. Thus, many of the gains we can attain by a single mutation in the simpler organism may already have been incorporated in the course of achieving our present longevities. In any event, applying a "life-extension factor" from one species to another assumes a linearity of response that is rarely found in biology.

Although we argue that ten-fold life extension, achieved in nematodes and yeast, may translate to a lesser effect in humans, evidence indicates that considerable potential exists for pharmacological intervention to increase the duration of healthy human life, far beyond what can be achieved through dietary restriction or IIS attenuation alone. This potential was first demonstrated in the nematode, and would have been difficult to glean from studies of mammals (or any metazoan organism in which adult survival requires ongoing cell division), because strong AKT signaling is essential for such proliferation. Maximal longevity may only be attainable by striking a balance between two conflicting cell states, proliferation and quiescence.

References

Apfeld J, Kenyon C (1999) Regulation of lifespan by sensory perception in *Caenorhabditis elegans*. Nature 402:804–809

Arantes-Oliveira N, Berman JR, Kenyon C (2003) Healthy animals with extreme longevity. Science 302:611

Ashrafi K, Lin SS, Manchester JK, Gordon JI (2000) Sip2p and its partner snf1p kinase affect aging in *S. cerevisiae*. Genes Dev 14:1872–1885

Ayyadevara S, Alla R, Thaden JJ, Shmookler Reis RJ (2008) Remarkable longevity and stress resistance of nematode PI3K-null mutants. Aging Cell 7:13–22

Ayyadevara S, Dandapat S, Singh S, Beneš H, Zimniak L, Shmookler Reis R, Zimniak P (2005) Lifespan extension in hypomorphic *daf-2* mutants of *Caenorhabditis elegans* is partially mediated by glutathione transferase CeGSTP2-2. Aging Cell 4:299–307

Ayyadevara S, Tazearslan C, Alla R, Bharill P, Siegel E, Shmookler Reis RJ (2009) C. elegans PI3K mutants reveal novel genes underlying exceptional lifespan and stress resistance. Aging Cell 8:706–725

Backer JM (2008) The regulation and function of Class III PI3Ks: novel roles for Vps34. Biochem J 410:1–17

Bartke A, Wright JC, Mattison JA, Ingram DK, Miller RA, Roth GS (2001) Extending the lifespan of long-lived mice. Nature 414:412

Berdichevsky A, Viswanathan M, Horvitz HR, Guarente L (2006) *C. elegans* SIR-2.1 interacts with 14-3-3 proteins to activate DAF-16 and extend life span. Cell 125:1165–1177

Bishop NA, Guarente L (2007) Genetic links between diet and lifespan: shared mechanisms from yeast to humans. Nat Rev Genet 8:835–844

Blakesley VA, Scrimgeour A, Esposito D, Le Roith D (1996) Signaling via the insulin-like growth factor-I receptor: does it differ from insulin receptor signaling? Cytokine Growth Factor Rev 7:153–159

Bluher M, Kahn BB, Kahn CR (2003) Extended longevity in mice lacking the insulin receptor in adipose tissue. Science 299:572–574

Bonkowski MS, Rocha JS, Masternak MM, Al Regaiey KA, Bartke A (2006) Targeted disruption of growth hormone receptor interferes with the beneficial actions of calorie restriction. Proc Natl Acad Sci USA 103:7901–7905

Burgering BM, Medema RH (2003) Decisions on life and death: FOXO Forkhead transcription factors are in command when PKB/Akt is off duty. J Leukoc Biol 73:689–701

Byron SA, Horwitz KB, Richer JK, Lange CA, Zhang X, Yee D (2006) Insulin receptor substrates mediate distinct biological responses to insulin-like growth factor receptor activation in breast cancer cells. Br J Cancer 95:1220–1228

Cameroni E, Hulo N, Roosen J, Winderickx J, De Virgilio C (2004) The novel yeast PAS kinase Rim 15 orchestrates G_0-associated antioxidant defense mechanisms. Cell Cycle 3:462–468

Chang JD et al. (2007) Deletion of the phosphoinositide 3-kinase p110gamma gene attenuates murine atherosclerosis. Proc Natl Acad Sci USA 104:8077–8082

Clancy DJ et al. (2001) Extension of life-span by loss of CHICO, a Drosophila insulin receptor substrate protein. Science 292:104–106

Curtis R, O'Connor G, DiStefano PS (2006) Aging networks in *Caenorhabditis elegans*: AMP-activated protein kinase (*aak-2*) links multiple aging and metabolism pathways. Aging Cell 5:119–126

Cypser JR, Johnson TE (1999) The *spe-10* mutant has longer life and increased stress resistance. Neurobiol Aging 20:503–512

D'Mello N P, Childress AM, Franklin DS, Kale SP, Pinswasdi C, Jazwinski SM (1994) Cloning and characterization of LAG1, a longevity-assurance gene in yeast. J Biol Chem 269:15451–15459

Dorman JB, Albinder B, Shroyer T, Kenyon C (1995) The *age-1* and *daf-2* genes function in a common pathway to control the lifespan of *Caenorhabditis elegans*. Genetics 141:1399–1406

Dragoi AM et al. (2005) DNA-PKcs, but not TLR9, is required for activation of Akt by CpG-DNA. Embo J 24:779–789

Easlon E et al. (2007) The dihydrolipoamide acetyltransferase is a novel metabolic longevity factor and is required for calorie restriction-mediated life span extension. J Biol Chem 282:6161–6171

Fabrizio P, Longo VD (2003) The chronological life span of *Saccharomyces cerevisiae*. Aging Cell 2:73–81

Friedman DB, Johnson TE (1988) Three mutants that extend both mean and maximum life span of the nematode, *Caenorhabditis elegans*, define the *age-1* gene. J Gerontol 43:B102–B109

Gami MS, Iser WB, Hanselman KB, Wolkow CA (2006) Activated AKT/PKB signaling in *C. elegans* uncouples temporally distinct outputs of DAF-2/insulin-like signaling. BMC Dev Biol 6:45

Gems D et al. (1998) Two pleiotropic classes of *daf-2* mutation affect larval arrest, adult behavior, reproduction and longevity in *Caenorhabditis elegans*. Genetics 150:129–155

Greer EL, Brunet A (2008) Signaling networks in aging. J Cell Sci 121:407–412

Greer EL et al. (2007a) An AMPK-FOXO pathway mediates longevity induced by a novel method of dietary restriction in *C. elegans*. Curr Biol 17:1646–1656

Greer EL et al. (2007b) The energy sensor AMP-activated protein kinase directly regulates the mammalian FOXO3 transcription factor. J Biol Chem 282:30107–30119

Guarente L, Kenyon C (2000) Genetic pathways that regulate ageing in model organisms. Nature 408:255–262

Hansen M, Hsu AL, Dillin A, Kenyon C (2005) New genes tied to endocrine, metabolic, and dietary regulation of lifespan from a *Caenorhabditis elegans* genomic RNAi screen. PLoS Genet 1:119–128

Hansen M, Taubert S, Crawford D, Libina N, Lee SJ, Kenyon C (2007) Lifespan extension by conditions that inhibit translation in *Caenorhabditis elegans*. Aging Cell 6:95–110

Haurie V, Perrot M, Mini T, Jeno P, Sagliocco F, Boucherie H (2001) The transcriptional activator Cat8p provides a major contribution to the reprogramming of carbon metabolism during the diauxic shift in *Saccharomyces cerevisiae*. J Biol Chem 276:76–85

Hawkins PT, Anderson KE, Davidson K, Stephens LR (2006) Signalling through Class I PI3Ks in mammalian cells. Biochem Soc Trans 34:647–662

Holzenberger M et al. (2003) IGF-1 receptor regulates lifespan and resistance to oxidative stress in mice. Nature 421:182–187

Houthoofd K, Braeckman BP, Johnson TE, Vanfleteren JR (2003) Life extension via dietary restriction is independent of the Ins/IGF-1 signalling pathway in *Caenorhabditis elegans*. Exp Gerontol 38:947–954

Jazwinski SM (1999) The RAS genes: a homeostatic device in *Saccharomyces cerevisiae* longevity. Neurobiol Aging 20:471–478

Johnson TE (1990) Increased life-span of *age-1* mutants in *Caenorhabditis elegans* and lower Gompertz rate of aging. Science 249:908–912

Kaeberlein M, McVey M, Guarente L (2001) Using yeast to discover the fountain of youth. Sci Aging Knowledge Environ 2001:pe1

Kaibuchi K, Miyajima A, Arai K, Matsumoto K (1986) Possible involvement of RAS-encoded proteins in glucose-induced inositolphospholipid turnover in *Saccharomyces cerevisiae*. Proc Natl Acad Sci USA 83:8172–8176

Kenyon C (2001) A conserved regulatory system for aging. Cell 105:165–168

Kenyon C, Chang J, Gensch E, Rudner A, Tabtiang R (1993) A *C. elegans* mutant that lives twice as long as wild type. Nature 366:461–464

Kimura KD, Tissenbaum HA, Liu Y, Ruvkun G (1997) *daf-2*, an insulin receptor-like gene that regulates longevity and diapause in *Caenorhabditis elegans*. Science 277:942–946

Klass MR (1983) A method for the isolation of longevity mutants in the nematode *Caenorhabditis elegans* and initial results. Mech Ageing Dev 22:279–286

Kondo M, Yanase S, Ishii T, Hartman PS, Matsumoto K, Ishii N (2005) The p38 signal transduction pathway participates in the oxidative stress-mediated translocation of DAF-16 to *Caenorhabditis elegans* nuclei. Mech Ageing Dev 126:642–647

Kulkarni RN (2002) Receptors for insulin and insulin-like growth factor-1 and insulin receptor substrate-1 mediate pathways that regulate islet function. Biochem Soc Trans 30:317–322

Kurosu H et al. (2005) Suppression of aging in mice by the hormone Klotho. Science 309: 1829–1833

Larsen PL, Albert PS, Riddle DL (1995) Genes that regulate both development and longevity in *Caenorhabditis elegans*. Genetics 139:1567–1583

Lin K, Dorman JB, Rodan A, Kenyon C (1997) *daf-16*: An HNF-3/forkhead family member that can function to double the life-span of *Caenorhabditis elegans*. Science 278:1319–1322

Lin K, Hsin H, Libina N, Kenyon C (2001) Regulation of the *Caenorhabditis elegans* longevity protein DAF-16 by insulin/IGF-1 and germline signaling. Nat Genet 28:139–145

Lin YJ, Seroude L, Benzer S (1998) Extended life-span and stress resistance in the Drosophila mutant methuselah. Science 282:943–946

Lithgow GJ, Walker GA. (2002) Stress resistance as a determinant of *C. elegans* lifespan. Mech Ageing Dev 123:765–71

Longo VD (2003) The Ras and Sch9 pathways regulate stress resistance and longevity. Exp Gerontol 38:807–811

Marden JH, Rogina B, Montooth KL, Helfand SL (2003) Conditional tradeoffs between aging and organismal performance of Indy long-lived mutant flies. Proc Natl Acad Sci USA 100: 3369–3373

Matsumoto M, Han S, Kitamura T, Accili D (2006) Dual role of transcription factor FoxO1 in controlling hepatic insulin sensitivity and lipid metabolism. J Clin Invest 116:2464–2472

Migliaccio E et al. (1999) The p66shc adaptor protein controls oxidative stress response and life span in mammals. Nature 402:309–313

Morris JZ, Tissenbaum HA, Ruvkun G (1996) A phosphatidylinositol-3-OH kinase family member regulating longevity and diapause in *Caenorhabditis elegans*. Nature 382:536–539

Murakami S (2006) Stress resistance in long-lived mouse models. Exp Gerontol 41:1014–1019

Murphy CT, Lee SJ, Kenyon C (2007) Tissue entrainment by feedback regulation of insulin gene expression in the endoderm of *Caenorhabditis elegans*. Proc Natl Acad Sci USA 104: 19046–19050

Ogg S, Ruvkun G (1998) The *C. elegans* PTEN homolog, DAF-18, acts in the insulin receptor-like metabolic signaling pathway. Mol Cell 2:887–893

Parkes TL, Elia AJ, Dickinson D, Hilliker AJ, Phillips JP, Boulianne GL (1998) Extension of Drosophila lifespan by overexpression of human SOD1 in motorneurons. Nat Genet 19: 171–174

Pettitt TR, Dove SK, Lubben A, Calaminus SD, Wakelam MJ (2006) Analysis of intact phosphoinositides in biological samples. J Lipid Res 47:1588–1596

Piper PW (2006) Long-lived yeast as a model for ageing research. Yeast 23:215–226

Remenyi A, Good MC, Lim WA (2006) Docking interactions in protein kinase and phosphatase networks. Curr Opin Struct Biol 16:676–685

Riddle DL, Swanson MM, Albert PS (1981) Interacting genes in nematode dauer larva formation. Nature 290:668–671

Roggo L et al. (2002) Membrane transport in *Caenorhabditis elegans*: an essential role for VPS34 at the nuclear membrane. Embo J 21:1673–1683

Rogina B, Reenan RA, Nilsen SP, Helfand SL (2000) Extended life-span conferred by cotransporter gene mutations in Drosophila. Science 290:2137–2140

Rulifson IC et al. (2007) Wnt signaling regulates pancreatic beta cell proliferation. Proc Natl Acad Sci USA 104:6247–6252

Samuelson AV, Carr CE, Ruvkun G. (2007) Gene activities that mediate increased life span of *C. elegans* insulin-like signaling mutants. Genes Dev 21:2976–2994

Sester DP et al. (2006) CpG DNA activates survival in murine macrophages through TLR9 and the phosphatidylinositol 3-kinase-Akt pathway. J Immunol 177:4473–4480

Shama S, Kirchman PA, Jiang JC, Jazwinski SM (1998) Role of RAS2 in recovery from chronic stress: effect on yeast life span. Exp Cell Res 245:368–378

Shmookler Reis RJ, Kang P, Ayyadevara S (2006) Quantitative trait loci define genes and pathways underlying genetic variation in longevity. Exp Gerontol 41:1046–1054

Steffen KK et al. (2008) Yeast life span extension by depletion of 60s ribosomal subunits is mediated by Gcn4. Cell 133:292–302

Taguchi A, Wartschow LM, White MF (2007) Brain IRS2 signaling coordinates life span and nutrient homeostasis. Science 317:369–372

Tatar M, Kopelman A, Epstein D, Tu MP, Yin CM, Garofalo RS (2001) A mutant Drosophila insulin receptor homolog that extends life-span and impairs neuroendocrine function. Science 292:107–110

Tazearslan C, Ayyadevara S, Bharill P, and Shmookler Reis RJ (2009) Positive feedback between transcriptional and kinase suppression in nematodes with extraordinary longevity and stress resistance. PLoS Genet 5:e1000452

Tirosh O, Schwartz B, Zusman I, Kossoy G, Yahav S, Miskin R (2004) Long-lived alpha MUPA transgenic mice exhibit increased mitochondrion-mediated apoptotic capacity. Ann N Y Acad Sci 1019:439–442

Tissenbaum HA, Ruvkun G (1998) An insulin-like signaling pathway affects both longevity and reproduction in *Caenorhabditis elegans*. Genetics 148:703–717

Tothova Z et al. (2007) FoxOs are critical mediators of hematopoietic stem cell resistance to physiologic oxidative stress. Cell 128:325–339

Troemel ER, Chu SW, Reinke V, Lee SS, Ausubel FM, Kim DH (2006) p38 MAPK regulates expression of immune response genes and contributes to longevity in *C. elegans*. PLoS Genet 2:e183

Urban J et al. (2007) Sch9 is a major target of TORC1 in *Saccharomyces cerevisiae*. Mol Cell 26:663–674

Vanhaesebroeck B et al. (2001) Synthesis and function of 3-phosphorylated inositol lipids. Annu Rev Biochem 70:535–602

Vieira OV et al. (2001) Distinct roles of class I and class III phosphatidylinositol 3-kinases in phagosome formation and maturation. J Cell Biol 155:19–25

Wang MC, Bohmann D, Jasper H (2005) JNK extends life span and limits growth by antagonizing cellular and organism-wide responses to insulin signaling. Cell 121:115–125

Wanke V et al. (2008) Caffeine extends yeast lifespan by targeting TORC1. Mol Microbiol 69: 277–285

Wei M et al. (2008) Life span extension by calorie restriction depends on Rim15 and transcription factors downstream of Ras/PKA, Tor, and Sch9. PLoS Genet 4:e13

Wolkow CA, Kimura KD, Lee MS, Ruvkun G (2000) Regulation of *C. elegans* life-span by insulinlike signaling in the nervous system. Science 290:147–150

Yamazaki S et al. (2006) Cytokine signals modulated via lipid rafts mimic niche signals and induce hibernation in hematopoietic stem cells. Embo J 25:3515–3523

Chapter 21
Reversing Age-Related DNA Damage Through Engineered DNA Repair

Clifford J. Steer and Betsy T. Kren

Contents

21.1 Introduction 641
    21.1.1 Overview of Gene Repair as a Concept 643
21.2 Homologous Pairing and DNA Strand Exchange 644
    21.2.1 In Eukaryotic Cells 644
    21.2.2 In Mammalian Cells 645
21.3 Ribozymes, Antisense, DNA Ribonucleases, and RNA Interference 646
    21.3.1 Ribozymes 646
    21.3.2 Antisense Oligonucleotides 647
    21.3.3 DNA Ribonucleases 648
    21.3.4 RNA Interference 649
21.4 Single Nucleotide Modification 650
    21.4.1 Chimeraplasty 650
    21.4.2 Single-Stranded Oligonucleotide Repair 653
    21.4.3 Studies in Animal Models 655
21.5 Triplex DNA 657
    21.5.1 Mechanisms and Pathways 657
    21.5.2 Potential Applications 659
21.6 What is in Our Future? 660
References 661

21.1 Introduction

Recent evidence has supported the possibility that mutations in nuclear DNA may account for some of the changes that have been observed in gene expression with aging (Lu et al. 2004; Niederhofer et al. 2006). However, advances in our

---

C.J. Steer (✉)
Department of Medicine, University of Minnesota Medical School, VFW Cancer Research Center, 406 Harvard Street SE, Minneapolis, MN, USA
e-mail: steer001@umn.edu

understanding of age- and disease-related alterations in gene expression, as well as endogenous pathways underlying these changes, have also prompted the development of countervailing new gene technologies for repairing mutated or damaged DNA. Important goals of gene therapy are to replace missing genes, correct over- or under-expression of normal genes, and to repair damaged or dysregulated genes (Richardson et al. 2002). In that vein, a number of studies have attempted to elucidate the endogenous mechanisms involved in DNA damage and repair. This work has resulted in the development of a number of exogenous approaches for leveraging and co-opting endogenous repair mechanisms to promote cell repair (Table 21.1). This has been done via the development, delivery, and integration of specific and selective modified DNA into precise locations in order to repair gene dysregulation and mutagen- and/or invader-induced cellular damage. In this chapter, we review these techniques and consider their pertinence to correcting age-related DNA damage.

Table 21.1 Methods for targeting permanent genetic change

Objective	Technique	Proposed mechanism(s)	Type of mutation(s) corrected
Targeted gene replacement	Zinc finger nucleases	Homologous pairing; DNA recombination	Replacement up to 7.7 kb of homologous genomic DNA
Single nucleotide modification	Chimeraplasty	Homologous pairing; DNA mismatch repair	Point mutations, deletions and/or insertions
	Single-stranded oligonucleotides	Homologous pairing; strand transfer and incorporation	Point mutations, deletions and/or insertions
Targeted repair or homologous recombination	Triplex DNA	Base specific interaction between bases in the third strand and purines of the target DNA duplex; nucleotide excision repair; transcription coupled repair; DNA recombination	Point mutations, small insertions or deletions

21.1.1 Overview of Gene Repair as a Concept

Gene repair has several advantages over other approaches (i.e., use of viral vectors or transgene expression) in terms of potential therapeutic application. First, it allows for the correction of single-point gene mutations without impacting functional regions of the gene. Notably, this approach allows for in situ, permanent, non-immunogenic, and heritable correction of the relevant aging- or disease-related point mutation, and allows for expression of the corrected gene under normal physiologic conditions. This approach can be applied to both dominant and recessive gene dysregulation, and does not require integration of a transgene. Importantly, use of the gene repair strategy is appealing from practical perspective in that it is inexpensive, and the targeting molecules are relatively easy to produce (Yin et al. 2004).

Table 21.2 Genetic modulation of gene expression

Technique	Intracellular site	Level of regulation	Molecular processes
Triple helix formation	Nuclear	Transcriptional	Inhibition of initiation, inhibition of transcript elongation
Chimeraplasty	Nuclear	Post-transcriptional	Modulation of pre mRNA splicing for exon skipping
Antisense forming oligonucleotides	Nuclear	Post-transcriptional	RNase H dependent transcript turnover with DNA antisense oligonucleotide
	Nuclear	Post-transcriptional	Modulation of pre mRNA splicing for exon skipping and/or inclusion
	Cytoplasm	Translational	Inhibition of translation initiation or protein elongation
Ribozymes	Cytoplasm	Post-transcriptional	Ribozyme mediated transcript cleavage
RNAi	Cytoplasm	Post-transcriptional	siRNA targeted RISC mediated mRNA cleavage

Nevertheless, it is noteworthy that gene repair as a potential therapeutic approach is currently limited by low levels of correction and by considerable variations in success.

It is well accepted that the extent and nature of modification in gene expression over the lifespan and/or as a result of disease is selective and specific, and thus requires targeted and specific repair approaches (Table 21.2). Several methods for exogenous gene repair have been studied; each has as a critical feature the delivery of nucleic acids, the maintenance of endogenous regulation of repaired genes, and reduction of nonspecific integration (Richardson et al. 2002). Although these various strategies for DNA repair may have some as yet unidentified common pathways, each is likely to have unique rate-limiting factors and pathways that will need to be characterized before they can be exploited for clinical use. This chapter will provide a review of the state of the science regarding exogenous strategies for DNA repair including approaches for gene modification and delivery. Its application to the aging process is cutting edge technology.

21.2 Homologous Pairing and DNA Strand Exchange

21.2.1 In Eukaryotic Cells

Biochemical analyses of purified RecA protein from *Escherichia coli* and of bacteriophage and yeast have elucidated a mechanism for homologous pairing and DNA strand exchange. Analyses conducted in eukaryotes ultimately led to the development of strategies for achieving similar ends in higher organisms, but not without the need to overcome the genetic contradictions and sparse biochemical data regarding appropriate specific targets for investigation.

The unexpected lack of illegitimate integration in yeast during plasmid transformations raises uncertainty regarding the ubiquity of this recombination process. In contrast, other data showing structural homologues of a number of yeast recombination proteins have been shown in higher eukaryotes (Aravind et al. 1999), supporting the notion that similar basic biological mechanisms are involved in this process at all levels of eukaryotic evolution. If this is indeed the case, key factors involved in successful gene targeting in lower eukaryotes may in fact have broad application. As a result of these and other advancements, it is now possible to successfully transfer a cloned, modified gene into the genome of the host organism (Thomas et al. 1986; Capecchi 1989); ideally these genes would be integrated only in their homologous location, thereby prompting specific and targeted chromosomal disjunction during meiosis (Thomas et al. 1986; Capecchi 1989).

Specific and targeted duplication of DNA is required for survival, and in this light it is not surprising that evolution of eukaryotic species has produced elaborate and redundant pathways by which the impact of mutagens and genetic invaders can be reduced. Despite this fact and a large number of studies designed to further elucidate DNA repair pathways in higher eukaryotes, homologous recombination as a means to provide targeted modification of genomic DNA has been relatively elusive (Yañez and Porter 1998). Before this process can be effectively applied to in vivo gene therapy, a number of barriers need to be overcome, including inefficient integration of exogenous DNA into the genome, integration of genomic DNA in the absence of sequence homology between the gene administered and the site of integration (Roth and Wilson 1985), and low efficiency of pairing between the introduced DNA and its target owing to sequestration of chromatin (Rubnitz and Subramani 1984). In contrast, more success has been realized for in vitro gene targeting (Cohen-Tannoudji and Babinet 1998; Müller 1999).

21.2.2 In Mammalian Cells

The growing body of literature elucidating the mechanisms of homologous recombination in mammalian cells has made the ability to selectively generate and integrate targeted DNA increasingly possible. For example, the results of a study in patients with cystic fibrosis showed a 1% frequency for successful use of a 488-nucleotide DNA fragment as a targeted replacement for the transmembrane conduction regulator associated with the 3-bp gene deletion (Goncz et al. 1998). Recently, in vivo gene repair using rAAV vectors as a source of single-stranded DNA for β-glucuronidase (Miller et al. 2006) and fumarylacetoacetate hydrolase (FAH) deficiency (Paulk et al. 2008) was achieved in livers of transgenic mice. Unfortunately, the success of selective integration occurs at the expense of other DNA repair proteins or pathways, which must be inactivated (Yañez and Porter 1998). Moreover, homologous recombination is determined by cell cycle, and is reduced outside of the S phase. Consequently, this approach cannot be applied to quiescent cell types (Wong and Capecchi 1987; Thyagarajan et al. 1996; Yamamoto et al. 1996).

A relatively new strategy for targeted gene replacement based on the use of zinc finger nucleases (ZFNs) resulted in improved efficiency of gene targeting. The technology is based on introducing DNA double-strand breaks in target genes, which then stimulate the cell's endogenous homologous recombination machinery (Porteus and Carroll 2005). Interestingly, the ZFNs can be engineered to specifically target unique DNA sequences providing the ability to modify almost any region of the genome (Porteus 2008). In fact, using engineered ZFNs, gene replacements as large 7.7 kb have been efficiently introduced into the host genome of mammalian cells (Moehle et al. 2007). In addition, ZFNs have been used for selective targeting and removal of mutated mitochondrial DNA in cultured cells (Minczuk et al. 2008). Thus, ZFN technology is powerful for its potential use in modulating genetic changes associated with aging in both nuclear and mitochondrial genomic DNA (Brunk and Terman 2002).

21.3 Ribozymes, Antisense, DNA Ribonucleases, and RNA Interference

21.3.1 Ribozymes

21.3.1.1 Mechanisms and Pathways

Ribozymes, RNA enzymes that catalyze endoribonucleic cleavage by binding to specific RNA substrates, are valuable tools for basic and translational research. This cleavage is involved in the processing of certain introns to form the mature RNA, and occurs under normal physiologic conditions.

Endoribonucleic cleavage, on the other hand, can also be developed as a *trans*-acting event. Under normal conditions, ribozymes undergo hybridization to complementary RNA sequences. During this process, the central portion of the RNA evolves into a secondary structure characterized by reactive groups in close proximity that direct the cleavage of target RNA. These moieties are functionally distinct from those that form base pairs with substrate RNA, allowing for circumscribed, substrate-specific ribozyme alteration that induces catalytic, trans-cleavage of specified sites within the target RNA. This approach has been used to target viral RNAs, oncogenes, and somatic mutations (Kruger et al. 1982; Symons 1992; Welch et al. 1998).

Ribozymes remain catalytically active for weeks in an intact organ following somatic gene transfer, and can be expressed in cells or synthesized and packaged for uptake. In these respects ribozymes are advantageous relative to antisense methods. Another significant advantage of the use of ribozymes relates to the relative lack of constraints associated with cleavage. Hairpin ribozymes, for example, only require the presence of a guanosine (G) residue immediately 3′ to the cleavage site; a GUC sequence is optimal. Even less constrained is the hammerhead ribozyme, which requires only a UN dinucleotide for cleavage, where N is either A, C, or U. The resulting RNA fragments are rapidly degraded, rendering the molecule nonfunctional.

21.3.1.2 Hepatitis Viral Gene Expression

Ribozymes have been designed for a number of purposes, including the cleavage of RNAs associated with human hepatitis viruses, although the absence of adequate tissue culture paradigms has limited advancement of this line of research. Nevertheless, available research has demonstrated the successful inhibition of viral production of both hepatitis B and hepatitis C by both hammerhead (Lieber et al. 1996; Sakamoto et al. 1996) and hairpin (Welch et al. 1996) ribozymes. When studied in primary human hepatocytes from patients with chronic hepatitis C infection or in cultured cells, the individually-expressed hammerhead ribozymes, as well as their combined expression, effectively reduced or eliminated the respective plus or minus strand RNAs (Lieber et al. 1996). In another study, however, cleavage-deficient ribozymes with a point mutation in the hammerhead domain had no significant effect (Sakamoto et al. 1996).

Another approach is to target a number of different highly-conserved hepatitis C viral RNA sequences in tandem with multiple ribozyme genes from a single vector (Welch et al. 1998). In theory, this type of approach could produce a constant and continuous supply of multiple intracellular ribozymes, which would reduce the likelihood of drug-resistant viral variants. In contrast to hepatitis C, hepatitis B is a partially double-stranded DNA virus that replicates through a pregenomic RNA intermediate. This process raises the possibility of a potential therapeutic target for novel antiviral gene therapy using ribozymal RNA cleavage. Indeed, hepatitis B viral replication can be disrupted by targeted delivery of hairpin ribozymes to pre-genomic RNA intermediates (Welch et al. 1997).

21.3.2 Antisense Oligonucleotides

21.3.2.1 Hepatitis Viral Gene Expression

Antisense oligonucleotides (ASOs) have been used to produce specific inhibition of hepatic C virus-directed cellular protein synthesis, when bound to asialoglycoprotein-polylysine complexes and targeted to HuH-7 (human hepatoma) cells by receptor-mediated endocytosis (Wu and Wu 1998). A similar approach was used to target antisense oligonucleotides to hepatitis B-infected cells via the asialoglycoprotein receptor, and to induce specific inhibition of viral protein synthesis and replication in vitro (Nakazono et al. 1996). A 21-mer phosphorothioate-linked oligonucleotide DNA complementary to the hepatitis B virus polyadenylation signal and 5′-upstream sequences, when coupled to a targetable DNA carrier of asialoglycoprotein/polylysine complex, showed significantly blocked viral gene expression and replication in HuH-7 cells transfected with B virus DNA. Importantly, an ASO is now marketed for the treatment of intraocular human immunodeficiency virus, suggesting its potential application in treating retinal degeneration or other age associated eye diseases (Yokota et al. 2009).

21.3.2.2 mRNA Splicing Modification

ASOs are also extensively used for modifying the splicing of pre-mRNAs to promote exon skipping (removal of specific exons from the final mRNA) (Yokota et al. 2009) and/or exon inclusion (mediating the inclusion of an exon normally spliced out) (Hua et al. 2007). The utility of exon-skipping was initially demonstrated in cultured muscle cells from the Duchenne muscular dystrophy (DMD) mouse (*mdx*) model (Dunckley et al. 1998). DMD is caused by mutations of the 79-exon dystrophin gene, usually associated with a frame shift and disruption of the codon reading frame of the mRNA and altered dystrophin protein production. In contrast, Becker muscular dystrophy (BMD) is a milder and varied disease in which the dystrophin gene mutation(s) usually preserves the translational open reading frame; thus the spliced mature BMD mRNA retains some ability to synthesize the dystrophin protein. The exclusion of frame-shifting exons by ASOs partially restores the dystrophin reading frame, changing a non-functional DMD to a Becker in-frame

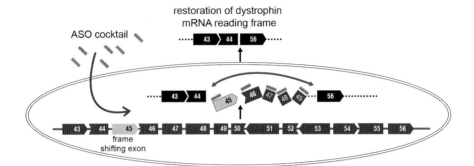

Fig. 21.1 Schematic of dystrophin gene exon skipping to correct a frame shift mutation. A portion of the 79 exon dystrophin gene pre-mRNA is shown depicting the removal of exons 45 through 56 via the ASO cocktail binding to the intervening exons splice sites and allowing them to be "skipped." The newly generated dystrophin mRNA isoform is exported into the cytosol carrying a corrected reading frame capable of producing a modified dystrophin protein in mature muscle fibers. Exon skipping results then in a more mild BMD phenotype versus the original severe (no dystrophin) DMD phenotype

transcript. Fortunately, most mutations in the dystrophin gene occur in regions that do not code for functionally essential regions of the protein. In fact, a single multi-exon skipping (exons 45 through 55) transcript of the dystrophin gene could rescue up to 63% of patients with DMD (Beroud et al. 2007) (Fig. 21.1). Impressive pre-clinical data in the mdx mouse for ASO mediated exon skipping has led to the first human clinical trials in DMD, with very promising results (van Ommen et al. 2008).

In addition to DMD, ASO mediated exon skipping has been used to correct IL-12Rα1 deficiency in T cells by removal of the mutant exon 2 during pre-mRNA splicing (van de Vosse et al. 2009). Moreover, this strategy for ASO modulation of pre-mRNA splicing has also been used for repairing other mutations by promoting exon inclusion (Hua et al. 2007), correcting aberrantly spliced mRNAs in ocular albinism (Vetrini et al. 2006) or generating novel isoforms of proteins such as ApoB in liver to alter lipoprotein metabolism (Khoo et al. 2007). Thus, ASO mediated modification of pre-mRNA splicing is functional in many different tissue types, and can mediate either inclusion or skipping of exons. Its meteoric path in the treatment of DMD from proof-of-principle to clinical trail in only 10 years suggests it may be a major approach for genetic engineering to ameliorate the aging process.

21.3.3 DNA Ribonucleases

21.3.3.1 Mechanism and Pathways

Like ribozymes, DNA ribonucleases are catalytic molecules; they are comprised of synthetic single-stranded DNA that causes specific cleavage of substrate RNA but with higher catalytic efficiency (Oketani et al. 1999). DNA ribonucleases can be prepared and delivered easily and have a higher threshold for chemical and enzymatic degradation. As is the case for ribozymes, DNA ribonucleases have three domains, including a catalytic domain of 15 nucleotides flanked by two substrate-recognition domains that bind target RNA through Watson-Crick base pairing.

21.3.3.2 Hepatitis Viral Gene Expression

When applied in the context of the hepatitis C viral genome, DNA ribonucleases can specifically cleave the targeted RNA. In a study of the inhibitory effects of DNA ribonucleases on the hepatitis C viral genome, those with point mutations in the catalytic domain had substantially lower, but not absent, inhibitory efficacy, suggesting some antisense contribution. DNA ribonucleases can be made to specifically cleave target hepatitis B viral RNA and substantially inhibit intracellular viral gene expression (Asahina et al. 1998).

21.3.4 RNA Interference

The most important recent discovery for genetic engineering applications is RNA interference (RNAi) (Martin and Caplen 2007), which is mediated by a novel class of RNAs (consisting of microRNAs and small interfering RNAs) that modulate gene expression rather than encoding proteins. These very small non-coding (nc) RNAs ~ 22 nt in length were first identified in plants, then nematodes and eventually humans. Depending on the type, they are derived from either perfect or imperfect complementary, stem-loop RNA precursors (Moulton 2005).

An improved understanding of RNAi has provided powerful tools for gene therapy for many acquired diseases, such as hepatocellular carcinoma, hepatitis B and C, and liver fibrosis (Arbuthnot et al. 2007; Arbuthnot and Thompson 2008; Chen et al. 2008; Cheng and Mahato 2007; Volarevic et al. 2007). The delivery of interfering RNA via hydrodynamic push to the liver (Lewis and Wolff 2007), has demonstrated a utility for modulation of hepatic gene expression in vivo. The rapid movement of the field suggests that RNAi based gene therapy will become a mainstay in the future for treatment of aging disorders resulting from deleterious protein expression.

21.3.4.1 microRNAs

In contrast to the nc tRNA genes, microRNA (miRNA) genes are transcribed by RNA polymerase II and are capped and polyadenylated similar to mRNAs (Tuschl 2003). These stem-loop configuration "primary miRNA" (pri-miRNA) transcripts are processed by the nuclear Drosha RNases, digesting the pri-miRNA into the small ~ 70 to 90 nt hairpin "precursor miRNA" (pre-miRNA) (Zeng 2006). The pre-miRNAs are translocated from the nucleus to the cytoplasm via the exportin-5 pathway. In the cytosol, the pre-miRNA are subjected to cleavage by the

RNase Dicer-1 to ~ 22 nt fragments. Mature miRNAs are paired with RISC (RNA Induced Silencing Complex), which aids in binding miRNA to the target mRNA, thereby acting as an active repressor of expression.

mRNAs that have been prevented from being translated by miRNAs are translocated to specialized processing (P) bodies located in the cell cytoplasm (Chu and Rana 2006). It should be noted, however, that miRNA mediated translational repression occurs independently of the movement of the miRNA-RISC:mRNA complex to the P-bodies (Eulalio et al. 2007). The mRNA has two fates within the P-body, which in part acts as a reservoir. It can either enter the mRNA decay pathway, or be returned to the cytosol to re-engage the translational machinery (Bhattacharyya et al. 2006, Parker and Sheth 2007, Sheth and Parker 2006).

21.3.4.2 Small Interfering RNAs

Small interfering RNAs (siRNAs) are processed much like miRNAs, in that they both require Dicer RNases for cleavage and shortening. In addition, siRNAs also require RISC to become active gene regulators, utilizing it to establish precise base pair targeting to mRNAs.

The fate of target mRNA is primarily dependent on the extent of base pairing to the ncRNA. Perfect target complementarity of the siRNA to the target mRNA will typically result in transcript turnover via cleavage. The presence of multiple, partially complementary target sites filled with miRNAs will inhibit translation without significantly affecting mRNA levels (Valencia-Sanchez et al. 2006).

The mechanism of siRNA-triggered mRNA degradation is not entirely known. However, it appears to be mediated in part by "Slicer" activity that is expressed by the human Argonaut (Ago2) protein. The cleavage mediated by Slicer activity is believed to mark the target mRNA for further degradation via deadenylation, decapping and $5'$-$3'$ exonuclease degradation in the cytosol.

21.4 Single Nucleotide Modification

21.4.1 Chimeraplasty

Another, more novel and controversial approach to gene therapy, known as chimeraplasty, leverages endogenous cellular repair mechanisms to correct mutations in single base pairs (Yoon 1999; Kmiec et al. 1999). This approach is predicated on the ability to correct genomic DNA point mutations using targeted chimeric RNA/DNA oligonucleotides, and has the advantage of generating targeted, site-specific and permanent correction, thereby maintaining endogenous gene regulation. The basis for chimeraplasty as an approach to gene alteration comes from studies showing greater efficiency of pairing of oligonucleotides ~50 bases in length and a genomic DNA target but only under limited circumstances, namely in the context of RNA replacement of DNA in a portion of the targeting oligonucleotides.

The development of chimeraplasty as a gene alteration approach arose from studies that elucidated the molecular aspects of DNA repair. The studies demonstrated that the efficiency of pairing increased, but only with RNA replacement of DNA in a portion of the targeting oligonucleotides (Kotani and Kmiec 1994a, b). In its initial iteration, the design used two single-stranded ends and consisted of the double-stranded region of the chimeric molecule flanked by unpaired nucleotide hairpin caps. Subsequent iterations of the hybrid chimeraplast showed improved stability and localization to mammalian cell genomic targets (Yoon et al. 1996).

21.4.1.1 Mechanism and Pathways

The juxtaposition and sequestration of the 5′ and 3′ ends, and the 2′-*O*-methylation of the RNA residues in part underlie the enhanced nuclease resistance of the chimeraplast, whereas oligonucleotide length determines the extent of homology of the chimeraplast to its intended genomic target.

One typical engineered mismatch, involving a 68-mer, includes a 25 base-pair region of homology with only one mismatch to the targeted gene sequence. The 68-mer initiates targeted genomic modification based on a chimeraplast template for DNA alteration (Fig. 21.2). It has been posited that the engineered mismatch between a chimeraplast and its target DNA signals an apparent base mutation, which activates endogenous enzymatic repair processes for natural and artificial DNA mutagens (Kmiec et al. 1999) and can be harnessed for the targeted modification of genes. The sequestered 5′ and 3′ ends minimize end-to-end ligation while the RNA

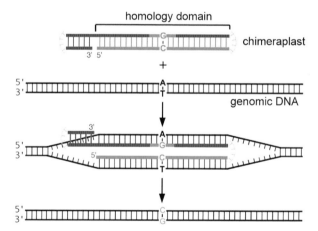

Fig. 21.2 Targeted correction of a single base pair mutation by chimeraplast. The schematic shows the homologous pairing of the target genomic DNA sequence with the homology domain of the chimeraplast, except for the engineered single base pair mismatch targeted for change. The alteration of the targeted nucleotides involves DNA mismatch repair pathways and is distinct from homologous recombination. The base changes can occur in a single strand or in both strands of a target locus as both DNA regions of the chimeraplast can be used as a template for repair. Black, RNA portions of the chimeraplast; gray, DNA portions of the chimeraplast; thin gray Ts, the unpaired thymines forming the hairpin loops of the chimeraplast structure

segments, the region of homology, and the "nick" are all essential for chimeraplast activity. The secondary structure and modified RNA confer the physical and enzymatic stability required for the survival of the chimeraplast en route to and within the cell.

Because the mechanisms underlying homologous recombination and DNA repair are highly conserved throughout evolution (Aravind et al. 1999), both bacterial and mammalian systems have been studied in order to elucidate important aspects of chimeraplasty and structural design. Plasmid-based selectable systems and colorimetric identification have been used to demonstrate chimeraplast-mediated gene conversion. The plasmid selectable system has been developed using neomycin phosphotransferase (*neo*) genes which confer either kanamycin (kanR) or tetracycline (tetR) resistance (Kren et al. 1999a; Cole-Strauss et al. 1999). The colorimetric assay utilizes the β-galactosidase enzymatic activity of the *lacZ* gene to cleave the synthetic substrate, X-gal, and produce the characteristic blue color (Igoucheva et al. 1999).

These bacterial systems have been used to identify key characteristics needed for improved targeting and nucleotide conversion (Kren et al. 1999a). An important factor appears to be homology length, which is correlated with gene repair activity. In one study, a chimeraplast with a 35 nucleotide homology region was 10 fold more active than one containing a 25 nucleotide region, which was ~40 fold more active than 15 nucleotides of homology (Andersen et al. 2002).

The chimeraplast is designed to be complementary to the Watson and Crick strands of the target DNA, and this raises the possibility that two distinct pairing events could occur, one involving an RNA/DNA hybrid strand that increases structural stability and pairing with target DNA, and the other a DNA/DNA duplex, each with a mismatched base pair. Research with 68-mer chimeraplasts containing only one mismatched strand with the *neo* gene revealed that the oligonucleotide with the mismatch on the "all-DNA" strand resulted in more efficient repair than the one with the mismatch on the "RNA/DNA" strand, or even the original "double mismatched" chimeraplast (Chen et al. 2001). These results were confirmed in a 2008 study published by Engstrom and Kmiec in which the investigators dissected the chimera to its functional parts. The results revealed that the gene repair reaction was initiated and regulated by the DNA strand of the oligonucleotide, and that the RNA strand had little if any functional role.

Based on these results, single-stranded DNA oligonucleotides (SSOs) varying in the number of phosphorothioate (PTO) linkages (3 or 6) were developed and evaluated (Yin et al. 2004, 2005). These modified oligonucleotides generated conversions 3- to 4-times more often than the chimera. Importantly, these modified oligonucleotide vectors robustly repaired mutant nucleotides.

21.4.1.2 Factors Affecting Gene Repair Activity

The various endogenous DNA repair pathways that contribute to chimeraplasty were investigated using *E. coli* strains deficient in specific repair proteins (Kren et al. 1999a). The results show that chimeraplasty resulted in either significantly reduced or undetectable gene modification in strains containing defects in the DNA pairing protein, RecA, or in the mismatch repair binding protein MutS. In contrast, other DNA and RNA modification enzymes involved in base excision repair, including adenine (*dam*) and cytosine (*dcm*) methylases, dUTPase (*dut*), and uracil N-glycosylase (*ung*) were dispensable. The fact that both RecA and MutS proteins appear to be required suggests a process involving both homologous recombination and mismatch repair processes. It has been theorized that the RecA-dependent pairing occurs first, followed by the mismatch and repair process. In addition, it appears that the RNA portion of the oligonucleotides serves an essential function in strand pairing based on the observation that control oligonucleotides comprised entirely of DNA with the same sequence and structure showed no activity. When the RNA/DNA strand of the chimeraplast was replaced with a strand comprised only of RNA, target conversion activity was improved, suggesting that the chimeric RNA/DNA duplex region is important for efficient pairing (Gamper et al. 2000a).

Both recombination and mismatch repair depend on evolutionarily conserved pathways, suggesting an application of chimeraplasty to mammalian systems (Aravind et al. 1999; Eisen 1998; Marra and Schar 1999). In support of this notion, HuH-7 cell-free extracts have been shown to support both dose-dependent conversion of the mutant *neo* gene and insertion of a deleted base pair into the tetracycline gene in vitro (Cole-Strauss et al. 1999). Conversely, when preparations either lacking the mismatch repair protein or containing an antibody to the protein were studied, little chimeraplast repair activity was detected.

Using the *lacZ* plasmid system, cell-free nuclear extracts from a variety of different cell lines efficiently catalyzed dose-dependent nucleotide conversion (Igoucheva et al. 1999). Furthermore, studies using the nuclear extract from a homozygous isogenic $p53^+/p53^+$ embryonic fibroblast cell line showed substantially reduced conversion relative to a p53 null extract, suggesting that wild-type p53 may inhibit the initial pairing step in chimeraplast repair. Taken together, data indicated that this mechanism of nucleotide conversion occurs in both prokaryotes and eukaryotes and is distinct from homologous recombination in its requirement for MutS or hMSH2 protein.

21.4.2 Single-Stranded Oligonucleotide Repair

In contrast to chimeraplasty, SSO-mediated repair is independent of the mismatch proteins MSH2 and MSH3 (Gamper et al. 2000a, b). In fact, the process appears to proceed by a mechanism of strand incorporation (Radecke et al. 2006); does not involve the ATM/ATR homologous recombination pathway used for double-stranded break repair (Wang et al. 2006); exhibits some strand bias in favor of targeting the non-transcribed strand (Liu et al. 2002); and the overall rate of targeted nucleotide replacement is enhanced > 10-fold by transcriptional activation of the targeted locus (Huen et al. 2007) (Fig. 21.3).

Studies have also investigated the impact of cell phase on single-strand oligonucleotide-induced gene repair. In one recent study, the mid- to late-S stage

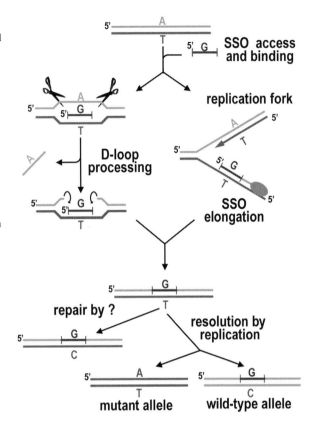

Fig. 21.3 Site-directed nucleotide exchange mediated by SSOs. This schematic shows strand transfer mediated repair of a point mutation mediated by SSOs. In contrast to chimeraplasty, the SSO-mediated repair process occurs via strand transfer and incorporation of the SSO carrying the desired base change, followed by alteration of the complementary genomic strand via endogenous DNA repair pathways or replication

was associated the most facile gene repair, which was further enhanced by the enrichment of the number of cells in these phases, regardless of the efficiency of transfection or the rate of replication (Engstrom and Kmiec 2008). An additional, but preliminary analysis of gene expression indicated the levels of cyclin G2, cyclin H, CDK12A and CDK12B were significantly elevated in cells that were most "amenable" to correction. Finally, the investigators observed delayed progression through the cell cycle in cells transfected with the oligonucleotide compared to those that were not treated.

Early cell culture studies found that chimeric RNA/DNA oligonucleotides also effect site-specific nucleotide conversion of episomal DNA (Yoon et al. 1996). Subsequently, it was reported that a targeted single nucleotide correction in cultured lymphoblastoid cells repaired the genomic DNA point mutation in patients with sickle cell anemia (Cole-Strauss et al. 1996).

This seminal paper led to further research that ultimately demonstrated the introduction of a missense mutation in genomic DNA in cultured HuH-7 cells by chimeric oligonucleotides (Kren et al. 1997). Although RNA/DNA hybrids actively promoted nucleotide exchange, DNA-only duplexes that were taken up were nevertheless inactive (Yoon et al. 1996; Cole-Strauss et al. 1996; Kren et al. 1997). These and other early studies underscored the need for improved delivery of chimeraplasts to target tissue. As an example, additional studies of gene repair in rat models of hepatic disease revealed that in order to be useful as a therapeutic intervention, not only targeted delivery of chimeraplasts but also efficient nuclear translocation and directed nucleotide exchange in quiescent cells were required.

To address this problem, chimeric oligonucleotides with an engineered nucleotide mismatch were designed to target both transcribed and nontranscribed rat factor IX genomic DNA strands. Thus, the nucleotide change at Ser^{365} would introduce a missense mutation in the rat genomic sequence resulting in the conversion of active site Ser^{365} to Arg365, characteristic of certain human factor IX mutations (Kren et al. 1998). Primary rat hepatocytes were transfected with chimeraplasts using polyethyleneimine (PEI) as a compacting agent and liposomal delivery systems, resulting in an A to C conversion efficiency of 25% at Ser^{365}, irrespective of the targeted strand. Similar results were obtained in the Chapel Hill strain of hemophilia B dogs (Evans et al. 1989).

In order to determine whether this approach could be used to correct a frameshift mutation, UDP-glucuronosyltransferase (*UGT1A1*)-deficient Gunn rats were tested in an animal model of Crigler-Najjar syndrome type I. Indeed, the simple insertion of one guanosine at position 1206 in the *UGT*1 coding sequence restored the proper reading frame of the *UGT1A1* mRNA, induced production of the UDP-glucurosyltransferase enzyme and re-established the wild-type *Bst*N I restriction site (Roy-Chowdhury et al. 1999). In contrast to previous studies, the correcting chimeraplast (CN3) used in this study included a 9-nucleotide DNA region with a central mismatched base pair flanked by the modified RNA. Nucleotide insertion determined by differential hybridization analysis of the wild-type corrected sequence (1206^G) and the 1206^A mutant sequence after transfection of isolated Gunn rat hepatocytes with CN3 chimeraplasts resulted in a dose-dependent frameshift correction frequency of about 25%.

21.4.3 Studies in Animal Models

21.4.3.1 Chimeraplast Mediated Repair

The successful use of chimeraplasty in the primary rat hepatocyte preparation led to the investigation of the potential of this approach for in vivo gene modification. A series of experiments was performed in male rats to determine whether chimeric oligonucleotides could induce mutation of factor IX in hepatocytes (Kren et al. 1998). Molecules were complexed to PEI and administered via injection into the tail vein. Liver tissue was collected, and genomic DNA was isolated for analysis several days after injection. PCR-amplified products analyzed by duplicate filter lift hybridization demonstrated dose-dependent conversion frequencies ranging from 15% to 40%. Similar results were obtained by RT-PCR analysis for RNA. Sequence analysis confirmed that the A to C conversion at Ser^{365} in both the cell culture and in vivo studies was specific.

In this study, the activated partial thromboplastin time (aPTT) assay was used to determine factor IX coagulant activity. The engineered rats were found to have a stable and long-lasting (72 week) significant reduction (~40%) in factor IX activity. Direct sequence analysis of the amplicons from random samples at 18 months confirmed the polymorphism at Ser^{365}, indicating that the site-directed conversion of the factor IX gene in intact liver is permanent and phenotypically stable. The replicative stability of the targeted nucleotide conversion was determined by performing a 70% partial hepatectomy, a surgical procedure that induces the liver remnant to undergo compensatory regeneration, after nucleotide modification. This process leads to 95% replication of hepatocytes in two synchronous waves (Higgins and Anderson 1931). Factor IX activity was determined by aPTT assays periodically between 9 and 78 weeks after intervention. The results showed that the genomic mutation was stable during hepatocyte replication.

Following these findings, additional experiments were done to establish the efficient replacement of deleted nucleotides by chimeric RNA/DNA oligonucleotides in vivo. Rats were given intravenous injections of CN3 chimeraplast designed to insert a single base pair in the mutated *UGT1A1* gene in Gunn rat hepatocytes using both delivery systems (Kren et al. 1999b). One week later, DNA was isolated from liver tissue and revealed a targeted G insertion rate of ~20%. However, there were no G insertions observed in the control animals. This genomic alteration persisted unchanged for 6 months, as confirmed by RFLP analysis as well as direct sequencing of the PCR amplified region of the *UG1A1* gene (Roy-Chowdhury et al. 1999). Southern blot analysis of genomic DNA from CN3-treated animals, but not control animals, showed partial restoration of the *Bst*N I restriction site.

Western blot analysis confirmed re-establishment of the 52 kDa UDP-glucuronosyltransferase type I microsomal hepatic enzyme expression. The effects of CN3 administration also were evident upon examination of serum bilirubin concentrations, which dropped below 50% of pretreatment levels in experimental but not in control rats. In fact, bilirubin levels increased in control animals. The effect of CN3 on bilirubin was sustained for more than one year post-treatment. Repeated administration of CN3 further reduced serum bilirubin. Finally, HPLC analysis of bile from the CN3-treated animals only revealed both mono- and diglucuronide-conjugated bilirubin, whereas analysis in the control animals showed only unconjugated bile. Collectively, the data from this series of studies of in models of liver infection and/or disease suggest that hepatic mismatch repair pathways may be sufficiently active to make this strategy feasible for other liver-related disorders resulting from single base pair mutations or deletions, including hemophilia B and α1-antitrypsin deficiency.

It is important to note that the use of chimeric oligonucleotides to modify genomic sequences has been successfully used in other model organ systems and disorders, including correction of the carbonic anhydrase II nonsense mutation in nude mouse primary kidney tubular cells (Lai et al. 1998, 1999), the missense mutation in the tyrosinase gene responsible for melanin production in albino mouse melanocytes (Alexeev and Yoon 1998; Alexeev et al. 2000) both in cell culture and in vivo, and the *mdx* point mutations underlying muscular dystrophy in both mice and dogs (Bartlett et al. 1998). Chimeraplasts have also been shown to produce efficient induction of targeted genomic DNA nucleotide changes in plants for improved herbicide resistance (Beetham et al. 1999; Zhu et al. 1999, 2000). Taken together, the results of these studies emphasize the fact that numerous cell types are capable of performing chimeraplast-mediated modifications of their genomic sequences. Interestingly, in addition to modulating gene expression by genomic alteration, chimeraplasty has also been used successfully for pre-mRNA splicing to promote exon skipping in the dystrophin gene in DMD resulting in production of normal muscle fibers both in vitro and in vivo (Bertoni and Rando 2002; Bertoni et al. 2003).

Targeted nucleotide exchange by SSOs has been successfully accomplished in numerous cell types using PTO 3′ modified SSOs. Increased rates of repair are associated with modulation of proteins and pathways involved in DNA repair and replication (Ciavatta et al. 2005; Dekker et al. 2006; Morozov and Wawrousek 2008; Schwartz and Kmiec 2007; Wu et al. 2005), particularly inhibition of MSH2 gene expression (Igoucheva et al. 2008). However, in repair efficiencies mediated by SSOs, 4.5% can be routinely achieved in replicating cells in culture by using 3' PTO modified SSOs with canonical 5′ phosphate (5′P) and 3′ hydroxyl (3′OH) termini (Olsen et al. 2005). Moreover, by coupling microarray gene expression analysis with oligonuclotide targeting studies (Igoucheva et al. 2006), it was shown that SSOs do not induce the DNA damage signaling and DNA repair genes, a potential advantage over double stranded DNA mediated repair.

21.4.3.2 Single-Stranded Oligonucleotide Repair

Successful in vivo application of SSOs has been reported in mice (Fan and Yoon 2003). More recently, SSO-mediated gene correction using polylysine conjugates targeted to the hepatocyte asialoglycoprotein receptor was successful in correcting the acid alpha-D-glucosidase gene as assessed by both PCR and phenotypic change (Lu et al. 2003). Using a 75-mer in transgenic mice expressing the mutant murine transthyretin (TTR) Val30Met gene, targeted nucleotide replacement leading to gene repair was achieved in 9% of the adult mouse hepatocytes, and phenotypic hepatic changes were also observed (Nakamura et al. 2004). Interestingly, 25-mer SSOs were shown to effectively repair the nonsense mutation in the *rd1* mouse retina responsible for retinal degeneration (Andrieu-Soler et al. 2007). Moreover, we have used 45-mer SSOs modified at their 3′ end with 3 phophorothioate residues and phosphorylated at their 5′ end and delivered to transgenic spf^{ash} ornithine transcarbamylase (OTC) deficient pups, resulting in restoration of enzymatic activity to 15% of wild-type, as reflected in the ~10 to 15% conversion of the targeted nucleotide from the mutant A to wild-type G (Kren et al. 2008).

21.5 Triplex DNA

21.5.1 Mechanisms and Pathways

Another approach to site-specific modification of genomic DNA involves the use of triplex DNA based on the formation of a three-stranded or triple-helical nucleic

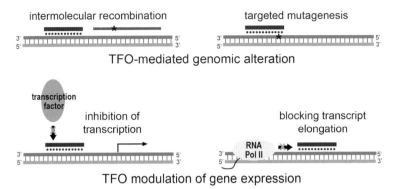

Fig. 21.4 Schematic of TFO-mediated processes applicable for genetic engineering. The *upper portion* indicates two TFO-mediated reactions that promote permanent genetic change. Although shown without "linking tethers", the strategy has also exploited non-homologous linkers to tether the TFO and the homologous DNA for the intermolecular recombination together. The *lower portion* shows TFO modulation of gene expression at the transcriptional level by preventing transcription initiation or blocking transcript elongation by RNA polymerase II (Pol II). The TFO is shown in black with the third strand bond formation represented by the black dots. * base targeted for change by mutagenesis or replacement by intermolecular recombination

acid structure (Chin and Glazer 2009). Use of this strategy provides the possibility of repairing more than single-base mutations, an obvious advantage over other methods aimed at targeted gene repair (Fig. 21.4). The basis of this approach is the binding of the exogenous third strand of nucleic acid in the major grove of a homopurine region of the DNA, resulting in the formation of Hoogsteen or reverse Hoogsteen hydrogen bonds with the purine base (Chan and Glazer 1997). Although this process can occur under physiologic conditions, some polypurine sites do not bind with high affinity; sufficient triplex formation requires guanine-rich polypurine regions that are 12 to 14 nucleotides in length.

The initial approach to the triple-helix site-directed modification of DNA used cross-linking agents, such as psoralen or other mutagens to form covalent bonds to the triplex-forming oligonucleotide (Chan and Glazer 1997). After intercalation of the psoralen at the target 5′ApT3′ site in the DNA, UV-irradiation results in cross-linking of the thymines in the two strands. This substrate is then repaired by the endogenous DNA repair activity in the cell producing the characteristic T:A to A:T transversions at a low frequency. Interestingly, the ability of the triplex-directed psoralen cross-linked DNA substrates to induce recombination is only partially dependent on functional nucleotide excision repair. This suggests that other endogenous DNA repair pathways are involved in processing the mutated DNAs into recombinagenic intermediates. However, it is now well established that triplex DNA-mediated recombination is not affected in mismatch repair-deficient cells.

An early strategy for producing triple-helix site-directed DNA modification used mutagens or cross-linking agents to form covalent bonds to the triplex-forming oligonucleotide (Chan and Glazer 1997). UV-radiation of the intercalated site results in cross-linked thymines in both strands, followed by endogenous DNA repair activity of the cell substrate that produces prototypic T:A to A:T transversions. This approach to in vitro episomal modification in mammalian cells has been successful, but limitations related to target sequence and the need for cross-linking have limited its use in vivo (Wang et al. 1995). Nevertheless, this approach has been used successfully to reactivate the hypoxanthine phosphoribosyl transferase gene in cultured cells after delivery of targeted gene knockouts. Unexpectedly, however, evidence suggests that the repair pathway engaged did not involve mismatch repair but rather the modification of DNA via nucleotide-induced promotion of target site-specific insertions and deletions without cross-linking agents (Majumdar et al. 1998; Vasquez et al. 1999). In the absence of cross-linking agents, these triple-helix forming oligonucleotides induce recombination via a nucleotide excision repair pathway (Faruqi et al. 2000), which in some cases is only partially reliant on functional nucleotide excision repair enzymes. Clearly, endogenous DNA repair pathways other than mismatch repair underlie triplex DNA-mediated recombination of intermediates from mutated DNA.

Bifunctional oligonucleotides, which contain both triple-helical structures and conventional Watson and Crick based pairs, have been used as a novel approach to addressing the limitations of sequence constraint for triple-helix formation, targeted single nucleotide, and homologous DNA-based gene correction (Broitman et al. 1999; Chan et al. 1999; Culver et al. 1999). Use of this approach has resulted in successful promotion of both site-specific nucleotide correction and gene targeting in both cultured cell and cell-free preparations. Triplex formation is inhibited under physiologic conditions by monovalent cations such as Na^+ and K^+, but modified bases can counteract this effect (Faruqi et al. 1997). For example, the thymidine purine analog 7-deaza-2'-deoxyxanthosine exhibits the ability to form triplex DNA at physiologic pH and salt conditions, as well as maintaining an all purine backbone motif (Faruqi et al. 1997).

21.5.2 Potential Applications

This technique has been used to modify episomal DNA in mammalian cells in vitro (Wang et al. 1995). The target sequence constraints and requirement for cross-linking agents have limited its application in living cells, but targeted gene knock-outs of the genomic hypoxanthine phosphoribosyl transferase gene in cultured cells were nevertheless successfully created using this approach (Majumdar et al. 1998). Interestingly, rather than the expected T to A transversions, the majority of the knock-outs resulted predominately from small deletions and some insertions, suggesting that the endogenous repair pathway involved was not that of mismatch repair.

Triple-helix-forming nucleotides can be used to promote insertions and deletions at target sites without cross-linking agents to promote modification of the DNA (Vasquez et al. 1999). These triple-helix forming oligonucleotides, in the absence of cross-linking agents, are able to induce recombination via a nucleotide excision repair pathway (Faruqi et al. 2000).

The problem of sequence constraint for triple-helix formation, targeted single nucleotide, and homologous DNA based gene correction has been overcome, in part, by the use of novel approaches such as bifunctional oligonucleotides (Broitman et al. 1999; Chan et al. 1999; Culver et al. 1999). These oligonucleotides contain regions that form triple-helical structures as well as conventional Watson and Crick base pairs. These modified oligonucleotides have been used successfully to promote site-specific nucleotide correction and gene targeting in both cell-free systems and cultured cells (Chin et al. 2008, Lonkar et al. 2009). Moreover, these triplex-forming oligonucleotides can be used to inhibit gene expression (Karympalis et al. 2004) and in such a capacity have been used successfully to modulate expression of genes involved in fibrosis (Cheng et al. 2005; Ye et al. 2007).

An intriguing aspect of the triple-helix forming oliognucleotides is the plethora of repair pathways and proteins involved in mediating the gene repair (Chin and Glazer 2009). This has led to an enhanced understanding of how these repair pathways are functionally regulated (Chin and Glazer 2009; Crosby et al. 2009), providing important information for enhancing all forms of targeted gene repair.

These exciting new advances using triple-helix forming oligonucleotides suggest that this form of genomic modification may have therapeutic potential in treating acquired and inherited genetic disorders implicated in aging.

21.6 What is in Our Future?

The advances in genetic engineering in the last decade hold great potential in treating acquired or inherited genetic disorders implicated in aging. Our ability to compensate for deficient or defective proteins by inducing permanent change at the genomic level or modulate gene expression at the transcriptional or post-transcriptional level provides the tools needed for modifying all aspects of nuclear gene function. Moreover, new uses of oligonucleotides impacting other cellular processes such as protein aggregation (Skogen et al. 2006) as a potential therapeutic for age-related inherited diseases such as Huntington's are being rapidly developed. Although the correction of aging-related mutations may require numerous interventions (Lu et al. 2004), this should be feasible in principle provided appropriate delivery systems for repair constructs can be developed.

Significant changes related to normal aging also occur in mitochondria, such as decreased respiration, increased production of reactive oxygen species, decreased degradation of defective mitochondria (mitophagy), increased molecular damage to mitochondrial DNA (mtDNA) and proteins, and increased mutations in mtDNA (Brunk and Terman 2002). While this is normally not an issue in embryogenesis as mtDNA undergo a purifying selection process in female germ line cells to eliminate mutant mtDNA (Stewart et al. 2008a, b; Fan et al. 2008), it may significantly impact aging. Thus, promoting our mtDNA to undergo this purifying selection, or other complementary approaches for ensuring the integrity of mtDNA over time (see Chapter 16), may be necessary to provide the ultimate benefit of nuclear genetic engineering for addressing the consequences of aging.

References

Alexeev V, Igoucheva O, Domashenko A, Cotsarelis G, Yoon K (2000) Localized in vivo genotypic and phenotypic correction of the albino mutation in skin by RNA-DNA oligonucleotide. Nature Biotechnol 18:43–47

Alexeev V, Yoon K (1998) Stable and inheritable changes in genotype and phenotype of albino melanocytes induced by an RNA/DNA oligonucleotide. Nature Biotechnol 16:1343–1346

Andersen MS, Sorensen CB, Bolund L, and Jensen TG (2002) Mechanisms underlying targeted gene correction using chimeric RNA/DNA and single-stranded DNA oligonucleotides. J Mol Med 80:770–781

Andrieu-Soler C, Halhal M, Boatright JH, Padove SA, Nickerson JM, Stodulkova E, Stewart RE, Ciavatta VT, Doat M, Jeanny JC, de Bizemont T, Sennlaub F, Courtois Y, Behar-Cohen F (2007) Single-stranded oligonucleotide-mediated in vivo gene repair in the rd1 retina. Mol Vision 13:692–706

Aravind L, Walker DR, Koonin EV (1999) Conserved domains in DNA repair proteins and evolution of repair systems. Nucleic Acids Res 27:1223–1242

Arbuthnot P, Longshaw V, Naidoo T, Weinberg MS (2007) Opportunities for treating chronic hepatitis B and C virus infection using RNA interference. J Viral Hepat 14:447–459

Arbuthnot P, Thompson LJ (2008) Harnessing the RNA interference pathway to advance treatment and prevention of hepatocellular carcinoma. World J Gastroenterol 14:1670–1681

Asahina Y, Ito Y, Wu CH, Wu GY (1998) DNA ribonucleases that are active against intracellular hepatitis B viral RNA targets. Hepatology 28:547–554

Bartlett RJ, Denis MM, Kornegay JN, et al. (1998) Can genetic surgery be used to revert muscular dystrophy mutations in live animals? In: Abstracts of the 1st Annual Meeting of the American Society of Gene Therapy. Seattle, WA, 153a

Beetham PR, Kipp RB, Sawycky XL, Arntzen CJ, May GD (1999) A tool for functional plant genomics: Chimeric RNA/DNA oligonucleotides cause in vivo gene-specific mutations. Proc Natl Acad Sci USA 96:8874–8778

Beroud C, Tuffery-Giraud S, Matsuo M, Hamroun D, Humbertclaude V, Monnier N, Moizard MP, Voelckel MA, Calemard LM, Boisseau P, Blayau M, Philippe C, Cossee M, Pages M, Rivier F, Danos O, Garcia L, Claustres M (2007) Multiexon skipping leading to an artificial DMD protein lacking amino acids from exons 45 through 55 could rescue up to 63% of patients with Duchenne muscular dystrophy. Hum Mutat 28:196–202

Bertoni C, Lau C, Rando TA (2003) Restoration of dystrophin expression in mdx muscle cells by chimeraplast-mediated exon skipping. Hum Mol Genet 12:1087–1099

Bertoni C, Rando TA (2002) Dystrophin gene repair in mdx muscle precursor cells in vitro and in vivo mediated by RNA-DNA chimeric oligonucleotides. Hum Gene Ther 13:707–718

Bhattacharyya SN, Habermacher R, Martine U, Closs EI, Filipowicz W (2006) Stress-induced reversal of microRNA repression and mRNA P-body localization in human cells. Cold Spring Harb Symp Quant Biol 71:513–521

Broitman S, Amosova O, Dolinnaya NG, Fresco JR (1999) Repairing the sickle cell mutation. I. Specific covalent binding of a photoreactive third strand to the mutated base pair. J Biol Chem 274:21763–21768

Brunk UT, Terman A (2002) The mitochondrial-lysosomal axis theory of aging: accumulation of damaged mitochondria as a result of imperfect autophagocytosis. Eur J Biochem 269:1996–2002

Capecchi MR (1989) Altering the genome by homologous recombination. Science 244:1288–1292

Chan PP, Glazer PM (1997) Triplex DNA: fundamentals, advances, and potential applications for gene therapy. J Mol Med 75:267–282

Chan PP, Lin M, Faruqi AF, Powell J, Seidman MM, Glazer PM (1999) Targeted correction of an episomal gene in mammalian cells by a short DNA fragment tethered to a triplex-forming oligonucleotide. J Biol Chem 274:11541–11548

Chen Y, Cheng G, Mahato RI (2008) RNAi for treating hepatitis B viral infection. Pharm Res 25:72–86

Chen Z, Felsheim R, Wong P, Augustin LB, Metz R, Kren BT, Steer CJ (2001) Mitochondria isolated from liver contain the essential factors required for RNA/DNA oligonucleotide-targeted gene repair. Biochem Biophys Res Commun 285:188–194

Cheng K, Mahato RI (2007) Gene modulation for treating liver fibrosis. Crit Rev Ther Drug Carrier Syst 24:93–146

Cheng K, Ye Z, Guntaka RV, Mahato RI (2005) Biodistribution and hepatic uptake of triplex-forming oligonucleotides against type $\alpha 1(I)$ collagen gene promoter in normal and fibrotic rats. Mol Pharm 2:206–217

Chin JY, Glazer PM (2009) Repair of DNA lesions associated with triplex-forming oligonucleotides. Mol Carcinog 48:389–399

Chin JY, Kuan JY, Lonkar PS, Krause DS, Seidman MM, Peterson KR, Nielsen PE, Kole R, Glazer PM (2008) Correction of a splice-site mutation in the β-globin gene stimulated by triplex-forming peptide nucleic acids. Proc Natl Acad Sci USA 105:13514–13519

Chu CY, Rana TM (2006) Translation repression in human cells by microRNA-induced gene silencing requires RCK/p54. PLoS Biol 4:e210

Ciavatta VT, Padove SA, Boatright JH, Nickerson JM (2005) Mouse retina has oligonucleotide-induced gene repair activity. Invest Ophthalmol Vis Sci 46:2291–2299

Crosby ME, Kulshreshtha R, Ivan M, Glazer PM (2009) MicroRNA regulation of DNA repair gene expression in hypoxic stress. Cancer Res 69:1221–1229

Cohen-Tannoudji M, Babinet C (1998) Beyond 'knock-out' mice: new perspectives for the programmed modification of the mammalian genome. Mol Hum Reprod 4:929–938

Cole-Strauss A, Gamper H, Holloman WK, Munoz M, Cheng N, Kmiec EB (1999) Targeted gene repair directed by the chimeric RNA/DNA oligonucleotide in a mammalian cell-free extract. Nucleic Acids Res 27:1323–1330

Cole-Strauss A, Yoon K, Xiang Y, Byrne BC, Rice MC, Gryn J, Holloman WK, Kmiec EB (1996) Correction of the mutation responsible for sickle cell anemia by an RNA-DNA oligonucleotide. Science 273:1386–1389

Culver KW, Hsieh W-T, Huyen Y, Chen V, Liu J, Khripine Y, Khorlin A (1999) Correction of chromosomal point mutations in human cells with bifunctional oligonucleotides. Nature Biotechnol 17:989–993

Dekker M, Brouwers C, Aarts M, van der Torre J, de Vries S, van de Vrugt H, te Riele H (2006) Effective oligonucleotide-mediated gene disruption in ES cells lacking the mismatch repair protein MSH3. Gene Ther 13:686–694

Dunckley MG, Villiet P, Eperon IC, Dickson G (1998) Modification of splicing in the dystrophin gene in cultured *Mdx* muscle cells by antisense oligoribonucleotides. Hum Mol Genet 7:1083–1090

Eisen JA (1998) A phylogenetic study of the MutS family of proteins. Nucleic Acids Res 26:4291–4300

Engstrom JU, Kmiec KB (2008) DNA replication, cell cycle progression, and the targeted gene repair reaction. Cell Cycle 7:1402–1414

Eulalio A, Behm-Ansmant I, Schweizer D, Izaurralde E (2007) P-body formation is a consequence, not the cause, of RNA-mediated gene silencing. Mol Cell Biol 27:3970–3981

Evans JP, Brinkhous KM, Brayer GD, Reisner HM, High KA (1989) Canine hemophilia B resulting from a point mutation with unusual consequences. Proc Natl Acad Sci USA 86:10095–10099

Fan W, Waymire KG, Narula N, Li P, Rocher C, Coskun PE, Vannan MA, Narula J, Macgregor GR, Wallace DC (2008) A mouse model of mitochondrial disease reveals germline selection against severe mtDNA mutations. Science 319:958–962

Fan W, Yoon K (2003) In vivo alteration of the keratin 17 gene in hair follicles by oligonucleotide-directed gene targeting. Exp Dermatol 12:832–842

Faruqi AF, Datta HJ, Carroll D, Seidman MM, Glazer PM (2000) Triple-helix formation induces recombination in mammalian cells via a nucleotide excision repair-dependent pathway. Mol Cell Biol 20:990–1000

Faruqi AF, Krawczyk SH, Matteucci MD, Glazer PM (1997) Potassium-resistant triple helix formation and improved intracellular gene targeting by oligodeoxyribonucleotides containing 7-deazaxanthine. Nucleic Acids Res 25:633–640

Gamper HB, Jr., Cole-Strauss A, Metz R, Parekh H, Kumar R, Kmiec EB (2000a) A plausible mechanism for gene correction by chimeric oligonucleotides. Biochemistry 39:5808–5816

Gamper HB, Parekh H, Rice MC, Bruner M, Youkey H, Kmiec EB (2000b) The DNA strand of chimeric RNA/DNA oligonucleotides can direct gene repair/conversion activity in mammalian and plant cell-free extracts. Nucleic Acids Res 28:4332–4339

Goncz KK, Kunzelmann K, Xu Z, Gruenert DC (1998) Targeted replacement of normal and mutant CFTR sequences in human airway epithelial cells using DNA fragments. Hum Mol Genet 7:1913–1919

Higgins GM, Anderson RM (1931) Experimental pathology of the liver. I. Restoration of the liver of the white rat following partial surgical removal. Arch Pathol 12:186–202

Hua Y, Vickers TA, Baker BF, Bennett CF, Krainer AR (2007) Enhancement of SMN2 exon 7 inclusion by antisense oligonucleotides targeting the exon. PLoS Biol 5:e73

Huen MS, Lu LY, Liu DP, Huang JD (2007) Active transcription promotes single-stranded oligonucleotide mediated gene repair. Biochem Biophys Res Commun 353:33–39

Igoucheva O, Alexeev V, Anni H, Rubin E (2008) Oligonucleotide-mediated gene targeting in human hepatocytes: implications of mismatch repair. Oligonucleotides 18:111–122

Igoucheva O, Alexeev V, Yoon K (2006) Differential cellular responses to exogenous DNA in mammalian cells and its effect on oligonucleotide-directed gene modification. Gene Ther 13:266–275

Igoucheva O, Peritz AE, Levy D, Yoon K (1999) A sequence-specific gene correction by an RNA-DNA oligonucleotide in a mammalian cells characterized by transfection and nuclear extract using a *lacZ* shuttle system. Gene Ther 6:1960–1971

Karympalis V, Kalopita K, Zarros A, Carageorgiou H (2004) Regulation of gene expression via triple helical formations. Biochemistry (Mosc) 69:855–860

Khoo B, Roca X, Chew SL, Krainer AR (2007) Antisense oligonucleotide-induced alternative splicing of the APOB mRNA generates a novel isoform of APOB. BMC Mol Biol 8:3

Kmiec EB, Kren BT, Steer CJ (1999) Targeted gene repair in mammalian cells using chimeric RNA/DNA oligonucleotides. In: Friedman T (ed) Development of human gene therapy. Cold Spring Harbor, Cold Spring Harbor Laboratory Press, New York, pp. 643–670

Kotani H, Kmiec EB (1994a) Transcription activates RecA-promoted homologous pairing of nucleosomal DNA. Mol Cell Biol 14:1949–1955

Kotani H, Kmiec EB (1994b) A role for RNA synthesis in homologous pairing events. Mol Cell Biol 14:6097–6106

Kren BT, Bandyopadhyay P, Steer CJ (1998) In vivo site-directed mutagenesis of the *factor IX* gene by chimeric RNA/DNA oligonucleotides. Nature Med 4:285–290

Kren BT, Cole-Strauss A, Kmiec EB, Steer CJ (1997) Targeted nucleotide exchange in the alkaline phosphatase gene of HuH-7 cells mediated by a chimeric RNA/DNA oligonucleotide. Hepatology 25:1462–1468

Kren BT, Metz R, Kumar R, Steer CJ (1999a) Gene repair using RNA/DNA oligonucleotides. Sem Liver Dis 19:93–104

Kren BT, Parashar B, Bandyopadhyay P, Chowdhury NR, Chowdhury JR, Steer CJ (1999b) Correction of the UDP-glucuronosyl-transferase gene defect in the Gunn rat model of Crigler-Najjar syndrome type I with a chimeric oligonucleotide. Proc Natl Acad Sci USA 96:10349–10354

Kren BT, Wong PY-P, Steer CJ (2008) Correction of the ornithine transcarbamylase point mutation in neonatal *spf^{ash}* mice using single-stranded oligonucleotides. Mol Ther 16:s323

Kruger K, Grabowski PJ, Zaug AJ, Gottschling DE, Cech TR (1982) Self-splicing RNA: Autoexcision and autocyclization of the ribosomal RNA intervening sequence of Tetrahymena. Cell 31:147–157

Lai L-W, Chau B, Lien Y-H (1999) In vivo gene targeting in carbonic anhydrase II deficient mice by chimeric RNA/DNA oligonucleotides. In: Abstracts of the 2nd Annual Meeting of the American Society of Gene Therapy. Washington DC, 236a

Lai L-W, O'Connor HM, Lien Y-HH (1999) Correction of carbonic anhydrase II mutation in renal tubular cells by chimeric RNA/DNA oligonucleotide. In: Abstracts of the 1st Annual Meeting of the American Society of Gene Therapy. Seattle, WA, 183a

Lewis DL, Wolff JA (2007) Systemic siRNA delivery via hydrodynamic intravascular injection. Adv Drug Deliv Rev 59:115–123

Lieber A, He CY, Polyak SJ, Gretch DR, Barr D, Kay MA (1996) Elimination of hepatitis C virus RNA in infected human hepatocytes by adenovirus-mediated expression of ribozymes. J Virol 70:8782–8791

Liu L, Rice MC, Drury M, Cheng S, Gamper H, Kmiec EB (2002) Strand bias in targeted gene repair is influenced by transcriptional activity. Mol Cell Biol 22:3852–3863

Lonkar P, Kim KH, Kuan JY, Chin JY, Rogers FA, Knauert MP, Kole R, Nielsen PE, Glazer PM (2009) Targeted correction of a thalassemia-associated β-globin mutation induced by pseudo-complementary peptide nucleic acids. Nucleic Acids Res 37: 3635–3644

Lu IL, Lin CY, Lin SB, Chen ST, Yeh LY, Yang FY, Au LC (2003) Correction/mutation of acid α-D-glucosidase gene by modified single-stranded oligonucleotides: in vitro and in vivo studies. Gene Ther 10:1910–1916

Lu T, Pan Y, Kao SY, Li C, Kohane I, Chan J, Yanker, B A (2004) Gene regulation and DNA damage in the ageing human brain. Nature 429:883–891

Majumdar A, Khorlin A, Dyatkina N, Lin FL, Powell J, Liu J, Fei Z, Khripine Y, Watanabe KA, George J, Glazer PM, Seidman MM (1998) Targeted gene knockout mediated by triple helix forming oligonucleotides. Nature Genet 20:212–214

Marra G, Schar P (1999) Recognition of DNA alterations by the mismatch repair system. Biochem J 338:1–13

Martin SE, Caplen NJ (2007) Applications of RNA interference in mammalian systems. Annu Rev Genomics Hum Genet 8:81–108

Miller DG, Wang PR, Petek LM, Hirata RK, Sands MS, Russell DW (2006) Gene targeting in vivo by adeno-associated virus vectors. Nat Biotechnol 24:1022–1026

Minczuk M, Papworth MA, Miller JC, Murphy MP, Klug A (2008) Development of a single-chain, quasi-dimeric zinc-finger nuclease for the selective degradation of mutated human mitochondrial DNA. Nucleic Acids Res 36:3926–3938

Moehle EA, Rock JM, Lee YL, Jouvenot Y, DeKelver RC, Gregory PD, Urnov FD, Holmes MC (2007) Targeted gene addition into a specified location in the human genome using designed zinc finger nucleases. Proc Natl Acad Sci USA 104:3055–3060

Morozov V, Wawrousek EF (2008) Single-strand DNA-mediated targeted mutagenesis of genomic DNA in early mouse embryos is stimulated by Rad51/54 and by Ku70/86 inhibition. Gene Ther 15:468–472

Moulton V (2005) Tracking down noncoding RNAs. Proc Natl Acad Sci USA 102:2269–2270

Müller U (1999) Ten years of gene targeting: targeted mouse mutants, from vector design to phenotype analysis. Mech Dev 82:3–21

Nakamura M, Ando Y, Nagahara S, Sano A, Ochiya T, Maeda S, Kawaji T, Ogawa M, Hirata A, Terazaki H, Haraoka K, Tanihara H, Ueda M, Uchino M, Yamamura K (2004) Targeted conversion of the transthyretin gene in vitro and in vivo. Gene Ther 11:838–846

Nakazono K, Ito Y, Wu CH, Wu GY (1996) Inhibition of hepatitis B virus replication by targeted pretreatment of complexed antisense DNA in vitro. Hepatology 23:1297–1303

Niedernhofer LJ et al. (2006) A new progeroid syndrome reveals that genotoxic stress suppresses the somatotroph axis. Nature 444:1038–1043

Oketani M, Asahina Y, Wu CH, Wu GY (1999) Inhibition of hepatitis C virus-directed gene expression by a DNA ribonuclease. J Hepatol 31:628–634

Olsen PA, Randol M, Krauss S (2005) Implications of cell cycle progression on functional sequence correction by short single-stranded DNA oligonucleotides. Gene Ther 12:546–551

Parker R, Sheth U (2007) P bodies and the control of mRNA translation and degradation. Mol Cell 25:635–646

Paulk N, Wursthorn K, Finegold M, Kay M, Grompe M (2008) AAV8-Mediated correction of a metabolic renal disease via gene repair in vivo. Mol Ther 16:s368

Porteus MH, Carroll D (2005) Gene targeting using zinc finger nucleases. Nat Biotechnol 23:967–973

Porteus M (2008) Design and testing of zinc finger nucleases for use in mammalian cells. Methods Mol Biol 435:47–61

Radecke S, Radecke F, Peter I, Schwarz K (2006) Physical incorporation of a single-stranded oligodeoxynucleotide during targeted repair of a human chromosomal locus. J Gene Med 8:217–228

Richardson PD, Kren BT, Steer CJ (2002) Gene repair in the new age of gene therapy. Hepatology 35:512–518

Roth DB, Wilson JH (1985) Relative rates of homologous and nonhomologous recombination in transfected DNA. Proc Natl Acad Sci USA 82:3355–3359

Roy-Chowdhury J, Huang TJ, Kesari K, Lederstein M, Arias IM, Roy-Chowdhury N (1999) Molecular basis for the lack of bilirubin-specific and 3-methylcholanthrene-inducible UDP-glucuronosyltransferase activities in Gunn rats. The two isoforms are encoded by distinct mRNA species that share an identical single base deletion. J Biol Chem 266:18294–18298

Rubnitz J, Subramani S (1984) The minimum amount of homology required for homologous recombination in mammalian cells. Mol Cell Biol 4:2253–2258

Sakamoto N, Wu CH, Wu GY (1996) Intracellular cleavage of hepatitis C virus RNA and inhibition of viral protein translation by hammerhead ribozymes. J Clin Invest 98:2720–2728

Schwartz TR, Kmiec EB (2007) Reduction of gene repair by selenomethionine with the use of single-stranded oligonucleotides. BMC Mol Biol 8:7

Sheth U, Parker R (2006) Targeting of aberrant mRNAs to cytoplasmic processing bodies. Cell 125:1095–1109

Skogen M, Roth J, Yerkes S, Parekh-Olmedo H, Kmiec E (2006) Short G-rich oligonucleotides as a potential therapeutic for Huntington's Disease. BMC Neurosci 7:65

Stewart JB, Freyer C, Elson JL, Larsson NG (2008a) Purifying selection of mtDNA and its implications for understanding evolution and mitochondrial disease. Nat Rev Genet, 9:657–662

Stewart JB, Freyer C, Elson JL, Wredenberg A, Cansu Z, Trifunovic A, Larsson NG (2008b) Strong purifying selection in transmission of mammalian mitochondrial DNA. PLoS Biol. 6:e10

Symons RH (1992) Small catalytic RNAs. Ann Rev Biochem 61:641–671

Thomas KR, Folger KR, Capecchi MR (1986) High frequency targeting of genes to specific sites in the mammalian genome. Cell 44:419–428

Thyagarajan B, Cruise JL, Campbell C (1996) Elevated levels of homologous recombination activity in the regenerating rat liver. Somatic Cell Mol Genet 22:31–39

Tuschl T (2003) Functional genomics: RNA sets the standard. Nature 421:220–221

Valencia-Sanchez MA, Liu J, Hannon GJ, Parker R (2006) Control of translation and mRNA degradation by miRNAs and siRNAs. Genes Dev 20:515–524

van de Vosse E, Verhard EM, de Paus RA, Platenburg GJ, van Deutekom JC, Aartsma-Rus A, van Dissel JT (2009) Antisense-mediated exon skipping to correct IL-12Rβ1 deficiency in T cells. Blood, 113:4548–4555

van Ommen GJ, van Deutekom J, Aartsma-Rus A (2008) The therapeutic potential of antisense-mediated exon skipping. Curr Opin Mol Ther 10:140–149

Vasquez KM, Wang G, Havre PA, Glazer PM (1999) Chromosomal mutations induced by triple helix-forming oligonucleotides in mammalian cells. Nucleic Acids Res 27:1176–1181

Vetrini F, Tammaro R, Bondanza S, Surace EM, Auricchio A, De Luca M, Ballabio A, Marigo V (2006) Aberrant splicing in the ocular albinism type 1 gene (OA1/GPR143) is corrected in vitro by morpholino antisense oligonucleotides. Hum Mutat 27:420–426

Volarevic M, Smolic R, Wu CH, Wu GY (2007) Potential role of RNAi in the treatment of HCV infection. Expert Rev Anti Infect Ther 5:823–831

Wang G, Levy DD, Seidman MM, Glazer PM (1995) Targeted mutagenesis in mammalian cells mediated by intracellular triple helix formation. Mol Cell Biol 15:1759–1768

Wang Z, Zhou ZJ, Liu DP, Huang JD (2006) Single-stranded oligonucleotide-mediated gene repair in mammalian cells has a mechanism distinct from homologous recombination repair. Biochem Biophys Res Commun 350:568–573

Welch PJ, Barber JR, Wong-Staal F (1998) Expression of ribozymes in gene transfer systems to modulate target RNA levels. Curr Opin Biotechnol 9:486–496

Welch PJ, Tritz R, Yei S, Barber J, Yu M (1997) Intracellular application of hairpin ribozyme genes against hepatitis B virus. Gene Ther 4:736–743

Welch PJ, Tritz R, Yei S, Leavitt M, Yu M, Barber J (1996) A potential therapeutic application of hairpin ribozymes: in vitro and in vivo studies of gene therapy for hepatitis C virus infection. Gene Ther 3:994–1001

Wong EA, Capecchi MR (1987) Homologous recombination between coinjected DNA sequences peaks in early to mid-S phase. Mol Cell Biol 7:2294–2295

Wu CH, Wu GY (1998) Targeted inhibition of hepatitis C virus-directed gene expression in human hepatoma cell lines. Gastroenterology 114:1304–1312

Wu XS, Xin L, Yin WX, Shang XY, Lu L, Watt RM, Cheah KS, Huang JD, Liu DP, Liang CC (2005) Increased efficiency of oligonucleotide-mediated gene repair through slowing replication fork progression. Proc Natl Acad Sci USA 102:2508–2513

Yamamoto A, Taki T, Yagi H, Habu T, Yoshida K, Yoshimura Y, Yamamoto K, Matsushiro A, Nishimune Y, Morita T (1996) Cell cycle-dependent expression of the mouse Rad51 gene in proliferating cells. Mol Gen Genet 251:1–12

Yañez RJ, Porter ACG (1998) Therapeutic gene targeting. Gene Ther 5:149–159

Ye Z, Houssein HS, Mahato RI (2007) Bioconjugation of oligonucleotides for treating liver fibrosis. Oligonucleotides 17:349–404

Yin W, Kren BT, Steer CJ (2004) Targeted gene repair: From RNA/DNA to single-stranded oligonucleotides In: Blum HE, Manns MP (eds) State-of-the-art of hepatology: molecular and cell biology, Falk Symposium 138, Kluwer Academic, pp. 172–195

Yin W, Kren BT, Steer CJ (2005) Site-specific base changes in the coding or promoter region of the human β- and γ-globin genes by single-stranded oligonucleotides. Biochem J 390:253–261

Yokota T, Takeda S, Lu QL, Partridge TA, Nakamura A, Hoffman EP (2009) A renaissance for antisense oligonucleotide drugs in neurology: exon skipping breaks new ground. Arch Neurol 66:32–38

Yoon K (1999) Single-base conversion of mammalian genes by an RNA-DNA oligonucleotide. Biogenic Amines 15:137–167

Yoon K, Cole-Strauss A, Kmiec EB (1996) Targeted gene correction of episomal DNA in mammalian cells mediated by a chimeric RNA·DNA oligonucleotide. Proc Natl Acad Sci USA 93:2071–2076

Zeng Y (2006) Principles of micro-RNA production and maturation. Oncogene 25:6156–6162

Zhu T, Mettenburg K, Peterson DJ, Tagliani L, Baszczynski CL (2000) Engineering herbicide-resistant maize using chimeric RNA/DNA oligonucleotides. Nature Biotechnol 18:555–558

Zhu T, Peterson DJ, Tagliani L, St. Clair G, Baszczynski CL, Bowen B (1999) Targeted manipulation of maize genes in vivo using chimeric RNA/DNA oligonucleotides. Proc Natl Acad Sci USA 96:8768–8773

Chapter 22
WILT: Necessity, Feasibility, Affordability

Aubrey D.N.J. de Grey

Contents

22.1	Cancer as a Component of Aging: The Challenge	667
	22.1.1 The Geriatric Illusion: Comprehensive Progress Presenting as Stasis	668
	22.1.2 SENS: A Plausible Route to Hugely Accelerated Progress	669
	22.1.3 The Unpalatable Prospect: Cancer as Our Main Killer	671
22.2	How can Brawn Reliably Defeat Brain?	671
	22.2.1 Cleverer than We Are – Much Cleverer	673
	22.2.2 Name Your Poison: One Blunderbuss or Ten Novice Marksmen	674
22.3	Feasibility of WILT: Selected Concerns	676
	22.3.1 Rendering Cells Telomere-Elongation-Incompetent In Situ	676
	22.3.2 Installing Ex Vivo-Modified Stem Cells	678
	22.3.3 Putative Roles of Telomerase Over and Above Telomere Maintenance	680
22.4	The Affordability of Such an Elaborate Therapy	681
	22.4.1 Relative Versus Absolute Benefits	681
	22.4.2 The Necessity of Regular Repeat Therapy	682
	22.4.3 Cost of Developing WILT in the First Place	682
22.5	Conclusion	682
References		683

22.1 Cancer as a Component of Aging: The Challenge

Those of a certain age, or a certain profession (oncology), will recall as if it were yesterday the words with which President Richard M. Nixon announced the launching of the War on Cancer (Nixon 1971). The centrepiece of his 1971 policy initiative

---

A.D.N.J. de Grey (✉)
SENS Foundation, PO Box 304, El Granada, CA 94018, USA
e-mail: aubrey@sens.org

was a sharp rise in the budget of the National Cancer Institute, and the motivation for it was the expert advice he had received to the effect that such an investment had a good chance of leading to a comprehensive cure for cancer within as little as five years.

As we all know, it didn't happen. The fact that public and philanthropic funding for cancer research have nonetheless been robustly sustained (National Institutes of Health 2006) is a cogent and telling rejoinder to those who claim that we must always lower expectations of the rate of biomedical progress for fear that dashed hopes will dissipate public support – but that is not my topic here. I wish to focus instead on *why* it didn't happen – and we can say with hindsight that there is one overwhelming reason, a reason that biologists have in fact known for a century: in Crick's succinct words (Dennett 1984), evolution is cleverer than you are. The more we discover, and subvert, particular tricks that cancer cells have employed to evade the assaults on them by the body's natural defences and by our medical efforts, the more we discover that, deprived of that trick, the cancer will simply turn up a new one.

22.1.1 The Geriatric Illusion: Comprehensive Progress Presenting as Stasis

But there's more. This depressing reality is bad enough today, but there is a very clear and present danger that it will get even worse – much worse – in the decades to come. Not because of increased exposure to carcinogens, but because our progress against *other* causes of death may well accelerate sharply.

It is incorrect to conclude that, because just as high a proportion of Americans die of cancer now as in 1971, the War on Cancer was a complete failure. There was a much cheaper way to cause a decline in cancer deaths, and one that would have been much surer of success: to close down research on cardiovascular disease, thus slowing our progress in postponing death from that cause and thereby "saving" huge numbers of people from living long enough to die of cancer. This policy was not adopted, even though death from a myocardial infarction occurs, as often as not, without a hint of prior symptoms – just the sort of "compression of morbidity" that so dominates the stated aspirations of influential geriatricians and gerontologists alike (Kalache et al. 2002). Lest you suspect that I feel we should indeed have rowed back on research into the postponement of death from heart attacks, let me reassure you that I do not: firstly because many people suffer severe symptoms from cardiovascular disease for much *longer* than the average time that one suffers from cancer, and secondly because such people generally seem rather to prefer still being alive, even in a diminished state of health, than being dead. But to return to my theme: the absence of a significant change in recent decades in the proportion of deaths from cancer versus other major age-related killers is a demonstration not that the war on cancer has failed, but simply that it has succeeded to roughly the same extent as have the less loudly declared wars on those other diseases.

Let us, then, extrapolate these achievements of the recent past into the future. And let us do so not by simple mathematical extrapolation, but by examining the precursor technologies, currently only being explored in the laboratory but with the potential for clinical application less far in the future than Nixon's speech is in our past, that may postpone the industrialised world's major killers other than cancer. For, just as in decades gone, it is our achievements in those fields that will define our goal in the cancer field: if progress against cancer lags behind, the proportion of deaths from cancer will rise.

22.1.2 SENS: A Plausible Route to Hugely Accelerated Progress

Unfortunately for any complacent oncologists out there, the prognosis on this point is (from their point of view) not good. The underlying molecular and cellular causes of cardiovascular disease, neurodegeneration and type 2 diabetes (to name just three) are still very incompletely understood at the *mechanistic* level, but they are really rather well understood at the *structural* level. That is to say: in each case there are clear molecular and/or cellular changes, occurring in the body throughout life, which are confidently understood to be (jointly) necessary and sufficient precursors of these diseases. What is less (albeit increasingly) well understood are the mechanisms whereby our normal metabolism ongoingly gives rise to these changes and those whereby the changes eventually give rise to the diseases. The same applies to the "susceptibilities" that we acquire in later life – changes such as loss of muscle mass and decline in immune function, which we do not honour with the status of "disease," but which play just as great a part in our increasing risk of imminent debilitation and death.

Your first reaction to the above may be to be unimpressed: after all, one oasis of knowledge flanked by two vast expanses of ignorance does not sound much like a basis for optimism. Closer inspection, however, leads to an altogether sunnier conclusion. Since – as for any machine – the body's function, including its likely remaining longevity, is determined wholly by its structure, we can make the following bold but undeniable statement: *reversal* of the ongoing molecular and cellular changes to which I have just referred, restoring the structure of a middle-aged adult's body to what it was at a younger age, would comprehensively uncouple the causes of those changes from their consequences. It would postpone pathology, simply by severing the chains of events that lead to that pathology. And it would do so whatever the (incompletely understood, as just noted) details of how those molecular and cellular changes either come about or give rise to pathology. In short, reversal of the structural changes occurring in the body throughout adulthood would thoroughly sidestep our ignorance of the detailed mechanistic basis of aging.

The above would, of course, be of purely academic interest if we did not in fact have any prospect of achieving this "structural rejuvenation" of the body – but we do. Some years ago I proposed (de Grey et al. 2002) that all such changes can be classified into seven major categories, for five of which a highly promising approach to reversal indeed exists. I refer the reader to my past publications for

Table 22.1 The seven major types of molecular and cellular damage that accumulate in aging, and foreseeable ways to repair five of them. For details, see de Grey et al. (2002)

Type of damage	Proposed repair strategy
Cell loss, cell atrophy	Stem cells, growth factors, exercise
Senescent/toxic cells	Ablation by immunisation or suicide gene therapy
Intracellular aggregates	Microbial hydrolases (transgenic or by enzyme therapy)
Extracellular aggregates	Immune-mediated phagocytosis
Extracellular crosslinks	AGE-breaking molecules or enzymes
Mitochondrial mutations	
Nuclear [epi]mutations	

details; Table 22.1 provides an overview. This classification appears to be standing the test of time: indeed, it could have been made a quarter of a century ago, since all seven types of "damage" were by 1982 the subject of widespread suspicion within biogerontology with regard to their contributions to aging and age-related pathology.

Five out of seven ain't bad, but one might be justified in the view that it also ain't good enough. My story is not over, however. The sixth category of damage, mitochondrial mutations, is in my view not likely to be amenable to outright repair (such as replacement of mutant mtDNA by non-mutant genomes), but there now seems to be a very strong prospect that such mutations can instead be *obviated*: made harmless, in other words, by the elimination of any pathways whereby they might contribute to age-related pathology. Obviation is a conceptually tricky business, since (as just mentioned) we know so little about how damage actually translates into pathology; but if an intervention targets the very first step that any such mechanism must involve, the prospects are good. Accordingly, the approach of "allotopic expression" – developing modified versions of the 13 protein-coding genes of the mtDNA, which can be introduced into the nuclear genome by gene therapy and complement any mutations that may occur in the mtDNA itself – seems highly likely to subvert any consequences of such mutations (de Grey 2000).

Six down, one to go. Mutations of the nuclear genome (or epimutations, defined as adventitious changes to the modifications of bases, histones or chromatin that determine gene expression) can of course cause cancer, and in principle they can also cause all manner of other failures of cellular function. If such changes result in the cell's death, or in its adoption of a distinctive state causing it to resist normal apoptotic signals, SENS already has them covered (see Table 22.1). In principle, however, such changes could be altogether more non-specific – degrading the performance of one cell in one way, the next in another, and so on. This might be daunting, or even impossible, to reverse. But here we have reason, paradoxically, to be grateful that we are susceptible to cancer. A single cell possessing the "right" constellation of mutations can kill us rather quickly by dividing uncontrollably, but dysfunction not related to cell cycle control can only be harmful if it affects a substantial proportion of the cells in a given tissue. This numerical disparity is only reduced, not eliminated, by the fact that we have many cancer-specific defence mechanisms. Thus, there is good reason to believe that cancer is the only consequence of nuclear mutations or epimutations that has pathological consequences in a currently normal lifetime (de Grey 2007).

22.1.3 The Unpalatable Prospect: Cancer as Our Main Killer

I can now, finally, return to the central topic of this section. The bottom-line conclusion of the foregoing analysis is that all aspects of age-related pathology other than cancer are – not certainly, of course, but quite probably – destined for comprehensive defeat within only a few decades. That is clearly a huge acceleration in progress when compared to the past several decades of medical advance against such scourges. Let us now suppose that, by contrast, progress in postponing the onset of and death from cancer continues only at a rate comparable to that seen in recent decades. I hardly need spell out the result that will befall us: a rather modest rise in life expectancy, combined with a calamitous rise in the proportion of our population who die of one of the most gruesome age-related diseases known to humanity.

It was the appreciation of this prospect that led me to acknowledge the need to explore, now, all conceivable avenues for the avoidance of such a future by the identification of anti-cancer modalities that might allow the postponement of cancer to keep pace with the postponement of the rest of aging. The result was WILT.

22.2 How can Brawn Reliably Defeat Brain?

Tumours are constantly evolving, and evolution is cleverer than we are – but we have many tools not available either to the cancer cell or to our inbuilt anti-cancer defences. How can our technological superiority overcome the cancer's greater ingenuity?

WILT is described in detail elsewhere (de Grey et al. 2004; de Grey 2005) so I provide only a brief summary here. The central idea is that mutations which alter the level of expression of genes are frequent, but the creation of entirely new genes is a vastly more intricate process – so intricate, indeed, that it will virtually never happen in a collection of "only" a trillion cells in a period of "only" a few decades. Thus, deletion from all our cells of genes necessary for extending telomeres would render tumours unable to transcend the Hayflick limit whatever genes they turned on or off, with the result that they would be initiated just as they naturally are – indeed, very probably more often than naturally, because critically short telomeres are highly mutagenic and oncogenic – but would wilt away before reaching a clinically problematic size. Selective targeting of cancerous or precancerous cells for this modification would be inadequate, because some such cells would already possess mutations that protected them from undergoing the modification; hence, it must be done pre-emptively to as many of our cells as possible (or, at least, the mitotically competent ones). This would clearly have the rather important side-effect of also rendering mortal the stem cells of our continuously renewing tissues (blood, gut, epidermis and lung) – but such problems could in principle be averted by replenishing these stem cell pools every decade or so with autologous stem cells that also lacked telomere-extension genes but had had their telomeres extended *ex vivo* with exogenous telomerase. This concept is depicted in Fig. 22.1.

The feasibility of WILT, even with the benefit of decades of future advances in gene therapy and stem cell therapy, is clearly very far from obvious. In the next section I will discuss some of the more daunting challenges to implementing it. By way of concentrating minds, however, in this prior section I critique the "one more push" anti-cancer philosophy: the idea that we need not explore such radical approaches as WILT because altogether less scary alternatives will almost certainly emerge just

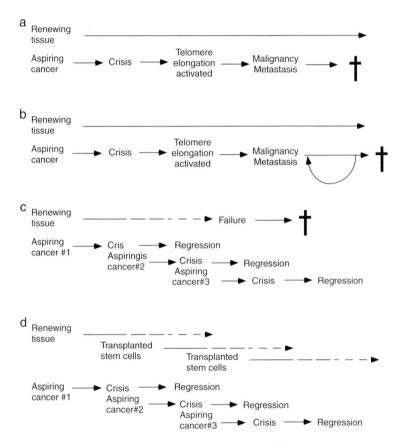

Fig. 22.1 The WILT concept. Metastatic cancers kill a typical untreated patient (**a**) rapidly and a beneficiary of contemporary treatments (**b**) only slightly less rapidly, on account of the genomic instability and inventiveness of such tumours. Uncompensated deletion of telomerase genes (**c**) prevents this but kills much sooner by preventing the maintenance of continuously renewing tissues such as the blood. In WILT, deletion of telomerase genes is combined with periodic stem cell therapy to maintain such tissues indefinitely, so that death from either cause is avoided (**d**)

as soon. Optimists on this point can be divided into two broad camps: those who pin their hopes on therapies that are truly clever enough to exhaust a cancer's options, and those who favour a "blunderbuss" attack that requires the cancer to solve a multitude of challenges simultaneously.

22.2.1 Cleverer than We Are – Much Cleverer

The "cleverer therapies" approach appeals strongly to the oncologist's self-esteem and is thus seductive – but a moment's reflection reveals that it is precisely the approach we have been following all along and with such modest success. Current thinking on such matters, when analysed dispassionately, reveals how frighteningly cancer research has departed from all normal levels of theoretical rigour in prioritising research directions. The most prominent current example of this is the mania over so-called "cancer stem cells" – cells that, though in a minority in a typical tumour, probably give rise to the bulk of that same tumour's cellular content a few years down the road because, just like normal stem cells, they can divide without losing much of their capacity to divide again (Lobo et al. 2007). Because a tumour's longevity relies on such cells, great effort is currently being directed at the development of therapies that specifically target them. Er... First of all, the fact that the cancer stem cell may be the required target is no reason to avoid killing the rest of the tumour. But more crucially, the whole reason why current anti-cancer therapies fail to kill cancer stem cells (and thus fail to kill cancers) is that cancer stem cells very closely resemble cells that we absolutely must not kill too many of with our therapies, namely our normal stem cells. As such, the contemporary focus on cancer stem cells is not really a new idea at all, but merely a restatement of the problem that oncologists have always faced, viz. that it's easy to kill a cancer cell, but not to kill it selectively.

A potential way out of the above might be thought to be provided by another increasingly fashionable "cleverer" approach in cancer research: the improved characterisation of individual tumours. If a therapy is chosen purely on the basis of where a tumour has arisen and how large it is (and perhaps one or two other gross histological features), there is much room for improvement: two tumours may arise in the same tissue and develop in essentially the same way as a result of quite different spectrums of mutations, and those differences may point rather clearly to differences in the likely relative efficacy of particular therapies. This has motivated the development of highly sophisticated molecular analyses of tumour gene expression, the results of which allow the subclassification of histologically hard-to-distinguish tumours into ones deemed suitable for one therapy or for another (Golub et al. 1999). While such work will indeed save lives, by addressing the genetic basis for the growth of *most* of the target tumour, the very fact that multiple alternative genetic misadventures can cause the same cellular phenotype is clear proof that in many cases a predominantly "type 1" tumour will simply regrow as a "type 2" tumour when the basis for type 1 growth is medically undercut. Because of this, treatment by improved characterisation can realistically provide only a modest postponement of cancer-related debilitation and death on average: quite probably not nearly enough to keep pace with therapies that combat the other major age-related killers. A particularly clear demonstration of this is provided by the drug Gleevec, which targets chronic myelogenous leukaemia on the basis of a particularly inviting mutation, the Philadelphia translocation. Gleevec's molecular target is a fusion protein that is not encoded in our non-mutant genome at all, yet is produced by every single cell of a cancer that has suffered this particular chromosomal rearrangement – and, indeed, patients with this type of cancer have been greatly helped by this drug. After several years of use, however, relapses began to be reported: the patients' cancers had returned, because of secondary mutations that rendered the drug ineffective (Kantarjian et al. 2006). The key point here is that this was a type of cancer that apparently did *not* have multiple subtypes distinguishable by additional molecular analysis, yet it *still* revealed itself to have multiple options at its disposal when forced to call upon them.

22.2.2 Name Your Poison: One Blunderbuss or Ten Novice Marksmen

So to the other mainstream approach to radically improving our current anti-cancer arsenal: multiple simultaneous therapies. This concept comes in a number of forms, Most simply, we just apply a variety of different drugs or vaccines that seem likely to undercut different defences that the cancer cell has erected against us. A second approach is to exploit the complexity of the immune system and target multiple cell cycle control pathways via antigens presented on the cancer cell surface. A third, which has been in clinical use for some time, is to "design" an immunotherapy specific to the individual cancer, by using cells taken from a surgically removed primary tumour as antigens for the production of antibodies against residual cancer cells (in particular, metastases) (Berd 2004). Other variations on these themes can be imagined.

Unfortunately, all such "blunderbuss" approaches possess two fatal (in my view) shortcomings.

The first has already been indirectly alluded to in the previous section. What makes a cancer hard to kill is not that the individual cancer cells are hard to kill, but rather that they (and especially the "cancer stem cells" that ensure the cancer's continued growth) are hard to distinguish from other cells on whose survival we rely for our own survival. Another way to state this problem is that cancer therapies have a rather low "therapeutic index" – range of dosage that is effective on the cancer cell but not on non-cancer cells. In reality, of course, a drug's therapeutic index is not defined at the level of the cell but at the clinical level – and the disparity between the two is different for cancer than for other diseases, because cancer is a disease of cell proliferation. Thus, in the case of cancer we will accept a much higher "false positive" rate (non-cancer cells killed) than we might for other diseases, so long as we have a really low "false negative" rate (cancer cells surviving). But this can only go so far, hence the need to avoid excessive doses of chemotherapy and such like. Now: the whole idea of applying multiple therapies to a cancer at the same time is that they will address distinct aspects of the cancer's methods for growth. This inevitably means that the individual therapies will incur different types of collateral damage – they will kill different (though typically overlapping) populations of non-cancer cells. Accordingly, each component therapy effectively reduces the therapeutic index of the other therapies with which it is co-administered.

The second difficulty with combination therapy is, in a sense, the mirror image of the first. It turns out that some of the most formidable defences that cancers employ in resisting our attacks are highly versatile. One is so versatile that it has even been named for that versatility: it is a membrane transport protein called the multi-drug resistance protein, and it works by simply shuttling out of the cell anything that it doesn't like the look of (Yasuhisa et al. 2007). (Why not just co-administer a drug that suppresses this transporter, along with the chemotoxic drug? Firstly, cancer cells will simply upregulate its expression to counter the suppression. Secondly and even worse, suppression of the transporter will render non-cancer stem cells more sensitive to the chemotherapy, whose therapeutic index will thus be reduced.) Similarly, though there are many ways to bring a rogue cell to the attention of the adaptive immune system, most of them rely on the target cell obligingly presenting antigens on its surface in the cleft of the major histocompatibility complex 1 (MHC1), and some cells just stop making MHC1. (This is such an obvious trick that we actually have cells, NK cells, to get rid of cells expressing no MHC1 – but cancers have counter-tricks, which exceed the scope of this chapter.)

It is perhaps worth looking closely at the specific anti-cancer modality that is conceptually the most similar to WILT: reversible, drug-based inhibition of telomerase (and of genes needed for ALT, as and when they are discovered). Perhaps because of its similarity to WILT, analysis of this approach reveals particularly starkly why something as ambitious as WILT will probably be needed if we are to control cancer indefinitely. First of all, in contrast to the permanent nature of genetic modification, any reversible suppression of expression of a given gene must be actively maintained by continued application of the suppressor (unless the reversal is effected by applying a second drug rather than just withdrawing the first, which is functionally equivalent to implementing WILT and then reintroducing telomerase by insertional gene therapy). This means that the cell constantly has the opportunity to mutate to exclude the drug. Secondly, the drug will inevitably inhibit telomerase in non-cancer cells where telomere elongation is needed (the stem cells of continuously renewing tissues), so it must be periodically withdrawn to allow those tissues to recover – and that will allow the cancer to recover too. Drug-based telomerase inhibition certainly has its place in the anti-cancer arsenal, as has been compellingly shown by recent results (Keith et al. 2007), but that place is short-term treatment, not permanent escape.

I hope, in this section, to have disabused you of any confidence you may have had that gradualist approaches to defeating cancer are definitely going to see us through. I do not claim that these approaches will definitely fail: all I claim is that their prognosis is inadequate to remove the incentive to look for something else. At the time of writing, WILT is the only "something else" on the table. Nothing would make me happier than the discovery that WILT can be consigned to the dustbin of unnecessarily elaborate ways to postpone aspects of aging – but nothing would make me unhappier than the failure to pursue WILT and the subsequent realisation that nothing else has turned up.

22.3 Feasibility of WILT: Selected Concerns

As noted earlier, WILT consists of two types of therapy:

– deletion of genes required for telomere maintenance; and
– regular replenishment of stem cell pools in thereby "mortalized" tissues.

Within each of these components lie many daunting challenges. I address a selection of these below.

22.3.1 Rendering Cells Telomere-Elongation-Incompetent In Situ

Our various cell types can be divided for present purposes into four classes:

1) postmitotic;
2) normally quiescent but mitotically competent: differentiated;
3) normally quiescent but mitotically competent: stem cells; and
4) rapidly dividing and/or short-lived.

Postmitotic cells, being constitutively unable to undergo mitosis, do not give rise to cancers and thus are not necessary targets either for telomere maintenance removal or for compensatory stem cell therapy. (I should stress that here I refer only to cells that have undergone gross structural alterations precluding the mechanics of mitosis, such as neurons or skeletal muscle fibres: I do not include cells that are non-dividing solely because they express proteins that suppress mitosis, as in the case of replicative senescence in vitro, since that suppression could theoretically break down.) Category 4 is also a low-priority target: such cells (either the short-lived terminally differentiated cells in continuously renewing tissues, such as keratinocytes or erythrocytes, or else the rapidly-dividing but developmentally committed transit amplifying cells that give rise to them) are virtually certain to die before they can accumulate the wealth of mutations necessary to turn them into a cancer, but even more conclusively they are the progeny of cells of type 3 (of which more in a moment) so are bereft of genes that have been removed from type-3 cells. Here I will, therefore, focus on cells of types 2 and 3. It should be noted that some cell types, such as hepatocytes, fall at the boundary between types 2 and 3 – they probably only divide on demand, but that demand is rather high on account of the high rate of cell death.

Differentiated cells that divide only "on demand" are present in many tissues. Most are covered by the collective term "mesenchymal cells" – fibroblasts, adipocytes, glia, muscle satellite cells, chondrocytes and such like. Certain epithelial cells also fall into this category, however: hepatocytes, for example. Such cells do not typically express detectable levels of the default telomere maintenance enzyme, telomerase; this probably explains why only around 50% of tumours arising from them use telomerase as their telomere maintenance mechanism, the remainder instead using ALT, "Alternative Lengthening of Telomeres" (Johnson and Broccoli 2007). They therefore present two main challenges. Firstly they must be genetically modified in situ rather than ex vivo, a procedure that deprives us of the ability to select post hoc for cells that have been modified in the desired way and no other. And secondly, since they so often use ALT as their telomere maintenance mechanism, we must delete genes both for telomerase and for ALT – and the latter remain obstinately undiscovered.

In situ gene therapy has had a rollercoaster ride over the past 20 years, as all readers will know. This has somewhat overshadowed the fact that it works really quite well in the laboratory, i.e. in contexts where safety is not an overriding issue. I have predicted (de Grey 2006) that the history of gene therapy, and most particularly the year-long suspension of essentially all gene therapy clinical trials in the wake of Jesse Gelsinger's death, will be the main trigger for society's flight from the dictum of "first do no harm" that served the medical profession so well since Hippocrates but has now, in my view, had its day – but I digress.

In the context of WILT the main requirement is not effective insertional gene therapy (introduction of a new gene in a more-or-less immaterial genomic location) but gene targeting, the alteration of specific pre-existing stretches of sequence, specifically those encoding one or both of the subunits of telomerase and its presumptive ALT counterpart. When WILT was first formulated, this seemed highly challenging: a few techniques were under development, but they all featured either a distinct lack of multi-laboratory reproducibility or an extremely low "efficiency" (proportion of cells appropriately modified) (Suzuki 2008). Moreover, their incidence of random modification of unintended sequences was high. Fortunately, this dismal scene is brightening apace.

Perhaps the most promising technology for in vivo gene targeting at present is zinc finger nucleases: fusions combining proteins that bind particular DNA sequences with high specificity and enzymes that create double-strand breaks at the site in question (Porteus et al. 2003). When such constructs are introduced into cells (using standard protein transduction techniques such as HIV-TAT, or by introducing the DNA encoding the protein using lentiviral vectors (Lombardo et al. 2007)) in concert with plasmids whose sequence is similar but appropriately modified, the cell's homologous recombination-based double-strand break repair machinery often uses the plasmid as template rather than the native chromosomal homologue, resulting in the desired modification of the target gene. This technology has already given highly promising results (Urnov et al. 2005; Lombardo et al. 2007). And it is not alone: an unrelated approach involving in vitro evolution of bacteriophage integrases also shows great promise (Sclimenti et al. 2001; Held et al. 2005).

One point that bears stressing is that, even though any mitotically competent cell in the body can potentially turn cancerous and kill us, most of them do not: hence, the postponement of cancer is a numbers game in which we (on average) gain some life from modifying a proportion of cells, more from modifying more, and so on. Thus, therapies that suffer from diminishing returns in terms of the difficulty of reaching a given proportion of cells do not so suffer in terms of the life extension that they confer. Suppose that an intervention that deletes telomere maintenance genes successfully modifies 90% of our cells. In the idealised situation where all cells were equally accessible, doubling the dose, or frequency, or potency of the intervention would be expected to modify 99% of the cells – a cell would, in essence, need to "get lucky" (be among the unaffected 10%) twice in order to retain telomere-elongation potential. In practice, however, some cells will be harder to modify than others, so maybe the double-strength therapy will only result in 95% of cells being modified. But that will still give twice the amount of postponement of death from cancer that modifying 90% of the cells would give. In this regard, a heartening feature of "gene targeting" technologies such as those mentioned above is that, because a successfully modified gene will not be further altered if targeted again, the efficiency of modification is not a major issue: so long as the fidelity of the method (avoidance of random modifications elsewhere in the genome) can be made extremely good, tissues can be treated arbitrarily often.

Our ignorance of the genetic basis of ALT remains a major Achilles' heel of WILT, but it too may be about to succumb to research efforts. It has long been known that ALT works by recombination, but this leaves many alternatives, which is the main reason why a "candidate gene" approach to characterising ALT has thus far failed. A somewhat inconspicuous result of telomere lengthening in telomerase knockout mice, reported in the discussion of a paper whose main focus was elsewhere (Herrera et al. 2000), has suggested to me that the genes encoding class switching in the spleen's germinal centres may be the long-sought mediators of ALT. Recently this idea has been powerfully reinforced by a report (Liu et al. 2007) of the same phenomenon in the early embryo, where a gene central to class switching, AID, is also expressed (Morgan et al. 2004). I have argued elsewhere that ALT must, like telomerase, be encoded by genes that are firmly suppressed in most cell types, and this narrows the field of candidates dramatically.

22.3.2 Installing Ex Vivo-Modified Stem Cells

When we consider cells of "type 2" above, i.e. the stem cells of continuously renewing tissues such as the blood and the gut, we encounter a very different set of challenges. Since ALT is so rare in tumours derived from such tissues, we may be able to get away for longer (though not forever!) with only targeting telomerase.

Secondly, the actual ex vivo genetic modification of the requisite cells is vastly easier, because techniques can be used that achieve the desired modification only in a very small proportion of treated cells, just so long as the modification is designed to allow the suitably modified cells to be distinguished and separated from those that have been inappropriately (if at all) altered – not a great challenge with today's molecular techniques – and then expanded to the required numbers (which, in the case of the blood, may be very small numbers (Spangrude et al. 1988)). Thirdly, in decades to come some epithelial tissues into which stem cells may not be able to migrate in sufficient numbers from the blood stream may be relatively easily maintained by wholesale transplantation of tissue-engineered organs: this might be a preferred option for the liver or kidneys, for example. But the exploitation of this opportunity requires a few additional obstacles to be overcome.

The main one is to eliminate our natural, telomerase-positive stem cells. Stem cells in the bone marrow and other continuously renewing tissues occupy "niches" – locations that maintain their milieu in a state conducive to the cells' retention of "stemness." Engineered stem cells injected (or carried by the circulation) into the vicinity may be unable to invade such niches in adequate numbers unless the niches have been for some reason vacated by their original residents (though a resident cell's resistance to replacement may be less than was once thought (Quesenberry et al. 2001)). An individual that had been rendered wholly telomerase-negative would probably be a fairly straightforward recipient of such treatment, because the uncompensated loss of telomeres of that person's stem cells would progressively cause them to differentiate and/or apoptose, conveniently vacating their niches for occupation by the cells introduced in the next round of stem cell replenishment. But what about the first time a patient is treated, when their niches are full of cells with perfectly functioning telomerase that may not so readily relinquish their nest?

In the case of the blood, a recent study (Czechowicz et al. 2007) reports a remarkably simple approach to this: introduction of an antibody that inactivates a cell-surface protein expressed by haematopoietic stem cells and thereby causes them to be lost from their niches (and, it seems, to die). After a carefully chosen interval (nine days in the case of mice), the replacement stem cells were injected. They engrafted rapidly enough to prevent any hematological problems, and analysis of the mice's blood some time later revealed that over 90% of the stem cells had been replaced.

A more elaborate, but possibly even more attractive, solution to this problem is to exploit a technique that has been in development for some years, being promising for anti-cancer therapies even independent of WILT. The limited efficacy of existing chemotherapeutic agents has two main causes: firstly the "therapeutic index" problem noted earlier, and secondly the ubiquitous "evolution is cleverer than you are" problem whereby the cancer cell may mutate to resist the therapy. Some examples of the latter have been molecularly characterised, and this has allowed researchers to turn cancer's ingenuity on itself (Hobin and Fairbairn 2002).

Specifically: consider a normal, non-cancer stem cell that has (by whatever means, such as by our own hand) acquired such a chemoresistance mutation. Now suppose that the chemotherapy in question is applied. At standard dose it kills a lot of the cancer, but not all, and it also kills quite a few stem cells, but not our engineered stem cell. Now let's raise the dose. Virtually all the cancer is killed. Unfortunately, most stem cells are also killed, which – other things being equal – would kill the patient. But wait! – the patient has some chemoresistant stem cells, which happily survive even this higher dose and duly repopulate the patient. Success!

The point for WILT, of course, is that this should work just as well if the chemoresistant stem cell has *also* been engineered to lack genes for telomerase and ALT. If such cells are injected just prior to high-dose chemotherapy, a double whammy will result: any cancer (whether or not already clinically identified) will be eliminated, and stem cell niches will be divested of their resident cells and thereby made available for the circulating engineered cells. The pre-injection part is probably appropriate only for the blood; for gut, lung and skin a post-chemotherapy approach may be favoured. If you doubt that this would be possible in practice, consider that in the case of the skin this is not materially different from standard contemporary burns therapy or cosmetic surgery (Roh and Lyle 2006) and that for the gut it has already been demonstrated in rodents (Tait et al. 1994) and dogs (Stelzner and Chen 2006).

It should also be taken into account that there is no need to restrict the in situ gene targeting approach to mesenchymal tissues. While the blood will almost certainly be easier to render incapable of telomere elongation using ex vivo manipulation than by in situ methods, the skin may well be treatable without actually removing and replacing the epidermal layer, and so may the gut.

22.3.3 *Putative Roles of Telomerase Over and Above Telomere Maintenance*

In recent years, a number of groups have reported evidence from mice that telomerase has roles in cell survival and/or proliferation in addition to its function in maintaining telomere length (reviewed in Sung et al. 2005). At this point, however, this work does not give grounds for appreciable concern that such roles will impact the feasibility of WILT, for several reasons:

- most, if not all, of the reported data can be explained in terms of the balance between cell division-independent telomere shortening (especially via oxidative damage) and telomerase-mediated telomere elongation;
- mouse data may be misleading, in that organisms that express telomerase very sparingly, such as humans, will have experienced selective pressure to evolve alternative methods to achieve whatever functions telomerase may perform in telomerase-profligate organisms such as mice;
- telomerase can be inactivated by deleting either the gene for its catalytic subunit or that for its template subunit; deletion of the latter would not affect functions

performed by the catalytic subunit on its own, which are what has been reported in most cases;
– altering six amino acids of the telomerase catalytic subunit disrupts its targeting to the telomere, and thereby abolishes its telomere-elongation activity in cellulo, without an effect on its biochemical function (Banik et al. 2002). Accordingly, this type of modification – which is easily complex enough never to be reverted by spontaneous mutation – is available as a "plan B" if entirely deleting the gene proves problematic.

22.4 The Affordability of Such an Elaborate Therapy

I am a British citizen, and as such I have grown up with the idea that health care is naturally free at the point of delivery. I am well aware that the US citizens among my audience are not in the same boat. I will therefore begin this section by giving you the bottom line: I believe that WILT – just like the rest of SENS – will be available to all those old enough to need it, both in nations that favour a public health care system and in those that do not. This is despite the fact that, in my view, WILT will probably be every bit as expensive as initial impressions of its complexity might suggest.

22.4.1 Relative Versus Absolute Benefits

My good friend Jay Olshansky's most prominent publications have been those that highlight how alarmingly little we can extend lifespan or healthspan by combating specific diseases, and thus how important it is to postpone their major "risk factor" (i.e., precursor), aging. One statistic he has derived is that the complete elimination of cancer would raise average age at death in the USA by less than two years (Olshansky et al. 1990). This might seem to demonstrate that the amount most of us would be willing to spend on avoiding death from cancer may be quite modest – far short of the likely cost of a therapy that involves so many state-of-the-art components as WILT does.

However, a proper analysis of this question must begin not from the life-extension benefit that we would enjoy *today* as a result of eliminating death from cancer, but from the benefit that would result at the time that WILT could become available. As noted in the first section of this chapter, it is quite possible that all the other causes of death that remain highly prevalent in today's industrialised world will have been (or, at least, will be in the process of being) brought under very comprehensive control by the time that WILT arrives. With these other killers removed, the difference in life expectancy that would be made by a truly complete elimination of death from cancer would be very great: decades if we consider the likely efficacy of these non-cancer therapies at that time, and centuries if we additionally take account of the rate at which those therapies will see subsequent improvement

(Phoenix and de Grey 2007). In such circumstances, the investment of a substantial proportion of an individual's – or of society's – wealth in a therapy that prevents cancer may seem altogether more attractive.

22.4.2 The Necessity of Regular Repeat Therapy

A feature of WILT that has intimidated many observers is the fact that the stem cell replenishment must be repeated every decade or so. The fear is that one may be putting oneself at risk of, for whatever reason, being unable to afford the therapy on schedule, with the result that even though one would not die of cancer, one would die of starvation and anaemia as one's gut and blood (respectively) lost the ability to self-renew.

This concern suffers, in my view, from the same flaw as the one I described above: it evaluates the importance of the treatment in the context of today's risk of death from other causes, rather than in the likely context that those causes have been essentially eliminated. Once society is faced with the choice of (a) having cancer as the only remaining age-related killer, or (b) spending quite a lot of money preventing it, it may exhibit rather less hesitation in choosing option (b) than it would today – even if there is a stringent requirement to make the prevention treatment very reliably available on schedule.

22.4.3 Cost of Developing WILT in the First Place

Suppose that pharmacological telomerase/ALT inhibition is 95% effective and costs 1000 times less than WILT, whereas WILT is 100% effective but costs 1000 times more. Is the greater marginal benefit of WILT sufficient to drive the economics of developing and using it? Or could it be that the need to develop stem cell therapies for aging in general will pay for the vast majority of the cost of preparing telomerase-deficient stem cells, which could be used instead of regular stem cells at little or no additional cost, and that the need to develop gene targeting technologies for reasons entirely unrelated to WILT will pay for the vast majority of the costs of deleting telomerase and ALT in vivo? Only time will tell, but my view on the basis of current evidence is that the latter is the case: most of the technologies upon which WILT depends are needed for other purposes too, which is why they are already being pursued. Moreover, when we have 99% effective treatments for all other causes of death, a 95% effective treatment for cancer may not seem quite so adequate as it may seem today.

22.5 Conclusion

WILT is a massively ambitious approach to defeating cancer. However, cancer is a hard problem – much harder than mainstream expert opinion had us believe back in 1971, and, I suspect, harder than quite a number of cancer researchers believe it to be even today. In particular, it may be a considerably harder problem than any of the other, also very hard, problems that comprise human aging. Thus, it is in grave danger of surviving as a threat to life long after those other killers have been swept aside. The way to minimise the risk of this unpalatable scenario is to think out of the box now, and identify radical new approaches, rather than to continue with the gradualist approach of attempting possibly over-simplistic refinements of therapies that have previously failed. WILT is such an approach. No one would be happier than I if it turns out to be unnecessary – but we will live (all too briefly) to regret it if we reject WILT as unnecessary until it is too late to save ourselves.

References

Banik SS, Guo C, Smith AC, Margolis SS, Richardson DA, Tirado CA, Counter CM (2002) C-terminal regions of the human telomerase catalytic subunit essential for in vivo enzyme activity. Mol Cell Biol 22(17):6234–6246

Berd D (2004) M-Vax: an autologous, hapten-modified vaccine for human cancer. Expert Rev Vaccines 3(5):521–527

Czechowicz A, Kraft D, Weissman IL, Bhattacharya D (2007) Efficient transplantation via antibody-based clearance of hematopoietic stem cell niches. Science 318(5854):1296–1299

de Grey ADNJ (2000) Mitochondrial gene therapy: an arena for the biomedical use of inteins. Trends Biotechnol 18(9):394–399

de Grey ADNJ (2005) Whole-body interdiction of lengthening of telomeres: a proposal for cancer prevention. Front Biosci 10:2420–2429

de Grey ADNJ (2006) Has Hippocrates had his day? Rejuvenation Res 9(3):371–373

de Grey ADNJ (2007) Protagonistic pleiotropy: why cancer may be the only pathogenic effect of accumulating nuclear mutations and epimutations in aging. Mech Ageing Dev 128(7–8): 456–459

de Grey ADNJ, Ames BN, Andersen JK, Bartke A, Campisi J, Heward CB, McCarter RJM, Stock G (2002) Time to talk SENS: critiquing the immutability of human aging. Annals NY Acad Sci 959:452–462

de Grey ADNJ, Campbell FC, Dokal I, Fairbairn LJ, Graham GJ, Jahoda CAB, Porter ACG (2004) Total deletion of in vivo telomere elongation capacity: an ambitious but possibly ultimate cure for all age-related human cancers. Annals NY Acad Sci 1019:147–170

Dennett DC (1984) Elbow room: the varieties of free will worth wanting. Clarendon Press, Oxford

Golub TR, Slonim DK, Tamayo P, Huard C, Gaasenbeek M, Mesirov JP, Coller H, Loh ML, Downing JR, Caligiuri MA, Bloomfield CD, Lander ES (1999) Molecular classification of cancer: class discovery and class prediction by gene expression monitoring. Science 286(5439):531–537

Held PK, Olivares EC, Aguilar CP, Finegold M, Calos MP, Grompe M (2005) In vivo correction of murine hereditary tyrosinemia type I by phiC31 integrase-mediated gene delivery. Mol Ther 11(3):399–408

Herrera E, Martinez-A C, Blasco MA (2000) Impaired germinal center reaction in mice with short telomeres. EMBO J 19(3):472–481

Hobin DA, Fairbairn LJ (2002) Genetic chemoprotection with mutant O6-alkylguanine-DNA-alkyltransferases. Curr Gene Ther 2(1):1–8

Johnson JE, Broccoli D (2007) Telomere maintenance in sarcomas. Curr Opin Oncol 19(4):377–382

Kalache A, Aboderin I, Hoskins I (2002) Compression of morbidity and active ageing: key priorities for public health policy in the 21st century. Bull World Health Organ 80(3):243–244

Kantarjian HM, Talpaz M, Giles F, O'Brien S, Cortes J (2006) New insights into the pathophysiology of chronic myeloid leukemia and imatinib resistance. Ann Intern Med 145(12):913–923

Keith WN, Thomson CM, Howcroft J, Maitland NJ, Shay JW (2007) Seeding drug discovery: integrating telomerase cancer biology and cellular senescence to uncover new therapeutic opportunities in targeting cancer stem cells. Drug Discov Today 12(15–16):611–621

Liu L, Bailey SM, Okuka M, Muñoz P, Li C, Zhou L, Wu C, Czerwiec E, Sandler L, Seyfang A, Blasco MA, Keefe DL (2007) Telomere lengthening early in development. Nat Cell Biol 9(12):1436–1441

Lobo NA, Shimono Y, Qian D, Clarke MF (2007) The biology of cancer stem cells. Annu Rev Cell Dev Biol 23:675–699

Lombardo A, Genovese P, Beausejour CM, Colleoni S, Lee YL, Kim KA, Ando D, Urnov FD, Galli C, Gregory PD, Holmes MC, Naldini L (2007) Gene editing in human stem cells using zinc finger nucleases and integrase-defective lentiviral vector delivery. Nat Biotechnol 25(11): 1298–1306

Morgan HD, Dean W, Coker HA, Reik W, Petersen-Mahrt SK (2004) Activation-induced cytidine deaminase deaminates 5-methylcytosine in DNA and is expressed in pluripotent tissues. J Biol Chem 279(50):52353–52360

National Institutes of Health. Appropriations (2006) NIH Publication No. 06-5:436–438

Nixon RM (1971) Annual Message to the Congress on the State of the Union. http://www.presidency.ucsb.edu/ws/index.php?pid=3110

Olshansky SJ, Carnes BA, Cassel C (1990) In search of Methuselah: estimating the upper limits to human longevity. Science 250(4981):634–640

Phoenix CR, de Grey ADNJ (2007) A model of aging as accumulated damage matches observed mortality patterns and predicts the life-extending effects of prospective interventions. AGE 29(4):133–189

Porteus MH, Cathomen T, Weitzman MD, Baltimore D (2003) Efficient gene targeting mediated by adeno-associated virus and DNA double-strand breaks. Mol Cell Biol 23(10):3558–3565

Quesenberry PJ, Stewart FM, Becker P, D'Hondt L, Frimberger A, Lambert JF, Colvin GA, Miller C, Heyes C, Abedi M, Dooner M, Carlson J, Reilly J, McAuliffe C, Stencel K, Ballen K, Emmons R, Doyle P, Zhong S, Wang H, Habibian H (2001) Stem cell engraftment strategies. Ann N Y Acad Sci 938:54–61

Roh C, Lyle S (2006) Cutaneous stem cells and wound healing. Pediatr Res 59(4 Pt 2):100R–103R

Sclimenti CR, Thyagarajan B, Calos MP (2001) Directed evolution of a recombinase for improved genomic integration at a native human sequence. Nucleic Acids Res 29(24):5044–5051

Spangrude GJ, Heimfeld S, Weissman IL (1988) Purification and characterization of mouse hematopoietic stem cells. Science 241(4861):58–62

Stelzner M, Chen DC (2006) To make a new intestinal mucosa. Rejuvenation Res 9(1):20–25

Sung YH, Choi YS, Cheong C, Lee HW (2005) The pleiotropy of telomerase against cell death. Mol Cells 19(3):303–309

Suzuki T (2008) Targeted gene modification by oligonucleotides and small DNA fragments in eukaryotes. Front Biosci 13:737–744

Tait IS, Evans GS, Flint N, Campbell FC (1994) Colonic mucosal replacement by syngeneic small intestinal stem cell transplantation. Am J Surg 167(1):67–72

Urnov FD, Miller JC, Lee YL, Beausejour CM, Rock JM, Augustus S, Jamieson AC, Porteus MH, Gregory PD, Holmes MC (2005) Highly efficient endogenous human gene correction using designed zinc-finger nucleases. Nature 435(7042):646–651

Yasuhisa K, Shin-Ya M, Michinori M, Kazumitsu U (2007) Mechanism of multidrug recognition by MDR1/ABCB1. Cancer Sci 98(9):1303–1310

Chapter 23
Comprehensive Nanorobotic Control of Human Morbidity and Aging

Robert A. Freitas Jr.

Contents

23.1	A Vision of Future Medicine	686
23.2	Nanotechnology, Nanomedicine and Medical Nanorobotics	687
23.3	Fundamentals of Medical Nanorobotics	689
	23.3.1 Nanobearings and Nanogears	690
	23.3.2 Nanomotors, Nanopumps, and Power Sources	694
	23.3.3 Nanomanipulators	697
	23.3.4 Nanosensors	699
	23.3.5 Nanocomputers	700
23.4	Manufacturing Medical Nanorobots	702
	23.4.1 Positional Assembly and Molecular Manufacturing	702
	23.4.2 Diamond Mechanosynthesis (DMS)	703
	23.4.3 Designing a Minimal Toolset for DMS	705
	23.4.4 Building the First Mechanosynthetic Tools	709
	23.4.5 Next Generation Tools and Components	709
	23.4.6 Strategies for Molecular Manufacturing	712
	23.4.7 R&D Timeline, Costs, and Market Value of Medical Nanorobots	713
23.5	Medical Nanorobot Biocompatibility	717
	23.5.1 Immune System Reactions	718
	23.5.2 Inflammation	719
	23.5.3 Phagocytosis	720
23.6	Control of Human Morbidity using Medical Nanorobots	721
	23.6.1 Advantages of Medical Nanorobots	722
	23.6.2 Curing Disease	726
	23.6.3 Reversing Trauma	736

R.A. Freitas Jr. (✉)
The Institute for Molecular Manufacturing, Palo Alto, CA, USA
e-mail: rfreitas@rfreitas.com

© Copyright 2007–8 Robert A. Freitas Jr. All Rights Reserved

	23.6.4 Cell Repair	751
23.7	Control of Human Senescence using Medical Nanorobots	764
	23.7.1 Nanomedically Engineered Negligible Senescence (NENS)	765
	23.7.2 Nanorobot-Mediated Rejuvenation	775
	23.7.3 Maximum Human Healthspan and the Hazard Function	780
23.8	Summary and Conclusions	782
References		783

23.1 A Vision of Future Medicine

Mankind is nearing the end of a historic journey. The 19th century saw the establishment of what we think of today as scientific medicine. But human health is fundamentally biological, and biology is fundamentally molecular. As a result, throughout the 20th century scientific medicine began its transformation from a merely rational basis to a fully molecular basis. First, antibiotics that interfered with pathogens at the molecular level were introduced. Next, the ongoing revolutions in genomics, proteomics and bioinformatics (Baxevanis and Ouellette 1998) began to provide detailed and precise knowledge of the workings of the human body at the molecular level. Our understanding of life advanced from organs, to tissues, to cells, and finally to molecules. By the end of the 20th century the entire human genome was finally mapped, inferentially incorporating a complete catalog of all human proteins, lipids, carbohydrates, nucleoproteins and other biomolecules.

By the early 21st century, this deep molecular familiarity with the human body, along with continuing nanotechnological engineering advances, has set the stage for a shift from present-day molecular scientific medicine in which fundamental new discoveries are constantly being made, to a future molecular technologic medicine in which the molecular basis of life, by then well-known, is manipulated to produce specific desired results. The comprehensive knowledge of human molecular structure so painstakingly acquired during the previous century will be extended and employed in this century to design medically-active microscopic machines. These machines, rather than being tasked primarily with voyages of pure discovery, will instead most often be sent on missions of cellular inspection, repair and reconstruction. The principal focus will shift from medical science to medical engineering. Nanomedicine (Freitas 1999, 2003) will involve designing and building a vast proliferation of incredibly efficacious molecular devices, and then deploying these devices in patients to establish and maintain a continuous state of human healthiness.

"Physicians aim to make tissues healthy," wrote one early pioneer (Drexler 1986) in medical nanorobotics, "but with drugs and surgery they can only encourage tissues to repair themselves. Molecular machines will allow more direct repairs, bringing a new era in medicine. Systems based on nanomachines will generally be more compact and capable than those found in nature. Natural systems show us only lower bounds to the possible, in cell repair as in everything else. By working along molecule by molecule and structure by structure, repair machines will be able to repair whole cells. By working along cell by cell and tissue by tissue, they (aided by larger devices, where need be) will be able to repair whole organs. By working through a [patient], organ by organ, they will restore health. Because molecular machines will be able to build molecules and cells from scratch, they will be able to repair even cells damaged to the point of complete inactivity. Thus, cell repair machines will bring a fundamental breakthrough: they will free medicine from reliance on self-repair as the only path to healing."

23.2 Nanotechnology, Nanomedicine and Medical Nanorobotics

The only important difference between the carbon atoms in a plain lump of coal and the carbon atoms in a stunning crystal of diamond is their molecular arrangement, relative to each other. Future technology currently envisioned will allow us to rearrange atoms the way we want them, consistent with natural laws, thus permitting the manufacture of artificial objects of surpassing beauty and strength that are far more valuable than bulk diamonds. This is the essence of **nanotechnology**: the control of the composition and structure of matter at the atomic level. The prefix "nano-" refers to the scale of these constructions. A nanometer is one-billionth of a meter, the width of about 5 carbon atoms nestled side by side.

Nanotechnology involves the engineering of molecularly precise structures and, ultimately, molecular machines. BCC Research (McWilliams, 2006) estimates the global market for nanotools and nanodevices was $1.3B in 2005 and $1.5B in 2006, projected to reach $8.6B by 2011 and rapidly gaining on the slower-growing nanomaterials market which is estimated at $8.1B (2005), $9.0B (2006) and $16.6B (2011). As distinct from nanoscale materials and today's simple nanotools and nanodevices having nanoscale features, molecular nanotechnology encompasses the concept of engineering functional mechanical systems at the molecular scale – that is, machines at the molecular scale designed and built to atomic precision. Molecular manufacturing (Section 23.4) would make use of positionally-controlled mechanosynthesis (mechanically-mediated chemistry) guided by molecular machine systems to build complex products, including additional nanomachines.

Nanomedicine (Freitas 1999, 2003) is the application of nanotechnology to medicine: the preservation and improvement of human health, using molecular tools and molecular knowledge of the human body. Nanomedicine encompasses at least three types of molecularly precise structures (Freitas 2005a): nonbiological nanomaterials, biotechnology materials and engineered organisms, and nonbiological devices including diamondoid nanorobotics. In the near term, the molecular tools of nanomedicine will include biologically active nanomaterials and nanoparticles having well-defined nanoscale features. In the mid-term (5–10 years), knowledge gained from genomics and proteomics will make possible new treatments tailored to specific individuals, new drugs targeting pathogens whose genomes have been decoded, and stem cell treatments. Genetic therapies, tissue engineering, and many other offshoots of biotechnology will become more common in therapeutic medical practice. We also may see biological robots derived from bacteria or other motile cells that have had their genomes re-engineered and re-programmed, along with artificial organic devices that incorporate biological motors or self-assembled DNA-based structures for a variety of useful medical purposes.

In the farther term (2020s and beyond), the first fruits of **medical nanorobotics** – the most powerful of the three classes of nanomedicine technology, though clinically the most distant and still mostly theoretical today – should begin to appear in the medical field. Nanotechnologists will learn how to build nanoscale molecular parts like gears, bearings, and ratchets. Each nanopart may comprise a few thousand precisely placed atoms. These mechanical nanoparts will then be assembled into larger working machines such as nanosensors, nanomanipulators, nanopumps, nanocomputers, and even complete nanorobots which may be micron-scale or larger. The presence of onboard computers is essential because in vivo medical nanorobots will be called upon to perform numerous complex behaviors which must be conditionally executed on at least a semiautonomous basis, guided by receipt of local sensor data and constrained by preprogrammed settings, activity scripts, and event clocking, and further limited by a variety of simultaneously executing real-time control protocols and by external instructions sent into the body by the physician during the course of treatment. With medical nanorobots in hand, doctors should be able to quickly cure most diseases that hobble and kill people today, rapidly repair most physical injuries our bodies can suffer, and significantly extend the human healthspan.

The early genesis of the concept of medical nanorobotics sprang from the visionary idea that tiny nanomachines could be designed, manufactured, and introduced into the human body to perform cellular repairs at the molecular level. Although the medical application of nanotechnology was later championed in the popular writings of Drexler (Drexler 1986; Drexler et al. 1991) in the 1980s and 1990s and in the technical writings of Freitas (Freitas 1999, 2003) in the 1990s and 2000s, the first scientist to voice the possibility was the late Nobel physicist Richard P. Feynman, who worked on the Manhattan Project at Los Alamos during World War II and later taught at CalTech for most of his professorial career.

In his prescient 1959 talk "There's Plenty of Room at the Bottom," Feynman proposed employing machine tools to make smaller machine tools, these to be used in turn to make still smaller machine tools, and so on all the way down to the atomic level (Feynman 1960). He prophetically concluded that this is "a development which I think cannot be avoided." After discussing his ideas with a colleague, Feynman offered the first known proposal for a medical nanorobotic procedure of any kind – in this instance, to cure heart disease: "A friend of mine (Albert R. Hibbs) suggests a very interesting possibility for relatively small machines. He says that, although it is a very wild idea, it would be interesting in surgery if you could swallow the surgeon. You put the mechanical surgeon inside the blood vessel and it goes into the heart and looks around. (Of course the information has to be fed out.) It finds out which valve is the faulty one and takes a little knife and slices it out. Other small machines might be permanently incorporated in the body to assist some inadequately functioning organ." Later in his historic 1959 lecture, Feynman urges us to consider the possibility, in connection with microscopic biological cells, "that we can manufacture an object that maneuvers at that level!" The field had progressed far enough by 2007, half a century after Feynman's speculations, to allow Martin Moskovits, Professor of Chemistry and Dean of Physical Science at UC Santa Barbara, to write (Moskovits 2007) that "the notion of an ultra-small robot that can, for example, navigate the bloodstream performing microsurgery or activating neurons so as to restore muscular activity, is not an unreasonable goal, and one that may be realized in the near future."

23.3 Fundamentals of Medical Nanorobotics

Many skeptical questions arise when one first encounters the idea of micron-scale nanorobots constructed of nanoscale components, operating inside the human body. At the most fundamental level, technical questions about the influence of quantum effects on molecular structures, friction and wear among nanomechanical components, radiation damage, other failure mechanisms, the influence of thermal noise on reliability, and the effects of Brownian bombardment on nanomachines have all been extensively discussed and resolved in the literature (Drexler 1992; Freitas 1999a). Molecular motors consisting of just 50–100 atoms have been demonstrated experimentally (e.g., see Section 23.3.2). Published discussions of technical issues of specific relevance to medical nanorobots include proposed methods for recognizing, sorting and pumping individual molecules (Drexler 1992a; Freitas 1999b), and theoretical designs for mechanical nanorobot sensors (Freitas 1999c), flexible hull surfaces (Freitas 1999d), power sources (Freitas 1999e), communications systems (Freitas 1999f), navigation systems (Freitas 1999g), manipulator mechanisms (Freitas 1999h), mobility mechanisms for travel through bloodstream, tissues and cells (Freitas 1999i), onboard clocks (Freitas 1999j), and nanocomputers (Drexler 1992b; Freitas 1999k), along with the full panoply of nanorobot biocompatibility issues (Freitas 2003) (see also Section 23.5).

The idea of placing semi-autonomous self-powered nanorobots inside of us might seem a bit odd, but the human body already teems with similar natural nanodevices. For instance, more than 40 trillion single-celled microbes swim through our colon, outnumbering our tissue cells almost ten to one (Freitas 1999m). Many bacteria move by whipping around a tiny tail, or flagellum, that is driven by a 30-nanometer biological ionic nanomotor powered by pH differences between the inside and the outside of the bacterial cell. Our bodies also maintain a population of more than a trillion motile biological nanodevices called fibroblasts and white cells such as neutrophils and lymphocytes, each measuring perhaps 10 microns in size (Freitas 1999m). These beneficial natural nanorobots are constantly crawling around inside us, repairing damaged tissues, attacking invading microbes, and gathering up foreign particles and transporting them to various organs for disposal from the body (Freitas 2003a).

The greatest power of nanomedicine will begin to emerge in a decade or two as we learn to design and construct complete artificial nanorobots using nanometer-scale parts and subsystems such as diamondoid bearings and gears (Section 23.3.1), nanomotors and pumps (Section 23.3.2), nanomanipulators (Section 23.3.3), nanosensors (Section 23.3.4), and nanocomputers (Section 23.3.5).

23.3.1 Nanobearings and Nanogears

In order to establish the foundations for molecular manufacturing and medical nanorobotics, it is first necessary to create and to analyze possible designs for nanoscale mechanical parts that could, in principle, be manufactured. Because these components cannot yet be physically built in 2009, such designs cannot be subjected to rigorous experimental testing and validation. Designers are forced instead to rely upon ab initio structural analysis and computer studies including molecular dynamics simulations. "Our ability to model molecular machines (systems and devices) of specific kinds, designed in part for ease of modeling, has far outrun our ability to make them," notes K. Eric Drexler (Drexler 1992). "Design calculations and computational experiments enable the theoretical studies of these devices, independent of the technologies needed to implement them."

Molecular bearings are perhaps the most convenient class of components to design because their structure and operation is fairly straightforward. One of the simplest classical examples is Drexler's early overlap-repulsion bearing design (Drexler 1992f), shown with end views and exploded views in Fig. 23.1 using both ball-and-stick and space-filling representations. This bearing has exactly 206 atoms including carbon, silicon, oxygen and hydrogen, and is comprised of a small shaft that rotates within a ring sleeve measuring 2.2 nm in diameter. The atoms of the shaft are arranged in a 6-fold symmetry, while the ring has 14-fold symmetry, a combination that provides low energy barriers to shaft rotation. Figure 23.2 shows an exploded view of a 2808-atom strained-shell sleeve bearing designed by Drexler and Merkle (Drexler 1992f) using molecular mechanics force fields to ensure that bond lengths, bond angles, van der Waals distances, and strain energies are reasonable. This 4.8-nm diameter bearing features an interlocking-groove interface which derives from a modified diamond (100) surface. Ridges on the shaft interlock with ridges on the sleeve, making a very stiff structure. Attempts to bob the shaft up or down, or rock it from side to side, or displace it in any direction (except axial rotation, wherein displacement is extremely smooth) encounter a very strong resistance (Drexler 1995). Whether these bearings would have to be assembled in unitary fashion, or instead could be assembled by inserting one part into the other without damaging either part, had not been extensively studied or modeled by 2009. There is some experimental evidence that these bearings, if and when they can be built, should work as expected: In 2000, John Cumings and Alex Zettl at U.C. Berkeley demonstrated experimentally that nested carbon nanotubes

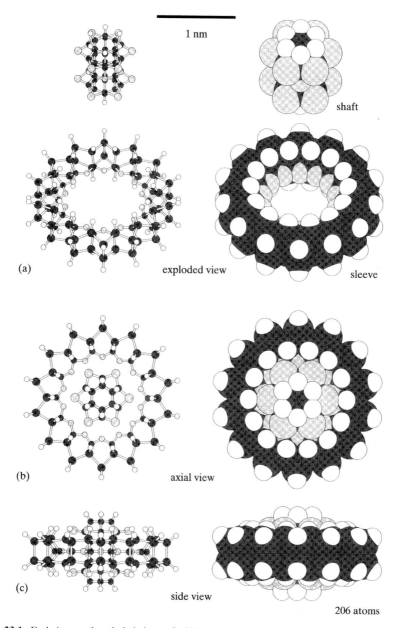

Fig. 23.1 End views and exploded views of a 206-atom overlap-repulsion bearing (Drexler 1992f). Image courtesy of K. Eric Drexler. ©1992 by John Wiley & Sons, Inc. Used with permission

Fig. 23.2 Exploded view of a 2808-atom strained-shell sleeve bearing (Drexler 1992f). Image courtesy of K. Eric Drexler. ©1992 by John Wiley & Sons, Inc. Used with permission

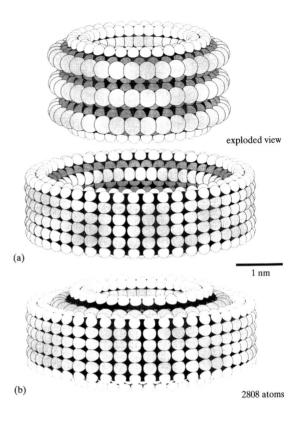

do indeed make exceptionally low-friction nanobearings (Cumings and Zettl 2000).

Molecular gears are another convenient component system for molecular manufacturing design-ahead. For example, in the 1990s Drexler and Merkle (Drexler 1992g) designed a 3557-atom planetary gear, shown in side, end, and exploded views in Fig. 23.3. The entire assembly has twelve moving parts and is 4.3 nm in diameter and 4.4 nm in length, with a molecular weight of 51,009.844 daltons and a molecular volume of 33.458 nm^3. An animation of the computer simulation shows the central shaft rotating rapidly and the peripheral output shaft rotating slowly as intended. The small planetary gears rotate around the central shaft, and they are surrounded by a ring gear that holds the planets in place and ensures that all of the components move in the proper fashion. The ring gear is a strained silicon shell with sulfur atom termination; the sun gear is a structure related to an oxygen-terminated diamond (100) surface; the planet gears resemble multiple hexasterane structures with oxygen rather than CH_2 bridges between the parallel rings; and the planet carrier is adapted from a Lomer dislocation (Lomer 1951) array created by R. Merkle and L. Balasubramaniam, and linked to the planet gears using C–C bonded bearings. View (c) retains the elastic deformations that are hidden in (a) – the gears are bowed.

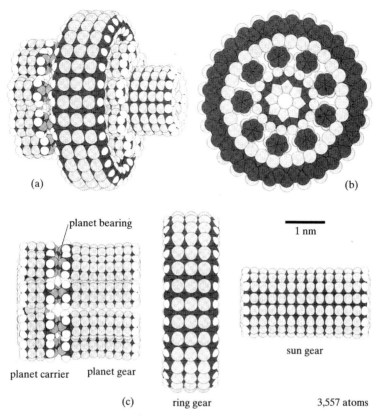

Fig. 23.3 End-, side-, and exploded-view of a 3557-atom planetary gear (Drexler 1992g). Image courtesy of K. Eric Drexler. ©1992 by John Wiley & Sons, Inc. Used with permission

In the macroscale world, planetary gears are used in automobiles and other machines where it is necessary to transform the speeds of rotating shafts.

Goddard and colleagues at CalTech (Goddard 1995; Cagin et al. 1998) performed a rotational impulse dynamics study of this "first-generation" planetary gear. At the normal operational rotation rates for which the component was designed (e.g., <1 GHz for <10 m/sec interfacial velocities), the gear worked as intended and did not overheat (Goddard 1995). However, when the gear was driven to ~100 GHz, significant instabilities appeared although the device still did not self-destruct (Goddard 1995). One run at ~80 GHz showed excess kinetic energy causing gear temperature to oscillate up to 450 K above baseline (Cagin et al. 1998). One animation of the simulation shows that the ring gear wiggles violently because it is rather thin. In an actual nanorobot incorporating numerous mechanical components of this type, the ring gear would be part of a larger wall that would hold it solidly in place and would eliminate these convulsive motions which, in any case, are seen in the simulation only at unrealistically high operating frequencies.

23.3.2 Nanomotors, Nanopumps, and Power Sources

Nanorobots need motors to provide motion, pumps to move materials, and power sources to drive mechanical activities. One important class of theoretical nanodevice that has been designed is a gas-powered molecular motor or pump (Drexler and Merkle 1996). The pump and chamber wall segment shown in Fig. 23.4 contain 6165 atoms with a molecular weight of 88,190.813 daltons and a molecular volume of 63.984 nm^3. The device could serve either as a pump for neon gas atoms or (if run backwards) as a motor to convert neon gas pressure into rotary power. The helical rotor has a grooved cylindrical bearing surface at each end, supporting a screw-threaded cylindrical segment in the middle. In operation, rotation of the shaft moves a helical groove past longitudinal grooves inside the pump housing. There is room enough for small gas molecules only where facing grooves cross, and these crossing points move from one side to the other as the shaft turns, moving the neon atoms along. Goddard (Cagin et al. 1998) reported that preliminary molecular dynamics simulations of the device showed that it could indeed function as a pump, although "structural deformations of the rotor can cause instabilities at low and high rotational frequencies. The forced translations show that at very low perpendicular forces due to pump action, the total energy rises significantly and again the structure deforms." The neon motor/pump is not very energy-efficient, but further refinement or extension of this crude design is clearly warranted. Almost all such design research in diamondoid nanorobotics is restricted to theory and computer simulation. This allows the design and testing of large structures or complete nanomachines and the compilation of growing libraries of molecular designs.

Although the neon pump cannot yet be built, proof-of-principle motors for nanoscale machines have already received a great deal of experimental attention

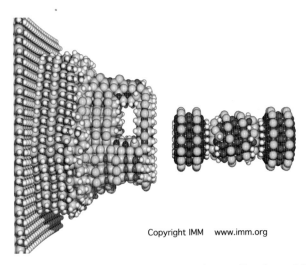

Fig. 23.4 Side views of a 6165-atom neon gas pump/motor (Drexler and Merkle, 1996). © Institute for Molecular Manufacturing (www.imm.org). Used with permission

including the 78-atom chemically-powered rotating nanomotor synthesized in 1999 by Kelly (Kelly et al. 1999), a chemically-powered rotaxane-based linear motor exerting ~100 pN of force with a 1.9 nm throw and a ~250 sec contraction cycle by Stoddart's group (Huang et al. 2003), a UV-driven catenane-based ring motor by Wong and Leigh (Leigh et al. 2003), an artificial 58-atom motor molecule that spins when illuminated by solar energy by Feringa (Koumura et al. 1999), and a great variety of additional synthetic molecular motor motifs as excellently reviewed by Browne and Feringa (Browne and Feringa 2006) and by Kay et al. (Kay et al. 2007). Zettl's group at U.C. Berkeley has experimentally demonstrated an essentially frictionless bearing made from two co-rotating nested nanotubes (Cumings and Zettl 2000), which can also serve as a mechanical spring because the inner nanotube "piston" feels a restoring force as it is extracted from the outer nanotube "jacket". Zettl's group then fabricated a nanomotor mounted on two of these nanotube bearings, demonstrating the first electrically powered nanoscale motor (Fennimore et al. 2003).

In 2005, Tour's group at Rice University reported (Shirai et al. 2005) constructing a tiny molecular "nanocar" measuring 3–4 nm across that consists of a chassis, two freely rotating axles made of well-defined rodlike acetylenic structures with a pivoting suspension, and wheels made of C_{60} buckyball (or, later, spherical carborane) molecules that can turn independently because the bond between them and the axle is freely rotatable (Fig. 23.5). Placed on a warmed gold surface held at 170°C, the nanocar spontaneously rolls on all four wheels, but only along its long axis in a direction perpendicular to its axles (a symmetrical three-wheeled variant just spins in place). When pulled with an STM tip, the nanocar cannot be towed sideways – the wheels dig in, rather than rolling. A larger, more functionalized version of the nanocar might carry other molecules along and dump them at will. Indeed, the Rice team (Shirai et al. 2006) has reportedly "followed up the nanocar work by designing a [motorized] light-driven nanocar and a nanotruck that's capable of carrying a payload" (Shirai et al. 2005).

Nanorobots working inside the body could most conveniently be powered by ambient glucose and oxygen found in the blood and tissues, which could be

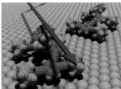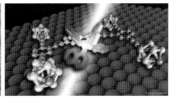

Fig. 23.5 Tour molecular nanocar (Shirai et al. 2005, 2006). *Left*: Original nanocar is depicted on gold surface. *Middle*: Two motorized nanocars are shown on a gold surface; each nanocar consists of a rigid chassis and four alkyne axles that spin freely and swivel independently of one another, with wheels made of p-carborane (spherical molecules of carbon, hydrogen and boron). *Right*: The nanocar's light-powered motor is attached mid-chassis; when struck by light, it rotates in one direction, pushing the car along like a paddlewheel. Used with permission.

converted to mechanical energy using a nanoengine (Freitas 1999aq) or to electrical energy using a nanoscale fuel cell (Freitas 1999ar). The first glucose-oxygen fuel cell was demonstrated experimentally by Nishizawa's group (Satoa et al. 2005) in 2005, who used a Vitamin K3-immobilized polymer with glucose dehydrogenase on one side as the anode and a polydimethylsiloxane-coated Pt cathode to yield an open circuit voltage of 0.62 volts and a maximum power density of 14.5 $\mu W/cm^2$ at 0.36 volts in an air-saturated phosphate buffered saline solution (pH 7.0) at 37°C containing 0.5 mM NADH and 10 mM glucose.

Another well-known proposal is for medical nanorobotic devices to receive all power (and some control) signals acoustically (Freitas 1999n; Drexler 1992c). Externally generated ultrasonic pressure waves would travel through the aqueous in vivo environment to the medical nanodevice, whereupon a piston on the device is driven back and forth in a well-defined manner, mechanically passing energy and information simultaneously into the device. Although an acoustically-actuated nanoscale piston has not yet been demonstrated experimentally, we know that pressure applied, then released, on carbon nanotubes causes fully reversible compression (Chesnokov et al. 1999), and experiments have shown very low frictional resistance between nested nanotubes that are externally forced in and out like pistons (Cumings and Zettl 2000). Masako Yudasaka, who studies C_{60} molecules trapped inside carbon nanotubes or "peapods" at NEC (Nippon Electric Corp.), expects that "the buckyball can act like a piston" (Schewe et al. 2001). In 2007, a prototype nanometer-scale generator that produces continuous direct-current electricity by harvesting mechanical energy from ultrasonic acoustic waves in the environment was demonstrated by Wang et al. (Wang et al. 2007).

Yet another important nanorobot component is the molecular sorting rotor (Freitas 1999o; Drexler 1992a) (see also Fig. 23.6a), which would provide an active means for pumping, say, individual gas molecules into, and out of, pressurized onboard microtanks, one molecule at a time. Sorting pumps are typically envisioned as ~1000 nm^3-size devices that can transfer ~10^6 molecules/sec and would be embedded in the hull of the nanorobot. Each pump employs reversible artificial binding sites (Freitas 1999p) mounted on a rotating structure that cycles between the interior and exterior of the nanorobot, allowing transport of a specific molecule even against concentration gradients up to ~20,000 atm. Sorting rotors are conceptually similar to the biological transporter pumps (Freitas 1999as) which are found in nature for conveying numerous ions (Gouaux and Mackinnon 2005), amino acids, sugars (Olson and Pessin 1996), and other small biomolecules (Sharom 1997) across cell membranes. The molecular structures of natural enzymatic binding sites for small molecules like oxygen, carbon dioxide, nitrogen, water and glucose have been known since the 1990s, and the design (Kapyla et al. 2007), simulation (Rohs et al. 2005), and fabrication (Bracci et al. 2002; Subat et al. 2004; Franke et al. 2007) of artificial binding sites for more complex molecules is an active field of research. Sequential cascades of sorting rotors (Fig. 23.6b) (Drexler 1992d; Freitas 1999o) could achieve high fidelity purification and a contaminant fraction of <10^{-15} for transporting small molecules of common types.

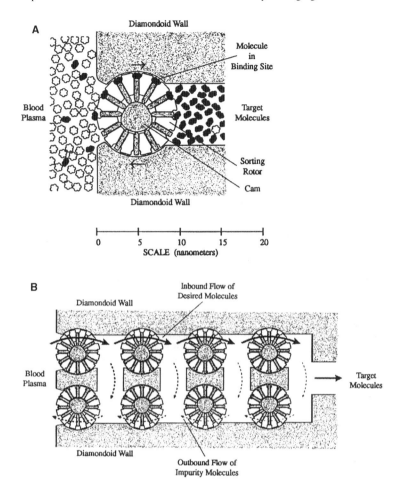

Fig. 23.6 (**a**) Individual sorting rotor and (**b**) a sorting rotor cascade, redrawn from Drexler (1992)

23.3.3 Nanomanipulators

Nanorobots require manipulators to perform grasping and manipulation tasks, and also to provide device mobility. One well-known telescoping nanomanipulator design (Fig. 23.7) features a central telescoping joint whose extension and retraction is controlled by a 1.5-nm diameter drive shaft (Drexler 1992e). The rapid rotation of this drive shaft (up to ~1 m/sec tangential velocity) forces a transmission gear to quickly execute a known number of turns, causing the telescoping joint to

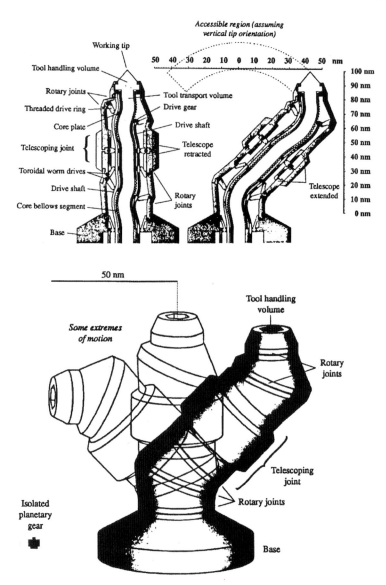

Fig. 23.7 Telescoping nanomanipulator design (Drexler 1992e). Image courtesy of K. Eric Drexler. ©1992 by John Wiley & Sons, Inc. Used with permission

slowly unscrew or screw in the axial direction, thus lengthening or contracting the manipulator. These shafts can be made to turn through a known number of rotations between locked states giving odometer-like control of manipulator joint rotations. Additionally, two pairs of canted rotary joints – one pair between the telescoping section and the base, the other pair between the telescoping section and the working tip – are controlled by toroidal worm drives. These joints enable a wide variety of complex angular motions and give full 6-DOF (degrees of freedom) access to the work envelope. The manipulator is approximately cylindrical in shape with an outside diameter of ~35 nm and an extensible length from 90 nm to 100 nm measured from top of base to working tip. The mechanism includes a hollow circular channel 7 nm in diameter to allow tool tips and materials to be moved from below the manipulator through the base up to the working tip. At the tip, a slightly larger region is reserved for a mechanism to allow positioning and locking of tool tips. This ~10^{-19} kg manipulator would be constructed of ~4×10^6 atoms excluding the base and external power and control structures, and is hermetically sealed to maintain a controlled internal environment while allowing leakproof operation in vivo.

Experimentally, a DNA-based robot arm has been inserted into a 2D array substrate by Seeman's group (Ding and Seeman 2006), and this simple rotary mechanism was then verified by atomic force microscopy to be a fully functional nanomechanical device.

23.3.4 Nanosensors

Medical nanorobots will need to acquire information from their environment to properly execute their assigned tasks. Such acquisition can be achieved using onboard nanoscale sensors, or nanosensors, of various types which are currently the subject of much experimental research (Nagahara et al. 2008). More advanced nanosensors to be used in medical nanorobots will allow monitoring environmental states including internal nanorobot states and local and global somatic states inside the human body. Theoretical designs for advanced nanosensors to detect chemical substances (Freitas 1999q), displacement and motion (Freitas 1999r), force and mass (Freitas 1999s), and acoustic (Freitas 1999t), thermal (Freitas 1999u), and electromagnetic (Freitas 1999v) stimuli have been described elsewhere.

For instance, medical nanorobots which employ onboard tankage will need various nanosensors to acquire external data essential in regulating gas loading and unloading operations, tank volume management, and other special protocols. Sorting rotors (Section 23.3.2) can be used to construct quantitative concentration sensors for any molecular species desired. One simple two-chamber design (Freitas 1999cj) uses an input sorting rotor running at 1% normal speed synchronized with a counting rotor (linked by rods and ratchets to the computer) to assay the number of molecules of the desired type that are present in a known volume of fluid. At typical blood concentrations, this sensor, which measures $45 \times 45 \times 10$ nm comprising ~500,000 atoms (~10^{-20} kg), should count, for example, ~100,000 molecules/sec of glucose, ~30,000 molecules/sec of arterial or venous CO_2, or ~2000 molecules/sec of arterial or venous O_2. It is also convenient to include internal pressure sensors to monitor O_2 and CO_2 gas tank loading, ullage (container fullness) sensors for ballast and glucose fuel tanks, and internal/external temperature sensors to help monitor and regulate total system energy output.

As another example of nanosensors, the attending physician could broadcast signals to nanorobotic systems deployed inside the human body most conveniently using modulated compressive pressure pulses received by mechanical transducers embedded in the surface of the nanorobot. Converting a pattern of pressure fluctuations into mechanical motions that can serve as input to a mechanical nanocomputer (Section 23.3.5) requires transducers that function as pressure-driven actuators (Drexler 1992c; Freitas 1999t, ck). Broadcast mechanisms similar to medical pulse-echo diagnostic ultrasound systems can transmit data into the body acoustically at ~10 MHz (~10^7 bits/sec) using peak-to-trough 10-atm pressure pulses that can be received onboard the nanorobot by nanosensors ~$(21 \text{ nm})^3$ in size comprising ~10^5 atoms. Such signals attenuate only ~10% per 1 cm of travel (Drexler 1992c), so whole-body broadcasts should be feasible even in emergency field situations. Pressure transducers will consume minimal power because the input signal drives the motion.

23.3.5 Nanocomputers

Many important medical nanorobotic tasks will require computation (Freitas 1999k) during the acquisition and processing of sensor data, the control of tools, manipulators, and motility systems, the execution of navigation and communication tasks, and during the coordination of collective activities with neighboring nanorobots, and also to allow a physician to properly monitor and control the work done by nanorobots. *Ex vivo* computation has few theoretical limits, but computation by in vivo nanorobots will be subject to a number of constraints such as physical size, power consumption, onboard memory and processing speed.

The memory required onboard a medical nanorobot will be strongly mission dependent. Simple missions involving basic process control with limited motility may require no more than ~10^5-10^6 bits of memory, comparable to an old Apple II computer (including RAM plus floppy disk drive). At the other extreme, a complex cell repair mission might require the onboard storage of the equivalent of a substantial fraction of the patient's genetic code, representing perhaps 10^9–10^{10} bits of memory which would be in the same range as the 1985 Cray-2 (2×10^{10} bits) or the 1989 Cray-3 (6×10^8 bits) supercomputers. Computational speed will also be strongly mission dependent. Simple process control systems in basic factory settings may only require speeds as slow as 10^4 bit/sec. At the other extreme, a processing speed of 10^9 bits/sec allows a ~10^9 bit genome-sized information store to be processed in ~1 sec, about the same as the small-molecule diffusion time across an average 20-micron wide cell.

Perhaps the best-characterized (though not yet built) mechanical nanocomputer is Drexler's rod logic design (Drexler 1992b). In this theoretical design, one sliding rod with a knob (Fig. 23.8a) intersects a second knobbed sliding rod at right angles to the first. Depending upon the position of the first rod, the second may be free to move, or unable to move. This simple blocking interaction serves as the basis for logical operations. One implementation of a nanomechanical Boolean NAND

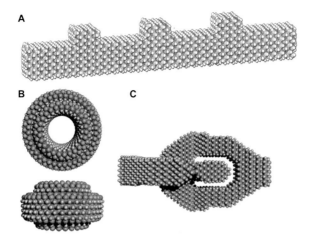

Fig. 23.8 (**a**) hydrocarbon logic rod, (**b**) hydrocarbon bearing, and (**c**) hydrocarbon universal joint (Nanofactory Collaboration 2007a)

"interlock" gate uses clock-driven input and output logic rods 1 nm wide which interact via knobs that prevent or enable motion, all encased in a housing, allowing ~16 nm^3/interlock. (Any logic function, no matter how complicated, can be built from NAND or NOR gates alone.) Similarly, a thermodynamically efficient class of register capable of mechanical data storage would use rods ~1 nm in width with 0.1-nanosec switching speeds, allowing ~40 nm^3/register. The benchmark mechanical nanocomputer design fits inside a 400 nm cube, consumes ~60 nW of power, and has 10^6 interlock gates, 10^5 logic rods, 10^4 registers, and an energy-buffering flywheel. Power dissipation is ~2×10^4 operations/sec-pW with a processing speed of ~10^9 operations/sec (~1 gigaflop), similar to a typical desktop PC in 2007.

Biocomputers (Freitas 1999ai; Guet et al. 2002; Yokobayashi et al. 2002; Basu et al. 2004, 2005) and both electronic (Heath 2000; Tseng and Ellenbogen 2001; Das et al. 2005) and mechanical (Blick et al. 2007) nanocomputers are active areas of current research and development. There has been progress toward nanotube- and nanorod-based molecular electronics (Collins et al. 2001; Reed and Lee 2003) and nanoscale-structured quantum computers (Stegner et al. 2006), possibly using diamond lattice (Dutt et al. 2007). A 160-kilobit memory device smaller than a white blood cell was fabricated by Stoddart's group in 2007 (Green et al. 2007) by laying down a series of perpendicular crossing nanowires with 400 bottom wires and 400 crossing top wires. (Sitting at each intersection of the tic-tac-toe structure and serving as the storage element were approximately 300 bistable rotaxane molecules that could be switched between two different states, and each junction of a crossbar could be addressed individually by controlling the voltages applied to the appropriate top and bottom crossing wires, forming a bit at each nanowire crossing.) A simple DNA-based molecular machine capable of translating "coded" information from one DNA strand to another, another basic nanocomputational activity, was demonstrated experimentally in 2007 by Seeman's group (Garibotti et al. 2007).

23.4 Manufacturing Medical Nanorobots

The development pathway for diamondoid medical nanorobots will be long and arduous. First, theoretical scaling studies (Freitas 1998, 2000a, b, 2005b, 2006a, 2007; Freitas and Phoenix 2002) are used to assess basic concept feasibility. These initial studies must then be followed by more detailed computational simulations of specific nanorobot components and assemblies, and ultimately full systems simulations, all thoroughly integrated with additional simulations of massively parallel manufacturing processes from start to finish consistent with a design-for-assembly engineering philosophy. Once nanofactories implementing molecular manufacturing capabilities become available, experimental efforts may progress from fabrication of components (from small-molecule or atomic precursors) and testing, to the assembly of components into nanomechanical devices and nanomachine systems, and finally to prototypes and mass manufacture of medical nanorobots, ultimately leading to clinical trials. By 2009 there was some limited experimental work with microscale-component microscopic microrobots (Ishiyama et al. 2002; Chrusch et al. 2002; Mathieu et al. 2005; Yesin et al. 2005; Monash University 2006) (see also Section "Endoscopic Nanosurgery and Surgical Nanorobots") but progress on nanoscale-component microscopic nanorobots today is largely at the concept feasibility and preliminary design stages and will remain so until experimentalists develop the capabilities required for molecular manufacturing, as reviewed below.

23.4.1 Positional Assembly and Molecular Manufacturing

Complex medical nanorobots probably cannot be manufactured using the conventional techniques of self-assembly. As noted in the final report (Committee 2006) of the 2006 Congressionally-mandated review of the U.S. National Nanotechnology Initiative by the National Research Council (NRC) of the National Academies and the National Materials Advisory Board (NMAB): "For the manufacture of more sophisticated materials and devices, including complex objects produced in large quantities, it is unlikely that simple self-assembly processes will yield the desired results. The reason is that the probability of an error occurring at some point in the process will increase with the complexity of the system and the number of parts that must interoperate."

The opposite of self-assembly processes is positionally controlled processes, in which the positions and trajectories of all components of intermediate and final product objects are controlled at every moment during fabrication and assembly. Positional processes should allow more complex products to be built with high quality and should enable rapid prototyping during product development. Positional assembly is the norm in conventional macroscale manufacturing (e.g., cars, appliances, houses) but is only recently (Kenny 2007; Nanofactory Collaboration 2007a) starting to be seriously investigated experimentally for nanoscale manufacturing. Of course, we already know that positional fabrication will work in the nanoscale realm. This is demonstrated in the biological world by ribosomes, which positionally assemble proteins in living cells by following a sequence of digitally encoded instructions (even though ribosomes themselves are self-assembled). Lacking this positional fabrication of proteins controlled by DNA-based software, large, complex, digitally-specified organisms would probably not be possible and biology as we know it would not exist.

The most important materials for positional assembly may be the rigid covalent or "diamondoid" solids, since these could potentially be used to build the most reliable and complex nanoscale machinery. Preliminary theoretical studies have suggested great promise for these materials in molecular manufacturing. The NMAB/NRC Review Committee recommended (Committee 2006) that experimental work aimed at establishing the technical feasibility of positional molecular manufacturing should be pursued and supported: "Experimentation leading to demonstrations supplying ground truth for abstract models is appropriate to better characterize the potential for use of bottom-up or molecular manufacturing systems that utilize processes more complex than self-assembly." Making complex nanorobotic systems requires manufacturing techniques that can build a molecular structure by positional assembly (Freitas 2005c). This will involve picking and placing molecular parts one by one, moving them along controlled trajectories much like the robot arms that manufacture cars on automobile assembly lines. The procedure is then repeated over and over with all the different parts until the final product, such as a medical nanorobot, is fully assembled using, say, a desktop nanofactory (see Fig. 23.17).

23.4.2 Diamond Mechanosynthesis (DMS)

Theorists believe that the most reliable and durable medical nanorobots will be built using diamondoid materials. What is diamondoid? First and foremost, diamondoid materials include pure diamond, the crystalline allotrope of carbon. Among other exceptional properties, diamond has extreme hardness, high thermal conductivity, low frictional coefficient, chemical inertness, a wide electronic bandgap, and is the strongest and stiffest material presently known at ordinary pressures. Diamondoid materials also may include any stiff covalent solid that is similar to diamond in strength, chemical inertness, or other important material properties, and possesses a dense three-dimensional network of bonds. Examples of such materials are carbon nanotubes and fullerenes, several strong covalent ceramics such as silicon carbide, silicon nitride, and boron nitride, and a few very stiff ionic ceramics such as sapphire (monocrystalline aluminum oxide) that can be covalently bonded to pure covalent structures such as diamond. Of course, large pure crystals of diamond are brittle and easily fractured. The intricate molecular structure of a diamondoid nanofactory macroscale product will more closely resemble a complex composite material, not a brittle solid crystal. Such products, and the nanofactories that build them, should be extremely durable in normal use.

Mechanosynthesis, involving molecular positional fabrication, is the formation of covalent chemical bonds using precisely applied mechanical forces to build, for example, diamondoid structures. Mechanosynthesis employs chemical reactions driven by the mechanically precise placement of extremely reactive chemical species in an ultra-high vacuum environment. Mechanosynthesis may be automated via computer control, enabling programmable molecular positional fabrication. Molecularly precise fabrication involves holding feedstock atoms or molecules, and a growing nanoscale workpiece, in the proper relative positions and orientations so that when they touch they will chemically bond in the desired manner. In this process, a mechanosynthetic tool is brought up to the surface of a workpiece. One or more transfer atoms are added to, or removed from, the workpiece by the tool. Then the tool is withdrawn and recharged. This process is repeated until the workpiece (e.g., a growing nanopart) is completely fabricated to molecular precision with each atom in exactly the right place. Note that the transfer atoms are under positional control at all times to prevent unwanted side reactions from occurring. Side reactions are also prevented using proper reaction design so that the reaction energetics help us avoid undesired pathological intermediate structures.

The positional assembly of diamondoid structures, some almost atom by atom, using molecular feedstock has been examined theoretically (Drexler 1992h; Merkle 1997; Merkle and Freitas 2003; Mann et al. 2004; Allis and Drexler 2005; Freitas 2005d; Peng et al. 2006; Temelso et al. 2006; Freitas et al. 2007; Temelso et al. 2007; Freitas and Merkle 2008) via computational models of diamond mechanosynthesis (DMS). DMS is the controlled addition of individual carbon atoms, carbon dimers (C_2), single methyl (CH_3) or like groups to the growth surface of a diamond crystal lattice workpiece in a vacuum manufacturing environment. Covalent chemical bonds are formed one by one as the result of positionally constrained mechanical forces applied at the tip of a scanning probe microscope (SPM) apparatus. Programmed sequences of carbon dimer placement on growing diamond surfaces *in vacuo* appear feasible in theory (Peng et al. 2006; Freitas and Merkle 2008), as illustrated by the hypothetical DCB6Ge tooltip which is shown depositing two carbon atoms on a diamond surface in Fig. 23.9.

The first experimental proof that individual atoms could be manipulated was obtained by IBM scientists in 1989 when they used a scanning tunneling microscope to precisely position 35 xenon atoms on a nickel surface to spell out the corporate logo "IBM" (Fig. 23.10). However, this feat did not involve the formation of covalent chemical bonds. One important step toward the practical realization of DMS was achieved in 1999 by Ho and Lee (Lee and Ho 1999), who achieved the first site-repeatable site-specific covalent bonding operation of two diatomic carbon-containing molecules (CO), one after the other, to the same atom of iron on a crystal surface, using an SPM. The first experimental demonstration of true mechanosynthesis, establishing covalent bonds using purely mechanical forces – albeit on silicon

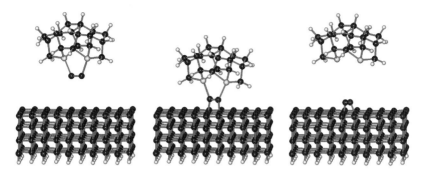

Fig. 23.9 DCB6Ge tooltip shown depositing two carbon atoms on a diamond surface (Nanofactory Collaboration 2007a)

Fig. 23.10 IBM logo spelled out using 35 xenon atoms arranged on a nickel surface by an STM (courtesy of IBM Research Division)

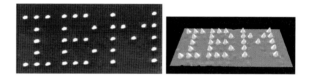

atoms, not carbon atoms – was reported in 2003 by Oyabu and colleagues (Oyabu et al. 2003) in the Custance group. In this landmark experiment, the researchers vertically manipulated single silicon atoms from the Si(111)–(7×7) surface, using a low-temperature near-contact atomic force microscope to demonstrate (1) removal of a selected silicon atom from its equilibrium position without perturbing the (7×7) unit cell and (2) the deposition of a single Si atom on a created vacancy, both via purely mechanical processes.

Following prior theoretical proposals (Freitas 2005d; Freitas and Merkle 2008) for experimental investigations, participants in the Nanofactory Collaboration (Nanofactory Collaboration 2007a) are now planning work designed to achieve DMS with carbon and hydrogen atoms using an SPM apparatus (Section 23.4.4).

23.4.3 Designing a Minimal Toolset for DMS

It is already possible to synthesize bulk diamond today. In a process somewhat reminiscent of spray painting, layer after layer of diamond can be built up by holding a cloud of reactive hydrogen atoms and hydrocarbon molecules over a deposition surface. When these molecules bump into the surface they change it by adding, removing, or rearranging atoms. By carefully controlling the pressure, temperature, and the exact composition of the gas in this process – called chemical vapor deposition or CVD – conditions can be created that favor the growth of diamond on the surface. But randomly bombarding a surface with reactive molecules does not offer fine control over the growth process and lacks atomic-level positional control. To achieve molecularly precise fabrication, the first challenge is to make sure that all chemical reactions will occur at precisely specified places on the surface. A second problem is how to make the diamond surface reactive at the particular spots where we want to add another atom or molecule. A diamond surface is normally covered with a layer of hydrogen atoms. Without this layer, the raw diamond surface would be highly reactive because it would be studded with unused (or "dangling") bonds from the topmost plane of carbon atoms. While hydrogenation prevents unwanted reactions, it also renders the entire surface inert, making it difficult to add carbon (or anything else) to it.

To overcome these problems, we're trying to use a set of molecular-scale tools that would, in a series of well-defined steps, prepare the surface and create hydrocarbon structures on a layer of diamond, atom by atom and molecule by molecule. A mechanosynthetic tool typically has two principal components – a chemically active tooltip and a chemically inert handle to which the tooltip is covalently bonded. The tooltip is the part of the tool where chemical reactions are forced to occur. The much larger handle structure is big enough to be grasped and positionally manipulated using an SPM or similar macroscale instrumentality. At least three types of basic mechanosynthetic tools (Fig. 23.11) have already received considerable theoretical (and some experimental) study and are likely among those required to build molecularly precise diamond via positional control:

(1) *Hydrogen Abstraction Tools.* The first step in the process of mechanosynthetic fabrication of diamond might be to remove a hydrogen atom from each of one or two specific adjacent spots on the diamond surface, leaving behind one or two reactive dangling bonds or a penetrable C=C double bond. This could be done using a hydrogen abstraction tool (Temelso et al. 2006) that has a high chemical affinity for hydrogen at one end but is elsewhere inert (Fig. 23.11a). The tool's unreactive region serves as a handle or handle attachment point. The

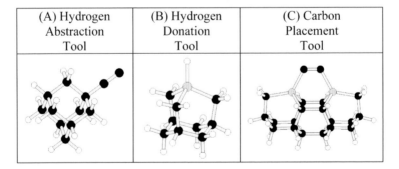

Fig. 23.11 Examples of three basic mechanosynthetic tooltypes that are required to build molecularly precise diamond via positional control (*black* = C atoms, *grey* = Ge atoms, *white* = H atoms) (Freitas and Merkle 2008)

tool would be held by a molecular positional device, initially perhaps a scanning probe microscope tip but ultimately a molecular robotic arm, and moved directly over particular hydrogen atoms on the surface. One suitable molecule for a hydrogen abstraction tool is the acetylene or "ethynyl" radical, comprised of two carbon atoms triply bonded together. One carbon of the two serves as the handle connection, and would bond to a nanoscale positioning tool through a much larger handle structure perhaps consisting of a lattice of adamantane cages as shown in Fig. 23.12. The other carbon of the two has a dangling bond where a hydrogen atom would normally be present in a molecule of ordinary acetylene (C_2H_2). The working environment around the tool would be inert (e.g., vacuum or a noble gas such as neon).

(2) *Hydrogen Donation Tools.* After a molecularly precise structure has been fabricated by a succession of hydrogen abstractions and carbon depositions, the fabricated structure must be hydrogen-terminated to prevent additional unplanned reactions. While the hydrogen abstraction tool is intended to make an inert structure reactive by creating a dangling bond, the hydrogen donation tool (Temelso et al. 2007) does the opposite. It makes a reactive structure inert by terminating a dangling bond. Such a tool would be used to stabilize reactive surfaces and help prevent the surface atoms from rearranging in unexpected and undesired ways. The key requirement for a hydrogen donation tool is that it include a weakly attached hydrogen atom. Many molecules fit that description, but the bond between hydrogen and germanium is sufficiently weak so that a Ge-based hydrogen donation tool (Fig. 23.11b) should be effective.

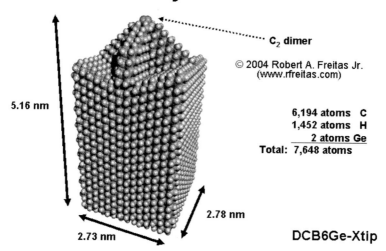

Fig. 23.12 Recyclable DCB6Ge tooltip with crossbar handle motif (Peng et al. 2006)

(3) *Carbon Placement Tools.* After the abstraction tool has created adjacent reactive spots by selectively removing hydrogen atoms from the diamond surface but before the surface is re-terminated by hydrogen, carbon placement tools may be used to deposit carbon atoms at the desired reactive surface sites. In this way a diamond structure would be built on the surface, molecule by molecule, according to plan. The first *complete* tool ever proposed for this carbon deposition function is the "DCB6Ge" dimer placement tool (Merkle and Freitas 2003) – in this example, a carbon (C_2) dimer having two carbon atoms connected by a triple bond with each carbon in the dimer connected to a larger unreactive handle structure through two germanium atoms (Fig. 23.11c). This dimer placement tool, also held by a molecular positional device, is brought close to the reactive spots along a particular trajectory, causing the two dangling surface bonds to react with the ends of the carbon dimer. The dimer placement tool would then withdraw, breaking the relatively weaker bonds between it and the C_2 dimer and transferring the carbon dimer from the tool to the surface, as illustrated in Fig. 23.9. A positionally controlled dimer could be bonded at many different sites on a growing diamondoid workpiece, in principle allowing the construction of a wide variety of useful nanopart shapes. As of 2009, the DCB6Ge dimer placement tool remains the most intensively studied of any mechanosynthetic tooltip to date (Merkle and Freitas 2003; Mann et al. 2004; Freitas 2005d; Peng et al. 2006; Freitas et al. 2007; Freitas and Merkle 2008), having had more than 150,000 CPU-hours of computation invested thus far in its analysis, and it remains the only DMS tooltip motif that has been successfully simulated and validated for its intended function on a full 200-atom diamond surface model (Peng et al. 2006). Other proposed dimer (and related carbon transfer) tooltip motifs (Drexler 1992h; Merkle 1997; Merkle and Freitas 2003; Allis and Drexler 2005; Freitas et al. 2007; Freitas and Merkle 2008) have received less extensive study but are also expected to perform well.

In 2007, Freitas and Merkle (Freitas and Merkle 2008) completed a three-year project to computationally analyze a comprehensive set of DMS reactions and an associated minimal set of tooltips that could be used to build basic diamond, graphene (e.g., carbon nanotubes), and all of the tools themselves including all necessary tool recharging reactions. The research defined 65 DMS reaction sequences incorporating 328 reaction steps, with 354 pathological side reactions analyzed and with 1,321 unique individual DFT-based (Density Functional Theory) quantum chemistry reaction energies reported. (These mechanosynthetic reaction sequences range in length from 1 to 13 reaction steps (typically 4) with 0–10 possible pathological side reactions or rearrangements (typically 3) reported per reaction step.) For the first time, this toolset provides clear developmental targets for a comprehensive near-term DMS implementation program (Nanofactory Collaboration 2007a).

23.4.4 Building the First Mechanosynthetic Tools

The first practical proposal for building a DMS tool experimentally was published by Freitas in 2005 and was the subject of the first mechanosynthesis patent ever filed (Freitas 2005d). According to this proposal, the manufacture of a complete DCB6Ge positional dimer placement tool would require four distinct steps: synthesizing a capped tooltip molecule, attaching it to a deposition surface, attaching a handle to it via CVD, then separating the tool from the deposition surface. The workability of the proposed process has already received valuable criticism from the scientific community and may be sufficiently viable to serve as a vital stepping-stone to more sophisticated DMS approaches.

An even simpler practical proposal for building DMS tools experimentally, also using only experimental methods available today, was published in 2008 by Freitas and Merkle as part of their minimal toolset work (Freitas and Merkle 2008) (see also Section 23.4.3). Processes are identified for the experimental fabrication of a hydrogen abstraction tool, a hydrogen donation tool, and two alternative carbon placement tools (other than DCB6Ge), and these processes and tools are part of the second mechanosynthesis patent ever filed and the first to be filed by the Nanofactory Collaboration (Nanofactory Collaboration 2007a). At this writing, Collaboration participants are undertaking preparatory steps (including equipment assessment and securing of funding) leading to direct experimental tests of these proposals.

Other practical proposals for building the first DMS tooltips, using existing technology, are eagerly sought by the Nanofactory Collaboration.

23.4.5 Next Generation Tools and Components

After the ability to fabricate the first primitive DMS tooltips has been demonstrated experimentally and repeatable sub-Angstrom positional placement accuracy for SPM tips has been developed, then-existing primitive tooltips could be manipulated to build the next generation of more precise, more easily rechargeable, and generally much improved mechanosynthetic tools. These more capable tools may include more stable handles of standardized dimensions, such as the rechargeable DCB6Ge dimer placement tool with the more reliable crossbar design (Peng et al. 2006) shown in Fig. 23.12, or tools with more complex handles incorporating moving components (Fig. 23.13). The end result of this iterative development process will be a mature set of efficient, positionally controlled mechanosynthetic tools that can reliably build molecularly precise diamondoid structures – including more DMS tools.

These more sophisticated tools also will be designed to allow building more complex components such as the all-hydrocarbon diamond logic rod (Fig. 23.8a), hydrocarbon bearing (Fig. 23.8b) and diamond universal joint (Fig. 23.8c), and related devices already described in Section 23.3. Once mechanosynthetic tooltips are developed for additional element types, a still wider variety of nanomachines

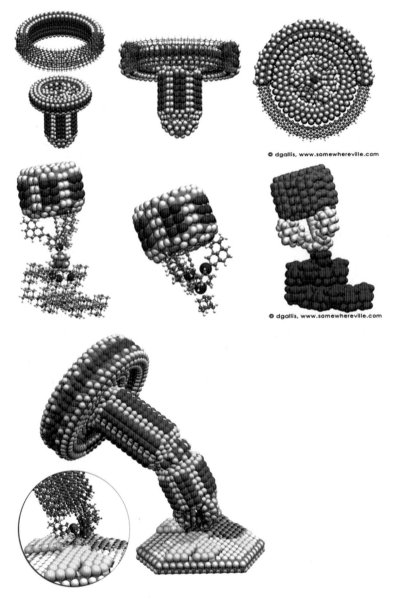

Fig. 23.13 Mechanosynthetic tooltip incorporating moving components (courtesy of Damian Allis). Used with permission

can be fabricated incorporating atoms other than hydrogen, carbon and germanium (e.g., silicon, oxygen, and sulfur). Examples of these diamondoid nanomachines include the speed reduction gear (Fig. 23.14a), in which the train of gears reduces the speed from the high-speed one on the left to the half-speed one on the right, and the differential gear (Fig. 23.14b) that smoothly converts mechanical rotation in one

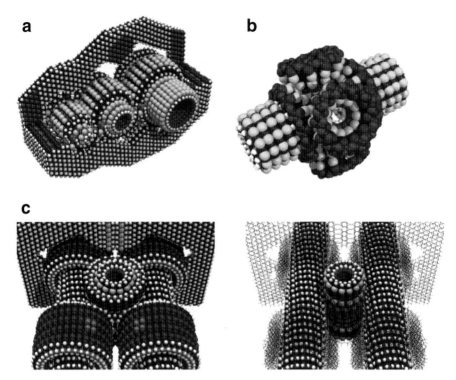

Fig. 23.14 (**a**) speed reduction gear, *above left*; (**b**) interior workings of differential gear, *above right*; (**c**) worm drive, *below* (*black* = silicon, *white* = hydrogen, *light grey* = sulfur, *dark grey* = oxygen). Images courtesy of Nanorex, used with permission

direction into mechanical rotation in the opposite direction. The largest molecular machine model that had been simulated as of 2009 using molecular dynamics was the worm drive assembly (Fig. 23.14c), consisting of 11 separate components and over 25,000 atoms; the two tubular worm gears progress in opposite directions, converting rotary into linear motion. Note that the magnitude of quantum effects is only ~10% of the classical (nonquantum) magnitudes for ~1 nm objects at 300 K, and even less significant for larger objects (Drexler 1992j).

Using computer-automated tooltips performing positionally-controlled DMS in lengthy programmed sequences of reaction steps, we should be able to fabricate simple diamondoid nanomechanical parts such as bearings, gears, struts, springs, logic rods and casings to atomic precision. Early tools would progress from single DMS tools manipulated by SPM-like mechanisms, to more complex multitip tools and jigs which the simple tools could initially fabricate, one at a time. In a factory production line (Fig. 23.15), individual DMS tooltips can be affixed to rigid moving support structures and guided through repeated contact events with workpieces, recharging stations, and other similarly-affixed apposable tooltips. These "molecular mills" could then perform repetitive fabrication steps using simple, efficient

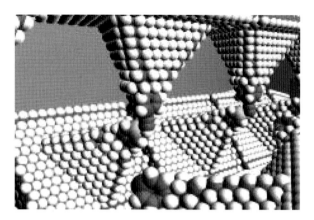

Fig. 23.15 Fabrication of nanoparts using DMS tooltips affixed to rigid moving support structures and guided through repeated contact events with workpieces under computer control in a nanofactory production line (courtesy of John Burch). Used with permission

mechanisms. Such mills can, in principle, be operated at high speeds – with positionally constrained mechanosynthetic encounters possibly occurring at up to megahertz frequencies.

The Nanofactory Collaboration has identified a large number of technical challenges (Nanofactory Collaboration 2007b) that must be solved before we can progress to building the kinds of complex nanoscale machinery described above. Among the theoretical and design challenges are: (1) nanopart gripper design, (2) nanopart manipulator actuator design, (3) design and simulation of nanopart feedstock presentation systems, (4) design and simulation of workpiece release surfaces, (5) design and simulation of nanopart assembly sequences, and (6) atomic rearrangements in juxtaposed nanoparts. Some experimental challenges include: (1) development of SPM technology to enable nanopart assembly work, (2) fabrication and testing of workpiece release surfaces, and (3) experimental proof-of-principle and early positional assembly demonstration benchmarks.

23.4.6 Strategies for Molecular Manufacturing

The ultimate goal of molecular nanotechnology is to develop a manufacturing technology able to inexpensively manufacture most arrangements of atoms that can be specified in molecular detail – including complex arrangements involving millions or billions of atoms per product object, as in the hypothesized medical nanorobots. This will provide the ultimate manufacturing technology in terms of precision, flexibility, and low cost. But to be practical, molecular manufacturing must also be able to assemble very large numbers of identical medical nanorobots very quickly. Two central technical objectives thus form the core of our current strategy for diamondoid molecular manufacturing: (1) programmable positional assembly including fabrication of diamondoid structures using molecular feedstock, as discussed above, and (2) massive parallelization of all fabrication and assembly processes, briefly described below.

Molecular manufacturing systems capable of massively parallel fabrication (Freitas and Merkle 2004a) might employ, at the lowest level, large arrays of DMS-enabled scanning probe tips all building similar diamondoid product structures in unison. Analogous approaches are found in present-day larger-scale systems. For example, simple mechanical ciliary arrays consisting of 10,000 independent microactuators on a 1 cm^2 chip have been made at the Cornell National Nanofabrication Laboratory for microscale parts transport applications, and similarly at IBM for mechanical data storage applications (Vettiger et al. 2002). Active probe arrays of 10,000 independently-actuated microscope tips have been developed by Mirkin's group at Northwestern University for dip-pen nanolithography (Bullen et al. 2002) using DNA-based "ink". Almost any desired 2D shape can be drawn using 10 tips in concert. Another microcantilever array manufactured by Protiveris Corp. has millions of interdigitated cantilevers on a single chip (Protiveris, 2003). Martel's group has investigated using fleets of independently mobile wireless instrumented microrobot manipulators called NanoWalkers to collectively form a nanofactory system that might be used for positional manufacturing operations (Martel and Hunter 2002).

Zyvex Corp. (www.zyvex.com) of Richardson TX received a $25 million, five-year, National Institute of Standards and Technology (NIST) contract to develop prototype microscale assemblers using microelectromechanical systems (Freitas and Merkle 2004d).

Eventually this research can lead to the design of production lines in a nanofactory, both for diamondoid mechanosynthesis and for component assembly operations. Ultimately, medical nanorobots will be manufactured in desktop nanofactories efficiently designed for this purpose. The nanofactory system will include a progression of fabrication and assembly mechanisms at several different physical scales (Fig. 23.16). At the smallest scale, molecular mills will manipulate individual molecules to fabricate successively larger submicron-scale building blocks. These are passed to larger block assemblers that assemble still larger microblocks, which are themselves passed to even larger product assemblers that put together the final product. The microblocks are placed in a specific pattern and sequence following construction blueprints created using a modern "design for assembly" philosophy. As plane after plane is completed, the product extrudes outward through the surface of the nanofactory output platform (Fig. 23.17).

23.4.7 R&D Timeline, Costs, and Market Value of Medical Nanorobots

The Nanofactory Collaboration (Nanofactory Collaboration 2007a) is establishing a combined experimental and theoretical program to explore the feasibility of nanoscale positional manufacturing techniques, starting with the positionally controlled mechanosynthesis of diamondoid structures using simple molecular

Fig. 23.16 Assembly of nanoparts into larger components and product structures using mechanical manipulators at various size scales on interconnected production lines inside a diamondoid nanofactory (courtesy of John Burch). Used with permission

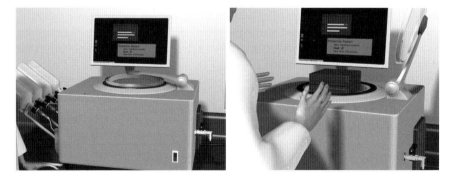

Fig. 23.17 Diamondoid desktop nanofactory (courtesy of John Burch). Used with permission

feedstock and progressing to the ultimate goal of a desktop nanofactory appliance able to manufacture macroscale quantities of molecularly precise product objects according to digitally-defined blueprints. The Collaboration was initiated by Freitas and Merkle in 2001 and has led to continuing efforts involving direct collaborations among 23 researchers and others, including 17 PhD's or PhD candidates, at 9 organizations in 4 countries – the U.S., U.K., Russia, and Belgium – as of 2009.

What will it cost to develop a nanofactory? Let's assume research funds are spent in a completely focused manner toward the goal of a primitive diamondoid nanofactory that could assemble rigid diamondoid structures involving carbon, hydrogen, and perhaps a few other elements. In this case, we estimate that an ideal research effort paced to make optimum use of available computational, experimental, and human resources would probably run at a $1–5 M/yr level for the first 5 years of the program, ramp up to $20–50 M/yr for the next 6 years, then finish off at a ~$100 M/yr rate culminating in a simple working desktop nanofactory appliance in year 16 of a ~$900 M effort. Of course the bulk of this work, after the initial 5 year period, would be performed by people, companies, and university groups recruited from outside the Nanofactory Collaboration. The key early milestone is to demonstrate positionally-controlled carbon placement on a diamond surface by the end of the initial 5 year period. We believe that successful completion of this key experimental milestone would make it easier to recruit significant additional financial and human resources to undertake the more costly later phases of the nanofactory development work.

Some additional costs would also be required to design, build, test, and obtain FDA approval for the many specific classes of nanorobots to be employed in various therapeutic medical applications (Sections 23.6 and 23.7). Medical nanorobots will certainly be among the first consumer products to be made by nanofactories because: (1) even relatively small (milligram/gram) quantities of medical nanorobots could be incredibly useful; (2) nanorobots can save lives and extend the human healthspan, thus will be in high demand once available; (3) manufacturers of such high value products (or of the nanofactories, depending on the economic model) can command a high price from healthcare providers, which means nanorobots should be worth building early, even though early-arriving nanomedical products are likely to be more expensive (in $/kg) than later-arriving products; and (4) the ability to extract, re-use and recycle nanorobots may allow the cost per treatment to the individual patient to be held lower than might be expected, with treatment costs also declining rapidly over time.

Is it worth spending billions of dollars to develop and begin deploying medical nanorobots? The billion-dollar R&D expense should be compared to the cost of doing nothing. Every year humanity suffers the death of ~55 million people, of which about 94% or 52 million of these deaths were not directly caused by human action – that is, not accidents, suicides, homicides or war – and thus all, in principle, are directly preventable by future nanomedical interventions (Section 23.6). We can crudely calculate the annual opportunity cost of a failure to intervene, as follows.

According to the Lasker Foundation (Lasker Foundation 2000), a dozen or so studies since the mid-1970s have found the value for human life is in the range of $3 to $7 million constant dollars, using many different methodologies. More recently, data from Murphy and Topel (Murphy and Topel 1999) at the University of Chicago, updated to Year 2000 dollars, show the value of human life at every age for white males (Fig. 23.18). It recognizes that fewer years remain to us at older ages. The chart in Fig. 23.19 gives an estimate of the number of people that died in the United States in the Year 2000, in each age cohort, year by year, again for white males. This estimate is computed by multiplying the estimated U.S. population of

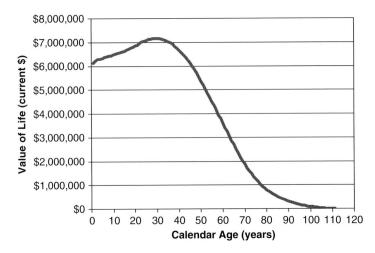

Fig. 23.18 U.S. Value of Human Life, by Age, for White Males in the Year 2000 (modified from Murphy and Topel 1999))

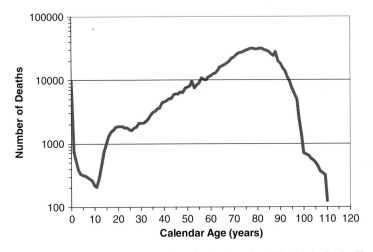

Fig. 23.19 U.S Number of Human Deaths in U.S., by Age, for White Males in the Year 2000 (values estimated using data from U.S. Census Bureau (Day 1993, Census Bureau 2001a) and from Vaupel et al. (1998))

while males (by age group, 0–110 years) (Day 1993) by the death rate by age for U.S. white males (ages 0–80 from Census Bureau (Census Bureau, 2001a), ages 81–110 estimated from Vaupel (Vaupel et al. 1998)). If we multiply the death rate at each age, from the chart in Fig. 23.19, by the dollar value at each age, from the previous chart in Fig. 23.18, we get the economic loss at each calendar age, due to human death. The sum of these economic losses divided by the total number of deaths gives the average economic value of a human life lost, across all the ages of a natural lifespan. The result is an average value of about $2.05 million dollars for each (white male) human life lost, with similar conclusions for either gender and for other races. If we assume that the population age structure, the age-specific mortality, and the value of human life is the same worldwide as in the United States, then the worldwide medically-preventable death toll of 52 million people in the Year 2000 represents an economic loss of about $104 trillion dollars per year, or ~$140T/year in 2007 dollars.

For comparison, taking Federal Reserve figures for the total tangible net wealth of the United States ($80.3T), which includes all household and business financial assets, all real estate, and all consumer durables, net of debt for 2007 (Federal Reserve System 2007), and applying the ratio (~29.4%, circa Year 2000) of U.S. GDP ($9.9T) (Census Bureau 2001b) to world GDP ($33.2T) (Department of Energy 1999) gives us a crude estimate of total global tangible net worth of $269 trillion dollars for the Year 2007. Thus every year, nanomedically preventable deaths deplete human capital by an amount exceeding half of the entire tangible wealth of the world. This ongoing *annual* capital loss is many of orders of magnitude greater than the entire likely *multi-decade* R&D expense of developing medical nanorobotics, whose deployment could spare us this great loss of human capital.

23.5 Medical Nanorobot Biocompatibility

The safety, effectiveness, and utility of medical nanorobotic devices will critically depend upon their biocompatibility with human organs, tissues, cells, and biochemical systems. An entire technical book published in 2003 (*Nanomedicine, Vol. IIA* (Freitas 2003)) describes the many biocompatibility issues surrounding the use of diamond-based nanorobots inside the human body, and broadens the definition of nanomedical biocompatibility to include all of the mechanical, physiological, immunological, cytological, and biochemical responses of the human body to the introduction of artificial medical nanodevices (Table 23.1). A large part of this work is an examination of the classical biocompatibility challenges including issues such as immune system reactions (Section 23.5.1), complement activation, inflammation (Section 23.5.2), thrombogenesis, and carcinogenesis that might be caused by medical nanorobots. But this study of classical challenges suggested a number of new biocompatibility issues that must also be addressed in medical nanorobotics including, most importantly, the areas of mechanocompatibility, particle biodynamics and distribution, and phagocyte avoidance protocols (Section 23.5.3). Readers interested

Table 23.1 Issues in nanorobot biocompatibility

Classical nanorobot biocompatibility issues	New nanorobot biocompatibility issues
• Adhesive Interactions with Nanorobot Surfaces	• Geometrical Trapping of Bloodborne Medical Nanorobots
• Nanorobot Immunoreactivity	• Phagocytosis of Bloodborne Microparticles
• Complement Activation	• Particle Clearance from Tissues or Lymphatics
• Immunosuppression, Tolerization, and Camouflage	• Phagocyte Avoidance and Escape
• Immune Privilege and Immune Evasion	• Nanorobotic Thermocompatibility and Electrocompatibility
• General and Nonspecific Inflammation	• Biofouling of Medical Nanorobots
• Coagulation and Thrombogenicity	• Biocompatibility of Nanorobot Effluents and Leachates
• Allergic and Other Sensitivity Reactions	• Biocompatibility of Nanorobot Fragments in vivo
• Sternutogenesis, Nauseogenesis and Emetogenesis	• Nanorobot Mechanocompatibility
• Nanoid Shock	• Mechanical Peristaltogenesis and Mucosacompatibility
• Nanopyrexia	• Nanorobotic Mechanical Vasculopathies
• Nanorobot Mutagenicity and Carcinogenicity	• Mechanocompatibility with Extracellular Matrix and Tissue Cells
• Protein Adsorption on Diamondoid Surfaces	• Mechanocompatibility with Nontissue Cells
• Cell Response to Diamondoid Surfaces	• Cytomembrane and Intracellular Mechanocompatibility
• Chemical Stability and Corrosion Degradation Effects	• Disruption of Molecular Motors and Vesicular Transport
• Nanorobot Hemolysis, Thrombocytolysis, and Leukocytolysis	• Mechanical Disruption of Intracellular Microzones
	• Mechanically-Induced Proteolysis, Apoptosis, or Prionosis

in biocompatibility issues not covered below can find a more comprehensive list of topics in the book *Nanomedicine, Vol. IIA* (Freitas 2003).

23.5.1 Immune System Reactions

Whether the human immune system can recognize medical nanorobots may depend largely upon the composition of the nanorobot exterior surfaces. Pure diamond is generally considered nonimmunogenic – e.g., chemical vapor deposition (CVD) diamond coatings for artificial joints are considered to have "low immunoreactivity", and as of 2009 there were no reports in the literature of antibodies having been raised to diamond. However, concerted experimental searches for antibodies to diamondoid materials have yet to be undertaken, and experimental failures rarely find their way into the literature. It is conceivable that different antibodies may recognize distinct faces of a crystal (possibly including diamond or sapphire crystal faces exposed at the surfaces of medical nanorobots) in an interaction similar to that of antibodies for repetitive epitopes present on protein surfaces – for instance, one monoclonal antibody (MAb) to 1,4-dinitrobenzene crystals was shown to specifically interact with the molecularly flat, aromatic, and polar (101) face of these crystals, but not with other faces of the same crystal (Kessler et al. 1999). Another concern is that antibodies may be raised against binding sites that are positioned on the nanorobot exterior, e.g., sorting rotor pockets (Freitas 1999o) which may be similar to traditional bioreceptors, and that these antibodies could then act as antagonists (Fauque et al. 1985; Wright et al. 2000) for such sites, since MAbs specific to biological binding sites are well known.

If antibodies to nanorobot exteriors are found to exist in the natural human antibody specificity repertoire, then to avoid immune recognition many techniques of immune evasion (Freitas 2003b) may be borrowed from biology, for example:

(1) *Camouflage.* Coat the nanorobot with a layer of "self" proteins and carbohydrate moieties resembling fibroblast, platelet, or even RBC (red blood cell) plasma membrane.
(2) *Chemical Inhibition.* Nanorobots may slowly secrete chemical substances into the perirobotic environment to make it difficult for Ig molecules to adhere to an otherwise immunogenic nanorobot surface.
(3) *Decoys.* Release a cloud of soluble nanorobot-epitope antigens in the vicinity of the nanorobot (though this method has limited utility because sending out decoys will only expand the number of attacking elements to overwhelm the decoys).
(4) *Active Neutralization.* Equip the nanorobot with molecular sorting rotors designed with binding sites similar or identical to the nanorobot epitopes that raised the target antibodies.
(5) *Tolerization.* Using only traditional methods, nanorobots introduced into a newborn may train the neonatal immune system to regard these foreign materials as "native," thus eliminating nanorobot-active antibodies via natural clonal deletion. However, it now appears possible to tolerize an adult to any antigen by regenerating the adult's thymus (the source of the newborn effect) and placing the antigen into the thymus where self-reactive clones are then deleted or anergized (Fahy 2003, 2007, 2010; Aspinall 2010).
(6) *Clonal Deletion.* Once the paratopes of antibodies that bind nanorobots are known, immunotoxin molecules can be engineered that display those paratopes, and upon injection into the patient, these targeted immunotoxins would bind to all T cell receptors that display this paratope, killing the nanorobot-sensitive T cells.

23.5.2 Inflammation

Could medical nanorobots trigger general inflammation in the human body? One early experiment (Royer et al. 1982) to determine the inflammatory effects of various implant substances placed subdermally into rat paws found that an injection of 2–10 mg/cm^3 (10- to 20-micron particles at 10^5–10^6 particles/cm^3) of natural diamond powder suspension caused a slight increase in volume of the treated paw relative to the control paw. However, the edematous effect subsided after 30–60 minutes at both concentrations of injected diamond powder that were tried, so this swelling could have been wholly caused by mechanical trauma of the injection and not the diamond powder. Another experiment (Delongeas et al. 1984) at the same laboratory found that intraarticulate injection of diamond powder was not phlogistic (i.e., no erythematous or edematous changes) in rabbit bone joints and produced no inflammation. Diamond particles are traditionally regarded as biologically inert and noninflammatory for neutrophils (Tse and Phelps 1970; Higson and Jones 1984; Hedenborg and Klockars 1989; Aspenberg et al. 1996) and are typically used as experimental null controls (Delongeas et al. 1984).

Since the general inflammatory reaction is chemically mediated, it should also be possible to employ nanorobot surface-deployed molecular sorting rotors to selectively absorb kinins or other soluble activation factors such as HMGB1 (High Mobility Group Box Protein 1) (Scaffidi et al. 2002), thus short-circuiting the inflammatory process. Active semaphores consisting of bound proteases such as gelatinase A could be deployed at the nanorobot surface to cleave and degrade monocyte chemoattractant molecules (McQuibban et al. 2000) or other chemokines, suppressing the cellular inflammatory response. Conversely, key inflammatory inhibitors could be locally released by nanorobots. For instance, Hageman factor contact activation inhibitors such as the 22.5-kD endothelial cell-secreted protein HMG-I (Donaldson et al. 1998), surface-immobilized unfractionated heparin (Elgue et al. 1993), and C1 inhibitor (Cameron et al. 1989) would probably require lower release dosages than for aspirin or steroids, and therapeutic blockade of factor XII activation has been demonstrated (Fuhrer et al. 1990). As yet another example, platelet activating factor (PAF) is a cytokine mediator of immediate hypersensitivity which produces inflammation. PAF is produced by many different kinds of stimulated cells such as basophils, endothelial cells, macrophages, monocytes, and neutrophils. It is 100–10,000 times more vasoactive than histamine and aggregates platelets at concentrations as low as 0.01 pmol/cm^3 (Mayes 1993). Various PAF antagonists and inhibitors are known (Freitas 2003c) – and these or related inhibitory molecules, if released or surface-displayed by medical nanorobots, may be useful in circumventing a general inflammatory response.

23.5.3 *Phagocytosis*

Invading microbes that readily attract phagocytes and are easily ingested and killed are generally unsuccessful as parasites. In contrast, most bacteria that are successful as parasites interfere to some extent with the activities of phagocytes or find some way to avoid their attention (Todar 2003). Bacterial pathogens have devised numerous diverse strategies to avoid phagocytic engulfment and killing. These strategies are mostly aimed at blocking one or more of the steps in phagocytosis, thereby halting the process (Todar 2003).

Similarly, phagocytic cells presented with any significant concentration of medical nanorobots may attempt to internalize these nanorobots. Virtually every medical nanorobot placed inside the human body will physically encounter phagocytic cells many times during its mission. Thus all nanorobots that are of a size capable of ingestion by phagocytic cells must incorporate physical mechanisms and operational protocols for avoiding and escaping from phagocytes (Freitas 2003d). Engulfment may require from many seconds to many minutes to go to completion (Freitas 2003e), depending upon the size of the particle to be internalized, so medical nanorobots should have plenty of time to detect, and to actively prevent, this process. Detection by a medical nanorobot that it is being engulfed by a phagocyte may be accomplished using (1) hull-mounted chemotactic sensor pads equipped with artificial binding sites that are specific to phagocyte coat molecules, (2) continuous monitoring of the flow rates of nanorobot nutrient ingestion or waste ejection mechanisms (e.g., blocked glucose or O_2 import), (3) acoustic techniques (Freitas 1999w), (4) direct measurement of mechanical forces on the hull, or (5) various other means.

The basic anti-phagocyte strategy is first to avoid phagocytic contact (Freitas 2003f), recognition (Freitas 2003g), or binding and activation (Freitas 2003h), and secondly, if this fails, then to inhibit phagocytic engulfment (Freitas 2003i) or enclosure and scission (Freitas 2003j) of the phagosome. If trapped, the medical nanorobot can induce exocytosis of the phagosomal vacuole in which it is lodged (Freitas 2003k) or inhibit both phagolysosomal fusion (Freitas 2003m) and phagosome metabolism (Freitas 2003n). In rare circumstances, it may be necessary to kill the phagocyte (Freitas 2003o) or to blockade the entire phagocytic system (Freitas 2003p). Of course, the most direct approach for a fully-functional medical nanorobot is to employ its motility mechanisms to locomote out of, or away from, the phagocytic cell that is attempting to engulf it. This may involve reverse cytopenetration (Freitas 1999x), which must be done cautiously (e.g., the rapid exit of nonenveloped viruses from cells can be cytotoxic (Oh 1985)). It is possible that frustrated phagocytosis may induce a localized compensatory granulomatous reaction. Medical nanorobots therefore may also need to employ simple but active defensive strategies to forestall granuloma formation (Freitas 2003q).

23.6 Control of Human Morbidity using Medical Nanorobots

Morbidity is the state of being unhealthy, sick, diseased, possessing genetic or anatomic pathologies or injuries, or experiencing physiological malfunctions, and also generally refers to conditions that are potentially medically treatable. Here we'll examine how human morbidity can be controlled and prevented by employing medical nanorobots to cure disease, reverse trauma, and repair individual cells. The descriptions of nanorobots suggested for each treatment are representative of the powerful new capabilities that are expected to be available some decades hence, but we do not provide an exhaustive summary of all devices that may be needed during each treatment as that would be beyond the scope of this chapter.

23.6.1 Advantages of Medical Nanorobots

Although biotechnology makes possible a greatly increased range and efficacy of treatment options compared to traditional approaches, with medical nanorobotics the range, efficacy, comfort and speed of possible medical treatments further expands enormously. Medical nanorobotics will be essential whenever the damage to the human body is extremely subtle, highly selective, or time-critical (as in head traumas, burns, or fast-spreading diseases), or when the damage is very massive, overwhelming the body's natural defenses and repair mechanisms – pathological conditions from which it is often difficult or impossible to recover at all using current or easily foreseeable biotechnological techniques.

While it is true that many classes of medical problems may be at least partially resolved using existing treatment alternatives, it is also true that as the chosen medical technology becomes more precise, active, and controllable, the range of options broadens and the quality of the options improves. Thus the question is not whether medical nanorobotics is absolutely required to accomplish a given medical objective. In many cases, it is not – though of course there are some things that only biotechnology and nanotechnology can do, and some other things that only nanotechnology can do. Rather, the important question is which approach offers a superior outcome for a given medical problem, using any reasonable metric of treatment efficacy. For virtually every class of medical challenge, a mature medical nanorobotics offers a wider and more effective range of treatment options than any other solution. A few of the most important advantages of medical nanorobotics over present-day and future biotechnology-based medical and surgical approaches include (Freitas 1999ct):

1. *Speed of Treatment.* Doctors may be surprised by the incredible quickness of nanorobotic action when compared to methods relying on self-repair. We expect that mechanical nanorobotic therapeutic systems can reach their targets up to ~1,000 times faster, all else equal, and treatments which require ~10^5 sec (~days) for biological systems to complete may require only ~10^2 sec (~minutes) using nanorobotic systems (Freitas 2005b).
2. *Control of Treatment.* Present-day biotechnological entities are not programmable and cannot easily be switched on and off conditionally (while following complex multidecision trees) during task execution. Even assuming that a digital biocomputer (Freitas 1999ai; Guet et al. 2002; Yokobayashi et al. 2002; Basu et al. 2004, 2005) could be installed in, for example, a fibroblast, and that appropriate effector mechanisms could be attached, such a biorobotic system would necessarily have slower clock cycles (Basu et al. 2004), less capacious memory per unit volume, and longer data access times, implying less diversity of action, poorer control, and less complex executable programs than would be available in diamondoid nanocomputer-controlled nanorobotic systems (Section 23.3.5). The mechanical or electronic nanocomputer approach (Freitas 1999k) emphasizes precise control of action (Freitas 2009), including control of physical placement, timing, strength, structure, and interactions with other (especially biological) entities.

3. *Verification of Treatment.* Nanorobotic-enabled endoscopic nanosurgery (Section "Endoscopic Nanosurgery and Surgical Nanorobots") will include comprehensive sensory feedback enabling full VR telepresence permitting real-time surgery into cellular and subcellular tissue volumes. Using a variety of communication modalities (Freitas 1999f), nanorobots will be able to report back to the attending physician, with digital precision and ~MHz bandwidth (Freitas 1999ah), a summary of diagnostically- or therapeutically-relevant data describing exactly what was found prior to treatment, what was done during treatment, and what problems were encountered after treatment, in every cell or tissue visited and treated by the nanorobot. A comparable biological-based approach relying primarily upon chemical messaging must necessarily be slow and have only limited signaling capacity and bandwidth.

4. *Minimal Side Effects.* Almost all drugs have significant side effects, such as conventional cancer chemotherapy which causes hair loss and vomiting, although computer-designed drugs can have higher specificity and fewer side effects than earlier drugs. Carefully tailored cancer vaccines under development starting in the late 1990s were expected unavoidably to affect some healthy cells. Even well-targeted drugs are distributed to unintended tissues and organs in low concentrations (Davis 1996), although some bacteria can target a few organs fairly reliably without being able to distinguish individual cells. By contrast, mechanical nanorobots may be targeted with virtually 100% accuracy to specific organs, tissues, or even individual cellular addresses within the human body (Freitas 1999g, 2006a). Such nanorobots should have few if any side effects, and will remain safe even in large dosages because their actions can be digitally self-regulated using rigorous control protocols (Freitas 2009) that affirmatively prohibit device activation unless all necessary preconditions have been met, and remain continuously satisfied. More than a decade ago, Fahy (1993) observed that these possibilities could transform "drugs" into "programmable machines with a range of sensory, decision-making, and effector capabilities [that] might avoid side effects and allergic reactions...attaining almost complete specificity of action.... Designed smart pharmaceuticals might activate themselves only when, where, and if needed." Additionally, nanorobots may be programmed to harmlessly remove themselves from the site of action, or conveniently excrete themselves from the body, after a treatment is completed. By contrast, spent biorobotic elements containing ingested foreign materials may have more limited post-treatment mobility, thus lingering at the worksite causing inflammation when naturally degraded in situ or removed. (It might be possible to design artificial eukaryotic biorobots having an apoptotic pathway (Freitas 1999ag) that could be activated to permit clean and natural self-destruction, but any indigestible foreign material that had been endocytosed by the biorobot could still cause inflammation in surrounding tissues when released).

5. *Faster and More Precise Diagnosis.* The analytic function of medical diagnosis requires rapid communication between the injected devices and the attending physician. If limited to chemical messaging, biotechnology-based devices such as biorobots will require minutes or hours to complete each diagnostic loop.

Nanomachines, with their more diverse set of input-output mechanisms, will be able to outmessage complete results (both aggregated and individual outliers) of in vivo reconnaissance or testing to the physician, literally in seconds (Freitas 1999f). Such nanomachines could also run more complex tests of greater variety in far less time. Nanomechanical nanoinstrumentation will make comprehensive rapid cell mapping and cell interaction analysis possible. For example, new instances of novel bacterial resistance could be assayed at the molecular level in real time, allowing new treatment agents to be quickly composed using an FDA-approved formulary, then manufactured and immediately deployed on the spot.

6. *More Sensitive Response Threshold for High-Speed Action.* Unlike natural systems, an entire population of nanorobotic devices could be triggered globally by just a single local detection of the target antigen or pathogen. The natural immune system takes $>10^5$ sec to become fully engaged after exposure to a systemic pathogen or other antigen-presenting intruder. A biotechnologically enhanced immune system that could employ the fastest natural unit replication time ($\sim 10^3$ sec for some bacteria) would thus require at least $\sim 10^4$ sec for full deployment post-exposure. By contrast, an artificial nanorobotic immune system (Freitas 2005b) could probably be fully engaged (though not finished) in at most two blood circulation times, or $\sim 10^2$ sec.

7. *More Reliable Operation.* Individual engineered macrophages would almost certainly operate less reliably than individual mechanical nanorobots. For example, many pathogens, such as *Listeria monocytogenes* and *Trypanosoma cruzi*, are known to be able to escape from phagocytic vacuoles into the cytoplasm (Stenger et al. 1998). While biotech drugs or cell manufactured proteins could be developed to prevent this (e.g., cold therapy drugs are entry-point blockers), nanorobotic trapping mechanisms could be more secure (Freitas 1999y, 2005b). Proteins assembled by natural ribosomes typically incorporate one error per $\sim 10^4$ amino acids placed; current gene and protein synthesizing machines utilizing biotechnological processes have similar error rates. A molecular nanotechnology approach should decrease these error rates by at least a millionfold (Drexler 1992i). Nanomechanical systems will also incorporate onboard sensors to determine if and when a particular task needs to be done, or when a task has been completed. Finally, and perhaps most importantly, it is highly unlikely that natural microorganisms will be able to infiltrate rigid watertight diamondoid nanorobots or to co-opt their functions. By contrast, a biotech-based biorobot more readily could be diverted or defeated by microbes that would piggyback on its metabolism, interfere with its normal workings, or even incorporate the device wholesale into their own structures, causing the engineered biomachine to perform some new or different – and possibly pathological – function that was not originally intended. There are many examples of such co-option in natural biological systems, including the protozoan mixotrichs found in the termite gut that have assimilated bacteria into their bodies for use as motive engines (Cleveland and Grimstone 1964; Tamm 1982), and the nudibranch mollusks (marine snails without shells) that steal nematocysts (stinging cells) away from

coelenterates such as jellyfish (i.e. a Portuguese man-of-war) and incorporate the stingers as defensive armaments in their own skins (Thompson and Bennett 1969) – a process which Vogel (Vogel 1998) calls "stealing loaded guns from the army."

8. *Nonbiodegradable Treatment Agents.* Diagnostic and therapeutic agents constructed of biomaterials generally are biodegradable in vivo, although there is a major branch of pharmacology devoted to designing drugs that are moderately non-biodegradable – e.g., anti-sense DNA analogs with unusual backbone linkages and peptide nucleic acids (PNAs) are difficult to break down. An engineered fibroblast may not stimulate an immune response when transplanted into a foreign host, but its biomolecules are subject to chemical attack in vivo by free radicals, acids, and enzymes. Even "mirror" biomolecules or "Doppelganger proteins" comprised exclusively of unnatural D-amino acids have a lifetime of only ~5 days inside the human body (Robson 1998). In contrast, suitably designed nanorobotic agents constructed of nonbiological materials will not be biodegradable. Nonbiological diamondoid materials are highly resistant to chemical breakdown or leukocytic degradation in vivo, and pathogenic biological entities cannot easily evolve useful attack strategies against these materials (Freitas 1999z). This means that medical nanorobots could be recovered intact from the patient and recycled, reducing life-cycle energy consumption and treatment costs.

9. *Superior Materials.* Typical biological materials have tensile failure strengths in the 10^6–10^7 N/m^2 range, with the strongest biological materials such as wet compact bone having a failure strength of ~10^8 N/m^2, all of which compare poorly to ~10^9 N/m^2 for good steel, ~10^{10} N/m^2 for sapphire, and ~10^{11} N/m^2 for diamond and carbon fullerenes (Freitas 1999aa), again showing a 10^3–10^5 fold strength advantage for mechanical systems that use nonbiological, and especially diamondoid, materials. Nonbiological materials can be much stiffer, permitting the application of higher forces with greater precision of movement, and they also tend to remain more stable over a larger range of relevant conditions including temperature, pressure, salinity and pH. Proteins are heat sensitive in part because much of the functionality of their structure derives from the noncovalent bonds involved in folding, which are broken more easily at higher temperatures. In diamond, sapphire, and many other rigid materials, structural shape is covalently fixed, hence is far more temperature-stable. Most proteins also tend to become dysfunctional at cryogenic temperatures, unlike diamond-based mechanical structures (Freitas 1999ab), so diamondoid nanorobots could more easily be used to repair frozen cells and tissues. Biomaterials are not ruled out for all nanomechanical systems, but they represent only a small subset of the full range of materials that can be employed in nanorobots. Nanorobotic systems may take advantage of a wider variety of atom types and molecular structures in their design and construction, making possible novel functional forms that might be difficult to implement in a purely biological-based system (e.g., steam engines (Freitas 1999ac) or nuclear power (Freitas 1999ad)). As another example, an application requiring the most effective bulk thermal conduction possible

should use diamond, the best conductor available, not some biomaterial having inferior thermal performance.

23.6.2 Curing Disease

Nanorobots should be able to cure most common diseases in a manner more akin to the directness and immediacy of surgery, fixing a given problem in minutes or hours, than to current treatment regimens for treating most disease conditions which typically involve (a) the injection or ingestion of slow-acting medications of relatively poor efficacy, (b) dietary and lifestyle changes, (c) psychological factors, and so forth, often taking weeks, months, or even years to provide what is sometimes only an incomplete cure. We focus here on the disease conditions that presently pose the greatest risk of death, most of which are, not surprisingly, presently associated with aging and include microbial infections (Section 23.6.2.1), cancer (Section 23.6.2.2), heart disease (Section 23.6.2.3), stroke (Section 23.6.2.4), and hormonal, metabolic and genetic disease (Section 23.6.2.5).

Bear in mind that each of the nanorobotic treatment devices described below has been subjected to a rigorous design and scaling study, most of which have been published in peer reviewed journals. Thus the proposed nanomachines are not cartoons or simplistic speculations but rather are genuine engineering constructs believed to be thoroughly feasible and likely to function as described once they (or similar devices) can be manufactured (Section 23.4).

23.6.2.1 Bacterial, Viral, and Other Parasitic Infection

Perhaps the most widely recognized form of disease is when the human body is under attack by invading viruses, bacteria, protozoa, or other microscopic parasites. One general class of medical nanorobot will serve as the first-line nanomedical treatment for pathogen-related disease. Called a "microbivore" (Fig. 23.20), this artificial nanorobotic white cell substitute, made of diamond and sapphire, would seek out and harmlessly digest unwanted bloodborne pathogens (Freitas 2005b). One main task of natural white cells is to phagocytose and kill microbial invaders in the bloodstream. Microbivore nanorobots would also perform the equivalent of phagocytosis and microbial killing, but would operate much faster, more reliably, and under human control.

The baseline microbivore is designed as an oblate spheroidal nanomedical device measuring 3.4 microns in diameter along its major axis and 2.0 microns in diameter along its minor axis, consisting of 610 billion precisely arranged structural atoms in a gross geometric volume of 12.1 micron^3 and a dry mass of 12.2 picograms. This size helps to ensure that the nanorobot can safely pass through even the narrowest of human capillaries and other tight spots in the spleen (e.g., the interendothelial splenofenestral slits (Freitas 2003r)) and elsewhere in the human body (Freitas 2003s). The microbivore has a mouth with an irising door, called the ingestion port, where microbes are fed in to be digested, which is large enough to

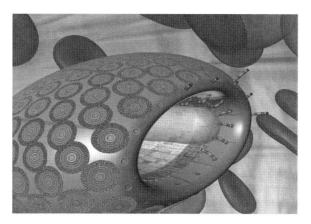

Fig. 23.20 An artificial white cell – the microbivore (Freitas 2005a). Designer Robert A. Freitas Jr., additional design Forrest Bishop. ©2001 Zyvex Corp. Used with permission

internalize a single microbe from virtually any major bacteremic species in a single gulp. The microbivore also has a rear end, or exhaust port, where the completely digested remains of the pathogen are harmlessly expelled from the device. The rear door opens between the main body of the microbivore and a tail-cone structure. According to this author's scaling study (Freitas 2005b), the device may consume up to 200 pW of continuous power (using bloodstream glucose and oxygen for energy) while completely digesting trapped microbes at a maximum throughput of 2 micron^3 of organic material per 30-second cycle. This "digest and discharge" protocol (Freitas 1999aj) is conceptually similar to the internalization and digestion process practiced by natural phagocytes, except that the artificial process should be much faster and cleaner. For example, it is well-known that macrophages release biologically active compounds during bacteriophagy (Fincher et al. 1996), whereas well-designed microbivores need only release biologically inactive effluent.

The first task for the bloodborne microbivore is to reliably acquire a pathogen to be digested. If the correct bacterium bumps into the nanorobot surface, reversible species-specific binding sites on the microbivore hull can recognize and weakly bind to the bacterium. A set of 9 distinct antigenic markers should be specific enough (Freitas 1999ak), since all 9 must register a positive binding event to confirm that a targeted microbe has been caught. There are 20,000 copies of these 9-marker receptor sets, distributed in 275 disk-shaped regions across the microbivore surface. Inside each receptor ring are more rotors to absorb ambient glucose and oxygen from the bloodstream to provide nanorobot power. At the center of each 150-nm diameter receptor disk is a grapple silo. Once a bacterium has been captured by the reversible receptors, telescoping robotic grapples (Freitas 1999am) rise up out of the microbivore surface and attach to the trapped bacterium, establishing secure anchorage to the microbe's cell wall, capsid, or plasma membrane (Fig. 23.21). The microbivore grapple arms are about 100 nm long and have various rotating and telescoping joints that allow them to change their position, angle, and length. After

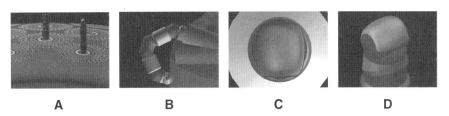

Fig. 23.21 Telescoping grapple manipulators for the microbivore (Freitas 2005a) help to capture and manipulate target pathogens into the interior of the device for digestion, and to assist in device mobility; (**a**) fully extended grapple, (**b**) grapple work envelope, (**c**) top view of grapple in silo with iris cover mechanism retracted, (**d**) grapple footpad covered by protective cowling. Images © 2001 Forrest Bishop, used with permission

rising out of its silo, a grapple arm could execute complex twisting motions, and adjacent grapple arms can physically reach each other, allowing them to hand off bound objects as small as a virus particle. Grapple handoff motions could transport a large rod-shaped bacterium from its original capture site forward to the ingestion port at the front of the device. The captive organism would be rotated into the proper orientation as it approaches the open microbivore mouth, where the pathogen is internalized into a 2 micron^3 morcellation chamber under continuous control of mouth grapples and an internal mooring mechanism.

There are two concentric cylinders inside the microbivore. The bacterium will be minced into nanoscale pieces in the morcellation chamber (Freitas 1999an), the smaller inner cylinder, then the remains are pistoned into a separate 2 micron^3 digestion chamber, a larger outer cylinder. In a preprogrammed sequence, ~40 different engineered digestive enzymes will be successively injected and extracted six times during a single digestion cycle, progressively reducing the morcellate to monoresidue amino acids, mononucleotides, glycerol, free fatty acids and simple sugars, using an appropriate array of molecular sorting rotors. These basic molecules are then harmlessly discharged back into the bloodstream through the exhaust port at the rear of the device, completing the 30-second digestion cycle. When treatment is finished, the doctor may transmit an ultrasound signal to tell the circulating microbivores that their work is done. The nanorobots then exit the body through the kidneys and are excreted with the urine in due course (Weatherbee and Freitas 2010).

A human neutrophil, the most common type of leukocyte or white cell, can capture and engulf a microbe in a minute or less, but complete digestion and excretion of the organism's remains can take an hour or longer. Our natural white cells – even when aided by antibiotics – can sometimes take weeks or months to completely clear bacteria from the bloodstream. By comparison, a single terabot (10^{12}-nanorobot) dose of microbivores should be able to fully eliminate bloodborne pathogens in just minutes, or hours in the case of locally dense infections. This is accomplished without increasing the risk of sepsis or septic shock because all bacterial components (including all cell-wall lipopolysaccharide) will be internalized and fully digested into harmless nonantigenic molecules prior to discharge from the device. And no matter that a bacterium has acquired multiple drug resistance to antibiotics or to any other traditional treatment – the microbivore will eat it anyway. Microbivores would be up to ~1000 times faster-acting than antibiotic-based cures which often need weeks or months to work. The nanorobots would digest ~100 times more microbial material than an equal volume of natural white cells could digest in any given time period, and would have far greater maximum lifetime capacity for phagocytosis than natural white blood cells.

Besides intravenous bacterial, viral, fungal, and parasitic scavenging, microbivores or related devices could also be used to help clear respiratory or cerebrospinal bacterial infections, or infections in other nonsanguinous fluid spaces such as pleural (Strange and Sahn 1999), synovial (Perez 1999), or urinary fluids. They could eliminate bacterial toxemias; eradicate viral, fungal, and parasitic infections; patrol tissues to remove pathological substances and organisms; disinfect surfaces, foodstuffs, or organic samples; and even help clean up biohazards and toxic chemicals.

Slightly modified microbivores are envisioned (Freitas 2005b) that could attack biofilms (Costerton et al. 1999) or small tumor masses (Section 23.6.2.2). A targeted cell-rich surface would be detected via receptor binding, then grapple arms would rotate the entire nanorobot perpendicular to the stationary biofilm. After positioning the nanorobot's mouth over the film and establishing watertight contact via lipophilic semaphores (Freitas 1999cu), operation of the vacuum piston draws biofilm contents into the morcellation chamber and the regular digestion cycle begins. The geometry of the nanorobot mouth can be altered (e.g., to square or hexagonal cross-section) to allow closer packing of a sufficient number of adjacent microbivores to avoid significant leakage of cell contents as the biofilm or tumor is planarly digested.

Bloodborne microbivores alone are not a complete solution to microbial disease – pathogens also accumulate in reservoirs inside organs, tissues, and even cells, and thus would need to be extirpated by more sophisticated tissue-mobile (Freitas 1999at) and even cytopenetrating (Freitas 1999x) microbivores. Similarly, viruses can insert alien genetic sequences into native DNA that must be rooted out using chromallocyte-class devices (Section 23.6.4.3), and so forth. But microbivore-class devices will be the foundation of our future first-line treatment against microbiological pathogens.

23.6.2.2 Cancer

A cell that has lost its normal control mechanisms and thus exhibits unregulated growth is called a cancer. Cancer cells can arise from normal cells in any tissue or organ, and during this process their genetic material undergoes change. As these cells grow and multiply, they form a mass of cancerous tissue that invades adjacent tissues and can metastasize around the body. Near-term alternatives to traditional chemotherapy (that kills not just cancer cells but healthy cells as well and causes fatigue, hair loss, nausea, depression, and other side effects) are being developed such as angiogenesis inhibitors (Bergers et al. 1999), autologous vaccines (Berd et al. 1998), and WILT (Section 23.7.1.7).

The healthy human body can use phagocytosis to dispose of many isolated cancer cells (Shankaran et al. 2001; Dunn et al. 2002; Street et al. 2004; Swann and Smyth 2007) before they can replicate and become established as a growing tumor – which happens more frequently in people with abnormally functioning immune systems (e.g., patients with autoimmune disease or on immunosuppressive drugs) – though some cancers can evade immune system surveillance even when that system is functioning normally. No such evasion is possible, however, if we use microbivore-class nanodevices, some with enhanced tissue mobility, that could patrol the bloodstream or body tissues, seeking out the clear antigenic signature of cancerous cells or tumors (see below) and then digesting these cancers into harmless effluvia, leaving healthy cells untouched. For example, active microbivores crowding on the exterior surface of a tumor mass could each excavate and digest the tumor mass beneath it at ~1 micron/min, requiring ~1 hour for ~4000 devices to digest a 100 micron diameter tumor mass or ~400,000 devices and ~10 hours for a 1 mm diameter tumor. Larger tumors could be infiltrated by tissue-mobile microbivores along numerous parallel strata or more tortuous vascular paths and then be rapidly consumed from multiple foci, from the inside out.

A more organized treatment protocol would begin with a comprehensive whole-body mapping of all tissue-borne cancer cells and tumors based on detection of specific biomarkers (Box 23.1) or other thermographic (Freitas 1999av) or chemographic (Freitas 1999aw) techniques. A trillion-nanorobot survey fleet that spends 100 seconds examining the chemical surface signatures of the plasma membranes of all ~10 trillion tissue cells in the body (Freitas 1999m) nominally would require ~1000 sec to complete the survey. Each device could reach the vicinity of most organs and tissues in the body in about one circulation time or ~60 sec, and could then reach most cells which lie well within ~40 microns (~2 cell widths) of a capillary exit point within ~40 sec even traveling at a very slow ~1 micron/sec through the tissues (comparable to leukocyte and fibroblast speeds (Freitas 1999au)). Adding this travel time increases survey time to ~1520 seconds for infusion, travel to 10 adjacent target cells, examination of target cells, and return to the bloodstream, so we should perhaps allow ~1 hour for the entire mapping process (Freitas 2007) which must also include ingress and egress of nanorobots from the body.

Once the locations of all cancerous cells in the body have been mapped, tissue-mobile microbivores can employ precise in vivo positional navigation (Freitas 1999cj) to return to the address of each isolated cancer cell or small cancer cell aggregate and destroy them. Tumor masses larger than 0.1–1 mm in diameter may be more practical to remove via endoscopic nanosurgery (Section "Endoscopic Nanosurgery and Surgical Nanorobots") in which a specialized nanomechanical probe instrument such as a nanosyringoscope (Section "Nanosyringoscopy") (whose intelligent tip is mobile and guided by continuous sensor readings to detect the perimeter of the cancerous region) would be inserted into the diseased tissue which is then excised and either digested or vacuumed out in a few minutes, much like fatty deposits during present-day atherectomies (Mureebe and McKinsey 2006), malignant tumors during endoscopic tumor microdebridement (Simoni et al. 2003), or less-precise laser-based tumor debulking procedures (Paleri et al. 2005). In principle, cell repair nanorobots called chromallocytes (Section 23.6.4.3) that are capable of in situ chromosome replacement therapy could be used to effect a complete genetic cure of diseased cancer cells, but this is not practical for large tumors (treatment time too long, cell population too protean) and because most tumors consist of surplus tissue that is more convenient to excise than to repair.

Box 23.1 Biochemical markers for cancer cell mapping

Cancer cells may display below-normal concentrations of β_4 integrins and above-normal concentrations of β_1 integrins, survivin, sialidase-sensitive cancer mucins, and leptin receptors such as galectin-3 (Dowling et al. 2007). Other cancer cell biomarkers are GM2 ganglioside, a glycolipid present on the surface of ~95% of melanoma cells with the carbohydrate portion of the molecule conveniently jutting out on the extracellular side of the melanoma cell membrane, and fucosyl GM1, which is only detected on small cell lung cancers (Zhang et al. 1997). GM2 and another ganglioside, GD2, are expressed on the surface of several types of cancer cells involved in small-cell lung, colon, and gastric cancer, sarcoma, lymphoma, and neuroblastoma (Zhang et al. 1997). Recognition of surface GM2 is the basis of anticancer vaccines currently under development (Livingston 1998; Knutson 2002). The membranes of cancer cells in gastrointestinal stromal tumors express CD117 (aka. Kit) 95% of the time, heavy caldesmon (80%) and CD34 (70%) (Miettinen and Lasota 2006). Other cancer cell membrane biomarkers include ERBB2 oncoprotein (aka. p185) in breast cancer (Wu 2002), DDR1 and CLDN3 in epithelial ovarian cancer (Heinzelmann-Schwarz et al. 2004), the cell surface glycoprotein CD147 (aka. emmprin) on the surface of malignant tumor cells (Yan et al. 2005), and the transmembrane protein EDFR in various human carcinomas (Normanno et al. 2006). Helpfully, expression of some cancer cell surface biomarkers is differentially related to tumor stage (Roemer et al. 2004) – for example, MMP-26 and TIMP-4 are strongly expressed in high-grade prostatic intraepithelial neoplasia but their expression significantly declines as the cancer progresses to invasive adenocarcinoma in human prostate (Lee et al. 2006). Many tumor-associated antigens are already known (Malyankar 2007) and the search for new membrane-bound (Liang et al. 2006, Alvarez-Chaver et al. 2007) and other (Zhang et al. 2007; Feng et al. 2007) cancer cell biomarkers is very active.

23.6.2.3 Heart and Vascular Disease

Perhaps the most common form of heart disease – and the leading cause of illness and death in most Western countries – is atherosclerosis, a condition in which the endothelial cell-lined artery wall becomes thicker and less elastic due to the presence of fatty-material-accumulating white cells under the inner lining of the arterial wall, creating a deposit called an atheroma. As the atheromas grow, the arterial lumens narrow. In time, the atheromas may collect calcium deposits, become brittle, and rupture, spilling their fatty contents and triggering the formation of a blood clot. The clot can further narrow or even occlude the artery, possibly leading to heart attack, or it may detach and float downstream, producing a vascular embolism.

By the era of nanomedicine in the 2020s and beyond, the incidence of heart disease in Western countries may be somewhat diminished compared to today because atherosclerosis is already partially reversible by controlling blood lipids (Wissler and Vesselinovitch 1990; Schell and Myers 1997; Grobbee and Bots 2004; Tardif et al. 2006), and future nanorobotic control of gene expression (Section 23.6.4.4) or stem cell treatments as already demonstrated in rodents (Lu et al. 2007) may prove even more effective as preventive measures. However, the regression of atherosclerotic plaque is generally accompanied by a decrease in total vessel size without an increase in luminal dimensions (Tardif et al. 2006), so restoring original luminal dimensions will likely still require a capability for direct vascular remodeling. Prevention is also likely to be underutilized by asymptomatic hyperlipidemic patients in wealthy countries, and may not be sufficiently available to less affluent patients or to patients in nonindustrialized countries.

The "vasculocyte" (Fig. 23.22) may be the nanorobotic treatment of choice for the limited vascular repair of primarily intimal arteriosclerotic lesions prior to complete arterial occlusion (Freitas 1996a). The device is designed as a squat, hexagonal-shaped nanorobot with rounded corners, measuring 2.7 microns across and 1 micron tall, that walks the inside surface of blood vessels atop telescoping

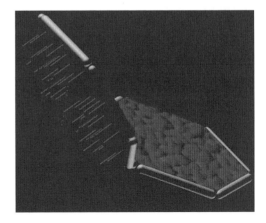

Fig. 23.22 The vasculocyte (Freitas 1996), nanorobotic treatment of choice for repairing atherosclerotic lesions on the vascular surface of arteries (courtesy of Forrest Bishop. Used with permission)

appendages arranged on its underbelly. Its 400-billion atom structure would weigh about 8 picograms. The machine is scaled so that its longest cross-body diagonal (Freitas 1999ax) is shorter than 4 microns, the diameter of the narrowest capillaries in the human body (Freitas 1999ay). The slightly-curved topmost surface will be almost completely tiled with 174,000 molecular sorting rotors (Section 23.3.2) to allow rapid exchange of specific molecules between the interior of the nanorobot and the patient's bloodstream.

On its six side walls the vasculocyte will be enveloped by an extensible "bumper" surface (Freitas 1999az) which cycles between 100 and 300 nm of thickness as internally-stored piston-pumped ballast fluid inflates and deflates the surface about once every second (Freitas 1999ba). This cycling will allow a nanorobot situated on an arterial wall to continuously adjust its girth by up to 15% to match the regular distensions of arterial wall circumference that occur during each systolic pulse of the heart (Freitas 1999bb), thus maintaining watertight contact with similarly-cycling neighboring devices all of which are stationkeeping over a particular section of vascular tissue.

On its bottom face, the vasculocyte will have 625 stubby telescoping appendages (Section 23.3.3), each capable of 1 cm/sec movements. Limbs are similar to those in the microbivore and are spaced out along a regular grid about 100 nm apart, with only 10% of them used at any one time both to preserve tenfold redundancy and to avoid any possibility of leg-leg collisions. Each leg walks on a "footpad" tool tip (Freitas 1999bc) that is 10 nm in diameter. Acting like a snowshoe, the footpad will distribute leg motion forces widely enough to avoid disrupting cell membranes (Freitas 1999bd).

Many different tool tips (Freitas 1999be) might be deployed up through the interior hollows of the 625 nanorobotic limbs. Appendages on the underbelly may be used as manipulator arms for blood clot and foam cell disassembly, endothelial cell herding, adhesive glycoprotein removal, and so forth. Syringe tips will allow suction or drug injection by penetrating the 10-nm thick cellular membranes over which the device is walking. Other specialized tips will be used for bulk tissue disposal (a rotating cutting annulus), molecular absorption (using binding sites keyed to the molecules that make up plaque), cell peeling (specialized grippers), and as sensors for biomarker detection, chemotactic mapping, and other physical measurements.

After injection, the vasculocytes will circulate freely in the patient's bloodstream for a few minutes, finally dropping out onto a capillary wall and beginning to crawl upstream (or downstream, if in the pulmonary bed) along the vessel surface. Each device moves past the precapillary sphincters, through the metarterioles to the wide end of the terminal arterioles, then up the terminal arterial branches (150 microns in diameter) and into the arteries, where it joins up with others, forming into traveling circumferential scanning rings consisting of millions of individual nanorobots walking side by side. Eventually these traveling bands (Lapidus and Schiller 1978) will enter the 25,000-micron diameter aorta, leading, ultimately, to the heart. Upon reaching the heart uneventfully, each device would release its grip on the arterial wall and return to the bloodstream, allowing removal from the body either by nanapheresis centrifugation (Freitas 1999bm) or by excretion through the kidneys (Weatherbee and Freitas 2010). Creeping along the arterial tree at a fairly modest speed of 100 microns/sec (Freitas 1999bf), a vasculocyte ring could travel the 70 cm mean distance from capillaries to heart in about 2 hours if uninterrupted.

However, if disease is present the nanorobots will detect sclerotic tissue based on surface plaque temperature heterogeneity (Stefanadis et al. 1999), directly sampled tissue biomarkers (Schönbeck and Libby 2001; Lipinski et al. 2004; Koenig and Khuseyinova 2007), observation of ultrastructural alterations in endothelial cell morphology (Walski et al. 2002), thinning of endothelial glycocalyx (Gouverneur et al. 2006) or other evidence of endothelial dysfunction (Hadi et al. 2005), and circumferential vasculometric variations. Upon such detection, enough vasculocytes would collect over the affected area to entirely cover the lesion. The nanorobots aggregate into a watertight arterial "bandage" by locking themselves together side by side through their inflatable bumpers, then establish mutual communications links (Freitas 1999bg) and anchor themselves securely to the underlying tissue to begin repair operations which may be externally supervised and directed by the physician in real time.

The total computational power inherent in each bandage would be fairly impressive: a 1 cm^2 patch of linked vasculocytes each running a tenfold-redundant 1 MB/sec nanocomputer having 5 MB of memory represents a 10-million nanorobot parallel computer with 100 terabit/sec processing capacity (crudely equivalent to the human brain) with 50 terabits of memory. Within each bandage, nanorobots would complete all repairs within 24 hours or less, faster than hemangioblast precursor cells derived from human stem cells that show robust reparative function of damaged rat/mouse vasculature in 24–48 hours (Lu et al. 2007). Repairs would occur in eight sequenced mission steps including: (1) reconnoiter, (2) clean the site, (3) strip the existing endothelial layer, (4) rebuild endothelial cell population, (5) remove lesions, (6) halt aberrant vascular muscle cell growth, (7) rebuild basement structure, and (8) reposition endothelial cells. A 1 cm^3 injection of 70 billion vasculocytes would be a large enough treatment dosage to entirely coat 50% of the entire human arterial luminal surface with these active, healing nanorobots. Supplemental endothelial cells may be manufactured exogenously (Section "Tissue Printers, Cell Mills and Organ Mills") and transported to active repair sites as required.

If complete arterial occlusion has occurred, the patient may require emergency endoscopic nanosurgery (Section "Endoscopic Nanosurgery and Surgical Nanorobots"), analogous to mechanical thrombectomy (Kasirajan et al. 2001) today, to quickly clear the obstruction, plus a local injection of respirocytes (Section 23.6.3.1) to reduce ischemic damage to the affected tissues; or, alternatively, a population of burrowing tissue-mobile microbivore-class devices could rapidly digest the embolus (Section 23.6.2.4). Nanorobotic devices can also be used to treat non-atheroma lesions of the vasculature, such as those caused by viral invaders that attack and damage the vascular endothelium (Sahni 2007), e.g., in viral hemorrhagic fevers (Marty et al. 2006).

23.6.2.4 Stroke and Cerebrovascular Disease

Strokes are the most common cause of disabling neurologic damage in the industrialized countries. In an ischemic stroke, a large fatty deposit (atheroma) can develop in a carotid artery, greatly reducing its blood flow feeding the brain. If fatty material breaks off from the carotid artery wall it can travel with the blood and become stuck in a smaller brain artery, blocking it completely. Also, a clot formed in the heart or on one of its valves can break loose, travel up through the arteries to the brain, and lodge there. When blood flow to the brain is disrupted, brain cells can die or become damaged from lack of oxygen. If the blood supply is not restored within a few hours, brain tissue dies, resulting in stroke. Insufficient blood supply to parts of the brain for brief periods causes transient ischemic attacks (TIAs), temporary disturbances in brain function, and brain cells can also be damaged if bleeding occurs in or around the brain, producing various cerebrovascular disorders. In a future nanomedical era the incidence of this form of disease should be somewhat reduced, but prevention may not be universally practiced or available for all patients.

Nanorobotic treatments might be applied as follows. First, in the case of partial occlusions of the carotid or lesser cranial arteries that are not immediately life threatening, vasculocytes (Section 23.6.2.3) could be employed to clear partial obstructions, repair the vascular walls, and to enlarge the vessel lumen to its normal diameter in a treatment lasting perhaps several hours. Second, in the case of small solid emboli blocking capillaries or small metarterioles, burrowing tissue-mobile microbivore-class devices could digest the obstructions in minutes (e.g., 8 minutes to clear an 8 micron diameter capillary, digesting an embolus at the ~1 micron/min rate; Section 23.6.2.2) with respirocytes added to the therapeutic cocktail to help maintain oxygenation of the affected tissues via diffusion from devices passing through adjacent capillaries. Finally, endoscopic nanosurgery (Section "Endoscopic Nanosurgery and Surgical Nanorobots") could be used to quickly clear a life-threatening total occlusion on an emergency basis, and intracranial hemorrhages may be dealt with using a combination of endoscopic nanosurgery (Section "Endoscopic Nanosurgery and Surgical Nanorobots"), vascular gates (Section "Vascular Gates") and clottocytes (Section 23.6.3.3).

23.6.2.5 Hormonal, Metabolic and Genetic Disease

Diabetes mellitus is a hormonal disorder that is the tenth leading cause of death in the United States and the leading cause of blindness with complications including kidney and nerve damage, cataracts, impairment of skin health and white cell function, and cardiovascular damage. In Type I diabetes, >90% of the insulin producing beta cells in the pancreas have been destroyed by the immune system, requiring regular insulin injections; in Type II, the pancreas continues to manufacture insulin but the body develops resistance to its effects, creating a relative insulin deficiency. Both forms have a genetic component. By the 2020s and beyond, as in the cases of heart disease (Section 23.6.2.3) and stroke (Section 23.6.2.4) conventional biotechnology-based cures for diabetes may exist. The immune disorder that causes type I diabetes might be eliminated by proper immunoengineering, perhaps using techniques that have already proven successful in animals, and the changes in gene expression with aging that give rise to type II diabetes occur not in the pancreas but in the tissues that normally use insulin but stop doing so with aging, and this may also be prevented.

Even if these methods prove unsuccessful or have drawbacks (e.g., side effects, excessive treatment time), in the era of medical nanorobotics cell repair devices called chromallocytes (Section 23.6.4.3) could permanently correct any genetic susceptibilities at their source, e.g., by rebuilding any missing pancreatic beta cells via genomic replacement in existing cells, creating healthy new beta cells that can be made more resistant to autoimmune destruction by editing out pancreatic antigens resembling those of the pancreas-destroying virus to which the immune system is responding, thus curing diabetes. Microbivore-class devices could also delete immune cells that recognize the self-antigens. Additional cell repair nanorobots could be used to correct aberrant or unreliable gene expression (Section 23.6.4.4) in tissue cells to eliminate any lingering insulin resistance effects. As a temporary stopgap measure, pharmacyte-class nanorobots (Section 23.6.3.2) or artificial implanted nanorobotic organs will comprehensively control serum levels of any small molecule such as insulin on a real time basis. Other endocrine disorders such as hypopituitarism, hyperthyroidism, and adrenal malfunction, metabolic diseases such as obesity, hyperlipidemia, and Tay-Sachs, and any of the thousands of known genetic diseases similarly could be permanently cured using chromallocytes (Section 23.6.4.3).

Another metabolic condition known as glycation, which may accumulate even when glucose levels are held in the normal range because glucose is chemically reactive and can combine with myelin and other biological components over time, may cause autoimmune conditions and other problems such as increased tissue stiffness. Unless the body already has adequate endogenous defenses against this problem that are not normally marshaled – not currently known one way or the other – glycation would eventually become serious enough to require attention. Nanorobotic deglycation of cell surfaces is briefly discussed in Section 23.7.1.2.

23.6.3 Reversing Trauma

Trauma is a physical injury or wound caused by external force or violence to the human body. In the United States, trauma is the leading cause of death between the ages of 1 and 38 years. The principal sources of trauma are motor vehicle accidents, suicide, homicide, falls, burns, and drowning, with most deaths occurring within the first several hours after the event. However, nanomedical interventions should be able to correct a great deal of the damage resulting from such events.

In this short Chapter we can only briefly summarize a few representative nanorobotic responses to some familiar situations involving traumatic injury, including suffocation and drowning (Section 23.6.3.1), poisoning (Section 23.6.3.2), hemostasis (Section 23.6.3.3), wound healing (Section 23.6.3.4), and internal injury requiring surgery (Section 23.6.3.5).

23.6.3.1 Suffocation and Drowning

The principal effect of a suffocation or drowning trauma is hypoxemic damage to tissues and organs. The first theoretical design study of a medical nanorobot ever published in a peer-reviewed medical journal (in 1998) described an artificial mechanical red blood cell or "respirocyte" (Freitas 1998) to be made of 18 billion precisely arranged atoms (Fig. 23.23) – a bloodborne spherical 1-micron diamondoid 1000-atmosphere pressure vessel (Freitas 1999bh) with active pumping (Freitas 1999o) powered by endogenous serum glucose (Freitas 1999bi), able to deliver 236 times more oxygen to the tissues per unit volume than natural red cells and to manage acidity caused by carbonic acid formation, controlled by gas concentration sensors (Freitas 1999bj) and an onboard nanocomputer (Drexler 1992b; Freitas 1999k). The basic operation of respirocytes will be straightforward. These nanorobots, still entirely theoretical, would mimic the action of the natural hemoglobin-filled red blood cells, while operating at 1000 atm vs. only 0.1–0.5 atm equivalent for natural Hb. In the tissues, oxygen will be pumped out of the device by the sorting rotors on one side. Carbon dioxide will be pumped into the device by the sorting rotors on the other side, one molecule at a time. Half a minute later, when the respirocyte reaches the patient's lungs in the normal course of the circulation of the blood, these same rotors may reverse their direction of rotation, recharging the device with fresh oxygen and dumping the stored CO_2, which diffuses into the

Fig. 23.23 The respirocyte (Freitas 1998), an artificial mechanical red cell. Designer Robert A. Freitas Jr. ©1999 Forrest Bishop. Used with permission

lungs and can then be exhaled by the patient. Each rotor requires little power, only ~0.03 pW to pump ~10^6 molecules/sec in continuous operation.

In the exemplar respirocyte design (Freitas 1998), onboard pressure tanks can hold up to 3 billion oxygen (O_2) and carbon dioxide (CO_2) molecules. Molecular sorting rotors (Section 23.3.2) are arranged on the surface to load and unload gases from the pressurized tanks. Tens of thousands of these individual pumps cover a large fraction of the hull surface of the respirocyte. Molecules of oxygen or carbon dioxide may drift into their respective binding sites on the exterior rotor surface and be carried into the respirocyte interior as the rotor turns in its casing. The sorting rotor array is organized into 12 identical pumping stations laid out around the equator of the respirocyte, with oxygen rotors on the left, carbon dioxide rotors on the right, and water rotors in the middle of each station. Temperature (Freitas 1999u) and concentration (Freitas 1999q) sensors tell the devices when to release or pick up gases. Each pumping station will have special pressure sensors (Freitas 1999t) to receive ultrasonic acoustic messages (Freitas 1999bk) so the physician can (a) tell the devices to turn on or off, or (b) change the operating parameters of the devices, while the nanorobots are inside a patient. The onboard nanocomputer enables complex device behaviors also remotely reprogrammable by the physician via externally applied ultrasound acoustic signals. Internal power will be transmitted mechanically or hydraulically using an appropriate working fluid, and can be distributed as required using rods and gear trains (Freitas 1999ao) or using pipes and mechanically operated valves, controlled by the nanocomputer. There is also a large internal void surrounding the nanocomputer which can be a vacuum, or can be filled with or emptied of water. This will allow the device to control its buoyancy very precisely and provides a crude but simple method for removing respirocytes from the body using a blood centrifuge, a future procedure now called nanapheresis (Freitas 1999bm).

A 5 cc therapeutic dose of 50% respirocyte saline suspension containing 5 trillion nanorobots would exactly replace the gas carrying capacity of the patient's entire 5.4 l of blood. If up to 1 l of respirocyte suspension can safely be added to the human bloodstream (Freitas 2003t), this could keep a patient's tissues safely oxygenated for up to 4 hours even if a heart attack caused the heart to stop beating, or if there was a complete absence of respiration or no external availability of oxygen. Primary medical applications of respirocytes would include emergency revival of victims of carbon monoxide suffocation at the scene of a fire, rescue of drowning victims, and transfusable preoxygenated blood substitution – respirocytes could serve as "instant blood" at an accident scene with no need for blood typing, and, thanks to the dramatically higher gas-transport efficiency of respirocytes over natural red cells, a mere 1 cm^3 infusion of the devices would provide the oxygen-carrying ability of a full liter of ordinary blood. Larger doses of respirocytes could also: (1) be used as a temporary treatment for anemia and various lung and perinatal/neonatal disorders, (2) enhance tumor therapies and diagnostics and improve outcomes for cardiovascular, neurovascular, or other surgical procedures, (3) help prevent asphyxia and permit artificial breathing (e.g., underwater, high altitude, etc.), and (4) have many additional applications in sports, veterinary medicine, military science, and space exploration.

23.6.3.2 Poisoning

Poisoning is the harmful effect that occurs when toxic substances are ingested, inhaled, or come into contact with the skin. To deliver antidote or to clear such substances from the bloodstream in the era of nanorobotic medicine, a modified respirocyte-class device called a "pharmacyte" (Freitas 2006a) could be used. The pharmacyte was originally designed as an ideal drug delivery vehicle with near-perfect targeting capability (Section 23.6.1(4)). In that capacity, the device would be targetable not just to specific tissues or organs, but to individual cellular addresses within a tissue or organ. Alternatively, it could be targetable to all individual cells within a given tissue or organ that possessed a particular characteristic (e.g., all cells showing evidence of a particular poison). It would be biocompatible and virtually 100% reliable, with all drug molecules being delivered only to the desired target cells and none being delivered elsewhere so that unwanted side effects are eliminated. (Sensors on the surface of the nanorobot would recognize the unique biochemical signature of specific vascular and cellular addresses (Freitas 1999ak), simultaneously testing encountered biological surfaces for a sufficiently reliable combination (at least 5–10 in number) of positive-pass and negative-pass molecular markers to ensure virtually 100% targeting accuracy.) It would remain under the continuous post-administration supervisory control of the supervising physician – even after the nanorobots had been injected into the body, the doctor would still be able to activate or inactivate them remotely, or alter their mode of action or operational parameters. Once treatment was completed, all of the devices could be removed intact from the body.

The exemplar 1–2 μm diameter pharmacyte would be capable of carrying up to ~1 μm^3 of pharmaceutical payload stored in onboard tanks that are mechanically offloaded using molecular sorting pumps (Section 23.3.2) mounted in the hull, operated under the proximate control of an onboard computer. Depending on mission requirements, the payload alternatively could be discharged into the proximate extracellular fluid (Freitas 1999bn) or delivered directly into the cytosol using a transmembrane injector mechanism (Freitas 1999bo, bp, bj, 2003u). If needed for a particular application, deployable mechanical cilia (Freitas 1999ae) and other locomotive systems (Freitas 1999i) could be added to the pharmacyte to permit transvascular (Freitas 1999bq) and transcellular (Freitas 1999x) mobility, thus allowing delivery of pharmaceutical molecules to specific cellular and even intracellular addresses.

Because sorting pumps can be operated reversibly, pharmacytes could just as easily be used to selectively extract specific molecules from targeted locations as well as deposit them. Thus in the case of poison control, these nanorobots might act in reverse to retrieve a specific chemical substance from the body, just as they can be used for targeted delivery of an antidote. Whole-body clearance rates for systemic poisons can be quite rapid. For example, a population of 10^{12} bloodborne pharmacytes having aggregate storage volume ~6 cm^3 could reduce serum alcohol from 0.2% in a seriously intoxicated 70 kg patient to 0.005% in ~1 second by prompt onboard sequestration, followed by catabolization of the entire inventory in ~10 minutes within a ~200 watt systemic caloric budget for waste heat production. Of course, continuing outflows from ethanol-soaked body tissues into the bloodstream and other factors complicate the process, e.g., such extremely rapid reduction of blood alcohol levels could be counterproductive because it might produce osmotic brain swelling in which water enters the brain (which still contains more alcohol than the blood) faster than alcohol can leave the brain.

23.6.3.3 Hemostasis

A major form of trauma occurs when the skin and underlying tissues are lacerated by violence, causing bleeding from broken capillaries or somewhat larger blood vessels. Total natural bleeding time, as experimentally measured from initial time of injury to cessation of blood flow, may range from 2–5 minutes (Kumar et al. 1978) up to 9–10 minutes (Hertzendorf et al. 1987; Lind 1995) if even small doses of anticoagulant aspirin are present (Ardekian et al. 2000), with 2–8 minutes being typical in clinical practice. Hemostasis is also a major challenge during surgery, as up to 50% of surgical time can be spent packing wounds to reduce or control bleeding and there are few effective methods to stop it without causing secondary damage. Modern surgical fibrin sealants (e.g., Crosseal, American Red Cross) composed of human clottable proteins and human thrombin can reduce mean hemostasis time to 282 seconds in clinical settings (Schwartz et al. 2004), and there is one report of a laboratory demonstration of artificial hemostasis in 15 seconds for multiple tissues and wound types in animal models using synthetic self-assembling peptides (Ellis-Behnke et al. 2007).

A medical nanorobot theoretical design study (Freitas 2000a) has described an artificial mechanical platelet or "clottocyte" that would allow complete hemostasis in ~1 second, even in moderately large wounds. This response time is on the order of 100–1000 times faster than the natural hemostatic system and 10–100 times faster than the best current artificial agents.

The baseline clottocyte is conceived as a serum oxyglucose-powered spherical nanorobot ~2 microns in diameter (~4 micron^3 volume) containing a fiber mesh that is compactly folded onboard. Upon command from its control computer, the device promptly unfurls its mesh packet (Fig. 23.24) in the immediate vicinity of an injured blood vessel – following, say, a cut through the skin. Soluble thin films coating certain parts of the mesh would dissolve upon contact with plasma water, revealing sticky sections (e.g., complementary to blood group antigens unique to red cell surfaces (Freitas 1999br)) in desired patterns. To stop flow, the net must be well anchored to avoid being swept along with the trapped red cells. A cut blood vessel has exposed collagen to which platelets normally adhere – the clottocyte netting may recognize collagen or even intact endothelial cells (or the junctions between endothelial cells) to provide the needed anchoring function. Blood cells

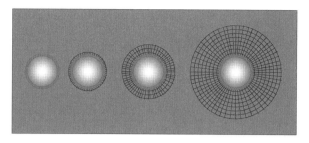

Fig. 23.24 The clottocyte (Freitas 2000a), an artificial mechanical platelet, rapidly unfurls its netting at the wound site, halting bleeding in ~1 second. Designer Robert A. Freitas Jr. © 2008 Robert A. Freitas Jr. (www.rfreitas.com). All Rights Reserved. Used with permission

are immediately trapped in the overlapping artificial nettings released by multiple neighboring activated clottocytes, and bleeding halts at once. The required blood concentration n_{bot} of clottocyte nanorobots required to stop capillary flow at velocity v_{cap} ~1 mm/sec (Freitas 1999bs) in a response time $t_{stop} = 1$ sec, assuming $n_{overlap} = 2$ fully overlapped nets each of area $A_{net} = 0.1$ mm^2, is n_{bot} ~$n_{overlap}$ / (A_{net} t_{stop} v_{cap}) = 20 mm^{-3}, or just ~110 million clottocytes in the entire 5.4-l human body blood volume possessing ~11 m^2 of total deployable mesh surface. This would be a total dose of ~0.4 mm^3 of clottocytes, producing a negligible serum nanocrit (nanorobot/blood volume ratio) (Freitas 1999bt) of ~0.00001%.

Clottocytes may perform a clotting function that is equivalent in its essentials to that performed by biological platelets – possibly including the release of vasoactive mediators, clotting factor cascade activators, etc. if needed – but at only ~0.01% of the bloodstream concentration of those cells. Hence clottocytes would be ~10,000 times more effective as clotting agents than an equal volume of natural platelets. While 1–300 platelets might be broken and still be insufficient to initiate a self-perpetuating clotting cascade, even a single clottocyte, upon reliably detecting a blood vessel break, could rapidly communicate this fact to its neighboring devices, immediately triggering a progressive controlled mesh-release cascade. Of course, onboard computerized control systems must ensure extremely safe and reliable operation (Freitas 2009).

23.6.3.4 Wound Healing

Once bleeding is stopped, the wound must be closed. Natural processes that rely solely upon wound self-repair often take months for completion and can leave unsightly, dense, shiny white fibrous scars – skin never heals into a condition that is "as good as new," and healed tissue is typically 15–20% weaker than the original tissue. There are a few notable counterexamples. Among mammals, the MRL/MpJ mouse displays accelerated healing and tissue regeneration with an extraordinary capacity to scarlessly heal ear-punch and other surgical wounds. Excision ear-punch wounds 2 mm wide close via regeneration after 30 days (Clark et al. 1998) with full re-epithelialization in just 5 days followed by blastema-like formation, dermal extension, blood vessel formation, chondrogenesis, folliculogenesis, and skeletal muscle and fat differentiation (Rajnoch et al. 2003); another MRL mouse study (Leferovich et al. 2001) found that even a severe cardiac wound healed in 60 days with reduced scarring and with full restoration of normal myocardium and function.

The goal in medical nanorobotics is to provide an equally effective alternative to wound healing that can work >1000 fold faster than the natural process, e.g., in minutes or hours. No comprehensive nanorobot design study has yet been published but a theoretical scaling study (Freitas 1996b) concentrating on nanomechanical activity requirements for minor dermal excision wound repair describes the dermal zipper or "zippocyte" as a roughly cubical nanorobotic device measuring $40 \times 40 \times 30$ microns in size. This study concludes that multipurpose nanorobotic manipulators 1-micron in length would cover five of the six faces of the device, forming a dense coating of ~7000 nanomechanical cilia of similar number density as might be found on the outer surface of a microbivore (Section 23.6.2.1) or the underside of a vasculocyte (Section 23.6.2.3). These utility appendages would serve many ancillary functions including sensing/mapping, wound debridement, individual locomotion, stationkeeping (by handholding with neighboring nanorobots), volume management of the collective, and binding to tissue walls. Actual repair work would be performed by larger manipulators on the sixth face located on the underbelly of each zippocyte, using derivatives of cell milling and tissue repair methods described elsewhere (Section "Tissue Printers, Cell Mills and Organ Mills"). The entire wound repair sequence, as seen from the viewpoint of a working dermal nanorobot, would occur in twelve sequenced mission steps including: (1) activation, (2) entry, (3) immobilization and anti-inflammation, (4) scan surface, (5) debridement, (6) muscle repair, (7) areolar (loose connective) tissue repair, (8) fatty tissue repair, (9) dermis repair, (10) germinative layer restoration, (11) corneum repair, and (12) exit and shutdown.

23.6.3.5 Internal Injury and Nanosurgery

Internal injuries are more serious and may include internal bleeding, crushed or damaged organs, electrical or burn injuries, and other serious physical traumas. This area of emergency medicine will demand some of the most sophisticated medical nanorobots available, with large numbers of devices of many different nanorobot types acting in concert under the most difficult conditions. In most cases some form of surgical intervention will be required, using nanorobotic surgical tools such as those described below.

Vascular Gates

Still only conceptual, the vascular gate (Freitas 2003r) will be a basic nanorobotic tool analogous to hemostats to allow surgeons to rapidly enable or disable free flow through whole sections of the vascular tree ranging from individual capillaries to entire capillary beds, all the way up to larger vessels the size of major arteries or veins. The most direct application in emergency medicine would be to allow the surgeon to quickly but reversibly seal off the open ends of hundreds or thousands of broken blood vessels simultaneously, to immediately stanch massive blood losses at the outset of trauma surgery and to provide reversible wide-area hemostasis in the surgical region by temporarily blockading all vessels. More complex vascular gates could also act as real-time content filters to choose which population of bloodborne objects can pass through any targeted section of the vasculature. For example, red cells could be allowed to pass but not platelets, or the gate might pass all formed blood elements but no nanorobots (or vice versa), or all fluids but no solid objects. Content filtration could be sensor-, time-, event-, or command-driven. In other non-emergency applications, gate nanorobots could be employed as intelligent embolic particles that would be directed to a specific organ or tumor within an organ, then triggered to halt flow in blood vessels supporting these structures for the purpose of selectively blocking organ perfusion or clogging tumor inputs.

The simplest gating nanorobots can be externally administered onto or into a wound area by the emergency surgeon; bloodborne ones not involved in emergency wound management can be internally administered like free-floating clottocytes (Section 23.6.3.3) or vasculomobile vasculocytes (Section 23.6.2.3). In wound scenarios the nanorobots can recognize cuts using chemotactic sensor pads (Freitas 1999cv) to detect the telltale molecular signatures of broken blood vessels much as occurs naturally by platelets (Cruz et al. 2005) and circulating vascular progenitor cells (Sata 2003), including but not limited to detection of exposed collagen (Cruz et al. 2005; O'Connor et al. 2006; Ichikawa et al. 2007) or elastin (Hinek 1997; Keane et al. 2007) from ruptured intima or media, or smooth muscle actin (Rishikof et al. 2006) and other molecular markers expressed on injured endothelial and smooth muscle cells (Takeuchi et al. 2007). Vessel recognition may be assisted by rapid advance nanoscopic mapping of the wound area with nanorobots subsequently proceeding to their assigned stations via informed cartotaxis (Freitas 1999cw). Once having arrived on site, internally administered gating nanorobots can use circumferential intraluminal pressure fit, analogous to the vasculoid design (Freitas and Phoenix 2002), to establish a leakproof seal that should easily withstand 1–2 atm exceeding the requisite maximum physiological backpressure. Externally administered nanorobots can employ a similar approach to apply anchoring pressure rings inside the undamaged section of a leaking vessel that lies nearest to the vessel's damaged terminus. Either process may be assisted by lipophilic semaphores (Freitas 1999cu) deployed on the nanorobot hull to help maintain noncovalently-bonded reversible leakproof seals to the plasma membranes of remaining undamaged endothelium or to subendothelial basement membrane.

A nanorobotic vascular gate installed across a large 6-mm diameter artery could be established using a sheet of ~10^7 micron-sized nanorobots each having a (~2 micron)^2 patrol area within the array, with each device stationkeeping in its patrol area by handholding with neighboring nanorobots (Section 23.6.3.4) and analogously as has been described elsewhere for "utility fog" (Hall 1993, 1996) (Fig. 23.25). A positive-pass gate might use contact sensor data to recognize impinging particulate matter that the physician desired to pass through, whereupon the gate would temporarily open wide enough to allow the desired particle to pass through, then quickly close again. A negative pass gate would normally allow everything

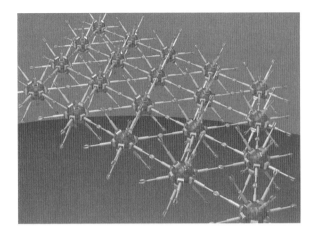

Fig. 23.25 A number of utility fog (Hall 1993) nanorobots hold hands with their neighbors, forming a strong, reconfigurable smart-matter array. Image © 1999, courtesy of J. Storrs Hall

to pass unless sensors detected an undesired particle, whereupon the gate would forcibly eject it and close up until the undesired particle had diffused away via osmotic dilution in the pulsed flow. If the vascular gate aggregate consists of vasculomobile nanorobots, then at mission's end the aggregate can disassemble itself and "walk away" much like vasculocytes (Section 23.6.2.3), then similarly be removed from the body.

Primitive but less effective analogs to vascular gates that are already in widespread surgical use include the inferior vena cava filter (Imberti et al. 2006; Dentali et al. 2006; Giannoudis et al. 2007; Patel and Patel 2007) for the prevention of pulmonary embolism as an alternative to anticoagulant therapy in high-risk patients (~100,000 cases annually in the U.S.), expandable net filters that are deployed while emplacing carotid artery stents in stroke patients (Henry et al. 2007), and the use of protective internal carotid artery flow reversal (Pipinos et al. 2006) during carotid angioplasty and stenting.

Tissue Printers, Cell Mills and Organ Mills

How can we restore severely injured organs that are too damaged to repair, or replace large chunks of missing tissue that has been excised from the body in a deep avulsive trauma? One answer is to manufacture new tissue from scratch (Mironov et al. 2003; Jakab et al. 2004). Tissue and organ printing is a very active area in biomedical research today (Box 23.2). With nanorobotic controlled precision and a massively parallel tip array, a future nanosurgical tissue printer might be used to squirt tissue matrix scaffold and tissue cells directly and accurately into a large immobilized wound, rebuilding missing tissues in situ. A sheet of finished tissue ~1 mm thick could be laid down every minute assuming a ~1 Hz scan rate and deposition layers one cell thick (~20 microns). Filling a 10 cm deep excision wound with fresh tissue would then require ~1.4 hours in an immobilized but stabilized patient. But MHC-compatible generic cells would have to be engineered for this purpose to avoid the need for immunosuppressive drugs. Alternatively, homologous cells of the patient's own type could be manufactured, using nanorobotic "organ mills" that will also allow surgeons to manufacture whole new fully-homologous organs, and then implant them, during the surgery.

Box 23.2 Tissue and organ printing

Boland's group at Clemson University has taken the first primitive steps towards 3D printing of complex tissues (Boland et al. 2006) and ultimately entire organs (Mironov et al. 2003). In one experiment (Xu et al. 2005), Boland's team used a modified Hewlett Packard 550C computer printer to print Chinese Hamster Ovary (CHO) and embryonic motoneuron cells into a pre-defined pattern using an "ink" of cells suspended in phosphate buffered saline solution. After deposition onto several "bio-papers" made from soy agar and collagen gel, the printed cells exhibited a healthy morphology with less than 8% cell lysis observed. In another experiment by the same group (Xu et al. 2006), complex cellular patterns and structures were created by automated and direct inkjet printing of primary embryonic hippocampal and cortical neurons – which maintained basic cellular properties and functions, including normal, healthy neuronal phenotypes and electrophysiological characteristics, after being printed through thermal inkjet nozzles. 3D cellular structures also were created by layering sheets of neural cells on each other (in a layer-by-layer process) by alternate inkjet printing of NT2 cells and fibrin gels (Xu et al. 2006). Cellular attachment and proliferation have been demonstrably controlled by precise, automated deposition of collagen (a biologically active protein) to create viable cellular patterns with 350-micron resolution (Roth et al. 2004). Boland defines his ultimate objective of "organ printing" as computer-aided, jet-based 3D tissue-engineering of living human organs involving three sequential steps: pre-processing or development of "blueprints" for organs, processing or actual organ printing, and postprocessing or organ conditioning and accelerated organ maturation (Mironov et al. 2003). Another group has printed rectangular tissue blocks of several hundred microns in thickness and tubular structures several millimeters in height (Jakab et al. 2006). Private companies are getting involved too: Therics (http://www.therics.com) is solid-printing resorbable implantable bone scaffolds that are already in use by surgeons, and Sciperio (http://www.sciperio.com) is developing an in vivo "Biological Architectural Tool" by which "clinicians and tissue engineers will be able to survey, diagnose, and construct new tissues via endoscopically manipulated vision, nonthermal tissue removal, and a direct-write tissue deposition apparatus."

In our vision of future nanomedicine, a desktop-type apparatus would accept as input the patient's DNA sequence, then manufacture large complex biological structures in a convergent assembly type process (Freitas and Merkle 2004b), as described in the following conceptual scenario.

The first module would synthesize copies of the patient's own homologous proteins and other relevant biomolecules, working from the patient's genome; as a proof of concept, functional copies of the human red cell anion exchanger, a proteinaceous transmembrane pump, have been self-assembled from sets of three, four or five complementary fragment "nanoparts" that were separately cloned in *Xenopus* oocytes (Groves et al. 1998). This process would include the manufacture of many duplicate copies of the patient's own DNA suitably methylated to match the expression pattern (e.g., the "methylome," "transcriptome," etc.) of the particular cells and organ being constructed, as described elsewhere in a lengthy technical paper (Freitas 2007). These fabricated biocomponents would then be fed to a second module which may positionally assemble them into bulk quantities of artificially fabricated organelles, membranes, vesicles, granules, and other key intracellular structures. Many such structures will self-assemble robustly (Marrink et al. 2001). As an experimental example of this, Golgi stacks (an important intracellular organelle) have been reassembled from isolated Golgi components including random assortments of vesicles, tubules, and cisternal remnants (Rabouille et al. 1995).

These mass-produced intracellular structures then serve as feedstock to the third production module, called a "cell mill," wherein these subcellular structures and materials would be assembled into complete cells of the requisite types, along with any extracellular matrix materials that might be required. This might be done using manufacturing systems analogous to 3D printing (Box 23.2). As long ago as 1970, an *Amoeba proteus* single-cell organism was reassembled from its major subcellular components – nucleus, cytoplasm, and cell membrane – taken from three different cells (Jeon et al. 1970), demonstrating the physical possibility of manually assembling living cells from more primitive parts. Others (Morowitz 1974) later reported that "cell fractions from four different animals can be injected into the eviscerated ghost of a fifth amoeba, and a living functioning organism results." Mammalian cells have also been assembled from separate nuclear and cytoplasmic parts (Veomett and Prescott 1976) and intracellular organelles have been individually manipulated both directly (Weber and Greulich 1992; Felgner et al. 1998; Bayoudh et al. 2001; Sacconi et al. 2005b) and nanosurgically (Section 23.6.4.2). Early cell assembly production systems might initially make partial use of more traditional biotechnologies such as cloning, stem cells, tissue engineering, animal cell reactors (Bliem et al. 1991; Nelson and Geyer 1991), transdifferentiation (Collas and Håkelien 2003) and nuclear reprogramming (Tada 2006).

In the fourth module, the manufactured cells are fed into a "tissue mill," which would mechanically assemble the cells into viable biological tissues using, again, positionally-controlled methods analogous to 3D printing (Box 23.2). 3D rapid prototyping has already created collagen scaffolds that can viably host human heart cells, the first step toward assembling an artificial heart valve (Taylor et al. 2006). Cells have also been manually assembled into larger artificial 3D structures such as chains, rings, and a pyramid-like tetrahedron using optical tweezers (Holmlin et al. 2000), a potentially automatable process in which different cell types could be linked together one at a time in precisely the order and the positions necessary to assemble new tissues and organs.

Finally, the manufactured tissues would be fed to the last module, the "organ mill," that assembles the tissues into working biological organs that could be surgically implanted (Section "Endoscopic Nanosurgery and Surgical Nanorobots"). Crude estimates suggest that throughput rates of materials in such nanorobotic-based assembly modules could be on the order of minutes, with an organ-build time on the order of a few hours. This is at least several orders of magnitude faster than growing organs from tissue-engineered organoids (Poznansky et al. 2000; Saito et al. 2006; McGuigan and Sefton 2006) or via homologous organ cloning (Wood and Prior 2001; Cui 2005) in a biotech reactor apparatus. This is also fast enough to fall within the time range of oxygenation and pH buffering provided by respirocytes, which could be supplied to the nascent vascular system as it is assembled along with local nutrients. Making organs under conditions of mild hypothermia would also reduce their metabolic demands until perfusion can be instituted, and, if need be, the growing organ could be perfused from time to time during the assembly process to keep the cells within the construct in good condition.

Endoscopic Nanosurgery and Surgical Nanorobots

From the hand saws of the 19th century to the powered drills and ultrasharp diamond cutting blades of the mid 20th century, surgical tools by the end of the last century had progressed to the concept of minimally invasive surgery (MIS). Rather than carving a giant 12 inch incision in a patient's abdomen and undertaking a marathon operation, the MIS surgeon could now open a few transcutaneous centimeter-sized holes, poke several rigid endoscopic tubes through the holes, then insert miniaturized surgical cutting, suturing, and visualization tools through the tubes, thus reducing both surgical intrusiveness and patient recovery time. Flexible catheters were also introduced that could be threaded through the largest blood vessels to install stents and to remove vascular blockages or arterial wall plaques (e.g., via mechanical debridement, laser ablation, or ultrasonic ablation).

In the first few decades of the 21st century, surgical endoscopes and catheters will become even smaller but also smarter, with sensors (temperature, pressure, chemical, mechanical, shear force, force feedback, etc.) and even computers installed initially near the tips but eventually along their entire working lengths. These devices will possess complex robotic manipulators at their business end, with the surgeon having the ability to change out multiple toolheads or inject nanoliter quantities of drugs in situ. Manipulators and sensors will become more numerous on each instrument, more densely packed and more information intensive. The earliest steps down this pathway involving microrobotics (Menciassi et al. 2007) are illustrated by the great variety of MEMS (microelectromechanical systems) -based miniaturized surgical tools (Salzberg et al. 2002) already coming into use. Examples include the "data knife" scalpel produced by Verimetra, Inc. (www.verimetra.com)

which incorporates pressure and strain sensors with cautery and ultrasonic cutting edges (Kristo et al. 2003), the MicroSyringe and Micro-Infusion Catheter systems of Mercator MedSystems (www.mercatormed.com) for site-specific perivascular injection, and the MEMS-based wireless implantable blood pressure biosensor from CardioMEMS (www.cardiomems.com) (Chaer et al. 2006).

Paralleling these developments is the emergence of "robotic surgery" and telesurgery systems that soon will include force reflection to allow the surgeon to feel what he's doing and thus achieve much better results (Rizun et al. 2006). With microscale sensors he can touch the patient with tiny micron-sized hands, feeling the smallest bumps and adhesions in the tissue he's working on. Telesurgery, telemedicine, microsurgical telemanipulator systems (Li et al. 2000; Knight et al. 2005; Katz et al. 2006) and even conventional laparoscopy are getting practitioners used to the idea of operating through a machine or computer interface, rather than traditional procedures involving more direct physical contact with the patient. This process of learning how to act through a machine intermediary will continue to progress, and eventually the surgeon will become comfortable using surgical robots that accept higher-level commands. For instance, simple autonomous action sequences such as surgical knot tying have already been demonstrated experimentally by surgical robots (Bauernschmitt et al. 2005). As the next developmental step, rather than repeatedly directing the manipulators to thread a suture at various sites, the surgeon may simply indicate the positions in the tissue where he wants a series of suture loops placed using guidance virtual fixtures (Kapoor et al. 2005) and the machine will then automatically go through the motions of placing all those sutures while he watches, without the surgeon having to actively direct each suture placement site. This capability for semi-autonomous robotic surgery (Rizun et al. 2004) is foreshadowed by present-day "offline robots" or "fixed path robots" which perform subtasks that are completely automated with pre-programmed motion planning based on pre-operative imaging studies where precise movements within set confines are carried out (Sim et al. 2006).

Another outcome of the growing machine intermediation is that the surgeon will gain the ability to easily control many more than one active surgical instrument or surgical task at a time (Zhijiang et al. 2005). For example, after he has ordered the suturing device to put a series of sutures along one line, while waiting for that task to finish he can direct another surgical tool to do something else somewhere else, or he can go check some sensor readings, or he can palpate a section of nearby tissue to test its strength, and so forth. This multitasking will speed the surgical process and increase the number of in vivo interventive foci to which the individual surgeon can simultaneously attend inside his patient. Immersive virtual reality interfaces will further extend the surgeon's ability to maintain proper control of a growing number of tools simultaneously, improve his efficiency and confidence in the multitasking situation, and generally allow him to work faster and safer while doing more. Collaborative robotic surgeries (Hanly et al. 2006) will also become more commonplace. In sum, the current trends in surgery are generally these: the tools will get smaller and more complex, and the surgeon will be working increasingly through a computerized intermediary in a rich sensory and control environment, while relying increasingly on the mechanized intermediary to carry out preprogrammed microtasks (enabling the surgeon to concentrate on the big picture and to guide the general course of the procedure) while being freed to multitask with an increasing number of tools and collaborators.

As the era of surgical nanorobotics arrives, these trends will accelerate and progress still further. Today's smallest millimeter diameter flexible catheters will shrink to 1–10 micron diameter bundles that can be steered (Glozman and Shoham 2006) through the tiniest blood vessels (including capillaries) or could even be inserted directly through the skin into organs without pain (Wang et al. 2005) or discomfort. Nanorobotic mechanisms embedded in the external surfaces of a nanocatheter or nanosyringe (Section "Nanosyringoscopy") will assist in actively propelling the telescoping apparatus gently through the tissues (Freitas 1999at), sampling the chemical environment (e.g., concentrations of oxygen, glucose, hormones, cytokines) along the way (Freitas 1999c), and providing a torrent of mechanical and optical sensory feedback along with precision positional metrology to allow the surgeon to know exactly where his tools are at all times, and also where his "virtual presence" is in relation to his targets. Internal hollow spaces inside the nanocatheter can be used to transport tools, sensors, fluids, drugs, or debridement detritus between patient and physician. The tip of the nanocatheter or nanosyringoscope may include a working head with thousands or millions of independent manipulators and sensors branching outward from the central trunk on retractile stalks, from which data can be encoded in real time and passed to external computers along an optical data bus located inside each nanocatheter. The endoscopic nanosurgeon's ability to multitask may extend to thousands of nanocatheters and millions or billions of simultaneously occurring mechanical and chemical processes during a single surgical procedure.

Populations of individual surgical nanorobots also could be introduced into the body through the vascular system or from the ends of catheters into various vessels and other cavities in the human body. Surgical microrobotics is already a thriving field of experimental research (Box 23.3). A future surgical nanorobot, programmed or guided by a human surgeon, would act as a semi-autonomous on-site surgeon inside the human body. Such devices could perform various functions such as searching for pathology and then diagnosing and correcting lesions by nanomanipulation, coordinated by an on-board computer while maintaining contact with the supervising surgeon via coded ultrasound signals.

Nanosyringoscopy

A common requirement for trauma treatment is foreign object removal. Carefully poking a needle-like 100-micron diameter nanosensor-tipped self-steering "nanosyringoscope" quickly through all intervening soft tissues to the immediate vicinity of a foreign object should cause minimal permanent damage, much like bloodless and painless microneedles (Cormier et al. 2004; Flemming et al. 2005; Coulman et al. 2006; Nordquist et al. 2007). After penetration, $\sim 10^{10}$ micron^3/sec of nanorobots flowing at 1 m/sec through the tube (typical syringe rate) could surround a cubic

1 cm³ target object to a coating thickness of 100 microns in ~10 seconds. The coating nanorobots would then dig out 1 micron wide grooves at a volumetric excavation rate of 1% nanorobot volume per second to partition the 1 cm³ object into 10⁶ 100-micron microcubes in ~300 seconds, after which the foreign object microcubes are transported out of the patient in single file at 1 m/sec through the nanosyringoscope in ~100 seconds, followed by the exiting nanorobots taking ~10 seconds, completing a ~7 minute object-removal nanosyringotomy procedure through a ~100-micron diameter hole.

Box 23.3 Experimental surgical microrobotics

There have already been early attempts to build less sophisticated stand-alone microrobots for near-term in vivo surgical use. For example, Ishiyama et al. (Ishiyama et al. 2002) at Tohoku University developed tiny magnetically-driven spinning screws intended to swim along veins and carry drugs to infected tissues or even to burrow into tumors and kill them with heat. Martel's group at the NanoRobotics Laboratory of Ecole Polytechnique in Montreal has used variable MRI magnetic fields to generate forces on an untethered microrobot containing ferromagnetic particles, developing sufficient propulsive power to direct the small device through the human body (Mathieu et al. 2005). In 2007 they reported injecting, guiding via computer control, and propelling at 10 cm/sec a prototype untethered microdevice (a ferromagnetic 1.5- millimeter-diameter sphere) within the carotid artery of a living animal placed inside a clinical magnetic resonance imaging (MRI) system (Martel et al. 2007) – the first time such in vivo mobility has been demonstrated. Nelson's team at the Swiss Federal Institute of Technology in Zurich has pursued a similar approach, in 2005 reporting (Yesin et al. 2005) the fabrication of a microscopic robot small enough (~200 μm) to be injected into the body through a syringe and which they hope might someday be used to perform minimally invasive eye surgery. Nelson's simple microrobot has successfully maneuvered through a watery maze using external energy from magnetic fields, with different frequencies able to vibrate different mechanical parts on the device to maintain selective control of different functions. Sitti's group at Carnegie Mellon's NanoRobotics Laboratory is developing (Behkam and Sitti 2007) a <100-micron swimming microrobot using biomimetic flagellar motors borrowed from *S. marcescens* bacteria "having the capability to swim to inaccessible areas in the human body and perform complicated user directed tasks." Friend's group in the Micro/Nanophysics Research Laboratory at Monash University in Australia is designing a 250-micron microrobot (Cole 2007) to perform minimally invasive microsurgeries in parts of the body outside the reach of existing catheter technology – such as delivering a payload of expandable glue to the site of a damaged cranial artery, a procedure typically fraught with risk because posterior human brain

> arteries lay behind a complicated set of bends at the base of the skull beyond the reach of all but the most flexible catheters. Friend's completed device, expected by 2009, will be inserted and extracted using a syringe and is driven by an artificial flagellar piezoelectric micromotor.

Tactile (Ku et al. 2003; Winter and Bouzit 2007), haptic (McColl et al. 2006) and other sensory feedback will allow emergency practitioners to steer the nanosyringoscope into a patient to remove a foreign object (Feichtinger et al. 2007), then to withdraw bloodlessly from the body. The nanosurgeon may control the procedure via hand-guided interfaces similar to various medical exoskeletal appliances (Fleischer et al. 2006; Cavallaro et al. 2006; Gordon and Ferris 2007), instrumented gloves (Castro and Cliquet 1997; Yun et al. 1997) and hand-held surgical robots (Tonet et al. 2006) that have been under development for several decades.

The nanosyringoscope could also rapidly and painlessly import macroscale quantities of cells to any location inside the body (Section 23.7.1.4).

23.6.4 Cell Repair

In suggesting the novel possibility of individual cell repair, Drexler (1986) drew inspiration from the cell's eye view to explain how medical nanorobotics could bring a fundamental breakthrough in medicine: "Surgeons have advanced from stitching wounds and amputating limbs to repairing hearts and reattaching limbs. Using microscopes and fine tools, they join delicate blood vessels and nerves. Yet even the best microsurgeon cannot cut and stitch finer tissue structures. Modern scalpels and sutures are simply too coarse for repairing capillaries, cells, and molecules. Consider 'delicate' surgery from a cell's perspective. A huge blade sweeps down, chopping blindly past and through the molecular machinery of a crowd of cells, slaughtering thousands. Later, a great obelisk plunges through the divided crowd, dragging a cable as wide as a freight train behind it to rope the crowd together again. From a cell's perspective, even the most delicate surgery, performed with exquisite knives and great skill, is still a butcher job. Only the ability of cells to abandon their dead, regroup, and multiply makes healing possible. Drug molecules are simple molecular devices [that] affect tissues at the molecular level, but they are too simple to sense, plan, and act. Molecular machines directed by nanocomputers will offer physicians another choice. They will combine sensors, programs, and molecular tools to form systems able to examine and repair the ultimate components of individual cells. They will bring surgical control to the molecular domain."

23.6.4.1 Mechanisms of Natural Cell Repair

Many so-called natural "cell repair" mechanisms are actually tissue repair mechanisms, some of which act by replacing, not repairing, existing cells. For instance, there are stem cells that can transform into other needed (differentiated) cell types at a site where a cell of the needed type has apoptosed or been phagocytosed and reabsorbed, in which case the stem cells are effectuating a fairly direct form of "repair by replacement" (Ahn et al. 2004; Ye et al. 2006; Gersh and Simari 2006; Reinders et al. 2006). Stem cells can also fuse with somatic cells and alter them (Padron Velazquez 2006). Other "cell repair" mechanisms include chondroblasts or fibroblasts that rebuild connective tissue by extruding collagen fibers and other ECM components, and oligodendrocyte progenitor cells that extracellularly remyelinate CNS cells that have been demyelinated by exposure to toxic chemicals (Armstrong et al. 2006) or viral infection (Frost et al. 2003).

As another example, injured or apoptosed cells can be replaced by new cells produced by the replication and division of neighboring cells of the same cytotype, as occurs in, for example, epithelial "cell repair" (actually tissue regeneration) of the gastrointestinal, kidney, lung, liver, skin, prostate and muscle tissues (Nony and Schnellmann 2003; Kawashima et al. 2006; Pogach et al. 2007). The predominant mode of "repair" in biology is probably turnover, a fairly robust process in which everything from molecules to whole cells is replaced with new molecules or new cells, with the old being discarded and not repaired. Most damaged molecules other than DNA are simply degraded and replaced, and all mRNAs and their precursors are degraded after limited use whether damaged or not. Typical protein turnover half-life is ~200,000 sec (Alberts et al. 1989; Becker and Deamer 1991), membrane phospholipid half-life averages ~10,000 sec (Becker and Deamer 1991) but plasma membrane turnover rate is ~1800 sec for macrophage (Lehrer and Ganz 1995) and ~5400 sec for fibroblast (Murray et al. 1993). Glycocalyx turnover in rat uterine epithelial cells is ~430,000 sec (Jones and Murphy 1994), and enterocyte glycocalyx is renewed in 14,000–22,000 sec as vesicles with adhered bacteria are expelled into the lumen of small and large intestine (Kilhamn 2003). Cell turnover rates are equally impressive. Neutrophil lifespan is ~11,000 sec in blood and ~260,000 sec in tissue (Black 1999); blood platelet lifespan is ~860,000 sec (Stein and Evatt 1992). Some mucosal surfaces may replace their entire luminal cell population every ~10^5 sec (~1 day): Cell turnover time is ~86,000 sec in gastric body, ~200,000 sec for duodenal epithelium, ~240,000 sec for ileal epithelium, and ~400,000 sec for gastric fundus (Peacock 1984). At the other extreme is the lens of the eye, where the rate of cell turnover and repair is very low (McNulty et al. 2004) and the lens crystalline is never subject to turnover or remodeling once formed (Lynnerup et al. 2008), and tooth enamel, dentine, and cementum (other biological structures that are preserved essentially without turnover; Boyde et al. 2006; Ubelaker et al. 2006).

There are at least six examples of true "cell repair" mechanisms. Most notable is eukaryotic DNA repair including excision repair (base excision repair and nucleotide excision repair), mismatch repair, repair of double-strand breaks, and cross-link repair (Sharova 2005). These repair processes boost the fidelity of DNA replication to error rates of ~10^{-11}.

Second, there is also a limited form of protein repair in which misfolding errors are corrected after protein synthesis or in response to pathological states, mediated by molecular chaperones (Craig et al. 2003) or heat shock proteins (Chow and Brown 2007).

Third, there is autophagy in which the stressed cell digests some of its own damaged components (e.g., long-lived proteins, cytomembranes and organelles) and then replaces these missing components with newly constructed ones (Bergamini et al. 2004; Malorni et al. 2007) – an activity whose failure appears linked to the process of aging (Bergamini et al. 2004; Bergamini 2006; Kaushik and Cuervo 2006; Donati 2006). This is replacement at the subcellular level but repair at the cellular level.

Fourth, there is cell membrane self-repair in which torn plasma membrane reseals with little loss of intracellular contents (Steinhardt et al. 1994; Bi et al. 1995). One or more internal membrane compartments accumulate at the disruption site and fuse there with the plasma membrane, resulting in the local addition of membrane to the surface of the mechanically wounded cell (Miyake and McNeil 1995) and activating repair-related gene expression inside the cell (Ellis et al. 2001). Plasma membrane disruptions are resealed by changes in the cellular cytoskeleton (partial disassembly) (Xie and Barrett 1991) and by an active molecular mechanism thought to be composed of, in part, kinesin, CaM kinase, snap-25, and synaptobrevin (Miyake and McNeil 1995), with vesicles of a variety of sizes rapidly (in seconds) accumulating in large numbers within the cytoplasm surrounding the disruption site, inducing a local exocytosis (Miyake and McNeil 1995). Intracellularly, torn Golgi membrane readily reconstitutes itself from a vesiculated state (Kano et al. 2000) and the nuclear membrane is reversibly disassembled and reassembled (a form of "repair") during mitosis (Georgatos and Theodoropoulos 1999).

Fifth, some limited forms of cytoskeletal self-repair exist, most notably the coordinated remodeling of plasma membrane-associated (cortical) cytoskeleton self-repair (Bement et al. 2007), autocatalytic microfilament actin polymerization (Pantaloni et al. 2001), and recovery from mechanical disruption of cross-bridged intermediate filament networks (Wagner et al. 2007).

Sixth, there is the lysosomal system (Walkley 2007) for recycling all major classes of biological macromolecules, with soluble products of this digestion able to cross the membrane, exit the organelle, and enter the cytosol for recycling into the cellular metabolism, and there is the proteasome/ubiquitin system (Wolf and Hilt 2004) for similarly recycling damaged proteins – both of which effect "repair by replacement" since the whole macromolecule is discarded and a new one is synthesized in its place.

Medical nanorobotics will make possible comprehensive true cell repair, including, most importantly, those repairs that the cell cannot make for itself when it is relying solely on natural self-repair processes.

23.6.4.2 Cell Nanosurgery

The earliest forms of cellular nanosurgery are already being explored today. Atomic force microscopes (AFMs) have been used to observe the movement of filaments beneath the plasma membrane of living eukaryotic (Parpura et al. 1993) and bacterial (Méndez-Vilas et al. 2006) cells. Microrobotic systems are being developed for single cell nanoscale probing, injection, imaging and surgery (Li and Xi 2004), and the differing effects of intracellular surgical nanoneedles having cylindrical or conical tips (Obataya et al. 2005), or DNA-functionalized tips (Han et al. 2005), have been explored experimentally. Optical tweezers and vortex traps (Jeffries et al. 2007) permit noncontact immobilization and manipulation of individual cells.

Basic individual cell manipulation is fairly commonplace in the laboratory. For more than four decades microbiologists have used nuclear transplantation (Gurdon 2006; Meissner and Jaenisch 2006) techniques to routinely extract or insert an entire nucleus into an enucleated cell using micropipettes without compromising cell viability. Direct microsurgical extraction of chromosomes from nuclei has been practiced since the 1970s (Korf and Diacumakos 1978, 1980; Frey et al. 1982; Maniotis et al. 1997), and microinjection of new DNA directly into the cell nuclei using a micropipette (pronuclear microinjection) is a common biotechnology procedure (Wall 2001) easily survived by the cell, though such injected DNA often eventually exits the nucleus (Shimizu et al. 2005). DNA microinjection into pronuclei of zygotes from various farm animal species has been practiced commercially since 1985 but has shown poor efficiency and involves a random integration process which may cause mosaicism, insertional mutations and varying expression due to position effects (Wolf et al. 2000).

Nanosurgery has been performed on individual whole cells by several means. For example, a rapidly vibrating (100 Hz) micropipette with a <1 micron tip diameter has been used to completely slice off dendrites from single neurons without damaging cell viability (Kirson and Yaari 2000), and individual cut nerve cells have been rejoined by microsuturing or fibrin glue welding (Zhang et al. 1998). Axotomy of roundworm neurons was performed by femtosecond laser (femtolaser) surgery, after which the axons functionally regenerated (Yanik et al. 2004). A femtolaser acts like a pair of "nano-scissors" by vaporizing tissue locally while leaving adjacent tissue unharmed. Femtolaser surgery has also performed localized nanosurgical ablation of focal adhesions adjoining live mammalian epithelial cells (Kohli et al. 2005). AFMs have dissected bacterial cell walls in situ in aqueous solution, with 26 nm thick twisted strands revealed inside the cell wall after mechanically peeling back large patches of the outer cell wall (Firtel et al. 2004). Maniotis et al. (Maniotis et al. 1997) has mechanically spooled and extracted chromatin from a nucleus, observing that "pulling a single nucleolus or chromosome out from interphase or mitotic cells resulted in sequential removal of the remaining nucleoli and chromosomes, interconnected by a continuous elastic thread."

Nanosurgery has also been reported on subcellular and even nanoscale structures deep inside individual living cells without killing them. For instance, femtolaser surgery has performed: (1) microtubule dissection inside live cells (Sacconi et al. 2005a, Colombelli et al. 2005, 2007), (2) severing a single microtubule without disrupting the neighboring microtubules less than 1 micron away (Heisterkamp et al. 2005), (3) altering depolymerization rate of cut microtubules by varying laser pulse duration (Wakida et al. 2007), (4) selective removal of sub-micron regions of the cytoskeleton and individual mitochondria without altering neighboring structures (Shen et al. 2005), (5) noninvasive intratissue nanodissection of plant cell walls and selective destruction of intracellular single plastids or selected parts of them

(Tirlapur and Konig 2002), and even (6) the nanosurgery of individual chromosomes (selectively knocking out genomic nanometer-sized regions within the nucleus of living Chinese hamster ovary cells) without perturbing the outer cell membrane (Konig et al. 1999). Zettl's group has demonstrated a nanoinjector consisting of an AFM-tip-attached carbon nanotube that can release injected quantum dots into cell cytosol, with which they plan to carry out organelle-specific nanoinjections (Chen et al. 2007). Gordon's group at the University of Manitoba has proposed magnetically-controlled "cytobots" and "karyobots" for performing wireless intracellular and intranuclear surgery. Future diamondoid medical nanorobots equipped with operating instruments and mobility will be able to perform precise and refined intracellular, intra-organelle, and nanometer-scale nanosurgical procedures which are well beyond current capabilities.

23.6.4.3 Chromosome Replacement Therapy

The chromallocyte (Freitas 2007) is a hypothetical mobile cell-repair nanorobot whose primary purpose will be to perform chromosome replacement therapy (CRT). In CRT, the entire chromatin content of the nucleus in a living cell will be extracted and promptly replaced with a new set of prefabricated chromosomes that have been artificially manufactured as defect-free copies of the originals.

The chromallocyte (Fig. 23.26) will be capable of limited vascular surface travel into the capillary bed of the targeted tissue or organ, followed by diapedesis (exiting a blood vessel into the tissues) (Freitas 1999bq), histonatation (locomotion through tissues) (Freitas 1999at), cytopenetration (entry into the cell interior) (Freitas 1999x), and complete chromatin replacement in the nucleus of the target cell. The CRT mission ends with a return to the vasculature and subsequent extraction of the nanodevice from the body at the original infusion site. This ~3 hour chromosome replacement process is expected to involve a 26-step sequence of distinct semi-autonomous sensor-driven activities, which are described at length in a comprehensive published technical paper on the subject (Freitas 2007) and in

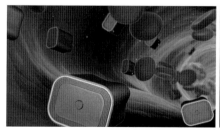

Fig. 23.26 Artist's conceptions of the basic chromallocyte (Freitas 2007) design: devices walking along luminal wall of blood vessel (*left*); schematic of telescoping funnel assembly and proboscis operation (*right*). Image © 2006 Stimulacra LLC (www.stimulacra.net) and Robert A. Freitas Jr. (www.rfreitas.com)

more detail below, and include: (1) injection, (2) extravasation, (3) ECM immigration, (4) cytopenetration, (5) inhibition of mechanotransduction (to avoid nanorobot mechanical actions triggering unwanted cell responses), (6) nuclear localization, (7) nucleopenetration, (8) blockade of apoptosis (to prevent misinterpretation of CRT processes as damage demanding cell suicide), (9) arrest of DNA repair (to prevent misinterpretation of CRT processes as damage demanding repair), (10) blockade of inflammatory signals, (11) deactivation of transcription, (12) detachment of chromatin from inner nuclear wall lamins (cortex proteins), (13) extension of the "Proboscis", (14) rotation of the Proboscis, (15) deployment of the chromosomal collection funnel, (16) digestion of stray chromatin, (17) dispensation of new chromatin, (18) decondensation of the new chromatin, (19) re-anchoring of the dispensed chromatin to inner nuclear wall lamins, (20) reactivation of transcription, (21) reactivation of DNA repair and other DNA-related maintenance and usage processes, (22) nuclear emigration, (23) cellular emigration, (24) ECM emigration, (25) return to original point of entry into the body, and (26) removal from the body. Treatment of an entire large human organ such as a liver, involving simultaneous CRT on all 250 billion hepatic tissue cells, might require the localized infusion of a ~1 terabot (10^{12} devices) or ~69 cm^3 chromallocyte dose in a 1-liter (7% v/v nanorobots) saline suspension during a ~7 hour course of therapy. This nanodevice population draws 100–200 watts which lies within estimated nanorobot thermogenic limits consistent with maintenance of constant body temperature (Freitas 1999 cm).

Replacement chromosomes would be manufactured in a desktop *ex vivo* chromosome sequencing and manufacturing facility, then loaded into the nanorobots for delivery to specific targeted cells during CRT. The new DNA is manufactured to incorporate proper methylation for the target cell type and other post-translational modifications constituting the "histone code" used by the cell to encrypt various chromatin conformations and gene expression states (Villar-Garea and Imhof 2006). A single fully-loaded lozenge-shaped 69 $micron^3$ chromallocyte will measure 4.18 microns and 3.28 microns along cross-sectional diameters and 5.05 microns in length, typically consuming 50–200 pW of power in normal operation and a maximum of 1000 pW in bursts during outmessaging, the most energy-intensive task. Onboard power can be provided acoustically from the outside in an operating-table scenario in which the patient is well-coupled to a medically-safe 1000 W/m^2 0.5 MHz ultrasound transverse-plane-wave transmitter throughout the procedure (Freitas 1999n) – the American Institute of Ultrasound in Medicine (AIUM) deems 10,000-sec exposures to 1000 W/m^2 ultrasound to be safe (Freitas 1999n). The chromallocyte design includes an extensible primary manipulator 4 microns long and 0.55 microns in diameter called the Proboscis that is used to spool up chromatin strands via slow rotation when inserted into the cell nucleus. After spooling, a segmented funnel assembly is extended around the spooled bolus of DNA, fully enclosing and sequestering the old genetic material. The new genetic material can then be discharged into the nucleus through the center of the Proboscis by pistoning from internal storage vaults, while the old chromatin that is sequestered inside the sealed leakproof funnel assembly is forced into the storage vaults as space is vacated by the new chromatin that is simultaneously being pumped out. The chromallocyte will employ a mobility system similar to the microbivore grapple system, possibly including a solvation wave drive (Freitas 1999bx) to help ensure smooth passage through cell plasma and nuclear membranes.

Modified procedures are proposed in the full technical description published elsewhere (Freitas 2007) for special cases including (1) proliferating, pathological, multinucleate, and karyolobate cells, (2) cells in locations where access is difficult such as brain, bone, or mobile cells, and (3) cells expressing genetic mosaicism, and also for alternative missions including (1) partial- or single-chromosome replacement, (2) single-cell and whole-body CRT, and (3) mitochondrial DNA replacement.

23.6.4.4 Modifying Cellular Controls and Cycles

Another important function of cell repair nanorobots (among those specialized for the task) would be the alteration of cellular control and metabolic cycle parameters. For example, the standard cell cycle for proliferating cells includes four rigidly-controlled and sequentially-executed phases, namely: entry G1 phase (cell expands in size), S phase (DNA synthesis), G2 phase (resting), and final brief M phase (mitosis or cell division, only ~4% of total cycle duration), with nonreplicating cells said to be in the quiescent G0 phase. Section 23.6.1 of the chromallocyte paper (Freitas 2007) reviews how a cell repair nanorobot might take complete control of a cell's mitotic cycle (Guardavaccaro and Pagano 2006), giving cell repair devices the ability to exercise nominal control over cell growth. Combined with control of cell development by changing the pattern of gene expression using CRT (Section 23.6.4.3), this provides a very general ability to replace cells. One clinical implication, for example, is that after a heart attack when scar tissue has replaced dead muscle, cell repair machines could stimulate unscarred regions of the heart to grow fresh muscle by resetting cellular control mechanisms, allowing the physician to guide the in situ self-healing of the heart.

The control of gene expression (e.g., via control of transcription factors, microRNAs, shRNAs, etc.) is paramount in the control of the cell. Important classes of cellular control modification – some having temporary, some having permanent, effects – might include direct intervention in protein synthesis (e.g., examining and editing extant mRNA tapes found in the cytosol, or fabricating and releasing supplemental natural or synthetic mRNA sequences, thus altering the rate of translation of specific protein sequences by the natural cellular machinery (Grudzien et al. 2004)); sequestration of key tRNA populations to sensitively influence the rate of protein synthesis (Delgado-Olivares et al. 2006); sequestration, augmentation, or chemical modification of key cell signaling molecules or ions to modulate internal signal pathways; artificial ubiquitination or de-ubiquitination (Johnston et al. 1999), or editing nuclear (Johnson et al. 2004) and cytoplasmic (Gomord et al. 1997) compartment localization sequences, on cytosolic proteins to assert control over trafficking; direct alteration of internal mitochondrial chemistry or internal lysosomal pH levels; artificially regulating normal cell functions including metabolism and secretion;

or cytocarriage (Freitas 1999ce) by nanorobotic "pilots" inside fibroblasts to direct the deposition and placement of collagen fibers by these mobile cells to rebuild extracellular matrix. Cellular controls and cycles could also be modified at their most upstream source by directly altering transcription (synthesis of RNA on a DNA template) in the nucleus, possibly by editing promoter sequences (Wray et al. 2003) in new replacement DNA that is installed by chromallocytes (Section 23.6.4.3), or by using sorting rotors to release or sequester inhibitors or transcription factors, since promoter activity is usually controlled by transcription factors that bind to the promoters or by inhibitors that inactivate the transcription factors.

23.6.4.5 Clearing Cytoplasm of Extraneous Materials and Devices

Cell repair nanorobots could restore and maintain cellular health by removing extraneous materials from the cytosol and other intracellular compartments. Perhaps the best-known example of such extraneous material is the insoluble age-pigment lysosomal granules called "lipofuscin" that collect in many of our cells, the accumulation starting as early in life as 11 years old and rising with age (Terman and Brunk 1998), activity level (Basson et al. 1982) and caloric intake (Moore et al. 1995), and varying with cell type (Brunk et al. 1992; Harman 1989). Clumps of these yellow-brown autofluorescent granules – typically 1–3 microns in diameter – may occupy up to 10% of the volume of heart muscle cells (Strehler et al. 1959), and from 20% of brainstem neuron volume at age 20 to as much as 50% of cell volume by age 90 (West 1979). Lipofuscin concentrations as high as 75% have been reported in Purkinje neurons of rats subjected to protein malnutrition (James and Sharma 1995). Elevated concentrations in heart cells appear not to increase the risk of heart attack (Strehler et al. 1959; Roffe 1998), nor to accelerate cellular aging processes in heart muscle or liver tissues (Blackett and Hall 1981), and brain cell lipofuscin is not associated with mental (West 1979; Drach et al. 1994) or motor (McHolm et al. 1984) abnormalities or other detrimental cellular function (Davies et al. 1983). However, hereditary ceroid lipofuscinosis (Shotelersuk and Gahl 1998) or neuronal ceroid-lipofuscinosis (NCL) diseases (Kida et al. 2001) can lead to premature death, though ceroid appears to be pathological only in neurons (Kida et al. 2001) or when loaded into human fibroblasts (Terman et al. 1999). There is also evidence that A2E, a hydrophobic fluorophore component of retinal pigment epithelial lipofuscin-like material, may contribute to age-related macular degeneration (De and Sakmar 2002). Lipofuscin is an indigestible lipid peroxidation product that cannot normally be excreted or metabolized by the cell, but which cytopenetrating microbivores that had entered the cell could readily detect and harmlessly digest – as the existence of various artificial lipofuscinolytic drugs (Totaro et al. 1985; James et al. 1992) and naturally occurring lipofuscinolytic bacteria (de Grey et al. 2005) attests is possible.

Other similarly inert intracellular pigments are known (Powell et al. 1996), along with a number of pathological intracellular storage diseases [e.g., of ER (Kim and Arvan 1998) and lysosomes (Winchester et al. 2000), etc.], including Fabrey's, Gaucher's, mannosidosis, Niemann-Pick (Simons and Gruenberg

2000), Tay-Sachs, Lewy bodies (Kosaka 2000) in Hallervorden-Spatz disease, and Hirano bodies (Yagishita et al. 1979). Neurofibrillary tangles (Mattson 2004) are pathological material found in neurons and are associated with Alzheimer's disease. Accumulation of lysosomal deposits of oxidized low-density lipoproteins or cholesterol crystals (Tangirala et al. 1994) in macrophage foam cells may contribute to atherosclerosis. Intracellular crystalloid bodies have been observed in the skeletal muscle cells of patients with hypothyroid myopathy (Ho 1987) and noninert amyloid deposits average ~12% of pancreatic islet cell volume in patients with maturity onset diabetes (Westermark and Wilander 1978). (See Section 23.7.1.1 for more on amyloidosis.) Excessive intracellular crystallization of drug molecules can lead to acute renal failure (Farge et al. 1986) and intracellular crystals have been found inside chondrocytes in certain crystal deposition diseases (Dijkgraaf et al. 1995). Other intracellular crystal deposition diseases are known such as mitochondrial crystalline inclusions (Farrants et al. 1988) and intermembrane inclusion bodies (O'Gorman et al. 1997), polyglucosan bodies (Matsumuro et al. 1993), and Fardeau-Engel bodies (Vital et al. 2002) that are involved in peripheral neuropathies. Several types of inorganic particles are highly toxic to phagocytes: just 0.05 µg of silica per 10^6 macrophages (Bateman et al. 1982), or 0.002% of cell volume assuming 1166 micron^3 per rat alveolar macrophage, is cytotoxic. Finally, heavy metals, radioactive ions, and metabolic poisons can also kill cells. All of these molecules, particles and deposits could either be digested to harmless effluents in situ by cytopenetrating microbivores (Section 23.6.2.1), or loaded into the large onboard storage tanks of chromallocyte-class nanorobots (Section 23.6.4.3) and transported intact out of the patient's body for external disposal.

Cell repair nanorobots may also remove extraneous nanodevices from the intracellular spaces. The most common of such devices would be natural biological nanomachines. For example, prions are the only known infectious intracellular pathogens that are devoid of nucleic acid (Prusiner 2001), and similarly viroids (Flores 2001) and viroid-like RNAs are intracellular pathogens lacking protein – and both are beyond the ability of current medicine to remove from infected cells (although antimisfolding agents to combat protein misfolding disorders like prions are under active study (Estrada et al. 2006)). Other biota that may live inside of cells include a variety of endosymbionts (Corsaro et al. 1999), viruses, and certain other entities involved in disease-associated emperipolesis (Freitas 2003y). In human cells, the tuberculosis bacterium enters the alveolar macrophage which transports the intruder into the blood, the lymphatic system, and elsewhere. Other intracellular microorganisms such as *Listeria* (~0.25 micron^3) and *Shigella* (~2 micron^3), once free in the cytoplasm, propel through the cytosol via continuous cytoskeleton-linked actin polymerization (Freitas 1999bz); macrophages infected with *Listeria* have been observed with ~2% of their volume co-opted by the microbes (~100 organisms) (Decatur and Portnoy 2000). Some motile intracellular parasites such as *Tyzzer* (Fujiwara et al. 1981) may cause disarrangement and depopulation of host cell organelles by the movement of their peritrichous (covering entire surface) flagella. Other motile intracellular parasites such as the spotted fever-group *Rickettsiae* (Hackstadt 1996) spread rapidly from cell to cell by actin-based movement but do not cause lysis of the host cell. Typhus-group rickettsiae (Hackstadt 1996) multiply in host cells to great numbers, though without profound damage until cell lysis finally occurs. Harmful pathogens such as malarial schizonts of *Plasmodium falciparum* may multiply to 50–70% of erythrocyte cytoplasmic volume before the red cell bursts, and other intracellular parasites have been observed at similar cytoplasmic volumetric fractions (Heydorn and Mehlhorn 1987; Abd-Al-Aal et al. 2000). Microbivore-class devices could remove all of these from intracellular spaces.

Beyond microbiological intruders, in a future era foreign nanorobots might be placed in a victim's body surreptitiously for unwanted or even malicious purposes. Chromallocyte-class personal cytosecurity nanorobots could be deployed having the ability to scavenge intracellular foreign nanodevices and either disable them in situ or transport them harmlessly out of the body, or to perform related sentinel or cytodefensive functions.

23.6.4.6 Organelle Testing, Replacement, or Repair

Another general class of cell repair machine would undertake the direct census and testing of intracellular organelle number and function, followed by appropriate corrective actions including the internal modification and repair or wholesale replacement of malfunctioning organelles. Excessive populations or damaged intracellular organelles could be removed using a cytopenetrating microbivore-class device; insufficient populations or volume of organelles can be addressed using a larger chromallocyte-class "delivery truck" type device (Section 23.6.4.3) to import supplemental organelles manufactured externally (Section "Tissue Printers, Cell Mills and Organ Mills").

Given the many thousands of unique biochemicals normally present within the cell, all having complex interactions, considerable R&D effort will be required to define the optimal testing regime for each organelle and is beyond the scope of this Chapter. However, simple tests are readily imagined as diagnostic indicators of correct organelle function. Cytosolic ATP concentrations, when combined with sensor readings for glucose and activity level indicators, can be diagnostic of proper metabolic function in the mitochondrial population. Organelle membrane breach (Freitas 2003z) is another concern – detection of free digestive enzymes in the cytosol may reveal a lysosomal or peroxisomal membrane breach, or the similar presence of cytochrome c may indicate mitochondrial wall breach. Neurons could be checked for proper ionic balance, ribosomes or mitochondria could be counted and inspected, and even vesicles, granules and vacuoles could be inventoried and sampled if deemed necessary or useful. A nonexhaustive list of general diagnostic mission classes (Freitas 1999ca) might include: (1) organelle counting, dimensional measuring, and general cytocartography (albeit somewhat ephemeral); (2) circumorganelle chemical assay; (3) organelle-specific surface membrane analysis or intracytoplasmic chemical assay; (4) dynamic functional or structural testing of cellular components; and (5) sampling, diagnosis, chemoinjection, replacement or repair operations to be performed upon an individual organelle or cytocomponent in a specific cell. Organelles could also be checked for organelle-specific storage diseases (Section 23.6.4.5) or for organelle-specific endosymbiont infestations such as mitochondrial mitophages (Sassera et al. 2006), and any unwanted foreign matter would be removed by cell repair nanorobots (Section 23.6.4.5).

The cell nucleus is the largest and most important intracellular organelle. Besides performing CRT (Section 23.6.4.3) on intranuclear genetic material, other activities that a cell repair machine might perform without entering the nucleus (Freitas 1999cb) could include: (1) physical mapping and compositional analysis of the nuclear envelope; (2) monitoring of nuclear pore traffic; (3) near-complete regulation of nuclear pore traffic using multiple manipulators or other devices; (4) monitoring, initiating, or modifying cytoskeletally-mediated mechanical signal transduction into the nuclear interior; and (5) injection of enzymes, RNA or DNA fragments, or other bioactive materials through nuclear pores using hollow nanoinjectors. The nucleus could also be checked for the presence of nucleus-specific intranuclear microbial parasites analogous to the tachyzoites of *Toxoplasma gondii* in mouse (which may enter the nucleus using its apical secretory organelle called the rhoptry) (Barbosa et al. 2005), *Nucleospora salmonis* (an intranuclear microsporidian parasite of marine and freshwater fish) (El Alaoui et al. 2006), the merogonic and gamogonic stages of *Eimeria* parasites in the goose (Pecka 1993), and MVM parvovirus (Cohen et al. 2006) – all of which could readily be detected and removed by medical nanorobots designed specifically for this task.

Cell membrane is a related compartment that suffers various molecular derangements that could be detected by medical nanorobots. For example, improper function of transmembrane glucose transporters could be detected by measuring interior glucose levels and comparing them to extracellular levels. Chemical testing could reveal toxins or poisons in cell receptors and transport channels, and related tests could be devised for other transmembrane pumps (e.g., the lack of pumps can be associated with disease (Chambers et al. 1999)), or to detect surface glycation (Section 23.7.1.2), and so forth. Upon detecting these conditions, nanorobots could appropriately edit or replace cell membrane components to repair all identified defects.

23.6.4.7 Cytostructural Testing, Replacement, or Repair

Besides testing for proper function, cell repair nanorobots could also examine cells for proper structure. For example, from outside the cell neural dendrites and other structural extensions could be checked for acceptable gross dimensions and appropriate connectivity (Freitas 1999co), adequate physical strength (Freitas 1999bd) and general health. Muscular dystrophy may be caused by disorganization of links between the intracellular cytoskeleton (e.g., dystrophin) and the ECM (Cohn and Campbell 2000), and the disruption of proper adhesive interactions with neighboring cells can lead to fatal defects in extracellular tissue architecture (Hagios et al. 1998). The cell plasma membrane and underlying cell cortex could be checked for mechanical integrity, and also to make sure that the correct surface receptors are in place and of the correct types and numbers (Freitas 1999cn). Cell membranes are ordinarily self-sealing after a puncture wound (Section 23.6.4.1), but a severely damaged membrane might need to be quickly patched using lipophilic materials dispensed from a repair nanorobot.

The cytoskeleton is normally self-repairing in a healthy cell – a fact that will help to allow nanorobots to transit the intracellular space without causing lasting damage. But some cells may experience "cytoskeletal disease" (Box 23.4) requiring nanorobotic repair. Direct nanorobotic intervention to repair these broken or inadequate cytoskeletal elements should be possible in all cases. However, the defect often will be widespread and caused by an underlying genetic (Section 23.6.4.3) or metabolic (Sections 23.6.2.5 and 23.6.4.4) pathology which should be directly corrected by the nanorobots, permanently curing the causative disease and allowing natural self-repair processes to resume their normal functions.

Box 23.4 Disorders of cytoskeletal architecture

Disorganization of the cytoskeletal architecture has been associated with diseases as diverse as heart failure (Hein et al. 2000; Lemler et al. 2000), rotavirus infection (Brunet et al. 2000), sickle cell anemia (Kuczera 1996), lissencephaly (Sapir et al. 1997), and Alzheimer's disease (Lee 1995), and a "collapse transition" of neurofilament sidearm domains may contribute to amyotrophic lateral sclerosis (ALS) and Parkinson's disease (Kumar et al. 2002). Cytoskeletal diseases most notably involve transmembrane linkage disruptions. For instance, breakage of major cytoskeletal attachments between the plasma membrane and peripheral myofibers in cardiac myocytes predisposes the cell to further mechanical damage from cell swelling or from ischemic contracture (Sage and Jennings 1988). Ellipocytosis (Liu et al. 1990) and other inherited hemolytic disorders (Delaunay 1995) are caused by disorganization of the subsurface spectrin-actin cell cortex in the erythrocyte (Zhang et al. 2001). Deeper inside the cell, perturbations in the architecture of the intermediate filament cytoskeleton in keratinocytes and in neurons can lead to degenerative diseases of the skin, muscle cells, and nervous system (Fuchs 1996). Tissues lacking intermediate filaments fall apart, are mechanically unstable, and cannot resist physical stress, which leads to cell degeneration (Galou et al. 1997). Perinuclear clumping of fragmented keratin intermediate filaments accompanies many keratin disorders of skin, hair, and nails (Sprecher et al. 2001). Impairment of normal axonal cytoskeletal organization in Charcot-Marie-Tooth disease results in distal axonal degeneration and fiber loss (Sahenk et al. 1999). A variety of human disorders are also associated with dysfunction of cytoskeleton-based molecular motors, including, for example: (1) the motor-based diseases involving defective cellular myosin motors (Keats and Corey 1999), e.g., implicated in Griscelli syndrome (Westbroek et al. 2001), hearing loss (Avraham 2002), hypertrophic cardiomyopathy (Rayment et al. 1995), and other myosin myopathies (Seidman and Seidman 2001); (2) spindle assembly- and function-related diseases (Mountain and Compton 2000) or kinesin- and dynein-related motor molecule diseases, e.g., implicated (Schliwa and Woehlke 2003) in Charcot-Marie-Tooth disease type 2A (Zhao et al. 2001), Kartagener syndrome (Marszalek et al. 1999) or primary ciliary dyskinesia (Olbrich et al. 2002), lissencephaly (Vallee et al. 2001), polycystic kidney disease (Qin et al. 2001), and retinitis pigmentosa (Williams 2002); and (3) other avenues for cellular malfunction (Schliwa and Woehlke 2003; Fischer 2000; Reilein et al. 2001; Schliwa 2003).

23.6.4.8 Intracellular Environmental Maintenance

Cell repair machines could also test, analyze, and restore a pathological cytoplasmic environment that has gotten too far from homeostatic equilibrium. This could be as simple as detecting and removing pathological proteins (or other damaged or unwanted biomolecules) from the cytosol, or it could involve actively manipulating intracellular pH, temperature, ionic balance, or metabolic inputs and byproduct concentrations. A nanorobot could straddle the plasma membrane of the cell, acting as a temporary artificial membrane transporter to pump out excess sodium, calcium, drug molecules, toxins, CO_2 and other waste products, or to pump in supplemental ions, O_2/glucose, or other nutrient molecules that are in short supply. These applications might require a pharmacyte-, microbivore-, or chromallocyte-class nanodevice, depending on circumstances.

As a simple example of the tremendous power of nanorobots to regulate the intracellular chemical environment, consider the Ca^{++} ion which serves as an intracellular mediator in a wide variety of cell responses including secretion, cell proliferation, neurotransmission, cellular metabolism (when complexed to calmodulin), and participates in signal cascade events that are regulated by calcium-calmodulin-dependent protein kinases and adenylate cyclases. The concentration of free Ca^{++} in the extracellular fluid or in the cell's internal calcium sequestering compartment (which is loaded with a binding protein called calsequestrin) is $\sim 10^{-3}$ ions/nm^3. However, in the cytosol, free Ca^{++} concentration varies from 6×10^{-8} ions/nm^3 for a resting cell up to 3×10^{-6} ions/nm^3 when the cell is activated by an extracellular signal; cytosolic levels $>10^{-5}$ ions/nm^3 may be toxic (Alberts et al. 1989), e.g., via apoptosis (Freitas 1999ag, cc).

To transmit an artificial Ca^{++} activation signal into a typical 20 micron cuboidal tissue cell in ~1 millisec, a single nanorobot stationed in the cytoplasm must promptly raise the cytosolic ion count from 480,000 Ca^{++} ions to 24 million Ca^{++} ions, a transfer rate of $\sim 2.4 \times 10^{10}$ ions/sec which may be accomplished using ~24,000 molecular sorting rotors (Freitas 1999o) operated in reverse, requiring a total nanorobot emission surface area of ~2.4 micron^2. Or, more compactly, pressurized venting or multiple ion diffusion nozzles may be employed (Freitas 1999 cd). Onboard storage volume of ~0.1 micron^3 can hold ~2 billion calcium atoms, enough to transmit ~100 artificial Ca^{++} signals into the cell (e.g., from $CaCl_2$) even assuming no ion recycling. In addition to the amplitude modulation (AM) of Ca^{++} signals noted above, De Koninck and Schulman (de Koninck and Schulman 1998) have discovered a mechanism (CaM kinase II) that transduces frequency-modulated (FM) Ca^{++} intracellular signals in the range of 0.1–10 Hz. Fine tuning of the kinase's activity by both AM and FM signals (either of which should be readily detected or generated by *in cyto* nanorobots) may occur as the molecule participates in the control of diverse cellular activities.

Similarly, high cytoplasmic calcium levels can destroy mitochondria by opening the mitochondrial "megapore" and activating destructive proteases (Dong et al. 2006), and elevated calcium levels are also expected under conditions of hypoxia, ischemia, and prolonged cold storage during cryopreservation. In such cases, the nanorobot described above can equally effectively extract excess calcium from the cytoplasm – dropping Ca^{++} cytosolic levels from a toxic 10^{-5} ions/nm^3 ($\sim 3 \times 10^6$ ions/cytosol) to a modest 10^{-7} ions/nm^3 resting-cell level ($\sim 3 \times 10^4$ ions/cytosol) in ~30 millisec, given a diffusion-limited ion current to the sorting rotor binding sites of ~10^8 ions/sec at 10^{-5} ions/nm^3 falling to ~10^6 ions/sec at 10^{-7} ions/nm^3 (Freitas 1999cp). The nanorobot can perform ~1000 such extractions before it must empty its tanks extracellularly.

23.7 Control of Human Senescence using Medical Nanorobots

Senescence is the process of growing old. Over the next few decades, it seems likely that a variety of purely biotechnological solutions to many of the major types of age-related damage will be found and will enter general therapeutic practice, for example, by following the illustrative SENS program (Section 23.7.1) developed by biogerontologist Aubrey de Grey, or by following other approaches (e.g., Fahy et al. 2010). De Grey's guarded expectation (de Grey 2005a) is that "all the major types of damage will be reversed, but only partly so. In several cases this incompleteness is because the category of damage in question is heterogeneous, consisting of a spectrum of variations on a theme, some of which are harder to repair than others. In the short term it's enough to repair only the easiest variants and thereby reduce the total damage load a fair amount, but in the longer term the harder variants will accumulate to levels that are problematic even if we're fixing the easy variants really thoroughly. Hence, we will have to improve these therapies over time in order to repair ever-trickier variants of these types of damage. I predict that nanotechnological solutions will eventually play a major role in these rejuvenation therapies."

In my view, nanotechnology will play a pivotal role in the solution to the problem of human aging. It is true that purely biotechnological solutions to many, if not most, of the major classes of age-related damage may be found, and even reach the clinic, by the 2020s. However, we have no guarantee that biotechnology will find solutions to *all* the major classes of age-related damage, especially in this timeframe. If treatments for any one of the numerous major sources of aging are not found, we will continue to age – albeit at a slower rate – and possibly with little or no substantial increase in the average human lifespan.

Medical nanorobotics, on the other hand, can undoubtedly offer convenient solutions to all known causes of age-related damage (Section 23.7.1) and other aspects of human senescence (Section 23.7.2), and most likely can also successfully address any new causes of senescence that remain undiscovered today. Medical nanorobotics is the ultimate "big hammer" in the anti-aging toolkit. Its development – as fast as humanly possible – is our insurance policy against the risk of a failure of biotechnology to provide a comprehensive solution to the problem of aging. Additionally, nanorobotic medicine, once developed, may offer superior treatments for aging, compared to the methods of biotechnology, as measured by a multitude of comparative performance metrics (Section 23.6.1). Finally, if we agree that a 16-year R&D effort costing a total of ~$1B launched today could result in a working nanofactory able to build medical nanorobots by the 2020s (Section 23.4.7), then it seems likely that by the late 2020s or early 2030s these powerful medical instrumentalities would begin to enter widespread clinical use, marking the beginning of the almost certain end to human aging (Section 23.7.1) while also providing cures for most other morbid afflictions (Section 23.6) of the human body.

23.7.1 Nanomedically Engineered Negligible Senescence (NENS)

According to Aubrey de Grey, SENS (Strategies for Engineered Negligible Senescence) (de Grey et al. 2002; de Grey 2006a, 2007a; de Grey and Rae 2007; Methuselah Foundation 2007) is a panel of proposed interventions in mammalian aging that "may be sufficiently feasible, comprehensive, and amenable to subsequent incremental refinement that it could prevent death from old age (at any age) within a time frame of decades." As explained in the foundational SENS paper (de Grey et al. 2002): "Aging is a three-stage process: metabolism, damage, and pathology. The biochemical processes that sustain life generate toxins as an intrinsic side effect. These toxins cause damage, of which a small proportion cannot be removed by any endogenous repair process and thus accumulates. This accumulating damage ultimately drives age-related degeneration. Interventions can be designed at all three stages. However, intervention in metabolism can only modestly postpone pathology, because production of toxins is so intrinsic a property of metabolic processes that greatly reducing that production would entail fundamental redesign of those processes. Similarly, intervention in pathology is a losing battle if the damage that drives it is accumulating unabated. By contrast, intervention to remove the accumulating damage would sever the link between metabolism and pathology, and so has the potential to postpone aging indefinitely. The term 'negligible senescence' (Finch 1990) was coined to denote the absence of a statistically detectable increase with organismal age in a species' mortality rate."

Seven major categories of such accumulative age-related damage have thus far been identified and targeted for anti-aging treatment within SENS. These include: removing extracellular aggregates (Section 23.7.1.1), removing extracellular crosslinks (Section 23.7.1.2), eliminating toxic death-resistant cells (Section 23.7.1.3), restoring essential lost or atrophied cells (Section 23.7.1.4), removing intracellular aggregates (Section 23.7.1.5), replacing mutant mitochondria (Section 23.7.1.6), and correcting nuclear mutations and epimutations (Section 23.7.1.7). As late as 2007 the prospective SENS treatment protocols (de Grey 2007a; de Grey and Rae 2007; Methuselah Foundation 2007) still lacked any serious discussion of future contributions from nanotechnology, an unfortunate omission which is corrected here by adding nanomedicine (medical nanorobotics) to SENS, obtaining "NENS".

23.7.1.1 Removing Extracellular Aggregates

Extracellular aggregates are biomaterials that have accumulated and aggregated into deposits outside of the cell. These biomaterials are biochemical byproducts with no useful physiological or structural function that have proven resistant to natural biological degradation and disposal. Two primary examples are relevant to the SENS agenda (de Grey 2003, 2006b).

First, there is the acellular lipid core of mature atherosclerotic plaques – which macrophages attempt to consume, but then die when they become full of the inert indigestible material, adding their necrotic mass to the growing plaques. One proposed SENS solution is to administer a bone marrow transplant of new bone marrow stem cells (cells that produce macrophages) that have been genetically reprogrammed to encode a new artificial macrophage phenotype that incorporates more robust intracellular degradation machinery. The resulting enhanced macrophages could then completely digest the resistant plaque material in the normal manner, though the full course of treatment would require months to run to completion and would likely yield only incomplete genetic substitution of stem cell genomes. Using NENS, vasculocytes (Section 23.6.2.3) would completely remove plaque deposits in less than a day, providing immediate vascular clearance and healing the vascular walls. For protection against future plaque development, chromallocytes (Section 23.6.4.3) could be targeted to the entire population of bone marrow stem cells to install the proposed more-robust macrophage phenotype using chromosome replacement therapy, in a thorough treatment also lasting less than a day.

Second, there are amyloid plaques that form as globules of indigestible material in small amounts in normal brain tissue but in large amounts in the brain of an Alzheimer's disease patient (Finder and Glockshuber 2007). Similar aggregates form in other tissues during aging and age-related diseases, such as the islet amyloid (Hull et al. 2004) in type 2 diabetes that crowds out the insulin-producing pancreatic beta cells, and in immunoglobulin amyloid (Solomon et al. 2003). Senile Systemic Amyloidosis or SSA (Tanskanen et al. 2006), caused by protein aggregation and precipitation in cells throughout the body, is apparently (Primmer 2006) a leading killer of people who live to the age of 110 and above (supercentenarians). One proposed SENS solution being pursued by Elan Pharmaceuticals to combat brain plaque is vaccination to stimulate the immune system (specifically, microglia) to engulf the plaque material, which would then be combined with the enhanced macrophages as previously described – although anti-amyloid immunization has not had great success experimentally (Schenk 2002; Patton et al. 2006). In NENS, amyloid binding sites could be installed on the external recognition modules of tissue-mobile microbivore-class scavenging nanorobots (Section 23.6.2.1), allowing them to quickly seek, bind, ingest, and fully digest existing plaques throughout the relevant tissues, in the manner of artificial mechanical macrophages. Chromallocytes could again be targeted to phagocyte progenitor cells to install the more robust macrophage phenotype to provide continuing protection against future plaque development.

Among the most promising investigational anti-amyloid therapies for Alzheimer's disease (Aisen 2005) is another potential SENS treatment for brain amyloid using anti-amyloid plaque peptides – one 5-residue peptide has already shown the ability, in lab rats, to prevent the formation of the abnormal protein plaques blamed for Alzheimer's and to break up plaques already formed (Soto et al. 1998), and to increase neuronal survival while decreasing brain inflammation in a transgenic mouse model (Permanne et al. 2002). However, a major challenge to the use of peptides as drugs in neurological diseases is their rapid metabolism by proteolytic enzymes and their poor blood-brain barrier (BBB) permeability (Adessi et al. 2003). In a NENS treatment model, a mobile pharmacyte-class nanorobot (Section 23.6.3.2) could steer itself through the BBB (Freitas 2003aa); release an appropriate engineered peptide antimisfolding agent (Estrada et al. 2006) in the immediate vicinity of encountered plaques so as to maintain a sufficiently high local concentration (Section 23.6.4.8) despite degradation; re-acquire the agents or their degradation products after the plaque dissolves; then exit the brain via the same entry route. Tissue-mobile microbivore-class devices could also be used to fully digest the plaques if it is deemed acceptable to ignore possible resultant localized deficits of normal soluble unaggregated amyloid-beta peptides. Nanorobots operating in the brain must be designed to accommodate the tight packing of axons and dendrites found there (Section 23.7.2(5)(a)).

23.7.1.2 Removing Extracellular Crosslinks

While intracellular proteins are regularly recycled to keep them in a generally undamaged state, many extracellular proteins are laid down early in life and are never, or only rarely, recycled. These long-lived proteins (mainly collagen and elastin) usually serve passive structural functions in the extracellular matrix and give tissue its elasticity (e.g., artery wall), transparency (e.g., eye lens), or high tensile strength (e.g., ligaments). Occasional chemical reactions with other molecules in the extracellular space may little affect these functions, but over time cumulative reactions can lead to random chemical bonding (crosslinks) between two nearby long-lived proteins that were previously unbonded and thus able to slide across or along each other (Methuselah Foundation 2007). Such crosslinking in artery walls makes them more rigid and contributes to high blood pressure.

In the SENS strategy (de Grey 2003, 2006b), it is theoretically possible to identify chemicals that can selectively dissociate crosslink bonds without breaking any other bonds, because many crosslink bonds have unusual chemical structures not found in proteins or other natural biomolecules. Some of these crosslink bonds may be unstable enough to be readily breakable by drugs, such as alagebrium chloride (aka. PMTC, ALT-711) which appeared to break one subset of glucose crosslinks (sugar-derived alpha-diketone bridges) in clinical trials (Bakris et al. 2004), but other crosslink bonds (e.g., acid-labile glucosepane (Lederer and Bühler 1999) and K2P (Cheng et al. 2004), and the highly stable pentosidine (Sell et al. 1991)) are probably too stable to be breakable by simple catalysis. SENS research proposals include: (1) finding new or synthetic deglycating enzymes that can couple the link-breakage to the hydrolysis of ATP to ADP (the most common power source inside cells), requiring the enzyme to shuttle back and forth across the cell membrane to acquire fresh ATP for each link-breakage cycle as there is very little ATP in the extracellular matrix; (2) engineering single-use link-breaking molecules analogous in action to the DNA repair protein MGMT which reacts with a stable molecule (DNA) but thereby inactivates itself (by transferring methyl and alkyl lesions from the O6 position of guanine on damaged DNA to a cysteine in its own structure (Pieper 1997)); or (3) increasing the rate of natural ECM turnover, taking care to avoid "dire side-effects such as hemorrhage from leaky blood vessels as collagen molecules are removed and replaced" (Furber 2006).

The NENS strategy proceeds similarly but more safely, using nanorobots as the delivery vehicle for the link-breaking molecules. In the first scenario, a population of $\sim 10^{12}$ (1 terabot) mobile pharmacytes would transverse the extracellular matrix in a grid pattern, releasing synthetic single-use deglycating enzymes (perhaps tethered (Craig et al. 2003; Holmbeck et al. 2004) to energy molecules, e.g., ATP) into the ECM to digest cross-linkages, then retrieving dispensed molecules before the nanorobot moves out of diffusive range. As an example, human skin and glomerular basement membrane (GBM) collagen has ~0.2 glucosepane (MW ~500 gm/mole) crosslinks per 100,000 kD strand of collagen in normally crosslinked aging tissue (Sell et al. 2005), indicating $\sim 2 \times 10^{18}$ glucosepane crosslinks in the entire human body which will require a very modest whole-body treatment chemical scission energy of ~0.2 joule per each ATP-ADP conversion event (~0.5 eV) required to energize cleavage of individual crosslink bonds. Each nanorobot would contain $\sim 2 \times 10^6$ enzyme molecules in a ~1 micron^3 onboard tank and would travel at ~3 micron/sec through ECM, releasing and retrieving enzymes in a ~10 micron wide diffusion cloud over a ~100 sec mission duration, with 10 successive terabot waves able to process all ~32,000 cm^3 of ECM tissue in the reference 70 kg adult male body in a total treatment time of ~1000 sec. Only 1 of every 10 enzymes released and retrieved are discharged by performing a crosslink bond scission; the rest are recovered unused. This treatment would likely be complete because full saturation of the targeted tissue volume can probably be achieved via diffusion, though some enzyme molecules may exit the diffusion cloud and become lost – lost molecules that must produce no side effects elsewhere or must be safely degradable via natural processes. In the second scenario, assuming $\sim 10^{19}$ collagen fibers in all ECM and allowing ~10 sec for a nanorobot to find and examine each fiber (thus removing one crosslink every ~50 sec), then $\sim 10^{14}$ nanorobots (~0.3% by volume of ECM tissue) using manipulators with enzymatic end-effectors could patrol ECM tissues, seeking out unwanted crosslink bonds and clipping them off, processing ~1 cm^3/min of crosslinked tissue and finishing the entire body in ~22 days. Enzymatically active components remain tethered and cannot be lost, reducing side effects to near-zero, but there may be some tight spaces that cannot easily be reached by the manipulator arms, possibly yielding an incomplete treatment. Further study is needed to determine the optimal combination of these two strategies.

23.7.1.3 Eliminating Toxic Death-Resistant Cells

A third source of age-related damage occurs from the accumulation of unwanted death-resistant cells that secrete substances toxic to other cells. These toxic cells are of several types: (1) fat cells (i.e., visceral adipocytes, which promote insulin resistance and lead to type 2 diabetes), (2) senescent cells (which accumulate in joint cartilage, skin, white blood cells, and atherosclerotic plaques, cannot divide when they should, and secrete abnormal amounts of certain proteins), (3) memory cytotoxic T cells (which can become too numerous, crowding out other immune cells from the useful immunological space, and which frequently become dysfunctional), (4) immune cells that have come to be hostile to endogenous antigens (autoimmune T and B cells), and (5) certain other types of immune cells which seem to become dysfunctional during aging (e.g., inability to divide, or immunosenescence) (de Grey 2006b; Methuselah Foundation 2007; Aspinall 2010).

There are several SENS strategies for reducing the number of senescent cells in a tissue: (1) conventional surgery, such as liposuction, wherein excess visceral fat tissue is simply cut out (e.g., eliminating pathology in diabetic rats (Barzilai et al. 1999)); (2) targeted apoptosis or cell suicide (Freitas 1999ag), in which only the chosen cells are induced to kill themselves in an orderly and non-necrotic manner (e.g., via immunotherapy in which immune cells would be sensitized to a diagnostic protein that is highly expressed only in the targeted cell type, or via somatic gene therapy (Campisi 2003) that would insert a suicide gene encoding a highly toxic protein controlled by a promoter that is activated only by the highly expressed diagnostic protein); and (3) de-senescing senescent cells by reversing the senescent phenotype (Beauséjour et al. 2003).

In NENS, tissue-mobile microbivore-class nanorobots (Section 23.6.2.1) would quickly and completely remove all unwanted cells, wherever located in the body, either by digesting them into harmless byproducts in situ or by sequestering their contents and transporting the compacted biomaterial out of the body for external disposal. Toxic cells could also be de-senesced using chromallocytes to wholly replace their nuclear genome with newly manufactured chromosomes (Section 23.6.4.3); alternatively, all DNA could be extracted from each toxic cell and the genome-free cell could then be flagged for natural macrophage removal (Freitas 1999cq) following the "neuter and release" protocol (Freitas 1999cr).

23.7.1.4 Restoring Essential Lost or Atrophied Cells

Cell depletion is another major source of age related damage (Methuselah Foundation 2007) that involves cell loss without equivalent replacement, most commonly in the heart, the brain, and in muscles. Missing cells leave gaps in tissues which may be filled by: (1) enlargement of adjacent similar cells (e.g., heart),

(2) invasion by dissimilar cells or fibrous acellular material (e.g., heart, brain), or (3) general tissue shrinkage (e.g., muscle).

Three SENS strategies to reverse cell depletion have been proposed (Methuselah Foundation 2007). The first two methods involve the natural stimulation of cell division by exercise (difficult in some muscles) or the injection of growth factors to artificially stimulate cell growth (Chen et al. 1995). Both methods may be of limited utility for normal dividing cells which may be robustly preprogrammed to avoid dividing excessively as a defense against cancer, but should be of greater utility in the case of stem cells, given that, for instance, marrow cells from older mice readily repopulate the irradiation-depleted marrow of young mice at least five times sequentially (Harrison and Astle, 1982) supporting the hypothesis that stem cells do not age, or age only very slowly. The third strategy would employ stem cell therapy to introduce new whole cells that have been engineered into a state where they will divide to fix the tissue even if cells already present in the body aren't doing so (Armstrong and Svendsen 2000).

The NENS approach starts with the manufacture of any needed replacement whole living cells, either very quickly with ideal quality control using external clinical cell mills (Section "Tissue Printers, Cell Mills and Organ Mills") or several orders of magnitude slower with inferior quality control using some variant of conventional mammalian cell reactors (Nelson and Geyer 1991). These replacement cells may include manufactured pluripotent stem cells. Nanosurgery is then employed to deliver the new cells to the repair site to assist the activities of vasculocytes (Section 23.6.2.3) and related nanorobots capable of controlled cell herding in vascular, ECM, or other cell-depleted tissue spaces. For example, a 1 cm^3 volume of 125 million 20-micron tissue cells, arranged in planar 10-cell slabs moving perpendicular to the slab plane through a tube, could be imported at ~1 m/sec through a 10-cm long nanosyringoscope (Section "Nanosyringoscopy") with a 100 micron inside diameter (possibly coated with mechanical cilia to facilitate efficient transport) to virtually anywhere inside the human body in ~250 sec (~4 min). A modest-sized array of 1000 safe and painless microneedles (Cormier et al. 2004; Flemming et al. 2005; Coulman et al. 2006; Nordquist et al. 2007) having a total ~10 mm^2 penetration cross-section for the array could transport ~1500 cm^3 of cells – the volume of the human liver, one of our largest organs – into the body during a ~6 minute transfer. A second nanosyringoscope can export a matching volume of body fluid or comminuted pathological tissue to precisely maintain conservation of volume/mass, if necessary. Arrival of conventionally vein-infused self-targeting stem cells at their designated destinations will take many orders of magnitude longer, and will not be 100% reliable and complete, as compared to nanosyringoscopy.

23.7.1.5 Removing Intracellular Aggregates

Intracellular aggregates are highly heterogeneous lipid and protein biomaterials that have accumulated and aggregated into clumps inside of the cell (Methuselah Foundation 2007). These biomaterials are normal intracellular molecules that have become chemically modified so that they no longer work and are resistant to the normal processes of degradation. Intracellular aggregates most commonly accumulate inside lysosomes, organelles that contain the most powerful degradation machinery in the cell. But if the lysosomes become congested and engorged, the cell will stop working properly – crudely analogous to a house whose toilets have all backed up. Cells in the heart and in the back of the eye, motor neurons and some other nerve cells, and white blood cells trapped within the artery wall appear most susceptible – intracellular aggregates have been associated with atherosclerosis (Brown et al. 2000) (the formation of plaques in the artery wall, which eventually occlude the vessel or calve material, causing heart attacks or strokes) and appear to be a contributing factor in several types of neurodegeneration (where the aggregates accumulate elsewhere than in the lysosome) and in macular degeneration (Reinboth et al. 1997) (the main cause of blindness in the old).

The proposed SENS strategy (de Grey et al. 2005; Methuselah Foundation 2007; de Grey 2006c) is to give all cells extra enzymes (such as microbial hydrolases found in natural soil bacteria and fungi) that can degrade the relevant biomaterial, or other accessory microbial proteins such as transporters to restore lysosomal acidity. The lack of such exogenous enzymes can be regarded as a genetic deficiency that results in pathological intracellular storage disease (Section 23.6.4.5), so the SENS treatment would be analogous to replacing a natural lysosomal enzyme for which patients are congenitally deficient as in enzyme replacement therapies (ERT). The ERT treatment can be directed to all cells as a complete whole-body gene therapy, or it can be directed only to modified stem cells via a bone marrow transplant that produces enhanced macrophages (Section 23.7.1.1), a stopgap approach that still allows the intracellular storage disease to progress to full senescence in somatic cells which are then removed and successfully digested by the enhanced macrophages. Possible difficulties with both approaches include: (1) inactivity or toxicity of microbial genes introduced into mammalian cells, (2) rapid degradation of the new microbial enzymes by lysosomal proteases whose normal function is to destroy other proteins, (3) immune rejection of microbial enzymes or proteins when cells expressing or containing them are attacked by lymphocytes, and (4) the inability of therapeutic enzymes in ERT to cross the blood-brain barrier in patients with cerebral neuropathies; though it is believed that further research can overcome all these problems (de Grey et al. 2005).

The proposed NENS strategy is twofold. First, storage-diseased lysosomes and other non-lysosomal intracellular aggregates could either be digested to harmless effluents in situ by cytopenetrating microbivores (Section 23.6.2.1) or by appropriate digestive enzymes temporarily injected into organelles, or could be loaded into onboard storage tanks of chromallocyte-class nanorobots and transported intact out of the patient's body for external disposal. This method could also effectuate cell-by-cell transplants of healthy lysosomes. Second, chromallocytes (Section 23.6.4.3) could install revised genomes in every cell in the human body, with the new chromosomes expressing the novel microbial-derived lysogenic enzymes and other requisite exogenous accessory proteins borrowed from the SENS program, assuming future research can validate the use of these or similar proteins.

23.7.1.6 Replacing Mutant Mitochondria

Mitochondria are the principal source of chemical energy in the cell, metabolizing oxygen and nutrients to carbon dioxide and water, producing energy-charged molecules of ATP that provide power for many important intracellular biochemical processes. Unlike other organelles, mitochondria have their own DNA that is susceptible to mutation, causing the mutated mitochondrion to malfunction leading to respiration-driven (i.e., oxidative damage-mediated) aging (Harman 1972; de Grey 1999, 2005b).

The principal SENS stopgap strategy (Methuselah Foundation 2007; de Grey 2000, 2005c) depends on the fact that of the ~1000 proteins present in the mitochondrion, only 13 (totaling under 4000 amino acids) are encoded by its own DNA. All the rest are encoded in the cell's nuclear DNA and are manufactured in the cytosol, then transported through the mitochondrial membrane wall by a complicated apparatus called the TIM/TOM complex (Rehling et al. 2001). By adding the genes encoding the unique 13 mitochondrial proteins to the better-protected nuclear chromosome content (Zullo et al. 2005), these proteins are anticipated to be produced when the mitochondria fail to do so and will be made to be imported through the organelle wall (Gearing and Nagley 1986), thus maintaining adequate energy-producing function even in mutated organelles. Nondividing cells such as muscle fibers and neurons accumulate mutant mitochondria most severely, so these cells most urgently need gene therapy to insert the supplementary genes. This is only a stopgap strategy because the mitochondria are not really "cured" of their pathology: new untreated cell pathologies hypothetically could appear if (1) the mutated mitochondrial DNA is left in place and the mutated DNA eventually comes to produce not just dysfunctional but actually harmful proteins (Baracca et al. 2007), or (2) the mutated mitochondrial DNA involves a dosage-sensitive gene with the disease phenotype resulting from multiple copies of a normal gene (Murakami et al. 1996). Other stopgap SENS strategies, also not amounting to complete or permanent cures, have been proposed, such as the injection of an antilipolytic agent to stimulate macroautophagy (the cell repair mechanism responsible for the disposal of excess or altered mitochondria under the inhibitory control of nutrition and insulin) in a presumably small number of the most severely injured mitochondria (Donati et al. 2006).

There are many possible NENS strategies for dealing with mutant mitochondria. First, chromallocytes (Section 23.6.4.3) could deliver into the nucleus of each cell in the human body a new set of manufactured chromosomes that incorporate genes encoding the 13 unique mitochondrial proteins, thus comprehensively effectuating the (incomplete) SENS proposal in a ~7 hour therapy for a single large organ such as liver or up to ~53 hours for a continuously-performed whole-body CRT procedure (Freitas 2007). Second, chromallocytes could employ a revised CRT treatment in which mitochondrial DNA is removed from each intracellular organelle in each cell and replaced with corrected versions of mtDNA (Freitas 2007), a more time-consuming approach. Third, replacement whole mitochondria containing non-mutated DNA could be manufactured in external clinical cell mills (Section "Tissue

Printers, Cell Mills and Organ Mills"), then delivered into the cytoplasmic compartment of target cells by chromallocyte-class nanorobots. Short-lifetime marker molecules (Freitas 2007) would distinguish new mitochondria from old, facilitating subsequent deportation of the old from the cell using exiting (now-empty) nanorobots, leaving behind only the new and also ensuring the removal of any mitophages (Sassera et al. 2006) that might be present, effectuating an all-cell mitochondrial transplant operation. Finally, replacement mitochondria re-engineered to contain no endogenous DNA could be installed in all cells by chromallocytes, after other chromallocytes have replaced nuclear DNA with new DNA containing the missing mitochondrial DNA, a treatment that would constitute a complete and permanent cure for inside-mitochondrion mutation. (Nuclear mutations continue to occur, and it has been claimed by some (Hayashi et al. 1994) that the mutation rate of genes encoding mitochondrial proteins might be higher in the nucleus than in the mitochondria, in which case the aforementioned strategy would be a way of greatly delaying but not permanently curing the problem of mitochondrial mutation.)

23.7.1.7 Correcting Cancer, Nuclear Mutations and Epimutations

Despite a sophisticated DNA self-repair system, chromosomes in the cell nucleus slowly acquire two types of irreversible age-related damage. First, there can be mutations, which are changes to the DNA sequence. Second, there can be epimutations, which are changes to the chemical decorations of the DNA molecule (e.g., DNA methylation) or to the histone modifications, that control DNA's propensity to be decoded into proteins, collectively representing the "epigenetic state" of the cell. (In a given patient, different cell types have the same DNA sequence but different epigenetic states.) When DNA damage of these types leads to uncontrolled rapid cell replication, the result is rapid tumor growth, aka. cancer (Section 23.6.2.2), and other loss of gene function unrelated to cancer can also occur. DNA damage and mutation may also be a significant cause of cell toxicity (Section 23.7.1.3) and cell depletion (Section 23.7.1.4) because cells can either commit suicide or go into a senescent non-dividing state as a pre-emptive response to DNA damage that stops it from developing into cancer (Methuselah Foundation 2007).

Traditional biotechnology knows no easy way to correct in situ large numbers of randomly occurring mutations or epimutations in the DNA of large numbers of randomly chosen cells. Consequently the SENS approach uses a stopgap strategy directed only at cancer (which is proposed to be the principal negative impact of mutated nuclear DNA on health and aging (de Grey 2007b)) via "Whole-body Interdiction of Lengthening of Telomeres" or WILT (de Grey et al. 2004; de Grey 2005d, 2010).

Here's how the SENS program of WILT would work. Telomerase (Autexier and Lue 2006) is a mainly nucleus-resident enzyme that acts to increase the length of telomeres, the endcaps of chromosomes, but is not normally expressed in most cells. Telomeres normally shorten at each cell division (accelerating after age 50 (Guan et al. 2007)), eventually resulting, after enough divisions, in chromosome dysfunction and cell senescence, a natural defense to runaway cancer. Cancer cells activate telomerase expression which removes this natural defense. WILT would forcibly reimpose the natural defense against cancer by totally eliminating the genes for telomerase and ALT (an alternative non-telomerase system for lengthening telomeres (Bryan et al. 1997)) from all cells that are able to divide. WILT would provide a permanent genetic alteration – not just a temporary improvement using drug-mediated telomerase inhibition as is currently being widely investigated (Cunningham et al. 2006) – by using gene deletion performed by comprehensive gene therapy (de Grey et al. 2002). WILT will require: (1) highly accurate gene targeting to delete the telomerase genes in tissues that don't rely on stem cells; (2) repopulating stem cells in the blood, gut, skin and any other tissues in which the stem cells divide a lot, with therapeutic infusions about once a decade (based on the apparent duration of the telomere reserve of neonatal stem cells judging (Methuselah Foundation 2007) from the age of onset of dyskeratosis congenita, a disease associated with inadequate telomere maintenance); and (3) growing engineered replacement stem cells whose telomeres have been restored in the laboratory, but which have no telomerase or ALT genes of their own. Also, cells already present in the body either must be destroyed without killing the engineered cells (in the case of stem cells for rapidly renewing tissues like the blood) or must have their telomerase and ALT genes deleted in situ (in the case of division-competent but normally quiescent cells, e.g., liver, glia) (Methuselah Foundation 2007). All this seems possible but represents a rather aggressive research agenda.

In NENS, chromallocytes (Section 23.6.4.3) could easily implement WILT, but why bother with a stopgap approach when nanorobots can fully address all nuclear mutations and epimutations, as well as cancer? Of course, cells can easily be killed by chromallocytes that extract nuclear DNA without replacing it (Freitas 1999cr), or by using cytocidal devices dramatically simpler than chromallocytes. But the optimal NENS solution to nuclear mutation and epimutation is to employ chromallocytes performing chromosome replacement therapy or CRT (Section 23.6.4.3) to replace all of the randomly damaged chromosomes with completely undamaged newly manufactured chromosome sets (Freitas 2007), in all cells of the body. As another benefit, CRT will automatically repair any somatic mutations in tumor-suppression genes, thus reinvigorating other components of the body's natural defenses against cancer – a repair that is wholly impractical using conventional biotechnology. As yet another benefit, the installed new chromosome sets can be manufactured with their telomeres re-extended to full neonatal reserve length, essentially "rolling back the clock" to birth on chromosome age and effectively implementing comprehensive cellular genetic rejuvenation (Section 23.7.2).

Because biology is highly complicated, the earliest implementations of nanorobotic CRT (perhaps in the 2030s) need not depend on knowing which DNA sequences and epigenetic states are "correct" (in the ideal functional sense), but merely on knowing which ones appear "normal" for a particular patient, with chromallocytes then reinstalling whatever is normal for each cell type. Normal can be measured by widespread sampling of DNA in the patient's native cells and statistically averaging out the observed random variations (Freitas 2007). In later implementations of CRT, we will know enough about the ideal epigenetic state of all cell types to be able to implement it just as precisely as we will be able to edit native DNA sequences or delete foreign sequences, using the same nanorobots.

23.7.2 Nanorobot-Mediated Rejuvenation

SENS or other fundamental approaches to the biology of aging, and more powerfully nanomedical implementations thereof, will give physicians the tools to eliminate all age-related damage (Section 23.7.1), and medical nanorobotics will provide comprehensive treatments for all common causes of human morbidity (Section 23.6). In many cases, a one-shot restoration of cells to their pristine undamaged state can re-establish the ability of those cells to maintain molecular homeostasis (Wiley 2005) and to resume normal self-healing activities in response to future cell damage that may occur. But there will still remain a residuum of ongoing cell damage that cells, tissues, and organs cannot heal on their own unless they are given novel capabilities for self-repair, or are given new engineered biochemical pathways that avoid creating the damage. perhaps by augmenting and reprogramming the human genome. Until and unless we implement these augmentations, injuries that the body is incapable of repairing on its own will resume their natural rate of accumulation, allowing natural aging to reappear. Periodic rejuvenative treatments will therefore be required to reverse this accumulating new damage to the body.

Important components of such periodic rejuvenative treatments may include, among other things:

(1) *All-Cell Genetic Renatalization*. Over time, new mutations and epimutations in the nuclear genome will continue to recur, and telomeres will resume growing shorter as cells continue normal division. Chromallocytes (Section 23.6.4.3) would deliver to all cells new mutation-free chromosome sets, thus periodically "rolling back the clock" to zero chromosome age (while leaving developmental controls in adult mode) and effectively implementing comprehensive genetic cellular rejuvenation (Section 23.7.1.7). The new error-free chromosome sets will be manufactured with their telomeres re-extended to full neonatal reserve length.
(2) *Whole-Body Cytological Maintenance*. Like an old house or car, cells and their immediate environs will need periodic maintenance to keep them in showroom condition. Primarily this would involve a NENS (Section 23.7.1) sweep of every cell in the body, eliminating intracellular aggregates and extracellular aggregates and crosslinks, and removing or replacing cells within tissues or organelles within cells as required to maintain optimal tissue and organ health. It would also include repairing errant or missing intercellular connections and other malformations of the extracellular matrix other than simple crosslinking, a category of tissue damage largely ignored by SENS, using a combination of fibroblast cytocarriage to lay down fresh fiber (Section 23.6.4.4) and surgical nanorobots (Section "Endoscopic Nanosurgery and Surgical Nanorobots")

with capabilities similar to dermal zippers (Section 23.6.3.3) to rebuild and reconstruct the ECM as needed. This kind of ECM damage may occur during scarring, burning or freezing injuries, which also may pull cells out of their proper positions thus requiring mechanical repositioning. Cell membranes could be edited to remove unwanted foreign molecules or mechanisms, and poisonous chemicals and heavy metals can be extracted. In this manner, cells and the matrix surrounding them could be restored to their ideal youthful state, effectively implementing comprehensive structural and functional cellular rejuvenation.

(3) *Whole-Body Anatomical Maintenance.* Patient anatomy could be mapped and recorded down to the cellular level, then compared to the ideal state desired by the patient (in consultation with his physician), then brought into compliance with the patient's wishes by the addition or removal of specific cells, tissue masses, or even organs via nanosurgery (Section 23.6.3.5). Many age-related cosmetically undesired changes in human appearance are completely non-pathological and reflect only an extension of normal cell growth processes that could be nanorobotically blocked or reversed, e.g., by cell removal (Section 23.7.1.3). Examples of such changes include the enlarged noses and ears in older people that arise from slow growth that proceeded unimpaired from birth until old age. Changes in chin prominence and other remodeling of the skull probably fall into the same category. Pathological anatomical damage must also be repaired. Physical trauma is an obvious source of new anatomical damage that could be repaired via medical nanorobotics (Section 23.6.3), and foreign-body granulomas, wherever situated, should also be excised. Comprehensive inspection and reconditioning of the human vascular tree by vasculocytes (Section 23.6.2.3) might be an important part of a periodic rejuvenation regimen, virtually eliminating all possibility of cardiovascular disease and brain damage due to stroke. Another age-related pathology of the ECM occurs when aging fibroblasts begin producing collagenase instead of collagen (Quan et al. 2006), tearing down the ECM and causing, for example, faces to wrinkle, sag, and become softer (because the ground substance that holds the face together is being torn apart), not stiffer as would be expected if facial aging was due to crosslinking. Rejuvenating an old face might therefore require in situ redeposition of collagen and elastin fibers unless this is found to occur automatically after aging fibroblasts have been removed (Section 23.7.1.3), new fibroblasts are installed (Section 23.7.1.4), and chromallocytes have reset the telomere lengths (Section 23.7.2(1)) of dermal cells and fibroblast precursor cells (Friedenstein et al. 1976).

(4) *Systemic Deparasitization.* Analogously to computer systems, human patients should be periodically "debugged" of unwanted parasitic entities present within the body. Parasitic entities may be present at all different levels of biological organization. At the molecular level, parasitic molecules such as prions and viroids should be eliminated (Section 23.6.4.5). At the genetic level, recent or ancient retroviral insertions into our DNA should be edited out using chromallocytes (Section 23.6.4.3), except for those known to have some beneficial effect

because we've adapted to their presence. We should also periodically clean out "transposable elements" or transposons (including retrotransposons) or "jumping genes" that may contribute to aging by inserting into the middle of other genes and deactivating them. Cancer cells, cancer cell microaggregates, and cancerous tumors are parasitic at the cellular level, and could be detected by periodic scans, and then excised (Section 23.6.2.2). A great variety of microbiological parasitic entities should be deleted from the body, most notably acute viral and bacterial infections (Section 23.6.2.1) but also including granuloma-encased tuberculosis bacteria and other latent biotic reservoirs such as those that produce periodic outbreaks of herpes, shingles, etc. later in life, and any nanorobotic intruders (Section 23.6.4.5). Nonsymptomatic infestations of commensal, amensal, or other endoparasites including protozoa and worms may also be removed using medical nanorobots.

(5) *Neural Restoration.* Adult neurons generally do not reproduce and cannot replace themselves once destroyed. Early workers in the 1950s (Brody 1955) attempted the first assessment of the long-term rate of natural attrition of brain cells. Losses ranged from none at all to very many in various parts of the organ, but the brainwide average loss was ~100,000 neurons per day, a rate consistent with loss of all brain cells (in some parts of the organ) over a period of about 250–350 years. More recent work (Lopez et al. 1997) has confirmed a similar ~3%/decade cell loss rate in some areas of the brain. A sufficient loss of neural connectivity or infrastructure from this source, or from physical brain trauma, would constitute effective creeping brain death. Several possible approaches to neural restoration have been identified.

(a) *Prevent or delay random cell death* within the neuronal network by using nanogerolytic treatments on individual cells, keeping each cell healthy and avoiding DNA mutations and microdeletions (Kamnasaran et al. 2003) via nanorobot-mediated CRT (Section 23.6.4.3). Note that the brain contains only 5% extracellular space and consists for the most part of densely-packed axons and dendrites with virtually no gaps between them, so neuron-targeted motile chromallocytes will often transit plasma membranes between neighboring cells rather than intercellular spaces. Because cell bodies containing the nucleus may be relatively far apart, these specialized nanorobots must be engineered either to migrate inside the larger-diameter axons without ruining neural function or external to the axons without disturbing the local ionic environment. This may require active nanorobotic monitoring and localized remediation of the ECM chemical environment (analogous to Section 23.6.4.8) during nanorobot locomotion, given that the minimal extracellular space in the brain controls the concentrations of extracellular ions that cross and re-enter the cell membrane during and after action potentials.

To effectuate neuronal CRT, one approach might be to block apoptosis to allow more time for DNA repair, then to osmotically expand the extracellular space on a local basis to allow relatively large nanorobotic devices to migrate wherever they need to go. Considerable expansion may be tolerable: Smith's classic hamster freezing experiments (Smith et al. 1954; Lovelock and Smith 1956; Smith 1965) showed that >60% of the water in the brain can be converted into extracellular ice without apparent brain damage, a distortion far in excess of what would be needed for nanorobot traffic. Recent unpublished observations by G. Fahy (personal communication, 2008) at 21st Century Medicine show that when ice forms in the brain even at low temperatures in the presence of cryoprotectants, neurons and nerve processes are neatly packaged and are not torn apart, supporting the idea that the extracellular space can be significantly locally expanded without lasting harm. The migration of newly-generated neurons through the brain provides additional evidence that the organ can tolerate significant local distortion of the extracellular space. For example, neurogenesis in the hippocampus is followed by neurons or their precursors migrating out of the hippocampus over large distances to other parts of the brain (Ehninger and Kempermann 2008), a mechanical process that is normal and apparently well tolerated, and microglial cells (the immune system phagocytes in the brain) have been observed (via two-photon imaging of mammalian neocortex) to have extremely motile processes and protrusions (Nimmerjahn et al. 2005).

(b) *Offset brain cell losses* by inducing compensatory regeneration and reproduction of existing neurons as in situ replacements, i.e., by stimulating endogenous neurogenesis (Tatebayashi et al. 2003). Successful neuron re-growth in response to growth factors, with associated cognitive benefits, has been reported in rats (Chen et al. 1995), and self-assembling peptide nanofiber scaffolds can create a permissive environment for axons to regenerate through the site of an acute injury and also to knit the brain tissue together, as demonstrated by the return of lost vision in one animal model (Ellis-Behnke et al. 2007).

(c) *Replace dysfunctional neurons,* by infusing stem cells, allowing normal memory-reinforcing cognitive processes to provide continuous network retraining. *Ex vivo*-cultured neural stem cells have been induced to differentiate and replace lost neurons after injection into the brain (Armstrong and Svendsen 2000) and dead neurons can be replaced by introducing stem or precursor cells that differentiate appropriately (Sugaya and Brannen 2001), but patterned neuronal networks, once thoroughly disrupted to the point of serious information loss, cannot be restored by stem cells or any other means as per SENS.

(d) *Rebuild neural tissue* using a combination of tissue mills (Section "Tissue Printers, Cell Mills and Organ Mills") and nanosurgery (Section "Endoscopic Nanosurgery and Surgical Nanorobots") following blueprints assembled from a complete brain state map acquired via comprehensive in vivo nanorobotic brain scans (Fig. 23.27). Maps of neural networks and neuron activity states could be produced by nanorobots positioned outside each neuron (Freitas 1999cf), after they have passed into the brain through the BBB (Freitas 2003aa), using tactile topographic scanning (Freitas 1999co) to infer connectivity along with noninvasive neuroelectric measurements (Freitas 1999cf) including, if necessary, direct synaptic monitoring and recording (Freitas 1999cs). The key challenge in making such scans feasible is obtaining the necessary bandwidth inside the body, which should be available using an in vivo optical fiber network (Freitas

Fig. 23.27 Artist's conception of a neuron inspection nanorobot; image courtesy of Philippe van Nedervelde, © 2005 E-spaces.com. Used with permission

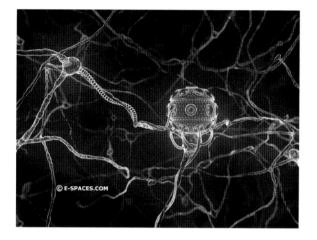

1999cg) distributed via nanocatheters (Section "Endoscopic Nanosurgery and Surgical Nanorobots"). Such a network could handle 10^{18} bits/sec of data traffic, capacious enough for real-time brain-state monitoring. The fiber network would have a 30 cm^3 volume and generate 4–6 watts of waste heat, both small enough for safe installation in a 1400 cm^3 25-watt human brain. Signals travel at most a few meters at nearly the speed of light, so transit time from signal origination at neuron sites inside the brain to the external computer system mediating the scanning process are ~0.00001 millisec which is considerably less than the minimum ~5 millisec neuron discharge cycle time. Neuron-monitoring chemical sensors (Freitas 1999ci) located on average ~2 microns apart can capture relevant chemical events occurring within a ~5 millisec time window, the approximate diffusion time (Freitas 1999ch) for, say, a small neuropeptide across a 2-micron distance. Thus human brain state monitoring can probably be "instantaneous", at least on the timescale of human neural response, in the sense of "nothing of significance was missed."

(e) *Avoid plethomnesia.* One theoretical additional health risk at very advanced calendar ages is that the total data storage capacity of the brain might eventually be reached. At this point, either no new memories could be stored or old memories would have to be overwritten and thus destroyed, giving rise to a hypothetical mental pathology involving forgetfulness most properly termed "plethomnesia" (from Gr. *plethos* (fullness, too full) + *mnasthai* (to remember, memory)). The data storage capacity of the human brain has been estimated using structural criteria to range from 10^{13}–10^{15} bits assuming ~1 bit per synapse (Cherniak 1990, Tipler 1994), or using functional criteria as 2.2×10^{18} bits for the information contained in a normal lifetime of experience (brain inputs) (Schwartz 1990; Tipler 1994; Baldi 2001) to ~10^{20} bits based on the accumulated total of all neural impulses conducted within the brain during a normal lifetime (von Neumann 1958). Given that experimental studies suggest a normal-lifetime limit for consciously recoverable data of only ~200 megabytes (~1.6×10^9 bits) (Landauer

1986), it appears that the human brain may have significant amounts of untapped reserve memory capacity. However, should plethomnesia occur it might most effectively be cured by employing nanomedicine via cognitive neural prosthetic implants (Berger and Glanzman 2005; Pesaran et al. 2006; Schwartz et al. 2006) linked through nanotechnology-based neural interfaces (Patolsky et al. 2006; Mazzatenta et al. 2007) to nanotechnology-based high-density read/write memory caches (Green et al. 2007; Blick et al. 2007), e.g., "brain chips".

Using perhaps annual nanorobot-mediated rejuvenative treatments such as the above, along with some occasional major repairs, it seems likely that all natural accumulative damage to the human body could be identified and eliminated on a regular basis. The net effect of these interventions will be the continuing arrest of all biological aging, along with the reduction of current biological age to whatever age-specific phenotype is deemed cosmetically desirable by the patient, severing forever the link between calendar time and biological health and appearance.

23.7.3 Maximum Human Healthspan and the Hazard Function

If all age-related causes of death and ill-health could be eliminated by medical nanorobotics and if the remaining non-medical causes of death are distributed randomly across all calendar ages, this gives a constant rate of death R_{mort} during any increment of time. The number of survivors $N(t)$ at time t, starting from an initial population N_{pop} at time t = 0, is estimated, using the standard exponential formula for an interval-constant decay rate, as $N(t) = N_{pop} \exp(-R_{mort} t)$. The median healthspan T_{half} is then given by a simple half-life formula (which is generally applicable to any process having a constant event rate): $T_{half} \sim \ln(2) / R_{mort}$, where R_{mort} is the cumulative death rate from all sources, in deaths/person-year. In this case, R_{mort} is the sum of five principal components:

(1) the fatal accident rate, including motor vehicle (presently 44% of the total) and all other causes (3.62×10^{-4} yr^{-1} for 1998 in the U.S. (Census Bureau 2001c));
(2) the suicide rate (1.13×10^{-4} yr^{-1} for 1998 in the U.S. (Census Bureau 2001c));
(3) the homicide rate (6.8×10^{-5} yr^{-1} for 1998 in the U.S. (Census Bureau 2001c));
(4) the combatant war casualty rate, deaths only, as a fraction of the general population (~3.2×10^{-5} yr^{-1} for all U.S. wars in the last 100 years, relative to average (~200 million) U.S population level during that period (Almanac 1994)); and
(5) the legal execution rate (2.27×10^{-7} yr^{-1} for 1998 in the U.S. (Death Penalty Information Center 1998)).

Summing these five items gives $R_{mort} \sim 5.75 \times 10^{-4}$ yr^{-1}, yielding a median healthspan of $T_{half} \sim 1200$ years. This is consistent with an independent estimate of $T_{half,10} \sim 5300$ years based upon the actuarial death rate of children in the 10-year-old cohort ($R_{mort,10} = 1.3 \times 10^{-4}$ yr^{-1} in 1998 in U.S. (Census Bureau 2001a)), whose death rate is the lowest for any age cohort and for whom the almost exclusive cause of death is accidents. (The death rates for children aged 1–15 is less than R_{mort} (that is, $<5.75 \times 10^{-4}$ yr^{-1}) (Census Bureau 2001a).) Note that in this model, T_{half} is the estimated median lifespan in a healthy non-aging state, with no part of that life spent in an infirm senescent state, hence the estimate reflects the anticipated length of healthy years, or *healthspan*, and not mere lifespan which today may include 25% or more time spent in a morbid condition.

It is worth pointing out (Freitas 2002) that the advances in medicine over the last two centuries have already effectively achieved a disease-related-mortality free condition for a few age cohorts of the human population in industrialized countries – our youngest children. Medical technology has had its greatest impact to date in preventing infant mortality, especially between the ages of 1 and 4. In the year 1865, a young child in this age cohort had a 6.86% probability of dying in the next year (Census Bureau 1989a), but by 1998 the probability of dying in the next year for these children had been slashed from 6.86 to 0.0345% (Census Bureau 2001a), a phenomenal 200-fold reduction. If we could keep our bodies in the same healthy condition that existed when we were young, we should have a median healthspan approaching $T_{half,10}$ ~5300 years as noted above. (This assumes the accident risks are roughly the same for adults (who drive cars, operate heavy machinery, etc.) as for children (who don't), which may seem improbable but is nonetheless approximately true: U.S. accident deaths for 1998 as a fraction of all deaths in each age cohort were 37% at 1–4 years, 42% at 5–14 years, 44% at 15–24 years, and 28% at 25–34 years (Census Bureau, 2001d).) Death would usually come from some form of non-medical accident, which is the leading cause of death up to the age range of 35–44 years (Census Bureau, 2001d). When future nanorobotic medicine is available as envisioned here, we shall extend this disease-related-mortality free condition to all age cohorts, not just to the children, and thus give all of us the potential to achieve $T_{half,10}$ ~5300 healthy years, or more.

The maximum likely healthspan in a world subject only to age-unrelated deaths can be estimated from the aforementioned exponential formula by taking N_{pop} ~6 billion, the current world population, and $N(t) = 1$, indicating the last survivor of this population, and a constant $R_{mort} = 5.75 \times 10^{-4}$ yr^{-1} as before, yielding t = T_{max} ~39,200 years, the maximum healthspan of the last random survivor from this cohort.

These projected healthspans seem incredibly long by current standards. Even so, it is safe to predict that people will desire more and will seek to reduce R_{mort} still further. The simplest way to reduce nonmedical hazards is to attack the largest source of them – the accident rate – by employing nanotechnology to create a safer and more hazard-free living environment. Motorized vehicles of all kinds (land, sea, air, and space) can be made more crash resistant, new forms of "airbags" can be designed to allow survival of high-speed impact forces from any direction, and the fallibility of human operators could be eliminated by switching to automated aircraft, cars, trucks, trains and ships. Buildings (including houses) can incorporate active safety devices. Extremely fine-grained simulations of the physical world could provide more accurate risk-prediction models, allowing potential dangers to be anticipated and avoided in advance. Implanted in vivo nanorobotic systems equivalent to respirocytes and clottocytes could greatly reduce accidental deaths from drowning and bleeding. Other basic augmentations to the human body could improve its durability and reduce its accident-proneness, including modifications to the human genome to engineer improved metabolism or increased intelligence, perhaps combined with more intrusive nanorobotic implants such as whole-body vascular replacement systems (Freitas and Phoenix 2002). Both homicide (inversely correlated with wealth and education) and suicide rates should fall as the spread of molecular manufacturing increases material prosperity (Freitas 2006b) and expands the diversity of life choices. These factors, along with greater access to knowledge, should also help to decrease the incidence of war.

The maximum speed at which R_{mort}, also known as the "hazard function," can be reduced is presently unknown but a conservative lower limit may be crudely estimated as follows. From 1933 (the first year reliable data became available) to 1998, annual accident rates fell by 50% from 71.9 per 10^5 (Census Bureau 1989b) to 36.2 (Census Bureau 2001c), suicide rates fell by 29% from 15.9 per 10^5 (Census Bureau 1989c) to 11.3 (Census Bureau 2001c), homicide rates fell by 30% from 9.7 per 10^5 (Census Bureau 1989c) to 6.8 (Census Bureau 2001c), and legal executions fell by 82% from 0.127 per 10^5 (Census Bureau 1989d) to 0.023 (Death Penalty Information Center 1998), a 65-year net decline of $\Delta R_{mort} = -0.863\%/yr$ in R_{mort}, from 1.01×10^{-3} yr^{-1} in 1933 to 5.75×10^{-4} yr^{-1} in 1998. Substituting $R(t) = R_{mort} (1-\Delta R_{mort})^{(t-1998)}$ for R_{mort} in our exponential formula to most simply represent this observed nonmedical death rate decline, T_{half} would rise from 1200 years in 1998 to 1300 years by 2009, 1500 years by 2029, and 2000 years by 2070. These figures appear conservative because if we can make our living environment as safe for adults as it currently is for our 10-year-olds, then T_{half} should more closely approximate $T_{half,10}$ ~5300 years, not 1300 years, for adults, in the present epoch.

23.8 Summary and Conclusions

This chapter has argued, I hope persuasively, that diamondoid medical nanorobotics can almost certainly achieve comprehensive control of human morbidity and aging. To the more limited extent that biotech-based instrumentalities can accomplish similar ends, nanorobot performance and safety will likely prove superior in comparison.

Some have averred that medical nanorobotics sounds like an "argument for infinity" because it appears to skeptical eyes to be a panacea that can do and cure anything. No such claim is advanced here or by any serious proponent of advanced nanomedicine. Nanorobots, no matter how capable, must always have very well-defined physical limitations. They are limited by mobility constraints, by the availability of energy, by mechanical and geometric constraints, by diffusion limits and biocompatibility requirements, and by numerous other constraints (Freitas 1999, 2003). Nanorobots cannot act instantly – they take time to effect their cure.

But because they will be constructed of superior building materials of surpassing strength and stiffness, diamondoid nanorobots will operate several orders of magnitude faster than analogous machinery built from biomaterials, and will be able to apply forces several orders of magnitude larger than those which may be applied by comparable biological- or biotech-based systems. Nanorobots will avoid almost all proximate side effects because they can operate under precise sensor-driven digital control, not drift aimlessly on the stochastic currents of the human body like nanoparticles and drug molecules. Nanorobots can be more reliable because they can report back to the physician what they are doing, both while they are doing it and after they've finished. They are safer because, unlike commonplace biotechnology-based approaches, a diamondoid nanomachine cannot be co-opted for hostile use by rapidly mutating microbes. And diamondoid nanorobots could incorporate biomaterials or biological components whenever necessary (e.g., in the design of exterior biocompatible coatings (Freitas 2003ab)), so hybrid bio-diamondoid nanorobots can assimilate any performance advantages of biotech as a subset of medical nanorobotics design.

Future clinical nanorobotic therapies will typically involve the administration of a cocktail of multiple nanorobot types, some performing the primary mission and others serving in a support role. After treatment is completed the nanorobots may be removed from the body, allowing human nature to resume its erratic but endlessly fascinating journey into the future.

Acknowledgments The author acknowledges private grant support for this work from the Life Extension Foundation, the Alcor Foundation, the Kurzweil Foundation, and the Institute for Molecular Manufacturing, and thanks Greg Fahy for extensive comments on an earlier version of this manuscript.

References

Abd-Al-Aal Z, Ramadan NF, Al-Hoot A (2000) Life-cycle of *Isospora mehlhornii* sp. Nov. (Apicomplexa: Eimeriidae), parasite of the Egyptian swallow *Hirundo rubicola savignii*. Parasitol Res 86:270–278

Adessi C, Frossard MJ, Boissard C et al. (2003) Pharmacological profiles of peptide drug candidates for the treatment of Alzheimer's disease. J Biol Chem 278:13905–13911; http://www.jbc.org/cgi/content/full/278/16/13905

Ahn JI, Terry Canale S, Butler SD et al. (2004) Stem cell repair of physeal cartilage. J Orthop Res 22:1215–1221

Aisen PS (2005) The development of anti-amyloid therapy for Alzheimer's disease: from secretase modulators to polymerisation inhibitors. CNS Drugs 19:989–996

Alberts B, Bray D, Lewis J et al. (1989) The Molecular Biology of the Cell, 2nd edn. Garland Publishing Inc., New York

Allis DG, Drexler KE (2005) Design and analysis of a molecular tool for carbon transfer in mechanosynthesis. J Comput Theor Nanosci 2:45–55; http://e-drexler.com/d/05/00/DC10C-mechanosynthesis.pdf

Almanac, U.S. Casualties in Major Wars: Total Deaths. The 1994 Information Please Almanac, Atlas and Yearbook, 47th edn. Houghton Mifflin Company, New York, p. 385

Alvarez-Chaver P, Rodríguez-Piñeiro AM, Rodríguez-Berrocal FJ et al. (2007) Identification of hydrophobic proteins as biomarker candidates for colorectal cancer. Int J Biochem Cell Biol 39:529–540

Ardekian L, Gaspar R, Peled M et al. (2000) Does low-dose aspirin therapy complicate oral surgical procedures? J Am Dent Assoc 131:331–335

Armstrong RJE, Svendsen CN (2000) Neural stem cells: from cell biology to cell replacement. Cell Transplantation 9:139–152

Armstrong RC, Le TQ, Flint NC et al. (2006) Endogenous cell repair of chronic demyelination. J Neuropathol Exp Neurol 65:245–56

Aspenberg P, Anttila A, Konttinen YT et al. (1996) Benign response to particles of diamond and SiC: bone chamber studies of new joint replacement coating materials in rabbits. Biomaterials 17:807–812

Aspinall R (2010) Chapter 15. Maintenance and Restoration of Immune System Function. In: Fahy GM et al. (eds) The Future of Aging: Pathways to Human Life Extension. Springer, Berlin Heidelberg New York, pp. 489–520

Autexier C, Lue NF (2006) The structure and function of telomerase reverse transcriptase. Annu Rev Biochem 75:493–517

Avraham KB (2002) The genetics of deafness: a model for genomic and biological complexity. Ernst Schering Res Found Workshop 36:271–297

Bakris GL, Bank AJ, Kass DA et al. (2004) Advanced glycation end-product cross-link breakers. A novel approach to cardiovascular pathologies related to the aging process. Am J Hypertens 17:23S–30S

Baldi P (2001) The Shattered Self: The End of Natural Evolution. MIT Press, Cambridge MA

Baracca A, Sgarbi G, Mattiazzi M et al. (2007) Biochemical phenotypes associated with the mitochondrial ATP6 gene mutations at nt8993. Biochim Biophys Acta 1767:913–919

Barbosa HS, Ferreira-Silva MF, Guimaraes EV et al. (2005) Absence of vacuolar membrane involving *Toxoplasma gondii* during its intranuclear localization. J Parasitol 91:182–184

Barzilai N, She L, Liu BQ et al. (1999) Surgical removal of visceral fat reverses hepatic insulin resistance. Diabetes 48:94–98

Basson AB, Terblanche SE, Oelofsen W (1982) A comparative study on the effects of ageing and training on the levels of lipofuscin in various tissues of the rat. Comp Biochem Physiol A 71:369–374

Basu S, Mehreja R, Thiberge S et al. (2004) Spatiotemporal control of gene expression with pulse-generating networks. PNAS 101:6355–6360; http://www.princeton.edu/~rweiss/papers/basu-pulse-2004.pdf

Basu S, Gerchman Y, Collins CH et al. (2005) A synthetic multicellular system for programmed pattern formation. Nature 434:1130–1134; http://www.princeton.edu/~rweiss/papers/basu-nature-2005.pdf

Bateman ED, Emerson RJ, Cole PJ (1982) A study of macrophage-mediated initiation of fibrosis by asbestos and silica using a diffusion chamber technique. Br J Exp Pathol 63:414–425

Bauernschmitt R, Schirmbeck EU, Knoll A et al. (2005) Towards robotic heart surgery: introduction of autonomous procedures into an experimental surgical telemanipulator system. Int J Med Robot 1:74–79

Baxevanis AD, Ouellette BF (eds) (1998) Bioinformatics: A Practical Guide to the Analysis of Genes and Proteins. Wiley-Interscience, New York

Bayoudh S, Mehta M, Rubinsztein-Dunlop H et al. (2001) Micromanipulation of chloroplasts using optical tweezers. J Microsc 203:214–222

Beauséjour CM, Krtolica A, Galimi F et al. (2003) Reversal of human cellular senescence: roles of the p53 and p16 pathways. EMBO J 22:4212–4222

Becker WM, Deamer DW (1991) The World of the Cell, 2nd edn. Benjamin/Cummings Publishing Company, Redwood City, CA

Behkam B, Sitti M (2007) Bacterial flagella-based propulsion and on/off motion control of microscale objects. Appl Phys Lett 90:1–3; http://nanolab.me.cmu.edu/publications/papers/Behkam-APL2007.pdf

Bement WM, Yu HY, Burkel BM et al. (2007) Rehabilitation and the single cell. Curr Opin Cell Biol 19:95–100

Berd D, Kairys J, Dunton C et al. (1998) Autologous, hapten-modified vaccine as a treatment for human cancers. Semin Oncol 25:646–653

Bergamini E (2006) Autophagy: a cell repair mechanism that retards ageing and age-associated diseases and can be intensified pharmacologically. Mol Aspects Med 27:403–410

Bergamini E, Cavallini G, Donati A et al. (2004) The role of macroautophagy in the ageing process, anti-ageing intervention and age-associated diseases. Int J Biochem Cell Biol 36:2392–2404

Berger TW, Glanzman DL (eds) (2005) Toward Replacement Parts for the Brain: Implantable Biomimetic Electronics as Neural Prostheses. MIT Press, Cambridge, MA

Bergers G, Javaherian K, Lo KM et al. (1999) Effects of angiogenesis inhibitors on multistage carcinogenesis in mice. Science 284:808–812

Bi G, Alderton JM, Steinhardt RA (1995) Calcium-regulated exocytosis is required for cell membrane resealing. J Cell Biol 131:1747–1758

Black J (1999) Biological Performance of Materials: Fundamentals of Biocompatibility, 3rd edn. Marcel Dekker, New York

Blackett AD, Hall DA (1981) Tissue vitamin E levels and lipofuscin accumulation with age in the mouse. J Gerontol 36:529–533

Blick RH, Qin H, Kim HS et al. (2007) A nanomechanical computer – exploring new avenues of computing. New J Phys 9:241; http://www.iop.org/EJ/article/1367-2630/9/7/241/njp7_7_241.html

Bliem R, Konopitzky K, Katinger H (1991) Industrial animal cell reactor systems: aspects of selection and evaluation. Adv Biochem Eng Biotechnol 44:1–26

Boland T, Xu T, Damon B et al. (2006) Application of inkjet printing to tissue engineering. Biotechnol J 1:910–917

Boyde A, Jones SJ, Robey PG (2006) Primer on the Metabolic Bone Diseases and Disorders of Mineral Metabolism. American Society for Bone and Mineral Research, Washington, DC; http://www.jbmronline.org/doi/xml/10.1359/prim.2006.0684

Bracci L, Lozzi L, Pini A et al. (2002) A branched peptide mimotope of the nicotinic receptor binding site is a potent synthetic antidote against the snake neurotoxin alpha-bungarotoxin. Biochemistry 41:10194–10199

Brody H (1955) Organization of the cerebral cortex, III. A study of aging in the human cerebral cortex. J Comp Neurol 102:511–556

Brown AJ, Mander EL, Gelissen IC et al. (2000) Cholesterol and oxysterol metabolism and subcellular distribution in macrophage foam cells. Accumulation of oxidized esters in lysosomes. J Lipid Res 41:226–237

Browne WR, Feringa BL (2006) Making molecular machines work. Nat Nanotechnol 1:25–35

Brunet JP, Jourdan N, Cotte-Laffitte J et al. (2000) Rotavirus infection induces cytoskeleton disorganization in human intestinal epithelial cells: implication of an increase in intracellular calcium concentration. J Virol 74:10801–10806

Brunk UT, Jones CB, Sohal RS (1992) A novel hypothesis of lipofuscinogenesis and cellular aging based on interactions between oxidative stress and autophagocytosis. Mutat Res 275: 395–403

Bryan TM, Englezou A, Dalla-Pozza L et al. (1997) Evidence for an alternative mechanism for maintaining telomere length in human tumors and tumor-derived cell lines. Nat Med 3: 1271–1274

Bullen D, Chung S, Wang X et al. (2002) Development of parallel dip pen nanolithography probe arrays for high throughput nanolithography. Symposium LL: Rapid Prototyping Technologies, Materials Research Society Fall Meeting; 2–6 Dec 2002; Boston, MA. Proc. MRS, Vol. 758, 2002; http://mass.micro.uiuc.edu/publications/papers/84.pdf

Cagin T, Jaramillo-Botero A, Gao G (1998) Molecular mechanics and molecular dynamics analysis of Drexler-Merkle gears and neon pump. Nanotechnology 9:143; http://www.wag.caltech.edu/foresight/foresight_1.html

Cameron CL, Fisslthaler B, Sherman A et al. (1989) Studies on contact activation: effects of surface and inhibitors. Med Prog Technol 15:53–62

Campisi J (2003) Consequences of cellular senescence and prospects for reversal. Biogerontology 4:13

Castro MC, Cliquet A (1997) A low-cost instrumented glove for monitoring forces during object manipulation. IEEE Trans Rehabil Eng 5:140–147

Cavallaro EE, Rosen J, Perry JC et al. (2006) Real-time myoprocessors for a neural controlled powered exoskeleton arm. IEEE Trans Biomed Eng 53:2387–2396

Census Bureau (1989a) Series B 201–213. Death Rate, by Age, for Massachusetts: 1865 to 1900. Vital Statistics, Historical Statistics of the United States: Colonial Times to 1970. U.S. Department of Commerce, Bureau of the Census, p. 63 (Series 203)

Census Bureau (1989b) Series B 149–166. Death Rate, for Selected Causes: 1900–1970. Vital Statistics, Historical Statistics of the United States: Colonial Times to 1970. U.S. Department of Commerce, Bureau of the Census, p. 58 (Series 163, 164, and 165)

Census Bureau (1989c) Series H 971–986. Homicides and Suicides: 1900–1970. Vital Statistics, Historical Statistics of the United States: Colonial Times to 1970. U.S. Department of Commerce, Bureau of the Census, p. 414 (Series 972 and 980)

Census Bureau (1989d) Series H 1155–1167. Prisoners Executed Under Civil Authority, by Race and Offense: 1900–1970. Vital Statistics, Historical Statistics of the United States: Colonial Times to 1970. U.S. Department of Commerce, Bureau of the Census, p. 422 (Series 1155); and Series A 6–8. Annual Population Estimates for the United States: 1790–1970. Vital Statistics, Historical Statistics of the United States: Colonial Times to 1970. U.S. Department of Commerce, Bureau of the Census, p. 8 (Series 7)

Census Bureau (2001a) Table 98. Expectation of Life and Expected Deaths by Race, Sex, and Age: 1998. Vital Statistics, Statistical Abstract of the United States: 2001. U.S. Census Bureau, p. 74

Census Bureau (2001b) The 2000 U.S. GDP was $9.896 trillion in constant year 2000 dollars, according to: Table 1340. Gross Domestic Product (GDP) by Country: 1995 to 2000. Comparative International Statistics, Statistical Abstract of the United States: 2001. U.S. Census Bureau, p. 841

Census Bureau (2001c) Table 105. Deaths and Death Rates by Selected Causes: 1990 to 1998. Vital Statistics, Statistical Abstract of the United States: 2001. U.S. Census Bureau, p. 79

Census Bureau (2001d) Table 107. Death by Selected Causes and Selected Characteristics: 1998. Vital Statistics, Statistical Abstract of the United States: 2001. U.S. Census Bureau, p. 81

Chaer RA, Trocciola S, DeRubertis B et al. (2006) Evaluation of the accuracy of a wireless pressure sensor in a canine model of retrograde-collateral (type II) endoleak and correlation with histologic analysis. J Vasc Surg 44:1306–1313

Chambers EJ, Bloomberg GB, Ring SM et al. (1999) Structural studies on the effects of the deletion in the red cell anion exchanger (band 3, AE1) associated with South East Asian ovalocytosis. J Mol Biol 285:1289–1307

Chen KS, Masliah E, Mallory M et al. (1995) Synaptic loss in cognitively impaired aged rats is ameliorated by chronic human nerve growth factor infusion. Neuroscience 68:19–27

Chen X, Kis A, Zettl A et al. (2007) A cell nanoinjector based on carbon nanotubes. Proc Natl Acad Sci USA 104:8218–8222

Cheng R, Feng Q, Argirov OK et al. (2004) Structure elucidation of a novel yellow chromophore from human lens protein. J Biol Chem 279:45441–45449

Cherniak C (1990) The bounded brain: toward quantitative neuroanatomy. J Cognitive Neurosci 2:58–68

Chesnokov SA, Nalimova VA, Rinzler AG et al. (1999) Mechanical energy storage in carbon nanotube springs. Phys Rev Lett 82:343–346

Chow AM, Brown IR (2007) Induction of heat shock proteins in differentiated human and rodent neurons by celastrol. Cell Stress Chaperones 12:237–244

Chrusch DD, Podaima BW, Gordon R (2002) Cytobots: intracellular robotic micromanipulators. In: Kinsner W, Sebak A (eds) Conference Proceedings, 2002 IEEE Canadian Conference on Electrical and Computer Engineering; 2002 May 12–15; Winnipeg, Canada,

IEEE, Winnipeg;http://www.umanitoba.ca/faculties/medicine/radiology/Dick_Gordon_papers/Chrusch,Podaima,Gordon2002

Clark LD, Clark RK, Heber-Katz E (1998) A new murine model for mammalian wound repair and regeneration. Clin Immunol Immunopathol 88:35–45

Cleveland LR, Grimstone AV (1964) The fine structure of the flagellate *Mixotricha* and its associated microorganisms. Proc Roy Soc London B 159:668–686

Cohen S, Behzad AR, Carroll JB et al. (2006) Parvoviral nuclear import: bypassing the host nuclear-transport machinery. J Gen Virol 87:3209–3213

Cohn RD, Campbell KP (2000) Molecular basis of muscular dystrophies. Muscle Nerve 23:1456–1471

Cole E (2007) Fantastic Voyage: Departure 2009. Wired Magazine, 18 January 2007; http://www.wired.com/medtech/health/news/2007/01/72448

Collas P, Håkelien AM (2003) Teaching cells new tricks. Trends Biotechnol 21:354–361

Collins PG, Arnold MS, Avouris P (2001) Engineering carbon nanotubes and nanotube circuits using electrical breakdown. Science 292:706–709

Colombelli J, Reynaud EG, Rietdorf J et al. (2005) In vivo selective cytoskeleton dynamics quantification in interphase cells induced by pulsed ultraviolet laser nanosurgery. Traffic 6:1093–1102

Colombelli J, Reynaud EG, Stelzer EH (2007) Investigating relaxation processes in cells and developing organisms: from cell ablation to cytoskeleton nanosurgery. Methods Cell Biol 82:267–291

Committee to Review the National Nanotechnology Initiative, National Materials Advisory Board (NMAB), National Research Council (NRC) (2006) A Matter of Size: Triennial Review of the National Nanotechnology Initiative. The National Academies Press, Washington, DC; http://www.nap.edu/catalog/11752.html#toc

Cormier M, Johnson B, Ameri M et al. (2004) Transdermal delivery of desmopressin using a coated microneedle array patch system. J Control Release 97:503–511

Corsaro D, Venditti D, Padula M et al. (1999) Intracellular life. Crit Rev Microbiol 25:39–79

Costerton JW, Stewart PS, Greenberg EP (1999) Bacterial biofilms: a common cause of persistent infections. Science 284:1318–1322

Coulman S, Allender C, Birchall J (2006) Microneedles and other physical methods for overcoming the stratum corneum barrier for cutaneous gene therapy. Crit Rev Ther Drug Carrier Syst 23:205–258

Craig EA, Eisenman HC, Hundley HA (2003) Ribosome-tethered molecular chaperones: the first line of defense against protein misfolding? Curr Opin Microbiol 6:157–162

Cruz MA, Chen J, Whitelock JL et al. (2005) The platelet glycoprotein Ib-von Willebrand factor interaction activates the collagen receptor alpha2beta1 to bind collagen: activation-dependent conformational change of the alpha2-I domain. Blood 105:1986–1991

Cui KH (2005) Three concepts of cloning in human beings. Reprod Biomed Online 11:16–17

Cumings J, Zettl A (2000) Low-friction nanoscale linear bearing realized from multiwall carbon nanotubes. Science 289:602–604

Cunningham AP, Love WK, Zhang RW et al. (2006) Telomerase inhibition in cancer therapeutics: molecular-based approaches. Curr Med Chem 13:2875–2888

Das S, Rose G, Ziegler MM et al. (2005) Architectures and simulations for nanoprocessor systems integrated on the molecular scale. In: Cuniberti G, Fagas G, Richter K (eds) Introducing Molecular Electronics. Springer, New York; http://www.mitre.org/work/tech_papers/tech_papers_05/05_0977/05_0977.pdf

Davies I, Fotheringham A, Roberts C (1983) The effect of lipofuscin on cellular function. Mech Ageing Dev 23:347–356

Davis S (1996) Chapter 16. Biomedical applications of particle engineering. In: Coombs RRH, Robinson DW (eds) Nanotechnology in Medicine and the Biosciences. Gordon & Breach Publishers, Netherlands, pp. 243–262

Day JC (1993) Table 2. Projections of the Population, by Age, Sex, Race, and Hispanic Origin, for the United States: 1993 to 2050 (Middle Series) – July 1, 2000, White Male. Population

Projections of the United States, by Age, Sex, Race, and Hispanic Origin: 1993 to 2050, Current Population Reports, Report No. P25-1104, Bureau of the Census, U.S. Department of Commerce, p. 26

De S, Sakmar TP (2002) Interaction of A2E with model membranes. Implications to the pathogenesis of age-related macular degeneration. J Gen Physiol 120:147–157

Death Penalty Information Center (1998) 1998 Year End Report: New Voices Raise Dissent, Executions Decline; http://www.deathpenaltyinfo.org/article.php?scid=45&did=542

Decatur AL, Portnoy DA (2000) A PEST-like sequence in listeriolysin O essential for *Listeria monocytogenes* pathogenicity. Science 290:992–995

de Grey ADNJ (1999) The Mitochondrial Free Radical Theory of Aging. Landes Bioscience, Austin, TX

de Grey ADNJ (2000) Mitochondrial gene therapy: an arena for the biomedical use of inteins. Trends Biotechnol 18:394–399; http://www.sens.org/allo.pdf

de Grey ADNJ (2003) Challenging but essential targets for genuine anti-ageing drugs. Exp Opin Therapeut Targets 37:1–5; http://www.sens.org/manu21.pdf

de Grey ADNJ (2005a) Developing biomedical tools to repair molecular and cellular damage. Foresight Nanotech Institute website; http://foresight.org/challenges/health002.html

de Grey ADNJ (2005b) Reactive oxygen species production in the mitochondrial matrix: implications for the mechanism of mitochondrial mutation accumulation. Rejuvenation Res 8:13–17; http://www.sens.org/AKDH.pdf

de Grey ADNJ (2005c) Forces maintaining organellar genomes: is any as strong as genetic code disparity or hydrophobicity? BioEssays 27:436–446; http://www.sens.org/HH-CDH-PP.pdf

de Grey ADNJ (2005d) Whole-body interdiction of lengthening of telomeres: a proposal for cancer prevention. Front Biosci 10:2420–2429; http://www.sens.org/WILT-FBS.pdf

de Grey ADNJ (2006a) Is SENS a farrago? Rejuvenation Res 9:436–439

de Grey ADNJ (2006b) Foreseeable pharmaceutical repair of age-related extracellular damage. Curr Drug Targets 7:1469–1477; http://www.sens.org/excellPP.pdf

de Grey ADNJ (2006c) Appropriating microbial catabolism: a proposal to treat and prevent neurodegeneration. Neurobiol Aging 27:589–595; http://www.sens.org/NBA-PP.pdf

de Grey ADNJ (2007a) The natural biogerontology portfolio: "defeating aging" as a multi-stage ultra-grand challenge. Ann NY Acad Sci 1100:409–423

de Grey ADNJ (2007b) Protagonistic pleiotropy: Why cancer may be the only pathogenic effect of accumulating nuclear mutations and epimutations in aging. Mech Ageing Dev 128:456–459; http://www.sens.org/nucmutPP.pdf

de Grey ADNJ (2010) Chapter 22. WILT: Necessity, Feasibility, Affordability. In: Fahy GM et al. (eds) The Future of Aging: Pathways to Human Life Extension. Springer, Berlin Heidelberg, New York, pp. 667–684

de Grey ADNJ, Rae M (2007) Ending Aging: The Rejuvenation Breakthroughs That Could Reverse Human Aging in Our Lifetime. St. Martin's Press, New York

de Grey ADNJ, Ames BN, Andersen JK et al. (2002) Time to talk SENS: critiquing the immutability of human aging. Ann NY Acad Sci 959:452–462

de Grey ADNJ, Campbell FC, Dokal I et al. (2004) Total deletion of in vivo telomere elongation capacity: an ambitious but possibly ultimate cure for all age-related human cancers. Annals NY Acad Sci 1019:147–170; http://www.sens.org/WILT.pdf

de Grey ADNJ, Alvarez PJJ, Brady RO et al. (2005) Medical bioremediation: Prospects for the application of microbial catabolic diversity to aging and several major age-related diseases. Ageing Res Rev 4:315–338; http://alvarez.rice.edu/emplibrary/63.pdf

De Koninck P, Schulman H (1998) Sensitivity of CaM kinase II to the frequency of Ca^{2+} oscillations. Science 279:227–230

Delaunay J (1995) Genetic disorders of the red cell membranes. FEBS Lett 369:34–37

Delgado-Olivares L, Zamora-Romo E, Guarneros G et al. (2006) Codon-specific and general inhibition of protein synthesis by the tRNA-sequestering minigenes. Biochimie 88:793–800

Delongeas JL, Netter P, Boz P et al. (1984) Experimental synovitis induced by aluminium phosphate in rabbits. Comparison of the changes produced in synovial tissue and in articular cartilage by aluminium phosphate, carrageenin, calcium hydrogen phosphate dihydrate, and natural diamond powder. Biomed Pharmacother 38:44–48

Dentali F, Ageno W, Imberti D (2006) Retrievable vena cava filters: clinical experience. Curr Opin Pulm Med 12:304–309

Department of Energy (1999) The 1999 GDP for 197 countries was $27.5769 trillion in constant 1990 dollars, according to: Table B2. World Gross Domestic Product at Market Exchange Rates, 1990–1999. Department of Energy (DOE); http://www.eia.doe.gov/emeu/iea/tableb2.html. Using Census Bureau data (Census Bureau, 2001b) to estimate the world GDP deflator, this figure for world GDP is estimated as equivalent to $33.7080 trillion in constant year 2000 dollars, for the year 2000.

Dijkgraaf LC, Liem RS, de Bont LG et al. (1995) Calcium pyrophosphate dihydrate crystal deposition disease. Osteoarthr Cartilage 3:35–45

Ding B, Seeman NC (2006) Operation of a DNA robot arm inserted into a 2D DNA crystalline substrate. Science 314:1583–1585

Donaldson VH, Wagner CJ, Mitchell BH et al. (1998) An HMG-I protein from human endothelial cells apparently is secreted and impairs activation of Hageman factor (factor XII). Proc Assoc Am Physicians 110:140–149

Donati A (2006) The involvement of macroautophagy in aging and anti-aging interventions. Mol Aspects Med 27:455–470

Donati A, Taddei M, Cavallini G et al. (2006) Stimulation of macroautophagy can rescue older cells from 8-OHdG mtDNA accumulation: a safe and easy way to meet goals in the SENS agenda. Rejuvenation Res 9:408–412

Dong Z, Saikumar P, Weinberg JM et al. (2006) Calcium in cell injury and death. Annu Rev Pathol 1:405–434

Dowling P, Meleady P, Dowd A et al. (2007) Proteomic analysis of isolated membrane fractions from superinvasive cancer cells. Biochim Biophys Acta 1774:93–101

Drach LM, Bohl J, Goebel HH (1994) The lipofuscin content of the inferior olivary nucleus in Alzheimer's disease. Dementia 5:234–239

Drexler KE (1986) Engines of Creation: The Coming Era of Nanotechnology. Anchor Press/Doubleday, New York

Drexler KE (1992) Nanosystems: Molecular Machinery, Manufacturing, and Computation. John Wiley & Sons, New York; (a) 13.2, (b) Ch. 12, (c) 16.3.2, (d) 13.2.2, (e) 13.4.1, (f) 10.4.7, (g) 10.7.8, (h) Ch. 8, (i) 8.3.4, (j) Fig. 5.9

Drexler KE (1995) Introduction to Nanotechnology. In: Krummenacker M, Lewis J (eds) Prospects in Nanotechnology: Toward Molecular Manufacturing. John Wiley & Sons, New York, pp. 1–19

Drexler KE, Merkle RC (1996) Simple pump selective for neon. Institute for Molecular Manufacturing website; http://www.imm.org/Parts/Parts1.html or http://www.imm.org/Images/pumpApartC.jpg

Drexler KE, Peterson C, Pergamit G (1991) Unbounding the Future: The Nanotechnology Revolution. William Morrow/Quill Books, New York

Dunn GP, Bruce AT, Ikeda H et al. (2002) Cancer immunoediting: from immunosurveillance to tumor escape. Nat Immunol 3:991–998

Dutt MVG, Childress L, Jiang L et al. (2007) Quantum register based on individual electronic and nuclear spin qubits in diamond. Science 316:1312–1316

Ehninger D, Kempermann G (2008) Neurogenesis in the adult hippocampus. Cell Tissue Res 331:243–250

El Alaoui H, Gresoviac SJ, Vivares CP (2006) Occurrence of the microsporidian parasite *Nucleospora salmonis* in four species of salmonids from the Massif Central of France. Folia Parasitol (Praha) 53:37–43

Elgue G, Sanchez J, Egberg N et al. (1993) Effect of surface-immobilized heparin on the activation of adsorbed factor XII. Artif Organs 17:721–726

Ellis PD, Hadfield KM, Pascall JC et al. (2001) Heparin-binding epidermal-growth-factor-like growth factor gene expression is induced by scrape-wounding epithelial cell monolayers: involvement of mitogen-activated protein kinase cascades. Biochem J 354:99–106

Ellis-Behnke RG, Liang YX, Tay DKC et al. (2007) Using nanotechnology to repair the body. Strategies for Engineered Negligible Senescence (SENS), Third Conference, Queens' College, Cambridge, England, 6–10 September 2007; http://www.sens.org/sens3/abs/Ellis-Behnke.htm

Estrada LD, Yowtak J, Soto C (2006) Protein misfolding disorders and rational design of antimisfolding agents. Methods Mol Biol 340:277–293

Fahy GM (1993) Molecular nanotechnology and its possible pharmaceutical implications. In: Bezold C, Halperin JA, Eng JL (eds) 2020 Visions: Health Care Information Standards and Technologies. U.S. Pharmacopeial Convention Inc., Rockville MD, pp. 152–159

Fahy GM (2003) Apparent induction of partial thymic regeneration in a normal human subject: a case report. J Anti Aging Med 6:219–227

Fahy GM (2007) Method for the prevention of transplant rejection. U.S. Patent No. 7,166,569

Fahy GM (2010) Chapter 6. Precedents for the biological control of aging: Experimental postponement, prevention, and reversal of aging processes. In: Fahy GM et al. (eds) The Future of Aging: Pathways to Human Life Extension. Springer, Berlin Heidelberg, New York, pp. 127–223

Farge D, Turner MW, Roy DR et al. (1986) Dyazide-induced reversible acute renal failure associated with intracellular crystal deposition. Am J Kidney Dis 8:445–449

Farrants GW, Hovmoller S, Stadhouders AM (1988) Two types of mitochondrial crystals in diseased human skeletal muscle fibers. Muscle Nerve 11:45–55

Fauque J, Borgna JL, Rochefort H (1985) A monoclonal antibody to the estrogen receptor inhibits in vitro criteria of receptor activation by an estrogen and an anti-estrogen. J Biol Chem 260:15547–15553

Federal Reserve System (2007) Flow of Funds Accounts of the United States, Flows and Outstandings, Third Quarter 2007, Federal Reserve Statistical Release Z.1, Board of Governors of the Federal Reserve System, Washington DC, 6 December 2007; http://www.federalreserve.gov/releases/z1/Current/z1.pdf. (Net worth 2007 QIII: Household $58.6T, Nonfarm Nonfinancial Corporate Business $15.5T, Nonfarm Noncorporate Business $6.2T.)

Feichtinger M, Zemann W, Kärcher H (2007) Removal of a pellet from the left orbital cavity by image-guided endoscopic navigation. Int J Oral Maxillofac Surg 36:358–361

Felgner H, Grolig F, Muller O et al. (1998) In vivo manipulation of internal cell organelles. Methods Cell Biol 55:195–203

Feng Y, Tian ZM, Wan MX et al. (2007) Protein profile of human hepatocarcinoma cell line SMMC-7721: identification and functional analysis. World J Gastroenterol 13:2608–2614

Fennimore AM, Yuzvinsky TD, Han WQ et al. (2003) Rotational actuators based on carbon nanotubes. Nature 424:408–410

Feynman RP (1960) There's plenty of room at the bottom. Eng Sci (CalTech) 23:22–36

Finch CE (1990) Longevity, Senescence and the Genome. University of Chicago Press, Chicago, IL

Fincher EF, Johannsen L, Kapas L et al. (1996) Microglia digest *Staphylococcus aureus* into low molecular weight biologically active compounds. Am J Physiol 271:R149–R156

Finder VH, Glockshuber R (2007) Amyloid-beta aggregation. Neurodegener Dis 4:13–27

Firtel M, Henderson G, Sokolov I (2004) Nanosurgery: observation of peptidoglycan strands in *Lactobacillus helveticus* cell walls. Ultramicroscopy 101:105–109

Fischer JA (2000) Molecular motors and developmental asymmetry. Curr Opin Genet Dev 10:489–496

Fleischer C, Wege A, Kondak K et al. (2006) Application of EMG signals for controlling exoskeleton robots. Biomed Tech (Berl) 51:314–319

Flemming JH, Ingersoll D, Schmidt C et al. (2005) In vivo electrochemical immunoassays using ElectroNeedles. 2005 NSTI Nanotechnology Conference; http://www.nsti.org/Nanotech2005/showabstract.html?absno=1112

Flores R (2001) A naked plant-specific RNA ten-fold smaller than the smallest known viral RNA: the viroid. CR Acad Sci III 324:943–952

Franke R, Hirsch T, Overwin H et al. (2007) Synthetic mimetics of the CD4 binding site of HIV-1 gp120 for the design of immunogens. Angew Chem Int Ed Engl 46:1253–1255

Freitas RA Jr (1996a) Dermal zippers. Institute for Molecular Manufacturing, internal design study

Freitas RA Jr (1996b) The End of Heart Disease. Institute for Molecular Manufacturing, internal design study

Freitas RA Jr (1998) Exploratory design in medical nanotechnology: A mechanical artificial red cell. Artif Cells Blood Subst Immobil Biotech 26:411–430; http://www.foresight.org/Nanomedicine/Respirocytes.html

Freitas RA Jr (1999) Nanomedicine, Volume I: Basic Capabilities. Landes Bioscience, Georgetown, TX; http://www.nanomedicine.com/NMI.htm; (a) 2.1, (b) Ch. 3, (c) Ch. 4, (d) Ch. 5, (e) Ch. 6, (f) Ch. 7, (g) Ch. 8, (h) 9.3, (i) 9.4, (j) 10.1, (k) 10.2, (m) 8.5.1, (n) 6.4.1, (o) 3.4.2, (p) 3.5, (q) 4.2, (r) 4.3, (s) 4.4, (t) 4.5, (u) 4.6, (v) 4.7, (w) 4.8.2, (x) 9.4.5, (y) 10.4.2, (z) 9.3.5.3.6, (aa) Table 9.3, (ab) 10.5, (ac) 6.3.1, (ad) 6.3.7, (ae) 9.3.1, (af) 5.3.1.4, (ag) 10.4.1.1, (ah) 7.3, (ai) 10.2.3, (aj) 10.4.2.4.2, (ak) 8.5.2.2, (am) 9.3.1.4, (an) 9.3.5.1, (ao) 6.4.3.4, (ap) 7.2.5.4, (aq) 6.3.4.4, (ar) 6.3.4.5, (as) 3.4.1, (at) 9.4.4, (au) 9.4.4.2, (av) 8.4.1.4, (aw) 8.4.3, (ax) 5.2.1, (ay) 8.2.1.2, (az) 5.4, (ba) 5.3.3, (bb) 5.2.4.2, (bc) 9.4.3, (bd) 9.4.3.2, (be) 9.3.2, (bf) 9.4.3.5, (bg) 5.4.2, (bh) 10.3, (bi) 6.3.4, (bj) 4.2.1, (bk) 7.2.2, (bm) 10.3.6, (bn) 9.2.6, (bo) 9.2.4, (bp) 9.2.5, (bq) 9.4.4.1, (br) 8.5.2.1, (bs) 8.2.1.1, (bt) 9.4.1.4, (bu) 5.2.5, (bv) 9.4.4.4, (bw) 9.4.2.6, (bx) 9.4.5.3, (by) 9.4.4.3, (bz) 9.4.6, (ca) 8.5.3.12, (cb) 8.5.4.7, (cc) 10.4.2.1, (cd) 9.2.7, (ce) 9.4.7, (cf) 4.8.6, (cg) 7.3.1, (ch) Table 3.4, (ci) 7.4.5.2, (cj) 4.2.3, (ck) 6.3.3, (cm) 6.5.2, (cn) 4.2.9, (co) 4.8.1, (cp) 4.2.5, (cq) 10.4.1.2, (cr) 10.4.2.5.2, (cs) 4.8.6.4, (ct) 1.3.3, (cu) 5.3.6, (cv) 4.2.8, (cw) 8.3.2

Freitas RA Jr (2000a) Clottocytes: artificial mechanical platelets. IMM Report No. 18, Foresight Update No. 41, pp. 9–11; http://www.imm.org/Reports/Rep018.html

Freitas RA Jr (2000b) Nanodentistry. J Amer Dent Assoc 131:1559–1566; http://www.rfreitas.com/Nano/Nanodentistry.htm

Freitas RA Jr (2002) "Death is an Outrage!" invited lecture, 5th Alcor Conference on Extreme Life Extension, 16 November 2002, Newport Beach, CA; http://www.rfreitas.com/Nano/DeathIsAnOutrage.htm

Freitas RA Jr (2003) Nanomedicine, Volume IIA: Biocompatibility. Landes Bioscience, Georgetown, TX, 2003; http://www.nanomedicine.com/NMIIA.htm; (a) 15.4.3.1, (b) 15.2.3.6, (c) 15.2.4, (d) 15.4.3.6, (e) 15.4.3.1, (f) 15.4.3.6.1, (g) 15.4.3.6.2, (h) 15.4.3.6.3, (i) 15.4.3.6.4, (j) 15.4.3.6.5, (k) 15.4.3.6.6, (m) 15.4.3.6.7, (n) 15.4.3.6.8, (o) 15.4.3.6.9, (p) 15.4.3.6.10, (q) 15.4.3.5, (r) 15.4.2.3, (s) 15.4.2, (t) 15.6.2, (u) 15.5.7.2.3, (v) 15.5.2.1, (w) 15.2.2.4, (x) 15.5.2.3, (y) 15.6.3.5, (z) 15.5.7.2.4, (aa) 15.3.6.5, (ab) 15.2.2

Freitas RA Jr (2005a) "What is Nanomedicine?" Nanomedicine: Nanotech Biol Med 1:2–9; http://www.nanomedicine.com/Papers/WhatIsNMMar05.pdf

Freitas RA Jr (2005b) Microbivores: Artificial mechanical phagocytes using digest and discharge protocol. J Evol Technol 14:1–52; http://www.jetpress.org/volume14/freitas.html

Freitas RA Jr (2005c) Current status of nanomedicine and medical nanorobotics. J Comput Theor Nanosci 2:1–25; http://www.nanomedicine.com/Papers/NMRevMar05.pdf

Freitas RA Jr (2005d) A simple tool for positional diamond mechanosynthesis, and its method of manufacture. U.S. Provisional Patent Application No. 60/543,802, filed 11 February 2004; U.S. Patent 7,687,146, issued 30 March 2010; http://www.MolecularAssembler.com/Papers/DMSToolbuildProvPat.htm

Freitas RA Jr (2006a) Pharmacytes: an ideal vehicle for targeted drug delivery. J Nanosci Nanotechnol 6:2769–2775; http://www.nanomedicine.com/Papers/JNNPharm06.pdf

Freitas RA Jr (2006b) Economic impact of the personal nanofactory. Nanotechnol Perceptions Rev Ultraprecision Eng Nanotechnol 2:111–126; http://www.rfreitas.com/Nano/NoninflationaryPN.pdf

Freitas RA Jr (2007) The ideal gene delivery vector: Chromallocytes, cell repair nanorobots for chromosome replacement therapy. J Evol Technol 16:1–97; http://jetpress.org/v16/freitas.pdf

Freitas RA Jr (2009) Computational tasks in medical nanorobotics. In: Eshaghian-Wilner MM (ed) Bio-inspired and Nano-scale Integrated Computing. Wiley, New York; http://www.nanomedicine.com/Papers/NanorobotControl2009.pdf

Freitas RA Jr, Merkle RC (2004) Kinematic Self-Replicating Machines. Landes Bioscience, Georgetown, TX; http://www.molecularassembler.com/KSRM.htm; (a) Section 5.7, (b) 5.9.4, (c) 3.8, (d) 4.20

Freitas RA Jr, Merkle RC (2008) A minimal toolset for positional diamond mechanosynthesis. J Comput Theor Nanosci 5: 760–861; http://www.molecularassembler.com/Papers/MinToolset.pdf

Freitas RA Jr, Phoenix CJ (2002) Vasculoid: A personal nanomedical appliance to replace human blood. J Evol Technol 11:1–139; http://www.jetpress.org/volume11/vasculoid.pdf

Freitas RA Jr, Allis DG, Merkle RC (2007) Horizontal Ge-substituted polymantane-based C_2 dimer placement tooltip motifs for diamond mechanosynthesis. J Comput Theor Nanosci 4:433–442; http://www.MolecularAssembler.com/Papers/DPTMotifs.pdf

Frey M, Koller T, Lezzi M (1982) Isolation of DNA from single microsurgically excised bands of polytene chromosomes of *Chironomus*. Chromosoma 84:493–503

Friedenstein AJ, Gorskaja JF, Kulagina NN (1976) Fibroblast precursors in normal and irradiated mouse hematopoietic organs. Exp Hematol 4:267–274

Frost EE, Nielsen JA, Le TQ et al. (2003) PDGF and FGF2 regulate oligodendrocyte progenitor responses to demyelination. J Neurobiol 54:457–472

Fuchs E (1996) The cytoskeleton and disease: genetic disorders of intermediate filaments. Annu Rev Genet 30:197–231

Fuhrer G, Gallimore MJ, Heller W (1990) FXII. Blut 61:258–266

Fujiwara K, Takahashi K, Shirota K et al. (1981) Fine pathology of mouse spinal cord infected with the Tyzzer organism. Jpn J Exp Med 51:171–178

Furber JD (2006) Extracellular glycation crosslinks: prospects for removal. Rejuvenation Res. 9:274–278

Galou M, Gao J, Humbert J (1997) The importance of intermediate filaments in the adaptation of tissues to mechanical stress: evidence from gene knockout studies. Biol Cell 89:85–97

Garibotti AV, Liao S, Seeman NC (2007) A simple DNA-based translation system. Nano Lett 7:480–483

Gearing DP, Nagley P (1986) Yeast mitochondrial ATPase subunit 8, normally a mitochondrial gene product, expressed in vitro and imported back into the organelle. EMBO J 5:3651–3655

Georgatos SD, Theodoropoulos PA (1999) Rules to remodel by: what drives nuclear envelope disassembly and reassembly during mitosis? Crit Rev Eukaryot Gene Expr 9:373–381

Gersh BJ, Simari RD (2006) Cardiac cell-repair therapy: clinical issues. Nat Clin Pract Cardiovasc Med 3:S105-S109

Giannoudis PV, Pountos I, Pape HC et al. (2007) Safety and efficacy of vena cava filters in trauma patients. Injury 38:7–18

Glozman D, Shoham M (2006) Flexible needle steering for percutaneous therapies. Comput Aided Surg 11:194–201

Goddard WA III (1995) Computational Chemistry and Nanotechnology. Presentation at the 4th Foresight Conference on Molecular Nanotechnology, November 1995

Gomord V, Denmat LA, Fitchette-Lainé AC et al. (1997) The C-terminal HDEL sequence is sufficient for retention of secretory proteins in the endoplasmic reticulum (ER) but promotes vacuolar targeting of proteins that escape the ER. Plant J 11:313–325

Gordon KE, Ferris DP (2007) Learning to walk with a robotic ankle exoskeleton. J Biomech 40:2636–2644

Gouaux E, Mackinnon R (2005) Principles of selective ion transport in channels and pumps. Science 310:1461–1465

Gouverneur M, Berg B, Nieuwdorp M et al. (2006) Vasculoprotective properties of the endothelial glycocalyx: effects of fluid shear stress. J Intern Med 259:393–400

Green JE, Choi JW, Boukai A et al. (2007) A 160-kilobit molecular electronic memory patterned at 10(11) bits per square centimetre. Nature 445:414–417

Grobbee DE, Bots ML (2004) Atherosclerotic disease regression with statins: studies using vascular markers. Int J Cardiol 96:447–459

Groves JD, Wang L, Tanner MJ (1998) Functional reassembly of the anion transport domain of human red cell band 3 (AE1) from multiple and non-complementary fragments. FEBS Lett 433:223–227

Grudzien E, Stepinski J, Jankowska-Anyszka M et al. (2004) Novel cap analogs for in vitro synthesis of mRNAs with high translational efficiency. RNA 10:1479–1487

Guan JZ, Maeda T, Sugano M et al. (2007) Change in the telomere length distribution with age in the Japanese population. Mol Cell Biochem 304:353–360

Guardavaccaro D, Pagano M (2006) Stabilizers and destabilizers controlling cell cycle oscillators. Mol Cell 22:1–4

Guet CC, Elowitz MB, Hsing W (2002) Combinatorial synthesis of genetic networks. Science 296:1466–1470; http://www.aph.caltech.edu/people/CombinatorialNetworks.pdf

Gurdon JB (2006) Nuclear transplantation in *Xenopus*. Methods Mol Biol 325:1–9

Hackstadt T (1996) The biology of rickettsiae. Infect Agents Dis 5:127–143

Hadi HA, Carr CS, Al Suwaidi J (2005) Endothelial dysfunction: cardiovascular risk factors, therapy, and outcome. Vasc Health Risk Manag 1:183–198

Hagios C, Lochter A, Bissell MJ (1998) Tissue architecture: the ultimate regulator of epithelial function? Philos Trans R Soc Lond B Biol Sci 353:857–870

Hall JS (1993) Utility fog: A universal physical substance. Vision-21: Interdisciplinary Science and Engineering in the Era of Cyberspace, Westlake OH. NASA Conference Publication CP-10129, pp. 115–126

Hall JS (1996) Utility fog: The stuff that dreams are made of. In: Crandall BC (ed) Nanotechnology: Molecular Speculations on Global Abundance. MIT Press, Cambridge, MA, pp. 161–184; http://www.kurzweilai.net/meme/frame.html?main=/articles/art0220.html?m%3D7

Han SW, Nakamura C, Obataya I et al. (2005) A molecular delivery system by using AFM and nanoneedle. Biosens Bioelectron 20:2120–2125

Hanly EJ, Miller BE, Kumar R et al. (2006) Mentoring console improves collaboration and teaching in surgical robotics. J Laparoendosc Adv Surg Tech A 16:445–451

Harman D (1972) The biologic clock: the mitochondria? J Am Geriatr Soc 20:145–147

Harman D (1989) Lipofuscin and ceroid formation: the cellular recycling system. Adv Exp Med Biol 266:3–15

Harrison DE, Astle CM (1982) Loss of stem cell repopulating ability upon transplantation. Effects of donor age, cell number, and transplantation procedure. J Exp Med 156:1767–1779

Hayashi JI, Ohtall S, Kagawall Y et al. (1994) Nuclear but not mitochondrial genome involvement in human age-related mitochondrial dysfunction: Functional integrity of mitochondrial DNA from aged subjects. J Biol Chem 269:6878–6883

Heath JR (2000) Wires, switches, and wiring. A route toward a chemically assembled electronic nanocomputer. Pure Appl Chem 72:11–20; http://www.iupac.org/publications/pac/2000/7201/7201pdf/2_heath.pdf

Hedenborg M, Klockars M (1989) Quartz-dust-induced production of reactive oxygen metabolites by human granulocytes. Lung 167:23–32

Hein S, Kostin S, Heling A et al. (2000) The role of the cytoskeleton in heart failure. Cardiovasc Res 45:273–278

Heinzelmann-Schwarz VA, Gardiner-Garden M, Henshall SM et al. (2004) Overexpression of the cell adhesion molecules DDR1, Claudin 3, and Ep-CAM in metaplastic ovarian epithelium and ovarian cancer. Clin Cancer Res 10:4427–4436

Heisterkamp A, Maxwell IZ, Mazur E et al. (2005) Pulse energy dependence of subcellular dissection by femtosecond laser pulses. Opt Express 13:3690–3696

Henry M, Polydorou A, Henry I et al. (2007) New distal embolic protection device the FiberNet 3 dimensional filter: first carotid human study. Catheter Cardiovasc Interv 69:1026–1035

Hertzendorf LR, Stehling L, Kurec AS et al. (1987) Comparison of bleeding times performed on the arm and the leg. Am J Clin Pathol 87:393–396

Heydorn AO, Mehlhorn H (1987) Fine structure of *Sarcocystis arieticanis* Heydorn, 1985 in its intermediate and final hosts (sheep and dog). Zentralbl Bakteriol Mikrobiol Hyg A 264: 353–362

Higson FK , Jones OT (1984) Oxygen radical production by horse and pig neutrophils induced by a range of crystals. J Rheumatol 11:735–740

Hinek A (1997) Elastin receptor and cell-matrix interactions in heart transplant-associated arteriosclerosis. Arch Immunol Ther Exp (Warsz) 45:15–29

Ho KL (1987) Crystalloid bodies in skeletal muscle of hypothyroid myopathy. Ultrastructural and histochemical studies. Acta Neuropathol (Berl) 74:22–32

Holmbeck K, Bianco P, Yamada S et al. (2004) MT1-MMP: a tethered collagenase. J Cell Physiol 200:11–19

Holmlin RE, Schiavoni M, Chen CY et al. (2000) Light-driven microfabrication: assembly of multicomponent, three-dimensional structures by using optical tweezers. Angew Chem Intl. Ed. 39:3503–3506

Huang TJ, Lu W, Tseng HR et al. (2003) Molecular shuttle switching in closely packed Langmuir films, 11th Foresight Conf. Mol. Nanotech., San Francisco, CA, 10–12 October 2003

Hull RL, Westermark GT, Westermark P et al. (2004) Islet amyloid: a critical entity in the pathogenesis of type 2 diabetes. J Clin Endocrinol Metab 89:3629–3643

Ichikawa O, Osawa M, Nishida N et al. (2007) Structural basis of the collagen-binding mode of discoidin domain receptor 2. EMBO J. 26:4168–4176

Imberti D, Ageno W, Carpenedo M (2006) Retrievable vena cava filters: a review. Curr Opin Hematol 13:351–356

Ishiyama K, Sendoh M, Arai KI (2002) Magnetic micromachines for medical applications. J Magnetism Magnetic Mater 242–245:1163–1165

Jakab K, Neagu A, Mironov V et al. (2004) Organ printing: fiction or science. Biorheology 41: 371–375

Jakab K, Damon B, Neagu A et al. (2006) Three-dimensional tissue constructs built by bioprinting. Biorheology 43:509–513

James TJ, Sharma SP (1995) Regional and lobular variation in neuronal lipofuscinosis in rat cerebellum: influence of age and protein malnourishment. Gerontology 41:213–228

James TJ, Sharma SP, Gupta SK et al. (1992) 'Dark' cell formation under protein malnutrition: process of conversion and concept of 'semi-dark' type Purkinje cells. Indian J Exp Biol 30: 470–473

Jeffries GD, Edgar JS, Zhao Y et al. (2007) Using polarization-shaped optical vortex traps for single-cell nanosurgery. Nano Lett 7:415–420

Jeon KW, Lorch IJ, Danielli JF (1970) Reassembly of living cells from dissociated components. Science 167:1626–1627

Johnson LR, Scott MG, Pitcher JA (2004) G protein-coupled receptor kinase 5 contains a DNA-binding nuclear localization sequence. Mol Cell Biol 24:10169–10179

Johnston SC, Riddle SM, Cohen RE et al. (1999) Structural basis for the specificity of ubiquitin C-terminal hydrolases. EMBO J 18:3877–3887

Jones BJ, Murphy CR (1994) A high resolution study of the glycocalyx of rat uterine epithelial cells during early pregnancy with the field emission gun scanning electron microscope. J Anat 185:443–446

Kamnasaran D, Muir WJ, Ferguson-Smith MA et al. (2003) Disruption of the neuronal PAS3 gene in a family affected with schizophrenia. J Med Genet 40:325–332

Kano F, Nagayama K, Murata M (2000) Reconstitution of the Golgi reassembly process in semi-intact MDCK cells. Biophys Chem 84:261–268

Kapoor A, Li M, Taylor RH (2005) Spatial motion constraints for robot assisted suturing using virtual fixtures. Med Image Comput Comput Assist Interv Int Conf Med Image Comput Comput Assist Interv 8:89–96

Kapyla J, Pentikainen OT, Nyronen T et al. (2007) Small molecule designed to target metal binding site in the alpha2I domain inhibits integrin function. J Med Chem 50:2742–2746

Kasirajan K, Marek JM, Langsfeld M (2001) Mechanical thrombectomy as a first-line treatment for arterial occlusion. Semin Vasc Surg 14:123–131

Katz RD, Taylor JA, Rosson GD et al. (2006) Robotics in plastic and reconstructive surgery: use of a telemanipulator slave robot to perform microvascular anastomoses. J Reconstr Microsurg 22:53–57

Kaushik S, Cuervo AM (2006) Autophagy as a cell-repair mechanism: activation of chaperone-mediated autophagy during oxidative stress. Mol Aspects Med 27:444–454

Kawashima R, Kawamura YI, Kato R et al. (2006) IL-13 receptor alpha2 promotes epithelial cell regeneration from radiation-induced small intestinal injury in mice. Gastroenterology 131: 130–141

Kay ER, Leigh DA, Zerbetto F (2007) Synthetic molecular motors and mechanical machines. Angew Chem Int Ed 46:72–191

Keane FM, Loughman A, Valtulina V et al. (2007) Fibrinogen and elastin bind to the same region within the A domain of fibronectin binding protein A, an MSCRAMM of *Staphylococcus aureus*. Mol Microbiol 63:711–723

Keats BJ, Corey DP (1999) The usher syndromes. Am J Med Genet 89:158–166

Kelly TR, De Silva H, Silva RA (1999) Unidirectional rotary motion in a molecular system. Nature 401:150–152

Kenny T (2007) Tip-Based Nanofabrication (TBN). Defense Advanced Research Projects Agency (DARPA)/Microsystems Technology Office (MTO), Broad Agency Announcement BAA 07–59; http://www.fbo.gov/spg/ODA/DARPA/CMO/BAA07-59/listing.html

Kessler N, Perl-Treves D, Addadi L et al. (1999) Structural and chemical complementarity between antibodies and the crystal surfaces they recognize. Proteins 34:383–394

Kida E, Golabek AA, Wisniewski KE (2001) Cellular pathology and pathogenic aspects of neuronal ceroid lipofuscinoses. Adv Genet 45:35–68

Kilhamn J (2003) The protective mucus and its mucins produced by enterocytes and goblet cells. Centrum for Gastroenterologisk Forskning, Goteborgs Universitet; http://www.cgf.gu.se/fouschema.html

Kim PS, Arvan P (1998) Endocrinopathies in the family of endoplasmic reticulum (ER) storage diseases: disorders of protein trafficking and the role of ER molecular chaperones. Endocr Rev 19:173–202

Kirson ED, Yaari Y (2000) A novel technique for micro-dissection of neuronal processes. J Neurosci Methods 98:119–122

Knight CG, Lorincz A, Cao A et al. (2005) Computer-assisted, robot-enhanced open microsurgery in an animal model. J Laparoendosc Adv Surg Tech A 15:182–185

Knutson KL (2002) GMK (Progenics Pharmaceuticals). Curr Opin Investig Drugs 3:159–164

Koenig W, Khuseyinova N (2007) Biomarkers of atherosclerotic plaque instability and rupture. Arterioscler Thromb Vasc Biol 27:15–26

Kohli V, Elezzabi AY, Acker JP (2005) Cell nanosurgery using ultrashort (femtosecond) laser pulses: Applications to membrane surgery and cell isolation. Lasers Surg Med 37:227–230

Konig K, Riemann I, Fischer P et al. (1999) Intracellular nanosurgery with near infrared femtosecond laser pulses. Cell Mol Biol 45:195–201

Korf BR, Diacumakos EG (1978) Microsurgically-extracted metaphase chromosomes of the Indian muntjac examined with phase contrast and scanning electron microscopy. Exp Cell Res 111: 83–93

Korf BR, Diacumakos EG (1980) Absence of true interchromosomal connectives in microsurgically isolated chromosomes. Exp Cell Res 130:377–385

Kosaka K (2000) Diffuse Lewy body disease. Neuropathology 20:S73–S78

Koumura N, Zijlstra RW, van Delden RA et al. (1999) Light-driven monodirectional molecular rotor. Nature 401:152–155

Kristo B, Liao JC, Neves HP et al. (2003) Microelectromechanical systems in urology. Urology 61:883–887; http://liaolab.stanford.edu/images/paper5_urology_2003.pdf

Ku J, Mraz R, Baker N et al. (2003) A data glove with tactile feedback for FMRI of virtual reality experiments. Cyberpsychol Behav 6:497–508

Kuczera K (1996) Free energy simulations of axial contacts in sickle-cell hemoglobin. Biopolymers 39:221–242

Kumar R, Ansell JE, Canoso RT et al. (1978) Clinical trial of a new bleeding-time device. Am J Clin Pathol 70:642–645

Kumar S, Yin X, Trapp BD et al. (2002) Role of long-range repulsive forces in organizing axonal neurofilament distributions: evidence from mice deficient in myelin-associated glycoprotein. J Neurosci Res 68:681–690; http://hohlab.bs.jhmi.edu/Hoh_lab_Media/neuro_JNR.pdf

Landauer TK (1986) How much do people remember? Some estimates of the quantity of learned information in long-term memory. Cognitive Sci 10:477–493

Lapidus IR, Schiller R (1978) A model for traveling bands of chemotactic bacteria. Biophys J 22:1–13

Lasker Foundation (2000) Exceptional Returns: The Economic Value of America's Investment in Medical Research. Funding First Reports, Lasker Medical Research Network, May 2000, p. 5; http://www.laskerfoundation.org/reports/pdf/exceptional.pdf

Lederer MO, Bühler HP (1999) Cross-linking of proteins by Maillard processes – characterization and detection of a lysine-arginine cross-link derived from D-glucose. Bioorg Med Chem 7:1081–1088

Lee HJ, Ho W (1999) Single bond formation and characterization with a scanning tunneling microscope. Science 286:1719–1722; http://www.physics.uci.edu/%7Ewilsonho/stm-iets.html

Lee S, Desai KK, Iczkowski KA et al. (2006) Coordinated peak expression of MMP-26 and TIMP-4 in preinvasive human prostate tumor. Cell Res 16:750–758

Lee VM (1995) Disruption of the cytoskeleton in Alzheimer's disease. Curr Opin Neurobiol 5:663–668

Leferovich JM, Bedelbaeva K, Samulewicz S et al. (2001) Heart regeneration in adult MRL mice. Proc Natl Acad Sci USA 98:9830–9835; http://www.pnas.org/cgi/content/full/98/17/9830

Lehrer RI, Ganz T (1995) Chapter 87. Biochemistry and function of monocytes and macrophages. In: William's Hematology, 5th edn. McGraw-Hill, New York, pp. 869–875

Leigh DA, Wong JKY, Dehez F et al. (2003) Unidirectional rotation in a mechanically interlocked molecular rotor. Nature 424:174–179

Lemler MS, Bies RD, Frid MG et al. (2000) Myocyte cytoskeletal disorganization and right heart failure in hypoxia-induced neonatal pulmonary hypertension. Am J Physiol Heart Circ Physiol 279:H1365–H1376

Li RA, Jensen J, Bowersox JC (2000) Microvascular anastomoses performed in rats using a microsurgical telemanipulator. Comput Aided Surg 5:326–332

Li WJ, Xi N (2004) Novel micro gripping, probing, and sensing devices for single-cell surgery. Conf Proc IEEE Eng Med Biol Soc 4:2591–2594

Liang X, Zhao J, Hajivandi M et al. (2006) Quantification of membrane and membrane-bound proteins in normal and malignant breast cancer cells isolated from the same patient with primary breast carcinoma. J Proteome Res 5:2632–2641

Lind SE (1995) Chapter 33. The Hemostatic System. In: Handin RI, Stossel TP, Lux SE (eds) Blood: Principles and Practice of Hematology. J.B. Lippincott Co., Philadelphia, PA, pp. 949–972

Lipinski MJ, Fuster V, Fisher EA et al. (2004) Technology insight: targeting of biological molecules for evaluation of high-risk atherosclerotic plaques with magnetic resonance imaging. Nat Clin Pract Cardiovasc Med 1:48–55

Liu SC, Derick LH, Agre P et al. (1990) Alteration of the erythrocyte membrane skeletal ultrastructure in hereditary spherocytosis, hereditary elliptocytosis, and pyropoikilocytosis. Blood 76:198–205

Livingston P (1998) Ganglioside vaccines with emphasis on GM2. Semin Oncol 25:636–645

Lomer WM (1951) A dislocation reaction in the face-centered cubic lattice. Phil Mag 42: 1327–1331

Lopez I, Honrubia V, Baloh RW (1997) Aging and the human vestibular nucleus. J Vestib Res 7:77–85

Lovelock JE, Smith AU (1956) Studies on golden hamsters during cooling to and rewarming from body temperatures below 0 degrees C. Proc R Soc Lond B Biol Sci 145:391–442

Lu SJ, Feng Q, Caballero S et al. (2007) Generation of functional hemangioblasts from human embryonic stem cells. Nature Methods 4:501–509

Lynnerup N, Kjeldsen H, Heegaard S et al. (2008) Radiocarbon dating of the human eye lens crystallines reveal proteins without carbon turnover throughout life. PLoS ONE 3:e1529; http://www.plosone.org/article/info:doi/10.1371/journal.pone.0001529

Malorni W, Matarrese P, Tinari A et al. (2007) Xeno-cannibalism: a survival "escamotage". Autophagy 3:75–77

Malyankar UM (2007) Tumor-associated antigens and biomarkers in cancer and immune therapy. Int Rev Immunol 26:223–247

Maniotis AJ, Bojanowski K, Ingber DE (1997) Mechanical continuity and reversible chromosome disassembly within intact genomes removed from living cells. J Cell Biochem 65:114–130

Mann DJ, Peng J, Freitas RA Jr et al. (2004) Theoretical analysis of diamond mechanosynthesis. Part II. C_2 mediated growth of diamond C(110) surface via Si/Ge-triadamantane dimer placement tools. J Comput Theor Nanosci 1:71–80; http://www.MolecularAssembler.com/JCTNMannMar04.pdf

Marrink SJ, Lindahl E, Edholm O et al. (2001) Simulation of the spontaneous aggregation of phospholipids into bilayers. J Am Chem Soc 123:8638–8639

Marszalek JR, Ruiz-Lozano P, Roberts E et al. (1999) Situs inversus and embryonic ciliary morphogenesis defects in mouse mutants lacking the KIF3A subunit of kinesin-II. Proc Natl Acad Sci USA 96:5043–5048; http://www.pnas.org/cgi/content/full/96/9/5043

Martel S, Hunter I (2002) Nanofactories based on a fleet of scientific instruments configured as miniature autonomous robots. Proc 3rd Intl Workshop on Microfactories; 16–18 Sep 2002; Minneapolis, MN, pp. 97–100

Martel S, Mathieu JB, Felfoul O et al. (2007) Automatic navigation of an untethered device in the artery of a living animal using a conventional clinical magnetic resonance imaging system. Appl Phys Lett 90:114105; http://wiki.polymtl.ca/nano/fr/images/1/14/J-2007-MRSUB-APL-Sylvain2.pdf

Marty AM, Jahrling PB, Geisbert TW (2006) Viral hemorrhagic fevers. Clin Lab Med 26:345–386

Mathieu JB, Martel S, Yahia L et al. (2005) MRI systems as a mean of propulsion for a microdevice in blood vessels. Biomed Mater Eng 15:367

Matsumuro K, Izumo S, Minauchi Y et al. (1993) Chronic demyelinating neuropathy and intra-axonal polyglucosan bodies. Acta Neuropathol (Berl) 86:95–99

Mattson MP (2004) Pathways towards and away from Alzheimer's disease. Nature 430:631–639

Mayes PA (1993) Metabolism of acylglycerols and sphingolipids. In: Murray RK, Granner DK, Mayes PA, Rodwell VW (eds) (1993) Harper's Biochemistry, 23rd edn. Appleton & Lange, Norwalk, CT, pp. 241–249

Mazzatenta A, Giugliano M, Campidelli S et al. (2007) Interfacing neurons with carbon nanotubes: electrical signal transfer and synaptic stimulation in cultured brain circuits. J Neurosci 27: 6931–6936

McColl R, Brown I, Seligman C et al. (2006) Haptic rendering for VR laparoscopic surgery simulation. Australas Phys Eng Sci Med 29:73–78

McGuigan AP, Sefton MV (2006) Vascularized organoid engineered by modular assembly enables blood perfusion. Proc Natl Acad Sci USA 103:11461–11466

McHolm GB, Aguilar MJ, Norris FH (1984) Lipofuscin in amyotrophic lateral sclerosis. Arch Neurol 41:1187–1188

McNulty R, Wang H, Mathias RT et al. (2004) Regulation of tissue oxygen levels in the mammalian lens. J Physiol 559:883–898; http://jp.physoc.org/cgi/content/full/559/3/883

McQuibban GA, Gong JH, Tam EM et al. (2000) Inflammation dampened by gelatinase A cleavage of monocyte chemoattractant protein-3. Science 289:1202–1206

McWilliams A (2006) Nanotechnology: A Realistic Market Assessment. BCC Research, July 2006; http://www.bccresearch.com/nan/NAN031B.asp

Meissner A, Jaenisch R (2006) Mammalian nuclear transfer. Dev Dyn 235:2460–2469

Menciassi A, Quirini M, Dario P (2007) Microrobotics for future gastrointestinal endoscopy. Minim Invasive Ther Allied Technol 16:91–100

Méndez-Vilas A, Gallardo-Moreno AM, González-Martín ML (2006) Nano-mechanical exploration of the surface and sub-surface of hydrated cells of *Staphylococcus epidermidis*. Antonie Van Leeuwenhoek 89:373–386

Merkle RC (1997) A proposed 'metabolism' for a hydrocarbon assembler. Nanotechnology 8: 149–162; http://www.zyvex.com/nanotech/hydroCarbonMetabolism.html

Merkle RC, Freitas RA Jr (2003) Theoretical analysis of a carbon-carbon dimer placement tool for diamond mechanosynthesis. J Nanosci Nanotechnol 3:319–324; http://www.rfreitas.com/Nano/JNNDimerTool.pdf

Methuselah Foundation (2007); http://www.sens.org

Miettinen M, Lasota J (2006) Gastrointestinal stromal tumors: review on morphology, molecular pathology, prognosis, and differential diagnosis. Arch Pathol Lab Med 130:1466–1478

Mironov V, Boland T, Trusk T et al. (2003) Organ printing: computer-aided jet-based 3-D tissue engineering. Trends Biotechnol 21:157–161

Miyake K, McNeil PL (1995) Vesicle accumulation and exocytosis at sites of plasma membrane disruption. J Cell Biol 131:1737–1745

Monash University (2006) Micro-robots take off as ARC announces funding. Press release, 11 October 2006; http://www.monash.edu.au/news/newsline/story/1038

Moore WA, Davey VA, Weindruch R et al. (1995) The effect of caloric restriction on lipofuscin accumulation in mouse brain with age. Gerontology 41:173–185

Morowitz HJ (1974) Manufacturing a living organism. Hospital Practice 9:210–215

Moskovits M (2007) Nanoassemblers: A likely threat? Nanotech. Law & Bus. 4:187–195

Mountain V, Compton DA (2000) Dissecting the role of molecular motors in the mitotic spindle. Anat Rec 261:14–24

Murakami T, Garcia CA, Reiter LT et al. (1996) Charcot-Marie-Tooth disease and related inherited neuropathies. Medicine (Baltimore) 75:233–250

Mureebe L, McKinsey JF (2006) Infrainguinal arterial intervention: is there a role for an atherectomy device? Vascular 14:313–318

Murphy KM, Topel R (1999) The economic value of medical research. Funding First Reports, Lasker Medical Research Network, Lasker Foundation, March 1998, revised September 1999; http://www.laskerfoundation.org/reports/pdf/economicvalue.pdf

Murray RK, Granner DK, Mayes PA, Rodwell VW (1993) Harper's Biochemistry, 23rd edn. Appleton & Lange, Norwalk, CT

Nagahara L, Tao N, Thundat T (2008) Introduction to Nanosensors. Springer, New York

Nanofactory Collaboration website (2007a); http://www.MolecularAssembler.com/Nanofactory

Nanofactory Collaboration website (2007b) Remaining technical challenges for achieving positional diamondoid molecular manufacturing and diamondoid nanofactories; http://www.MolecularAssembler.com/Nanofactory/Challenges.htm

Nelson KL, Geyer S (1991) Bioreactor and process design for large-scale mammalian cell culture manufacturing. Bioprocess Technol 13:112–143

Nimmerjahn A, Kirchhoff F, Helmchen F (2005) Resting microglial cells are highly dynamic surveillants of brain parenchyma in vivo. Science 308:1314–1318

Nony PA, Schnellmann RG (2003) Mechanisms of renal cell repair and regeneration after acute renal failure. J Pharmacol Exp Ther 304:905–912

Nordquist L, Roxhed N, Griss P et al. (2007) Novel microneedle patches for active insulin delivery are efficient in maintaining glycaemic control: an initial comparison with subcutaneous administration. Pharm Res 24:1381–1388

Normanno N, De Luca A, Bianco C et al. (2006) Epidermal growth factor receptor (EGFR) signaling in cancer. Gene 366:2–16

Obataya I, Nakamura C, Han S et al. (2005) Mechanical sensing of the penetration of various nanoneedles into a living cell using atomic force microscopy. Biosens Bioelectron 20: 1652–1655

O'Connor MN, Smethurst PA, Farndale RW et al. (2006) Gain- and loss-of-function mutants confirm the importance of apical residues to the primary interaction of human glycoprotein VI with collagen. J Thromb Haemost 4:869–873

O'Gorman E, Piendl T, Muller M et al. (1997) Mitochondrial intermembrane inclusion bodies: the common denominator between human mitochondrial myopathies and creatine depletion, due to impairment of cellular energetics. Mol Cell Biochem 174:283–289

Oh JO (1985) Immunology of viral infections. Int Ophthalmol Clin 25:107–116

Olbrich H, Haffner K, Kispert A et al. (2002) Mutations in DNAH5 cause primary ciliary dyskinesia and randomization of left-right asymmetry. Nature Genet 30:143–144

Olson AL, Pessin JE (1996) Structure, function, and regulation of the mammalian facilitative glucose transporter gene family. Annu Rev Nutr 16:235–256

Oyabu N, Custance O, Yi I et al. (2003) Mechanical vertical manipulation of selected single atoms by soft nanoindentation using near contact atomic force microscopy. Phys Rev Lett 90:176102; http://link.aps.org/abstract/PRL/v90/e176102

Padron Velazquez JL (2006) Stem cell fusion as an ultimate line of defense against xenobiotics. Med Hypotheses 67:383–387

Paleri V, Stafford FW, Sammut MS (2005) Laser debulking in malignant upper airway obstruction. Head Neck 27:296–301

Pantaloni D, Le Clainche C, Carlier MF (2001) Mechanism of actin-based motility. Science 292:1502–1506

Parpura V, Haydon PG, Henderson E (1993) Three-dimensional imaging of living neurons and glia with the atomic force microscope. J Cell Sci 104:427–432; http://jcs.biologists.org/cgi/reprint/104/2/427.pdf

Patel SH, Patel R (2007) Inferior vena cava filters for recurrent thrombosis: current evidence. Tex Heart Inst J 34:187–194

Patolsky F, Timko BP, Yu G et al. (2006) Detection, stimulation, and inhibition of neuronal signals with high-density nanowire transistor arrays. Science 313:1100–1104

Patton RL, Kalback WM, Esh CL et al. (2006) Amyloid-beta peptide remnants in AN-1792-immunized Alzheimer's disease patients: a biochemical analysis. Am J Pathol 169:1048–1063

Peacock EE Jr (1984) Wound Repair. WB Saunders Company, Philadelphia, PA

Pecka Z (1993) Intranuclear development of asexual and sexual generations of *Eimeria hermani* Farr, 1953, the coccidian parasite of geese. Zentralbl Bakteriol 278:570–576

Peng J, Freitas RA Jr., Merkle RC et al. (2006) Theoretical analysis of diamond mechanosynthesis. Part III. Positional C_2 deposition on diamond C(110) surface using Si/Ge/Sn-based dimer placement tools. J Comput Theor Nanosci 3:28–41; http://www.MolecularAssembler.com/Papers/JCTNPengFeb06.pdf

Perez LC (1999) Septic arthritis. Baillieres Best Pract Res Clin Rheumatol 13:37–58

Permanne B, Adessi C, Saborio GP et al. (2002) Reduction of amyloid load and cerebral damage in a transgenic mouse model of Alzheimer's disease by treatment with a beta-sheet breaker peptide. FASEB J 16:860–862

Pesaran B, Musallam S, Andersen RA (2006) Cognitive neural prosthetics. Curr Biol 16:R77–R80

Pieper RO (1997) Understanding and manipulating O6-methylguanine-DNA methyltransferase expression. Pharmacol Ther 74:285–297

Pipinos II, Bruzoni M, Johanning JM et al. (2006) Transcervical carotid stenting with flow reversal for neuroprotection: technique, results, advantages, and limitations. Vascular 14:245–255

Pogach MS, Cao Y, Millien G et al. (2007) Key developmental regulators change during hyperoxia-induced injury and recovery in adult mouse lung. J Cell Biochem 100:1415–1429

Powell JJ, Ainley CC, Harvey RS et al. (1996) Characterisation of inorganic microparticles in pigment cells of human gut associated lymphoid tissue. Gut 38:390–395

Poznansky MC, Evans RH, Foxall RB et al. (2000) Efficient generation of human T cells from a tissue-engineered thymic organoid. Nat Biotechnol 18:729–734

Primmer S (2006) First Public Meeting of the Supercentenarian Research Foundation: Meeting Report by Stan Primmer, President of the SRF. GRG Breaking News Items [2006], 4 June 2006; http://www.grg.org/breakingnews2006.htm

Protiveris Corp. (2003) Microcantilever arrays; http://www.protiveris.com/cantilever_tech/microcantileverarrays.html

Prusiner SB (2001) Shattuck lecture – neurodegenerative diseases and prions. New Engl J Med 344:1516–1526

Qin H, Rosenbaum JL, Barr MM (2001) An autosomal recessive polycystic kidney disease gene homolog is involved in intraflagellar transport in *C. elegans* ciliated sensory neurons. Curr Biol 11:457–461

Quan T, He T, Shao Y et al. (2006) Elevated cysteine-rich 61 mediates aberrant collagen homeostasis in chronologically aged and photoaged human skin. Am J Pathol 169:482–490; http://ajp.amjpathol.org/cgi/content/full/169/2/482

Rabouille C, Misteli T, Watson R et al. (1995) Reassembly of Golgi stacks from mitotic Golgi fragments in a cell-free system. J Cell Biol 129:605–618

Rajnoch C, Ferguson S, Metcalfe AD et al. (2003) Regeneration of the ear after wounding in different mouse strains is dependent on the severity of wound trauma. Dev Dyn 226:388–397; http://www3.interscience.wiley.com/cgi-bin/fulltext/102525230/PDFSTART

Rayment I, Holden HM, Sellers JR et al. (1995) Structural interpretation of the mutations in the beta-cardiac myosin that have been implanted in familial hypertrophic cardiomyopathy. Proc Natl Acad Sci USA 92:3864–3868

Reed MA, Lee T (eds) (2003) Molecular Nanoelectronics. American Scientific Publishers, Stevenson Ranch, CA

Rehling P, Wiedemann N, Pfanner N et al. (2001) The mitochondrial import machinery for preproteins. Crit Rev Biochem Mol Biol 36:291–336

Reilein AR, Rogers SL, Tuma MC et al. (2001) Regulation of molecular motor proteins. Int Rev Cytol 204:179–238

Reinboth JJ, Gautschi K, Munz K et al. (1997) Lipofuscin in the retina: quantitative assay for an unprecedented autofluorescent compound (pyridinium bis-retinoid, A2-E) of ocular age pigment. Exp Eye Res 65:639–643

Reinders ME, Rabelink TJ, Briscoe DM (2006) Angiogenesis and endothelial cell repair in renal disease and allograft rejection. J Am Soc Nephrol 17:932–942

Rishikof DC, Lucey EC, Kuang PP et al. (2006) Induction of the myofibroblast phenotype following elastolytic injury to mouse lung. Histochem Cell Biol 125:527–534

Rizun PR, McBeth PB, Louw DF et al. (2004) Robot-assisted neurosurgery. Semin Laparosc Surg 11:99–106

Rizun P, Gunn D, Cox B et al. (2006) Mechatronic design of haptic forceps for robotic surgery. Int J Med Robot 2:341–349

Robson B (1998) Doppelganger proteins as drug leads. Nature Biotechnol 14:892–893

Robson B, Pseudoproteins: Non-protein protein-like machines. Paper presented at 6th Foresight Conference on Molecular Nanotechnology, November 1998

Roemer A, Schwettmann L, Jung M et al. (2004) The membrane proteases adams and hepsin are differentially expressed in renal cell carcinoma. Are they potential tumor markers? J Urol 172:2162–2166

Roffe C (1998) Ageing of the heart. Br J Biomed Sci 55:136–148

Rohs R, Bloch I, Sklenar H et al. (2005) Molecular flexibility in *ab initio* drug docking to DNA: binding-site and binding-mode transitions in all-atom Monte Carlo simulations. Nucleic Acids Res 33:7048–7057

Roth EA, Xu T, Das M et al. (2004) Inkjet printing for high-throughput cell patterning. Biomaterials 25:3707–3715

Royer RJ, Delongeas JL, Netter P et al. (1982) Inflammatory effect of aluminium phosphate on rat paws. Pathol Biol (Paris) 30:211–215

Sacconi L, Tolic-Norrelykke IM, Antolini R et al. (2005a) Combined intracellular three-dimensional imaging and selective nanosurgery by a nonlinear microscope. J Biomed Opt 10:14002

Sacconi L, Tolić-Nørrelykke IM, Stringari C et al. (2005b) Optical micromanipulations inside yeast cells. Appl Opt 44:2001–2007

Sage MD, Jennings RB (1988) Cytoskeletal injury and subsarcolemmal bleb formation in dog heart during in vitro total ischemia. Am J Pathol 133:327–337

Sahenk Z, Chen L, Mendell JR (1999) Effects of PMP22 duplication and deletions on the axonal cytoskeleton. Ann Neurol 45:16–24

Sahni SK (2007) Endothelial cell infection and hemostasis. Thromb Res 119:531–549

Saito M, Matsuura T, Masaki T et al. (2006) Reconstruction of liver organoid using a bioreactor. World J Gastroenterol 12:1881–1888

Salzberg AD, Bloom MB, Mourlas NJ et al. (2002) Microelectrical mechanical systems in surgery and medicine. J Am Coll Surg 194:463–476

Sapir T, Elbaum M, Reiner O (1997) Reduction of microtubule catastrophe events by LIS1, platelet-activating factor acetylhydrolase subunit. EMBO J 16:6977–6984; http://emboj.oupjournals.org/cgi/content/full/16/23/6977

Sassera D, Beninati T, Bandi C et al. (2006) '*Candidatus Midichloria mitochondrii*', an endosymbiont of the tick *Ixodes ricinus* with a unique intramitochondrial lifestyle. Int J Syst Evol Microbiol 56:2535–2540

Sata M (2003) Circulating vascular progenitor cells contribute to vascular repair, remodeling, and lesion formation. Trends Cardiovasc Med 13:249–253

Satoa F, Togoa M, Islamb MK et al. (2005) Enzyme-based glucose fuel cell using Vitamin K3-immobilized polymer as an electron mediator. Electrochem Commun 7:643–647

Scaffidi P, Misteli T, Bianchi ME (2002) Release of chromatin protein HMGB1 by necrotic cells triggers inflammation. Nature 418:191–195

Schell WD, Myers JN (1997) Regression of atherosclerosis: a review. Prog Cardiovasc Dis 39:483–496

Schenk D (2002) Amyloid-beta immunotherapy for Alzheimer's disease: the end of the beginning. Nat Rev Neurosci 3:824–828

Schewe PF, Stein B, Riordon J (2001) A carbon nanotube integrated circuit. American Institute of Physics Bulletin of Physics News No. 531, 22 March 2001; http://newton.ex.ac.uk/aip/physnews.531.html

Schliwa M (ed) (2003) Molecular Motors. VCH-Wiley, Weinheim

Schliwa M, Woehlke G (2003) Molecular motors. Nature 422:759–765

Schönbeck U, Libby P (2001) CD40 signaling and plaque instability. Circ Res 89:1092–1103

Schwartz AB, Cui XT, Weber DJ et al. (2006) Brain-controlled interfaces: movement restoration with neural prosthetics. Neuron 52:205–220

Schwartz JT (1990) The new connectionism: Developing relationships between neuroscience and artificial intelligence. In: Graubard SR (ed) The Artificial Intelligence Debate. MIT Press, Cambridge, MA

Schwartz M, Madariaga J, Hirose R et al. (2004) Comparison of a new fibrin sealant with standard topical hemostatic agents. Arch Surg 139:1148–1154

Seidman JG, Seidman C (2001) The genetic basis for cardiomyopathy: from mutation identification to mechanistic paradigms. Cell 104:557–567

Sell DR, Nagaraj RH, Grandhee SK et al. (1991) Pentosidine: a molecular marker for the cumulative damage to proteins in diabetes, aging, and uremia. Diabetes Metab Rev 7:239–251

Sell DR, Biemel KM, Reihl O et al. (2005) Glucosepane is a major protein cross-link of the senescent human extracellular matrix. Relationship with diabetes. J Biol Chem 280:12310–12315; http://www.jbc.org/cgi/content/full/280/13/12310

Shankaran V, Ikeda H, Bruce AT et al. (2001) IFNgamma and lymphocytes prevent primary tumour development and shape tumour immunogenicity. Nature 410:1107–1111

Sharom FJ (1997) The P-glycoprotein efflux pump: how does it transport drugs? J Membr Biol 160:161–175

Sharova NP (2005) How does a cell repair damaged DNA? Biochemistry (Mosc) 70:275–291

Shen N, Datta D, Schaffer CB et al. (2005) Ablation of cytoskeletal filaments and mitochondria in live cells using a femtosecond laser nanoscissor. Mech Chem Biosyst 2:17–25

Shimizu N, Kamezaki F, Shigematsu S (2005) Tracking of microinjected DNA in live cells reveals the intracellular behavior and elimination of extrachromosomal genetic material. Nucleic Acids Res 33:6296–6307

Shirai Y, Osgood AJ, Zhao Y et al. (2005) Directional control in thermally driven single-molecule nanocars. Nano Lett 5:2330–2334. See images at: "Rice scientists build world's first single-molecule car," 20 October 2005; http://www.media.rice.edu/media/NewsBot.asp?MODE=VIEW&ID=7850

Shirai Y, Morin JF, Sasaki T et al. (2006) Recent progress on nanovehicles. Chem Soc Rev 35:1043–1055

Shotelersuk V, Gahl WA (1998) Hermansky-Pudlak syndrome: models for intracellular vesicle formation. Mol Genet Metab 65:85–96

Sim HG, Yip SK, Cheng CW (2006) Equipment and technology in surgical robotics. World J Urol 24:128–135

Simoni P, Peters GE, Magnuson JS (2003) Use of the endoscopic microdebrider in the management of airway obstruction from laryngotracheal carcinoma. Ann Otol Rhinol Laryngol 112:11–13

Simons K, Gruenberg J (2000) Jamming the endosomal system: lipid rafts and lysosomal storage diseases. Trends Cell Biol 10:459–462

Smith AU (1965) Problems in freezing organs and their component cells and tissues. Fed Proc 24:S196–S203

Smith AU, Lovelock JE, Parkes AS (1954) Resuscitation of hamsters after supercooling or partial crystallization at body temperature below 0 degrees C. Nature 173:1136–1137

Solomon A, Weiss DT, Wall JS (2003) Immunotherapy in systemic primary (AL) amyloidosis using amyloid-reactive monoclonal antibodies. Cancer Biother Radiopharm 18:853–860

Soto C, Sigurdsson EM, Morelli L et al. (1998) Beta-sheet breaker peptides inhibit fibrillogenesis in a rat brain model of amyloidosis: implications for Alzheimer's therapy. Nat Med 4:822–826

Sprecher E, Ishida-Yamamoto A, Becker OM et al. (2001) Evidence for novel functions of the keratin tail emerging from a mutation causing ichthyosis hystrix. J Invest Dermatol 116:511–519

Stefanadis C, Diamantopoulos L, Vlachopoulos C et al. (1999) Thermal heterogeneity within human atherosclerotic coronary arteries detected in vivo: A new method of detection by application of a special thermography catheter. Circulation 99:1965–1971

Stegner AR, Boehme C, Huebl H et al. (2006) Electrical detection of coherent ^{31}P spin quantum states. Nature Phys 2:835–838

Stein SF, Evatt B (1992) Chapter 13-2. Thrombocytopenia. In: Hurst JW, Medicine for the Practicing Physician, 3rd edn. Butterworth-Heinemann, Boston, MA, pp. 769–771

Steinhardt RA, Bi G, Alderton JM (1994) Cell membrane resealing by a vesicular mechanism similar to neurotransmitter release. Science 263:390–393

Stenger S, Hanson DA, Teitelbaum R et al. (1998) An antimicrobial activity of cytolytic T cells mediated by granulysin. Science 282:121–125

Strange C, Sahn SA (1999) The definitions and epidemiology of pleural space infection. Semin Respir Infect 14:3–8

Street SE, Hayakawa Y, Zhan Y et al. (2004) Innate immune surveillance of spontaneous B cell lymphomas by natural killer cells and gammadelta T cells. J Exp Med 199:879–884

Strehler BL, Mark DD, Mildvan AS et al. (1959) Rate and magnitude of age pigment accumulation in the human myocardium. J Gerontol 14:430–439

Subat M, Borovik AS, Konig B (2004) Synthetic creatinine receptor: imprinting of a Lewis acidic zinc(II)cyclen binding site to shape its molecular recognition selectivity. J Am Chem Soc 126:3185–3190

Sugaya K, Brannen CL (2001) Stem cell strategies for neuroreplacement therapy in Alzheimer's disease. Med Hypotheses 57:697–700

Swann JB, Smyth MJ (2007) Immune surveillance of tumors. J Clin Invest 117:1137–1146

Tada T (2006) Nuclear reprogramming: an overview. Methods Mol Biol 348:227–236

Takeuchi T, Adachi Y, Ohtsuki Y et al. (2007) Adiponectin receptors, with special focus on the role of the third receptor, T-cadherin, in vascular disease. Med Mol Morphol 40:115–120

Tamm SL (1982) Flagellated ectosymbiotic bacteria propel a eukaryotic cell. J Cell Biol 94:697–709

Tangirala RK, Jerome WG, Jones NL et al. (1994) Formation of cholesterol monohydrate crystals in macrophage-derived foam cells. J Lipid Res 35:93–104

Tanskanen M, Kiuru-Enari S, Tienari P et al. (2006) Senile systemic amyloidosis, cerebral amyloid angiopathy, and dementia in a very old Finnish population. Amyloid 13:164–169

Tardif JC, Gregoire J, L'Allier PL et al. (2006) Effect of atherosclerotic regression on total luminal size of coronary arteries as determined by intravascular ultrasound. Am J Cardiol 98:23–27

Tatebayashi Y, Lee MH, Li L et al. (2003) The dentate gyrus neurogenesis: a therapeutic target for Alzheimer's disease. Acta Neuropathol (Berl) 105:225–232

Taylor PM, Sachlos E, Dreger SA et al. (2006) Interaction of human valve interstitial cells with collagen matrices manufactured using rapid prototyping. Biomaterials 27:2733–2737

Temelso B, Sherrill CD, Merkle RC, Freitas RA Jr (2006) High-level *ab initio* studies of hydrogen abstraction from prototype hydrocarbon systems. J Phys Chem A 110:11160–11173; http://www.MolecularAssembler.com/Papers/TemelsoHAbst.pdf

Temelso B, Sherrill CD, Merkle RC, Freitas RA Jr (2007) *Ab initio* thermochemistry of the hydrogenation of hydrocarbon radicals using silicon, germanium, tin and lead substituted methane and isobutane. J Phys Chem A 111:8677–8688; http://www.MolecularAssembler.com/Papers/TemelsoHDon.pdf

Terman A, Brunk UT (1998) Lipofuscin: mechanisms of formation and increase with age. APMIS 106:265–276

Terman A, Dalen H, Brunk UT (1999) Ceroid/lipofuscin-loaded human fibroblasts show decreased survival time and diminished autophagocytosis during amino acid starvation. Exp Gerontol 34:943–957

Thompson TE, Bennett I (1969) Physalia nematocysts utilized by mollusks for defense. Science 166:1532–1533

Tipler FJ (1994) The Physics of Immortality. Doubleday, New York

Tirlapur UK, Konig K (2002) Femtosecond near-infrared laser pulses as a versatile non-invasive tool for intra-tissue nanoprocessing in plants without compromising viability. Plant J 31:365–374

Todar K (2003) Evasion of host phagocytic defenses. University of Wisconsin-Madison website; http://www.bact.wisc.edu/microtextbook/disease/evadephago.html

Tonet O, Focacci F, Piccigallo M et al. (2006) Comparison of control modes of a hand-held robot for laparoscopic surgery. Int Conf Med Image Comput Comput Assist Interv 9:429–436

Totaro EA, Pisanti FA, Continillo A et al. (1985) Morphological evaluation of the lipofuscinolytic effect of acetylhomocysteine thiolactone. Arch Gerontol Geriatr 4:67–72

Tse RL, Phelps P (1970) Polymorphonuclear leukocyte motility in vitro. V. Release of chemotactic activity following phagocytosis of calcium pyrophosphate crystals, diamond dust, and urate crystals. J Lab Clin Med 76:403–415

Tseng GY, Ellenbogen JC (2001) Toward nanocomputers. Science 294:1293–1294

Ubelaker DH, Buchholz BA, Stewart JEB (2006) Analysis of artificial radiocarbon in different skeletal and dental tissue types to evaluate date of death. J Forensic Sci 51:484–488

Vallee RB, Tai C, Faulkner NE (2001) LIS1: cellular function of a disease-causing gene. Trends Cell Biol 11:155–160

Vaupel JW, Carey JR, Christensen K et al. (1998) Biodemographic trajectories of longevity. Science 280:855–860

Veomett G, Prescott DM (1976) Reconstruction of cultured mammalian cells from nuclear and cytoplasmic parts. Methods Cell Biol 13:7–14

Vettiger P, Cross G, Despont M et al. (2002) The Millipede – nanotechnology entering data storage. IEEE Trans Nanotechnol 1:39–55

Villar-Garea A, Imhof A (2006) The analysis of histone modifications. Biochim Biophys Acta 1764:1932–1939

Vital C, Bouillot S, Canron MH et al. (2002) Schwannian crystalline-like inclusions bodies (Fardeau-Engel bodies) revisited in peripheral neuropathies. Ultrastruct Pathol 26:9–13

Vogel S (1998) Cats' Paws and Catapults: Mechanical Worlds of Nature and People. W.W. Norton and Company, New York

von Neumann J (1958) The Computer and the Brain. Yale University Press, New Haven CT

Wagner OI, Rammensee S, Korde N et al. (2007) Softness, strength and self-repair in intermediate filament networks. Exp Cell Res 313:2228–2235

Wakida NM, Lee CS, Botvinick ET et al. (2007) Laser nanosurgery of single microtubules reveals location-dependent depolymerization rates. J Biomed Opt 12:024022

Wall RJ (2001) Pronuclear microinjection. Cloning Stem Cells 3:209–220

Walski M, Chlopicki S, Celary-Walska R et al. (2002) Ultrastructural alterations of endothelium covering advanced atherosclerotic plaque in human carotid artery visualised by scanning electron microscope. J Physiol Pharmacol 53:713–723

Walkley SU (2007) Pathogenic mechanisms in lysosomal disease: a reappraisal of the role of the lysosome. Acta Paediatr Suppl 96:26–32

Wang PM, Cornwell M, Prausnitz MR (2005) Minimally invasive extraction of dermal interstitial fluid for glucose monitoring using microneedles. Diabetes Technol Ther 7:131–141

Wang X, Song J, Liu J et al. (2007) Direct-current nanogenerator driven by ultrasonic waves. Science 316:102–105

Weatherbee CW, Freitas RA Jr (2010) The structure of the human kidney as applicable to nanoscale robot navigation. In preparation.

Weber G, Greulich KO (1992) Manipulation of cells, organelles, and genomes by laser microbeam and optical trap. Int Rev Cytol 133:1–41

West C (1979) A quantitative study of lipofuscin accumulation with age in normals and individuals with Down's syndrome, phenylketonuria, progeria and transneuronal atrophy. J Comp Neurol 186:109–116

Westbroek W, Lambert J, Naeyaert JM (2001) The dilute locus and Griscelli syndrome: gateways towards a better understanding of melanosome transport. Pigment Cell Res 14:320–327

Westermark P, Wilander E (1978) The influence of amyloid deposits on the islet volume in maturity onset diabetes mellitus. Diabetologia 15:417–421

Wiley C (2005) Nanotechnology and molecular homeostasis. J Amer Geriatrics Soc 53:S295–S298

Williams DS (2002) Transport to the photoreceptor outer segment by myosin VIIa and kinesin II. Vision Res 42:455–462

Winchester B, Vellodi A, Young E (2000) The molecular basis of lysosomal storage diseases and their treatment. Biochem Soc Trans 28:150–154

Winter SH, Bouzit M (2007) Use of magnetorheological fluid in a force feedback glove. IEEE Trans Neural Syst Rehabil Eng 15:2–8

Wissler RW, Vesselinovitch D (1990) Can atherosclerotic plaques regress? Anatomic and biochemical evidence from nonhuman animal models. Am J Cardiol 65:33F–40F

Wolf DH, Hilt W (2004) The proteasome: a proteolytic nanomachine of cell regulation and waste disposal. Biochim Biophys Acta 1695:19–31

Wolf E, Schernthaner W, Zakhartchenko V et al. (2000) Transgenic technology in farm animals – progress and perspectives. Exp Physiol 85:615–625

Wood KJ, Prior TG (2001) Gene therapy in transplantation. Curr Opin Mol Ther 3:390–398

Wray GA, Hahn MW, Abouheif E et al. (2003) The evolution of transcriptional regulation in eukaryotes. Mol Biol Evol 20:1377–1419; http://mbe.oxfordjournals.org/cgi/content/full/20/9/1377

Wright LM, Brzozowski AM, Hubbard RE et al. (2000) Structure of fab hGR-2 F6, a competitive antagonist of the glucagon receptor. Acta Crystallogr D Biol Crystallogr 56:573–580

Wu JT (2002) C-erbB2 oncoprotein and its soluble ectodomain: a new potential tumor marker for prognosis early detection and monitoring patients undergoing Herceptin treatment. Clin Chim Acta 322:11–19

Xie XY, Barrett JN (1991) Membrane resealing in cultured rat septal neurons after neurite transection: evidence for enhancement by Ca(2+)-triggered protease activity and cytoskeletal disassembly. J Neurosci 11:3257–3267

Xu T, Jin J, Gregory C et al. (2005) Inkjet printing of viable mammalian cells. Biomaterials 26:93–99

Xu T, Gregory CA, Molnar P et al. (2006) Viability and electrophysiology of neural cell structures generated by the inkjet printing method. Biomaterials 27:3580–3588

Yagishita S, Itoh Y, Nakano T et al. (1979) Crystalloid inclusions reminiscent of Hirano bodies in autolyzed peripheral nerve of normal Wistar rats. Acta Neuropathol (Berl) 47:231–236

Yan L, Zucker S, Toole BP (2005) Roles of the multifunctional glycoprotein, emmprin (basigin; CD147), in tumour progression. Thromb Haemost 93:199–204

Ye L, Haider HK, Sim EK (2006) Adult stem cells for cardiac repair: a choice between skeletal myoblasts and bone marrow stem cells. Exp Biol Med 231:8–19

Yanik MF, Cinar H, Cinar HN et al. (2004) Neurosurgery: functional regeneration after laser axotomy. Nature 432:822

Yesin KB, Exner P, Vollmers K et al. (2005) Biomedical micro-robotic system. 8th Intl. Conf. on Medical Image Computing and Computer Assisted Intervention (MICCAI 2005 / www.miccai2005.org), Palm Springs CA, 26–29 October 2005, p. 819

Yokobayashi Y, Weiss R, Arnold FH (2002) Directed evolution of a genetic circuit. Proc Natl Acad Sci USA 99:16587–16591; http://www.pnas.org/cgi/content/abstract/99/26/16587

Yun MH, Cannon D, Freivalds A et al. (1997) An instrumented glove for grasp specification in virtual-reality-based point-and-direct telerobotics. IEEE Trans Syst Man Cybern B Cybern 27:835–846

Zhang C, Gu Y, Chen L (1998) Study on early frozen section after nerve repair with cell surgery technique. Chin J Traumatol 1:45–48

Zhang H, Li N, Chen Y et al. (2007) Protein profile of human lung squamous carcinoma cell line NCI-H226. Biomed Environ Sci 20:24–32

Zhang S, Cordon-Cardo C, Zhang HS et al. (1997) Selection of tumor antigens as targets for immune attack using immunohistochemistry: I. Focus on gangliosides. Int J Cancer 73:42–49

Zhang Z, Weed SA, Gallagher PG et al. (2001) Dynamic molecular modeling of pathogenic mutations in the spectrin self-association domain. Blood 98:1645–1653

Zhao C, Takita J, Tanaka Y et al. (2001) Charcot-Marie-Tooth disease type 2A caused by mutation in a microtubule motor KIF1Bbeta. Cell 105:587–597

Zhijiang D, Zhiheng J, Minxiu K (2005) Virtual reality-based telesurgery via teleprogramming scheme combined with semi-autonomous control. Conf Proc IEEE Eng Med Biol Soc 2:2153–2156

Zullo SJ, Parks WT, Chloupkova M et al. (2005) Stable transformation of CHO Cells and human NARP cybrids confers oligomycin resistance (oli(r)) following transfer of a mitochondrial DNA-encoded oli(r) ATPase6 gene to the nuclear genome: a model system for mtDNA gene therapy. Rejuvenation Res 8:18–28

Appendices: Two Unusual Potential Sources of Funding for Longevity Research

Editorial Note

The following appendices (A and B) provide information that may be useful to investigators seeking funding for experimental types of aging intervention that may be difficult to fund through more traditional sources. The Editors make no representations about the ability of the presenting organizations to fund research or about the veracity of the presenting organizations' views about the biology of aging or of the prospects for intervening into aging in the ways envisioned by either of these organizations. Our intention here is simply to inform readers about additional possible sources of support for their research.

– The Editors

Appendix A
SENS Foundation: Accelerating Progress Toward Biomedical Rejuvenation

Michael Rae

A.1 Invitation to Submit Research Proposals

SENS Foundation (website: http://www.sens.org/) is positioned as the most effective philanthropic organization investing in biomedical gerontological research today: a California-based biomedical nonprofit[1] committed to critical-path research within a comprehensive panel of therapeutics that is aimed at curing age-related disease, disability, suffering, and death.

This mission most obviously distinguishes the Foundation from the large number of nonprofit organizations devoted to treatments for specific age-related diseases. However, the Foundation's unique strategic approach to advancing progress toward the cure of biological aging is also unique even within the small field of organizations devoted to biogerontological research. The Foundation's focus is on the development of biomedical interventions rather than on descriptive studies – and on interventions that will not simply *retard* the rate of aging, but *restore* the aged body's original biomolecular- and cellular-level fidelity – and through it, replace the frailty of biological age with the robust homeostatic resilience of biological youth.

To this end, SENS Foundation directly funds research projects aimed at the ultimate development of interventions to remove, repair, replace, or render harmless molecular and cellular damage prevalent in the body in situ late in life.

We are now inviting all qualified researchers in the biological or medical sciences to submit proposals aimed at molecular- and cellular-level structural 'rejuvenation' of the aging body. Established fundable program areas and Requests for Proposals (RFPs) for specific projects are outlined below.

M. Rae (✉)
SENS Foundation, 1230 Bordeaux Drive, Sunnyvale, CA 94089, USA
e-mail: michael.rae@sens.org

[1] SENS Foundation is a California nonprofit organization, and is in the process of applying for recognition of exemption from taxation under section 501(c)(3) of the Internal Revenue Code. The organization expects to receive tax-exempt status effective as of the date of incorporation.

A.2 Funding Priorities and Principles

SENS Foundation's research funding priorities emerge from critical-path analysis of areas of high biomedical importance within the "Strategies for Engineered Negligible Senescence" (SENS) platform proposed by Dr. Aubrey de Grey to develop a comprehensive panel of robust anti-aging interventions. As explained in detail elsewhere (de Grey 2003; de Grey et al. 2002; de Grey and Rae, 2007), SENS is based on an "engineering" heuristic for the development of therapeutics targeting biological aging, in contrast to the so-called "gerontological" approach that underlies most past and present attempts to do so.

SENS Foundation is the only nonprofit organization whose research funding is devoted to biomedical gerontology, but whose remit allows it to avoid the diversion of funds into the inefficient "gerontological" approach. Instead, SENS Foundation's research program targets funding to research that advances the development of the specific biotechnologies identified through the "engineering" approach to anti-aging biomedicine: in essence, the extension of regenerative medicine principles to aging.

I will now outline these two approaches, and the reasons why the Foundation's strategy of targeted investments in the former rather than the latter can be expected to yield anti-aging interventions of dramatically greater benefit.

A.2.1 The "Gerontological" Approach: Modulation of Metabolic Pathways Contributing to Biological Aging

There is broad-based consensus in biogerontology (e.g., Kirkwood, 2008; Hayflick, 2007; Holliday, 2004; de Grey et al. 2002) that aging is the result of accumulating stochastic damage to the body's cellular and molecular structures, as a result of unintended biochemical side-effects of metabolism such as reactive oxygen species (ROS), nonenzymatic glycation, and errors in DNA replication and epigenetics. Most biogerontology research, therefore, is based on enhancing our understanding of those metabolic processes and their mechanistic links to aging damage, in hopes that those pathways can be modulated in ways that reduce the rate at which such damage forms and accumulates, leading in turn to a slower loss of physiological integrity and retarded organismal aging.

Within this framework, putative interventions might downregulate pathways responsible for ROS generation, sequester reactive intermediates of glycolysis, upregulate repair mechanisms, interfere in the action of anabolic pathways responsible for cellular proliferation, or otherwise antagonize or cushion such pathways. Examples will come readily to mind, and pervade the choice of interventions selected for inclusion in the NIA's *Interventions Testing Program* (ITP (Nadon et al. 2008; National Institute on Aging 2008).

This way of pursuing anti-aging intervention is grounded in the logic of basic research science: analyze a phenomenon in progressively finer detail to discover its underlying basis, and then use the results of such studies to modify hypotheses, thereby generating new questions in a recursive process of scientific progress.

The same strategy is also widely employed as part of conventional drug development. Researchers begin by analyzing and characterizing the various components of a metabolic pathway that is defective in the disease state, identifying the normal function of each. They then compare the function of those components in health and disease, thereby revealing the dysfunctional steps in the process that form the molecular basis of the illness. Such components then become therapeutic targets, subject to manipulation to normalize function, whether by small molecules or other means (such as gene therapy or prosthesis).

Unfortunately, this strategy is ill-suited to the near-term development of interventions against the biological aging process, for the overarching reason that aging damage is not the result of the de novo *dysfunction* of metabolic pathways, but of the undesirable biochemical side-reactions of *normal metabolic activity*. This is precisely why aging is a *universal* disease, and why many physicians and even biogerontologists are reluctant to characterize it as a "disease" at all: biological aging is the *pathological result* of *perfectly-functioning*, "healthy" – but necessarily imperfect – metabolic processes. Hence Dr. de Grey's quip that "aging is a side-effect of being alive." In other words, any "gerontological" anti-aging intervention would of necessity be grounded upon interfering with the biochemical processes that sustain our very lives – and doing so day in and day out, from the day that a "patient" first begins therapy until his or her death.

The pathways whose reactive products contribute to aging are *there for a reason*, even though they have long-term downsides, and we begin dialing them up or down at our peril. Even the reactive products that link metabolism to aging damage are often indispensable, evolution having learned to harness them for metabolic purposes, such as when ROS are used as signaling molecules (D'Autréaux and Toledano, 2007). Indeed, the body's metabolic processes are *already* embedded within hopelessly complex regulatory pathways and interlocking feedback loops, precisely in order to ensure that they continue to operate in strict coordination with one another, and within a range that natural selection has 'learned' to be optimal for sustaining fitness.

This fact is illustrated in the very phenomena that advocates of "gerontological" intervention seek to mimic pharmacologically: animals that age slowly because of either genetic mutations or the imposition of calorie restriction. Slow-aging organisms are vanishingly rare in natural populations for a reason: not that evolution "wants us to age" as is often mistakenly said, but because the metabolic abnormalities that are responsible for such animals' slow-aging phenotype also ensure that they can only survive in the sheltered conditions of the laboratory. In the wild, these organisms would be unfit relative to their normally-aging cohorts for most of their natural lifespans, and the more aggressively the interventions are applied to enhance the anti-aging effect, the more severe the phenotype would become.

An additional, and equally crucial, limitation to "gerontological" intervention is that no matter how much the accumulation of further aging damage is *slowed* by such means, it cannot be entirely *stopped* without turning off the underlying metabolic processes entirely. Nor can such interventions do anything about the *pre-existing* burden of molecular and cellular lesions. Therefore, the ultimate benefits of "gerontological" interventions will by definition be inversely proportional to the degree of medical need of the recipient.

This weakness of all interventions whose expected effect is to decelerate aging changes has been acknowledged as a significant impediment to a realistic program of human clinical testing in a recent review by current NIA Director Richard Hodes, past Biology of Aging program director Huber Warner, and their colleagues (Hadley et al. 2005). It also limits the potential benefits of such therapies even supposing they were developed, tested clinically, and widely disseminated. For example, while calorie restriction (CR) imposed from weaning extends total, largely healthy lifespan by approximately 30%, the total lifespan extension declines when CR is imposed for the first time in adulthood or in early seniority. Based on a direct extrapolation from the available rodent data, (Rae, 2004) a small-molecule mimetic of severe CR beginning at weaning, at age 35, or at age 54, would extend human life expectancy to 110.5, 100, or 94.3 years, respectively. A 54-year-old would gain only 9.3 years compared to the 85 year life expectancy that would otherwise result from "aging as usual."

A.2.2 The SENS Heuristic: Regenerative Biomedical Gerontology

In contrast to pure science, pioneering engineering is a highly empirical endeavor, oriented not so much to detailed description of reality but to its practical manipulation. While informed by available scientific knowledge, it is very much a second-order discipline, utilizing proven rules of thumb in the absence of fully fleshed-out theoretical understanding of the phenomena within whose parameters it advances. The subject of concern is not whether a phenomenon is fully *understood*, but whether it can be reliably and advantageously *manipulated* to achieve some outcome.

Thus, for example, Edison and others were able to harness the power of electricity to great practical effect decades before the discovery of the electron was made, and when indeed the mysterious electrical "fluid" was mistakenly believed to flow from the cathode to the anode, rather than vice-versa. In terms of public health, the earliest (and most dramatic) advances were made on the same basis: contrast the weakness of the understanding of microbiology possessed by the likes of John Snow, Jenner, or Pasteur with the effectiveness of the solutions they pioneered. Even today, the very *names* of such high-tech fields as tissue "engineering" and biomedical "engineering" accurately reflect their highly empirical basis in replicating the key structural-functional characteristics of native tissues, even when the materials and processes used are highly remote from their biological originals.

But while the engineering heuristic is well-established in other fields of biomedicine, its use in biomedical gerontology is a recent, dramatic, and even disruptive departure from the "gerontological" approach that has dominated thinking on the subject to date – a genuine paradigm shift. Dr. de Grey is responsible for first explaining the distinction between the two strategies as they apply to intervention in aging (de Grey 2003; de Grey et al. 2002) and then using its heuristic principles (in collaboration with experts in widely diverse relevant fields) to develop a specific suite of emerging or foreseeable biotechnologies that comprise a platform of comprehensive rejuvenation biotechnology: the SENS program.

A.2.2.1 Principles of "Engineered" Rejuvenation

The engineering paradigm begins with the same understanding of the ultimate cause of biological aging as the "gerontological" paradigm: that the progressive increase in frailty that characterizes aging is the result of the accumulation of cellular and molecular damage that ensues from the normal operation of metabolic processes. Also shared with (but rarely emphasized in) "gerontological" thinking is the fact that, while the formation and accumulation of aging damage is an ongoing process that begins with life itself and continues throughout the lifespan as a result of normal metabolism, neither the metabolic processes that cause the accumulation, nor the accumulation process per se, have a significant impact on functionality for the first three to four decades of life. This is because of the great redundancy of the body's systems. Thus, there is little difference in the physical performance or risk of death between reasonably health-conscious people aged twenty, thirty, or forty, despite the fact that the burden of aging damage progressively accumulates over each additional decade. It is only once the total burden of such damage reaches a critical "threshold of pathology" that aging begins to impact our health and increase our risk of imminent death.

Granted these shared premises, the key strategic departure separating the engineering heuristic from that of "gerontology" is the identification of the cellular and molecular damage of aging *itself* as the preferred therapeutic target, rather than the metabolic pathways that contribute to the formation of that damage. By removing, repairing, replacing, or rendering irrelevant damaged cells and biomolecules, the body's structural fidelity can, logically, be maintained at – or, indeed, restored to – that of a young adult. And just as the degradation of such structural fidelity causes the progressive loss of function that characterizes biological aging, so the restoration of such fidelity necessarily must lead to the restoration of youthful functionality – to *rejuvenation*, in a word.

This shift in therapeutic targets yields several practical advantages to engineering interventions. First, it escapes the core dilemma of "gerontological" interventions: the risk to normal functionality entailed in perturbing normal metabolic processes in hopes of reducing the production of their unintended molecular and cellular side-effects. Instead, those metabolic processes can be allowed to proceed unimpeded, targeting only their deleterious consequences.

Second: because aging damage is initially inert, its removal should be intrinsically benign (again, something that is not the case for the manipulation of metabolism).

Third: because such damage takes so long (several decades) to accumulate to pathological levels, aging damage can be allowed to accumulate to levels within the range encountered over the two decades or so between maturity and early middle age before reaching the 'threshold of pathology' and requiring a cycle of biomedical removal to restore youthful molecular fidelity. Such a periodic schedule of intervention affords an extended period between cycles, during which the body can rest and recuperate from the rigors of treatment and engage its own intrinsic capacity for repair and regeneration; it also reduces the risk that a subtle side-effect of therapy can gradually lead to long-term, cumulative negative consequences. This is again in contrast to "gerontological" intervention, which requires chronic dosing because the metabolic pathways that it targets are continually active and the ensuing accumulation of damage proceeds on an unremitting basis.

Fourth: a focus on aging damage as the therapeutic target greatly reduces the number of such targets required to impact aging globally. While the metabolic processes that generate aging damage are hopelessly complex, multifarious, and interlocking, their direct consequences – aging damage itself– consists of a relatively small number of classes of lesion which can be directly targeted for removal (see next section).

Fifth, interventions designed to *reverse* aging changes can be tested on much more rapid timescales than can those designed to *retard* them: the clinical testing of any intervention whose effect is to "*decelerate* aging changes ... *must be initiated relatively early in life and sustained for decades* before clinical effects occur, [making] the logistical challenges of conducting a trial ... extremely formidable ... [By contrast,] If the intervention is expected to *reverse* in older persons adverse aging changes that increase their risk of clinical outcomes, it may be possible to design a study to measure the effect of the intervention on these outcomes directly" (Hadley et al. 2005) (my emphasis).

Finally, the magnitude of the benefits of such therapies to recipients – in delayed or escaped age-related disease, disability, dementia, dependence, and ultimately, premature death – can similarly be expected to be correspondingly greater when they can undo, rather than merely hold back, the molecular and cellular ravages of decades of normal metabolism. This implication is elaborated and modeled in (ii) *Prospects*, below.

A.2.2.2 Specific Engineering Targets

Decades of exhaustive characterization of the aging of mammals indicate there are no more than seven major classes of cellular and molecular aging lesions (see Table A.1). We can be confident this listing is exhaustive, not least because of

Table A.1 Classes of aging damage, dates of identification, and foreseeable solutions

Aging damage	Identified	SENS biotechnology
Intracellular aggregates	1959	Novel lysosomal hydrolases ("LysoSENS")
Mitochondrial mutations	1972	Allotopic expression ("MitoSENS")
Extracellular aggregates	1907	Immunotherapeutic clearance ("AmyloSENS")
Nuclear [epi]mutations	1959, 1982	Interdiction of telomere-lengthening ("OncoSENS")
Death-resistant cells	1965	Targeted ablation ("ApoptoSENS")
Loss of elasticity	1958, 1981	De-stiffening ("GlycoSENS"); tissue engineering
Cell loss, tissue atrophy	1955	Stem cells and tissue engineering ("RepleniSENS")

the *lack* of any progress in identifying new ones in nearly a generation of ongoing research, despite the ever-increasing number of detailed descriptive studies, the increasingly-sophisticated molecular-level probes and tools deployed, and the discovery of ever more layers of complexity in metabolic processes.

As Table A.1 also indicates, the path ahead to repairing these forms of damage is also discernable: for each major aging lesion, biotechnology for its removal or repair either already exists in prototype form, or is clearly foreseeable from existing scientific developments. Each of the biotechnologies in the SENS panel (Table A.1) is intended to *remove* (or, in two cases, obviate) aging damage central to the etiology and pathogenesis of major age-related diseases, and so should individually constitute actual *cures* for distinct therapeutic indications, in addition to collectively contributing to the global rejuvenation of the body. This fact will greatly ease the regulatory hurdles facing each individual intervention and provide incentives for private industry financing to advance their development.

A.3 Comparative Effectiveness Modeling

SENS provides a detailed roadmap of the way ahead for biogerontological engineering, based on our existing knowledge of aging damage and on the existing or foreseeable biotechnologies whose maturation will be needed for us to repair it. The first iterations of these technologies will repair the forms of aging damage that contribute most to age-related frailty, disease, and death within a currently-normal lifetime; thereafter, normal biomedical progress will iteratively refine and expand the SENS platform, repairing increasingly subtle or slowly-accumulating forms of aging damage. This will allow us to live progressively longer – and healthier – lives, until these technologies can clear all aging damage and our rejuvenation is complete.

This prospect is qualitatively illustrated in Fig. A.1. In principle, damage-removal protocols could become sufficiently comprehensive to place their beneficiaries permanently in a condition similar to those of healthy twenty- to thirty-year olds, having negligible age-related mortality risk and possessing a healthy life expectancy that is literally indefinite. A model of the impact on successive generations of progressively improved engineering interventions for various cohorts (Phoenix and de Grey 2007) is also illustrated in Fig. A.2 below.

Even a mature "gerontological," damage-prevention strategy cannot address patients' pre-existing burden of aging damage and can only slow down the ongoing progression of aging. By contrast, engineering interventions can potentially *remove* already existing aging damage, thereby *reducing* biological age in the face of increasing chronological age and leading to a progressive reduction in age-related mortality risk, suffering, and resulting medical costs. Therefore, continuing on the traditional road has a large opportunity cost, entailing the consignment of 100,000 people each and every day, indefinitely, to death from aging, and of many hundreds of times this number of person-days to progressively-increasing frailty, disability,

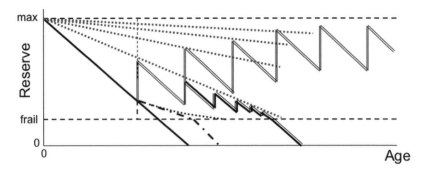

Fig. A.1 Comparison of the effects of various interventions on a middle-age person's arrival at a "frailty threshold" (loss of homeostatic reserve). "Normal" aging (*solid line*); "gerontological" intervention that halves the rate of further damage accumulation (*dashed dot*) and its iterative improvement every 7 years (the typical human mortality-rate doubling time) (*dotted line*); an "engineering" therapy that removes half the damage (*thick double line*) and its iterative improvement every 20 years (*thin double line*). (Adapted from Phoenix and de Grey 2007)

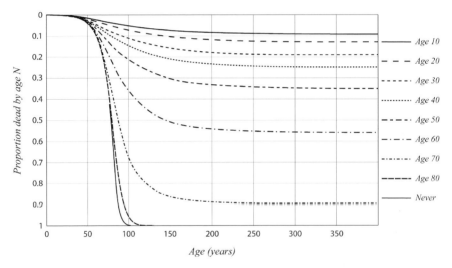

Fig. A.2 Survival curves for "normal aging" (*leftmost solid line*) and for cohorts that are aged 10, 20, ... 80 when the first therapy arrives, assuming a 42-year cycle of doubling of the efficacy of damage removal. (Adapted from Phoenix and de Grey 2007)

disease, and dementia – not forgetting the enormous cost to friends, loved ones, government entitlement budgets, and economic productivity. SENS Foundation believes we know enough *today* – both about aging damage, and about the path toward its repair – to start advancing toward a cure for aging using engineering principles, with expected success on a foreseeable timescale, subject only to the magnitude of investment in this research.

A.4 SENS Foundation-Funded Research

The clear advantages of the engineering paradigm make hastening progress in the development of the SENS platform the focus of the Foundation's core venture: direct investment in the development of SENS's constituent biotechnologies (Table A.1). SENS Foundation has continued and expanded the SENS research projects formerly under the aegis of the Methuselah Foundation, and ongoing fundraising continues to expand our research budget, and the number and size of our investments.

SENS Foundation is based in California, but has global reach through its staff, funded researchers, and wider network of contacts. The day-to-day operations of the Foundation are overseen by its senior management team, who themselves report to the SENS Foundation Board. Research priorities are identified by the Chief Science Officer and Research Advisory Board (RAB), based on their ability to advance the progress of the SENS platform. Funding priorities are identified on a critical-path basis. While a complete panel of anti-aging biotechnologies will be required in order to implement a comprehensive rejuvenation program, progress toward clinical implementation of those interventions stands at widely disparate stages of development; the Foundation allocates funding to the foreseeable bottlenecks in progress of the platform as a whole.

Thus, some SENS-platform interventions are not subject to Foundation funding because they have progressed to the very rim of the therapeutics pipeline independently, rendering additional Foundation funding superfluous. A notable example of this is immunotherapy for the removal of the accumulating beta-amyloid protein (Solomon 2008); the recent FDA approval of the first human clinical trial of embryonic stem cell-based therapy in patients with acute spinal cord injury (Geron 2009) is another. In other cases, research is still at a relatively preliminary stage, but the level of scientific interest and private and government funding is sufficient to give confidence in the developmental trajectory, leaving the Foundation in a similar position as with interventions already in late testing.

However, many of the key components of a comprehensive panel of rejuvenating biotechnologies have been the subject of little or no development to date, despite their biomedical importance. In these "bottleneck" areas, seed funding from the Foundation can have a very large effect on the rate of progress of that plank of the platform, and thus toward the ultimate goal of a comprehensive panel of rejuvenation biotechnologies.

The SENS Foundation Research Center (SENSF-RC) is the Foundation's intramural facility, and undertakes mid-level translational research between early cellular models and work in mammalian systems within the SENS platform. Originally based in Tempe, AZ, the SENSF-RC has relocated to Sunnyvale, CA, to facilitate collaboration with the strong life science/biotech industry in the Bay Area.

Extramural research grants are directed to academic centers with specialized expertise for the relevant projects, and with consideration for the availability of matching grants and/or existing infrastructure to leverage Foundation dollars to produce research deliverables.

The small size and highly focused remit of the Foundation allows for *a relatively rapid review of proposals and dispersions of research investments. Foundation* *grants are typically for approximately US $100,000 per year per researcher (typically a graduate student or postdoc) per annum, tailored to the needs and available resources of the principal investigator.*

A brief discussion of active and projected Foundation research activities – *including areas where the Foundation is actively inviting proposals from extramural researchers* – follows.

A.4.1 LysoSENS: Medical Bioremediation

The initial goal of the "LysoSENS" project is to identify microbial enzymes capable of breaking down the metabolic waste products that accumulate in aging cells, impair or destroy their function, and thereby contribute to a range of age-related diseases, including atherosclerosis, Alzheimer's disease, age-related macular degeneration (primarily the "dry" form) (de Grey et al. 2005), and other age-related disorders. Once identified, suitably-modified versions of such enzymes would be used to fortify our cells' own enzymatic armamentarium, thereby allowing the removal of such wastes and restoring normal cell function.

Currently, three Foundation-funded labs (the SENSF-RC, and labs at Rice and Arizona State Universities) are employing complementary approaches to pursue this benchmark. Significant progress has been made in identifying and characterizing enzymes for the degradation of *7-ketocholesterol*, an oxidation product of serum cholesterol widely thought to be central in the progression of atherosclerotic plaque (Rittmann and Schloendorn 2007; Mathieu et al. 2008; Mathieu et al. 2009).

Additionally, the ASU team has (with outside collaboration) identified other enzymes that efficiently degrade *A2E*, a recalcitrant metabolic byproduct whose accumulation is central to most age-related macular degeneration. The team is now collaborating with Dr. Janet Sparrow of Columbia University, who has repeated and confirmed the ASU group's results in A2E-loaded *retinal pigment epithelial cells* (RPE), the population vulnerable in macular degeneration (Rittman et al. 2008), and is now being funded to lend her expertise to the further characterization and development of the candidate enzymes. After identifying the products of A2E degradation by the enzymes, her lab will perform preliminary safety and efficacy screens in RPE. If the cell culture results are promising, the Foundation will fund the testing of potentially viable enzymes in a mouse model of macular degeneration similar to *Stargardt's disease*, a human congenital form of macular degeneration.

As the Foundation's research budget continues to expand, *we will expand these intramural and extramural research efforts* in the identification, characterization, and testing of important microbial enzymes.

A.4.2 MitoSENS: Obviation of Mitochondrial Mutations

The "MitoSENS" project aims to address another form of aging damage: mutations in mitochondrial DNA. Although most of the genes that encode the proteins used in the mitochondrial energy-harvesting process are safely housed in the cell nucleus, 13 such proteins are instead encoded by genes situated within the mitochondria themselves. These genes are extremely vulnerable to mutations resulting from the ongoing production of reactive waste products during energy production. While the *mechanisms* linking mitochondrial mutations to aging and age-related pathology continue to be debated (de Grey 2002, 2004; Khrapko et al. 2004, 2006; Smigrodzki and Khan 2005), such mutations (and especially large mitochondrial DNA deletions) accumulate in individual postmitotic cells with aging and are widely held to play an important role in human aging and age-related disease.

The technical hurdles of repairing such mutations are not generally thought to be surmountable with foreseeable biotechnologies. There are, however, a small number of potentially viable solutions to the problem of mitochondrial mutations. The most promising is *allotopic expression* (AE), the placement of "backup copies" of the 13 vulnerable genes into the nuclear genome, suitably modified to ensure the delivery of their encoded proteins from their remote site of synthesis in the cell body to their appropriate location within the mitochondria. Even if the native mitochondrial DNA is irreparably damaged, the existence of this alternative source of functional energy-harvesting proteins would prevent any dysfunctional metabolic consequences, effectively rendering any such mutations harmless.

Foundation-funded work on AE is centered in the Quinze-Vingts National Center of Ophthalmology in Paris, to support ongoing work based on the extremely promising early fruits of a novel AE strategy developed by Dr. Marisol Corral-Debrinski (Kaltimbacher et al. 2006), extending most recently to the initiation and reversal of an animal model of the mitochondriopathy *Leber hereditary optic neuropathy* (LHON) (Ellouze et al. 2008). With further Foundation support, Dr. Corral-Debrinski's team is working to further validate and systematize the technique, and if successful, to extend it to obviation of other mitochondrial mutations in cellular and laboratory animal models of mitochondrial disease.

A.4.3 AmyloSENS: Removal of Pathological Extracellular Aggregates

A range of bodily proteins suffer damage that causes them to aggregate into small, soluble oligomers and larger fibrillar formations; these accumulate with age and are evidently pathological. The most prominent of these is *beta-amyloid*, widely thought to be a key pathogenic species in Alzheimer's disease; others of concern are aggregated *islet amyloid polypeptide (IAPP)*, apparently responsible for beta-cell death in diabetes; *alpha-synuclein* in Parkinson's disease and other so-called synucleinopathies; and wild-type *transthyretin (TTR)*, deposition of which underlies senile cardiac amyloidosis (SCA) which increasingly impairs cardiac function with age and appears to be a significant contributor to death in centenarians (Tanskanen et al. 2008; Bernstein et al. 2004; Hashizume et al. 1999; Steve Coles, Los Angeles Gerontology Research Group (LA-GRG), personal communication).

Multiple pharmaceutical companies and academic labs are actively pursuing immunological removal of beta-amyloid, and there is early work with similar strategies for human IAPP (Lin et al. 2007) and alpha-synuclein (Masliah et al. 2005) aggregates in transgenic mice. To our knowledge, however, *no group is yet pursuing immunotherapeutic removal of wild-type human TTR. The Foundation is now inviting proposals for projects aimed at the development and subsequent testing in transgenic mice of such an intervention.*

A.4.4 OncoSENS: Containment of All Cancers

SENS Foundation believes that a comprehensive program for the indefinite extension of healthy human life requires a comprehensive cure for all cancers. This is a very challenging requirement, because of cancer's uniquely evolutionary nature: due to its rapidly proliferating, genetically unstable, highly heterogeneous subpopulations, a malignancy is an engine of rapid evolution, generating mutations that ultimately elude the mechanism of any given intervention that targets a particular feature of the original cancer. This perspective and the WILT (*"Whole-body Interdiction of Lengthening of Telomeres"*) plan for dealing with it have been described elsewhere (de Grey 2005; de Grey et al. 2004) and in Chapter 22 of the present volume (de Grey 2010), and the reader is referred to this chapter for additional discussion. WILT is the most complex and technically challenging part of the SENS platform – but, as a roundtable report including recognized experts in all aspects of this intervention has concluded (de Grey et al. 2004), no aspect is theoretically insurmountable; in fact, all of the required biotechnology is foreseeable from current developments, positioning WILT firmly within the timescales of the rest of the SENS platform.

One concern surrounding WILT is that some reports suggest that the telomere maintenance machinery may have indispensable physiological functions beyond the actual lengthening of telomeres (Fauce et al. 2008; Ju et al. 2007; Passos et al. 2007; Flores et al. 2005; Sarin et al. 2005; Liu et al. 2004). The evidence in all cases is ambiguous, but SENS Foundation is now funding research by Dr. Zhenyu Ju (formerly of Dr. K. Lenhard Rudolph's laboratory and now at the Chinese Academy of Medical Sciences) to help resolve the issue by monitoring the effects of transplanting telomerase-deficient but *ex vivo* telomere-extended bone marrow into mice. Additionally, a few genes have recently been identified that appear to be involved in noncanonical, non-cancer telomere maintenance, and SENS Foundation is now funding work at SENS-RC and extramurally to test their involvement in ALT cancers.

Finally, there is the question of the intuitive, widely-accepted, but theoretically and empirically debatable (de Grey, 2007), notion that age-related, somatic nuclear (epi)mutations other than those leading to cancer accumulate to a degree that meaningfully limits a currently-normal lifespan. For some time now, this important question has been the chief research interest and occupation of Dr. Jan Vijg, an innovative and productive researcher in this area, who is enthusiastic to further test this hypothesis. The Foundation has now funded a project developed with Dr. Vijg for a more rigorous test of this idea in the brains of aging mice.

A.4.5 ApoptoSENS: Ablation of Maladaptive Cells

During aging, a variety of cells undergo maladaptive changes causing them to enter into abnormal metabolic states that appear to be highly detrimental to the organism as a whole, and which simultaneously render them resistant to cell death programs that are designed to ensure the removal of such cells. Identified cases exhibiting this problem include visceral adipose tissue macrophages (thought responsible for the metabolic syndrome associated with abdominal obesity), "senescent" cells (which secrete factors that create a permissive environment for cancer), and "anergic" CD8+ and possibly CD4+ T-cells (which appear to be major contributors to immunological senescence) (de Grey 2006). ApoptoSENS is dedicated to the selective removal of such cells, employing techniques similar to those used in modern targeted cancer therapies.

SENS Foundation is currently funding an experimental protocol for restoration of youthful immune function in aged mice involving clearance of anergic T-cells – combined with efforts aimed at restoring the declining function and cellularity of the thymus gland – in the lab of prominent immunosenescence researcher Dr. Janko Nikolich-Zugich. Separately, the SENSF-RC recently completed a proof-of-concept study, in which putative anergic T-cells were "scrubbed" from the blood of aged mice ex vivo using magnetic nanoparticles coated with antibodies to target cell surface markers, reducing cell count 7.3-fold (Rebo et al. 2010). *We would also be open to proposals and preliminary discussions for similar projects for the depletion of age-related accumulations of other dysfunctional cell types.*

A.4.6 GlycoSENS: Cleavage of Extracellular Protein Crosslinks

Long-lived structural proteins undergo occasional nonenzymatic crosslinking by advanced glycation endproducts (AGE). Because such crosslinks tend to accumulate with age, such tissues progressively lose their elasticity and other functionality. In the cardiovascular system, this leads to age-related diastolic dysfunction in the heart and the systemic rise in systolic blood pressure and decline in resilience in the vasculature, resulting in stroke, kidney damage, and other pathologies of major organs.

Because nonenzymatic AGE crosslinks are structurally distinct from physiologic proteins (including the products of enzymatic glycosylation reactions) they can in principle be selectively targeted for cleavage by appropriately-designed pharmaceuticals. Indeed, the lead member of one class of such drugs (*Alagebrium*/ALT-711) has been documented to restore elasticity and improve cardiovascular functions and outcomes in a range of animal models, and advanced into limited human clinical trials (Zieman et al. 2007; Thohan et al. 2006; Little et al. 2005; Bakris et al. 2004; Kass et al. 2001). However, the results of those trials have been disappointing, likely because the class of crosslinks that Alagebrium severs (alpha-diketone bridges) is relatively rare in humans; indeed, current evidence suggests that the unrelated *glucosepane* is by far the most important AGE crosslink in human collagen.

The development of a pharmaceutically-acceptable glucosepane-cleaving agent is therefore the clear priority within this arm of SENS, and the Foundation invites preliminary discussion and/or proposals from investigators with relevant medicinal chemistry or other expertise to engage in a research project to identify, modify, or design such an agent.

Another likely contributor to age-related cardiovascular stiffening is mechanical fatigue of the large arteries, mediated by the fraying of the anchored lamellar structures responsible for the cushioning of pulse pressure in the arterial wall (O'Rourke, 2007). If such mechanical fatigue proves to be the dominant contributor to arterial stiffening during the course of a currently-normal lifetime, then a solution based on advanced tissue engineering to restore lamellar connectivity may abrogate or greatly forestall the need for a pharmaceutical AGE-cleaver. Work on such a solution is in progress, albeit only in very early stages at this time (e.g. Zavan et al. 2008; Gao et al. 2008; Mendelson et al. 2007).

A.4.7 RepleniSENS: Restoration of Lost Cellularity

As humans age, cells are continuously lost to mechanical damage, internal and external toxins, accumulation of metabolic damage, and active removal by cell death and senescence programs. In long-lived tissues in which cell replacement is limited, this leads to gradual loss of tissue function, contributing to global aging and a range of age-related diseases. These facts are already widely appreciated, and thus nearly all of the relevant clinical indications are already under vigorous pursuit by biotech companies and academic institutions. Since the marginal gains that might be reaped by additional investment from the Foundation at this stage are therefore likely negligible, we have no plans for RFPs in this area.

The principal exception to this generalization is thymic regeneration, which (perhaps because it is not linked to a narrowly-defined "disease" indication) has been the subject of remarkably little work; moreover, much of the work that has been done has focused on enhancing residual organ function using 'gerontological' strategies, rather than restoring cellularity (Chidgey et al. 2007). As noted above (Section A.4.5), SENS Foundation is currently testing a combination immunorejuvenation protocol that includes thymic cell therapy in aging mice; if results are positive, we anticipate the advancement of this protocol for testing in a nonhuman primate model.

A.5 The Past, the Present, and the New Future of Aging Research

As most biogerontologists will readily admit, a great deal of current research investment aimed at alleviating the suffering and loss of life expectancy caused by the "diseases of aging" is ultimately misplaced. Conventional medical charities, and even the National Institutes of Health, continue to pursue the piecemeal treatment of age-related diseases in isolation from the underlying biological aging process. This approach has been successful in alleviating much of the *early* mortality resulting from poor social conditions and lifestyle choices, but is increasingly ill-suited to a population in which increasingly large numbers of people reach the limits of a 'normal' lifespan and the synergistic, exponential increase in frailty which comes with aging becomes the dominant reason for disability, death, and disease.

A few organizations – primarily the Buck Institute, the Ellison Medical Foundation, and the NIA's Division of Aging Biology – recognize this fact, and have focused some of their energies and resources on research whose ultimate goal is intervention in the basic processes of aging. However, all of these organizations continue to pursue their new goal within a framework inappropriately modeled on conventional disease research science and pharmacology: the "gerontological" paradigm.

SENS Foundation's strategy is therefore unique, having pioneered and embraced the new "engineering" approach in biomedical gerontology and initiated targeted, critical-path research to advance its core rejuvenation biotechnologies. Such research investments can be predicted to yield more rapid and more robust medical advances in the prevention of age-related disease, disability, dementia, and ultimately risk of death, delivering larger expansions of youthful, healthy lifespan than those focused on conventional disease-oriented medicine or even traditional biogerontological research.

As funding continues to expand, SENS Foundation will accelerate the movement of core SENS science through the therapeutic pipeline and on into the clinic, reversing aging damage and restoring youthful health and vigor.

We invite investigators to submit preliminary sketches of research project proposals for funding, or via email to Dr. de Grey <aubrey@sens.org>

References

Bakris GL, Bank AJ, Kass DA et al. (2004) Advanced glycation end-product cross-link breakers. A novel approach to cardiovascular pathologies related to the aging process. Am J Hypertens 17(12 Pt 2):23S–30S

Bernstein AM, Willcox BJ, Tamaki H et al. (2004) First autopsy study of an Okinawan centenarian: absence of many age-related diseases. J Gerontol A Biol Sci Med Sci 59(11):1195–1199

Chidgey A, Dudakov J, Seach N, Boyd R (2007) Impact of niche aging on thymic regeneration and immune reconstitution. Semin Immunol 19(5):331–340

Ellouze S, Augustin S, Bouaita A et al. (2008) Optimized allotopic expression of the human mitochondrial ND4 prevents blindness in a rat model of mitochondrial dysfunction. Am J Hum Genet 83(3):373–387

D'Autréaux B, Toledano MB (2007) ROS as signalling molecules: mechanisms that generate specificity in ROS homeostasis. Nat Rev Mol Cell Biol 8(10):813–824

de Grey AD (2002) The reductive hotspot hypothesis of mammalian aging: membrane metabolism magnifies mutant mitochondrial mischief. Eur J Biochem 269(8):2003–2009

de Grey AD (2003) An engineer's approach to the development of real anti-aging medicine. Sci Aging Knowledge Environ 2003(1):VP1

de Grey AD (2004) Mitochondria in homeotherm aging: will detailed mechanisms consistent with the evidence now receive attention? Aging Cell 3(2):77

de Grey AD (2005) Whole-body interdiction of lengthening of telomeres: a proposal for cancer prevention. Front Biosci 10:2420–2429

de Grey AD (2006) Foreseeable pharmaceutical repair of age-related extracellular damage. Curr Drug Targets 7(11):1469–1477

de Grey AD (2007) Protagonistic pleiotropy: Why cancer may be the only pathogenic effect of accumulating nuclear mutations and epimutations in aging. Mech Ageing Dev 128(7–8): 456–459

de Grey A (2010) WILT: Necessity, feasibility, affordability. In: Fahy GM, West M, Coles LS, Harris SB (eds) The Future of Aging: Pathways to Human Life Extension. Springer, Berlin Heidelberg New York, 667–684

de Grey AD, Alvarez PJ, Brady RO et al. (2005) Medical bioremediation: prospects for the application of microbial catabolic diversity to aging and several major age-related diseases. Ageing Res Rev 4(3):315–338

de Grey AD, Ames BN, Andersen JK et al. (2002) Time to talk SENS: critiquing the immutability of human aging. Ann NY Acad Sci 959:452–462

de Grey AD, Campbell FC, Dokal I et al. (2004) Total deletion of in vivo telomere elongation capacity: an ambitious but possibly ultimate cure for all age-related human cancers. Ann N Y Acad Sci 1019:147–170

de Grey AD, Rae MJ (2007) Ending Aging: The Rejuvenation Breakthroughs that Could Reverse Human Aging in Our Lifetime. St. Martin's Press, New York, NY (ISBN 0-312-36706-6)

Fauce SR, Jamieson BD, Chin AC, et al. (2008) Telomerase-based pharmacologic enhancement of antiviral function of human CD8+ T lymphocytes. J Immunol 181(10):7400–7406

Flores I, Cayuela ML, Blasco MA (2005) Effects of telomerase and telomere length on epidermal stem cell behavior. Science 309(5738):1253–1256

Gao J, Crapo P, Nerem R, Wang Y (2008) Co-expression of elastin and collagen leads to highly compliant engineered blood vessels. J Biomed Mater Res A 85(4):1120–1128

Geron (Press Release) (2009) Geron Receives FDA Clearance to Begin World's First Human Clinical Trial of Embryonic Stem Cell-Based Therapy. Available online at http://www.geron.com/media/pressview.aspx?id=1148. Accessed 2009-07-01

Hadley EC, Lakatta EG, Morrison-Bogorad M et al. (2005) The future of aging therapies. Cell 120(4):557–567

Hashizume Y, Wang Y, Yoshida M (1999) Neuropathological study in the central nervous system of centenarians. In: Tauchi H, Sato T, Watanabe T (eds) Japanese Centenarians: Medical Research for the Final Stages of Human Aging. Institute for Medical Science of Aging, Aichi, Japan, 137–154

Hayflick L (2007) Biological aging is no longer an unsolved problem. Ann NY Acad Sci 1100:1–13

Holliday R (2004) The close relationship between biological aging and age-associated pathologies in humans. J Gerontol A Biol Sci Med Sci 59(6):B543–B546

Ju Z, Jiang H, Jaworski M et al. (2007) Telomere dysfunction induces environmental alterations limiting hematopoietic stem cell function and engraftment. Nat Med 13(6):742–747

Kaltimbacher V, Bonnet C, Lecoeuvre G et al. (2006) mRNA localization to the mitochondrial surface allows the efficient translocation inside the organelle of a nuclear recoded ATP6 protein. RNA 12(7):1408–1417

Kass DA, Shapiro EP, Kawaguchi M et al. (2001) Improved arterial compliance by a novel advanced glycation end-product crosslink breaker. Circulation 104(13):1464–1470

Khrapko K, Ebralidse K, Kraytsberg Y (2004) Where and when do somatic mtDNA mutations occur? Ann NY Acad Sci 1019:240–244

Khrapko K, Kraytsberg Y, de Grey AD et al. (2006) Does premature aging of the mtDNA mutator mouse prove that mtDNA mutations are involved in natural aging? Aging Cell 5(3):279–282

Kirkwood TB (2008) A systematic look at an old problem. Nature 451(7179):644–647

Lin CY, Gurlo T, Kayed R et al. (2007) Toxic human islet amyloid polypeptide (h-IAPP) oligomers are intracellular, and vaccination to induce anti-toxic oligomer antibodies does not prevent h-IAPP-induced beta-cell apoptosis in h-IAPP transgenic mice. Diabetes 56(5):1324–1332

Little WC, Zile MR, Kitzman DW et al. (2005) The effect of alagebrium chloride (ALT-711), a novel glucose cross-link breaker, in the treatment of elderly patients with diastolic heart failure. J Card Fail 2005;11(3):191–195

Liu L, DiGirolamo CM, Navarro PA et al. (2004) Telomerase deficiency impairs differentiation of mesenchymal stem cells. Exp Cell Res 294(1):1–8

Masliah E, Rockenstein E, Adame A et al. (2005) Effects of alpha-synuclein immunization in a mouse model of Parkinson's disease. Neuron 46(6):857–868

Mathieu J, Schloendorn J, Rittmann BE, Alvarez PJ (2008) Microbial degradation of 7-ketocholesterol. Biodegradation. 19(6):807–813

Mathieu JM, Schloendorn J, Rittmann BE, Alvarez PJ (2009) Medical bioremediation of age-related diseases. Microb Cell Fact 8:21

Mendelson K, Aikawa E, Mettler BA et al. (2007) Healing and remodeling of bioengineered pulmonary artery patches implanted in sheep. Cardiovasc Pathol 16(5):277–282

Nadon NL, Strong R, Miller RA et al. (2008) Design of aging intervention studies: the NIA interventions testing program. AGE 30(4):187–99

National Institute on Aging. Interventions Testing Program (ITP) (2008) Online at http://www.nia.nih.gov/ResearchInformation/ScientificResources/InterventionsTestingProgram.htm. Accessed 2008-05-21

O'Rourke MF (2007) Arterial aging: pathophysiological principles. Vasc Med 12(4):329–341

Passos JF, Saretzki G, von Zglinicki T (2007) DNA damage in telomeres and mitochondria during cellular senescence: is there a connection? Nucleic Acids Res 35(22):7505–7513

Phoenix CR, de Grey AD (2007) A model of aging as accumulated damage matches observed mortality patterns and predicts the life-extending effects of prospective interventions. AGE 29(4):133–189

Rae M (2004). It's never too late: calorie restriction is effective in older mammals. Rejuvenation Res 7(1):3–8

Rebo J, Causey K, Zealley B, Webb T, Hamalainen M, Cook B, Schloendorn J (2010) Whole-animal senescent cytotoxic T cell removal using antibodies linked to magnetic nanoparticles. Rejuvenation Res 13(2–3):298–300

Rittman BE, Kemmish K, Schloendorn J, Jiang L (2008) Cleaning out the junk with medical bioremediation. Understanding Aging: Biomedical and Bioengineering Approaches. Conference Program and Abstract Book:12

Rittmann BE, Schloendorn J (2007) Engineering away lysosomal junk: medical bioremediation. Rejuvenation Res 10(3):359–365.33

Sarin KY, Cheung P, Gilison D et al. (2005) Conditional telomerase induction causes proliferation of hair follicle stem cells. Nature 436(7053):1048–1052

SENS Foundation (2010) Scientific Publications. Online at http://www.sens.org/publications. Accessed online 10-06-14

SENS Foundation Website: http://www.sens.org

Smigrodzki RM, Khan SM (2005) Mitochondrial microheteroplasmy and a theory of aging and age-related disease. Rejuvenation Res 8(3):172–198

Solomon B (2008) Immunological approaches for amyloid-beta clearance toward treatment for Alzheimer's disease. Rejuvenation Res 11(2):349–357

Tanskanen M, Peuralinna T, Polvikoski T et al. (2008) Senile systemic amyloidosis affects 25% of the very aged and associates with genetic variation in alpha2-macroglobulin and tau: a population-based autopsy study. Ann Med 40(3):232–239

Thohan V, Koerner MM, Pratt CM, Torre G (2006) Structural, hemodynamic and clinical improvements among patients with advanced heart failure treated with Alagebrium (a novel oral advanced glycation end-product crosslink breaker). J Heart Lung Transplant 25(2 Suppl 1):151

Zavan B, Vindigni V, Lepidi S et al. (2008) Neoarteries grown in vivo using a tissue-engineered hyaluronan-based scaffold. FASEB J 22(8):2853–2861

Zieman SJ, Melenovsky V, Clattenburg L et al. (2007) Advanced glycation endproduct crosslink breaker (alagebrium) improves endothelial function in patients with isolated systolic hypertension. J Hypertens 25(3):577–583

Appendix B
The Manhattan Beach Project

David Kekich

B.1 Introduction: The Problem(s) and the Need

Traditional venture capital funding for companies focused on life extension technologies has generally been difficult to acquire. Such technologies are perceived as being too risky, having exit horizons that are too far into the future, and even lacking credibility.

The Maximum Life Foundation (MaxLife), a 501(c)3 corporation, was formed in 1999 to address this problem. It currently raises funds as a not-for-profit corporation to support both basic and applied research. The Foundation may support some of the MaxLife Capital projects described in this chapter as well as new projects that it will review over time. The Foundation has only raised a few hundred thousand dollars to date, and has invested most of it in promising stem cell technologies. Smaller amounts have gone to support kindred organizations as well as three scientific conferences. Aggressive fund raising campaigns will start in the last quarter of 2010, and the author welcomes suggestions as to how to make them more fruitful. You can find more information at www.MaxLife.org.

MaxLife strongly believes in the promise of anti-aging therapeutics and has created a unique funding mechanism tailored to the development of such treatments on the one hand and to the exceptional reward of investors on the other. Because of this it is believed that the MaxLife approach may be a particularly attractive way for ideas such as those described in this book to be converted into anti-aging treatments.

Historically, significant financing for biomedical technologies in the U.S. has come from government-related and other private non-profit sources. Over time, the changing political economy has led to a shortfall of these traditional funding sources.[2] This is especially true for life extension technologies targeted at maximizing longevity while maintaining a high quality of living.

D. Kekich (✉)
Maximum Life Foundation, Huntington Beach, CA 92648, USA
e-mail: kekich@maxlife.org

[2]This is not a world-wide trend. The governments of Australia, China, Singapore and Taiwan, for example, have launched expansive programs of aggressive government-based financial and infrastructure support to capitalize on the financing difficulties occurring in the U.S.

For a while, private equity venture capital fund managers perceived this trend as an opportunity for venture capital placements. Some prominent successes have been achieved. However, this biomedical sector generally could not keep pace with the high returns obtained in ever decreasing time periods in other non-regulated (or less regulated) technology sectors.

Most venture funds have become increasingly impatient to harvest returns from their investments, many expecting an "exit" within only 2–4 years. This trend has been adverse for many biomedical technologies, since they generally have a much longer gestation cycle than communications, electronics or information technologies. In addition, technologies requiring regulatory approvals further increase development costs and lengthen the investment cycle. Since most biotech investments will require FDA approval, compliance is a hurdle that must be overcome before most of the developments will reach the US and most other markets.

Other forms of non-currency financial underpinnings have also begun to dry up for the biomedical sector. The 1990s were marked by a rapid expansion of technology incubators (both public and private) that provided infrastructure and mentoring to technologists. Many of these incubators evolved into real estate projects, whereby a building or business park is developed to be inhabited by promising technology companies. Too frequently, the incubator primarily became simply a vehicle for cheap subsidized rent for entrepreneurial ventures without sufficient process or controls to direct successful outcomes. And, many incubators found it was very difficult to fill their projects with clients, or to evict those clients that weren't making progress. The incubator managers have been under continuous economic pressure to fill space in their project buildings, and consequently many end up lowering selection criteria to increase occupancy. As a result, incubator outcomes generally have failed to produce many successful enterprises.

Because of the preceding events, a clear picture of needs has begun to develop:

- Provision of long-term "patient" capital to commercialize promising biomedical technologies over time frames of up to ten years or longer
- Supplemental backing in terms of a developmental infrastructure provided by premium services providers and managerial oversight
- Reasonable assurance of good financial returns to investors

I outline here strategies developed by MaxLife that can in principle overcome these and most other barriers to raising the necessary capital. One particular strategy has the advantages of combining the benefits of capital preservation, liquidity, annual dividends and attractive venture capital returns.

In fact, this strategy, known as LIVES^{TM}, may actually be able to give investors a higher than average return on investment with lower than average risk compared to traditional investment mechanisms. LIVES^{TM}, as well as a more traditional funding approach, could provide the capital required to fund much of the research described in this book.

These funding strategies are designed to maximize health and longevity in minimum time ... and maximize investment returns while minimizing risk. It

Appendix B: The Manhattan Beach Project

is MaxLife's intent to generate significant funds from current associates and to be in the position to fund some of the more important emerging life extension technologies.

B.1.1 Visualizing the Need for Investment in Life Extension

An exercise that is useful in bringing home the need for investing now in life extension technologies begins with constructing a diagram that represents the length of the investor's life. In Fig. B.1, each square represents one month, and each row represents two years for a total of 80 years.

Ten years ago, the author drew this chart for himself on his 55th birthday. You'll notice some of the squares are filled in, and some are empty. The gray squares represented the months I already lived. Actuarial tables tell us men 55 years of age can expect to live to about 80 years of age (statistically, half die before the age of 80 and half after). The empty squares therefore represented the number of months I could have expected to have left.

Based on this diagram, I noticed a serious problem: I had already used up most of my years... and the most healthy, vibrant and productive ones at that!

So after some thought, I came up with the solution shown in Fig. B.2: I needed to add more squares!

And that's how Maximum Life Foundation was born.
This idea is at the heart of the strategy:
"Live long enough to live as long as you want."

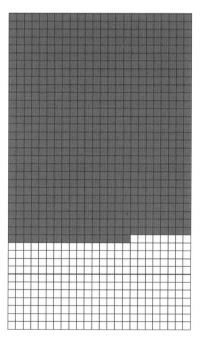

Fig. B.1 Diagram of a typical human life today. Each row equals 24 months, each filled *box* is a year already lived, and each *white box* is a year of remaining life

Fig. B.2 Diagram of a modified human life

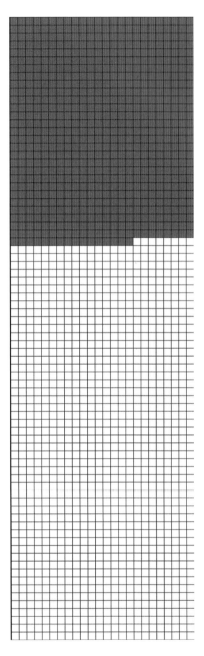

MaxLife intends on fast tracking key research and making the benefits available to most people alive today.

B.1.2 The New Timescale for Progress in Biology

MaxLife intends to achieve extreme life extension in the intermediate future. What could justify such an aim, especially in light of the fact that we haven't even been able to cure cancer after spending hundreds of billions over several decades?

In the modern era, our knowledge has been advancing by *quantum leaps* compared to most of human history. For instance, scientific knowledge doubled from the year 1 to 1500 A.D. But by 1967, it doubled *five more times* ... and each time, faster than before. Now it doubles in less than 5 years.

Supercomputers, like the kind now being used in bioinformatics, are part of the reason. These computers can do some experiments in *15 seconds* that used to take years.

Also, for example, newly developed research tools, gene chips, can do some tissue studies in *hours* – or even *minutes* – that used to take *years* of animal studies (or that couldn't be done at all).

Ray Kurzweil made perhaps the most profound observation (Kurzweil 2005). He observed that the rate of change itself is accelerating. This means the past is no longer a reliable guide to the future. The 20th century was not 100 years of progress at *today's* rate but, rather, was equivalent to about 20 years at the rate of change in the year 2000 because we've been speeding up to current rates of change. And we'll make another 20 years of progress at that rate, equivalent to that of the entire 20th century, by 2014. Then we'll do it again in just 7 years, and so on.

Because of this exponential growth, the 21st century should equal *20,000* years of progress at year 2000s rate of progress (Kurzweil 2005) – 1,000 times greater than what we witnessed in the 20th century, an amazingly progressive century in itself.

These are some of the reasons why we foresee dramatic interventions in the aging process in the next ten to twenty years. We're making progress faster, better and cheaper, and at a continually accelerating rate (Kurzweil and Grossman 2010).

But why not make progress faster still? To further accelerate progress, MaxLife has positioned itself directly in the path of this major trend.

B.1.3 Aging Intervention as a Major Emerging Growth Industry

The human life-extension segment of the life sciences industry, commonly known as "Anti-Aging Medicine", is still in its infancy, but some think it will grow to become a trillion-dollar industry (Pilzer 2002), and it offers profit opportunities today. As stated above, the current venture capital community is not designed to take advantage of many of the most promising opportunities in this area.

The largest segment of the US population, the Baby Boomer generation, is entering the anti-aging marketplace right now. Baby Boomers are already interested in longer quality of life, looking and feeling younger, enjoying a longer period of sexual health, tissue regeneration, memory enhancement, curing the very diseases they will face shortly, extending the healthy human life span, and so on.

Tomorrow's pursuits are more ambitious. Sights are set on maximum intervention of the aging process – reversing the biological effects of aging. Throughout the last half of the 20th century, the international medical community invested heavily in interventions that treat aging-related diseases and conditions. As a result, significant advancements designed to delay the onset of aging-dependent health challenges are emerging, and research is proceeding that may gradually lead to the prevention or cure of most age-dependent chronic diseases.

B.2 A Brief History of MaxLife Activities

Over the past ten years, MaxLife has sponsored four life extension scientific conferences. About a dozen world-class research scientists attended each working weekend. The first two were brainstorm sessions to better understand how to intervene in the aging process. The objective was to develop a cross disciplinary approach which by the year 2030 would dramatically intervene in human aging, and allow us to repair much of the damage done by the aging process – and to identify specific market opportunities for such technologies in the interim.

The results of these dialogs were spectacular, and somewhat surprising. The meetings brought to light several interesting things, including:

- How to get around two of the biggest obstacles of turning good science into marketable products.
- Most of the best independent and university researchers in this field spend valuable time teaching, writing grant proposals and performing administrative duties rather than thinking creatively and doing research. Funding will enable them to maximize their talents and productivity.

We also concluded, in June of 2000, that the upcoming convergence among biotechnology, information technology and nanotechnology will change the world as we know it. This conclusion later received validation by the National Science Foundation's 450-page report titled, *Converging Technologies for Improving Human Performance* (Roco and Bainbridge 2002). The likely imminence of such a convergence has further impelled us to fast-track the MaxLife mission.

B.2.1 The Manhattan Beach Project

Two years after the author's "squares exercise," MaxLife launched the "Manhattan Beach Project". This focused and targeted all-out assault on the human aging process was spawned during MaxLife's first international scientific conference on June 24th and 25th, 2000 in Manhattan Beach, California.

The plan is to assemble some of the world's leading anti-aging researchers in a focused effort, with definite deadlines and ambitious goals, much like the "Manhattan Project" that ended World War II. Unlike its namesake, the Manhattan "Beach" Project's goal is life building rather than life threatening. Another difference is that the Manhattan Beach Project will be a commercial enterprise driven by the private sector. It marries a scientific roadmap developed by our non-profit foundation to intervene in the human aging process with a for-profit enterprise, called MaxLife Capital. MaxLife Capital has developed a traditional funding strategy as well as LIVES^{TM}, an optional, reduced-risk, financial model to fund the research and development. It will carry out these goals by funding and developing technologies having growth or value creation potential in the fields of anti-aging, medical nanotechnology, and artificial intelligence.

The Project's core team includes many of the world's leading doctors and scientists in the fields of genetics, genomics, stem cell biology, gerontology, nanotechnology and artificial intelligence. In addition to these leading scientists, MaxLife's Advisory Boards also include top business, marketing and finance minds.

Starting with the inaugural scientific conference, MaxLife's scientific advisors and others formulated a scientific roadmap to control and reverse the aging process. Now that the science is being tackled by some of the world's leading scientists, the author decided to concentrate not on the scientific mysteries related to aging, but on the problem of overcoming the financial roadblocks standing in the way of funding the research. Since investors typically want safety and a healthy return, the primary focus was to establish a funding strategy that eliminates the reasons investors don't invest in an area that is perceived as risky and uncertain.

FDA approval is one such issue. MaxLife plans on helping take companies through Phase I or even possibly Phase II where necessary. Some technologies may be developed and tested under FDA guidelines, but preferably in countries where they may be commercialized with fewer regulatory hurdles. And finally, some technologies may be developed into nutraceuticals that need little or no regulatory approval in the United States.

The author also wanted to make it possible for investors to be among the first to enjoy the health benefits the funding might provide. Our plans for accomplishing these goals are outlined below and will be found also in more detail and in continually updated form at www.manhattanbeachproject.com and www.maxlife.org/mbp.pdf.

B.3 Specifics of the Manhattan Beach Project

Most of the Project's resources will be dedicated to developing specific technologies in the fields of Molecular Biology; Stem Cell Therapy; Organ and Tissue

Regeneration; Whole Genome Reengineering[3]; Gene Therapy; Nutraceuticals; Pharmaceuticals; and Therapeutic and Diagnostic Devices.

Some of the focus will be in the field of nanotechnology, broken down into various types of devices and therapies, including nanorobots (cell-sized robots with molecular sized components) acting as artificial blood cells, repairing DNA, providing energy to cells, and interfacing with extracellular devices.

In addition, MaxLife intends to support companies that are striving to create human-level artificial general intelligence (AGI), which it believes can be harnessed to accelerate anti-aging research at all levels.

Table B.1 outlines potential general technologies MaxLife plans to help develop and when they might become available to benefit investors, principals, and members of the general public. Although the major technologies are projected to mature beyond typical venture capital time horizons, the reader will see that some profitable interventions may be possible in the near future. Projections are hypothetical and are best-case estimates of when the author believes they could be developed once they are fully funded.

Table B.1 Gant chart for the development of exemplary milestones

Longevity technology	Years from initiation of funding													
	1	2	3	4	5	6	7	8	9	10	12	15	20	25
Reprogram Biochemistry[1]	I P													
Nutraceuticals[2]	I	P												
Genetics[3]		I	P											
Genome Reengineering[4]							I				P			
Escape Velocity[5]											I	P		
Nanomedicine[6]													I	P

[1] *Supplementation Activity Lifestyle Anti-aging medicine Diet Stress reduction.* See report at http://maxlife.org/salads.pdf for key present strategies for human lifespan extension.
[2] New effective products are available now and others will be introduced in 2010–2011
[3] Oral supplements and diagnostic tools based on current genetic research will be available to the general public and clinics by 2011.
[4] Early versions of technology may be clinically proven by 2015 and available by 2020.
[5] "Escape Velocity" means to add 1 year for every additional year lived – best case scenario.
[6] Nanomedicine refers to aging reversal – best case scenario.
I: When technologies may be available to investors, researchers, management, large donors.
P: When technologies may be available to the general public.

---

[3] "Whole Genome Reengineering" refers to the process of gradually reconstructing genomes so they can be reengineered and become easier to reengineer over time. Viruses, e.g. CMV, HSV, HIV, do this over time, in that in various ways they corrupt the human genome. Scientists now argue that we can now take control of the process.

Table B.2 Hypothetical project areas for investment by MaxLife and projected investment requirements for each project area

Industry	Business	Annual investment
Genetics[a]	Pharmaceutical/Nutraceutical	$7,000,000
Molecular Biology	Genome Reengineering[b]	$5,000,000
Nanotechnology	Nanomedicine[c]	$2,000,000
IT	Artificial General Intelligence[d]	$4,000,000
Regenerative Medicine	Stem Cell Technologies	$3,500,000
Genomics[e]	Pharmaceutical/Nutraceutical	$1,000,000
Various	SENS[f]	$8,000,000
	Misc.	$6,000,000
	Annual Budget Years 1–3	$36,500,000
	Total 1st Year	$36,500,000

[a] "Genetics" includes modulation of gene expression.
[b] "Genome Reengineering" includes key gene therapies related to aging mitigation.
[c] "Nanomedicine" includes short-term projects that build toward the kind of capabilities indicated in Chapter 23 (by Freitas) but produce useful and profitable results in the short run.
[d] The investment in "artificial general intelligence" is based on unique ongoing research known to MaxLife.
[e] "Genomics" refers to gene chip studies targeted to the testing and production of useful new age-mitigating agents.
[f] "SENS" refers to strategically targeted areas of de Grey's "strategies for engineered negligible senescence."

Funding will be allocated among at least five synergistic industries and seven business categories. In addition, opportunities within Dr. de Gray's SENS project (de Grey and Rae 2007) and other industries and businesses falling in or outside these groups may be identified during the life of the Manhattan Beach Project.

Table B.2 breaks down the presently-targeted average annual budget for the first three years. It covers key life extension technologies. MaxLife is investigating other promising technologies as well. The budget indicated is our targeted budget and is presented for purposes of illustration. It can be ramped up to accelerate progress beyond projections or scaled back in at least some disciplines with minimal delay in projections. Beyond year 3, a substantial increase in the genome reengineering and nanomedicine budgets is anticipated. Milestones have been worked out for many viable technologies, and more technologies will be investigated as they are incorporated into the Maxlife strategic planning process.

One hypothetical overall scenario illustrating our objectives is shown in Fig. B.3.

B.3.1 Initial Vision

One key to our ambitious undertaking is to achieve at least one breakthrough that will make the world take notice. One goal is to rejuvenate lab animals (turning old mice into biologically "young" mice) while growing profitable companies in the process. This goal is illustrated by Dr. Aubrey de Grey's development of SENS, a

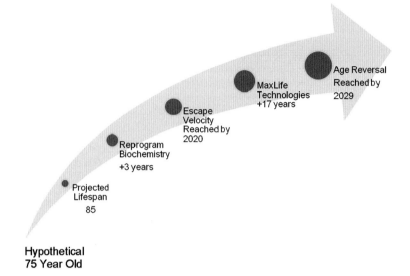

Fig. B.3 According to statistics, a 75 year old has a 50% chance of dying within 10 years (about half the 75 year olds will die before age 85 and half after; 65 year olds can expect to die by age 81 without intervention). However, if the Manhattan Beach Project is successful, humans might potentially enjoy a series of life-extension breakthroughs leading ultimately to an open-ended and youthful future. MaxLife believes technologies developed prior to 2029 could add approximately 17 years to the subject's life

seven-pronged plan to accomplish rejuvenation based on already accepted scientific disciplines (de Grey and Rae 2007). Dr. de Grey has proposed that seven fundamental kinds of damage account for aging (generating such side effects as heart disease, arthritis, diabetes, cancer, etc.). MaxLife believes some of the SENS allied technologies could eventually lead to practical solutions for many of them.

MaxLife may develop one or more equivalent proofs of concept. Once rejuvenation is proven in animals, the investing public should catch the vision and grasp the investment opportunities. Then all the necessary money could become available to fine tune intervention in human aging. MaxLife plans on being one of the catalysts to that next step.

MaxLife will consider funding private companies, as well as limited basic research, in some of these seven disciplines. It plans on developing other technologies as well, including select nanotechnology and artificial intelligence projects, in addition to shorter term projects such as promising nutraceuticals, cosmeceuticals and diagnostics, some of which are aptly described in this book.

MaxLife believes average human age can be extended by a few healthy decades in the intermediate future. In fact, MaxLife also has good reason to believe the Manhattan Beach Project might be able to see more than one year added to nearly everyone's expected lifespan every calendar year in about fifteen years. This may greatly improve the reader's odds of surviving long enough to take advantage of

Appendix B: The Manhattan Beach Project 837

more extreme technologies. Further advances will then be made that could ratchet up the human lifespan dramatically. MaxLife's Manhattan Beach Project spells out how a concerted effort over the next 19 years might result in *full age-reversal* capability for as little as $1.9 billion in addition to the $1 billion SENS project.

B.3.2 The LIVES^{TM} Funding Mechanism

The strategy called "LIVES^{TM}" was designed to give investors a higher than average return on investment with lower than average risk. The strategy will finance the research needed to profitably develop life extending technologies. The LIVES^{TM} acronym stands for:

1. Liquidity. Invested funds or "deposits" are redeemable, much like a bank CD.
2. Income. Depositors earn projected annual dividends.
3. Venture capital returns. Depositors earn an above average projected annual internal rate of return (IRR) over the 10 year life of the fund plus residuals from the 11th to 17th years or beyond.
4. Equity. Depositors get positions in potentially some of the world's bigger and more profitable companies. The projected returns do not reflect this potentiality.
5. Safety. Efforts will be made to insure deposits are never exposed to more than a 10% drawdown.

In the LIVES^{TM} plan, a low-risk Renewal Account is set up to generate both immediate returns for investors and funding for the sponsored research entities through a separate Accelerator Fund without relying on investments to mature. Over the past 40 years, institutional venture capital has developed an effective model for due diligence, oversight, governance and investment liquidation that can be applied to successfully manage an Accelerator Fund. Through this mechanism, investors in MaxLife Capital can expect to achieve financial returns approximately equal to venture capital rates of return.

The management goals are to give depositors liquidity, safety, annual dividends equal to certificates of deposit plus potential venture capital returns.

LIVES^{TM} improves upon the traditional venture fund application in seven fundamental ways:

1. Instead of investing or donating, Partners make refundable deposits.
2. While reducing market risk, MaxLife's institutional portfolio management team generates cash flow from the deposited funds to pay Partners annual dividends and to fund portfolio companies.
3. This model helps build in insurance against losing money in today's rapidly changing markets and technologies and is designed to put a floor under potential losses.

4. The targeted return plus residuals is based on refundable deposits, not on invested cash and includes dividends as well as carried interest.
5. After the first two years, deposits may be withdrawn with a 90 day notice.
6. The project has a ten-year term, with regular distributions, and with provisions for extensions.
7. If Partners are health care, biotech or technology companies, the Fund may joint venture with them by matching the portfolio companies with their strategic initiatives to add an immediate impact to their earnings and/or business plans.

These relationships should also accelerate the acceptance of the portfolio companies in the marketplace.

B.3.3 Supporting both Investor Needs and the MaxLife Companies

We have structured the MaxLife Capital Fund such that investments into various technologies could become profitable enough after the ninth year or sooner to self-fund two major technologies which we feel are needed to reverse aging, Genome Reengineering and Nanomedicine.

The Fund will provide intensive hands-on participation with portfolio technologies and companies. This involves a structured role in portfolio selection, development of intellectual properties, organizational and personnel development, financial oversight and product and service commercialization. It also involves continuous participation of specialty professional service providers that work with the portfolio companies throughout their life cycle. In essence, it frees the innovators up to focus on the science while business and investment professionals manage the companies.

MaxLife will manage its funds through a Fund Management team comprised of Managing Directors with approximately 100 person-years of venture capital experience in "finding and minding" successful opportunities that achieve prominence in their marketplaces, as well as healthy financial returns to investors.

It will also associate with Consortium Innovation Centers (CIC) in Southern California, which has been derived from two of the pre-eminent technology accelerators in the U.S. CIC has contracted with premium service providers such as Price-Waterhouse Coopers, Marsh & McLennan, Blaine Group, and Staubach Realty Advisors, etc. to mitigate problems before capital is invested.

MaxLife has established an investment strategy with supporting tactics that emphasizes the following criteria:

- Industry emphasis will incorporate a focus on proprietary technologies that have potential for meeting optimum life extension objectives.
- A tech-transfer emphasis, leveraging the Manager's global network to identify the best qualifying opportunities, and then, if it is advantageous, to relocate their central operations to appropriate sites

- A preference for companies that can rationally demonstrate the existence of a model of or a working proof-of-concept product(s) or process(es) that guide their R&D and/or commercialization
- Preliminary screening and on-going grooming of investment opportunities by referral to CIC's Success Formulas® process. This augments each opportunity with a virtual team of prominent management expertise by its *pro bono* sponsors in the areas of Big-4 accounting, legal services and governance, executive recruitment, risk management, etc.
- Initial individual investments ranging from $1.5 to $10 million with co-investment objectives to provide up to an additional $10 million in capital availability (some of which will be derived from members of CIC's Capitalist Club)
- Provision for follow-on investing for up to 8 years, provided predetermined progress benchmarks are achieved
- A requirement that all investees have, or will establish, an administrative headquarters located in the U.S. within 2 hours commuting time of a Fund (or affiliate) office.

B.3.4 Summary: Advantages of the MaxLife Approach

Although a number of venture capital funds invest in companies that have products to treat diseases of aging, it is believed MaxLife Capital plans the only Fund that exclusively focuses on life extension. Also, LIVES^{TM} may be the only funding mechanism capable of long-term exit strategy horizons of 7, 10 or more years, while still delivering an annual return to the investors.

Although this funding strategy is in an early stage, MaxLife is currently in discussions with several funding sources and hopes to secure a partnership or partnerships in a reasonable amount of time.

The advantageous features of the MaxLife approach are summarized in Table B.3.

Table B.3 The MaxLife value proposition

- √ Unique Niche
- √ Optional Refundable Deposits Instead of Investments or Donations
- √ Annual Dividend on Deposits with LIVES^{TM}
- √ Numerous Funding Opportunities
- √ Socially and Personally Desirable Outcomes
- √ Societal Need
- √ Huge Markets
- √ Committed Management
- √ Strong Potential for Large Return with Managed Risk
- √ Synergistic advantages of fund focus and cross-pollination of portfolio companies as a result of shared technical, market and scientific data
- √ With LIVES^{TM}, the only private venture funding structure capable of long term, 7+ year exit horizons

B.4 Parting Thoughts

The Manhattan Beach Project, as envisioned by MaxLife Capital, offers an alternative to many popular investments that may be ruining our health and shortening our lives. Examples are popular investments in such areas as fast foods, processed foods, alcoholic and soft drinks and tobacco. MaxLife Capital encourages investors to commit a portion of their portfolios to investments that cure diseases, promote wellness and extend healthy life, anywhere in the world and with any viable investment group or with any emerging company. Over time, we expect to see more and more opportunities for both institutional and private investors. Not only would adding more healthy years to people's lives be a positive and humanitarian thing to do, it could also represent one of the more important business and investment opportunities of all time.

Consider. If you extended your healthy lifespan by just 15 years, your net worth would quadruple if it compounded at 10% per year.

Consider. The human death toll in the Year 2001 from all 227 nations on Earth was nearly 55 million people. 52 million of these were "natural" deaths, and 37 million were aging related.

As identified by Robert Freitas, each one of us carries within us a unique and complex universe of knowledge, life experience, and human relationships. Almost all this rich treasury of information is forever lost when we die. If the vast content of each person's life can be summarized in just one book, then every year, natural death robs us of 52 million books, worldwide. So each year, we allow a destruction of knowledge equivalent to three Libraries of Congress.

Consider. Natural death also destroys wealth on a grand scale, an average value of about $2 million for each human life lost in developed countries, or an economic loss of about $104 trillion if that value were assigned to every life – every year. That equals the tangible net worth of the entire world.

Consider. If we can speed up developing dramatic life extending technologies by just a few years, we could ultimately save more lives than were lost in every war since the beginning of recorded history.

Consider. We can do this. And better yet, we can do it with moderate financial risk and with potentially massive financial rewards.

Can you think of a nobler legacy?

Our life spans depend mostly on how soon real anti-aging medicine is developed and commercialized. Supporting this research could be our surest path to extreme health and longevity.

Every life form fights for survival. That's how each evolved. Therefore . . . . The Purpose of Life is, and always has been. . . . To Delay and Avoid Death.

References

de Grey A, Rae M (2007) Ending Aging: The Rejuvenation Breakthroughs that Could Reverse Human Aging in Our Lifetime. St. Martin's Press, New York

Kurzweil R (2005) The Singularity is Near. Penguin Group, New York

Kurzweil R, Grossman T (2010) Bridges to Life. In: Fahy GM, West MD, Coles LS, Harris SB (eds) The Future of Aging: Pathways to Human Life Extension. Springer, Berlin, Heidelberg, New York, pp. 3–22, Dordrecht

Pilzer PZ (2002) The New Wellness Revolution: How to Make a Fortune in the New Wellness Industry. John Wiley & Sons, New York

Roco MC, Bainbridge WS (2002) Converging Technologies for Improving Human Performance. National Science Foundation, Arlington

Index

A
ABCA1, 279, 287
Accident, 7, 738, 780–782
Accumulated damage, 89–92, 130
Accumulated mutation theory, 94–95
Acetylcarnitine, 158, 180, 184–186, 190
Active neutralization, 719
Active regulator of SIRT1 (AROS), 334–335
Active safety devices, 780
Acute phase response, 298
Adaptation, 96–97, 101, 103–106, 108–114, 268, 357–358, 361, 364, 376
Adaptive immunity/adaptive immune (response, system), 95–96, 99, 103–105, 108–109, 145, 314, 322, 338–339, 376–377, 395, 524–526, 675
Adaptive pleiotropy, 103–104
Adaptive theories of aging, 104–114
Additive genetic variance, 97, 105
Adducts, glycation, 597–599, 603, 606
Adenylate cyclase, 629, 763
Adipocytes, 6, 336, 504, 590, 677, 769
Adipose and the health effects of CR, 405–406
Administration of DHEA, 240
  beneficial effects, 170–172, 240
Administration of melatonin, 247
  prolonged life span in rodents and, 247
Adrenal hyperplasia, 144
"Adrenal involution", 170, 172, 174
Advanced glycation endproducts, 588, 591–594, 596, 598–603, 605–606, 821
Advanced lipoxidation endproducts, 588, 592, 599, 602–603, 605
Aerobic, 7–8, 107, 172, 186, 262–264, 268, 270, 284–285, 350, 628
AGE-1, 100, 152, 624–628, 632–634

Age-associated diseases, 150, 589, 600
Age and CR responsiveness, 374–375
Age-related fat gain, 170
AGEs (advanced glycation endproducts), 588, 591–594, 596, 598–603, 605–606, 821
aggregation, 348, 380, 393, 458, 548, 600, 660, 766
Aging
  as a basic "ground rule" of life history "design", 169
  biological control of, 127–202
  inescapable, 97–98, 137, 172
Aging program, 88–89, 93, 152, 812
Aging-related diseases, 574–575, 577–580, 832
Airbags, 781
AKT-1, 627–628, 631, 634
Alagebrium (ALT-711), *3-(2-phenyl-2-oxoethyl)-4, 5-dimethylthiazolium*, 607, 767–768, 821
Albumin, 320
Alcor Foundation, 783
Aldehyde, 593
ALEs (advanced lipoxidation endproducts), 588, 592, 599, 602–603, 605
Allogeneic use, 323
Allotopic expression, 531, 670, 814, 819
ALT-711 (alagebrium), 607, 767, 821
ALT (Alternative Lengthening of Telomeres), 459, 675, 677–678, 680, 820
Alteon Pharmaceuticals, 607
Alternative Lengthening of Telomeres(ALT), 459, 675, 677–678, 680, 820
ALT pathway, 459, 470
Altruistic death, 145
Alzheimer's dementia, 589

843

Alzheimer's disease (AD), 12–13, 17, 32, 96, 117, 348, 396, 525–526, 529, 592, 601, 610, 759, 762, 766–767, 818–819
Amadoriase, 606–607
Amadori ene-dione, 594
Amadori product, 593, 606–607
Ambystoma mexicanum, 612
American Academy of Anti-Aging Medicine (A4M), 25, 27–30
Ames dwarf mouse, 153–154
Aminoguanidine, 603–604
Amish families, 168
AMP-activated protein kinase (AMPK), 158–159, 186, 349, 398, 400–403, 411, 631–633
Amphibians, 612
AMPK, 158–159, 186, 349, 398, 400–403, 411, 631–633
Amyloid, 600–601, 610
Amyloid-beta protein, 525–526
Amyloidosis, extracellular, 600–601
Amyotrophic lateral sclerosis (ALS), 348–349, 385, 396, 476, 762
Anabolic, 93, 169, 172, 174, 230, 243, 267, 390–391, 810
Anabolic enzymes, 93
Anabolism, 91–92, 170, 174
Anaerobic, 321
Anatomical maintenance, 776
Anemia, pernicious, 509, 604–605
Aneurysm, 283, 559
Angiogenic agents, 564
Annual plants, 106, 145
Ansell's mole rats (*Cryptomys ansellí*), 175
Antagonistic pleiotropy, 99, 103–104, 136, 138, 463
Anti-aging interventions, 26, 30, 810
  See also Pharmacological and nutritional interventions, in mammals
Anti-angiogenic agents, 564–565
Antibiotics, 18, 29, 320, 361, 575, 686, 728–729
Anticancer activity, potential, 343
Anticancer effects of CR in rodents, 378, 387–391, 411
Antigenic response, 613
Antigen-specific therapies, 322
Antimicrobial effectors, 319
Antimisfolding agent, 759, 767
Antioxidants, 27, 161, 177, 236, 247–248, 253, 339, 526, 604
  defenses, 154, 169, 172–174, 525

enzymes, 149, 160–161, 169, 173, 177, 190, 299, 301
Antisense oligonucleotides (ASOs), 335, 647–648
  against Duchenne muscular dystrophy, 647
  against hepatitis, 647
  against human immunodeficiency virus, 647
Antler fly, 94
Aortic disease, 283
Aβ peptide, 349
Apheresis, 323, 325, 734, 738
apoC III, 150
Apo E (Apolipoprotein E), 13, 301
Apolipoprotein B (apoB), 286, 648
Apolipoprotein A-I (apoA-I), 300–301
Apolipoprotein A-I mimetic peptide (D-4F), 293–300
  action of, 294–295
  and atherosclerosis, 295–296
  and brain arteriole inflammation, 297
  and cardiovascular complications, 297–298
  and endothelial health, 296–297
  and influenza A infection, 298
  Phase 1 clinical trial, 300
  and vasodilation, 296–297
  and vessel wall thickness, 296–297
Apolipoprotein J (apoJ), 292, 299
Apoptosis, 96, 105, 115–117, 158, 183, 313, 319, 332, 335–337, 339, 343–345, 376, 380, 387–390, 392, 394, 397, 400, 404, 411, 443, 522, 524, 526, 528, 579, 600, 718, 756, 763, 769, 777
  and anticancer effects of CR, 388–389
  of cancer cells, 344
Arginine, 80, 171, 319, 590, 594, 597–598
Argument for infinity, 782
Arithmetic, 573–581
AROS (active regulator of SIRT1), 334–335
Arteries, 13, 258, 283, 297, 300, 564, 589, 597, 607, 732–733, 735, 742, 751, 822
Arthritis, 117, 286, 411, 509, 588, 609, 836
Artificial DNA repair technologies, 199, 641–661
Artificial heart, 746
Artificial Intelligence, 4, 16–18, 833, 836
Ascites, 310–312
Asparagine, 595
Aspartate, 396, 591, 595, 598
Asphyxia, 738
Aspirin, 171, 285, 603–604, 740
Associative learning, 349

Atherosclerosis, 16, 117, 155, 238, 240, 252, 282, 285, 287, 293–296, 298–301, 373, 453, 462, 476, 578, 598, 732, 759, 771, 818
 and P-selectin, 285
 and serum amyloid A (SAA), 285
 and tumor necrosis factor alpha, 285
 and vascular cell adhesion molecule (VCAM-1), 280, 285
Atherosclerotic plaque, 732, 818
Athymic mice, 314
Atomic force microscope (AFM), 705, 753
ATP, 152, 186, 285, 287, 289, 320–321, 336, 399–400, 524–528, 531, 606–607, 626, 760, 768, 772
Atrophy of organs
 reversed by growth hormone, 195–197
Augmentation, 757, 775, 782
Autoimmunity, 240, 395, 509
Autologous cells
 cavernosal, 555
 chondrocyte, 547, 560–562
 endothelial, 555, 559, 564–565
 muscle, 548, 553–556, 558–559, 563–564
 urothelial, 545, 554
Automated aircraft, 781–782
Autophagy, 152, 161, 165, 178, 183, 186–187, 190, 380–381, 391–393, 627, 753
Autopsy, 65, 76, 266, 382, 600
Autoxidation, 603–604
Avon Products, Inc, 609
Axolotl, 612

B
Bacteria, live, anticancer effect of, 320–321
Bakers yeast, 370, 623
Balance and motor coordination
 improved, 350
BALB/c mice, 311, 343, 504
Banking, 68, 325, 475
Base excision repair (BER), 181–183, 186, 337, 653, 752
Basophil, 322–323
Bats, 93, 236
Bax, 335, 348
Benfotiamine, 603, 605
Benzoic acid, 603–604
Beta-amyloid accumulation
 and deprenyl, 160
Beta amyloid plaques, 592, 601, 610, 766
Beta galactosidase, 464, 652
Bifunctional oligonucleotides, 659–660
Bim, 338
Biocompatibility, 546, 689, 717–718, 782

Biocomputer, 701, 722
Bioethics, 27, 32, 41, 66, 78–79, 82, 368
Biofilm, 729
Biogenesis, mitochondrial, 158, 186, 336, 348–351, 399–401, 525–526
Biological constraints, 574–575
Biological evolution, 15, 103
Biologically controlled aging
 in iteroparous organisms, 146–159
Biological robots, 688
Biological warranty period, 575
Biology, 4, 14–16, 18–19, 22, 167, 307, 576, 578–579, 599, 812, 823, 831, 833–835
Biomarker, 149, 233, 249, 290, 733
Biomaterials, 546–548, 552, 559, 725, 766, 770, 783
 acellular matrices, 546
 alginate, 546–547, 561–562, 564
 collagen, 544, 546–547, 552–556
 naturally derived, 546–547
 synthetic polymers, 546–548
Biorobotic, 722–723
Biotechnology (biotechnological) revolution, 4, 18–19
Bladder
 engineered tissue, 546, 553, 564
 gastrointestinal segments in repair of, 554
 replacement in dogs, 554
 replacement in humans, 561
 subtotal cystectomy, 554
 urinary, 597
Blastocyst, 64, 67–71, 76–77, 452, 466, 469, 472, 548, 551, 556
Blastomere, 70
Blindness, 14, 474, 534, 735, 771
Blood-brain barrier (BBB), 767, 771, 778
Blood vessel, 70, 144, 239, 282–283, 296, 558–559, 589, 611, 688, 732, 740–743, 747, 749, 751, 755, 768
 replacement with autologous graft, 553
 replacement with synthetics, 556
B lymphocytes, 309, 579
Body mass index (BMI), 269–270, 293, 372–374, 387, 406–408
 and all-cause mortality, 263, 269, 406–407
 and smokers, 269
Boiling food, 602
Bone
 marrow, 68, 296, 323, 337, 395, 404, 469, 476, 494, 506, 510–511, 545, 559, 561, 579–580, 679, 766, 771
 stem cells, 510, 559, 611, 766

Bowhead whale, 137, 173, 199
Brain state map, 778
Brain-state monitoring, 779
Bridge one (life extension), 4–14
Bridge two (biotechnology/biotechnological revolution), 15–16
Bridge three [nanotechnology-AI (Artificial Intelligence) revolution], 16–18
Broiling food, 602
Brown adipose tissue, 334, 338–339
Bush rat (*Rattus fuscipes*), 144
Butterflies, 153

C

Caenorhabditis elegans, 4, 95, 100, 105, 107, 129–132, 138, 147–152, 154, 165–167, 170, 175, 369, 380, 383–385, 393, 398, 401, 403, 623–628, 630–633
Caloric density, 6, 375
Caloric energy, 102
Caloric restriction (CR), 4, 7, 16, 26, 88, 99, 102, 107, 115, 117–118, 261–262, 334, 369, 374–375
  age of onset and, 138–139
  cardiovascular effects, 373, 394–395
  duration and lifespan, 411
  effect, evolutionary origin of, 383
  exercise and, 404–405
  genic effects of, 397
  in animals, 7
  in humans, 5–7, 371–374
  hypothermia and, 403–404
  immunological effects of, 395
  intensity and lifespan, 375
  mimetics, 7, 115, 117–118, 159
  neurological effects of, 396–397
  physiological "memory" of, 377
  reproductive effects of
    humans, 387
    rodents, 385
Calorie blockers, 6–7
Calories
  dietary intake of, 262
  food, definition of, 251
  and pounds of extra body fat, 269
  and weight loss, 6
Camouflage, 718–719
Cancer
  breast, 337, 343
  cells, 335, 343–344
  immunological killing of, 160
  killing by neutrophils (granulocytes), 313, 317, 320–321
  lung, 335, 343
  prostate, 343
  resistance/resistant, 188, 311–312, 314–317
  -specific antigens, 322
  surveillance, 163, 173, 309–310, 314, 316, 318, 322, 324
  WILT and, 667–683
Cancer-killing activity (CKA), 316
Carbon placement, 708–709, 715
Carbonyl, 375, 594, 604
Carboxyethyllysine, 594
Carboxymethyllysine, 591
Cardiac computed tomography (CT), 195, 247, 285
Cardiomyocytes, 337, 348, 394, 559–560
Cardiomyopathies, 265, 337, 378, 394, 527, 535, 589, 762
Cardiovascular aging, 279–302
  of heart, 282–285, 298
  of vasculature, 284
Cardiovascular disease (CVD), 7–8, 12–13, 31–32, 117, 150, 166–167, 240, 252, 254, 264–265, 269, 281, 284, 349, 369, 371, 373, 394, 399, 406–408, 577, 668–669, 676
Cardiovascular disease (CVD) risk, 283–284
  factors, novel, 285
  factors, traditional, 283–284
    diabetes, 282
    modification, 284–285
    saturated fat diet, 283–284
    smoking, 284
Cardiovascular effects of CR, 373, 394–395
Carnitines, total
  decline with age in brain and muscle, 186, 190
Carnosine, 603, 605
Carotenoids, 253, 603–604
Carotid artery stenosis, 283
Carrel, Alexis, 454
Cartilage, 199, 244, 548, 560–562, 589, 591, 598–600, 611, 769
  articular cartilage, 560–561, 591, 599, 611
  chondrocytes, 561–562
  fetal tissue engineering, 545
  regeneration of, in humans, 199
  trachea, 560–561
Cartotaxis, 743
Cassette transporter A1 (ABCA1), 279, 287
Castration
  and lifespan in humans, 144, 166–167, 176
  and lifespan in outbred male cats, 166

and lifespan in salmon, 144
and likelihood of death by infection, 166
Catalase, 92, 247
Cataracts, 137, 350, 462, 588–589, 735
β-Catenin, 336, 631
Cation peptides, and recognition of cancer cells and bacteria, 319–320
CD28, 503, 579–580
CD28 antigen, 579
CD4+ cells, 579
CD4:CD8 ratio, 503
CD8+ cells, 512, 579
CDK1-cyclin B, 337
Cell(s)
  cycle, 70, 335, 338, 402, 454–456, 461, 471, 576, 621, 629, 634, 645, 654, 674, 757
    checkpoint activation, 576
  disassembly, 733
  herding, 733, 770
  mill, 734, 742, 744, 746, 760, 770, 772–773, 778
  morphology, 734
  nanosurgery, 753–755
  repair, 642, 687, 700, 731, 736, 751–753, 755, 757–761, 763, 772
  replication, 577, 773
  survival, 297, 332, 335, 400, 680
  turnover, 381, 528, 752
  viability, 357, 754
Cellular injection therapy
  bulking agents, 561–563
  with chondrocytes, 561
  with hepatocytes, 560
  for incontinence, 561–563
  with muscle cells, 563
  for vesicoureteral reflux, 561–562
Cellular rejuvenation, 775–776
Cellular reprogramming
  human fibroblasts, 550–551
  induced pluripotent state, 80–82, 548, 551
  mouse embryonic fibroblasts, 548, 551
  retroviral integration, 552
Cellular senescence, 96, 105, 154, 442, 474
  See also Replicative senescence
CEL (N-epsilon-carboxyethyllysine), 594–595, 602
Centenarians, 5, 150, 228, 300, 315, 369, 372, 464, 766, 819
  See also Supercentenarians
Centrophenoxine, 161, 177–180, 189–190
CETP, 150, 279, 287, 292
Charge-neutral, 320

Charges, 319–321, 594
Charlesworth, B., 358, 508
Chelating, 603
Chemical inhibition, 719
Chemical vapor deposition (CVD), 705–706, 718
Chemo-attractant/attractive, 312
Chemotactic sensor pad, 721, 743
Chemotaxis, 292, 312, 319
Childhood, 133, 172, 575, 578
Chimeraplasty, 642, 643, 650–655, 657
Chimeric RNA/DNA oligonucleotides, 650, 654, 656
Chloride, 252–253
Cholesterol
  HDL, 255, 257–258, 279, 287, 381
  LDL, 240, 255, 257–258, 270, 279, 283–284, 373, 381
Cholesterol ester transfer protein (CETP), 150, 279, 287, 292
Choline
  and brain structure, 199
  and centrophenoxine, 161, 199
  and DMAE, 161
Chondrocytes, 453, 547, 560–562, 590–591, 598, 611, 677, 759
Chromallocyte, 727, 755–757, 759–760, 763, 771, 773
Chromium picolinate, 159, 243
Chromosomal linkage, 314
Chromosome deletions
  and increased human lifespan, 167–168
Chromosome replacement therapy (CRT), 731, 755, 766, 774
Chronic inflammatory response, 284
Chronological aging, 231, 628
Cicada, 133
Cigarette smoking, 284, 298, 315, 348, 577
Circulation, 9, 70, 294, 301, 316, 321–323, 325, 444–445, 559, 579, 589, 597, 602, 679, 724, 730, 737
"Classical HDACs", 332
Clinical trials, 28, 117, 300, 302, 321, 475, 562, 565, 580, 604, 607–608, 610, 648, 677, 702, 768, 821
CLK-1, 95
Clonal deletion, 509, 719
Cloning
  nuclear, 549
  reproductive, 549
  therapeutic, 544–545, 549–550, 556–558
  See also Somatic cell nuclear transfer
Clottocyte, 735, 740–741, 743, 782

CML (N-epsilon-carboxymethyllysine), 591, 594–595, 599, 602
Cochlear hair cells
regenerated by activating Atoh-1, 199
Coenzyme Q, 161–163, 186, 525
See also Ubiquinone
Cognition, 50–51, 76–78, 361
Coley's toxins, 321
Coley, William, 320–321
Collagen, 5, 232, 267, 384, 394, 544, 546–547, 552–556, 589–592, 594–598, 601–602, 605–608, 610–612, 740, 743, 745–746, 752, 758, 767–768, 776, 821
Collagenase, 242, 461, 776
Colony size distribution, 464
Common sense diet, 261
Complex trait, 576
Compressed morbidity, 575
Comte de St Germain, 490–491, 493
Conatus, 47–55, 57–58
Conditional semelparity, 144
Congestive heart failure (CHF), 265, 279, 281, 589
Conserved genes for aging, 96
Contact zone, 313, 320
Cooking, 602
Copper, 526, 594, 603
Coral, golden, 1800 year lifespan of individuals of, 137
Coronary artery disease, 11, 144, 156, 269, 288
Coronary calcium, 285
Coronary heart disease (CHD), 13, 150, 279, 281, 288–290, 348, 373
Corpora allata
surgical ablation of, 153
Cortisol, 144, 172, 230, 239, 317
CRC2, 349
C-reactive protein, 285, 371, 373, 397
Crisis, 97, 459–460, 672
Crosslinks
alpha-diketone, 768, 821
dilysine, 590
glucosepane, 594, 605, 609–610, 613, 768, 821–822
glycation, 591, 598, 605–607
tetralysine, 591
"CR state"
rapidity and reversibility, 376–377
Cryonic suspension, 68
Crystal deposition disease, 759
Cultural construction of aging, 31
Cyclin dependent kinase (CDK)

inhibitor, 442–443, 445
Cysteine, 165, 319, 594, 603, 605, 768
Cystic fibrosis, engineered correction of, 645
Cytocarriage, 758, 775–776
Cytochrome *c*, 185, 334, 339, 526, 760
Cytocidal, 774
Cytokines, 285, 287, 319, 400, 443, 501, 591, 599, 749
Cytological maintenance, 775
Cytolysis, 313, 319, 718
Cytomegalovirus (HHV-5), 497–498, 502
Cytopenetration, 721, 755–756
Cytoskeletal disease, 762
Cytoskeleton, 753–754, 759, 761–762
Cytostructural testing, 761–763
Cytotoxic granules, 322–323

D

D-4FD-4F, *see* Apolipoprotein A-I mimetic peptide (D-4F)
DAF, 100–101
Daf-2, 101, 146–148, 150, 152, 154, 393, 624–625, 632–633
Daf-12, 147, 625
Daf-16, 147, 149–151, 393, 624, 627, 629, 631–632
Daphnia pulex, 100, 140
Data storage capacity of the human brain, 779
Dauer, 129–132, 148, 624, 779
DCB6Ge, 704–705, 708–709
Deamidation, asparagine, 595
Death rates for children, 781
Death-resistant cells, 765, 769, 814
Decoys, 719
Deficiencies
of nutrients, 251–252
of vitamins, 10, 270
Deglycation/deglycating, 606, 736, 768
Degranulation, 318–320
de Grey, Aubrey, 26–27, 29–30, 32–34, 89, 135, 174, 201, 360, 371–372, 533, 667–683, 758, 764–767, 769, 771–774, 810–812, 815–816, 818–821, 835–836
Dehydroepiandrosterone (DHEA), 117, 170–172, 239–240, 372–373
administration, beneficial effects of, 170–172, 240, 244
deficiency relative to GH/IGF-1, 170–171
and exercise capacity, 158, 171–172
and GH-induced hyperinsulinemia, 170
and LDL levels, 240, 257
side effects of, 172, 240

Index

Delayed maturation, 131
  effect on lifespan, 129–134
Dell'Orco, Robert, 454
Deme (definition), 109
Dementia, 11, 14, 117, 464, 577, 589, 814, 816, 823
Demographic homeostasis, 101, 110–111
Demographic theory of aging, 110
Dentine, 752
Denuded endothelium, 578
Deparasitization, 776–777
Deprenyl, 160–161, 177, 190, 194, 409
Dermal zipper, 742, 776
"Developmental program"
  ELT-5, ELT-6 and, 152
Developmental theories of aging, 201
dFOXO, 100
DHA (docosahexaenoic acid), 7, 11
DHEA, 117, 170–172, 239, 240, 372
DHEA sulfate (DHEA-S), 170–171, 239, 373
DHF, 597, 609
Diabetes, 5–7, 12, 16–17, 150, 170–171, 240, 252, 258, 265, 269, 279, 282, 301, 332, 339, 349, 351, 371, 381, 402, 405–406, 408, 462, 464, 477, 514, 522, 535, 589, 601, 605, 669, 735, 766, 769, 819, 836
Diabetes mellitus (DM), 279, 282, 291, 296, 339, 349, 735
Diabetic complications, 588
Diamond/diamondoid, 142, 144, 171, 490, 562, 687, 690, 694, 701–715, 717–718, 720, 722, 724–726, 737, 747, 755, 782–783
Diamond particle, 720
Diapause, 134, 153
Diastolic dysfunction, 589, 821
Diastolic heart failure, 597, 609
Dideoxyosone, 594
Diet, 250–262, 408, 601–602, 834
Dietary composition and lifespan, 377–383
Dietary restriction, 237, 369, 388, 625, 629–630, 634–635
Diets
  common sense, 261
  comparison of, 261
Differential gear, 710–711
Diffusible molecules, 313
Dimer placement tool, 708–709
Dimethylaminoethanol (DMAE), 161, 190
Diseases, 5, 7–8, 10–12, 14, 19, 32, 64, 115, 117, 150, 200, 230–231, 233, 245, 251, 257, 262, 265, 281–285, 291, 295–298, 300, 302, 320, 322, 332, 336, 340, 345, 349, 360, 364, 369–370, 372, 384, 394, 396, 407, 411, 462, 474–475, 493, 503, 509, 513–514, 522, 525, 544, 550, 574–575, 577–581, 588–589, 599–600, 614, 647, 649, 660, 668–669, 671, 674, 681, 688, 722, 726, 736, 758–760, 762–763, 766–767, 809, 815, 818, 822, 832, 839–840
"Disorganized development", 134
Disposable soma, 91–92, 97–104
Dissipative structure, 58
DMAE, lifespan extension by, 161
DMS, 703–709, 711–713
DNA, 623–635, 641–661
DNA damage, 10, 93, 181–184, 199–201, 315, 335, 374–375, 394, 446, 455, 458–459, 462, 465, 501, 576, 641–661, 773
  response, 335, 576
DNA polymerase, 181–183, 455–456, 523, 527, 533
DNA polymerase alpha
  A1 and A2 forms, 182
DNA polymerase beta, 181
DNA polymerase gamma, 182–183, 523
DNA repair, 5, 91, 149, 154, 169, 172–174, 181–182, 199–200, 332, 335, 337, 398, 523, 641–661, 752, 756, 768, 777
  technologies, artificial, 199, 641–661
DNA ribonucleases, 646–650
DNA segments
  block replacement of, *in vivo*, 199
DNA stability, 346
DNA strand exchange, 644–645
Dogs, 160, 192, 411, 554, 609, 611, 655–656, 680
Double-strand breaks (DSBs), 645, 677, 752
Drexler, K. Eric, 690–693, 698
Drosophila, 4, 99, 134, 148, 153, 161, 175, 235, 338, 360, 370, 378, 382–386, 398, 401, 403, 623, 625, 630, 635
Drosophila melanogaster, 99, 161, 370
Drowning, 736–738, 782
Drug resistance, 551–552, 675, 729
DS (disposable soma), 91–92, 97–104
Duchenne muscular dystrophy (DMD), 647–648, 657
  ASO mediated exon skipping clinical trial, 648

Dysdifferentiation, 235–236
Dyskeratosis congenital, 469, 580, 774
Dysnutrition, 252

E
e1368 mutation, 146
Ecdysone, 153
ECM (extracellular matrix), 81–82, 461, 544–547, 588–596, 598, 601–602, 606, 610–614, 752, 756, 761, 768, 770, 776–777
Ectoderm, 70, 466
Effector mechanism, 318–320
Effect, Warburg, 321
Elastin, 589–591, 595–598, 600, 608, 743, 767, 776
Elastin-laminin receptor, 597
Electromagnetic energy, 610
Electrophiles, 594
Elongation factor 1a (EF1a), 179
ELT-5, ELT-6, 152
　　programming of aging by, 152
　　inhibition of elt-3 by, 152
Embryo, 64–65, 69–78, 80, 466, 471–474, 494, 545, 548–549, 678
Embryonic progenitor cells, 467
Embryonic stem cells, 63–83, 340, 444–445, 451–477, 544–545, 548, 551, 559
Emergence, 48, 56–58, 93, 459, 513, 579, 748
Emotional stress, 316–317
Endocrine disorder, 736
Endocrine replacement
　　cell encapsulation, 564
　　Leydig cells for, 564
　　testosterone, 563–564
Endoderm, 70, 467, 548
Endoparasites, 777
Endoscopic nanosurgery, 702, 723, 730, 734–735, 747, 775, 778–779
Endothelial
　　dysfunction, 283–285, 296–297
　　progenitor cells, 296
　　sloughing, 297
　　vasodilation, 297
Endothelial cells, 284, 294, 296–297, 300, 348, 476, 548, 555, 564, 590, 597, 599, 720, 734, 740
Endurance, 8, 157–158, 186, 263–265, 350, 373
Engineered gene repair, 645
Engineered Negligible Senescence (ENS), 32, 686, 765, 835
English Sweat (sudor anglicus), 493

eNOS, 279, 283, 296, 297, 599, 631
Entelechy, 65
Entropy, 52–55, 90–91, 371, 452
Environmental factors, 12, 231, 349, 573, 576, 579
Enzyme replacement therapies, 771
Enzymes, bioengineered, 601
Enzymes, fungal, 606
Eosinophil, 322
EPA (Eicosapentanenoic acid), 7, 11
Epidemiology, 578
Epigenetic, 149, 234, 551, 579, 773–775
Epigenetic regulation of gene expression, 331
Epigenetic state, 551, 773–774
Epimutation, 670–671, 765–766, 773–775
Epithelial progenitor cells, 578
Epstein Barr virus (HHV-4), 458, 502–503
Erectile dysfunction, 554, 589, 597, 600, 609
Error-free chromosome, 775
Escape velocity, 29, 834
E-selectin, 285
Essential fatty acids, 256–257
EST mutants, 456
Estrogen replacement therapy, 237
Estrogen, transdermal, 245
Ethanol, 628, 740
Eusocial mammals, 175
Every other day calorie restriction
　　and IGF-1, 173
Every-other-day feeding and lifespan, 381
Evolution, 14, 88–89
Evolution of aging, 89–90
Evolutionary origin
　　of lifespan and health benefits of CR, 383
Excution, 700, 722, 700
Exercise
　　bone density loss in older adults offsetting, 267
　　and CR, 404–405
　　decreased intra-abdominal fat and, 265
　　favorable responses to, 262–263
　　how much?, 263
　　lower arterial stiffness and, 264
　　lower blood pressure and, 265
　　lower death rate and, 263
　　lowered all-cause mortality and, 263
　　lower plasma insulin levels and, 265
　　two to threefold increases in muscle strength and, 267
　　VO_2max and, 264–265
　　weight loss interventions and, 267
Exponential growth, 15, 831

Index 851

Extended lifespans
   in age-1 mutants, 100, 152, 624–626, 628, 632–634
   due to pharmacological and nutritional interventions, in mammals, 159–165
   in genetically altered mice, 153–159
   in humans, 166–168
   in mice with delayed maturation, 129
   prior to sexual maturation, 129–134, 138–139
Extracellular matrix (ECM), 81–82, 461, 544–547, 588–596, 598, 601–602, 606, 610–614, 752, 756, 761, 768, 770, 776–777
Extracellular senescence control by a single gene, 154
Extrachromosomal circles, 460, 462
Extra-uterine life, 577

F

Fasting, 118, 133, 242–243, 254, 265, 292, 373, 381, 385, 387, 392, 397, 401, 601–603
Fat accumulation
   decreased, 336
Fat blockers, 6–7
Fat insulin receptor, 6, 15–16, 155
Fat mobilization, 336, 349
Fatty acids
   essential, 256–257
   MUFAs, 256–258
   omega-3 and omega-6, 257
   PUFAs, 256–258
   saturated (SFAs), 256–258
   trans (TFAs), 257–258
FDA, 82, 255, 323, 508, 547, 608, 715, 724, 817, 828, 833
Fecundity, 100–101, 108, 132, 137, 158, 174, 201, 363, 369, 385, 624
Femtolaser surgery, 754
Fertility, 90, 92–93, 96–102, 107, 109–111, 115, 153–156, 160, 173, 363, 369, 383, 385, 527
Fetus, 65, 69, 71, 74–76, 79, 182, 184, 548–549
Feynman, Richard P., 688
Fiber, 6, 173, 251, 254–255, 260, 270, 293, 399, 610–611, 740, 762, 768, 775, 778–779
   hyperinsulinemia and, 170, 255, 261
Fibrinogen, 255, 285
Fibroblast, 160, 182, 184, 460, 508, 558, 590–591, 596, 611, 653, 719, 722, 725, 752, 775–776

Fibronectin, 506, 589, 592, 596–598, 600
Fibrosis, 297, 373, 394, 395, 397, 580, 611, 645, 649
Finch, 117, 129, 136, 142, 151, 246, 599
FIRKO (Fat-specific Insulin Receptor Knock Out) mice, 6, 155, 405–406
Fish oil, 7, 10–11
Flaxseed oil, 11
FLCs (fibroblast-lineage cells), 590, 592, 596, 610–612
Flexibility
   decline with age, 234
Flies, 95, 97, 100, 107, 118, 140, 153, 161, 165, 179, 332, 359–360, 364, 370, 377, 385, 574, 632–633
Foam cell, 285–286, 599, 733, 759
Force of natural selection, 93, 136, 139, 358
FOXO, 100, 120, 150–151, 155, 172–173, 334, 338, 393, 398, 624, 627, 629, 631–632
FOXO1 (human analog of daf-16), 150
FOXO3, 335, 338, 631
Free energy, 53–54, 90–91
Free radical damage, 14, 247
Free radicals, 5, 234, 246, 253, 357, 523, 594, 599, 725
Freitas, Robert A., Jr, 17, 174, 200–201, 685–783, 835, 840
French Paradox, 348
Fruit flies, 95, 99–100, 359, 363–364, 383
Fruit and vegetable intake
   and risk of cancer, 253
Frying food, 602
Full genome sequencing, 314–315
Funding, 25–26, 30, 79, 82, 363, 668, 709, 810–815, 817, 820–821, 823, 827–828, 832–839
Fusitriton oregonensis, 131
Fuzzy logic, 71–72
Fuzzy set, 64, 71–73

G

G_0, 629, 634, 757
Gastrulation, 70
G-CSF or Neupogen, 323
Gene expression, 7, 128, 149–152, 159, 163, 174, 180–181, 200–201, 331, 349–350, 359, 362, 364, 374–376, 390, 394, 397–398, 403, 410, 441–443, 447, 458, 461, 467, 469, 471, 473, 551, 590, 598, 628, 642–644, 646–647, 670, 673, 732, 736, 753, 756

Gene repair
  cystic fibrosis, 645
  engineered, 641–661
  Huntington's disease, 660
  in vivo, 645, 649, 655–657, 659
Genes associated with extraordinary ages in humans, 95, 150, 394, 399
Genescient, Inc., 359–364
Gene targeting, 477, 644–645, 659–660, 677–678, 680, 682, 774
Gene therapy, 19, 446, 465, 531–535, 611, 613, 642, 645, 647, 649–650, 670, 672, 675, 677, 769, 771–772, 774, 811, 834
  See also In vivo gene repair
Genetic, 12–14, 27, 54, 79–81, 93, 95, 97–98, 100–101, 103–108, 110, 114–116, 128, 138, 140, 143, 145, 148, 150–154, 166–169, 176, 197, 200, 231–236, 282, 311–318, 324, 359, 440
Genetic disease, 10, 477, 726, 735–736
Genetic engineering, 27, 81, 648–649, 658, 660–661
Genetic exchange, 104
Genetic pathways that shorten lifespan, 146
Genetic program, 90, 95, 97, 106, 143, 154, 235, 399
Genetic rejuvenation, 774
Genetic renatalization, 775
Genetic tradeoff, 98–101, 103
Genic effects of CR, 397
Genome, 11–12, 14, 70, 95–96, 104, 116, 128, 150–151, 186, 198–200, 235, 314–315, 322, 331, 359–360, 384, 398, 410, 446, 470, 472, 477, 507, 521–525, 529–535, 558, 612, 644–645, 649, 670, 674, 678, 686, 688, 700, 746, 766, 769, 771, 775, 782, 819, 834–835, 838
Genomics, 11–14, 19, 314–315, 357–364, 686, 688, 833, 835
  stability, maintenance of, 337
  testing, 11–14
Germ line, 68, 103–104, 146–148, 154, 167, 170, 314, 358, 452–470, 472–473, 528, 551, 660
  cells, negative effect of on lifespan, 147
  transmission, 466
GH/IGF/insulin system
  dual role as anabolic system and downregulator of antioxidant enzymes and DNA repair, 172

Ghrelin, 245, 504–505
GH secretagogues (GHS)
  improvements in muscle strength and function and, 244–245
Gilgamesh, 489, 490
GIS1, 629
Gleevec
  and hair repigmentation, 197
Gliomas, 335
Glomerular basement membrane, 589, 591, 599, 768
Glucocorticoids, 391, 393–394, 411
Glucose, 6, 158–159, 175, 186, 239, 241–244, 254, 265, 292, 320–321, 339, 349–350, 373, 380–381, 385, 389–392, 398–403, 557, 593–594, 601, 603–605, 628–630, 695–696, 699, 721, 727, 736–737, 749, 760–761, 763, 767
Glucosepane, 594, 605, 609–610, 613, 768, 821–822
Glutathione, 165, 247, 380, 603, 605–606
Glycation
  inhibitors, 602–603, 607, 613–614
  pathways, 592, 595, 606
Glycolysis, 320–321, 397, 524, 529, 630, 810
Glycoproteins, 589, 592
Glycosylation, nonenzymatic, 588
Glycoxidation, 592, 594, 601, 608
Gonadectomy, 141–143, 168
Granulocyte donation, 323
Granulocyte mobilization, 323, 325
Grasshoppers, 153
Grossman, Terry, M.D., 3–19, 28, 831
Growth factors, 150, 153, 155, 160, 169, 188, 199, 243, 246, 298, 343, 370, 373, 384, 389–390, 468, 504, 508, 514, 546, 560, 564, 602, 611–612, 670, 770, 778
Growth hormone (GH)
  and acromegalics, 246
  administration of, 244–245
  and body composition, 244, 249
  diabetogenic effects of, 170
  and estrogen, 237–240, 245, 253
  and prostate cancer, 245–246, 253
  and reversal of age-related atrophy, 196
  and "somatic involution", 169, 171–174
Growth hormone receptor knockout mice, 155, 157, 358
Growth hormone secretagogue, 244–245, 504–505

Index 853

Guanido group, 594, 597
Guppies, 94–95, 370
GWAS: Genome-Wide Association Studies, 360

H
H_2O_2, 96
Hair repigmentation, 197
Hamiltonian gerontology, 358–359
Hamilton's Rule, 109
Hamilton, W.D., 91, 358, 505
Hayflick, Leonard, 82, 188, 454
Hayflick limit, 188, 454, 457, 460, 576, 671
Hazard function, 780–782
Health maintenance processes, 234–235
Healthspan, 139, 159, 200, 228, 236, 361–365, 514, 574, 681, 686, 688, 715, 780–782
Healthspan extension, 139, 159, 361–363, 365, 514, 681, 688, 715
Heart
    decellularized cadaveric heart, 559
    disease, 7, 11, 13–14, 16, 19, 116–117, 150, 233, 252, 265, 283, 288, 348, 372–373, 559, 688, 726, 732, 736, 836
    failure
        congestive, 265, 279, 281, 589
        diastolic, 597
    injectable cells for repair, 547
    patch repair, 559
Heat-killed bacteria, 321
Heavy metal, 361, 759, 776
HeLa cells, 180, 184, 309, 316
HeLa cells fused with enucleated old fibroblasts
    reduced mitochondrial protein synthesis eliminated by, 184
Hemangioblasts, 474, 476
Hematopoietic stem cells, 464, 467–468, 476, 577
Hematopoietic system, 476, 576
Hemeoxygenase-1 (HO-1)ATP-binding, 296
Hemostasis, 737, 740–741, 743
Heritable, 130, 149, 579, 643
Heritable differences in gene expression, 149
Herpes virus, 502
Heterochronic, 443–444, 473–476
Heterochronic parabiosis, 187–188, 444
Heteroplasmy, 522–523, 529
High-calorie diets, 350
High-density lipoprotein (HDL)
    anti-inflammatory, 287–289, 291–296, 298–301
    composition, 286–288, 292–293
    dysfunctional, 290–293
    function, 282–288, 290–293, 296–298, 301–302
    inflammatory index, 288–290, 293, 295, 301
    life cycle, 286
    mature, 287, 294–296, 319, 324, 398
    particle size, 281, 299–300, 302
    pre-beta, 287, 294–295, 298–301
    pro-inflammatory, 285, 288–295, 297–301
    serum level (HDL-C), 288–290, 293, 295, 299
High throughput screening (HTS), 347, 617
Histidine, 255, 594, 603, 605
Histone deacetylases, 331–332, 335
Histone H3K9, 337
Histones, 332–335, 375, 670
History, 28–31, 47–50, 72–74, 229–230, 390–391, 832–833
HNP (human neutrophil peptides), 319
Hobbes, 48–50, 52–53, 58, 83
Homicide, 64, 715, 736, 780, 782
Hominization, 65
Homocysteine, 10, 285
Homologous genes, 95, 642
Homologous recombination, 458–459, 642, 645, 651–653, 677
Hormesis, 96, 107–108, 380
hsp 16.2, 149
hTERT, 512
Human, 5–7, 63–83, 107, 135, 166–168, 195–197, 227–271, 315–318, 320–325, 336–339, 367–412, 451–477, 502, 508, 532, 548, 573–581, 610, 685–783
Human BMI, morbidity, and mortality, 406–408
Human embryonic stem cells (hESCs), 63–83, 545, 548
Human immunodeficiency virus (HIV), 340, 345, 465, 504, 647, 677, 834
Human neutrophil peptides (HNP), 319
Huntington's disease (HD), 336, 348–349, 393, 396, 399, 402
Hutchinson-Gilford syndrome, 461–462, 477
Hydrogen abstraction, 706–707, 709
Hydrogen donation, 707, 709
8-Hydroxy-2'-deoxyguanosine, 181
Hyperadrenocorticism, 142–143
Hypercortisolemia, 144
Hyperinsulinemia, 170, 255, 261

Hypertension (HTN), 16, 150, 254, 265, 269, 279, 281–282, 300, 399, 464, 476, 529, 564, 589, 609
Hypertrophy, left ventricular, 589
Hypophysectomy, 141–142, 237
Hypothermia and CR, 403–404

I

Ibuprofen, 603–604
IFN signaling pathway, 314
IGF-1, 155, 160, 169–173, 196, 243–247, 602, 611, 624, 630, 632–634
    levels, inverse correlation with disability and institutionalization, 172
    signaling, 155, 630, 633–634
IGFI and insulin receptor signaling, 389–390
IGF pathway, 95
IL-6, 230, 285, 298, 602
IL-7, 170, 495, 504–508
    and T cell production, 506–507
    and thymic involution, 170
Imatinib (Gleevec)
    and hair repigmentation, 197
Immortal, 92, 137, 452–453, 455–460, 466, 468, 477, 489, 491, 551
Immortality, 19, 29, 34, 59, 135, 452–455, 467–470, 477, 489–491, 513
Immortalization, 454–460
Immune risk phenotype, 503
Immune suppressor, 310, 316–317
Immune system, 118, 160, 174, 191, 198, 240–241, 309–310, 312, 318, 322, 324–325, 384, 395, 489–514, 528, 579, 674–675, 718–719, 724, 730, 735–736, 766, 778
    rejuvenation of, 241
Immunity, 247, 311, 314, 318, 322, 324–325, 378, 498, 501, 510, 513–514, 633
Immunological effects of CR, 395
Immunosenescence, 170, 190, 769, 821
    and risk of death in humans, 170
Immunotherapies, against cancer, 309–310, 322, 324
Implantation, 69, 71, 74, 77–78, 143, 191, 194, 281, 476, 544, 547, 550, 555, 557, 559, 564
Incontinence
    urinary, 589
    See also Cellular injection therapy
Increased intelligence, 782
Individual fitness, 93–98, 108–110, 112
Induced pluripotent stem cells, 80–81, 472–473, 551–552

INDY, 100
Infant mortality, 781
Infections, 72, 107–108, 115–117, 166–167, 230, 232, 239, 286, 292, 298, 300, 314, 318, 320–321, 323, 325, 340, 345, 395, 458, 469, 474, 491–493, 497–498, 502–504, 511–514, 558, 560, 588, 599, 646, 656, 726–729, 752, 762, 777
Infectious diseases, 513
Infiltration, 312, 323, 504, 547, 553
Inflammation, 10–11, 115, 117, 170, 253, 285–293, 295, 297–301, 332, 348, 351, 373, 394, 397, 399, 411, 461, 471, 508, 546, 580, 588, 592, 595, 598–600, 602–603, 633, 717–720, 723, 742, 767
Influenza, 285, 292, 395, 497–498, 502, 507, 514
Information processes, 14–16, 18–19
Innate immunity, 633
Inositol, 186, 190, 603–604, 626–629
Institute for Molecular Manufacturing, 694
Insulin/IGF1-like signaling, 627–628
Insulin-like growth factor-1 (IGF-1), 155, 243, 247, 602, 611
    and acromegalics, 246
    and binding protein balance, 173, 246
    and breast cancer, 238, 245, 253
    and IGFBP3, 173
    and prostate cancer, 173, 245–246, 253
    receptor knockout mouse, 155
Insulin-like signaling pathways, 153
Insulin and/or IGFI signaling, 370, 383–386, 411
Insulin resistance, 7, 16, 242–243, 255, 336, 349–350, 373, 399, 401, 578, 736, 769
    and chromium deficiency, 243
Insulin secretion, 242, 254, 336, 339, 349, 631
Insulin sensitivity, 118, 155–156, 158, 240, 243, 262, 265, 349–350, 371, 379–380, 389, 401, 405, 410
Integrin receptor, 589–590, 598
Intercellular adhesion molecule-1 (ICAM-1), 285, 292–293
Interferon (INF), 314
Interleukin-7, 495, 504–508, 514
    receptor, 506
Interleukin-6 (IL-6), 230, 285, 298, 300, 602
Intermittent fasting and lifespan, 381
Interventions Testing Program, 165, 810

Intracellular aggregates, 670, 765, 770–771, 775, 814
   See also Lipofuscin
Intracellular parasite, 759–760
Intracellular storage disease, 758, 771
In vitro, 16, 177, 182, 187–188, 243, 295, 300–301, 313, 323, 337–338, 340–341, 344–347, 349, 362, 401–402, 444, 452, 454, 458, 460, 462, 466–469, 475–477, 495, 502–503, 510–513, 533, 545, 548, 551, 554, 556, 558, 559–561, 577, 579, 594, 598, 600, 605–607, 610, 613, 627, 645, 647, 653, 657, 659, 676, 678
In vitro assay, 316, 342
In vivo, 17, 187, 189, 194, 199–200, 295, 301, 316, 321, 323, 335, 340–341, 344, 346, 348, 350, 388, 391, 401, 443–444, 452, 455, 459, 463–466, 502, 510, 512, 523, 532–533, 545–548, 551, 556, 558, 560–561, 563–564, 577–580, 599–600, 605–606, 608, 610, 627, 649, 655–657, 659, 677, 682, 688, 696, 699–700, 718, 724–725, 730, 745, 748, 750, 778, 782
In vivo gene repair, 199, 645
iPS Cells, see Induced pluripotent stem cells
IRS-1, 631, 633
IRS-2, 384, 633
Isoaspartate, 595
Isochronic, 444
Isomerization, 588, 590, 595
   amino acid, 595
Isonicotinamide, 340

J
JNK1, 631, 633
Juvenile hormone, 153, 175

K
Kekich, David, 32, 827–840
Keratinocyte growth factor, 504, 508, 514
Kidney, 165, 181, 195, 197, 244, 291–292, 334–335, 338, 387, 392, 406, 476, 534, 556–558, 560, 564, 589–591, 598–600, 605, 608–609, 656, 735, 752, 763
   bovine model, 556
   extracorporeal renal unit, 556
   rejection, 557–558
   therapeutic cloning of, 556–558
   tissue engineering of, 556

Kidney disease, 291–292, 476, 556, 589, 763
Kidney epithelial cells, 599
Kilocalories, 251
Klotho, 155–156, 384, 389
   polymorphisms in humans, 156
Krebs cycle, 92, 339, 391
kri-1, 147
Ku70, 334–335, 348
Kurzweil Foundation, 783
Kurzweil, Ray, 3–19, 28–29, 831

L
Laboratory selection, 115
Lactate (lactic acid), 157, 320–321
Lagging strand DNA synthesis, 456–457
Lamins, 756
Laminin, 589, 597–598
Lampreys, 140–142
Laron dwarf mouse, 157
Laser, 100, 146–147, 610, 731, 747, 754
   gonad ablation, 146
Lasker Foundation, 715–716
LAT1, 630
Law of accelerating returns, 12, 18–19
Lead compound (drug development), 613
Lecithin cholesterol acyltransferase (LCAT), 287, 292, 298–299
Leishmaniosis, 340, 345
Lens, 26, 58, 137, 199, 591, 752, 767
LESPs (long-lasting extracellular structural proteins), 585, 588–590, 592, 597, 607, 610–611
Leukocyte activation, 314
Levels of selection, 103
Life expectancy, 4–6, 15, 18–19, 134, 228, 233, 240, 315, 368, 377, 574–575, 671, 681, 812, 815, 822
Life-extending mutants, 153, 625
Life extension, 4–5, 16, 29, 39–59, 89, 99–102, 107–108, 116, 118, 128, 133, 139–176, 188, 201, 228–231, 237, 240, 243, 270, 308, 325, 544–565, 625, 630, 632–635, 678, 827, 829, 831–832, 835–836, 838–839
   ethics, 42
   by reproduction, 175
Life Extension Foundation, 783
Life history (Life-history), 72, 98, 169, 229, 372, 390, 575
   and anticancer effects of CR, 390–391
   "design", and aging, 169
   evolution, 98

Lifespan
   cost for eating excess calories, 261–262
   extension, 133, 141, 146, 151, 155, 165, 167–168, 229, 370, 374–376, 378, 380–381, 393, 398, 407
   fly, shortening by smelling food, 100
   increase, 131, 164–165, 173, 398, 624
   maximum, 4–5, 131–132, 142–143, 146, 148, 152–156, 159–161, 163–166, 168, 175–176, 182, 228, 263, 337, 368–369, 370, 372, 374, 377–379, 383, 388, 402, 404–405, 625, 630
   mean, 132, 146, 153, 155, 156, 161, 164–165, 175–176, 182, 191, 194, 381, 404, 409, 574
   and methionine restriction, 164–165, 379–380
   and protein restriction, 378–379
   and protein synthesis, 151–152
   and specific nutrients, 381–383
   and tryptophan restriction, 163–165, 379
LIFT (leukocyte infusion therapy), 324–325
Ligaments, 8–9, 267, 589, 767
Limbo, 70, 88
Lin-4 microRNA, 149
Lipid hydroperoxides (LOOH), 286, 291, 293–295, 298–301
Lipids, 19, 118, 149, 161, 190, 234, 240, 254, 262, 265, 281, 283, 285–287, 292, 294–302, 319, 350, 371, 380–381, 391–392, 397–399, 401, 411, 522, 588–589, 592, 594–597, 601–605, 633, 686, 732, 758, 766, 770
Lipofuscin, 180, 186, 189–190, 283, 758
Lipofuscinolysis, 189–190
Lipoprotein (a), 285
Lipotoxicity, 158, 170
Lipoxidation, 588, 592, 594–595, 602–606
Liquid nitrogen, 68
"Little mouse/mice", the, 155, 384
Liver
   bioartificial device, 560
   hepatocyte injection, 560
Lizards, 118
Logic, fuzzy, 71–72
Longevity, 18–19, 149, 377, 383–385, 408–411, 573–581, 623–635, 834
Longevity assurance, 624, 629
   genes, 624
Longevity determinant genes (LDGs), 234–236
Longevity mutants, 146, 149, 169, 361–362

Longevity mutations, 159, 176, 361, 383–385, 408
   and molecular mechanisms of CR, 383–385
Longevity therapeutics, 408–411
Lotka, A.J., 93–94, 110–113
Low-density lipoprotein (LDL)
   minimally-modified (MM-LDL), 286, 293
   native, 296
   oxidized (OX-LDL), 285–286, 296
   particle size, 299–300
   serum level (LDL-C), 283–285, 293, 373, 381
Low socio-economic status, 577
Lung cancer, 315, 335, 343, 731
Lungs, 16, 70, 162, 252, 315, 334–335, 343–344, 388, 392, 404, 411, 589, 597, 600, 612, 672, 680, 731, 737–738, 751–752
Lymphocytes, 241, 309, 313, 389, 393, 395, 463–465, 469, 492–493, 497, 500, 507, 510, 512, 558, 579, 689, 771
Lymphocytic, 313–314
Lymphoid, 503, 578
Lysine, 197, 255, 319, 331–332, 335, 340, 593–595, 597, 605, 607, 647
Lysosomes, 590, 758, 771

M

Macrophages, 241, 285–286, 291, 293, 297, 300–301, 312–316, 395, 494, 588, 590–591, 595, 598–599, 633, 720, 724, 727, 752, 759, 766–767, 769, 771, 821
Macular degeneration, 475, 529, 771, 818
Macular degeneration, age-related, 14, 453, 474, 475, 758, 818
Malnutrition
   deaths attributable to, 252
Malthusian Parameter, 115
Mammalian analog of daf-16, see FOXO; FOXO1; FOXO3
Manganese superoxide dismutase, 155, 247
MAPK, 187–188, 631–633
Margarines containing plant stanols and sterols, 258
Marginotomy, 454
Marsupial mice, 144–145
Martin, G.M., 176, 466
Masterpiece of nature, 104
Matrix metalloproteinase (MMP), 547, 591–592, 597
Maximal lifespan, 575
Maximum healthspan, 781

Index
857

Maximum lifespan, 4–5, 131–132, 142–143, 146, 148, 152–156, 159–160, 161, 163–166, 168, 175–176, 182, 228, 263, 337, 368–370, 372, 374, 377–379, 383, 388, 402, 404–405, 625, 630
Maximum tolerated dose (MTD), 160, 312
MCF7 cells, 316
MCLK1, 95
Meal frequency and lifespan, 377–383
Mean lifespan, 132, 146, 153, 155–156, 161, 164–165, 175–176, 182, 191, 194, 381–382, 404, 409, 574
Mechanical senescence, 198–199
Mechanisms of lifespan extension
   by protein, methionine and tryptophan restriction, 380–381
Mechanosynthesis, 687, 703–705, 709, 713–714
Meclofenoxate, 161
Median lifespan, 155–157, 159–160, 164–166, 188, 404, 463, 781
Medical nanorobotics, 685–783
Mediterranean Agaves, 145
Melatonin, 164, 247, 371, 409
   biological activities of, 247
   prolonged life span in rodents and, 247
Membrane pores, 320
Memory cells, 395, 493, 497, 500–501, 503, 506, 579
Memory T cell, 497, 499, 501
MEMS, 747–748
Men, 8, 50, 70–72, 144, 166–168, 172, 195–197, 238–240, 242–245, 247, 255, 258, 264–268, 288, 373–374, 406–408, 505, 510, 564, 577–579, 609, 829
Mendellian mutation, 314
Merkle, Ralph C., 690, 692, 694, 704–706, 708–709, 713, 715, 746
Mesenchymal, 199, 508, 552, 559, 561, 590, 611–612, 677, 680
Mesenchymal stem cells, 199, 559, 590
Mesoderm, 70, 467, 548
Metabolic disease, 265, 349–351, 736
Metabolic effects of aging and CR, 391–392
Metabolic hibernation, 361
Metabolic pattern, 320, 324
Metabolic syndrome, 286, 291, 293, 300, 336, 349–350, 399, 402, 410, 529, 821
Metabolic tradeoff, 98, 100, 102, 107
Metaethics, 50–52

Metformin, 7, 159, 401–402, 410, 604–605, 607
Methionine, 10, 164–165, 255, 379–380, 410
Methionine restriction and lifespan, 164–165, 379–380
Methuselah Flies, 359, 361, 364
Methuselah Mice, 359
Methuselah Mouse Prize, 26
Methylation, 10, 149, 345, 380, 473, 551, 651, 756, 773
Mice/mouse, 7, 26, 107, 144–145, 153–159, 182, 186, 310–316, 323–324, 336, 350, 359, 384, 398, 551, 632–633
Microarray, 314, 384, 397, 402, 657
Microbivore, 726–730, 733–736, 742, 757–760, 763, 766–767, 769, 771
Microdebridement, 731
Microdeletions, 777
Microglobulin, 320
Microheteroplasmy, 522, 529
Microneedle, 749, 770
MicroRNA (miRNA), 149–150, 649–650
Microrobot/microrobotics, 17, 702, 713, 747, 749–751, 753
Microsurgery/microsurgical, 689, 750
Microtubules, 337–338, 343, 754
Migration, 32, 106, 288, 314, 390, 443, 461, 494, 508, 546, 589, 598, 612, 756, 778
Milk, pasteurized, 602
Mineral, 9–11, 244, 248–249, 251–253, 267, 374, 377, 382
Miniaturization, 16–18, 747
Mitochondria/mitochondrial, 182–187, 334, 338–339, 521–535, 670, 772–773, 814, 818–819
   aging, reversal of, 182–187
   and autophagy, 186–187
   protein synthesis decline with age, human, 180, 184, 399–400
   rejuvenation of, 180, 184–187, 530–535
Mitochondrial biogenesis, 158, 186, 336, 348–351, 399–401, 411, 525–526
Mitochondrial DNA
   cloning, 530, 532
   damage, and down-regulation of BER, 183
   mutated, removal in cultured cells mtDNA, 645
   point mutations in, 182, 527
   purifying selection process in female germ line cells, 660
   transfection, 186, 533

Mitochonrial DNA mutations, *see* MtDNA mutations
Mitochondrial gene therapy, 532–535
Mitochondrial number (number of mitochondria), 186, 349, 399–401, 525
Mitochondrial oxidative damage
 and aging, 182, 523, 525–526
Mitochondrial protection, 165
Mitochondrial protein synthesis
 increased by acetylcarnitine, 180
 restored by fusion with HeLa cells, 180, 184
Mitochondrial size and/or number
 increases in, 186, 349, 400
Mitochondrial theory of aging, 521–530
Mitochondrial transcription
 restored by acetylcarnitine, 184–185
Mitosis, 104, 337, 579, 611, 676, 753, 757
MMP (matrix metalloproteinase), 547, 591–592, 597
Model organism, 359, 574, 624
Moieties, glycoxidation, 601
Molecular bearing, 690
Molecular manufacturing, 687, 690, 692, 694, 702–703, 712–713, 782
Molecular motor, 689, 694–695, 718, 762
Monoamine oxidase, 160–161, 194
Monocarpic plants, 145, 148
Monocyte chemoattractant protein-1 (MCP-1), 286, 293, 297–299
Monocyte chemotactic activity (MCA), 288, 293–295
Monozygotic, 148–149, 469
Monozygotic twins, 149
Morbidity, 27, 117, 196, 230, 302, 336, 372, 402, 406–408, 512, 514, 553, 561, 575, 668, 685–783
Mortality rate, 97, 109, 134–137, 150, 166, 318, 363, 372, 498, 765, 816
 constant with age, 135–137, 765
 declining with age, 134
 reduced, 150, 166
Mosaic, 74, 100–101, 106, 754, 757
MSN2, 629
MSN4, 629
mtDNA, 155, 157, 182–184, 186, 374, 399, 470, 522–533, 535, 557–558, 645, 660, 670, 757, 772–773, 818–819
mtDNA deletion(s), 186, 522–523, 527, 819

mtDNA mutations, 182–183, 186, 522–530, 532–535, 660, 670, 722, 773, 814, 818–819
Multiple vitamin/mineral formulation, 10
Multipotent, 67–68, 494–495, 549, 552
Muscle aerobic capacity
 increased, 157–158, 350
Muscle regeneration
 and fibroblast growth factor 2, 188
 and soluble frizzled-related protein 3 (sFRP3), 187
 and TGF-beta, 187
 and Wnt3A, 187
Mutant mitochondria, 765, 772
Mutational load, 93, 95
Myelocyte-specific promoter, 315
Myelocytic, 313–314
Myeloid, 197, 337, 503, 578
Myoblast, 440–441, 444, 563–564
Myofiber, 440–441, 443, 563, 762
Myofibroblasts, 599

N

N-acetylcysteine (NAC), 605
NAD^+-binding domain, 339
Naïve T cell, 395, 497, 499, 501, 503, 513
Naked mole rats, 93, 137–138, 236
Nanapheresis, 734, 738
Nanobearing, 690–693
Nanobot, 613
Nanocar (Tour "nanocar"), 695
Nanocatheter, 749, 779
Nanocomputer, 688–690, 700–702, 722, 734, 737–738, 751
Nanocrit, 741
Nanodissection, 754
Nanofactory/nanofactories, 702–704, 712–715, 765
Nanofactory Collaboration, 701, 703, 705, 708–709, 712–713, 715
Nanogear, 690–693
Nanogerolytic treatments, 777
Nanoinjector, 755, 761
Nanomachine, 686–689, 694, 702, 709–710, 724, 726, 759, 783
Nanomanipulator, 688, 690, 697–699
Nanomechanical, 689, 699, 700–702, 711, 724–725, 730, 742
Nanomedical, 715, 717, 726, 735–737, 765–775
Nanomedically engineered negligible senescence (NENS), 765–775
Nanomedicine, 686–690, 717–718, 732, 745, 766, 780, 782, 834–835, 838

Nanomotor, 689–690, 694–697
Nanopump, 688, 694–697
Nanorobot/nanorobotics, 17, 685–783, 834
Nanorobotic brain scans, 778
Nanorobotic red blood cell, 17
Nanosensor, 688, 690, 699–700, 749
Nanosurgery, 723, 730, 734–735, 742–751, 753–755, 770, 776, 778
Nanosurgically, 746
Nanosyringe, 749
Nanosyringoscopy/nanosyringoscope, 730, 749–751, 770
Nanosyringotomy, 750
Nanotechnology, 4, 11, 16–18, 687–689, 702, 712, 722, 724, 764, 766, 780–781, 832–836
Nanotruck, 695
National Institute on Aging (NIA), 29, 231, 812
Nature of value, 41–43, 45–46, 58
Negatively charged, 319–320, 595
"Negative reproductive costs", 174–176
Negligible senescence, negligible aging, 129, 137, 765
Nematode, 129, 345, 359, 369, 574, 623–625, 627, 629, 633–635, 649
NENS, 765–775
Neonatal reserve length, 774–775
Nervous signaling, 105–106
Neural interfaces, 780
Neural prosthetic implants, 780
Neural restoration, 199, 777
Neurodegeneration, 336, 349, 669, 771
Neurogenesis, 396, 778
Neurological disease, 396, 767
Neurological effects of CR, 396–397
Neuronal CRT, 777–778
Neuropathy, 525, 589, 599, 605, 609
Neurotoxicity, 601
Neutrophils, 313–315, 317, 319–324, 395, 494, 689, 720, 728, 752
Newborns, 577, 579, 719
Newts, 118, 475
NF-κB, 334, 336–337, 398
Niche, cell, 445, 680
Niche exhaustion, 194
Niche(s), 395, 440, 443–447, 476, 500–501, 588, 679–680
Niche, stem cell, 445, 680
Nicotinamide concentrations, 340
Nitration, 588

Nitric oxide (NO), 280, 283, 284, 286, 296, 300, 597, 599, 600, 603
Nitric oxide synthase, 599, 603–604
Nonimmunogenic, 718
Nonmedical hazards, 781
NOS (nitric oxide synthase), 603–604
Notch, 187–188, 439–447, 556
Nous, 65
NSAID, 117
Nuclear DNA mutations, 5, 172, 182, 199, 232, 308, 314, 387, 390, 399, 477, 525, 641–643, 646, 650, 660, 670–671, 766, 773–775, 777, 820
Nuclear mutation, 671, 765, 773–775
Nuclear transfer, 79, 471–472, 545, 548–552, 556, 558
Nucleophiles, 605
Nucleophilic, 604, 607
Nude mice, 556, 563–564
Nutrient
  absorption, 10, 254
  deprivation, 337, 387, 392, 403
  sensing, 629–630, 632
Nutrigenomics, 357–365
Nutritional supplementation/supplements, 6–7, 9–11, 13, 362–364, 605–606

O

Obesity, 12, 16, 156–157, 239–240, 244, 251, 255, 257, 259, 265, 268–269, 332, 349–351, 397, 406–408, 577, 580, 633, 736, 821
Octopus, 106, 140–141
Octopus hummelincki, 140–141
Odds of living to 100, 150
Offspring, 39, 58, 91, 93–94, 109, 111, 114–115, 130, 137, 144–145, 150, 175, 299–300, 311, 379, 383, 462, 552, 579, 635
Okinawa, 5–6, 372
Olovnikov, Alexey, 454–455, 457
Omega-3 fatty acids, 11, 256, 259
Omega-3 to omega-6 fatty acids
  imbalance of, 243
Optic glands, 106, 140–141
Orchiectomized men
  lower mortality rate of, 166
Organelle testing, 760–761
Organ mill, 734, 742, 744–747, 760, 770, 773, 778
Organogenesis, 70, 74, 76–77, 441
Organ printing, 744–745

Organs, 81, 104, 118, 139, 144, 162–164, 169, 171, 183, 195–197, 234, 237, 240–241, 283, 297, 376, 386–387, 389, 392, 439–447, 453–454, 505, 507, 543–565, 589, 602, 607, 646, 656, 679, 686–687, 689–690, 717, 723, 729–730, 734, 736–737, 739, 742–747, 749, 755–756, 760, 770, 772–773, 775–778, 821–822, 833–834
Orgel, Leslie
   second law and, 88
Ornithine, 594, 598, 657
Orthologs, 332–333, 359–360, 398, 401, 627, 630
Osteoarthritis, 453, 589, 591, 599
Osteoblasts, 393, 590
Oxidation, 154, 177, 253, 285–286, 291–294, 298–301, 357, 398–401, 524, 588, 592, 594–595, 599, 818
Oxidative damage, 93, 172, 182, 371, 378, 460, 576, 607, 633, 680, 772
Oxidative phosphorylation, 163, 320–321, 350, 399, 410, 521–522, 524–526
Oxidative stress (OS), 155–156, 182–183, 235–236, 243, 249, 286, 291, 293, 295, 298, 333, 335, 337, 393, 396, 442, 576–577, 580, 602
Oxidized-LDL, 285–286
Oxidized lipid, 285–294, 296, 299
Oxidized phospholipid, 288, 294, 298
Oxoaldehydes, 594
Oxygen, 17, 130, 156, 186, 234, 321, 403, 409, 460, 464, 524–525, 592, 613, 690, 692, 695–696, 710–711, 727, 735, 737–738, 749, 772, 810

P
p15, 442–443
p16, 443, 463
p21, 442–443
p27, 443
p38 kinase, 631
p53, 334–337, 343–344, 390, 398, 459, 463, 653
p66^{shc} gene knockout, 155
PABA, 604
Parabiosis, 187–188, 444
Paradigm shift rate, 30
Parameters, 110, 115, 228, 238, 245, 248, 321, 372, 409, 412, 472, 575, 738–739, 757, 812
Paramutation, 314

Paraoxonase (PON), 287, 292, 295, 298–301
Parkinson's disease (PD), 338, 340, 348–349, 370, 393, 396, 453, 474–475, 529, 762, 819
Parthenogenesis, 471
Paternal age, 579
Pathological anatomical damage, 776
Pathologies, age-associated, 588, 614
PCMT1, 595
PDK-1, 624, 627–628, 631
Penis
   cavernosal cells, 555
   erectile dysfunction, 554
   rabbit model of injury, 555
   reconstruction of, 554–555
PEPCK-C (mus) mice, 157–159, 186
Peripheral T cell pool, 496–502, 504, 509, 513
Peritoneal cavity, 310, 313
Perls, T., 150, 575
Permeability, 187, 319, 767
Personhood, 43–47, 55, 64–66, 71–79, 83
Petromyzon marinus, 142
PGC-1α, 158, 186, 334, 336, 349, 397–401, 403, 411
Phagocytes, 313, 318, 717–718, 720–721, 727, 759, 767, 778
Phagocytosis, 547, 590, 599–600, 670, 718, 720–721, 726, 729–730
Pharmacological and nutritional interventions, in mammals, 159–165
Pharmacyte, 736, 739–740, 763, 767–768
Phenformin, 159, 402
Phenotype, cancer killing, 316
Phestilla sibogae, 131, 138
Phosphatidylinositide, 626
Phospholipid, 286–288, 292, 294, 298–299, 592, 629, 752
Phospholipid transfer protein (PLTP), 292
Photoaging, 588
Phylogenic conservation, 369–370
Physical activity
   favorable responses to, 262
Physiological theory of aging
   central hypothesis, 201
Phytochemicals, 253–254
PI3 kinase, 624, 626
Pimagedine, 603–604
PIMT, 595
Pineal gland, 174, 236, 247
   and hypothalamic sensitivity to feedback, 247
   involution of, 174, 247
Pineal polypeptide extracts, 247

pKa, 605
Planaria, 118
Planetary gear, 692
Plaque, atherosclerotic, 285, 732, 766, 769, 818
Plaques, beta-amyloid, 592, 601, 610, 767
Platelet activating factor acetylhydrolase (PAF-AH), 287, 292–293, 298–299
Pleckstrin homology (PH), 627–628
Pleiotropic, 88, 94, 96–101, 136, 378–379, 443, 505
Pleiotropy, 96–104, 107, 136, 138, 463
  as adaptation, 103–104
Plethomnesia, 779–780
Pluripotent, 68–70, 80, 467, 472–473, 544, 548–551, 770
Poisoning, 737, 739–740
Polymerase gamma, 183, 523, 527
Polymorphisms, 10, 13, 156, 339, 529, 656
Polymorphonuclear cells or PMN, 322
Popper, K.R., 358
Popular diets, comparison of, 260
Population dynamics, 95, 110–114
Population genetics, 90, 109
Population genetic theory, 97, 105, 115
Positional assembly, 702–704, 712–713
Positional navigation, 730
Postmenopausal, 237–238, 578
Postmitotic, 148, 181, 381, 390–391, 397, 523, 576, 628–629, 634, 676, 819
PPAR-γ, 334, 336, 349
PPARδ, 158, 186
Predator-prey dynamics, 110, 114
Predictions/predicting, 23–34, 43, 54, 88, 95, 100, 115, 134, 236, 288, 410, 613, 781–782
Predictive genomics, 11–14
Preembryo, 65, 70
Preventive settings, 322
Prions, 759, 776
Proapoptotic genes, 344
Probucol, 604
Procollagen, 590
Progenitor cell, 188, 439–441, 444, 453, 467–468, 494–495, 527, 560–561, 743, 752, 767
Progeria, *see* Hutchinson-Gilford syndrome
Programmed aging, 89, 141, 152
Programmed death, 96, 105–109, 145, 453
Programmed senescence, 107, 139, 232, 246
Prolongevity factors, 332

Prophet of Pit-1 (Prop-1) mutation, 153–154
Prostate cancer, 245–246, 253, 324, 343, 388, 505, 514
Proteases, 189, 319, 392, 461, 592, 595–597, 600, 720, 764, 771
Protein(s)
  Biological Value (BV) of, 256
  extracellular, 320, 471, 588–589, 592, 597, 614, 767, 821–822
  misfolded, 600
  recommended intake of, 255
  restriction, and lifespan, 378–379
  strand breaks, 588, 595–597
  synthesis, 145, 151–152, 164–165, 177–180, 184, 244, 380, 390, 392, 394, 400, 403, 530, 647, 724, 752, 757
Protein kinase A, 629
Proteoglycans, 589, 591–592
Protons, 336, 524
Pseudomys fumeus, 144
PTB, *3-(2-phenyl-2-oxoethyl)-thiazolium bromide*, 607–608
PTEN, 324, 626–627
PTEN knockout, LIFT treatment of malignancy resulting from, 324
Pulmonary fibrosis, 580
Puya raimondii, 145
Pyridoxamine, 604–605
Pyruvate dehydrogenase, 630

Q
Quackery, 28, 30–31
Quantum computer, 701
Quickening, 64

R
Rabbits, wild
  negative reproductive costs in, 175–176
Racemization, amino acid, 595
Radioactive, 323, 562, 759
RAGE (receptor of AGEs), 598–599, 601, 608
Random damage, 89, 129,131, 135, 148–149, 176–177
Rapamycin, 165, 403, 629–631
Rapid death of plants
  hormones and, 145
RAS1, 629
RAS2, 629, 633
Rational drug design, 16, 601, 612–613
Rattus fuscipes, 144
Raw food, 251, 602

Reactive oxygen species (ROS), 92–93, 95, 170, 286, 336, 338, 348–349, 523–529, 600–601, 603, 660, 810–811
Recent thymic emigrants, 505, 509
Receptors, 6, 15–16, 115, 150, 153, 155, 157, 175, 186–187, 197–198, 243, 245, 285–287, 294, 301, 314, 336, 375, 384–385, 389–390, 396–398, 400–401, 403, 405, 441–442, 492–494, 496–497, 499–500, 504, 506–508, 511, 589–590, 592, 596–600, 608, 611, 624–625, 627, 631–633, 647, 657, 719, 727, 729, 731, 761
Recombination and mismatch repair, 653
Red cell, 323, 737–738, 740, 743, 746, 760
Red Queen hypothesis, 114
Reductionism, 43, 56–59
Regeneration, 98, 115, 118–119, 143, 187, 190, 192–195, 201, 439–445, 473–476, 552, 558, 563, 596, 611–612, 656, 741, 752, 778, 814, 822, 832, 834
Regenerative competence
  restored in old livers, 188
  restored in old muscle, 187–188
Regenerative medicine, 451–477, 544–545, 810, 835
  tissue engineering in, 544–545, 559
Rejuvenation/rejuvenating, 27, 187, 241, 444–447, 504–511, 513–514, 606, 612, 669, 764, 774–780, 809–823
Remodeling of ECM, 590
Repair by replacement, 199, 752–753
Replacement mitochondria, 186, 773
Replicative lifespan, 370, 398, 454, 458–459, 461, 463–464, 477, 501, 512, 611, 628–629
Replicative senescence, 188, 456, 462, 464, 573, 576–580, 676
Replicometer, 454
Reproductive costs
  negative, 174–176
Reproductive cycling
  reversal of cessation of, 160, 194–195, 386
Reproductive effects of CR
  in humans, 387
  in rodents, 385–386
Respirocyte, 17, 734–735, 737–739, 747, 782
Resveratrol (RSV), 7, 118, 158–159, 336, 340, 346–351, 382, 604
  and metabolites of, 158, 336
  See also Sirtuin activators

Retinopathy, 589
Retrotransposons, 777
Retroviral components, 315
Retrovirus, 551
Reverse cholesterol transport, 286
Reversed development, 133
Reverse-engineer (the brain), 17–18
RGD sequence, 590
Rheumatoid arthritis (RA), 286, 291–292, 509, 589, 599
Ribozymes, 643, 646–649
RIM15, 629–630
Risk of dying, 150
Risk-prediction models, 781–782
RNA interference, 16, 19, 101, 146–147, 336, 646–650
Robotic surgeries, 748
Rosettes, 313
ROS (reactive oxygen species), 92–93, 95, 170, 286, 336, 338, 348–349, 523–529, 600–601, 603, 660, 810–811
Rupture, cell, 319

S
S180, 310–312, 324
S6 kinase, 380, 630, 634
Saccharomyces cerevisiae, 370, 397, 623–624, 628–630
Salicylic acid, 604
Salmon, 106, 140, 142–145
Sapience, 66, 76–79
SAP protein, 600
Satellite cells, 187–188, 440–444, 563, 677
Scanning probe microscope (SPM), 704–707, 709, 711–712
Scanning rings, 733
Scar formation, 592, 611–612
Scavenger receptor B1 (SR-B1), 287
SCH9, 629–630, 634
Schiff base, 593, 605
SCNT, 68–69, 71, 470–473, 476, 549–550, 552
Sea anemones, 136, 357
Seasonality, of cancer killing effect, 317–318
Second law of thermodynamics, 52–54, 90
α-Secretase activity, 349
Sedentary lifestyle, 7, 282, 577–578
"Segmental aging reversal", 176–198
Self-assembly, 702–703
'Self-destruct' system, 117, 693, 723
Semaphores, 720, 729, 743
Semelparous, 106–107, 137, 139–146, 148, 167–169
Senescence, physiological, 588

α-Secretase activity, 349
α-Synuclein, 338, 345, 396, 819–820
Senescent cells, 454, 461, 464, 471–472, 769, 821
SENS, 27, 669–671, 681, 764–773, 775, 778, 809–823, 835–837
SENSE: Strategies for Engineering Negligible Senescence Evolutionarily, 358–359
Senstatic activation, 461
Sentience, 66, 76–78
Serum proteins, 320, 592
Sex hormones
    negative effects on immune function, 167
Sexual maturation
    and aging phenotypes, 129–134, 169–174
    prevention of, 142
SGK-1, 632
Sickle cell anemia
    correction of, 81
Sickle cell disease, 284, 300
Side effects, 77, 88, 90, 98, 108, 133, 136, 155, 164, 200–201, 234–235, 240, 308, 325, 361, 507–508, 512–514, 544, 559, 604, 608, 613, 671, 723, 729, 736, 739, 765, 768, 783, 810, 813–814, 836
Sigmoid function, 72, 74
Signaling, hormonal, 100, 400
Signaling pathways, 95, 149, 153, 314, 335, 350, 383–384, 390, 402, 441, 599, 611, 630, 632–634
Silent inflammation, 10–11
Silent Information Regulator, 332–335, 340–341, 398, 630, 632
Sinclair, D.A., 117–118, 159, 332–333, 336, 370
Single nucleotide modification
    engineered, 650–657
Single nucleotide polymorphisms (SNPs), 12–13, 339
Single-stranded oligonucleotides (SSOs), 642, 653–655, 657
SIR2, 332–335, 340–341, 398, 630, 632
SIRT1, 186, 332–336, 338–351, 397–399, 411
    active regulator of, 334
SIRT3 variant
    in males older than 90 years, 338
Sirtuin, 7, 158–159, 164, 174, 331–351, 398, 630
Sirtuin activation, 158, 174, 346
Sirtuin activators
    butein, 346
    piceatannol, 346
    quercetin, 346
    resveratrol, 340, 346–347
    SRT1460, 348
    SRT1720, 340, 347–350
    SRT2183, 348
Sirtuin-based therapies, 332
Sirtuin classes, 333–334
    Class I, 333
    Class II, 333
    Class III, 333
    Class IV, 333
Sirtuin inhibitors
    AC-93253, 344
    AGK2, 345
    cambinol, 341–343
    in cancer therapy, 343–345
    chemotherapeutic properties, 343
    HR73, 345
    in other diseases, 345
    salermide, 340, 342, 344
    sirtinol, 340–341, 343, 345
    splitomicin, 340–341, 343, 345
    suramin, 340, 342, 345
    tenovin, 340, 344
Skeletal muscle, 157–158, 187–188, 262, 334–337, 371, 375, 392, 399–401, 403, 440–442, 467, 559, 576, 591, 602, 676, 742, 759
Skin, 53, 67, 81, 143, 171, 180, 184, 196, 230, 232, 239, 249, 297, 346, 387, 393, 462, 476–477, 490, 497, 508, 510–511, 545, 548, 552, 556, 558, 563, 588–589, 591–592, 596, 608–609, 612, 680, 725, 735, 739–741, 749, 752, 762, 768–769, 774
Smad, 442
Small interfering RNAs (siRNAs), 337, 339, 643, 649–650
Smith, Kirby, 167–168
Smokers, 269, 315, 578
Smokey mouse (*Pseudomys fumeus*), 144
Snell dwarf mouse, 154
Social distance, 492
SOD, 92, 182, 385, 396, 629, 633
Somatic cell nuclear transfer, 68–69, 71, 470–473, 476, 549–550, 552
Somatic cells, 67, 104, 116, 308, 452–464, 466, 469–470, 472, 549–551, 577, 752, 771
Somatic involution, 170–174
Sorting rotor, 696–697, 699, 719–720, 728, 733, 737–738, 758, 763–764

Specific nutrients and lifespan, 381–383
Speed reduction gear, 710–711
Spinoza, 48–52, 55, 57–58
Spontaneous regression, 311–312
"Squaring the curve", 228
SR/CR mice, 311–316, 323–324
SRT1720, 340, 347–350
   See also Sirtuin activators
Stanols and sterols, 258
Starch blockers, 6
Starfish, 118
Statin, 289, 608
Stem cells
   adult, 68, 199, 439–440, 444, 446, 548
   amniotic fluid, 68, 548, 559
   differentiation of, 466–467
   embryonic, 63–83, 340, 344–345, 451–477, 544–545, 548–549, 559, 817
   fetal, 545
   mesenchymal, 199, 559, 590
   placental, 548–549
Stem cell therapy, 64, 79, 118, 199, 558, 672, 676, 682, 770, 833
Steric hindrance, 606–607, 610, 612
Stochasticity of aging, 91, 128, 135, 148–151, 154, 166, 458, 464, 576, 810
Strength (resistance) training, 8–9, 266–267
Stress hormones, 316–318
Stress induced premature senescence (SIPS), 460
Stress-mediated cell death, 348
Stretching, 9, 268, 597
Stroke, 7, 11, 17, 117, 150, 156, 233, 252, 269, 281, 349, 396, 464, 475–476, 522, 597, 726, 735–736, 744, 771, 776, 821
Strong inference, 358
Suffocation, 337–739
Sugar, 7, 118, 251, 254, 261, 514, 588–589, 592–595, 601–602, 605, 607, 696, 728, 768
Suicide, 42, 46–47, 142, 174, 252, 670, 715, 736, 756, 769, 773, 780, 782
Sulfur, 253, 605, 692, 710–711
Sunburn, 596
Supercentenarians, 369, 766
   age-independent mortality rate of, 135
Superoxide dismutase, 155, 247, 301, 338, 629
Surface recognition, 313, 320–321, 324
Surgical microrobotics, 749–751
Surgical nanorobot/surgical nanorobotics, 685–783
Surgical robot, 748, 751

Survival, 161, 228–231, 626, 816
Survival time, 140–141, 161–163, 165, 402
SV40 virus, 458–459
Swelling, 27, 319, 720, 740, 762
Synaptic monitoring and recording, 778
α-Synuclein, 338, 345, 396, 819–820
Synvista Therapeutics, 607
Systemic lupus erythematosus (SLE), 286, 291–292
Systemic sclerosis (SSc), or scleroderma, 292, 297

T

T cells, 118, 241, 371, 395, 463, 492, 494–495, 497–507, 509–513, 557–558, 579–580, 648, 719, 769, 821
Tachigalia versicolor, 145
Target cells, 246, 316, 319–320, 530, 611, 675, 730, 739, 755–756, 773
Target of rapamycin, 403, 629–631
Target structure (drug development), 612–613
Technological evolution, 15
Telemedicine, 748
Telesurgery, 748
Telomerase, 96, 116, 188, 446, 455–460, 462–465, 467–470, 472, 477, 512, 577, 580, 672, 675, 677–682, 773–774
   activation, 456, 458–459, 462–463, 465, 467–469, 472
Telomeres
   attrition, 464, 576–579
   dynamics, 454–460, 470–472, 573, 576–580
   erosion, 576–577, 579–580
   extension in mice, 477
      40% extension of median lifespan by, 159, 188
   length, 116, 452, 455, 459, 462, 464–465, 467, 470, 472–473, 477, 573–574, 576–580, 680, 776
   shortening, 116, 134, 455–458, 460, 468, 578, 580, 680
   in vivo, 577–580
Telomeric aging, 96, 115, 454–460
Telomeric erosion, 577
Telomeric G triplets, 576
Telosome, 458, 472–473
Tendons, 8–9, 267, 589, 591, 602, 608, 614
TERT, 458, 460, 462–463, 465, 469, 473, 611
Testosterone
   and reduced mortality from all causes, 144, 167

Index   865

Tetrapeptides, 299
  FREL, 299
  KERS, 299
  KRES, 299
TGF-beta, 187, 439–447, 602
Theories of aging, 231–232
  accumulated damage, 90–93
  adaptive theories, 104–114
  antagonistic pleiotropy, 98–101, 103–104
  developmental, 201
  disposable soma, 101–103
  evolutionary, 358
  mutation accumulation, 93, 97
  other, 232
Therapeutic modalities, 564, 575
Therapeutic settings, 322–323
Therapeutic window, 325
Thermodynamics, 52–55, 90, 262, 268
Thermogenesis, 170, 338–339, 386
Thiamine (vitamin B1), 605, 607
Thiazolium, 607–609, 614
Three dimensional (3D) printing, 745–746
Thymic epithelium, 505, 508–510
Thymic hyperplasia, 191–192
Thymic involution
  and aging, 170
  lack of, in hyperthyroid individuals, 191
  reversal of, 191–194
Thymic regeneration, 191–194, 504–510
Thymocyte, 463, 494–496, 499, 507, 509
Thymosins, 241
Thymulin, 191–192
Thymus
  as a controller of many aging processes, 198
  rejuvenation of, 191–194, 504–510
  transplants, 170, 198
    lives extended by, 198
Thyroxine, 191–192, 238
Tissue(s)
  mill, 746, 778
  printer, 734, 742, 744–747, 760, 770, 778
t loops, 457–459, 462
T lymphocytes, 309, 313, 469, 497, 507, 558, 579
TNFα, 336
Tolerization, 718–719
Tooth replacement
  continuous throughout life, 199
Tooth wear
  as limiting for lifespan, 199
TOR, 151–152, 165, 180, 380, 629–632, 634
Torpor, 579

Totipotent, 67, 70, 465–466
Toxic cells, 670, 769
Tradeoffs, 97–104, 138, 145, 157, 176, 463
  evolutionary, 90
Tragedy of the Commons, 110
Transcriptional integrator p300, 335
Transcriptional programs
  in aged tissues, 337
Transforming growth factor beta, 187, 373, 612
Transglycation, 604–606
Transient ischemic attack (TIA), 735
Transition metals, 594
Transmembrane potential, 319
Transposons, 314, 777
Transthyretin (TTR), 600–601, 657, 819–820
Trauma, 81, 170, 198, 239, 255, 318, 345, 461, 471, 475, 554, 720–722, 736–751, 776–777
TREC (T cell receptor excision circle), 505, 507
TRF (telomere restriction fragment), 455–456, 462, 464, 469, 471, 477
Triglycerides, 11, 13, 157–158, 255, 257–258, 265, 287, 292, 371, 373
Triplex DNA
  homologous pairing of, 657–660
Trolox, 605
Trophoblast, 67–68, 70
Tropism, 77
Tryptophan, 163–165, 237, 255, 379–380, 411
  deficiency, 163–165
Tryptophan-deficient diets, 163–165, 237
Tryptophan restriction and lifespan, 163–165, 379–381
TTR (transthyretin), 600–601, 657, 819–820
Tumor(s)
  cells, 241, 335, 338, 343, 388–390, 410, 456, 731
  promoter, 335
  suppressor, 338
Turnover, collagen, 591, 608, 611
Twins, monozygotic, 149
Type 2 diabetes (type II diabetes), 7, 12, 19, 170, 240, 339, 349, 408, 411, 462, 477, 535, 605, 669, 736, 766, 769

U

Ubiquinone, 92, 95, 162
Ultrasound, 71, 700, 728, 738, 749, 756
Uncoupling protein (UCP), 186, 336, 404
United Therapeutics, 16

Urethra
    hypospadias repair, 553
    onlay replacement, 553–554
    replacement of, 553–554
    stricture disease, 553–554
    tubularized repair, 553–554
Urinary bladder, 597
Urine, 181, 247, 249, 557, 590, 605, 728
Uterus, 64, 68, 70, 77–78, 334–335, 549, 555
Utility fog, 743–744

V

Vaccination, 322, 395, 498, 507, 512, 514, 766
Vagina, 556
Value of life, 39–59, 716
Valvular heart disease
    aortic valve, 283
Varicella Zoster Virus (HHV-3), 502
Vascular dysfunction, 284, 350, 476, 578
Vascular gate, 735, 742–744
Vascular repair, 732
Vasculocyte, 732–735, 742–744, 766, 770, 776
Vaupel, 92, 103, 134–135, 321, 368, 574, 716–717
Vena cava filter, 744
Ventricular hypertrophy, 589, 609
Vesicoureteral reflux, *see* Cellular injection therapy
Vicious cycle, 588, 591, 597–598, 600
Viroids, 759, 776
Vitamin
    vitamin B1 (thiamine), 604–605
    vitamin C, 243, 253, 603
    vitamin E, 190, 253, 382, 604
Vitellogenin, 175
Vitronectin, 589, 598
VPS34, 627

W

War, 52–53, 228, 668, 688, 715, 725, 780, 782, 833, 840
Warburg effect, 321
Warburg, Otto, 321
Watson, James, 12, 54, 455
WBC's, 17, 241, 292, 316, 404, 574, 577–580, 701, 729, 769, 771
Wear-and-tear, 92, 232, 234, 246, 252
Webster, George, 177–179
Weismann, August, 92, 108, 453–454, 461, 463, 466
Werner syndrome, 456, 461–462, 464–465, 477
West Nile Virus, 497
White blood cells, 17, 241, 292, 316, 404, 574, 577–580, 701, 729, 769, 771

Whole-body Interdiction of Lengthening of Telomeres, 667–683, 730, 773–774, 820
Wild rabbits
    negative reproductive costs in, 176
WILT, 667–683, 730, 773–774, 820
Winter, effect of on cancer killing, 317–318
Wnt, 187, 441, 631
Women, 8, 13, 64, 68, 166, 168, 195, 197, 230, 237–238, 243–245, 247, 255, 258, 264–268, 288, 368, 373–374, 406–408, 509–510, 577–578
Worm drive assembly, 711
Worms, 95–96, 100–101, 106–107, 118, 129–130, 138, 146–149, 152–153, 155, 159, 163, 165, 332, 574, 625–627, 632–633, 635, 777
Wound healing, 119, 232, 463, 588–589, 592, 612, 737, 741–742
Wright, Woodring, 454, 459–460, 465, 469, 576–577

Y

Y-chromosome deletions
    human lifespan and, 166–168
Yeast, 95–96, 100, 118, 145, 153, 159, 165, 332–334, 340–341, 346, 370, 378, 380, 398, 401, 456, 623, 627–630, 633–635, 644

Z

Zero aging
    in box huckleberry clones, 137
    in Bristlecone pine, 137
    in *C. elegans*, 129–131
    in fish, 129
    in *Fusitriton oregonensis*, 131
    in giant fungus (*Armillaria bulbosa*), 137
    in golden corals, 137
    in hydras, 136
    in naked mole rat, 137
    in *Phestilla sibogae*, 131
    in POSCH-2 mouse, 134
    in queens (insect), 136
    in sea anemones, 136
    in sea urchins (red), 136
    in Sequoia, 137
    in *Trogoderma glabrum*, 132–133
    in *Trogoderma tarsale*, 133
    in tubeworms, 136
    in turtles, 137–138
Zinc finger nucleases (ZFNs), 642, 645, 677
Zippocyte, 742
Zona reticularis, 170
Zygote, 65, 69, 73, 79, 528, 754